2019 International Aegean Conference on Electrical Machines and Power Electronics (ACEMP 2019) & 2019 International Conference on Optimization of Electrical and Electronic Equipment (OPTIM 2019)

Istanbul, Turkey
27 – 29 August 2019

IEEE Catalog Number: CFP1922D-POD
ISBN: 978-1-5386-7688-2

**Copyright © 2019 by the Institute of Electrical and Electronics Engineers, Inc.
All Rights Reserved**

Copyright and Reprint Permissions: Abstracting is permitted with credit to the source. Libraries are permitted to photocopy beyond the limit of U.S. copyright law for private use of patrons those articles in this volume that carry a code at the bottom of the first page, provided the per-copy fee indicated in the code is paid through Copyright Clearance Center, 222 Rosewood Drive, Danvers, MA 01923.

For other copying, reprint or republication permission, write to IEEE Copyrights Manager, IEEE Service Center, 445 Hoes Lane, Piscataway, NJ 08854. All rights reserved.

*** This is a print representation of what appears in the IEEE Digital Library. Some format issues inherent in the e-media version may also appear in this print version.

IEEE Catalog Number:	CFP1922D-POD
ISBN (Print-On-Demand):	978-1-5386-7688-2
ISBN (Online):	978-1-5386-7687-5
ISSN:	1842-0133

Additional Copies of This Publication Are Available From:

Curran Associates, Inc
57 Morehouse Lane
Red Hook, NY 12571 USA
Phone: (845) 758-0400
Fax: (845) 758-2633
E-mail: curran@proceedings.com
Web: www.proceedings.com

TABLE OF CONTENTS

PLENARY SESSIONS

ACEMP-OPTIM 2019 OPENING SPEECH...5
H. Bülent Ertan

FUTURE OF EV CHARGING...8
Pavol Bauer

IN MEMORIAM - PROFESSOR ROBERT D. LORENZ 10
Tom Jahns

ELECTRIFIED AVIATION: ARE MOTOR DRIVES READY TO SPREAD THEIR WINGS AND FLY?..16
Thomas M. Jahns

REVIEW OF ELECTRIC VEHICLE POWERTRAIN TECHNOLOGIES WITH OEM PERSPECTIVE 18
Mustafa Karamuk

LARGE VARIABLE SPEED GENERATORS DESIGN AND THEIR CONTROL: A REVISIT IN 2019 29
Ion Boldea ; Lucian Tutelea ; Ana Popa

REQUIRED TECHNOLOGY FOR UPGRADING EFFICIENCY OF HIGH-SPEED MOTOR WITH HIGH POWER DENSITY...41
Masato Enokizono

DEVELOPMENTS AND TRENDS IN THE ADJUSTABLE SPEED DRIVES INDUSTRY 51
Norbert Hanigovszki

SYNCHRONOUS RELUCTANCE MOTOR DRIVES: STILL A NICHE TECHNOLOGY?.................57
Gianmario Pellegrino

DESIGN OF ELECTRICAL HIGH SPEED DRIVES FOR VEHICLE DRIVETRAINS...............58
Markus Henke

AUTOMOTIVE POWER ELECTRONICS: CURRENT STATUS AND FUTURE TRENDS...................60
Bülent Sarlioglu

TECHNICAL TRACK ON ELECTRICAL MACHINES, INDUSTRIAL DRIVES AND CONTROL

10MW, 10RPM, 10HZ DIRECTLY-DRIVEN CAGE ROTOR INDUCTION GENERATOR (CRIG): PRELIMINARY DESIGN WITH KEY FEM VALIDATIONS...65
I. Boldea ; L. N. Tutelea ; I. Torac ; F. Marignetti

A DIRECT-DRIVE, LINEAR ACTUATOR OF A HYBRID STRUCTURE 71
Ning Zhang ; Michael Collins ; Michael Rae ; Howard Locatt

A NOVEL APPROACH TO PLCS BASED SYSTEMS UTILIZED IN ELECTRIC DRIVES 77
Adrian Daniel Martin ; Lucian Tutelea ; Radu Babau ; Ion Boldea

A NOVEL KURTOGRAM-BASED HEALTH INDEX FOR INDUCTION MOTOR FAULT DIAGNOSIS...85
Azadeh Gholaminejad ; Farzaneh Sabbaghian Bidgoli ; Javad Poshtan ; Majid Poshtan

A NOVEL SURFACE IMPEDANCE BASED CLAMPING PLATE LOSS MODEL FOR LARGE SYNCHRONOUS GENERATORS..93
Torben Fricke ; Babette Schwarz ; Bernd Ponick

ADAPTIVE ALGORITHM TO REDUCE ACOUSTIC NOISE AND TORQUE RIPPLE IN LOW-COST PM MOTORS..100
Martin Sumega ; Simon Zossak ; Patrik Varecha ; Pavol Rafajdus ; Marek Stulrajter

ANALYSIS OF THE THERMAL INFLUENCE ON THE VIBRATIONAL BEHAVIOR OF THE STATOR END-WINDING REGION 108
Sebastian Lange ; Martin Pfost

ASSESSMENT OF 5 KW INDUCTION MOTOR FINITE ELEMENT COMPUTATIONS WITH A COMMERCIAL AND AN OPEN-SOURCE SOFTWARE...114
M. Zaheer ; P. Lindh ; L. Aarniovuori ; J. Pyrhönen

AUTOMATIC SUPERIMPOSED DROOP FREQUENCY CONTROL SCHEME FOR DC MICROGRIDS 120
Mohammad Jafari Matehkolaei ; Hossein Mokhtari ; Majid Poshtan

BEHAVIOR OF A FIVE-PHASE PENTACLE CONNECTED IM OPERATED UNDER ONE-PHASE FAULT 126
Pavel Zaskalicky

CALCULATING FREQUENCY RESPONSES OF SYNCHRONOUS MACHINES USING MIMO TRANSFER FUNCTIONS 132
Matthias Kalla ; Olga Korolova ; Alexander Neufeld ; Lutz Hofmann ; Bernd Ponick

COMPARISON OF PI AND FOPI BASED VOLTAGE AND CURRENT CONTROLLED DC MOTOR DRIVE SYSTEM 139
Hafiz M. Usman ; Abdel Gafoor Haddad ; Habibur Rehman ; Shayok Mukhopadhyay

COMPARISON OF TIME-DOMAIN AND TIME-SCALE DATA IN BEARING FAULT DETECTION 143
I. Halil Ozcan ; Levent Eren ; Turker Ince ; Bulent Bilir ; Murat Askar

COMPARISON OF TWO-LEVEL AND THREE-LEVEL NPC INVERTER TOPOLOGIES FOR A PMSM DRIVE FOR ELECTRIC VEHICLE APPLICATIONS 147
Alican Madan ; Emine Bostanci

CURRENT DATA FUSION THROUGH KALMAN FILTERING FOR FAULT DETECTION AND SENSOR VALIDATION OF AN ELECTRIC MOTOR 155
Sadra Mousavi ; Duygu Bayram ; Serhat Seker

CURRENT HARMONIC SUPPRESSION FOR PERMANENT MAGNET SYNCHRONOUS MOTORS 161
Ngoc-Tu Trinh ; Fabien Vidal-Naquet

CUTTING TECHNOLOGIES INFLUENCE ON MAGNETIC PROPERTIES OF ELECTRICAL STEELS USED IN HIGH-EFFICIENCY MOTORS MANUFACTURING 166
Gheorghe Paltanea ; Veronica Manescu Paltanea ; Elena Helerea ; Iosif-Vasile Nemoianu ; Marius Daniel Calin

DESIGN OF CONICAL ROTOR FLUX-SWITCHING PERMANENT MAGNET MACHINE WITH IMPROVED FLUX-WEAKENING CAPABILITY FOR TRACTION APPLICATIONS 172
Hao Ding ; Mingda Liu ; Bulent Sarlioglu

DEVELOPMENT OF HIGH EFFICIENCY AND HIGH-SPEED MOTOR WITH HIGH POWER DENSITY 178
Masato Enokizono ; Naoya Soda ; Daisuke Wakabayashi ; Shohei Ueno ; Yuji Tsuchida

DIAGNOSIS OF DIFFERENT ECCENTRICITY FAULTS IN INDUCTION MOTORS BASED ON ELECTRICAL AND MAGNETIC SIGNATURES AND UNBALANCED MAGNETIC PULL 186
T. A. Sarikaya ; A. Polat ; L. T. Ergene

DIRECT RADIAL AND CIRCUMFERENTIAL ANALYTICAL AIR-GAP FIELD CALCULATION FOR ELECTRICAL MACHINES 191
Jan Andresen ; Bernd Ponick ; Axel Mertens

EARLY DETECTION OF TURN-TO-TURN FAULTS IN POWER TRANSFORMER WINDING: AN EXPERIMENTAL STUDY 199
Arash Moradzadeh ; Kazem Pourhossein

ECCENTRICITY FAULT DIAGNOSIS IN PMSM USING MOTOR CURRENT SIGNATURE ANALYSIS 205
Zakaria Gherabi ; Noureddine Benouzza ; Djilali Toumi ; Azeddine Bendiabdellah

HIGHLY EFFICIENT MULTI-JUNCTION SOLAR CELLS PERFORMANCE IMPROVEMENT FOR AC INDUCTION MOTOR CONTROL USING THE DSPIC30F MICROCONTROLLER ... 211
Abdelkader Hadj Dida ; Mohamed Bourahla ; H. Bülent Ertan

IMPROVED DIAGNOSIS OF INDUCTION MOTOR'S ROTOR FAULTS USING THE PAPOULIS WINDOW 216
Mohamed Boudiaf Koura ; Ahmed Hamida Boudinar ; Ameur Fethi Aimer

MODEL PREDICTIVE CURRENT AND CAPACITOR VOLTAGE CONTROL OF POST-FAULT THREE-LEVEL NPC INVERTER-FED SYNCHRONOUS RELUCTANCE MOTOR DRIVES ... 221
Yong-Chao Liu ; Salah Laghrouche ; Abdoul N'Diaye ; Maurizio Cirrincione

MODELING AND DIAGNOSIS OF STATOR WINDING FAULTS IN PMSM USING MOTOR CURRENT SIGNATURE ANALYSIS 227
Zakaria Gherabi ; Djilali Toumi ; Noureddine Benouzza ; Noureddine Henini

MOTOR EFFICIENCY DETERMINATION OF SYNRM AND MEASUREMENT UNCERTAINTY ... 233
Youn-Hwan Kim ; Hee-Deuk Jun ; Jae-Won Moon ; Rae-Eun Kim ; Se-Hyun Rhyu ; Sang-Young Jung

NUMBER OF TURNS INFLUENCE ON THE PARAMETERS OF HIGH SPEED SWITCHED RELUCTANCE MOTOR 240
S. Kocan ; P. Rafajdus ; P. Makys ; R. Bastovansky

NVH-SIMULATION OF SALIENT-POLE SYNCHRONOUS MACHINES FOR TRACTION APPLICATIONS 246

Stephan-Akash Vip ; Jan Hollmann ; Bernd Ponick

OUTER RACE FAULT DIAGNOSIS BY COMPARISON BETWEEN THE POWER SPECTRAL DENSITY AND THE KURTOGRAM 254

Mohammed-El-Amine Khodja ; Ahmed Hamida Boudinar ; Ameur Fethi Aimer ; Azeddine Bendiabdellah

SENSORLESS SYNCHRONOUS RELUCTANCE GENERATOR CONTROL BASED ON Q AXIS ESTIMATED CURRENT 260

Liviu-Danut Vitan ; Lucian Tutelea ; Nicolae Muntean ; Ion Boldea

SHORT CIRCUIT LOCATION IN TRANSFORMER WINDING USING DEEP LEARNING OF ITS FREQUENCY RESPONSES..268

Arash Moradzadeh ; Kazem Pourhossein

TEMPORAL ENVELOPE ESTIMATION OF STATOR CURRENT BY PEAKS DETECTION FOR IM FAULT DIAGNOSIS 274

Hamid Khelfi ; Samir Hamdani ; Youcef Chibani

TORQUE ERROR REDUCTION OF INTERIOR PERMANENT MAGNET SYNCHRONOUS MOTOR DRIVES USING A STATOR FLUX LINKAGE OBSERVER 280

Sungmin Choi ; Seung-Hwan Lee ; Jae Suk Lee

TECHNICAL TRACK ON POWER ELECTRONICS AND POWER CONVERSION

A BIDIRECTIONAL HYBRID SWITCHED-CAPACITOR DC-DC CONVERTER WITH A HIGH VOLTAGE GAIN 289

Dan Hulea ; Nicolae Muntean ; Mihaita Gireada ; Octavian Cornea

A COMPARATIVE STUDY OF CAPACITIVE AND INDUCTIVE PULSED POWER SUPPLY TOPOLOGIES FOR ELECTROMAGNETIC LAUNCHER APPLICATIONS...................................297

Doga Ceylan ; Siamak Pourkeivannour ; Ozan Keysan

A FAMILY OF QUADRATIC DC/DC CONVERTERS WITH ONE LOW-SIDE SWITCH AND A TAPPED INDUCTOR AT THE OUTPUT SIDE 304

Felix A. Himmelstoss ; Helmut L. Votzi

A FRAMEWORK FOR FAST SIMULATION OF POWER ELECTRONIC CIRCUITS 310

Hadhiq Khan ; Mohammad Abid Bazaz ; Shahkar Ahmad Nahvi

A FULL SOFT SWITCHED BRIDGELESS POWER FACTOR CORRECTED AC-DC CONVERTER..315

Sevilay Cetin ; Veli Yenil

A METHOD FOR ACCELERATING FPGA BASED DIGITAL CONTROL OF SWITCHED MODE POWER SUPPLIES 322

Tudor Gherman ; Dorin Petreus ; Remus Teodorescu

A REAL TIME SIMULATOR OF A PEV'S ON BOARD BATTERY CHARGER 329

Tudor Gherman ; Dorin Petreus ; Remus Teodorescu

A SINGLE-SWITCH ZCS BOOST CONVERTER WITH LOW CONDUCTED EMI 336

Mohammad Rouhollah Yazdani ; Mohammad Pahalvandust

A TLBO ALGORITHM FOR DESIGN OPTIMIZATION OF DVRS IN AN INTERLINE DVR (IDVR) 341

Mahdi Jabbari ; Majid Moradlou ; Mehdi Bigdeli

AN APPROACH FOR SPACE VECTOR PWM TO REDUCE HARMONICS IN LOW SWITCHING FREQUENCY APPLICATIONS 347

Ali Bakbak ; Erkan Mese

ANALYSIS OF CURRENT-FEEDBACK PWM PROCEDURES BASED ON HYSTERESIS AND CURRENT-CARRIER-WAVE CONTROL FOR VSI-FED INDUCTION MOTOR DRIVE 351

Csaba Szabo ; Eniko Szoke ; Norbert Szekely ; Vlad Zacharias ; Maria Imecs

CAPACITOR VOLTAGE BALANCE ON NPC MULTILEVEL CONVERTER 359

Juan Diego Nieto Cardona ; Fabio Gómez-Estern Aguilar ; Francisco Gordillo

CASCADED FUZZY CONTROLLER FOR ELECTRIC VEHICLE TRACTION SYSTEM BATTERY ENERGY MANAGEMENT 366

Ahmed Sayed Abdelaal Abdelaziz ; Habib-Ur Rehman ; Shayok Mukhopadhyay

CONTROL STRATEGY FOR FLYWHEEL ENERGY STORAGE SYSTEMS ON A THREE-LEVEL THREE-PHASE BACK-TO-BACK CONVERTER 372

M. Di Benedetto ; A. Lidozzi ; D. M. Kumar ; H. K. Mudaliar ; M. Cirrincione

DIGITAL HYBRID CURRENT MODE CONTROL WITH ASYMMETRIC SLOPE COMPENSATION FOR THREE-LEVEL FLYING CAPACITOR BUCK CONVERTER..................377
Abdulkerim Ugur ; Murat Yilmaz

DUAL-MODE OPERATION OF 3-LEVEL 4-LEG AT-NPC INVERTER FOR MICROGRIDS..........383
Emre Avci ; Mehmet Ucar

IMPROVING THE MODULAR LAYER METHOD TO REPRESENT THE CAPACITIVE EFFECTS OF OVERLAPPING LAYERS IN PLANAR TRANSFORMERS..................389
Ismail Onur Loraz ; M. Timur Aydemir

INVESTIGATION OF THE EFFECTS OF SWITCHING TECHNIQUE ON THE PERFORMANCE OF FOUR SWITCH BUCK-BOOST BIDIRECTIONAL DC/DC CONVERTERS..................395
Ibrahim Koçak ; Hulusi Bülent Ertan

LIFETIME ESTIMATION AND RELIABILITY OF PV INVERTER WITH MULTI-TIMESCALE THERMAL STRESS ANALYSIS..................402
Sara Bouguerra ; Kamel Agroui ; Oussama Gassab ; Ariya Sangwongwanich ; Frede Blaabjerg

MULTILOOP PR+P CONTROLLER OF INTEGRATED BESS-DVR FOR POWER QUALITY IMPROVEMENT..................409
Abdul Muiz Sufianto ; Jaeho Choi ; Nanang Hariyanto ; Arwindra Rizqiawan

OPTIMAL LOW-PASS BUTTERWORTH FILTER DESIGN BY AN ENHANCED ACO ALGORITHM..................417
Bachir Benhala

OPTIMIZATION OF EFFICIENCY AND HARMONICS FOR GBIT FLASH MEMORY BASED PWM INVERTERS..................423
Dorin O. Neacsu

POWER LOSS ANALYSIS IN MODULAR MULTILEVEL CONVERTERS..................431
Ahmed Eshwiage ; Suleiman M. Sharkh ; Sara Bouguerra

THREE-PHASE MODIFIED Z-SOURCE THREE-LEVEL T-TYPE INVERTERS WITH CONTINUOUS SOURCE CURRENT..................439
Anh-Vu Ho ; Anh-Tuan Huynh ; Tae-Won Chun

TECHNICAL TRACK ON RENEWABLE ELECTRIC ENERGY CONVERSION, PROCESSING AND STORAGE

A SUPERIMPOSED FREQUENCY METHOD WITH AN ADAPTIVE DROOP CHARACTERISTIC FOR DC MICROGRIDS..................447
Mohammad Jafari Matehkolaei ; Hossein Mokhtari

ANALYSIS OF MMC HVDC SYSTEM USING SYMMETRIC COORDINATE METHOD..................453
Chan-Ki Kim ; Soo-Yeon Sim ; Sang-Min Kim ; Kyeon Hur

CONTROL OF MULTI-SOURCES ENERGY PV/FUEL CELL AND BATTERY BASED MULTI-LEVEL INVERTER FOR AC LOAD..................459
Mostefa Koulali ; Bachir Boumediene ; Karim Negadi ; Siamak Pourkeivannour ; Mohamed Mankour ; Attalah Smaili

CONTROL STRATEGY FOR OPTIMIZING ENERGY MANAGEMENT IN MICROGRID SYSTEM USING ADAPTIVE CONTROL..................466
R Dimas Dityagraha ; Jaeho Choi ; Nanang Hariyanto

HYBRID STORAGE SYSTEM ASSOCIATED WITH A GRID-CONNECTED WIND GENERATOR..................473
Karima Boulaam ; Akkila Boukhelifa

IDENTIFYING INTERNAL DEFECTS OF PHOTOVOLTAIC PANELS USING SWEEP FREQUENCY RESPONSE ANALYSIS..................481
Kazem Pourhossein ; Meysam Asadi

INTEGRATION OF OFFSHORE WIND FARM PLANTS TO THE POWER GRID USING AN HVDC LINE TRANSMISSION..................486
Abderrahmane Berkani ; Siamak Pourkeivannour ; Karim Negadi ; Bachir Boumediene ; Tayeb Allaoui ; H. Bülent Ertan

MATLAB/SIMULINK MODEL FOR HVDC FAULT CALCULATIONS..................493
Ahmad Mustapha Usman ; Mahir Kutay ; Tuncay Ercan

MODEL COMPARISON AND PARAMETER ESTIMATION OF POLYMER EXCHANGE MEMBRANE (PEM) FUEL CELL BASED ON NONLINEAR LEAST SQUARES METHOD..................500
Krishnil R Ram ; Karteek Naidu ; Ravinesh Kumar ; Maurizio Cirrincione ; Ali Mohammadi

MODELING AND ANALYSIS OF A RENEWABLE-ENERGY-POWERED GREENHOUSE..................506
Yerbol Akhmetov ; Mehdi Bagheri ; G. B. Gharehpetian

MODELING AND SITING OF WIND FARMS USING SUPPORT VECTOR REGRESSION (SVR) 511
Meysam Asadi ; Kazem Pourhossein
**MPPT BASED ADAPTIVE CONTROL ALGORITHM FOR SMALL SCALE WIND ENERGY
CONVERSION SYSTEMS WITH PMSG** 517
M. C. Akkaya ; A. Polat ; L. T. Ergene

TECHNICAL TRACK ON MECHATRONICS, INDUSTRIAL AUTOMATION AND CONTROL

A NOVEL DEVELOPMENT OF ACOUSTIC SLAM 525
Joseph O'Reilly ; Silvia Cirstea ; Marcian Cirstea ; Jin Zhang
ALTERNATIVE APPROXIMATION METHOD FOR TIME DELAYS IN AN IMC SCHEME 532
Cristina I. Muresan ; Isabela R. Birs ; Cosmin Darab ; Ovidiu Prodan ; Robin De Keyser
DESIGN OF PROGRAMMABLE, HIGH-FIDELITY HAPTIC PADDLE 540
Seyit Yigit Sizlayan ; Mustafa Mert Ankarali
MICROCONTROLLER-BASED MOTION CONTROL FOR DC MOTOR DRIVEN ROBOT LINK 547
Mustafa M. Mustafa ; Ibrahim Hamarash
**PATIENT-SPECIFIC IMAGINARY MOTOR MOVEMENT CLASSIFICATION OF EEG
SIGNALS AND CONTROL OF ROBOTIC ARM** 553
Özer Can Devecioglu ; Burak Yaman ; Özle Mesekoparan ; Can Çakir ; Türker Ince
**SOUTH AFRICAN POWER DISTRIBUTION NETWORK LOAD FORECASTING USING
HYBRID AI TECHNIQUES: ANFIS AND OP-ELM** 557
Sibonelo Motepe ; Ali N. Hasan ; Bhekisipho Twala ; Riaan Stopforth ; Nancy Alajarmeh
**SUPERCAPACITOR PARAMETER IDENTIFICATION USING GREY WOLF OPTIMIZATION
AND ITS COMPARISON TO CONVENTIONAL TRUST REGION REFLECTION
OPTIMIZATION** 563
Ravneel Prasad ; Utkal Mehta ; Kajal Kothari ; Maurizio Cirrincione ; Ali Mohammadi
**TUNING PID CONTROLLER USING HYBRID GENETIC ALGORITHM PARTICLE SWARM
OPTIMIZATION METHOD FOR AVR SYSTEM** 570
Faouzi Aboura
**USING DEEP LEARNING TECHNIQUES FOR SOUTH AFRICAN POWER DISTRIBUTION
NETWORKS LOAD FORECASTING** 575
Sibonelo Motepe ; Ali N. Hasan ; Bhekisipho Twala ; Riaan Stopforth

AUTOMOTIVE POWER CONVERSION – MOTORS, POWER ELECTRONICS, BATTERIES, AND CHARGERS

**DESIGN OF A CONTROLLER FOR TORSIONAL VIBRATIONS OF AN ELECTRIC VEHICLE
POWERTRAIN** 583
Mustafa Karamuk ; Salih Baris Ozturk
**DEVELOPMENT OF FUZZY LOGIC BASED ENERGY MANAGEMENT CONTROL
ALGORITHM FOR A PLUG-IN HEV WITH FIXED ROUTED** 590
Hazal Sölek ; Kenan Müderrisoglu ; Cem Armutlu ; Murat Yilmaz
**ELECTRIC MULTIPURPOSE VEHICLE POWER TAKE-OFF: OVERVIEW, LOAD CYCLES
AND ACTUATION VIA SYNCHRONOUS RELUCTANCE MACHINE** 596
Branko Ban ; Stjepan Stipetic
**STATE-OF-CHARGE ESTIMATION OF LI-ION BATTERY CELL USING SUPPORT VECTOR
REGRESSION AND GRADIENT BOOSTING TECHNIQUES** 604
Eymen Ipek ; M. Kerem Eren ; Murat Yilmaz
**TRACKING CONTROLLER DESIGN OF A RF MATCHING BOX WITH PLASMA LOAD
VARYING** 610
Yen-Fang Li ; Ming-Heng Hsieh ; Ren-Sian Liou
Author Index

Proceedings

2019 International Aegean Conference on Electrical Machines and Power Electronics (ACEMP) & 2019 International Conference on Optimization of Electrical and Electronic Equipment (OPTIM)

BAHÇEŞEHİR UNIVERSITY
Istanbul, Turkey
27 - 29 August, 2019

Organized by

Middle East Technical University
Atilim University
Bahcesehir
University
Transilvania University of Brasov
University Politehnica of Timisoara
Technical
University Cluj-Napoca

Sponsored by

Institute of Electrical and Electronics Engineers (IEEE)
IEEE Industrial Electronics
Society (IES)

Co-sponsored by

IEEE IAS
IEEE PES

978-1-5386-7688-2/19 $31.00 © 2019 IEEE

978-1-5386-7688-2/19 $31.00 © 2019 IEEE

Plenary Sessions

978-1-5386-7688-2/19 $31.00 © 2019 IEEE

ACEMP-OPTIM 2019 Opening Speech

H. Bülent Ertan,
Atılım University and Middle East Technical University of Ankara, Turkey

Ladies and gentlemen

Welcome to the 2019 ACEMP-OPTIM Joint International Conference: Aegean Conference on Electrical Machines and Power Electronics (ACEMP) and Optimization of Electrical & Electronic Equipment (OPTIM). These two conferences which date back to the early 1990s are jointly organized since 2015, with a view to increasing the quality, participation, international recognition and reputation of both conferences.

The third edition of the conference is organized by Atılım University and Middle East Technical University of Ankara and is hosted by Bahçeşehir University of Istanbul, between 27-29 August.

The conference is technically co-sponsored by the Institute of Electrical and Electronics Engineers (IEEE) through three of its societies: Industrial Applications (IAS), Industrial Electronics (IES) and Power Electronics (PELS). I would like to thank Professors Ion Boldea, Marcian Cirstea and Mihai Cernat, who helped us with the IEEE arrangements.

We are holding this meeting in Istanbul at a beautiful location by the Bosphorus, at the campus of Bahçeşehir University. Thanks go to Bahçeşehir University administration for offering us to use their facilities.

I am glad to inform you that we have participants coming from 5 continents and 27 countries. During this conference we shall have five plenary sessions and 15 oral sessions. All 94 papers will be presented orally. In the usual tradition of ACEMP and OPTIM conferences, in the plenary sessions we have presentations by prominent scientists. In the last name order, we shall have the pleasure of listening to the following keynote speeches:

Pavol Bauer	TU Delft, Netherlands	The Future of EV Charging
Ion Boldea	University Politehnica of Timisoara, Romania	Renewable energy (hydro and wind) large power systems: a revisit in 2019
Masato Enokizono	Oita University, Japan	Required Technology for Upgrading Efficiency of High-Speed Motor with High Power Density
Norbert Hanigovszki	Danfoss A/S, Denmark	Developments and trends in the adjustable speed drive industry
Markus Henke	Braunschweig Technical University, Germany	Design of electrical high-speed drives for vehicle drivetrains
Thomas M. Jahns	University of Wisconsin-Madison, USA	Electrified Aviation: Are Motor Drives Ready to Spread Their Wings and Fly?
Mustafa Karamuk	Ford-Otosan, Turkey	Electrified Powertrain Engineering in Ford-Otosan: Review of Electric Vehicle Powertrain Technologies with OEM Perspective
Gianmario Pellegrino	Politecnico di Torino, Italy	Synchronous Reluctance Motor Drives: Still a niche Technology?
Bülent Şarlıoglu	University of Wisconsin-Madison, USA.	Automotive Power Electronics: Current Status and Future Trends

We gratefully acknowledge the contribution of the keynote speakers to this meeting.

Yesterday we had three tutorials presented by authors coming from industry. These were:

Vladimir Blasko, United Technologies Research Center USA
State of the Art and Practical Aspects of the Development of Power Converters and Control for Electrical Drives

Reza Rajabi Moghaddam, ABB Sweden
Eddy current in electrical machines and stray losses, an introduction to investigation methods

Mircea Popescu and David Stanton, Motor Design Ltd. UK
Practical design aspects for electrical machines in power traction applications for Electrical Vehicles

I would like to thank them, for their effort and generously contributing to ACEMP-OPTIM 2019.

I would like to bring to your attention that we have a special session for two of our colleagues who passed away recently, Prof. Lazslo Szentirmai from the University of Miskolc, Hungary and Prof. Robert D. Lorenz from the University of Wisconsin-Madison, USA. We acknowledge their great contribution to science and to our community.

I would like to note that there will be a boat tour along the historic and beautiful shores of Bosphorous this evening after the last session today. Gala dinner is on 28th August; with music and dances from Turkey.

I would like to finish my speech by thanking our students who helped with the organization and many others who contributed to the organization of this event. Special thanks go to Prof. Mihai Cernat, Prof. Marcian Cirstea, Prof. Ion Boldea and Prof. Bülent Sarlioglu for their help in organizing this conference. And of course greatest thanks go to you, participants, for coming here and sharing your ideas during this meeting.

I hope we all will depart home with new ideas and motivation for furthering our research and studies.

Conference Chairs: H. Bülent ERTAN
Conference Vice-Chairs: Marcian CIRSTEA and Ion BOLDEA
Conference Program Chair: Mihai CERNAT and Carmen GERIGAN
Conference Secretary: Ozan KEYSAN

Local Organizing Committee :

Ali Güngör
Bahçeşehir University, Istanbul, Turkey
Bülent Şarlıoğlu
University of Wisconsin-Madison, USA
Erhan Akın
Fırat University, Elazığ , Turkey
Emine Bostancı
Middle East Technical University, Ankara, Turkey
Ercan Ertürk
Bahçeşehir University, Istanbul, Turkey
Eyüp Akpınar
Dokuz Eylül University, Izmir, Turkey
Lale Tükenmez Ergene
İstanbul Technical University, Turkey
Levent Eren
İzmir University of Economics, Turkey
Murat Barut
Niğde Halis Demir University, Nevşehir,Turkey
Muammer Ermiş
Middle East Technical University, Ankara, Turkey
Ozan Keysan
Middle East Technical University, Ankara, Turkey
Sedat Sunter
Fırat University, Elazığ , Turkey
Timur Aydemir
Gazi University, Ankara,Turkey

Future of EV Charging

Prof.Dr. eng. Pavol Bauer, Delft University of Technology, The Netherlands

Abstract

Charging infrastructure for electric vehicles will be the key factor for ensuring a smooth transition to e-mobility. It is here, that five technologies will play a vital role in the EV charging infrastructure: smart charging (including vehicle-to-grid V2G technology), charging of EVs from photovoltaic panels, (ultra)fast charging, contactless charging and on-road charging of EVs. With the use of smart charging, the EV charging power and direction can be continuously controlled. Smart charging of EVs can provide several benefits to the EV owner and to the providers of the EV charging infrastructure like reduced peak demand on the grid and reduced cost.

In order to ensure that the use of electric vehicles results in net zero CO_2 emissions, it is important that the charging infrastructure derives all/majority of its power from renewable energy sources. It is here that the falling costs of photovoltaic panels (PV) over the years and the ease of integrating into the distribution network play a key role. Workplaces like office buildings and industrial areas are ideal to facilitate solar EV charging where the rooftops and car parks can be installed with PV panels. There are several additional advantages of charging EVs from photovoltaic panels: EV battery can be used as an energy storage for the PV and reduced energy and peak power demand on the grid as the EV charging power is locally generated from PV.

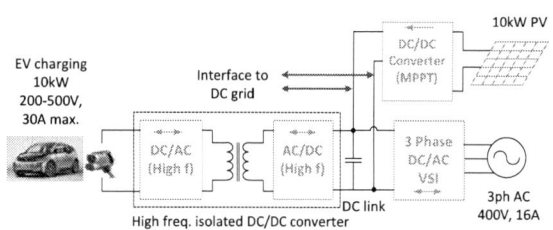 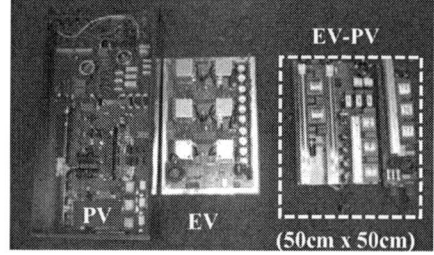

Figure 1:Left: Topology of power converter; Right: 10kW prototype of SiC based converter developed with PRE (Power Research Electronics) compared to a conventional PV inverter and EV charger of 10kW

With respect to (ultra)fast charging, new EVs are designed to withstand high power and for the EU market new standards are being developed with 350 kW charging. Research on what is the fast charger architecture and power electronic components which gives the most competitive advantage considering, the product development (cost, manufacturing, operability, compactness, power efficiency, etc) is conducted. New research question is how to maximize the utilization of installed power of the EV charger. Therefore a concept of a multiport, flexible and intelligent fast charger which features multiple output charging spots through the implementation of multiplexing techniques, scheduling and simultaneous charging is developed (see Figure 2).

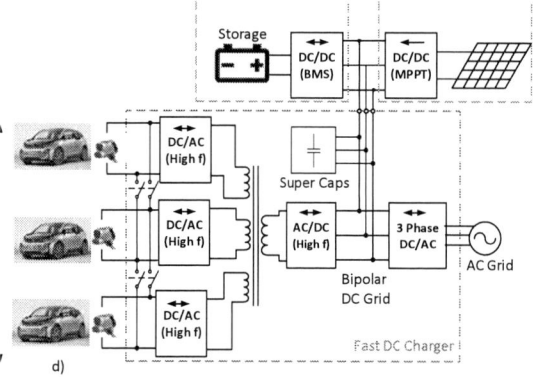

Figure 2:Modular multiport fast DC charger for simultaneous charging of multiple EVs

Contactless charging of EVs using Inductive Power Transfer (IPT) and on road charging is a technology that is increasingly becoming acceptable as an important feature of autonomous charging and key element enabling autonomous driving of EVs. This technology uses electromagnetic energy transfer between loosely coupled charge-pads which are placed with an air-gap in between. A block diagram of such a system is shown in Figure 3.

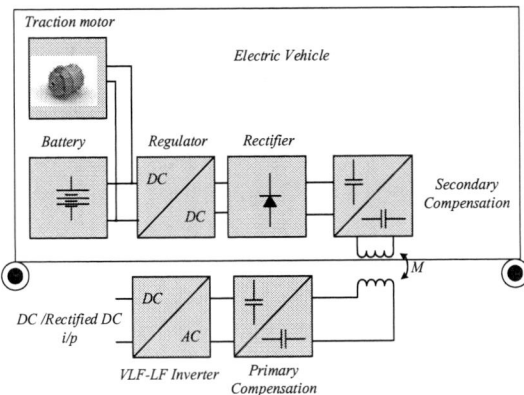

Figure 3: Block diagram of an EV IPT based system highlighting the various power conversion stages.

Prof.Dr. eng. Pavol Bauer

Delft University of Technology

Pavol Bauer is currently a full Professor with the Department of Electrical Sustainable Energy of Delft University of Technology and head of DC Systems, Energy Conversion and Storage group. He received Masters in Electrical Engineering at the Technical University of Kosice ('85), Ph.D. from Delft University of Technology ('95) and title prof. from the president of Czech Republic at the Brno University of Technology (2008) and Delft University of Technology (2016). He is also honorary professor at Politehnica University Timisioira in Romania. From 2002 to 2003 he was working partially at KEMA (DNV GL, Arnhem) on different projects related to power electronics applications in power systems. He published over 100 journal and over 350 conference papers in his field (with H factor Google scholar 37, Web of science 24), he is an author or co-author of 8 books, holds 6 international patents and organized several tutorials at the international conferences. He has worked on many projects for industry concerning wind and wave energy, power electronic applications for power systems such as Smarttrafo; HVDC systems, projects for smart cities such as PV charging of electric vehicles, PV and storage integration, contactless charging; and he participated in several Leonardo da Vinci and H2020 EU projects as project partner (ELINA, INETELE, E-Pragmatic, Smart Charging, Metrology for Inductive Charging, Trolley 2.0) and coordinator (PEMCWebLab.com-Edipe, SustEner, Eranet DCMICRO). He is a Senior Member of the IEEE ('97), former chairman of Benelux IEEE Joint Industry Applications Society, Power Electronics and Power Engineering Society chapter, chairman of the Power Electronics and Motion Control (PEMC) council, member of the Executive Committee of European Power Electronics Association (EPE) and also member of international steering committee at numerous conferences.

978-1-5386-7688-2/19 $31.00 © 2019 IEEE

In Memoriam - Professor Robert D. Lorenz

Robert Donald "Bob" Lorenz, distinguished IAS member and IEEE Life Fellow, passed away on January 27, at age 72. He received the B.S., M.S., and Ph.D. degrees from the University of Wisconsin, Madison, and the M.B.A. degree from the University of Rochester, Rochester, NY. From 1969 to 1970, he did his Master thesis research in adaptive control of machine tools at the Technical University of Aachen. From 1972 to 1982, he was a member of the research staff with the Gleason Works, Rochester, NY, working on high-performance drives and synchronized motion control. Since 1984 he was with the faculty with the University of Wisconsin, Madison, where in 1999 he was named the Mead Witter Foundation Consolidated Papers Professor of Controls Engineering in both the Department of Mechanical Engineering and the Department of Electrical and Computer Engineering. In 2016, he was named Elmer R. and Janet A. Kaiser Chair. Prof. Lorenz was Codirector of the Wisconsin Electric Machines and Power Electronics Consortium, the world's largest industrial research consortium on motor drives and power electronics. He was the Research Leader for control and sensor integration and integrated modular motor drives with the National Science Foundation Engineering Research Center for Power Electronic Systems. Prof. Lorenz was the Industry Applications Society (IAS) President in 2001, an IAS Distinguished Lecturer in 2000/2001, the IEEE Division II Director in 2005/2006, past Chair of the IAS Awards Department, and past Chair of the IAS Industrial Drives Committee. In addition to his distinguished service to IAS, he was active in the IEEE Power Electronics Society, the IEEE Sensor Council, the American Society of Mechanical Engineers, the Instrument Society of America, the International Society for Optical Engineers, and the European Power Electronics (EPE) Committee. He authored over 300 published technical papers and held 26 patents. His awards included the IAS Outstanding Achievement Award, the EPE Power Electronics and Motion Control Outstanding Achievement Award, the IAS Distinguished Service Award, the IEEE Richard Harold Kaufmann Technical Field Award, the EPE Outstanding Achievement Award, and 34 IEEE prize paper awards.

Robert D. Lorenz

Prof. Robert (Bob) Lorenz passed away on Sunday, Jan. 27, 2019.

Bob was a pioneer of self-sensing control of electrical drives, an unmatched example of technical excellence and a friend of us.

He was keynote speaker at SLED past editions many times, lastly in 2017 in Catania, with an energizing lecture entitle **The Path to Widespreead Use of Self-Sensing**. At the time he passed away, he was a member of the Steering Committee of SLED 2019.

SLED 2019 will remember Prof. Lorenz through special dedicated moments, including a dedicated Special Session.

More information about Bob's life and legacy can be found on WEMPEC and UW Madison dedicated webpages. Other details are on Prof. Lorenz's Caring Bridge site.

https://wempec.wisc.edu/in-memory-of-prof-bob-lorenz/

In Memory of Prof. Bob Lorenz

1946 – 2019

Delivered at the Celebration of Life Service at Covenant Presbyterian Church

Madison, WI, February 9, 2019

by Prof. Tom Jahns

How does one find the right words to capture the essence of Bob Lorenz? *"Special"*, *"unique"*, and *"one-of-a-kind"* are certainly among the terms that come to mind, but they don't come anywhere close to capturing the breadth and depth of the very singular man that all of us knew. I have had the privilege of knowing Bob for almost all of my professional life. I first met him over 35 years ago soon after he had joined the mechanical engineering faculty here at UW-Madison. He had become actively involved in the then-new university-industry consortium named the Wisconsin Electric Machines and Power Electronics Consortium (WEMPEC for short). I was working in industry at the time, but it didn't take long for me to tell that Bob was no ordinary young faculty member, and that he was destined to leave a special mark in our field, which is exactly what he did.

My close association with Bob began 20 years ago when I left industry and joined the electrical engineering faculty here at UW and I, too, became heavily involved in the WEMPEC Consortium. I have had the special privilege of working very closely with Bob as a fellow co-director of WEMPEC for the past 12 years during which we developed a strong technical partnership and a close personal friendship. Bob was absolutely and completely dedicated to WEMPEC and I, like all of the other WEMPEC-affiliated faculty members, were in awe of the energy and diligence that Bob invested in maintaining and strengthening WEMPEC's international reputation as one of the strongest university programs in our field anywhere in the world. Bob and WEMPEC were almost synonymous.

Rather than trying to describe Bob's many technical accomplishments, I would like to share my observations on Bob's role and reputation at the university and, more specifically, in WEMPEC. Bob was passionate about his guiding principle that WEMPEC is more than a center of technical excellence; WEMPEC is a *family* whose members really care for each other. Bob was always the one to remind us of this fact and to set the example for the rest of us. Accordingly, he reveled in the large numbers of US and international students that WEMPEC attracted to Madison from the four corners of the earth, and he went to great lengths to make certain that they all felt welcome and appreciated as members of our special community, even if they were thousands of miles from their homes.

Bob always had a large group of graduate students that he supervised, often numbering 25 or more, which is about 4 times higher than the average for a faculty member in our College of Engineering. Despite this heavy load, Bob always found the time to get to know each of his students and their families on a personal basis. The annual holiday parties that Bob and Sally held every December were legendary and filled their home to the brim with laughter and holiday cheer. Despite his failing health, Bob, with Sally's help, managed to hold a modified version of his holiday party as recently as two months ago, a fitting tribute to the depth of commitment he had to his students. In fact, as Sally knows, it was only the onset of a major snowstorm that prevented Bob from attending and leading a PhD preliminary exam for one of his grad students only 4 days before he left this world. That was classic Bob right to the end.

In the classroom, Bob was a natural-born proselytizer. If there was an undergraduate sitting in the back of the room staring at his cell because he really didn't want to be taking his class, Bob took it as his *mission* and *personal challenge* to convince that student that electric machines and controls were the most important technologies ever invented! He didn't always succeed, but sometime he did. It's a fact that some of his best graduate students were those he first identified and nurtured when they were taking his class. Bob was similarly single-minded and persuasive when faced with the challenge of convincing skeptical sponsors about the importance and potential of his latest technical breakthroughs. Here again, he didn't always succeed, but let it be said that Bob's track record for winning over the skeptics was impressive to the rest of us WEMPEC faculty members by any standards.

Even though I am purposely not dwelling on Bob's specific academic accomplishments, I want to make it clear that Bob had an incredibly productive career as an engineering scholar and innovator. He published over 400 papers, supervised more than 200 graduate students, and was granted more than 40 patents. Lest there be any doubt about Bob's stellar academic credentials, I am pleased to report that just two days ago it was announced that Bob has been elected to the US National Academy of Engineering, one of the highest honors that an engineer can receive in the United States. Of course, it is sad and bittersweet that this very special honor was announced only 11 days after Bob left us, but I am still very pleased and touched that Bob's legacy has been recognized so publicly, even if he can't be here to enjoy the recognition he so richly deserves for his accomplishments.

Yet, despite the impressiveness of his technical achievements, I feel certain that Bob, if he were still with us, would take even greater pride and satisfaction in the number of lives that he influenced and changed for the better during his time here on earth. During the past two weeks, I have had the privilege of receiving e-mail messages from many of his former students and professional colleagues who have borne testimony to the tremendous impact that Bob had on their lives. *"Inspiring"*, *"dynamic"*, *"enthusiastic"*, *"passionate"*, and *"life-changing"* were among the phrases that appeared repeatedly in their heartfelt messages. Although Bob can't be here to bask in the touching warmth of the spontaneous tributes and memorials that appeared in these messages, all of us here in this room, and particularly those of us who are educators, can and should be inspired by the shining example that Bob set for us throughout his highly active career in academia.

I can attest that Bob had a major impact on *my* life, inspiring me to strive to be a better teacher, a better research supervisor, and a better person. I will always miss Bob, but his memory and his legacy will live on with me for the rest of my life, as it will with so many of you in the audience and around the world. So, thank you, Bob, for sharing your uniquely productive and inspiring life so generously with all of us; yours was truly a life well-lived to its fullest. Now, my friend, rest in peace. You succeeded in your life-long quest to make the world a better place by your being here. Now it's for us, the living, to find our own way to do the same.

In Memoriam Prof. László Szentirmai

László Szentirmai was born in Nagykanizsa, Hungary on February 15, 1930. After completing high school with honors, he graduated-at the Budapest University of Technology with a major in electrical power engineering in 1952. He obtained his PhD degree (at that time it was called Candidate of technical sciences) in 1976. In 1977 he became Doctor of technical sciences at Budapest University of Technology.

Next, to his main profession of an engineer at Ganz Electricity Works and the Mining Design Institute, he was a lecturer at Technical High School. As a recognized expert, professor Szentirmai contributed to establishing a new Technical University in Algeria (Algeria 1970-72). During multiple visits in Egypt in the period between 1968 and 1973 he helped developing research and education programs. As a leading consultant for UNESCO, he assisted in setting up a new technical university and educated lecturers in Iran (1974-77 and 1991-92).

From 1979 to 1995 he was a professor at the University of Miskolc and head of the Department of Electrical Engineering. In 2000 he became professor emeritus. He coordinated 6 projects of European Union in microelectronics, computer science, electric drives, energy technology, mechatronics. Additionally, Laszlo coordinated PhD programs in English and technology transfer science with 15 technical universities from 10 different countries (between 1991-2002). Furthermore, he was visiting Professor, session chairman at International Conferences, and invited and guest lecturer in many countries on four continents. László Szentirmai was also head of the Automation and Informatics Committee of the Miskolc Academic Committee of the Hungarian Academy of Science (1985-1997), Chairman of the Editorial Board of the Journal of Electrical Engineering (1995-2010), Member of the Board of Directors of the European Society of Engineering Training (SEFI) (from 1990). He was also chair of a dissertation opponent committee and examiner of numerous PhDs. He wrote more than 200 International publications such as books, university notes, and papers in English, in the fields of electrical machinery,

978-1-5386-7688-2/19 $31.00 © 2019 IEEE

drives, metrology, and renewable energy. László Szentirmai is a recipient of Gold Medal at the World Exhibition for Research and Industrial Innovation for the New Development of Renewable Energy Sources (Brussels, 1994); Honorary Medal of the Signum Aureum Universitatis (1993) and Albert Szent-Györgyi Prize (1997)

Electrified Aviation: Are Motor Drives Ready to Spread Their Wings and Fly?

Thomas M. Jahns

Grainger Professor of Power Electronics and Electric Machines, University of Wisconsin – Madison
t.jahns@ieee.org

Abstract

Aviation is in the early stages of a technological revolution that is leading to progressively increasing levels of electrification of the propulsion systems for aircraft of all types and sizes. This trend is being driven by the increasingly urgent need to dramatically reduce the greenhouse gas emissions of commercial aircraft that are currently responsible for 2.0 to 2.5% of total global annual CO_2 emissions. State-of-the-art commercial aircraft such as the Boeing 787 and Airbus A380 have taken important initial steps towards achieving significant efficiency improvements by installing much higher power electrical systems to displace conventional hydraulic and pneumatic secondary power systems and to electrify the climate control system (i.e., bleed air elimination).

Longer-term success of this electrification initiative depends on the ability of future aircraft propulsion systems to dramatically reduce their dependence on fossil fuels by displacing conventional jet engines with lightweight electrical machines

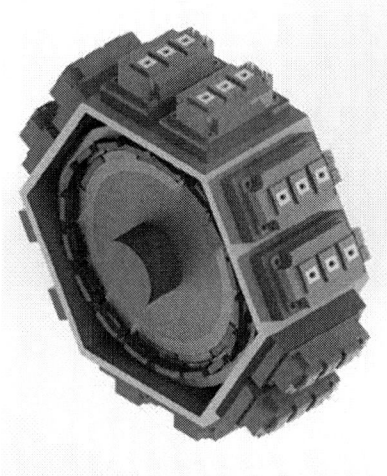

Integrated machine drive concept

and inverters that draw their energy from batteries and other types of advanced electrical energy storage devices. Similar to the history of land-based electric vehicles, early effort is focused on the development of partially-electrified aircraft propulsion systems. More specifically, there is high interest in the development of hybrid propulsion configurations combining electric machines and smaller optimized jet engines that are designed to operate at significantly higher efficiencies, lowering the required jet fuel consumption.

There are also major development activities under way around the world that are pushing the limits of all-electric propulsion systems for small aircraft. Rapidly growing interest in urban air mobility aircraft (i.e., air taxis) is helping to accelerate this technology thrust. High-performance demonstration aircraft such as the battery-powered two-passenger Siemens 330LE have attracted considerable international attention to what can be achieved within the limits of currently-available technology. Researchers in industry and academia are competing to raise the specific power density values of electric machines to values exceeding 15 kW/kg (including housing). Serious development efforts are also under way to combine electric machines and power electronics into tightly integrated drive systems sharing the same housing.

Major technology advances are critical to achieving the ambitious long-term objectives of this electrified aviation vision. First and foremost, breakthroughs in electrical energy storage technology are key to achieving dramatic improvements in energy density that will enable progressively longer ranges of all-electric aircraft. Wide-bandgap power semiconductors and additive manufacturing are additional examples of rapidly-evolving technologies that have much to contribute to reducing the mass and volume of future electric aircraft propulsion systems. In addition, major improvements in the reliability and fault tolerance of future electric drive systems will be essential to meeting the demanding requirements of future aircraft.

Thomas M. Jahns

Biography

Prof. Thomas M. Jahns received his bachelors, masters, and doctoral degrees from MIT, all in electrical engineering.

Dr. Jahns joined the faculty of the University of Wisconsin-Madison in 1998 as a Grainger Professor of Power Electronics and Electric Machines in the Department of Electrical and Computer Engineering. He is the Director of the Wisconsin Electric Machines and Power Electronics Consortium (WEMPEC), a university/industry consortium with over 85 international sponsors.

Prior to coming to UW-Madison, Dr. Jahns worked at GE Corporate Research and Development in upstate New York for 15 years, where he pursued new power electronics and motor drive technology in a variety of research and management positions. His current research interests at UW-Madison include integrated electric machines, electric vehicle propulsion, and distributed energy systems.

Dr. Jahns is a Fellow of IEEE. He received the 2005 IEEE Nikola Tesla Technical Field Award "for pioneering contributions to the design and application of AC permanent magnet machines". Dr. Jahns is a Past President of the IEEE Power Electronics Society and the recipient of the 2011 Outstanding Achievement Award presented by the IEEE Industry Applications Society. He was elected to the US National Academy of Engineering in 2015.

Review of Electric Vehicle Powertrain Technologies with OEM Perspective

Mustafa Karamuk
Advanced Powertrain Department
Ford Otosan
Istanbul, Turkey
mkaramuk@ford.com.tr

Abstract— Abstract— **Electric vehicle technologies have rapid growth by the development of the multitude solutions at component and subsystem level. Compared to technology level in year 2010, automotive technologies have entered a new era similar to industry 4.0. On the other hand, understanding the benefits and trade-offs of the new technologies have crucial importance for the development and life cycle of the electric vehicle. Beside the development challenges, automotive manufactures need to manage the complexity of the solutions and find the optimal technologies that fulfil the requirements for the target market vs. development cost. In this study, the recent technologies and system integration issues of electric vehicles are reviewed. Electric vehicle control unit and application software modules are reviewed with respect to vehicle performance. System level problems including the functional safety, motor insulation aging, bearing currents and electric powertrain vibrations are reviewed with application aspects. Electric axle topologies and trends in non-rare earth and reduced rare earth motors are also discussed. Technology trends in traction inverters and motors are compared with trends in industrial inverters and motors. Future technology trends in automotive inverters and motors are estimated. Importance of post-crash diagnostic analysis of electric vehicle components is also highlighted for aftersales services of electric vehicles.**

Keywords—electric powertrain, technology trends, vehicle control unit, electric vehicle system level problems, electric axles, trends in traction motors and inverters, after sales.

I. INTRODUCTION

Electric vehicle (EV) market has rapid momentum since 2010 [1-2]. Despite the incentives and developed new technologies, the potential customers of EV have still hesitations on driving range, cost, lack of charging infrastructures and safety of EV [1]. Meanwhile, EV technologies provide multiple solutions at each subsystem. Some recent technological improvements can be listed such as higher efficiency power electronic units and motors, increase of specific power of traction motors, increase of range by high density batteries, improved thermal management and range at cold temperatures by waste heat recovery systems, compact and integrated electric axle solutions. Considering the multitude solutions, development of an EV for an automotive manufacturers (Original Equipment Manufacturer-OEM) also becomes a challenge. Because, the optimization between the target market and development cost of the required components, evaluation of the technical benefits and trade-offs of the available solutions require multi-parameter optimization.

Power sizing of an electric motors and inverter require a duty cycle to characterize the electrical and thermal load of the motor as function of its mechanical load. Drive cycles, as being the counterpart of industrial term "duty cycle", enables the characterization of mechanical load and hence definition of electrical and thermal limits of electrical components such as high voltage (HV) battery, traction and inverter and motor. However, drive cycle is specific to applications. Therefore, the target drive cycles are needed for the right sizing of the HV battery, traction motor and inverter.

In order to select and size the components, definition of system topologies and subsystems should be defined first as below:

- Definition of vehicle performance metrics such as acceleration time, maximum uphill driving, estimated range.

- Powertrain topology and suitable motor topology in terms of dimensions and power. Powertrain topology includes the number of the drive axles and motors for each axle.

- Thermal system that fulfils the thermal requirements of the electric powertrain components.

- HV system topology and safety concepts.

- CAN network topology.

- Vehicle control unit (VCU) software architecture, control algorithms and functional safety concepts.

EV has more system level and multidisciplinary problems than combustion engine vehicles. Because the control signal interfaces of HV components with VCU, powertrain topologies and their control problems, interactions between thermal, electrical and mechanical power loops, grid integration of charging stations and power quality issues increases the system engineering effort of EVs. Realization of differences between EV and combustion engine vehicles are essential to the success of EV projects.

II. ELECTRIC VEHICLE MARKET IN A FEEDBACK LOOP

In order to solve the range problem, multitude of solutions are developed to increase the system efficiency and range, reduce the dimensions of the components. As illustrated in Fig.1, EV market can be modelled in terms of a control loop where the reference is generated by competition and trend makers mostly [3].

While the technology progress into direction of compact, high power density and efficient components, the developed solutions need more application feedback such as variation of range, thermal management performance at different weather conditions, drive cycles, and durability performance at high vibration levels etc. Integration of every new technology bring new engineering problems as will be discussed in Section IV. As the feedback loop returns new results from different applications, these technologies need to be improved further.

978-1-5386-7688-2/19 $31.00 © 2019 IEEE

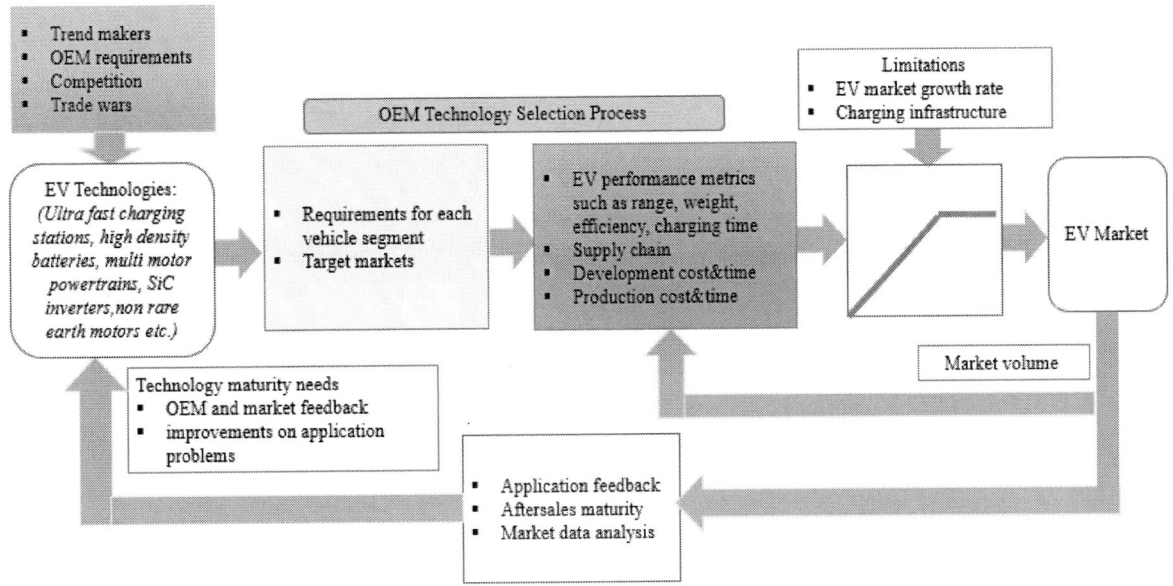

Fig. 1. EV technologies and market in a feedback loop.

TABLE I.   SUMMARY OF TRENDS IN ELECTRIC VEHICLE TECHNOLOGY [4-17]

Efficiency, performance, diagnostics, charging infrastructure	EV Technologies
Component technologies	• High speed (n > 10.000 rpm) radial flux electric motors to increase specific power. • Axial flux motors to increase specific power and enable compact sizes for electric axle topologies. • Improvements in non-rare earth motors to reduce material cost and avoid supply problems of rare earth elements [4-5]. • High efficiency SiC MOSFET inverters (up to 98-99 % inverter efficiency) [6-7]. • Multifunctional power electronic components (charger and inverter) to reduce component sizes and cost [8]. • Development of high energy battery chemistries like lithium air and solid state batteries to have higher range [9].
Energy and thermal management	• Improved energy management by route based algorithms [10]. • Improved thermal management by waste heat recovery systems [11].
Electric powertrain topologies and control	• Electric axle topologies to reduce powertrain sizes and speed up system integration [12]. • Dual axle driven topologies to reduce motor sizes and optimize powertrain efficiency [13]. • 2-speed automatic transmission to increase powertrain efficiency and downsize traction motor and inverter [14]. • Torque vectoring for improved lateral vehicle dynamics [15].
Service and maintenance	• Remote software updates for vehicle control units by over the air (OTA) technologies [16].
Grid technologies	• Smart grid, Vehicle to Grid (V2G) technologies to stabilize and support the grid for charging [17].

EV technologies will be subject to evolution in terms of cost, efficiency, charging infrastructure, life time, durability and aftersales issues. The recent EV technologies are summarized in Table I [4-17].

The first generation of EVs was developed as the conversion of combustion engine vehicles into battery electric vehicles. The powertrain topology consists of a single motor installed between the driven wheels and a single-speed transmission including a differential. This topology is known as central drive topology. In order to improve the vehicle's lateral and longitudinal dynamics, individually controlled electric powertrains, in-wheel motors, multi-motor topologies up to four electric motors are being developed. Topologies have different performance in terms of vehicle dynamics behaviour and energy efficiency [15].

EV applications can be divided into three categories in terms of vehicle segment, traction power and HV range [18, 19]:

1) Passenger car in B segment: 300-500 V DC battery voltage range, up to 120 kW peak power range.

2) Passenger car in C and D segment, light commercial electric vehicles, medium duty trucks: 400-600 V DC battery voltage range, up to 200-230 kW peak power range.

3) Heavy duty (HD) electric vehicles in electric bus and truck segment: 600-800 V DC battery voltage range, up to 300-350 kW peak power range.

978-1-5386-7688-2/19 $31.00 © 2019 IEEE

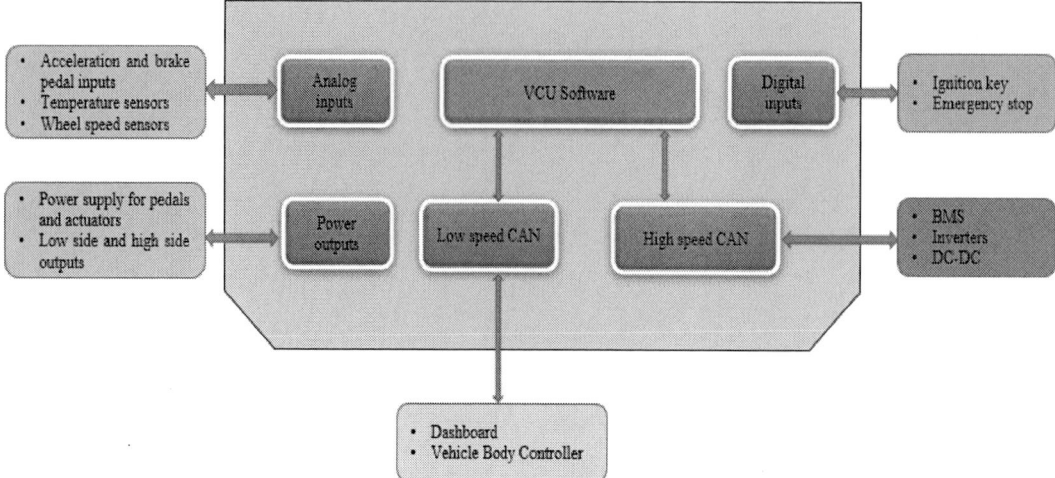

Fig. 2. Generic VCU interface of an EV [22].

The ratio between the peak to continuous torque is typically 2:1. There is also growing trend in light electric vehicle applications supplied by 48 V battery [20, 21]. The advantage of 48 V application is that it is a safer voltage level and easier to implement as it is below 60 V DC. Each of these applications has different requirements, design constraints in terms of vehicle segment, target market, vehicle performance metrics such as acceleration and maximum uphill driving capability, maximum weight, dimensions of components, component layout etc. To fulfil these requirements, EV component and system technologies are being developed as given in Table I.

III. ELECTRIC VEHICLE PERFORMANCE AND VEHICLE CONTROL UNIT

Electric powertrain efficiency is dependent on the efficiencies of traction motor, inverter and transmission. But, the system efficiency is also dependent on the application control algorithms of VCU, as they set the operating points and the system limits. In this section, VCU functions are reviewed briefly. Input-output interfaces of a generic VCU is shown in Fig. 2 [22]. The most important modules and their functions are listed below [22, 23]:

1) *State Machine (Software State Controller)*
2) *Energy Management*
3) *Torque Management*
4) *Thermal Management*
5) *Diagnostic Management*
6) *Safety Management*

State machine module defines the states of the EV e.g. charge, drive, normal shutdown or emergency shutdown, and ensures that the vehicle subsystems and components are activated or de-activated in that state safely. It is a supervisory module that coordinates the sequences of the software and EV system functions [22, 23].

A. Functions of VCU

An engine controller unit (ECU) and VCU have similar hardware whereas the functions are completely different. Summary of the VCU functions are listed below:

1) VCU is a system level controller: Unlike ECU, VCU is not the direct controller of the electrical motor. It produces the torque reference based on the vehicle driver's torque request. It monitors the voltage, current and temperature limits of HV battery via battery management system (BMS), traction motor and inverter. In other words, it does not have a closed loop current and torque controller of traction motor as this task is performed by the traction inverter. However, powertrain controllers like torque vectoring is implemented in VCU as it outputs different torque reference for each traction inverter based on vehicle lateral dynamics [15].

2) VCU coordinates the control inputs and outputs of multiple components: VCU has interfaces over CAN network with all subsystems such as BMS, electric powertrain, charger system and instrument cluster of electric vehicle.

3) VCU is the supervisory diagnostic manager: It evaluates the drive enable or shutdown condition by interpreting the diagnostic messages of all subsystems.

4) VCU performs the functional safety tasks: VCU is responsible for monitoring of safety related states of EV and performs the defined tasks based on hazard analysis and risks assessment (HARA) [24].

B. The Relationship between VCU, EV Range and Efficiency

VCU application modules have impact on EV range as they determine the system operating points.

Considering the efficiency of an EV from HV system to the wheels, there are two level of efficiency:

1) Component level efficiencies: Traction inverter and motor efficiencies, efficiencies of auxiliary inverters and motors, transmission efficiency.

2) System level efficiencies: Control algorithms of VCU determine the operating points of the electrical loads on the HV battery. In that sense, energy, torque and thermal management system have impact on component and system efficiency.

C. Energy Management

For an optimum energy management in an EV, VCU needs to know the drive cycle and grade information. In that way, energy management module (EMM) in VCU can predict the required energy and distribute the available HV battery energy between the auxiliary loads and traction system. By using the state of charge (SOC) of the HV battery, road traffic and route through global positioning system (GPS) system and grade information, EMM can predict an optimum acceleration and economy mode driving for energy saving [10, 25-26]. In [25], a real time drive cycle and road grade prediction method is proposed for EMM.

In case of low SOC, an EMM can minimize the power consumption of auxiliary loads and enable higher regenerative power depending on the regenerative brake method. In case of high SOC and downhill driving condition, EMM can maximize the consumption of auxiliary loads to support the brake resistor operation to waste more regenerative power [27].

D. Torque Management

Torque management produces the torque reference based on the torque request of the vehicle driver and limits the torque reference within battery and motor-inverter system limits. Electric motor torque reference should be limited and filtered to improve driveability and reduce drivetrain vibrations.

VCU sends single torque reference to traction inverter. Depending on the driving states of EV, torque reference has the following operation modes. Torque Management module should produce the torque reference considering these operation modes:

1) Acceleration torque: Acceleration torque is mostly peak torque of EM (electrical motor) which is also limited by motor thermal condition and battery voltage.

2) Regenerative braking torque: The ratio of the total braking torque is distributed into the mechanical brake torque and regenerative brake by brake management module. Series or parallel braking methods are applied in EV applications [28].

3) Coast down braking torque: As in combustion engine vehicles, coast down effect can be generated by VCU torque management. Based on vehicle's longitudinal dynamics model, and coast down deceleration profile, regenerative torque reference can be set in coast down mode. Implementation examples are given in [28-29].

E. Thermal Management

Thermal Management of a VCU controls the temperature reference of the cooling and heating loops and keeps the temperature of HV battery, traction and auxiliary motors and power electronic units within specified temperature limits. Typically all motors and power electronic units require upper temperature limits at 60-65 °C level whereas a HV battery requires the upper limit around 35 to 40 °C. HV battery also requires heating at minus temperatures and therefore needs a heating loop. Including the cabin heating, there are three thermal loops in an EV [30]:

- HV battery heating and cooling loop.

- Cabin air conditioning loop.

- Cooling loops for power electronic units such as DC-DC converter, charger, traction and auxiliary system inverters and motors.

Range of EV can be reduced by thermal subsystem loads such as HV battery and cabin heating by as much as 45 % below −10 °C of ambient temperatures. Heat pump can recover the waste heat back to the battery and it can improve the range at cold temperatures. [30].

Definition of the coolant temperature and flow rate is critical for available torque of the traction motor. Peak and continuous torque of the traction motor vary depending on its thermal limits. Traction inverter is also subject to thermal cycling and therefore it can be in power de-rating in case of overheating.

Energy Management and Thermal Management have direct impact on EV range and vehicle performance.

IV. SYSTEM LEVEL PROBLEMS OF ELECTRIC VEHICLES

An EV can be modelled in such a way that the HV battery is a variable and limited energy source supplying electrical loads that have thermal and electromechanical energy conversion loops. Therefore, design parameters such as drive cycles, target range, vehicle performance requirements at various climates and geographies should be defined at concept design phase to ensure that the EV system runs within the electrical and thermal limits.

To this end, system level engineering problems should be analysed for the selected powertrain and all subsystems. A summary of common problems are listed in Table II [6, 14, 31-57]. It should be noted that every EV application has specific topologies and subsystems. The Table II is given to provide an overview. Some of the selected problems are discussed below:

1) Functional safety: The road vehicles functional safety standard ISO 26262 defines the functional safety of automotive electrical and electronics (E/E) parts. ISO 26262 identifies four critical levels denoted by automotive safety integrity levels (ASILs) grouped as A, B, C, and D where ASIL D refers to the highest safety level. ISO 26262 requires that all potantial risks should be evaluated and the necessary measures should be implemented [31-33].

By using the HARA analysis, functional safety of an EV should be implemented in V-curve process at every stage of development. Risks and safety goals need to be defined for each component and subsystem of EV such as VCU, brake system, HV battery, traction inverter and motor, charger unit, electric steering, HV system, and CAN network [31-39].

The most critical safety problems for EV are given below [31-35]:

- Unintended vehicle acceleration or deceleration.

- Increased vehicle stopping distance.

- Vehicle instability.

- Exposure to hazardous voltages or substances such as toxic or flammable substances.

TABLE II. SYSTEM LEVEL PROBLEMS OF ELECTRIC VEHICLES [6,14, 31-57]

System	Development Issues
▪ Functional safety	▪ HARA analysis and implementation of safety measures in VCU, traction inverter, HV battery, brake system, electric steering system, CAN network [31-39].
▪ Electric powertrain	▪ Variation of available torque between continuous and peak torque region due to thermal limitation and HV battery change [40-42]. ▪ Control of regenerative energy [27, 28]. ▪ Drive cycle effect on life time of IGBT power modules of traction inverter. Life time estimation of the traction inverter [43]. ▪ Optimization of electric powertrain efficiency by 2-speed transmission [14]. ▪ Insulation aging of traction motors due to the vibration, ambient temperature, high PWM switching frequency. ▪ Bearing currents and premature failure of bearings of traction and auxiliary motors [6, 44].
▪ HV system	▪ Detection of crash and disconnection of HV system from the HV battery, discharge of HV capacitors within 5 s ▪ Reinforcement of HV battery casing and HV wirings ▪ Component layout with respect to vehicle crash zones [45-46] ▪ Limitation of regenerative energy at full battery during downhill driving [27, 47]. ▪ Prevention of over-voltage transients [47, 48]. ▪ EMC/EMI problems [49].
▪ Interaction between power electronic system and HV battery	▪ High frequency current ripples caused by power electronic systems and evaluation of risks of degradation and capacity fade of battery cells [50, 51].
▪ Thermal system	▪ Range improvements below minus temperatures [30]. ▪ Thermal optimization of traction system and HV battery to increase the available traction power [52].
▪ Electric NVH (noise, vibration, harshness)	▪ Filtering of the noise of high frequency PWM emitted from power electronic units, and noise of auxiliary units [53]. ▪ Damping of electric powertrain vibrations during tip in and tip outs [54].
▪ Charging system and grid	▪ Harmonics, voltage drop and overload impact of charging stations on grid and distribution transformers [55-57].

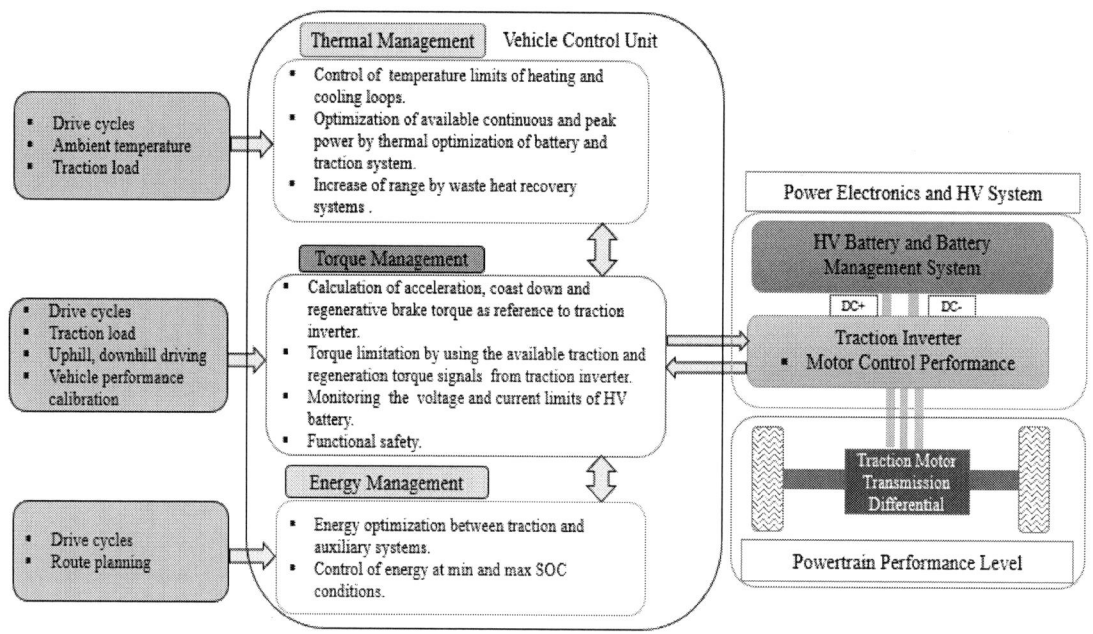

Fig. 3. VCU and electric powertrain performance parameters.

Functional descriptions of VCU control modules and relationship with the vehicle performance and electric powertrain system is illustrated in Fig.3. As seen in the figure, drive cycle is the common input parameters for Thermal Management, Torque Management and Energy Management.

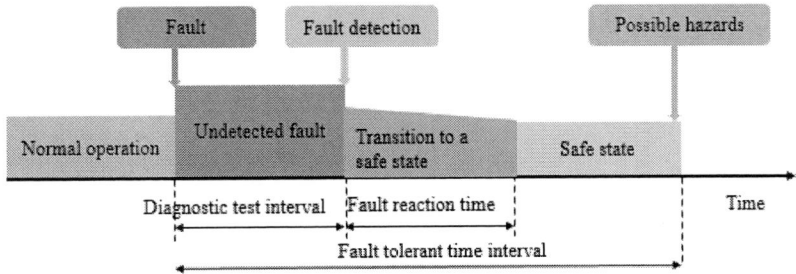

Fig. 4. Transition between normal operation, fault detection and safe state in an EV [33-34].

Regarding the unintended acceleration or deceleration, implementation of torque monitoring is the critical safety measure in VCU hardware and software, and also in traction inverter. Implementation examples are given in [31-39]. To define the technical safety requirements (TSR), failure mode effect analysis (FMEA) and fault tree analysis (FTA) methods are used. In [32], detailed analysis is given for ASIL C level safety of a traction motor contol system. To fulfil the safety requirements of ASIL C in ISO 26262, redundancy measures are required to avoid or reduce the occurance of safety failures. Information redundancy in CAN network, hardware redundancy (sensor hardware) and software redundancy in VCU should be implemented [32, 36, 38].

Regarding the software redundancy, application software in a VCU can be implemented in three level as given in [36]. The vehicle control functions are executed at Level-1. Level-2 executes the redundant software independent from Level-1. In case of any discrepancy between Level-1 and Level-2, Level-2 forces the critical safety task by setting the torque reference of the traction inverter to neutral state. Level-3 is the low level monitoring of the software and hardware [36]. ASIL D level VCU given in [36] has dual core microcontroller such as CPU0 and CPU1 and they will constantly crosscheck each other. If the check fails, VCU should shut off the actuator outputs. For example, the traction inverter torque reference can be set to zero to avoid hazardous situations.

As depicted in Fig.4, specifications of functional safety concept includes fail safe or fault tolerant behaviour, allowable time to detect and react to a fault, warning strategy to inform the presence of faults to the driver [33-34].

2) Variation of available continuous and peak torque:

a) HV battery voltage change: As the state of charge of the HV battery reduces to minimum level, battery voltage also reduces to minimum level. The required torque speed curve of the traction motor should be designed or selected for the specified minimum HV battery condition. Application examples are available in [40].

b) Motor thermal condition: Unlike ICE, power of an electrical motor derates at higher temperature and traction inverter informs the VCU for the actual available torque. This derating is needed to protect the motor against thermal overload [41-42]. Motor supplier can provide a thermal analysis including the stator winding and rotor temperature analysis for the target drive cycles and required coolant temperature. Typically 60-65 °C is the upper limit of coolant temperature.

3) Interaction between traction inverter and traction motor:

a) Bearing currents: Electrical motors driven by inverters have capacitive coupling between stator winding and rotor shaft. Summation of the output voltages of the three phase PWM inverter is not zero. Therefore, the common mode voltages (CMV) occurs. CMV is the voltage between neutral point to ground of the motor-inverter system. CMV cause high frequency current path from stator winding to rotor shaft and then to rotor bearings. The problem is already known in industrial applications. As the high switching frequency applications (10 kHz and higher) are used in EV applications, which increases the severity of the problem, bearing failure issue becomes a raising concern in EVs [6, 44, 58-61].

The most effective solutions are to use insulated bearings, ceramic ball bearings and shaft grounding rings [60-61]. Grounding resistance between the housings of motor, inverter and vehicle chassis must be as low as possible. Shielded cables must be used between motor and inverters. As stated by UNECE-R100, grounding resistance should be less than 0.1 Ohm. These precautions should be implemented for traction and also auxiliary inverters and motors used for steering assist, air brake compressor, and power take off systems as well.

b) Motor insulation aging vs. inverter switching frequency : In order to increase the specific power (kW/kg ratio) and operate the traction motor at higher efficiency area, high speed motors (above 10.000 rpm) coupled with 2-speed transmissions are being developed [14]. High speed operation requires high bandwidth current controllers and PWM switching frequency as well. Silicon carbide (SiC) MOSFET inverters are being developed to increase the inverter efficiency at high PWM switching frequency (10 kHz and higher) as their switching behavior leads to lower switching losses [6].

Regarding the simulation study in [7], SiC MOSFETs and SiC diodes increase the inverter efficiency up to 97.7 % at 10 kHz, 400 V DC voltage for New European Drive Cycle (NEDC), whereas inverter efficiency is 95,6 % at 10 kHz, 400 V DC with Si-IGBT and Si-diodes at same cycle. While the SiC MOSFET inverters enable using high PWM switching frequency and reducing the thermal load of the inverters, high PWM switching frequency increases the voltage gradient dV/dt at the motor terminals which accelerates the aging of motor winding insulation [6, 44].

Use of dV/dt filter in the inverter, reinforce the motor winding insulation are possible solutions [44]. Motor condition monitoring sensors to sense the aging of the motor windings is also a solution to improve the quality processes of the motor manufacturing and enable predictive maintenance as well [62]. From OEM side, life time estimation and periodical insulation diagnosis of the traction motor is critical.

4) Active short circuit protection function for permanent magnet synchronous motors: Traction inverter of a permanent magnet synchronous motor (PMSM) or permanent magnet assisted synchronous reluctance motor (PMaSyRM) must have an active short circuit protection function. In case of vehicle crash, this function should be activated to prevent any torque production and regeneration effect of the back-emf of the PMSM or PMaSyRM motor. In this way, it is ensured that the high voltage (HV) system is below 60 V DC, which is the safe condition [63, 64].

5) High voltage system design

a) Interaction between power electronic systems and HV battery: Power electronic units such as charger, DC-DC converter, auxiliary and traction inverters generate high frequency current ripples on HV DC bus of EV. The current ripples cause increase of cell impedance and capacity fade. As the frequency of the current ripples increase, this effect increases as well [50-51].

For current ripple evaluation, EV should be tested both in slow and fast charging modes as the chargers produce current ripples as well. Another test condition can be to drive the vehicle at top speed to increase the current of traction inverter and activate auxiliary power electronic units to increase the current ripples. Evaluation of the measured current ripples can be done by battery supplier.

b) Limitation of regenerative energy at downhill driving at full battery condition: In this driving condition, a brake resistor should be connected to HV DC bus. It increases the safety of HV system. In case of full battery or heating of hydraulic brakes, brake resistor waste the regenerative energy into heat [47, 65]. Brake resistors are equipped with its chopper unit [47]. Another benefit of brake resistor is limitation of any transient overvoltage condition that may occur during sudden changes of the electrical loads connected to the HV DC bus [27, 47].

6) Electric powertrain vibrations during tip in and tip out: This problem already exists in combustion engine powertrains. However, electric powertrains do not have clutch or torque converter. Torque rise time of an electric motor is in milliseconds range which is much faster than internal combustion engine (ICE). It is smoothed by filters in VCU. Weak damping of the electric powertrain, sudden changes of the vehicle driver's torque request, road disturbances are causes of torsional vibrations occurring at motor shaft speed and torque.

Oscillation damping controllers are developed and implemented in traction inverters [54]. The damping torque is applied to the torque reference sent from VCU. In order to improve the NVH (noise, vibration, harshness) and driveability performance, this function needs to be activated and tuned in traction inverter.

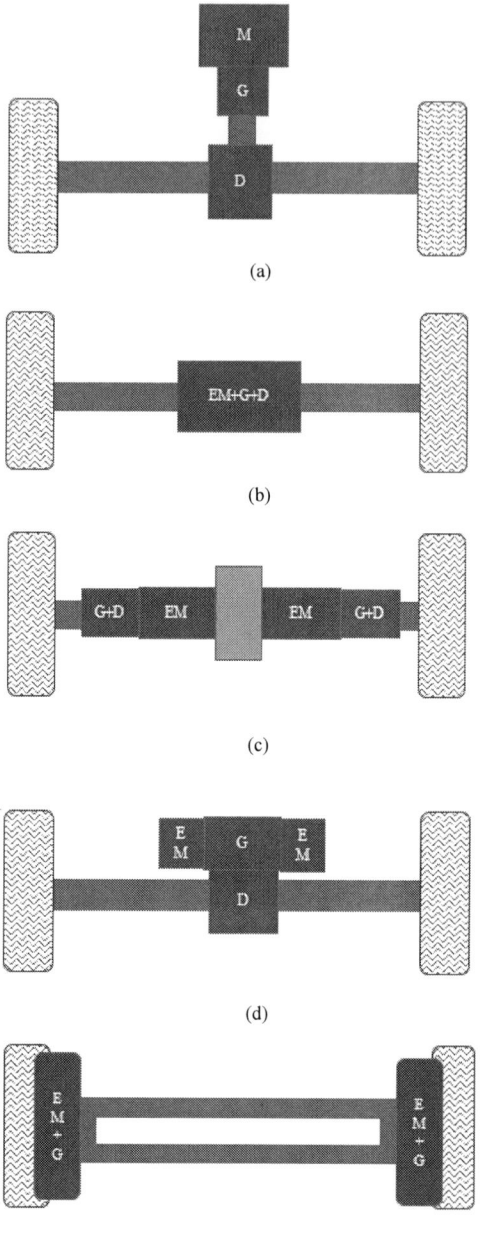

Fig. 5. a) Central drive topology with propeller shaft b) Single motor electric axle c) Dual motor near-wheel electric axle d) Dual motor center aligned electric axle e) In-wheel topology.

V. STATE OF ART POWERTRAIN TOPOLOGIES, TRACTION INVERTERS AND MOTORS

A. Evaluation of Electric Powertrain Topologies

Development of an electric powertrain has geometrical integration issues such as finding the optimum dimensions of transmission, traction motor and inverter that match with the requirements of EV project. In-wheel and near-wheel topologies, single or dual motors integrated into the axle center are available solutions as given in Fig.5 [12, 66-71]. Central drive topology is not a compact topology as it requires more space because of the propeller shaft.

By using electric axle topologies, the propeller shaft can be removed. This provides more space especially for electric bus and truck applications. To solve the geometrical integration and speed up the development time, electric axles are being developed by suppliers and OEMs. Compared to the in-wheel topology, near-wheel electric axle topology does not increase the mass of the wheel which is an advantage in terms of driving comfort and handling [70].

Electric axle development requires high specific power (kW/kg) ratio due to the size limitations. As shown in benchmark in [4], a high speed radial flux motor has 15.000 rpm maximum speed, 120 kW peak power and 4.6 kW/kg specific peak power ratio, and an axial flux motor has 3250 rpm maximum speed, 199 kW peak power and 6 kW/kg specific peak power ratio. High speed motors have reduced motor diameter which is advantage for near-wheel dual motor electric axles. Axial flux motors has reduced length compared to radial flux motors, which is also advantage for single or dual motor topology integrated into the center of the axle as shown in Fig.5(d). Use of axial flux motors and high speed radial flux motors in electric axles will be increased especially in HD segment as this applications require high specific motor power and compact motor topologies.

B. A Case Study for Comparison of 1-and 2-Speed Transmission

Electric city bus and refuse trucks have low speed stop and go cycles and therefore, the traction motor runs at low efficiency area in these applications. High speed motors coupled with 2-speed transmission systems can be controlled to run at high efficiency area to give the required motor torque [14, 72-74]. Instead of using a higher rated power motor and running it at low efficiency area by 1-speed transmission, a motor with lower rated power can be run at higher efficiency area by 2-speed transmission to give the same torque. In that way, it is possible to downsize the motor torque, inverter current, and motor diameter which is important for electric axle [14].

In [14], a comparative evaluation of 1 and 2-speed transmission systems are given. The simulation study in [14] is performed In NEDC cycle and it is shown that motor torque and inverter current can be reduced as given in Table III.

Some critical application issues of 2-speed transmission are given below [72-74]:
- Gear shifting should not be allowed during regeneration to avoid torque transients [73]. Otherwise, the change of regenerative torque can cause an overvoltage on the HV battery terminals.

- Gear shifting should be as fast as possible. In [73], shifting time is implemented in less than one second, with 100 ms shifting actuator rise time.

- Gear shifting performance is dependent on control dynamics and inertia of traction motor, shifting actautor rise time [73].

- Gear shifting should be optimized for the drive cycle for energy optimization. Drive cycle can be predicted or assumed to be known.

TABLE III. PERFORMANCE VALUS OF 1-AND 2-SPEED TRANSMISSONS [14]

Transmission	Max. motor torque	Max. inverter current	Gear ratio	Motor losses in NEDC cycle
1-speed	196 Nm	325 A rms	7.73	1.35 kWh/100 km
2-speed	143 Nm	280 A rms	12.91/ 6.63	1.09 kWh/100 km

C. State of the Art Traction Inverters and Motors

In industrial applications, inverters and motors are evolving towards the system solutions rather than single component alone. Use of non-rare earth motors like synchronous reluctance motors in high efficiency IE4 or IE5 class motors are increasing and combined efficiency data including motor and inverter is given [75, 76]. As part of this trend, combined efficiency of motor and inverter is expected be given in automotive as in industrial applications. Regarding the industry 4.0 concept, motor condition monitoring sensors and internet of things (IoT) applications are also increasing to enable predictive maintenance and remote monitoring [75]. The development trends in industrial and automotive inverters and motors are summarized in Table IV [3].

Condition monitoring sensors are used to detect bearing faults, stator or rotor electrical faults, and rotor eccentricity and imbalance faults. ABB has commercialized the condition monitoring sensors for low voltage industrial motors [75]. Studies have shown that the failure occurrence rates of electrical motors are mostly stator and bearing faults. Fault rates are distributed below [77]:

- Bearing faults occur at 40% ~50 % rate.

- Open or short circuit faults in stator occur at 40% ~50% rate.

- Eccentricity related faults occur at 5% ~10% rate.

- Rotor related faults occur at 5%-10% rate.

Current state of art EV technology has only state of health (SOH) estimation for HV battery. It is expected that condition monitoring systems will be either an integrated system to the traction motor and/or an after sales maintenance tools for periodic check of EV. Condition monitoring sensors developed for EVs [62] can also change the diagnostic modelling between VCU and electric powertrain system and make it more intelligent.

Cost reduction, higher specific power and efficiency will be the high priority for the traction and auxiliary inverters. Power sizing configurator tools for auxiliary motors and inverters will help to reduce the costs as such tools can help to optimize the required power size.

Speed sensorless control is expected to be developed both for traction and auxiliary inverters. From functional safety aspect, in case of any sensor failure, speed sensorless control can be activated in limp home mode or redundant control mode for traction and also steering inverter systems.

EV applications require zero and low speed operation in traction and regeneration modes at full load. Electrical parameters of traction motors drift due to the temperature

978-1-5386-7688-2/19 $31.00 © 2019 IEEE

TABLE IV.   DEVELOPMENT TRENDS IN INDUSTRIAL AND AUTOMOTIVE   INVERTERS AND MOTORS [3]

1990-2005	*2005-2019*
Industry	
• For IM applications: general purpose stand-alone inverters. • For PM motors: servo inverters.	• Motor integrated inverters. • Compact sized stand-alone inverters. • Inverters with integrated PLC functions. • Application optimized inverter and motor sets (pump, HVAC, crane). • Commissioning wizards, application configurators. • More applications of speed sensorless control. • Premium efficiency motors in IE3-IE4-IE5 efficiency classes. • System efficiency requirement including motor, inverter and load machine such as pump. • IoT technologies for inverters. • Motor condition monitoring sensors. • Premium efficiency SynRM as alternative to PM motor. • Process efficiency analysers.
Automotive	
2005-2019	*2020-2030*
• Stand-alone inverters, integrated motor inverters. • Application specific functions: active short circuit, vibration damping. • Alternative to PM motors: PMaSyRM, Hybrid PMaSyRM, Axial Flux motors, IM • Electric axle as system solution.	• Premium efficiency SiC inverters and high speed motors. • More applications of multispeed transmissions. • System efficiency data for electric powertrain including traction motor, inverter and transmission for a reference drive cycle. • Multi-functional power electronic units. • Electric axle including a configurable VCU as system solution. • Application optimized auxiliary motor and inverter sets. • Motor condition monitoring sensors. • Speed sensorless control as redundant mode and limp home mode for traction and steering inverters. • Self-check functions for inverters, user friendly diagnostic tools. • Drive cycle efficiency analysers.

change. Moreover, there is no steady-state operation in an EV. These factors create difficulties for applications of speed sensorless control in EVs. State of the art review of speed sensorless control in EV applications is given in [78]. In [79], speed sensorless auxiliary inverter and motor solutions are given for automotive applications.

D. Trends in Non-Rare Earth and Reduced Rare Earth Motors

OEMs and suppliers continue to develop non rare earth (NRE) or reduced rare earth (RRE) motors as alternative to interior permanent magnet (IPM) synchronous motors. It should be noted that rare earth elements are also needed to develop permanent magnet motors in other industries such as industrial automation, consumer electronics and defence industry. Therefore, the need for NRE and RRE motors will continue.

Regarding some EV applications, Danfoss Editron [80] has developed permanent magnet assisted synchronous reluctance motor (PMaSyRM), BMW has developed a hybrid PMaSyRM [81]. Magnets of a hybrid PMaSyRM contain both ferrite and rare earth magnets. ZF has developed induction motor (IM) and applied in electric axle models [82]. Audi has developed an IM and applied in an electric axle system [13]. Benchmark in [5] shows the comparative evaluation of Prius IPM motors with other motor types including the PMaSyRM, Hybrid PMaSyRM as well. In terms of efficiency and power for the same dimensions as Prius IPM motor, PMaSyRM and Hybrid PMaSyRM have promising results. In NEDC cycle, Prius IPM motor has 86.94 % efficiency, whereas Hybrid PMaSyRM 88.74 % and NdFeB SynRM have 88.34 % efficiency.

VI.   POST- CRASH DIAGNOSTICS OF ELECTRIC VEHICLES

After-sales service and post-crash diagnostics of an EV is critical issues for growth of EV market and for EV customers as well. EV will absolutely change the after-sales services in automotive industry. Beside sales price, range and charging stations, growth of EV market is also dependent on the spare part logistics and after sales services as well.

Test procedures are needed for diagnosis of the system and component failures after a crash. Considering the service costs after a crash,   technical requirements to decide the re-usability of EV components should be defined. A post-crash component analysis is given in [83]. Critical issue is the definition of the conditions for re-usability of EV components. EV technologies have different powertrain and motor topologies, which also increase the functional complexity and safety requirements. Some of the component check points after a crash are listed below:

- Component housings.
- Integrity of signal and HV wirings and connectors.
- Grounding resistance between grounding points of HV components and vehicle chassis.
- Insulation resistance of all motors and HV power units.
- IP protection of the impacted area.

- Rotor imbalance, eccentricity of all motors.
- Position sensors of motors.
- Any leakage in thermal system.

The list can be improved by the case studies and application feedbacks. It is intended to make a focus on post-crash check points and do not cover all details for every component.

Referring to the crash analysis and timings of rear crash in [83], there is a delay between the instants of impact on the components, crash detection and discharge of HV circuit. Therefore, it is critical to make HARA analysis for the safety functions like active short circuit (ASC) function of PMSM or PMaSyRM motors considering any motor wiring damage that can occur before the activation of the ASC function. Because, ASC function short circuits the motor phase terminals by switching on IGBTs either on top or bottom legs of the traction inverter [37, 63]. Wiring and connector damages shown in [46, 83] may occur before the activation of ASC function. Precautions should be taken in case the PM motor rotates freely and a damaged HV wiring between motor and inverter creates conductive path to the vehicle chassis when the ASC function is activated. In case of separate installation of traction motor and inverter, HV wiring protection materials should be used between traction motor and inverter. In [84], some HV wiring protection materials are given.

VII. CONCLUSION

In this study, EV technologies have been reviewed with system perspective and future trends. Almost all available EV technologies have some trade-offs and therefore, OEMs should evaluate the technology selection with all long terms benefits and engineering effort for system integration as well.

Beside HV battery and electric powertrain technologies, growth of EV market is also dependent on charging stations that comply with power quality standards [55-57], and after sales network that can provide cost effective solutions for maintenance or repair after a crash.

In terms of application case studies and product features, industrial applications of electrical motors and inverters can provide useful feedback for automotive applications. It can be predicted that both industrial segments will have more synergy in near future.

VIII. REFERENCES

[1] "New Markets, New Entrants, New Challenges, Battery Electric Vehicle," 2019, www.deloitte.com.

[2] "Electric Vehicle Outlook 2019," BloombergNEF, about.bnef.com.

[3] M. Karamuk, "Development of heavy duty electric powertrain focusing on cost, time, efficiency and range," IQPC Conference on Powertrain Electrification for Medium and Heavy Duty Vehicles, June 2019.

[4] A. EL-Refaie and M.Osama, "High specific power electrical machines: a system perspective," CES Transactions on Electrical Machines and Systems, vol. 3, March 2019.

[5] Z. Q. Zhu, W. Q. Chu, and Y. Guan, "Performance of electrical machines for HEVs/EVs," CES Transactions on Electrical Machines and Systems, vol. 1, March 2017.

[6] K.Vogel and A. Brodt, "Benefits and challenges of new semiconductor solutions in AC-Drives," 18th European Conference on Power Electronics and Applications , September 2016 .

[7] M. Hofmann, "Evaluation of potentials for Infineon SiC-MOSFETs in automotive inverter applications," Fraunhofer IISB.

[8] www.tm4.com.

[9] "Research progress: next generation secondary batteries," www.toyota.com.

[10] F. Aymen and C. Mahmoudi, "A novel energy optimization approach for electrical vehicles in a smart city," Energies, vol.12, pp.1-22, February 2019.

[11] J. Meyer et al., "Range extension opportunities while heating a battery electric vehicle," SAE Technical Paper, April 2018.

[12] M. Milbrandt, "Skalierbare Module aus Antrieb und Achse für die Elektromobilitaet," publica.fraunhofer.de, 2016.

[13] J. Doerr, T. Attensperger, L. Wittmann, and T. Enzinger, "The new electric axle drives from Audi," MTZ Worldwide, vol.79, pp 18–25, June 2018

[14] A.Schönknecht, A. Babik, and V. Rill, "Electric powertrain system design of BEV and HEV appling a multi objective optimization methodology," 6th Transport Research Arena, April, 2016, Elsevier

[15] L. D. Novellis et al., "Torque vectoring for electric vehicles with individually controlled motors: state-of-the-art and future developments," EVS26 Los Angeles, California, May 6-9, 2012

[16] www.bosch-mobility-solutions.com

[17] Y. Zhou and X. Li, "Vehicle to grid technology: A review," 34th Chinese Control Conference (CCC) , 2015

[18] T.Hulshorst, "Future of e-mobility next generation electric vehicles," 2nd International Automotive Engineering Conference, November 2017.

[19] Marktubersicht Elektrobusse, www.mobilitatsmanagement.at, Januar 2018.

[20] www.gemmotors.si.

[21] www.ashwoodselectricmotors.com.

[22] M.Karamuk, M.Cepni and S.Gur, "Electric vehicle powertrain development-conceptual design and implementation," International Conference on Automotive and Vehicle Technologies, AVTECH 2013, Yildiz Technical University Istanbul.

[23] O. Spinka, M.Rezac, and J. Rathousky, "Evolution of a powertrain manager for electric and hybrid-electric vehicles," www.porsche engineering.com.

[24] G. Xie, Y. Chen, Y. Liu, R. Li, K. Li, "Minimizing development cost with reliability goal for automotive functional safety during design phase," IEEE Transactions on Reliability, vol.67, March 2018.

[25] J.J. Valera, B. Heriz, G. Lux, J. Caus, B. Bader, " Driving cycle and road grade on-board predictions for the optimal energy management in EV-PHEVs," EVS 27, 2013.

[26] F. Zhang, X. Hu, R. Langari, D. Cao, "Energy management strategies of connected HEVs and PHEVs: Recent progress and outlook," Elsevier, vol 73. , July 2019.

[27] P. Marx, "Dauerbremse (Bremshilfe) für Elektrofahrzeuge," German Patent, DE102013013258, publication date: December 2015.

[28] B. Balasubramanian and A. C. Huzefa, "Development of regeneration braking model for electric vehicle range improvement," IEEE Transportation Electrification Conference (ITEC-India), December 2017.

[29] M. Sway-Tin, M. Arcori, K. L. Cartwright, T. Roterman, and J. McCoy, "Intelligent coast down algorithm for electric vehicle," US Patent 6364434B1, April, 2002.

[30] S. Chowdhury, L. Leitzel, M. Zima, M. Santacesaria et al., " Total thermal management of battery electric vehicles (BEVs)," SAE Technical Paper, 2018-37-0026, May 2018.

[31] A. Watson, "Lessons Learned Developing ISO 26262 Supervisory controllers for EV/HEV," www.pi-innovo.com, June 2018.

[32] H-P. Li, Y-W. Li, "The research of electric vehicle's MCU system based on ISO 26262," 2nd Asia-Pacific Conference on Intelligent Robot Systems, June 2017.

[33] D. Ward, "Practical experiences in applying "concept phase" of ISO 26262," MIRA, November 2012.

[34] A. Nardi, A. Armato, "Functional Safety Methodologies for Automotive Applications," www.cadence.com.

[35] S. Christiaens, J. Ogrzewalla and S. Pischinger, " Functional safety for hybrid and electric vehicles," SAE Techical Paper, 2012-01-0032.

[36] EV2274A datasheet, www.ecotrons.com.

[37] S. Batchu, "Functional safety in inverter hardware," SAE Technical Paper, 2016-28-0166.

[38] Z.Wu, K. Lu, Y. Zhu, X. Lei et al., "Functional safety and secure CAN in motor control system design for electric vehicles," SAE Technical Paper, 2017-01-1255.

[39] J. Zhang, A. Amodio, B. A. Guvenc, G. Rizzoni, and P. Pisu, "Investigation of torque security problems in electrified vehicles," Proceedings of the ASME 2015, Dynamic Systems and Control Conferences, October, 2015.

[40] HVH series motor datasheets, www.borgwarner.com.

[41] Z. X. Fu, "Real-time prediction of torque availability of an IPM synchronous machine drive for hybrid electric vehicles," IEEE Conference on Electrical Machines and Drives, May 2005.

[42] A. Specht and J. Boecker, "Observers for the rotor temperature of IPMSM," 14th International Power Electronics and Motion Control Conference (EPE/PEMC), 2010.

[43] Y. Sang et al., "Analysis on multiple factors influencing the lifetime of IGBTs of electric vehicles converters," 43rd Annual Conference of the IEEE Industrial Electronics Society, IECON October 2017.

[44] K. Vogel and A. J. Rossa, "Improving efficiency in AC drives: Comparison of topologies and device technologies," PCIM Europe, May 2014.

[45] M. Wisch, J. Ott, R. Thomson, and M. Abert, "Recommendations for safe handling of electric vehicles after severe road traffic accidents," www-esv.nhtsa.dot.gov, 2015.

[46] R. Justen and R. Schoeneburg, "Crash safety of hybrid- and battery electric vehicles," 22nd International Technical Conference on the Enhanced Safety of Vehicles (ESV), June 2011.

[47] PowerMELA_BC brake chopper data sheet, www.stw-technic.com.

[48] H.Wen, W.Xiao, and H.Li,X.Wen, "Analysis and minimisation of DC bus surge voltage for electric vehicle applications," IET Electrical Systems in Transportation vol. 2, June 2012, pp.68-76.

[49] A. Kishore, C. Patki, M. Anwar, W. Ivan, and M. Teimor, "Investigation of common mode noise in electric propulsion system high voltage components in an electrified vehicle," IEEE Transportation Electrification Conference and Expo, June 2016.

[50] K. Uddin, A. D. Moore, A. Barai, and J. Marco, "The effects of high frequency current ripple on electric vehicle battery performance," Elsevier, June 2016.

[51] A. Bessman, "Interactions between battery and power electronics in an electric vehicle drivetrain," Doctoral Thesis, KTH, Sweden, 2018.

[52] M. Stellato, L. Bergianti, and J. Batteh, "Powertrain and Thermal System Simulation Models of a High Performance Electric Road Vehicle," Proc. of the 12th International Modelica Conference May 2017.

[53] J. G. Cherng, "ME 570 Powertrain NVH of Electrified Vehicles," Mechanical Engineering Department University of Michigan Dearborn.

[54] N. Amann, J. Boecker,and F. Prenner, "Active damping of drivetrain oscillations for an electrically driven vehicle," IEEE/ASME Transactions on Mechatronics, vol.9, December 2004.

[55] A. Lucas, F. Bonavitacola, E. Kotsakis, and G. Fulli, "Grid harmonic impact of multiple electric fast charging", Elsevier, June 2015.

[56] R. Carter, A. Cruden, A. Roscoe, D. Densley, and T. Nicklin, " Impacts of harmonics distortion from charging electric vehicles on low voltage networks," EVS 26, May 2012.

[57] A. Lucas, G. Trentadue, H. Scholz and M. Otura, "Power quality performance of fast charging under extreme temperature conditions", Energies, vol.11, October 2018.

[58] www.abb.com, External drive hardware-selection and application

[59] www.skf.com.

[60] www.est-aegis.com.

[61] T. Hadden et al., "A review of shaft voltages and bearing currents in EV and HEV motors," 42nd Annual Conference of the IEEE Industrial Electronics Society, October 2016.

[62] K.Watkins and C.P.Wong, "Condition monitoring sensor for electric vehicle motor and generator insulation systems," EVS26, May 2012.

[63] A. Birk, M. Schwab, "System zum aktiven Kurzschliessen von Phasen eines Wechselrichters und Kraftfahrzeugantrieb," ZF Friedrichshafen AG, German Patent DE102016207195, November 2017.

[64] VDE-Kompendium , Elektromobilitaet, www.dke.de.

[65] L. Yutong, Z.Junzhi, L. Chen, K. Decong, and H. Chengkun, "Research of regenerative braking system for electrified buses equipped with a brake resistor," IEEE Vehicle Power and Propulsion Conference, October 2013.

[66] J. Brousek and R. Vozenilek, "Gearbox design for a dual motor drive system," Proced. of the 7th International Conference on Mechanics and Materials in Design, June 2017.

[67] www.axletech.com.

[68] www.danaelectrified.com.

[69] www.bpw.de.

[70] B. Wang, D. Hung, J. Zhong, and K-Y. Teh, " Energy consumption analysis of different BEV powertrain topologies by design optimization," International Journal of Automotive Technology, vol. 19, pp. 907-914, October 2018.

[71] www.ziehl-abegg.com.

[72] U.Knödel, A. Strube, U. C. Blessing, and S. Klostermann "Design and implementation of requirement-driven electric drives," ATZ Worldwide, vol.112, pp.56-60, June 2010.

[73] M. Allende, P. Prieto, B. Heriz, J. M. Cubert, and T. Gassman, "Advanced shifting control of a two speed gearbox for an electric vehicle," EVS28, May 2015.

[74] F. Viotto, "A novel seamless 2-speed transmission system for electric vehicles: principles and simulation results," Electronic Systems for Vehicle Propulsion Symposium, SAE Technical Paper, 2011-37-0022.

[75] www.abb.com.

[76] www.ksb.com.

[77] F.Lin, K.T.Chau, C.C. Chan and C.Liu, "Faults diagnosis of power components in electric vehicles," Journal of Asian Electric Vehicles, vol.11, December 2013.

[78] S. Rind, Y. Ren, and L. Jiang, "Traction Motors and Speed Estimation Techniques for Sensorless Control of Electric Vehicles: A Review," 49th International Universities Power Engineering Conference (UPEC), September 2014.

[79] www.keb.com.

[80] www.danfoss.com.

[81] J. Merweth, "The hybrid synchronous machine of the new BMW i3&i8," BMW Group, 2014.

[82] www.zf.com.

[83] Y.Lian, D. Zeng, S. Ye, B. Zhao and H. Wei, "High-voltage safety improvement design for electric vehicle in rear impact," Automotive Innovation, pp.211-225, July 2018.

[84] P. Marks, "Electric vehicle wiring design challenges," ATZ Elektronik, 0512012, Volume 7.

Large variable speed generators design and their control: a revisit in 2019

Ion Boldea
Electrical Engineering
Politehnica University Timisoara
Romanian Academy
Timisoara, Romania
ion.boldea@upt.ro

Lucian Tutelea
Electrical Engineering
Politehnica University Timisoara
Timisoara, Romania
lucian.tutelea@upt.ro

Ana Popa
Electrical Engineering
Politehnica University Timisoara
Timisoara, Romania
ana.popa@upt.ro

Abstract—**The paper presents a panoramic view of various large power variable speed generators design and their control in hydro and wind energy conversions. Recent systems that reached industrial stage and new developments are discussed in terms of their topologies, merits, demerits and some case study optimal designs of generators and their PWM converters reliable control.**

Keywords— *variable speed generators design, doubly fed induction generator (DFIG,) cage rotor induction generators, wind energy conversion.*

I. INTRODUCTION

Hydro power is currently the largest renewable source for electric power generators in the world. It produced 3431TWh meeting about 16% of global electricity needs in 2010 and will remain no. 1 in renewables (Fig. 1).

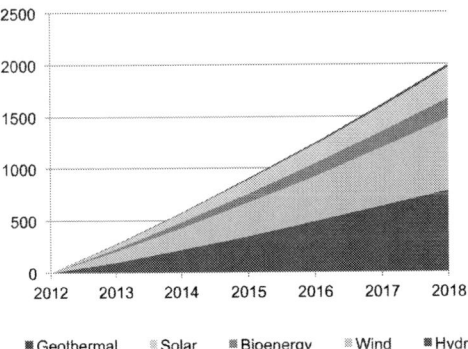

Fig. 1, Forecast cumulative additions of power resources (TWh)

On the other hand, wind energy penetration in electric power system in 2018 is about 500GW (about 2.5% of all installed power in the world); however the potential for growth rate here is even larger than for hydro, where most important site are exploited already (especially in developed countries). Large variable speed generators are required in hydro (especially in pump storage) – in the hundreds of MW and in offshore wind (10+MW) energy conversion to tap better the primary energy at site and offer flexibility to control the flow of energy [1-30]. Additional ac-dc-ac PWM converters, two level or multilevel type for low (up to 690V) or medium (3.6+ kV) voltage generators are required for the scope of handling grid connected, standalone connected, asymmetric grid voltage sags and voltage and current harmonics reduction in the generators and in the power grid; all at reasonable volume and cost of the converters of high efficiency (above 97% in general).

This paper aims to offer a panoramic realistic view of existing (commercial) and newly proposed high power wide speed electric generator systems for hydro (especially pump storage) and wind energy conversion, in terms of topologies, merits and demerits, optimal design with case study results, more reliable PWM converters with advanced control under normal and under representative voltage sags conditions.

The paper continues with: variable speed hydro generator systems (section2), wind energy large power generators state of the art (section 3), doubly fed induction generator (DFIG) control (section 4), cage rotor induction generators (CRIG) and its control (section 5), PMSG standard and new proposed) optimal design and control (section6), dc excited standard and S.C-SGs optimal design and control (section 7) and conclusion (section 8).

II. VARIABLE SPEED HYDRO-GENERATOR SYSTEMS

Besides the efficiency increase (Fig. 2) the main benefits of variable speed in hydro-generators are the following:
- Larger operation time of hydropower system due to reduced noise and cavitation problems,
- Increased flexibility in site solution,
- Relaxation of parameter requirements on electric generator design,
- Fast power response for grid systems with integrated wind/solar power,
- Less environmental impact.

An example of such a hydropower plant connected to HVDC transmission system is described in Table 1 and in Figs 3, 4. It has dc excited variable speed synchronous generators a.c. connected in parallel ac. And then, through 2 transformers, to a bipolar HVDC transmission line whose sending end uses two full power controlled rectifiers (fig.4).

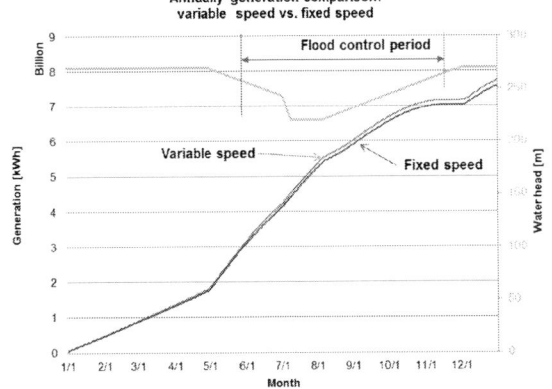

Fig. 2. Comparison of the hydropower generated between fixed speed operation and variable speed operation.

978-1-5386-7688-2/19 $31.00 © 2019 IEEE

TABLE 1 MAIN PARAMETERS OF XIAOLANGDI HYDROPOWER PLANT

Nominal power of turbines	6×306 MW
Nominal power of generators	6×300 MW
Pole pairs number	28
Generator synchronous speed	107.1 r/min
Efficiency of generators	0.98
Nominal diameter of turbines	6 m
Maximum water head	139.2 m
Weighted average water head	119.9 m
Optimized unit rotating speed	62 r/min

As visible in Fig. 3 the energy yield per year in not increased by much (2%) but it counts and the flexibility and other advantages lined-up above to justify the practicality of the solution. In contrast, DFIG is used also for variable speed replace pump storage system as its PWM bidirectional converter is designed in general to less than 10-15% kVA rating thus lowering the inverter cost of the effective system. An example is shown in Table 2 and Fig. 5 while Fig. 6-9 present:

- 50% voltage drop and speed transients (Fig. 6,7),
- instantaneous power injection by lowering temporally the generator/turbine speed (Fig. 8); that is exploiting the inertia energy available in the system to treat grid power transients.

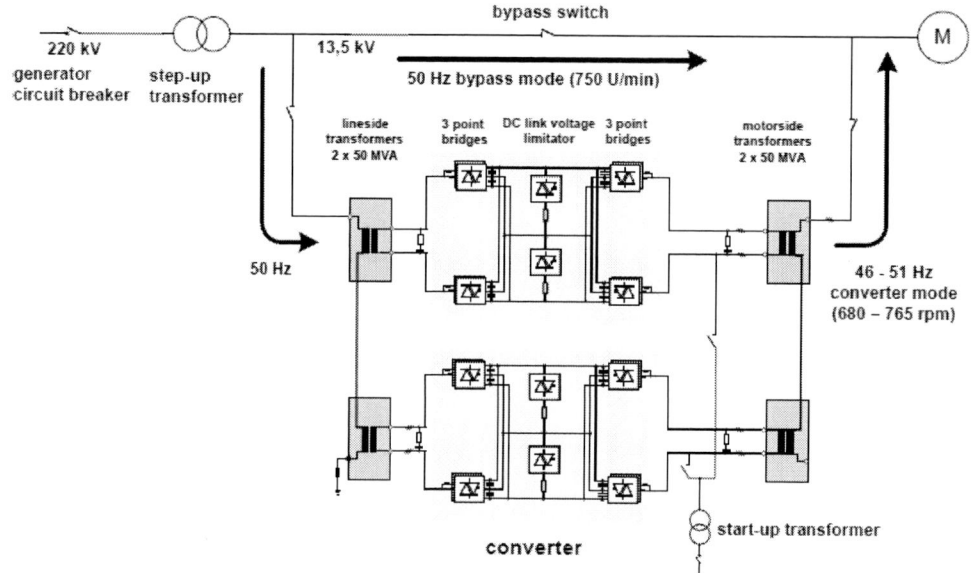

Fig.3. Twin Converters: full power Synchronous Generator.

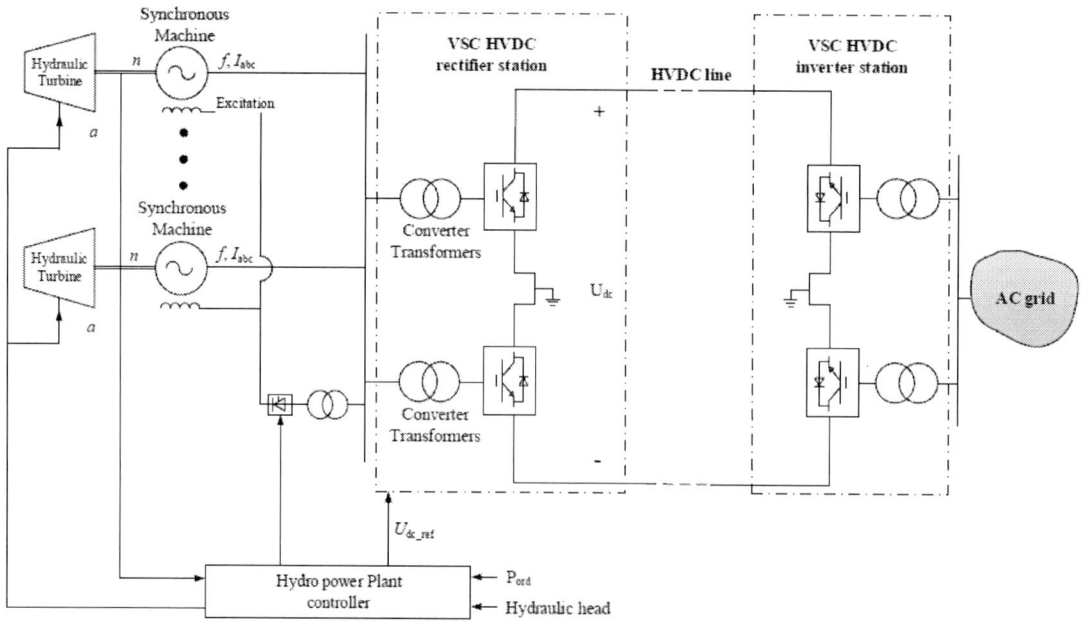

Fig.4. Direct connection of hydro generator with VSC HVDC.

TABLE 2 HYDRO DOUBLY FED ASYNCHRONOUS MACHINE (DFIG) - MACHINE AND GRID PARAMETERS

Rated apparent power:	MVA	230
Synchronous speed:	rpm	333.33
Speed range:	rpm	300-366
Rated machine voltage:	kV	15.75
Grid voltage:	kV	245
Grid short circuit power:	MVA	1120
Frequency:	Hz	50
Shaft line inertia:	tm²	1410

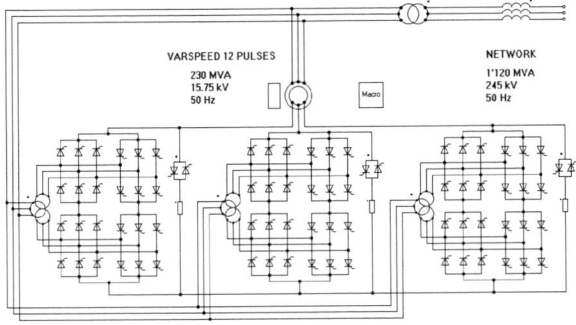

Fig. 5. Diagram of a DFIG - VARSPEED hydro-installation.

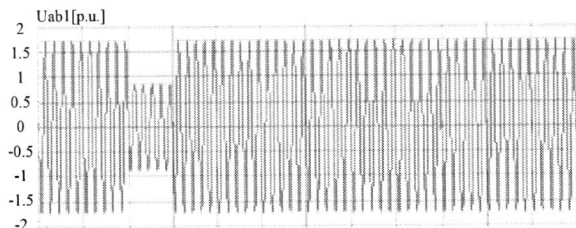

Fig.6.Hydro–DFIG 50% voltage drop on grid's side(100 ms).

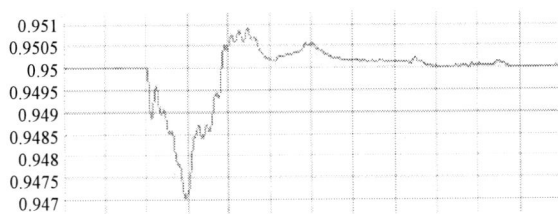

Fig. 7 Hydro – DFIG: Speed during the voltage drop.

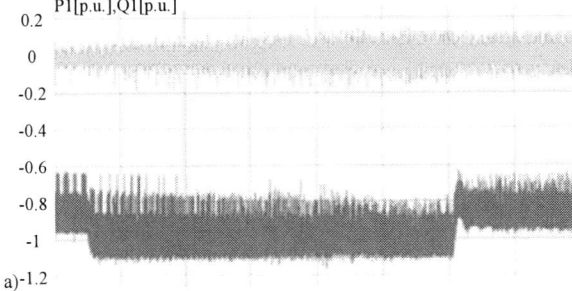

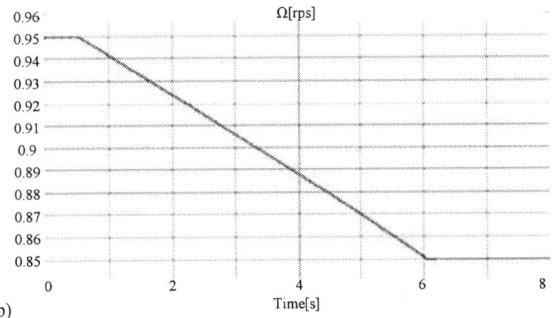

Fig. 8. Hydro – DFIG: a) Instantaneous active/reactive power injection into the grid; b) mechanical speed of the machine.

Note. Through only a small fraction of hydro-plant use variable speed systems their scale is expected to grow when renovation of old power plants takes place and for new systems, based largely on the technologies in wind generator systems, notwithstanding that, for hydro power, the speed is roughly 10 times larger than in wind generators.

III. LARGE WIND ELECTRICAL ENERGY GENERATIONS

Large wind generators power per unit increased steadily in the last two decades using widely the wind tower concept (Fig.9) up to power envisaged at 10-20 MW, but imposing an increasingly tall wind towers with a heavy nacelle on top. To reduce the tower height and costs, INVELOX.com has proposed recently a new wind energy collector system with a probably flexible (lighter) tower, a Venturi tube to accelerate the air 4-5 times, with the wind turbine generator in the tube, on the ground (or just above to water level), Fig.10. This new system, experimented so far at few hundred kW is credited with 30% lower overall costs and more power yield per site, Fig.11. Aggressive R&D efforts are needed to prove INVELOX practicality in the 10+MW wind turbine generator systems. They should cover: in tube wind turbine design, air-flow around the generator, exhaust air usage and noise reduction, optimal design and the control of the system, etc.

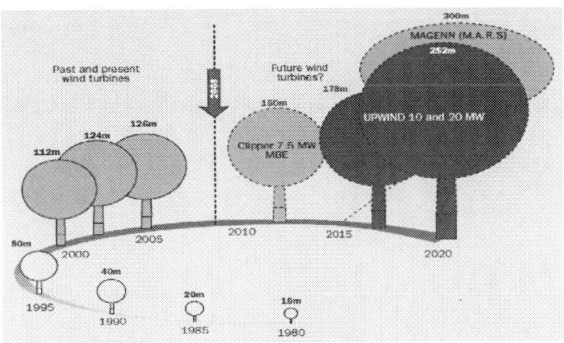

Fig. 9. Evolution of wind turbine power /unit (after [25]).

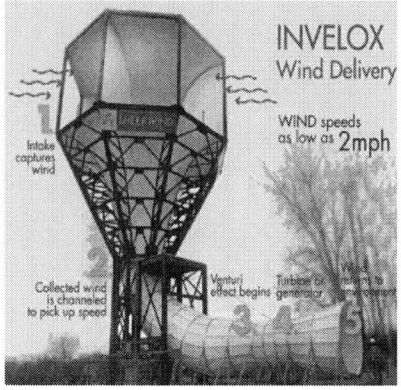

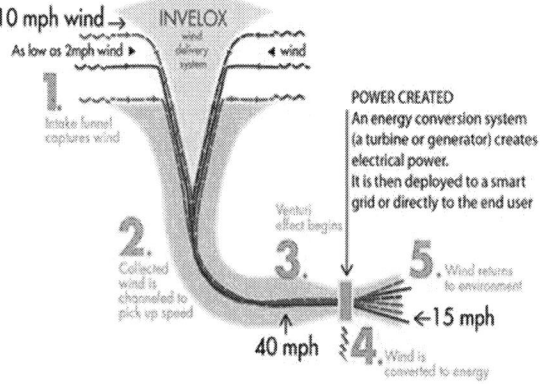

Fig. 10. INVELOX wind energy conversion (after [5]).

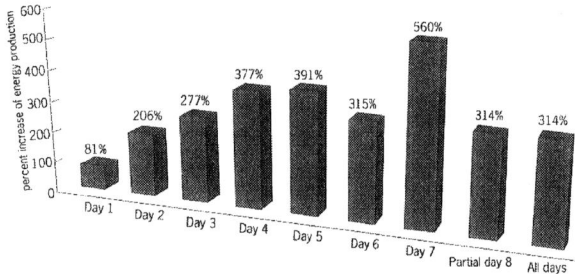

Fig. 11. Energy production implemented by INVELOX (after [5]).

Coming back to the wind tower topologies, a typical initial cost by source breakdown is shown in Fig. 12.

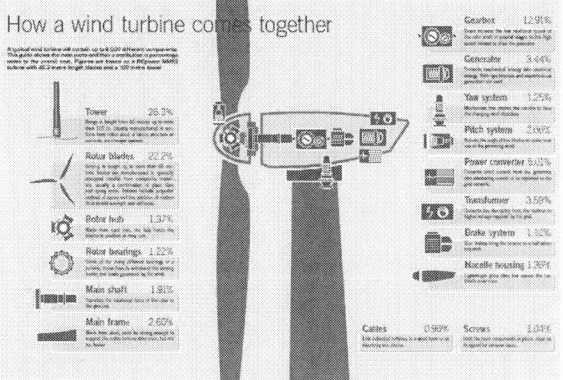

Fig. 12. Generic wind tower power plant (after [26]).

The cost breakdown shows that the total cost of the wind generator, transformer, transmission and power electronics

represents less than 25% of the total, Fig.12. This means that its optimum design would impede its volume and weight influence on the nacelle volume/weight and cost.

The largest envisaged wind-tower generator systems so far are shown in Fig 13 and Fig 14:

- the 8 MW, 480 rpm PMSG with 3 phase, 3 slots per pole by Vestas (Fig. 13),
- the 7.6 MW, 9.1 rpm (direct driven) dc excitation synchronous generator by Energon, Fig. 14.
- A 12 MW, 9 rpm, wind generator is due soon for commissioning by GE.

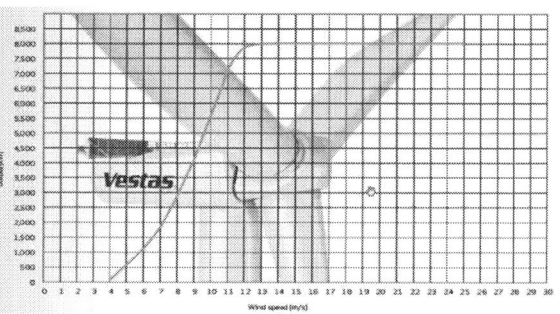

Fig. 13. 8 MW, 490 rpm PMSG, Power Curve for V164-8.0 MW.

Cut-in wind speed is 4 m/s, operation rotor speed: 4.8-12.1 rpm, rated rotor speed: 10.5 rpm, rotor diameter: 164 m, frequency: 50Hz, converter type–full scale. Nominal voltage is 33-35 and 66 kV, blade length: 80 m, blade chord: 5.4 m, blade weight: 35 tones, nacelle including hub weight: 390 tones, power regulation–pitch regulated with variable speed (after [6]).

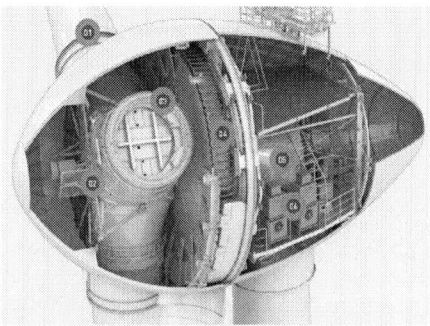

01 - Rotor blade   04 - Annular generator
02 - Rotor hub   05 - Main carrier
03 - Hub adapter   06 - Yaw drive

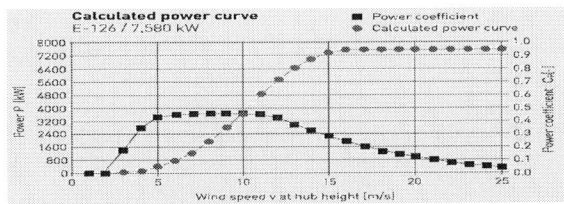

Fig. 14. 7.5 MW, 10 rpm, Enercon E-126, rotor diameter 127 m, d.c. excited synchronous generator (after [7]).

Besides PMSG and dc-excited SGs, cage rotor induction generators (CRIGs), all with full power PWM converters,

DFIGs (with partial (15- 50)% PWM converters rating, are also commercial up to 5-6MW per unit in transmission-topologies. The two (690V) or multilevel (3-6kV) dual ac-dc-ac PWM converters used practically in industrial systems are shown in fig. 15. The SIG PWM Converters (based on 3.3 kV, 30 A devices) might lead to two level multiMW implementation due to their increased switching frequency and lower losses. After this short state of the art briefing let us go into some design and control details of the main 4 types wind generators.

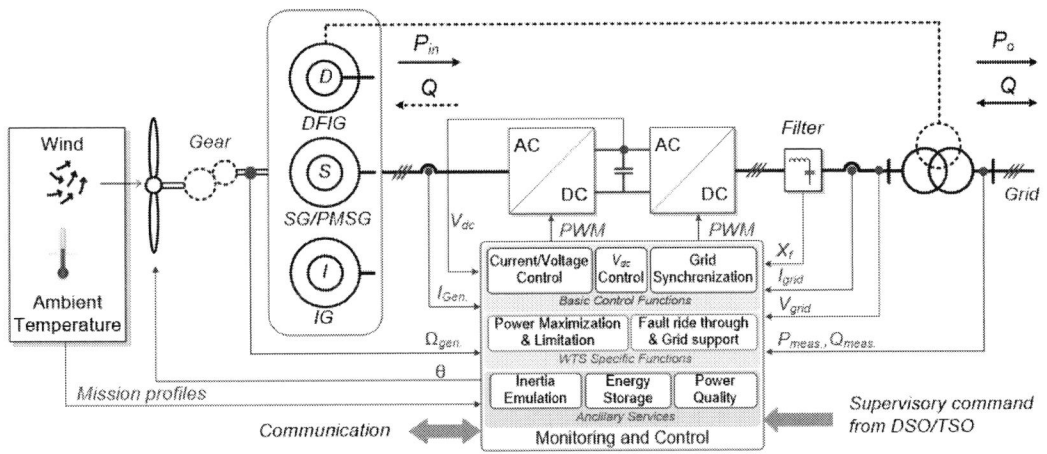

Fig. 15. Generic wind generators control system by ac-dc-ac bidirectional static power converters (after [8]).

IV. DFIG WIND GENERATOR SYSTEMS

DFIG represents more than 50% of all installed wind power system and is preferred mainly because of the converter reduced kVA(15-30%) and costs. Its main anatomy rotor side and grid side converters field oriented control are illustrated in Figs. 16-18.

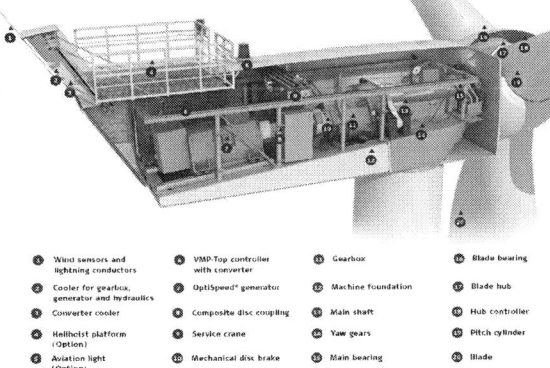

Fig. 16. 4.5MVA DFIG [after 12] 50Hz, 6000V – Vestas V 120, technical data: maximum average wind speed 10m/s, maximum reference turbulence 12%, rotor diameter 120m, operation speed 9.9-14.9rpm, power regulation Pitch/OptiSpeed, cut-in wind speed 4m/s, cut-out wind speed 25m/s, rated speed (4.5MW) 12m/s, gearbox 3G (two planetary and one parallel stage), nacelle weight 145t, rotor weight 75t, (after [12]).

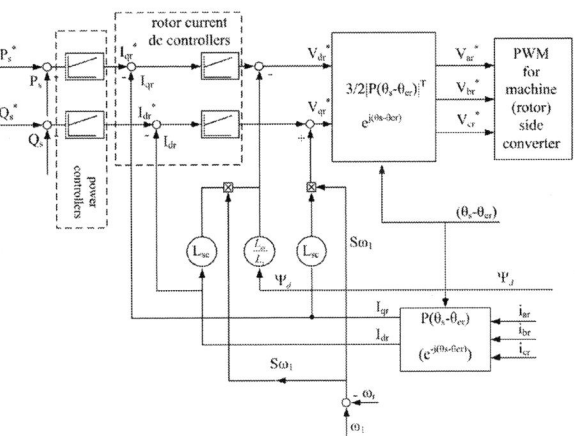

Fig.17. P_s and Q_s FOC of DFIG side ac-dc PWM converter.

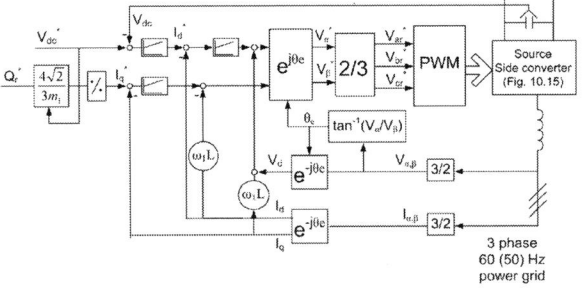

Fig. 18. V_{dc} and Q_r FOC of grid side converter of DFIG.

978-1-5386-7688-2/19 $31.00 © 2019 IEEE

Rotor position is needed for FOC (field orientated control) but not for direct P_S and Q_S control, and it is either measured by an encoder or estimated by an observer (or the latter may be used for redundancy).

Typical short circuit transients on a 2MW DFIG, with encoder-less control are shows in Fig. 19. There is no ballast load and thus the speed increases notably, though it comes back steadily after fault clearing.

Numerous methods to limit rotor voltages and currents during symmetric and asymmetric voltage sags have been proposed [32].

Recent design attempts extend the power range of DFIG to 10MW in direct drives (10rpm), but with a flexible frame to allow an airgap of 1.5-2mm for a 10-12m air-gap diameter [13].

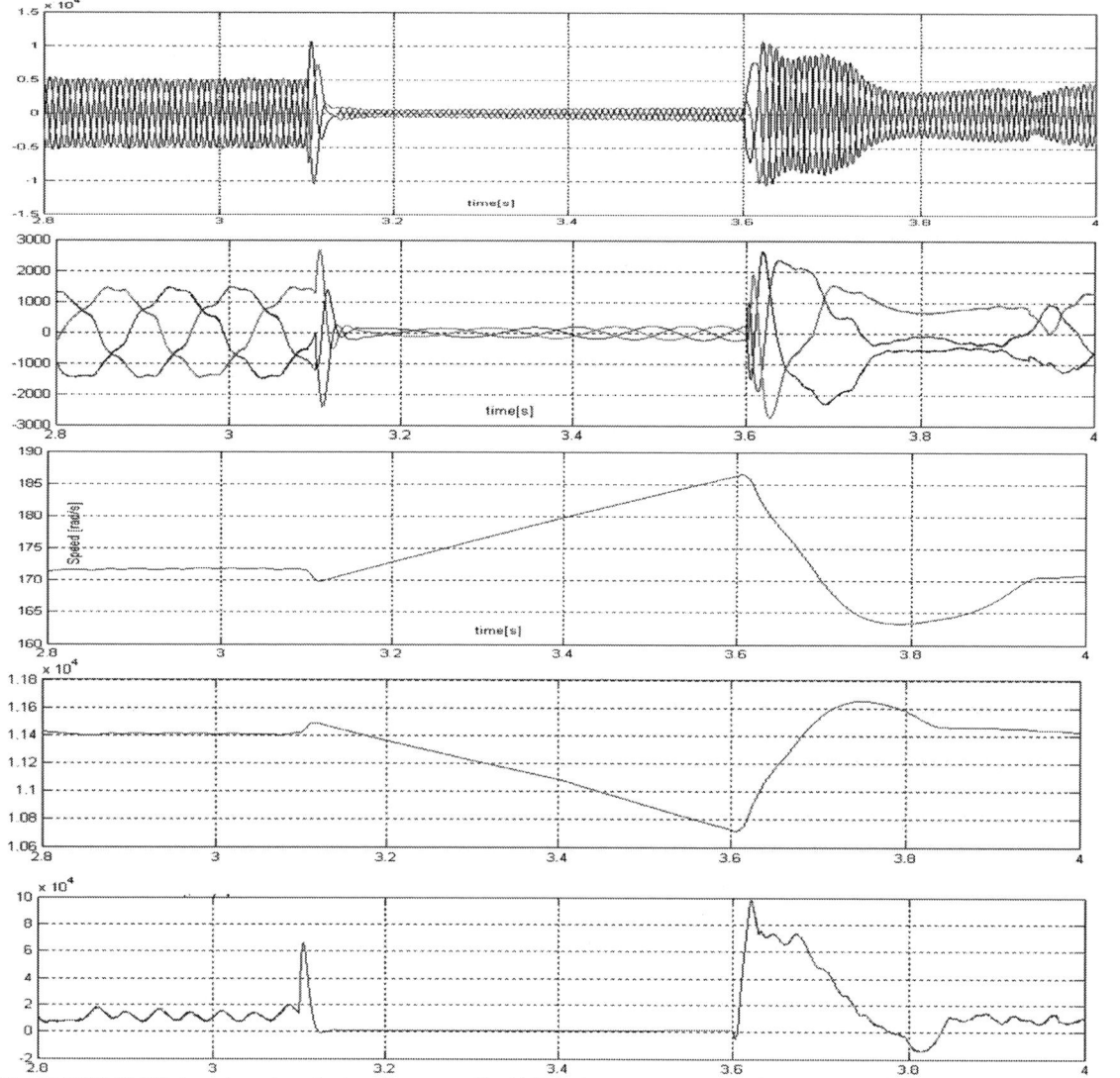

Fig. 19. DFIG (2MW) three phase short circuit on the power grid with limitations on rotor currents: a)stator currents, b)rotor currents, c)speed, d) turbine torque [Nm] & DFIG electromagnetic torque.

As all reactive power generated to the grid by DFIG "springs" from the dc link capacitor, most of the time DFIG operates at unity power factor in the stator, thus playing a small role in voltage regulation in the power grid. Also, recently, variable frequency not only in the rotor but also in the stator of DFIG was proposed either to reduce the PWM converter kVA or to allow only a diode rectifier output as sending end of a MHVDC transmission line from the wind farm (avoiding the full power controlled rectifier at sending lend of M(H)VDC transmission line).The presence of the copper rings and brushes to collect/transmit (15-30%) of power from/to the rotor is still a severe limitation of DFIG.

V. VARIABLE SPEED CRIG WIND GENERATORS

CRIG is brushless, rugged, but implies a full power dual ac-dc-ac converter while allows for generous contribution to grid voltage regulation. A typical realization at 5.6 MW is shown in its anatomy in Fig. 20, with its typical control in Fig. 21and typical grid voltage and current waveforms in Fig.22. Needless to say that encoder-less control is performed but not yet widely applied in industrial systems, together with more robust (sliding mode) state observers and controllers. The solution is robust with acceptably good efficiency and thus it is credited with stronger penetration in power grids in the near future.

978-1-5386-7688-2/19 $31.00 © 2019 IEEE

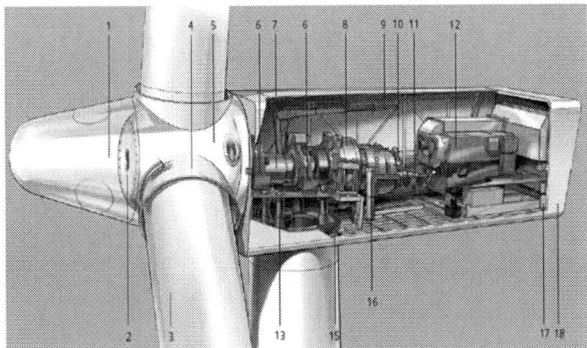

Fig.20. CRIG (Siemens SWT-3.6-120, S_n=3.6MW , n=5 to 13rpm , diameter 120m, gear box ratio 1:119, cut-in wind speed 3-5m/s, cut-out speed 25m/s, rated speed 12-13m/s, rotor weights 100t, nacelle weight 140t, after [16]).

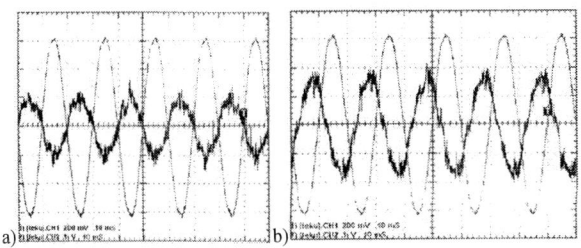

Fig. 21 Indirect or direct FOC of CRIG with emf compensation

a) b)

Fig. 22. Steady state grid voltage and current for generator at full (100%) torque (active power) and 1500rpm (11kW CRIG): a) zero reactive (Qg=0), b) 50% reactive power delivery (after [4] vol.2).

VI. VARIABLE SPEED WIND PM SYNCHRONOUS GENERATORS (PMSGS)

Surface PM, radial air-gap SGs with 3 slots/pole have been so far used for direct drive PMSG up to 3MW, 15 rpm and up to 8MW at 490rpm (q>1 distributed windings) in three phase configuration. The surface PM distributed

winding (q≥1) topology was used, in our view, mainly because of its total small synchronous reactance Xs which implies lower voltage regulation that leads to lower dc link capacitor rating. Sample results from our optimal design of 8MW, at 480 rpm (and 1500 rpm) PMSG are shows in Fig. 23 and in Table 3.

TABLE 3 8MW, 3600V, 1500RPM AND 480 RPM PMSG.

Generator	Efficiency	Total weight (kg)	Total initial cost (USD)	Outer stator diameter (m)	Stack length (m)	Frequency (Hz)	Number of poles	PM weight (kg)
8MW, 1500rpm	0.98296	8 560	197 027	1.252	3.5	75	6	629
8MW, 480rpm	0.98475	9 799	249 376	2.177	3	56	14	875

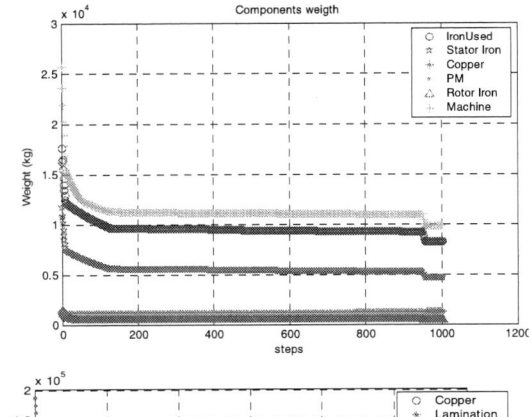

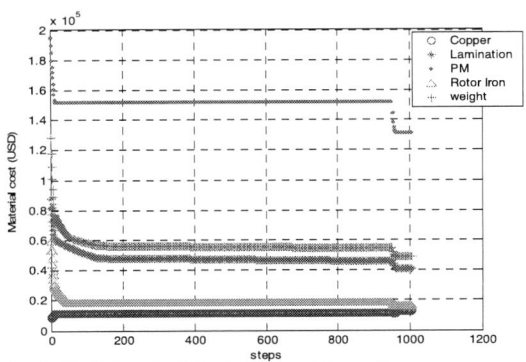

Fig. 23. Evolution of weight, a) and material cost, b)

The performance (efficiency, weight) is good but the mandatory transmission with its weight and cost should be added to the picture too.

Typical FOC of PMSG full power side converter (Fig. 24) in an encoder-less implementation shows satisfactory performance in asymmetric voltage sags (Fig. 25, 26). The full power ac-dc-ac converter also uses a filter for the voltage and current in the power grid or to filter directly and symmetrize stator & rotor currents; finally it has to handle automatically the connection and disconnection (with ballast load) to/from the power grid (Fig. 26).

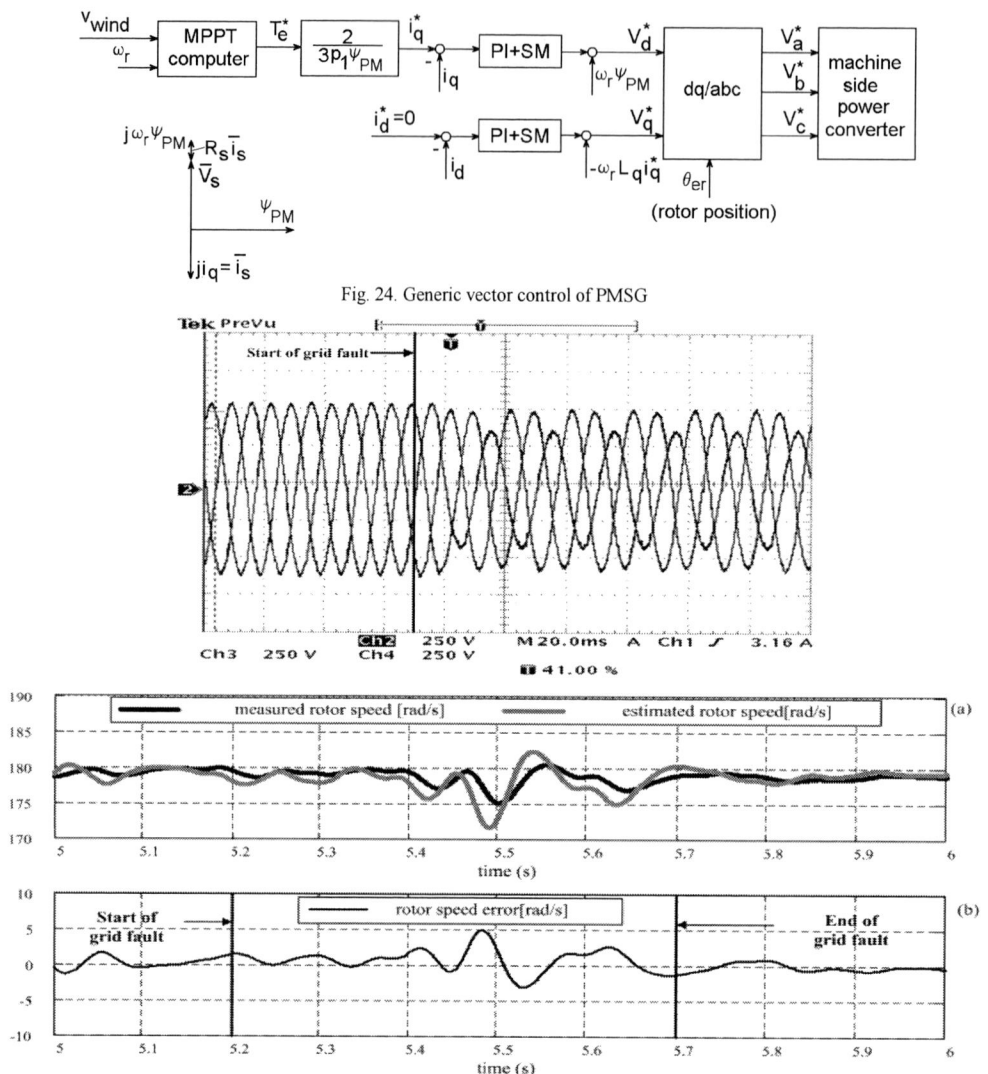

Fig. 24. Generic vector control of PMSG

Fig. 25. Single phase voltage sag operation of PMSG: a) grid side voltage, b) estimated, measured speed and estimated speed in rad/s (after [22]).

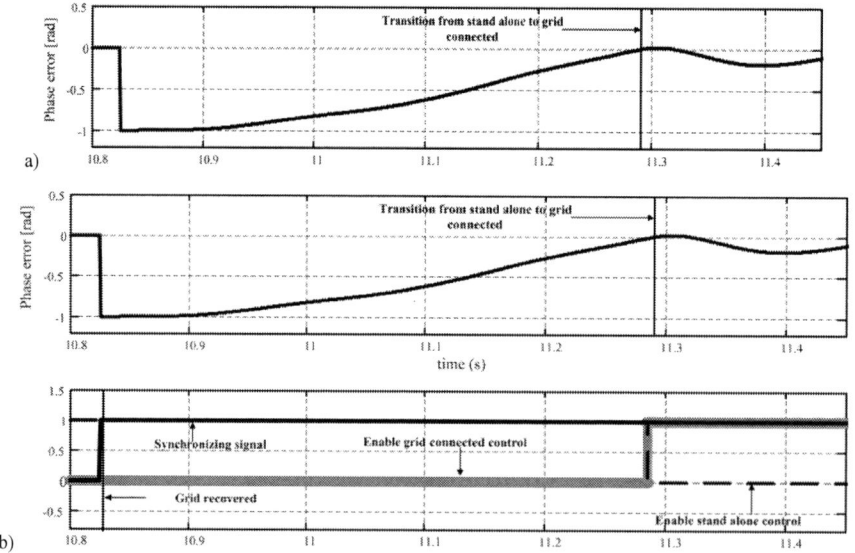

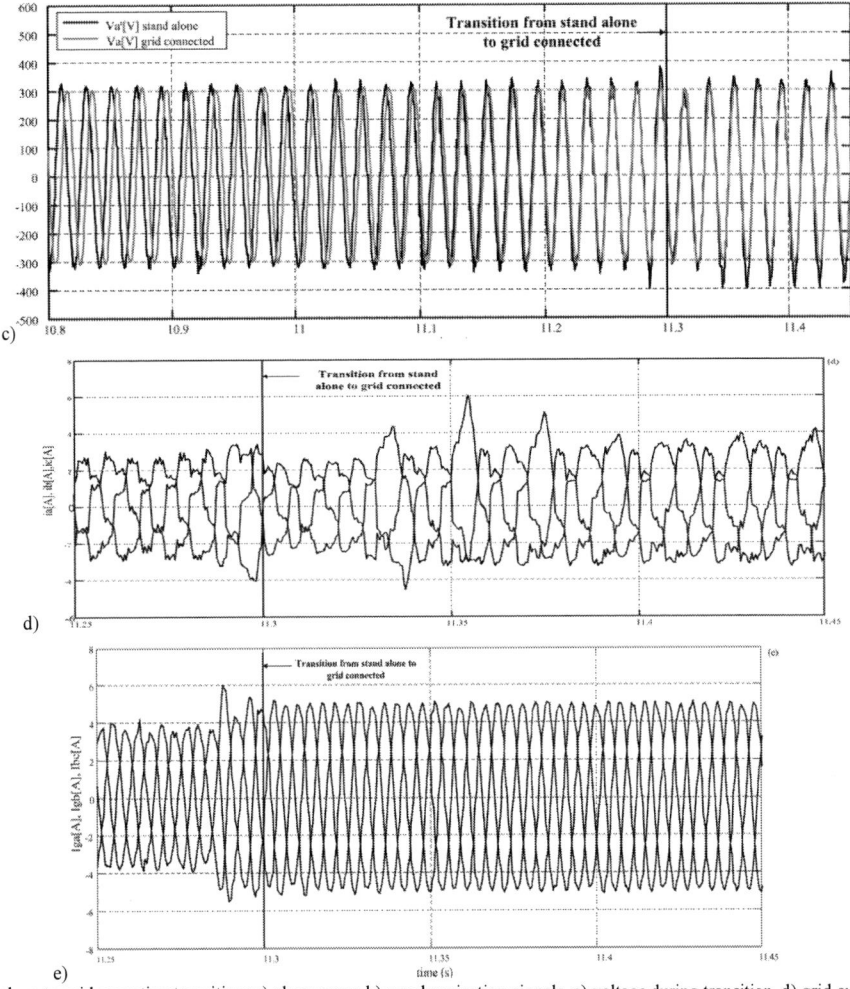

c)

d)

e)

Fig. 26. PMSG stand alone to grid operation transition: a) phase error, b) synchronization signals, c) voltage during transition, d) grid currents, e) generator currents (after [22]).

Rather recently other PMSGs have been proposed but did not reach yet wide industrial stage:
- flux reversal tooth wound PMSG, Fig. 27, with PM–rotor
- flux reversal dual stator PMSG, Fig. 28, with PM–rotor
- BLDC-MRM (multiphase reluctance generator),Fig. 29a
- Vernier PMSG, Fig. 29b,
- Magnet geared PMSG, Fig. 30,
- Transverse flux axial airgap PMSG, Fig. 32.

Each of them has merits and demerits. There are many other radial or axial airgap alternative configurations in recent literature [36], but they are more or less similar to the types illustrated here.

The TF- axial airgap SG (Fig. 31) is treated here via optimal design in a 3 MW, 11 rpm case study as it turned out to need only 1.73 tons of NdFeB for a total 9300kg of active weight only. Sample optimal design which uses a 3Dnonlinear magnetic circuit model of the machine, was performed with sample results given in Table 4.

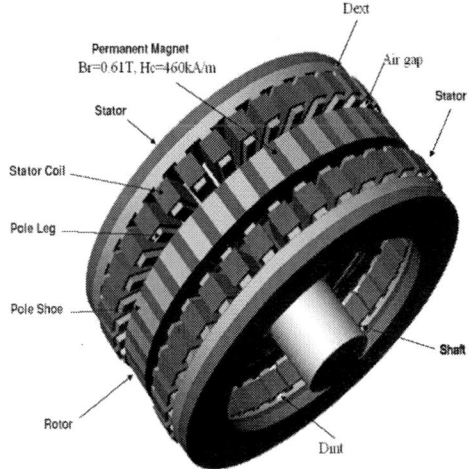

Fig. 27. Flux reversal tooth–wound coil PMSG with PM rotor.

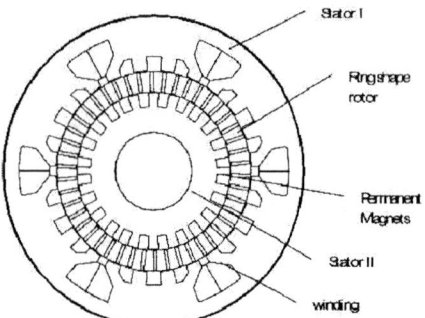

Fig. 28. Flux reversal dual stator PMSG with PM rotor.

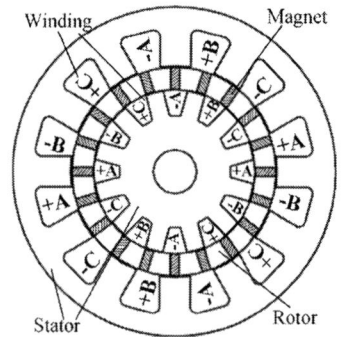

Fig. 29 Vernier PMSG (with dual stator).

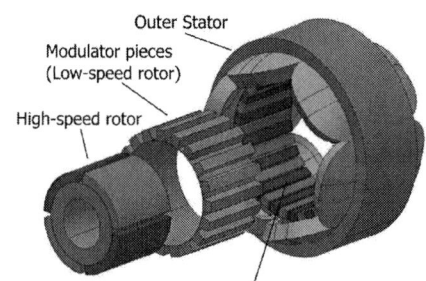

Fig. 30 Typical structure of an OS-MGPM machine (after [34]).

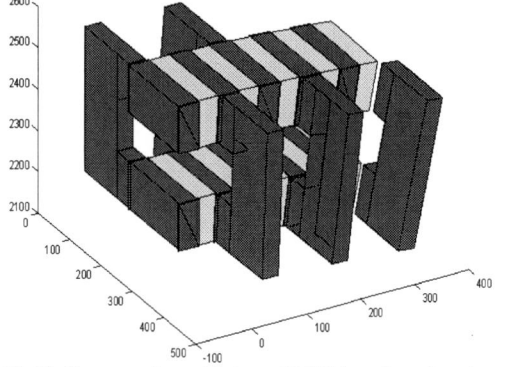

Fig 31. Transverse flux axial air-gap PMSG (one phase shown).

TABLE 4. TF-AXIAL AIRGAP PMSG (3 MW, 11 RPM) COST PARAMETERS. ANALYTICAL MODEL. OPTIMIZED ANALYTICAL MODEL [33].

No.	Poles	Outer Diameter (m)	Efficiency (%)	Weight (tons)	Comments
1	220	14.18	96.53	70.67	Better efficiency
2	220	13	92.81	66.81	Lowest weight
3	140	10	94.68	132.9	Lowest diameter
4	200	13	96.32	77.02	Optimum 1
5	180	12	94.54	80.0	Optimum 2

VII. DC EXCITED SG

A typical multipole SG with dc excitation configuration (Fig. 32) may be benefit by using inter-pole rotor magnet to reduce the excitation flux leakage between rotor poles to increase power by up to 20% and increase efficiency but at least 1%. As this type of generator has been used mainly as a direct drive up to 7.5 MW at 10 rpm we present here some optimal design results for such a case study (in table 5 and in Fig. 33).

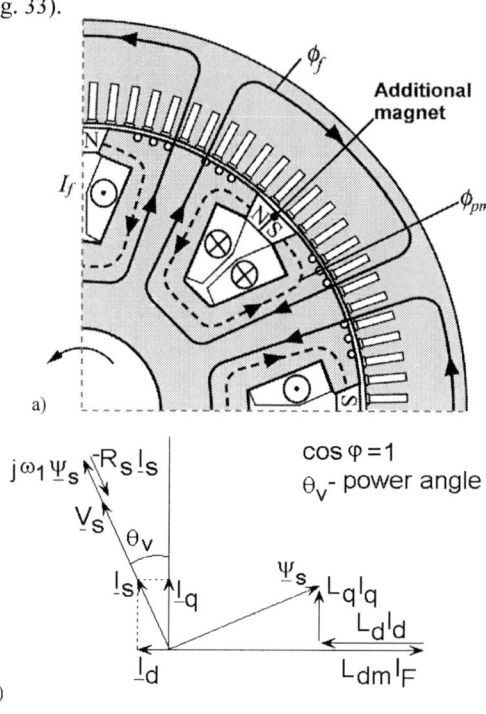

Fig. 32. DC-E SG with assisting interpole PMs: a) cross – section, b) phasor diagram at $\cos\varphi_1=1$.

TABLE 5. 7.6 MW, 11RPM, DCE-SG.

Cost parameters	Analytical model value	Optimized analytical value
Total copper cost: Cu_c	8018 [USD]	19140 [USD]
Total iron lamination cost: lam_c	35419 [USD]	28474 [USD]
Total active material cost: i_cost	144684 [USD]	134424 [USD]
Total passive material cost: pmw_c	69373 [USD]	65416 [USD]
Inverter cost: inverter_c	157842.6 [USD]	112781 [USD]
Energy cost: energy_c	267197 [USD]	153977 [USD]
Total generator costs: t_cost	639100 [USD]	466600 [USD]

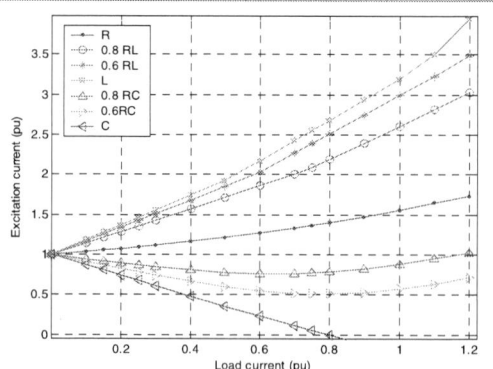

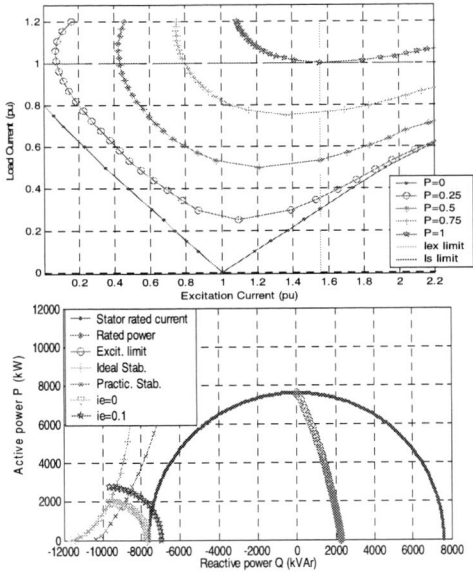

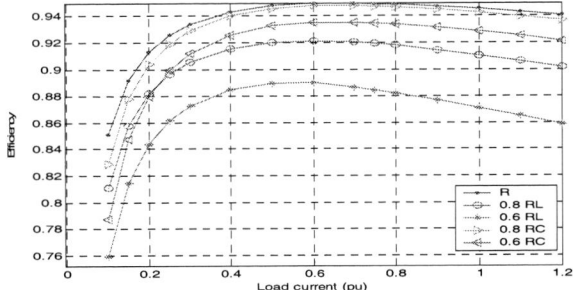

Fig. 33. Optimal design of DCE-SG, 7.6 MW, 3.6 kV, n=11 rpm

A typical FOC for dc excited generator with full power ac-dc-ac PWM converter is shown in Fig. 34.

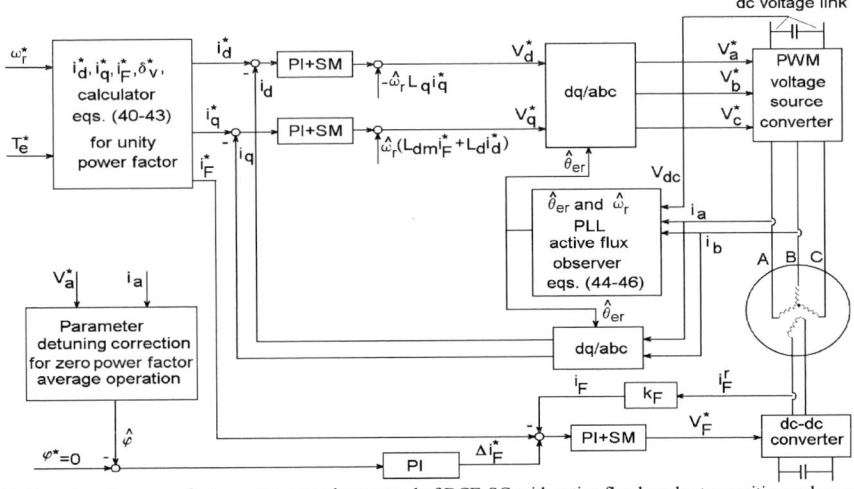

Fig. 34. Generic unity power factor vector sensorless control of DCE-SG with active flux based rotor position and speed observer.

The above synthetic knowledge of dc. excited leads to remarks such as:

- For the case in unit (7.5 MW, 11 rpm) it is possible to reach a machine efficiency of 96.32% for 77 tons of active (table 5) at unity power factor with 100 pole pairs and an airgap diameter D_{airgap}=13 m.
- The machine is capable, as expected to deliver large reactive power to grid. (Fig. 32)
- Yes, the machine seems heaving, but it is a direct drive so the transmission weight, volume, cost are eliminated.
- In direct drives, with all generator types, the larger diameter rotor may be suspended magnetically by dividing the stator winding in 4 parts with 4 (25% each) machine side ac-dc converter whose magnetic flux will control the airgap (like in MAGLEVs).
- Additionally we present here two superconducting SG recent designs (Fig.35) whose active weight seems to be about 20tons (with air core at 10MW 10 rpm, 0.833Hz, 0.964 efficiency.
- The very small fundamental frequency rules out any standard ac-dc-ac PWM converter and thus, apparently, a formidable R&D effort is required to

develop a performance cost competitive such converter. Also the refrigerator for the rotor requires a sizeable additional volume (weight) fabrication effort, and additional costs. The small fundamental frequency stems from the rather large pole pitch required to provide enough large Super Conducting (SC) flux density in the ac stator coils (the SC flux density reduces 2.8 times at a radial distance of τ/π).

- Probably a dual stator central rotor axial-airgap configuration would allow lower pole pitches and thus may allow a larger fundamental frequency to yield hopefully, cost competitive PWNM converters for the scope.

VIII. DIRECT DRIVE CRIG AT 10 MW?

Finally we explored via an elaborated analytical preliminary design a 10MW, 10rpm, 10Hz CRIG generator with copper cage. Synthetic output data are as in Table 6, which includes also direct drive PMSG and DFIG design data at 10MW. Again, as for DFIG, a flexible frame is required to secure a 2mm airgap at 10-12m airgap diameter (at small less than 6m/s peripheral rotor speed such an airgap is feasible.

978-1-5386-7688-2/19 $31.00 © 2019 IEEE

TABLE 6. CRIG: 10MW, 10 RPM, 10 Hz PRELIMINARY DESIGN RESULTS VERSUS DD – PMSG AND DFIG

		10 MW, 10 rpm	DD-PMSG
Number of phases m=6 (2x3) Pole pairs: 60 Torque: 10.37 MNm Lstack: 1.2 m Dos: 12 m! Rated slip: 0.02 Efficiency: 0.936 Power factor: 0.866! Active weight: 62 tons	Stator inner diameter(m)		5
	Stack length (m)		1.6
	Airgap(mm)		10
	Pole pair p1/q1		160/1
	Material and cost		
	Iron (ton) (4 $/kg)		47
	Copper (ton) (20 $/kg)		12
	PM (ton) (33 $/kg)		6
	Construction (ton)		260
	Total (ton)		325
	Total losses (GWh)		2.87
	Energy yield (GWh)		48.4
	Total cost		$3.325*10^6$

It should be noted, that the performance in terms of active weight and efficiency are almost similar for CRIG, PMDG, DFIG 10MW, 10rps/ direct driven. The power factor of CRIG is rather good (0.866) and thus the inverter is less expensive. It has a smaller number of poles (smaller frequency, 10 Hz. The 6 phase configuration should reduce somewhat the strain on the dc capacitor of the 7MW converter.

IX. CONCLUSION

* Large dc excited synchronous variable speed hydro generators up to 100MW with full power converter ac-dc-ac PWM converters have been introduced in industry for 2-3% energy yield addition and for flexibility in the power grid.

* Large (up to 400MVA) pump storage hydro DFIGs have been successfully introduced in industry. PMSG at 8MW, 480rpm and dc excited SG at 10 rpm with full power ac-dc-ac converter have been commissioned recently.

* CRIG with transmission wind generators up to 4.5-6.5 MW are used in most wind power plants.

* Increasing the power to 10 MW in CRIGS and DEFIGs direct drives at 10 rpm seems feasible but the design so far have to go through "the strains" of full scale prototyping etc. before they may reach the markets.

The TF axial airgap PMSG and superconducting SG at 10MW, 10 rpm seems to show unusually small active weight, but are still burdened by respectively 5 tons of PMs or respectively very low frequency converters.

As the speed for given power in hydro sites it about 10 times larger than foe wind generators, the later technologies may be extended rather easily to hydro-systems with variable speed generators.

PWM converter performance and reliability are on the rise with the cost still going down. The use of DFIG with voltage boost transformer and diode rectifier output and variable stator (not only rotor) frequency to provide the sending end of H(M)VDC transmissions lines may also represent a way of the future.

REFERENCES

[1] B. Wu, Y. Lang, N. Zargari, S. Kouro, "Power Conversion and Control of Wind Energy Systems, Wiley-IEEE Press eBook Chapters", 2011, pp. 449 - 453.

[2] G. Abad, J. Lopez, M. Rodriguez, L. Marroyo, G. Iwanski, "Doubly Fed Induction Machine: Modeling and Control for Wind Energy Generation", Wiley, 2011, ISBN: 978-0-470-76865-5.

[3] MG Simoes, FA Farret, "Renewable energy systems: design and analysis with induction generators", CRC Press Taylor and Francis Group, New-York, 2004.

[4] I. Boldea, "Electric Generators Handbook", Vol.1 and 2, book, CRC Press Taylor and Francis Group, New-York, (2nd edition 2016).

[5] McGraw-Hill 2015, "Yearbook of science & Technology", book, pp.203-207.

[6] http://nozebra.ipapercms.dk/Vestas/Communication/Productbrochure/V16480MW/V16480MW/

[7] http://www.enercon.de/en/products/ep-8/e-126/

[8] K. Ma, L. Tutelea, I. Boldea, D. Ionel F. Blabjerg, "Power Electronic Drives, Controls, and Electric Generators for Large Wind Turbines-An Overview", EPCS Journal, Vol. 43 no.12, pp. 1406-1421.

[9] H. Polinder, F.F.A. Van Der Pijl, G.-J. De Vilder,P.J. Tavner, "Comparison of direct-drive and geared generator concepts for wind turbines", Record of IEEE-IEMDC, 2005, pp. 543-550.

[10] H. Polinder, D.J. Bang, R.P.J.M. Van Rooij, A. S. McDonald, M.A. Mueller, "10 MW Wind Turbine Direct-Drive Generator Design with Pitch or Active Speed Stall Control", Record of IEEE-IEMDC, 2007, pp. 1390-1395.

[11] Li F. Hui, Z. Chen, "Design optimization and evaluation of different wind generator systems", Record of ICEMS 2008, pp. 2396-2401.

[12] www.nrg-systems.hu/dok/en/V120_UK.pdf

[13] V. Delli Colli, F. Marignetti, C. Attaianese, "Analytical and Multiphysics Approach to the Optimal Design of a 10-MW DFIG for Direct-Drive Wind Turbines", IEEE Trans. on Industrial Electronics, Vol. IE-59, no 7, 2012, pp. 2791-2799.

[14] R. A. McMahon, P.C. Roberts, X. Wang, P.J. Tavner, "Performance of BDFM as generator and motor", Electric Power Applications, IEE Proc. Vol. EPA-153, No. 2, 2006, pp. 289-299.

[15] L. Mihet-Popa, I. Boldea, "Variable speed wind turbines using induction generators connected to the grid: Digital simulation versus test results", OPTIM 2004, Vol. II: pp. 287-294.

[16] http://www.energy.siemens.com/co/en/renewable-energy/wind-power/platforms/g4-platform/wind-turbine-swt-3-6-120.htm#content=Design

[17] I. Boldea, S. A. Nasar, "Electric drives"–book, 3rd edition, 2016, CRC Press Taylor &Francis Group, New York.

[18] I. Boldea, L. Tutelea,"Electric machines", (chapter 14-15), 2010, CRC Press Taylor & Francis Group, New York.

[19] B. Novakovic, Y. Duan, M. Solveson, A. Nasiri, D. M. Ionel, "Comprehensive modeling of turbine systems from wind to electric grid", Record of IEEE-ECCE, 2013, pp. 2627 - 2634.

[20] A. Fatemi, D. M. Ionel, N.A.O. Demerdash, T.W. Nehl, "Fast multi-objective CMODE-type optimization of electric machines for multicore desktop computers", IEEE-ECCE, 2015, pp. 5593 – 5600.

[21] L. Tutelea, I.Boldea, "Surface permanent magnet synchronous motor optimization design: Hooke Jeeves method versus genetic algorithms", Record of IEEE-ISIE, 2010, pp. 1504 - 1509.

[22] M. Fatu, F. Blaabjerg, I.Boldea, "Grid to Standalone Transition Motion-Sensorless Dual-Inverter Control of PMSG with Asymmetrical Grid Voltage Sags and Harmonics Filtering", IEEE Trans. Vol. PE-29, no. 7, 2014, pp. 3463- 3472.

[23] I. Boldea, M.C. Paicu, G. Andreescu, "Active Flux Concept for Motion-Sensorless Unified AC Drives", IEEE Trans. Vol. PE- Mo.5, 2008, pp. 2612-2618.

[24] I. Boldea, G.D. Andreescu, C. Rossi, A. Pilati, D. Casadei, "Active flux based motion-sensorless vector control of DC-excited synchronous machines", Record of IEEE-ECCE 2009, pp. 2496-2503.

[25] Khron S(editor). Poul-Erik Mort horst P-E, Awerbuch S., "The Economics of wind Energy", A report by the European Wind Energy Association, March, 2009.

[26] http://www.power-technology.com/projects/havoygavlen/havoygavlen5.html

[27] http://www.4coffshore.com/windfarms/turbine-neg-micon--nm-72-2000-tid29.html

Required Technology for Upgrading Efficiency of High-Speed Motor with High Power Density

Masato ENOKIZONO

Vector Magnetic Characteristic Technical Laboratory, Usa, Japan,
Nippon Bunri University, Oita, Japan
enoki@oita-u.ac.jp

Abstract—This paper describes the necessary technical issues for the development of high-speed motors. Especially, it aims at high efficiency by reducing core loss. Its key point is the necessary magnetic properties of materials for high speed motor iron core, its vector magnetic characteristics and magnetic characteristics analysis of the magnetic circuit design.

Keywords—vector magnetic characteristic, core loss. Ultra-thin electrical steel sheet, high efficiency, high speed motor

I. INTRODUCTION

Research on improving the efficiency of the motor is very much,, but research on low loss of the motor iron core is very few. The major research report concerns drive and control of the motor. [1] - [4] Although this is effective for power saving technology, it is different from the technique for reducing the core loss of the motor alone. It seems that efficiency improvement by control is confused with high quality by low loss design of the motor. In this paper, a method for realizing efficiency upgrading by the designing the motor with low loss is described. In order to develop the low loss and high efficiency motor, it is necessary to review the evaluation method of the motor core material and the motor analytic technology. [5][6]

For down-sizing and high power output of the motor, high speed drive and multiple excitation are more promising, but its development is very difficult. This is because the driving excitation frequency must be increased to increase the output power, so that the core loss increases greatly and the motor efficiency decreases. At this time, it is necessary to select the suitable high frequency core material for high speed motors as the primary point required. Useful materials for high frequency cores in the electric power machine are electrical steel sheet, amorphous, 6.5% silicon steel plate, etc.

Table 1 Comparison of high frequency power magnetic materials (<1 kHz)

	Ultra-thin Electrical Steel Steel Sheet	Amorphous and Nano Material Ribbon	6.5% Silicon Steel Sheet
Saturated Magnetic Flux Density	◎	△	△
Magnetic Permeability and Core Loss	○	◎	○
Insulate Coating	◎	×	○
Processability.	◎	×	△
Building Factor Magnetostriction	○	△	◎

Table 1 shows a comparison from the practical point of view of magnetic materials listed as candidates for power high frequency core material. We focused on the ultra-thin electrical steel sheet (50 - 80 µm thick) developed by Nippon Kinzoku Co., Ltd., and an examined the development of technology to effectively utilize it, and decided to suggest

guidelines for the development design of high speed motor. Low loss and high efficiency of the motor are not easy enough to achieve by simply replacing the core material. It is necessary to design and develop according to the magnetic properties of the material. It is based on the detailed grasp of the vector magnetic characteristics of the structural design of the magnetic circuit. In addition, it is also necessary to construct various other necessary technologies, which are difficult to obtain from the conventional techniques.

II. RESEARCH PROBLEM

A. Difficulty of Loss Reduction and High Efficiency of High Speed Motor with High Power Density

The difficulty of development target of realizing the motor is listed below and the research subjects are clarified.

(1) In order to develop the downsizing, high output power of high speed motor, that is, to develop high power density motor, high magnetic flux density design of core material at high frequency is required, but the loss rapidly increases and the efficiency downs remarkably. Also, temperature rise due to heat generation from increased loss becomes a problem.

(2) In particular, eddy current loss increases in proportion to the square of velocity. In addition, countermeasures against an increase in mechanical loss (friction loss) are necessary.

(3) The core structure design (magnetic loading) for suppressing the increase in the hysteresis loss, and the design of the electrical loading will align the bolt.

(4) Improve the difficulty of assembly and manufacturing.

(5) Consistency with motor performance by reexamination of the drive system.

B. Research Subject of Low Loss and High Efficiency of High Power Density Motor

In order to solve the difficult problems in the previous section, the following research subjects are set.

(1) Employ ultra-thin electrical steel sheet (80 µm thick) developed by Nippon Metal Co., Ltd. As high-speed motor core material and grasp the frequency characteristics of vector magnetic characteristics of structure design.

(2) To clarify the iron stator core evaluation technology (building factor) by the development of laminated iron core technology of ultra-thin electromagnetic steel sheet (by Yoshikawa Industry Co., Ltd.) and to improve the performance.

(3) Vector magnetic characteristic analysis for high power density and low loss design of ultra-thin electromagnetic steel sheet (Mutec Corporation). Designing a low loss

structure by design (magnetic / electric loaded design) using μ-E & S vector magnetic characteristic analysis software.

(4) Production and assembly of high speed and high power density motor will be commercialized under the technology of ShinMeiwa Industry Co., Ltd.

(5) During the load test of the motor, the magnetic characteristics in the iron core are grasped while evaluating the magnetic characteristics and the core loss.

(6) As technologies for development, we will update the frequency characteristics of magnetic property, two-dimensional magnetostriction, secondary processing, and improvement of building factors.

(7) Development of drive drivers suppressing high frequency noise, development of bearing evaluation characteristics technology for suppressing mechanical loss.

III. REQUIREMENTS FOR LOW LOSS AND HIGH EFFICIENCY OF HIGH SPEED MOTOR

A. Frequency Characteristic of Vector Magnetic Property.

A-1. Basic Properties of Ultra-Thin Electrical Steel Sheet

Fig. 1 shows the magnetic characteristics of the power magnetic material shown in Table 1 at 50 Hz, the magnetic characteristics of the ultra-thin steel sheet shows almost the same magnetic characteristics as the ordinary non-oriented electrical steel sheet (NO). This improves on the conventional guess that the magnetic properties deteriorate as the thickness of the sheet is decreased. In addition, since the composition is 3.5% Si content, the saturation magnetic flux density is high, enabling high magnetic flux density design for high power density. [7][8]

Furthermore, we have succeeded in suppressing the occurrence of rolling magnetic anisotropy during strong cold rolling for ultra-thin steel sheeting. [9] [10]

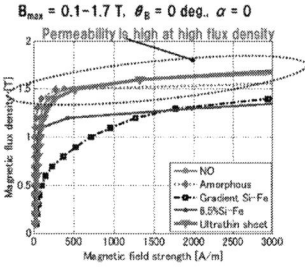

Fig. 1 Basic characteristic of ultra-thin steel sheet

A-2. Vector Magnetic Characteristic

Fig2 shows the duality of electrical and magnetic properties of the motor. The concept of vector magnetic characteristics, including the phase angle θ_{BH} satisfies the duality instead of the scalar magnetic characteristics expressed only by the magnitudes of the conventional magnetic flux density **B** and the magnetic field strength **H** was shown. The magnetic characteristics indicated by the conventional general magnetic characteristic catalog data do not provide accurate information for estimating the performance of the motor. In particular, attention to the

phase angle θ_{BH} between vectors is also important on the magnetic characteristics as well as the power factor in the electrical characteristics.

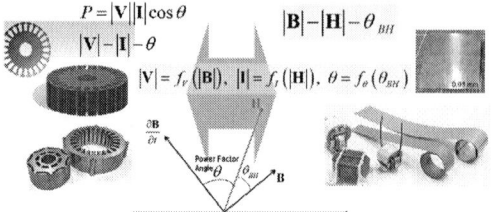

Fig. 2 Concept of vector magnetic characteristics.

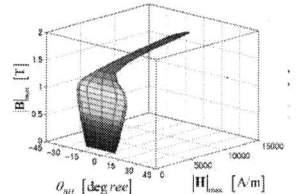

(a) $|\mathbf{B}|$-θ_{BH}-$|\mathbf{H}|$ vector characteristic

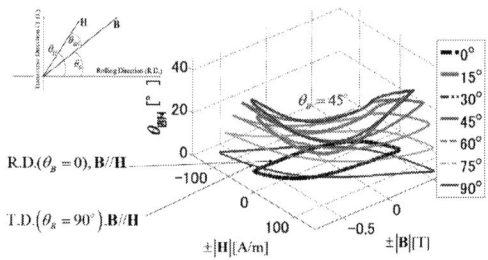

(b) Vector magnetic hysteresis characteristic.
Fig. 3 Vector magnetic characteristics

Figs. 3 (a) and (b) show the result of vector magnetic characteristics in arbitrary directions. Furthermore, its vector magnetic hysteresis loop is shown in Fig. 3 (b). In a direction different from the rolling direction, the vector magnetic hysteresis loop shows a curved shape.

The core loss characteristics in arbitrary direction considering these points are shown in Fig. 4 as W - θ_B - $|\mathbf{B}|_{max}$ characteristics. Furthermore, the vector magnetic properties provide θ_{BH} -θ_B - $|\mathbf{B}|_{max}$ and $|\mathbf{H}|_{max}$ -θ_B - $|\mathbf{B}|_{max}$ characteristics shown in Figs. 5 (a) and (b) as new findings, respect every. These magnetic characteristic data are useful for designing low loss and high efficiency of electric machines, and development technology utilizing them is required. Although the magnetic characteristics in arbitrary direction from the rolling direction of the electromagnetic steel sheet are expressed as vector magnetic characteristics, the grasp of the magnetic field strength necessary for obtaining the magnetic flux density in an arbitrary direction and the spatial phase difference angle θ_{BH} between the **H** vector and the **B** vector is an important point in motor design. The vector magnetic characteristics cannot be obtained from conventional measurement and evaluation methods.

978-1-5386-7688-2/19 $31.00 © 2019 IEEE

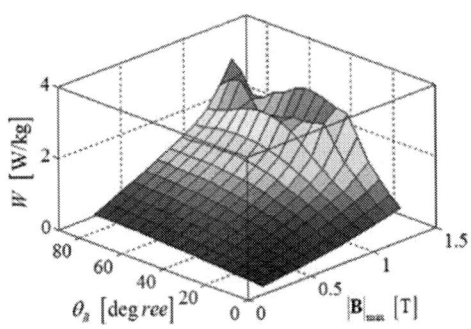

Fig, 4 W-θ_B-$|\mathbf{B}|_{\max}$ characteristic

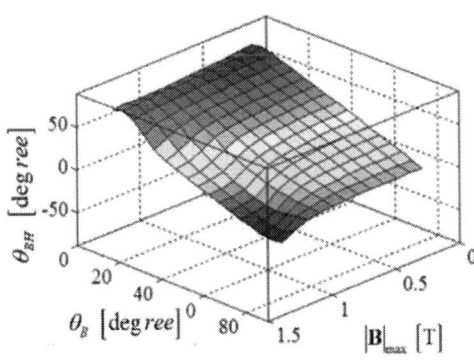

(a) θ_{BH}-θ_B-$|\mathbf{B}|_{\max}$ characteristic

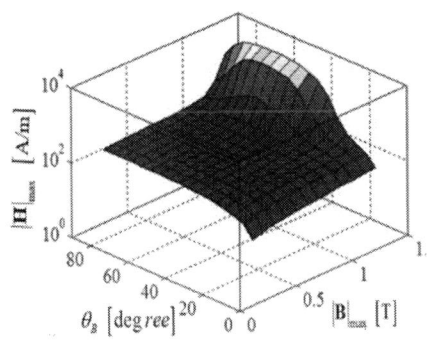

(b) $|\mathbf{H}|_{\max}$-θ_B-$|\mathbf{B}|_{\max}$ characteristic

Fig, 5 θ_{BH}-θ_B-$|\mathbf{B}|_{\max}$ characteristic

A-3. Frequency Characteristic

For the design and development of high-speed motors, the frequency characteristics of the magnetic material are required, and then the frequency characteristics are shown for the vector magnetic characteristics shown in Fig. 6. In order to clarify the characteristics of the ultra-thin electrical

steel sheet applied to this development, the $|\mathbf{B}|_{\max}$-θ_{BH}-$|\mathbf{H}|_{\max}$ characteristic is shown compared to the conventional non-oriented electrical steel sheet. In ordinary electrical steel sheets, as shown in (a), the spread of θ_{BH} becomes smaller, which is a factor of decreasing the power factor. On the other hand, in ultra-thin electrical steel sheets, as shown in (b), no widening of θ_{BH} is observed, and the characteristics of the low frequency region are retained.

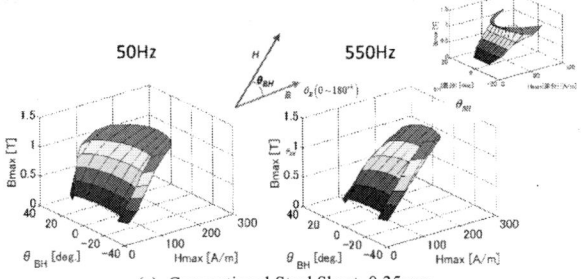

(a) Conventional Steel Sheet, 0.35mm

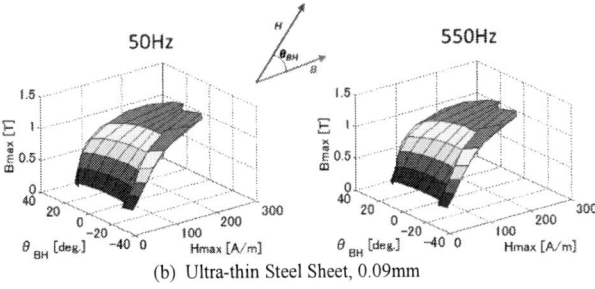

(b) Ultra-thin Steel Sheet, 0.09mm

Fig. 6 $|\mathbf{B}|_{\max}$-θ_{BH} -$|\mathbf{H}|_{\max}$ characteristic

Furthermore, the individual magnetic frequency characteristics are shown in Fig. 7 (a) - (c). These results show the characteristics of the inclination angle of 45 ° from the rolling direction. The frequency dependence of the magnetic permeability constituting the inductance of the motor is larger than that of the conventional non-oriented electrical steel sheet, and a large difference occurs in the high frequency region. This suggests a decrease in core cross section and downsize of the motor. Furthermore, when looking at the change of the magnetic field strength $\mathbf{H}$, it shows an almost constant value, giving insight on suppression of the necessary exciting current and design of electrical loading. Furthermore, the change of θ_{BH} is similarly small, it is useful for design consideration for power factor, $\cos\theta$ and design of the motor core structure, leading to suppression of copper loss. Core loss characteristics will affect the excitation current of the motor and will provide insight into the design of the electrical load.

978-1-5386-7688-2/19 $31.00 © 2019 IEEE

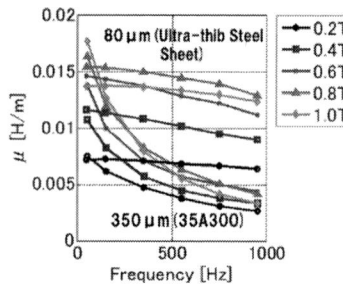

(a) Permeability characteristic

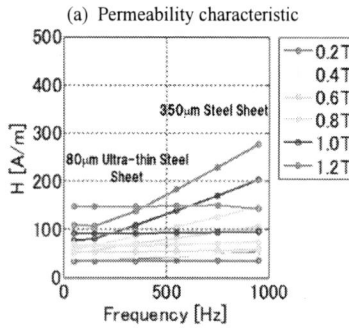

(b) $|\mathbf{H}|_{max}$ characteristic

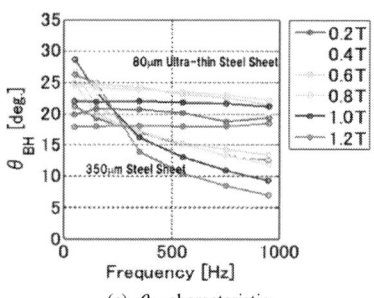

(c) θ_{BH} characteristic

Fig. 7 Frequency characteristic of magnetic property for basic design

IV. EXPECTED MOTOR ELECTRICITY - MAGNETIC PROPERTIES

A. Electrical Loading

By looking at the effect on the performance of the high-speed motor when using ultra-thin electrical steel sheet as the motor core, consider the insight for actual high-speed motor development and discuss the design can direction toward low loss and high efficiency.

Fig. 8 shows the estimated effect of core replace of various electric and magnetic properties in the high speed range. If the conventional material is replaced with the ultra-thin electrical steel sheet and used as the motor core, since the inductance at high speed rotation becomes large, the exciting current under a constant voltage drive becomes difficult to follow. That is, in order to utilize ultra-thin electrical steel sheets, it is necessary to review magnetic loading and design of electric loading taking advantage of the characteristics by greatly suppressing eddy current loss. Particularly, since the magnetic permeability is high in the high speed range, it is necessary to design and review the electric load by changing the size and structure.

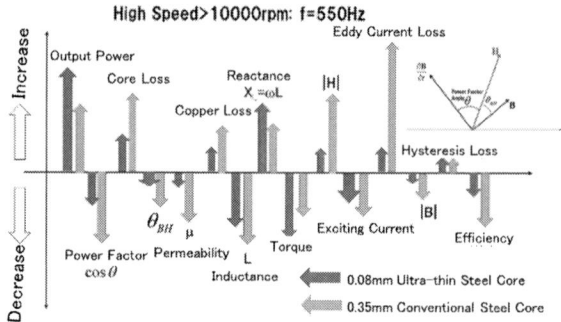

Fig. 8 Estimation of motor by steel sheet performance.

B. Loss reduction design by vector magnetic characteristic analytical Technology

B-1. Benefit of Vector Magnetic Characteristic Analysis

Although computer simulation technology has been used for motor development and design for a long time, it is difficult to grasp the magnetic characteristics in detail.

Fig. 9 compares magnetic field analysis method with conventional general-purpose (commercial packaged software program) with magnetic characteristics analysis proposed by the author. In the conventional method, the core loss as a function of magnetic flux density $|\mathbf{B}|$ characteristic obtained from the catalog data of the electrical steel sheet is used for the estimation of the core loss by using a mathematical approximation function. In the analysis of the conventional method, since only the behavior of vector $\mathbf{B}$ can be solved, there is a fundamental problem that it cannot be directly calculated by using expressions constituting core loss.

In other words, the conventional method represents the core loss distribution depending on the obtained $|\mathbf{B}|$ distribution. It suggested that the magnetic flux density level only has to be lowered for the decreasing core loss motor. The conventional method has been effective in the design mainly on the torque output of the motor. That is because the electromagnetic force depends only on the vector $\mathbf{B}$ as seen in the Maxwell stress equation. In the vector magnetic characteristic analysis, $\mathbf{B}$ vector and $\mathbf{H}$ vector can be solved at the same time, and each vector locus is obtained from the result and it is calculated by assigning it to the iron loss formula. Core loss data are unnecessary for this analysis, and relational data between $\mathbf{B}$ vector and $\mathbf{H}$ vector is required.

This analysis method provides the trajectory (locus) of each vector, the hysteresis loop (B_x-H_x, B_y-H_y), and the characteristic value concerning the core loss from the analysis results of the behaviors of vector $\mathbf{B}$ and $\mathbf{H}$ vectors. In addition, θ_B, θ_{BH}, α, etc. Concerning the behavior of the vector can also be obtained. [7]

978-1-5386-7688-2/19 $31.00 © 2019 IEEE

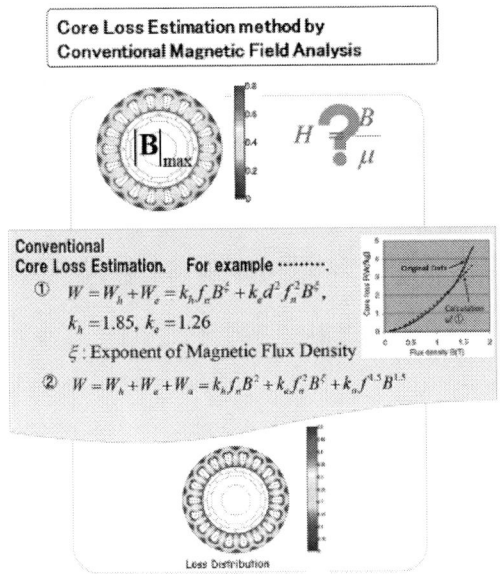

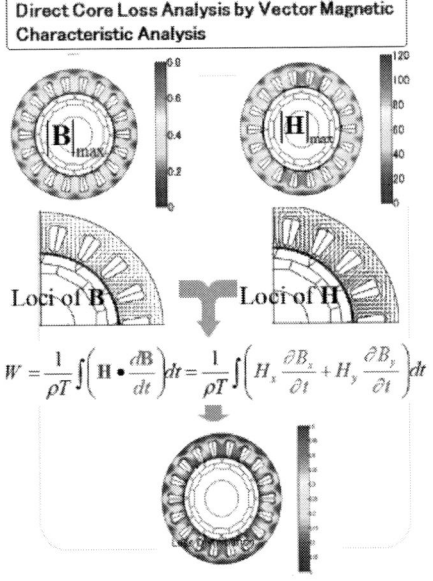

(a) Estimation of core loss by conventional magnetic field analysis

(b) Magnetic characteristic analysis considering the vector magnetic property

Fig. 9 Magnetic characteristic analysis for in design core loss reduction

B-2. Vector Magnetic Characteristic Analysis

Fig. 10 shows the results of (a) maximum magnetic field strength $|\mathbf{H}|_{max}$ distribution and (b) core loss distribution, respectively, due to the difference in sheet thickness of the electrical steel sheet, using vector magnetic characteristic analysis (dynamic E&S model).[7] The magnetic characteristic analysis results obtained under a constant voltage focus on the behavior of the vector $\mathbf{H}$.

Although there is no change in $\mathbf{H}$ and core loss distribution in the low rotational speed region, changes in each distribution are appearing as the rotation speed increases.

The tendency is that $|\mathbf{H}|_{max}$ and core loss are smaller as the thickness of the electromagnetic steel sheet becomes thinner. It would also be possible to see the threshold value around 20000 [rpm] as an effective range of 80μm thicknesses.

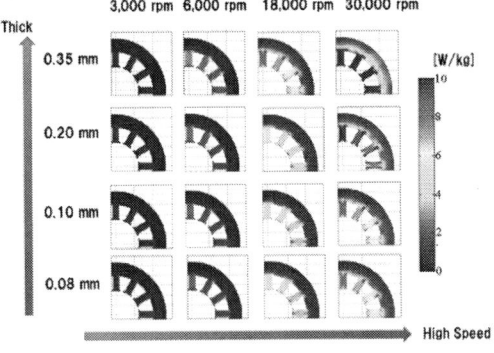

(a) Distribution of $|\mathbf{H}|_{max}$

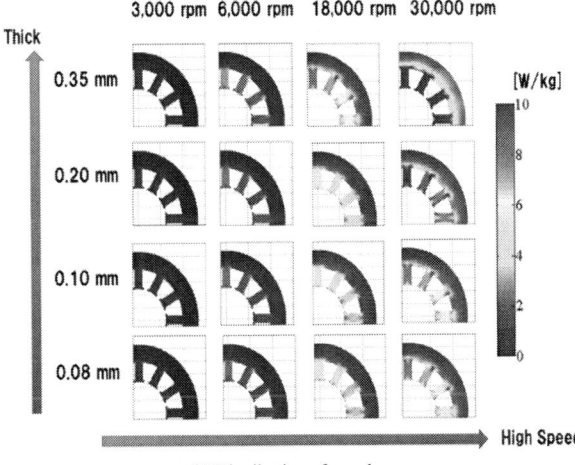

(b) Distribution of core loss

Fig. 10 Frequency characteristic analysis of magnetic field distribution and core loss distribution

V. DESIGN GUIDELINES FOR LOW LOSS AND HIGH EFFICIENCY

In general, the point of view of designing the shape of the motor core with high efficiency due to the low loss of the motor can be determined from the following equation and the core loss calculation formula shown in- (1),

$$ W = \frac{1}{\rho T} \int \left(\mathbf{H} \cdot \frac{\partial \mathbf{B}}{\partial t} \right) dt = \frac{1}{\rho T} \sum_{t=1}^{n} \left\{ |\mathbf{H}|_t \frac{\partial |\mathbf{B}|_t}{\partial t} \cos\left(\frac{\pi}{2} - \theta_{BH} \right)_t \right\} \quad (1) $$

where $(\pi/2 - \theta_{BH})$ is due to the power factor angle. Therefore, the characteristic corresponding to this angle is important for the design of a low loss motor since it cannot be obtained from the conventional magnetic characteristic.

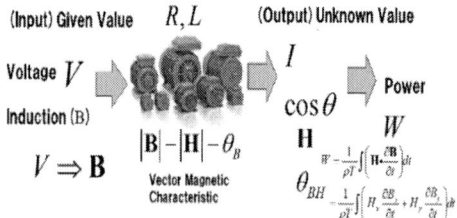

Fig. 11 Given values and unknown value in magnetic characteristic analysis.

Fig.11 shows the contents of the vector magnetic characteristics analysis, basically giving a voltage and solving the **H** vector and θ_{BH}. In other words, the vector hysteresis characteristic should be solved as unknown. The vector magnetic characteristic analysis basically becomes an inverse problem analysis. [11] Fig. 12 shows the matching relationship between electrical loading and magnetic loading in the core loss calculation formula. Matching of iron loss and copper loss is required to obtain a high efficiency motor.

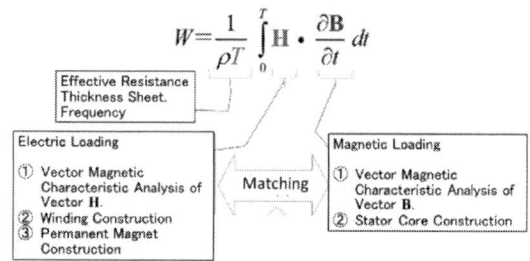

Fig. 12 Procedure of motor design for core loss analysis.

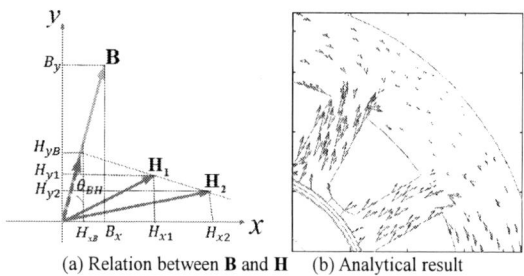

(a) Relation between **B** and **H**   (b) Analytical result

Fig. 13 Behavior of vector magnetic characteristic.

The electrical characteristics and power factor of the motor can be obtained by magnetic characteristic analysis under constant voltage analysis, and loss characteristics and efficiency excluding mechanical loss can be estimated.

By vector magnetic characteristic analysis, it is possible to solve the relation among the **B** vector, the **H** vector and the phase difference angle θ_{BH} as shown in Fig. 13 (b). As a result from definition of Fig. 13 (a), the core loss factors related to this equation are |**H**|, θ_{BH}, α (**B** and **H** vector axis ratio in the trajectory of one cycle = short axis/long axis), and the inclination angle θ_B from the rolling direction.

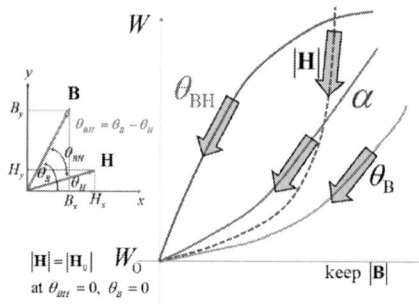

Fig. 14 Parameter for core loss reduction.

Fig. 14 shows the change of core loss with each parameter of the vector magnetic property that affects the increase and decrease of core loss. The core loss value indicates the minimum value (W_0: θ_{BH} = 0, **H**//**B**) in the rolling direction. However, it is difficult to uniformly reduce all of these factors, so it is desirable to make a judgment based on an intensive approach. For example, due to changes in core shape, the θ_{BH} parameter may decrease, but |**H**| and α may increase, The overall judgment is necessary by adopting the control of the parameter that contributes a lot to the reduction of the core loss relatively.

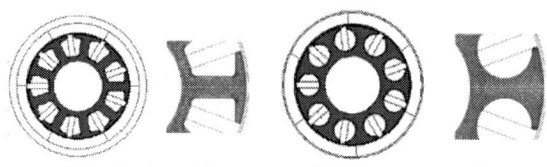

(a)   Basic model   (b) Improvement model

Fig. 15 Analytical models for core loss reduction.

A specific application example of the present technology will be shown and described. The development results of the high speed motor, which is outer rotor typed motor are discussed with the optimum shape obtained by focusing on the above parameters in order to reduce the core loss from the analysis model shown in Fig. 15. Fig. 15 (a) is the basic model designed by conventional technique, and (b) is an improved model designed the core loss minimization model by using the vector magnetic characteristic technique.

In order to reduce core loss from the basic analysis model shown in Fig. 15(a), the result of obtained the optimum shape as shown in Fig. 15 (b) focusing on the above parameters is shown in Figs. 16 (a) - (b). Each distribution is contrasted to investigate the effect of optimization.

Fig. 16 (a) compares the core loss distribution and shows the difference in the total core loss. Core loss is reduced by 18% and improved significantly. This result means the improvement of magnetic loading. Figs. 16 (b) - (d) show the factors that reduced the core loss. A slight decrease is observed despite the design keeping the magnetic flux density level. On the other hand, the magnetic field strength is improved by 7%, resulting in the reduction of the excitation current and the reduction in the copper loss. Overall improvement can be confirmed from the change of θ_{BH} related to the power factor.

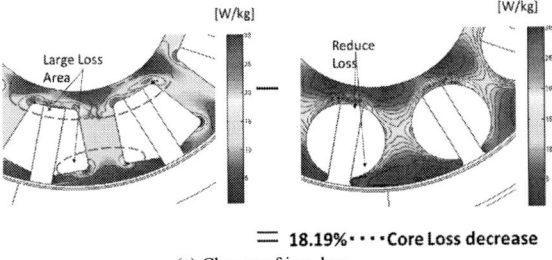

= 18.19% ····Core Loss decrease

(a) Change of iron loss

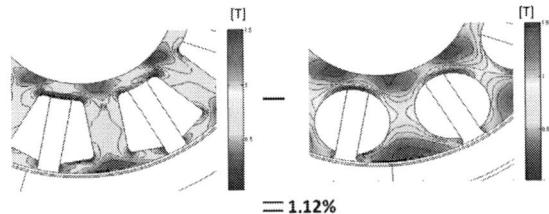

= 1.12%

(b) Change of magnetic flux density |**B**|

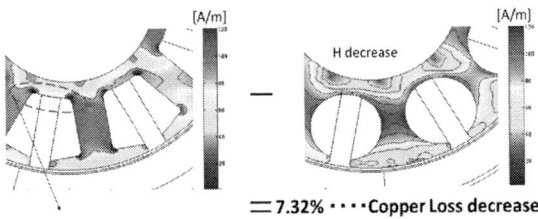

= 7.32% ····Copper Loss decrease

(c) Change of magnetic field strength |**H**|

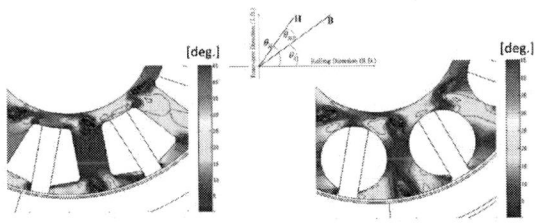

(d) Change of spatial phase angle θ_B

Fig. 16 Change of core loss distribution and θ_{BH}, |**H**| and |**B**| by optimization of motor core shape.

By focusing on the parameters of vector magnetic characteristics that control core loss, it is possible to gain the knowledge on the core shape optimization for core loss reduction.

The changes in magnetic properties were investigated in more detail for each vector locus distribution (rotational flux and rotational magnetic field) and iron loss distribution.

Figs. 17 (a) and (b) show the behavior between vector B and vector H. As shown in the figure, vector B and vector H are not parallel in most places.

The relation of these vectors locus and core loss as shown in Figs. 18 (a) and (b). Naturally vector B and vector H have a phase difference Relation between vector locus and core loss are shown in Figs. 18 (a) and (b). At the portion where the core loss decreases, the axis ratio of magnetic flux density vector trajectory decreases, and the direction of the magnetic field strength vector approaches the flow of the magnetic flux density and becomes smooth. This

phenomenon causes changes in the tips and corners of the teeth, as shown in Figs. 18 (a) and (b). From this result, it is not preferable that the shape of the magnetic flux changes rapidly.

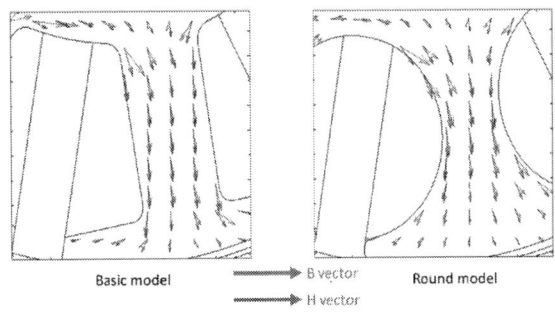

(a) Basic model   (b) Improved model

Fig. 17 Distribution of behavior of vector **B** and vector **H**

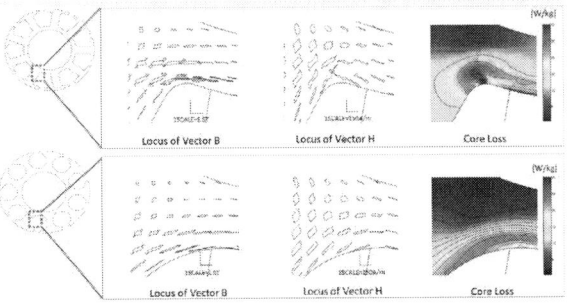

(a)   Region of teeth back core

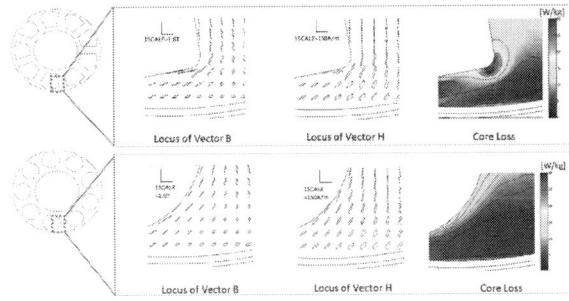

(b)   Region of teeth top core.

Fig. 18 Vector behavior and relation between vector locus and core loss.

VI. Requirement of Analysis - Design Procedure

A. Increase factor of motor loss

According to the design procedure of motor development, the design policy will be determined from the output. However, the design with low loss is possible only by solving the B vector, the H vector, and the phase difference angle θ_{BH} between the vectors under the constant voltage input condition, based on the vector magnetic characteristics of the core material.

We will investigate shape deformation by trial and error while paying attention to the iron loss increase factor shown in Fig. 14. Under the constant voltage condition, the magnetic flux density of the moto core is kept constant, but its direction is unknown. Then, while grasping the motor performance, we examine the output and efficiency from current and power factor.

Fig. 19 increases the core loss of the motor, thereby increasing the building factor. This increase factor can be classified into material selection, magnetic circuit design (magnetic loading), excitation structure (electric loading), and stress effect of manufacturing assembly. These factors influence each other, but the magnetic properties of the material are influenced. In particular, two-dimensional magnetostriction and stress magnetic properties based on vector magnetic properties are the main elements. Residual stress due to punching of steel sheet deteriorates the magnetic characteristics of the iron core. Such deterioration of the motor performance due to deterioration of the magnetic characteristics is a serious matter. Degradation of the magnetic properties of the core means an increase in the building factor. It is strongly required to establish a technology for diagnosing the deterioration of the magnetic properties of the iron core of the motor production process. In the future, development of a building factor evaluation technology should be emphasized in order to ensure high reliability due to the high quality of the motor.

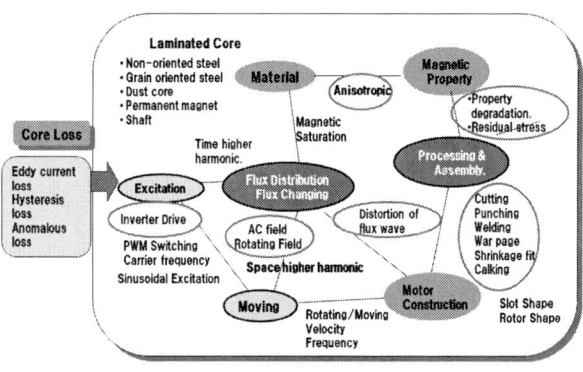

Fig. 19 Increase factor of core loss.

B. Motor load - magnetic characteristic measurement evaluation system

Fig. 20 shows the proposal of the procedure for designing the low loss of the motor. Vector magnetic characteristic analysis method is mainly considered, and the magnetic characteristics of the motor iron core are monitored. Following traditional design techniques to obtain the basic structure, vector magnetic characteristics technology has been introduced for low loss of the obtained structure. In order to suppress deterioration of the magnetic characteristics of the core, a secondary processing technique is used. The state of the magnetic properties of the iron core must be grasped at each stage of development. And it is necessary to be reflected in improvement of motor performance. This process can be repeated several times to reach the final stage.

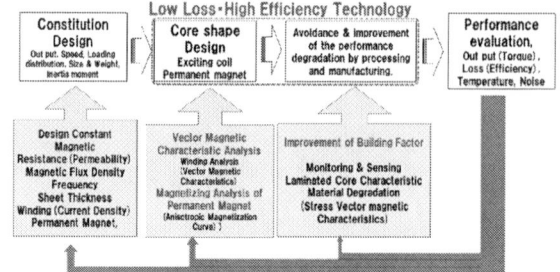

Fig. 20 Suggestion of design process

C. Motor load - magnetic characteristic measurement evaluation

For the development of motors, it is important to utilize computer simulation and actual experimental results like the two wheels of a car. Evaluation of high speed motors is technically very difficult compared to conventional low speed motors. The most difficult points are the load system that can cope with high speed rotation and the determination of the true torque by removing the influence of the rotational moment of inertia. Since it is difficult to separate the motor loss from the mechanical loss, an accurate evaluation method for iron loss has not been developed.

For that purpose, conventional motor testers are not enough, and it is necessary to capture the motor from the viewpoint of magnetic characteristics of motor core material.

Fig. 21 shows the outline of the proposed motor load-magnetic characteristic measurement system.

Measurement must be performed under voltage sinusoidal drive via a power amplifier. Apply a probe coil for magnetic flux density measurement to the motor iron core and grasp the state of the magnetic flux density waveform inside the iron core. Furthermore, by installing a temperature sensor in the iron core, thermal local iron loss can be estimated. From the result under the constant magnetic flux density condition by feedback control, the performance of the motor can be evaluated from the magnetic test.

Especially at high speed driving, the influence of the moment of inertia of the rotor on the output affects the result, so its evaluation needs attention.

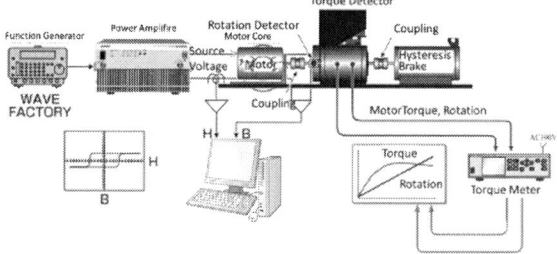

Fig. 21 Motor-torque-speed-magnetic characteristic evaluation system.

VII. MAGNETIC EVALUATION OF MOTOR PERFORMANCE

It is necessary to evaluate from the same viewpoint as comparative evaluation of magnetic material in order to judge high efficiency by decreasing core loss of the motor. By evaluating the motor performance under the constant magnetic flux density condition, it is possible to judge the effective utilization degree of the material. Fig. 22 shows load test results of core loss of motor under constant voltage condition (10 [V]) and constant flux density level condition (B=1.2 [T]), respectively. Comparing the difference in the obtained results, the characteristics of the core material are not sufficiently reflected in the load test under the constant voltage condition. Although the result of the performance evaluation of the motor is sufficient, the information for effective utilization of the material is insufficient. It is necessary to change the design of magnetic loading and electric loading from the result of the difference in loss characteristics due to replacement of the core material. For that change the magnetic properties of the result must be analyzed. Therefore, in order to know the utilization situation of materials, it is necessary to evaluate the load characteristics with respect to core loss and copper loss under the condition where the magnetic flux density level is constant.

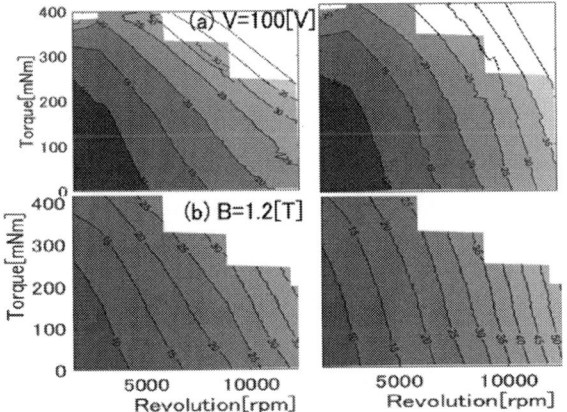

Ultra-thin steel (80μm) core   Conventional steel (350μm) core

Fig. 22 Comparison by measurement condition

Fig. 23 (a) shows the iron loss versus frequency characteristic of the material, and (b) shows the iron loss versus frequency characteristic of the motor. By comparing the frequency characteristics of (a) and (b), it is possible to investigate the building factor of material deterioration. Focusing on 500 Hz, the 350 μm steel sheet increased from 50 [W/kg] to 70 [W/kg], the 80μm ultra-thin steel sheet increased from 20 [W/kg] to 40 [W/kg], and the building factors were 1.4 and 2.0, respectively. As the steel sheet becomes thinner, the influence of residual stress applied to processing and manufacturing process increases. Therefore, secondary treatment for removal of residual stress is important.

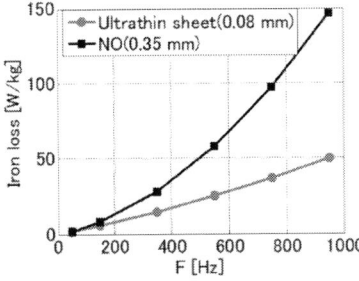

(a) Characteristic of material

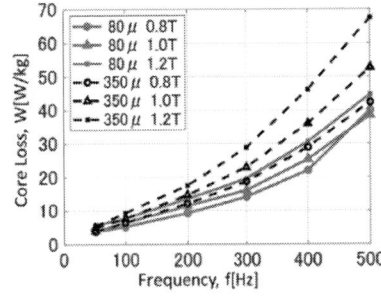

(b) Characteristic of motor

Fig. 23 Evaluation of building factor

Fig. 24 shows the frequency characteristics of copper loss. By using ultra-thin electrical steel core in the case of fixed the magnetic flux density level B=1.2 [T], excitation current decreases and copper loss decreases. It can be understood that the resistance magnetic field due to the eddy current decreases and the exciting current is caused by the ultra-thin electrical steel core.

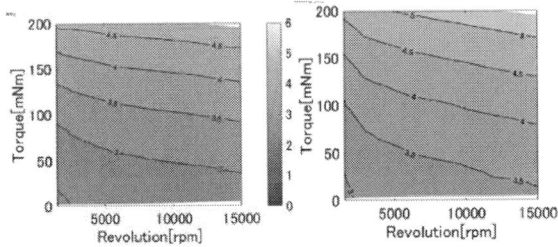

Ultra-thin steel (80μm) core   Conventional steel (350μm) core

Fig. 24 Frequency characteristic of copper loss

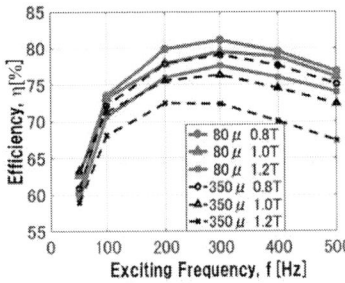

Fig. 25 Frequency characteristic of motor efficiency.

Fig. 25 shows the frequency characteristic of efficiency. The results show that the efficiency is highest around 300 [Hz]. This means that the copper loss and the iron loss are aligned in this vicinity. Here, the effect of replacing the iron core of the same structure with the material was investigated, but it is understood that it is not reflected in the improvement of the efficiency. Fig. 26 shows the appearance and performance of JAXA motor designed by appropriate means. This motor is for space exploration vehicles, and a special specification was imposed. This motor was compact and required high output and low loss. The target specification of this motor is more15000 [rpm] , and it is the high power density of 25 [g] weight and 50 [W] output. Furthermore, because it is the motor for space rover, small size, small weight and lower loss are required. Since a cooling device can not be used, the temperature rise due to loss must be minimized.

Voltage : 40 V
Weight : 25 g
Output : 50 W
Speed : 24000 rpm
Torque : 20 mN
Efficiency ; 85 %
Thickness : 6 mm

Fig. 26 the appearance and performance of JAXA motor

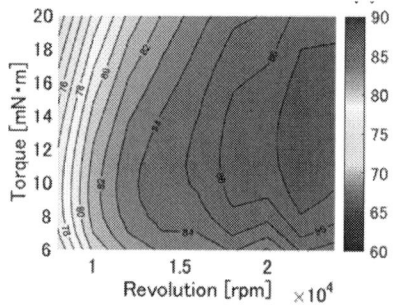

(a) Ultra-thin steel (80μm) core motor

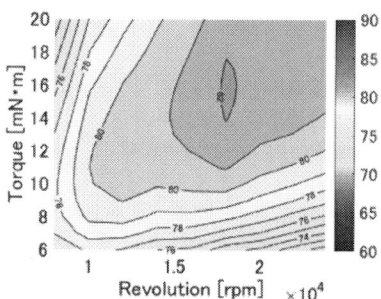

Conventional steel (350μm) core motor

Fig. 27 Efficiency characteristic of high speed motor with high power density. 25g, 50W, 24000rpm

Fig. 27 shows the efficiency characteristics compared with the motor using the conventional electrical steel sheet 350 [μm] core. We could develop the highly efficient and high power density motor by designing the motor iron core shape of ultra-thin electrical steel sheet by vector magnetic characteristics analysis considering vector magnetic characteristics.

VIII. CONCLUSION

The technologies necessary for low loss and high efficiency of a high power density, high speed motor are summarized as follows.

(1) For high-speed drive motors, the ultra-thin electrical steel sheets that can suppress an increase in eddy current loss are useful. Furthermore, if the hysteresis loss of the steel sheet is small, the effect of suppressing an increase in iron loss increases. Being able to suppress the generation of eddy currents reduces the excitation current of the motor and leads to the improvement of copper loss.

(2) By analyzing the iron core of the motor with the vector magnetic characteristics analysis using the vector magnetic characteristics of the ultra-thin magnetic steel sheet, it is possible to obtain an optimum core shape focusing on the movement of the $\mathbf{H}$ vector. The attention of the $\mathbf{H}$ vectors different from the conventional viewpoint has been neglected so far, but it plays an important role in improving the efficiency of the motor.

(3) It is necessary to change the shape of the iron core so that θ_{BH} or α obtained by the movement of the $\mathbf{H}$ vector becomes small. It is important to pay attention to the suppression of hysteresis loss in decreasing core loss design. Therefore, in order to investigate these important factors, a magnetic property analysis method incorporating the vector magnetic property is required.

(4) To obtain high output, inductance design that matches, electric loading and magnetic loading is required. As the number of magnetic poles increases, high power density can be obtained by high frequency excitation.

(5) The improvement of the motor's building factor needs to be emphasized in order to obtain high quality and high reliability of the motor.

REFERENCES

[1] M. Popescu, D. G. Dorrell, D. M. Ionel, "A study of the engineering calculations for iron losses in 3-phase AC motor models", *Proc. 33rd IEEE IECON*, pp. 169-174, Nov. 5-8, 2007.
[2] A. H. Bonnett, "An update on AC induction motor efficiency", *IEEE Trans. Ind. Applicat.*, vol. 30, pp. 1362-1372, Sept./Oct. 1994.
[3] A. H. Bonnett, "An update on AC induction motor efficiency", *IEEE Trans. Ind. Applicat.*, vol. 30, pp. 1362-1372, Sept./Oct. 1994.
[4] W.-J. Wang, C.-C. Wang, "Speed and efficiency control of an induction motor with inputoutput linearization", *IEEE Trans. Energy Conv.*, vol. 14, pp. 373-378, Sept. 1999.
[5] Fu Fengli, "International developing tendency and researching direction of small & middle size three-phase asynchronous motors," *Electrical Machinery Technology*, no. 11, pp. 3-8, 2005.
[6] Liu Chengyu, Zhang Wei, "Calculation of losses and efficiency of squirrel-cage asynchronous motor driven by an inverter," *Electric Machines & Control Application*, no. 3, pp. 2-8, 1991.
[7] S. Urata, H. Shimoji, T. Todaka, and M. Enokizono, "Measurement of two-dimensional vector magnetic properties on frequency dependence of electrical steel sheet," *Int. J. Appl. Electomagn. Mech.*, vol. 20, pp. 155–162, 2004.
[8] R. M. Bozorth, *Ferromagnetism*, Wiley-IEEE Press, New York, pp. 145-149, 1998.
[9] M. Yamagashira, S. Ueno, D. Wakabayashi, and M. Enokizono, "Vector magnetic properties and two-dimensional magnetostriction of various soft magnetic materials," *International Journal of Applied Electromagnetics and Mechanics.*, vol. 44, no. 3,4, pp. 387–400, 2014.
[10] S. Ueno, M. Enokizono, Y. Mori, K. Yamazaki, "Vector Magnetic Characteristics of Ultrathin Electrical Steel Sheet for Development of High Efficiency High Speed Motor" *IEEE Trans. On Magnetics*, Vol.53, no.11. Nov.2017. Art.no.6300604.
[11] Information on http://www.vector-magtec.ne.jp

Developments and Trends in the Adjustable Speed Drives Industry

Norbert Hanigovszki
Danfoss Drives A/S
Gråsten, Denmark
norbert@danfoss.com

Abstract—The paper explores the impact of global mega-trends on the adjustable speed drive industry. Two current topics are detailed: energy efficiency and intelligent drives.

Keywords—adjustable speed drive, energy efficiency, digitalization, Industry 4.0

I. INTRODUCTION

Until a couple of years ago, the industrial adjustable speed drives market seemed to have matured. Across industry, the two-level voltage source inverter (VSI) established itself as the dominant topology. Alternative topologies such as matrix converters and three-level inverters remained niche applications. In general, the developments were incremental. The dynamics were limited to advances in microelectronics (more powerful DSP chips) and field bus communications. Concerns were voiced that the drives market will follow the fate of the motor market and become commoditized.

During the last decade, some global mega-trends have taken shape, and this led to interesting developments impacting the drives technology and market. These mega-trends are:

- Digitalization
- Electrification
- Urbanization
- Food Supply
- Climate Change

The current paper is exploring the impact of these mega-trends on the developments in the industrial adjustable speed market.

II. GLOBAL MEGA-TRENDS

In this section it will be outlined how the five global mega-trends affect the adjustable speed drive industry.

A. Digitalization

"Digitalization" means the use of digital technologies to change business models and produce new revenue sources. This term is related to *Industry 4.0* that is a generic term, suggesting a fourth industrial revolution which can be characterized by networking (following the first industrial revolution – mechanization, the second – electrification and the third – automation). Although the term is somewhat vague, a possible definition could be "*Industry 4.0 describes the intelligent networking of people, things and systems by utilizing all the possibilities of digitalization across the entire value chain.*"

As Industry 4.0 deals with networking, the role of adjustable speed drives changes. Additional to being power processors, drives become elements in the information chain, being increasingly used as smart sensors or controllers.

B. Electrification

"Electrification" is nothing new – we have experienced it for more than 150 years. The current mega-trend relates to the replacement of traditional combustion engine or hydraulic drives with electric drives. It impacts all sectors, from the ubiquitous electrical automobile, to hybrid and all-electric vessels in the transportation sector, replacement of hydraulic and pneumatic drives with electric drives in machines, and the use of electrically driven heat pumps instead of fossil fuels for heating.

Adjustable speed drives are a key enabler for electrification. In some applications (for example marine), industrial drives can be used directly, or with some minor adaptation. In other applications, the construction of the drives is impacted by the space, cooling and reliability requirements, resulting in radical design changes.

C. Urbanization

Back in 1960, two thirds of the population were living in rural areas and one third in urban. The share of urban population increased during the second half of the 20th century, not by the reduction of rural population, but by increase of urban population. In 2007 the urban population surpassed the rural one. According to the UN World Urbanization Prospects [1], by 2050 it is projected that 68 % of the world's population will live in urban areas.

Urbanization requires building infrastructure, which translates to heating, ventilation and air condition systems, elevators, escalators, off-highway vehicle and other means of personal mobility. It also means infrastructure such as water and wastewater networks. All these rely heavily on variable speed drives for achieving comfort and efficiency. Buildings and buildings construction sectors combined are responsible for 36% of global final energy consumption, with heating ventilation and air conditioning in buildings accounting for 42% of the energy used [3]. Increased installation density of power electronics also leads to power quality related challenges, mainly harmonics.

D. Food Supply

Increased population and urbanization require changes in the food supply, mainly in terms of efficiency. This applies both in the production phase (increase of agricultural efficiency through "mechanization", including use of electrical vehicles and autonomous vehicles) and in the entire food chain by ensuring the continuity of the cold chain – needed for preserving the aliments. There is a huge improvement potential, as 30% of the produced food is wasted [2].

Under this section the supply of fresh water needs also to be mentioned. According to the International Energy Agency [5], many countries face water scarcity. It is reported that one billion people live in areas with water stress and this figure is expected to triple by 2025.

In this context, variable speed driven water pumps are an important solution to counteract water stress. According to [6], a study gathering experience from 112 systems in 10 countries, pressure management leads to 20 – 40 % reductions in energy consumption, typical 38% water leakage reduction, breaks reduced by 53% and extended asset lifetime.

E. Climate change

For reducing the effect of human activity on the climate, it is necessary to reduce the use of resources. One way to achieve this is energy efficiency. Indeed, energy efficiency was proclaimed as "the new renewable" [7]. The attention for increased energy efficiency is both market-driven (as consumers select energy efficient solutions which lead to reduced operating costs) and policy driven. For example, the revised EU Energy Efficiency Directive of 2018 targets an efficiency improvement of at least 32,5% by 2030. This objective seems ambitious, but not unrealistic, as the improvement potential is tremendous. 20% of the world's energy is electrical energy. And 50% of electrical energy is used by motors. 80 % of the motors are driving fans, pumps and compressors – all applications which can benefit by the advantages of variable speed drives [4]. Drives typically save 15 – 40% of the consumed energy. The potential of variable speed drives alone is to save 8% of the global electricity consumption by 2040.

Out of the above-mentioned global mega-trends and their effect on drives, we have selected two topics which will be explored in the present paper. The selection was made considering the importance and impact of the topics. The two topics are energy efficiency and intelligent drives.

III. ENERGY EFFICIENCY

Reducing energy use was traditionally the main reason for using adjustable speed dries. During the last years, an increased interest in energy efficiency is noticed. There are two main drivers: market demand and policy.

A. Market demand

The market demand is restrained to a certain extent by relatively low energy cost in the last five years. Figure 1 shows the development of crude oil price during the last ten years. It can be observed that the price topped in 2011 and dropped during 2014. This leads to a rather moderate interest in energy saving solutions.

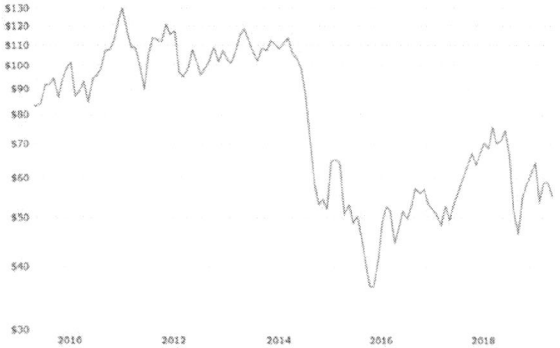

Figure 1 Crude oil price development (source: macrotrends.net)

The forces acting in the opposite direction are changes in business models, transitioning from selling physical products to selling uptime or services. There is also an increased focus on operating performance, aiming the reduction of operating cost.

A trend in the motor market can also be observed where motors with efficiency classes above the policy imposed minimum requirement (IE4 and IE5) start to gain market share and are often used for marketing purposes. According to the European Committee of Manufacturers of Electrical Machines and Power Electronics (CEMEP), by 2020 the share of IE4 motors will be 4% (Figure 2).

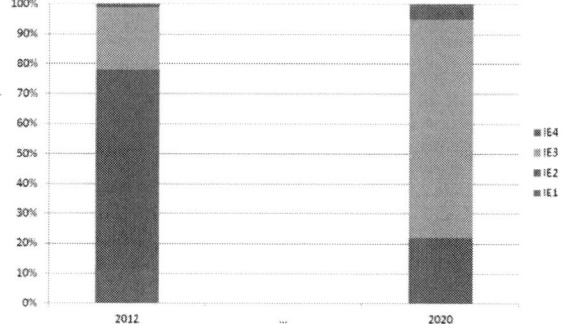

Figure 2 Market share prediction for Europe for efficiency levels IE1 – IE4

B. Policy

Policy remains the biggest driver in component level efficiency improvements. For motors there are minimum efficiency performance standards (MEPS) around the world, including all major economies like USA, China, EU, Japan. The requirements are evolving in time, and the effect on the market is visible. Figure 3 shows the historical evolution of the market share of IE classes in Europe.

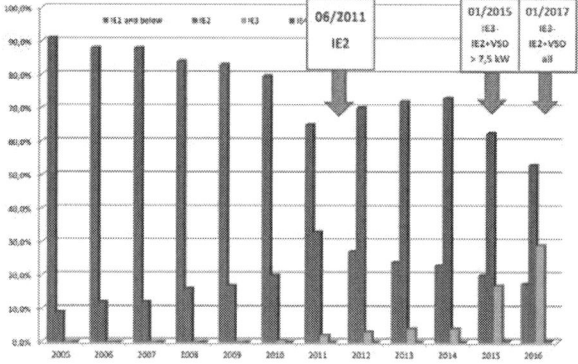

Figure 3 Market share development of efficiency classes IE1 – IE4 in Europe

In the European Union a new regulation is on the way, at the current date (August 2019) still in draft. A summary of the requirements is shown in Table 1. This regulation increases the efficiency requirements for motors. It is important to note, that the new regulation goes beyond the scope of previous regulations and it introduces two new elements: minimum efficiency requirements for drives and requirements for publishing partial load (both torque and speed) efficiency data for motors and drives.

Table 1 Overview of the MEPS requirements in EU to be published in 2019 (EuP Lot 30)

Start date	Motor (2,4, 6, 8 pole)		VSD	
	Power	*MEPS*	*Power*	*MEPS*
1.7.2021	3~ 0,12 – 0,75 kW	IE2	3~ 0,12 – 1000 kW	IE2
	3~ 0,75 – 1000 kW	IE3		
1.7.2023	1~ ≥0,12 kW	IE2		
	2, 4, 6 pole			
	3~ 75 – 200 kW	IE4		
	EX env. 2, 4, 6, 8 pole			
	3~0,12 – 1000 kW	IE2		

Traditionally, market regulations are addressing the component level efficiency and their effect is relatively low. The benefits of system level optimization are much higher, as explained in the following section. However, it needs to be commended that the newest EU regulation requires publication of part load efficiency data. This is a key element in determining the system efficiency in variable speed applications. Although, fulfilling this requirement is a huge effort for manufacturers, especially for motor manufacturers who need to produce all this data. It is also an opportunity for academia for proposing analytic and simulation-based solutions which can accelerate the generation of data.

C. Component, variable speed or system and process optimization?

The potential of energy savings from the motor drive system is 40% of the energy used, with 10 % from increased efficiency of components and 30 % from variable speed operation. The biggest potential of 60% lays within optimizations at system and process level [6]. This picture points out the huge savings potential which is currently not addressed. An important enabler is a cross-disciplinary approach. Electric energy savings can not be addressed by electrical engineers alone, and the participation of process engineers, mechanical engineers and data scientists is needed.

D. Energy efficiency driven trends in motor technologies

Motor technologies which once have been limited to niche applications, are now entering the mainstream market – driven by efforts to increase energy efficiency. More than a decade ago the choice was between *induction motor* and *permanent magnet motor*. But the surge in raw material prices in 2011 (Figure 4) changed this paradigm.

The first move was to explore motor technologies which do not require magnets. This put the *synchronous reluctance motor* (SynRM) in the limelight. Major motor producers have launched quite extensive power ranges of SynRM motors. But the disadvantage of the poor power factor of the SynRM has led to many situations where it was necessary to increase the size of the variable speed drive. Manufacturers adopted strategies to alleviate this disadvantage, such as re-calibrating the drive line-up, but the root cause persisted.

Because of the power factor disadvantage of the SynRM, and partially also because of the aim to achieve even higher efficiencies, the attention shifted towards the *permanent magnet assisted SynRM* (PMaSynRM). After all, this motor is a good compromise: good power factor, small amounts of magnets, possibility of using ferrite magnets and excellent efficiency. Most SynRM manufacturers are migrating towards PMaSynRM. Even manufacturers of interior permanent magnet (IPM) motors opt for the PMaSynRM as a cost optimization solution. There have been early adopters, but the growth still needs to unfold. The main barriers to overcome are related to manufacturing technology, availability of drives with appropriate control capability and availability of second source.

The availability of a broad range of motor technologies can be a challenge both to the users and to drive manufacturers. For the users it is somewhat difficult to navigate through the broad offering and selecting the optimal solution. Drive manufacturers need motor control technology for all motor types, especially manufacturers who don't bundle drives and motors. The theoretical fundament for universal motor control exists. Still, practical implementation aspects should not be underestimated.

E. Drives efficiency

When looking at the efficiency of VSDs, the picture is very different from the one for motors. The reason is that the physics of power electronics is quite different from the physics of electric machines. The cost-efficiency dependency is different, illustrated in Figure 5. For electric machines in order to increase efficiency it is necessary to use more expensive materials, or to increase the volume of material. This means that a more efficient motor is more expensive than a less efficient motor. In drives this trend is almost opposite. A less efficient drive is a drive with more losses, which requires more cooling, a larger heatsink and a bigger enclosure. Drive manufacturers have a natural incitement to produce as efficient drives as technically feasible, because these lead to greater profits.

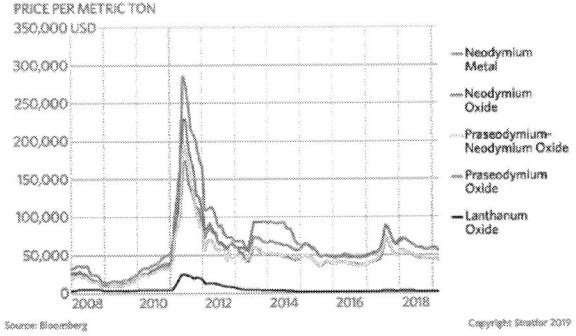

Figure 4 Evolution of rare earths price. Source: Bloomberg

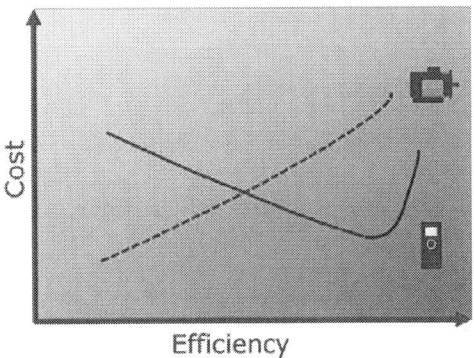

Figure 5 Cost-efficiency dependency of motors and drives

978-1-5386-7688-2/19 $31.00 © 2019 IEEE

The efficiency – price curve for drives has an upper limitation. Basically, the efficiency of most drives can be increased (slightly) at no cost, by re-designing the gate circuit and increasing the dV/dt. Still, this is practically not possible. The limitation comes from a number of factors:

- The motor isolation is stressed by high dV/dt values.

- Drives are required to accommodate motor cable lengths sometime in excess of 100 – 200 meters.

- EMC – higher dV/dt leads to higher emissions, and there are strict requirements for both radiated and conducted emissions.

These factors are also a challenge for SiC based inverters for motor drives (along with cost related concerns). The fast switching of SiC leads to the need of passive filtering which introduces additional losses and complexity. Of course, when an output filter is needed anyway, SiC devices can be used. For the front-end of low-harmonic or regenerative drives, the above mentioned issues are not present, because of the inherent presence of the input filter (typically LCL filter). The solution which is able to push these limitations is the closer integration of motor and drive.

If the drive efficiency is analyzed from a system perspective instead of a component perspective, new possibilities arise, such as using the heat dissipated by the active and passive components to feed other sections of the process. An example is shown in Figure 6. A liquid-cooled drive has an integrated pump and heat exchanger that make possible the use of the heat dissipated by the inductors in the LCL filter, motor sine-wave filter, and the semiconductors in the grid converter and motor inverter section. The heat is re-used in the production process. Judging the efficiency of this drive only from a component perspective is a disadvantage, because of the addition of the pump, but judging from a system perspective shows the advantages of the solution.

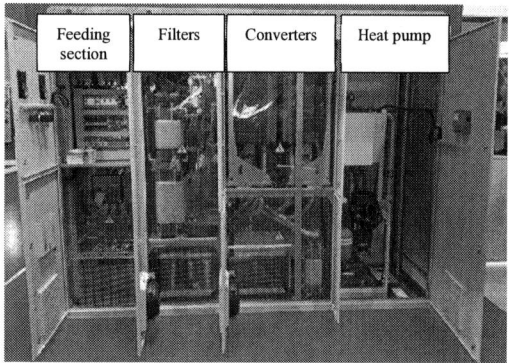

Figure 6 Liquid-cooled active front-end drive with heat recuperation

Going even further, application integral solutions can offer in some situations the highest efficiency levels. In this case, not only the motor and drive are integrated, but also the application is integrated. An example is shown in Figure 7 where a turbocompressor is integrated with a high-efficiency motor and drive – all cooled by the refrigerant.

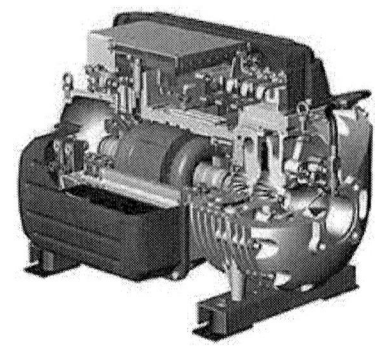

Figure 7 Turbocompressor with integrated motor and drive (Danfoss Turbocor)

IV. THE INTELLIGENT DRIVE

Since the introduction of microprocessors to control the drives, additional functionality has been added to the original function – which is that of a power processor. For example, drives are able to perform motion control, to control several pumps in a cascade system in water pumping applications or can by-pass certain frequencies to avoid resonances.

The advance of Industry 4.0 has given an additional boost to these auxiliary functions. As Industry 4.0 deals with information and networking, we start using the drives as smart and networked sensors.

A. Industry 4.0 in motor and drive systems

The impact of this trend on motor systems is a migration from what is known as "automation pyramid" (Figure 8) to networked systems (Figure 9). This means that the various elements of the system, such as motors, drives, sensors and controls, get interconnected and also connected to a cloud – where data is stored, processed, analyzed and decisions are made.

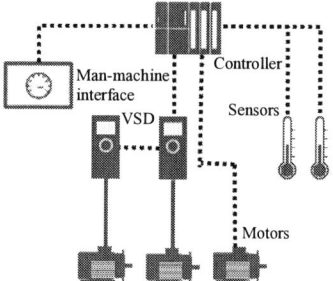

Figure 8 Automation pyramid

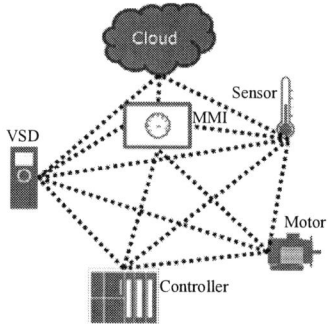

Figure 9 Automation network

B. The Industry 4.0 drive

In the Industry 4.0 network, the drive plays an important role and is characterized by some enabling features (Figure 10):

- *Secure connectivity* to other elements in the network (drives, PLCs, sensors) and cloud

- *Drive as a sensor* – using motor current and voltage signature analysis for sensing the motor and application

- *Sensor hub* – using the drive for data acquisition from external sensors related to the process controlled by the drive

- *Drive as a controller*, for replacing the PLC whenever the application constraints allow that

- *Wireless connectivity* to smart devices (smartphone, tablet) – the "bring your own device" concept

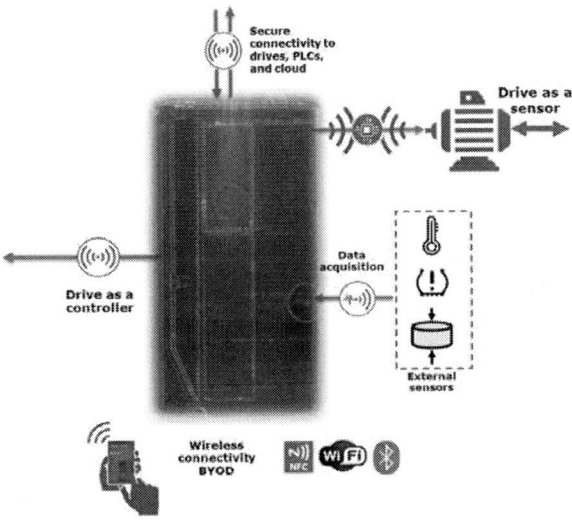

Figure 10 The Industry 4.0 Intelligent Drive

C. The drive as a sensor

The availability of microprocessors in the drive and bus communication options, combined with current and voltage sensors opens new opportunities. Moreover, additional sensors (such as vibration and pressure sensors) can be connected to the drive at almost no cost. This allows the drive to be used as a smart sensor (Figure 11). The available information offers various use cases, e.g. system optimization, energy efficiency optimization, and condition-based maintenance.

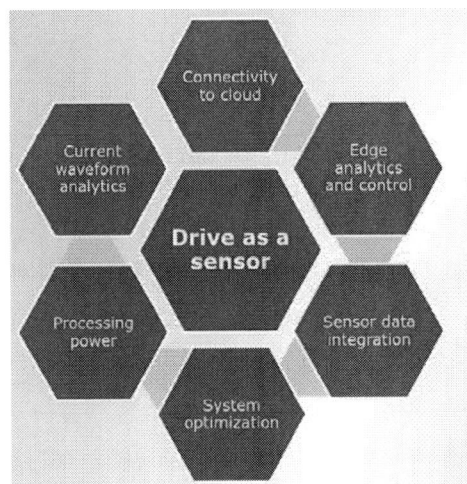

Figure 11 Drive as a smart sensor

The information available from the drive can be structured on three levels: instantaneous signals, processed signals and analyzed information (Figure 12). In the following section, the use of the drive as a sensor for condition monitoring will be described, as an application example.

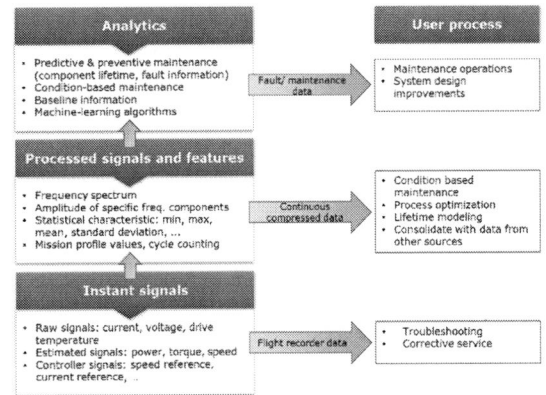

Figure 12 Information layers in the drive as a smart sensor

D. Condition monitoring

Condition monitoring is a technique to monitor the health of equipment in service. For this purpose, key parameters need to be selected as indicators for developing faults. The equipment condition typically degrades over time. Figure 13 shows a typical degradation pattern, also known as PF-curve. The point of functional failure is when the equipment fails to provide the intended function. The idea of condition-based maintenance is to detect the potential failure before the actual failure occurs. In this case, maintenance actions can be planned before functional failure, with advantages such as: reduction of downtime, elimination of unexpected production stops, maintenance optimization, reduction of spare part stock, and others.

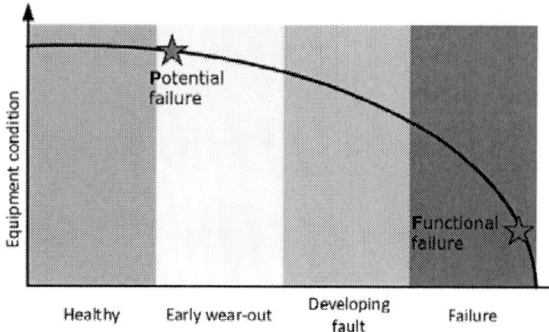

Figure 13 P-F curve representing the condition of a component until functional failure

The condition of the motor and application can also be monitored through electrical signature analysis. This technique has been under research for many years. The early studies have addressed direct online machines, and later variable speed drive applications have been investigated too[9][10][11]. With the available processing power and memory in today's drives, these techniques are now integrated into products as product features.

Figure 14 illustrates the basic concept. Fault condition indicators can be extracted from the motor currents and voltage signals. Frequency components of currents and voltages can be related to motor or application faults, e.g. shaft misalignment or stator winding faults. The current and voltage sensors are essential components of drives anyway. They provide the necessary signals for controlling the motor.

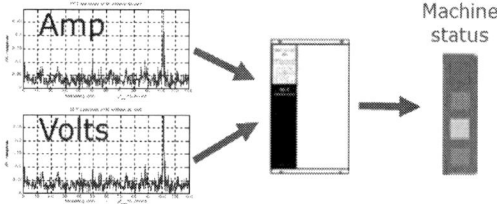

Figure 14 Motor current and voltage signature analysis

These signals can be used for monitoring purpose. Thus, no extra sensor costs are added. Signal processing and analytic techniques play an important role in this context.

The drive being the controller of the motor can correlate the monitoring values, e.g. specific current harmonics, with other available information inside the drive. Knowing the controller state for instance, the drive knows when meaningful spectrum calculations can be performed. The monitored values can be correlated with motor speed, load, and other relevant process data (e.g. pressure in water pipes) to get more accurate fault information.

V. CONCLUSION

Global megatrends (digitalization, electrification, urbanization, food supply and climate change) have an impact also on the adjustable speed drives industry. Drives have been typically used as power processors for achieving energy savings. On one hand this role is accentuated on the other hand it is augmented by using the drive as an intelligent element in the information chain – both as a sensor and as a controller. This leads to the need of developing new technologies and will change the feature set expected from an adjustable speed drive.

REFERENCES

[1] United Nations, *World Urbanization Prospects 2018*, 2018

[2] Food and Agriculture Organization of the United Nations, *Food Loss and Food Waste*, 2018

[3] International Energy Agency, *Energy Efficency: Buildings*, [Online] https://www.iea.org/topics/energyefficiency/buildings/. Accessed: 19/02/2019

[4] International Energy Agency, *World Energy Outlook 2018*, 2018

[5] International Energy Agency, *World Energy Outlook 2016*, Chapter 9: Water-energy nexus, 2016, Paris, France

[6] Thornton and Lambert, Water Service Association of Australia Asset Management 2007

[7] International Energy Agency, *World Energy Outlook 2018*

[8] Eichhammer, Wolfgang & Fleiter, Tobias, *Energy efficiency in electric motor systems: Technology, saving potentials and policy options for developing countries*, 2012

[9] Hamid A. Toliyat, Subhasis Nandi, Seungdeog Choi, Homayoun Meshgin-Kelk: Electric Machines: Modeling, Condition Monitoring, and Fault Diagnosis, CRC Press, 2013

[10] Howard P. Penrose: Electrical Motor Diagnostics, Success By Design; 2nd ed. edition (2008)

[11] Sanjeet Kumar Dwivedi, Jorg Dannehl: Modeling and simulation of stator and rotor faults of induction motor and their experimental comparison, 2017 IEEE 11th International Symposium on Diagnostics for Electrical Machines, Power Electronics and Drives (SDEMPED)

Energy Department

Synchronous Reluctance Motor Drives:
Still a Niche Technology?

Gianmario Pellegrino

Politecnico di Torino - gianmario.pellegrino@polito.it

Abstract – **The Synchronous Reluctance machine is becoming increasingly popular under the push of new efficiency standards and the high cost of rare-earth permanent magnets. This technology has been known as far as since 1923 and reached kind of technical maturity after the introduction of variable speed drives and digital vector control by the end of 1990s. Although the growing market share will foreseeably continue to grow even faster in the future, to date this technology is still mastered by a relatively small number of companies and academic research groups. This is true both for the machine design, where flux barriers design and torque ripple minimization are the main challenges, and the digital control, which is known to be parameter-dependent and hard to calibrate. The lecture will briefly review the state of the art of Synchronous Reluctance motor drives, from the early years to the most recent advancements in machine design, sensorless control and motor commissioning techniques. Strength and weaknesses, as well as the key controversial aspects that relented the momentum of this technology will be summarized and associated to existing or foreseeable solutions. The open-source platform SyR-e will be presented as an easy-to-use tool for initial machine design and finite-element identification.**

Gianmario Pellegrino, Ph.D., is an Associate Professor of Electrical Machines and Drives at the Politecnico di Torino, Turin, Italy. Dr. Pellegrino is engaged in several research projects with the industry, and one of the authors of the open-source project SyR-e for the design of electrical motors. He was a visiting fellow at Aalborg University, Denmark, the University of Nottingham, UK, and the University of Wisconsin-Madison, USA. Dr. Pellegrino is an Associate Editor for the IEEE Transactions on Industry Applications and an IEEE Senior Member. He has 40+ IEEE journal papers, one patent and seven Best Paper Awards. He is one of the founding members of the PEIC, the Power Electronics Interdepartmental Laboratory established in 2017 at the Politecnico di Torino, and a member of the Advisory Board of PCIM Europe. He is currently the Vice President of the CMAEL Association, representing the field of Power Converters, Electrical Machines and Drives in Italy, and the Advisor to the Rector of Politecnico di Torino for the implementation of interdepartmental centers.

POWER ELECTRONICS INNOVATION CENTER
Politecnico di Torino   Corso Duca degli Abruzzi, 24 – 10129 Torino – Italia
PEIC@polito.it   www.peic.polito.it

Design of electrical high speed drives for vehicle drivetrains

Markus Henke, Technische Universität Braunschweig, Germany

Abstract

In many applications, electric drives with very high power density are needed to fulfill requirements of low weight and low volume. These range from industrial and vehicle drives to aerospace applications. An increase in power can be realized by increasing the rotor speed due to the proportionality of power and speed. Higher speeds result in higher performance or smaller volumes. However, the high speed presents the engine with other challenges. The cooling of the machine becomes very important, since significantly more power loss must be dissipated in a smaller space. An important target of machine design must therefore be loss minimization and sufficient cooling concepts. In addition to strength requirements, rotor dynamic effects occur, which must be taken into account in the drive design process.

In this presentation the main design rules and conception considerations for high speed drives are outlined and some data of an experimental 24.000/31.000 rpm machine just under test are given.

Prof. Dr.-Ing. Markus Henke received the Dipl.-Ing. degree in electrical engineering from the University of Paderborn. In 1996 he became a research assistant at the University of Paderborn designing linear electrical machines in mechatronic railway applications. In 2002 he received his Dr.-Ing. in the field of electrical machine design and control. In 2003, he joined Volkswagen AG Wolfsburg, Germany, where in 2007 he became head of the research department for electrical drives. Since 2012, he is University Professor for electrical drive systems at the Technische Universität Braunschweig, and there he is the director of the Institute for Electrical Machines, Traction and Drives (IMAB)

Gap in pagination due to formatting issues.

Page 59

Automotive Power Electronics: Current Status and Future Trends

Bülent Şarlioğlu

The Wisconsin Electric Machines and Power Electronics Consortium (WEMPEC)
Dept. of Electrical and Computer Engineering, University of Wisconsin – Madison, USA
sarlioglu@wisc.edu

Abstract

In the electric automotive industry, traction drive with high power is necessary for both Hybrid Electric Vehicle (HEV) and Battery Electric Vehicle (BEV) applications. In the case of HEV, industries are moving from micro-HEV to full-HEV due to a higher fuel economy. However, this technical shift calls for increased power from the battery while reducing power from the conventional engine. In the case of light-duty BEV, the peak power of 100~150 kW is required for vehicles such as Nissan LEAF, BMW i3, Hyundai IONIQ, and Chevrolet BOLT [1].

Integrated Modular Motor Drive (IMMD) concept [2]

To achieve a higher power rating, the traction drives must have higher voltage and/or current ratings. Although most of the Li-ion batteries in EV/HEV have a nominal voltage around 300~400 V, it is expected that the battery voltage will go higher than 800 V. As a result, power devices with a voltage rating of 1.2 kV or higher are expected to become dominant for the next generation traction drive system [2]. The Si-IGBT has been dominantly used in this voltage level. However, due to the advent of SiC MOSFET with a voltage rating of over 1.2 kV, it is possible to use MOSFET instead of IGBT for traction inverter design. However, SiC MOSFET suffers from a low current rating compared to Si IGBT. It is inevitable to parallel discrete SiC MOSFET devices for BEV and PHEV applications at this moment, but this can change in the near future due to rapid growth.

Automotive industries are moving towards an integrated motor drive (IMD). In the case of the Toyota Prius, inverter peak power density increased roughly from 4.5 kW/L to 11.5 kW/L as the technology evolved from 2004 to 2017. Chevrolet improved the power density of "single power inverter module" (SPIM) from roughly 10.7 kVA/L to 19.6 kVA/L when they developed the motor drive system for Chevrolet Bolt EV [3]. Although technical details for each vehicle may vary, manufacturers are making the motor drive system more and more compact by integrating the traction drive system, which includes traction inverter, and cooling system as close as possible to the traction motor. A special type of IMD known as an integrated modular motor drive (IMMD) [2]. This high level of inverter and motor modularization offers benefits for manufacturing scalability as well as enhanced fault tolerance of the traction drive system.

Multiple automotive power electronics topologies have been proposed to increase the performance of the motor drives. Segmented motor drive systems are proposed to reduce capacitor ripple current for VSI [4]. By using the 90 degrees phase-shifted PWM method, the capacitor rms ripple current can be reduced about 55~75 %. Segmented inverter topology is equivalent to having two identical motor systems each with half of the power rating. As a result, even if one traction drive system fails, the other system can still drive the vehicle. Multi-level inverters are also proposed to use high-performance devices with low voltage ratings and reduce the THD of the output phase voltage. Current source inverters (CSI) in [5] are drawing more attention in improving the overall performance of a motor drive system in terms of increasing the constant power speed ratio (CPSR) as well as reducing the PWM-induced motor iron losses.

[1] K. Namiki, K. Murota, and M. Shoji, "High Performance Motor and Inverter System for a Newly Developed Electric Vehicle," presented at the WCX World Congress Experience, 2018, pp. 2018-01–0461.

[2] C. Jung, "Power Up with 800-V Systems: The benefits of upgrading voltage power for battery-electric passenger vehicles," *IEEE Electrific. Mag.*, vol. 5, no. 1, pp. 53–58, Mar. 2017.

[3] J. Liu., "Design of the Chevrolet Bolt EV Propulsion System," *SAE Int. J. Alt. Power.*, vol. 5, no. 1, pp. 79–86, Apr. 2016.

[4] G. Su and L. Tang, "A Segmented Traction Drive System with a Small Dc Bus Capacitor." *In IEEE Energy Conversion Congress and Exposition (ECCE)*, 2847–53. Raleigh, NC, USA: IEEE, 2012.

[5] G. Su and L. Tang, "A Current Source inverter based motor drive for EV/HEV applications," *SAE 2011 World Congress*, Detroit, MI, 2011.

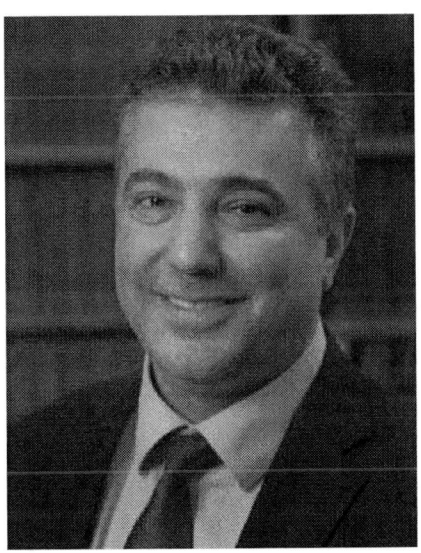

Bülent Şarlıoglu is a Professor at University of Wisconsin–Madison, and Associate Director, Wisconsin Electric Machines and Power Electronics Consortium (WEMPEC). Dr. Şarlıoglu spent more than ten years at Honeywell International Inc.'s aerospace division, most recently as a staff system engineer, earning Honeywell's technical achievement award in 2003 and an outstanding engineer award in 2011. Dr. Şarlıoglu contributed to multiple programs where high-speed electric machines and drives are used mainly for aerospace and ground vehicle applications. Dr. Şarlıoglu is the inventor or co-inventor of 19 US patents and many other international patents. His research areas are motors and drives including high-speed electric machines, novel electric machines, and application of wide bandgap devices to power electronics to increase efficiency and power density. He received the NSF CAREER Award in 2016 and the 4th Grand Nagamori Award from Nagamori Foundation, Japan in 2019. Dr. Şarlıoglu became IEEE IAS Distinguished Lecturer in 2019. He is the technical program co-chair for ECCE 2019 and was the general chair for ITEC 2018.

978-1-5386-7688-2/19 $31.00 © 2019 IEEE

Technical track on Electrical Machines, Industrial Drives and Control

978-1-5386-7688-2/19 $31.00 © 2019 IEEE

10MW, 10rpm, 10Hz directly-driven cage rotor induction generator (CRIG): preliminary design with key FEM validations

I.Boldea
Dept of Electrical Engineering
Politehnica University
Timisoara
Romanian Academy Timisoara Branch
Timisoara
Romania
email: ion.boldea@upt.ro

L. N. Tutelea
Dept of Electrical Engineering
Politehnica University
Timisoara
Romanian Academy Timisoara Branch
Timisoara
Romania
email:lucian.tutelea @upt.ro

I. Torac
Romanian Academy Timisoara Branch
Timisoara
Romania
email: ileana_torac@yahoo.com

F. Marignetti
Dept of Electrical Engineering
University of Cassino
Italy
email:marignetti@unicas.it

Abstract—**With two 8MW wind generators in the commercial stage – both synchronous types: one with permanent magnets at about 480rpm (with transmission) and one directly driven d.c. excited (at 9rpm) and 10MW designs for directly driven at 10 rpm doubly fed induction generators (at 1.5-2 mm air gap for air gap diameters of 12m, grace to flexible framing allowable by the small speed / centrifugal force), the present paper tries to investigate if at 10MW, 10 rpm 10Hz; the cage rotor induction generator (CRIG) is feasible. Preliminary design and key FEM validations, which constitute the core of the paper yield a CRIG characterized by an efficiency of 0.92, power factor 0.86, active materials (core and copper-stator and rotor) weight of 82 tons, in an outer rotor topology with air gap diameter of 12m and an air gap of 2mm (a flexible frame is required).**

Keywords—directly driven cage rotor induction generator, FEM validation

I. Introduction

Large power/unit wind generators are required for offshore large wind parks [1]. Two 8MW generators have recently reached the markets: both synchronous, both with full power ac-dc-ac converter interface to power grid. One of them is directly driven at around 9 rpm, with dc-excited rotor, while the other, with permanent magnet rotor, at around 480rpm (with 50/1 transmission).

10MW directly driven generator (10rpm) designs have been proposed, both with PM-rotors [2] or with doubly fed induction generators. The directly-driven 10MW permanent magnet synchronous generator (PMSG) avoids the multistage transmission with its bulky volume and maintenance problems but it requires a large quantity of PMs [4] if designed for limited active materials weight (around 65tons).

A 10MW, 9rpm DFIG was designed, at around 80 tons of active materials but with a flexible frame to admit a 2mm air gap [3].

Design attempts to design transverse flux PMSGs (TF-PMSGs) [5] for high power wind generators led to lower PM weight per KW, at good efficiency but at 0.65-0.7 power factor which impedes realistically through on ac-dc-ac converter ratings and costs.

In [5] at 3MW, 11rpm, f1 around 58Hz, 1800kg of PMs weight, efficiency is 0.965, power factor 0.67 and the active materials weight 90tons, it may be possible to optimally design a 10MW 10rpm TF-PMSG for 30tons of active materials out of which around 54 tons will be high energy PMs at the cost of lower power factor and thus higher (60%) inverter cost. Still the inverter cost is smaller than the directly driven generator active and passive materials costs [5].

In an effort to promote PM-less large wind generators besides the DFIG, the cage rotor induction generators (CRIGs) have been introduced and they have been commercialized with full power ac-dc-ac converters up to 5MW/unit but at speeds above 1000rpm (with multistage transmissions) [6,7].

To reduce the total wind generator weight in directly driven designs, active magnetic suspension of rotor has been proposed [2] based on the MAGLEV vehicles principles [9], but at a much lower rotor peripheral speed (less than 10m/s). If INVELOX wind energy harvesting system [10] through a Venturi tube is going to be practical, the speed of collected wind air will be magnified 5 times, say from 10rpm to 50rpm in the generator/turbine system to be placed in the tube, 10MW/unit wind generators will be still directly driven but much smaller in size.

In view of the above, the present paper aims at a realistic preliminary design methodology with key FEM validation for a 10MW, 10rpm, 10Hz directly driven CRIG with outer copper-cage rotor and full power ac-dc-ac converters for the purpose. The obtained 0.92 efficiency at 0.86 power factor for around 82 tons of active materials is considered to have promising performance, worthy of further optimal design investigation in the near future.

978-1-5386-7688-2/19 $31.00 © 2019 IEEE

The paper continues from here with a preliminary design methodology by a case study (section II), key electromagnetic FEM validation and discussion.

II. PRELIMINARY DESIGN METHODOLOGY

2.1 Specifications:

P_n=10MW, n_n=10rpm, f_{1n}=10Hz, m_1=6 phases (2x3phase), copper cage outer rotor, d.c. link rated voltage V_{dc}=4.6kVdc (rated phase voltage VphRMS=3.6kV/$\sqrt{3}$, two stator connection of phases, with two ac-dc-ac converters in parallel), air-gap g=2mm (for flexible generator framing [3]) air gap inner-stator outer diameter D_{os}=12m

2.2 Torque production

The number of pole pairs is simply:
$$p_1 = \frac{f_{1n}}{n_n} = (10/10) \times 60 = 60 \text{ pole pairs} \quad (1)$$

and thus the pole τ pitch is:
$$\tau = \pi \times D_{os} / (2p_1) = \pi \times 12 / 120 = 0.314m \quad (2)$$

Adopting a 2x3 phase stator winding with 30^0 phase shift (asymmetric 6 phase winding) and q_1=1 slot/pole/phase, the stator slot pole pitch
$$\tau_s = \tau /(m_1 \times q_1) = 0.313 /(6 \times 1) = 0.05233 \text{ m}$$

Adopting a rectangular slot width b_{ss}=27mm, the average tooth width close to the air gap diameter b_{ts}=26.33mm.
With an assumed efficiency, $\eta_n \approx 0.94$, a rated slip s_n=0.02 (both to be recalculated in the design) the generator torque T_{en} is:
$$T_{en} = P_n \times \frac{p_1}{\omega_1} \times \frac{1}{(1-s_n) \times \eta} = $$
$$= \frac{10 \times 10^6 \times 60}{2 \times \pi \times 10 \times (1 \quad 0.02) \times 0.94} \approx 10.37 \times 10^6 \, Nm \quad (3)$$

Now the d-axis phase mmf required to produce the rated air gap flux density B_{g1}=0.75T is calculated as (peak value) [11]:
$$w_1 I_{1d} = \frac{B_{g1} \times \pi \times p_1 \times g \times k_c (1+k_s)}{m_1 \times k_{w1} \times \mu_0} = $$
$$= \frac{0.75 \times \pi \times 60 \times 0.002 \times 1.15(1+1.4)}{6 \times 0.955 \times 4 \times \pi \times 10^{-7}} = 63,219 \, Aturns / phase \quad (4)$$

kw1=1, winding factor as through q=1 we use two 3 phase windings electrically dephased by 30^0, to reduce harmonics. The magnetization inductance L_m is [11]:
$$L_m = \frac{2}{\pi} B_{g1} \frac{\tau \times l_{stack} \times w_1^2 \times k_{w1}}{w_1 \times I_{1d}} \times K_{red} = $$
$$= \frac{2}{\pi} \frac{0.75 \times 0.314 \times l_{stack} \times w_1^2 \times 0.955}{63219} \times K_{red} = $$
$$= 3.927 \times 10^{-6} \times l_{stack} \times w_1^2 \quad (5)$$

K_{red} accounts for unconsidered nonlinearities $K_{red} = 1.3$.
With l_{stack}- stack length, w_1-turns per phase, the torque is simply (as in rotor flux oriented control):
$$T_{en} = \frac{m_1}{2} p_1 (L_m - L_{rl}) I_{1d} I_{1q} * K_{red} \quad (6)$$

I_{1q} torque current component (I_{1d} – field current components) and L_{rl}-rotor leakage inductance.
Assuming very conservatively L_{rl}=0.05L_m, from (3),(5),(6) the stator laminated stack length l_{stack} is:
$$l_{stack} = \frac{T_{en}}{3p_1(L_m - L_{rl}) \times (I_{1q}/I_{1d}) \times (w_1 I_{1d})^2 * K_{red}} = $$
$$= \frac{10.37 \times 10^6}{3 \times 60 \times 0.95 \times 3.927 \times 10^{-6} \times 4.5 \times (63219)^2 \times 10^6 \times 1.3} = $$
$$\approx 1.6m \quad (7)$$

The ratio between the torque and field current component (on rotor flux orientation) was assumed $I_{1q}/I_{1d} = 4.5$ to secure a reasonable power factor and torque density.

2.3. Stator slotting

We assume open stator slots with magnetic wedges (μ_{rel}=5) to secure preformed coils (Fig.1).

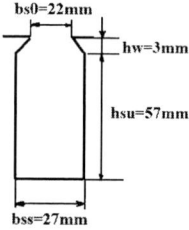

Fig.1. Stator slot (6 slots/pole)

The stator yoke height is:
$$h_{ys} = \frac{\tau}{\pi} \times \frac{B_{g1}}{B_{ys}} = \frac{0.314}{\pi} \times \frac{0.75}{1.5} \approx 0.05m$$

Now the stator slot useful area A_{slotu} is:
$$A_{slotu} = \frac{w_1 I_{1d} \sqrt{1 + (I_{1q}/I_{1d})^2}}{\sqrt{2} \times k_{fills} \times j_{con} \times p_1 \times q_1} = $$
$$= \frac{63.219 \times \sqrt{1 + (4.5)^2}}{\sqrt{2} \times 0.5 \times 4.5 \times 60 \times 1} = 1530mm^2 \quad (8)$$

We have adopted $j_{con} = 4.5 A / mm^2$ and $k_{fills} = 0.5$ (preformed coils in rectangular slots).

2.4. Rotor current I_r, resistance R_r, copper losses $p_{co \, rotor}$ and slotting

Let us consider $L_{sc} = 2L_{rl} = 0.1L_m$ (as before in the design). For zero rotor flux in axis q:
$$L_r I_r + L_m I_{1q} = 0$$
$$\text{So } I_r = -\frac{L_m}{L_r} I_{1q} = 0.95 \times 4.5 \times I_{1d} \quad (9)$$

Still $I_{1q}/I_{1d} = 4.5$ (for full torque).
Considering that a small speed machine with the need to reduce rotor copper cage losses (rotor cooling is not easy to perform at low speeds), the cross section total rotor slot area per total stator slot area is considered here 0.8 with 7 rotor bars per pole (there are 6 stator slots/pole), to reduce space harmonics and parasitic torques; the rotor/stator resistances

R_r/R_s (with short end coils $l_{stack}/\tau \geq 4/1$) is (for same current density):

$$R_r \approx R_s \times \frac{1}{0.8 \times 2} \qquad (10)$$

while

$$I_r / I_s = \frac{0.95 \times 4.5 \times I_{1d}}{I_{1d}\sqrt{1 + 4.5^2}} = 0.927 \qquad (11)$$

The factor 2 in (10) comes from the fact that the stator slot fill factor is $k_{fills} = 0.5$ while for the rotor bars it is $k_{fillr} \approx 1$.

Consequently, the rotor copper cage to stator copper losses ratio may be considered as:

$$\frac{p_{corotor}}{p_{costator}} = \frac{6R_r I_r^2}{6R_s I_s^2} \approx \frac{1}{1.6} \times (0.97)^2 = 0.537 \qquad (12)$$

With the rotor slot area as 80% of stator one approximately and the rotor slot height $h_{ru} = h_{su} \times 0.8 = 57 \times 0.8 = 45.6mm$, the rotor rectangular slot average width

$$b_{sr} = \frac{6}{7} b_{ss} = \frac{6}{7} \times 27 = 23.14mm \qquad (13)$$

And the rotor yoke depth $h_{yr} = h_{ys} = 0.05m$.
The rotor slot geometry is given in Fig.2.

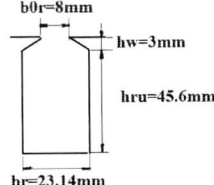

b0r=8mm
hw=3mm
hru=45.6mm
br=23.14mm

Fig.2. Rotor slot (7 slots/pole)

2.5. Losses and efficiency

The losses in the CRIG comprise the stator copper losses $p_{costator}$, the rotor copper losses $p_{corotor}$, the stator core losses p_{iron}, the mechanical losses p_{mec} and the auxiliary losses (stray load losses and the harmonic losses due to the inverter supply) p_{add}.

As the fundamental frequency is small ($f_1 = 10Hz$) the skin effect will be insignificant for the fundamental frequency and the core losses will be small.

So the stator copper losses are:

$$p_{costator} = m_1 \times \rho_{co80C} \times l_{coil} \times \frac{w_1}{\sqrt{2}} \times \sqrt{1 + (I_{1q}/I_{1d})^2} \times j_{con} \qquad (14)$$

With stator coil length l_{coil}:

$$l_{coil} \approx 2l_{stack} + 0.1 + 2 \times 1.5 \times \tau \approx 4.24m \qquad (15)$$

So

$$p_{costator} = 6 \times 2.3 \times 10^8 \times 4.24 \times \frac{63.219}{\sqrt{2}} \times \sqrt{1 + 4.5^2} \times 4.5 \times 10^6 = \qquad (16)$$

$$= 543kW$$

The rotor copper cage losses, on the other hand (12) are:

$$p_{corotor} \approx p_{costator} \times 0.537 = 543kW \times 0.537 \approx 292kW \qquad (17)$$

The stator core weight G_{siron}:

$$G_{siron} \approx [\frac{\pi(D_{os}^2 - (D_{os} - 2(h_{su} + h_w) - 2h_{ys})^2}{4} - m_1 \times 2p_1 \times A_{slots}] \times l_{stack} \times \gamma_{iron} =$$

$$= [\frac{\pi(12^2 - (12 - 2(0.06 + 0.05))^2}{4} - 6 \times 120 \times 1530 \times 10^{-6}] \times 1.6 \times 7600 =$$

$$= 36.8tons \qquad (18)$$

Considering a 0.4W/kg at 10 Hz and 1.5T for regular 0.5mm silicon laminations the stator core losses are:

$$p_{irons} = G_{siron} \times 0.4W/kg = 36.8 \times 0.4kW = 14.72kW \qquad (19)$$

With $p_{mec} = 0.1\% P_n = 10kW$ and additional losses equal to $p_{irons} \approx 15kW$ the machine total rated losses would be:

$$\sum p_n = p_{costator} + p_{corotor} + p_{itons} + p_{mec} + p_{add} =$$

$$= (543 + 292 + 14.72 \times 2 + 10)kW \approx 875kW \qquad (20)$$

So the rated efficiency η_n is:

$$\eta_n = \frac{P_n}{P_n + \sum p_n} = \frac{10000kW}{10000kW + 875kW} = 0.919545 \approx 0.92 \qquad (21)$$

As η_n is close to the targeted efficiency ($\eta_n = 0.94$), the design is considered to hold.

As active weights we consider here the stator and rotor cores and windings weights: G_{siron}, G_{riron}, $G_{costator}$, $G_{corotor}$:

$$G_{riron} \approx G_{siron} \times 0.778 = 36.8tons \times 0.778 = 28.63tons \qquad (22)$$

$$G_{wstator} = m_1 \times l_{coil} \times \frac{w_1 I_s}{j_{con}} \times \gamma_{co} =$$

$$= 6 \times 4.24 \times \frac{63.219 \times \sqrt{1 + 4.5^2}}{\sqrt{2} \times 4.5 \times 10^6} \approx 10,400kg. \qquad (23)$$

Let us consider the rotor copper cage weight proportional to the ratio of copper losses (12)

$$G_{corotor} \approx G_{costator} \times 0.537 \approx 5,585kg \qquad (24)$$

So the active weight G_a is:

$$G_a = G_{siron} + G_{riron} + G_{costator} + G_{corotor} =$$
$$= (36.8 + 28.63 + 10.4 + 5.585)tons \approx 82tons$$

Note: This active weight value is considered rather competitive with existing large generators and with recent designs. With more active weight the efficiency may be improved further as cooling is a strong challenge, though the copper windings may sustain higher temperatures.

2.6. Power factor

The space phasor diagram of CRIG (Fig. 3) helps into calculating power factor even before calculating the number of turns (to suit the d.c. link voltage of the dual converters two 3 phase such converters in parallel).

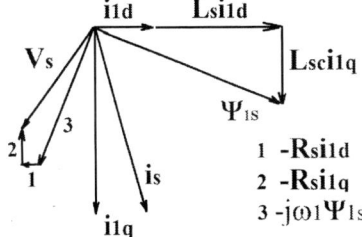

i1d   Lsi1d
Vs
Lsci1q
Ψ1s
2 3
1
is
i1q

1 -Rsi1d
2 -Rsi1q
3 -jω1Ψ1s

Fig.3.Space phasor diagram for flux oriented control of CRIG

The power factor angle
$\varphi_{1s} \cong \tan^{-1}(I_{1d}/I_{1q}) + \tan^{-1}([(L_{sc}i_{1q}\omega 1 + R_s i_{1d})/(L_s i_{1d}\omega_1 - R_s I_{1q})] \qquad (26)$
Now $L_s = L_m + L_{sl}$ with L_{sl} - stator leakage inductance

Standard calculations for L_{sl} lead in our case to

$$L_{sl} \approx 0.035 L_m \qquad (28)$$

With $L_{sl} \approx L_{rl}$, $L_{sc} \approx 0.07 L_m$

Thus we have all the data to calculate the power factor angle and power factor $\cos\varphi_{1s}$.

$$\cos\varphi_{1s} \cong \cos[\tan^{-1}(1/45)+\tan^{-1}(0.07/4.5)]=0.866 \qquad (29)$$

This is a very reasonable value that the full power ac-dc-ac converter can handle without unreasonable larger costs.

2.7 Stator winding connections, number of turns per coil (slot) and wire diameter d_{co}.

The two × three phase windings with q=1 which make the 6 phase machine (Fig.4) are connected to two twin inverters in parallel (Fig.4.b), to offer redundancy or even operation with one inverter at low load (after carefully balancing the radial forces for this operation mode).

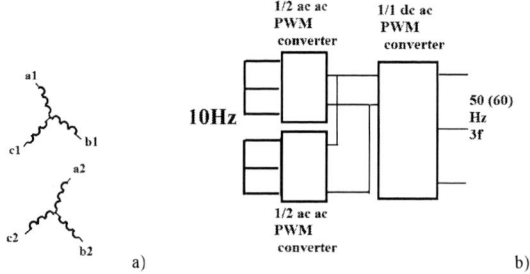

Fig.4. Two×three phase stator windings (electrically dephased by 30^0) a) and windings to converter connections, b)

With $V_{phRMSideal}=3600V/\sqrt{3}=2080V$ (30)

we need first the stator flux (for rotor flux orientation control):

$$\Psi_{1s} = \sqrt{(L_s i_{1d})^2 + (L_{sc} i_{1q})^2} \approx \qquad (31)$$
$$\approx L_m i_{1d}\sqrt{1.035^2 + (2\times0.035\times4.5)^2} \approx L_m \times 1.047 \times i_{1d}$$

But

$$V_{phRMS} \approx \frac{1}{\sqrt{2}} L_{md} i_{1d} \times 1.047 \times \omega_1 = \qquad (32)$$
$$\frac{1}{\sqrt{2}} \times 3.021 \times 1.2 \times w_1^2 \times i_{1d} \times 1.047 \times 2\pi \times 10 \approx 10.75 w_1$$

Leaving a margin of 5% in stator voltage, the number of turns per phase for one current path would be:

$$(w_1)_{onepath} = \frac{V_{phRMS} \times 0.95}{10.75} = \frac{2080\times0.95}{10.75} \approx 183.8 turn/phase \qquad (33)$$

With single layer windings the minimum number of turns should be a multiple of pole pairs p_1=60; which is $(w1)_{onepath}$=180 turns per phase with 3 turns per coil (slot); if convenient, twisted multiple strand bars may be used to make the winding.

However, noticing the large number of pole pairs p_1=60, at least a=20 current path could be adopted; in this case the number of turns per coil (slot) would be 3×20=60 turns/coil (slot).

The total phase current I_{sn} is:

$$I_{snRMS} = \frac{P_n}{6 \times V_{ph} \times \cos\varphi_{1s}} = \frac{10\times10^6}{6\times2080\times0.95}/0.866 = 974A \qquad (34)$$

with a=20 current path in parallel the coil (path) current I_{coil} is:

$$I_{coil} = \frac{I_{sn}}{a} = \frac{974}{20} = 48.7 A/path \qquad (35)$$

Even in such a case the coil current requires a copper area A_{co}:

$$A_{co} = \frac{I_{coil}}{j_{co}} = \frac{48.7}{4.5} = 10.822 mm^2 \qquad (36)$$

Six 1.55 mm diameter copper wires in parallel would do.
Note: Alternatively we may adopt 60 current path (a=p_1=60) and in this case we may use a single conductor of 2.2 mm diameter (3 ×60=180 turns/coil (slot)).
But the parallel connection of 60 coils/phase may pose challenges, too.

2.8. Rated slip and rotor skin effect

The rated slip S_n may easily be calculated for rotor flux oriented control as:

$$S_n\omega_{1n} = \frac{I_{1q}}{I_{1d}} \times \frac{R_r}{L_r} = 4.5 \times \frac{R_r}{1.035 L_m} = \qquad (37)$$
$$= \frac{4.5\times0.625\times1.689\times10^{-6}\times w_1^2}{1.035\times3.637\times w_1^2} = 1.262$$
$$S_n = 1.262/(2\pi\times10) \approx 0.02 \qquad (38)$$

Note: This is not a very small value but it is justified by the low frequency and limited weight target.
With the rotor 46.5 mm deep slots (copper bars), to check for skin effect at rated slip frequency $s_n\omega_{1n} = 1.262$ is required.
The depth of penetration δ_{co} is:

$$\delta_{co} = \sqrt{\frac{2\rho_{co}}{\mu_0 \omega_{1n} s_n}} \approx 0.174 m \qquad (39)$$

which is much larger than the rotor slot height h_{ru}=46.5mm. At least for the fundamental rotor frequency the skin effect may be neglected.

III. FEM KEY VALIDATION ATTEMPTS AN DISCUSSION

2D FEM software was used to verify the essential results of analytical design. Fig. 5 shows the flux density distribution on machine cross section for no load case and I_{s0}=750A, 60 turns per coil (slot) (q1=1).

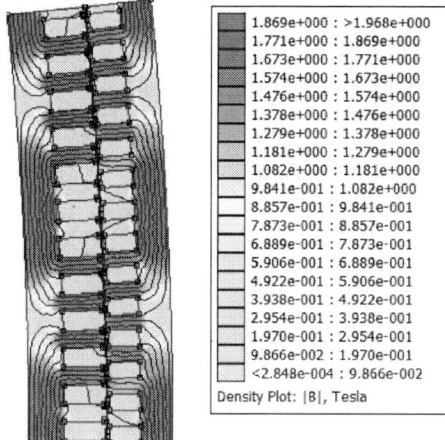

Fig.5. No load flux density distribution in saturated conditions (I_s=750A; 0.77 p.u.)

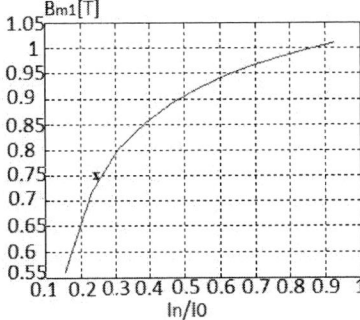

Fig 6. No load air gap flux density fundamental versus p.u current (x analytical)

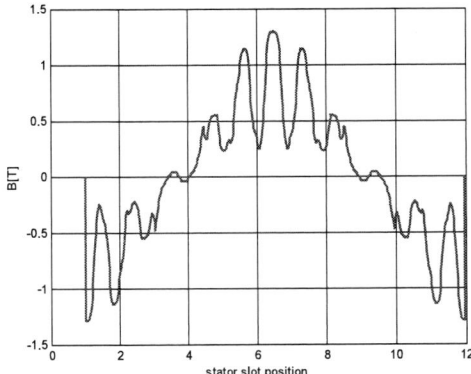

Fig.7.No load air-gap flux density distribution in saturated conditions (Is_0=250A)

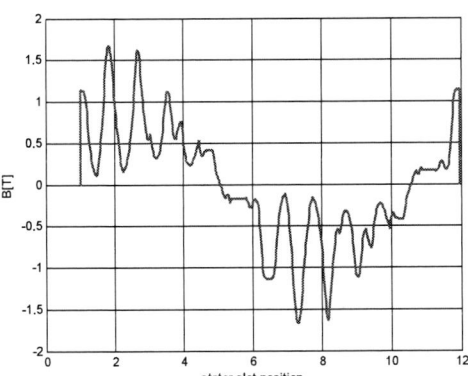

Fig.8. Air-gap flux density distribution on load case (Isn=974A, Irn=Isn*cosφ*6/7=700A, cosφ=0.84)

The obtained torque variation versus rotor position for rated sinusoidal currents and slip conditions (analytical calculated) is presented in Fig. 9.

The obtained average torque value is Te=9.9918e+006, only almost 3.4% less than the analytical value for rated sinusoidal currents and slip conditions (analytically calculated).

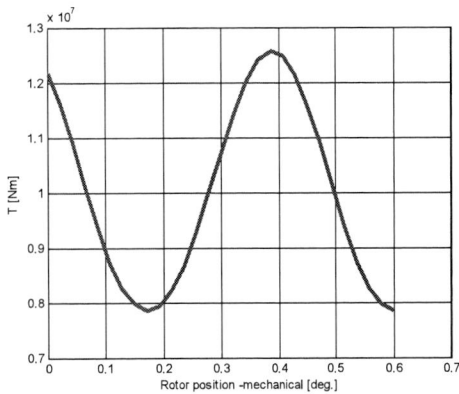

Fig.9. The torque depending on rotor position (on load case (Isn=974A, Irn=Isn*cosφ*6/7=700A, cosφ=0.84)

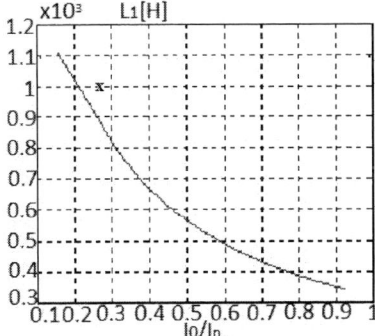

Fig. 10. FEM inductance on no load L_1 versus p.u. current (x analytical)

The results in the paper may be summarized as:

-The nonlinear analytical conservative design was validated by 2D FEM in terms of air gap flux density fundamental, average, torque and no load inductance for analytically calculated currents and losses.

-The torque pulsations are still rather large and either higher μ wedges or semiopen stator slots should be used to reduce them further (6slots/pole).

IV. CONCLUSION

The 2mm air gap, 12m air gap (inner stator outer) diameter 1.6 stack length, 82 tons active weight, 0.92 efficiency, 0.866 power factor CRIG design at 10Hz for n_n=10rpm is considered feasible, with a flexible frame.

Thermal and mechanical FEM validation should follow together with optimal design.

ACKNOWLEDGMENT

This work was also supported by a grant of the Romanian Ministry of Research and Innovation, project number 10PFE/16.10.2018, PERFORM-TECH-UPT - *The increasing of the institutional performance of the Polytechnic University of Timişoara by strengthening the research, development and technological transfer capacity in the field of "Energy, Environment and Climate Change"*, within Program 1 - Development of the national system of Research and Development, Subprogram 1.2 - Institutional Performance - Institutional Development Projects - Excellence Funding Projects in RDI, PNCDI III.

REFERENCES

[1] I. Boldea, L. Tutelea, F. Blaabjerg High power wind generator design with less or no PMs: an overview ECPS journal

[2] H. Polinder, D. J. Bang, R.P.J.M. Van Rooi, A.C.Mc Donald, M.A.Mueller "10MW wind turbine direct drive generator design with pitch and active speed still control" Record of IEEE IEMD2007, pp1390-1395

[3] V. Delli Colli , F. Marignetti, C. Attaianese, " Analytical and multiphisics approach to an optimal design of a 10MW DFIG for direct drive wind turbine" IEEE Trans vol IE −59, no7, 2012, pp2791-2799

[4] A. Penzkofer, K. Atallah "Analytical modeling and optimization of pseudodirect drive PM machine for large wind turbines" IEEE Trans vol Mag-51,no12,2015, pp8700814

[5] O. F. Andonie, L.N. Tutelea, Ana Popa (Moldovan), I. Boldea "Improved transverse flux directly driven wind PM generator: optimal design with key FEM validation" Record of ICEM 2018

[6] A. Bensaleh, M. A. Benhamida, G. Barakat," Large wind turbine generators: a review" ICEM 2018 pp 2205-2221

[7] I. Boldea, L. Tutelea, F. Blaabjerg "Large power wind generators: an overview"Record of ICEMS 2014-IEEE Explore

[8] I. Boldea, L.N. Tutelea, D. Ursu, "BLDC multiphase reluctance machines for wide speed range applications: a revival attempt", Record of EPE-PEMC-2014(IEEExplore)

[9] I. Boldea, "Linear electric machines, drives and MAGLEV Handbook, CRC press, Taylor and Francis Group, New York, 2016.

[10] . "New wind power technology INVELOX" ,Mc Graw-Hill Year book of science and technology 2015" pp203-207

[11] I. Boldea, L. Tutelea, "Electric machines: steady state transients and design with MathLab", book, CRC Press, Taylor&Francis, New York, 2010

A Direct-Drive, Linear Actuator of a Hybrid Structure

Ning Zhang
Manufacturing Flagship
CSIRO
Sydney, Australia
ning.zhang@csiro.au

Michael Collins
Energy
CSIRO
Newcastle, Australia
mike.collins@csiro.au

Michael Rae
Energy
CSIRO
Newcastle, Australia
michael.rae@csiro.au

Howard Locatt
Manufacturing Flagship
CSIRO
Sydney, Australia
howard.lovatt@csiro.au

Abstract -- A novel, direct drive, linear actuator suitable for sun tracking applications, heliostats, was proposed and investigated. The proposed actuator has a hybrid structure which combines a permanent magnet excited component (PEC) and a current excited component (CEC). The PEC provides the force to hold the load still when power is switched off; the CEC together with the PEC provides the drive force to move the attached load when power is switched on. First, FE models were benchmarked using actuators of simple magnetic structures, and good agreement was observed between predictions by FE models and experimental measurements. Next, 2D axisymmetric and 3D full size FE models were developed to simulate the proposed actuator. Investigation shows that, with a stator of a length of 1 m and a radius of 49.5 mm, the proposed actuator is able to move and hold still a load as high as 200 N. Replacing the ferrite magnets with neodymium magnets further improves the load carrying ability of the actuator to approximately 380 N.

Keywords— finite element analysis, magnetic flux, ferrites, neodymium, actuators

I. Introduction (Heading 1)

Concentrating solar power plants use fields of heliostats to reflect sunlight towards predetermined targets. This often requires the orientations of the mirrors of the heliostats to be precisely controlled by a computer system and stepping motors. Linear stepping actuators can provide the mechanical movement and torque necessary for maneuvering the heliostat mirrors, and are popular choices for heliostat application [1]. However; most commonly used are rotary-to-linear actuators, which integrate a rotary motor with a lead screw, lead nut, and lubricant in a dust sealed housing to convert the rotational motion into linear motion [2]. However, rotary-to-linear actuators are prone to accumulative mechanical abrasion from the daily operation. They are also susceptible to unexpected strong wind load which may cause irreversible damage to the lead screw and lead nut [3]. The maintenance and replacement of the damaged actuators are expensive and can lead to an unscheduled shut-down of heliostat plants. The actuators are also self-locking, meaning that the heliostat frame and support needs to be designed to withstand worst-case wind loading without damage; this makes them expensive [4].

The prospective of designing a direct drive, linear actuator for heliostat application in investigated in [4]. The investigated actuator does not involve converting rotary motion to linear motion and is designed to slip at excessive load. This brings two benefits. First, by removing the lead screw and lead nut, the stator and mover of the actuator become contact free. The actuator becomes more robust and durable which reduce the need for maintenance and replacement significantly. Second, it also reduces the heliostat frame and support cost since these are not required to withstand the worst-case wind loading any longer. However, the linear actuator in [4] undesirably requires energization to generate sufficiently large holding force to hold the load still, which is not energy efficient.

In this paper, a direct drive, linear actuator of a hybrid structure which has a permanent magnet excited component (PEC) and a current excited component (CEC) is proposed. The proposed actuator is investigated using Finite Element (FE) models. In Section II, two benchmark actuators of simple magnetic structures were built. The FE models are benchmarked by the experimental measurements of two benchmark actuators. Section III presents the design of the proposed actuator and evaluates its potential performance using FE models. Section IV provides a summary of the presented work and suggests future research directions.

II. Benchmarking FE Models

A. Benchmark Actuators

To benchmark the FE models, two actuators of simple magnetic structure were constructed. The two benchmark actuators are the foundations of the actuator proposed in Section III. The axisymmetric view of the first benchmark actuator, which is refer to as a current excited linear actuator (CEA), is shown in Fig. 1. The magnetic field was generated by the current carrying winding. The second benchmark actuator, hereafter referred to as the permanent-magnet excited linear actuator (PEA), generates the magnetic field by a magnet ring. Its axisymmetric view is shown in Fig. 2. The mover and stator of both the CEA and the PEA were made from 0.65 mm thick electrical-steel (Lycore 140) sheets. The winding of the CEA was made from 0.55 mm thick copper sheets. The permanent magnet ring of the PEA was ferrite magnet (YB30BT). It had an inner diameter of 40 mm, an outer diameter of 80 mm and a thickness of 15 mm. Although the structure of PEA and CEA are only simplified version of the hybrid structure presented in Section III, providing experimental validation to the FE models of the PEA and

CEA provides extra confidence in the setting of the FEM software.

B. FE Models

FE analysis is an effective tool for analyzing problems involving non-linear material and complicated structures. [5]-[7]. The FE software used here was Comsol Multiphysics™ (Comsol Inc., Burlington, MA, USA). Electromagnetic forces can be calculated using Virtual Work (VW) method [8] and Maxwell Stress Tensor (MST) method [9]. FE models were built using dimensions of the PEA and the CEA described in Fig. 1 and Fig. 2.

C. Experimental Setup

Fig. 3 illustrates the experimental setup to measure the magnetic force acting on the mover of the PEA and the CEA as functions of the mover position. A milling machine was used to position the mover in the stator and also measure its vertical displacement (the mill was not rotating!). The stator of the benchmark actuator to be measured was attached firmly to the table of a milling machine. The stainless-steel shaft of the mover was attached to a load cell. The load cell was fixed to the spindle of the milling machine via a stainless-steel rod. The table of the milling machine can move the stator vertically with a resolution of 1 μm. Note that for measurement of the PEA the devices in the dotted box were not required.

D. Results and Discussion

The benchmarking of the FE model of the CEA has been presented in [4]. For readers' convenience, the results for magnetic motive forces (MMFs) of 370 ampere-turns and 515 ampere-turns are reproduced in Fig. 4 and Fig. 5. For the PEA, the mover was moved 6 mm away from the 0 position upwards and downwards. The predicted force by the FE model is compared to the measurement in Fig. 6. Good agreement is observed. This demonstrates that the FE models developed are reliable tools for predicting the force of linear actuators excited by currents and permanent magnets.

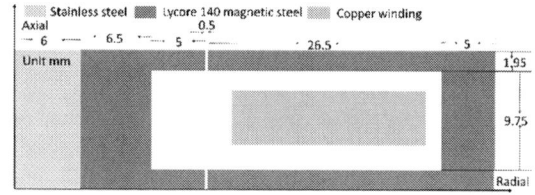

Figure 1. Axisymmetric view of the CEA. Mover on left, stator on right, and symmetry along left edge. Units: mm.

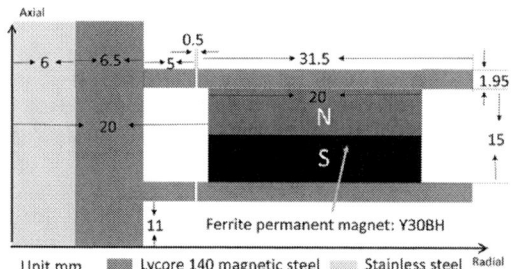

Figure 2. Axisymmetric view of the PEA. Mover on left, stator on right,

and symmetry along left edge. Units:mm.

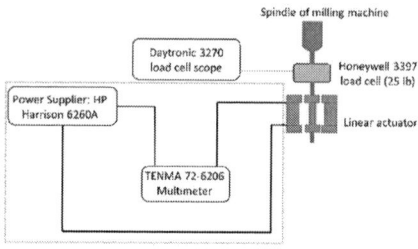

Figure 3. Experimental setup to measure the magnetic force of the benchmark actuators. The devices in the blue dotted box are not required for measurements of the PEA.

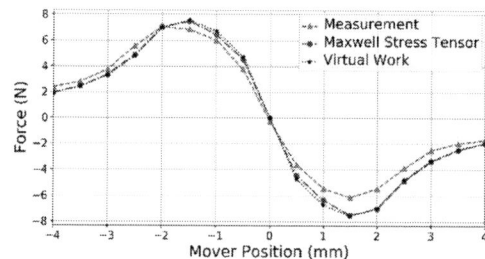

Figure 4. Measured axial magnetic force of the CEA with a MMF of 370 ampere-turns compared to the prediction of the FE model using MST method and VW method.

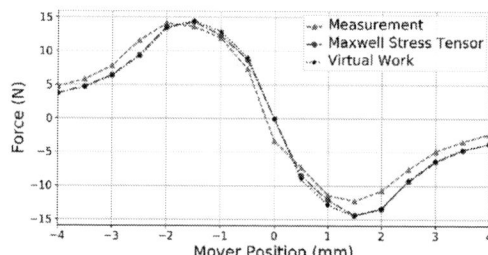

Figure 5. Measured axial magnetic force of the CEA with a MMF of 515 ampere-turns compared to the prediction of the FE model using MST method and VW method.

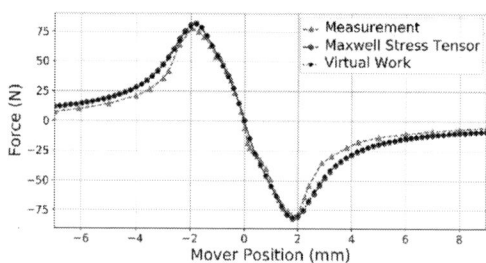

Figure 6. Measured axial magnetic force of the PEA (ferrite magnets) compared to the prediction of the FE model using MST method and VW method.

III. DIRECT DRIVE, LINEAR ACTUATOR OF HYBRID STRUCTURE

A. The proposed hybrid structure

The proposed linear actuator of a hybrid structure, referred to as the hybrid actuator hereafter, is illustrated in Fig. 7 and Fig. 8. Fig. 7 shows the hybrid actuator in 3D and Fig. 8 shows the cross-section view of the hybrid actuator in 2D. As shown the hybrid actuator consists of a permanent magnet excited component (PEC) and a current excited component (CEC), separated by Air Separations. The space of the Air Separations are used for accommodating the windings of the CEC. The windings of the CEC, the permanent magnets of the PEC, and half of the stator are omitted in Fig. 7 for better illustration of the internal structure. The size of the PEC, the CEC and the Air Separations are denoted by α_{pec}, α_{cec} and α_{as} respectively as shown in Fig. 8. The following relationship always hold: $\alpha_{pec}+\alpha_{cec}+\alpha_{as}=360°$.

The CEC is based on the CEA in Section II and its axisymmetric view is shown in Fig 9. It is also similar to the linear actuator in [4] which has a step size of 1.8 mm. Fully-pitched winding is employed with three phase sinusoidal currents corresponds to a peak MMF of 705 ampere-turn. Further increase of the MMF causes saturation of the magnetic flux in the electrical steel. 2D axisymmetric and 3D FE models of the CEC were built, and the force-density were simulated with MST and VW methods. The results are compared in Fig. 10. Good consistency between the results of 2D and 3D FE models and different force computation methods are observed. Note that for the 3D FE model only VW method was used because the MST method requires a much higher density mesh than the VW method, making the FE simulation computationally unaffordable.

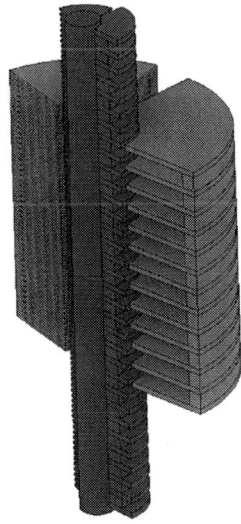

Figure 7. Illustration of the hybrid actuator in 3D. For better illustration, magnets, windings and half of the stator are omitted. The stator part of the CEC and the PEC are highlighted in red and green respectively. The mover is highlighted in blue and the stainless-steel shaft is in grey.

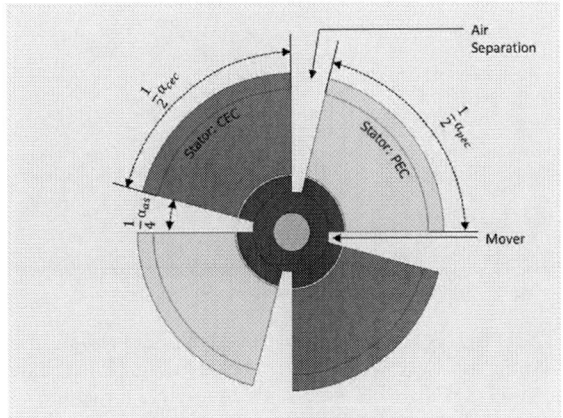

Figure 8. The cross-section view of the hybrid actuator in 2D. The size of the PEC, the CEC and the Air Separation are denoted by α_{pec}, α_{cec} and α_{as} respectively. The stator part of the CEC and the PEC are highlighted in red and green respectively. The mover is highlighted in blue and the stainless-steel shaft is in grey.

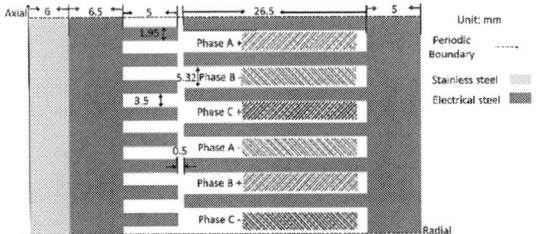

Figure 9. Axisymmetric view of one periodic structure of the CEC.

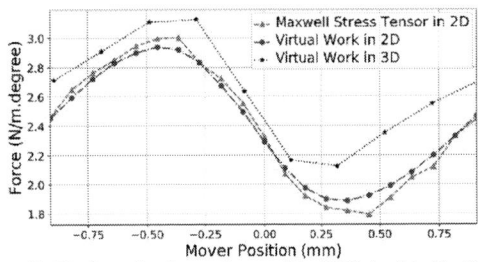

Figure 10. The force-density per degree of the CEC simulated by 2D and 3D FE models using the MST and VW methods.

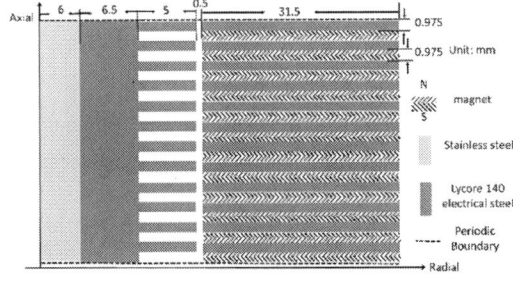

Figure 11. Axisymmetric view of 6 periodic structure of the PEC. The axial length of the shown PEC corresponds to half of the CEC shown in Fig. 9.

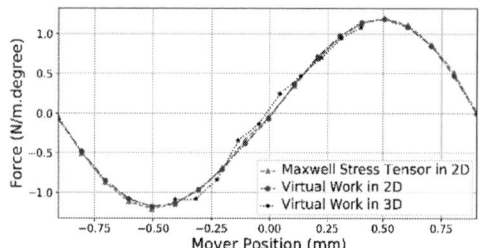

Figure 12. The force-density per degree of the PEC simulated by 2D and 3D FE models using the MST and VW methods.

The PEC is based on the PEA in Section II. The axisymmetric view of the PEC is shown in Fig 11. The axial dimensions of the PEC were adjusted to yield a step size of 1.8 mm. Its radial dimensions were adjusted to match the radial dimensions of the CEC. When the power is off, the PEC is designed to provide a holding force to hold the load still. 2D asymmetric and 3D FE models were built and the MST and VW methods were used to estimate the force-density, which are compared in Fig. 11. Good agreement is observed.

The load carrying ability of the hybrid actuator is characterized by the maximum load the hybrid actuator can move and hold still with a backlash no more than half of a step size. The hybrid actuator is designed to function at low speed and therefore the momentum of the mover can be neglected. The maximum load the hybrid actuator can move is determined by the minima of the force as a function of the mover position when the power is switched on, i.e., the combined force of the PEC and the CEC. The maximum load the actuator can hold still is determined by the maxima of the force when the power is switched off, i.e., the force of the PEC. As shown in Fig. 10 and Fig 12, the force-density of the CEC and the PEC as functions of the mover position are both approximately sinusoidal. The principle of the proposed hybrid actuator is to align the CEC and the PEC in the axial direction by an appropriate offset such that the maxima of force-density of the CEC coincides with the minima of the PEC, and vice versa. This way the variation in the combined force-density of the hybrid actuator is smoothed, and minima of the combined force-density is increased, which increases the load the hybrid actuator can move.

B. Simulation

The force-density of the hybrid actuator can be simulated with two methods. The first method is building a FE model in 3D for the complete hybrid actuator including the CEC, the PEC and the Air Separations. It has the disadvantage of large computational overhead because of 3D FE models are involved, making a parameter sweep computationally very expensive. The second method is adding the contributions of the PEC and the CEC simulated by 2D FE models using the following equation:

$$F_{hybrid} = \alpha_{pec} f_{pec} + \alpha_{cec} f_{cec}$$

where f_{pec} and f_{cec} are the force-densities of the PEC and the CEC per unit angle respectively. Method 2 has a smaller computational overhead because 2D FE models are used instead of a 3D FE model. However, it assumes that the magnetic linkage via the Air Separations between the CEC and the PEC, and the magnetic loss in the Air Separations are

small. To quantify the effect of this assumption, the force-density of a hybrid actuator with α_{cec}=140°, α_{pec}=160° and α_{as}=60° was simulated using both methods. The results are compared in Fig. 13 and are in good agreement. There are discrepancies between the predicted results by method 1 and method 2 at some mover positions; method 2 overestimates the force-density by approximately 50 N/m compared to method 1. This is possibly due to the loss of the magnetic flux to the Air Separations as well as the error introduced by the coarser mesh used in the 3D FE model of method 1. Nonetheless the discrepancy is small relative to the overall magnitude of the force-density, and method 2 will be used to accelerate the parameter sweep in the subsequent investigation.

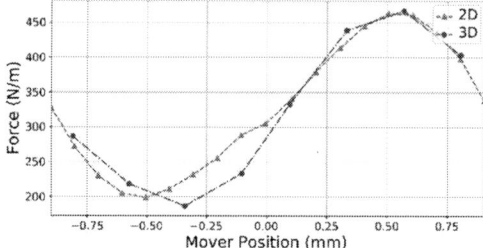

Figure 13. Simulated force-density of the hybrid actuator (ferrite magnets) with $\alpha_{cec} = 140°$, $\alpha_{pec} = 160°$ and $\alpha_{as} = 60°$ from 2D and 3D FE models.

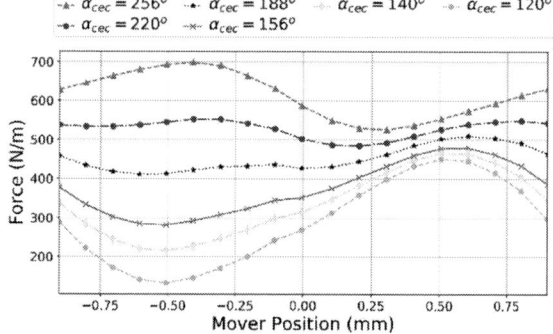

Figure 14. Simulated force-density of the hybrid actuators (ferrite magnets) with varied α_{cec} when power is switched *on*.

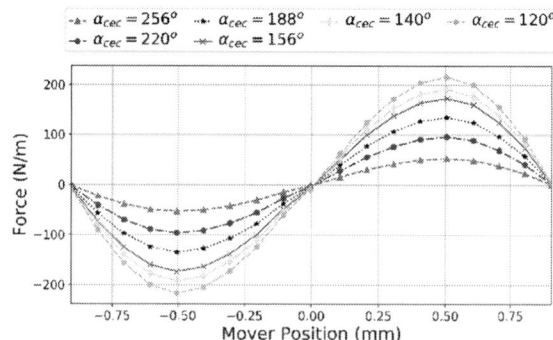

Figure 15. Simulated force-density of the hybrid actuator (ferrite magnets) with varied α_{cec} when power is switched *off*.

C. Optimization

In this section, the size of the Air Separation was fixed at α_{as}=60° which is enough space to accommodate the copper windings of the CEC. α_{cec} was swept from 156° to 256° while α_{pec} was varied from 144° to 44°. The parametric sweep was done using method 2 discussed in Section IIIB. Fig 14 shows the force-density when the power is on and Fig 15 shows the linear-force-density when the power is off. As shown in Fig. 14, the minima of the linear-force-density was increased from 120 N/m (yellow) to 550 N/m (red) with the increasing size of the CEC when the power is on. On the other hand, the maxima of the force-density is reduced from 200 N/m (yellow) to 120 N/m (red) when the power is off. The peak to peak variation in the force-density as a function of the mover position was reduced from 300 N/m (yellow) to a minimum of 100 N/m (black) with the increasing size of the CEC as shown in Fig. 14. Further increases in the size of the CEC increases the variation to 180 N/m (red) when α_{cec}=256⁰. For a hybrid actuator with a 1 m long stator with α_{cec}=140⁰ (cyan curves in Fig 14 and Fig 15), the maximum load the hybrid actuator can move when power is on and hold still when power is off are both approximately 200 N, which is the load carrying ability of the hybrid actuator.

D. Hybrid actuator with neodymium magnet

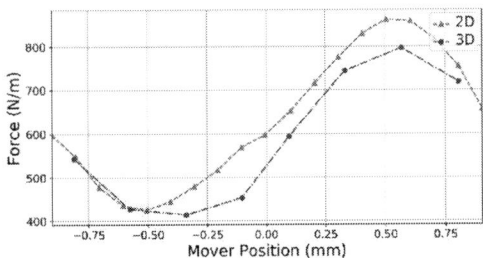

Figure 16. Simulated force-density of the hybrid actuator (neodymium magnets) with α_{cec} = 264⁰, α_{pec} = 36⁰ and α_{as} = 60⁰, simulated by 2D FE models compared to the 3D FE model

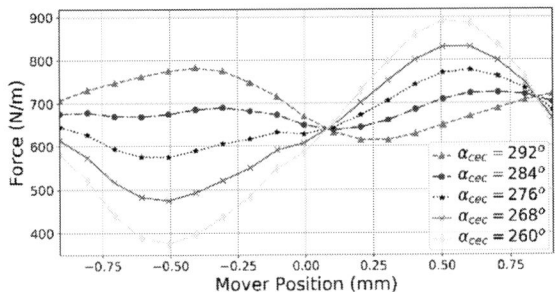

Figure 17. Simulated force-density of the hybrid actuator (neodymium magnets) with varied α_{cec} when power is switched *on*.

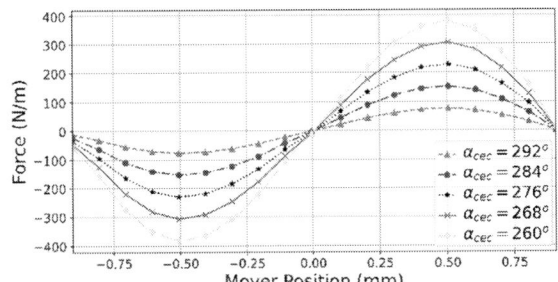

Figure 18. Simulated force-density of the hybrid actuator (neodymium magnets) with varied α_{cec} when power is switched *off*.

Replacing the ferrite magnets in the PEC with stronger magnets will improve the force-density of the PEC. This allows for a PEC of smaller size, subsequently leaves more space for the CEC, and increases load carrying ability of the hybrid actuator. To assess the effect of stronger magnets, ferrite magnets were replaced by neodymium magnet (grade N33 and Br = 1.1 T), and the analysis in Section IIIC was repeated.

A 3D FE model (method 1 in Section IIIB) and 2D FE models (method 2 in Section IIIB) of the hybrid actuator with neodymium magnets were built with α_{pec}=36⁰, α_{cec}=266⁰ and α_{as}=60⁰. The simulated force-densities were compared in Fig 16. Good agreement was observed with discrepancies similar to what was described and discussed in Section IIIB. This is possibly due to the loss of the magnetic flux to the Air Separations as well as the error introduced by the coarser mesh used in the 3D FE model of method 1. Then α_{pec} was swept from 40⁰ to 8⁰ and α_{cec} from 260⁰ to 292⁰. The resultant force-density with power on and off are shown in Fig 17 and Fig. 18 respectively. When the power is on, the minima of the force-density was increased to 600 N/m (red) when α_{cec}=292⁰ from 380 N/m (cyan) when α_{cec}=260⁰. When the power is switched off, the maxima of the force-density was decreased from 400 N/m (cyan) to 100 N/m (red). The variation in the force-density is minimized when α_{cec}=284⁰ (blue). For a hybrid actuator with 1 m long stator, the maximum load the actuator can move when power is on and hold still when power is off are both approximately 380 N. That is to say, replacing the ferrite magnets with neodymium magnets has approximately doubled the load carrying ability of the hybrid actuator.

IV. CONCLUSION

In this paper, a novel, direct drive, linear actuator was proposed. The hybrid actuator involves two fundamental components: a PEC to provide the holding force when the power is switched off; and a CEC to provide the required force to move the load together with the PEC when the power is switched on. The proposed actuator was assessed by FE models which were benchmarked by experimental measurements. It has been shown that, by adjusting the relative size of the CEC and the PEC, the maximum load the hybrid actuator can move and hold still can be maximized. A 200 N load carrying ability is achieved for a hybrid actuator with a 1 m long stator and ferrite magnets. Replacing the ferrite magnets with stronger neodymium magnets has almost doubled the load carrying ability.

In the future, in order to further improve the load carrying ability of the proposed hybrid actuator without increasing its size, current waveforms driving the CEC will be tailed to produce a force-density of greater maxima.

The proposed hybrid actuator is potentially advantageous in heliostats application. With the feature of being contact free between stator and mover and being able to slip when experiencing excessive wind load, the hybrid actuator may significantly reduce the maintenance and replacement cost of linear steppers used of heliostat power plants.

ACKNOWLEDGMENT

The project is funded by Australia Renewable Energy Agency.

REFERENCES

[1] G. Prinsloo and R. Dobson, Solar Energy Harvesting, Trough, Pinpointing and Heliostat Solar Collecting Systems, Stellenbosch: Solar Books, 2015.

[2] T. Owada and M. Kagami, "Moterized Linear Module for Tracking System of Solar Light/Solar Heat Power Genration," *NTN Technical Review,* vol. 80, no. 1, pp. 19-22, 2012.

[3] J. Coventry, J. Campbell, Y. P. Xue, C. Hall, J. S. Kim, J. Pye, G. Burgess, D. Lewis, G. Nathan, M. Arjomandi, W. Stein, M. Blanco, J. Barry, M. Doolan, W. Lpinski and A. Beath, "Heliostat Cost Down Scoping Study Final Report," Australian Solar Thermal Research Initiative, 2013.

[4] N. Zhang, M. Collins and H. Lovatt, "A Direct-Drve, Linear Actuator for a Heliostat Tracking System," in *2017 20th International Conference on Electrical Machines and Systems (ICEMS),* Sydney, 2017.

[5] A. Hellany, M. Nagrial, W. Aljaism and J. Rizk, "Analysis Design Optimisation and Simulation of Switched Reluctance Motors," in *Conference on Energy Conversion,* Johor Bahru, 2014.

[6] L. Jolly, M. A. Jabbar and Q. Liu, "Design Optimization of Permanent Magnet Motors Using Response Surface Methodology and Genetic Algorithm," *IEEE Transactions on Magnetics,* vol. 41, no. 10, pp. 3928-3930, 2005.

[7] S. Kocman, P. Pecinka and T. Hruby, "Induction Motor Modeling Using Comsol Multiphysics," in *International Scientific Conference on Electric Power Engineering,* Prague, 2016.

[8] K. J. Binns and P. J. Lawrenson, Analysis and Computation of Electric and Magnetic Field Problems, Hungary: Pergamon Press, 1973.

[9] F. Henrotte and K. Hameyer, "Computation of Electromagnetic Force Densities: Maxwell Stress Tensor vs. Virtual Work Principle," *Journal of Computational and Applied Mathematics,* vol. 168, no. 1-2, pp. 235-243, 2004.

A novel approach to PLCs based systems utilized in electric drives

Adrian Daniel Martin
Electrical Engineering
University Politehnica of Timisoara
Timisoara, Romania
daniel.martin@student.upt.ro

Lucian Tutelea
Electrical Engineering
University Politehnica of Timisoara
Timisoara, Romania
lucian.tutelea@upt.ro

Radu Babau
Electrical Engineering
SC. Beespeed Automatizari SRL
Timisoara, Romania
rbabau@beespeed.ro

Ion Boldea
Electrical Enegineering
University Politehnica Timisoara
Timisoara, Romania
ion.boldea@upt.ro

Abstract—**This paper studies the possibilities of extending the usage range of classical PLCs, and proposes the minimum requirements for these in electric drives. Direct torque and closed-loop critical computation methods are implemented in two different PLC types. For both PLC types, simulation results are compared with experimental results, in order to validate the accuracy of the system.**

Keywords—PLC, Induction Machine, Torque estimator, Electric drives.

I. INTRODUCTION

Nowadays, many grid connected induction motors are part of more complex systems monitored by PLCs (Programmable Logic Controller). In some of the pulsating torque damping applications, it is useful to estimate the electromagnetic torque directly with the existing equipment. Several recent such methods can be found in the literature [1], [2]. Adequate technical prerequisites should be taken in order to implement such methods on PLC platforms.

This article reviews and emphasises the advantage of using existing medium performance industrial automation equipment to identify or prevent possible malfunctions in electric drive systems, without installing additional specialised equipment. For induction motors fed directly from main grid, detailed information about the state of operation of the mechanical systems [3] can be obtained. by estimating the torque in a PLC-based system. This way, the overall performance of the drive system can be improved using the existing equipment, relevant changes being required in PLC software.

The currently used PLC platforms have a large industrial deployment, are stable against temperature, dust, vibration and electromagnetic pollution, are easy to maintain, have standardized communication protocols, and can be programmed using standardized programming practices [4]. The PLCs have a wide range of utilization such as drinking water and wastewater applications, heating, cooling and ventilation, research and education process, control and management of power generation and manufacturing processes [5]. In the literature, we can find power applications where PLCs are used as main controllers [6]. In [7], the PLC controls the opening angle of a valve to improve a compressor testing efficiency.

In this paper two torque estimation methods are implemented on PLC based systems. The PLC is used as a main engine for flux and torque estimation. The electrical parameters of the machine are already known, their determination not being the subject of this paper. Methods regarding induction machine losses can be found in [8].

II. THE SYSTEM LAYOUT OF PLC BASED DATA ACQUISITION AND CONTROL SYSTEM (PLCb-DACS)

Our testing stand consists of a pair of coupled IMs. One of the machines (11 kW, 1000 rpm) will be used as a Main Motor (MM), connected directly to the grid. The other one (15kW, 1500 rpm) is driven by a 4Q variable frequency converter (VFC) and will be used as an Emulated Load Machine (ELM). This ELM set is therefore able to impose various speed and torque load types to the MM. The Direct Torque Controlled based VFC receives a torque reference from the PLC in digital format (Profibus), at a 2 msec. rate. The electric machines are rigidly coupled, so the speed of the common shaft depends on how the grid connected MM machine reacts to the torque applied by the ELM set. This MM+ELM test arrangement has been proposed in order to develop and calibrate a torque estimator of the grid connected MM machine, based on the measured phase voltages, (U) and currents, (I). The U, I data acquisition is implemented in the PLCb-DACS (Fig.1), along with the ELM torque reference capabilities.

During the development of the MM torque estimator techniques and procedures, the U, I data acquired by the PLC were transferred to a PC, for off-line data processing using Simulink. Once a certain ELM torque shape and amplitude was referenced by the PLCb-DACS, the calculated MM torque estimator output should match the referenced torque. The Simulink torque estimator structure and parameters were developed and refined in order to minimise the difference between the estimated torque $\left(T_{Simulink}\right)$ and the VFC referenced torque $\left(T^*\right)$ for various puslating torque shapes (sinus, ramps, etc.).

$$T_{Simulink} - T^* \to 0 \qquad (1)$$

Some of the developed MM torque estimator models have been already implemented in the PLC (chapter IV), while in future research stages (not in the scope of this paper), the best MM torque estimator will be established and implemented in the PLC software as for online torque calculation and monitoring, as a relevant protection feature   for the industrial pulsating torque applications.

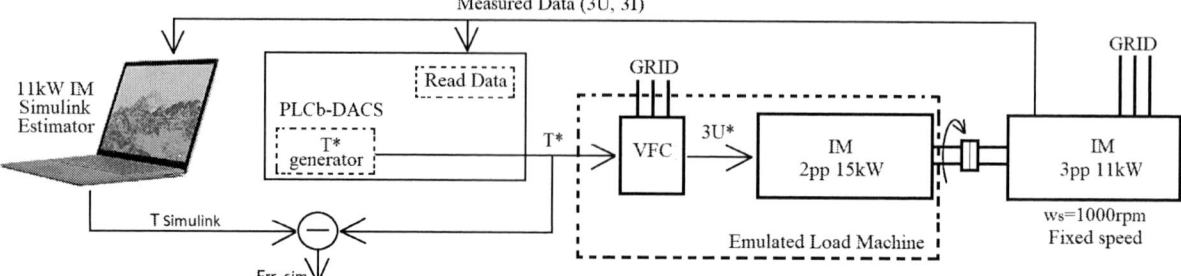

Fig.1   Experimental setup

In Fig.1 the utilization of the PLC as a data acquisition and control system is presented. In our paper, the torque estimator is implemented in Simulink / PC, while the PLC covers the data acquisition and provides the ELM torque reference. According to the computation speed and the system structure, two different PLCb-DACS topologies are studied. These are based on two different PLCs:

A. Low speed data acquisition system (Fig.2):

This is a decentralized system, without buffered data. The analog to digital converter (ADC) is a 16bit converter (including sign). The voltages and currents are measured via an analog input module (AIM), and then transmitted to the PLC via an Ethernet bus. The communication between the 8xAI Module and the PLC is established using an isochronous real time (IRT) channel. The 8 channel data are read, digitally converted, sent to the PLC processor and then torque computed at each 500 µs (minimum. acquisition rate using this PLC type).

B. High speed data acquisition system (Fig.3):

This is a centralized system, with buffered data. The analog to digital converter (ADC) is a 16 bit converter (including sign). The 8xAI Module is connected to the PLC via a high-speed bus. The 3 voltages and 3 currents are synchronously acquired at each 25 µs (max. acquisition rate). Two different modes are available for this system: Oscilloscope Mode, or FIFO Mode:

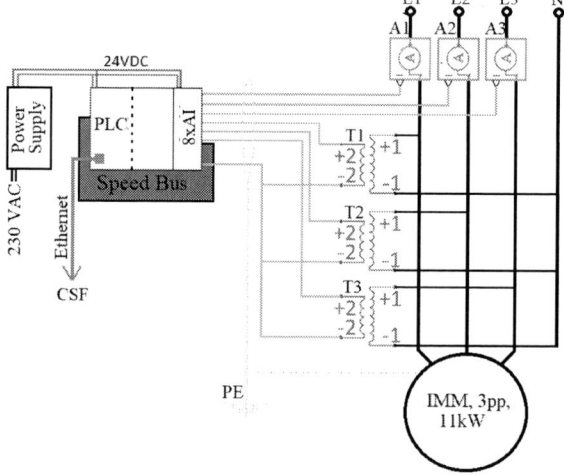

Fig.3   High speed data acquisition system

Our High speed system uses the Oscilloscope Mode for data acquisition. Once the internal buffer of the module is full, the PLC can access and process the available data. Therefore, this system can only perform offline computation cycles (i.e 1 sec. cycle = 200 msec – data acquisition + 800 msec- data processing).

III. ESTIMATOR'S STRUCTURES-SIMULINK IMPLEMENTATION

Mathematical models are used to analyze the induction machine steady state and transient processes. These models are commonly used in literature to estimate the electromagnetic torque both for the induction machine (IM) and synchronous machine (SM) [9],[10]. Tools that study electromagnetic fields from either PMIG or PMSG, based on analytical methods [11] can be found in literature.

The torque estimator simulations were performed through the Matlab & Simulink environment, Fig.4. The article investigates the way in which the complexity level of the newly proposed software structures influences the performance of PLCs. Thus the electromagnetic torque is estimated by two different methods. These were done in discrete time, the sampling time of the model and the integration time was set according to the program cycles of the above mentioned two PLCb-DACS topologies (500[µs], 25[µs]). The used induction machine model (Induction Machine Block) is fed with the voltages acquired with PLCb-DACS. The Torque Estimator receives on the one hand the currents obtained from the simulation of the IM

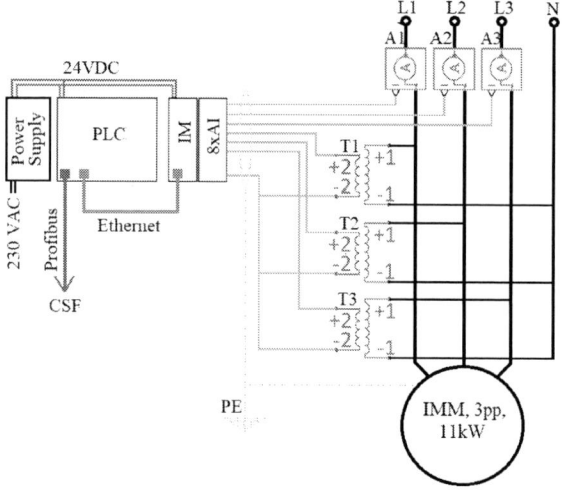

Fig.2   Low speed data acquisition system

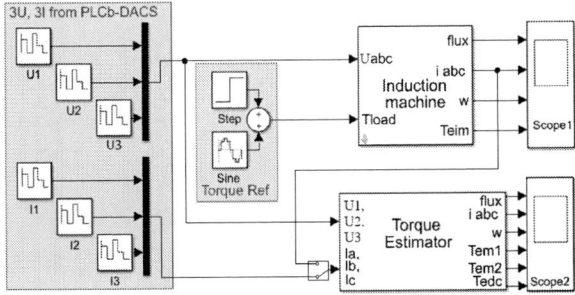

Fig.4   General simulation model

and on the other hand the currents acquired by the acquisition system. Two methods of calculating the electromagnetic torque are presented, as follows in Fig.5, Fig.6.

A. Torque estimation based on induction machine dynamic model

The torque estimator is based on the IMM 11kW dynamic model in orthogonal stator coordinates. The load torque is obtained by means of a PI regulator, based on the weighted errors between the model currents and measured currents as it is shown in Fig.5. The fluxes from the quadrature axes are the weight factors. The equations from block diagram are presented briefly:

$$U_{1s} = U_\alpha - a_{11} \cdot \Psi_{1s} + a_{12} \cdot \Psi_{1r} \qquad (2)$$

$$U_{2s} = U_\beta - a_{11} \cdot \Psi_{2s} + a_{12} \cdot \Psi_{2r} \qquad (3)$$

$$U_{1r} = a_{21} \cdot \Psi_{1s} - a_{22} \cdot \Psi_{1r} - p \cdot \omega \cdot \Psi_{2r} \qquad (4)$$

$$U_{2r} = a_{21} \cdot \Psi_{2s} - a_{22} \cdot \Psi_{2r} + p \cdot \omega \cdot \Psi_{1r} \qquad (5)$$

$$I_{\alpha e} = c_1 \cdot \Psi_{1s} + c_2 \cdot \Psi_{1r} \qquad (6)$$

$$I_{\beta e} = c_1 \cdot \Psi_{2s} + c_2 \cdot \Psi_{2r} \qquad (7)$$

$$\sigma = 1 - \frac{L_m^2}{L_s \cdot L_r}; \qquad (8)$$

$$a_{11} = \frac{R_s}{\sigma \cdot L_s}; \qquad (9)$$

$$a_{12} = R_s \cdot \frac{1-\sigma}{\sigma \cdot L_m}; \qquad (10)$$

$$a_{21} = R_r \cdot \frac{1-\sigma}{\sigma \cdot L_m}; \qquad (11)$$

$$a_{22} = \frac{R_r}{\sigma \cdot L_s}; \qquad (12)$$

$$c_1 = \frac{1}{\sigma \cdot L_s}; \qquad (13)$$

$$c_2 = \frac{\sigma-1}{\sigma \cdot L_m}; \qquad (14)$$

where: $u(1), u(2), u(3), u(4)$ are stator and rotor α, β-axis flux components, $u(5)$ is the shaft speed, $u(6), u(7)$ are stator α, β-axis calculated voltages, p is the number of pole pairs, R_s, R_r are the stator and rotor resistance, L_m, L_s, L_r are magnetizing, stator, respectively rotor inductances, $U_{1s}, U_{2s}, U_{1r}, U_{2r}$ are the stator and rotor voltages, $\Psi_{1s}, \Psi_{2s}, \Psi_{1r}, \Psi_{2r}$ are the stator and rotor fluxes, U_α, U_β are the stator α, β-axis voltage components, $I_{\alpha e}, I_{\beta e}$ are the stator α, β-axis estimated current components $a_{11}, a_{12}, a_{21}, a_{22}, c_1, c_2, \sigma$ are coefficients.

Table I presents the induction machine parameters values.

TABLE I.   INDUCTION MACHINE PARAMETERS

R_s	R_r	L_m	L_s	L_r
0.34[Ω]	0.33[Ω]	76.39[mH]	78.39[mH]	80.023[mH]

B. Direct computation method

The simulation layout of the direct torque computation method is presented in Fig.6.

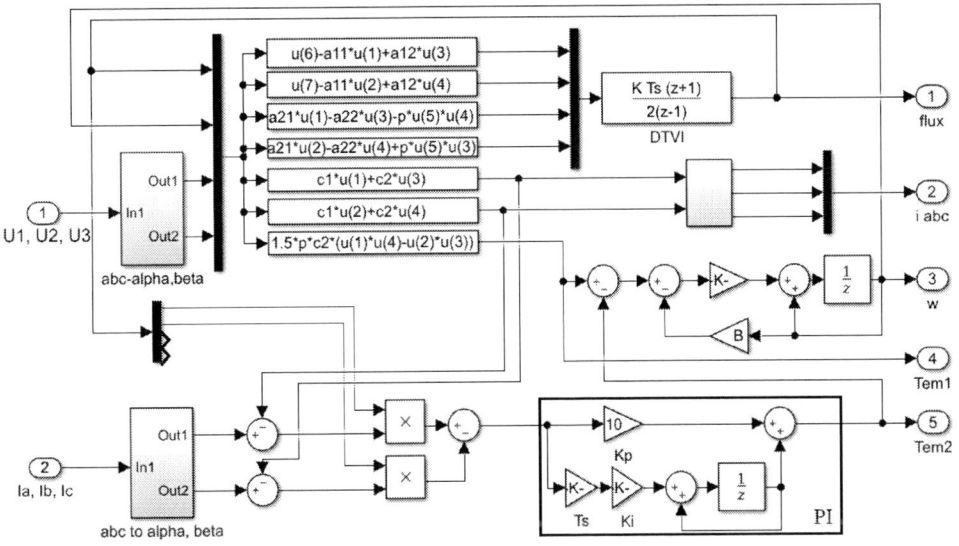

Fig.5   Torque estimator based on dynamic model – block diagram

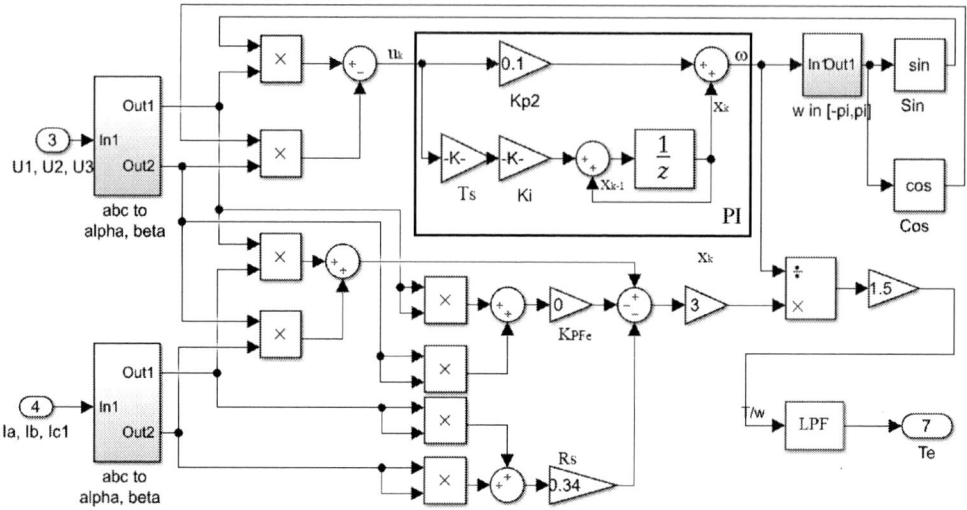

Fig.6   Direct torque estimator – block diagram

$$P = \frac{3}{2} \cdot \left(U_\alpha \cdot I_\alpha + U_\beta \cdot I_\beta \right) \qquad (15)$$

$$P_{Cu} = \frac{3}{2} \cdot R_s \cdot \left(I_\alpha^2 + I_\beta^2 \right) \qquad (16)$$

$$P_{Fe} = 0.5\% \cdot P_N \qquad (17)$$

$$T_{elm} = \frac{P - P_{Cu} - P_{Fe}}{\omega} \qquad (18)$$

The torque estimation is based on the power computation [1]. One of the benefits of using this estimator is that it can be easily implemented on both PLCb-DACS, but this method is more sensitive to measured noise. The dynamic model doesn't take into account the iron losses. In order to compare the results from both models the iron losses from direct computation method should be 0, $\left(K_{PFe} = 0 \right)$. The rotational speed is determined by means of a phase detector (PLL). Many schemes of sensorless speed determination are reported in literature [12], [13].

System (19) presents a simple way to implement the PLL on PLC:

$$\begin{cases} x_k = x_{k-1} + k_i \cdot T_s \cdot \left(U_\alpha \cdot \sin \theta_{t-1} - U_\beta \cdot \cos \theta_{t-1} \right) \\ \omega = x_k + k_{P2} \cdot u_k \end{cases} \qquad (19)$$

where: x_k is the state value at moment k, x_{k-1} is the state value at moment k-1, k_i is the integrator coefficient, T_s is the sample time, θ_{t-1} is the rotary angle at moment t-1, ω is the speed (the transfer function output), k_{P2} is the proportional coefficient and the u_k is the transfer function input at moment k.

The response time of this estimator is clearly superior to the torque estimator based on dynamic model.

Different modeling and discrete control practices can be found in [14].

IV. SIMULATED AND EXPERIMENTAL SETUP

Expanding the range of PLC utilization raises a number of technical issues. Generally speaking, torque estimation requires integration of voltages (dynamic model).

$$\Psi(t) = \int_0^t (U - I \cdot R) \cdot dt + \Psi(0) \qquad (20)$$

where: $\Psi(0)$ is the flux at moment 0 and could be considered 0 fort IM.

The Simulink layout from Fig.5 was implemented in PLC, as a Synchronous function, totally separated from the main program which is the VFC torque reference generator. The DTVI block represents the integrator. The sampling time of the integrator is given by the synchronous function cycle. The results are stored in an array, in which the position of the value multiplied by the sampling rate, generates the time of the signals.

In Fig.8 the estimated torque using the dynamic model is presented, utilizing the Euler integration method.

This way, the global error at a certain time is directly proportional to the sample time (cycle program time). Equations (21) and (22) present how this method is implemented in PLCb-DACS.

$$t_{n+1} = t_n + h \qquad (21)$$

$$y_{n+1} = y_n + h \cdot f(t_n, y_n) \qquad (22)$$

where: t_n , t_{n+1} represent the time at moment n, respectively, at moment $n+1$, h is the sample time, y is the output and f represents the output y at moment n.

For the low speed data acquisition system ($500\mu s$ sampling rate, Fig.2), improved results can be obtained by using a higher-performance integration method.

Given the limitation by the PLC processor, the sampling time couldn't be decreased. Two different types of discrete integration methods were studied in order to achieve better performance on PLC. Runge-Kutta (RK) method is generally more accurate. The approximate solution converged faster to the exact solution compared to the Euler method. In many applications this integration method can be found in [15]. In Fig.7 a block diagram of this method, for one simple function can be seen.

$$y_{n+1} = y_n + \frac{1}{6} \cdot \left(k_1 + 2 \cdot k_2 + 2 \cdot k_3 + k_4 \right) \quad (23)$$

$$k_1 = h \cdot f\left(x_n, y_n \right) \quad (24)$$

$$k_2 = h \cdot f\left(x_n + \frac{h}{2}, y_n + \frac{k_1}{2} \right) \quad (25)$$

$$k_3 = h \cdot f\left(x_n + \frac{h}{2}, y_n + \frac{k_2}{2} \right) \quad (26)$$

$$k_4 = h \cdot f\left(x_n + h, y_n + k_3 \right) \quad (27)$$

This way the final value is a weighted average of all four intermediate values: k_1, k_2, k_3, k_4. To compute the first value (k_1) Euler's Method is used. Next, two values (k_2, k_3) use estimates of the slope of the solution at the midpoint. The last intermediate value (k_4) uses an estimate of the slope at the right end-point. In Fig.8, $Tref\_PLC$ represents the torque reference prescribed by the torque generator from PLCb-DACS, $Tem1\_PLC$, represents the electromagnetic torque obtained from PLC with real voltages and currents, and $Tem1\_sim$ represents the electromagnetic torque obtained from simulation for the same acquired data set. The $Tem1\_PLC$ doesn't appear on the graph because is perfectly overlapped over the simulated electromagnetic torque. This means that the simulated estimator and the PLC works the same. The estimated torque approaches the prescribed reference after approximately 0.1s after the reference was given to the VFC.

In Fig.9, the results of the dynamic model implemented on the low speed PLC which uses the Runge-Kutta fourth order integration method are presented. Also the steady state

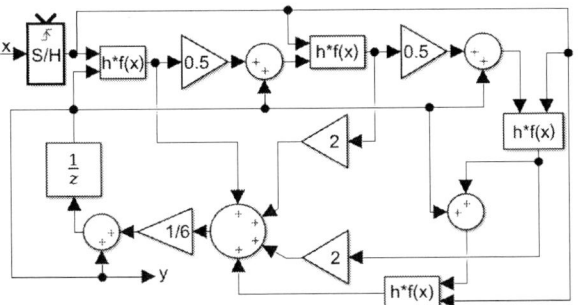

Fig.7   Simulink implementation of Runge-Kutta fourth order method

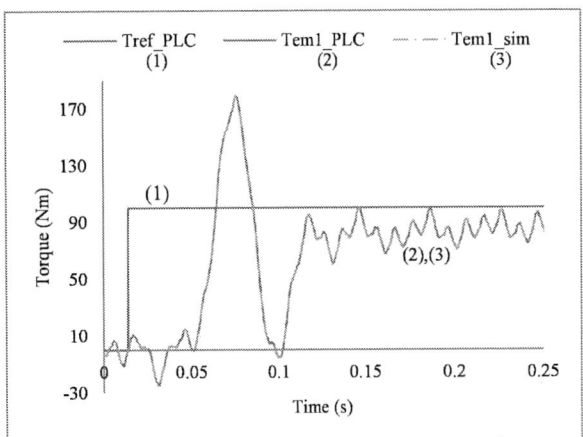

Fig.8   Low speed PLC – dynamic model – Euler method

torque error seems to be larger when the Euler method was used. Fig.10 and Fig.11 present the results obtained using the direct calculation method implemented on the low speed PLC. Tdc_PLC represent the estimated torque produced by PLC when the direct computation method is utilized. The PLC implementation of the layout from Fig.6 involves a smaller effort, and thus fewer hardware resources used. Even if we utilized the Euler method or the RK no. 4 method, the results present considerable deviation from the reference.

The behavior of the both models are illustrated in Fig. 12, where the induction machine and the torque estimator were simulated considering a real input voltage acquired with the high speed system (including higher voltage harmonics and measurements errors) and an input load $Tload$:

$$T_{load} = T_c + T_a \cdot \sin\left(\omega \cdot t \right) = 100 + 25 \cdot \sin\left(50 \cdot t \right) \quad (28)$$

It could be noticed that estimated load with dynamic model, $Tem2$, follows the input load $Tload$ while electromagnetically estimated torque from dynamic model, $Tem1$ follows the electromechanical torque of induction machine $Teim$. The estimated torque by direct method, $Tedc$ follows the electromagnetically torque even closer than $Tem1$ but the direct method is not able to estimate the load torque. The electromagnetic torque ripples are produced by the higher harmonics from the measured supply voltage.

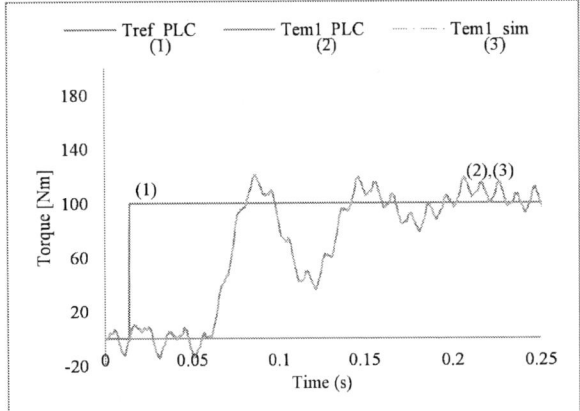

Fig.9   Low speed PLC–dynamic model - Runge Kutta no. 4 method

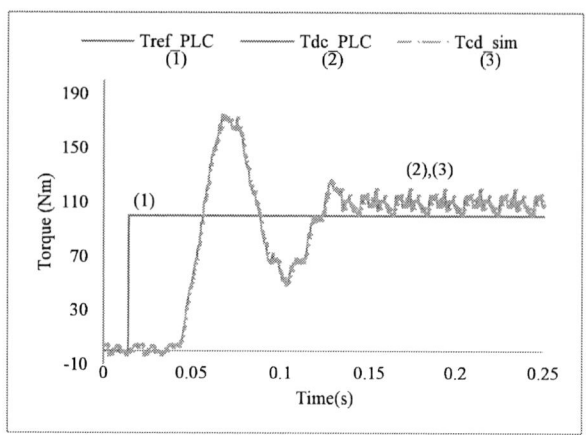

Fig.10 Low speed PLC–direct computation method

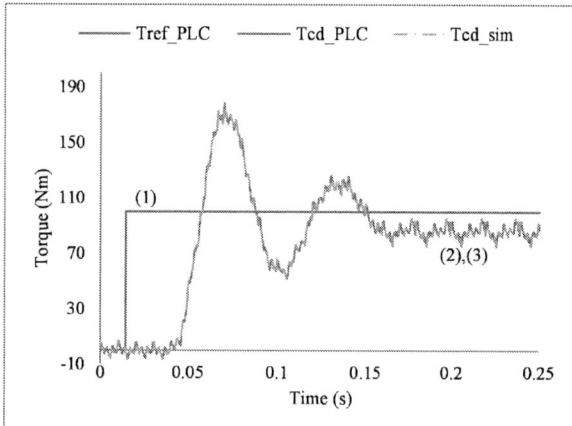

Fig.11 High speed PLC - Direct computation method

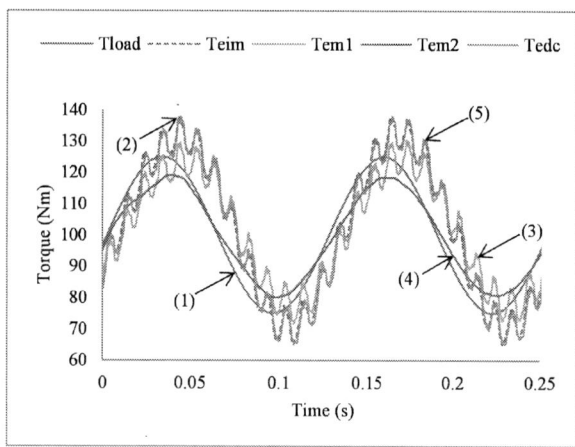

Fig.12 Dynamic and direct torque model – comparison results

V. ESTIMATOR SENSITIVITY ON INDUCTION MACHINE

Correct estimation of the induction machine torque and speed is strongly influenced by the change of parameters. Induction machine sesitivity analisys is done by changing next 5 parameters one at the time: R_s, R_r, L_s, L_r, L_m.

In Fig. 13 – Fig. 18 are presented the simulation results by reporting the real torque value (obtained from the

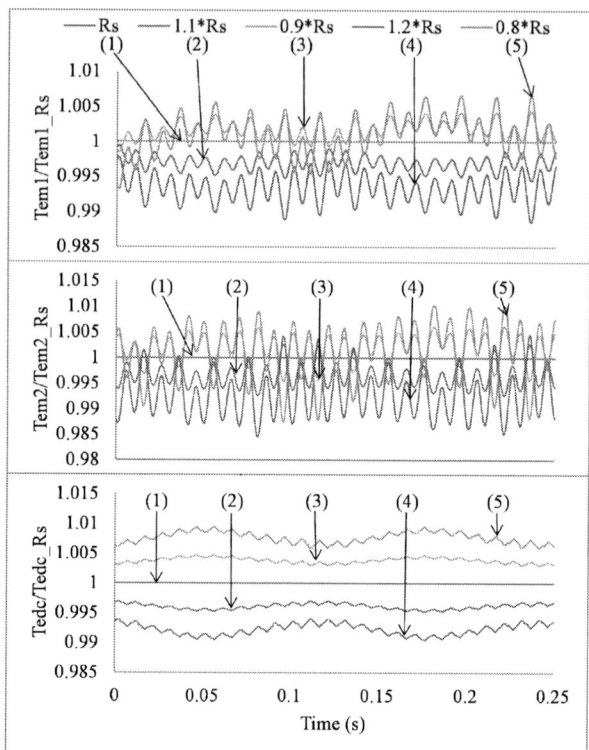

Fig.13 Reported response of the estimator by modifing the Rs parameter

estimator with unchanged parameters) and the modified torque value (obtained from the estimator with 1 single parameter changed).

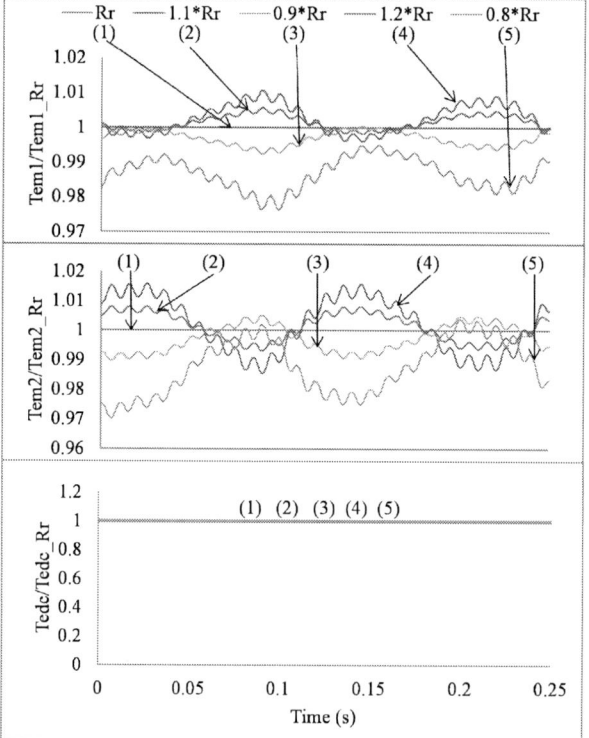

Fig.14 Reported response of the estimator by modifing the Rr parameter

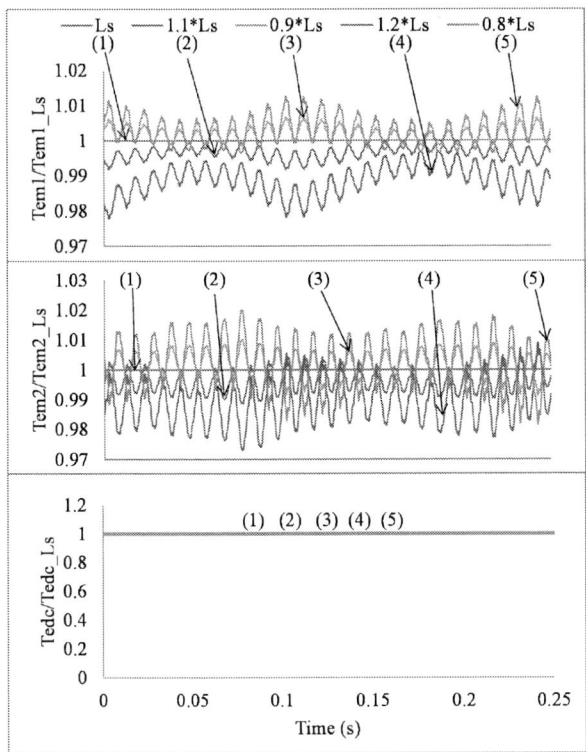

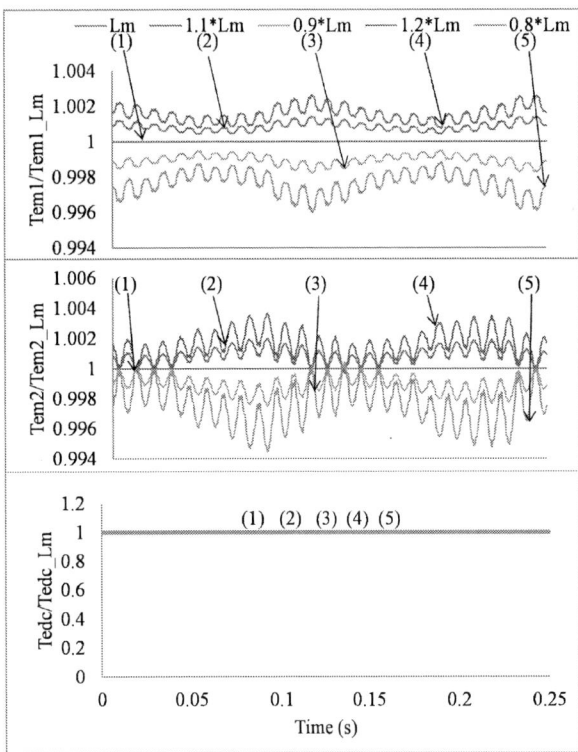

Fig.15 Reported response of the estimator by modifing the Ls parameter

Fig.17 Reported response of the estimator by modifing the Lm parameter

The reported values of the estimated torques without changing any parameter were noted with (1). For a better understanding, the studied parameter appears as a suffix in the y-axis name on previous graphs.

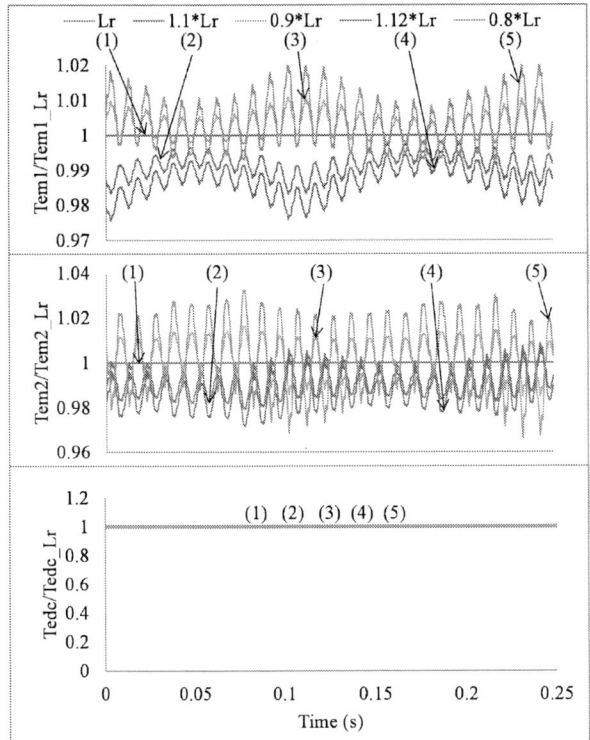

Fig.16 Reported response of the estimator by modifing the Lr parameter

The estimated electromagnetic torque obtained with the direct method T_{edc} is influenced only by the changing of R_s. The variation range of L_r is $[0.8,\ 1.12]$. Passing above the upper limit of 12% of L_r, the electromagnetic torque estimated with the dynamic method $T_{em1}\_L_r$ presents great deviations.

VI. CONCLUSION

The torque estimation methods were implemented on PLC and also offline simulated on Simulink with the same data set. Simulink model represents a good solution to troubleshoot the torque estimator implemented on PLC, because it allows an easy step-by-step monitoring of each state variable.

From previous graphs it can be observed that using the same data set there is no significant difference between the online results (PLC based systems) and offline (Simulink based) results. The direct method implemented on both systems produces about the same result. The Simulink implementation could be used as a tool to check and debug the PLC code.

The direct torque computation method offers a smaller average error in steady state compared to the dynamic model that uses the Euler integration method. The dynamic torque estimation method implemented on the low speed system produced better results if the Euler integration method was replaced by the Runge-Kutta 4th order integration method. It could be noticed that torque ripple computed by Euler method are larger than those computed with Runge-Kutta method, which is, however, harder to implement. At the same time the overshoot of the Euler method is significantly bigger that of RK no. 4 method. Direct computation method is more stable in case of an accidental change of a parameter during operation.

Regardless of the modified parameter, both estimators follow the load torque frequency of 50 rad/s, but present significant phase shift compare to the results based on the real parameters estimator. For a positive deviation of the machine's parameters ($[0\% - 20\%]$ from nominal value), the electromagnetic and mechanical torques estimated by the dynamic model, show a lower average value than those estimated with nominal parameters.

Although the use of high acquisition speed PLCs would increase the estimator performance, the ratio between data acquisition and processing times has to be carefully chosen in relation with the PLC hardware structure and drive system requirements.

We conclude that a reasonably accurate and rather simple torque estimation can be achieved for any grid connected motor, by choosing the proper combination of PLC hardware & torque estimator. The above proposed structures can be used on a wide range of PLC types, and will be very useful in pulsating torque damping applications.

ACKNOWLEDGMENT

This work was supported by a grant of the Romanian Ministry of Research and Innovation, project number10PFE/16.10.2018,PERFORM-TECH-UPT-*The increasing of the institutional performance of the Polytechnic University of Timișoara by strengthening the research, development and technological transfer capacity in the field of "Energy, Environment and Climate Change"*, within Program 1 - Development of the national system of Research and Development, Subprogram 1.2 - Institutional Performance - Institutional Development Projects - Excellence Funding Projects in RDI, PNCDI III". The test benches required to fulfill this work was provided by BEESPEED Automatizari SRL, Timisoara, Romania.

REFERENCES

[1] Marcelo M. Stopa, Marcos A. Saldanha, Alex-Sander A. Luiz, Lane M. R. Baccarini, George A. M. Lacerda, "A simple Torque Estimation for In-Service Efficiency Determination of Induction Motors", 2017 IEEE Industry Applications Society Annual Meeting, 1-5 Oct. 2017, Cincinnati, OH, USA;

[2] Toshie Kikuchi, Yasushi Matsumoto, Akira Chiba, „Fast Initial Speed Estimation for Induction Motors in the Low-Speed Range", IEEE Transactions on Industry Applications, vol 54 , issue: 4 , July-Aug. 2018;

[3] Konstantinos N. Gyftakis, Dionysios V. Spyropoulos, Joya C. Kappatou, Epaminondas D. Mitronikas, „A Novel Approach for Broken Bar Fault Diagnosis in Induction Motors Through Torque Monitoring", IEEE Transaction on Energy Conversion, vol. 28, no. 2, June 2013.

[4] Karl-Heinz John, Michael Tiegelkamp, „IEC 61131-3:Programming Industrial Automation System. Concepts and Programming Languages, Requirements for Programming Systems, Aids to Decision-Making Tools".

[5] Ephrem Ryan Alphonsus, Mohammad Omar Abdullah, "A review on the applications of programmable logic controllers (PLCs)", Renewable and Sustainable Energy Reviews, vol. 60, pp. 1185-1205 July 2016,;

[6] Nordin Saad , M. Arrofiq, „A PLC-based modified-fuzzy controller for PWM-driven induction motor drive with constant V/Hz ratio control", Robotics and Computer-Integrated Manufacturing, Volume 28, Issue 2, pp 95-112, April 2012;

[7] Yituan HE, „Proportional-Integral-Differential(PLC) control system design of compressor performance test bench", International Conference on Electrical and Control Engineering, 25-27 June 2010, Wuhan, China.

[8] „112-1996 - IEEE Standard Test Procedure for Polyphase Induction Motors and Generators", 8 May 1997;

[9] S. R. Holm, H. Polinder, J. A. Ferreira, „Analytical Modeling of a Permanent-Magnet Synchronous Machine in a Flywheel", IEEE Transaction on Magnetics, vol. 43, Issue 5, May 2007;

[10] G. von Pfingsten, M. Nell, K. Hameyer, „Hybrid simulation methods for induction machine calculation reduction of simulation effort by coupling static FEA with transient FEA and analytic formulations", 18th International Symposium on Electromagnetic Fields in Mechatronics, Electrical and Electronic Engineering (ISEF) Book of Abstracts, 14-16 Sept. 2017, Lodz, Poland;

[11] Andre Mrad, Ziad Noun, Mohamad Arnaout, „SIMUPMSAM – An analytical modeling tool for permanent magnet synchronous and asynchronous machines", International Conference on Computer and Applications (ICCA), 25-26 Aug. 2018, Beirut, Lebanon.

[12] M. A. Asha Rani, Chilakapati Nagaman, G. Saravana Ilango, „An Improved Rotor PLL (R-PLL) for Enhanced Operation of Doubly Fed Induction Machine", IEEE Transactions on Sustainable Energy, vol. 8, Issue 1, Jan. 2017;

[13] Kyoung-Jun Lee, Jong-Pil Lee, Dongsul Shin, Dong-Wook Yoo, Hee-Je Kim, „A Novel Grid Synchronization PLL Method Based on Adaptive Low-Pass Notch Filter for Grid-Connected PCS", IEEE Transactions on Industrial Electronics, vol: 61 , Issue 1 , Jan. 2014;

[14] Jorge Rivera Dominguez, „Discrete-Time Modeling and Control of Induction Motors by Means of Variational Integrators and Sliding Modes–Part I: Mathematical Modeling", IEEE Transactions on Industrieal Electronics, vol. 62, Issue 9, Sept. 2015.

[15] Marius Brehler, Malte Schirwon, Dominik Göddeke, Peter M. Krummrich,"A GPU-Accelerated Fourth-Order Runge–Kutta in the Interaction Picture Method for the Simulation of Nonlinear Signal Propagation in Multimode Fibers", Journal of Lightwave Technology, vol 35, Issue 17, Sept.1, 2017.

A novel kurtogram-based health index for induction motor fault diagnosis

Azadeh Gholaminejad
Electrical Engineering department
Iran University of Science and Technology
Tehran, Iran
gholaminejad.a@gmail.com

Farzaneh Sabbaghian Bidgoli
Electrical Engineering department
Iran University of Science and Technology
Tehran, Iran
sabbaghianf@yahoo.com

Javad Poshtan
Electrical Engineering department
Iran University of Science and Technology
Tehran, Iran
jposhtan@iust.ac.ir

Majid Poshtan
Electrical engineering department
California polytechnic state university
San Luis Obispo, CA, USA
mposhtan@calpoly.edu

Abstract— **Nowadays fault diagnosis is a necessary part in all manufacturing processes considering economical and safety aspects. One of the main applications of fault diagnosis is condition-based maintenance which plays a key role in modern industries. Vibration-based techniques are among the most famous methods in this area. They are especially successful in mechanical faults diagnosis. In this paper, first some of the main approaches for fault diagnosis of induction motors using vibration data are briefly reviewed. Mostly the practical points are highlighted and the results of a laboratory setup is presented. At the end, a novel approach for induction motor fault diagnosis using vibration signals based on kurtosis and FFT is proposed.**

Keywords— fault diagnosis, vibration signal, induction motor, kurtosis

I. INTRODUCTION

Maintenance has been always a challenging issue in all industries, and numerous strategies have been proposed and applied in this field. One of the main applications of fault diagnosis (FD) methods is in Preventive Maintenance (PM) strategy. This need has caused a drastically fast evolution in different methods of fault diagnosis. PM is an alternative to the Corrective Maintenance (CM) strategy. The concept of PM includes the performance of maintenance prior to the failure of equipments. In Table I the improvements due to monitoring systems implementation is demonstrated quantitatively. Condition-based maintenance (CBM) is the most popular technique. It is applied to maximize the effectiveness of the PM decision making. In this method, the system condition is monitored constantly using real-time data. Vibration signals are the most popular and effective signals used for FD in rotary machines. Vibration prevention is practically impossible. There are several sources of vibration in equipments. It can be a result of dynamic effects such as production change, gaps, rotary or frictional connections between machine parts and also out-of-balance forces in rotating and vacillating parts. Even an unremarkable small vibration may excite resonance frequencies of components and convert to a main source of noise and vibration. Especially resonances in machine's working speed range can be catastrophic, because each time passing the resonance frequency amplifies the vibration amplitude hazardously [1] [2] [3].

TABLE I. RELIABILITY IMPROVEMENT USING A MONITORING SYSTEM

Maintenance costs	Reduction of 50%–80%
Equipment damages	Reduction of 50%–60%
Extra hours expenses	Reduction of 20%–50%
Machine life expectancy	Increase of 50%–60%
Total productivity	Increase of 20% to 30%

Depending on selected features, diagnosis can be performed through analysis in time domain (average, kurtosis, variance…), frequency domain (envelope analysis, harmonic analysis…), time-frequency domain (short-time Fourier transform, Hilbert-Hung transform, wavelet transform…), cepstral and nonlinear (semi-phase portraits, correlation dimensions, information entropy…) forms [3].

On the other side, electrical motors are the most common energy converters in industry. Their malfunction may cause production loss and affect the entire system. Therefore, motor fault diagnosis is an important issue which has been the subject of many recent researches. Early detection of faults before a complete failure of the motor can make planned maintenance possible without loss in production. Fault detection in these motors are performed through signal processing in three main underparts: stator, rotor and bearing. Fig. 1 shows the dispersion of most common faults in induction motors. Among various methods proposed in this field, vibration-based approaches have proven to be effective especially in mechanical fault diagnosis [4] [5] [6] [7].

There is a huge source for literature review in the field of vibration-based fault diagnosis and also its application to rotary machines. In the sequel, some of the main and most relevant ones to our research are reviewed.

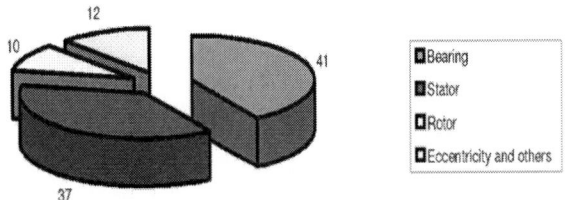

Fig. 1. Induction motors fault dispersion [7]

978-1-5386-7688-2/19 $31.00 © 2019 IEEE

In [8] the economic impact of vibration-based maintenance is investigated and applied on a paper mill in Sweden. It is shown that there is an average yearly of 3.8 million SEK profit and also an average potential saving of 30 million SEK due to economic losses reduction. In [9] vibration signal monitoring is used for pump cavitation fault diagnosis. It is demonstrated that vibration-based techniques are a powerful tool for accurate detection of cavitation in real time applications. In [10] an envelope demodulation method is proposed. It is a short time fractal dimension (STFD) transform which extracts features from vibration signals for bearing fault diagnosis. This research claims that the lower the inference frequency, the less effect the inference has on STFD presentation. Therefore, a kurtosis-based peak search algorithm (KPSA) is also proposed. The STFD-KPSA method is examined and compared to other envelope methods using experimental data. In [11] operational deflection shape and kurtosis of vibration response are used for crack identification in a rotor-bearing system. Also kurtosis deviation curve and amplitude deviation curves are investigated for shaft crack detection. The results show successful diagnosis of single cracks and also two cracks. In [12] vibration signals are used for finding frequencies of tapered roller bearings, gear mesh and belt drive in a lathe. It is shown that experimental prediction of defects is correct and accurate. In [13] two real-time vibration-based monitoring case studies are presented. Spectral analysis is used as a means of failures early stage identification. In [7] dynamic and static eccentricity faults of a rotor are investigated using vibration signals. It is shown that the static eccentricity causes change in vibration while dynamic eccentricity changes the vibration order and frequency. It is concluded experimentally that static eccentricity increases the vibration of the original frequency and dynamic eccentricity leads to sideband frequency and little changes in the vibration of the original frequency. In [14] three fault feature extraction methods are investigated. They include Fourier transform frequency spectrum, envelopment analysis and local mean decomposition. Their mathematical theory is developed and the vibration signal of common machinery fault is designed. The successful performance of methods is also proven using experimental vibration signals. In [15] bearing fault diagnosis using vibration signal is studied. Various techniques in time, frequency and time-frequency domains are explored. Experimental examination results show accurate bearing fault diagnosis. [16] investigated vibrational resonance method for filtering and denoising in order to detect weak bearing fault. The method is validated using simulation data and it is shown that the method is a practical solution for minor faults of bearing. In [17] the issue of distributed bearing fault diagnosis is studied. A procedure for these faults detection employing vibration signal analysis is proposed. It is concluded that fault diagnosis is possible due to differences in features extracted from vibration signals in healthy, localized and distributed fault conditions.

In this paper a novel approach for induction motor monitoring is proposed employing FFT and kurtosis concepts. First the issue of vibration-based fault diagnosis especially in rotary machines is investigated and some of the most famous and effective methods are briefly reviewed with an emphasis on the practical points. They are also applied on experimental vibration data collected from an electro-pump system and the results are presented.

In the remainder of this paper, first in section two the issue of vibration-based fault diagnosis and some of the main analysis used in this area are explained. In section three the experimental results of these methods are presented. Also a heuristic method is proposed and experimental application results are shown. And the paper ends in section four by conclusion and future works offers.

II. FAULT DIAGNOSIS METHODS USING VIBRATION SIGNAL

Usually for general judgement about machine condition, a feature or index is extracted from vibration signal. Then it is monitored and compared to historical previous measurement records, validated standards suggested value or manufacturing company value. Methods applied on vibration signals may be in time, frequency, cepstral, time-frequency or nonlinear domain [3].

Some of these approaches are studied below from a practical point of view especially for implementation on an induction motor fault diagnosis system.

A. Time Domain Analysis

Vibration time waveform or time signal is the first output of the sensor. It is a direct demonstration of machine dynamic behavior and its component condition. Especially faults which cause pulses and changes in common sinusoidal form of signals are trackable by time domain analysis. Localized faults in ball bearings, a crack in inner or outer race of a bearing are some of the main faults that can be identified through analysis of the time domain signal.

B. Spectrum Analysis

Another approach in rotary machines fault diagnosis using vibration signals is spectrum analysis. Shifting from time domain to frequency domain is performed through fast Fourier transform (FFT) and thus the frequency domain signal is the result. This approach is the most important and common method for various faults diagnosis. Time domain signal has several frequencies. FFT extracts these frequencies and demonstrates each frequency with the respective amplitude. It detects harmonics and sidebands in the signal [3].

There are some parameters in this method needed to be considered and adjusted correctly in order to result in a good performance. They are average method, number of averages, window and filter selection.

Machine behavior is not uniform in time. There are some irregularities meanwhile performing a frequency analysis like changes in load or speed, impulses or adjacent machines vibration disturbance. These disorders affect analysis results. Recording a single spectrum is in fact a coincidental record of machine vibrational behavior. Averaging uses a series of records in definite time intervals which result in machine real behavior and condition representation. It increases results repeatability and facilitates noisy and complicated signals explanation [18].

Averaging methods are various. Linear averaging is the common traditional one. Another approach is peak hold averaging. In this method the peak value in each analysis is saved and displayed. In fact, it forms an envelope of the highest amplitude of spectral line. It is useful for transient conditions like coast downs. In exponential averaging the last spectra are considered more important. It is a weighted averaging with higher weights for the most recent spectra. It is useful in slow processes with respect to sampling time.

978-1-5386-7688-2/19 $31.00 © 2019 IEEE

Synchronous time method uses a synchronizing signal from monitoring machine. It is used in time domain. It is effective for machines with many rotational parts with different speeds. In this way the vibrations which are synchronous with the synchronizing signal are given priority [18].

The number of averages parameter determines the number of spectrums used in averaging. The higher the number of spectrums is, the softer the noise impulses become. It results in more accurate spectrum peaks. It needs to be noted that as the number of averages increases more data should be collected, thus more time is needed for calculating the average spectrum.

The collected data is not usually used directly for spectrum calculating. Regarding FFT method limitations, first some corrections are applied on the data. In other words, after digitalizing the signal using an analog to digital transformer, in the next step the windowing function is applied and then the FFT algorithm is used. Since FFT algorithm considers discontinuity as modulating frequencies and displays them in the form of sidebands in the spectrum. While in fact these frequencies do not exist in the signal. And also it affects the capability of adjacent frequencies analysis while maintaining the amplitude accuracy. Thus windowing is necessary in spectrum analysis. It is often carried out through multiplication to a correction window. It prevents signal leakage and manipulation. Windowing function selection is an issue. Window passing is multiplying signal samples by a windowing function with the same length. In this way data discontinuity is filled by nullifying sampled data at the beginning and the end of the sampling period. Rectangular, flat top and hanning are the most common windows. Rectangular or uniform window does not apply any window. It is useful for transition signal resulted from a momentary shock. In comparison to rectangular window, with using flat top window a broader peak in FFT is observed. Hanning window has the same effect but less than the flat top one. Recognizing adjacent frequencies is very difficult with broad peaks. When identifying a part of a signal (peak) in a special frequency is desired, applying rectangular window is a better choice. But if the amplitude of the peak is important, flat top window is definitely the best one. Hanning window has a high frequency resolution which makes it a good option for periodic monitoring. Exponential window is used in bump test. It is not recommended for application in machine stable working condition. It is clear that the best window selection depends on the application [18] [19].

Another issue is filter selection. There are two main choices in frequency analysis of vibration signal. They are constant bandwidth and constant percentage bandwidth filters. Constant bandwidth filter has a constant resolution. In constant percentage bandwidth filter the filter bandwidth is a constant percentage of central frequency. Therefore, as the frequency increases the bandwidth also increases. Bandwidth and number of lines parameter determine the speed of data acquisition. A spectrum with higher number of lines is more informative. It means as the number of lines increases more data is collected and of course it will need more time and memory. Spectrum precision (resolution) is directly affected by the number of lines. For instance, if there are vibrational parts which their distance is less than the distance between frequency lines, they will not be distinguished. It should be noted again that the time of data acquisition has an inverse relation with the distance between frequency lines. Bandwidth

filter selection which means the narrowness of passing band also regulates frequency analysis resolution. Using a narrow bandwidth result in more details and makes possible the isolation of vibration spectrum peaks. But the downside is that the time needed to achieve a certain accuracy increases remarkably by narrowing the bandwidth of the filter. Thus usually an elementary analysis is performed by a broad bandwidth filter. In this way the important parts of frequency spectrum are detected. Then the analyzer is changed to a one with narrow bandwidth for more details. The best option for bandwidth selection is mostly the one which provides a sufficient resolution in the entire frequency range and enables analysis in the shortest possible time. Bandwidth can be calculated by dividing frequency range into number of lines and then multiplying the result by the window function. Resolution is twice the bandwidth. Sampling time is achieved by dividing number of lines into frequency range. Increase in F_{max} makes measurement faster, bandwidth broader and more lines are needed in order to do not miss the details. It does not cause raise in number of collected data. But spreads data in a broader range of frequency. Generally, for rotary machines like pumps, fans and motors F_{max} is selected 20 to 40 times of operation speed. For gearboxes it is at least 3 times of tooth involvement frequency. For the first analysis of a machine it is recommended to begin with two spectrums: one in 10 times of operation speed and another one in 100 times of operation speed. Then one will make sure that no important frequency (in high and low frequency ranges) is missed [18] [19].

C. Cepstrum Analysis

Cepstrum analysis is a technique consisting functions in the form of "spectrum of logarithmic spectrum". This method has been used effectively since 1990s in the field of fault diagnosis. Rotary machines spectrum may be so complex with several harmonics or sidebands due to frequency modulations. Cepstrum brings back the spectrum to time domain. Therefore, it has picks in the period of common frequency distances in spectrum. In other words cepstrum reveals the periodicity of the frequency spectrum and the patterns of harmonics and sidebands in it [18] [29] [3].

D. Envelope Analysis

Envelope analysis of the frequency spectrum is the FFT of the modulating signal. It is mainly used for bearings and gearboxes fault detection. Envelope analysis or amplitude demodulation is a technique in which the modulating signal is extracted from an amplitude modulated signal. The user may also use some filters for focusing the signal processing on a definite frequency zone. The result is the time history of the modulating signal. It can be studied in time or frequency domains. It is a well-known technique to extract the periodic impulses from machine vibration signal. It can perform this even for the impulses with low level of energy and hidden with other vibration signals. Therefore, it is an applicable method for maintenance people. For instance, bearing fault detection is performed through this method very fast. For ball bearings, when the rotating part with a local fault meets the inner or outer race produces an impulse. These impulses modulate a signal in bearing passing frequencies. A common approach for envelope generation is to calculate the FFT of the time domain signal and each bin frequency (for example the division of sampling rate to time series length). The window indexes are also needed to be calculated. Then the Fourier coefficients of desired window are copied into the baseband indexes (Heterodyne function) and remaining Fourier coefficients are made zero. The last step is taking the absolute value of inverse FFT Fourier coefficients [21] [22] [18].

One of the main considerations in this method is correct filter placement. Averaging is also necessary to improve signal quality and signal to noise ratio [23].

E. Kurtosis Coefficient and Kurtogram

As mentioned in the envelope analysis explanations, choosing a correct filter is an issue in envelope analysis and also in many other applications. One of the methods proposed for this purpose is spectral kurtosis. Kurtosis is a non-dimensional parameter which measures the "peakedness" of a probability distribution of a random variable. In fact, it is used to check whether the data set is peaked or flat in comparison to the normal distribution. High kurtosis value shows high variance due to infrequent extreme deviations. In other words, it is the fourth standard moment and is defined as below

$$kurtosis = \frac{\sum_{i=1}^{n}(x_i - \bar{x})^4}{(n-1)\sigma^4} - 3 \qquad (1)$$

This definition gives the kurtosis of a standard normal distribution as zero [24] [11].

If the selected band pass filter is truly adjusted, the kurtosis coefficient of the signal will increase due to the fault. Generally the kurtosis coefficient of the envelope of the filtered signal is more liable for fault detection.

Spectral kurtosis is a technique for finding the best band pass in envelope analysis. It is a statistical parameter that shows signal impulsiveness changes due to frequency. Reference [25] has investigated this approach thoroughly. In [26] kurtogram is developed. it is a map that demonstrates the optimum central frequency and band width. The area with the highest kurtosis is selected as the range showing bearing fault the best.

F. CPB

Constant percentage bandwidth analysis is a kind of frequency spectra in which the whole frequency range is divided into some limited number of bands. It is a tool in vibration analysis for fault diagnosis which has a good repeatability and precision. It has a smart filter and focuses on reducing the number of false alarms. Each band width is a constant percentage of the central frequency. It means that frequency resolution in lower frequencies is relatively higher than higher frequencies. It is good for a reliable primary fault diagnosis. the number of bands is selected by the user. Higher number of bands increases the fault diagnosis system ability and also the need of memory. CPB spectra is usually logarithmic that allows a higher range of frequency demonstration [27] [18] [22].

III. EXPERIMENTAL INVESTIGATION

In this section a fault diagnosis interface development is explained. The introduced methods are used in this interface. The methods are applied on experimental vibrational signals collected from the laboratory system. The laboratory system consists of a 3-phase 2-pole 3000-rpm 3-KWatt squirrel cage induction motor. This motor is a driver for a 3-stage centrifugal pump which is usable for maximal output of 74 meters. The system is shown in Fig. 2.

All the algorithms are coded in MATLAB. But considering the Lab-View user friendly environment, it is used as an interface between MATLAB and the use and the adjustments by user and results demonstration are performed in Lab-View. Math script block is used for the connection between MATLAB and Lab-VIEW. The Lab-View main page environment is shown in Fig. 3.

Vibration signals collected from the electropump system are used for methods evaluation. Data acquisition is performed using a magnetic accelerometer sensor AS-065. It is located on the coupling between motor and pump as it is shown in Fig. 2..

There are three kinds of contact vibrometers: accelerometers, velocity meters and displacement meters. The accelerometers are the best. They show the speed of the changes in speed of vibrating part. The piezo-electric accelerometers are the most applicable ones. They have a broad, linear and dynamic frequency range. They are relatively robust and reliable. It means that their characteristics is stable in time. They are also self-generator. Therefore, they do not need a power supply. Another advantage is that they do not have any moving part [2] [19].

The developed user interface system for fault diagnosis enables data acquisition in three modes. Measurement can be performed based on a pre-determined time, pre-determined speed (frequency) and through selecting the stop button. When the speed mode is selected, the user should also choose the machine condition from two options of coast down or start up. Data demonstration is possible in the linear and logarithmic units. In Fig. 3. to Fig. 6. the user interface environment and the online time domain signal, average time domain signal, signal FFT and the adjustments in these three modes are showed.

After selecting the data acquisition mode, several analyses are possible.

In Fig. 7. the signal in time domain and also its frequency spectrum are showed. The number of lines are adjustable.

Fig. 2. Lab elecrtopump system

978-1-5386-7688-2/19 $31.00 © 2019 IEEE

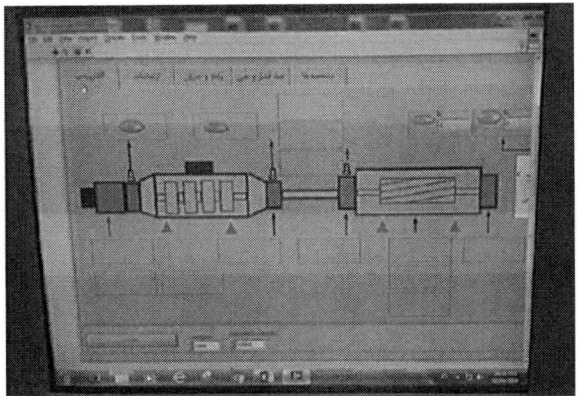

Fig. 3. Lab-View environment

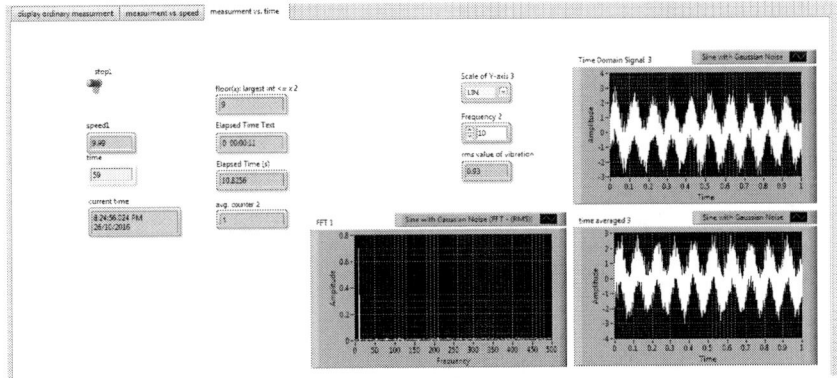

Fig. 4. Pre-determined-time data acquisition

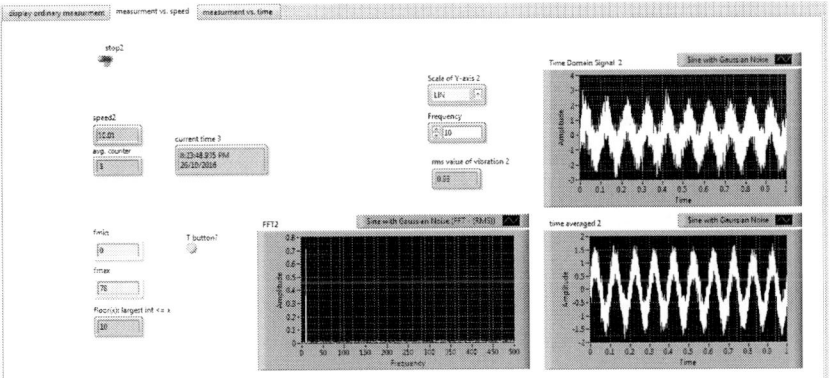

Fig. 5. Pre-determined-speed data acquisition

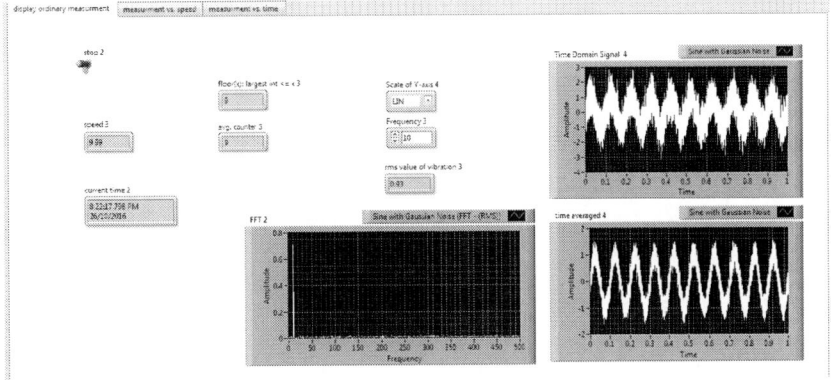

Fig. 6. Ordinary data acquisition

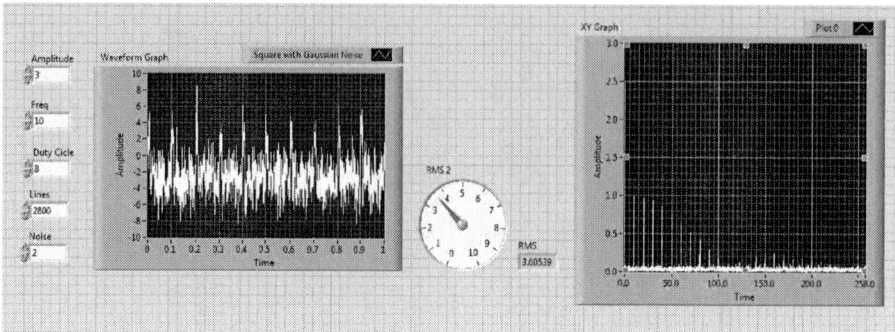

Fig. 7. Time domain and FFT of vibration signal

The next possible analysis is envelope analysis. Kurtogram is used for filter selection. In Fig. 8 vibration signal kurtogram is showed. It shows that the f_c=4687.5 Hz and 3125Hz for bandwidth are appropriate choices.

In Fig. 9 the time domain signal and its envelope using the filter determined through kurtogram are demonstrated.

It is also possible in the program to adjust the filter by user. Therefore, by choosing the Envelope SED, low and high frequency of filter are asked from the user.

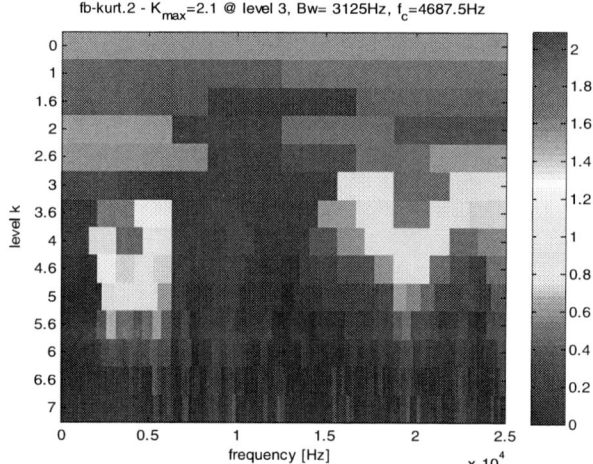

Fig. 8. Vibration signal kurtogram

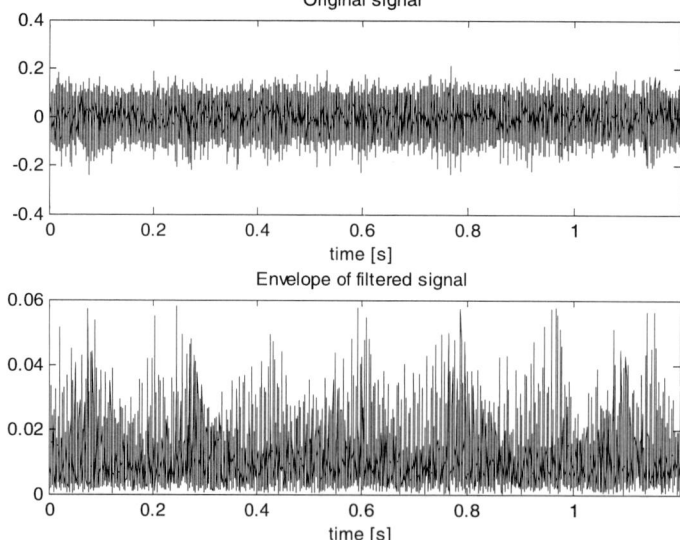

Fig. 9. Time domain vibration signal and its envelope using kurtogram-based filter

978-1-5386-7688-2/19 $31.00 © 2019 IEEE

To check CPB algorithm performance following signal is used. It consists of two sinusoidal terms with the frequencies of 50 Hz and 150 Hz and amplitudes of 100 and 1 respectively. A white noise with the variance of 1 is also added.

$$x(t) = 100 \sin(100\pi t) + \sin(300\pi t) + n(t) \qquad (2)$$

In Fig. 10., time domain signal, its FFT and CPB are demonstrated. It shows that both FFT and PCB could identify the base frequencies.

In the next step, the amplitude of the sinusoidal term with the frequency of 150 Hz is multiplied by 3 (let's call it y(t)). The result is shown in Fig. 11.. comparing these two results it is clear that the changes in amplitude in the frequency of 150 Hz in CPB is more than FFT. Therefore, by using CPB changes in amplitude of small vibrational signals are detected better which may lead to a better fault diagnosis.

A. Novel suggested Health Index

A novel method is also developed and a novel Health Index (HI) is suggested in this research. Fig. 12. shows the kurtogram of a collected vibration signal in the frequency of 48.5 Hz. As it is clear in the kurtogram the highest values of kurtosis are in the frequency range of 312.5-1562.5 Hz. Therefore, regarding the explanations presented for the kurtosis, this range is a good choice to be selected for fault diagnosis. In the next step the single-sided amplitude spectrum is shown in Fig. 13. and then the power of Fourier spectrum is calculated in the selected range which is demonstrated. It is considered as HI and is equal to 8.0866×10^{-7}. This value and procedure can be used as a method for fault diagnosis. Process condition monitoring may be performed by monitoring the HI through recording this value in time and comparing it with the value of faulty modes.

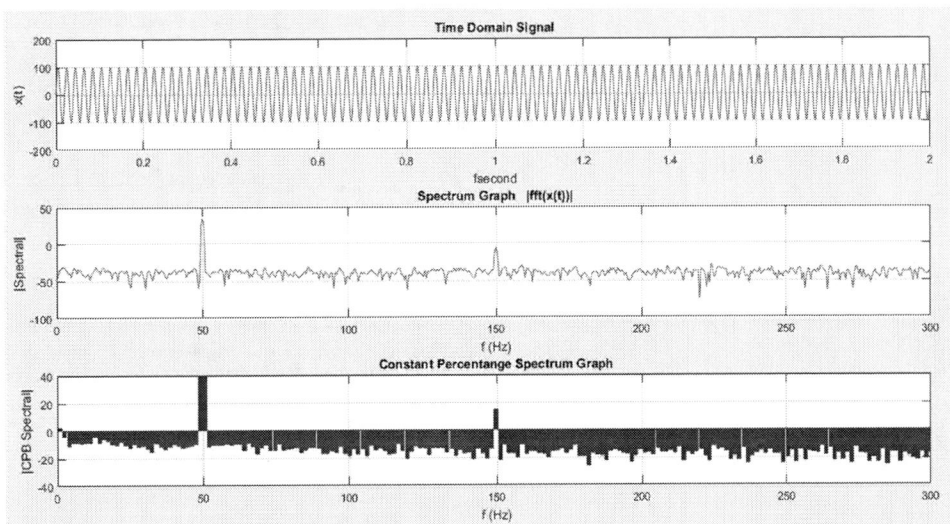

Fig. 10. X signal, its time domain, FFT and CPB

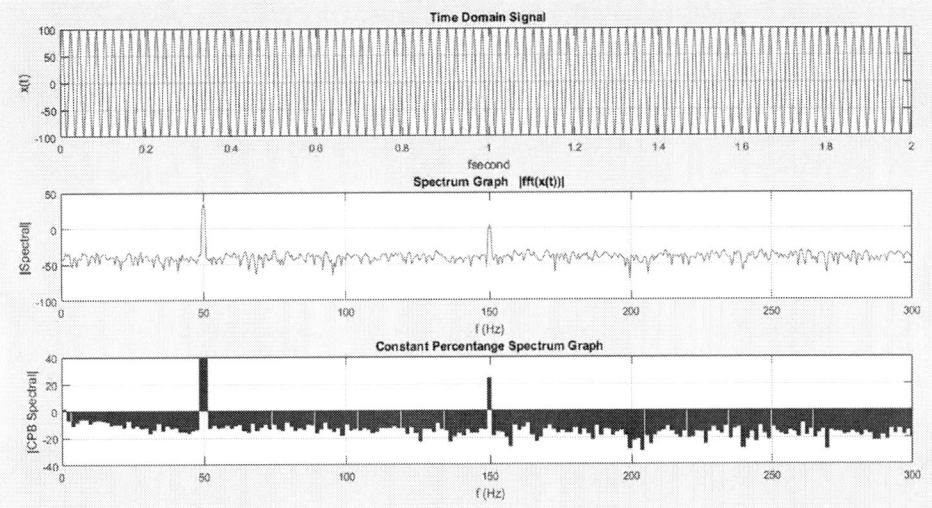

Fig. 11. Y signal, its time domain, FFT and CPB

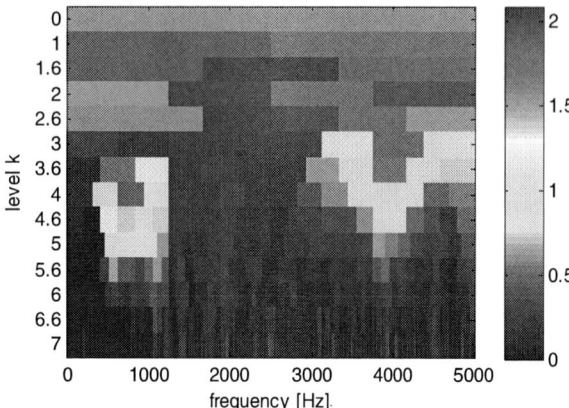

Fig. 12. Kurtogram of the vibration signal with the frequency of 48.5Hz

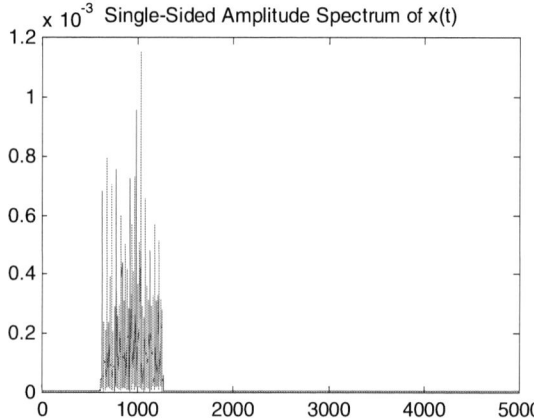

Fig. 13. Power spectrum of the signal

IV. CONCLUSION AND FUTURE DEVELOPMENT SUGGESTIONS

Fault diagnosis and equipment condition monitoring is an important part of industries. On the other hand, vibration-based methods have proven to present a high efficiency in this subject. In this research a fault diagnosis interface using some of the most reliable methods is developed. Regarding practical implementation of the methods on a real electro-pump system, the practical points of the subject are thoroughly investigated. At the end the results of algorithms implementations on real collected vibration signals are demonstrated. Also a novel method is proposed for fault diagnosis using kurtogram and FFT. In the future works, the developed interface can be used as a tool for various faults diagnosis and the results of different methods can be compared.

REFERENCES

[1] R. Ahmad and S. Kamaruddin, "An overview of time-based and condition-based maintenance in industrial application," *Comput. Ind. Eng.*, vol. 63, no. 1, pp. 135–149, 2012.

[2] Brüel & Kjær, "measuring vibration," 1982.

[3] P. Henriquez, J. B. Alonso, M. A. Ferrer, and C. M. Travieso, "Review of automatic fault diagnosis systems using audio and vibration signals," *IEEE Trans. Syst. Man, Cybern. Syst.*, vol. 44, no. 5, pp. 642–652, 2014.

[4] S. Ergin, A. Uzuntas, and M. B. Gulmezoglu, "Detection of stator, bearing and rotor faults in induction motors," *Procedia Eng.*, vol. 30, no. 2011, pp. 1103–1109, 2012.

[5] M. E. H. Benbouzid, "A Review of Induction Motors Signature Analysis," *Power*, vol. 47, no. 5, pp. 1950–1955, 1998.

[6] S. H. Chetwani, M. K. Shah, and M. Ramamoorty, "Online condition monitoring of induction motors through signal processing," *2005 Int. Conf. Electr. Mach. Syst.*, vol. 3, no. 6, pp. 2175–2179, 2005.

[7] Ç. Ferhat, "Detection of eccentricity fault based on vibration in the PMSM," vol. 10, no. January, pp. 760–765, 2018.

[8] B. Al-Najjar and I. Alsyouf, "Enhancing a company's profitability and competitiveness using integrated vibration-based maintenance: A case study," *Eur. J. Oper. Res.*, vol. 157, no. 3, pp. 643–657, 2004.

[9] D. Siano and M. A. Panza, "Diagnostic method by using vibration analysis for pump fault detection," *Energy Procedia*, vol. 148, no. Ati, pp. 10–17, 2018.

[10] J. Shi and M. Liang, "A fractal-dimension-based envelope demodulation for rolling element bearing fault feature extraction from vibration signals," *Proc. Inst. Mech. Eng. Part C J. Mech. Eng. Sci.*, vol. 230, no. 18, pp. 3194–3211, 2016.

[11] K. Saravanan and A. S. Sekhar, "Crack detection in a rotor by operational deflection shape and kurtosis using laser vibrometer measurements," *JVC/Journal Vib. Control*, vol. 19, no. 8, pp. 1227–1239, 2013.

[12] M. Senthilkumar, M. Vikram, and B. Pradeep, "Vibration monitoring for defect diagnosis on a machine tool: A Comprehensive case study," *Int. J. Acoust. Vib.*, vol. 20, no. 1, pp. 4–9, 2015.

[13] S. Orhan, N. Akturk, and V. Celik, "Vibration monitoring for defect diagnosis of rolling element bearings as a predictive maintenance tool: Comprehensive case studies," *NDT E Int.*, vol. 39, no. 4, pp. 293–298, 2006.

[14] C. Chu, Z. Zuo-xi, K. Xin-rong, and G. Yun-zhi, "The Research of Machinery Fault Feature Extraction Methods Based on Vibration Signal," *IFAC-PapersOnLine*, vol. 51, no. 17, pp. 346–352, 2018.

[15] A. Khadersab and S. Shivakumar, "Vibration Analysis Techniques for Rotating Machinery and its effect on Bearing Faults," *Procedia Manuf.*, vol. 20, pp. 247–252, 2018.

[16] L. Xiao, X. Zhang, S. Lu, T. Xia, and L. Xi, "A novel weak-fault detection technique for rolling element bearing based on vibrational resonance," *J. Sound Vib.*, vol. 438, pp. 490–505, 2019.

[17] B. Dolenc, P. Boškoski, and Đ. J. Juričić, "Distributed bearing fault diagnosis based on vibration analysis," 2015.

[18] Paresh Girdhar, "Practical Machinery vibration Analysis & Predictive maintenance," p. 264, 2004.

[19] "How is Vibration Measured? - Reliabilityweb_ A Culture of Reliability," *From the Beginner's Guide to Machine Vibration.* 2006.

[20] R. B. Randall, "Cepstrum Analysis and Gearbox Fault Diagnosis.," *Maint. Manag. Int.*, vol. 3, no. 3, pp. 183–208, 1982.

[21] E. Bechhoefer, "A QUICK INTRODUCTION TO BEARING ENVELOPE ANALYSIS," *Green Power Monit. Syst.*

[22] "Envelope Analysis _ Brüel & Kjær Vibro." .

[23] H. Konstantin-Hansen, "Envelope analysis for diagnostics of local faults in rolling element bearings," *Brüel & Kjær, Denmark*, 2003.

[24] S. Karmakar and M. Mitra, *Induction Motor Fault Diagnosis.* Springer, 2016.

[25] R. B. Randall, *Vibration-based Condition Monitoring: Industrial, Aerospace and Automotive Applications.* 2010.

[26] J. Antoni, "Fast computation of the kurtogram for the detection of transient faults," *Mech. Syst. Signal Process.*, vol. 21, no. 1, pp. 108–124, 2007.

[27] B&K company, "Using Selective Envelope Detection (SED) for the early detection of faults in rolling-element bearings," no. BPT0046-EN-11, pp. 1–15.

A Novel Surface Impedance Based Clamping Plate Loss Model for Large Synchronous Generators

Torben Fricke
*Institute for Drive Systems
and Power Electronics
Leibniz University Hannover*
Hannover, Germany
torben.fricke@ial.uni-hannover.de

Babette Schwarz
*Electromagnetics
Voith Hydro Holding GmbH & Co. KG*
Heidenheim, Germany
babette.schwarz@voith.com

Bernd Ponick
*Institute for Drive Systems
and Power Electronics
Leibniz University Hannover*
Hannover, Germany
ponick@ial.uni-hannover.de

Abstract—This paper presents a clamping plate loss model based on surface impedance boundary conditions. The clamping plate surface is discretized using a single loop of nodes. By making a clever choice of boundary conditions instead of modeling an entire pole pitch and only considering the fundamental temporal and spacial harmonic of the field normal to the clamping plate surface, the number of nodes (and elements) required is reduced to around 100, making the proposed method computationally inexpensive. However, the presented loss model requires the knowledge of the flux density normal to the clamping plate, making it only useful in situations where a fast method of obtaining the normal flux density distribution is available. In this paper, three-dimensional time step FEA is used to both validate the proposed loss model and assess the impact of assumptions made in its derivation. For the geometry investigated in this paper, only 61 nodes were required leading to a computing time of 20 ms with an error of 3 % when compared to time step 3D FEA.

Index Terms—Clamping System, End Region Losses, Hydroelectric Power Generation, Surface Impedance Boundary Conditions, Synchronous Generator

TABLE I
LIST OF SYMBOLS

A_e	area of an element
A_{node}	area of a node
$\underline{\vec{B}}_e$	element flux density as a 2D (rad-ax/tan) vector
$\underline{B}_{e,rms}$	rms value of the element flux density
$\underline{B}_n$	flux density normal to the clamping plate surface
$\underline{B}_s$	surface flux density amplitude
$\underline{E}_s$	surface electric field strength
$\underline{H}_s$	surface magnetic field strength amplitude
l	edge length
P_{cp}	clamping plate losses for the entire generator
P_e	element losses
p_s	surface loss density
p	number of pole pairs
$\underline{V}_{m,i}$	magnetic potential (nodal potential) of a node with index i
w	width
$\underline{Y}a,b$	edge admittance between two adjacent nodes a and b
$\underline{Z}_s$	surface impedance
δ	skin depth
γ_{el}	electric boundary pitch
γ_{mech}	mechanical boundary pitch
μ	permeability
$\underline{\Phi}_{n,i}$	flux normal to a node i
$\underline{\Phi}_{a,b}$	edge flux between two adjacent nodes a and b
σ	electric conductivity

Fig. 1. Rendering of a generic stator end region including the clamping system

I. INTRODUCTION

Due to their size, turbo and large hydro generators require an extensive clamping system to apply enough clamping force to the stator core, to prevent damage to the stator winding insulation resulting from vibrating laminations. Fig. 1 shows a rendering of a generic clamping system, consisting of bolts applying force onto the clamping plate through large spring washers. Pressure fingers (usually made of non-magnetic austenitic steel) apply the clamping force to the stator teeth. Cost constraints often require the solid clamping plate to be made of magnetic steel, leading to increased eddy current losses. Accurately predicting these losses during the design stage is crucial for a correct dimensioning of the cooling system and an accurate prediction of the generator efficiency. End region losses in general and clamping plate losses in particular have thus been extensively studied in the past. Mecrow et al. [1] introduced the use of 3D and quasi 3D FEA to end region loss calculation in 1989. Silva et al. [2] were the first to use surface impedance boundary conditions to predict stator end region losses. Most modern literature [3], [4] still relies on these loss modeling and field calculation approaches, due to the complex, three-dimensional nature of end region effects. Traxler [5] devised an analytical approach to calculate clamping plate losses instead, greatly reducing computational cost, but sacrificing accuracy in the process.

978-1-5386-7688-2/19 $31.00 © 2019 IEEE

The method proposed in this paper is numerical in nature and based on surface impedances like Silva's [2] approach but aims at reducing the model order to the point where the computational cost becomes negligible. Instead of hundreds of thousands or even millions of elements required for 3D FEA, only around 100 nodes are needed. One disadvantage of the proposed method compared to 3D FEA, however, is the requirement of a fast and reliable way of obtaining the flux density distribution normal to the clamping plate surface. Therefore, the presented approach should by no means be seen as a stand-alone tool or a replacement for 3D FEA but rather as one building block on the path towards a less computationally expensive clamping plate loss calculation tool chain.

II. SURFACE IMPEDANCE BOUNDARY CONDITIONS

Due to the high permeability ($\mu_r \approx 900$) and conductivity ($\sigma \approx 5$ MSm^{-1}) of ferritic non-alloy steel, fields do not penetrate deep into the material, even at grid frequencies. Modeling eddy effects accurately using traditional FEA requires the mesh size normal to the surface to be very small. Usually around 3 mesh layers within the first skin depth

$$\delta = \frac{1}{\sqrt{\pi f \sigma \mu_r \mu_0}} \tag{1}$$

are considered adequate, though this is highly dependent on implementation details of the FEA software used. Given the conductivity and permeability from above and a grid frequency of 50 Hz, the skin depth becomes $\delta = 1.1$ mm. Surface impedance boundary conditions present a neat alternative to modeling the entire conductive volume using volumetric elements. Instead, the ratio between the electric field $\underline{E}_s$ and the magnetic field $\underline{H}_s$ (orthogonal to $\underline{E}_s$) on the surface

$$\underline{Z}_s = \frac{\underline{E}_s}{\underline{H}_s} = \frac{1+j}{\sigma\delta} \tag{2}$$

is used as a boundary condition [6]. The derivation of the surface impedance $\underline{Z}_s$ requires some assumptions to be made:

- All relevant material properties (μ_r, σ) must be constant,
- only one frequency can be considered and
- all geometric features must be large compared to the skin depth δ.

None of these assumptions presents a significant challenge when approaching clamping plate losses. As will be shown later, comparing linear and non-linear B(H) characteristics in FEA shows practically no difference (see fig. 9). Eddy currents and other field effects are governed by the grid frequency as will also be shown later (see fig. 8). Some error can be expected at the sharp edges of the clamping plate, as their edge radius is essentially zero and thus smaller than the skin depth δ. The impact of this error is gauged by comparing the overall losses to those calculated using FEA.

The eddy current surface loss density

$$p_s = \frac{1}{2} \left| \underline{H}_s \right|^2 \cdot \Re(\underline{Z}_s) \tag{3}$$

depends only on the surface impedance $\underline{Z}_s$ and the magnetic field amplitude parallel to the surface $\underline{H}_s$ [2].

III. NODE LOOP SURFACE IMPEDANCE LOSS MODEL

The underlying idea of the proposed loss model is generally not dissimilar to the approach employed by Silva [2]. However, instead of modeling the clamping plate surface along an entire pole pitch, only a singe loop of nodes around the clamping plate profile, as seen in fig. 2, is required. This is where the name *node loop surface impedance loss model* originates from. The following subsections will outline the approach starting with chosen boundary conditions. From there, edge impedances and a system matrix are determined. Using the flux density normal to the surface, nodal potentials can be calculated and used to in turn calculate the fluxes passing between adjacent nodes. Lastly, these can be used to determine the element flux density and losses.

A. Boundaries

Central to the node loop surface impedance loss model is the idea of modeling only one loop of nodes around the clamping plate profile. This forces some constraints. For one thing, the proposed model needs to be a frequency-domain model due to the use of surface impedances. Therefore, only the fundamental temporal harmonic of any clamping plate field effect can be considered. Magnetic voltages $\underline{V}_{m,i}$ are calculated as amplitude phasors at each node. Each node is connected to two other nodes along the clamping plate profile.

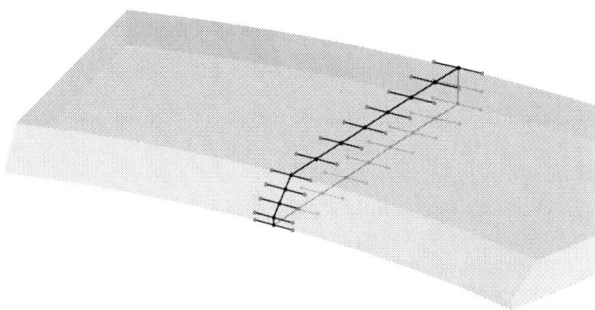

Fig. 2. Single node loop around clamping plate

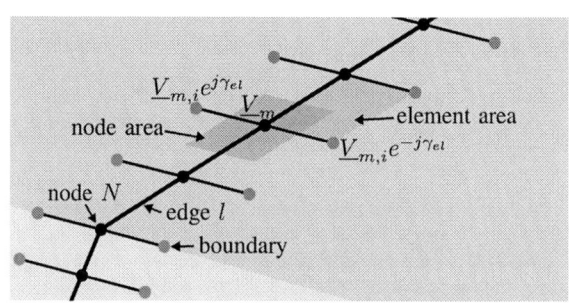

Fig. 3. Nodes, boundaries and nomenclature in a zoomed in view of a node loop

Adjacent to each node in clockwise and counterclockwise direction, a boundary condition is placed. Only considering the fundamental spacial harmonic of the flux density normal to the clamping plate surface allows the boundary condition to be devised. For a boundary offset by a mechanical angle γ_{mech} from a node with the magnetic voltage $\underline{V}_{m,i}$, this boundary condition becomes

$$V_{m,boundary} = V_m e^{j\gamma_{el}}, \tag{4}$$

as seen in fig. 3, where $\gamma_{el} = p \cdot \gamma_{mech}$ is the electric boundary pitch. Analogously, the mechanical angle between a boundary and its corresponding node γ_{mech} is referred to as mechanical boundary pitch. The electric boundary pitch can be chosen arbitrarily. It should generally be chosen small enough to ensure a fine effective spatial mesh resolution in tangential direction, but large enough to prevent numerical inaccuracies. A reasonable value for the electric boundary pitch was found to be $\gamma_{el} = 2\pi/100$.

B. Edge impedances

Before outlining how edge impedances are determined, some nomenclature is introduced. Using only square elements allows adjacent nodes (and boundaries) to be referred to as top, bottom, left and right neighbors, simplifying both implementation and explanation of the proposed loss model. Neighboring nodes in radial-axial direction (i.e. in direction of the node loop) are referred to as top or bottom neighbors, while neighboring nodes or boundaries in tangential direction are referred to as left and right neighbors.

In this context edge impedance refers to the magnetic impedance represented by edges connecting nodes to one another and to boundaries. The impedance along the edge l_{top} between a node N_i and its top neighbor $N_{i,top}$ can now be calculated as

$$\underline{Z}_{i,top} = \underline{Z}_s \cdot \frac{l_{i,top}}{w} = \underline{Z}_s \cdot \frac{l_{i,top}}{l_{i,left}/2 + l_{i,right}/2}, \tag{5}$$

where the edge $l_{top} = \left| \vec{N}_{i,top} - \vec{N}_i \right|$ simply corresponds to the distance between the node and its top neighbor. The width w (seen in fig. 5) depends on the distance to the right $l_{i,right}$ and left neighbor $l_{i,left}$ (lengths orthogonal to the direction of the edge impedance). Impedances to all other neighbors, including boundaries, are calculated analogously.

C. Normal flux

Having found mathematical descriptions for both boundary conditions and edge impedances, the only remaining step before assembling the system matrix is finding the flux entering the clamping plate at each node. For this, the area of a node, shown in fig. 3, can be calculated as

$$A_{node} = h \cdot w = (l_{top} + l_{bottom})/2 \cdot (l_{left} + l_{right})/2 \tag{6}$$

by referring to the previously mentioned concept of neighbors. Given the flux density normal to the surface at a node $\underline{B}_n$

through some field calculation approach (in this paper FEA is used), the flux entering the node area is calculated as

$$\begin{aligned} \underline{\Phi}_n &= \underline{B}_n \cdot h \cdot 2\sin(\gamma_{el}/2)\, r \\ &\approx \underline{B}_n \cdot A_{node}, \end{aligned} \tag{7}$$

making use of the sinusoidal field distribution in tangential direction, where r is the nodes' radius. For sufficiently small electric boundary pitches γ_{el}, the simplified approximation can be used. This step highlights the key limitation of the proposed loss model: A reliable way of obtaining the flux density normal to the clamping plate surface $\underline{B}_n$ needs to be devised. For this paper, to ensure comparability, the field is extracted from time step FEA used to validate the loss model.

D. System matrix

Nodal analysis is employed to calculate the magnetic voltages V_m at each node. The flux entering at each node is represented by the right side of the equation system

$$\mathbf{A} \cdot \begin{pmatrix} \underline{V}_{m,1} \\ \underline{V}_{m,2} \\ \underline{V}_{m,3} \\ \vdots \\ \underline{V}_{m,k} \end{pmatrix} = \begin{pmatrix} \underline{\Phi}_{n,1} \\ \underline{\Phi}_{n,2} \\ \underline{\Phi}_{n,3} \\ \vdots \\ \underline{\Phi}_{n,k} \end{pmatrix} \tag{8}$$

To assemble the system matrix $\mathbf{A}$, all nodes need to be uniquely numbered, as seen in fig. 4. The system matrix for k nodes

$$\mathbf{A} = \begin{pmatrix} \underline{Y}_1 & -\underline{Y}_{1,2} & 0 & \cdots & -\underline{Y}_{1,k} \\ -\underline{Y}_{1,2} & \underline{Y}_2 & -\underline{Y}_{2,3} & \cdots & 0 \\ \vdots & \vdots & \vdots & \ddots & \vdots \\ -\underline{Y}_{1,k} & 0 & 0 & \cdots & \underline{Y}_k \end{pmatrix}. \tag{9}$$

becomes (k, k) in size. The system matrix consists of edge admittances $\underline{Y} = 1/\underline{Z}$. The edge admittance between node N_1 and N_2 can be referred to in multiple ways

$$\underline{Y}_{1,2} = \underline{Y}_{1,top} = \underline{Y}_{2,bottom}. \tag{10}$$

The concept of neighbors is translated into the implementation developed for this paper. It allows for a single loop over all nodes to assemble the entire system matrix without having to determine neighboring nodes during each loop iteration, saving

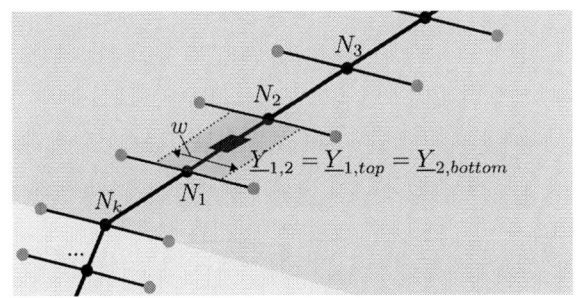

Fig. 4. Nomenclature around assembling the system matrix

computing time at the cost of memory. The diagonal elements contain both, the admittances to all neighboring nodes as well as the adjacent boundary conditions. Assuming the left and right neighbors of a node N_i are boundary conditions, while the top and bottom neighbors are nodes, the diagonal element of the system matrix for node N_i becomes

$$\underline{Y}_i = \underline{Y}_{1,top} + \underline{Y}_{1,bottom} \\ - \underline{Y}_{1,left} \cdot e^{j\gamma_{el}} - \underline{Y}_{1,right} \cdot e^{-j\gamma_{el}}. \tag{11}$$

This step highlights the key advantage of the node loop surface impedance loss model: The number of required nodes and accompanying computational cost can be greatly reduced by directly incorporating these boundaries into the diagonal elements of the system matrix and only modeling a single node loop.

Having assembled the system matrix $\mathbf{A}$ and calculated the flux entering each node $\underline{\Phi}_n$, the equation system can be solved for the nodal voltages $\underline{V}_m$ using any solver of choice.

E. Surface losses

Having calculated the magnetic voltage at each node V_m leaves only a few post-processing steps to obtaining the clamping plate losses. Losses are not calculated at each node, but rather at each element as seen in fig. 5. First, the flux along each edge is calculated

$$\underline{\Phi}_{a,b} = \underline{Y}_{a,b} \cdot \left(\underline{V}_{m,a} - \underline{V}_{m,b} \right), \tag{12}$$

where a and b refer to arbitrary neighboring node indices. As the loss model is similar in nature to a magnetic equivalent circuit, the flux along the surface is discretized as edge fluxes. For the flux between node N_2 and N_3 in fig. 5, this becomes

$$\underline{\Phi}_{2,3} = \underline{\Phi}_{2,top} = \underline{Y}_{2,top} \cdot \left(\underline{V}_{m,2} - \underline{V}_{m,2,top} \right), \tag{13}$$

again, utilizing the naming convention of neighbors to simplify implementation. The flux between a node and a boundary can be calculated analogously. For the example shown in fig. 5, this yields

$$\underline{\Phi}_{3,right} = \underline{Y}_{3,right} \cdot \left(\underline{V}_{m,3} - \underline{V}_{m,3} e^{-j\gamma_{el}} \right). \tag{14}$$

To quantify the surface losses of an element, the average element flux density needs to be calculated. But beforehand, a correlation between the edge flux and the surface flux density is derived by integrating the surface flux density given in [6]

$$\underline{\Phi} = w \int_{-\infty}^{0} \underline{B}_s e^{-\frac{1+j}{\delta}x} \, dx = w \frac{\delta}{1+j} \underline{B}_s, \tag{15}$$

in the direction normal to the surface dx, taking into account the effective width w of the flux path represented by the edge. This equation is rearranged for the surface flux density

$$\underline{B}_s = \frac{1+j}{w\delta} \underline{\Phi}. \tag{16}$$

The width w is calculated as before, when calculating the edge impedances (see equation (5)). The width w is again determined using the length of adjacent edges **orthogonal** to the flux direction. If for instance the flux flowing from node N_i in top direction is to be calculated, equation (16) results in

$$\underline{B}_{s,i,top} = \frac{1+j}{\delta \left(l_{i,left}/2 + l_{i,right}/2 \right)} \underline{\Phi}_{i,top} \tag{17}$$

Returning to the example shown in fig. 5, equation (16) can now be used to calculate the element surface flux density in tangential direction $\underline{B}_{s,tan}$ as the mean

$$\underline{B}_{s,tan} = \frac{1}{2} \left(\underline{B}_{s,2,right} + \underline{B}_{s,3,right} \right) \tag{18}$$

of both tangential flux densities ($\underline{B}_{s,2,right}$ and $\underline{B}_{s,3,right}$) adjacent to the element. The same is done for the radial-axial flux

$$\underline{B}_{s,rax} = \frac{1}{2} \underline{B}_{s,2,top} \left(1 + e^{-j\gamma_{el}} \right) \tag{19}$$

making use of the boundary condition. The resulting element flux density is a two-dimensional vector of amplitude phasors

$$\underline{\vec{B}}_e = \begin{pmatrix} \underline{B}_{s,rax} \\ \underline{B}_{s,tan} \end{pmatrix}, \tag{20}$$

The element type is intrinsically two-dimensional, despite the mesh being wrapped around the surface of a three-dimensional body. The root mean square (rms) of the element flux density

$$B_{e,rms} = \sqrt{\frac{1}{2\pi} \int_0^{2\pi} \left| \Re(\underline{\vec{B}}_e e^{j\varphi}) \right| \, d\varphi} \tag{21}$$

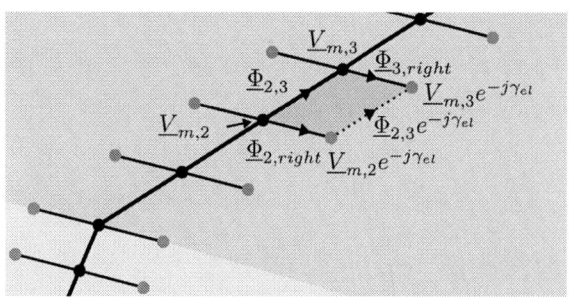

Fig. 5. Element area and its adjacent nodal potentials as well as edge fluxes

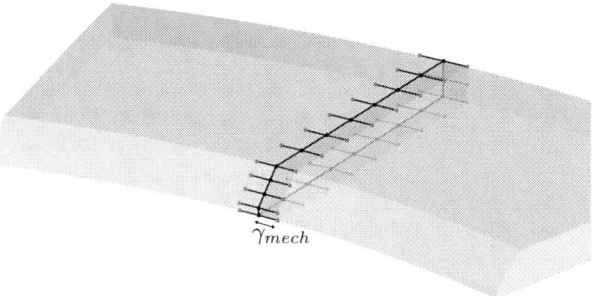

Fig. 6. Smallest symmetry unit losses can be calculated for and applied to the entire clamping plate

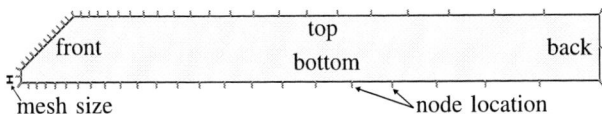

Fig. 7. Mesh for node loop surface impedance loss model. This mesh, containing 61 nodes, is used for later comparisons to FEA.

is calculated from the time value of the element flux density over one period. Inserting this value into equation (3) yields the surface loss density

$$p_s = \frac{1}{2} \left| \frac{\sqrt{2}B_{e,rms}}{\mu_r \mu_0} \right|^2 \cdot \Re(\underline{Z}_s) = \left| \frac{B_{e,rms}}{\mu_r \mu_0} \right|^2 \cdot \Re(\underline{Z}_s) \quad (22)$$

The surface loss density and surface area of each element is multiplied

$$P_e = p_s \cdot A \quad (23)$$

to obtain the element losses. Adding up losses for all elements and applying the symmetry finally yields the clamping plate eddy current losses for the entire generator

$$P_{cp} = 2\frac{2\pi}{\gamma_{mech}} \sum_i P_{e,i}. \quad (24)$$

The factor of 2 accounts for the fact that there are two clamping plates (drive and non-drive end).

To make best use of the symmetry, element losses are only calculated for elements towards one side of the node loop as seen in fig. 6. If losses were to be calculated for both sides, equation (24) would need to be adapted.

F. Mesh

Due to the partially primitive ways lengths, widths and areas are calculated in the proposed loss model, large differences in size between adjacent elements can cause significant errors, especially when the normal flux is large. To prevent this from impacting results, special care was taken of the mesh. Fig. 7 shows the mesh in a two-dimensional representation. Nodes in the front of the clamping plate are being equidistantly spaced at intervals of 2.5 mm. As most losses occur in that part of the clamping plate, the mesh size increases exponentially with a base of 1.1 towards the back. This procedure ensures that no two adjacent node spacings differ by more than 10 %, reducing the number of nodes while keeping errors low.

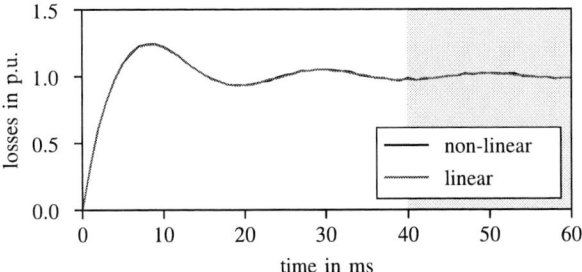

Fig. 9. Clamping plate losses from 3D FEA over 3 electrical periods.

IV. FEA SIMULATIONS

To validate the node loop surface impedance loss model, the geometry of a 600 kVA test generator was modeled using 3D FEA. To reduce the test stand power requirements is very short, still allowing to study end region effects. Table II shows the key generator dimensions. Pressure fingers were not modeled.

TABLE II
KEY GENERATOR DIMENSIONS

description	value
number of poles	16
stator frequency	50 Hz
inner stator diameter	1.3 m
radial clamping plate height	25 cm
axial clamping plate length	2.5 cm

A time-step simulation with 20 time steps per electric period was set up and solved for 3 electric periods in open-circuit operation. Measurements on the actual generator hardware were used to determine the open-circuit field current used in the FEA. Close attention was paid to the clamping plate mesh generation, taking findings by Waldhart [7] into account. Wherever possible, mosaic and hexahedral elements were used, with three surface mesh layers at intervals of 0.4 mm to accurately represent the skin effect. Fig. 9 shows the clamping plate losses for each calculated time step, with the last electric period being highlighted. For later comparisons, losses, averaged over the last electric period are used. Fig. 9 also shows effectively no difference between FEA with non-linear and linear clamping plate B(H) characteristics. This validates the

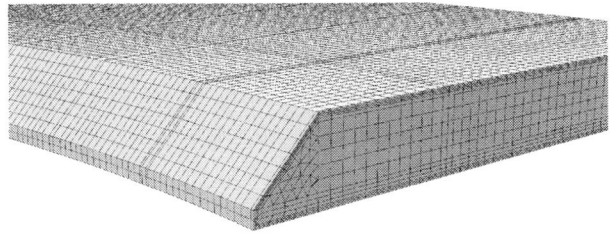

Fig. 8. Clamping plate mesh of the FEA model used.

Fig. 10. Test generator (right) coupled with the load machine.

3D FEA

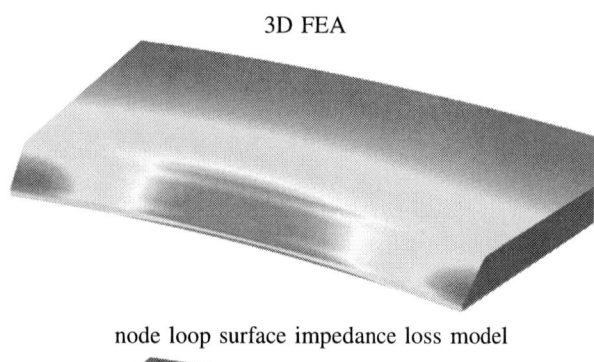

node loop surface impedance loss model

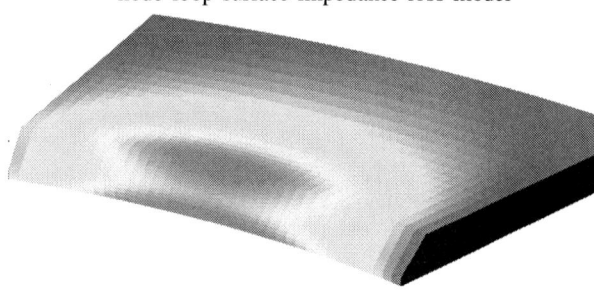

Fig. 11. Loss distribution calculated using 3D FEA and the node loop surface impedance loss model.

assumption of linear material properties, necessary for surface impedance boundary conditions.

A. Extracting the field normal to the clamping plate surface

To accurately compare the node loop surface impedance loss model to FEA, the field normal to the clamping plate surface needs to be extracted from the same FEM simulation. To accomplish this, the field normal to the surface is evaluated along the top, front and bottom edges (see fig. 7 for nomenclature) for each time step over the last electric period, disregarding the back side, as the normal field there is essentially non-existent. A discrete Fourier transform for each point along the clamping plate profile gives the temporal harmonics shown in fig. 13. Evidently, the harmonic distortion is very low (THD < 0.5 %), validating the second assumption enforced by surface impedances: Only one frequency can be considered. The fundamental temporal harmonic is then used in the node loop surface impedance loss model.

V. RESULTS

The most important measure when evaluating the performance of the node loop surface impedance loss model are the overall clamping plate losses. Directly comparing the losses calculated using the proposed loss model to those calculated using 3D FEA yields only a 3 % discrepancy between both methods (see table III). Fig. 11 illustrates how these values were gathered. Considering the challenging conditions posed by the small skin depth of 1 mm, this is within the realm of the expected numerical inaccuracies of the FEA.

Besides comparing the overall losses, fig. 11 shows the loss density distribution for both calculation methods. Despite

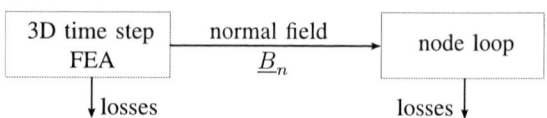

Fig. 12. Flow chart highlighting how losses are calculated for the comparison seen in fig. 11.

having paid close attention to the clamping plate mesh, some mesh artifacts can still be observed in the FEA. Most notably, lines in tangential direction indicate some numerical inaccuracies. Results cannot be compared quantitatively, as the FEA computes losses per volume, whereas the node loop surface impedance loss model calculates losses per surface area. A quantitative comparison of the overall losses can be found in table III. Given this caveat, the general loss distribution can still be qualitatively judged. The proposed loss model is generally able to calculate hot spots (areas of high loss densities) on the clamping plate surface with a reasonably degree of accuracy. This ability can be very useful when designing the cooling system of a new generator.

TABLE III
LOSSES CALCULATED USING 3D FEA AND THE NODE LOOP SURFACE
IMPEDANCE LOSS MODEL

calculation method	losses in p.u.
3D FEA	100 %
3D FEA linear materials	100 %
node loop surface impedance model	103.1 %

A. Mesh impact

To quantify the mesh impact on the clamping plate loss calculation in the proposed loss model, the mesh size was varied between 1 μm and 5 mm. Fig. 14 shows this comparison. Distinct steps in the loss variance $\Delta P = \frac{P}{P(2.5\ \text{mm})} - 1$ are caused by the actual mesh size being internally adjusted to be a divisor of the length of the smallest geometric feature (5 mm).

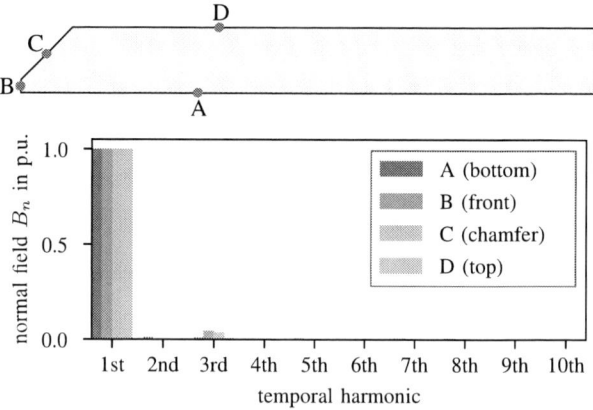

Fig. 13. Temporal harmonics of the field normal to the clamping plate surface B_n at four exemplary points along the clamping plate profile (amplitudes normalized to the fundamental harmonic).

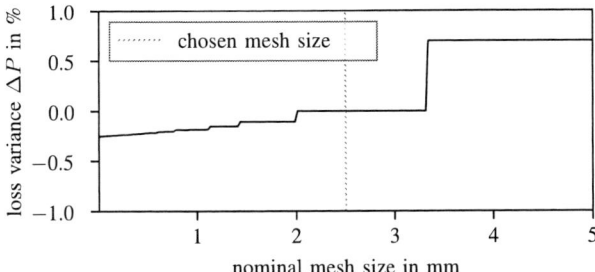

Fig. 14. Impact of mesh size on the losses calculated by the node loop surface impedance loss model, normalized to the losses at the chosen nominal mesh size of 2.5 mm $\Delta P = \frac{P}{P(2.5\,\text{mm})} - 1$.

The small overall loss variance is likely to be dwarfed by other errors in a comprehensive clamping plate loss calculation tool chain. The chosen nominal mesh size represents a reasonable compromise between accuracy and computational cost.

B. Computational cost

Simulating 60 time steps of the 3D FEM model in open-circuit operation required around a day to finish on current desktop computer hardware, not accounting for post-processing. The python implementation of the node loop surface impedance loss model only requires around 20 ms to compute. This represents a difference in computing time by a factor 10^6.

However, it is important to reiterate that the proposed loss model is by no means a replacement for 3D FEA, but rather a single building block of a more comprehensive end region loss calculation tool chain. The proposed loss model is only useful, when paired with an accurate, yet computationally inexpensive field calculation approach.

VI. CONCLUSION

The node loop surface impedance loss model manages to very accurately calculate clamping plate losses with one major caveat: The magnetic flux density normal to the surface needs to be known. This limits the loss model's use case to circumstances where a fast and accurate end region field calculation approach is available. Finding such an approach is a central part of current and future work.

Besides evaluating overall clamping plate losses, some of the assumptions required to devise the node loop surface impedance loss model were tested individually. Comparing non-linear and linear clamping plate materials using 3D FEA showed the non-linear clamping plate B(H) characteristic to have no impact. Evaluating the temporal flux density harmonics normal to the surface revealed everything but the fundamental harmonic to be barely discernible, validating the use of a frequency domain model. With an overall error of 3 % in losses compared to FEA and a reduction in computational cost by a factor of 10^6, the proposed loss model allows for parameter variations, which would be infeasible using only 3D FEA.

Besides finding a fast yet accurate field calculation approach, future work includes

- using the available test generator to provide measurement validation for both, the FEA and node loop surface loss model and
- potentially incorporating flux shielding effects due to the electrically connected pressure fingers and the clamping plate by adjusting the surface impedance at nodes interfacing pressure fingers.

REFERENCES

[1] B. C. Mecrow, A. G. Jack, and C. S. Cross, "Electromagnetic design of turbogenerator stator end regions," *Transmission and Distribution IEE Proceedings C - Generation*, vol. 136, no. 6, pp. 361–372, Nov. 1989.

[2] V. C. Silva, Y. Marechal, and A. Foggia, "Surface impedance method applied to the prediction of eddy currents in hydrogenerator stator end regions," *IEEE Transactions on Magnetics*, vol. 31, no. 3, pp. 2072–2075, May 1995.

[3] S. Li, N. A. Gallandat, J. R. Mayor, T. G. Habetler, and R. G. Harley, "Calculating the Electromagnetic Field and Losses in the End Region of a Large Synchronous Generator Under Different Operating Conditions With 3-D Transient Finite-Element Analysis," *IEEE Transactions on Industry Applications*, vol. 54, no. 4, pp. 3281–3293, Jul. 2018.

[4] B. Marcusson, "Magnetic Leakage Fields and End Region Eddy Current Power Losses in Synchronous Generators," *DIVA*, 2017. [Online]. Available: http://urn.kb.se/resolve?urn=urn:nbn:se:uu:diva-331182

[5] G. Traxler-Samek, *Zusatzverluste im Stirnraum von Hydrogeneratoren mit Roebelstabwicklung*. Technische Universität Wien, 2003.

[6] S. V. Yuferev and N. Ida, *Surface Impedance Boundary Conditions: A Comprehensive Approach*. CRC Press, Sep. 2009.

[7] F. J. Waldhart, J. P. Bacher, and G. Maier, "Modeling eddy current losses in the clamping plate of large synchronous generators using the finite element method," in *Automation and Motion International Symposium on Power Electronics Power Electronics, Electrical Drives*, Jun. 2012, pp. 1468–1473.

Adaptive algorithm to reduce acoustic noise and torque ripple in low-cost PM motors

Martin Sumega, Simon Zossak, Patrik Varecha, Pavol Rafajdus
University of Zilina
Zilina, Slovak Republic
martin.sumega@fel.uniza.sk

Marek Stulrajter
NXP Semiconductors
Roznov pod Radhostem, Czech Republic
marek.stulrajter@nxp.com

Abstract—This paper presents a method for torque ripple, acoustic noise and vibrations reduction in three-phase permanent magnet (PM) motors. Field oriented control (FOC) has been extended by compensation voltage in *q*-axis to minimize the ripple in angular velocity. Amplitude and phase of injected *q*-axis voltage harmonic are adjusted online by least mean squares (LMS) learning rule, based on the ripple in angular velocity. Proposed method significantly reduces impact of cogging torque and non-sinusoidal Back-EMF voltage on produced torque ripple, acoustic noise and vibrations. Effectiveness of proposed algorithm is experimentally verified by sound FFT analysis. Conventional techniques require a map of the cogging torque and Back-EMF shapes, while proposed method calculates the compensation *q*-axis voltage directly out of the ripple in angular velocity, which is great advantage in comparison to conventional methods.

Index Terms—torque ripple, cogging torque, noise, vibrations, BLDC, PMSM, torque ripple reduction, adaptive control

Nomenclature

ϑ_r	mechanical angular rotor position
ϑ_e	electrical angular rotor position
f_e	stator electrical frequency
ω_e	electrical angular velocity
ω_r	mechanical angular velocity
T_e	electromagnetic torque
T_{cogg}	cogging torque
e_a, e_b, e_c	phase Back-EMF voltages
i_a, i_b, i_c	phase currents

I. Introduction

Nowadays, there is a strong trend of product cost and size reduction clearly seen in the industry. This leads to more frequent use of low-cost motors or sensorless control techniques. However, the usage of these low-cost motors with simple and poor quality construction often leads to torque ripple, vibrations and acoustic noise [1]. Magnetic radial forces, torque ripple and the cogging torque are the main electromagnetic sources of vibrations and acoustic noise in PM motors with low and medium power [2], [3].

More expensive motors have higher quality constructions, which is essential to better performance in terms of torque ripple, acoustic noise and vibrations. There have been proposed several approaches in the past to suppress the torque ripple. One of these techniques is a proper modification in the

motor construction which yields to minimizing the cogging torque. Most conventional technique is stator slots or rotor PM segments skewing, which significantly reduces cogging torque [4], [5], [6]. Also the magnet pole arc can have a great impact on the magnitude of the cogging torque [7], [4], [8]. Notches in the stator teeth can reduce main harmonic component of cogging torque [4]. Magnet shifting and also fractional number of slots per pole lead to reducing of the overall cogging torque [7], [4], [6]. All of these construction changes have great success in minimization of the cogging torque, but also have a higher cost and much more complicated manufacturing. These construction features however absent in the low-cost PM motors thus these motors are characterized by rippled torque operation. Due to this fact, advance compensation control techniques seem to be the only option how to achieve the smooth torque operation.

Majority of the approaches for torque ripple minimization require ripple source identification and mapping of its characteristic. Based on the ripple map the *dq* current harmonic components are adjusted to compensate torque ripple sources to achieve smoother operation of the motor [9], [10], [11]. This basic principle is applicable only at lower frequencies, because required current trajectories have several times higher frequencies than stator electrical frequency. Bandwidth of current controllers may be too low and the compensation becomes insufficient. Frequency of required current trajectories for reduction of influence of non-sinusoidal Back-EMF has mainly 6-times higher frequency than electrical frequency f_e. Required current trajectories for reduction of cogging torque depend on its harmonic components. For precise tracking of *dq* current trajectories, what is crucial in torque ripple reduction algorithms, it is necessary to use an advanced control techniques.

Control techniques for characterization and minimization of cogging torque in low-cost BLDC drives were introduced in [1]. Authors inserted a position based cogging torque map into the control and achieved higher performance in robotic servo applications, compared to more expensive BLDC motors.

High-bandwidth current control to reduce influence of non-sinusoidal Back-EMF was introduced by use of dead-beat controller, which was extended by current predictor to achieve stable regulation. Inputs to this algorithm were torque ripple functions, which had to be precisely tracked. This control

978-1-5386-7688-2/19 $31.00 © 2019 IEEE

technique was compared to conventional current PI controller extended with feed-forward compensation [12]. Same authors also used a self-commissioning algorithm for identification of torque ripple functions [13].

Repetitive current control is able to remove periodic disturbances in regulation error. This method was used to precisely track the current trajectories to eliminate impact of non-sinusoidal Back-EMF on torque ripple [14].

Multiple reference frame algorithm was used for precise tracking of current harmonics, which was controlled separately by synchronous regulator [15], [16].

Modified field oriented control uses extended Park transformation and extended model of three-phase PM motor to calculate correct compensation voltages to minimize ripple in electromagnetic torque [17].

Resonant controller was successfully used for compensation of various sources of torque ripple including non-sinusoidal Back-EMF and cogging torque [18].

Disadvantage of methods described above is in required identification of the sources of torque ripple and in forming of required current trajectories. A simple single layer artificial neural network solved the problem with identification of torque ripple sources and forming of current trajectories by utilizing the ripple in speed or in estimated torque. Authors used neural speed/torque and neural current controllers together and achieved great results [19]. Same authors used similar technique of torque ripple minimization for synchronous reluctance motor [20].

In this article simplified and effective solution was introduced, which is based on method above [19], [20]. Proposed adaptive algorithm extends speed controller in classic FOC and calculates adaptation of injected q-axis voltage online by LMS learning rule, based on speed ripple connected with 6-th harmonic component (6-multiply of f_e). This is great advantage in comparison to conventional methods, because there is no need to identify cogging torque or Back-EMF.

II. Sources of torque ripple in 3-phase PM motor

Construction of three-phase permanent magnet (PM) motor has a huge influence on the torque ripple, but also defines its specific features. The main torque ripple sources in three-phase PM motor due to its construction are:

- Cogging torque
- Interaction between Back-EMF and current excitation
- Mechanical and electrical faults

The main torque ripple sources in three-phase PM motor due to its control are:

- Waveform of phase currents, which depends on the type of control technique
- Commutation of phase currents in six-step control
- Dead-time and zero current clamping effect

A. Cogging torque

Cogging torque represents the interaction between rotor permanent magnets and stator teeth or poles independent of any stator current. In three-phase PM motors it causes ripple in

produced torque and in rotor speed, mainly in low speed area. Cogging torque also causes vibrations and acoustic noise not only at low speeds, because harmonic magnetic forces due to cogging torque generate stator radial vibrations which result in acoustic noise. The effect of the cogging torque on the speed ripple depends on the moment of inertia which behaves as a filter. A mathematical equation for cogging torque is:

$$T_{\text{cogg}} = -\frac{1}{2}\phi_\delta^2 \frac{dR}{d\vartheta_r} \tag{1}$$

where ϕ_δ describes PM flux crossing the air-gap, R describes total reluctance through which the flux passes. If the total reluctance R doesn't vary as the rotor rotates, the derivation in (1) is zero and also the cogging torque is equal to zero [21], [22], [7].

Cogging torque can be mathematically described also by Fourier series:

$$T_{\text{cogg}} = \sum_{k=1}^{\infty}(T_{\text{ksin}}sin(kLCM(Q,2p)\vartheta_r)+ \\ +T_{\text{kcos}}cos(kLCM(Q,2p)\vartheta_r)) \tag{2}$$

T_{ksin} and T_{kcos} describes amplitude of k-th harmonic component for sine or cosine part. $LCM(Q,2p)$ represents least common multiple of number of slots Q and the number of poles $2p$ [7].

Fig. 1 shows construction of BLDC motor Linix 45ZWN24-40, which is used for experimental verification. Motor is controlled by FOC. Motor has number of slots $Q = 6$ and number of poles $2p = 4$. According to (2) cogging torque has 12 periods in one mechanical revolution.

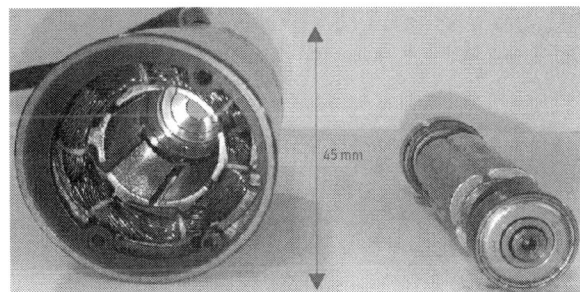

Fig. 1: Construction of BLDC motor Linix 45ZWN24-40

TABLE I: Parameters of used BLDC motor Linix 45ZWN24-40

Nominal power	$P_N[W]$	40
Nominal current	$I_N[A]$	2,3
Nominal voltage	$U_N[V]$	24
Nominap speed	$n_N[rpm]$	4000
Nominal torque	$T_N[Nm]$	0,097
Number of poles	$2p[-]$	4
Stator resistance	$R_s[\Omega]$	0,5255
D-axis inductance	$L_d[H]$	0,000472
Q-axis inductance	$L_q[H]$	0,000496
Moment of inertia	$J[kgm^{-2}]$	0,000009
Voltage constant	$K_e[\frac{Vs}{rad}]$	0,013715

978-1-5386-7688-2/19 $31.00 © 2019 IEEE

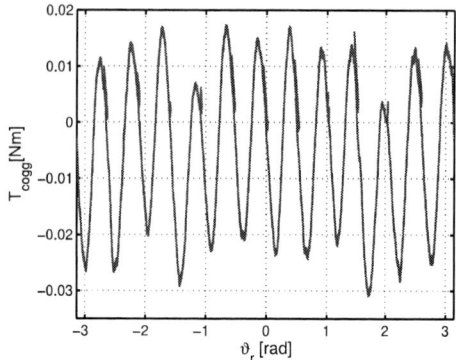

Fig. 2: Measured cogging torque

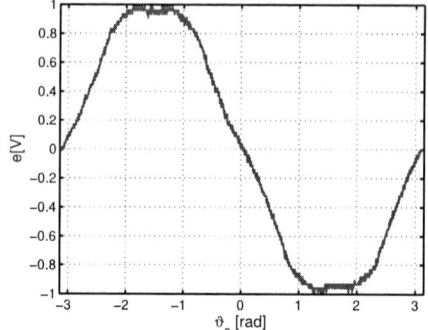

Fig. 4: Measured phase Back-EMF voltage

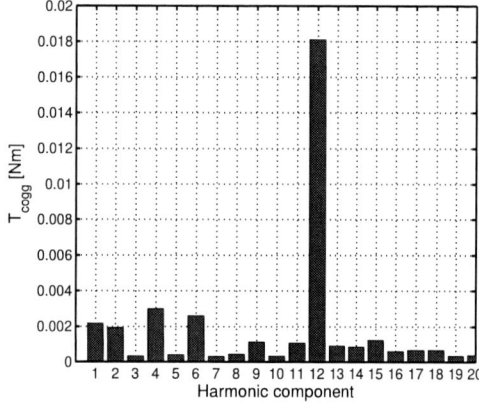

Fig. 3: FFT analysis of measured cogging torque (fig. 2)

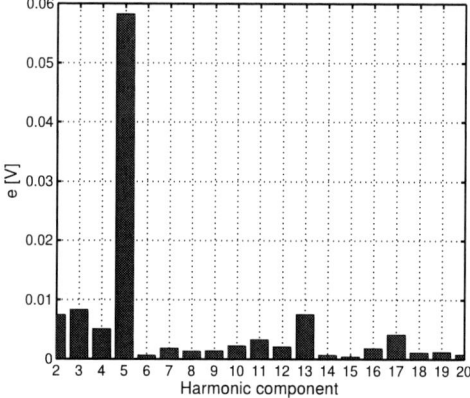

Fig. 5: FFT analysis of measured phase Back-EMF voltage (fig. 4)

Fig. 2 shows measured cogging torque map for one mechanical revolution of used PM motor and fig. 3 shows FFT analysis of measured cogging torque. Dominant harmonic component of the measured cogging torque represent 18,5% of the rated torque T_N. Cogging torque leads to torque ripple, vibrations and noise with frequency, which is 12-multiply of mechanical frequency or 6-multiply of electrical frequency f_e for the Linix motor, because number of pole pairs $p = 2$.

B. Interaction between Back-EMF and current excitation

Interaction between stator current excitation and Back-EMF is very important factor in terms of torque ripple and can be analyzed through the shape of the motor phase currents and Back-EMF voltages. Back-EMF voltages and their shape is defined by the permanent magnets, their polarization and positioning on the rotor. Shape of the phase currents can be determined by type of control technique. Assuming an ideal case, the FOC technique produce sinusoidal currents, while six-step (SS) commutation control leads to a rectangular shape of the currents. If the motor with trapezoidal Back-EMF voltage (BLDC motor) is controlled by FOC, or sinusoidal Back-EMF motor (PMSM) is controlled by SS control, then

a ripple in the electromagnetic torque is expected. In other words, the harmonic spectra misalignment between the currents and Back-EMF leads to torque ripple. Because of that, it is very important to ensure correct interaction of Back-EMF voltages and phase currents. This results from equation for electromagnetic torque [23], [22]:

$$T_e = \frac{i_a e_a + i_b e_b + i_c e_c}{\omega_r} \qquad (3)$$

Fig. 4 shows waveform of Back-EMF voltage of used PM motor measured in generator mode. Fig. 5 shows FFT analysis of measured phase Back-EMF voltage. There is dominant 5th harmonic component, but also less significant 11th, 13th and 17th harmonic components. This will lead to ripple in electromagnetic torque T_e, if field oriented control with sinusoidal currents is used.

5th, 7th, 11th and 13th harmonic components in *abc* Back-EMF voltage will lead to the ripple in electromagnetic torque as a 6th and 12th harmonic components. Fig. 6 shows calculated electromagnetic torque according to (3) and its FFT analysis, which confirms 6th and 12th harmonic content.

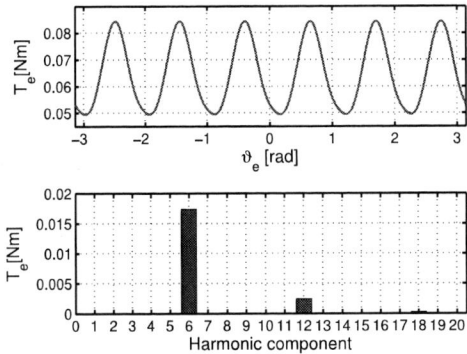

Fig. 6: Electromagnetic torque and its FFT

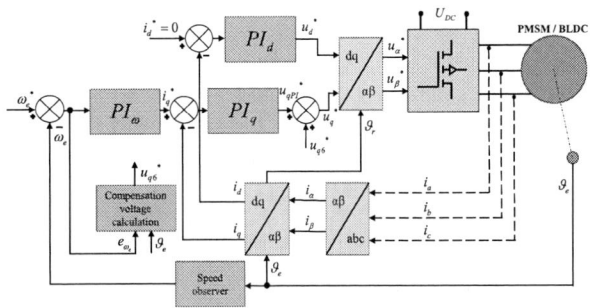

Fig. 7: Proposed modification of FOC

III. PROPOSED ALGORITHM

As mentioned above, main sources of torque ripple for the tested motor are reflected mostly with 6-multiply and 12-multiply of stator electrical frequency f_e. This includes non-sinusoidal Back-EMF interacting with sinusoidal currents (FOC), cogging torque and also dead-time. However dead-time has very small effect in this electric drive, so it is not considered. It means, that if the electrical frequency is 50 Hz, torque ripple, noise and vibrations will be mainly produced at frequency of 300Hz, and also little part will be seen in 600 Hz. Based on that, there is an assumption that injection of 6th q-axis voltage harmonic with correct phase and amplitude will lead to q-axis current harmonic, which will compensate sources of torque ripple, noise and vibrations mentioned above. For the effective compensation a precise q-axis current trajectory tracking must be achieved without amplitude attenuation or phase shift. This can be a challenge, because required q-axis current trajectory will have the frequency 6 and 12 times higher than fundamental frequency f_e.

Proposed algorithm just injects the 6th q-axis voltage harmonic. Amplitude and phase of this injected voltage are adapted online, based on the ripple in angular velocity. This leads to correct q-axis current harmonic, which acts against the sources of ripple. Fig. 7 shows extension of FOC scheme with proposed calculation and injection of the q-axis voltage.

Input to this compensation algorithm is an error in electrical angular velocity and electrical position of the rotor. Amplitude x_{q6} and phase ϕ of compensation voltage are adjusted during operation, based on change in angular velocity error. This voltage can be described as:

$$u_{q6}^* = x_{q6}sin(6\vartheta_e + \phi) =$$
$$= x_{q6}[sin(6\vartheta_e)cos(\phi) + sin(\phi)cos(6\vartheta_e)] \quad (4)$$

Voltage can be separated into sine and cosine part:

$$x_{q6}sin(6\vartheta_e + \phi) = w_{6sin}sin(6\vartheta_e) + w_{6cos}cos(6\vartheta_e) \quad (5)$$

w_{6sin} is weight coefficient of sine part and w_{6cos} is weight coefficient of cosine part. Variations in these weight coefficients

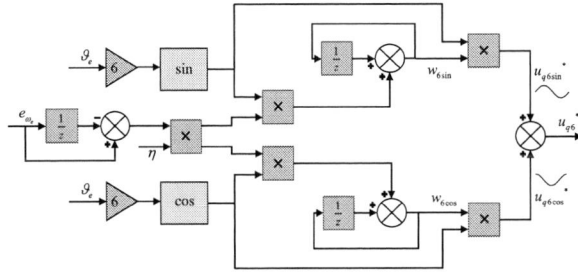

Fig. 8: Adaptive algorithm for calculation of compensation q-axis voltage

will lead to variations in amplitude and phase of injected q-axis voltage. w_{6sin} and w_{6cos} are DC values in steady-state.

For calculation of these weight coefficients an iterative least mean squares learning algorithm (Widrow-Hoff learning method) is used, to achieve minimum value of 6th harmonic component in angular velocity ω_e[24], [19], [20]:

$$w_{6sin}(k+1) = w_{6sin}(k) + \eta\Delta e_{\omega_e}sin(6\vartheta_e)$$
$$w_{6cos}(k+1) = w_{6cos}(k) + \eta\Delta e_{\omega_e}cos(6\vartheta_e) \quad (6)$$

k represent step, Δe_{ω_e} describes change in error of angular velocity ω_e between two samples. η is learning rate (chosen between 0 and 1), which determines speed of convergence of weight coefficients to correct values. If the value of η is too high, it will lead to instability of the system, due to missing of the searched local minimum. Too small value of η will lead to very slow convergence to correct values of weight coefficients. Fig. 8 shows algorithm for calculation of q-axis compensation voltage.

The advantage of this method is in the fact, that it is not necessary to identify any of the above mentioned sources of torque ripple and store them in look-up table. Algorithm is able to find itself the correct compensation q-axis voltage during operation to suppress torque ripple, noise and vibrations based on the ripple in angular velocity.

978-1-5386-7688-2/19 $31.00 © 2019 IEEE

IV. EVALUATION OF PRODUCED ACOUSTIC NOISE

For evaluation of produced acoustic noise software Spectroid was used, which makes online sound FFT analysis from microphone signal and visualizes data as a spectrogram. In this case, microphone was placed 10 cm from the motor. Bottom part of Fig. 9, 13, 14, 12 shows time dependent FFT analysis of sound (one FFT is calculated every 20 ms during 9 seconds interval). The time axis is on the right side of the figures. Amplitude of noise in *dB* is determined based on the color scale on the left side of the figures. Yellow waveform in upper part of these figures shows the last FFT of measured sound.

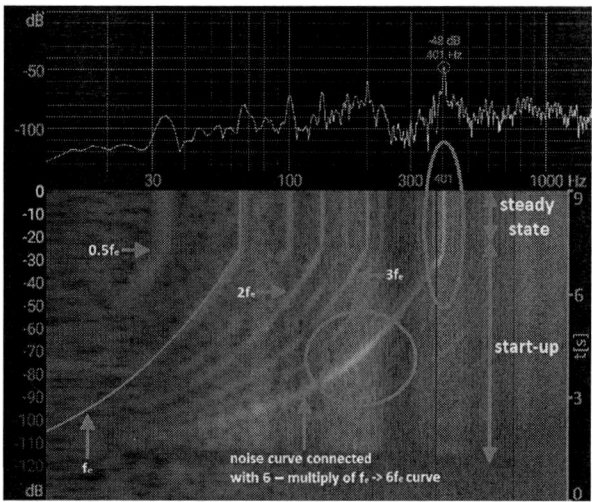

Fig. 9: Sound FFT analysis of the start-up sequence for ω_e from 0 to 420 $\frac{rad}{s}$. Speed ramp 300 $\frac{rpm}{s}$

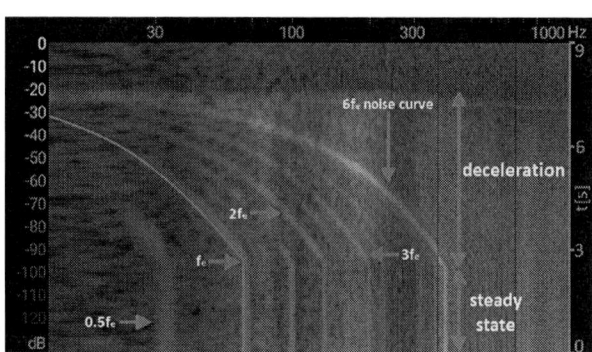

Fig. 10: Sound FFT analysis of slowing-down sequence, deceleration for ω_e from 420 $\frac{rad}{s}$ to 0. Speed ramp 300 $\frac{rpm}{s}$

Fig. 9 shows analysis of the motor start-up sequence under conventional FOC technique. Electrical angular velocity ω_e increased from 0 to 420 $\frac{rad}{s}$ (stator electrical frequency f_e = 66,8 Hz; mechanical speed n = 2000 rpm). Gray color indicates fundamental frequency f_e curve in all figures below. The complete acoustic noise spectra is represented by grid of curves that are an integer multiply of the fundamental

frequency f_e. As seen in Fig. 9, the most visible noise values are at 6 multiply of f_e, which confirms the theoretical assumption. Two peak noise values are identified at 180 Hz and 401 Hz, which corresponds to fundamental frequency 30 Hz and 66.8 Hz respectively. These two areas are in red circles in the Fig. 9.

Fig. 10 shows the same noise measurement results recorded during deceleration from 420 $\frac{rad}{s}$ to 0. This FFT noise analysis confirms the theoretical assumption, that the main sources of torque ripple for Linix motor are reflected mainly in frequency, which is 6-multiply of f_e.

V. EXPERIMENTAL RESULTS

Above proposed algorithm for torque ripple, acoustic noise, and vibration reduction by injection of 6th harmonic component of q-axis voltage was implemented and experimentally verified on real PM drive. Learning rate has been set to $\eta = 0,001$. Fig. 11 shows experimental setup with microcontroller NXP MPC5643L, 3-phase power inverter and PM motor.

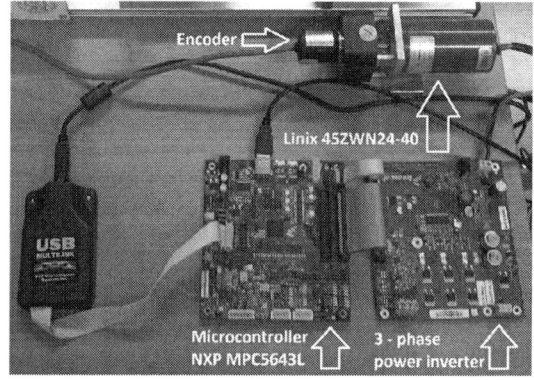

Fig. 11: Experimental setup

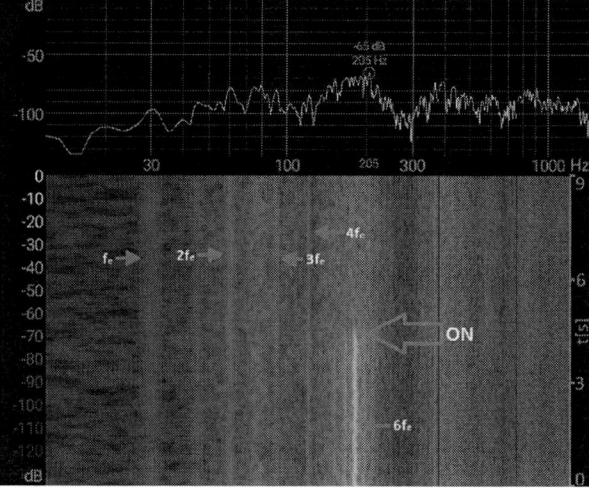

Fig. 12: Sound FFT analysis of steady-state at $\omega_e = 190\frac{rad}{s}$ with applied compensation algorithm

The experimental results with compensation algorithm show a significant improvement in the acoustic noise reduction. Fig. 12 shows steady-state case at $\omega_e = 190\frac{rad}{s}$. Red arrow indicates turning on of the algorithm. After that, noise connected with 6-multiply of f_e is completely reduced (around -30 dB). Fig. 13 shows greatly minimized noise during the start-up for ω_e from 0 to 420 $\frac{rad}{s}$. Improved noise curve is highlighted with red circles. Fig. 14 shows deceleration.

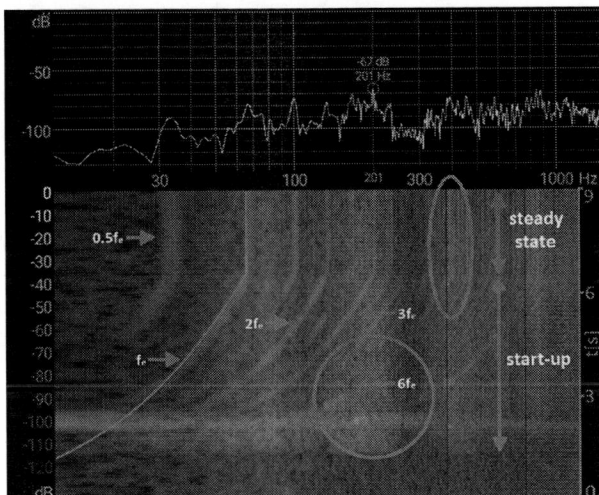

Fig. 13: Sound FFT analysis of the start-up sequence for ω_e from 0 to 420 $\frac{rad}{s}$ with applied compensation algorithm. Speed ramp 300 $\frac{rpm}{s}$

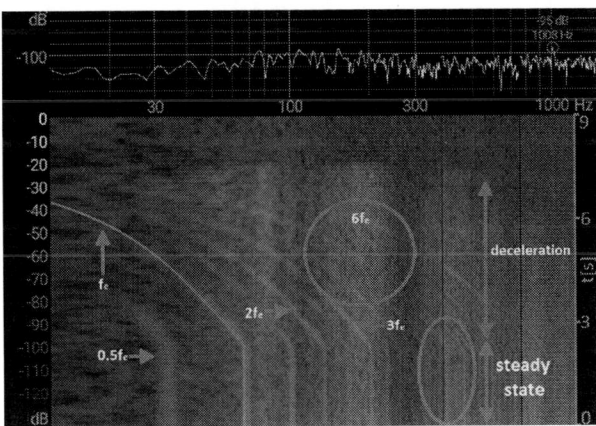

Fig. 14: Sound FFT analysis of deceleration for ω_e from 420 $\frac{rad}{s}$ to 0 with applied compensation algorithm. Speed ramp 300 $\frac{rpm}{s}$

Fig. 15 shows graph of acoustic noise connected with 6-multiply of f_e ($6f_e$ curve in figures above) vs. mechanical speed n_r expressed in *rpm*. Measurement of noise was done in steady state for lot of speeds. Application of adaptive algorithm reduce acoustic noise connected with 6th harmonic

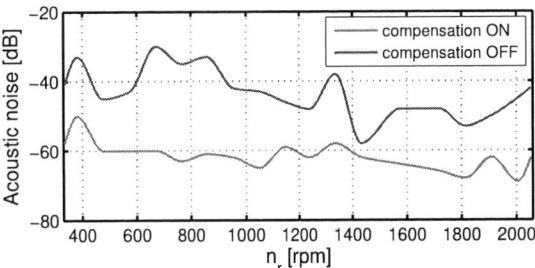

Fig. 15: Acoustic noise connected with 6-multiply of f_e vs. mechanical speed n_r

in whole measured speed range significantly (at least -10 *dB*, what is 10 times lower noise).

Fig. 16 shows significantly reduced ripple in speed connected with 6th harmonic component in one electrical revolution, when the compensation adaptive algorithm is turned on. Fig. 17 shows q-axis currents and voltages before and after compensation.

Fig. 18 shows transient state in speed ripple and in weight coefficients of the compensation voltage after turning on the proposed compensation algorithm. Weight coefficients w_{6sin} and w_{6cos} converge to final steady-state values approximately in 1 *s*. Frequency of control loop is 10 *kHz*.

These figures confirm the correct assumption, that the appropriate 6th q-axis current harmonic injected into the used PM motor, is able to greatly reduce acoustic noise and speed ripple. Acoustic noise and speed behavior improvement give us the confidence, that torque ripple and vibrations were improved too even their were not measured directly.

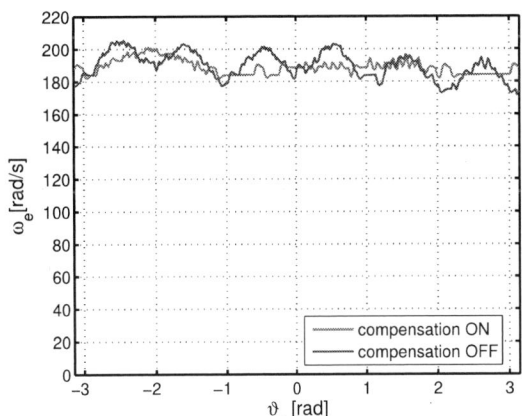

Fig. 16: Speed ripple in one electrical revolution at $\omega_e = 190\frac{rad}{s}$ before and after compensation

VI. CONCLUSION

This article brings up a new adaptive compensation algorithm for minimization of torque ripple, acoustic noise and

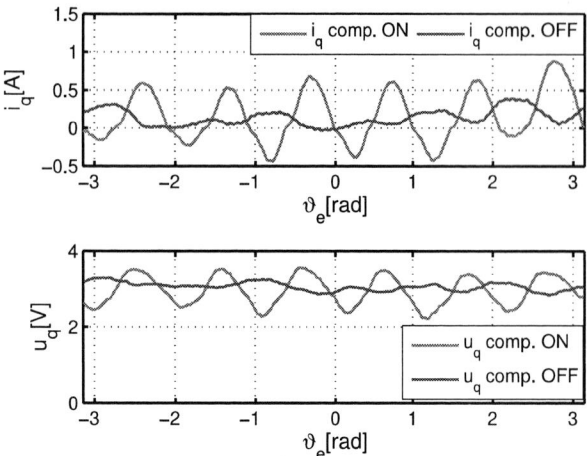

Fig. 17: q-axis currents and voltages before and after compensation at $\omega_e = 190 \frac{rad}{s}$

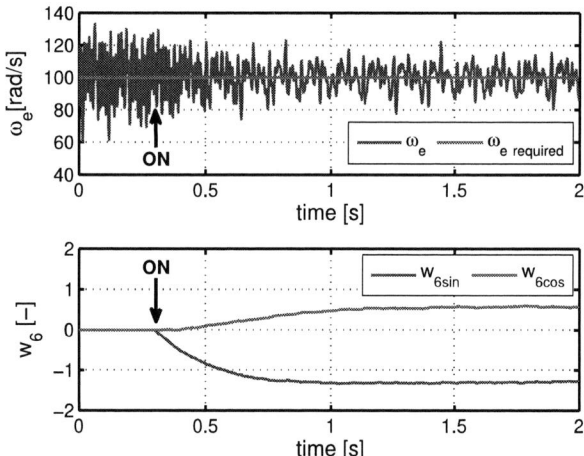

Fig. 18: Speed ripple and weight coefficients after turning on the compensation at $\omega_e = 100 \frac{rad}{s}$

vibrations. As an evaluation method online sound FFT analysis of produced acoustic noise was used, which can be considered as a unusual approach. Sound FFT analysis offers very simple and cheap method for verification of proper functionality of proposed adaptive algorithm for torque ripple, acoustic noise and vibrations minimization.

It has been proven that proposed compensation technique significantly reduce acoustic noise and also ripple in speed connected with 6-multiply of f_e. Algorithm works perfectly in a steady-state and also through slow start-up and deceleration. Based on ripple in speed, a correct 6-th harmonic of q-axis voltage has been calculated, which has been injected to the low-cost PM motor. This leads to increasing its performance.

Described method doesn't need identification of sources of torque ripple like non-sinusoidal Back-EMF voltage or the cogging torque, because ripple in speed is used for forming of the compensation current harmonics. This is a great advantage over the other conventional methods, which require this identification.

Proposed adaptive algorithm has also some limits. Angular velocity is calculated in a speed observer from position and its bandwidth can lead to damping and phase shift in observed speed at higher frequencies. This can lead to instability of the weight coefficients calculation. Ripple in speed must be large enough for correct functionality of adaptive algorithm.

Further work will be focused on the instable area of proposed algorithm. Moreover, this adaptive algorithm can be used in current control loop running in parallel to PI controller. The aim is to offload the main PI controller hence the loop bandwidth can be wider. With proper mapping of the cogging torque distribution or the Back-EMF voltage shape, proposed adaptive algorithm can be used in the current loop to achieve even better performance also in higher speed region. Other conventional methods achieved success mainly in lower speed region.

ACKNOWLEDGMENT

This work was supported by Slovak Scientific Grant Agency VEGA No. 1/0615/19, and by projects ITMS:26220120046, cofounded from EU sources and European Regional Development Fund.

REFERENCES

[1] M. Piccoli and M. Yim, "Anticogging: torque ripple suppression, modeling, and parameter selection," The international journal of robotics research, vol. 35, iss. 1-3, pp. 148-160, 2016

[2] R. Islam, I. Husain, "Analytical Model for Predicting Noise and Vibration in Permanent-Magnet Synchronous Motors", *IEEE Transactions on Industry Applications*, VOL. 46, NO. 6, August 2010, pp. 2346-2354, ISSN: 0093-9994, DOI: 10.1109/TIA.2010.2070473

[3] J. Le Besnerais, "Acoustic noise & vibrations due to magnetic forces in permanent magnet synchronous machines for traction application", *Eomys Engineering*, October 2016

[4] N. Bianchi, S. Bolognani, "Design Techniques for Reducing the Cogging Torque in Surface-Mounted PM Motors", *IEEE Transactions on Industry Applications*, VOL. 38, NO. 5, October 2002, pp. 1259-1265, ISSN: 0093-9994, DOI: 10.1109/TIA.2002.802989

[5] R. Islam, I. Husain, A. Fardoun, K. McLaughlin, "Permanent-Magnet Synchronous Motor Magnet Designs With Skewing for Torque Ripple and Cogging Torque Reduction", *IEEE Transactions on Industry Applications*, VOL. 45, NO. 1, February 2009, pp. 152-160, ISSN: 0093-9994, DOI: 10.1109/TIA.2008.2009653

[6] D. Q. Zhu, D. Howe, "Influence of Design Parameters on Cogging Torque in Permanent Magnet Machines", *IEEE Transactions on energy conversion*, VOL. 15, NO. 4, December 2000, pp. 407-412, ISSN: 0885-8969, DOI: 10.1109/60.900501

[7] L. Dosiek, P. Pillay, "Cogging Torque Reduction in Permanent Magnet Machines", *IEEE Transactions on Industry Applications*, VOL. 43, NO. 6, December 2007, pp. 1565-1571, ISSN: 0093-9994, DOI: 10.1109/TIA.2007.908160

[8] T. Li, G. Slemon, "Reduction of cogging torque in permanent magnet motors", *IEEE Transactions on Magnetics*, VOL. 24, Issue 6, November 1988, pp. 2901 - 2903, ISSN: 0018-9464, DOI: 10.1109/20.92282

[9] H. W. Park, S. J. Park, Y. W. Lee, S. Song, Ch. Kim, "Reference frame approach for torque ripple minimization of BLDCM over wide speed range including cogging torque", *IEEE Transactions on industrial electronics*, VOL. 47, NO. 1, February 2000, pp. 637-642, ISBN: 0-7803-7090-2 , DOI: 10.1109/ISIE.2001.931869

[10] D. Hanselman, "Minimum Torque Ripple, Maximum Efficiency Excitation of Brushless Permanent Magnet Motors ", *IEEE Transactions on Industrial Electronics*, VOL. 41, NO. 3, June 1994, pp. 292-300, ISSN: 0278-0046, DOI: 10.1109/41.293899

[11] S. J. Park, H. W. Park, M. H. Lee, F. Harashima, "A New Approach for Minimum-Torque-Ripple Maximum-Efficiency Control of BLDC Motor", *IEEE Transactions on industrial electronics*, VOL. 47, NO. 1, February 2000, pp. 109-114, ISSN: 0278-0046, DOI: 10.1109/41.824132

[12] L. Springob, J. Holtz, "High-Bandwidth Current Control for Torque-Ripple Compensation in PM Synchronous Machines", *IEEE Transactions on industrial electronics*, VOL. 45, NO. 5, October 1998, pp. 713-721, ISSN: 0278-0046, DOI: 10.1109/41.720327

[13] J. Holtz, L. Springob, "Identification and Compensation of Torque Ripple in High-Precision Permanent Magnet Motor Drives", *IEEE Transactions on industrial electronics*, VOL. 43, NO. 2, April 1996, pp. 309-320, ISSN: 0278-0046 , DOI: 10.1109/41.491355

[14] P. Mattavelli, L. Tubina, M. Zigliotto, "Torque-Ripple Reduction in PM Synchronous Motor Drives Using Repetitive Current Control", *IEEE Transactions on power electronics*, VOL. 20, Issue 6, November 2005, pp. 1423-1431, ISSN: 0885-8993, DOI: 10.1109/TPEL.2005.857559

[15] P. L. Chapman, S. D. Sudhoff, "A Multiple Reference Frame Synchronous Estimator/Regulator", *IEEE Transactions on energy conversion*, VOL. 15, NO. 1, June 2000, pp. 197-202, ISSN: 0885-8969, DOI: 10.1109/60.867000

[16] M. Musak, M. Stulrajter, V. Hrabovcova, M. Cacciato, G. Scarcella, G. Scelba, "Suppression of Low-order Current Harmonics in AC Motor Drives via Multiple Reference Frames Based Control Algorithm", *Electric Power Components and Systems*, VOL. 43, Issue 18, September 2015, pp. 2059-2068, ISSN: 1532-5016

[17] M. Lazor, M. Stulrajter, "Modified Field Oriented Control for Smooth Torque Operation of a BLDC Motor", *10th International Conference ELEKTRO 2014*, May 2014, pp. 180-195, ISBN: 978-1-4799-3720-2

[18] Ch. Xia, B. Ji, Y. Yan, "Smooth Speed Control for Low Speed High Torque Permanent Magnet Synchronous Motor Using Proportional Integral Resonant Controller", *IEEE Transactions on industrial electronics*, VOL. 62, Issue 4, April 2015, pp. 2123-2134, ISSN: 0278-0046, DOI: 10.1109/TIE.2014.2354593

[19] D. Flieller, N. K. Nguyen, P. Wira, G. Strutzer, D. O. Abdeslam, J. Merckle, "A Self-Learning Solution for Torque Ripple Reduction for Nonsinusoidal Permanent-Magnet Motor Drives Based on Artificial Neural Networks", *IEEE Transactions on industrial electronics*, VOL. 61, Issue 2, February 2014, pp. 655-666, ISSN: 0278-0046 , DOI: 10.1109/TIE.2013.2257136

[20] P. H. Truong, D. Flieller, N. K. Nguyen, J. Merckle, G. Strutzer, "An Investigation of Adaline for Torque Ripple Minimization in Non-Sinusoidal Synchronous Reluctance Motors", 2013, pp. 2602-2607, ISBN: 978-1-4799-0224

[21] R. Krishnan, "Permanent Magnet Synchronous and Brushless DC Motor Drives", *Electrical and Computer Engineering Department, Virginia Tech, Blacksburg, Virginia, U.S.A.*, CRC Press 2010, ISBN: 978-0-8247-5384-9

[22] D. Hanselman, "Brushless Permanent Magnet Motor Design. Second Edition", *Lebanon : Magna Physics Publishing*, 2006. pp. 392. ISBN: 1-881855-15-5

[23] S. Wang, "BLDC Ripple Torque Reduction via Modified Sinusoidal PWM", *FAIRCHILD SEMICONDUCTOR POWER SEMINAR 2008 - 2009*, 2009, s. 10.

[24] D. Nguyen, B. Widrow, "Neural Networks for Self-Learning Control Systems", *IEEE Control Systems Magazine*, Volume: 10 , Issue: 3 , April 1990, pp. 18-23, DOI: 10.1109/37.55119, ISSN: 2374-9385

Analysis of the Thermal Influence on the Vibrational Behavior of the Stator End-Winding Region

Sebastian Lange
Chair of Energy Conversion
TU Dortmund University
Dortmund, Germany
sebastian2.lange@tu-dortmund.de

Martin Pfost
Chair of Energy Conversion
TU Dortmund University
Dortmund, Germany
martin.pfost@tu-dortmund.de

Abstract—In this study, the influence of the thermal conditions on the vibrational behavior of the stator end-winding region at the excitation end of turbogenerators is investigated. As a result of the flexible operation and volatile grid feed-in, caused by an increasing share of renewable energies, the operating point of electrical machines shifts from a time-invariant to a dynamic operation. Influenced by this, the temperature distribution in electrical machines changes more frequently. The varying thermal load leads to thermomechanical stress due to thermal expansion. Furthermore, the mechanical properties of composite materials are dependent of the temperature, which leads to lower stiffness of the system. In order to analyze the vibrational behavior in consideration of these mechanisms, the thermal expansion and the vibrations during the transient event of a 3-pole short circuit are analyzed for different thermal conditions.

Index Terms—3-pole short circuit, electrical machines, temperature, thermal expansion, thermomechanical strain, turbogenerator, vibration

I. INTRODUCTION

The fluctuation of the available power due to renewable energies in modern power grids asks for a more flexible operation of turbogenerators in conventional power plants. The operating power has higher gradients, which are occurring more frequently [1]. These dynamics are influencing the operating current of turbogenerators. The operating current is the dominant influencing factor on the Lorentz forces that acts on the stator end-winding region and on the losses, which have a major impact on the temperature profile [2]. A thermal load causes thermal expansion of the components, which adds to the displacement due to the Lorentz forces. Further, several materials, especially glass fiber reinforced materials, show a decreasing stiffness with increasing temperature. During glass transition, the Young's Moduli of those materials are going down to a level of 20-50% of the value at room temperature. Consequently, the system stiffness varies with the temperature distribution.

The influence of temperature effects on eigenfrequencies are experimentally shown in [3] via the experimental modal analysis. An increase of the ambient temperature from 24°C to 90°C reduces the eigenfrequencies to about 90% of the initial value for the investigated 2-pole machine.

A shift of eigenfrequencies could also be explained by aging. [4]–[6] present aging phenomena, which could cause a change of Young's Moduli, shear Moduli, and Poisson's ratio. For instance, abrasion between stator bars and the support ring, the loss of spacing constructions between the stator bars, and broken ties are considered. Each of these damages can easily be explained by vibrations and by a loosened linkage of the components.

A semi-analytical approach for calculating the vibrations of a 202 MVA turbogenerator is used in [7]. The study describes a good agreement with experimental results. But this method does not support the calculation of strains and stresses at material interfaces (e.g. between cured copper and electrical insulation system).

[8]–[11] use 3D finite element analysis (FEA) for estimating the vibrational behavior. These studies are mainly focused on the calculation of vibrations and eigenfrequencies. For that, simplified geometries of main components such as the stator bars and spacing constructions are used. Further, these studies neglect any influence of the temperature.

However, the influence of different temperature distributions is of major interest for an accurate stress estimation for flexible operation. The results are also important for diagnosis and optimization of turbogenerators in conventional power plants. Therefore, this study is focused on the estimation of the vibrational behavior of the stator end-winding region under thermal influence. To take this into account, a 3D FEA is used to estimate the deformations, strains and stresses at room temperature (RT) and under thermal load, considering that the system is stress-free at curing temperature.

The thermal expansion corresponds to a steady-state stress on the system, which gets overlaid by transient Lorentz forces. In this study the thermal expansion is analyzed and critical positions are highlighted. Further, at the same time the impact of temperature-dependent material properties on vibrations due to transient Lorentz forces is introduced.

978-1-5386-7688-2/19 $31.00 © 2019 IEEE

II. STATOR END-WINDING REGION MODEL

A. Temperature Dependency of Mechanical Properties

The stator end-winding region consists of several glass fiber reinforced materials, which are impregnated with epoxy resin. A characteristic of those materials and components is the glass transition at a certain temperature which is mainly dependent on the share of epoxy resin in the material, the concentration of nanoparticle filler (if used) and the hardening temperature. This causes a temperature-dependent stiffness. All materials have a characteristic loss of stiffness with increasing temperature and a highly increased loss factor tan δ around the glass transition temperature in common. A schematic representation is given in Fig. 1. For each of the materials, temperature-dependent properties are assumed, cf. Fig. 1. The parameters are adapted with a material-specific fitted trend and an individual glass transition temperature.

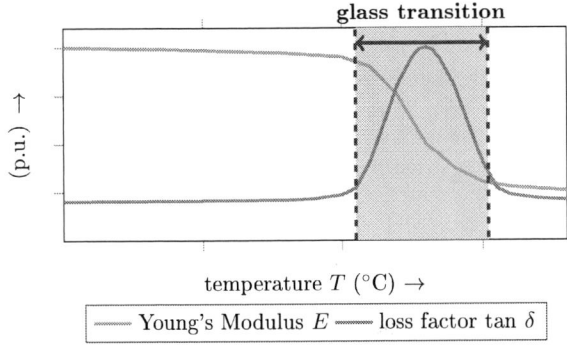

Fig. 1. Typical temperature dependency of the Young's Modulus E (green) and of the loss factor tan δ (orange). The glass transition is in the gray marked range.

B. 3D Finite Element Model and Loads

Fig. 2a) shows the evaluated stator end-winding region model of the excitation end of a turbogenerator. The model is characterized by a separate modeling of important components such as the stator bars, consisting of the cured copper bar and the electrical insulation system (EIS), cf. Fig. 2b), and the inter stator-bar fixation, consisting of a sandwich of two swell mats and a diamond spacer, cf. Fig. 2c). This separation leads to a more realistic stiffness and mass distribution. As shown in [12] it is possible to calculate the vibrations for the estimated operating conditions of no-load and continuous short circuit with an accuracy of more than 90%. Further, the model has the ability to calculate the vibrational behavior under consideration of thermal expansion, transient events and thermally dependent material properties.

The forces are calculated with the tool developed in [13], which is based on the Biot-Savart law. The calculation of the forces is dependent on the currents and the given geometry of the stator end-winding bars, which are represented as lines. The results match well to 3D FEA [13].

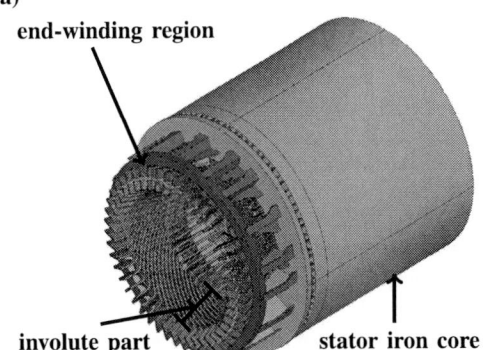

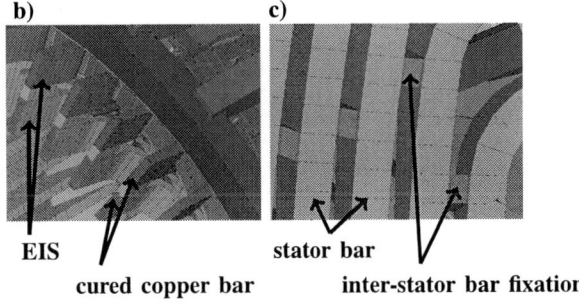

Fig. 2. a) 3D FE model of the stator end-winding region, b) detail of the geometry of a cured copper bar (red, at the bottom layer; pink, at the top layer) and EIS (blue, at the bottom layer; green, at the top layer), c) zoom into the inter-stator bar fixation structure consisting of different insulation materials (red and blue)

III. THERMAL EXPANSION AND THERMOMECHANICAL STRAIN

A. Assumptions

The current-carrying copper windings produce losses. Because of this, the cured copper bar is the heat source and the hot spot in the electrical machine. The temperature of the cured copper bar is the main influencing factor of the temperature profile.

The cured copper bar provides a good thermal conductivity. Contrary to that, the EIS is a poor thermal conductor and impairs the heat dissipation from the cured copper. The process of heating up leads to thermal expansion which is the dominating aging factor on the EIS. Because of different material properties, both components need a different duration for heating up. Hence, there is a temperature difference and a gradient at the surface of these components. Further, both materials have coefficients for thermal expansion that differ strongly. As consequence, the stator bars are undergoing large strains and stresses due to the temperature distribution and the corresponding inhomogeneous thermal expansion.

For the estimation of thermal expansion, a reference temperature has to be defined. For this temperature, the system is assumed to be free of strains and stresses. In this case, the reference temperature is set to the curing temperature of the turbogenerator. Before impregnation, the machine is heated up to this temperature. During impregnation all strains and stresses are eliminated. Nevertheless, the stator bars are undergoing strain for temperatures above the curing temperature or compression for temperatures below the curing temperature [14].

The strain is a stress parameter of huge importance to estimate the influence of a load on the aging behavior. Twelve dedicated points at the cross-sectional area of the stator bar are chosen for the analysis, see Fig. 3. The points 1-8 estimate the strain at the center of the narrow side of the stator bar, with consideration of the interface between cured copper bar and EIS. The positions A-D estimate the strain at the center of the broadsides of the stator bar.

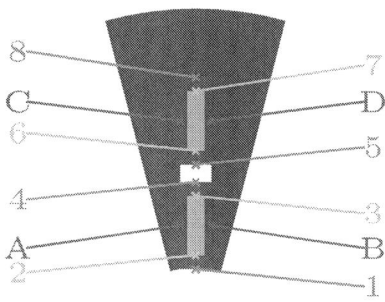

Fig. 3. Schematic view on a slot with marked evaluation points. The green points (1,4,5,8) are the mid-side nodes of the narrow side of the bar, the orange points (A,B,C,D) are the mid-side nodes of the broadside of the bar, the lime points (2,3,6,7) are the mid-side nodes of the narrow side of the interface between the cured copper bar and the EIS. The bar nearer to the air gap is the top layer bar, the other bar is the bottom layer bar.

For estimation of the strain the sum of all acting strains above each of the $n = 48$ stator bars per layer and each of the $m = 80$ axial FEA-elements per bar is calculated using (1), for each of the 12 analyzed points (cf. Fig. 3).

$$\varepsilon_{\text{sum}} = \sum_{m=0}^{80} \sum_{n=0}^{48} \varepsilon_{\text{n,m}} \qquad (1)$$

B. Steady-State Thermal Expansion - Homogeneous Temperature Distribution

The worst case for stresses resulting from thermal expansion are large differences between the reference temperature and the steady-state temperature. Therefore, in this study the worst case is when the machine is homogeneously cooled down from reference temperature (curing temperature) to room temperature. The thermal conditions are in a steady-state for this analysis.

The displacement results for the steady-state thermomechanical calculation at RT are shown in Fig. 4. In three-phase electrical machines the gap between two stator bars

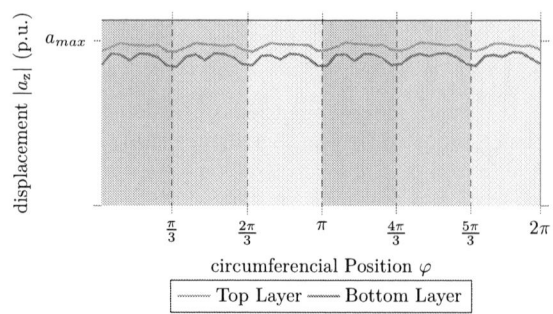

Fig. 4. Detailed view on the axial displacement a_z at the location of the series connection in dependency of the circumference (stator bar) for the top layer bars and the bottom layer bars. The dashed lines mark the position of a phase shift between two stator bars of different electrical phase

of a different electrical phase is a significant point for observing vibrations. Because of the large Lorentz forces at this position, there are additional diamond spacers and swell mats for an increased stiffness. This characteristic leads to the axial displacements shown in Fig. 4 calculated by FEA. The displacement of the stator bars at the series connections is shown for the top layer bars and the bottom layer bars for the center of each of these bars. The series connections are located at the front of the stator end-winding region, where the top layer and bottom layer are connected.

The displacement is periodically distributed along the circumference. The minimal displacement occurs at the shift between two adjacent phases, the maximum arises at the stator bars located at middle of the phase. The stator bars located at the phase shift are up to 1.6% less displaced than stator bars at the middle of the phase. A possible reason is the higher number of diamond spacers in the phase shift. Further, the top layer bars are undergoing a larger displacement than the bottom layer bars. The deviation between the maximal displacement of the top layer bars and the bottom layer bars amounts to 2.1%. This deviation seems plausible because of the connection between the bottom layer bars and components of iron stator core by glass fiber reinforced components and bandages.

Fig. 5 shows the axial distribution of the displacement a_z into axial direction. Within the slot of the stator iron core the stator bars are expanding straightly into axial z-direction, as well as at the series connections where the stator bars are not curved. Between these two sections of the stator bars, the stator bars are curved in an involute form. The global z-direction is not equal to the local z-direction of the stator bar surface. During thermal expansion, the straight part pushes into the involute part. Consequently, the curvature into the involute part is of interest for analysis. The involute part is less displaced into z-direction. The gradient of the displacement in the involute part is about 3.7 times lower than in the straight part.

The resulting strain distribution is shown in Fig. 6. All values are normalized on the maximal $\varepsilon_{\text{sum,max}}$, which is cal-

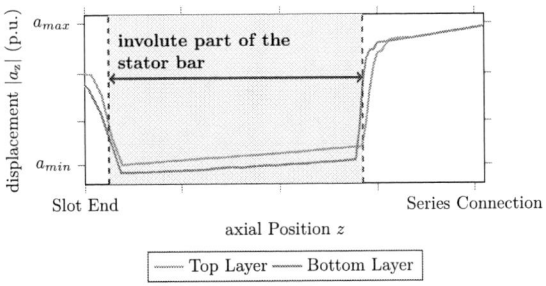

Fig. 5. Axial displacement a_z in dependency of the axial position z for the top layer bar (green) and bottom layer bar (orange)

culated according to (1), from the twelve analyzed positions.

The highest summed up strain occurs at the narrow side point of the top layer bars at surface of the EIS (point 1). This point is the nearest to the air gap, cf. Fig. 3. With increasing distance to the air gap, the strain is decreasing for the corresponding points on the surface of the EIS at the narrow side of the stator bars. The interface between the cured copper bar and the EIS is less strained, the corresponding FEA nodes have a roughly 50% lower value for $\varepsilon_{\mathrm{sum}}$. The strain distribution at the broadsides of the stator bars shows a characteristic pattern. The points B and C are more stressed than the points A and D, cf. Fig. 6. Considering the geometry of the stator end-winding region, the nodes B and C are both directed into positive axial z-direction and are facing towards the series connections. These nodes are pushing forward due to thermal expansion, which could be a possible reason for the occurrence of higher strain at these positions. The points A and D are on the reverse side of the bar, where less stress is occurring.

In [12], large strain values are observed at axial positions which are fixated with diamond spacers and at the curvatures into and out of the involute part. To analyze this axial dependence, for each axial position the circumferential average of the local strain ε_{i} is calculated and depicted in Fig. 7. It is focused on the points 1 and B, which show the largest sum of strains.

For point 1 (narrow side of the stator bar, node nearest to the air gap) and point B (broadside, node facing into axial direction) the peak strain values are located within the involute part of the stator bar and stand out at the axial strain distribution in Fig. 7. Each peak is close to the curvature into or out of the involute part. Further, at these positions a fixation via diamond spacers and swell mats is located. Accordingly, the first and the last inter-bar fixation seem to be a characteristic point for a peak value of the strain. In addition to point 1, for point B other positions with diamond spacers and swell mats are noticeable in the strain distribution.

These results have to be evaluated under consideration of the characteristics of the finite element model. Positions with stiffening constructions are in the FEA modeled as contact elements with a large stiffness. This could lead to an overestimation of the observed strain.

This case considered a cooling-down from reference temperature to room temperature. Therefore, the system is mainly stressed by thermal compression of the stator end-winding region. The results of this analysis are similar to the reverse case, when room temperature is the reference and the system is heated up to the curing temperature. The characteristic axial positions and the typical peaks remain the same. Similarly, the points B and C (cf. Fig. 3) are highly stressed.

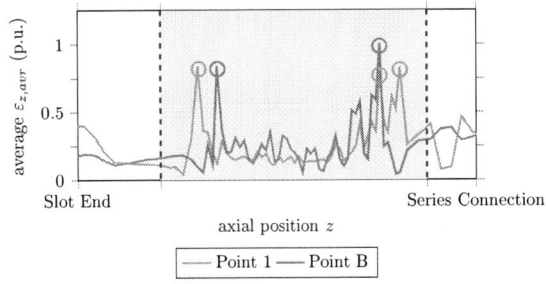

Fig. 7. Average strain $\varepsilon_{\mathrm{z,avr}}$ along the axial position z for point 1 (mid-side node of the narrow side of the top layer bars) and point B (mid-side node of the broadside of the top layer bars). All values are normalized to the maximum

IV. VIBRATIONAL BEHAVIOR UNDER THERMAL LOAD

A. Eigenfrequencies

Eigenfrequencies are strongly correlated to the system stiffness. In case of a decreasing stiffness, the eigenfrequencies decrease linearly with the square root of the system stiffness [15].

This investigation is focused on the 1-lobe mode, the 2-lobe mode, and the 3-lobe mode. The 1-lobe mode is characterized by one sinusoidal period of the deformation and two corresponding nodes along the circumference. Similarly to this, the 2-lobe mode (3-lobe mode) shows two (three) sinusoidal periods of the displacement and four (six) nodes along the circumference. These mode shapes are always occurring pairwise with a deviation of less than 3%. Thereby, the mode shapes of these eigenfrequency pairs are circumferentially shifted with a mode-shape-specific angle [16].

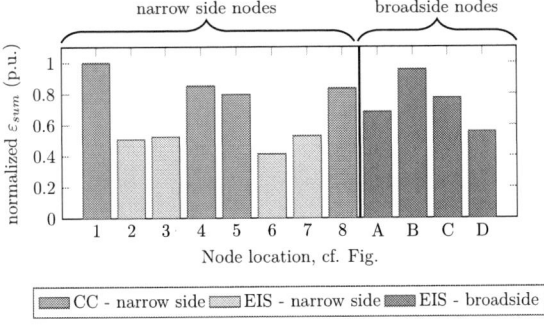

Fig. 6. Summed strain $\varepsilon_{\mathrm{sum}}$ for the evaluated points, cf. Fig. 3

978-1-5386-7688-2/19 $31.00 © 2019 IEEE   111

The influence of three homogeneous temperatures (20°C, 90°C and curing temperature) is depicted in Fig. 8. As expected, the eigenfrequencies decrease with increasing temperature because of the loss of stiffness with increasing temperature. Unlike to [3], where a decrease of the eigenfrequency $f_{4N,2-lobe}$ of about 10% is observed for the temperature step from 20°C to 90°C, the eigenfrequency $f_{4N,2-lobe}$ decreases in this case by only 3.5%. Here, in this temperature range the analyzed system stiffness is not very sensitive to temperature changes. Also, the other eigenfrequencies show a similar trend for 90°C. Contrary to this, at curing temperature a significant decrease of the eigenfrequencies is visible. At this temperature level some materials are entering into the glass transition or have passed through it. The stiffness of these materials behaves as shown in Fig. 1. Finally, the eigenfrequencies are reducing to 82 - 88% of the starting value. But the eigenfrequencies stay out of the forbidden eigenfrequency range around twice of the grid frequency.

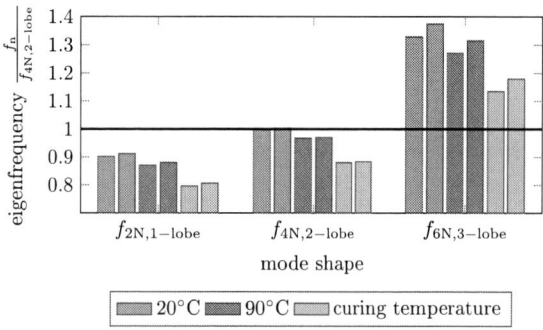

Fig. 8. Overview of the first three eigenfrequencies. Values are normalized to $f_{4N,2-lobe,1}$

B. Thermally Dependent Vibrations

For analyzing the vibrational behavior, the 3-pole short circuit is chosen as a transient event. The results from the Lorentz force calculation tool from [13] are used for simulation. The resulting vibrations are depicted in Fig. 9 focusing on the circumferential 4N (2-lobe)-share of the radial displacement. In this Figure, the time-dependent displacement of the series connections, where the maximal displacement is located, is shown for the three evaluated temperature distributions.

A growing displacement due to increasing temperature is visible, which is due to the decreasing stiffness caused by the increasing temperature. The relation between the maximal displacement at RT and at 90°C is $a_{max,90°C} = 1.067 \cdot a_{max,20°C}$. The shape of both deformations is similar. The slight change of the eigenfrequencies between 20°C and 90°C indicates the minor influence of this temperature change on the displacement. But it has to be considered that only the vibrations due to Lorentz forces are reviewed for this investigation. The superimposing thermomechanical stress is much lower at 90°C than at RT.

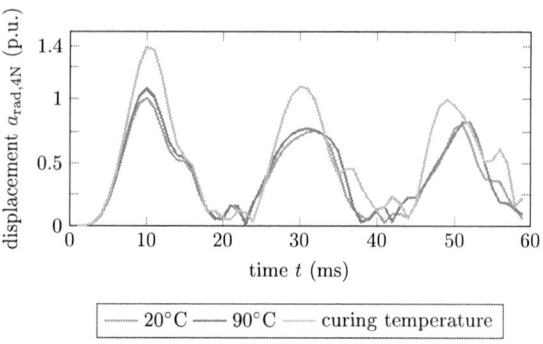

Fig. 9. Time-dependent displacement at the series connection of a phase A bar (point of maximal displacement) for the three evaluated temperature distributions

The impact of curing temperature is much more visible. The displacement level is increasing up to 141% of the maximal displacement at RT. In the same manner, as the eigenfrequencies are changing, the vibrations are changing when the curing temperature is exceeded. At this level some components are less stiff because the temperature range of their glass transition is reached.

C. Superposition

Under real conditions, vibrations and thermal expansion are acting at the same time. A superposition of both mechanisms leads to the real stress of the stator end-windings. The results of this superposition for the von-Mises strain ε_{EQV} are shown in Fig. 10.

Here, the maximal strain around the circumference for each of the axial positions are evaluated. The lower side of the green-marked range represents the von-Mises strain ε_{EQV} at 20°C without considering thermal expansion. The upper side marks the von-Mises strain ε_{EQV} at 20°C with consideration of thermal expansion. The green-marked corridor represents

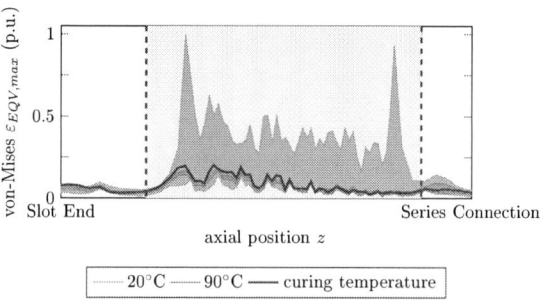

Fig. 10. Axial distribution of the maximal circumferential von-Mises strain $\varepsilon_{EQV,max}$ at point 1 (cf. Fig. 3). The black line marks the strain at curing temperature, the green range marks the superposition of strain due to vibrations (lower boundary of the range) and due to superposition with thermal expansion (upper boundary of the range) for 20°C and the orange range marks the trend for 90°C. The gray marked range marks the involute part of the stator bar. All values are normalized to the maximum.

the strain, which superimposes the strain from vibrations. For 90°C the results are visualized in the same manner, the trend is marked in orange. The black line marks the von-Mises strain ε_{EQV} at curing temperature, where no thermal expansion is present and the strain is solely due to vibrations.

Fig. 10 shows that thermal expansion is the dominating stressing factor for the case considered here. The stress is mainly concentrated on the involute part of the stator bars. Positions with spacers and a resulting increased local stiffness are more sensitive for stress due to thermomechanical strain than from vibrations alone. At 20° the superpositions of vibrations and thermal expansion leads to the largest value for the von-Mises strain $\varepsilon_{EQV,max}$. With increasing temperature, the share of the vibrations on the von-Mises strain $\varepsilon_{EQV,max}$ increases, but the peak value decreases. The von-Mises strain at curing temperature, where no thermal expansion is existent, and at 90°C (superposition of vibrations and thermal expansion) are more or less comparable. The influence of an increasing strain due to increasing temperature (increasing vibration level, cf. Fig. 9) is nearly compensated due to the thermal expansion at 90°C. This shows that the thermal load and the temperature distribution are a factor with high impact on the stress distribution.

V. CONCLUSION

In this study the thermal influence on the vibrational behavior of the stator end-winding region is shown. Under thermal load the structure is undergoing thermal expansion, which stresses the materials. The analyzed cases show peak loads at the curvature into and out of the involute part. In particular, the first and the last sandwich of swell mats and a diamond spacer are stressed. At the cross-sectional area, the point nearest to the air gap on the narrow side of the top layer bar is undergoing the largest load. On the broadside the points facing into positive axial z-direction are remarkable, they see high loads as well.

Under thermal loading, the eigenfrequencies are shifting to lower values and therefore the shape under load (due to Lorentz forces) is varying, too. The 2-lobe-vibration level increases up to 141% for the temperature step from room temperature to curing temperature because of the decreasing stiffness of glass fiber reinforced materials with increasing temperature. The system is getting weaker while the forces are staying the same. As an additional consequence of the eigenfrequency shift, the shape of the deformation is varying as well. It is getting a higher 1-lobe share, but the 2-lobe share remains dominant.

The subsequent analysis of the thermal influence on vibrations and the superposition of vibrations and thermal expansion shows the dominating influence of thermal expansion as stressing factor. With increasing temperature, the strain due to

vibrations is increasing, but simultaneously the maximal strain decreases because of lower stress due to thermal expansion. The resulting strain due to the superposition of these mechanisms is decreasing with increasing temperature.

ACKNOWLEDGMENT

This work was supported by the German Federal Ministry of Economic Affairs and Energy under grant 03ET7078D. The authors are responsible for the contents of this publication.

REFERENCES

[1] A. Joswig, H. Steins, J. R. Weidner, "Impact of new flexible load operation and grid codes on turbine generators with a focus on end windings," in *Proc. Power Gen Europe*, Amsterdam, Netherlands, 9-11 June 2015.

[2] S. GC, E. Boulter, and I. Culbert, *Electrical Insulation for Rotating Machines : Design, Evaluation, Aging, Testing, and Repair*, 02 2004.

[3] J. Letal and V. Warren, "Optimize stator endwinding vibration monitoring with impact testing," in *The Premier Electrical Maintenance and Safety Event*, 2017.

[4] M. S. J. Kapler, J. Letal and G. C. Stone, "Recent endwinding vibration problems in air-cooled turbine generators," in *Cigre*, 2014.

[5] G. C. Stone and R. Wu, "Examples of stator winding insulation deterioration in new generators," in *2009 IEEE 9th International Conference on the Properties and Applications of Dielectric Materials*, July 2009, pp. 180–185.

[6] A. Ttreault and Z. Zhengping, "End-winding vibration monitoring: Pivotal in preventing major damage on a large turbo-generator," in *2013 IEEE Electrical Insulation Conference (EIC)*, June 2013, pp. 1–6.

[7] H. Molki, P. Bahemmat, and A. A. Kharamani, "Theoretical study on vibration behavior of stator of 202 MVA large turbo-generator," in *The 3rd Conference on Thermal Power Plants*, Oct 2011, pp. 1–5.

[8] K. Senke, S. Kulig, J. Hauhoff, D. Wünsch, "Vibrational behaviour of the turbogenerator stator end winding in case of electrical failures," in *Cigre 97*, Yokohama, Japan, 29 October 1997.

[9] Y. Zhao, B. Yan, C. Chen, J. Deng, and Q. Zhou, "Parametric study on dynamic characteristics of turbogenerator stator end winding," *IEEE Transactions on Energy Conversion*, vol. 29, no. 1, pp. 129–137, March 2014.

[10] B. Schlegl, F. Schönleitner, A. Marn, F. Neumayer, and F. Heitmeir, "Analytical determination of the orthotropic material behavior of stator bars in the range of the end windings and determination of the material characteristics of the orthotropic composite space brackets via experimental modal analysis and FE-calculation," in *Proc. ICEM 2012*, Sept 2012, pp. 1948–1956.

[11] S. Exnowski, "Excitability of different modes of vibration of stator end windings," in *Proc. IECON 2012*, Oct 2012, pp. 1781–1785.

[12] S. Lange, M. Pfost, "Validation and verification of a structural mechanical stator end-winding region model," in *in Proc. IEMDC 2019*, San Diego, USA, May 2019.

[13] A. Gruening, "Electromechanical behaviour of stator end-winding region of large synchrounous machines during steady-state and electrical failure: (in german)," Ph.D. dissertation, TU Dortmund University, Dortmund, December 2006.

[14] O. P. Chabra, "Residual technological stresses in stator bar insulation of large machines," in *Conference Record of the 1992 IEEE International Symposium on Electrical Insulation*, June 1992, pp. 495–498.

[15] W. Weaver, S. Timoshenko, and D. Young, *Vibration Problems in Engineering*, ser. A Wiley-Interscience publication. John Wiley & Sons, 1990.

[16] M. Humer, R. Vogel, and S. Kulig, "Monitoring of generator end winding vibrations," in *in Proc. ICEM*, Sep. 2008, pp. 1–5.

Assessment of 5 kW Induction Motor Finite element computations with a Commercial and an Open-source software

M. Zaheer
LUT University
Lappeenranta, Finland

P. Lindh, *SM, IEEE*
LUT University
Lappeenranta, Finland

L. Aarniovuori
Aston University
Birmingham, UK

J. Pyrhönen, *SM, IEEE*
LUT University
Lappeenranta, Finland

Abstract – **A 5 kW high efficiency induction motor is analyzed using an open-source platform and the results are compared with measured data and commercial software. Comparison of the results shows that the open-source platform gives good results that are close to measured and commercial software computed values – especially for torque, current and losses. The open-source method required less computational time than the commercial software. It is concluded that for 2D applications, open-source software can be used on industrial-scale problems with acceptable results.**

Index Terms-- **electrical machine, finite element analysis, finite element method, induction machine, IEC-loss components iron losses, rotor joule losses, stator joule losses.**

I. INTRODUCTION

THe primary aim this research work is to benchmark a multi-physical open-source platform against well-known commercial FEA software to assess whether it can be used in electrical machine computations. Current, slip, electromagnetic torque and associated losses calculated using the open-source platform are compared with the commercial finite element method and measured data of the real motor. There are several reasons to favor an open-source platform. Firstly, the platform enables flexible modification of the modelling features. Secondly, it is possible to include attributes that do not exist in commercial software, for example, the open-source platform supporting alternatives for pre-processing and post-processing. Lastly, use of an open-source platform permits cost-efficient massive computation [1]. Additionally, open-source software is transparent; it is thus possible to check the source code, investigate the functionalities implemented, and judge their reliability [1].

The utilized open-source finite element method (FEM) was developed by the Finnish company CSC – IT Center for Science Ltd. Both the commercial and open-source FEM software use Maxwell's equations and the Bertotti model to solve the electromagnetic problems and the iron losses. For inter-process communication and parallel processing, the open-source software utilizes the standardized message passing interface MPI, which makes it possible to run analyses in multi-core as well as multi-processor environments [2]. The commercial FEA software also supports parallel computing but the computation time of the commercial FEA computations depends on the number of cores selected and the RAM of the operating machine [3]. For the open-source system there is no pre-requisite to have an efficient operating machine because it can utilize supercomputing resources and access CSC servers employing up to 672 cores for robust results. The open-source FEM uses mesh generation tools for complex geometries [4]. Importing the mesh from GMSH (open-source software) to open-source FEM could result in dislocations of mesh nodes, which would affect the accuracy of the results. In [5], where open-source meshing tools such as OpenFOAM, SALOME and GMSH are studied to analyze their computational meshing capabilities, it is concluded that open-source tools are less user-friendly, that there is complexity in different tool interactions and data incompatibility in various process phases [5]. Commercial tools are generally based on integrated modelling and the in-built advanced meshing capability of commercial FEA software allows it to calculate more precise results due to the absence of dislocations in the mesh nodes.

An induction motor was chosen as a test case because it is is widely used in the industrial sector. They are easy to design, control and operate safely within a wide field-weakening range [6]. Induction motors are often named the workhorses of the industry, and they are widely used in variable speed applications such as conveyor belts, robots, electrical vehicles, cranes, elevators and home appliances [7]. The 5 kW motor in this work is operated with a frequency converter and is specially designed to power industrial lifting equipment. The design and main parameters of the 5 kW prototype IM are introduced in section II. Sections III and IV compare the current and torque results from the open-source and commercial FEA software. Section V presents the losses obtained with the two software platforms and the real measured losses obtained with the IEC losses segregation method for the 5 kW prototype machine. The computation time of the open-source FEM is investigated in section VI.

II. MOTOR DESIGN AND ANALYTICAL CALCULATIONS

The studied 5 kW motor is a 4-pole squirrel-cage machine. Because of symmetry, it is sufficient to analyze one pole in the FEM. One fourth of the FEA model of the machine having one pole with ten aluminum rotor bars and twelve slots in the stator is shown in Fig. 1. The main data of the 5 kW IM rated values are given in Table I. The motor is designed for lifting purposes operating mainly at 75% torque. To deliberately increase the share of iron losses and facilitate examination of the iron losses, M800-65A material steel sheets with high specific loss value of 8 W/kg at 50 Hz are used in construction of the rotor and stator stacks. The dimensions of the machine, stator circuit configuration and details of the winding configuration can be found in [8]. The rotor has 8.5 degrees skew, so the magnetic flux is non-isotropic in the axial direction, and with 2D analysis, some difference to measured values is to be expected. In the open-source software simulations, the free CAD tool GMSH (open-source software) is used for modelling, as GMSH has excellent meshing capabilities for 2D problems, due to its efficient meshing algorithms, and scripting can be done for effective construction of geometry. The conductivity of aluminum and copper is 24 MS/m and 48 MS/m, respectively when there is temperature rise in the winding.

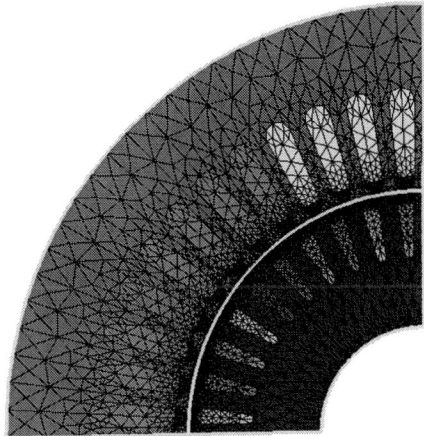

Fig. 1. One fourth of the 5kW induction machine geometry and the used mesh in open source FEM.

The stator slot and rotor bar dimensions are shown in Fig. 2.

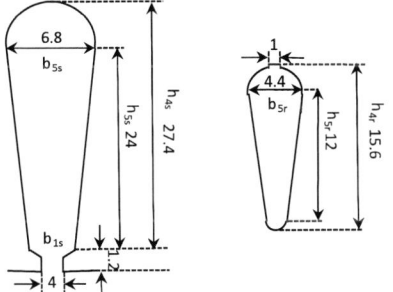

Fig. 2. Stator slot and rotor bar dimensions.

TABLE I. INDUCTION MACHINE RATED VALUES

Parameter	Value
Stator stack length	160 mm
Stator core outer diameter	220 mm
Stator core inner diameter	125 mm
Air gap	0.5 mm
Number of winding turns in one phase	128
Winding configuration	Y
Rated voltage	400 V
Rated frequency	50 Hz
Rated power	5 kW
Rated speed	1467 min-1
Rated current	10.4 A

In design of induction motors, the equivalent circuit parameters offer indispensable information, and they can be obtained analytically or by FEA [9]. The circuit parameters utilized in our research were calculated analytically. The 2D-model cannot take into account three-dimensional effects and therefore they are included in an external circuit model. The motor end effects are divided into stator end-winding inductance and rotor end ring resistance and inductance [10]. The rotor circuit parameters have an effect on the slip and the related torque production of IMs. Lombard and Zidat [11] computed the impedance of a machine using 2D and 3D computation and studied the effect of the end ring on the machine performance. They found that the computation results for an induction machine are very sensitive to the end ring parameters. In this study the measured phase resistance information is used in the FEA, and the resistance value is corrected to the corresponding winding temperature at the operating point studied.

A. End ring resistance
The end ring resistance is given by [12],[13]

$$R_{\text{ring}} = \frac{2\pi D_{\text{r,o}} \rho K_{\text{ring}}}{h_{\text{ring}}(D_{\text{r,o}} - D_{\text{r,i}})} \qquad (1)$$

where h_{ring} represents the height of the ring, $D_{\text{r,o}}$ is the rotor outer diameter, $D_{\text{r,i}}$ is the end ring internal diameter, $K_{\text{ring}} = 1.05$ is a correction factor and ρ is the resistivity of the end ring conductor [12]. The skin effect can be taken into account in (1) by varying the useful height of the rotor bar and the end-ring thickness. In some cases, the current does not penetrate completely into the bar (height), and therefore, the resistance should be computed for a larger diameter using a corrected rotor bar height and end-ring thickness [14].

B. End ring leakage inductance
End ring leakage inductance is given by [15]

$$L_{\text{ring}} = \mu_0 \frac{Q_r}{2m_s p^2}\left[\frac{2}{3}(l_{\text{bar}} - l'_r) + \text{U}\frac{\pi D'_r}{2p}\right] \qquad (2)$$

where Q_r is the number of rotor slots, m_s is the number of stator phases, p represents the number of pole pairs, l_{bar} is the rotor bar length, l'_r is the rotor core length, factor $\text{U} = 0.18$, D'_r is the average diameter of the short-circuit ring and μ_0 is the vacuum permeability.

C. End winding leakage inductance

When applying transient 2D finite element tools, the main interest in the stator leakage components lies in the end-winding leakage connected to the FEA circuit model. The end-winding leakage inductance represents the flux distribution in the end region produced by currents flowing outside the stator iron stack. Analytical calculation of the leakage fluxes is challenging because of the complex multilayered end-winding areas. End-winding leakage inductance is normally given by [15].

$$L_{\text{w}} = \frac{4m}{Q_s} q N^2 \mu_0 l_{\text{w}} \lambda_{\text{w}} \qquad (3)$$

where Q_s is the number of stator slots, N is the number of coil turns in the phase winding, l_{w} is the average end winding length. λ_{w} is permeance factors, whose product is given as

978-1-5386-7688-2/19 $31.00 © 2019 IEEE

$$l_w \lambda_w = 2l_{ew}\lambda_{lew} + W_{ew}\lambda_{Wew} \qquad (4)$$

where l_{ew} is the axial length of the end winding measured from the end of the stack, and W_{ew} is the coil span. The permeance factor $\lambda_{lew} = 0.5$ and λ_{wew} is 0.2 for a three-phase two-layer stator winding and rotor cage winding type [15]. The IM circuit parameters (1)-(4) is given in Table II.

TABLE II. MODEL PARAMETERS

Parameter	Value
Phase resistance (θ_w=102°C)	0.582 Ω
End winding leakage inductance	6 mH
End ring resistance	1.4 μΩ
End ring inductance	44 nH

D. Iron losses

Much research has studied losses in a three-phase induction motor. Based on a 2D FEM, Belahcen, Rasilo and Arkkio [17] segregated the iron losses into eddy, hysteresis and excess losses. The iron losses estimated analytically were less precise than calculated. Saheb and Ali [18] introduced an approach to estimate induction motor iron and copper losses by finite element analysis considering stator winding end effects, rotor end ring end effects and rotor bars skewing.

In the open-source FEA in this study, steady-state iron losses estimation is done using the Steinmetz equation and a Bertotti model in which the total iron losses are described as a contribution of hysteresis, eddy current and excess losses [8].

$$P_{Fe} = \Sigma P_k = C_1 f_k^{a1} B_k^{b1} + C_2 f_k^{a2} B_k^{b2} + C_3 f_k^{a3} B_k^{b3} \qquad (5)$$

where P_k is a harmonic power loss component associated with the kth flux density harmonic, B_k is the peak flux density value of the kth B harmonic, and f_k is the kth harmonic frequency [8]. The values of parameters a, b and C are obtained from curve fitting of the measured data for various frequencies and flux densities. In this research, the values for harmonic loss frequency exponents $a1$, $a2$, and $a3$ are 1.0, 2.0, and 1.5. Harmonic loss field exponents $b1$, $b2$, and $b3$ are 1.776, 2.0 and 1.5, respectively [8]. The first, second and third terms in (5) represents hysteresis, eddy current and excess losses respectively.

III. CURRENT EVALUATION USING FEA AND MEASUREMENTS

At the rated slip value, the current waveforms were analyzed using both the commercial and the open-source FEM. The measured current waveform and current amplitudes given by the open-source and the commercial FEA are shown in Fig. 3

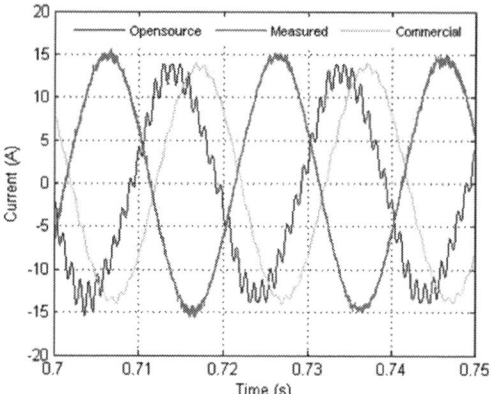

Fig. 3. Open-source, commercial FEA and measured currents waveform at 50 Hz on rated torque.

The stator current peak value obtained with the commercial FEA is 13.4 A, the corresponding measured value is 14.7 A, and the open-source software gave phase peak current of 13.3 A. The FEA computed results might differ from the measured values for one obvious reason: the 29th and 31st voltage harmonics are produced by a generator supply unit. Fig. 4 compares the open-source, the commercial FEA and the measured current harmonics and it can be seen that the 31st harmonic is present in the measured spectrum but negligible in the FEA computed waveforms as it originates from the generator voltage. From the 17th to the 23rd harmonic, there is a significant difference between the amplitude of the measured and FEA computed waveform because the harmonics originate from the rotor slotting. In the simulations, the higher amplitudes of the current harmonics might be a consequence of a non-modelled skewing effect [16].

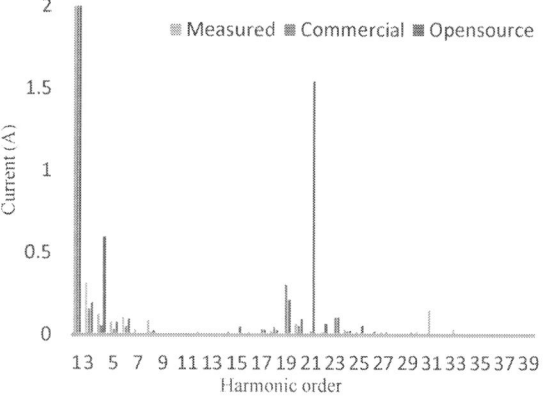

Fig. 4. Open-source and commercial FEA and measured fast fourier transform (FFT) analysis for current at 50 Hz.

IV. TORQUE EVALUATION USING FEA AND MEASUREMENTS

The finite element analyses were done at the rated slip and the rated load. Computations were performed in the Altair Flux 12.3 finite element program and in the open-source FEM program Elmer to get torque waveforms at the rated load (time stepping computation with sinusoidal voltage supply). The FEA model had 120,000 nodes and a time step of 50 μs. Torque waveforms as a function of time for both software are illustrated in Fig. 5 for 5 ms because the fundamental waveform period is 20 ms and, due to symmetry, one fourth is taken into account. The torque ripple is large because rotor skewing is not taken into account in this 2D model [8].

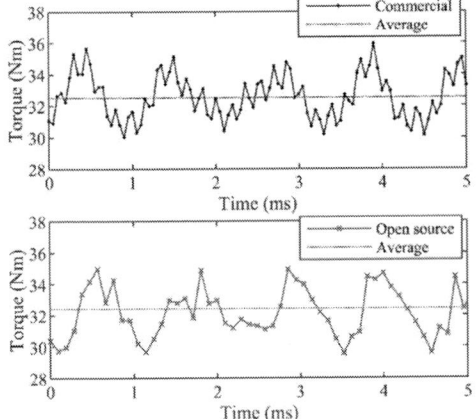

Fig. 5.   Open-source and commercial FEA torque waveform as a function of time at 50 Hz.

Torque fast Fourier transform results are depicted in Fig. 6 for both software.

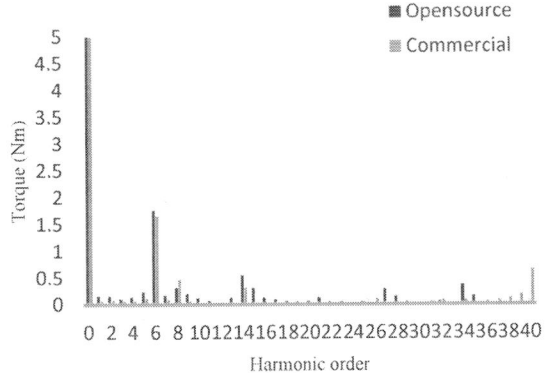

Fig. 6.   Open-source and commercial FEA torque FFT analysis at 50 Hz.

Comparison of the FFT results shows that both software show torque ripple at the 6th harmonic resulting from 5th and 7th current linkage harmonics. The back-to-back test bench used in this study cannot be used to analyze the torque harmonics of the tested machine since the load machine and the test bench mechanics affect the results.

The results obtained from the open-source FEM were compared with the commercial software at different slip values. In Fig. 7, it can be seen that for smaller slip values both software show almost the same behavior. As the slip values increase, the open-source FEM gives smaller values for the torque compared to the commercial FEA. At unity slip, the start-up-torque value is practically the same. The peak torque value for the commercial FEA is higher compared to the open-source FEA results.

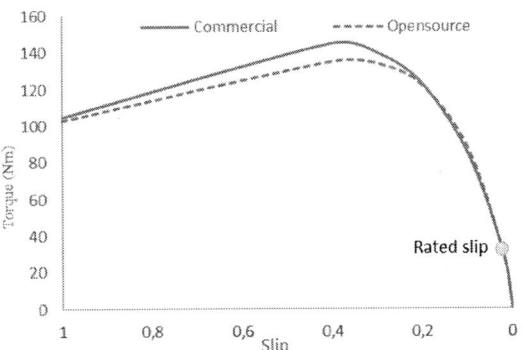

Fig. 7.   Torque as a function of slip from Open-source and commercial FEA.

V.   LOSSES FROM DYNAMIC STATE FEA COMPUTATIONS AND MEASUREMENTS

In this research, pure sinusoidal voltage was supplied without considering time harmonics. Time step FEA computations were performed to analyze losses at four different frequencies. Measured results were determined at four different frequencies of 50 Hz, 37.5 Hz, 25 Hz and 12.5 Hz with sinusoidal generator supply and constant U/f ratio, i.e. the line-to-line voltages were 400 V, 300 V, 200 V and 100V, respectively. The IEC [16] standard loss segregation method 2-1-1B was used to measure different motor loss components. Three tests were used to obtain the data for the loss segregation method. The stator and rotor Joule losses were determined from the rated load heat run test; the no-load voltage curve test was used to determine the mechanical losses and the iron losses; and the additional stray load losses were determined from the load curve test [16].

In this research the special attention is paid to determine the iron losses. The measured results used to calculate the iron losses as a function of supply frequency are shown in Fig. 8. Obtaining the iron losses accurately from the no-load test is difficult because the machine is running with a low power factor, which results in a low power measurement accuracy. Further, reducing the fundamental waveform frequency will reduce the accuracy of the power measurement. At 12.5 Hz, the iron losses do not behave logically. It should be noted that the iron losses in Fig. 8 are measured with a constant voltage frequency ratio to keep the flux linkage level constant, similar to when the machine is driven with a frequency converter.

The measured iron losses are determined at the voltage point U_i, where the effect of the stator voltage drop is removed, that is:

$$U_i = \sqrt{\left(U - \frac{\sqrt{3}}{2}IR\cos\varphi\right)^2 + \left(\frac{\sqrt{3}}{2}IR\sqrt{1-\cos^2\varphi}\right)^2}, \quad (6)$$

where U is the measured stator voltage, I is the measured stator current, R is the stator resistance at the temperature of the corresponding measurement point, and $\cos\varphi$ is the power factor at that specific point.

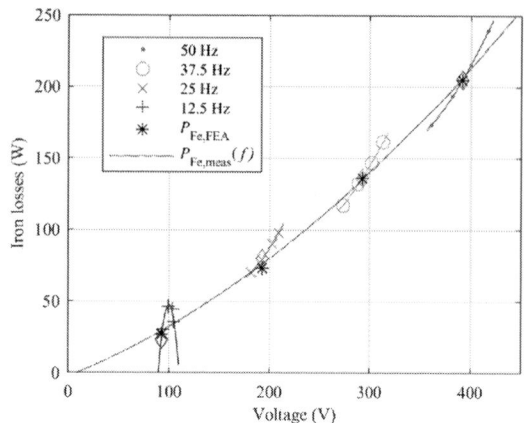

Fig. 8. Measured iron losses as a function of stator voltage. The diamond sign (◊) is used to indicate the iron loss determination point where the stator inner voltage drop has been taken into account.

The measurement results differ from an ideal FEA computation since the supply unit generates some voltage harmonics. Consequently, it can be expected that also the simulation results will deviate slightly from the measurement results. The difference in current amplitude is determined by the mechanical losses. Analytical values for the iron losses at different frequencies and line-to-line voltage are compared with the FEA computed results in Fig. 9.

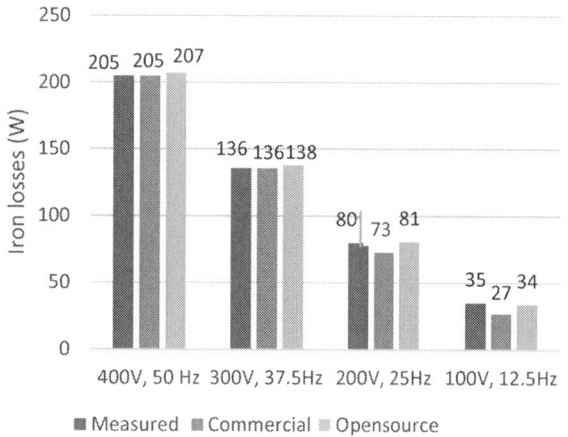

Fig. 9. Iron losses comparison for measured and FEA computed values for loss at 32.5 Nm load with end-ring resistance 1.36 µΩ for the studied 5 kW IM.

Compared to the measured iron losses values, the commercial FEA tool showed higher percentage error for low frequencies than the open-source FEA. At 12.5 Hz, the commercial FEA showed more than 20% error, whereas the open-source tool error percentage is below 2%.

Simulated results from the open-source FEA are given in Table III. Table IV lists the results from the commercial FEA. Current RMS value for the open-source FEA is 9.78 A, while the commercial software showed 9.87 A at 32.5 Nm at 50 Hz at 0.023 slip value.

TABLE III. OPEN-SOURCE COMPUTATIONAL RESULTS FOR DIFFERENT FREQUNCIES AT RATED LOAD 32.5 NM.

Parameter	Values	Values	Values	Values
Frequency, Hz	12.5	25	37.5	50
Iron Losses, W	34	81	138	207
Rotor Joule Losses, W	130.6	124.4	122	125
Stator Joule Losses, W	154.6	152.6	149	151
Total Electrical Losses, W	319.2	358	409	483

TABLE IV. COMMERICIAL FEA COMPUTATIONAL RESULTS FOR DIFFERENT FREQUNCIES AT RATED LOAD 32.5 NM.

Parameter	Values	Values	Values	Values
Frequency, Hz	12.5	25	37.5	50
Iron Losses, W	27	73	136	205
Rotor Joule Losses, W	137.4	129.7	128.2	131
Stator Joule Losses, W	163	161	156	157
Total Electrical losses, W	327.4	363.7	420.2	493

The measured values obtained using the IEC-segregation of losses method for losses at different frequencies are presented in Fig. 10.

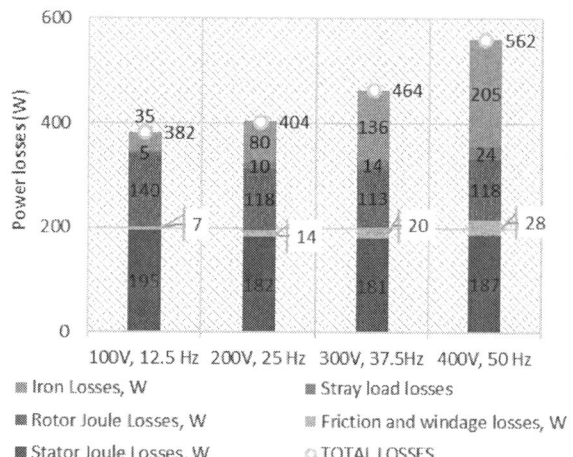

Fig. 10. Measured losses values at 32.5 Nm load and 1.36 µΩ end-ring resistance for 5kW IM

Table V shows the impact on iron, stator and rotor losses when changing the end ring bar to bar resistance and keeping all other parameters the same. It can be seen that the change in the resistance value has a small effect on the rotor and stator Joule losses. Iron losses are almost the same for all end ring resistance values.

TABLE V. OPEN-SOURCE COMPUTATIONAL RESULTS FOR DIFFERENT END RING RESISTANCE VALUES AT 50 HZ.

Parameter	Values	Values	Values	Values
Resistance between bars, µΩ	0.7	1	1.36	1.7
Iron Losses, W	205	207	207	207
Rotor Joule Losses, W	122.5	124	125	125
Stator Joule Losses, W	150.8	150	151	149.8
Total Electrical Losses, W	478.3	481	483	481.8

VI. OPEN-SOURCE FEA COMPUTATIONAL TIME

In iron loss computation at 50 Hz supply, the commercial and open-source FEA time step size was 50 μs and the number of mesh nodes was around 120,000. Table VI shows the relationship between the number of time step intervals, job type, number of processor cores and computational time for the open-source FEM when solving the problem. It can be observed that the computational time decreases more than 50% when the number of cores is increased from 14 to 26 for 16,000 intervals. The relation of computation time to number of cores is not linear. The computational time could be further reduced by using more parallelization.

TABLE VI. OPEN-SOURCE FEA COMPUTATIONAL TIME FOR 50 Hz INDUCTION MACHINE.

Number of intervals	Time step	Job type	Number of cores	Simulated time
16000	50 μs	serial	1	8280 min
16000	50 μs	parallel	14	1280 min
16000	50 μs	parallel	26	520 min
60000	50 μs	parallel	6	10080 min
60000	50 μs	parallel	22	1860 min
60000	50 μs	parallel	24	1580 min

VII. CONCLUSION

The aim of this study was the assessment of induction machine by FEA and the results are compared with measured data. 2D FEM analysis was performed for a 5 kW induction motor with commercial software Flux 2D Altair and then an open-source software was used to analyze the same induction motor. The machine electromagnetic state remained the same during the whole analysis. FEA results found from both open-source and commercial tool were compared with measured values. At 50 Hz, the commercial FEA showed more than 8% error compared with the measured fundamental wave value of the stator current, while the open-source FEA result deviated less than 4% from the measured current value. The experimentally determined value for iron losses at 50 Hz is 205 W, while the open-source and commercial FEA computed values are 207 W and 205 W respectively. At 12.5 Hz, the iron losses for the commercial FEA showed more than 20% error, while the open-source error percentage was below 2% compared to the measured value. It can be seen that the open-source software shows acceptable results in comparison with the commercial software. Furthermore, the computational time can be significantly reduced by using a higher number of cores. Therefore, it is concluded that for 2D applications, it is possible to utilize the open-source platform in industrial-scale applications to get acceptable results without incurring license fees. This study did not consider the skewing effect in the models, which may have affected the accuracy of the results. A 3D model of the 5 kW induction motor will be studied in further work along with analysis of the skewing effect.

REFERENCES

[1] P. Råback, "Benefits of open source software for industry," CSC – IT Center for Science, 2013.

[2] "CSC - IT Center for Science (CSC)," [Online]. Available: https://www.csc.fi/web/elmer. [Date Accessed 7.3.2019].

[3] "Flux™ 2018 Release Notes," Altair Engineering, Inc., 2018.

[4] M. Lyly, "ElmerGUI manual," CSC – IT Center for Science, 2015.

[5] J. Kortelainen, "Meshing Tools for Open Source CFD – A Practical Point of View," VTT Technical Research Centre of Finland LTD, 2009.

[6] U. Schuffenhauer, S. Miersch, N. Michalke, T. Schuhmann and A. Bárdos, "Modeling and practical investigation of the efficiency and operational behavior of induction machines with die-cast copper rotor," *2017 International Symposium on Electrical Machines (SME)*, Naleczow, 2017, pp. 1-6.

[7] I. Daut, K. Anayet, M. Irwanto, N. Gomesh, M. Muzhar, M. Asri and Syatirah. , "Parameters Calculation of 5 HP AC Induction Motor," Proceedings of International Conference on Applications and Design in Mechanical Engineering (ICADME), pp. 12B-1- 12B-5, 2009.

[8] P. Ponomarev, "Elmer FEM Induction Machine Tutorial," VTT Technical Research Centre of Finland LTD, 2017.

[9] K. Yamazaki, "An efficient procedure to calculate equivalent circuit parameter of induction motor using 3-D nonlinear time-stepping finite-element method," in *IEEE Transactions on Magnetics*, vol. 38, no. 2, pp. 1281-1284, March 2002.

[10] R. De Weerdt, K. Hameyer and R. Belmans, "End ring inductance of a squirrel-cage induction motor using 2D and 3D finite element methods," Conference Record of the 1995 IEEE Industry Applications Conference Thirtieth IAS Annual Meeting (IAS '95), Orlando, USA, Vol. 1, pp. 515-522, 1995.

[11] P. Lombard and F. Zidat, "Determining end ring resistance and inductance of squirrel cage for induction motor with 2D and 3D computations," 2016 XXII International Conference on Electrical Machines (ICEM), Lausanne, pp. 266-271, 2016.

[12] P. H. TRICKEY, "Induction motor resistance ring width," *IEEE*, Vol. 55, pp. 144-150, 1936.

[13] J. F. Fuller, E. F. Fuchs and D. J. Roesler, "Influence of harmonics on power distribution system protection," in *IEEE Transactions on Power Delivery*, Vol. 3, no. 2, pp. 549-557, April 1988.

[14] J. Langheim, "Modelling of rotorbars with skin effect for dynamic simulation of induction machines", Conference Record of the IEEE Industry Applications Society Annual Meeting (IAS), Vol.1 pp. 38 - 44, 1989

[15] J. Pyrhönen, T. Jokinen and V. Hrabovcova, Design of Rotating Electrical Machines, John Wiley & Sons, Ltd, 2008.

[16] P. Lindh, L. Aarniovuori, H. Kärkkäinen, M. Niemelä and J. Pyrhönen, "IM Loss Evaluation using FEA and Measurements," International Conference on Electrical Machines (ICEM), 2018.

[17] A. Belahcen, P. Rasilo and A. Arkkio, "Segregation of Iron Losses From Rotational Field Measurements and Application to Electrical Machine," in *IEEE Transactions on Magnetics*, vol. 50, no. 2, pp. 893-896, Feb. 2014, Art no. 7022104.

[18] A. N. Abdul Saheb and A. M. Ali, "Estimation of Copper and Iron Losses in a Three-Phase Induction Motor using Finite Element Analysis," *2018 2nd International Conference for Engineering, Technology and Sciences of Al-Kitab (ICETS)*, Karkuk, Iraq, 2018, pp. 6-10.

978-1-5386-7688-2/19 $31.00 © 2019 IEEE

Automatic Superimposed Droop Frequency Control Scheme for DC Microgrids

Mohammad Jafari Matehkolaei
The Center of Excellence in Power
System Management & Control
Department of electrical engineering
Sharif University of technology
Tehran, Iran
m.jfr.m@ee.sharif.edu

Hossein Mokhtari
The Center of Excellence in Power
System Management & Control
Department of electrical engineering
Sharif University of Technology
Tehran, Iran
mokhtari@sharif.edu

Majid Poshtan
Department of electrical engineering
California Polytechnic State University
San Luis Obispo, USA
mposhtan@calpoly.edu

Abstract— This paper proposes a control strategy to improve the superimposed droop frequency control scheme in DC microgrids. Conventional droop-based schemes for DC microgrids control suffer from issues such as inaccurate power sharing among the sources, voltage control with insufficient quality, and adverse effect of the line resistance on the droop characteristic. To overcome these challenges, the superimposed droop frequency method has been introduced. However, the stable operation in terms of the load variations and improper voltage quality are still the challenges with this prior-art method due to the injected AC voltages. In this paper, a method is proposed to solve the stability problem and alleviate the overall voltage quality in the superimposed droop frequency scheme. The performance of the proposed method is verified by different simulation studies in MATLAB/SIMULINK environment.

Keywords— DC microgrids, droop-based control method, superimposed frequency scheme, power-sharing, microgrid stability, voltage quality, transmission line resistance effect.

I. INTRODUCTION

Renewable energy sources (RES) mitigate limitations in energy resources and environmental issues caused by fossil fuels [1]. Despite these advantages, integration of RES to the main grid results in some major problems such as harmonic injection, inverse power flow and voltage deviations which adversely affects the overall voltage quality and power system protection process [2]- [4]. Widespread integration of the RES to the main grid entails the centralized power system to move forward to a more distributed platform, the so-called concept of distributed generation (DG) [5]- [7]. Microgrids provide the best mean to overcome mentioned challenges and integrate DGs into the main power system which also increase overall system reliability and add controllability to DGs [8]- [10].

Recently, DC microgrids have gained more attention by providing higher efficiency, reliability and stability compared to the AC microgrids [11]. Also, most of the RESs and the loads are DC in nature which is another motivation to use more DC microgrids. Further, DC microgrids efficiently utilize the transmission lines capacity since the reactive current concept doesn't exist in these systems. Considering the mentioned advantages as well as recent advances in power electronic technologies, make DC microgrids as a proper alternative for power distribution [11].

Conventionally, sources in DC microgrids are controlled via droop-based methods. In droop-based schemes, current/power set-point of each source is determined by its terminal voltage level via droop characteristic. The droop

gains have the major effect on the power/current sharing and system overall performance [12]- [15], and then, must be selected carefully. In this respect, using low droop gains yields to a perfect voltage regulation. However, transmission line resistances adversely affect the equivalent droop characteristic and correspondingly the power sharing among the sources. This is more challenging at high load conditions which may overload some sources [11]. On the other hand, using high droop gains yield to an accurate power sharing among the sources. However, the voltage regulation is a concern for high droop gains [11].

To mitigate the improper voltage regulation of high droop gains, a secondary controller-based method is proposed by [11], [16], [17]. Secondary controller adaptively increases voltage reference when the load increases which logically compensates the voltage drops at high load conditions. However, secondary controller-based methods require a communication network to collect the data from all the lines which affects the system overall reliability considering the communication system failures.

To solve these issues, a control strategy based on the superimposed frequency is proposed in [18] and [19]. In this method, a small AC voltage with a constant amplitude is injected to the DC microgrid by each source in the system. The frequency of the injected AC voltages is adjusted by a frequency/current droop characteristic. Also, DC voltages of sources are regulated by a voltage/reactive power droop characteristic [18]. Since the frequency is a global parameter in the system, the DC current of all the sources is proportional to their frequency droop gain at steady state operation of the grid [18]- [19]. Therefore, the superimposed frequency scheme provides an accurate power sharing among the sources in the system. The average of buses DC voltages is equal to the microgrid reference voltage which is so desirable compared to the conventional control methods [18].

However, the superimposed droop frequency method has two major issues. First, the system stability is not guaranteed in terms of the load variations [20]. Second, the overall voltage quality is affected by the injection of AC voltages. With a fixed amplitude of the injected AC voltage, the maximum transferred reactive power is limited which limits the system loading [18]. Injection of high amplitude of the AC voltages deteriorates the voltage quality, and on the other hand, the system stable operation is not guaranteed at low load levels [18]- [21]. In this paper, a new method is proposed to determine the amplitude of the injected AC voltages. This method ensures the stable operation of the grid at high load levels and enhances the stability and the voltage quality at low load levels.

The paper is structured as follows. The impacts of the injected AC voltage amplitude on the maximum load of the system is explained in section II. Section III presents the proposed method. In section IV, the dynamic behavior of the system is analyzed with small signal stability analysis. In section V, the performance of the proposed method is investigated by simulation results. Finally, section VI concludes the paper.

II. EFFECT OF THE INJECTED AC VOLTAGE ON THE MAXIMUM LOAD OF THE SYSTEM

The test system of Fig. 1 with the given parameters in Table I is considered to show the impacts of the injected AC voltage amplitude on the maximum loading of the system. The test system consists of two sources and a local load connected to the point of common coupling (PCC).

With refer to [18], the transferred reactive power for supplying a special load level is as:

$$Q = \frac{V_{ref}(d_{f1}r_2 - d_{f2}r_1)}{R_{load}(d_1 + d_2)(d_{f1} + d_{f2}) + r_1 d_2 d_{f2} + r_2 d_1 d_{f1}} \quad (1)$$

where R_{load} models the load, d_1 and d_2 are the voltage droop gains, r_1 and r_2 are the transmission line resistances, and d_{f1} and d_{f2} are the frequency droop gains.

Variations of the transferred reactive power in terms of the load changes is depicted in Fig. 2. As shown in this figure, increasing the load level leads to an increase in the transferred reactive power.

Considering the fixed amplitude of the injected AC voltage, the maximum transferred reactive power limits the microgrid maximum loading. The injected reactive power and the maximum transferred reactive power are as given in (2) and (3), respectively.

$$\begin{cases} Q_1 = -\frac{A^2}{2(r_1+r_2)}\sin(\delta) \\ Q_2 = \frac{A^2}{2(r_1+r_2)}\sin(\delta) \end{cases} \quad (2)$$

$$Q_{max} = \frac{A^2}{2(r_1 + r_2)} \quad (3)$$

In these equations, A is the amplitude of the injected AC voltage, $\delta = \delta_1 - \delta_2$ where δ_1 and δ_2 are the angle of the injected AC voltage of source1 and 2, respectively.

TABLE I. PARAMETERS OF THE STUDY MICROGRID [18]

Quantity	Value
d_{f1}, d_{f2}	0.15 , 0.15
r_1, r_2	2 , 4 Ω
Voltage reference	700 V
d_1, d_2	2.5 , 2.5
Reference Frequency	50 Hz
Filter Cutoff Frequency (w_c)	35 Hz

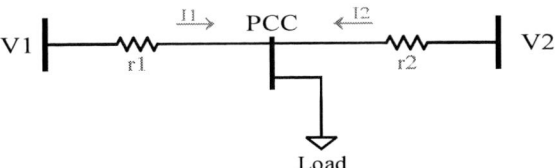

Fig. 1. Typical DC Microgrid (V_1 and V_2 are the two power sources)

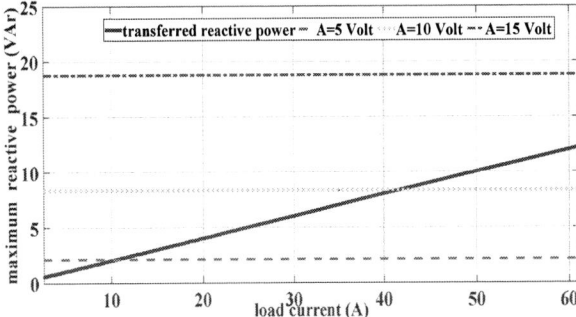

Fig. 2. The maximum transferrable reactive power at different amplitudes of the injected AC voltage

The higher the amplitude of the injected AC voltage, the higher the microgrid maximum load level. Fig. 2 demonstrates the microgrid maximum load level at different amplitudes of the injected AC voltage. As shown in this figure, injection of AC voltage with high amplitude increases the maximum load of the system due to increase in the maximum transferred reactive power. However, it deteriorates the overall voltage quality of the DC microgrid. Besides, large imaginary parts of the dominant poles with the small damping ratio at low loads, may result in instability. This subject is further discussed in section IV and V.

III. THE PROPOSED SUPERIMPOSED DROOP FREQUENCY CONTROL METHOD

As mentioned in section II, AC voltages with low amplitude are desirable at low load levels to ensure the system stable operation and enhance the overall voltage quality. As the load increases, the transferred reactive power increases, and the amplitude of the injected AC voltage should be increased accordingly to guarantee stable operation of the system. This is the base for the proposed method in this paper.

To perform the proposed method, an indicator is required to evaluate the overall load of the system. System frequency is a perfect choice because; it is available in the whole system at steady state operation of the microgrid. In the superimposed droop frequency method, the frequency and source DC currents are related as (4), hence both the frequency and DC current of the sources can be used to determine the system overall load level.

$$f = f_{ref} - d_{fi} I_i. \quad (4)$$

where, I_i and d_{fi} are the DC current and the frequency droop gain of the ith source.

In the proposed method, the amplitude of the injected AC voltage is set as:

$$A = A_{min} + \alpha_A (f_{ref} - f). \quad (5)$$

where, α_A [Volt/Hz] is the rate of change of voltage over the frequency which is calculated as (7). A_{min} and A_{max} are the lowest and highest amplitudes of the injected AC voltage which corresponds to no load and full load conditions, respectively

Increasing the load, decreases the system frequency, and according to (5), higher amplitude of the AC voltage is injected to the system. Substituting (4) into (5) yields to:

$$A = A_{min} + \alpha_A d_{fi} I_i . \qquad (6)$$

Variation of the injected AC voltage amplitude of the sources in terms of load variations is depicted in Fig. 3, and control block diagram of the system is as Fig. 4. The main idea of the superimposed droop frequency scheme is to change the DC voltage levels of the sources using the injected ac voltages. As shown in Fig.4, the injected reactive power is the parameter used for changing the DC voltage level [18-20]. The main idea of the proposed control method is to change the injected AC voltage amplitude to ensure the system stability when the load changes.

$$\alpha_A = \frac{A_{max} - A_{min}}{f_{ref} - f_{min}} = \frac{A_{max} - A_{min}}{d_{fi} I_{i,max}} \qquad (7)$$

To determine A_{min} and A_{max} values, the voltage quality and the stability requirement must be considered. In fact, A_{min} and A_{max} are the lowest and highest amplitudes of the injected AC voltages that keeps the system stable at no load and high load conditions, respectively. It should be noted that A_{min} should not be too small (e.g. lower than 5V in 700V system) due to the difficulties in detecting such small AC voltage in a medium voltage DC microgrid

Using the proposed method, the injected AC voltage amplitude can be set in a wide range which covers both low and high load conditions. Therefore, compared to the conventional superimposed droop frequency method, the overall voltage quality is enhanced at low loads, and the stable operation of the system is guaranteed by generating the higher amplitudes of the AC voltage. Also, with the proposed method, the system maximum loading is increased, and then, the system is able to supply the loads in a wide range. Hence, the proposed method solves the stability problem of conventional superimposed droop frequency method and enhances the overall voltage quality of the system.

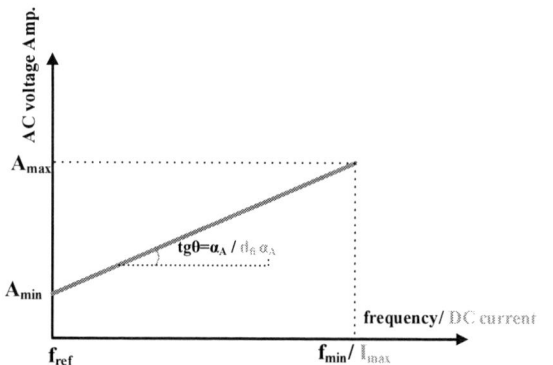

Fig. 3. Variation of injected AC voltage amplitude in terms of load change

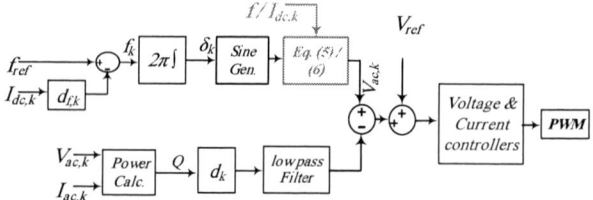

Fig. 4. Control block diagram of the kth source in the proposed control method

IV. SMALL SIGNAL STABILITY ANALYSIS

In this section, the dynamic behavior of the system is investigated with the small signal stability analysis. Substituting (6) into (2), the linear models of the equivalent equations are as:

$$\begin{cases} \Delta Q_1 = -k_1 d_{f1} \Delta I_1 - k_2 \Delta \delta \\ \Delta Q_2 = k_2 d_{f2} \Delta I_2 + k_2 \Delta \delta \end{cases} \qquad (8)$$

$$\text{where,} \begin{cases} k_1 = \dfrac{A_0 \alpha_A}{r_1 + r_2} \sin \delta_0 \\ k_2 = \dfrac{A_0^2}{2(r_1 + r_2)} \cos \delta_0 \end{cases} \qquad (9)$$

In this equation, A_0 is the amplitude of the injected AC voltage at the operating point which is calculated by (6).

Considering Fig.1, in the superimposed droop frequency method, equation (10)- (13) are satisfied at steady state operation of the system as [18]:

$$\delta = \frac{2\pi}{S}(d_{f2} I_2 - d_{f1} I_1) \qquad (10)$$

$$V_i = V_{ref} - d_i Q_i ; i = 1,2 \qquad (11)$$

$$V_i = V_{PCC} + r_i I_i ; i = 1,2 \qquad (12)$$

$$V_{PCC} = R_{load} (I_1 + I_2) \qquad (13)$$

Substituting (8) into the linear model of (11)- (13) yields ΔI_1 and ΔI_2. Then, substituting ΔI_1 and ΔI_2 in the linear model of (10) yields to the system characteristic equation as:

$$s^2 + w_c s + \frac{\beta}{g} = 0 \qquad (14)$$

where,

$$\begin{aligned} g =\; & r_1 r_2 + R_{load} (r_1 + r_2) + k_1 d_2 d_{f2} (R_{load} + r_1) \\ & - k_1 d_1 d_{f1} (R_{load} + r_2) - k_1^2 d_1 d_2 d_{f1} d_{f2} \end{aligned} \qquad (15)$$

and

$$\beta = 2\pi R_{load} k_2 w_c \left[\begin{array}{l} d_{f1}\left(d_1 \times \dfrac{(R_{load}+r_2+k_1 d_2 d_{f2})}{R_{load}}+d_2\right) \\ +d_{f2}\left(d_1+d_2 \times \dfrac{(R_{load}+r_1-k_1 d_1 d_{f1})}{R_{load}}\right) \end{array} \right] .(16)$$

Also, w_c is the cutoff frequency of the applied low pass filter [22-24]. For the system of Fig. 1 with the parameters given in table I, the dominant pole location of the system characteristic equation and its variations in terms of load change is depicted in Fig. 5. As this figure shows, the system is unstable at high load levels since one of the system poles moves toward the imaginary axis. Also, for low load levels, the imaginary part of system dominant poles is too high and the system damping ratio is too small, which makes the system operation undesirable. Furthermore, the effect of increasing the injected AC voltage amplitude for a constant load is depicted in Fig. 6. As shown in this figure, increasing the amplitude of the injected AC voltage mitigates the effect of the load increase. However, high amplitude of the injected AC voltage deteriorates the damping ratio of the poles at low load levels, and low amplitude of injected AC voltage causes instability at high loads.

Using the proposed automatic injection of the AC voltage in this paper, the system dominant poles are kept in an acceptable area for wide range of load variations. Dominant pole placement for the system of Fig. 1 is depicted in Fig. 7 for different load levels. As shown in this figure, dominant poles are placed on real axis which yields a desirable damping condition at low load levels. Also, at high loads, dominant poles have an acceptable distance from the imaginary axis which guarantees the system stability.

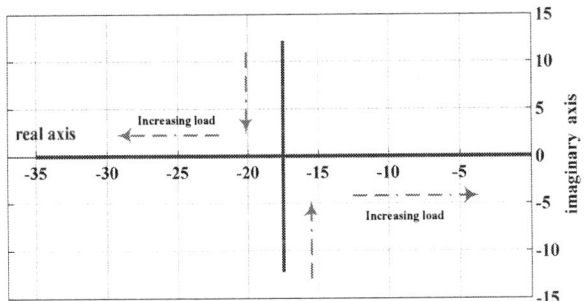

Fig. 5. Effect of increasing the load on the system dominant pole placement for convetional method (A=10V, 15.3<R_{load}<120 Ω)

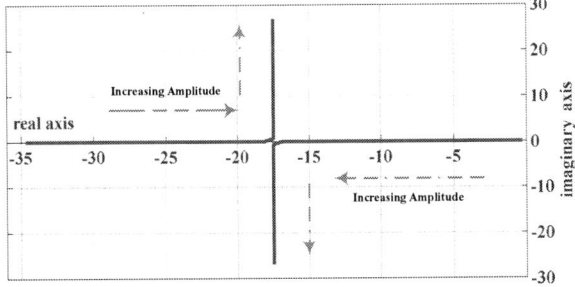

Fig. 6. Effect of increasing the injected AC voltage amplitude on the system dominat pole placement at constant load (R_{load}=66 Ω , α_A=1 , 5<A<15 Volt)

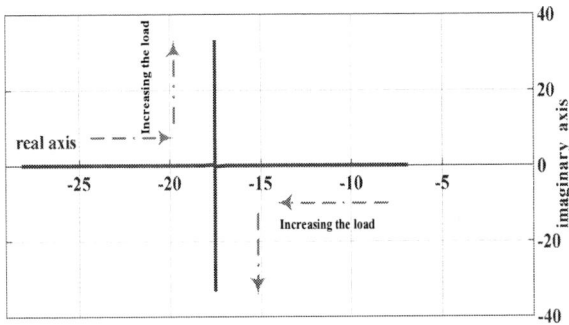

Fig. 7. Dominant pole placement for the proposed method in termd of load variations (A_{min}=5.5 V, α_A=2.7 Volt/Hz , 10<R_{load}<120 Ω)

V. SIMULATIONS

In this section, the performance of the proposed automatic voltage injection method is investigated with simulation results. Fig. 2 shows the test system with the parameters given in Table I. The superimposed droop frequency method is simulated for three different cases as:

Case 1: low amplitude of AC voltage is injected,

Case 2: high amplitude of AC voltage is injected, and

Case 3: the proposed automatic injection of AC voltage is considered.

A. Case1

In this case, injecting low amplitude of AC voltage is analyzed at high load levels to show the unstable operation of the system in this condition. Output current of the sources are depicted in Fig. 8. As this figure shows, system is not stable.

B. Case2

In this case, injecting high amplitude of AC voltage is considered to analyze the system performance at low load levels. Simulation result is provided in Fig. 9. As this figure shows, system does not reach stability.

C. Case3

The performance of the proposed automatic injection of AC voltage method is investigated in this section (A_{min}=5.5 Volt, α_A=2.7 Volt/Hz). In order to validate the performance of the proposed method, the system load is changed from low loads to high loads. Load current profile is shown in Fig. 10. The DC current of the sources are depicted in Fig. 11. The same frequency droop gains are considered to share the load equally among the sources.

The frequency of the injected AC voltages is depicted in Fig. 12 for the two sources. As this figure shows, the injected frequencies of the two sources converges to a constant value. Also, the stable frequency response at different load levels verifies the performance of the proposed method.

The amplitude of the injected AC voltages are depicted in Fig. 13 for the two sources. As shown in this figure, the amplitude of the injected AC voltages increase/decrease as the load increases/decreases.

DC voltage of the sources and their average are depicted in Fig. 14. The average of DC voltages is equal to the microgrid reference DC voltage at different load levels, and the proposed scheme provides a desirable voltage regulation.

978-1-5386-7688-2/19 $31.00 © 2019 IEEE

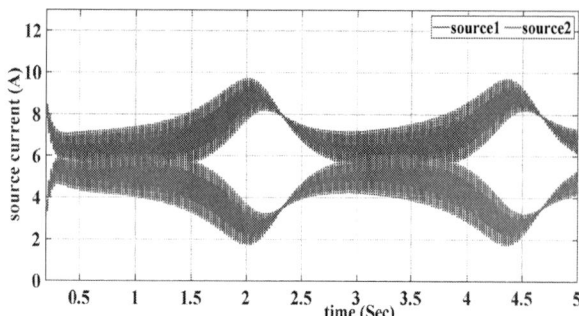

Fig. 8. Instability of the grid with low amplitude of the injected AC voltage at high loads (A=5 V, R_{load}=60 Ω)

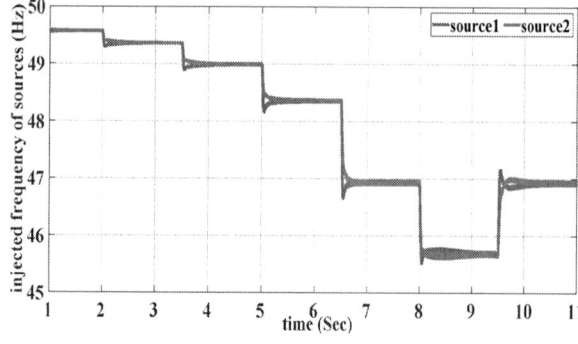

Fig. 12. Source injected frequency

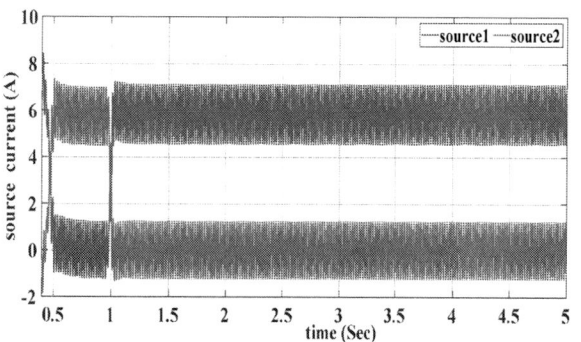

Fig. 9. Undesired operation of the grid by injectig high amplitude of AC voltage at low loads (A=12V , R_{load}=120Ω)

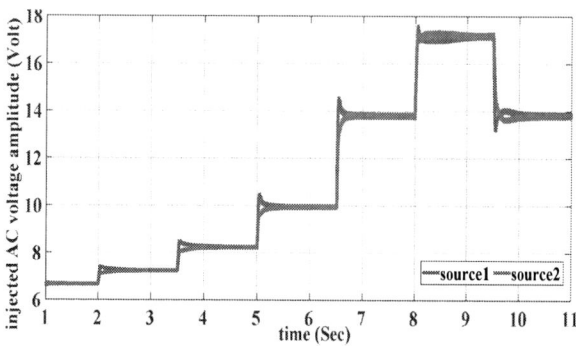

Fig. 13. Amplitude of the injected AC voltage of sources

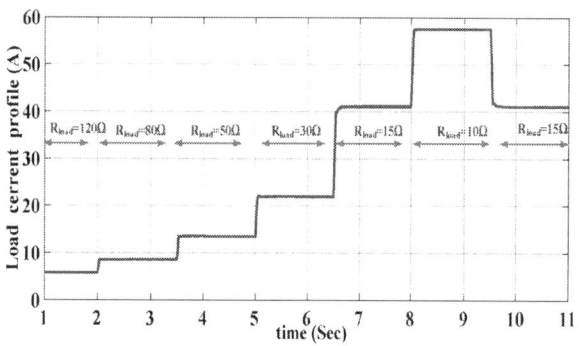

Fig. 10. Load current profile

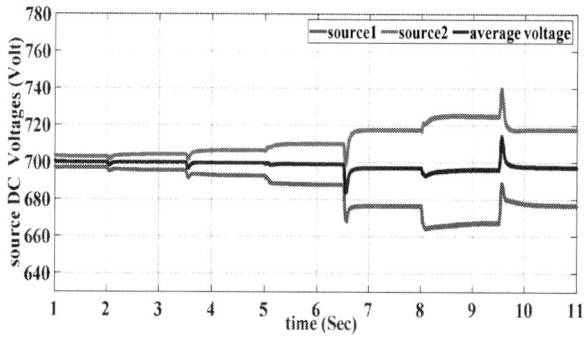

Fig. 14. Microgrid bus DC voltages and average bus voltage

VI. CONCLUSION

This paper proposed an automatic control for regulating the amplitude of the injected AC voltage in order to enhance the performance of the superimposed droop frequency method in DC microgrids. Considering the limited transferable reactive power, the conventional superimposed droop frequency method covers only limited range of load variations. In this paper, the proposed method automatically determines the amplitude of the injected AC voltages proportional to the grid overall load level. Using this method, the system voltage quality and the stability are improved for the wide range of load variations. Also, by increasing the amplitude of the injected AC voltage, the system maximum load is increased accordingly. The proposed method locally controls the sources, and it does not require any

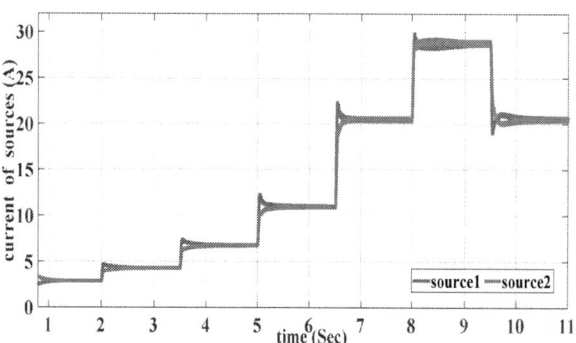

Fig. 11. Source DC currents (A_{min}=5.5 V, $α_A$=2.7 Volt/Hz)

978-1-5386-7688-2/19 $31.00 © 2019 IEEE

communication infrastructure. The performance of the proposed method is verified by different simulation studies in MATLAB/SIMULINK environment, and compared with that of the existing methods in the literature.

REFERENCES

[1] H. M. Xiao, S. X. Ling, L. S. Bo, and Z. Z. Kui, "Transient analysis and control for microgrid stability controller," in Proc. IEEE Grenoble Power Tech., 2013, pp. 1–6.

[2] Yang Han, Hong Li, Pan Shen, Ernane Antˆonio Alves Coelho, and Josep M. Guerrero, "Review of Active and Reactive Power Sharing Strategies in Hierarchical Controlled Microgrids," IEEE Trans. Power Electron., VOL. 32, NO. 3, MARCH 2017.

[3] N. Hatziargyriou, H. Asano, R. Iravani, and C. Marnay, "Microgrids," IEEE Power and Energy Magazine, vol. 5, no. 4, pp. 78–94, 2007.

[4] F. Katiraei, R. Iravani, N. Hatziargyriou, and A. Dimeas, "Microgrids Management," IEEE Power Energy Mag., vol. 6, no. 3, pp. 54–65, 2008.

[5] J. M. Guerrero, J. C. Vasquez, J. Matas, L. G. de Vicuna, and M. Castilla, "Hierarchical control of droop-controlled AC and DC microgrids—A general approach toward standardization," IEEE Trans. Ind. Electron., vol. 58, no. 1, pp. 158–172, 2011.

[6] J. A. P. Lopes, C. L. Moreira, and A. G. Madureira, "Defining control strategies for microgrids islanded operation," IEEE Trans. Power Syst., vol. 21, no. 2, pp. 916–924, 2006.

[7] D. Olivares, A. Mehrizi-Sani, A. Etemadi, C. A. Cañizares, R. Iravani, M. Kazerani, A. H. Hajimiragha, O. Gomis-Bellmunt, M. Saeedifard, R. Palma-Behnke, G. A. Jimenez-Estevez, and N. D. Hatziargyriou, "Trends in microgrid control," IEEE Trans. Smart Grid, vol. 5, no. 4, pp. 1905–1919, 2014.

[8] Seyed Fariborz Zarei, Mohammad Amin Ghasemi, Hossein Mokhtari, Frede Blaabjerg, (2019). "Performance Improvement of AC-DC Power Converters under Unbalanced Conditions". Scientia Iranica, early access, doi: 10.24200/sci.2019.53337.3254.

[9] F. Li, Z. Lin, Z. Qian and J. Wu, "Active DC bus signaling control method for coordinating multiple energy storage devices in DC microgrid," 2017 IEEE Second International Conference on DC Microgrids (ICDCM), Nuremburg, 2017, pp. 221-226, doi: 10.1109/ICDCM.2017.8001048 .

[10] D. Boroyevich, I. Cvetkovic, R. Burgos and D. Dong, "Intergrid: A Future Electronic Energy Network?," in IEEE Journal of Emerging and Selected Topics in Power Electronics, vol. 1, no. 3, pp. 127-138, Sept.2013, doi: 10.1109/JESTPE.2013.2276937.

[11] S. Peyghami, H. Mokhtari, and F. Blaabjerg, "Hierarchical Power Sharing Control in DC Microgrids," in Microgrid: Advanced Control Methods and Renewable Energy System Integration, First Edition, Magdi S Mahmoud, Elsevier Science & Technology, 2017, pp. 63–100.

[12] J. Schonbergerschonberger, R. Duke and S. D. Round, "DC-Bus Signaling: A Distributed Control Strategy for a Hybrid Renewable Nanogrid," in IEEE Transactions on Industrial Electronics, vol. 53, no. 5, pp. 1453-1460, Oct. 2006, doi: 10.1109/TIE.2006.882012.

[13] D. Chen and L. Xu, "Autonomous DC Voltage Control of a DC Microgrid With Multiple Slack Terminals," in IEEE Transactions on Power Systems, vol. 27, no. 4, pp. 1897-1905, Nov. 2012. doi: 10.1109/TPWRS.2012.2189441.

[14] K. Sun, L. Zhang, Y. Xing and J. M. Guerrero, "A Distributed Control Strategy Based on DC Bus Signaling for Modular Photovoltaic Generation Systems With Battery Energy Storage," in IEEE Transactions on Power Electronics, vol. 26, no. 10, pp. 3032-3045, Oct. 2011, doi: 10.1109/TPEL.2011.2127488.

[15] D. Chen and L. Xu, "DC microgrid with variable generations and energy storage," IET Conference on Renewable Power Generation (RPG 2011), Edinburgh, 2011, pp. 1-6, doi: 10.1049/cp.2011.0167.

[16] J. M. Guerrero, J. C. Vasquez, J. Matas, L. G. de Vicuna and M. Castilla, "Hierarchical Control of Droop-Controlled AC and DC Microgrids—A General Approach Toward Standardization," in IEEE Transactions on Industrial Electronics, vol. 58, no. 1, pp. 158-172, Jan. 2011, doi: 10.1109/TIE.2010.2066534 .

[17] X. Lu, J. M. Guerrero, K. Sun and J. C. Vasquez, "An Improved Droop Control Method for DC Microgrids Based on Low Bandwidth Communication With DC Bus Voltage Restoration and Enhanced Current Sharing Accuracy," in IEEE Transactions on Power Electronics, vol. 29, no. 4, pp. 1800-1812, April 2014, doi: 10.1109/TPEL.2013.2266419.

[18] S. Peyghami, H. Mokhtari, P. C. Loh, P. Davari and F. Blaabjerg, "Distributed Primary and Secondary Power Sharing in a Droop-Controlled LVDC Microgrid with Merged AC and DC Characteristics," IEEE Trans. Smart Grid, vol. 9, no. 3, pp. 2284-2294, May 2018 , doi: 10.1109/TSG.2016.2609854.

[19] S. Peyghami, H. Mokhtari, and F. Blaabjerg, " Autonomous Power Management in LVDC Microgrids based on a Superimposed Frequency Droop," IEEE Trans. Power Electron., vol. 33, no. 6, pp. 5341-5350, June 2018, doi: 10.1109/TPEL.2017.2731785.

[20] S. Peyghami, H. Mokhtari and F. Blaabjerg, "Decentralized Load Sharing in a Low-Voltage Direct Current Microgrid With an Adaptive Droop Approach Based on a Superimposed Frequency," IEEE Trans. Power Electron., vol. 5, no. 3, pp. 1205-1215, Sept. 2017, doi: 10.1109/JESTPE.2017.2674300.

[21] S. B. Bashir and A. R. Beig, "An improved voltage balancing algorithm for grid connected MMC for medium voltage energy conversion," International Journal of Electrical Power & Energy Systems, vol. 95, pp. 550-560, 2018.

[22] A. R. Dekka, A. R. Beig, and M. Poshtan, "Comparison of passive and active power filters in oil drilling rigs," in Electrical Power Quality and Utilisation (EPQU), 2011 11th International Conference on, 2011, pp. 1-6, doi: 10.1109/EPQU.2011.6128815.

[23] R. S. R. Chilipi, N. Al Sayari, K. Al Hosani, and A. R. Beig, "Adaptive notch filter based multipurpose control scheme for grid-interfaced three-phase four-wire DG inverter," in 2016 IEEE Industry Applications Society Annual Meeting, 2016, pp. 1-8.

[24] W. Taha, A. R. Beig, and I. Boiko, "Quasi optimum PI controller tuning rules for a grid-connected three phase AC to DC PWM rectifier," International Journal of Electrical Power & Energy Systems, vol. 96, pp. 74-85, 2018.

978-1-5386-7688-2/19 $31.00 © 2019 IEEE

Behavior of a Five-Phase Pentacle Connected IM Operated under One-Phase Fault

Pavel Zaskalicky

Faculty of Electrical Engineering and Informatics
Technical University of Kosice
Letna 9, 042 00 Kosice, Slovakia
Email: pavel.zaskalicky@tuke.sk

Abstract—**The presented paper deals with the five-phase induction motor (IM) having pentacle connected stator winding, which is working under one phase supply failure. Computation of the motor electromagnetic quantities are made using the space vector theory in the complex plane. Analysis is done assuming the motor is supplied by a pulse width modulation (PWM) controlled inverter with sufficiently high modulation frequency. Only the first stator voltage harmonics is taken into consideration. On the base of measured the IM parameters, trajectories of stator and rotor current space vectors were investigated. On their basis, the motor electromagnetic torque ripple waveform for failure supply mode is derived. Finally a possibility to reduce torque ripple in failure state is shown.**

Index Terms—**Complex Fourier series, Five phase inverter, Induction machine, Torque ripple, Pentacle supply voltage, One phase fault, Three phase operation**

I. INTRODUCTION

Five phase asynchronous motors have been used recently in various applications of drive technology. Compared to three phase asynchronous motors they have several advantages. Firstly they have a lower phase current for the same output motor power. This is very advantageous in case of a high power asynchronous motor supplied by a power semiconductor converters. It is known, that five phase motors have a smoother run, comparing to three phase one. This makes them as an ideal drive for a residential home elevators. Another their great advantage is, that they are able to continue operation even in case of one phase supply failure. This often occurs with frequency controlled drives at the sudden overload in the semiconductor devices of the inverter. This feature brings a great advantage for the elevator drives in the high buildings or hospitals, where the failure of elevator drive can cause considerable problems.[14], [15], [16]

Stator winding of the five phase motors can be connected in three ways:[9],

- Star connection.
- Pentagon connection.
- Pentacle connection (pentagram).

Each of the connections has its advantages and disadvantages. In the case of one phase supply failure, a star connection is the most advantageous. On the other side, the pentacle connection gives the highest value of phase voltage, for the same DC inverter input supply.

In the majority cases, supply of the five phase induction motors in the majority cases is provided by a voltage source inverter (VSI). Number of pulse width modulation techniques (PWM) are available to control VSI output voltage. The space vector PWM technique has become the most popular in this days because of its easy digital implementation and better DC utilization compared to the ramp comparison sinusoidal PWM method.

Output voltage control in the semiconductor converters has a significant effect on the electromagnetic torque ripple of the motor. A specific task in multiphase drive system is that generation of certain low-order harmonics in the VSI output voltage can lead to large stator current harmonics since these are restricted only by leakage impedance. [4], [11]

In case of a five phase IM machine with sinusoidal winding distribution which is supplied by voltages containing harmonic components of the third and seventh order, corresponding stator current harmonics flow freely through the machine and their magnitude is be limited by a stator leakage impedance only. [6], [10], [12]

On the contrary, five phase IM with concentrated windings requires and it is desirable to utilize the third harmonic stator current injection to enhance the torque production.

II. FIVE PHASE VSI INVERTER MODELLING

The five phase motors need for their operation a supply from semiconductor converter, which is able to generate a five phase supply voltages. The power circuit topology of the five phase voltage source inverter (VSI) is shown in Fig.1. [18], [19],

VSI consists of parallel connection of five transistor legs. It is supplied by a constant voltage source consisting of an isolated DC-source and a capacitive DC-link. Each leg contains two IGBT transistors with anti-parallel connected freewheeling diodes used to ensure a negative current path through the switches. Inverters output terminals are numbered as $1 - 2 - 3 - 4 - 5$. [13]

Assume that the inverter DC input voltage is constant for any supply current.

Next, let's suppose that inverters output voltage is controlled by a PWM with very high modulation frequency ($f_m > 10\,\mathrm{kHz}$). On the base of this assumption we can suppose that the inverter output voltages is ideal sinusoidal.[3]

978-1-5386-7688-2/19 $31.00 © 2019 IEEE

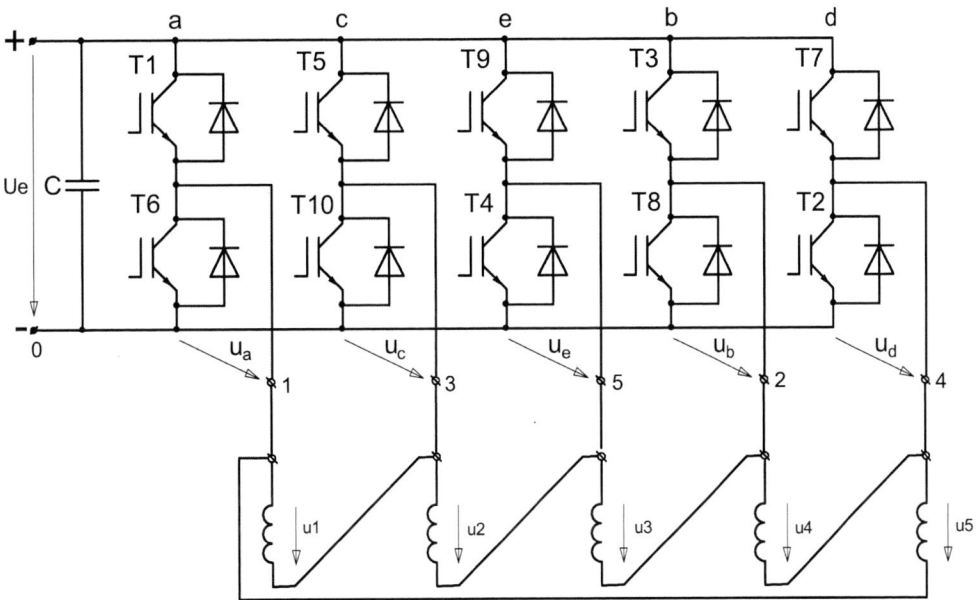

Fig. 1. Five-phase bridge connected VSI

The transistors in the first inverters leg is controlled to obtain output leg voltage in the form:[5], [8],

$$\mathbf{u}_a = \mathbf{u}_{dc} + \mathbf{u}_{01} = \frac{U_e}{2} + r\frac{U_e}{2}e^{j\omega t} \qquad (1)$$

where, $\mathbf{u}_{dc}$ is DC and $\mathbf{u}_{01}$ is AC voltage component, r is voltage control coefficient $r \in \langle 0,1 \rangle$ and $\omega = 2\pi f$ is angular frequency.

Leg voltages are measured between the leg nodes and negative rail of the DC supply.

The AC voltage components in other legs are shifted gradually by $\frac{2\pi}{5}$. This shift is mathematically expressed by the complex shifting factor $\mathbf{a} = e^{j\frac{2\pi}{5}}$

$$\mathbf{u}_{02} = \mathbf{a}\,\mathbf{u}_{01} \qquad \mathbf{u}_{03} = \mathbf{a}^2\,\mathbf{u}_{01}$$

$$\mathbf{u}_{04} = \mathbf{a}^3\,\mathbf{u}_{01} \qquad \mathbf{u}_{05} = \mathbf{a}^4\,\mathbf{u}_{01} \qquad (2)$$

The motor phase voltages of the pentacle connected winding in the stator are given by a difference between two legs voltages as follows

$$\mathbf{u}_1 = \mathbf{u}_a - \mathbf{u}_c = \mathbf{u}_{01} - \mathbf{u}_{03} = (1 - \mathbf{a}^2)\mathbf{u}_{01}$$
$$\mathbf{u}_2 = \mathbf{u}_c - \mathbf{u}_e = \mathbf{u}_{03} - \mathbf{u}_{05} = (\mathbf{a}^2 - \mathbf{a}^4)\mathbf{u}_{01}$$
$$\mathbf{u}_3 = \mathbf{u}_e - \mathbf{u}_b = \mathbf{u}_{05} - \mathbf{u}_{02} = (\mathbf{a}^4 - \mathbf{a})\mathbf{u}_{01} \qquad (3)$$
$$\mathbf{u}_4 = \mathbf{u}_b - \mathbf{u}_d = \mathbf{u}_{02} - \mathbf{u}_{04} = (\mathbf{a}^2 - \mathbf{a}^3)\mathbf{u}_{01}$$
$$\mathbf{u}_5 = \mathbf{u}_d - \mathbf{u}_a = \mathbf{u}_{04} - \mathbf{u}_{01} = (\mathbf{a}^3 - 1)\mathbf{u}_{01}$$

Fig. 2 depicts the phase voltages formed in the pentacle stator winding connection.

To simplify the calculation of AC motor quantities, it is very advantageous to employ space vectors theory. It's utilization

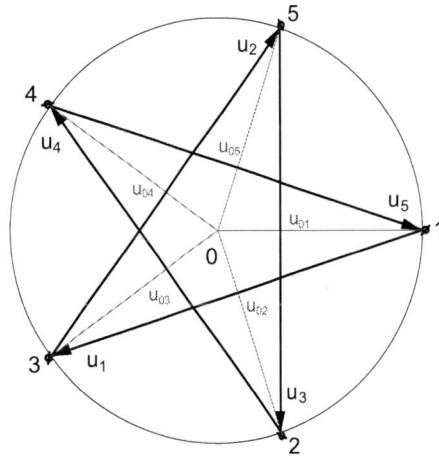

Fig. 2. Pentacle connection voltage phasors

very simplifies analyzes of multi-phase electric systems too. The term "space" originally stands for the two-dimensional complex plane, in which the multi-phase quantities are transformed.

The transformation of space space vector can be directly derived from the sum of voltage phasors. Based on the equation (3), voltage space vector transformation is thus defined as: [7]

978-1-5386-7688-2/19 $31.00 © 2019 IEEE

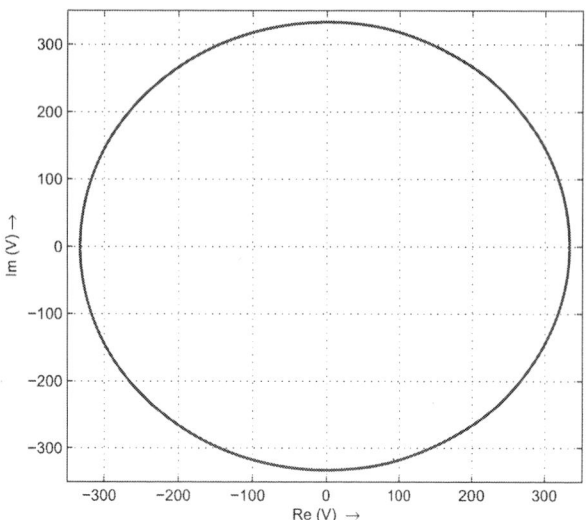

Fig. 3. Voltage space vector trajectory

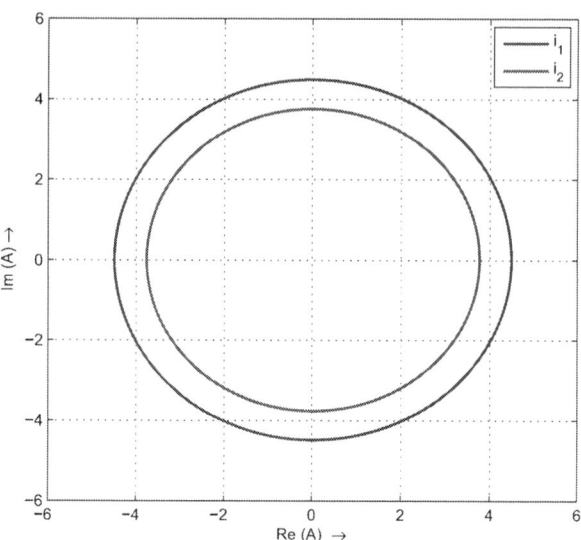

Fig. 5. Stator and rotor current space vector trajectories

$$\underline{u} = \frac{2}{5}\Big(\mathrm{Re}(\mathbf{u}_1) + \mathbf{a}_1\,\mathrm{Re}(\mathbf{u}_2) + \\ \mathbf{a}_1^2\,\mathrm{Re}(\mathbf{u}_3) + \mathbf{a}_1^3\,\mathrm{Re}(\mathbf{u}_4) + \mathbf{a}_1^4\,\mathrm{Re}(\mathbf{u}_5)\Big) \tag{4}$$

where, $\mathbf{a}_1 = e^{j2\frac{2\pi}{5}}$ is space shifting factor. The pentacle stator winding connection forms the second order multi phase system.

Coefficient $\frac{2}{5}$ keeps constant the magnitude of phasors during transformation.

Fig. 3 shows the space vector trajectory of the five phase voltage system. The trajectory was calculated on the base of equation (4) for the inverter's DC input value 350 V.

For a symmetric five phase harmonic voltage system, the space vector trajectory is a circle as shown in Fig. 3.

III. Current Space Vector Calculation

For stator and rotor current space vectors calculation, a classical one phase equivalent circuit of induction machine is advantageously used (Fig. 4).[1], [17]

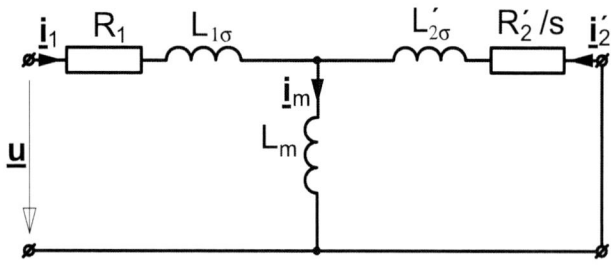

Fig. 4. One phase IM equivalent circuit

Motor parameters used for next calculation are listed in Appendix.

Referred to the equivalent circuit above, the following equation for the stator current space vector is valid

$$\underline{i}_1 = \frac{\underline{u}}{R_1 + \left[j\omega L_{1\sigma} + \frac{j\omega L_m(R_2'/s + j\omega L_{2\sigma}')}{R_2'/s + j\omega(L_m + L_{2\sigma}')}\right]} \tag{5}$$

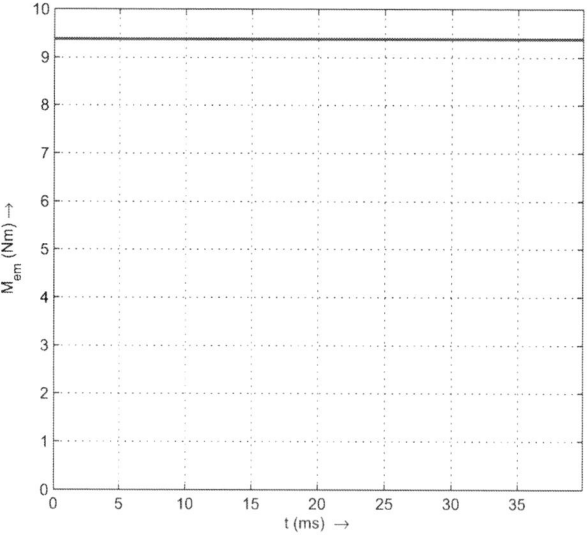

Fig. 6. Electromagnetic torque waveform

where, $s = \frac{\omega - p\omega_m}{\omega}$ is the motor slip and $\omega = 2\pi f$ is an angular frequency, ω_m is a motor speed.

The rotor current space vector can by determined using Thevenin theorem. For the Thevenin voltage the following equation applies

$$U_{eth} = U_e \frac{\omega L_m}{\sqrt{R_1^2 + (\omega L_{1\sigma} + \omega L_m)^2}} \qquad (6)$$

The Thevenin impedance is given by

$$\mathbf{Z}_{th} = R_{th} + j\omega L_{th} = \frac{j\omega L_m(R_1 + j\omega L_{1\sigma})}{R_1 + j\omega(L_{1\sigma} + L_m)} \qquad (7)$$

Because $L_m \gg L_{1\sigma}$ and $\omega(L_m + L_{1\sigma}) \gg R_1$, the Thevenin resistance and inductance are approximately given by

$$R_{th} \approx R_1 \left(\frac{L_m}{L_{1\sigma} + L_m} \right) \qquad (8)$$

$$L_{th} \approx L_{1\sigma}$$

Then for the rotor current space vector following equation is valid

$$\mathbf{i}_2' = - \frac{\frac{U_{th}}{U_e} \mathbf{u}}{R_2'/s + R_{th} + j\omega(L_{th} + L_{2\sigma}')} \qquad (9)$$

Assume, the motor operates at rating speed, the sleep ($s = 0.03$) and frequency ($f = 50\,\text{Hz}$). Fig. 5 depicts the stator and rotor current space vector trajectories which were calculated on the base of equations (5) and (9).

The electromagnetic torque is calculated on the base of the following equation

$$M_{em} = \frac{5}{2} p L_m \, Im \left(\mathbf{i}_1 \mathbf{i}_2^* \right) \qquad (10)$$

In the Fig. 6 there is shown the time course of electromagnetic torque, calculated on the base of equation (10). In case of a failure-free state the motor electromagnetic torque is constant at all times.

IV. ONE PHASE FAILURE OPERATION

Let us assume fault on the terminal "2" - the transistors ($T3 - T8$) ruptured and $\mathbf{u}_{02} = 0$. Subsequently across the phases "3" and "4" there appears the voltage equal to the difference between leg voltages $\mathbf{u}_{04}$ and $\mathbf{u}_{05}$, as seen in the Fig. 7.

Assuming symmetry windings in each phase, the voltage is evenly distributed between the phases "3" and "4".

$$\mathbf{u}_3 = \mathbf{u}_4 = (\mathbf{a}^4 - \mathbf{a}^3) r \frac{U_e}{4} e^{j\omega t} \qquad (11)$$

On the base of equation (4) there is possible to calculate voltage space vector trajectory. This trajectory is no longer circular, but elliptical, as depicted in Fig. 8.

From equations (5) and (9) the trajectories of stator and rotor current space vector were calculated. They are shown in Fig. 9. The calculation was made for supply frequency $f = 50\,\text{Hz}$ and nominal motor torque $M_p = 9.5\,\text{Nm}$. Motor speed has dropped approximatively by $100\,\text{rev/min}$ to the value $2814\,\text{rev/min}$, corresponding to $s = 0.062$.

Fig. 10 shows calculated electromagnetic torque waveform. This one is strong pulse with second harmonic of the supply voltages frequency. In the calculated waveform, the electromagnetic moment changes within the range $9,5 \pm 8\,\text{Nm}$.

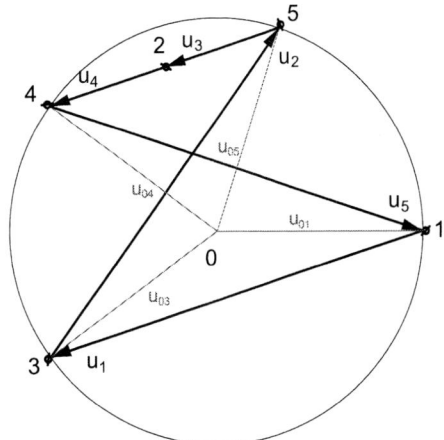

Fig. 7. One phase rupture voltage phasors

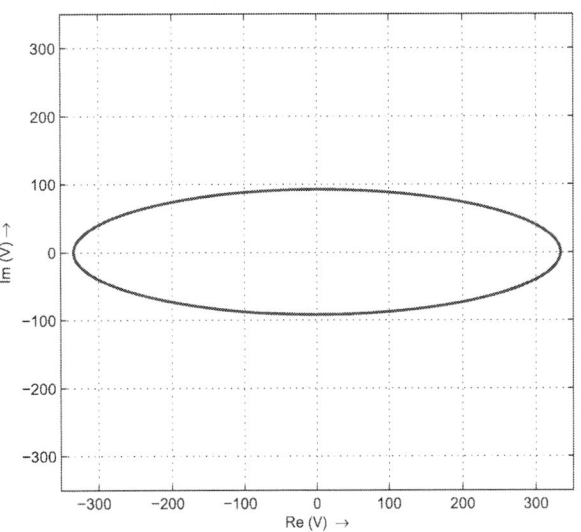

Fig. 8. Voltage space vector trajectory in case of one phase rupture

V. THREE PHASE OPERATION

In case of one phase failure, as shown above, there is a possibility to smooth the disturbed waveform of the electromagnetic torque. For this, it is necessary to eliminate the failure in phases "3" and "4". This can be accomplished by shifting leg voltages u_{04} and u_{05}, as shown in Fig. 11. In order to get the trajectory of voltage space vector circular, we must shift the voltages u_{01} and u_{03}.

Let us define the fault shifting factor

$$\mathbf{a}_f = e^{\frac{j2\pi}{10}} \qquad (12)$$

then the shifted leg voltages take the form

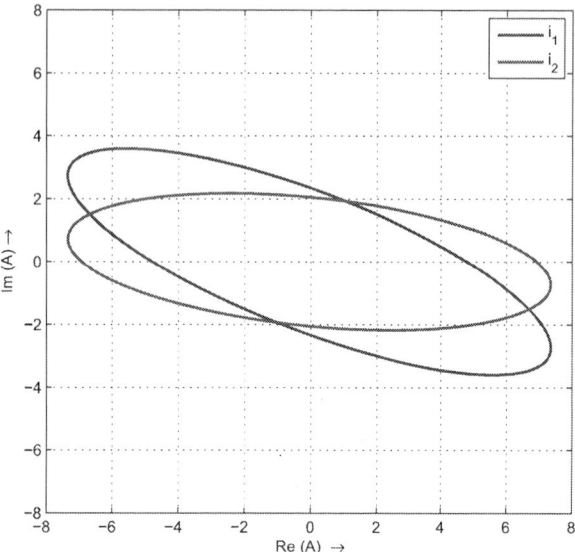

Fig. 9. Current space vectors trajectories in case of one phase rupture

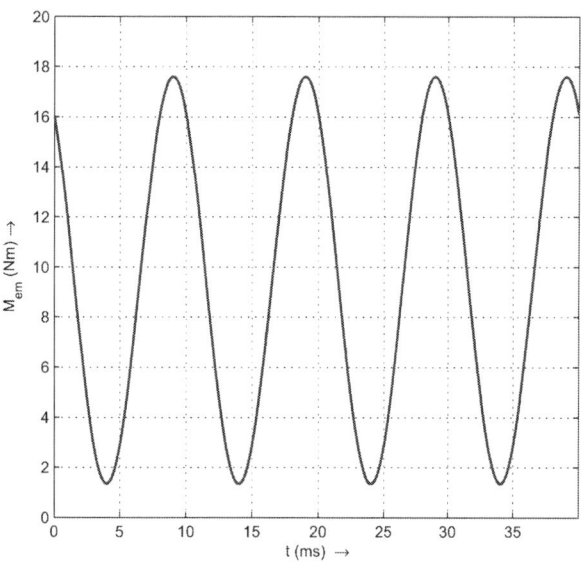

Fig. 10. Electromagnetic torque waveform in case of one phase rupture

$$\mathbf{u}_{01}' = \mathbf{a}_f^{-1}\,\mathbf{u}_{01} \qquad \mathbf{u}_{03}' = \mathbf{a}_f\,\mathbf{u}_{03}$$
$$\mathbf{u}_{04}' = \mathbf{a}_f^{-1}\,\mathbf{u}_{04} \qquad \mathbf{u}_{05}' = \mathbf{a}_f\,\mathbf{u}_{05} \qquad (13)$$

Then, for the motor phase voltages we can write equations

$$\mathbf{u}_1 = \mathbf{u}_{01}' - \mathbf{u}_{03}' \qquad \mathbf{u}_2 = \mathbf{u}_{03}' - \mathbf{u}_{05}'$$
$$\mathbf{u}_5 = \mathbf{u}_{04}' - \mathbf{u}_{01}' \qquad (14)$$

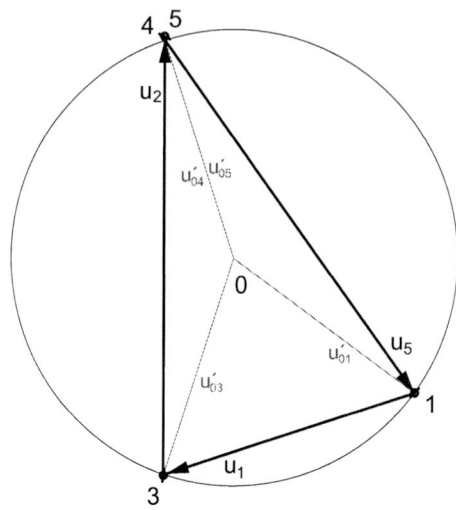

Fig. 11. Three phase voltage phasors

The voltage space vector for three phase operation is

$$\underline{\mathbf{u}} = \frac{2}{3}\mathrm{Re}(\mathbf{u}_1) + \mathbf{a}_1\,\mathrm{Re}(\mathbf{u}_2) + \mathbf{a}_1^4\,\mathrm{Re}(\mathbf{u}_5) \qquad (15)$$

Fig. 12 shows calculated voltage space vector trajectory for three phase operation of the five phase IM. This is again circular, but with reduced voltage.

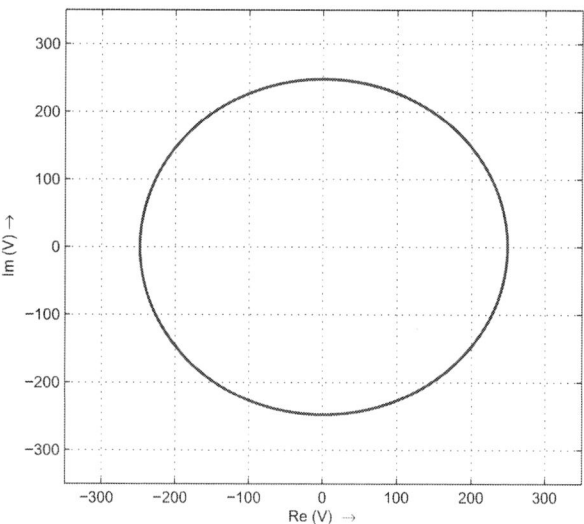

Fig. 12. Three phase voltage space vector trajectory

By eliminating two motor phases and by switching to three phase operation the motor lost 40 % of power capacity. In addition, the motor operates with reduced excitation (three phases work only). If the motor has to operate at a rating torque, it is necessary to reduce the speed due to current overload.

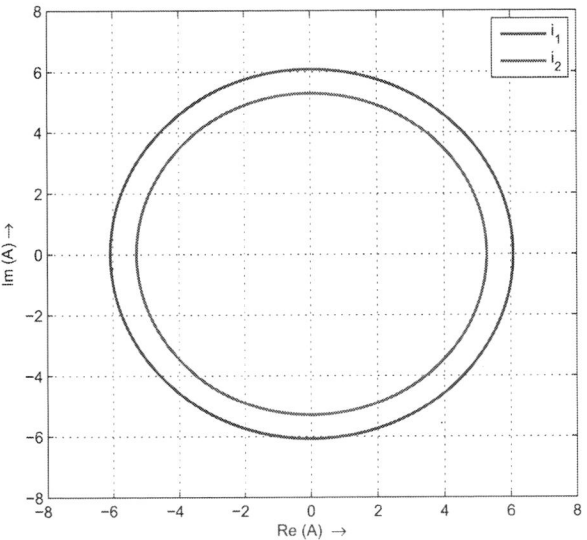

Fig. 13. Three phase current space vector trajectories

Fig. 13 shows stator and rotor space vector trajectories. The motor operates in three phase mode with nominal load (M_p = 9.5 Nm). Supply voltage frequency is reduced to f = 30 Hz and the motor is working at speed n = 1690 rev/min. Fig. 13 also shows that the motor is slightly overloaded. The time course of the electromagnetic moment is again smooth, as shown in Fig. 6.

VI. CONCLUSION

The paper deals with the five-phase pentacle connected winding of an induction motor (IM), which is working under one phase supply failure. Computation of the motor electromagnetic quantities are made using the space vector theory in the complex plane. Assuming, the motor is supplied by a PWM controlled inverter with a sufficiently high modulation frequency, only the first stator voltage harmonics are taken in consideration. On the base of measured IM parameters, the trajectories of a stator and rotor current space vectors were investigated. Based of them, the motor electromagnetic torque ripple waveform for faulty supply working mode has been derived. It is shown, that in case of the fault state there is possible switch motor to a three-phase operation state without electromagnetic torque ripple. In any case, the motor loses its original performance power, but it is capable to work in smooth operation.

APPENDIX

Five phase induction motor parameters:
P_n = 3 kW; U_n = 5 × 230 V/50 Hz;
n_n = 2910 rev/min.; p = 1;
R_1 = 3.778 Ω; R_2' = 2.498 Ω;
L_m = 0.436 H; $L_{1\sigma}$ = 6.83 mH; $L_{2\sigma}'$ = 11.88 mH;

ACKNOWLEDGMENT

The author would like to thank Slovak Research and Development Agency for financial support. Paper was made under the contract No: APVV-16-0270.

REFERENCES

[1] J. Chatelain, *Machines électriques*, Vol. X, Edition Georgi, 1983
[2] J. Machowski, W. Bialek, J.R. Bumby, *Power System Dynamics: Stability and Control*. John Wiley & Sons, 2011
[3] P. Špánik, B. Dobrucký, M. Frivaldský, et al. *Measurement of swtitching losses in power transistor structure*, Elektronika ir Elektrotechnika, No. 2, pp. 75-78, 2008
[4] P. Zhao, T.A. Lipo, *Space vector PWM control of dual-phase induction machine using vector space decomposition*, IEEE Trans. on Industry Applications, Vol. 31, No. 5, pp. 1100-1109, 1995
[5] M. Zaskalicka, P. Zaskalicky, M. Benova, M.A.R. Abdalmula, B. Dobrucky, *Analysis of complex time function of converter output quantities using complex Fourier transform/series*, Communications - Scientific Letters of the University of Zilina, Vol. 12, No. 1, pp. 23-30, 2010
[6] A. Iqbal, E. Levi, *Space Vector PWM Techniques for Sinusoidal Output Voltage Generation with Five-Phase Voltage Source Inverter*, Electric Power Components and Systems, No. 34, pp. 119-140, 2006
[7] H.M. Ryu, S.K. Sul, *Multiple d-q spaces concept for multi-phase AC motor drive*, ICPE04, pp. 670-674, 2004
[8] B. Dobrucký, O.V. Chernoyarov, M. Marčoková, *Computation of the total harmonic distortion of impulse system quantities using infinite series*, APLIMAT 2015, 14th conference on applied mathematics, Proceedings pp. 213-220, 2015
[9] M. Chomat, L. Schreier, *Effect of Stator Winding Configuration on Operation of Converter Fed Five-Phase Induction Machine*, Electric Drives and Power Electronics (EDPE), The High Tatras, pp. 488-496, 21-23 Sept. 2015
[10] H.M. Kim, N.H. Kim, W.S. Baik, *A Five-Phase Induction Motor Speed Control System Excluding Effects of 3rd Current Harmonics Component*, Journal of Power Electronics, Vol. 11, No. 3, pp. 294-303, 2011
[11] P. Brandstetter, M. Kuchar, O.Skuta, *Implementation of RBF Neural Network in Vector Control Structure of Induction Motor*, International Review of Electrical Engineering, Vol. 9, No. 4, 2014
[12] E. Levi, *Multiphase electric machines for variable-speed applications*, IEEE Trans. on Ind. Electronics, Vol. 55, No. 5, pp. 1893-1909, 2008
[13] P. Zaskalicky, *Mathematical Model of a Five-phase Voltage-source PWM Controlled Inverter*, Electrical Engineering - Archiv für Elektrotechnik, SPRINGER, No. 99, pp. 1179-1184, 2017
[14] M. Trabelsi, K.N. Ngac, E. Semail, *Real-Time Switches Fault Diagnosis Based on Typical Operating Characteristics of Five-Phase Permanent-Magnetic Synchronous Machines*, IEEE Trans. on Ind. Electronics, Vol. 63, No. 8, pp. 4683-4694, 2016
[15] J.O. Estima, A.J. Marques Cardoso, *A new algorithm for real-time multiple open-circuit fault diagnostic in voltage-fed PWM motor drives by the reference currents errors*, IEEE Trans. on Ind. Electronics, Vol. 28, No. 5, pp. 3496-3505, 2013
[16] B. Dobrucký, M. Marčoková, M. Pokorný, R. Šul, *Using Orthogonal and Discrete Transform for Single-Phase PES Transients - A New Approach*, Proceedings of the 27th IASTED international conference on modeling identification and control. Innsbruck, 2008
[17] D. Balara, J. Timko, J. Žilková, M. Lešo, *Neural networks application for mechanical parameters identification of asynchronous motor*, Neural Network World, Vol. 27, No. 3, pp. 259-315, 2017
[18] M. Frivaldský, B. Dobrucký, M. Praženica, J. Koscelník, *Multi-tank resonant topologies as key design factors for reliability improvement of power converter for power energy applications*, Electrical Engineering - Archiv für Elektrotechnik, SPRINGER, Vol. 97, No. 4, pp. 287-302, 2015
[19] M. Frivaldský, B. Dobrucký, M. Pridala, *Analysis of LCLC DC-DC resonant converter in steady state operation*, 42nd Annual Conference of the IEEE-Industrial-Electronics-Society (IECON), Firenze, October 24-27, 2016

Calculating frequency responses of synchronous machines using MIMO transfer functions

Matthias Kalla*, Olga Korolova*, Alexander Neufeld[†], Lutz Hofmann[†], Bernd Ponick*

* *Institute for Drive Systems and Power Electronics, Leibniz University Hannover*, Hannover, Germany

[†]*Institute of Electric Power Systems, Leibniz University Hannover*, Hannover, Germany

Abstract—This paper addresses the issue of synchronous machines performing at an operating point with small oscillations of the supply voltage. When the stator voltage spectrum consists of more than one frequency component, the electrical machine has a response with this distorting frequency. To calculate the behaviour of the machine, a Multiple Input Multiple Output (MIMO) transfer function is proposed, which also considers oscillations of the torque and the field winding voltage for electrically excited synchronous machines. With this transfer function, possible resonance effects can be studied. This is suitable for harmonic studies in power grids, where a frequency-dependent model of electrical generators is needed. In addition, oscillations of the torque are analysed and a link to the complex torque coefficient is discussed.

Index Terms—transfer functions, linearisation, synchronous machines, resonances

I. NOMENCLATURE

A. Formal conventions

a	Real instantaneous value
A	Root mean square (RMS) value
$\mathbf{a}, \mathbf{A}$	Matrix
$\vec{a}, \vec{A}$	Vector
$\underline{A}$	Complex RMS value

B. Formula symbols

$\mathbf{E}$	Identity matrix
f	Frequency
$G(s)$	Common transfer function
I	Current
J	Moment of inertia
L	Inductance
m	Torque
M	Mutual inductance
n	rotational speed
p	Number of pole pairs
R	Resistance
S	Apparent power
U	Voltage
δ	Load angle
ω	Angular frequency
Ω	Angular velocity

C. Indices

1	Stator value
2	Field winding value
3	Damper cage value
0	0 component
d	d component
q	q component
σ	Leakage value
m	Mechanical
N	Rated value
OP	Operating point

II. INTRODUCTION

When analysing power grids in search of resonant frequencies, electrical machines like synchronous generators have to be considered. Integrating generator models in the model of a power grid for methods like the Resonance Mode Analysis (RMA) from [1], the equations of the generator model have to be transformed into a system that is compatible with the power grid model. Usually, the power grid model is formulated in symmetrical components and consists of a grid impedance or admittance matrix, as in [2] and [3], where a node-oriented representation is proposed which leads to an admittance matrix. The admittances are frequency-dependent and represent the positive sequence components. The task at hand is to implement the generator model in this representation considering the frequency dependency. For this purpose, the response of a generator to a given frequency has to be calculated. Solving the differential equations for a voltage input with the main grid frequency and a second, distorting frequency in the time domain would yield the results needed. But solving in the time domain results in long simulation durations and generates only one time domain result for the given frequency. Therefore, the aim should be a solution in the frequency domain, requiring a transfer function, which needs to be linearised. In [4], a possible linearised model is given, but not in a generalized form. The advantage of the form presented in this paper is that it can be easily expanded with further linear effects, e.g. in order to consider the skin effect in the damper cage or current displacement.

The transfer function will represent the generator as a MIMO system, where the input variables are the stator voltages, the drive torque and the dc rotor field voltage. The outputs of the transfer function are the angular velocity and the currents, which are given as complex values that describe the amplitude and the phase of the oscillations with the frequency of the input. In this way, it is possible to calculate a positive-sequence current as phasor or as a function of time. Furthermore, all output signals are easily accessible. This procedure needs only an evaluation of a transfer function at a given frequency and not a time-consuming solving of differential equations in the time domain. This method can further be used for the calculation of the complex torque coefficient from [5]. Another application is the estimation of power losses in the damper cage.

The generator model used as an example is that of an electrically excited synchronous machine. Due to the nature of how the model is formulated, other machine types can be calculated, as well.

III. ACQUIRING THE STATE-SPACE MODEL

A. Modelling the synchronous machine

The foundation of the transfer function required consists of a linearised non-linear model of the synchronous machine in the time domain. The non-linear model used here is derived from [6] and describes the differential equations in the rotating d/q reference frame and covers transient effects. The model considers only the spatial fundamental field. Saturation and current displacement are currently not taken into account. These effects are not considered in this paper as the main goal is to present the general applicability of the method.

The linearised model is acquired by the first-order Taylor expansion around the rated point of operation. Though there are infinite eligible points to linearise at, the rated point is the best choice for the validation of the method used. The values of the operating point can be derived from the phasor diagram from [7].

The linearised electrical equation results in

$$\dot{\vec{\Delta i}} = \mathbf{L}^{-1}(-\mathbf{R}\vec{\Delta i} - \mathbf{L}_{\mathrm{rot}}\omega_{\mathrm{OP}}\vec{\Delta i} - \\ \mathbf{L}_{\mathrm{rot}}\vec{i}_{\mathrm{OP}}\Delta\omega + p\vec{u}_{\mathrm{OP}}\Delta\varphi + \vec{\Delta u}_{\mathrm{in}}), \tag{1}$$

where $\mathbf{L}$ and $\mathbf{R}$ have the dimension 6x6 for the considered generator. After linearisation, resistance matrix $\mathbf{R}$ and inductance matrices $\mathbf{L}$ and $\mathbf{L}_{\mathrm{rot}}$ remain the same and can be found in the appendix. The equation of motion results in

$$\Delta\dot{\Omega}_{\mathrm{m}} = \frac{1}{J}\left(\frac{3}{2}(\vec{i}_{\mathrm{OP}}^{\mathrm{T}}\mathbf{L}_{\mathrm{rot}}\vec{\Delta i} + \vec{\Delta i}^{\mathrm{T}}\mathbf{L}_{\mathrm{rot}}\vec{i}_{\mathrm{OP}}) - \Delta m_{\mathrm{drive}}\right), \tag{2}$$

where Ω is the derivative of the mechanical rotor angle

$$\Delta\dot{\varphi}_{\mathrm{m}} = \Delta\Omega_{\mathrm{m}}. \tag{3}$$

Having linearised the non-linear equations of the transient model, the prerequisites for a state-space model are met. The first step is to formulate the vector of the state variables[1] as

$$\vec{x} = \begin{bmatrix} i_{\mathrm{1d}} & i_{\mathrm{1q}} & i_{10} & i_2 & i_{\mathrm{3d}} & i_{\mathrm{3q}} & \Omega_{\mathrm{m}} & \varphi_{\mathrm{m}} \end{bmatrix}^{\mathrm{T}}. \tag{4}$$

They consist of the stator d/q and zero current components, the dc current of the field winding, the d/q components of

[1]For the purpose of readability, the Δ that should stand before the variables is left out from this point on.

the damper cage current, the rotor angular frequency and the rotor angle. As mentioned earlier, the model will be a MIMO system, so the input is a vector. The possible inputs are the stator d/q and zero voltages, the dc field winding voltage and the drive torque. Consequently, the input vector is

$$\vec{u} = \begin{bmatrix} u_{\mathrm{1d}} & u_{\mathrm{1q}} & u_{10} & u_2 & m_{\mathrm{drive}} \end{bmatrix}^{\mathrm{T}}. \tag{5}$$

Having set the input and the state variables, the state-space equation

$$\dot{\vec{x}} = \mathbf{A}\vec{x} + \mathbf{B}\vec{u} \tag{6}$$

can be formulated. The system matrix $\mathbf{A}$ can be derived from equations (1), (2) and (3). This results in the 8x8 dimensional matrix

$$\mathbf{A} = \begin{bmatrix} \mathbf{L}^{-1}(-\mathbf{R} - \mathbf{L}_{\mathrm{rot}}\omega_{\mathrm{OP}}) & \mathbf{L}^{-1}\mathbf{L}_{\mathrm{rot}}\vec{i}_{\mathrm{OP}} & p\mathbf{L}^{-1}\vec{u}_{\mathrm{OP}} \\ \frac{3}{2J}\left((\mathbf{L}_{\mathrm{rot}}\vec{i}_{\mathrm{OP}})^T + \vec{i}_{\mathrm{OP}}^T\mathbf{L}_{\mathrm{rot}}\right) & 0 & 0 \\ 0 & 1 & 0 \end{bmatrix}. \tag{7}$$

The input matrix $\mathbf{B}$ is calculated as

$$\mathbf{B} = \begin{bmatrix} \mathbf{L}^{-1} \cdot \begin{bmatrix} 1 & 0 & 0 & 0 & 0 \\ 0 & 1 & 0 & 0 & 0 \\ 0 & 0 & 1 & 0 & 0 \\ 0 & 0 & 0 & 1 & 0 \\ 0 & 0 & 0 & 0 & 0 \\ 0 & 0 & 0 & 0 & 0 \end{bmatrix} \\ 0 & 0 & 0 & 0 & \frac{1}{J} \\ 0 & 0 & 0 & 0 & 0 \end{bmatrix} \tag{8}$$

with a dimension of 8x5, completing the state-space equation. With the goal to build a transfer function, the output equation

$$\vec{y} = \mathbf{C}\vec{x} + \mathbf{D}\vec{u} \tag{9}$$

has to be calculated. The variables performing as outputs are the state variables. knowing this, the output matrix $\mathbf{C}$ becomes an 8x8 identity matrix and the feed through matrix $\mathbf{D}$ becomes an 8x5 zero matrix.

The last step is obtaining the transfer function out of the state-space model. The conversion of the state-space model into a transfer function is given in [8] and yields

$$\mathbf{G}(s) = \mathbf{C}(s\mathbf{E} - \mathbf{A})^{-1}\mathbf{B}. \tag{10}$$

B. Using the transfer function

In order to use the derived transfer function for the estimation of the generator state-space variable responses, the following recommendations should be considered.

Because the basic non-linear model is formulated in the rotating d/q reference frame, so is the transfer function. Hence, if the input signal is *not* in the used grid reference frame, the signal has to be converted into this reference frame. This applies to the stator voltage distortion, but not the torque and the field winding voltage, because they both originate from the rotating reference frame already, while the stator voltage distortion originates from the fundamental frequency reference frame. The input signal has to be a harmonic oscillation, that can be transformed into a phasor. If not, the signal has to

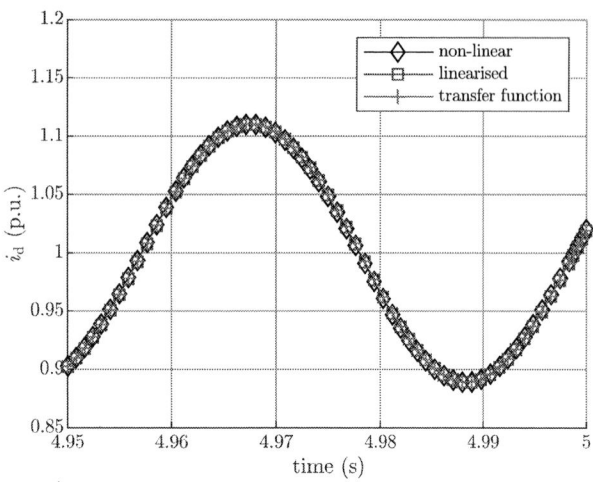

Fig. 1. i_d of generator A for a distorting frequency of 26 Hz

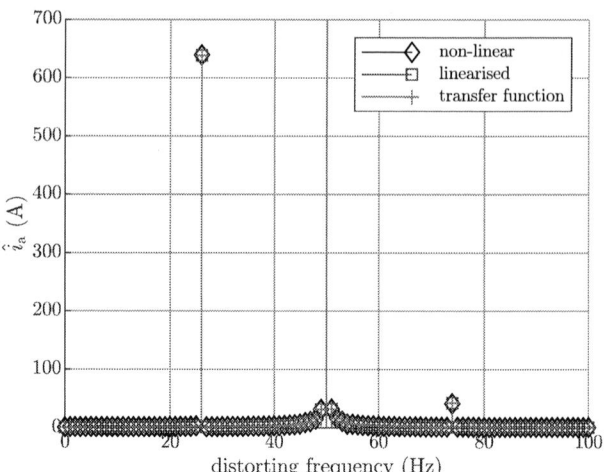

Fig. 2. Fourier analyses of the current i_a of generator A for a distorting frequency of 26 Hz

be deconstructed into its Fourier components. In the case of multiple frequencies, the transfer function has to be evaluated for each frequency with the respective amplitude and phase, but can be added for the full response afterwards, because the principle of superposition is eligible for the MIMO transfer function.

Like the input signal, the output signal is in the form of a vector of phasors. To create a signal in the time domain, the real part has to be calculated as

$$\vec{x}(t) = \mathrm{Re}\{\underline{\mathbf{G}}(\mathrm{j}\omega) \cdot \underline{\vec{u}}(t)\}. \tag{11}$$

When an external oscillation is applied to the system, the rotor angle, which is used for the d/q transformation of the stator voltages, oscillates, too. The rotor angle has the form

$$\varphi(t) = \omega t \cos(\xi(t)). \tag{12}$$

In [9], a method is shown how to expand (12) into a Bessel series using the Jacobi-Anger expansion in order to decouple the expression. In this paper, only small oscillations are taken into account, so that this effect can be neglected.

IV. Validation of the MIMO transfer function

A. Model comparison

In this section, the time domain results of the presented models are compared using parameters of different generators. Two synchronous generators were chosen, a salient-pole machine (generator B) and a cylindrical rotor machine (generator A). Both operate at the same voltage and power factor with a rated apparent power of 15 MVA and 80 MVA, respectively. Generator A was chosen, because it is directly connected to the power grid via a transformer and not via power electronics. This makes the generator vulnerable to oscillations of the stator voltage, which would otherwise be significantly dampened by the power electronics. The other generator is built for being operated in an isolated network and is powered by a combustion machine. Therefore, it is

susceptible to oscillations of the torque because of the combustion processes. They both have either a damper winding or a solid rotor which are capable of dampening the oscillations, what makes both machines particularly interesting for this simulation. The parameters for the machines can be found in the appendix.

For the tests, a voltage component is added with a different frequency than the fundamental grid frequency, which in this case is 50 Hz. The models are all implemented in the software Matlab and the numerical solver used for the differential equations is ODE45. As a reference for the validation, the non-linear model is used. The goal is to observe the same steady-state response as the non-linear model gives. To neglect transients completely, the response is therefore calculated for five seconds and only the last second is analysed.

The voltage distortion used for the simulation consists of an RMS of 100 V (line-to-line) and a frequency of 26 Hz. This distorting voltage is 0.95 % of the fundamental stator voltage. The response to the described distortion can be seen in Figs. 1 to 4, where the oscillation of the d component of the stator current can be seen with the frequency spectrum of the respective transformed stator current for two different generators. Looking at generator A in Fig. 1, it can be seen that the d current describes a sine wave with a main frequency of 24 Hz, an amplitude of roughly 0.1 p.u. and a mean value of 1 p.u. . The linear model and the MIMO transfer function model yield similar results. In Fig. 2, the frequency spectrum of the signal, transformed into natural coordinates, from Fig. 1 is shown. In natural coordinates, the constant component observed is at 50 Hz, as expected. The anticipated harmonic of 26 Hz can be observed, while all three models show the same value at this frequency. This is a strong indicator that the MIMO transfer function is eligible in terms of calculating the steady-state system response. There is a third, unanticipated frequency at 74 Hz, shown by all three models with the same

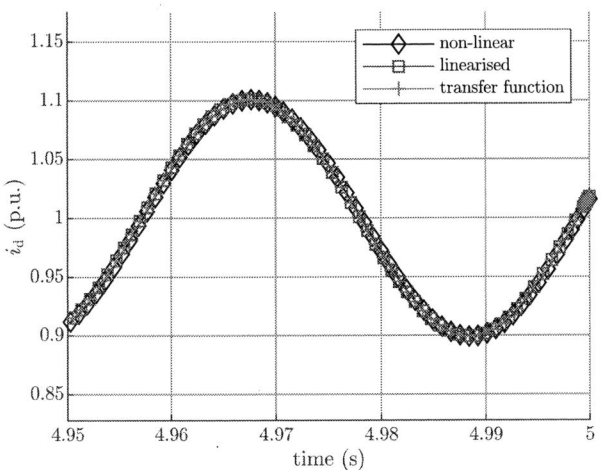

Fig. 3. i_d of generator B for a distorting frequency of 26 Hz

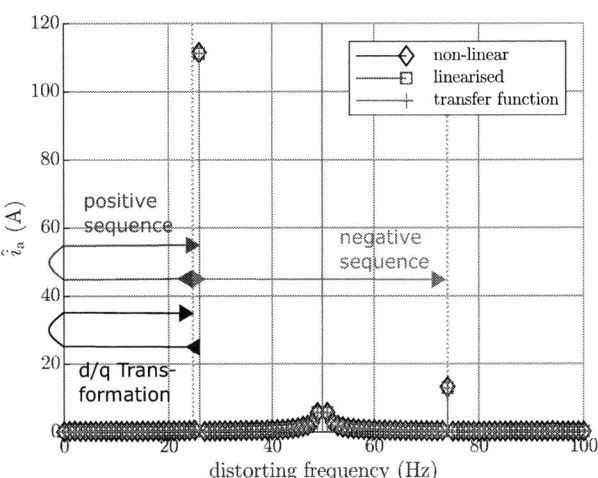

Fig. 4. Fourier analyses of the current i_a of generator B for a distorting frequency of 26 Hz

amplitude as well.

This third component is a negative-sequence component. When a signal is transformed into the rotating reference frame, its frequency is subtracted by the rotating frequency. In this case, the 26 Hz in natural coordinates would be -24 Hz in the 50 Hz rotating coordinates. The negative-sequence component has a 'negative' frequency, which through the double negative would be 24 Hz.

Now, the ± 24 Hz are added to the 50 Hz, which results in the already known 26 Hz and the newly observed 74 Hz. This behaviour for the frequency of the negative-sequence component is described by

$$f_- = | -(f_\mathrm{dis} - f_\mathrm{dq}) + f_\mathrm{dq}|. \qquad (13)$$

With this equation, the frequency of the negative-sequence component in natural coordinates can be determined, where f_dis represents the frequency of the distortion and f_dq is the frequency of the d/q reference frame. This issue is further discussed later in this paper.

As for the second generator B, the results are rather similar to the already observed effects. Figure 3 shows the d current for generator B, which is a sine wave as well, and the amplitude is a bit smaller than for generator A . The linearised and the MIMO transfer function model show the same results again. The frequency spectrum in Fig. 4 shows that for generator B the 26 Hz component is smaller in relation to the 74 Hz component than for Generator A. For this generator, all three models generate the same results, as it is the case for generator A. Therefore, it can be stated, that the MIMO transfer function is eligible to calculate the steady-state frequency response of a synchronous generator.

B. Analysing the symmetrical components

As stated in the previous section, the additional frequency is caused by a negative-sequence component. In this section, the symmetrical components are analysed further. In references

[3], [10] and [11] a method is proposed to calculate an equivalent of the symmetrical components especially for rotating d/q values. The positive-sequence component's equivalent is calculated as

$$\begin{bmatrix} \underline{g}_1^\mathrm{R} \\ \underline{g}_1^\mathrm{R}* \\ \underline{g}_\mathrm{h} \end{bmatrix} = \begin{bmatrix} 1 & \mathrm{j} & 0 \\ 1 & -\mathrm{j} & 0 \\ 0 & 0 & 2 \end{bmatrix} \begin{bmatrix} g_\mathrm{d} \\ g_\mathrm{q} \\ g_0 \end{bmatrix}, \qquad (14)$$

where $\underline{g}_1^\mathrm{R}$ is the positive-sequence component of the rotating d/q stator value g and $\underline{g}_1^\mathrm{R}*$, the complex conjugate of $\underline{g}_1^\mathrm{R}$, is the negative-sequence component. While in [3] the d/q values are real numbers, in this calculation the complex representation

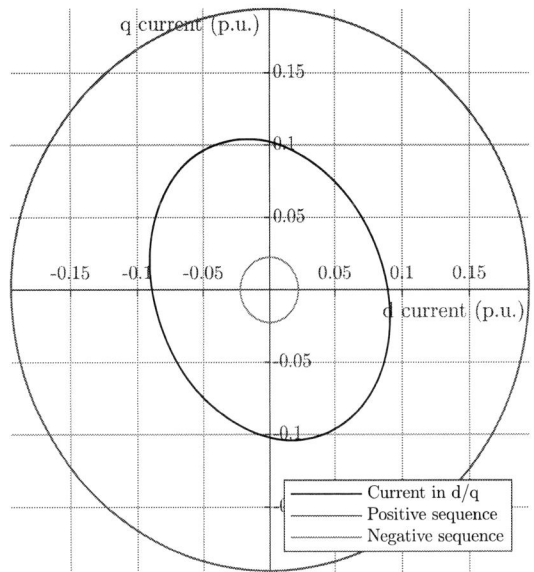

Fig. 5. Phase portrait of the stator current and positive- and it's negative-sequence components for Generator B

978-1-5386-7688-2/19 \$31.00 © 2019 IEEE

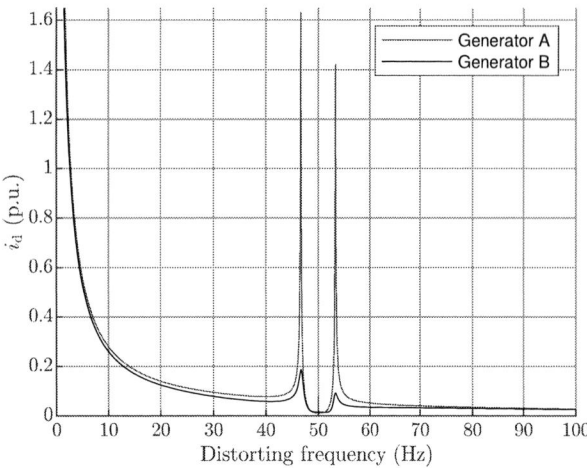

Fig. 6. d-axis current as a function of the distorting frequency with 100 V RMS line-to-line voltage

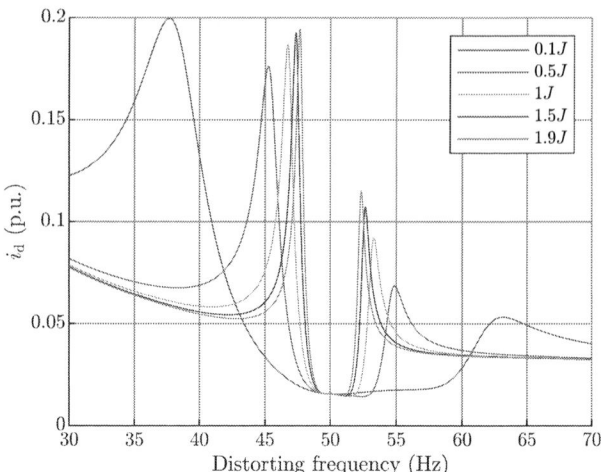

Fig. 7. Influence of the inertia on the d-axis current of generator B

was used. In this equivalent of the symmetrical components, the values have the double magnitude, as can be observed for the zero component. With this in mind, (14) can be applied to the MIMO transfer function response of the generator. This is done exemplarily for generator B, which can be seen in Fig. 5. The elliptical-shaped phase portrait of the stator current is depicted in black. Without a negative-sequence component, its shape would be a circle. The circle in blue represents the positive-sequence component, the negative sequence is displayed in red, which is also a circle. This illustrates the existence of the negative-sequence components.

V. STUDIES IN THE FREQUENCY DOMAIN

Having validated the method, the next step is to study the generator in the frequency domain. For this purpose, the MIMO transfer function is evaluated at a broad frequency range. Figure 6 shows the d current of both generators from 0 Hz to 100 Hz. It can be seen that low frequencies lead to a very high magnitude in the d current. This is because the frequency-depended component of the impedance decreases with decreasing frequency. At the point of 0 Hz, only the resistive component is active, which for both generators is smaller than $0.1\,\Omega$, this leading to very high currents.

The next points of interest lie around 50 Hz. Both machines have peaks in this area, especially generator A with peaks over 1 p.u.. This is a resonance effect of the coupling between the electro-magnetic and the mechanical parts of the machine, because both generators possess either a damper cage or a solid rotor, whose eddy currents have a similar effect like a damper cage. An equivalent mechanical system of rotational spring, dampener and mass can be applied. For this, [12] and [13] propose a mechanical equation, where the equivalent mechanical system is applied and calculated as

$$m(t) = \frac{J}{p}\frac{\mathrm{d}^2}{\mathrm{d}t^2}\delta_{\mathrm{p}} + \underbrace{\frac{M_{\mathrm{i}}}{s_{\mathrm{i}}\cdot 2\pi f_1}}_{D}\frac{\mathrm{d}}{\mathrm{d}t}(\delta_{\mathrm{p}} - \delta_{\mathrm{n}}) + \underbrace{M_{\mathrm{break}}\cos(\delta_0)}_{E}(\delta_{\mathrm{p}} - \delta_{\mathrm{n}}). \tag{15}$$

This equation has three parts, where the first part is related to the inertia of the rotor. The second part originates from the asynchronous damper torque of the damper cage, which represents the dampening of the oscillation. The third part consists of the breaking torque multiplied with the cosine of the steady-state load angle. As for a mechanical oscillation, the eigenfrequencies can be calculated as in [14]. For the non-rotating natural coordinates, the frequencies are

$$f_{\mathrm{e}} = 50\,\mathrm{Hz} \pm \frac{1}{2\pi}\sqrt{p\frac{E}{J} - \left(\frac{p\cdot D}{2\cdot J}\right)^2} \tag{16}$$

for an excitation of the oscillations originating from the stator reference frame. This is indicated by the 50 Hz component. To investigate whether the method developed shows the same behaviour as equation (16), the dependency on the moment of inertia is tested. In case the moment of inertia is increased, the frequencies are expected to be shifted closer to 50 Hz, and for a decrease they should shift away from 50 Hz. The simulation results are shown in Fig. 7, where the moment of inertia is varied from 0.1 to 1.9 of the true value. It can be seen, that the peaks are not only symmetrical to 50 Hz, but the previously stated behaviour to the variation of the moment of inertia can be observed. We conclude, that the observed resonance phenomenon is a result of the mechanical oscillation of the rotor, which is damped by the asynchronous torque of the damper cage.

Figures 6 and 9 also show that the peaks at the resonance frequencies of generator A are sharper and higher than for generator B. Especially in Fig. 6, there is a huge difference

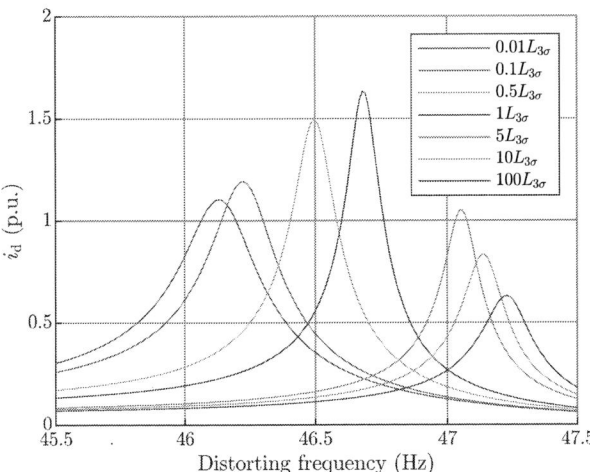

Fig. 8. Variation of the leakage inductance of the damper cage for generator A

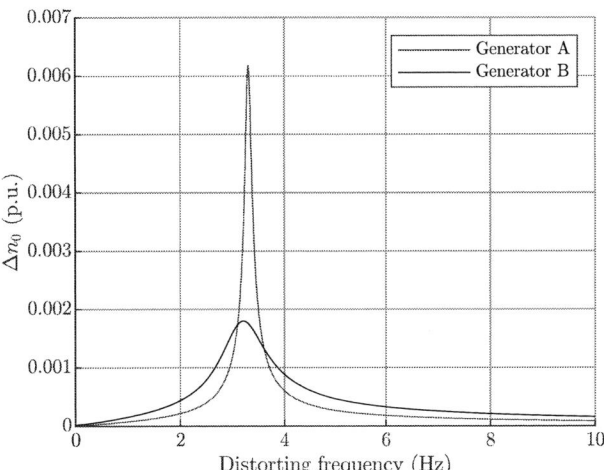

Fig. 9. Impact of an oscillating torque with an amplitude of 2 % of the rated torque on the rotational speed of the rotor for both machines

when comparing 20 % of the rated current to 160 % for the first peak. A method for optimizing the behaviour at the resonance frequency could be a variation of the leakage inductance of the damper cage. In Fig. 8, such a variation was performed for Generator A. The impact on the d current is shown in the frequency domain for different leakage inductances of the damper cage. The amplitude starts for small inductances by about 1.1 p.u. and grows bigger up to the already seen 1.6 p.u. and falls after that to 0.6. p.u. for a 100 times bigger leakage inductance. Also the resonance frequency shifts closer to 50 Hz with growing leakage inductance. This is in accordance with equations (15) and (16). When the leakage inductance is increased, the torque of the damper cage gets smaller. Therefore, the frequencies shift closer to 50 Hz. With the corresponding boundary conditions, design decisions can be taken.

When the excitation of the oscillations is caused by the torque or the rotor dc voltage, the natural frequencies are at ± 0 Hz, because both excitations originate already from the rotating reference frame. An example of a frequency response for this case is given in Fig. 9, where the change of the rotational speed is calculated for an oscillating torque. It becomes apparent, that for an amplitude of 2 % of the rated torque, the rotational speed deviates 0.6 % from the rated rotational speed for generator A and less than 0.2 % for generator B at their respective resonance frequency. The resonance frequency is around 3 Hz and not at 47 Hz or 53 Hz like in Fig. 6. However, the resonance originates from the same effects stated earlier. This result is especially interesting for generators that are powered by internal combustion engines (ICE). Reference [15] illustrates, that the oscillation of the torque of the ICE is related to the rotational speed of the crank shaft and the amount of ignitions in one cycle. This is the reason why four-stroke engines, where one ignition per cylinder happens in two revolutions of the crank shaft, are capable of generating

frequency components of the 0.5th order. These orders were marked critical in [16] for oscillating synchronous machines, as they result in frequencies in the range from 0.5 to 10 Hz, depending on the rotational speed. For generator B with a rotational speed of $750\,\mathrm{min}^{-1}$ (at 50 Hz stator frequency), the 0.5th order has the frequency 6.25 Hz, which is close to the resonance frequency of 3 Hz.

For the purpose of studying synchronous machines coupled with a source of mechanical energy, like a ICE or a water turbine, the "complex torque coefficient"[2] from [5] and [13] is used. The problem with this method is that the calculation and the analysis of this complex torque coefficient takes effort, but it can indicate the resonance frequencies for torque oscillations. With the MIMO transfer function, these resonance frequencies can be determined just by looking at the frequency spectrum of the machine, making it the more convenient method.

VI. CONCLUSION

The MIMO transfer function, which was derived from a non-linear simulation model using means of linearisation, was validated for calculating the steady-state response of synchronous generators with an oscillating operating point. It was observed, that the MIMO transfer function is eligible for calculating the symmetrical components and identifying possible resonance frequencies of the generator. Furthermore, the resonance effects were explained and the influence of the damper cage was estimated.

It was shown that, with the MIMO transfer function, the influence of single parameters, like the moment of inertia and the leakage inductance of the damper cage, can be studied and visualized in a convenient way. With the generalized matrix form used to describe the equations, the method presented

[2]In European technical literature also known as "complex synchronizing coefficient" [17]

in this paper can be used for many applications in conjunction with synchronous machines, such as the calculation of the complex torque coefficient, which was presented in this paper. The information about resonance frequencies of torque excitations can easily be extracted from the transfer function, while it is also applicable to stator voltage oscillations.

The frequency dependent-behaviour of the damper cage can be further investigated with this method, for example the active and reactive power consumption. Another application could be the prediction of current amplitudes for harmonics, when the induced voltages of the harmonics and their order are known.

APPENDIX

A. Parameters of the generators studied

TABLE I
PARAMETERS OF THE USED GENERATORS

	Generator A	Generator B
S_N	80 MVA	15 MVA
U_N	10.5 kV	10.5 kV
p	1	4
$\cos(\varphi)$	0.8	0.8
R_1 in p.u.	0.0014	0.0034
R_2 in p.u.	$0.943 \cdot 10^{-3}$	$0.991 \cdot 10^{-3}$
R_{3d} in p.u.	0.0729	0.074
R_{3q} in p.u.	0.0729	0.0376
L_{1d} in p.u.	2.7796	1.666
L_{1q} in p.u.	2.667	1.0703
L_{10} in p.u.	0.253	0.2401
L_2 in p.u.	2.77	1.69
L_{3d} in p.u.	2.72	1.63
L_{3q} in p.u.	2.61	0.996

B. Resistance and inductance matrices

$$\mathbf{R} = \begin{bmatrix} R_1 & 0 & 0 & 0 & 0 & 0 \\ 0 & R_1 & 0 & 0 & 0 & 0 \\ 0 & 0 & R_1 & 0 & 0 & 0 \\ 0 & 0 & 0 & R_2 & 0 & 0 \\ 0 & 0 & 0 & 0 & R_{3d} & 0 \\ 0 & 0 & 0 & 0 & 0 & R_{3q} \end{bmatrix}$$

$$\mathbf{L} = \begin{bmatrix} L_{1d} & 0 & 0 & M_{12} & M_{13d} & 0 \\ 0 & L_{1q} & 0 & 0 & 0 & M_{13q} \\ 0 & 0 & L_{10} & 0 & 0 & 0 \\ \frac{3}{2}M_{12} & 0 & 0 & L_2 & M_{23} & 0 \\ \frac{3}{2}M_{13d} & 0 & 0 & M_{23} & L_{3d} & 0 \\ 0 & \frac{3}{2}M_{13q} & 0 & 0 & 0 & L_{3q} \end{bmatrix}$$

$$\mathbf{L}_{rot} = p \cdot \begin{bmatrix} 0 & -L_{1q} & 0 & 0 & 0 & -M_{13q} \\ L_{1d} & 0 & 0 & M_{12} & M_{13d} & 0 \\ 0 & 0 & 0 & 0 & 0 & 0 \\ 0 & 0 & 0 & 0 & 0 & 0 \\ 0 & 0 & 0 & 0 & 0 & 0 \\ 0 & 0 & 0 & 0 & 0 & 0 \end{bmatrix}$$

REFERENCES

[1] C. Amornvipas, L. Hofmann, "Resonance analyses in transmission systems: Experience in Germany", InPower and Energy Society General Meeting, 2010 IEEE 2010 Jul 25 (pp. 1-8), IEEE.

[2] Xi-Fan Wang, Yonghua Song, Malcolm Irving, "Modern power systems analysis", Springer Science & Business Media, 2010.

[3] B. Oswald "Berechnung von Drehstromnetzen"(in German), Springer, 2009.

[4] P. C. Krause, O. Wasynczuk, S. Sudhoff, and S. Pekarek, "Analysis of Electric Machinery and Drive Systems", 3rd ed. Hoboken, NJ, USA: John Wiley & Sons, Inc., 2013.

[5] G. Müller, B. Ponick, "Theorie elektrischer Maschinen" (in German), Wiley-VCH, 2009.

[6] R. H. Park, "Two reaction theory of synchronous machines generalized method of analysis-part I", AIEE Transactions Vol. 48, 1929, p. 716.

[7] H. O. Seinsch, "Grundlagen elektrischer Maschinen und Antriebe"(in German) Springer, 1993.

[8] L. C. Westphal, "Sourcebook of control systems engineering", Springer Science & Business Media, 2012.

[9] G. Holmes, T. A. Lipo, "Pulse Width Modulation for Power Converters: Principles and Practice", Wiley-IEEE Press, 2003.

[10] Lyon W.V., "Transient Analysis of Alternating-current Machinery: An Application of Method of Symmetrical Components", Technology Press of Massachusetts Institute of Technology, and Wiley, New York, 1954.

[11] Gerardus C. Paap, "Symmetrical components in the time domain and their application to power network calculations", IEEE Transactions on power systems, 2000, 15. Jg., Nr. 2, S. 522-528.

[12] E. Arnold, J. L. La Cour, "Die synchronen Wechselstrommaschinen. Generatoren, Motoren und Umformer. Ihre Theorie, Konstruktion, Berechnung und Arbeitsweise"(in German), Manuldruck 1923, Springer-Verlag, 2013

[13] K. Bonfert, "Betriebsverhalten der Synchronmaschine"(in German), Vol. 122. Berlin, Springer, 1962.

[14] T.L. Schmitz, K. S. Smith, "Machining dynamics", Springer Publishing Co., New York, 2008.

[15] M. Beuschel, "Neuronale Netze zur Diagnose und Tilgung von Drehmomentschwingungen am Verbrennungsmotor"(in German), Doctoral dissertation, Technische Universität München, 2000

[16] L. A. Kilgore,E. C. Whitney, "Spring and damping coefficients of synchronous machines and their application", Transactions of the American Institute of Electrical Engineers 69.1 (1950): 226-230.

[17] I. M. Canay, "A novel approach to the torsional interaction and electrical damping of the synchronous machine Part I: Theory", IEEE Transactions on Power Apparatus and Systems 10 (1982): 3630-3638.

Comparison of PI and FOPI Based Voltage and Current Controlled DC Motor Drive System

Hafiz M. Usman[1], Abdel Gafoor Haddad[2], Habibur Rehman[3], Shayok Mukhopadhyay[4]

[1,3,4]*Dept. of Electrical Engineering, American University of Sharjah, UAE*
[2]*Dept. of Electrical and Computer Engineering, Khalifa University, Abu Dhabi, UAE*
[1,3,4]{b00071330, rhabib, smukhopadhyay}@aus.edu, [2] 100049699@ku.ac.ae

Abstract—This paper aims to improve the performance of DC motor drive system by utilizing a fractional-order proportional-integral (FOPI) speed regulator. Available tuning guidelines to design the FOPI speed regulator for a DC motor drive system are also presented. Two closed loop speed control strategies, namely DC motor armature voltage control and armature current control, are proposed for comparative investigation. Numerical simulations and quantitative analysis are performed to evaluate the efficacy of FOPI speed regulator-based DC motor drive system. Simulation results show that the FOPI speed regulator-based voltage and current control strategies are robust against external disturbances and are effective for DC motor drive system.

Index Terms—DC motor, voltage control, current control, fractional-order PI controller

I. INTRODUCTION

DC motors are being widely used in numerous applications related to portable electronics, home automation, and robotics. The popularity of a DC motor has led many researchers to develop, improve, and enhance robustness of its speed control strategies against external disturbances. Different approaches have been proposed in the literature to control DC motors, especially as the motor characteristic curves have a nonlinear region of operation that requires a sophisticated control method to ensure that the system operates over an extensive range of operating points.

A sensorless passivity-based approach is proposed in [1] for the control of DC motor. The control signal is fed to a buck-boost converter-inverter system to control the speed and bidirectional movement of a DC motor. A generalized proportional-integral observer based multivariable control approach is developed in [2]. The proposed strategy in [2] is combined with parallel DC buck converters for better armature current of DC motor and to enhance the robustness against external disturbances. The authors in [3] present a neural network-based control scheme. A neural network is used to predict the behavior of a DC motor using motor armature current. A fuzzy-logic-based adaptive PID control strategy is reported in [4] for speed control of DC motor. The combination of DC motor and buck converter transfer functions are utilized for initial settings of a PID controller, which is achieved by Ziegler-Nichols tuning method. Afterward, a fuzzy-logic-based controller is employed for online updates of PID controller settings. A comprehensive comparison between a regular PID controller and a fuzzy-logic-based PID

controller is also performed in [4]. The simulation results exhibit that the fuzzy-logic-based PID control strategy is robust to external disturbances and has minimal overshoot. The authors in [5] present a recursive nonlinear adaptive controller based on Lyapunov functions for speed control of a DC motor. Moreover, the system parameters are estimated using an adaptation law, and the estimated parameters are incorporated in the Lyapunov functions to enhance the robustness and effectiveness of the designed controller. The comparison of the system's performance with an adaptive backstepping controller in [5] illustrates its effectiveness in terms of speed tracking and system stability.

In this paper, fractional-order proportional-integral (FOPI) based speed control strategies are proposed for DC motor drive system. This paper evaluates the performance of two candidate FOPI-based speed control schemes, i.e., armature voltage control and armature current control, against trial-and-error (TEPI) tuned PI speed controller schemes. Unlike [1] and [2], the proposed strategy effectively controls the speed and is robust to external disturbances. Furthermore, the proposed strategy is relatively less computationally intensive and very simple to implement compared to [3], [4], and [5]. Although some research attempts have been reported in the literature, e.g. [6], [7], and [8], for the speed control of DC motor system, this work presents a comprehensive comparison between FOPI-based voltage and current control schemes for DC motor drive system which has not been explored to date. In our previous works, a comparative analysis among various integer-order PI controllers and an FOPI controller is presented for field-oriented induction motor drive system [9], [10].

This paper is organized as follows. Section-II describes the voltage and current control strategies and provides the tuning guidelines of FOPI controller. The comparison of TEPI and FOPI control strategies and the detailed performance analysis is presented in Section-III. Finally, the concluding remarks are given in Section-IV.

II. METHODOLOGY

The armature voltage-controlled speed regulation strategy for a DC motor is shown in Fig. 1. In this technique, the FOPI speed regulator ensures the actual speed of the DC motor is equal to the reference speed at steady state. The FOPI speed regulator generates the armature voltage command which is then converted into a PWM signal by the chopper circuit. The

978-1-5386-7688-2/19 $31.00 © 2019 IEEE

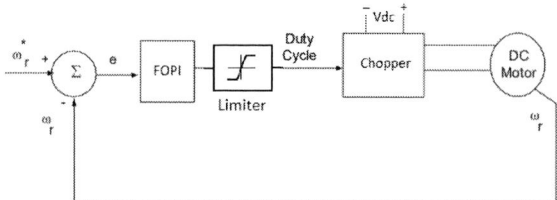

Fig. 1: DC motor speed control using armature voltage control

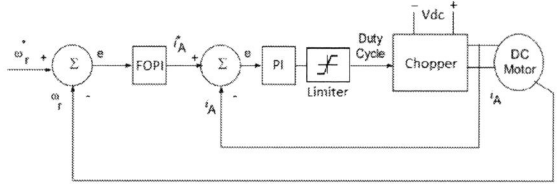

Fig. 2: DC motor speed control using armature current control

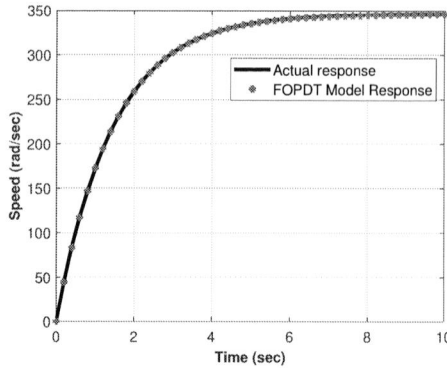

Fig. 3: Step response of DC motor and approximated FOPDT model

cascaded armature current-controlled speed regulation strategy for a DC motor is illustrated in Fig. 2. In this technique, the outer FOPI speed regulator generates the reference current command, which is fed into the inner armature current regulator to generate the armature voltage command for the chopper circuit. The guidelines to tune the gains of FOPI speed regulator are provided as follows.

The dynamic equations of a DC motor model are presented by (1)-(2) [1].

$$L_a \frac{di_a(t)}{dt} = V_{dc}(t) - R_a i_a(t) - K_b \omega(t) \qquad (1)$$

$$J \frac{d\omega(t)}{dt} = K_t i_a(t) - b\omega(t) - \tau_L(t). \qquad (2)$$

Here V_{dc} is the input voltage (V), R_a is the equivalent armature circuit resistance (Ohm), i_a represents the armature circuit current (A), K_b denotes the back EMF constant $(V.s/rad)$, L_a is the armature inductance (H), J represents the rotor moment of inertia $(Kg.m^2)$, $\omega(t)$ denotes the speed (rad/s), K_t represents the torque constant $(N.m/A)$, b is the coefficient of viscous friction $(N.m.s/rad)$, and τ_L represents the load torque $(N.m)$. The input DC voltage to the chopper circuit is 24 V. The DC motor parameters used in this work are as follows: $R_a = 0.5$, $K_b = 0.06$, $L_a = 1e-3$, $J = 2.52e-4$, $K_t = 0.06$, $b = 1.73e-04$.

Equation (2) is employed to design the FOPI speed regulator at no-load condition. The open-loop transfer function using (2) is described by (3).

$$G_p(s) = \frac{\omega}{i_a} = \frac{K_t}{Js + b}. \qquad (3)$$

The first step toward the design of FOPI controller involves the first order plus dead time (FOPDT) approximation of (3).

The general form of FOPDT model is described by (4). In (4), the parameters K, T, and L denote process gain (rad/sec), time constant (sec), and dead time (sec), respectively. The MATLAB PID tuner toolbox tunes K, T, and L parameters to approximate (3) by (4), and provides $K = 346.271$, $T = 1.45$, and $L = 0.0001$. The step response of (3) and approximated FOPDT model response of (4) is shown in Fig. 3. Next, the values of K, T, and L are utilized to tune K_p, K_i, and α gains of an FOPI controller.

$$G_{approx} = \frac{K}{Ts + 1} e^{-Ls}. \qquad (4)$$

The general form of an FOPI controller is described by (5). The D operator calculates the integral of any order, i.e. α, using (6).

$$u(t) = K_p e(t) + K_i \big({}_\alpha D_t^\alpha e(t) \big), \qquad (5)$$

$${}_\alpha D_t^\alpha f(t) = \frac{1}{\Gamma(n - \alpha)} \int_\alpha^t \frac{f^{(n)}(\tau)}{(t - \tau)^{\alpha + 1 - n}} d\tau$$
$$(n - 1 < \alpha < n), (n \in \mathbb{R}). \qquad (6)$$

Theoretical details related to tuning rules of FOPI controller are provided in [9], [11], [10]. The vital steps to evaluate the K_p, K_i, and α gains of FOPI controller are given by (7)-(8).

$$\tau = \frac{L}{T + L}, \alpha = \left\{ \begin{array}{ll} 1.1, & \tau \geq 0.6 \\ 1, & 0.4 \leq \tau < 0.6 \\ 0.9, & 0.1 \leq \tau < 0.4 \\ 0.7, & \tau < 0.1 \end{array} \right\} \qquad (7)$$

$$\left. \begin{array}{l} K_p = \frac{1}{K} \left(\frac{0.2978}{L + 0.000307} \right), \\ T_i^* = T \left(\frac{0.8578}{L^2 - 3.402L + 2.405} \right), \\ K_i = \frac{K_p}{T_i^*}. \end{array} \right\} \qquad (8)$$

The values of K_p, K_i, and α obtained by solving (7)-(8) are given in Table I. MATLAB packages such as 'Ninteger toolbox' are readily available [12] for FOPI controller implementation. The initial guesses for K_p, and K_i gains for TEPI controller are also provided in Table I. In the next section, the Table I gain values are employed to perform the numerical simulations.

TABLE I: Controller Gains

Parameters	Voltage-Controlled TEPI	Current-Controlled TEPI	FOPI
α	1	1	0.7
K_p	3.2	3.2	2.2875
K_i	3.2	3.2	4.4216

III. SIMULATION RESULTS

In this section, the armature voltage-controlled and armature current-controlled FOPI speed regulation strategies are investigated and are compared with their TEPI controller counterparts. The simulation results of speed and armature current are analyzed on a reference speed command at 1000 RPM with an external step disturbance of 0.6 Nm after 0.1 seconds. The results of speed tracking and armature current of voltage-controlled speed regulation strategy is shown in Fig. 4a and Fig. 4c, respectively. The results of TEPI controller as shown in Fig. 4b and Fig. 4d are not desirable as they exhibit higher steady state speed error and higher ripple amplitude in the armature current, respectively. On the other hand, the FOPI speed regulator-based voltage-controlled strategy shows better performance in terms of speed tracking and armature current ripples. The quantitative analysis of maximum overshoot,

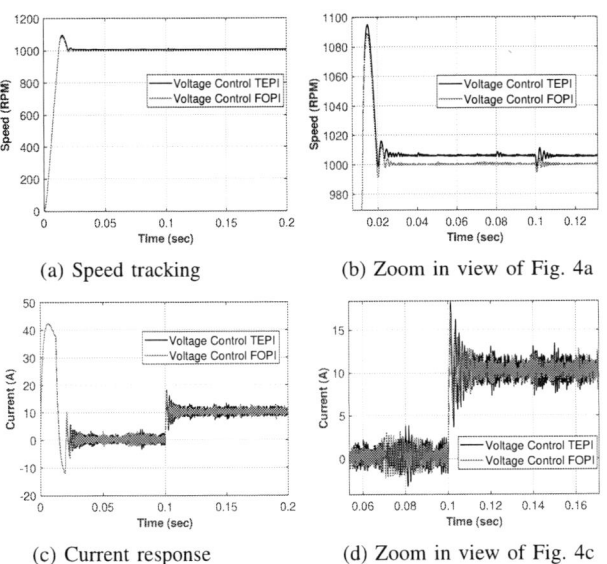

(a) Speed tracking   (b) Zoom in view of Fig. 4a

(c) Current response   (d) Zoom in view of Fig. 4c

Fig. 4: Armature voltage-controlled scheme results for TEPI and FOPI controllers

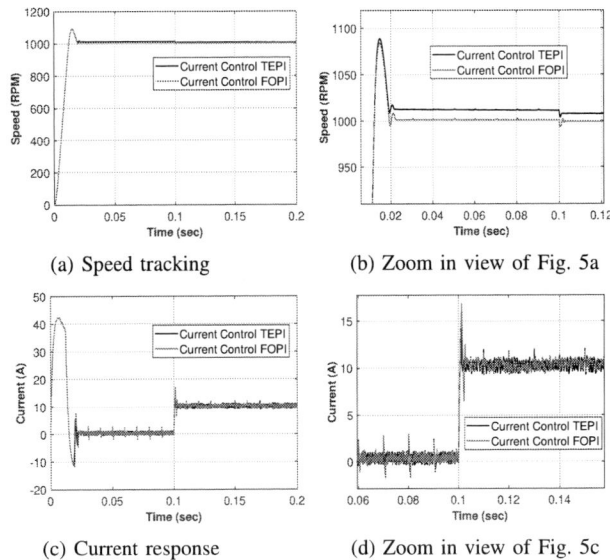

(a) Speed tracking   (b) Zoom in view of Fig. 5a

(c) Current response   (d) Zoom in view of Fig. 5c

Fig. 5: Armature current-controlled scheme results for TEPI and FOPI controllers

settling time, and steady state error of Fig. 4 is presented in Table II.

The results of speed tracking and armature current of current-controlled speed regulation strategy is shown in Fig. 5. Similar to FOPI controller-based voltage-controlled strategy results, the FOPI controller-based current-controlled scheme outperforms its TEPI counterpart as shown in Fig. 5. The quantitative analysis of maximum overshoot, settling time, and steady state error of Fig. 5 is described in Table II. It can be inferred from Table II results that the current-controlled FOPI scheme performs better than its TEPI counterpart. Furthermore, the FOPI-based current-controlled scheme causes lesser ripples in armature current compared to its voltage-controlled FOPI counterpart as shown in Fig. 6. Thus, FOPI-

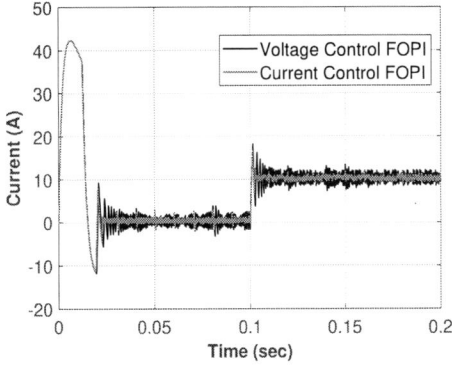

Fig. 6: Comparison of armature current response for FOPI-based voltage-controlled and current-controlled schemes

TABLE II: Step-response characteristics for voltage-controlled and current-controlled TEPI-based and FOPI-based speed regulation schemes for DC motor drive system

Step-Response Characteristics	Voltage-Controlled TEPI	Voltage-Controlled FOPI	Current-Controlled TEPI	Current-Controlled FOPI
Maximum Overshoot (%)	9.5	8.8	8.7	8.3
Settling Time (sec)	0.0191	0.0188	0.0187	0.0187
Steady State Error (RPM)	~ 7	~ 5	~ 6	~ 1

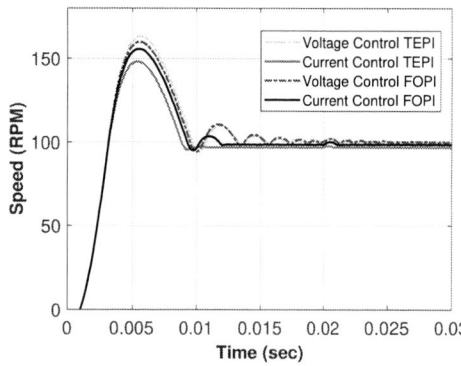

Fig. 7: Low speed tracking comparison for TEPI-based and FOPI-based armature voltage-controlled and current-controlled schemes

based current-controlled strategy exhibit better speed tracking and current response and effectively rejects the effects of applied external disturbances.

The effectiveness of FOPI-based current-controlled speed regulation strategy is also illustrated at low speed tracking test. The DC motor drive system is commanded to operate at 100 RPM with 0.6 N.m load. The comparison of voltage-controlled and current-controlled TEPI-based and FOPI-based schemes is shown in Fig. 7. Although current-controlled TEPI-based strategy has lesser overshoot compared to other schemes, it does not nullify the effect of 0.6 N.m load as it has steady state error of ~ 4 RPM. On the other hand, FOPI-based current-controlled strategy completely cancels out the effect of 0.6 N.m load and exhibits no steady state error. Compared to a normal PI controller-based voltage or current controlled DC motor drive systems, FOPI-based current-controlled strategy performs better and may increase the battery state of heath and hence the battery lifetime, especially in robotics applications.

IV. CONCLUSION

In this paper, two closed loop FOPI controller-based speed regulation strategies for DC motor drive system are compared. The armature voltage-controlled FOPI speed regulator scheme is simple and exhibit good performance in terms of speed tracking and armature current ripple contents. However, the armature current-controlled FOPI speed regulator scheme

performs better by showing comparatively lesser overshoot, settling time, and steady state error in speed tracking and lesser ripples in the armature current. Furthermore, the low speed tracking results demonstrate that the FOPI-based control strategies perform better compared to their TEPI-based counterparts.

REFERENCES

[1] E. Hernández-Márquez, R. Silva-Ortigoza, J. R. García-Sánchez, M. Marcelino-Aranda, and G. Saldana-Gonzalez, "A DC/DC buck-boost converter–inverter–DC motor system: Sensorless passivity-based control," *IEEE Access*, vol. 6, pp. 31 486–31 492, 2018.

[2] E. Guerrero, J. Linares, E. Guzman, H. Sira, G. Guerrero, and A. Martinez, "DC motor speed control through parallel DC/DC buck converters," *IEEE Latin America Transactions*, vol. 15, no. 5, pp. 819–826, 2017.

[3] B. Eskandari, H. V. Haghi, M. T. Bina, and M. Golkar, "An experimental prototype of buck converter fed series DC motor implementing speed and current controls," in *2010 International Conference on Computer Applications and Industrial Electronics*. IEEE, 2010, pp. 606–609.

[4] R. Abhinav and S. Sheel, "An adaptive, robust control of DC motor using fuzzy-PID controller," in *2012 IEEE International Conference on Power Electronics, Drives and Energy Systems (PEDES)*. IEEE, 2012, pp. 1–5.

[5] T. Roy, L. Paul, M. Sarkar, M. Pervej, and F. Tumpa, "Adaptive controller design for speed control of DC motors driven by a DC-DC buck converter," in *2017 International Conference on Electrical, Computer and Communication Engineering (ECCE)*. IEEE, 2017, pp. 100–105.

[6] S. W. Khubalkar, A. S. Junghare, M. V. Aware, A. S. Chopade, and S. Das, "Demonstrative fractional order–PID controller based DC motor drive on digital platform," *ISA transactions*, 2017.

[7] R. V. Jain, M. Aware, and A. Junghare, "Tuning of fractional order PID controller using particle swarm optimization technique for DC motor speed control," in *2016 IEEE 1st International Conference on Power Electronics, Intelligent Control and Energy Systems (ICPEICES)*. IEEE, 2016, pp. 1–4.

[8] A. Rajasekhar, S. Das, and A. Abraham, "Fractional order PID controller design for speed control of chopper fed DC motor drive using artificial bee colony algorithm," in *2013 World Congress on Nature and Biologically Inspired Computing*. IEEE, 2013, pp. 259–266.

[9] A. Khurram, H. Rehman, S. Mukhopadhyay, and D. Ali, "Comparative analysis of integer-order and fractional-order proportional integral speed controllers for induction motor drive systems," *Journal of Power Electronics*, vol. 18, no. 3, pp. 723–735, 2018.

[10] H. M. Usman, S. Mukhopadhyay, and H. Rehman, "Electric vehicle traction system performance enhancement using FO-PI controller," in *14th IEEE Vehicle Power and Propulsion Conference (VPPC)*. IEEE, 2018, pp. 1–5.

[11] Y. Chen, T. Bhaskaran, and D. Xue, "Practical tuning rule development for fractional order proportional and integral controllers," *Journal of Computational and Nonlinear Dynamics*, vol. 3, no. 2, p. 021403, 2008.

[12] D. Valerio, "Ninteger v. 2.3 fractional control toolbox for Matlab." [Online]. Available: https://www.mathworks.com/matlabcentral/fileexchange/8312-ninteger

978-1-5386-7688-2/19 $31.00 © 2019 IEEE

Comparison of time-domain and time-scale data in bearing fault detection

I. Halil Ozcan
Electrical and Electronics Engr. Dept.
Izmir University of Economics
Izmir, TURKEY

Levent Eren
Electrical and Electronics Engr. Dept.
Izmir University of Economics
Izmir, TURKEY

Turker Ince
Electrical and Electronics Engr. Dept.
Izmir University of Economics
Izmir, TURKEY

Bulent Bilir
Electrical and Electronics Engr. Dept.
Izmir University of Economics
Izmir, TURKEY

Murat Askar
Electrical and Electronics Engr. Dept.
Izmir University of Economics
Izmir, TURKEY

Abstract— Recently various machine learning techniques have been applied as a solution to the problem of timely and accurate detection of motor bearing faults. Conventional decision support systems consist of feature extraction and classification phases which require manual design and optimization of each block separately with increased computational cost. In our work, one dimensional Convolutional Neural Networks (1D CNN) with an inherent adaptive design are applied to efficiently combine the feature extraction and classification into a single learning body. Wavelet packet decomposition (WPD) is applied to represent the bearing vibration signals in time-scale domain efficiently. We then compare performance of raw time-domain vibration input and time-scale transformed input signals for classification of bearing faults using real experimental data.

Keywords—bearing fault, 1D CNN, wavelet packet transform

I. INTRODUCTION

Data-driven deep learning models are becoming widely used in bearing fault detection and diagnosis problems lately. Deep learning methods come with the advantage of automatic learning of good features from raw input data through training [1-4]. But, they require large (labeled) training datasets and have much higher computational complexity. A shallow adaptive 1D Convolutional Neural Network (1D CNN) classifier can be used to overcome these limitations [5-8]. In this work, an adaptive 1D CNN classifier is applied to real-time detection of bearing faults from both raw time-domain and time-scale data.

The network is capable of learning filters in a data-driven fashion for given inputs to extract features automatically. The back-propagating classification error is used to optimize the convolutional filters of the CNN during the supervised training phase helping learn highly discriminative features from the input data. As a result, the system can optimally perform feature extraction and classification after the network is properly trained using bearing vibration dataset. In this study, for optimization of network parameters (filter coefficients, multi-layer perceptron weights and biases), the back-propagation (BP) algorithm that iteratively searches using gradient-descent method is used. In our recent works, the 1D CNNs have been successfully applied for feature extraction as well as classification of the raw electrocardiogram (ECG) data [5] and raw motor current and vibration data [6-8]. In [10], signal-based approaches based on time-frequency analysis of the measured signals to extract

information about the health of the systems are reviewed. In [12], a wavelet kernel local fisher discriminant analysis optimized by a particle swarm optimization based classifier is employed for bearing defect classification. An improved wavelet packet transform (WPT) and distance evaluation based feature selection is applied to a support vector machine (SVM) ensemble for detecting faults in [13]. In [14], a novel framework for adaptive feature extraction based on higher order cumulants (HOCs) and wavelet transform (WT) is proposed as input to a *K*-nearest neighbor classifier for categorizing the fault types. In another study [15], SVM is used along with continuous WT to analyze frame vibrations during start-up for condition monitoring of an induction motor. [16] proposes combination of wavelet packet decomposition (WPD) and empirical mode decomposition (EMD) to extract fault feature frequencies and applies them to a neural network for rotating machinery early fault diagnosis.

The 1D CNNs have been successfully applied for feature extraction as well as classification of the raw motor current and vibration data [6-8]. The shallow 1D CNN classifier has advantages such as low complexity architecture, cost effective and practical real-time hardware implementation, and ability to work with limited size of training data set. In this study, we show that shallow 1D CNNs performing 1D convolutions of input bearing vibration signals is able to learn to detect bearing faults better with time-scale data than time-domain data. The time-scale data provides higher time resolution for higher frequencies. As a result, the time-scale data input is less susceptible to the sub sampling process in 1D CNN structure.

The adaptive 1D CNN classifier structure and its training is described in Section 2. Wavelet packet decomposition is briefly covered in Section 3. Bearing fault related vibration frequencies are introduced in the following section. In Section 5, the performance of the proposed system is tested under two different input conditions. Finally, conclusions are presented in Section 6.

II. ADAPTIVE 1D CNN

The Adaptive 1D CNN architecture of the proposed bearing fault detection system is depicted in Figure 1. The raw time-domain or time-scale domain vibration data is preprocessed by filtering, decimation and normalization before being input to the adaptive 1D CNN classifier for testing.

978-1-5386-7688-2/19 $31.00 © 2019 IEEE

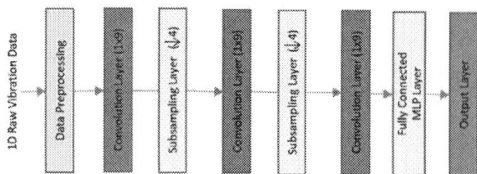

Fig. 1. Overview of a sample conventional 1D CNN

The 1D forward propagation from convolution layer *l-1* to the input of a neuron in layer *l* is expressed as,

$$x_k^l = b_k^l + \sum_{i=1}^{N_{l-1}} conv1D(w_{ik}^{l-1}, s_i^{l-1}) \qquad (1)$$

where the scalar bias of the kth neuron b_k^l, the output of the ith neuron at layer *l-1* s_i^{l-1}, and the kernel from the ith neuron at layer *l-1* to the kth neuron at layer *l* w_{ik}^{l-1} are used to determine the input x_k^l at layer *l*.

The intermediate output of the neuron, y_k^l, is a function of the input, x_k^l, and the output of the neuron s_k^l at layer *l* is a sub-sampled version of y_k^l as,

$$y_k^l = f(x_k^l) \ and \ s_k^l = y_k^l \downarrow ss \qquad (2)$$

The feature extraction and classification are merged into a single 1D CNN learner as described in [5]. Both can be optimized by using the BP algorithm during training to maximize the classification performance. This CNN topology could adapt to any input layer dimension without changing of parameters manually. Furthermore, within the adaptive CNN architecture the hidden neurons of the convolution layers (so called CNN layers) perform both convolution and subsampling operations by redesigning each neuron operations as shown in Figure 2. Here, 1D convolution of 1D signals with kernel filters are performed as opposed to multiplication with a scalar in a regular MLP classifier. Further details about adaptive 1D CNNs can be found in [5].

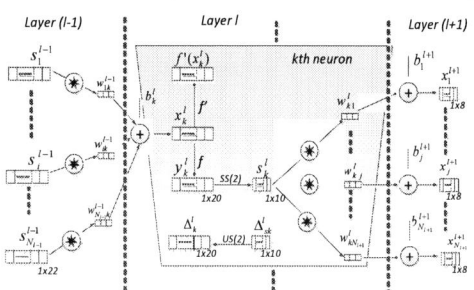

Fig. 2. The specialized neuron structure of the adaptive 1D CNN architecture

III. WAVELET PACKET DECOMPOSITION

Wavelet transform techniques have been applied to motor fault detection applications successfully [6-16]. WPD is a very useful signal processing tool in analysis of both motor current and vibration data for condition monitoring. The frequency resolution obtained by decomposing input vibration signal three levels with half-band filters is displayed in Figure 3.

Here, f_0 is frequency band resolution obtained by WPD analysis.

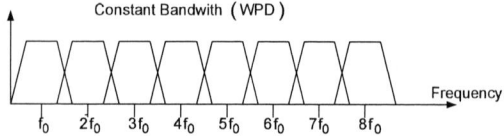

Fig. 3. WPD frequency separation

The poly-phase decomposition of any input data results in more efficient filtering implementation computationally. Here a filter *H(z)* can be written as

$$H(z) = \sum_{k=-\infty}^{\infty} h(k)z^{-k} \qquad (3)$$

Defining poly-phase components $P_r(z)$ as

$$P_r(z) = \sum_{l=-\infty}^{\infty} h(r + lM)z^{-l} \qquad (4)$$

where r=0,1,...,M-1. Then, the filter becomes

$$H(z) = \sum_{r=0}^{M-1} z^{-r} P_r(z^M) \qquad (5)$$

For M=2, the prototype low-pass filter $H_0(z)$ then has 2 poly-phase parts.

$$H_0(z) = H(z) = P_0(z^2) + P_1(z^2)z^{-1} \qquad (6)$$

The poly-phase parts of the prototype filter are then used to obtain $H_1(z)$, the high-pass filter as shown in equation 7.

$$H_1(z) = P_0(z^2) - P_1(z^2)z^{-1} \qquad (7)$$

The filter structure for two levels of WPD using half-band filters is depicted in Figure 4.

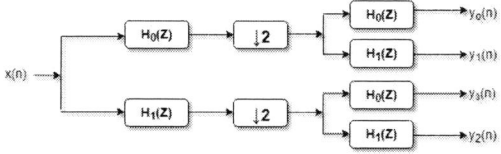

Fig. 4. Filter structure of WPD

Here, the input signal is split into four equally spaced frequency bands. The lower two bands ($y_3(n)$ and $y_2(n)$) in the figure will have higher time resolution compared to higher two bands. It also should be noted that $y_3(n)$ and $y_2(n)$ are in reversed order. This is due to natural, or Paley, order produced by phase reversal in high-pass filter.

IV. BEARING FAULT FREQUENCIES

Bearing faults cause vibration at frequencies associated with the developed fault types. If the bearing geometry and speed of rotation are known, the characteristic fault frequencies can be determined analytically [17]. The typical ball bearing geometry is depicted in figure 5. The bearing geometry information is usually provided in datasheets.

978-1-5386-7688-2/19 $31.00 © 2019 IEEE

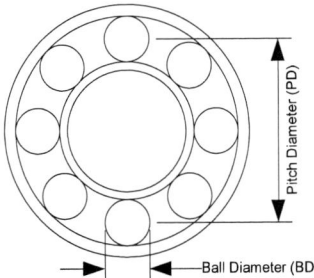

Fig. 5. Ball bearing geometry

Outer race defect frequency, f_{OD}, is expressed as

$$f_{OD} = \frac{n}{2} f_{rm}(1 - \frac{BD}{PD}\cos\phi) \qquad (8)$$

where f_{rm} is the rotor speed in revolutions per second, n is the number of balls, and the angle φ is the contact angle which is zero for ball bearings.

Inner race defect frequency f_{ID}, is given by

$$f_{ID} = \frac{n}{2} f_{rm}(1 + \frac{BD}{PD}\cos\phi) \qquad (9)$$

Ball defect frequency f_{BD}, is determined from

$$f_{BD} = \frac{PD}{2BD} f_{rm}(1 - (\frac{BD}{PD})^2 \cos^2\phi) \qquad (10)$$

V. DATA ANALYSIS

The Intelligent Maintenance Systems (IMS) bearing dataset, generated and provided by the University of Cincinnati Center for IMS, is used as a benchmark in this study [18]. The rotation speed was fixed at 2000 rpm and a radial load of 6000 lbs is applied onto the shaft and bearing by a spring mechanism. There was one accelerometer installed on the bearing housing for each bearing for the dataset. Data is collected until the bearings worked for over the designed lifetime of more than 100 million revolutions and outer race failure occurred (test-to-failure experiments). Data collection was made using NI DAQ card 6062E with sampling rate of 20 kHz. Each recording consisted of 20,480 sample points.

The proposed bearing fault detection algorithm is applied to the data collected from a bearing which develops an outer race defect. Rexnord ZA-2115 bearings with a pitch diameter of 2.815 inches, roller diameter of 0.331 inches, and a tapered contact angle of 15.17° are used in the test rig. Equation (8) would yield the outer race fundamental vibration frequency of 236 Hz at rotational speed of 2000 RPM.

The raw input vibration signal is decimated by 8 to provide reduction in CNN configuration for the system implementation. The frequencies up to 10 kHz could be detected in original raw vibration data since the sampling was done at 20 kHz. Decimating the original vibration data by 8 would result in the reduced detectible bandwidth of 1250Hz. The fundamental bearing fault frequency is at 236Hz for the dataset used in the analysis and the first five integer multiples of this fault frequency would fall in the frequency band of 0-1180 Hz. The decimated data was used as time-domain input whereas it was decomposed two levels as shown in figure 4 for generating time-scale data. In time-scale case, each band contains 312.5 Hz. Since the fundamental frequency (236 Hz) is in the lowest band, the data for band 1, $y_0(n)$, is used in the evaluations.

The input to the 1D CNN is 240 (time-domain) samples of the bearing vibration data and the output is MLP layer with size of 2 indicating faulty or healthy classes. True positive (TP), false negative (FN), true negative (TN), and false positive (FP) are used as hit counters to express the confusion matrix. The code utilizes 10-fold cross-validation technique to prevent overfitting and improve generalization in training the 1D CNNs.

Three different CNN structures were tested in comparing the performance of time-scale data with time-domain data as input. Three convolution layers have 20, 20, and 10 neurons respectively while the fully connected MLP layer has 10 neurons in the first configuration. Three convolution layers have 40, 40, and 10 neurons respectively while the fully connected MLP layer has 10 neurons in the second configuration. Three convolution layers have 80, 60, and 40 neurons respectively while the fully connected MLP layer has 20 neurons in the last configuration. The total parameters to be determined during training is 5762, 22292, and 66562 for three configurations respectively. The confusion matrix obtained from all (10) test runs of the proposed time-domain input of case three is presented in Table 1.

TABLE I. CONFUSION MATRIX FOR TIME DOMAIN

		Classification Result	
		H	F
Ground Truth	H	992 (TN)	45 (FN)
	F	8 (FP)	955 (TP)

The confusion matrix for time-scale input of case three is given in table 2.

TABLE II. CONFUSION MATRIX FOR TIME SCALE

		Classification Result	
		H	F
Ground Truth	H	1000 (TN)	12 (FN)
	F	0 (FP)	988 (TP)

The standard performance metric accuracy is calculated from confusion matrices. The results for all three configurations are given in table 3.

TABLE III. THREE CONFIGURATIONS

Case	Comparison of time domain and time scale data		
	CNN Structure	Time Domain	Time Scale
1	(0 20 20 10 10 0)	0,9895	0,9930
2	(0 40 40 20 10 0)	0.9765	0.9920
3	(0 80 60 40 20 0)	0.9735	0.9940

In all three configurations, applying time-scale values as input results in better performance. The best results for time-scale input are obtained in the third configuration. Whereas, the best results are given in the first configuration with simplest architecture for the time-domain input. Lower performance with higher complexity architecture indicates over training. Since the time-scale results are all very close in all three configurations, the first case would be selected for the implementation.

VI. CONCLUSION

In this study, it was shown that shallow 1D CNNs performing 1D convolutions of input bearing vibration signals is able to learn to detect bearing faults better with time-scale data than time-domain data. The time-scale data provides higher time resolution for higher frequencies. As a result, the time-scale data input is less susceptible to the sub sampling process in 1D CNN structure and the features of time-scale data are better preserved in the process.

Three different CNN structures were tested in comparing the performance of time-scale data with time-domain data as input. In all three configurations, applying time-scale values as input resulted in improved fault detection performance. Only a single band for time-scale data including outer race fundamental fault frequency was used in the analysis. The future work could include multi-band input for the 1D CNN classifier to explore the possibility of further improving the detection rates.

REFERENCES

[1] R. Zhang, Z. Peng, L. Wu, B. Yao, and Y. Guan, "Fault diagnosis from raw sensor data using deep neural networks considering temporal coherence," Sensors, doi:10.3390/s17030549, pp. 549-565, 2017.

[2] D.C. Ciresan, U. Meier, L.M. Gambardella, and J. Schmidhuber, "Deep big simple neural nets for handwritten digit recognition," Neural Comput., vol. 22, pp. 3207-3220, 2010.

[3] D. Scherer, A. Muller, and S. Behnke, "Evaluation of pooling operations in convolutional architectures for object recognition," In Proceedings of the Int. Conf. on Artificial Neural Networks (ICANN), Thessaloniki, Greece, pp. 92-101, Sept. 2010.

[4] A. Krizhevsky, I. Sutskever, and G. Hinton, "Imagenet classification with deep convolutional neural networks," In Proceedings of the Advances in Neural Information Processing Systems (NIPS), Lake Tahoe, pp. 1097-1105, Dec. 2012.

[5] S. Kiranyaz, T. Ince, and M. Gabbouj, "Real-Time Patient-Specific ECG Classification by 1D Convolutional Neural Networks," IEEE Trans. Biomed. Eng., vol. 63, pp. 664-674, 2015.

[6] T. Ince, S. Kiranyaz, L. Eren, M. Askar, and M. Gabbouj, " Real-Time Motor Fault Detection by 1D Convolutional Neural Networks," IEEE Trans. Ind. Electron., vol. 63, pp. 7067-7075, 2016.

[7] L. Eren, "Bearing Fault Detection by One-Dimensional Convolutional Neural Networks," Mathematical Problems in Engineering, pp. 1-9, 2017.

[8] L.Eren, T.Ince, and S.Kiranyaz, "A Generic Intelligent Bearing Fault Diagnosis System Using Compact Adaptive 1D CNN Classifier," Journal of Signal Processing Systems, vol. 91, pp. 179-189, 2019.

[9] L. Eren, and M.J. Devaney, "Bearing Damage Detection via Wavelet Packet Decomposition of the Stator Current," IEEE Trans. Instrum. Meas. vol. 53, pp. 431–436, 2004.

[10] R. Yan, R.X. Gao, and X. Chen, "Wavelets for fault diagnosis of rotary machines: A review with applications," Signal Process, vol 96, pp. 1-15, 2014.

[11] K. Kim, and A.G. Parlos, "Induction motor fault diagnosis based on neuropredictors and wavelet signal processing," IEEE/ASME Trans. Mechatron, vol. 7, pp. 201-219, 2002.

[12] M. Van, and H.-J. Kang, "Bearing defect classification based on individual wavelet local fisher discriminant analysis with particle swarm optimization," IEEE Trans. Ind. Informat., vol.12, pp.124-135, 2016.

[13] H. Qiao, H. Zhengjia, Z. Zhousuo, and Z. Yanyang, "Fault diagnosis of rotating machinery based on improved wavelet package transform and SVMs ensemble," Mech. Syst. Signal Process, vol. 21, pp. 688-705, 2007.

[14] M.F. Yaqub, I. Gondal, and J. Kamruzzaman, "Inchoate fault detection framework: Adaptive selection of wavelet nodes and cumulant orders," IEEE Trans. Instrum. Meas., vol. 61, pp. 685-695, 2012.

[15] P. Konar, and P. Chattopadhyay, "Bearing fault detection of induction motor using wavelet and support vector machines (SVMs)," Appl. Soft Comput., vol. 11, pp. 4203–4211, 2011.

[16] G.F. Bin, J.J. Gao, X.J. Li, and B.S. Dhillon, "Early fault diagnosis of rotating machinery based on wavelet packets – empirical mode decomposition feature extraction and neural network," Mech. Syst. Signal Process, vol.27, pp. 696-711, 2012.

[17] V. Wowk, Machinery Vibration, Measurement and Analysis, McGraw-Hill, 1991.

[18] Case Western Reserve University Bearing Data Center Website ⟨http://csegroups.case.edu/bearingdatacenter/home⟩.

Comparison of Two-Level and Three-Level NPC Inverter Topologies for a PMSM Drive for Electric Vehicle Applications

Alican Madan
Electrical and Electronics Engineering Department
Middle East Technical University
Ankara, Turkey
alican.madan@metu.edu.tr

Emine Bostanci
Electrical and Electronics Engineering Department
Middle East Technical University
Ankara,Turkey
emineb@metu.edu.tr

Abstract— **Multidimensional comparison of two-level and three-level DC/AC converters for a 120 kW permanent magnet synchronous machine (PMSM) drive is carried out in this study. Comparison of two topologies with Sinusoidal Pulse Width Modulation (SPWM) and Space Vector Pulse Width Modulation (SVPWM) by means of output current THD, conduction and switching losses, thermal stresses on semiconductors and switching frequency limitation are investigated at various operating points. Electro-thermal simulation of both topologies are achieved using PLECS tool with Silicon IGBT and diode pairs from SEMIKRON. Results of these analysis show the limitations of two-level inverter topology in terms of the switching frequency and point out the advantage of three-level inverter in high speed PMSM drives.**

Keywords—electric vehicle, motor drive, PMSM, three-level inverter, two-level inverter

I. INTRODUCTION

Permanent Magnet Synchronous Machines (PMSMs) are commonly used in vehicle traction applications and there are numerous studies focusing on increasing the performance and reliability of their drive systems. It is evident that the trend in electric machine design is to increase the rotational speed aiming a higher torque and power density of the drive system. As fundamental frequency of the drive system increases, operation at higher switching frequencies and if possible at higher DC bus voltage levels are required to assure a low output voltage and current THD as well as a good dynamic control performance.

Two-level voltage source inverter (VSI) is the commercial converter type used in PMSM drives. As semiconductor switches, conventional Si IGBT and diode pairs are used for driving PMSMs in traction applications. However, in recent years, SiC MOSFETs and SiC diodes have been involved in the motor drive market to account for the need in higher switching frequency modulation [1]. As stated in [2], switching frequency should be at least 9-12 times of the maximum fundamental frequency. However, conventional Si IGBT diode pairs suffer from high switching losses in two-level VSI topology. SiC MOSFET and Si based IGBT diode pair comparison is done for an 80 kW electric motor PMSM drive in [1]. Usage of SiC MOSFETs instead of Si IGBT results in 5% range extension for the electric car. SiC Mosfets are now preferable with two-level topology due to low switching losses of semiconductors and high switching frequency requirement of PMSM in maximum speed operation.

Three-level topology is expected to be advantageous at higher switching frequency operations with conventional Si IGBT diode pairs due to lower switching losses. The reasons are as follows: First, blocking voltage requirement of the main switch is halved in three-level topology. Second, switching losses do not linearly change with blocking voltage, for example an exponent of 1.3-1.4 is used to account for the effects of the blocking voltage in [3]. That means switching losses per switch are expected to drop below 50% in a 3-level topology, giving us the opportunity to increase the switching frequency.

In the literature, advantages of three-level inverters over two-level inverters are discussed for PMSM drive applications. Three-level and two-level topologies are compared in terms of output voltage THD, efficiency, and fault tolerance in [4]. In all aspects, three-level topology has a superior performance. Similarly, three-level NPC topology is shown to be more compatible with PMSM drives for traction applications in terms of current THD, torque ripple, and switching losses aspects in [5]. Besides, according to the comparative cost analysis of two-level and three-level topologies presented in [2], three-level NPC type topology has a slightly higher initial cost but a lower operational cost due to its higher efficiency. Although superior performance stated in the literature, higher number of switches and floating capacitor voltage are main design challenges of the three-level NPC topology. To address the floating capacitor voltage issue, various balancing techniques are presented in [6-8].

Considering the increase in DC bus voltage levels, topology selection is also getting as important as switch selection. Commercially available semiconductor switches have discrete steps of blocking voltage levels. In DC bus voltage levels around 700-900 V, switches with 1200 V voltage blocking capability are required in 2-level VSIs. At this point, semiconductor technologies with lower switching losses are the choice of developers. On the other hand, a multi-level VSI may provide a solution with conventional semiconductor switches thanks to the lower turn-on and turn-off energies compared to 2-level topologies. This study aims to explore the operating limits of a 3-level NPC VSI by investigating its output characteristics and thermal limits at various switching frequencies.

A multidimensional comparison of two-level and three-level inverter topologies for a 120 kW PMSM drive is made in this paper. Comparison of these topologies with two major switching techniques namely Sinusoidal Pulse Width Modulation (SPWM) and Space Vector Pulse Width Modulation (SVPWM) are conducted using SEMIKRON conventional Si IGBT diode pairs. DC Bus voltage of the motor drive is selected as 850 V. Therefore; 650 V IGBT diode pair is used in three-level topology while1200 V IGBT is selected for two-level topology. In order to make a

978-1-5386-7688-2/19 $31.00 © 2019 IEEE

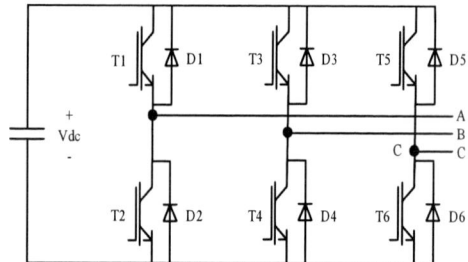

Fig. 1. Two-level topology

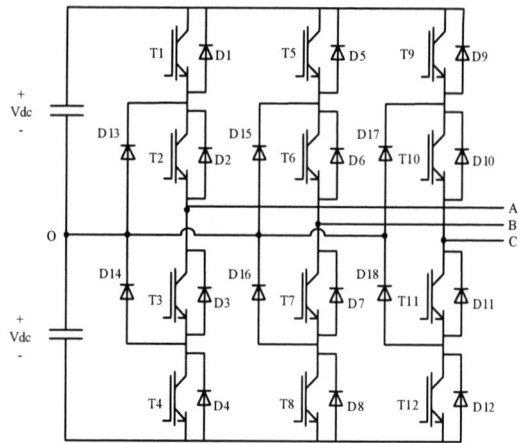

Fig. 2. Three-level NPC topology

TABLE I.   SELECTED MODULES

Topology	IGBT and Diode Module	VCE(max)(V)	Ic(A)
Two-level	SKiiP39GB12E4V1_HPTP	1200	388
Three-level NPC	SkiM401MLI07E4	650	317

reasonable comparison, same heatsinks are assumed in both topologies.

The following four cases are analyzed; two-level topology with SPWM, two-level topology with SVPWM, three-level NPC topology with SPWM, and three-level topology with SVPWM. Comparisons are done in terms of output voltage and current quality, junction temperatures, and switching frequency limitation aspects. First of all, quality of the output voltage and current waveforms are compared at 30% of rated torque and at various rotational speeds. These operation points are selected to account for a daily drive cycle. This is followed by thermal analysis in Plexim/PLECS simulation platform performed at rated torque and rated speed to investigate thermal behavior of the inverter at full load. Junction temperatures of each semiconductor device are calculated and switching frequency limits of each case are explored by limiting the maximum thermal junction temperature of semiconductors to 130ºC.

II. TOPOLOGIES AND SWITCHING TECHNIQUES

Two-level three phase topology consists of six switches as shown in Fig. 1. T1 and T2, T3 and T4, T5 and T6 work complementarily in order to avoid cross-conduction. Three-level NPC topology has four main switches and two clamping diodes in each leg as shown in Fig. 2. T1 and T3, T2 and T4 work complementarily. In the other inverter legs, corresponding main switches work complementarily as well. Bi-directional main switches should be used in both topologies. MOSFET, SiC MOSFET, GaNFET, IGBT-Diode pairs could be used as the main switch depending on the current, blocking voltage and switching frequency requirements. For the selected power level, conventional IGBT-Diode pairs are widely used in industry. Therefore; IGBT-Diode pairs from SEMIKRON are selected for the analysis in this paper and the two IGBT-Diode modules used in each topology are given in Table I.

A. Two-Level Inverter with SPWM

In SPWM applied for two-level three phase inverter, three reference sine waves with 120ºdeg phase shifts wrt. each other are used as reference signals for each leg. By comparing reference sine waves with carrier triangle waveforms, gate signals are obtained for upper and lower switches. Exemplary carrier signal, three reference signals and node voltages with respect to DC bus ground of each leg are shown in Fig. 3.

B. Two-Level Inverter with SVPWM

SVPWM technique is widely used in many inverter applications. The main advantage of SVPWM is the degree of freedom of selecting voltage vectors in a switching period compared with SPWM method. SVPWM also provides a better DC bus voltage utilization than SPWM. For a two-level

inverter topology, 61.2% DC bus utilization can be achieved with SPWM and 70.7% with SVPWM. However; SVPWM technique needs higher computational power due to its nature. In recent years, microcontrollers and DSPs allow designers to implement SVPWM algorithms easily.

SVPWM technique is based on the selection of optimum voltage vectors in each switching period. In order to achieve this, three phase reference vectors are transformed to α-β axis (Clarke's transformation) and a rotating reference vector is formed as a combination of $2^3 = 8$ voltage vectors, from which two are zero vectors that are 000 and 111 vectors. In each switching cycle, components of the reference vector and dwell times are calculated by software.

Switch logic signals for corresponding space vectors are given in Table II. Space vectors of two-level topology and defined sectors are shown in Fig. 4 and Table III, respectively. Each region corresponds to a 60ºdeg rotation of reference vector V_r as shown in Fig. 4.

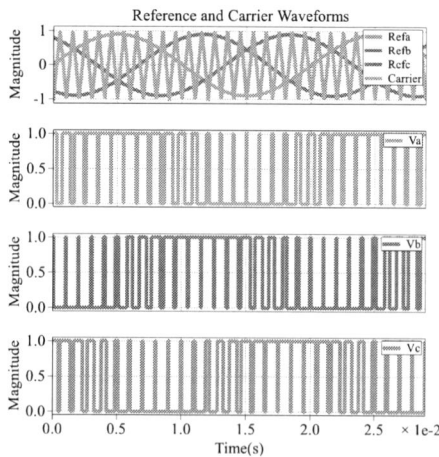

Fig.3.   Reference and carrier waveforms of two-level SPWM.

TABLE II. SWITCH LOGIC FOR CORRESPONDING SPACE VECTORS

		UPPER SWITCHES			LOWER SWITCHES		
Vectors		T1	T3	T5	T2	T4	T6
U0	000	1	1	1	0	0	0
U1	100	0	1	1	1	0	0
U2	110	0	0	1	1	1	0
U3	010	1	0	1	0	1	0
U4	011	1	0	0	0	1	1
U5	001	1	1	0	0	0	1
U6	101	0	1	0	1	0	1
U7	111	0	0	0	1	1	1

Clark's transformation is done as follows:

$$\begin{bmatrix} V_\alpha \\ V_\beta \end{bmatrix} = \frac{2}{3} \begin{bmatrix} 1 & -\frac{1}{2} & -\frac{1}{2} \\ 0 & \frac{\sqrt{3}}{2} & -\frac{\sqrt{3}}{2} \end{bmatrix} \begin{bmatrix} V_a \\ V_b \\ V_c \end{bmatrix} \qquad (1)$$

Then reference vector V_r is defined as:

$$\vec{V_r} = V_\alpha + jV_\beta \qquad (2)$$

In SVPWM algorithm, amplitude and phase of the reference vector V_r are required. These quantities are determined as follows:

$$\alpha = \tan^{-1}\left(\frac{V_\beta}{V_\alpha}\right) \text{ and } |\vec{V_r}| = \sqrt{V_\alpha^2 + V_\beta^2} \qquad (3)$$

After determining the magnitude and the phase of V_r, the space vectors and their on-times must be calculated. In order to calculate on-times for each main vector, volt-seconds law is applied. On-time calculation of vectors in sector 1 can be done as follows:

$$\int_0^{T_s} \vec{V_r} = \int_0^{T_1} \vec{V_1} dt + \int_0^{T_2} \vec{V_2} dt + \int_0^{T_0} \vec{V_0} dt \qquad (4)$$

$$\vec{V_r} = |V_r| e^{j\alpha}, \vec{V_0} = 0, \vec{V_1} = \frac{2}{3}V_{dc}, \vec{V_2} = \frac{2}{3}V_{dc}e^{\frac{j\pi}{3}}, \vec{V_7} = 0 \qquad (5)$$

$$m_a = \frac{3V_r}{2V_{dc}} \qquad (6)$$

TABLE III. SECTOR AND POSITION OF REFERENCE VECTOR

Sector	Position of $\vec{V_r}$
1	0°<ωt<60°
2	60°<ωt<120°
3	120°<ωt<180°
4	180°<ωt<240°
5	240°<ωt<300°
6	300°<ωt<360°

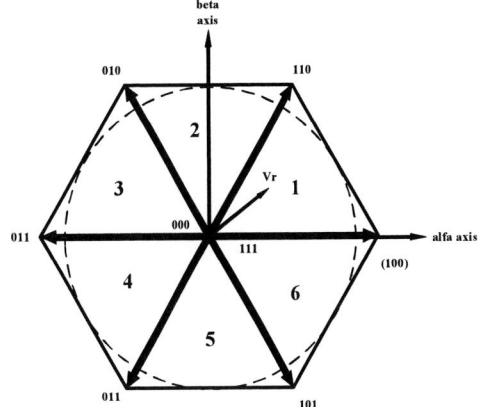

Fig. 4. Two-level space vectors

TABLE IV. ON-TIME OF SEMICONDUCTORS IN EACH REGION

R	T_1	T_2	T_0
1	$T_s m_a \sin(\frac{\pi}{3} - \alpha)$	$T_s m_a \sin(\alpha)$	$T_0 = T_s - T_1 - T_2$
2	$T_s m_a \sin(\frac{2\pi}{3} - \alpha)$	$T_s m_a \sin(\alpha - \frac{\pi}{3})$	$T_0 = T_s - T_1 - T_2$
3	$T_s m_a \sin(\pi - \alpha)$	$T_s m_a \sin(\alpha - \frac{2\pi}{3})$	$T_0 = T_s - T_1 - T_2$
4	$T_s m_a \sin(\frac{4\pi}{3} - \alpha)$	$T_s m_a \sin(\alpha - \pi)$	$T_0 = T_s - T_1 - T_2$
5	$T_s m_a \sin(\frac{5\pi}{3} - \alpha)$	$T_s m_a \sin(\alpha - \frac{4\pi}{3})$	$T_0 = T_s - T_1 - T_2$
6	$T_s m_a \sin(2\pi - \alpha)$	$T_s m_a \sin(\alpha - \frac{5\pi}{3})$	$T_0 = T_s - T_1 - T_2$

From (5) and (6), on-time calculations of space vectors in each sector are derived as in Table IV. T_1 is the on-time of the first main space vector $\vec{V_1}$ in the active state, T_2 is the on-time of the second main space vector $\vec{V_2}$. T_0 time is shared equally among two zero vectors $\vec{V_0}$ and $\vec{V_7}$.

C. Three-Level NPC Topology with SPWM

In three-level NPC type three phase inverter topology, SPWM is applied with two different carrier waveforms separated with a DC offset from each other. Carrier waveforms and reference waveforms are shown in Fig. 5. carrier1 is used for the top switches of the upper side of each leg that are T_1, T_5, and T_9 and carrier2 is for bottom switches of the upper side that are T_2, T_6, and T_{10}. Similar to a two level topology, Phase B and Phase C legs have shifted reference sine waves by 120°deg wrt. each other. The lower side switches in each leg works complementarily with top side corresponding switch. Switch logic signals, reference and carrier waveforms are shown in Fig. 6.

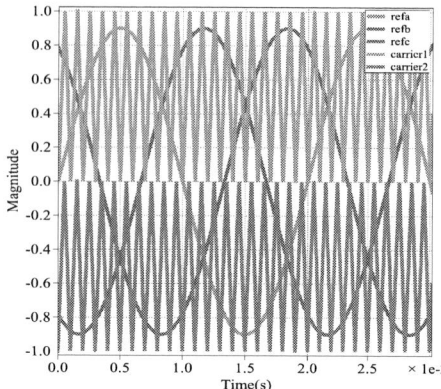

Fig. 5. Three-level SPWM three-phase references and carriers

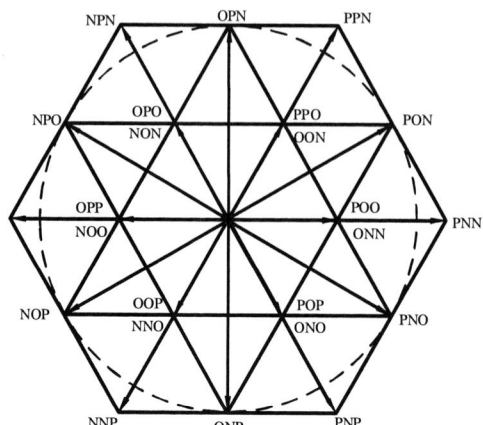

Fig. 7. Three-level space vectors

D. Three-Level NPC Topology with SVPWM

In three level topology, there are in total of 3^3=27 voltage vectors. 24 of them are active vectors, from which 12 are short vectors, 6 are medium vectors and 6 are long vectors. Other 3 vectors are zero vectors located in the center. In three level inverter, six main sectors are the same as in two-level case and each sector consists of four sub-sectors. Fig. 7 and 8 show space vectors and four sub-sectors in the first main sector, respectively. Due to the additional degree of freedom, there are various modulation strategies to manage space vectors to generate PWM signals. In this study, symmetrical SVPWM method that is commonly used to achieve low current THD at the output is used [9].

V_{dc-cap} is defined as the voltage of one of the floating capacitors as can be seen in Fig. 2. The capacitor voltages are assumed to be equal in this study that is $V_{dc-cap}= V_{dc}/2$. Each connection point can attain three different voltage levels in a 3-level VSI; P represents V_{dc-cap}, O represents zero potential and N represents -V_{dc-cap} at the switching node of a phase leg as listed in Table V. Small vectors (POO, ONN, etc.) have magnitude of $2V_{dc-cap}/3$. Medium vectors (PON, OPN, etc.) have magnitude of $2\sqrt{3} V_{dc-cap}/3$. Large vectors (PNN, PPN, etc..) have magnitude of $4V_{dc-cap}/3$.

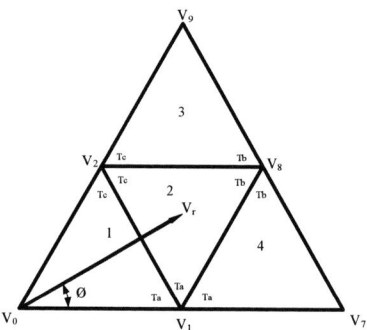

Fig. 8. Sub-sectors in three-level SVPWM

TABLE V.   SYMBOLS AND CORRESPONDING SWITCH STATES & PHASE VOLTAGES

Symbol	T1	T2	T3	T4	Voltage
P	1	1	0	0	V_{dc}
O	0	1	1	0	0
N	0	0	1	1	-V_{dcs}

Main sector detection is done the same way as in two-level inverter. Sub–sector detection is done with the angle and magnitude of V_r vector. One example of on-time calculation in main sector 1 and sub-sector 1 is as follows:

$$\vec{V_1}T_a + \vec{V_2}T_c = \vec{V_r}T_s \tag{7}$$

$$\frac{2V_{dc}}{3} e^{j0}T_a + \frac{2V_{dc}}{3} e^{\frac{j\pi}{3}}T_c = |\vec{V_r}|T_s \tag{8}$$

k is defined as follows:

$$k = \frac{\sqrt{3}}{2}\left(\frac{|\vec{V_r}|}{V_{dc-cap}}\right) \tag{9}$$

Where,

$$V_\alpha = V_{aref} - 0.5V_{bref} - 0.5V_{cref} \tag{10}$$

$$V_\beta = (V_{bref} - V_{cref})(\frac{\sqrt{3}}{2}) \tag{11}$$

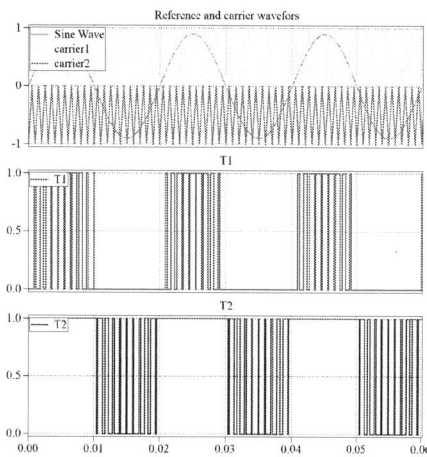

Fig. 6. Three-level SPWM phase A switching signals

978-1-5386-7688-2/19 $31.00 © 2019 IEEE

TABLE VI. CALCULATED TIMES AND SUB-SECTORS

T \ S	T_a	T_b	T_c
1	$2k\sin(\frac{\pi}{3}-\varnothing)$	$T_s-2k\sin(\frac{\pi}{3}+\varnothing)$	$2k\sin(\varnothing)$
2	$T_s-2k\sin(\varnothing)$	$2k\sin\left(\frac{\pi}{3}+\varnothing\right)-T_s$	$T_s-2k\sin(\frac{\pi}{3}-\varnothing)$
3	$2k\sin(\varnothing)-T_s$	$2k\sin(\frac{\pi}{3}-\varnothing)$	$2T_s-2k\sin(\frac{\pi}{3}+\varnothing)$
4	$2T_s-2k\sin(\frac{\pi}{3}+\varnothing)$	$2k\sin(\varnothing)$	$2k\sin(\frac{\pi}{3}-\varnothing)-T_s$

$$\varnothing = \tan^{-1}(\frac{V_\beta}{V_\alpha}) \qquad (12)$$

Then T_a, T_b, and T_c is calculated for each sub-sector as in Table VI, where T_s is switching period, T_a, T_b and T_c are on-time of each vector as shown in Fig. 8.

III. SIMULATION RESULTS AND DISCUSSIONS

A. Electrical Simulations

In this part, a 120 kW PMSM drive is modelled in PLECS software with the motor and driver parameters given in Table VII. Two-level VSI and three-level NPC VSI topologies are simulated with both SPWM and SVPWM, so in total four cases are analyzed. In order to conduct a fair comparison procedure, controller performances of the motor drives are kept same for all cases. Moreover, the same heatsinks are chosen for both topologies to avoid influence of the size of the cooling system. Analyses are done with different switching frequencies (carrier frequencies) varying between 8-12 kHz. Dead time is not included into simulations. Minimum switching frequency is selected as 8 kHz by considering maximum speed of the electric machine, that is 12000 rpm resulting in 600 Hz fundamental stator frequency. Maximum switching frequency is selected

TABLE VII. MOTOR PARAMETERS AND INVERTER SPECIFICATIONS

Parameter	Value
V_{dc}(V)	850
R_{stator}(mΩ)	6.6
L_d, L_q(uH)	600
Flux induced (V.s)	0.222
Nominal speed (rpm)	4583
Maximum speed (rpm)	12000
Pole pair number	3
$I_{phase,rms}$(A)	177
F_{sw} (kHz)	8-12
T_{max} (Nm)	250

by considering the thermal limitation of the two-level topology by setting the maximum junction temperature to 130 ℃.

As a controller structure, field oriented control (FOC) technique is implemented using outer speed and inner current loops. In order to compare controller performances, speed is kept constant at rated speed and step load torque change is applied to all four cases. An exemplary controller performance of motor drive is shown in Fig. 9. Same controller performances, which are speed regulation and dynamic response with a step load change, are achieved for each topology and each switching technique. I_d and I_q currents that are Park's transformation outputs are regulated by the inner current control loop. In FOC technique, I_q current supplies electrical torque to sustain load torque in a surface mount PMSM, whereas I_d current reference is zero until rated speed is exceeded. In the field weakening region, I_d current reference is adjusted to be able to reach higher speed with a limited DC bus voltage.

Current THD analyses are done with constant torque at 75 Nm (30% of the rated torque). At different operating speeds, current THDs are observed with different switching frequencies. The most distinct operation between analyzed cases is observed at rated speed. Therefore, current THD values with rated and half of the rated speed are given in Fig. 10 and Fig. 11, respectively. Difference between current THDs gets more distinct as speed gets close to rated speed since average on time of semiconductors increase with the motor speed. Therefore; current ripple increases for higher speeds.

Three-level NPC topology with SVPWM has superior performance for all motor speeds and switching frequency cases. Three-level topology generates voltage pulses of half of the amplitude of the two-level topology at the output. Therefore; current THD of the three-level topology is

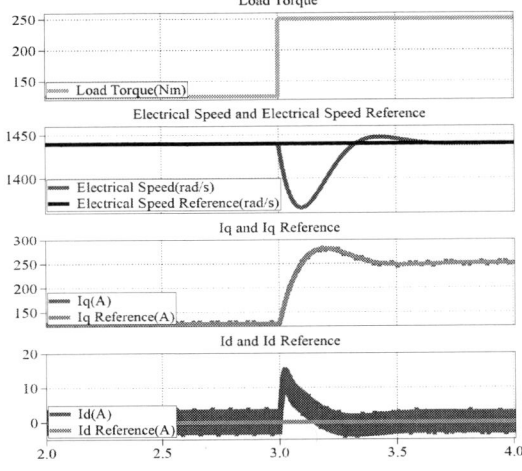

Fig. 9. Dynamic control performance from half load to full load

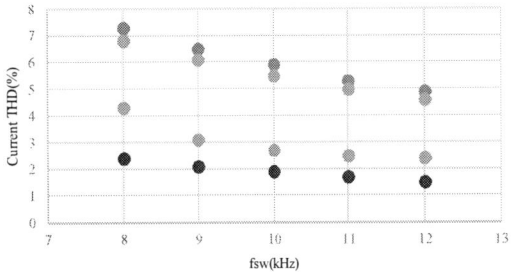

Fig. 10. Current THD (%) vs. switching frequency at 4583 rpm

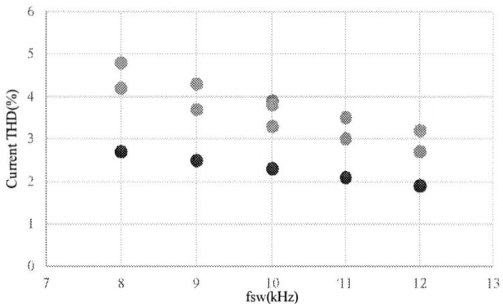

Fig. 11. Current THD (%) vs. switching frequency at 2291 rpm

● Two-Level SPWM ● Two-Level SVPWM ● Three-Level SPWM ● Three-Level SVPWM

expected to be lower than the two level topology. Moreover, SVPWM has better performance than SPWM in terms of current THD in both topologies. Whereas, the performance difference is more noticeable in the three-level topology. Current THD in all cases decreases as switching frequency increases and decrease in output current THD is expected to result in lower iron losses in the machine [10].

B. Electro-thermal Simulations

Electro-thermal simulations are done at rated speed and rated torque. Both semiconductor modules are assumed to be directly connected to the heatsink without any thermal interface material. PLECS thermal models calculate loss of a semiconductor by summing up the energy losses of semiconductors within a switching period. Then, averaging these in a fundamental cycle of output voltage and current, average loss values of each module are calculated. Loss calculation of the IGBTs and the diode of the selected modules are carried out regarding to the given information in [3].

Electro-thermal simulation models can be seen in Fig. 12 and Fig. 13. Same heatsink thermal parameters are used for both topologies having R_{s-a}=0.023 °C/W as sink-ambient thermal resistance. Since both modules are connected to the

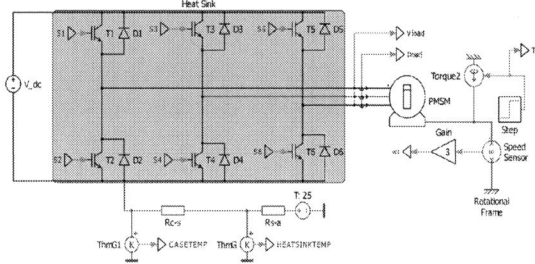

Fig. 12. Two-level topology electro-thermal model

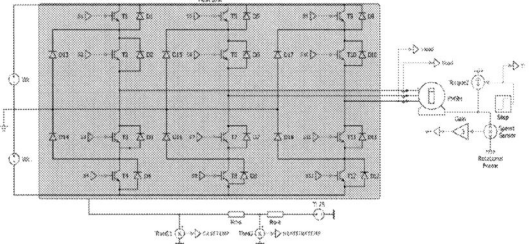

Fig. 13. Three-level topology electro-thermal model

heatsink directly, R_{c-s} case-sink thermal resistances are assumed to be zero. Case-heatsink and heatsink-ambient thermal resistances are implemented in the electro-thermal simulation model as depicted in the circuit schematics given Fig. 12 and Fig. 13. IGBT and diode junction-case thermal resistances are derived from datasheet and SEMISEL design tool values. SEMIKRON's WP16_280 heatsink module is used in both designs. Following assumptions are made for the thermal calculations:

- Water cooling is assumed with a flow rate of 6 l/min.
- Liquid inside the heatsink is assumed to be composed of 50% water and 50% glycol.
- Ambient temperature is set to 25 °C.

1) IGBT Conduction Losses

Collector-emitter voltage V_{CE} vs. collector current I_C curve is embedded into PLECS IGBT thermal model. Temperature dependency of conduction losses are not integrated into software due to its negligible effect on losses. Conduction losses is calculated as follows:

$$P_{con,IGBT} = \frac{1}{T_1} \int_0^{T_1} I_{C(t)} V_{CE(t)} \, D_{IGBT}(t) dt \quad (13)$$

$$V_{CE(t)} = V_{CE0} + I_o(t) r_{ce} \quad (14)$$

V_{CE0}: Zero crossing of linearized I_{CE} vs. V_{CE} curve

r_{ce}: Inverse slope of I_{CE} vs. V_{CE} curve

By implementing collector-emitter voltage by curve fitting, PLECS software calculates the average conduction loss via (13) and (14).

2) IGBT Switching Losses

Switching losses are calculated via given turn-on energy E_{on} and turn-off energy E_{off} graphs in the datasheet of each semiconductor. Turn-on and turn-off energies are calculated for the given collector emitter current and the blocking voltage of the semiconductor switch. Mathematical equations that are (15), (16) and (17) are implemented into the software. By summing up the turn-on and turn-off energies in each switching cycle, total switching loss energy of a IGBT is calculated. By averaging this in one fundamental frequency, switching loss of a IGBT is determined.

K_v in (16) and (17) is a manufacturer dependent coefficient that is selected as 1.3 or 1.4 for IGBTs of SEMIKRON [3].

$$P_{sw,IGBT} = \frac{1}{T_1} \sum_{i=1}^{T_1 f_{sw}} E_{on}(t_i) + E_{off}(t_i) \quad (15)$$

$$E_{on} = E_{on}(I_c) \left(\frac{v_{in}}{v_{ref}}\right)^{K_v} \left(1 + TC_{Esw}\left(T_j - T_{ref}\right)\right) \quad (16)$$

$$E_{off} = E_{off}(I_c) \left(\frac{v_{in}}{v_{ref}}\right)^{K_v} \left(1 + TC_{Esw}\left(T_j - T_{ref}\right)\right) \quad (17)$$

V_{in} : V_{CE} blocking voltage of IGBT

$E_{on,off}(I_c)$: E_{on} and E_{off} value for corresponding current

K_v : Exponents for voltage dependency of switching losses

TC_{Esw} : Temperature coefficient of IGBT switching losses

3) Diode Conduction Losses

$$P_{con,diode} = \frac{1}{T_1} \int_0^{T_1} I_{F(t)} V_{F(t)} D_{diode}(t) dt \qquad (18)$$

$$V_F(t) = V_{F0} + I_o(t) r_d \qquad (19)$$

V_{F0}: Zero crossing of linearized I_F vs. V_F curve of diode

r_d: Inverse slope of I_F vs. V_F curve of diode

4) Diode Switching Losses

Switching loss calculation of diode is done with the same procedure as for the IGBT using (20) and (21).

$$P_{sw,IGBT} = \frac{1}{T_1} \sum_{i=1}^{T_1 f_{sw}} E_{rr}(t_i) \qquad (20)$$

$$E_{rr} = E_{rr} \left(\frac{I_{out}(t)}{I_{ref}}\right)^{K_i} \left(\frac{V_{in}}{V_{ref}}\right)^{K_v} \left(1 + TC_{Err}\left(T_j - T_{ref}\right)\right) (21)$$

E_{rr}: Reverse recovery energy of diode

TC_{ERR}: Temperature coefficient of diode switching loss

Loss distribution results for analyzed cases are shown in Fig. 14 for switching frequencies of 8-12 kHz. SVPWM and SPWM does not have significant effect on thermal performance in both topologies that is conduction losses and switching losses are almost the same in two switching techniques. Switching loss decrease is significant from two-level to three-level topology. Since three-level topology has more switches than two-level topology, conduction losses are dominant in the three-level topology. Conduction losses are almost doubled in three-level topology.

Due to high fundamental frequency requirement in the field weakening region, high frequency switching is required in PMSM drives. However; in two-level topology, main switches are operating against double V_{CE} voltage stress compared with three-level topology. Therefore; switching losses are dominant in two-level topology. Another reason of high switching losses in two-level topology is as follows: Relation between switching losses and blocking voltage is not linear. Dominance of switching losses can be seen in Fig. 14.

At 12 kHz switching frequency, maximum junction temperature constraint is reached for two-level topology as can be seen in Fig. 15. On the other hand, three-level topology reaches a maximum junction temperature of 130 °C at 22 kHz switching frequency. That is with the given semiconductors, three-level topology could be used even at 22 kHz although two-level topology is limited to 12 kHz switching frequency. As expected, 22 kHz switching frequency will result in a better output current THD and less torque ripple at the output.

IV. CONCLUSION

A multidimensional comparison of two VSI topologies is done with two different switching techniques. The three-level NPC VSI topology is superior in both thermal performance and output quality point of views. Conduction losses are dominant in three-level topology. However; due to lower collector-emitter blocking voltage requirement, 650 V IGBT products can be used in three-level design. Therefore; switching losses have minor effect in three-level topology for the same switching frequencies. For the given semiconductors, two-level topology is limited to 12 kHz switching frequency. On the other hand, three-level topology

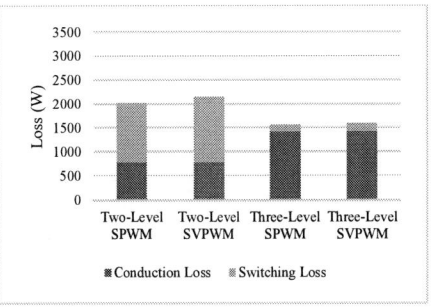

a) 8 kHz switching frequency

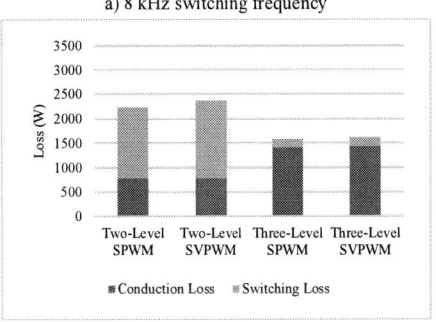

b) 9 kHz Switching Frequency

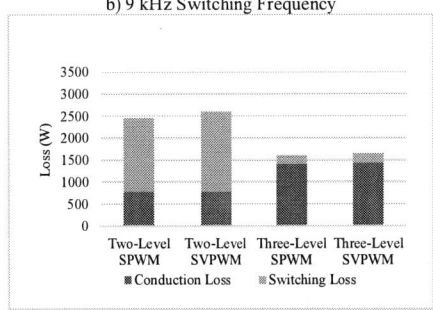

c) 10 kHz Switching Frequency

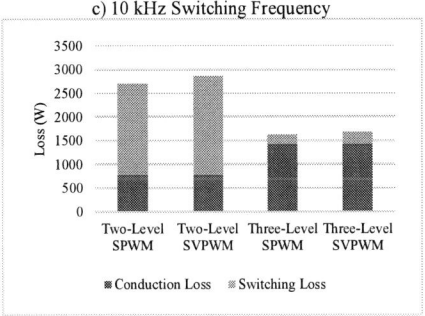

d) 11 kHz Switching Frequency

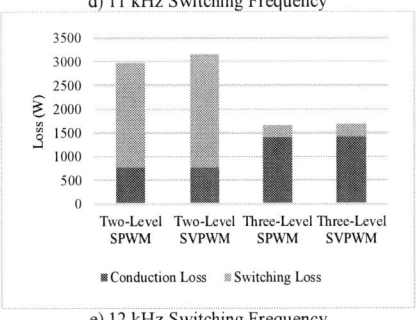

e) 12 kHz Switching Frequency

Fig. 14. Loss distribution of topologies and switching techniques

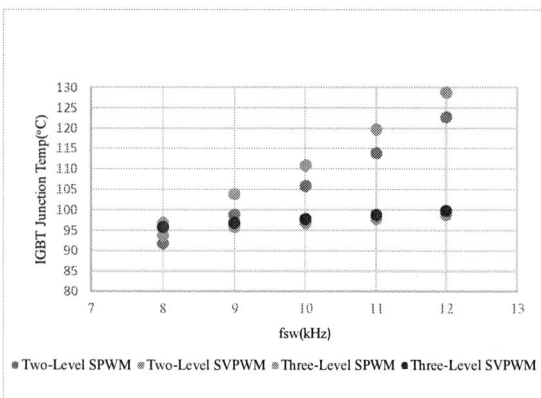

Fig. 15. Junction temperatures of IGBTs in each topology and switching technique

can reach up to 22 kHz for the same maximum junction temperature limitation. Therefore; three-level topology is superior for applications such that high fundamental frequency is needed. However; this inference is valid for conventional Si IGBT diode pairs. With the recent developments in semiconductor technologies, higher switching frequency operation in such power levels with two-level topology is possible as SiC MOSFETs have better switching performance than conventional Si IGBTs [1].

Thermal performance effect of switching techniques is not significant in both topologies. This can be seen from the efficiency comparison results given in Table VIII for a power factor of 0.83. On the other hand, three-level NPC topology has a better efficiency performance in whole switching frequency range. SVPWM has a better output current THD performance compared with SPWM in both topologies. Therefore; SVPWM technique can be preferable due to low THD at the output. Additionally, due to better DC bus utilization, SVPWM can also be preferable in order to decrease DC bus voltage level and switching losses. Although three-level SVPWM is able to achieve a superior performance in all dimensions, topology has few drawbacks. Floating capacitor voltage and manipulation of higher number of switches make the three-level topology more challenging compared to the two-level topology. Also the high number of switches increases initial cost.

TABLE VIII. EFFICIENCY RESULTS FOR TWO TOPOLOGIES AND SWITCHING TECHNIQUES

Topology	Switching Technique	Efficiency (%) with f_{sw} (kHz)					P.F.
		8	9	10	11	12	
Two-level	SPWM	98.3	98.1	97.9	97.7	97.5	0.83
Two-level	SVPWM	98.2	98.0	97.8	97.6	97.4	0.83
Three-level NPC	SPWM	98.7	98.7	98.7	98.6	98.6	0.83
Three-level NPC	SVPWM	98.7	98.6	98.6	98.6	98.6	0.83

REFERENCES

[1] Kumar, K.; Bertoluzzo, M.; Buja, G., "Impact of SiC MOSFET traction inverters on compact-class electric car range," in Power Electronics, Drives and Energy Systems (PEDES), 2014 IEEE International Conference on , vol., no., pp.1-6, 16-19 Dec. 2014.

[2] R. Teichmann and S. Bernet, "A comparison of three-level converters versus two-level converters for low-voltage drives, traction, and utility applications," Industry Applications, IEEE Transactions on, vol. 41, no. 3, pp. 855 – 865, 2005.

[3] A. Wintrich, U. Nicolai, and T. Reimann, "Semikron application manual," 2011.

[4] A. Bubert, S.-H. Lim, and R. W. De Doncker, "Comparison of 2-level B6C and 3-level NPC inverter topologies for electric vehicles," in Proc. IEEE Southern Power Electron. Conf. (SPEC), Puerto Varas, Chile, 2017, pp. 1–6

[5] A. Choudhury, P. Pillay, M. Amar and Sheldon. S. Williamson, "Performance comparison study of two and three-level inverter for electric vehicle application," in Proc. on IEEE Transportation Electrification Conf. and Expo, Dearborn, USA, June 2014, pp. 1-6.

[6] Peng, F.Z.: 'A generalized multilevel inverter topology with self voltage balancing', IEEE Trans. Ind. Appl., 2001, 37, pp. 611–618

[7] J. Pou, D. Boroyevich, and R. Pindado, "New feedforward space-vector PWM method to obtain balanced ac output voltages in a three-level neutral-point-clamped converter," IEEE Trans. Power Electron., vol. 49, pp. 1026–1034, Oct. 2002

[8] A. Choudhury, P. Pillay, and S. Williamson, "DC-link voltage balancing for a three-level electric vehicle traction inverter using an innovative switching sequence control scheme," IEEE J. Emerging Selected Topics Power Electron., vol. 2, no. 2, pp. 296-307, June 2014.

[9] Haibing Hu; Wenxi Yao; Zhengyu Lu, "Design and Implementation of Three-Level Space Vector PWM IP Core for FPGAs," Power Electronics, IEEE Transactions on , vol.22, no.6, pp.2234-2244, Nov. 2007

[10] G. Mademlis, Y. Liu, J. Zhao, "Comparative Study of the Torque Ripple and Iron Losses of a Permanent Magnet Synchronous Generator Driven by Multilevel Converters," XIII International Conference on Electrical Machines (ICEM), pp. 1406-1412, 2018

Current Data Fusion Through Kalman Filtering for Fault Detection and Sensor Validation of an Electric Motor

Sadra Mousavi
Department of Electrical Engineering
Istanbul Technical University
Istanbul, Turkey
sadramousavi@itu.edu.tr

Duygu Bayram
Department of Electrical Engineering
Istanbul Technical University
Istanbul, Turkey
bayramd@itu.edu.tr

Serhat Seker
Department of Electrical Engineering
Istanbul Technical University
Istanbul, Turkey
sekers@itu.edu.tr

Abstract— In this paper, a data fusion method through Kalman filtering for condition monitoring (CM) and fault detection (FD) of electrical motors (EM) is proposed. Moreover, sensor validation (SV) and tracking the fault source (either the sensor or the process) are possible through this approach. A current signal, obtained from different sensors, is used for the case study. A fused current information is calculated through Kalman filtering. Afterwards, the effects of the measurement and process noises on the fused signal, are discussed, respectively. Then, it is noticed distinctive features by the comparison of the fused and original signal in terms of spectral and statistical properties. In addition, Kalman gain is monitored to investigate the impact of the process noise and measurement noise to perform SV. The proposed method is developed on the artificial data and then tested on the real data collected from an Induction Motor (IM).

Keywords—Kalman filtering, Kalman gain, fault detection, condition monitoring, current signal, data fusion, sensor validation, electrical motors.

I. INTRODUCTION

In 1960, Rudolf Kalman presented a recursive algorithm to solve the discrete data linear filtering problem. The method was then developed by Kalman and Bucy in 1961. Nowadays, this method is known as Kalman filtering which has been using in various applications widely. Kalman filtering is employed to approximate the state of a process through recursive equations, which provide minimum error. The strongest ability of the filter is providing accurate estimation of the past, present and future states of a system, even when the model is not precise enough [1-3]. Many manuscripts have been published about Kalman filtering application from different point of views.

Since 1970s, FD has attracted the scientists' attention, because of its vital role in security and reliability of the complex systems. Fault detection utilizing Kalman filtering is one of the popular methods. Kalman filtering method is significantly efficient while the system is disrupted by random noises, because it can deal easily with disturbances or process noises [4, 5]. In paper [4], the authors presented a FD method based on Kalman filtering for a linear dual motor system considering influence of sensor faults, actuator faults and random noises. In this method, the residual signal is generated based on Kalman filter and a threshold is defined to determine the occurrence of faults. In paper [5] a structure of FD using Kalman filter is presented and applied to DC motor. Through Kalman filter, uncertainties are eliminated, as a result, error in FD is reduced [5]. The authors in paper [6] proposed an approach based on Extended Kalman Filter (EKF) to estimate the disturbance of the stator phase currents in an IM. The extended model of the IM under healthy conditions containing disturbances is applied in the approach.

In paper [7], an electrical reduced-order model for an IM considering iron loss, is first presented to achieve thermal monitoring in real time. Then, different model extensions are applied to estimate the rotor and stator temperatures through EKF. In paper [8] a model-based CM method, which is tested in real time, is presented to detect and diagnose any fault in the stator windings of the brushless wound-field synchronous generator. EKF is employed as a state and parameter estimation technique for the presented model. The estimated rotor current, a fault parameter and the harmonic content of the damper and field currents are the examples of state and parameters which are obtained by means of EKF.

Kalman filtering is also one of the most popular methods for data fusion applications such as object tracking using multiple cameras [9], enhancement of FD and diagnosis [10] and etc.

The fault in machinery systems can be originated by the sensor or the process. To the best of authors knowledge, there is not any example to distinguish the fault source through a data fusion method empowered by Kalman filtering, in the literature. In this study, a data fusion application through Kalman filtering for CM and FD purposes is proposed. In this application, current signal of an IM is acquired from two different current sensors disturbed by the measurement and the process noises. These collected signals are fused together to evaluate the impact and effectiveness of disturbing noises. The origination of the noise is estimated, whether it is sensors or process. Moreover, Kalman gain is monitored to peruse the process noise effect and achieve SV. As a result, Kalman gain is proposed as a compact measure to diagnose the changes in the system behavior by its ability of providing information about the error covariance matrix and the state vector. Also, in literature [11] Kalman gain is used to demonstrate any alteration and fault in the system.

The study gives a brief mathematical background of the Kalman filtering in Section II. In Section III, data acquisition system (DAQ) is presented. In Section IV, a flow chart of the study is presented, the effects of measurement and process noises are discussed. Fault source determination by means of the Kalman gain and the effect of the different noises on the Kalman filter gain is studied on Section V. Finally, the paper is concluded at Section VI.

II. MATHEMATICAL BACKGROUND

A. Kalman method

In this section, a brief mathematical background for Kalman filtering and data fusion is represented [12-14]. The Kalman filter model assumes that the state of a system at a time k evolved from the prior state at time $k-1$ according to the equation below;

978-1-5386-7688-2/19 $31.00 © 2019 IEEE

$$x_k = Fx_{k-1} + Gu_{k-1} + w_{k-1} \qquad (1)$$

where x is the state vector, F is the state matrix, G is the input matrix and w is the process noise vector. The output vector of the system can also be computed, according to the model;

$$y_k = Hx_k + v_k \qquad (2)$$

where y is the output vector, H is the observation matrix, v is the measurement noise vector. The variables in (1) and (2) are defined in the table below:

TABLE I.   DIMENTIONS AND DEFINITIONS OF VARIABLES [12].

Variable	Description	Dimension
x	State vector	$n_x \times 1$
y	Output vector	$n_y \times 1$
u	Input vector	$n_u \times 1$
w	Process noise vector	$n_x \times 1$
v	Measurement noise vector	$n_y \times 1$
F	System matrix – state	$n_x \times n_x$
G	System matrix – input	$n_x \times n_u$
H	Observation matrix	$n_y \times n_x$

Note that in most of the applications the input matrix G and the observation matrix H, do not need to be square matrices. Output vector y can be obtained by measurement and it should be expressed as a function of the state vector x.

w_k contains the process noise terms for the state vector. The process noise is assumed to be a normal distribution Gaussian noise with covariance matrix Q [14].

v_k contains the measurement noise terms for each observation in the measurement vector. Like the process noise, the measurement noise is assumed to be zero mean Gaussian noise with covariance R [14].

The Kalman filter algorithm involves two stages; prediction and measurement updates. The standard Kalman filter equations for the prediction stage are given in (3) and (4) [12-14];

$$\hat{x}_{k|k-1} = F\hat{x}_{k-1} + Gu_{k-1} \qquad (3)$$

$$P_{k|k-1} = FP_{k-1}F^T + Q_t \qquad (4)$$

And measurement update equations are presented as follow;

$$K_k = P_{k|k-1}H^T(HP_{k|k-1}H^T + R_k)^{-1} \qquad (5)$$

$$\hat{x}_k = \hat{x}_{k|k-1} + K_k(\hat{y}_k - H\hat{x}_{k|k-1}) \qquad (6)$$

$$P_k = (I - K_k H)P_{k|k-1} \qquad (7)$$

In these equation $\hat{x}_k$ is predicted state vector, P_k is initial state error covariance matrix, K_k is Kalman gain matrix, $\hat{y}_k$ is measurement output and I is identity matrix. In order to start the loop, the initial state estimation matrix $\hat{x}_0$ and the initial state error covariance matrix P_0 are just needed.

First, state vector can be predicted using (3). In the next step, the state error covariance matrix can also be predicted using (4). Once the predicted values are obtained, the Kalman Gain matrix K is calculated by (5). In order to make a better estimation, predicted state vector is updated through (6). As the last step, the state error covariance matrix is restored using (7).

B. Data fusion through Kalman filtering

In fusion algorithm, assignment of the observation matrix H is very important. Observation matrix H maps the state vector into the output vector. Hence, in linear Kalman filter H matrix provides a linear transformation. H matrix indicates that which state variables are included and which are not. In addition, it illustrates dependence of the output vector to the state vector. For example, if the first and second states of a 3-dimensional state vector, are measured, H matrix should be selected as [12]:

$$H = \begin{bmatrix} 1 & 0 & 0 \\ 0 & 1 & 0 \\ 0 & 0 & 0 \end{bmatrix}; \qquad (8)$$

$$y_k = Hx_k \Rightarrow y_k = \begin{bmatrix} 1 & 0 & 0 \\ 0 & 1 & 0 \\ 0 & 0 & 0 \end{bmatrix}\begin{bmatrix} x_1 \\ x_2 \\ x_3 \end{bmatrix} = \begin{bmatrix} x_1 \\ x_2 \end{bmatrix} \qquad (9)$$

Furthermore, H matrix can be used for state combination which serves as a fusion tool. For example, in a system which the measured output depends on two states, the observation matrix is assigned as seen in (10), providing the linear combination of input states [12]:

$$H = \begin{bmatrix} 1 & 1 \end{bmatrix}; \qquad (10)$$

$$y_k = Hx_k \Rightarrow y_k = \begin{bmatrix} 1 & 1 \end{bmatrix}\begin{bmatrix} x_1 \\ x_2 \end{bmatrix} = x_1 + x_2 \qquad (11)$$

III. DATA ACQUISITION SYSTEM

The study is developed on artificial data then it is tested on real data collected from an IM. The data, which are acquired through NISCXI signal conditioning interface, are sampled with 12 kHz sampling rate. The antialiasing filter which is utilized in the DAQ system is NISCXI-1142 eight-order elliptic low pass filter with 4 kHz cutoff frequency, -80 dB stop band and 135 dB/octave roll-off. In order to obtain the current signal through DAQ system, the current sensor is installed. The test motor also, is loaded by a 3 kW, 1800 rpm dynamometer, excited by 0-2 A, 0-200 V dc and loaded with 21.63 Ω resistance [15].

IV. SENSOR FUSION APPLICATION THROUGH KALMAN FILTERING TO DISTINGUISH PROCESS AND SENSOR FAULT

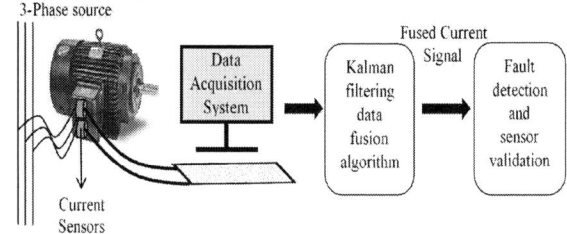

Fig. 1.   Schematic for sensor fusion through Kalman filter.

Sensor fusion through Kalman filter is utilized for FD and SV in this paper. The results of the analyses is presented in the following sections. Fig. 1 and Fig. 2 represent the proposed method. The authors assume that there is a second sensor having the same information with a deformation like an offset or phase difference. In addition, the collected current signal is duplicated with biases for the Kalman filtering application. Each signal includes measurement noise.

According to the proposed algorithm, the outputs of the sensors, x_{t1}, x_{t2} are fused together through Kalman filtering data fusion algorithm. Thereupon, fused signal x_f and the sensor outputs are compared for the evaluation of fusion algorihm. Basic statistical and spectral analyses, Kalman Gain calculation, errors between fused and actual data are considered. Finally, in order to detect the origin of the fault and validate the sensors, according to the results of the analyses, the effects of the process noise and measurement noise on fused signal are discussed.

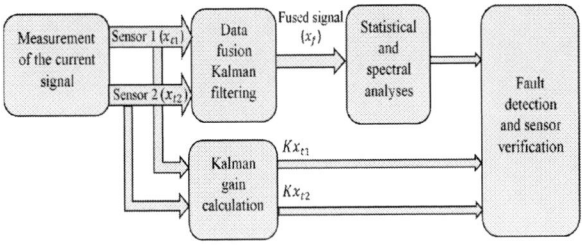

Fig. 2. Process of fault detection and sensor validation.

A. Sensor validation

In this section, effect of different measurement noises on the fused signal is negotiated. The statistical and spectral results is presented.

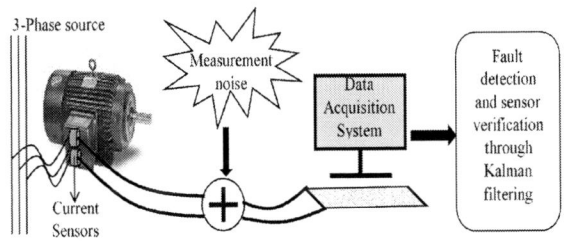

Fig. 3. Sensor validation while the sensors are influenced by the noises.

Table II-III represent the statistical analysis of the real current signals. It is supposed that, the sensors include measurement noise and biases with $+b$ and $-b$ quantities for Sensor 1 and Sensor 2, respectively. Then, the fused signal x_f which estimates the original signal is obtained through the Kalman filtering technique. In these tables, Q is the process noise, R_1 and R_2 are measurement noises which associates Sensor 1 and Sensor 2, respectively.

Moreover, the following equations are used for error calculations;

$$MSE_1 = (x_f - x_{t1}) = \frac{1}{n}\sum_{i=1}^{n}(x_{fi} - x_{t1i})^2 \qquad (12)$$

$$MSE_2 = (x_f - x_{t2}) = \frac{1}{n}\sum_{i=1}^{n}(x_{fi} - x_{t2i})^2 \qquad (13)$$

$$RMSE_1 = (x_f - x_{t1}) = \sqrt{\frac{1}{n}\sum_{i=1}^{n}(x_{fi} - x_{t1i})^2} \qquad (14)$$

$$RMSE_2 = (x_f - x_{t2}) = \sqrt{\frac{1}{n}\sum_{i=1}^{n}(x_{fi} - x_{t2i})^2} \qquad (15)$$

where MSE_1 and MSE_2 are mean square errors, $RMSE_1$ and $RMSE_2$ are root mean square error between the fused signal and the output of the sensor 1 and 2, respectively. In the tables below, std x_{t1}, std x_{t2}, std x_f are standard deviations of x_{t1}, x_{t2} and x_f; M x_{t1}, M x_{t2}, M x_f are mean values of x_{t1}, x_{t2} and x_f; Kurt x_{t1}, Kurt x_{t2}, Kurt x_f are kurtoses of x_{t1}, x_{t2} and x_f signals, respectively.

In the Table II, the measurement noises of two sensors are increased equally, hence, the error values are raised. As a result of the added noises, standard deviation of the Sensor 1 and Sensor 2 grows. However, standard deviation of the fused signal decreases. It is an important finding, because data fusion through kalman filtering method diminishes the noise of the fused signal and cancels the noise of the sensors. In Table III, it is assumed that one of the sensors is faulty and the noise rises gradually. Thus, the statistical parameters for the faulty sensor deviates while the fused signal's statistical parameters converges to the healthy signal. In addition, by observing the spectral results which are presented in Figs.4-6, more supporting features can be acquired.

Fig. 4, Fig.5 which are related to the Table II and Fig. 6 which is relevant to the Table III, demonstrate the frequency domain analyses. According to the Figs.4-6, it is clearly seen that, the faulty sensor is distinguishable, because the amplitude of the power spectral density (PSD) of the fused signal remains constant at high frequency region. Fig. 5 demonstrates that due to the faulty sensors, the fundamental frequency of the original signal is not estimated precisely. As regards, when one of the sensors is damaged or faulty (Fig.6), hence, amplitude of the PSD related to the faulty sensor increases. However, the fused signal, which converges to the healthy signal, can approximate the fundamental frequency of the original signal accurately. As a consequence, in spite of the existence of the faulty sensors, the source of the faults (either sensor or process) can be easily identified. Even if the fundamental frequency is roughly estimated, sensor verification can be performed. The outputs of the spectral analyses and the statistical results justify each other.

TABLE II. EFFECT OF THE MEASUREMENT NOISE INCREMENT ON THE STATISTICAL VALUES OF THE CURRENT DATA WHILE THE MEASUREMENT NOISES ARE THE SAME.

Test	R1	R2	Q	MSE 1	MSE 2	RMSE 1	RMSE 2	Std x_{t1}	Std x_{t2}	Std x_f	M x_{t1}	M x_{t2}	M x_f	Kurt x_{t1}	Kurt x_{t2}	Kurt x_f
#1	0.06	0.06	0.04	25	25	5	5	12.3	12.3	12.2	5	-5	0	1.5	1.5	1.5
#2	5	5	0.04	37.4	37.3	6.1	6.1	12.5	12.5	12	5	-5	0	1.6	1.6	1.5
#3	10	10	0.04	49.7	49.6	7	7	12.7	12.7	11.6	5	-5	0	1.7	1.7	1.5
#4	50	50	0.04	128	128	11.3	11.3	14.2	14.2	9.7	5	-5	0	2.2	2.2	1.5
#5	100	100	0.04	203	203	14.3	14.3	15.8	15.8	8.3	5	-5	0	2.5	2.5	1.5
#6	150	150	0.04	269	269	16.5	16.4	17.4	17.4	7.3	5	-5	0	2.6	2.6	1.5

TABLE III. EFFECT OF THE MEASUREMENT NOISE INCREMENT ON THE STATISTICAL VALUES OF THE REAL CURRENT SIGNAL WHILE THE MEASUREMENT NOISES ARE DIFFERENT.

Test	R1	R2	Q	MSE 1	MSE 2	RMSE 1	RMSE 2	Std x_{t1}	Std x_{t2}	Std x_f	M x_{t1}	M x_{t2}	M x_f	Kurt x_{t1}	Kurt x_{t2}	Kurt x_f
#1	0.06	0.01	0.04	73.5	2.04	8.6	1.4	12.3	12.3	12.3	5	-5	-3.3	1.52	1.51	1.51
#2	0.06	1	0.04	0.45	89.9	0.67	9.48	12.3	12.3	12.3	5	-5	4.7	1.51	1.53	1.51
#3	1	0.06	0.04	89.9	0.45	9.5	0.67	12.3	12.3	12.3	5	-5	-4.6	1.53	1.51	1.51
#4	5	0.01	0.04	104	0.01	10.2	0.09	12.5	12.3	12.3	5	-5	-4.7	1.6	1.5	1.5
#5	20	0.01	0.04	119	0.009	10.9	0.09	13	12.3	12.3	5	-5	-4.9	1.84	1.5	1.5

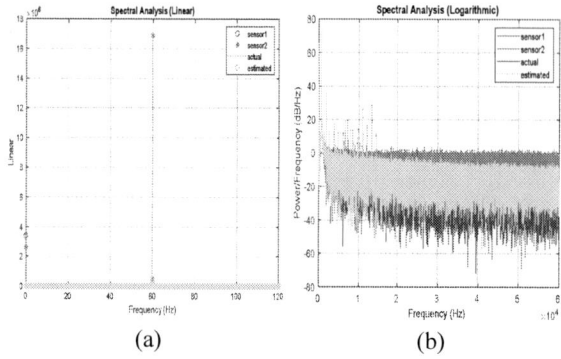

(a)   (b)

Fig. 4. Spectral results related to Table II test1; (a) Linear, (b) Logarithmic.

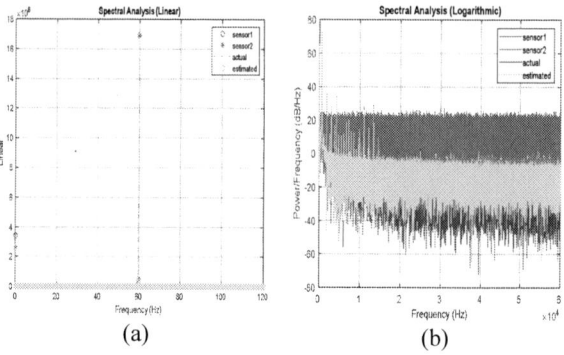

(a)   (b)

Fig. 6. Spectral results related to Table III test 5; (a)Linear, (b) Logarithmic.

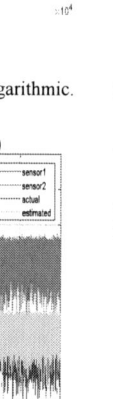

(a)   (b)

Fig. 5. Spectral results related to Table II test 6; (a) Linear, (b) Logarithmic.

B. Effect of the process noise

In this section, the influence of the process noise variations during the data fusion through Kalman filtering procedure is investigated.

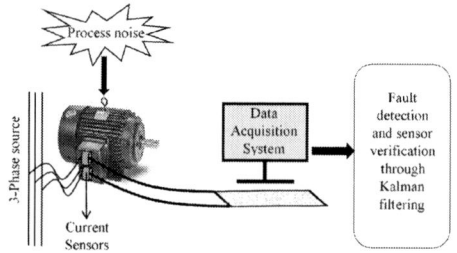

Fig. 7. Kalman filtering data fusion algorithm while the process is noisy.

TABLE IV. EFFECT OF THE PROCESS NOISE INCREMENT ON THE STATISTICAL VALUES OF THE REAL SIGNAL.

Test	R1	R2	Q	MSE 1	MSE 2	RMSE 1	RMSE 2	Std x_{t1}	Std x_{t2}	Std x_f	M x_{t1}	M x_{t2}	M x_f	Kurt x_{t1}	Kurt x_{t2}	Kurt x_f
#1	0.06	0.06	0.01	25	25	5	5	12.3	12.3	12.3	5	-5	0	1.5	1.5	1.5
#2	0.06	0.06	5	25	25	5	5	12.5	12.5	12.5	5	-5	0	1.6	1.6	1.6
#3	0.06	0.06	10	25	25	5	5	12.7	12.7	12.7	5	-5	0	1.7	1.7	1.7
#4	0.06	0.06	50	25	25	5	5	14.2	14.2	14.2	5	-5	0	2.2	2.2	2.2
#5	0.06	0.06	100	25	25	5	5	15.8	15.8	15.8	5	-5	0	2.5	2.5	2.5
#6	0.06	0.06	150	25	25	5	5	17.4	17.4	17.4	5	-5	0	2.6	2.6	2.6

Table IV and Fig. 8 indicate that the process noise (Q) fully affects the measured signals and the fused signal. By increasing the process noise, the fused signal diverges from original one and the fused signal resembles the measured signals. However, fundamental frequency is estimated precisely. Faulty process influences similarly on the both sensors and the fused signal which is output of data fusion through Kalman filtering algorithm. Consequently, the error remains constant and the standard deviation increases equally. Note that the kurtosis of a signal completely depends on the noise rate. The noises that are used in this work are normal distributed Gaussian noise and kurtosis of the noise is equal to 3. As a consequence, by increasing the rate of the noise, kurtosis of the current signal converges to 3.

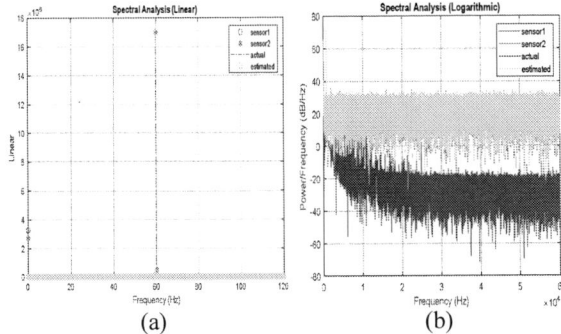

(a)   (b)

Fig. 8. Spectral results related to Table IV Test 6;(a)Linear,(b) Logarithmic.

V. FAULT SOURCE DETERMINATION THROUGH KALMAN FUSION

Kalman gain monitoring is one of the most effective methods to investigate the measurement noises and process noise effects, which is totally new for the literature. Otherwise speaking, the Kalman gain is an important factor for diagnosing the changes in the system behavior. In this section also, the authors have worked on the both real and artificial data in order to have better comprehension and to validate the method.

TABLE V.   EFFECT OF THE PROCESS NOISE INCREMENT ON THE KALMAN GAINS OF THE SENSORS.

Test	R1	R2	Q	K_{xt1}	K_{xt2}
#1	0.06	0.06	0.01	0.22	0.22
#2	0.06	0.06	5	0.5	0.5
#3	0.06	0.06	10	0.5	0.5
#4	0.06	0.06	50	0.5	0.5
#5	0.06	0.06	100	0.5	0.5
#6	0.06	0.06	150	0.5	0.5

TABLE VI.   EFFECT OF THE MEASUREMENT NOISE VARIATION ON THE KALMAN GAINS OF THE SENSORS.

Test	R1	R2	Q	K_{xt1}	K_{xt2}
#1	0.06	0.06	0.04	0.3	0.3
#2	5	5	0.04	0.059	0.059
#3	10	10	0.04	0.043	0.043
#4	50	50	0.04	0.0196	0.0196
#5	100	100	0.04	0.0139	0.0139
#6	150	150	0.04	0.0114	0.0114
Test	R1	R2	Q	K_{xt1}	K_{xt2}
#1	0.06	0.01	0.04	0.121	0.725
#2	0.06	1	0.04	0.53	0.0316
#3	1	0.06	0.04	0.0316	0.527
#4	5	0.01	0.04	0.0017	0.827
#5	20	0.01	0.04	0.0004	0.83

Table V and VI illustrate the influence of the process noise and the measurement noises on the Kalman gains of the sensors, which are obtained from sensor fusion algorithm empowered by Kalman filtering. In these tables, K_{xt1} and K_{xt2} are Kalman gain of the first sensor and the second sensor, respectively. Through the Kalman gain tracking, the fault source can be indicated as well. Whenever the process is

faulty, the Kalman gain saturates. However, any increment of the measurement noise causes Kalman gain to reduce. Afterwards, if either of the sensors are faulty, the Kalman gain of the corresponding sensor decreases. Accordingly, the healthy sensor can be validated and the faulty source is distinguishable.

VI. CONCLUSION

The sensor fusion approach empowered by Kalman filtering is developed for IM data collection system to determine the origin of the fault (either the process or sensors) and validate the sensors. First, the statistical and spectral analyses are achieved on current data for both the measurement and process noise interferences. Next, the Kalman gains of the sensors for all scenarios are monitored. The results which are presented in this study, is concluded as follow:

- When one of the sensors is faulty, the fused signal precisely estimates the healthy one by converging to it. As a consequence, the faulty sensor become distinguishable.

- Through interpretation of the spectral analysis, damaged sensor or sensors can be identified. Since, the fused signal diverges from the defective sensors at the high frequency regions.

- When the process noise is dominant, the PSD amplitude of the sensor outputs as well as the fused signal raise. In spite of the process noise increment, the algorithm accurately estimates the fundamental frequency.

- Either one of the sensors or all of them are faulty, Kalman gain of the corresponding sensor decreases.

- Kalman gain saturates when the process noise increases.

Furthermore, the Kalman gain results and the outputs which are obtained from statistical and spectral analyses support each other. Accordingly, the outputs which are presented in this paper, prove that the data fusion through Kalman filtering is an impressive method and an original idea which is extremely reliable for FD and SV purposes. Moreover, this method can be generalized and employed for various applications in industry.

REFERENCES

[1] R.E. Kalman, "A new approach to linear filtering and prediction problems," Transaction of the ASME—Journal of Basic Engineering, pp. 35-45, March 1960.

[2] Welch, G. and G. Bishop, "An Introduction to the Kalman Filter," Technical Report TR 95-041, University of North Carolina, Department of Computer Science, 1995.

[3] R.E. Kalman, R.S. Bucy, "New results in linear filtering and prediction theory," Transaction of the ASME—Journal of Basic Engineering, vol 83, pp. 95-108, December 1961.

[4] G. Fumin, R. Xuemei, L. Zhijun and H. Cunwu, "Kalman filter based fault detection of dual motor systems," 2017 36th Chinese Control Conference (CCC), Dalian, 2017, pp. 7133-7137.

[5] T. Park and K. Park, "Kalman filter-based fault detection and isolation of direct current motor: Robustness and applications," 2008 International Conference on Control, Automation and Systems, Seoul, 2008, pp. 933-936.

[6] A. J. Fernandez Gomez, V. H. Jaramillo and J. R. Ottewill, "Fault detection in electric motors by means of the extended Kalman Filter as disturbance estimator," 2014 UKACC International Conference on Control (CONTROL), Loughborough, 2014, pp. 432-437.

978-1-5386-7688-2/19 $31.00 © 2019 IEEE

[7] E. Foulon, C. Guibert and L. Loron, "Reduced-order electrical model extensions for induction machine temperature monitoring with extended Kalman filter," 2005 European Conference on Power Electronics and Applications, Dresden, 2005, pp. 10 pp.-P.10.

[8] S. Nadarajan, S. K. Panda, B. Bhangu and A. K. Gupta, "Online Model-Based Condition Monitoring for Brushless Wound-Field Synchronous Generator to Detect and Diagnose Stator Windings Turn-to-Turn Shorts Using Extended Kalman Filter," in IEEE Transactions on Industrial Electronics, vol. 63, no. 5, pp. 3228-3241, May 2016.

[9] Chin-Wen Wu, Y. Chung and Pau-Choo Chung, "A Kalman filtering based data fusion for object tracking," 2010 5th IEEE Conference on Industrial Electronics and Applications, Taichung, 2010, pp. 2291-2295.

[10] M. Mosallaei and K. Salahshoor, "Sensor Fault Detection using Adaptive Modified Extended Kalman Filter Based on Data Fusion Technique," 2008 4th International Conference on Information and Automation for Sustainability, Colombo, 2008, pp. 513-518.

[11] S. C. Stubberud and K. A. Kramer, "Monitoring the Kalman Gain Behavior for Maneuver Detection," 2017 25th International Conference on Systems Engineering (ICSEng), Las Vegas, NV, 2017, pp. 39-44.

[12] Rhudy, Matthew & A Salguero, Roger & Holappa, Keaton. (2017), "A Kalman Filtering Tutorial for Undergraduate Students," International Journal of Computer Science & Engineering Survey, 08. 01-18. 10.5121/ijcses.2017.8101.

[13] L. Jinfang and D. Zili, "Information fusion Kalman filter for two-sensor system with time-delayed measurements," Procedia Engineering, vol. 29, pp. 630–636, 2012.

[14] Wichit, Nattawut and Anant Choksuriwong, "Multi-sensor Data Fusion Model Based Kalman Filter Using Fuzzy Logic for Human Activity Detection," International Journal of Information and Electronics Engineering, Vol. 5, No. 6, November 2015.

[15] D. Bayram and S. Seker, "Redundancy-Based Predictive Fault Detection on Electric Motors by Stationary Wavelet Transform," IEEE Transaction on Industry Applications, vol. 53, no. 3, MAY/JUNE 2017.

Current harmonic suppression for permanent magnet synchronous motors

Ngoc-Tu Trinh
Research engineer in Digital Science
and Technology Direction,
IFP Energies Nouvelles,
Rueil Malmaison, France
ngoc-tu.trinh@ifpen.fr

Fabien Vidal-Naquet
Research engineer in Mobility and
Systems Direction,
IFP Energies Nouvelles,
Rueil Malmaison, France.
fabien.vidal-naquet@ifpen.fr

Abstract—This paper deals with problem of current harmonic suppression in permanent magnet synchronous motors (PMSM). First, the current harmonic model is considered in details in the *dq* coordinate frame. Then the current harmonic observer design is proposed to estimate online all harmonic components. Based on the harmonic observer, current harmonics are suppressed to obtain accurate currents. Finally, we present some simulations and experiments to validate the algorithm performance.

Keywords—*current harmonic, harmonic suppression, PMSM, rotor dq coordinate frame, harmonic observer.*

I. INTRODUCTION

Permanent magnet synchronous motors (PMSM) are nowadays increasingly used in industrial applications because of their high efficiency, compactness, and high speed operation capacity. However, one of its important issues is current harmonics which are often available in the rotating *dq* frame with the 6th , 12th , 18th … components or in the *abc* frame and the αβ coordinate frame with 5th ,7th , 11th , 13th , … ones, see in [1], [2], [3]. These harmonics caused from machine structure, inverters, dead-time, and other sources lead to torque ripple, joule loss, and decreased performance in motor control. Therefore, current harmonic suppression in PMSM becomes an attracted topic of researchers.

The problem of eliminating current harmonics has led to numerous studies in the literature, for example [4], [5], [6] to cite but a few. Many results apply control algorithms to compensate current harmonics such as repetitive control in [1], [6]; propotional integral controllers and resonanant controllers in [7], [8]. In other works, current harmonic suppression is based on harmonic observers which estimate current harmonics, and then remove them in the control parts. The observers used can be a multiple reference frame synchronous estimator (MRFSE) in [5], the zero-sequence estimator in [4], or multive adaptive feed-forward algorithm in [9] and [10].

In this paper, our objective is to suppress all current harmonic components to obtain more accurate currents by using an online observer which detects dynamically all current harmonic components in the rotor cordinate *dq*.

The paper is organized as follows. Section II is devoted to present current harmonic analysis, harmonic observer design, and harmonic suppression structure. In Section III, some numerical simulations and experiments are carried out to validate effectiveness of our algorithm. Finally, conclusions are given in Section IV.

II. CURRENT HAMONIC SUPRESSION METHOD

A. Current harmonic analysis

In the rotor coordinate system *dq*, the currents of PMSM include principal harmonics of 6th , 12th and 18th order components (see in [1], [2]) :

$$
\begin{aligned}
I_d = I_{d0} &+ I_{d6} \sin(6\theta + \phi_{d6}) + I_{d12} \sin(12\theta + \phi_{d12}) \\
&+ I_{d18} \sin(18\theta + \phi_{d18}) \\
I_q = I_{q0} &+ I_{q6} \sin(6\theta + \phi_{q6}) + I_{q12} \sin(12\theta + \phi_{q12}) \\
&+ I_{q18} \sin(18\theta + \phi_{q18})
\end{aligned}
\tag{1}
$$

where I_d, I_q are currents in axes d and q, θ is the electrical angle. I_{dk}, I_{qk}, ϕ_{dk}, ϕ_{qk} ($k=6,12,18$) represent respectively amplitudes and phase angles of kth harmonics to be determined. Note that the electrical angle $\theta=\omega t$ where ω denotes fundamental frequency.

It is equivalent to rewrite currents I_d and I_q from (1) in the following form :

$$
\begin{aligned}
I_d = I_{d0} &+ I_{d6a} \sin(6\theta) + I_{d6b} \cos(6\theta) + I_{d12a} \sin(12\theta) \\
&+ I_{d12b} \cos(12\theta) + I_{d18a} \sin(18\theta) + I_{d18b} \cos(18\theta) \\
I_q = I_{q0} &+ I_{q6a} \sin(6\theta) + I_{q6b} \cos(6\theta) + I_{q12a} \sin(12\theta) \\
&+ I_{q12b} \cos(12\theta) + I_{q18a} \sin(18\theta) + I_{q18b} \cos(18\theta)
\end{aligned}
\tag{2}
$$

where

$I_{dka} = I_{dk} \cos(\phi_{dk})$, $I_{dkb} = I_{dk} \sin(\phi_{dk})$,

$I_{qka} = I_{qk} \cos(\phi_{qk})$, $I_{qkb} = I_{qk} \sin(\phi_{qk})$.

Hence, in order to estimate online all harmonic amplitudes and all harmonic phase angles, it is necessary to determine all parameter I_{dka}, I_{dkb}, I_{qka}, I_{qkb}. In the following, we present the observer design to estimate all fundamental amplitudes (I_{d0}, I_{q0}) and above harmonic parameters.

B. Harmonic observers

The current harmonic observer design presented in this section is to estimate fundamental values (I_{d0}, I_{q0}) and all harmonic components of 6th , 12th and 18th orders containing in currents I_d, I_q from (2) (i.e. all parameters I_{dka}, I_{dkb}, I_{qka}, I_{qkb} for $k=6,12,18$).

Let denote $\hat{I}_{d0}, \hat{I}_{q0}, \hat{I}_{dka}, \hat{I}_{dkb}, \hat{I}_{qka}, \hat{I}_{qkb}$ be respectively estimations of $I_{d0}, I_{q0}, I_{dka}, I_{dkb}, I_{qka}, I_{qkb}$. In addition, let define the following current estimations $\hat{I}_d$ and $\hat{I}_q$:

978-1-5386-7688-2/19 $31.00 © 2019 IEEE

$$\hat{I}_d = \hat{I}_{d0} + \hat{I}_{d6a} \sin(6\theta) + \hat{I}_{d6b} \cos(6\theta) + \hat{I}_{d12a} \sin(12\theta)$$
$$+ \hat{I}_{d12b} \cos(12\theta) + \hat{I}_{d18a} \sin(18\theta) + \hat{I}_{d18b} \cos(18\theta)$$
$$\hat{I}_q = \hat{I}_{q0} + \hat{I}_{q6a} \sin(6\theta) + \hat{I}_{q6b} \cos(6\theta) + \hat{I}_{q12a} \sin(12\theta)$$
$$+ \hat{I}_{q12b} \cos(12\theta) + \hat{I}_{q18a} \sin(18\theta) + \hat{I}_{q18b} \cos(18\theta)$$

The harmonic observer for the *d*-axis is described as follows :

$$\dot{\hat{I}_{d0}} = \alpha(I_d - \hat{I}_d)$$
$$\dot{\hat{I}_{dka}} = \alpha.\sin(k\theta)(I_d - \hat{I}_d) \qquad (3)$$
$$\dot{\hat{I}_{dkb}} = \alpha.\cos(k\theta)(I_d - \hat{I}_d)$$

and similarly for the *q*-axe :

$$\dot{\hat{I}_{q0}} = \alpha(I_q - \hat{I}_q)$$
$$\dot{\hat{I}_{qka}} = \alpha.\sin(k\theta)(I_q - \hat{I}_q) \qquad (4)$$
$$\dot{\hat{I}_{qkb}} = \alpha.\cos(k\theta)(I_q - \hat{I}_q)$$

where $\alpha > 0$ is tuning parameter of the observer.

The harmonic observer design is shown in Fig. 1.

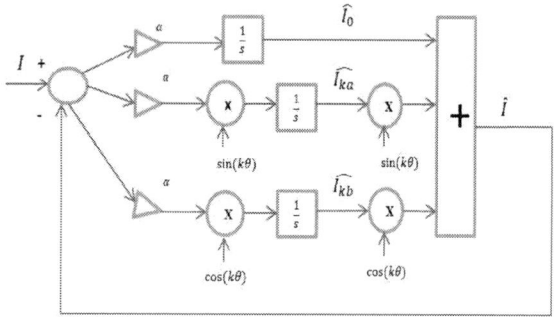

Figure 1. Harmonic observer structure

The proof of the above observer convergence is given in Appendix A.

Note that the bigger α is, the faster the current observer convergence is, but the more the observer state is oscillatory; and a big value of α leads to the instability of the observer.

C. Harmonic suppression structure

Following to the harmonic analysis and the harmonic observer mentioned in Sections II.A and II.B, the current harmonics are removed by the following :

$$I_d^f = I_d - H_{6d} - H_{12d} - H_{18d}$$
$$I_q^f = I_q - H_{6q} - H_{12q} - H_{18q}$$

where I_d^f, I_q^f are filtered currents, H_{dk} and H_{qk} denotes respectively k^{th} harmonic components in axes *d* and *q* defined by :

$$H_{dk} = \hat{I}_{dka} \sin(k\theta) + \hat{I}_{dkb} \cos(k\theta)$$
$$H_{qk} = \hat{I}_{qka} \sin(k\theta) + \hat{I}_{qkb} \cos(k\theta).$$

The harmonic suppression diagram is given in Figure 2. Note that all harmonic parameters $\hat{I}_{dka}, \hat{I}_{dkb}, \hat{I}_{qka}, \hat{I}_{qkb}$ are obtained dynamically from observers (3) and (4), therefore the current harmonic suppression is carried out online.

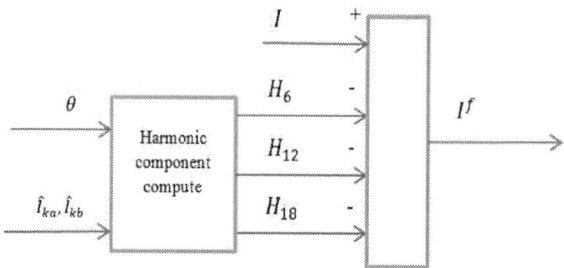

Figure 2. Harmonic suppression diagram

III. SIMULATIONS AND EXPERIMENTATION

A. Simulations

In the simulation, we apply the proposed method to consider the harmonic suppression in current signal I_q. The data used are the following : $I_{q0} = 10\ A$, $I_{q6a} = 0.1\ A$, $I_{q6b} = -0.15\ A$,

$I_{q12a} = 0.14\ A$, $I_{q12b} = 0.09\ A$, $I_{q18a} = -0.05\ A$, $I_{q18b} = -0.2\ A$. The electrical angle $\theta = \omega t$ (rad) where $\omega = 200\ rad/s$. Note that in the simulation, the harmonic observer is active from $t = 1s$, and the tuning parameter α for harmonic observer is chosen by 10.

The simulation results are firstly shown in Figure 3 which gives the estimations of I_{qka}, I_{qkb} from the current harmonic observe. The convergence of observer parameters to real harmonic coefficients is verified, and the observer gives therefore precise estimations of I_{qka}, I_{qkb}.

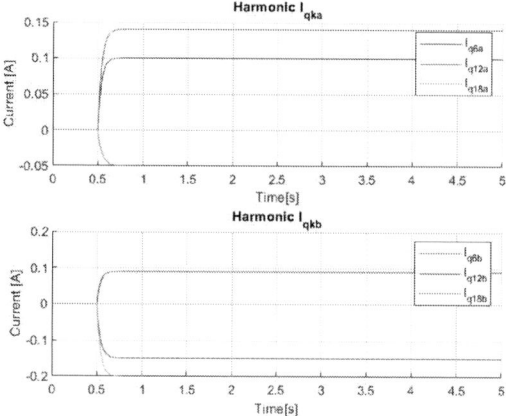

Figure 3. Estimation of I_{qka}, I_{qkb} from harmonic observer

Figure 4 presents performance of harmonic suppression method for current I_q. In fact, a comparison between current I_q without harmonic suppression (figure below) and the one with proposed method (figure above) is given. Obviously, by proposed method, current harmonic components are eliminated and we obtain the current I_q more accurate.

Figure 5 shows FFT analysis of current I_q without harmonic suppression (blue line) and the one with proposed harmonic suppression approach (red line). It is clearly that all current harmonic components (6th , 12th and 18th orders) are totally removed with the proposed approach.

978-1-5386-7688-2/19 $31.00 © 2019 IEEE

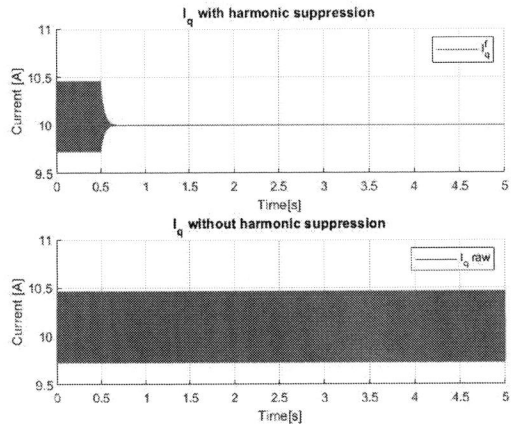

Figure 4. Current I_q with and without proposed method

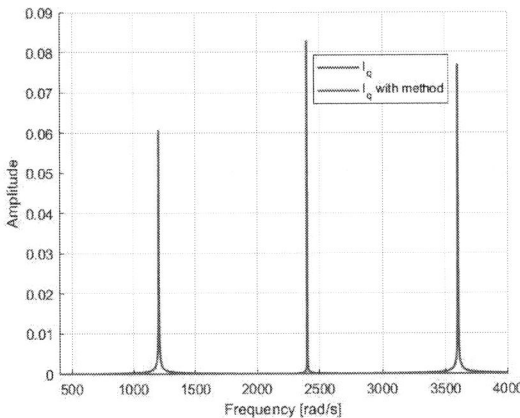

Figure 5. FFTs of I_q with and without proposed method

In Figure 6, we consider the performance of the proposed algorithm in transient zone in which the fundamental frequency ω of the electrical angle changes from 20 to 200 *rad/s*. The simulation shows that in spite of changing frequency, the current I_q obtained from the proposed method is accurate by suppressing harmonic components.

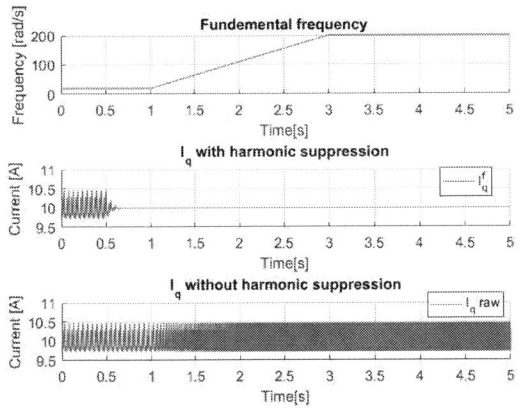

Figure 6. Simulation results with frequency change from 20 to 200 rad/s

B. Experimental results

In this Section, we present the experimental results to verify the algorithm. The parameters of PMSM used in the experiment are found in Table 1.

Tableau I   PMSM PARAMETERS

Parameters	Value
Phase number	3
Number of pole pairs	4
Stator resistance	0.0082 Ω
Magnetic flux	0.0216 Wb
Voltage (rated)	350 V
Current (rated)	300 A
Speed (rated)	7500 rpm
Torque max	100 Nm

Figure 7 shows the evolution of current I_q in the experiment with and without proposed method. Compared to the signal I_q without processing (blue line), the proposed processing method gives a better result (red line) in which the harmonic currents are filtered out.

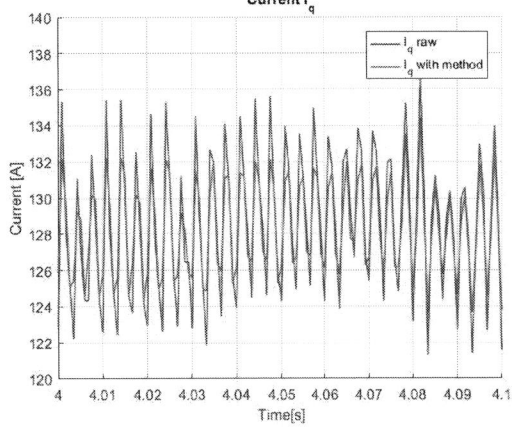

Figure 7. Current I_q in experiments

In Figure 8, the FFT (Fast Fourier Transform) analysis of current signal I_q is presented to consider the performance of the studied method. It is clearly to see that 6th, 12th and 18th harmonic components are all removed in the processed current signal I_q with the proposed method (red line) while these harmonic components appear in the raw signal (blue line). In addition, Figure 9 gives the online estimated value of harmonic parameters $I_{q6a}, I_{q12a}, I_{q18a}$ and $I_{q6b}, I_{q12b}, I_{q18b}$ in the current signal I_q.

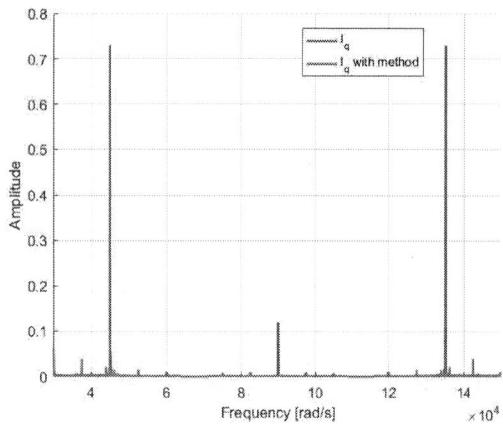

Figure 8. FFTs of I_q with and without proposed method

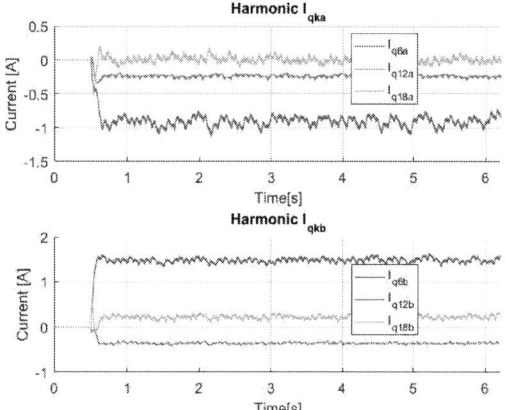

Figure 9. Estimation of I_{qka}, I_{qkb} in experiments

IV. CONCLUSIONS

In the paper, the problem of current harmonic suppression for PMSM has been studied. The current models which are considered in the dq coordinate frame contain harmonics of 6th, 12th and 18th order components. Two harmonic observers are designed for axe d and axe q to detect online all current harmonic components. Based on harmonic estimations received from two harmonic observers, we can remove all harmonic components in current signals to obtain accurate currents. The numerical simulations and experimental results are presented to verify performance of the proposed method. In addition, it should be remarked that the proposed approach can be easily adapted to work with signals which contain various types of harmonics.

V. APPENDIX

A. Proof of observer convergence

Because of the similarity in observer design of d-axe and q-axe, we present therefore the proof of convergence only for d-axe.

It is known that $\theta = \omega t$ where ω denotes electrical speed which is also fundamental frequency of current signals.

Let denote for each harmonic components

$$S_k(t) = \hat{I}_{dka}(t) \sin(k\omega t) + \hat{I}_{dkb}(t) \cos(k\omega t),$$

and for fundamental part $S_0(t) = \hat{I}_{d0}(t)$.

Hence $\hat{I}_d = S_0 + S_6 + S_{12} + S_{18}$.

Inspired of frequency analysis idea in [11] with Laplace transform, we obtain the following :

$$S_0(s) = \left(I_d(s) - \hat{I}_d(s)\right) \frac{\alpha}{s},$$

$$S_k(s) = \left(I_d(s) - \hat{I}_d(s)\right) \frac{\alpha s}{s^2 + (k\omega)^2}$$

Furthermore, it can be deduced that

$$\hat{I}_d(s) = S_0(s) + S_6(s) + S_{12}(s) + S_{18}(s)$$

$$= \left(I_d(s) - \hat{I}_d(s)\right) \left(\frac{\alpha}{s} + \sum_{k=6,12,18} \frac{\alpha s}{s^2 + (k\omega)^2} \right)$$

Therefore,

$$H(s) = \frac{\hat{I}_d(s)}{I_d(s)} = \frac{\dfrac{\alpha}{s} + \sum_{k=6,12,18} \dfrac{\alpha s}{s^2 + (k\omega)^2}}{1 + \dfrac{\alpha}{s} + \sum_{k=6,12,18} \dfrac{\alpha s}{s^2 + (k\omega)^2}}$$

Considering the function $H(s)$: its gain is one at $s=0$ and $s=jk\omega$, and almost zero if not. It implies the convergence $\hat{I}_d(t) \rightarrow I_d$, and other convergences of harmonic observer components.

It should be mentioned that the observer convergence can be proven by other way, i.e. by a time-domain method in [9] and [12].

REFERENCES

[1] T. Nakai and H. Fujimoto, "Harmonic Current Suppression Method of PMSM Based on Repetitive Perfect Tracking Control," in *IECON 2007 - 33rd Annual Conference of the IEEE Industrial Electronics Society*, pp. 1049-1054, 2007.

[2] X. Wang, W. Zhou and R. Dou, "Analysis of Harmonic Current in Permanent Magnet Synchronous Motor and Its Effect on Motor Torque," *Journal of Electromagnetic Analysis and Applications*, vol. 4, pp. 15-20, 2012.

[3] M. Morimoto, T. Zanma, I. Muneaki, G. Pipeleers and J. Swevers, "Suppression of harmonic current for IPMSM using generalized repetitive control," in *IEEE Annual Conference on Industrial Electronics Society*, pp. 2070-2075, 2000.

[4] J. Hwang and H. Wei, "The Current Harmonics Elimination Control Strategy for Six-Leg Three-Phase Permanent Magnet Synchronous Motor Drives," *IEEE Transactions on Power Electronics*, pp. 3032-3040, 2014.

[5] M. Musak, M. Stulrajter, V. Hrabovcova, M. Cacciato, G. Scarcella and G. Scelba, "Suppression of Low-order Current Harmonics in AC Motor Drives via Multiple Reference Frames Based Control Algorithm," *Electric Power Components and Systems*, vol. 43, no. 18, pp. 2059-2068, 2015.

[6] J. Kim, S. Doki and M. Ishida, "Suppression of Harmonic Current in Vector Control for IPMSM by Utilizing Repetitive Control," in *IEEE International Confernce on Industrial Technology*, pp. 264-267, 2002.

[7] C. Liu, F. Blaabjerg, W. Chen and D. Xu, "Stator current harmonic control with resonant controller for doubly fed

induction generator," *IEEE Transactions on Power Electronics,* vol. 27, no. 7, pp. 3207-3220, 2012.

[8] C. Lascu, L. Asiminoaei, I. Boldea and F. Blaabjerg, "High Performance Current Controller for Selective Harmonic Compensation in Active Power Filters," *IEEE Transactions on Power Electronics,* vol. 22, no. 5, pp. 1826 - 1835, 2007.

[9] S. Leng, W. Liu, I.-Y. Chung and D. Cartes, "Active Power Filter for Three-Phase Current Harmonic Cancellation and Reactive Power Compensation," in *IEEE American Control Conference,* pp. 2140-2147, 2009.

[10] L. Qian, D. Cartes and Q. Zhang, "Three-Phase Harmonic Selective Active Filter Using Multiple Adaptive Feed Forward Cancellation Method," in *2006 CES/IEEE 5th International Power Electronics and Motion Control Conference,* vol. 2, pp. 1-5, 2006.

[11] S.-Y. Jung and K. Nam, "PMSM Control Based on Edge-Field Hall Sensor Signals Through ANF-PLL Processing," *IEEE Transactions on Industrial Electronics,* vol. 58, no. 11, pp. 5121-5129, 2011.

[12] S. Osowski, "Neural network for estimation of harmonic components in a power system," in *IEEE Generation, Transmission and Distribution,* vol. 139, pp. 129-135, 1992.

978-1-5386-7688-2/19 $31.00 © 2019 IEEE

Cutting Technologies Influence on Magnetic Properties of Electrical Steels used in High-Efficiency Motors Manufacturing

Gheorghe Paltanea
Electrical Engineering Department
University Politehnica of Bucharest
Bucharest, Romania
paltanea03@yahoo.com,
0000-0003-3251-7146

Veronica Manescu (Paltanea)
Electrical Engineering Department
University Politehnica of Bucharest
Bucharest, Romania
m1vera2@yahoo.com,
0000-0002-2535-5698

Elena Helerea
Department of Electrical Engineering
Transilvania University of Brasov
Brasov, Romania
helerea@unitbv.ro,
0000-0001-7145-8762

Iosif-Vasile Nemoianu
Electrical Engineering Department
University Politehnica of Bucharest
Bucharest, Romania
inemoianu@yahoo.com,
0000-0001-5610-8318

Marius Daniel Calin
Department of Electrical Engineering
Transilvania University of Brasov
Brasov, Romania
marius.calin@unitbv.ro

Abstract—Aiming to obtain high efficiency electrical machines, the manufacturers pay attention at every aspect, which may lead to the energy loss reduction. The cutting technology is very important, because the classical punching method induces mechanical stresses that affect the total energy losses. In this paper, the magnetic behavior of non-oriented M300-35A electrical steel, cut through punching and water-jet technologies, was investigated. The strips with a width variating between 5 mm and 60 mm and assembled side by side 1, 2, 4, 6, 8, 10 and 12 samples have been prepared. The total energy losses, the magnetic permeability with its components (real and imaginary parts) and the coercive field have been determined as depending on the peak magnetic polarization, with a laboratory Single Strip Tester, at industrial frequency of 50 Hz. The normal magnetization curves were determined by extracting from the experimental measurements the tip points of the nested symmetric minor hysteresis cycles. The study shows that the water-jet cutting technology has a lower impact on the magnetic properties' deterioration.

Keywords— *high quality non-oriented electrical steel, mechanical cutting, water-jet cutting technology, energy efficiency of electric motors, relative magnetic permeability, coercive field, normal magnetization curve.*

I. INTRODUCTION

Starting in 2017, there has been an accelerated rise in global energy consumption and environmental issues. Recent studies show that by 2040 global energy demand will increase by 30% [1].

In order to ensure future energy demand and to limit the environmental impact, the most direct approach is to produce improved energy efficiency equipment using materials with appropriate properties.

Regarding the production of electric machines, the study of the International Energy Agency shows that 50% of the world's electricity consumption is used by electric motors, although in some European countries this demand is estimated at about 70%. Thus, increasing the energy efficiency of electric motors can be a clean and cheaper way to save nearly 300 TWh per year of total energy consumption in 2030, avoiding the release of about 200 Mt of CO_2 [1], [2]. The regulations, as IEC 60034-30-1:2014 [3],

categorized the electrical motors in terms of their efficiency, and a set of limit efficiency values for electric motors in operation, and manufactured with different technologies is imposed. Thus, to increase the energy efficiency of electric motors, new tasks have emerged for researchers as well as for designers and manufacturers.

The researchers have included in their studies tasks such as: developing new schemes, new algorithms, new models and design methods. But not many studies focus on high-efficiency induction motors with respect to the requirements of IEC 60034-30. The reason could be that the requirements of this standard are highly applied through technological solutions [4].

The tasks of the designers and manufacturers of electric motors are related to the developing and implementing new and advanced technologies to meet these requirements. But, a common task for researchers, designers and manufacturers is to find solutions for implementing the new high-quality materials, and new technologies, with an adequate price of manufacturing and friendly with environment [5].

In this respect, many researches are directed to the including the influence of the manufacturing effects on the material properties, and in modeling and simulation the operation of electric motors.

Regarding the magnetic materials, the common materials used for rotor and stator packages of electric machines are non-oriented (NO) silicon iron alloy, with properties specified in standards IEC 60404-8-2 and 60404-8-4, and manufactured in different grades. By a proper choice of composition and suitable metallurgical and thermal treatments [6], more efficient NO silicon iron magnetic materials can be obtained [5].

Besides an improved design for electric motors, the use of better material grades and the deep knowledge of the influence of manufacturing processes (cutting, pressing during stacking, welding etc.) on the magnetic material properties can sustain the high efficiency task. Knowing the effects of cutting processes of silicone steel sheets on magnetic properties is important for designing electric machines, in terms of magnetic field and loss calculations.

978-1-5386-7688-2/19 $31.00 © 2019 IEEE

Many researchers analyzed the effect of mechanical cutting on the magnetic properties of the electrical steel sheets [6]-[11]. Others investigated the effect of laser cutting [12]-[14]. Many papers proposed comparative analyzes of the effects of different cutting procedures (punching, laser, abrasive water jet, electric discharge machining) [15]-[21].

All these studies have shown that cutting processes for magnetic core manufacture influence the properties of magnetic materials: magnetic permeability decreases, and magnetic losses increase. The degradation mechanisms are related to the change of grain size, disorientations within the grains, plastic deformation and residual, thermal or mechanical stresses induced in proximity of the cut edge.

Some studies have considered reducing the effect of mechanical and thermal stresses by applying specific annealing processes [22]-[24].

New models for magnetization curves have been proposed that take into account the degradation effect due to manufacturing processes, completing the catalog data of FeSi alloy suppliers [11], [25], [26]. The results of the studies allowed the introduction of correction parameters in the modeling, simulation and design of the magnetic circuit of the electric motor [7], [12], [13], [25], [27]-[31].

However, there are not many studies to investigate the dimensional effect of cutting on the magnetic properties [11], [15], [18], and new theoretical and experimental researches are required to clarify the dependence of magnetic properties on the geometry of parts at cutting.

In this paper the dimensional effect at cutting is investigated. There is shown that the cutting perimeter of the M300-35A high quality non-oriented steel samples obtained with traditional mechanical and water-jet cutting technologies influences the magnetic permeability, the magnetic losses and the normal magnetization curve. This effect needs to be taken into account in the design and realization of high energy efficient electric motors.

II. MATERIALS AND METHODS

The magnetic measurements were conducted at the industrial frequency of 50 Hz, at a given magnetic polarization set between 20 mT and 1700 mT, utilizing a standardized Single Strip Tester (SST), which permits the magnetic characterization of samples with a maximum area of 300 mm × 60 mm. Aiming to analyze the influence of the cutting procedure on the magnetic behavior of non-oriented electrical steel, the strips were prepared with the same length of 300 mm and different widths, varying from 5 mm to 60 mm. A number of 12, 10, 8, 6, 4, 2 and 1 strips were placed side by side, to reconstruct a 60 mm width sample. In this way the cutting perimeter is increased, and the influence on the normal magnetization curve, the total energy losses, the relative magnetic permeability and coercive field is investigated. For the SST measurements were prepared 7 samples in the case of each cutting technology.

The strips were obtained through mechanical punching procedure and water-jet cutting technology.

For the mechanical machining a Trumpf Trumatic 500 stamping & punching press was utilized, and the strips were cut very fast, with the help of a computer program that executes automatically, the cutting steps. In the case of

water-jet technology a Maxiem 1530 machine was used which has a high-pressure pump, able to provide a maximum pressure of 413.64 MPa by using as abrasive particles a mixture of Garnet 80 mesh.

TABLE I. ELEMENT WIDTH AND MASS OF M300-35A SAMPLES OF 60 mm WIDTH

Structure of sample	Number of strips	Element width [mm]	Mass [g]	
			Punching	Water-jet
1x60 mm	1	60	46.79	46.62
2x30 mm	2	30	46.74	46.65
4x15 mm	4	15	47.01	46.65
6x10 mm	6	10	46.18	46.64
8x7.5 mm	8	7.5	47.71	46.68
10x6 mm	10	6	45.51	46.55
12x5 mm	12	5	46.83	46.51

A double C laminated yoke with two concentric solenoids of 240 mm magnetic path length was used to characterize the samples. As magnetic field source a 288 turns coil is used and a 250 turns pickup coil for the secondary voltage is placed near the tested sample.

Energy losses and magnetic permeability were determined by controlling with an Agilent 33210A arbitrary function generator, the sinusoidal waveform of the secondary voltage according to the IEC 60404-3 standard [3], using a 12 bit 500 MHz HDO4054 LeCroy oscilloscope, for the signals' acquisition.

The current through the magnetizing coil was measure with a non-inductive 10 Ω manganine shunt resistor.

This precision resistor has a low temperature coefficient in the domain of 20°C to 50°C, with a parabolic shape of the resistance-temperature dependence. When this special resistor is used for precision measurements its temperature should not exceed 60 °C.

Concerning M300-35A non-oriented steel the catalog data provides two values of 1.10 W/kg and 2.62 W/kg for the specific power losses at 50 Hz, determined at 1 T and 1.5 T peak magnetic polarizations. Other useful mechanical properties are: yield strength 370 N/mm², tensile strength 490 N/mm², rolling direction Young's modulus 185000 N/mm² and hardness 180 HV5 [21]. The material average mass density is 7.65 g/cm³, the electrical resistivity is 50× 10^{-8} Ωm.

The element width and mass of M300-35A samples with thickness of 0.35 mm and total width of 60 mm are given in Table I.

III. EXPERIMENTAL MEASUREMENTS

Symmetrical hysteresis loops were measured for several peak magnetic polarization values, starting with 20 mT and then 40 mT, 50 mT, 100 mT, 200 mT, 400 mT, 500 mT, 800 mT, 1000 mT, 1200 mT, 1400 mT, 1500 mT, 1600 mT and 1700 mT at industrial frequency f = 50 Hz.

The maximum positive values of the magnetic field strength H_p and the magnetic polarization J_p were extracted, and then the normal magnetization curve was plotted for each independent sample prepared using mechanical or water-jet cutting technologies. The normal magnetization curves specific to M300-35A samples are shown in Fig. 1 and Fig. 2.

In Fig. 3 and Fig. 4 the total energy losses as a function of the magnetic polarization are displayed. It can be noticed that the increase of the cutting perimeter determines an increase of the total energy losses for both types of cutting technology.

Fig. 5 and Fig. 6 show the variation of magnetic permeability as a dependence of the peak magnetic polarization. Lower magnetic permeability values function of the cutting perimeter is being noticed in both cases, due do the magnetic properties deterioration.

The real part of relative magnetic permeability is shown in Fig. 7 and Fig. 8 as a dependence of the peak magnetic

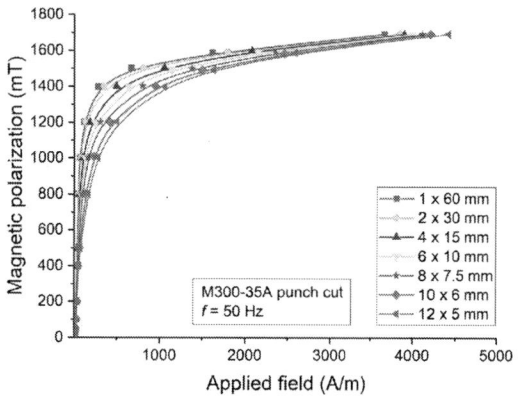

Fig. 1. Normal magnetization curves plotted in the case of M300-35A non-oriented strips generated using punching technology.

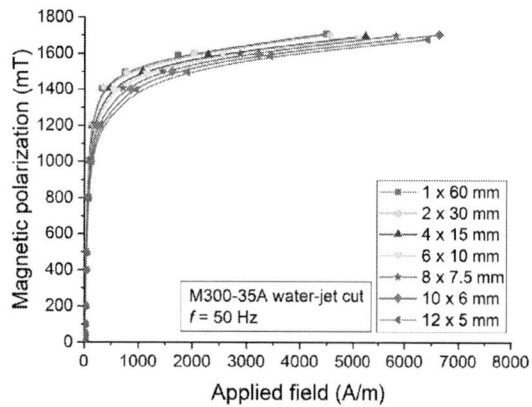

Fig. 2. Normal magnetization curves plotted in the case of M300-35A non-oriented strips generated using water-jet technology.

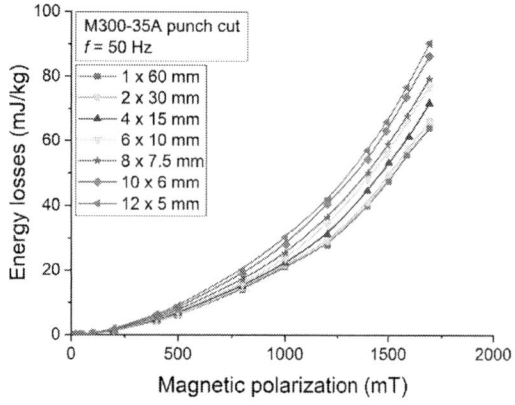

Fig. 3. Total energy losses versus magnetic polarization dependencies, obtained in the case of M300-35A non-oriented strips generated using punching technology.

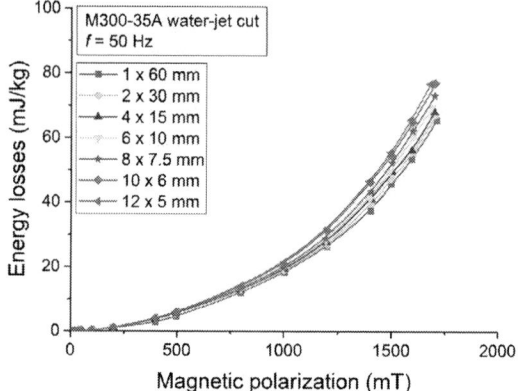

Fig. 4. Total energy losses versus magnetic polarization dependencies, obtained in the case of M300-35A non-oriented strips generated using water-jet technology.

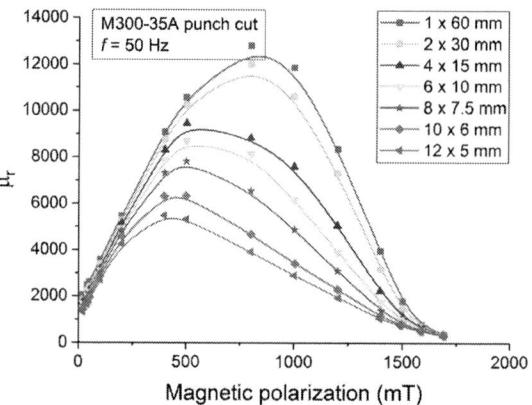

Fig. 5. Relative magnetic permeability versus magnetic polarization dependencies, obtained in the case of M300-35A non-oriented strips generated using punching technology.

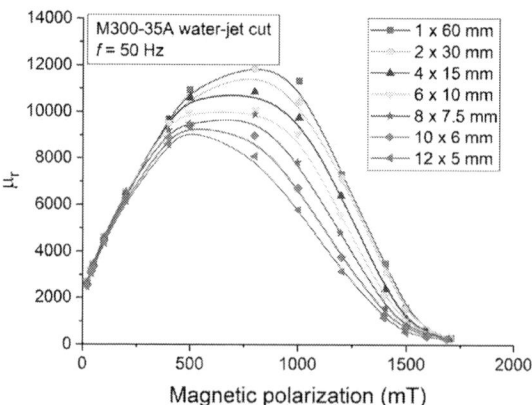

Fig. 6. Relative magnetic permeability versus magnetic polarization dependencies, obtained in the case of M300-35A non-oriented strips generated using water-jet technology.

978-1-5386-7688-2/19 $31.00 © 2019 IEEE

polarization, for each cutting technology. The imaginary part of the relative magnetic permeability is linked to the energy losses.

In Fig. 9 and Fig. 10 the variation of the imaginary part versus peak magnetic polarization is presented for each cutting technology.

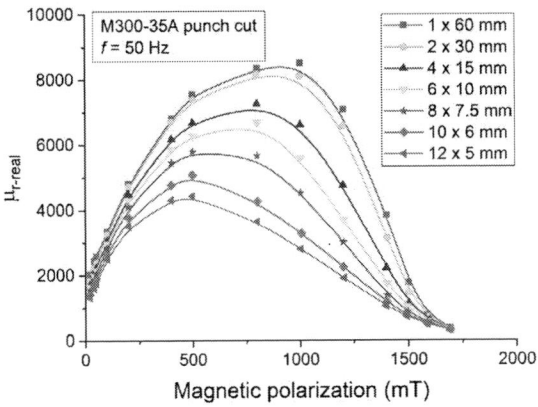

Fig. 7. Real part of the relative magnetic permeability versus magnetic polarization dependencies, obtained in the case of M300-35A non-oriented strips cut through punching technology.

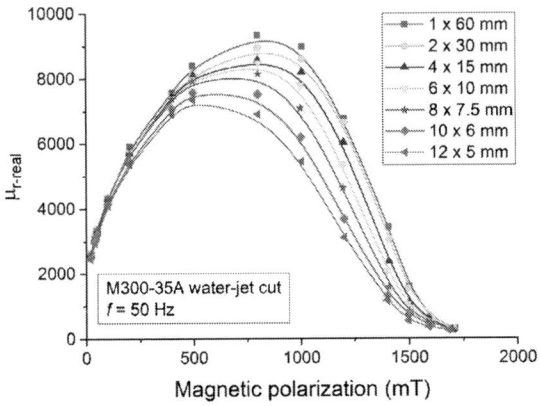

Fig. 8. Real part of the relative magnetic permeability versus magnetic polarization dependencies, obtained in the case of M300-35A non-oriented strips generated using water-jet technology.

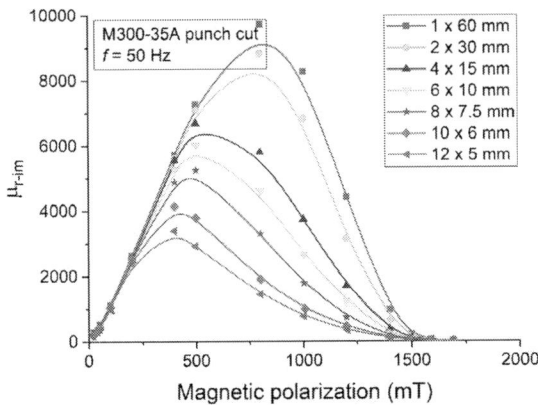

Fig. 9. Imaginary part of the relative magnetic permeability versus magnetic polarization dependencies, obtained in the case of M300-35A non-oriented strips generated using punching technology.

The value of coercive field in electrical steel is directly connected with the entire magnetization process that is made through domain wall displacement.

The influence of the cutting technology on the coercive field is presented in Fig. 11 and Fig. 12.

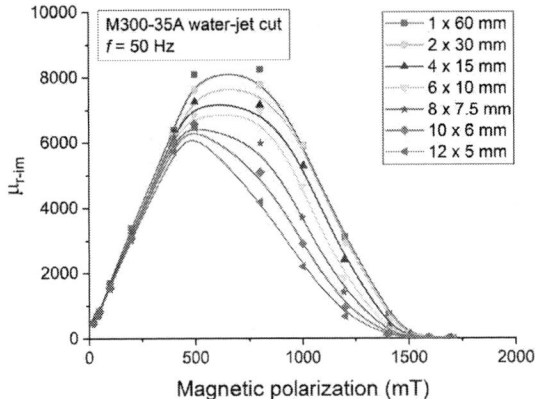

Fig. 10. Imaginary part of the relative magnetic permeability versus magnetic polarization dependencies, obtained in the case of M300-35A non-oriented strips generated using water-jet technology.

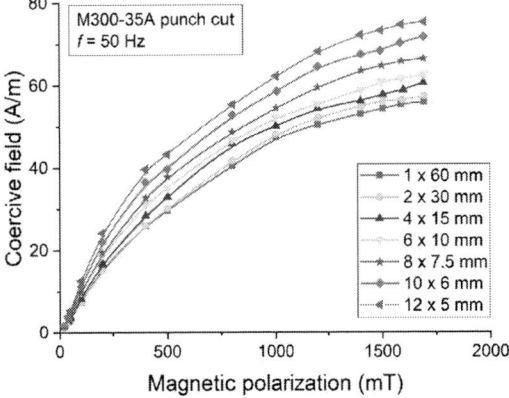

Fig. 11. Coercive field versus magnetic polarization dependencies, obtained in the case of M300-35A non-oriented strips generated using punching technology.

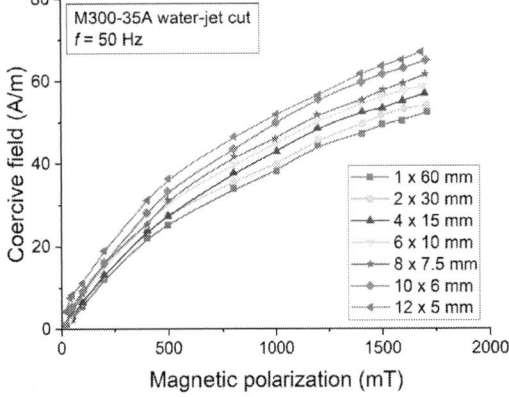

Fig. 12. Coercive field versus magnetic polarization dependencies, obtained in the case of M300-35A non-oriented strips generated using water-jet technology.

978-1-5386-7688-2/19 $31.00 © 2019 IEEE

IV. DISCUSSIONS

Analyzing the normal magnetization curves, presented in Figs. 1 and 2, it can be observed that the cutting perimeter increasing determines a harder magnetization process of the samples. In order to obtain the value of the saturation magnetic polarization it is necessary to apply a higher value of the magnetic field intensity.

The M300-35A electrical steel sheets used for electric machine magnetic cores are relative thin, so that the effect of the cutting technology is difficult to highlight. However, the measurements made in the magnetic polarization range from 1000 mT to 1500 mT shown a small improvement in the case of water-jet cut sample. This fact is due to the reduced number of pinning sites, placed near cut edge, that limit the domain wall movement [32], [33].

Regarding the efficiency of applications in which the high-quality electrical steel M300-35A alloy is used, it is expected to obtain lower energy losses, with positive impact on the reduction of global pollution and on the resource's conservation. The investigated alloy has 3 % Si and presents an important magneto-crystalline anisotropy that has a significant influence on the magnetization phenomenon, being characterized by an intermediary magnetic behavior between non-oriented and grain-oriented steels.

In the case of mechanical punching the influence of the cutting technology on the energy losses versus peak magnetic polarization is more pronounced, because during cutting process mechanical stresses are induced and they determine an increase of the total energy losses (Fig. 3). For water-jet cutting technology, the differences between energy loss dependences are more reduced, because this method is a less invasive one and the deterioration of the material, due to cutting procedure, is minimal (Fig. 4). The magnetic properties of a given material depend on different factors as frequency, temperature, linearity, homogeneity and isotropy. The dispersive behavior of non-oriented silicon iron alloys is represented by the complex magnetic permeability (Figs. 5 and 6).

For a linear approximation, when a given magnetic field strength is applied, the magnetic material response is entirely characterized by the complex magnetic permeability concept [32], [33]. Figures 7 and 8 show that the capacity to store magnetic energy exhibited by the investigated material is strongly affected by the increase of the cutting perimeter, since the value of the real part of relative magnetic permeability decreases from 8000 to 4000 in the case of punching technology, and from 9000 to 7000, for the water jet method. Also, can be observed a shift to the left of the maximum value of the magnetic permeability, when the cutting perimeter increases.

The effect of the cutting technology on the energy losses is better observed from the imaginary part of magnetic permeability as a function of the magnetic polarization dependencies, as shown in Figs. 9 and 10. For the low polarization region (0 to 500 mT) it can be observed that the magnetic domain movements is not yet influenced by the pinning sites, but between 500 mT and 1200 mT in the knee zone of the normal magnetization curve the variation of the magnetic permeability imaginary part is strongly influenced by the cutting technology, fact that suggest an expansion of the magnetic domains through a pseudo-mono-domanial configuration.

TABLE II. ENERGY LOSSES AND RELATIVE PERMEABILITY VALUES MEASURED AT PEAK MAGNETIC POLARIZATION $J_p = 1000$ mT

Sample	Cutting Technology			
	Punching		Water-Jet	
	Energy losses [mJ/kg]	Relative magnetic permeability	Energy losses [mJ/kg]	Relative magnetic permeability
1 × 60 mm	21.255	11339	18.335	11868
12 × 5 mm	30.336	2901	21.506	5789

TABLE III. ENERGY LOSSES AND RELATIVE PERMEABILITY VALUES MEASURED AT PEAK MAGNETIC POLARIZATION $J_p = 1500$ mT

Sample	Cutting Technology			
	Punching		Water-Jet	
	Energy losses [mJ/kg]	Relative magnetic permeability	Energy losses [mJ/kg]	Relative magnetic permeability
1 × 60 mm	47.402	1589	45.774	1781
12 × 5 mm	66.008	542	55.626	729

The soft magnetic materials have a coercive force in the order of the Earth's magnetic field or lower [32]-[35]. Figs. 11 and 12 show that, in the case of M300-35A non-oriented soft alloy, the maximum value of the coercive magnetic field strength is almost 50 A/m at a peak value of 1700 mT for the peak magnetic polarization. The influence of the cutting technology is more obvious in the case of the coercive field. Indeed, depending on the punching method, one can notice higher experimental coercive field values starting from 50 A/m (in the case of 60 mm width sample) up to 75 A/m (for 5 mm samples). In the case of water-jet technology a 65 A/m coercive field is obtained for the 5 mm width sample.

To make a direct comparison between the two cutting technologies it was chosen to present the total energy losses and the relative magnetic permeability values at two peak magnetic polarizations of 1000 mT (Table II) and 1500 mT (Table III), in the case of samples 1 × 60 mm and 12 × 5 mm. As a general observation, the water jet cutting technology leads to lower total energy losses, in the case of both peak magnetic polarizations and to higher values of the magnetic permeability. The mechanical cutting method determines a higher grade of magnetic properties deterioration.

V. CONCLUSIONS

The mechanical cutting procedure has a negative impact on the magnetic characteristics of M300-35A non-oriented electrical steel. This fact was put in evidence in the paper, by analyzing the evolution of the normal magnetization curve, relative magnetic permeability and its components – real and imaginary parts, energy losses and coercive field.

In the water-jet cutting technology case, a lower deterioration of the above-mentioned physical quantities was noticed. This is due to the fact that during punching mechanical stresses are induced in the material. In contrast water-jet cutting method is considered to be a non-invasive cutting technology.

The classical punching is used worldwide, in the electrical motor mass production, because it is a fast and simple method. The water-jet technology is applied only in the case of special motors, which are produced in a smaller amount, due to their increased manufacture costs.

In order to choose a proper cutting technology, the producers must analyze carefully the material quality and to decide what cutting technology is the most adequate for the electrical equipment.

ACKNOWLEDGMENT

The work was partially supported by the University POLITEHNICA of Bucharest under the PubArt program. All the measurements, reported in the paper, have been performed at Instituto Nazionale di Ricerca Metrologica, Turin, Italy. Dr. Fausto Fiorillo and Dr. Enzo Ferrara are kindly acknowledged for the fruitful discussions and suggestions concerning the experiments and the obtained results.

REFERENCES

[1] *** International Energy Agency statistics (2006), available at http://www.iea.org/stats.

[2] *** Commission regulation (EC) no. 640/2009 of 22 July (2009) implementing directive 2005/32/EC of the European parliament and of the council with regard to ecodesign requirements for electric motors, Official Journal of the European Union, vol. 52.

[3] *** IEC standard 60034-30-1, 2014: Rotating electrical machines – Efficiency classes of line operated AC motors.

[4] C.M. Gheorghe and S. Piperca, "The induction machine in Eastern Europe: A research agenda," Rev. Roum. Sci. Techn. – Électrotechn. et Énerg., Vol. 63, 4, 2018, pp. 371–378.

[5] V. Manescu (Paltanea), G. Paltanea, and H. Gavrila, "Non-oriented silicon iron alloys – State of the art and challenges," Rev. Roum. Sci. Techn. – Electrotechn. et Energ., vol. 59, no. 4, 2014, pp. 371-380.

[6] Y. Kurosaki, H. Mogi, H. Fujii, T. Kubota, and M. Shiozaki, "Importance of punching and workability in non-oriented electrical steel sheets," Journal of Magnetism and Magnetic Materials, vol. 320, 2008, pp. 2474–2480.

[7] M. Bali, H. De Gersem, and A. Muetze, "Finite-Element modeling of magnetic material degradation due to punching," IEEE Transactions on Magnetics, Vol. 50, No. 2, 2014, 7018404.

[8] B. Augustyniak, L. Piotrowski, P. Jasnoch, and M. Chmielewski, "Impact of tensile and compressive stress on classical and acoustic Barkhausen effects in grain-oriented electrical steel," IEEE Transactions on Magnetics, Vol. 50, No. 4, 2014, 6100304.

[9] H.M.S. Harstick, M. Ritter, and W. Riehemann, "Influence of punching and tool wear on the magnetic properties of nonoriented electrical steel," IEEE Transactions on Magnetics, Vol. 50, No. 4, 2014, 6200304.

[10] V. Manescu-Paltanea, G. Paltanea, and I.V. Nemoianu, "Influence of edge mechanical stress on the 50 Hz magnetic properties of thin electrical steel," Proceedings of International Conference on Optimization of Electrical and Electronic Equipment (OPTIM-ACEMP), May 2017, pp. 450–455.

[11] V. Manescu-Paltanea, G. Paltanea, and I.V. Nemoianu, "Degradation of static and dynamic magnetic properties of non-oriented steel sheets by cutting," IEEE Transactions on Magnetics, Vol. 54, No. 11, 2018, 2001705.

[12] G. Crevecoeur, P. Sergeant, L. Dupre, L. Vandenbossche, and R. Van de Walle, "Local identification of magnetic hysteresis properties near cutting edges of electrical steel sheets," IEEE Transactions on Magnetics, Vol. 44, No. 6, 2008, pp. 3173-3176.

[13] A. Belhadj, P. Baudouin, and Y. Houbaert, "Simulation of the HAZ and magnetic properties of laser cut non-oriented electrical steels," Journal of Magnetism and Magnetic Materials, vol. 248, 2002, pp. 34–44.

[14] V. Puchy, F. Kovač, L. Falat, I. Petryshynets, a.o., "The effects of CO2 laser and thulium-doped fibre laser scribing on magnetic domains structure, coercivity, and nanohardness of Fe-3.2Si grain-oriented electrical steel sheets," Kovove Materialy, vol. 56, no. 6, 2018, pp. 389–395.

[15] E.G. Araujo, J. Schneider, K. Verbeken, G. Pasquarella, and Y. Houbaert, "Dimensional effects on magnetic properties of Fe–Si steels due to laser and mechanical cutting," IEEE Transactions on Magnetics, Vol. 46, No. 2, 2010, pp. 213-216.

[16] G. Loisos and A.J. Moses, "Effect of mechanical and Nd:YAG laser cutting on magnetic flux distribution near the cut edge of non-oriented steels," Journal of Materials Processing Technology, vol. 161, 2005, pp. 151–155.

[17] A.J. Moses, N. Derebasi, G. Loisos, and A. Shoppa, "Aspects of the cut-edge effect stress on the power loss and flux density distribution in electrical steel sheets," Journal of Magnetism and Magnetic Materials, 10.1016/S0304-8853(00)00260-2, 2000, pp. 690-692.

[18] R. Siebert, J. Schneider, and E. Beyer, "Laser cutting and mechanical cutting of electrical steels and its effect on the magnetic properties," IEEE Transactions on Magnetics, Vol. 50, No.4, 2014, 2001904.

[19] S. Bayraktar and Y. Turgut, "Effects of different cutting methods for electrical steel sheets on performance of induction motors," Proc IMechE Part B: J. Engineering Manufacture, vol. 232, no. 7, 2016, pp. 1287–1294.

[20] M. Emura, F.J.G. Landgraf, W. Ross, and J.R. Barreta, "The influence of cutting technique on the magnetic properties of electrical steels," Journal of Magnetism and Magnetic Materials, 254–255, 2003, pp. 358–360.

[21] G. Paltanea, V. Manescu - Paltanea, H. Gavrila, A. Nicolaide, and B. Dumitrescu, "Comparison between magnetic industrial frequency properties of non-oriented FeSi alloys, cut by mechanical and water jet technologies," Rev. Roum. Sci. Techn. - Série Électrotechnique et Énergétique, vol. 61, no. 1, 2016, pp. 26-31.

[22] C.C. Chiang, A.M. Kni, M.F. Hsieh, M.G. Tsai, a.o "Effects of annealing on magnetic properties of electrical steel and performances of SRM after punching," IEEE Transactions on Magnetics, Vol. 50, No. 11, 2014, 8203904.

[23] Z. Wang, S. Li, R. Cui, X. Wang, and B Wang, „Influence of grain size and blanking clearance on magnetic properties deterioration of non-oriented electrical steel," IEEE Transactions on Magnetics, Vol. 54, No. 5, 2018, 2000607.

[24] G. Paltanea, V. Paltanea, and I.V. Nemoianu, "Magnetic properties of non-oriented silicon iron sheets in case of external applied thermal treatments," Rev. Roum. Sci. Techn. – Electrotechn. et Energ., vol. 55, no. 4, 2010, pp. 357-364.

[25] G. Goldbeck, M. Cossale, M. Kitzberger1, G. Bramerdorfer, a.o. "Numerical implementation of local degradation profiles in soft magnetic materials," The XIII International Conference on Electrical Machines (ICEM), 2018, pp. 1037-1043.

[26] D. Singh, P. Rasilo, F. Martin, A. Belahcen, and A. Arkkio, "Effect of mechanical stress on excess loss of electrical steel sheets," IEEE Transactions on Magnetics, Vol. 51, No. 11, 2015, 1001204.

[27] E. Cardelli, E. della Torre, and E. Pinzaglia, "Using the reduced Preisach vector model to predict the cut angle influence in Si-Fe steels," IEEE Transactions on Magnetics, Vol. 41, No. 5, 2005, pp. 1560-1563.

[28] L.R. Dupré, R. Van Kerr, and J.A.A. Melkebeek, "A study of laser cutting and punching on the electromagnetic behavior of electrical steel sheets using a combined finite element dynamic Preisach model," Proc. 4th Int. Workshop of Electric and Magnetic Field, Marseilles, France, May, 1998, pp. 195-200.

[29] B. Vaseghi, S.A. Rahman, and A.M. Knight, „Influence of steel manufacturing on J-A model parameters and magnetic properties," IEEE Transactions on Magnetics, Vol. 49, No. 5, 2013, pp. 1961-1964.

[30] B. Koprivica, M. Šućurović, and A. Milovanović," Calibration of AC induction magnetometer," Facta Universitatis - Series: Electronics and Energetics, vol. 31, no. 4, 2018, pp. 613-626.

[31] S. Zurek, P. Borowik, and K. Chwastek, "Anisotropy of loss density of chosen electrical steel sheets," Przeglad Elektrotechniczny, Vol: 94, no: 2, pp: 96-99, 2018.

[32] F. Fiorillo, Measurement and Characterization of Magnetic Materials, Elsevier Academic Press, 2004.

[33] G. Bertotti, Hysteresis in Magnetism, Elsevier Academic Press, 1998.

[34] A. Nicolaide and S. Öner, "Considerations on the magnetization characteristics of soft magnetic materials," Rev. Roum. Sci. Techn. Ser. Electrotechnique et Energ., vol. 56, no. 4, pp. 349-358, 2011.

[35] A. Nicolaide and S. Öner, "Determination of the hysteresis loop and losses by the DC tests and programming facilities," Rev. Roum. Sci. Techn. Ser. Electrotechnique et Energ., vol. 56, no. 1, pp. 25-35, 2011.

978-1-5386-7688-2/19 $31.00 © 2019 IEEE

Design of Conical Rotor Flux-Switching Permanent Magnet Machine with Improved Flux-Weakening Capability for Traction Applications

Hao Ding, Mingda Liu, and Bulent Sarlioglu
Wisconsin Electric Machines and Power Electronics Consortium (WEMPEC)
University of Wisconsin – Madison
Madison, WI, USA
sarlioglu@wisc.edu

Abstract—The purpose of this paper is to propose a novel flux-switching permanent magnet machine (FSPM) with a conical rotor. The conical rotor has two degree-of-freedom movement, including axial and radial movements. The axial movement is realized by controlling the axial force generated by the *d*-axis current. During the positive axial movements, the air gap length is reduced and leads to higher air gap flux density to increase the torque production capability. During the negative axial movements, the air gap flux density is reduced due to the increase of air gap length, which creates an opportunity for the improved flux-weakening compared with normal FSPM machines. In this paper, the design parameters of the conical FSPM are provided. The axial force and torque production capability are investigated by analytical equations and 3-D finite element analysis (FEA). Flux-weakening is also evaluated at various axial positions and *d*-axis currents.

Keywords—axial force, conical machine, flux-switching permanent magnet machine, flux-weakening

I. INTRODUCTION

Electric machines with conical shape stator/rotor have been investigated for two-degree-of-freedom movements for the presence of axial force created by the axial components of the normal force on the rotor. Magnetic bearings can be realized by conical machines which provide supporting force against both axial and radial displacements. The radial and axial force exerted on the conical rotor surface are used to control the position of the rotor [1]-[2]. In reference [3], two conical PM machines are put back-to-back to form a five-axis bearingless motor. The two conical motors are controlled in such a way that they produce both torque and radial/axial force. In reference [4], the axial force generated by a conical induction motor is used for actuator engagement/disengagement for aircraft wheel, which eliminates the need for an extra axial actuator. Flux-switching permanent magnet (FSPM) machine has been widely investigated in the past few decades [5]–[8]. The FSPM machine has the permanent magnet in the stator and can be easily cooled by a water jacket. Another advantage of FSPM machine is the robust rotor structure, which has less mechanical issues at high-speed condition compared to other PM machines with permanent

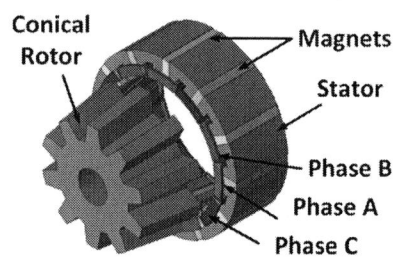

Fig. 1. Exploded view of CR-FSPM machine

magnets in the rotor [9]. However, no research has been reported to investigate the FSPM machine with a conical rotor.

Wide speed range is one of the requirements for electric machines designed for traction application. Flux-weakening is usually applied to the motor when operating beyond the base speed to keep the motor terminal voltage under the limit of the inverter. In PM machines, the negative *d*-axis current is used to counteract the magnetic flux from the permanent magnet. Depending on the inverter current limit and the machine's characteristic current, the constant power speed range (CPSR) of the machine varies [10]-[12]. However, the applied negative *d*-axis current introduces copper losses in the armature winding and does not contribute to the output power. Therefore, it is hard to achieve high efficiency beyond the base speed, especially in the deep flux-weakening region.

Adjusting the machine's air gap field density by changing the effective stack length has been investigated in [13]. A DC motor and an axial position sensor are used to adjust the alignment of stator and rotor. The machine is able to achieve flux-weakening without changing the angle of the current vector. However, there is a significant increase in system complexity. Flux-weakening performance of a conical surface PM machine is studied in [14]. It is shown that the air gap length and the stack length can be controlled by *d*-axis current and effective flux-weakening can be achieved. However, no research has been reported to investigate how the flux-weakening is related to the axial forces of a conical rotor FSPM (CR-FSPM)

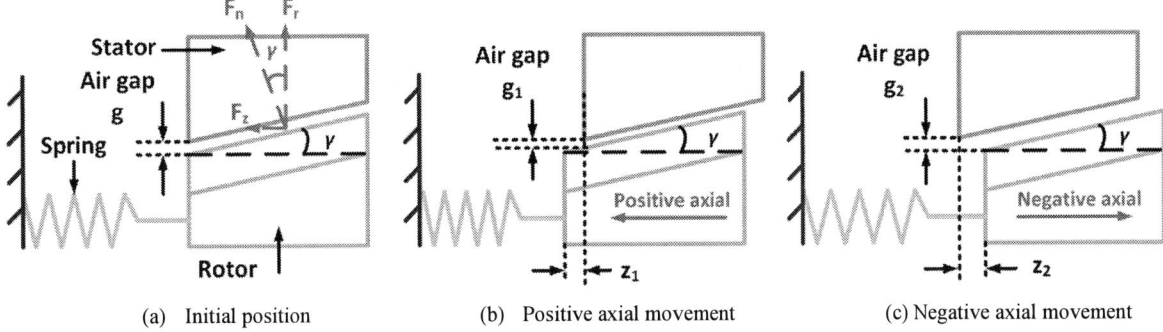

(a) Initial position   (b) Positive axial movement   (c) Negative axial movement

Fig. 2. Demonstration the axial forces and axial movements of the CR-FSPM machine

machine. This paper proposes an FSPM machine with conical shape rotor shown in Fig. 1. The proposed CR-FSPM machine has two-degree-of-freedom movements including axial and radial movements. The positive axial movements lead to shorter air gap length and higher torque production capability. The negative movements allows an effective field weakening realized by a combination of negative d-axis current and controlled air gap enlargement.

The contribution of this paper is to propose a novel FSPM machine with a conical shape rotor, which is able to achieve bi-directional axial movements and deeper flux-weakening operation than the conventional FSPM machine. The structure and operating principles of the CR-FSPM machine are presented in Section II. Section III shows the performance evaluation of the CR-FSPM machine. Conclusions are given in Section IV.

II. DESIGN AND OPERATING PRINCIPLES

A. Operating Principles

In the electric machines with a cylindrical air gap, there are no axial components of the normal forces on the cylindrical air gap. However, for the conical machines, the air gap is no longer cylindrical, and a cone angle γ is introduced. Therefore, as is shown in Fig. 2 (a), the force normal to the rotor pole surface F_n is divided into the radial force F_r and the axial force F_z. The force density can be calculated using Maxwell's stress tensor and the normal and tangential magnetic stress can be calculated based on the normal and tangential air gap flux density. The force density can be simplified by ignoring the tangential flux density when calculating the normal force F_n [3], [4]. For the conical shape of the rotor, the mid-plane of the rotor can be used to estimate the radius of the air gap and calculate the effective space between the rotor and stator. Therefore, the axial force can be calculated with (1)

$$F_z = \frac{\pi r_{mid} \dfrac{L_{is}}{\cos\gamma} \sum B_{ni}^{2}}{2\mu_0} \sin\gamma \tag{1}$$

where L_{is} is the motor effective stack length, r_{mid} is the rotor mid-plane air gap radius, γ is the cone angle, B_n is the air gap flux density in the normal direction, and the subscription i refers to the number of harmonics orders.

In (1), the axial force is directly related to the field strength in the air gap. Therefore, the axial movement can be controlled by controlling the air gap flux density. Controlling d-axis current or change the air gap length can increase or decrease the air gap flux density to control the axial force. By controlling the axial force to increase or decrease, the proposed CR-FSPM machine can achieve positive and negative axial movements shown in Fig.2. In Fig. 2, the CR-FSPM machine is connected to a spring with a constant factor k as well as a linear bushing for combined linear and rotary motions. In Fig. 2 (a), at the initial position, the spring needs to be preloaded due to the inherent PM flux linkage of the FSPM machine at no load condition which produces a positive F_z exerting on the spring. In order to realize the positive movements, increasing d-axis current leads to a larger axial force, which further compresses the spring until a balance of spring force F_s with rotor axial force F_z is reached. As is shown in Fig. 2 (b), the axial movements lead to the reduction of air gap length, which further increase the air gap flux density besides the increase of d-axis current. The negative axial movement is shown in Fig. 2 (c). By decreasing the d-axis current, the axial force F_z is decreased and the spring force F_s, therefore, is larger than the axial force initiating a negative movement. While the rotor starts the negative axial movement, the air gap increases due to the conical shape of the rotor which further reduce the field strength and axial force until a balance between the axial force and spring force is reached. The increased air gap length also creates a "free" opportunity for the flux-weakening operation, and it will be discussed in the later section.

B. Design of the CR-FSPM Machine

The calculation of the air gap length at various axial position is shown in (2) where z is the distance for axial movement, g is the air gap length at the initial position, and γ is the cone angle.

$$g_c = -z \tan\gamma + g \tag{2}$$

Based on (2), the positive axial movement has a limitation as the air gap is decreasing with the increase of positive movement and the cone angle. Therefore, the cone angle selection and the control of axial force need carefully consider the reliability issues. Hence, the minimum air gap length limitation is set to be 3mm.

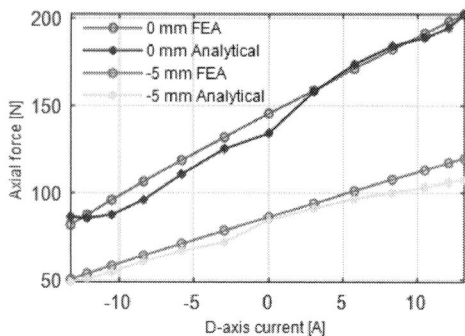

Fig. 3. *d*-axis current vs axial force by analytical and FEA

FSPM machine is known for high flux density due to the flux intensify effect. 12-slot, 10-pole FSPM machine has been proved to produce high power density and has good potential for high-speed applications [9]. Besides, sinusoidal back-EMF of 12/10 FSPM machine reduces the complexity of motor control. Therefore, 12/10 is selected for the slot and pole combinations for the CR-FSPM machine. In an FSPM machine, the permanent magnets are sandwiched between the stator cores and creates strong PM flux linkage with high torque. The strong magnetic field in the air gap leads to a high axial force, which requires a spring with high spring constant k to balance the axial force from the rotor. According to the axial force equation in (1), the cone angle should be set relatively small than the induction machines with conical rotors to avoid extremely large axial forces and k [4]. In this design, the cone angle is set to be 8 degrees, considering the air gap length limit and the limitation of the spring. The air gap length at the initial position is selected to be 1 mm, and the axial movement is investigated from -5 mm to 5 mm. The magnets are selected to be ferrite to reduce the magnet losses and to avoid high axial forces. In addition, using ferrite magnets prevents the heavy saturation in the laminations during the large positive axial movements (small air gap length).

The design parameters are shown in Table. I. Fig. 3 shows the axial force with respect to *d*-axis current at the initial position and -5 mm axial movement by analytical methods and 3-D finite element analysis (FEA). As can be seen in Fig. 3, the analytical calculations match the FEA results, and the axial force increases linearly with the increasing of d-axis current.

In (1), as the axial force is directly related to the air gap flux density, and the air gap flux density is affected by the magnets and the *d*-axis current. The axial force F_z can be separated into the axial force caused by the magnets F_m and the axial force generated by the d-axis current F_d. With analytical calculations and 3-D FEA, F_m is calculated to be 131.2 N. In order to control the axial position of the rotor, the spring is compressed by a length x. The spring force $F_s(x)$ equals to F_m at the initial position shown in (3). If the rotor moves a distance of z, the spring force $F_s(x+z)$ changes correspondingly until the rotor axial force equals to the spring force, which is shown in (4).

$$F_s(x) = F_m = kx \qquad (3)$$

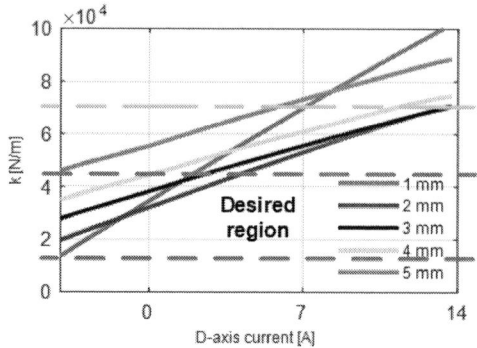

Fig. 4. *d*-axis current vs spring constant at positive positions

$$F_s(x+z) = k(x+z) = F_m + F_d(z) \qquad (4)$$

Substituting (3) into (4), the spring constant can be calculated in (5) for each axial position. Fig. 4 shows the spring constant k versus *d*-axis current at positive axial positions. As the rotor is connected to the same spring at each location, the spring constant should be the same at each location. Therefore, in Fig. 4, the spring constant for the up to 5 mm axial movement is between the range of 46000 N/m to 70000 N/m. It is noted that, with shorter axial movement, the range of the spring constant is enlarged. For example, at 4 mm, the spring constant is from 35000 N/m to 75000 N/m, and at 1 mm, the spring constant is from 13500 N/m to 100000 N/m.

$$k = \frac{F_d(z)}{z} \qquad (5)$$

In Fig. 5, the spring constant k versus *d*-axis current at negative axial positions are plotted. The desired k for the -1 mm to -5 mm axial movement is much lower than the positive movement due to the lower requirement of axial force. This is because when increasing the air gap length in the negative axial movements, the air gap magnetic field is getting weaker with respect to the positive axial movement. Therefore, the desired spring constant for -5 mm movement is from 2100 N/m to 17000 N/m, for -4 mm movement, the range is from 0 N/m to 20000 N/m. It is noted that in order to have bi-directional axial

TABLE I. DESIGN PARAMETERS

Parameters	Value
Rotor speed n [rpm]	3,000
Current density [A/mm²]	7.5
Magnet remanence [T]	0.4
Winding turns per coil	20
Air gap length [mm]	1
Cone angle [deg]	8
Front rotor OD [mm]	70
Rear rotor OD [mm]	80
Stator stack length [mm]	71
Stator OD [mm]	200
Positive movements [mm]	5
Negative movements [mm]	-5
L_d [mH]	1.9
L_q [mH]	1.9
Loaded Torque [Nm]	7.16
Output power [W]	2500

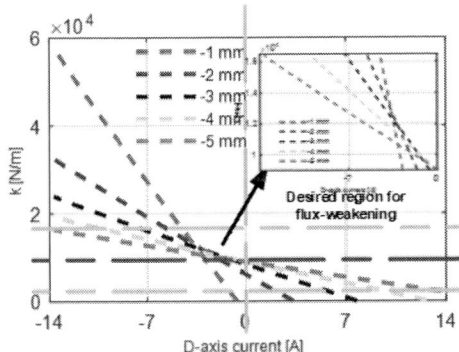

Fig. 5. *d*-axis current vs spring constant at negative positions considering filed weakening operations

movement, the distance for axial movement needs to be small. The smaller $|z|$, the larger overlap region of k for positive movements and negative movements, for example, from -5 mm to 1 mm with k from 13500 N/m to 17000 N/m. If -5 mm to 4 mm bi-directional movement needs to be achieved, a hybrid spring with adjustable spring constant can be applied.

C. Opportunities for Flux-Weakening

With the negative axial movements, the air gap length increases, leading to a weakened field. Thus, a "natural" flux-weakening is induced as a negative *d*-axis current offset compared with the cylindrical type machine with a constant air gap. As a result, in the CPSR region, the CR-FSPM machine requires much less *d*-axis current to achieve the same effect of flux-weakening compared with cylindrical type machine. In addition, the CR-FSPM machine has a lower inductance than the cylindrical FSPM machine due to a larger air gap. Therefore, the power factor of the CR-FSPM machine is higher than the cylindrical type machine during field weakening. In addition, due to the limitation of the inverter, the CR-FSPM machine has more room to apply negative *d*-axis current than the cylindrical type machine to further weaken the field and reach a larger CPSR range. Therefore, in Fig. 5, considering the negative *d*-axis current requirements for flux-weakening at the negative axial positions up to -5 mm, the desired range of k for flux-weakening is from 9000 N/m to 17000 N/m.

Characteristics current I_{ch} is also used as a metric for evaluating the performance of flux-weakening. The calculation of I_{ch} is shown in (6) where Ψ_m is the rms PM flux linkage, L_d is the *d*-axis inductance. With the increases of the length of the air gap, the PM flux linkage decreases much faster than the *d*-axis inductance and results in a smaller characteristic current.

$$I_{ch} = \frac{\Psi_m}{L_d} \qquad (6)$$

III. PERFORMANCE COMPARISON

A. Electromagnetic Performance at Various Axial Positions

3-D FEA is implemented for evaluating the electromagnetic performance of the CR-FSPM machine at positive and negative axial positions. Fig. 6 shows the flux density distribution of CR-FSPM machine at no load condition at -5mm, 0 mm, and 4 mm axial positions. In Fig. 6, it can be observed, the flux density in the materials increases from negative movements to positive movements. However, at 4 mm position with 0.44 mm air gap, the material is not saturated due to the usage of ferrite magnets.

Fig. 7 shows the air gap flux density at -5 mm, 0 mm, and 4 mm positions. The 4th, 6th, 14th, and 16th orders of harmonics are the dominant components in the CR-FSPM machine. Fig. 8 and Fig. 9 show the no-load flux linkage and back-EMF comparison at -5 mm, 0 mm, and 4 mm. The PM flux linkage decreases from the positive movements to the negative movements. The back-EMF of -5 mm position is reduced to 50.1 V_{rms}.

Fig. 10 and Fig. 11 shows the average loaded torque and axial force with respect to the *q*-axis current and *d*-axis current from -5 mm to 4 mm axial position, respectively. Using the data from Fig. 10 and Fig. 11, a LUT can be created for investigation the capabilities of flux-weakening of the CR-FSPM machine.

B. Performance Evaluation of the Flux-Weakening of the CR-FSPM Machine

The flux-weakening of the CR-FSPM machine is described in the following procedures. At the initial position, the spring is compressed, and the spring force is balanced with the CR-FSPM machine at no load condition at 3000 rpm. With 0 A *d*-axis current, -13.7 A *q*-axis current maximum torque can be achieved as 7.16 Nm at the initial position. Then the machine starts to go into the CPSR region. By decreasing the d-axis current, the rotor begins to move in the negative axial direction. By controlling the *d*-axis current with the method proposed in Fig. 5 together with the LUT in Fig. 10 and Fig. 11, the force can be balanced at -1 mm position and the terminal voltage is the same with the terminal voltage at the initial position. Similarly, in the CPSR region, for each given spring constant *k* from the desired region, by controlling the *d*-axis current, an optimal speed can be obtained at each location, and the axial force is also balanced at that location.

Fig. 12 shows the CPSR region of the torque-speed curve for the CR-FSPM machine from -1 to -5 mm positions with various spring constant *k* of 9000 N/m, 12000 N/m, 13500 N/m, and 16000 N/m. The optimal speed at each location increases with the increase of the spring constant. This is for the reason that with larger spring constant, from Fig. 5, a larger *d*-axis current is required for the negative axial force, which also weakens the field. For example, when *k* equals to 13500 N/m, the optimal speed at -5 mm position is 6000 rpm. When k equals to 16000 N/m, the optimal speed at -5 mm is 13000 rpm.

Fig. 13 shows the optimal speed at each location for a given spring constant *k* versus the d-axis current. It can be observed that when *k* is small, e.g., 9000 N/m and 12000 N/m, in order to obtain an optimum speed at each location to balance the axial force with the spring force, the *d*-axis current needs to be increased. In the cases with small *k*, the rotor-spring force balance condition does not need that much negative axial force created by the large negative *d*-axis current. As the field is already weakened at negative axial positions, a small amount of

Fig. 6. No load flux density distribution of the CR-FSPM machine at -5 mm, 0 mm, and 4 mm axial positions

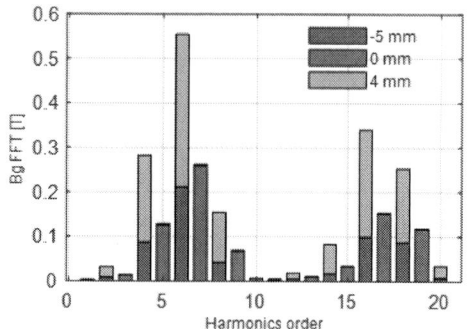

Fig. 7. Air gap flux density of the CR-FSPM machine at various axial positions

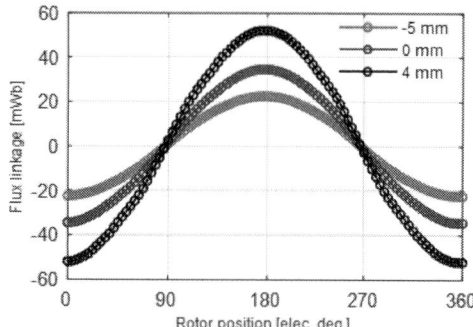

Fig. 8. No load flux linkages of the CR-FSPM machine at various axial positions

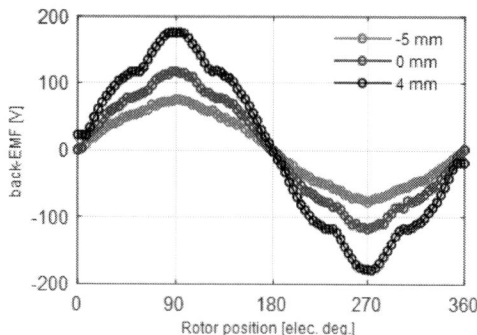

Fig. 9. Back-EMF of the CR-FSPM machine at various axial positions

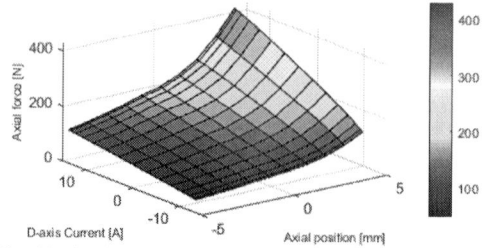

Fig. 10. d-axis current vs axial force at various positions

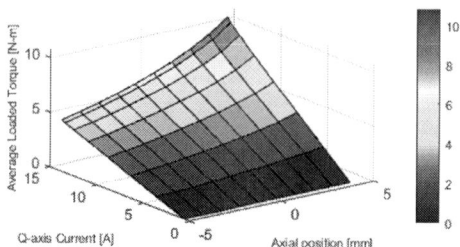

Fig. 11. q-axis current vs average torque at various positions

negative d-axis current is enough to balance the force and achieve the flux-weakening operation. However, in the cases with large k, a large amount of negative d-axis current is needed to balance the axial force, and in the meanwhile flux-weakening operation benefits in the large negative d-axis current.

C. Comparison of the Flux-Weakening Capability of the CR-FSPM Machine with a Normal FSPM Machine

The flux-weakening capability of a normal cylindrical type of FSPM machine is compared with the proposed CR-FSPM machine. The two designs have the same stack length. The size of the cylindrical FSPM machine equals to the mid-plane of the CR-FSPM machine at the initial position. With 7.5 A/mm² current density and 20 turns, 13.7 A_{pk} current is applied in the windings for both machines.

Characteristics current can be used to compare the flux-weakening performance of the proposed CR-FSPM machine with a normal FSPM machine. For the CR-FSPM machine at the initial position, the rms value of PM flux linkage equals to 24.4 mWb. The d-axis inductance equals to 1.9 mH, and I_{ch0} is calculated to be 9.08 A_{rms} which close to the rated current 9.7 A_{rms}. At -5 mm position, the rms value of PM flux linkage is 15.9 mWb, d-axis inductance is 1.62 mH, and I_{ch-5} is calculated to be 9.81 A which is very close to the rated current. However, for the normal FSPM machine, the rms value of PM flux linkage is 24.4 mWb, the d-axis inductance is 1.79 mWb, and I_{ch-n} is calculated to be 9.64 A_{rms} which is larger than the characteristics current of the CR-FSPM machine.

In addition, by comparing the flux linkage of the normal PM machine with the CR-FSPM machine at -5 mm position, the normal FSPM machine needs to apply -6 A d-axis current to get the same flux linkage. When k equals to 13500 N/m, optimal speed is 6000 rpm, the CR-FSPM machine requires -8.2 A d-axis current and 11 A q-axis current. While for the normal FSPM

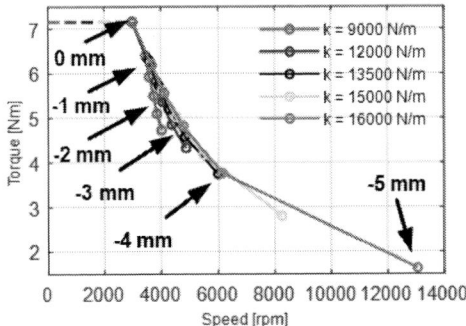

Fig. 12. Torque-speed curve at CPSR region with different spring constant at various negative axial positions

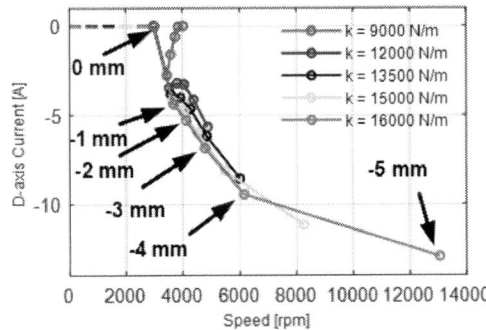

Fig. 13. *d*-axis current-speed curve at CPSR region with different spring constant at various negative axial positions

machine with the same speed and terminal voltage, -11 A *d*-axis current and 6.9 A *q*-axis current are required. In this case, the CR-FSPM machine can apply 2.8 A more *d*-axis current than the normal FSPM machine to achieve a wider CPSR range considering the inverter current limit.

IV. CONCLUSION

This paper design and provide a novel CR-FSPM machine with two degree-of-freedom movements. The axial movements are controlled by the *d*-axis current and a strong spring. The axial forces and torque production capability of the CR-FSPM are investigated for positive and negative axial movements from -5 mm to 4 mm. Reducing the distance of the positive axial movements enables a wider overlapped range of a fixed spring constant for both the positive and negative movements. In this design, the spring constant with 13500 N/m is selected for the bi-directional axial movement from -5 mm to 1 mm with improved flux-weakening capability at negative axial positions.

The negative movement of the CR-FSPM machine increases the air gap length, which enables a wider CPSR range for the flux-weakening operations. The magnetic field of -5 mm position of CR-FSPM machine is equivalents to a normal FSPM machine applying -6 A *d*-axis current. The characteristics current of the CR-FSPM machine is much lower and closer to the rated current with respect to the normal FSPM machine indicating a better flux-weakening capability of the CR-FSPM machine.

REFERENCE

[1] Shilei Xu and Jiancheng Fang, "A novel conical active magnetic bearing with claw structure," *IEEE Trans. Magn.*, vol. 50, no. 5, pp. 1–8, May 2014.

[2] EA Fisher and E Richter - US Patent 5,233,254, 1993.

[3] G. Munteanu, A. Binder, and S. Dewenter, "Five-axis magnetic suspension with two conical air gap bearingless PM synchronous half-motors," in *International Symposium on Power Electronics Power Electronics, Electrical Drives, Automation and Motion*, 2012.

[4] S. Roggia, F. Cupertino, C. Gerada, and M. Galea, "A two-degree-of-freedom system for wheel traction applications," *IEEE Trans. Ind. Electron.*, vol. 65, no. 6, pp. 4483–4491, Jun. 2018.

[5] J. T. Chen and Z. Q. Zhu, "Winding configurations and optimal stator and rotor pole combination of flux-switching PM brushless AC machines," *IEEE Trans. Energy Convers.*, vol. 25, no. 2, pp. 293–302, Jun. 2010.

[6] H. Ding, Y. Li, D. Han, M. Liu and B. Sarlioglu, "Design of a novel integrated motor-compressor machine with GaN-based inverters," in *IEEE Power Electronics and Applications (EPE'17 ECCE Europe).*, Warsaw, Poland, 2017.

[7] Y. Li, J. H. Kim, R. Leuzzi, M. Liu, and B. Sarlioglu, "Novel 6-slot 4-pole dual-stator flux-switching permanent magnet machine comparison studies for high-speed applications," in *IEEE Energy Conversion Congress and Exposition (ECCE)*, Milwaukee, WI, 2016.

[8] H. Ding, W. Sixel, M. Liu, Y. Li, and B. Sarlioglu, "Influence of rotor pole thickness on optimal combination of stator slot and rotor pole numbers in integrated flux-switching motor-compressor," in *IEEE Energy Conversion Congress and Exposition (ECCE)*, Portland, OR, 2018.

[9] A. S. Thomas, Z. Q. Zhu, and G. W. Jewell, "Comparison of flux switching and surface mounted permanent magnet generators for high-speed applications," *IET Electr. Syst. Transp.*, vol. 1, no. 3, p. 111, 2011.

[10] A. M. EL-Refaie, "Fractional-Slot Concentrated-Windings Synchronous Permanent Magnet Machines: Opportunities and Challenges," *IEEE Trans. Ind. Electron.*, vol. 57, no. 1, pp. 107–121, Jan. 2010.

[11] H. Dai, R. A. Torres, T. M. Jahns, and B. Sarlioglu, " Characterization and implementation of hybrid reverse-voltage-blocking and bidirectional switches using WBG devices in emerging motor drive applications," in *IEEE Applied Power Electronics Conference and Exposition (APEC)*, Anaheim, CA, USA, 2019.

[12] L. Chang, T. M. Jahns, and R. Blissenbach, "Characterization and modeling of soft magnetic materials for improved estimation of PWM-induced iron loss," in *IEEE Energy Conversion Congress and Exposition (ECCE)*, Portland, OR, 2018.

[13] K.-C. Kim, "A novel magnetic flux weakening method of permanent magnet synchronous motor for electric vehicles," *IEEE Trans. Magn.*, vol. 48, no. 11, pp. 4042–4045, Nov. 2012.

[14] F. Chai, K. Zhao, Z. Li, and L. Gan, "Flux weakening performance of permanent magnet synchronous motor with a conical rotor," *IEEE Trans. Magn.*, vol. 53, no. 11, pp. 1–6, Nov. 2017.

Development of High Efficiency and High-Speed Motor with High Power Density

Masato ENOKIZONO
Vector Magnetic Characteristic Technical Laboratory
Usa, Japany
enoki@oita-u.ac.jp

Naoya SODA
Department of Electrical and Electronic Engineering
Ibaraki University
Hitachi, Japan
naoya.soda.magtec@vc.ibaraki.ac.jp

Daisuke WAKABAYASHI
School of Engineering,Mechanical and Electrical Engineering
Nippon Bunri University
Oita, Japan
wakabayashids@nbu.ac.jp

Shohei UENO
Faculty of Science and Technology
Oita University
Oita, Japan
ueno-shohei@oita-u.ac.jp

Yuji TSUCHIDA
Faculty of Science and Technology
Oita University
Oita, Japan
tsuchida@oita-u.ac.jp

Abstract—**This paper describes the development of the high speed motor of 50 [W] at 25 [g] as a space probe rover motor. Furthermore, this motor was required to have high power density. Ultra-thin electrical steel sheet with the thickness of 80μ was used for the iron core of this motor. The iron core shape was designed by vector magnetic characteristic analysis, introducing vector magnetic characteristics representing the magnetic characteristics in arbitrary direction of this electrical steel sheet.**

Keywords—vector magnetic characteristic, core loss. Ultra-thin electrical steel sheet, high efficiency, high speed motor, high power density.

I. INTRODUCTION

Recently the high-speed motor has been noticed as the power of the moving object. This reason is due to following equation that depends on speed of motor.

$$\text{Output}: P \propto \overbrace{D^2 \times L}^{} \times B \times AC \times N \quad (1)$$

On the other hand, this equation also shows the relationship between magnetic loading and electrical loading Furthermore, output power depends on the number of pole pairs (=number of pole/2), and in order to power-up the increment of pole pairs are obtained but the rotational speed decreases. Therefore, it becomes necessary to increase driving frequency. As a result, power loss will be increased with increasing frequency excitation. In order to solve this problem, it is necessary to reduce the thickness of the steel sheet of the core material. This paper describes the advantages of the 80 μm-thick ultra-thin electrical steel sheet developed as an iron core material for high-speed motors and its application technology. The main technology is a vector magnetic characteristic technology that evaluates the magnetic property in arbitrary directions. And vector magnetic characteristics of this material are introduced [1]-[7]. Furthermore, (2) shows the relation between output power

and loss. The kind of loss is a magnetic power loss, copper loss and mechanical loss. Naturally, each loss must be decreased in order to improve an output of motor. By linking the copper loss with core loss, it is possible to decrease the copper loss. Especially, it is important to reduce the core loss of motor, because the contribution ratio of core loss increases in the high speed rotation area of the motor.

$$P_{out} = \omega T_{out} = \omega T_0 - \begin{cases} W_{mag} = W_{hys}f + W_{eddy}f^2 \\ W_{copper} = I_{exc}^2 R_{wind} + I_{eddy}R_{sheet} \\ W_{mech} \end{cases} \quad (2)$$

where W_{mag} is core loss, W_{hys} is hysteresis loss, W_{eddy} is eddy current loss, W_{copper} is copper loss, I_{eddy} is depend eddy current field of core, and W_{mech} is mechanical loss. T_0 is torque by synchronous watt, T_{out} is output torque and P_{out} output power.

In space utilization, it is almost impossible to expect heat dissipation due to convection on the vacuum state of the moon's surface or the surface of the atmosphere, Mars, so it is very important for the motor with high efficiency and little heat generated to be mounted on space exploration equipment. As commercialization of the ground applications such as drone in which the motor mass is directly linked to the cruising time, joints of the robot, driving precision measuring instruments that want to avoid temperature change, etc. are conceivable.

II. REPLACEMENT EFFECT OF ULTRA-THIN ELECTRICALSTEEL SHEET

A. Basic Magnetic Characteristics

It is necessary to select the ultra-thin electrical steel sheet as an iron core material conforming to the specifications of a high-speed motor and to examine its frequency magnetic characteristics. Figs. 1 (a) and (b) show the comparison with conventional non-oriented electrical steel sheet with 350 μm thickness and ultra-thin steel sheet with 80 μm thickness about the frequency change of core loss, and eddy current loss is improved. It is indicated that the copper loss in high speed can be decreased this. Therefore, it is necessary to design the motor in order to reflect these characteristics.

978-1-5386-7688-2/19 $31.00 © 2019 IEEE

These characteristics give suggestion the design construction of winding coil as the design of electrical loading.

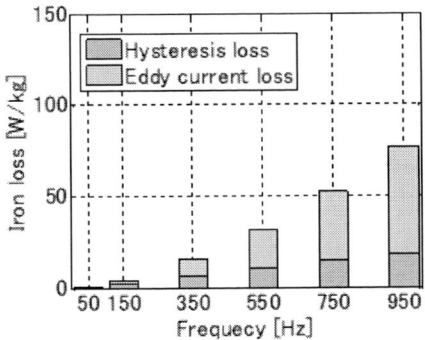

(a) Conventional non-oriented electrical steel sheet (35A300), 350μm thick.

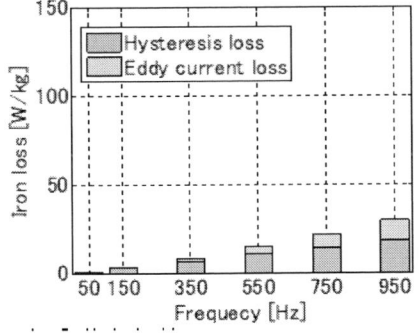

(b) Ultra-thin electrical steel sheet, 80μm thick.

Fig. 1. Core losses depending on the frequency [$\mathbf{B}_{max}$ = 1.0, θ_B = 450 deg.].

B. Basic characteristics

Fig. 2 shows the fundamental frequency characteristics of the motor iron core. For this reason, it is necessary to examine the evaluation of the motor iron core from the viewpoint of the characteristic evaluation of magnetic material. Generally by fixing the magnetic flux density level, the evaluation of the different magnetic material must measure it. Furthermore, it is required that it is executed under the sinusoidal voltage wave excitation in order to evaluate the high frequency property of the iron core.

As shown in Fig. 2, core loss of the motor core of ultra-thin steel sheet increases rectilinear in comparison with the motor core of conventional material. This tendency of the core loss can be inferred with the increase in the hysteresis loss only, since the eddy current loss is very small as shown in Fig. 2 (a). Fig. 2 (b) shows the temperature rise characteristic at exciting frequency 550 Hz.

The temperature rise of motor core was suppressed at 25%, and the gradient of the temperature rise is small. The gradient of the temperature rise is proportional to the core loss as following equation.

$$W = c \frac{\partial T}{\partial t} \qquad (3)$$

where c is the specific heat constant. The core loss estimate has reduced about 60%. On the other hand, this core loss

estimate is not accurate, but a qualitative assessment is possible.

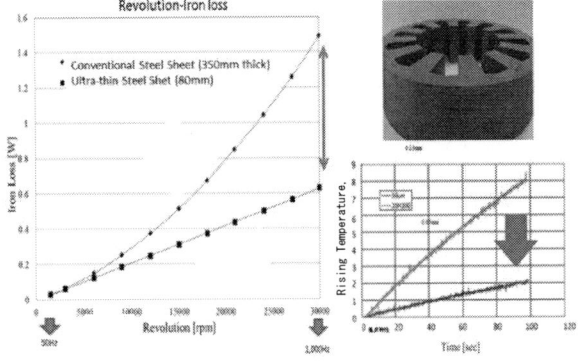

(a) Core loss dependence of motor speed.

(b) temperature rise characteristic.

Fig. 2 Fundamental characteristics of stator c

Fig. 3 shows the surface temperature distribution of conventional electric steel sheet core and ultra-thin electrical steel sheet core observed by using the thermography camera at 10,000 rpm.

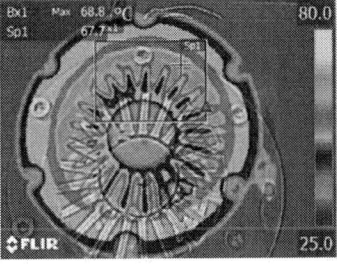

(a) Conventional electrical steel sheet core motor.

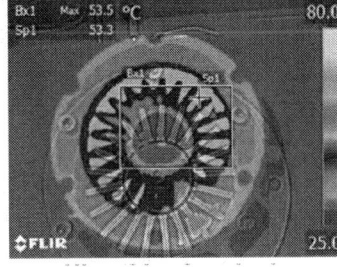

(b) Ultra-thin electrical steel sheet core motor.

Fig. 3. Temperature rise distribution at 10,000 rpm.

III. DESIGN GUIDELINES FOR LOW LOSS AND HIGH EFFICIENC

Fig. 4 compares magnetic field analysis by conventional general-purpose software with magnetic characteristics analysis proposed by the author. In the conventional method, the core loss as a function of magnetic flux density |$\mathbf{B}$|

characteristic obtained from the catalog data of the electrical steel sheet is used for the estimation of the core loss by using a mathematical approximation function. In the analysis of the conventional method, since only B can be solved, there is a fundamental problem that it cannot be directly calculated by using expressions constituting core loss. In other words, the conventional method represents the core loss distribution depending on the obtained by vector distribution. It suggested that the magnetic flux density level only has to be lowered for the decreasing core loss motor. In the vector magnetic characteristic analysis, **B** vector and **H** vector can be solved at the same time, and each vector locus is obtained from the result and it is calculated by assigning it to the core loss formula. Core loss data are unnecessary for this analysis, and relational data only between **B** vector and **H** vector is required. This analysis method can provide the trajectory (locus) of each vector, the hysteresis loop (Bx-Hx, By-Hy), and the characteristic value concerning the core loss from the analysis results of **B** and **H** vectors. In addition, the inclination angle of **B** vector θ_B, the spatial difference phase angle between **B** vector and **H** vector θ_{BH}, the axis ratio α, etc. concerning the behavior of the vector can also be obtained. [7] Vector magnetic characteristics analysis is used to adjust the factor that dominates core loss, and motor iron core shape is required to obtain low loss and high power density. As adjusting the factor, the behavior of vector **H** and its amplitude |**H**| and spatial phase angle θ_{BH} between vector **H** and vector **B**.[8]-[10]

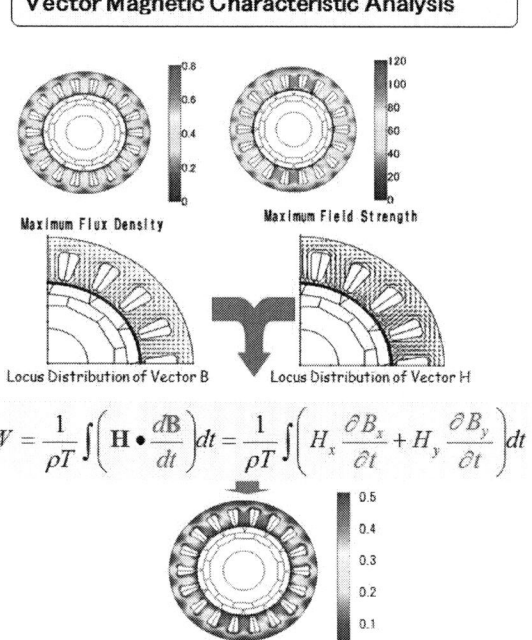

Fig. 4 Magnetic characteristic analysis for in design core loss reduction.

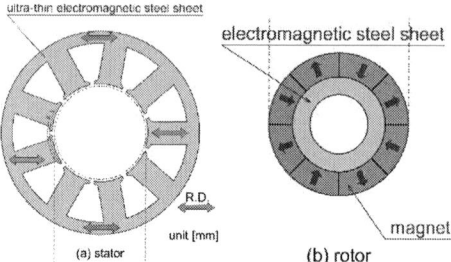

Fig. 5 Magnetic characteristic analysis for in design core loss reduction.

In this research, the shape of a small motor with 8 poles and 9 slots, which is the base model, is shown in Fig. 5. Ultra-thin electrical steel sheet was used for the stator core, and the thickness of the laminated core was fixed at 12.5 mm. Compare core loss and power density when inputting the same three-phase sinusoidal voltages to all motor models when the shape of the stator core is changed and rotating at 10,005 rpm (excitation frequency 667 Hz). However, if the shape of the iron core is changed, it is expected that the induced voltage value will change because the slot shape is also changed. Therefore, in order to compare and investigate the motors whose iron core shape was changed, the analysis conditions of the various motors are made consistent according to the following procedure.

(1) By changing the number of windings, the effective value of the induced voltage waveform of the motor whose stator shape is changed is made uniform.

(2) Change winding resistance according to winding number ratio.

(3) Analyze by inputting the same three-phase sinusoidal Therefore, it possible to compare and examine motors with different structures.

Therefore, it is possible to compare and examine motors with different structures. In this study, we analyzed the small motor by the finite element method which introduced the integral type E&S model to investigate the core loss distribution inside the stator core. This vector magnetic characteristic analysis can analyze directly core loss distribution as shown in Fig. 4.[7]-[9]

$$H_k(\tau) = \nu_{kr} B_k(\tau) + \nu_{ki} \int B_k(\tau) d\tau \qquad (4)$$

Here, the subscript k is x or y, and ν_{kr} and ν_{ki} are the magnetic resistivity coefficient and the magnetic hysteresis coefficient, respectively. Moreover, in this study, in order to analyze when applying a three-phase sinusoidal voltage to a small motor and rotating it at 10,005 rpm, the two-dimensional field governing equation introducing (4) and the circuit equation of the following equation are co-established.

$$V_{0n} = \frac{\partial}{\partial t} \int_c A ds + R_{mn} I_{mn} \qquad (5)$$

Here, V_{0n} ($n = 1$ to 3) is the terminal voltage, R_{mn} is the winding resistance, and I_{mn} is the excitation current. Exciting current was obtained as analysis results. And the copper loss of the small motor can be obtained from the winding resistance. Furthermore, in the analysis using the integral E&S model, the core loss distribution in the stator core can be directly obtained from the magnetic flux density vector **B**

978-1-5386-7688-2/19 $31.00 © 2019 IEEE

and the magnetic field intensity vector **H** of the analysis result using the following equation.

$$P_i = \frac{1}{\rho T} \int_0^T \left(H_x \frac{dB_x}{dt} + H_y \frac{dB_y}{dt} \right) dt \qquad (6)$$

Here, ρ is the material density, and T is the excitation period. By obtaining the sum of the core loss distribution in the stator core, the core loss of the small motor can be obtained. The torque was calculated by Maxwell's stress method. In addition, losses of small motors ignored mechanical losses and considered a core loss and copper loss only.

A. Constitution's design

In physique design, in order to determine the iron core size effective for reducing core loss of small motors, as shown in Fig. 5 the dimensions of the back yoke width B, the tooth length T, and the tooth width W are set to be We changed and analyzed. The dimensions of the base model are T = 6.2 mm, W = 3.0 mm, B = 1.5 mm. In this paper, we report the case where only the back yoke width B is changed by fixing T = 6.2 mm and W = 3.0 mm.

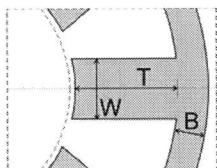

Fig. 6 Parameters T, W and B which definition of stator core.

When changing the back yoke width B, the outer diameter of the back yoke was changed so that the tooth length T did not change. The back yoke width B was changed from 0.9 mm to 2.5 mm for the reference model size of 1.5 mm. Fig. 7 shows the core loss distribution when only the back yoke width B is changed. When the width of the back yoke is changed, narrowing the width B increases the magnetic flux density in the back yoke and the core loss increases, and if it is expanded, the magnetic flux density decreases and the core loss decrease. However, there is no change in the core loss distribution in the teeth.

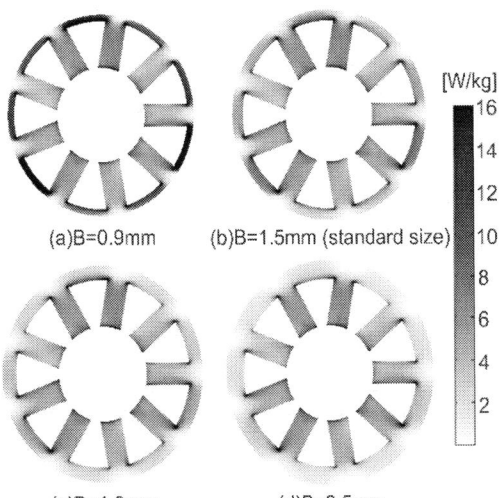

Fig. 7 Core loss distribution for various width B.

Next, Fig. 8 shows the change rate from the base model of the average magnetic flux density distribution, core loss, and copper loss when the back yoke width B is changed from 0.9 mm to 2.5 mm. When only the back yoke width B is changed, since the outer diameter of the stator is increased, the coil region is not changed and there is almost no influence on the coil. Therefore, even if the back yoke width is changed, the copper loss is almost constant. Core loss decreases as back yoke width B increases. In order to reduce the core loss, it is not so much as to expand the back yoke width B further. The rate of change of the standard deviation of the magnetic flux density distribution when the back yoke width B is changed is shown in Fig. 9. As shown in Fig. 9, the standard deviation of the magnetic flux density becomes large even if it is too narrow or too wide, and it is minimum at the width of 1.7 mm. This standard deviation represents the fluctuation of the magnetic flux density, and when T = 6.2 mm and W = 3.0 mm, B = 1.7 mm is optimal, and it shows that the region where magnetic flux does not pass increases even if it is expanded further.

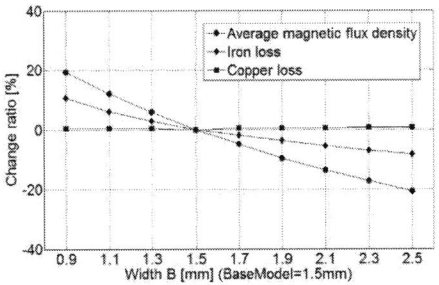

Fig. 8 Relation between the width B and the change ratio of average magnetic flux density and so on.

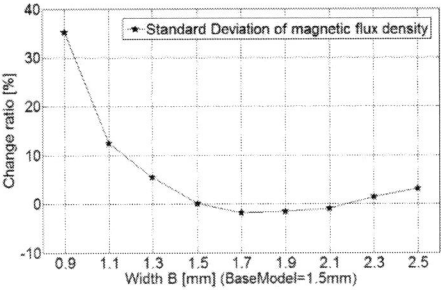

Fig. 9 Relation between the width B and the change ratio of standard deviation of magnetic flux density.

B. Stator Shape Design

In the shape design, attaching a radius to the corner of the teeth tip and the teeth root portion makes the magnetic flux flow smoother to reduce the core loss. In order to reduce the core loss, it is possible to suppress the concentration of the magnetic flux by making the slot shape along the flow of the magnetic flux line distribution. Furthermore, due to the relationship between the vector magnetic characteristics and the core loss, it is necessary to reduce the spatial phase difference angle θ_{BH} between the **B** vector and the **H** vector.

The core shape was examined, paying attention to the above points. The radius of the radius to be attached to the corner of the teeth collar is defined R_t as shown in Fig. 10. Fig. 11 shows the rate of change from the base model in the average of magnetic flux density, core loss, and copper loss when radius R_t is changed. The rate of change was also small because of the small flange portion, but the core loss was reduced as the radius R_t was increased, and the core loss reduction effect was confirmed. Next, the radius R_b of the radius to be attached to the root of the tooth is defined as shown in Fig.12. Fig. 13 shows the average magnetic flux density, core loss, core loss, and copper loss from the base model when the radius R_b is changed. As the radius R_b was increased, the core loss decreased, and the effect of reducing the core loss could be confirmed even when the rounded portion of the tooth was rounded. In addition, from Fig. 11 and Fig. 13, even if the radii are the same value, the range of the radius of the teeth becomes wider than the corner of the flange portion, so the core loss reduction effect becomes large.

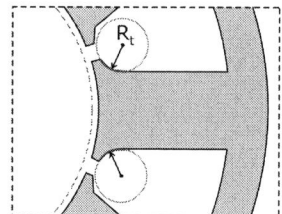

Fig. 10 Definitions of radius R_t at the teeth collar of motor stator for reduction of core loss.

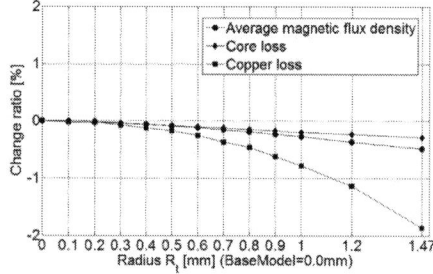

Fig. 11 Relation between the radius R_t and the change ratio of average magnetic flux density and so on.

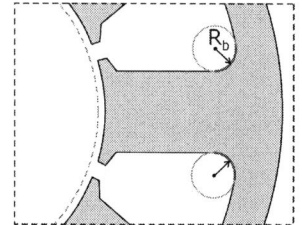

Fig. 12 Definitions of radius R_b at the teeth root of motor stator for reduction of core loss.

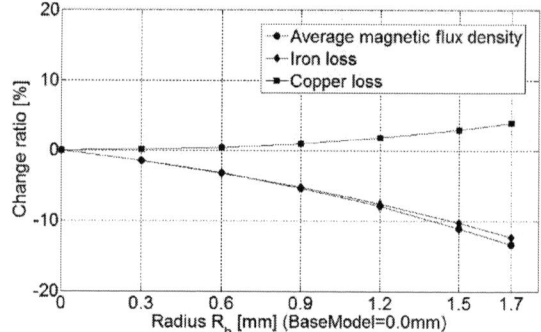

Fig. 13 Relation between the radius R_b and the change ratio of average magnetic flux density and so on.

When compared with the base model, the low core loss model has a lower magnetic flux density in the back yoke portion surrounded by a broken line. This is because the corners of the tooth root were made smooth, so that magnetic flux concentration to the corner was suppressed and no magnetic flux protruded to the back yoke side. Even if this low magnetic flux density region in the back yoke is cut out.

C. Improved Design of Power Density

A comparison of the magnetic flux density distributions of the base model and the low core loss model with the shape of smooth corners of the teeth root part is shown in Fig. 14. When compared with the base model, the low core loss model has a lower magnetic flux density in the back yoke portion surrounded by a broken line. This is because the corners of the tooth root region were made smooth, so that magnetic flux concentration to the corner was suppressed and no magnetic flux protruded to the back yoke side. Even if this low magnetic flux density region in the back yoke is cut out, it does not have a significant effect on the magnetic circuit and it is considered not to lead to a reduction in torque. As a result, it is possible to reduce only the iron core weight and improve the power density.

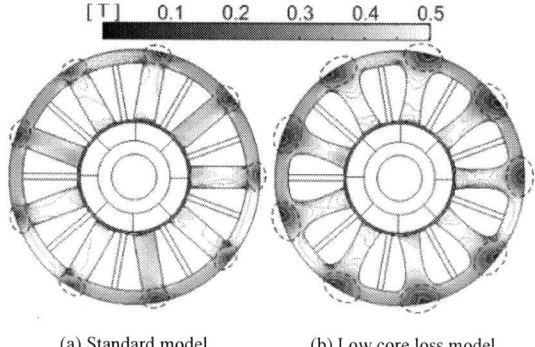

(a) Standard model   (b) Low core loss model

Fig. 14 Magnetic flux density distribution of Standard model and the Low core loss model shown by low magnetic flux density level.

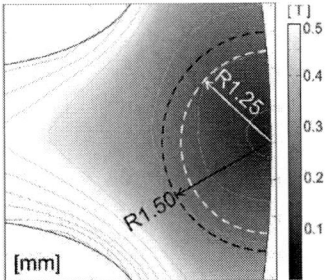

Fig. 15 Relation between the width B and the change ratio of standard deviation of magnetic flux density.

The area of the back yoke cut from the low core loss model is shown in Fig.15. Fig. 15 is an enlarged view of the back yoke portion surrounded by the broken line in Fig. 14 (b). In this paper, the range of 0.2 to 0.3 T is considered to be low magnetic flux density region, and the range to cut out was set to two circles with a radius of 1.25 mm and a circle of 1.50 mm so that it falls within the range.

Fig. 16 shows the core loss distribution of low core loss model, R1.25 cutout model, and R1.50 cutout model. As shown in Fig. 16, core loss distributions in the teeth of R1.25 cutout model and R1.50 cutout model are almost the same as that of a low core loss model. The reason for this is believed that even if the low magnetic flux density region of the back yoke is cut out, there is no influence on the magnetic circuit and there is no increase or decrease in the torque.

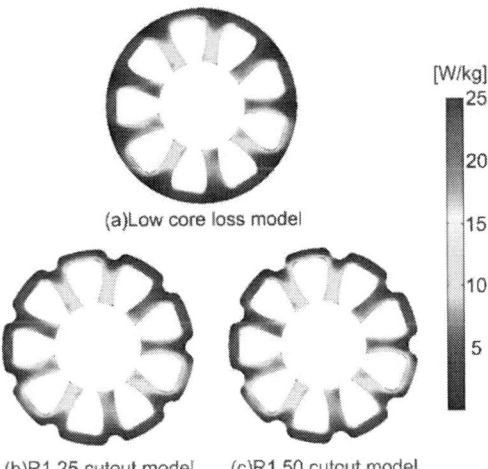

Fig. 16 Core loss distribution of various shape core models.

IV. PERFORMANCE OF HIGH EFFICIENCY HIGH SPEED MOTOR WITH HIGH POWER DENSITY

The stator core of the motor obtained by vector magnetic property analysis was created and assembled to the high speed motor. The motor was 8 poles and 9 slots and the permanent magnet arrangement was affixed to the surface of the rotor core. In accordance with the specifications designed in the previous section, Yoshikawa Industries Co., Ltd. created laminated iron core, manufactured by ShinMaywa Industry Co., Ltd.

Fig. 17 shows the structure of the developed motor and its performance. Since this motor is a specification as the space exploration rover motor, the emphasis is placed on the applied voltage and its magnitude.

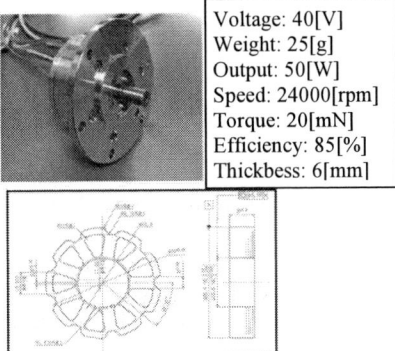

Fig. 17 Structure of the developed motor and its performance.

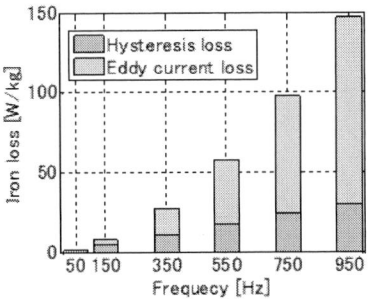

(a) Non-oriented electrical steel sheet.

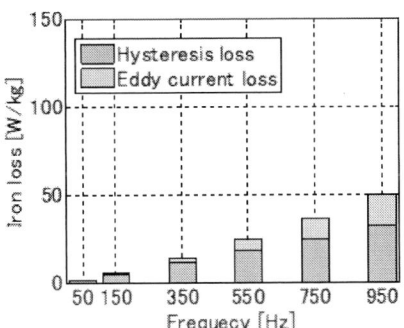

(b) Ultra-thin electrical steel sheet.

Fig.18 Core losses depending on frequency under the rotational flux condition at $\mathbf{B}_{max} = 1.0$, *at* 1.0.

Fig. 18 shows the frequency characteristic of the core loss under the rotational magnetic flux condition obtained from evaluating the magnetic characteristics of the iron core of the motor. The core loss under rotational magnetic flux is called the rotational core loss and is larger than the core loss under alternating magnetic flux. This is due to the increase of the eddy current loss caused by the generation of the domain wall and the annihilation process.

978-1-5386-7688-2/19 $31.00 © 2019 IEEE

In order to investigate the performance of this motor, the load test was conducted by changing the rotation speed of the motor with constant torque. In this load test, compared with the conventional iron core motor of the same size structure. Fig. 19 (a) and (b) shows the core loss map characteristics measured from this load test apparatus, respectively. From this figure, the core loss of this motor can be seen to be reduced by more than 60% compared with the case of using the conventional material.

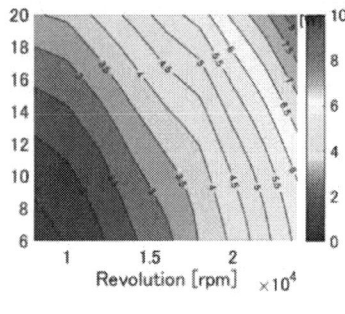

(a) 350 μm core motor.

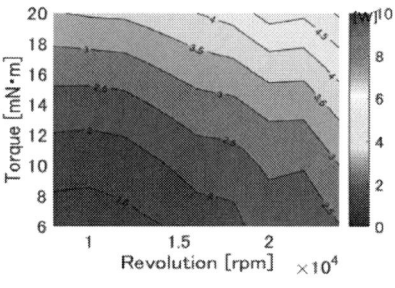

(b) 80 μm core motor.

Fig. 19 Core loss of torque vs. speed characteristics.

Fig. 20 shows the rotational speed characteristics of core loss obtained from no-load test about both motors. As can be seen from the no-load test results, the difference between the both motors increase as the rotation speed increases, and the utilization effect of the ultra-thin electrical steel sheet is exhibited.

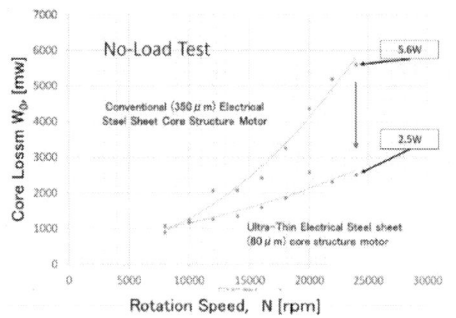

Fig. 20 Core loss vs frequency characteristic by no-load test.

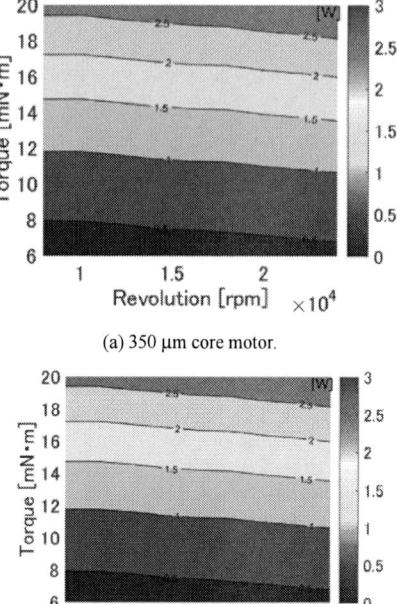

(a) 350 μm core motor.

(b) 80 μm core motor.

Fig. 21 Copper loss of torque vs. speed characteristics.

Fig. 21 (a) and (b) shows the copper loss map characteristics measured from this load test apparatus, respectively. From this figure, the copper loss of this motor could not be reduced. This reason is due to copper machine.

Fig. 22 shows the efficiency map characteristics of the motor. The efficiency has improved drastically from the conventional 60% level to 85%. This developed motor can achieve high efficiency at high speed rotation. The temperature rise of the motor due to the various losses is shown in Fig. 23. The temperature rise of the motor made of the ultra-thin electrical steel sheet is suppressed compared to the conventional product. Fig. 24 shows the frequency characteristics of the iron loss when the magnetic material density and the magnetic flux density level in the iron core are kept constant by this measuring system. From this, it is possible to evaluate the motor performance by reflecting the magnetic characteristics of the core material and to obtain the design clues corresponding to the characteristics. In other words, it is strongly insisted that the design of the motor must be changed if the material is changed.

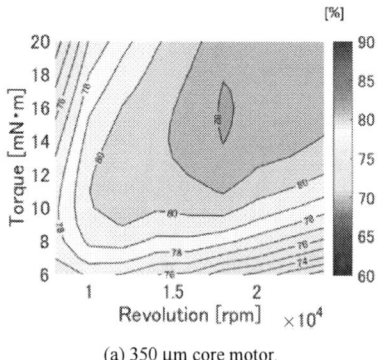

(a) 350 μm core motor.

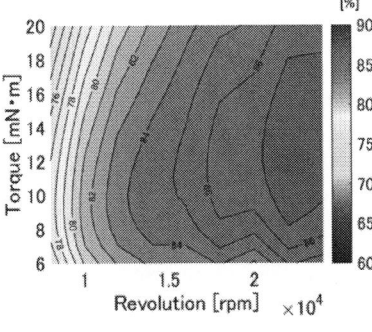

(b) 80 μm core motor

Fig. 22 Efficiency loss of torque vs. speed characteristics

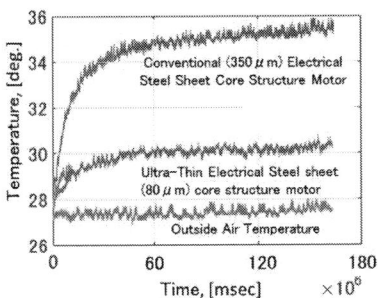

Fig. 23 Temperature rise characteristics.

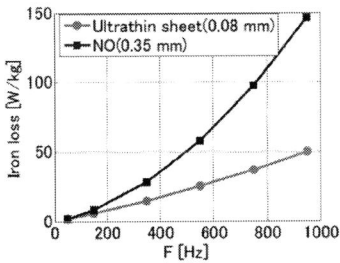

(a) Core loss of electrical steel sheet

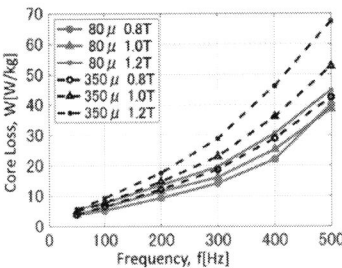

(b) Core loss of motor core

Fig. 24 Evaluation of building factor

Focusing on the frequency 550 Hz shown in the figure, the building factor can be evaluated by comparing the core loss value between the material and the motor. It is thought that the laminated core made of the ultra-thin electrical steel sheet requires about 5 times the number of the conventional material, so that the magnetic characteristics were

deteriorated due to much influence of stress in the assembly process of the motor.

V. CONCLUSION

In this paper, the development process of high-speed motor with high efficiency and high power density aiming at high efficiency and high output of high speed motor was described. The results and knowledge obtained below will be summarized.

It was shown the usefulness of ultra-thin electrical steel sheet of 80 μm thickness as a core material for high speed motor.

(1) An optimum shape of stator core that suppresses core loss by using ultrasound electromagnetic steel sheet is required by the vector magnetic characteristic analysis.

(2) As a compact, high speed motor for space exploration rover, the high power density motor has been developed.

(3) The developed high-speed motor achieved an efficiency of 85% at an applied voltage of 40 V, weight of 25 g, output of 50 W, output torque of 20 mNm.

On the other hand, as the findings obtained, the reduction of core loss also reduces copper loss. However, an increase in the number of laminated iron cores increases the building factor. In the future, it will be desirable to study the optimum number of laminated sheets and the number of windings.

REFERENCES

[1] J. Pyrhonen, J. Nerg, P. Kurronen, U. Lauber, "High Speed High-Output Solid-Rotor Induction-Motor Technology Gas Compression", IEEE Transactions on Industrial Electronics, Jan. 2010, vol. 57, no. 1, pp. 272-280.

[2] C. Bailey, D. Saban and P. Guedes-Pinto, "Design of high-speed, direct-connected, permanent-magnet motors and generators for the Petrochemical Industry", IEEE PCIC Conference Record, Sep 2007.

[3] K. Weeber, C. Stephens, J. Vandam, A. Gravame, J. Yagielski and D. Messervey, "High-speed permanent-magnet Motors for Oil & Gas Industry," Proceedings of GT 2007 ASME Turbo Expo 2007: Power for Land and Air, May 14-17, 2007, Montreal, Canada.

[4] F. Luise, A. Tessarolo, S. Pieri, P. Raffin, M. Di Chiara, F. Agnolet, M. Scalabrin, "Design and technology solutions for high-efficiency high-speed motors", 2012 XX International Conference on Electrical Machines (ICEM), pp.157-163, 2-5 Sept. 2012.

[5] X. Jannot, J.-C. Vannier, C. Marchand, M. Gabsi, J.Saint-Michel, D. Sadarnac, D., "Multiphysic Modeling of a High-Speed Interior Permanent-Magnet Synchronous Machine for a Multiobj active Optimal Design", IEEE Transactions on Energy Conversion, vol. 26, no. 2, pp. 457,467, June 2011.

[6] A.S. Bornschlegell, J. Pelle, S. Harmand, A. Fasque Ile, J.-P. Curio, "Thermal Optimization of a High-Power Salient-Pole Electrical Machine", IEEE Transactions on Industrial Electronics, Vol. 60, no. 5, pp. 1734-1746, May 2013.

[7] E.P. Wiechmann, P. Aqueveque, R. Burgos, J. Rodriguez, "On the Efficiency of Voltage Source and Current Source Inverters for High-Power Drives", IEEE Transactions on Industrial Electronics, Vol. 55, no. 4, pp. 1771-1782, April 2008.

[8] M. Yamagashira, S. Ueno, D. Wakabayashi, and M. Enokizono, "Vector magnetic properties and two-dimensional magnetostriction of various soft magnetic materials," International Journal of Applied Electromagnetics and Mechanics., Vol. 44, no. 3,4, pp. 387–400, 2014

[9] S. Ueno, M. Enokizono, Y. Mori, K. Yamazaki, "Vector Magnetic Characteristics of Ultrathin Electrical Steel Sheet for Development of High Efficiency High Speed Motor" IEEE Trans. On Magnetics, Vol. 53, no. 11, Nov. 2017, Art. no. 6300604.

[10] Information on http://www.vector-magtec.jp

Diagnosis of Different Eccentricity Faults in Induction Motors Based on Electrical and Magnetic Signatures and Unbalanced Magnetic Pull

T. A. SARIKAYA
Department of Electrical Engineering
Istanbul Technical University
34469, Istanbul, Turkey
sarikayatu @itu.edu.tr

A. POLAT
Department of Electrical Engineering
Istanbul Technical University
34469, Istanbul, Turkey
polata@itu.edu.tr

L. T. ERGENE
Department of Electrical Engineering
Istanbul Technical University
34469, Istanbul, Turkey
ergenel@itu.edu.tr

Abstract— In induction motors, eccentricity faults are very common. If it is not detected at early stages, these faults would develop more serious problems. In this paper static, dynamic and mixed eccentricity are modeled for understanding the effects of this type of faults. The effects are examined in terms of alterations in stator current, torque characteristic, radial flux, and magnetic pull. Results are compared with each other and with a healthy machine.

Keywords— *induction motor, fault diagnosis, eccentricity, harmonic components, unbalanced magnetic pull*

I. INTRODUCTION

Three-phase induction motors are being frequently used in the industry because of their easy maintenance, sturdy and simple structure and low cost. Their robust and reliable structure let induction motor to be used for not just for general purposes also for hazardous and harsh environments. Due to their vast area of use, induction machines are exposing to diverse environmental stresses. Environmental effects along with the aging of the induction machines, make the machines susceptible to faults [1]. With the increasing demand for reliability and desire to minimize earning losses in the industry, fault detection is gaining an interest [2].

Faults in an induction motor can occur either internally or externally. When the fault source is taken as a reference, a fault may be caused from electrical or mechanical problem. Depending on fault location, fault can be categorized as a stator fault or rotor fault. When individually examined, induction motor faults can be generally categorized as bearing failures, stator-winding faults, rotor faults, air gap eccentricities, mechanical vibrations, etc. [3]. Occurrence rates of the different faults are shown in Fig.1.

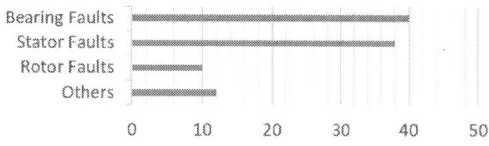

Fig. 1. Fault occurrence distribution [3]

Bearing faults are the most common fault type with 40% rate. Bearings consist of two rings with a set of balls between them. Defects on the balls or waviness on the rings generate mechanical vibrations and alter rotation characteristic of the machine. Stator faults are relatively common. These type of faults are thought initiate as a small turn to turn faults which

develop into major phase to phase or phase to ground faults [4]. Rotor faults are infrequent with 10% occurrence rate. Rotor faults can occur from mechanical or electrical failures. Thermal increase due to electrical failure leads to rotor bar defect or breakage. Mechanical problems such as load unbalance or shaft misalignment can damage bearings and bearing housing [3]. These type of problems along with bearing faults are also related to air gap eccentricity [5].

Eccentricity is the case when the air gap between stator and rotor show non-uniform characteristic. Thus rotor is affected by an unbalanced electromagnetic pull. There are three types of eccentricity: static, dynamic, and mixed eccentricity which is a combination of static and dynamic eccentricity.

In static eccentricity, center and rotation axes of the rotor does not intersect with the center of the stator. Hence air gap width is narrower at one side of the machine and location of that independent from rotor motion. Static eccentricity strongly connected to the manufacturing process and may be caused by oval stator or wrong placement of the rotor [6-7].

In dynamic eccentricity, the rotation axis of the rotor is same with stator axis and rotor center axis does not intersect with them. Therefore, non-uniform and time-dependent air gap occurs. This fault can be caused by a bent rotor shaft, bearing faults, mechanical imbalances. Undetected dynamic eccentricity on a machine can lead to a total breakdown [8]. Cross-section view of static and dynamic eccentricity can be seen in Fig. 2.

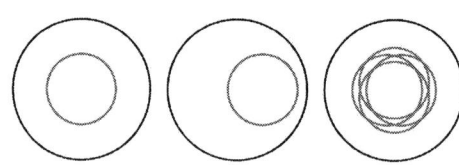

Fig. 2: Eccentricity types (left) healthy machine, (middle) static eccentric machine, (right) dynamic eccentric machine.

Machine with eccentricity shows that changes in parameters: increase in torque ripple, decrease in average torque, reduction in efficiency, increase in magnitudes of air gap flux harmonics, etc.

For detection of the eccentricity faults, many methods have been developed and proposed by researchers. Most common one of those methods is motor current signature analysis (MCSA). Cause of eccentricity, new harmonic components appear on harmonic spectrum of air gap flux.

978-1-5386-7688-2/19 $31.00 © 2019 IEEE

These components also occur in stator current as harmonics. When the current signal is processed, increase in the magnitude of high order spectral components of the current can be seen [9]. On the other hand, the harmonic spectrum of the air gap flux density can be directly examined for detection of the faults. Each eccentricity fault alters the harmonic spectrum differently. Fault detection can be done by comparing the healthy and faulty machine harmonic spectrums [10-11]. Further, the vibration of the motor can be analysed for fault detection but in practice vibration is generally used for specifying the fault magnitude. That is performed in this way, because of parasitic environmental vibrations affects vibration measurement of the motor [12]. In healthy machines magnetic pull that affecting the rotor is uniform. In case of non-uniform air gap width, magnetic pull increases in narrow widths and decreases in wide widths. Non-symmetrical stator currents can also be a reason for unbalanced magnetic pull. If unbalanced magnetic pull is analysed, faults especially eccentricity ones could be detected. However, measuring the magnetic pull is more complicated than other induction motor signals [13]. Recently, artificial intelligence based methods have been used in induction machine fault diagnosis. These methods can analyse any signal of the induction machine and almost obtain the machine status instantaneously. Their classification ability let them find and decide the fault that occurs in the machine [2, 4, 14-15].

This paper concerns the dynamic model of the squirrel cage induction motor and presents the diversion in the magnetic flux, current, torque values and unbalanced magnetic pull. Faulty conditions are formed for static, dynamic and mixed eccentricity types. The air gap flux density, current, and torque characteristics are examined for healthy and asymmetric operational conditions. Changes in magnetic pull are derived and compared finally.

II. ANALYSIS TECHNIQUE

A 2.2 kW, 4 poles, 50 Hz, delta connected squirrel cage induction motor is forms a basis for simulations in this paper. Nominal voltage and current of the motor are 400V, 5A and nominal speed of the rotor is 1440 rpm.

First of all, induction motor parameters are obtained and the induction motor model is created for the finite element analysis. Consistencies of simulation results are confirmed by comparing them to the test results of the motor. Motor geometry is given in Fig. 3.

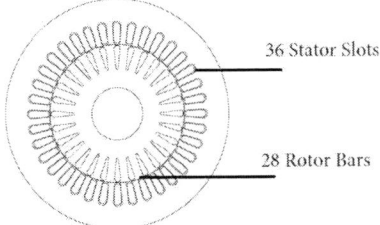

Fig. 3: Cross-section of the induction motor

Finite element method (FEM) is a numerical technique which is being used to solving and optimizing engineering problems in different diciplines. This method gives a unique solution for each partial differential equation that bounded with boundary values. FEM divides the area that inside of the model to group of smaller parts that is called as mesh, and solves the equations for each of them. Final result obtained via summing all of these results. Increasing the mesh number enhances solutions accuracy with the drawback of increasing the computing time.

For obtaining accurate solutions mesh density must be optimized and mesh formation in the air gap region should be chosen denser. In Fig. 4, higher mesh density over air gap region can be seen.

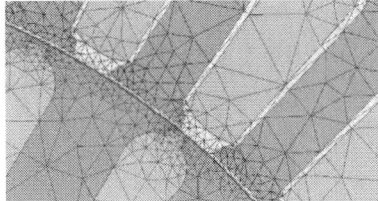

Fig. 4: Mesh topology

III. DYNAMIC MODEL OF INDUCTION MOTOR

Dynamic models for the induction motor were created for healthy motor and static, dynamic and mixed eccentricity conditions. Radial air gap flux densities, torque ripples, and current characteristics were examined and magnetic pulls affecting the rotor were compared.

A. Healthy Condition

Simulation for healthy motor created for rated values of the chosen induction motor. Air gap region is symmetrically formed. Torque and current results for healthy machine obtained from simulation and given in Fig. 5 and 6, respectively.

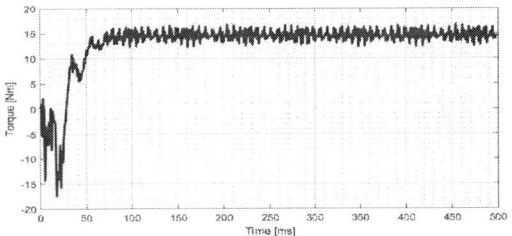

Fig. 5: Torque profile of the healthy induction motor

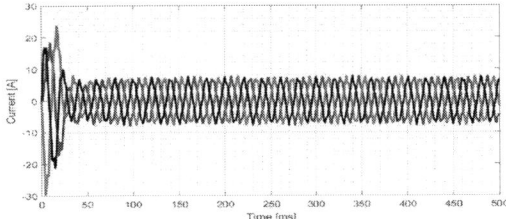

Fig. 6: Phase currents of the healthy induction motor

The radial flux density of the motor obtained from the center of the air gap as shown in Fig.7. Harmonics of the air gap flux obtained via using FFT. In healthy induction machine when stator slots are considered, air gap widths are narrower on stator teeth. That is the other way round at slot openings. Due to the magnetic reluctance, flux densities are stronger on stator teeth.

978-1-5386-7688-2/19 $31.00 © 2019 IEEE

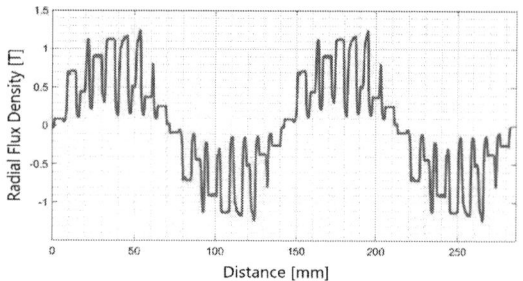

Fig. 7: Radial air gap flux

In Fig. 7, impulse locations on the graph, are corresponding to stator teeth. So the total impulse number is equal to the stator slot number.

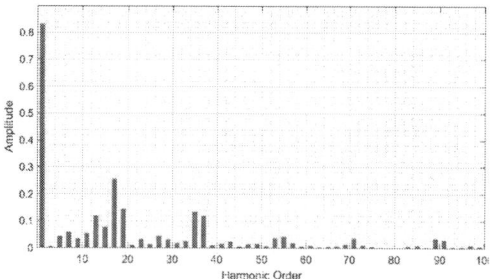

Fig. 8: Harmonic order of air gap flux

Harmonic spectrum of heathy machine radial flux density is given in Fig. 8. In harmonic distribution of the air gap flux density, due to symmetry of the system even harmonic orders not exist. Distinct magnitudes of higher order harmonics are caused by stator slots and rotor bars. Effect of slotting, special harmonics can be observed at 35th and 37th harmonics and this repeat itself on multiples of 36 which is number of stator slots. These harmonics called as slot harmonics and rotor bars also cause similar harmonic formation.

B. Faulty Conditions

In this part implementation of the eccentricity failures will be expressed. Faulty motors were created for static dynamic and mixed eccentricity faults. Fault degree of simulations were adjusted by moving rotor, stator and rotation center over x-axis for various configurations.

1) Static Eccentricity

Static eccentricity (SE) formed by moving rotor center and rotor rotation center together to 30%, 50%, 70% of the air gap width over the x-axis. That way, non-uniform air gap that independent from time is obtained. Static eccentricity change is shown in Fig. 9.

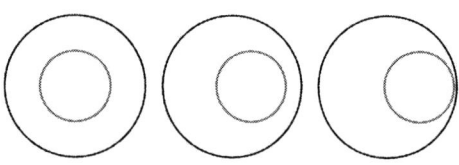

Fig. 9: Healthy motor and static eccentricity

2) Dynamic Eccentricity

Dynamic eccentricity (DE) formed by moving the rotor center to 30%, 50%, 70% of the air gap width over the x-axis. In this case, rotor rotation center aligns with the stator center. That way, air gap width that altering with the rotation of rotor is obtained. Dynamic eccentricity change is shown in Fig. 10.

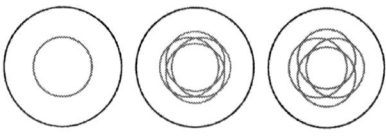

Fig. 10: Healthy motor and dynamic eccentricity

3) Mixed Eccentricity

Mixed eccentricity (DE+SE) is formed by combining dynamic and static eccentricity. Constant static eccentricity added over to dynamic eccentricity. With increasing the distance between rotor rotation center and rotor center dynamic eccentricity applied at different levels. Air gap width changes along with rotor motion but due to static eccentricity difference between minimum and maximum air gap widths is higher than dynamic eccentricity. Mixed eccentricity change is shown in Fig. 11.

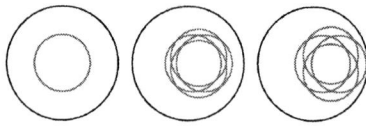

Fig. 11: Healthy motor and mixed eccentricity

IV. SIMULATION RESULTS AND COMPARISON

Obtained faulty and healthy induction machine simulation results will be compared in this section. Comparison and evaluation of the results will be done in terms of torque ripple, average torque, stator current, air gap flux density and harmonics of air gap flux.

In induction motors when eccentricity fault is present, torque ripple and peak torque levels increases significantly. A decrease in the average torque can also be observed. Comparison of the torque characteristic of the two machines are given in Fig. 12.

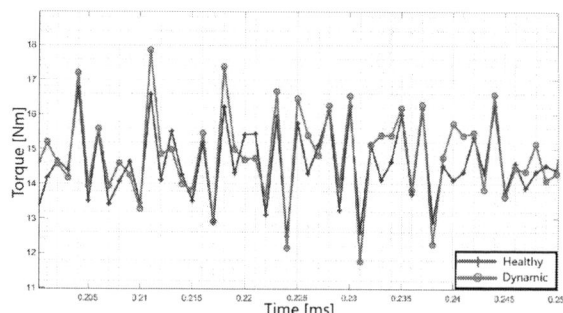

Fig. 12: Torque comparison of healthy and %70 dynamic eccentric motors

Torque results for healthy and eccentric machines are given in Table 1.

978-1-5386-7688-2/19 $31.00 © 2019 IEEE

TABLE I. TORQUE RESULTS

	Peak to Peak Torque [Nm]	Peak to Peak Error (%)	Average Torque [Nm]
Healthy	4.4272	-	14.5912
70% SE	5.1852	17.12	14.5657
70% DE	6.1428	38.75	14.7904
70% DE + 10% SE	6.1996	40.03	14.7881

Although average torque changes are not noticeable, increases in the peak to peak torque is significant.

Air gap flux density at any point takes values that based on magnetic reluctance at that point. Air gap flux density values takes lower values at wide air gap widths and higher values at narrow air gap widths. Magnetic flux density color spectrums of healthy and static eccentric motors are given in Fig. 13 for the comparison.

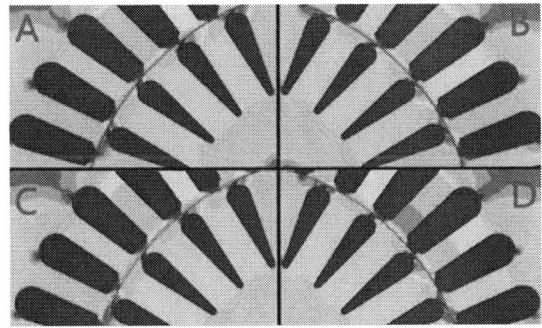

Fig. 13: Colors spectrum of flux density:
(a, b) left-right upper quadrant of healthy motor respectively,
(c, d) left-right quadrant of static eccentric motor

When harmonics of the air gap flux densities are examined and the faulty and healthy machine results are compared, a decrease in the magnitude of the fundamental frequency is obvious. There are rises in the magnitude of the higher order harmonics. Due the simulation results, increase in the stator slot harmonics is approximately 20% in the static eccentricity. Dynamic eccentricity does not show any significant increase in the slot harmonics. Increase in 3rd, 5th harmonics (negative sequence) are 21% in the dynamic eccentricity case while they reach maximum of 8% in the static eccentricity. The harmonic spectrum of air gap fluxes of the healthy machine and mixed eccentricity that contains 70% dynamic eccentricity and 10% static eccentricity are compared as given in Fig. 14.

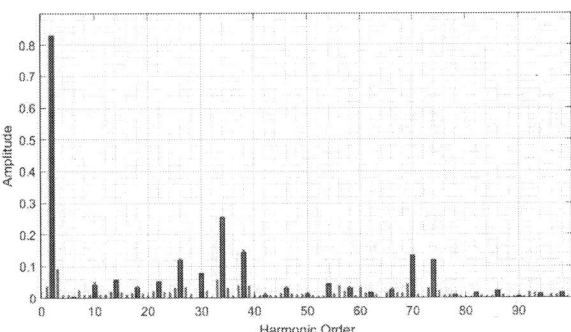

Fig 14: Comparison of the harmonic spectrums of the healthy and mixed eccentric machine

Even harmonic components exist as shown in Fig.14. This is caused by nonsymmetrical motor structure due to eccentricity fault.

Effective and peak values of the stator phase currents increase when eccentricity failure exist. In simulation results average change in peak stator currents are almost 1%. Effective values of phase currents according to eccentricity types are given in Table 2.

TABLE II. CURRENT RESULTS

	Phase A RMS [A]	Phase B RMS [A]	Phase C RMS [A]
Healthy	4.6205	4.6263	4.6131
70% SE	4.6768	4.6754	4.6700
70% DE	4.7789	4.7737	4.7767
70% DE + 10% SE	4.7660	4.7646	4.7647

Induction motor rotor affected by a balanced magnetic pull in normal conditions. However, due to non-uniform air gap in eccentricity failures, unbalanced magnetic pull (UMP) occurs. Related to eccentricity type, UMP shows different characteristics over x and y axes. In dynamic eccentricity x and y components oscillates over zero axis with phase difference. In static eccentricity x and y components oscillates over constant values that related fixed position of the rotor in air gap. UMP characteristics of static, dynamic and mixed eccentricities can be observed in Fig. 15.

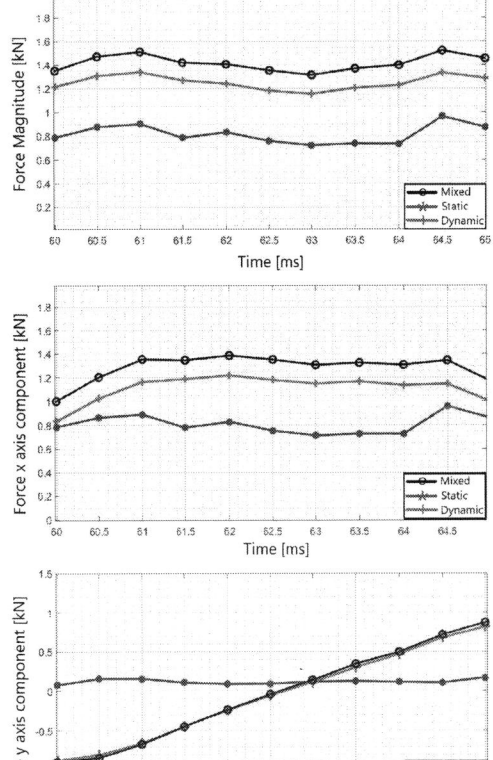

Fig. 15: Ump comparison:
(top) magnitude, (middle) x component, (bottom) y component

978-1-5386-7688-2/19 $31.00 © 2019 IEEE

Value of UMP is calculated from vectorial summation of x and y axes components. Therefore, maximum value of UMP takes higher values in dynamic eccentricity than static eccentricity because of both components can reach high values at close times. Maximum values of UMP that obtained from simulations for increasing levels of static, dynamic, and mixed eccentricity are given in Fig. 16.

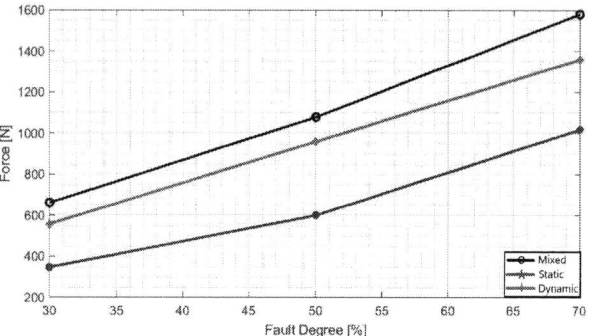

Fig 16: Maximum obtained UMP values for static, dynamic and mixed eccentricity.

These results obtained from 30%, 50%, 70% static and dynamic eccentricity and mixed eccentricity that contains 30%, 50%, 70% dynamic eccentricity with 10% static eccentricity. As expected UMP reaches higher values in dynamic eccentricity than static eccentricity. In mixed eccentricity static eccentricity act like an added constant over dynamic eccentricity that is explains why highest reached value in mixed eccentricity.

V. Conclusion

In conclusion, in this paper induction motor is modelled and simulated for healthy and static, dynamic and mixed eccentricity conditions. Healthy motor results are verified for nameplate values. After the verification, static dynamic and mixed eccentric machines derivated from the healthy machine. Alteration in the motor currents, torque characteristics, radial air gap flux densities and magnetic pulls, showed for eccentric conditions and results are investigated. For all faulty conditions following statements are obtained. Torque ripples - especially in dynamic eccentric conditions - are increased. Phase current rms values slightly increased. In harmonic spectrum of radial air gap flux density, magnitudes of higher order harmonics are increased. Increase in slot harmonics for static eccentricity and 3rd, 5th harmonics for dynamic eccentricity are more distinctively observed. Inequal magnetic pulls are occurred over faulty machine rotors. Fault detection methods of eccentricity are studied in terms of electrical, mechanical and magnetic parameters.

References

[1] Chow, M. (1990). Artificial neural network methodology in real-time incipient fault detection of rotating machines. In Proceedings of National Science Foundation workshop on artificial neural network methodology in power systems engineering

[2] Kanika Gupta1, Arunpreet Kaur2 (2012) A Review on Fault Diagnosis of Induction Motor using Artificial Neural Networks

[3] Bhowmik1 P. S., Pradhan S. and Prakash M. (2013) Fault Diagnısis and Monitoring Methods: a Review

[4] Khireddine, M.S. & Slimane, N & Abdessemed, Yassine & Makhloufi, M.T. (2014). Fault detection and diagnosis in induction motor using artificial intelligence technique. MATEC Web of Conferences. 16. 10.1051/matecconf/20141610004

[5] M. Ojaghi and M. Mohammadi, "Unified Modeling Technique for Axially Uniform and Nonuniform Eccentricity Faults in Three-Phase Squirrel Cage Induction Motors," in IEEE Transactions on Industrial Electronics, vol. 65, no. 7, pp. 5292-5301, July 2018. doi: 10.1109/TIE.2017.2760280

[6] K. N. Gyftakis and J. C. Kappatou, "A Novel and Effective Method of Static Eccentricity Diagnosis in Three-Phase PSH Induction Motors," in IEEE Transactions on Energy Conversion, vol. 28, no. 2, pp. 405-412, June 2013. doi: 10.1109/TEC.2013.2246867

[7] G. Mirzaeva and K. I. Saad, "Advanced Diagnosis of Stator Turn-to-Turn Faults and Static Eccentricity in Induction Motors Based on Internal Flux Measurement," in IEEE Transactions on Industry Applications, vol. 54, no. 4, pp. 3961-3970, July-Aug. 2018. doi: 10.1109/TIA.2018.2821098

[8] A. Barbour andW. T. Thomson, "Finite element study of rotor slot designs with respect to current monitoring for detecting static airgap eccentricity in squirrel-cage induction motor," in Proc. IEEE Ind. Appl. Soc. Annu. Meeting Conf., New Orleans, LA, USA, Oct. 5–8, 1997, pp. 112–119.

[9] M. E. H. Benbouzid, M. Vieira and C. Theys, "Induction motors' faults detection and localization using stator current advanced signal processing techniques," in IEEE Transactions on Power Electronics, vol. 14, no. 1, pp. 14-22, Jan. 1999. doi: 10.1109/63.737588

[10] Siddique, Arfat & Yadava, G.s & Singh, Bhim. (2005). A Review of Stator Fault Monitoring Techniques of Induction Motors. Energy Conversion, IEEE Transactions on. 20. 106 - 114. 10.1109/TEC.2004.837304.

[11] A. A. Salah, D. G. Dorrell and Y. Guo, "A Review of the Monitoring and Damping Unbalanced Magnetic Pull in Induction Machines Due to Rotor Eccentricity," in IEEE Transactions on Industry Applications, vol. 55, no. 3, pp. 2569-2580, May-June 2019.doi: 10.1109/TIA.2019.2892259

[12] Y. Gritli, A. Bellini, C. Rossi, D. Casadei, F. Filippetti and G. Capolino, "Condition monitoring of mechanical faults in induction machines from electrical signatures: Review of different techniques," 2017 IEEE 11th International Symposium on Diagnostics for Electrical Machines, Power Electronics and Drives (SDEMPED), Tinos, 2017, pp. 77-84. doi: 10.1109/DEMPED.2017.8062337

[13] C. Di, X. Bao, H. Wang, Q. Lv and Y. He, "Modeling and Analysis of Unbalanced Magnetic Pull in Cage Induction Motors With Curved Dynamic Eccentricity," in IEEE Transactions on Magnetics, vol. 51, no. 8, pp. 1-7, Aug. 2015, Art no. 8106507. doi: 10.1109/TMAG.2015.2412911

[14] Jose, Greety & Jose, Victor. (2013). Induction Motor Fault Diagnosis Methods: A Comparative Study.

[15] Vilas N. Ghate, Sanjay V. Dudul, Optimal MLP neural network classifier for fault detection of three phase induction motor, Expert Systems with Applications, Volume 37, Issue 4, 2010, Pages 3468-3481

Direct Radial and Circumferential Analytical Air-Gap Field Calculation for Electrical Machines

Jan Andresen
*Institute for Drive Systems
and Power Electronics
Leibniz University Hannover*
Hannover, Germany
jan.andresen@ial.uni-hannover.de

Bernd Ponick
*Institute for Drive Systems
and Power Electronics
Leibniz University Hannover*
Hannover, Germany
ponick@ial.uni-hannover.de

Axel Mertens
*Institute for Drive Systems
and Power Electronics
Leibniz University Hannover*
Hannover, Germany
mertens@ial.uni-hannover.de

Abstract—This paper introduces a new approach for directly calculating the radial flux density in a slotted electrical machine. Through being based on the subdomain approach, it is capable of calculating the air-gap field more accurately than traditional one-dimensional approaches. Contrary to existing subdomain models and finite element analysis, the new approach has the advantage that the cause of certain spatial harmonics of the radial flux density is directly visible. The approach is based on the separation of variables of the vector potential inside the air gap. The radial flux density is calculated at both, the inner and the outer side of the air gap. Using the radial flux density, the circumferential flux density can be calculated. This enables the calculation of the electromagnetic forces including the torque. The approach also encompasses effects due to rotation by assigning each spatial harmonic to a rotational order. This eliminates the need to calculate multiple time steps. Furthermore, the approach provides an extensibility similar to traditional one-dimensional air-gap conductivity based methods.

Index Terms—magnetic fields, magnetic flux density, magnetic forces, torque

I. INTRODUCTION

The purpose of this paper is to introduce a new approach of analytical field calculation for electrical machines that is a valuable tool for analysing the behaviour of machines as well as important parasitic effects such as torque pulsation or acoustic noise emissions. In order to be valuable, the approach has to fulfil two purposes. On the one hand, it is used to quantify the behaviour of a machine by determining flux density, torque and radial forces. On the other hand, it is supposed to explain the machine's behaviour by linking an observed effect to its cause.

Traditional one-dimensional analytical field analysis modelling, as summarized in [1]–[3], involves two steps: First, the MMF is calculated. Then, it is multiplied by a permeability function. The permeability is mostly determined by the low permability of the air gap which is additionally modulated by the influence of slots and saliency. While being intuitive, this approach is a help to understand the behaviour of numerous electrical machines, especially induction motors. However, through being one-dimensional, only the radial flux density is directly calculated in this approach. The circumferential flux density is not calculated. This makes the

calculation of torque and forces more challenging.

Besides conformal mapping e.g. [4] and magnetic equivalent circuits e.g. [5], subdomain models represent a more advanced approach of electrical machine modelling which has attracted considerable attention over the past decades. An overview of the broad spectrum of publications can be found in [6] and [7]. Most publications focus on permanent magnet machines, which are modelled in great detail in [8] and [9]. Some recent advances have been made for induction machines [10] and [11]. The underlying principle of the subdomain model is that the magnetic potential, magnetic vector potential, or the flux density [12] is described in multiple subdomains as a function of a set of coefficients. Through border conditions between these subdomains, a linear equation system can be derived, which can be numerically solved. Using the determined potential, the flux density can be derived. Since a system of numerical equations has to be solved, the impact of the individual parts on the air-gap field is not directly visible. Most papers only consider the field solution at a particular instant of time. However, it has been shown [13] how introducing two coefficients for both rotational orders can be used to directly calculate the time dependency of the field. By just considering particular orders [14], the computation effort can be significantly reduced [15]. Combinations of a subdomain model with other modelling techniques, e.g. with magnetic equivalent circuits [16] or homogenization techniques [17] are studied in literature as well, enabling the coupling of a smooth air gap with other types of field representations. The comprehensive research in this has enabled to precisely calculate the machine behaviour in many applications. In existing subdomain models, the equations are, however, not intended to be interpreted but are solely derived to be solved numerically. This makes it difficult to explain the machine behaviour based on the derived equations.

In contrast to existing subdomain models, the proposed approach makes the machine behaviour interpretable since it highlights the cause of each spatial harmonic of the radial flux density. This is achieved by separately deriving the influence of the electric loading and the slotting. The approach is

978-1-5386-7688-2/19 $31.00 © 2019 IEEE

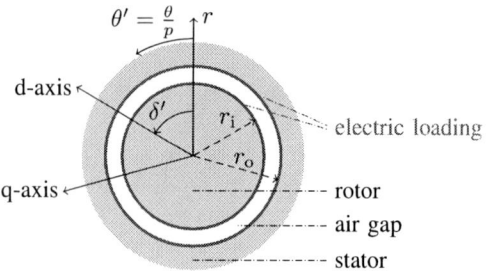

Fig. 1. Configuration with a smooth air gap showing both, the electric loading and the coordinate systems used for a four-pole electrical machine.

based on the subdomain model approach and shows how the flux density can be directly calculated, if only the air-gap subdomain excited by an electric loading is considered. Both the radial and circumferential flux density are calculated. Furthermore, it is shown how an additional subdomain, in this case the stator slots, can be incorporated. The approach of only considering dedicated circumferential and rotational orders is used throughout the paper.

To derive the new approach, the field in the case of a smooth air gap is calculated and discussed in Section II. This section also includes the derivation of the radial and circumferential spatial harmonics and the calculation of the resulting torque. Given the results for a smooth air gap, a field calculation approach that includes the effect of stator slot openings is derived in Section III. The results are validated through a comparison to a numerical model solved using the Finite Element method (FEM) in Section IV. Using an example, the interpretability of the new approach is shown in Section V. Conclusions are drawn in Section VI.

II. CALCULATION OF THE AIR-GAP FLUX DENSITY FOR SMOOTH STATOR AND ROTOR

In this section, the flux density for a smooth air gap is determined. The field is assumed to be excited by an electric loading. Fig. 1 shows the configuration considered in this section.

A. Representation of Field Quantities

In this paper, field quantities are represented as spatial harmonics. An example of a spatial harmonic of the flux density (B) is

$$
\begin{aligned}
B =& \hat{B}_{\cos,n'_1,k'_1} \cdot \cos\left(n'_1\theta' - k'_1\delta'\right) + \\
& \hat{B}_{\sin,n'_1,k'_1} \cdot \sin\left(n'_1\theta' - k'_1\delta'\right),
\end{aligned}
\tag{1}
$$

where θ' is the mechanical circumferential angle, n' is the circumferential order, δ' is the mechanical angle between d-axis and the fixed coordinate system and k' is the rotational order. The d-axis is assumed to be fixed to the rotating part of the machine. Since the rotor position is accounted for by $k'\delta'$,

an explicit time dependency is not necessary. The amplitude of a spatial harmonic is represented as a vector

$$
\hat{\underline{B}}_{n'_1,k'_1} = \begin{bmatrix} \hat{B}_{\cos,n'_1,k'_1} \\ \hat{B}_{\sin,n'_1,k'_1} \end{bmatrix}.
\tag{2}
$$

A vector of spatial harmonics is written as

$$
\vec{\hat{B}} = \begin{bmatrix} \hat{\underline{B}}_{n'_1,k'_1} \\ \hat{\underline{B}}_{n'_2,k'_2} \end{bmatrix}.
\tag{3}
$$

A matrix which describes the influence of one spatial harmonic to another is written as $\underline{M}$, whereas $\vec{\underline{M}}$ denotes a matrix which couples vectors of spatial harmonics.

B. Representation of the Current

As introduced in [3], [10], [18]–[20], the current is represented as the electric loading at the surface towards the surface of the respective part, where the winding is located. A speciality is that the electric loading is directly represented in the spatial harmonic representation used throughout this paper. Additionally, the currents are represented in a dq-reference frame, which is fixed to the rotor using the transformation matrices ($\mathbf{T}$)

$$
\mathbf{T}_{abc,\alpha\beta} = \begin{bmatrix} \frac{2}{3} & -\frac{1}{3} & -\frac{1}{3} \\ 0 & \frac{1}{\sqrt{3}} & -\frac{1}{\sqrt{3}} \end{bmatrix},
\tag{4}
$$

mapping the phase currents to the $\alpha\beta$-coordinate system and

$$
\mathbf{T}_{\alpha\beta,dq} = \begin{bmatrix} \cos\left(p\delta'\right) & \sin\left(p\delta'\right) \\ -\sin\left(p\delta'\right) & \cos\left(p\delta'\right) \end{bmatrix}
\tag{5}
$$

further mapping them to the dq-coordinate system, where p is the number of pole pairs. Using this approach, the electric loading J of an m-phase winding can be described as

$$
\begin{aligned}
J =& -\sum_{n'} \hat{J}_{d,n'} \cdot \sin\left(n'\theta' - p\delta'\right) \\
& -\sum_{n'} \hat{J}_{q,n'} \cdot \sin\left(n'\theta' - p\delta' - \frac{\pi}{2}\right)
\end{aligned}
\tag{6}
$$

with

$$
\begin{aligned}
\hat{J}_{d,n'} &= i_d \xi_{n'} \frac{m \cdot w}{\pi \cdot r_o} \\
\hat{J}_{q,n'} &= i_q \xi_{n'} \frac{m \cdot w}{\pi \cdot r_o},
\end{aligned}
\tag{7}
$$

$$
n' = p\left(1 + 2mg\right) \quad g \in \mathbb{Z},
\tag{8}
$$

and

$$
\xi_{n'} = \frac{\sin\left(n'q\frac{\pi}{N}\right)}{q\sin\left(n'\frac{\pi}{N}\right)} \sin\left(\frac{n'}{p}\frac{y}{y_\varnothing}\frac{\pi}{2}\right) \frac{\sin\left(n'\frac{b_S}{2R}\right)}{n'\frac{b_S}{2R}},
\tag{9}
$$

where w is the number of turns per phase, ξ is the winding factor, r_o is the outer air-gap radius (r), b_S is the width of a slot, $y_\varnothing$ is the pole pitch, y is the winding pitch, N is the number of slots and q the number of slots per pole and per phase. The zero component of the currents is assumed to be zero.

C. Representation of the Air-Gap Flux Density

Throughout this paper, the air-gap flux density is assumed to be two-dimensional: Radial and circumferential components are considered, whereas the axial component of the flux density is neglected. As a result, only the axial component of the magnetic vector potential is unequal to zero. The vector potential can be derived based on the principle of separable partial differential equations to be

$$A_{n',z} = \sum_{n'} \left(r^{n'} + K_{a,n'} \cdot r^{-n'} \right) \cdot \tag{10}$$
$$\left(K_{b,n'} \cdot \cos\left(n'\theta'\right) + K_{c,n'} \cdot \sin\left(n'\theta'\right) \right),$$

where $K_{a,n'}, K_{b,n'}$ and $K_{c,n'}$ are constants determined by the magnetic field strength. Using the curl of the vector potential, the circumferential flux density can be calculated as

$$B_{\theta'} = -\frac{d}{dr}A \tag{11}$$

and the radial flux density can be calculated as

$$B_r = \frac{1}{r}\frac{d}{d\theta'}A. \tag{12}$$

D. Field due to a Known Circumferential Field

The next step is to calculate the radial field strength at a radius r_2, assuming that the magnetic field strength at the radius r_1 ($\underline{\hat{B}}_{\theta',n'}(r_1)$) is known and the circumferential field at the radius r_0 is zero. r_0, r_1 and r_2 are arbitrary radii. Specific radii will be used later on. Using the property

$$B_{\theta'}(r_0) = 0, \tag{13}$$

the vector potential can be simplified to

$$A_{n',z} = \left(r^{n'} + r_0^{2n'} r^{-n'} \right) \cdot \tag{14}$$
$$\left(K_{1,n'} \cdot \cos\left(n'\theta'\right) + K_{2,n'} \cdot \sin\left(n'\theta'\right) \right).$$

The constants $K_{1,n'}$ and $K_{2,n'}$ can be determined using $\underline{\hat{B}}_{\theta',n'}(r_1)$. Using the vector potential, any circumferential or radial field within the boundaries of the air gap can be calculated. For future reference, two symbols for frequently used ratios are introduced. We introduce a symbol (χ) for the ratio between a circumferential spatial harmonic and a radial spatial harmonic of the same orders n' and k'

$$\underline{\hat{B}}_{n',r}(r_2) = \begin{bmatrix} 0 & -1 \\ 1 & 0 \end{bmatrix} \frac{\left(r_2^{n'-1} + r_0^{2n'} r_2^{-n'-1} \right)}{\left(r_1^{n'-1} - r_0^{2n'} r_1^{-n'-1} \right)} \underline{\hat{B}}_{n',\theta'}(r_1)$$
$$= \underline{\underline{\chi}}_{n'}(r_0, r_1, r_2)\underline{\hat{B}}_{n',\theta'}(r_1). \tag{15}$$

with

$$\chi_{n'}(r_0, r_1, r_2) = \frac{\left(r_2^{n'-1} + r_0^{2n'} r_2^{-n'-1} \right)}{\left(r_1^{n'-1} - r_0^{2n'} r_1^{-n'-1} \right)} \tag{16}$$

and a symbol (ζ) for the ratio between the amplitudes of the same radial spatial harmonic at two different radial positions

$$\underline{\hat{B}}_{n',r}(r_2) = \frac{r_2^{n'-1} + r_0^{2n'} r_2^{-n'-1}}{r_1^{n'-1} + r_0^{2n'} r_1^{-n'-1}} \begin{bmatrix} 1 & 0 \\ 0 & 1 \end{bmatrix} \underline{\hat{B}}_{n',r}(r_1)$$
$$= \underline{\underline{\zeta}}_{n'}(r_0, r_1, r_2)\underline{\hat{B}}_{n',r}(r_1). \tag{17}$$

E. Relationship between Electric Loading and Circumferential Field

It is assumed that the electric loading is an infinitesimally thin film either at the outer air-gap radius (r_o) or at the inner air-gap radius (r_i). These assumptions are also explained in [21] and [22]. In case the electric loading is at the outer radius, r_o^- and r_o^+ are the radii right below and above this radius. Since an ideal magnetic conductor is assumed to start at the radius r_o^+, the circumferential magnetic field strength (H) at this radius is zero

$$H_{\theta'}(r_o^+) = 0. \tag{18}$$

$\oint H ds = i$ with i as the current yields

$$H_{\theta'}(r_o^-) = -J(r_o). \tag{19}$$

If the electric loading is located at r_i the result is

$$H_{\theta'}(r_i^+) = J(r_i). \tag{20}$$

Since the radii r_i^+ and r_o^- are within the air gap, the flux density can be calculated as

$$B_{\theta'}(r_o^-) = \mu_0 H_{\theta'}(r_o^-), \tag{21}$$

where μ_0 is the permeability of vacuum.

F. Representation of the Final Result

As an example, the radial flux density of the order n' at the radius r_o is calculated. No electric loading is present at r_i ($B_{\theta'}(r_i) = 0$), while an electric loading located at r_o is exciting the air-gap field,

$$\underline{\hat{B}}_{n',r}(r_o) = \underline{\underline{\chi}}_{n'}(r_i, r_o, r_o)\underline{\hat{B}}_{n',\theta'}(r_o^-)$$
$$= -\mu_0 \underline{\underline{\chi}}_{n'}(r_i, r_o, r_o)\underline{\hat{J}}_{n'}(r_o). \tag{22}$$

The flux density at both surfaces is particularly important in the next steps. Both flux densities can be calculated based on the loading at both surfaces through superposition as

$$\vec{\underline{B}}_r = \begin{bmatrix} \underline{\hat{B}}_{n',r}(r_i) \\ \underline{\hat{B}}_{n',r}(r_o) \end{bmatrix}$$
$$= \mu_0 \begin{bmatrix} \underline{\underline{\chi}}_{n'}(r_o, r_i, r_i) & \underline{\underline{\chi}}_{n'}(r_i, r_o, r_i) \\ \underline{\underline{\chi}}_{n'}(r_o, r_i, r_o) & \underline{\underline{\chi}}_{n'}(r_i, r_o, r_o) \end{bmatrix} \begin{bmatrix} \underline{\hat{J}}_{n'}(r_i) \\ -\underline{\hat{J}}_{n'}(r_o) \end{bmatrix} \tag{23}$$
$$= \mu_0 \underline{\underline{\vec{\chi}}} \vec{\underline{J}}.$$

The circumferential spatial harmonic is not directly visible. However, it can be calculated at any radius inside the air gap (r_3) based on the vector potential as

$$\underline{\hat{B}}_{n',\theta'}(r_3) = \frac{\frac{r_3^{n'-1}}{r_2^{n'+1}} + \frac{r_2^{n'-1}}{r_3^{n'+1}}}{\frac{r_1^{n'-1}}{r_2^{n'+1}} - \frac{r_2^{n'-1}}{r_1^{n'+1}}} \begin{bmatrix} 0 & 1 \\ -1 & 0 \end{bmatrix} \underline{\hat{B}}_{n',r}(r_1)$$
$$- \frac{\frac{r_3^{n'-1}}{r_1^{n'+1}} + \frac{r_1^{n'-1}}{r_3^{n'+1}}}{\frac{r_1^{n'-1}}{r_2^{n'+1}} - \frac{r_2^{n'-1}}{r_1^{n'+1}}} \begin{bmatrix} 0 & 1 \\ -1 & 0 \end{bmatrix} \underline{\hat{B}}_{n',r}(r_2). \tag{24}$$

G. Example Flux Distribution in the Air Gap

Fig. 2 shows the field distribution inside an air gap for different radial positions of the electric loading in the case of a smooth air gap. If the field lines are assumed to behave like 'elastic straps', it becomes obvious that only the field caused by two electric loadings close to the stator and rotor surface, respectively, generates the resulting torque. The resulting torque (T) can be calculated as [21]

$$T = r_o l \int_0^{2\pi} \frac{B_r B_{\theta'}}{\mu_0} r_o \mathrm{d}\theta', \tag{25}$$

where l is the length of the stator and rotor core. As (25) shows, knowledge of the circumferential and the radial field components enables the direct calculation of the torque. Since both field components can be calculated by using this theory, the torque can always be calculated directly.

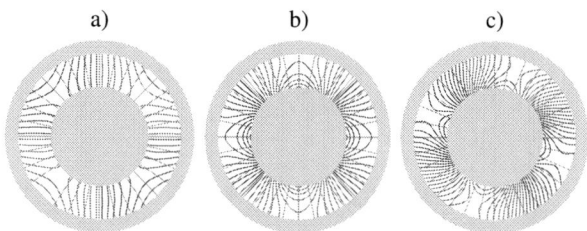

Fig. 2. Field distribution inside the air gap caused by : a) one electric loading close to the stator surface, b) one electric loading close to the rotor surface, and c) two electric loadings close to the stator and rotor surface respectively, with the same orders n' and k'.

III. INFLUENCE OF THE STATIONARY SLOTS

This section covers the analytical calculation of the influence of stator slot openings. The slot opening has the width of τ_N and is located at the angle θ'_v as shown in Fig. 3. Since the field inside the whole slot is calculated, a very simple slot geometry was chosen. The shaded area is assumed to be an ideal magnetic conductor. The fundamental idea is that at the slot opening, the radial component of the air-gap flux density has to be equal to the radial component of the flux density inside the slot

$$B_r(r_1^+) = B_r(r_1^-) \quad \text{for} \quad \theta'_v - \frac{\tau_N}{2} < \theta' < \theta'_v + \frac{\tau_N}{2}. \tag{26}$$

Based on this equality, the circumferential field component inside the slot is calculated. The link between the radial and the circumferential field component is again derived from the vector potential. This in turn influences the field inside the air gap.

A. Fourier Series of the Air-Gap Field

The field inside the slot is again represented by a Fourier series. However, a spatial decomposition is necessary, as described below. The first step is to decompose the field into an even and an odd fraction. Fig. 4 shows the decomposition. For the even fraction, a period of τ_N is considered as the fundamental period. For the odd fraction, a period of $2\tau_N$ is

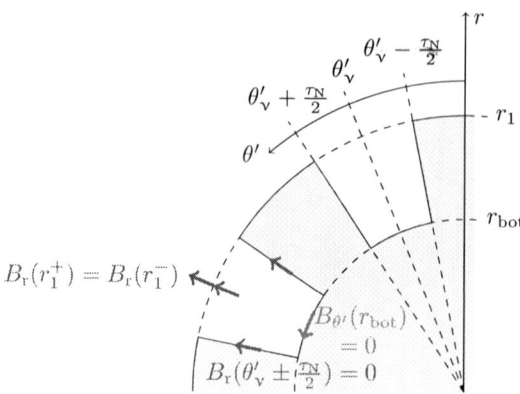

Fig. 3. Schematic view of a slot.

taken as the fundamental period, where the second half is equal to the first half, shifted by half a period and multiplied by minus one. Due to the symmetry, this function only contains harmonics that have the period $2\tau_N/g$ with $g \in \mathbb{Z}$. As a result of the decomposition, the overall Fourier series of the field in slot n can be written as

$$\hat{B}_{\text{slot,r,v}} = \sum_{\mu'=1+2g} \hat{B}_{\text{slot},\mu',v} \cos\left(\mu'(\theta'-\theta'_v)\frac{\pi}{\tau_N}\right)$$
$$+ \sum_{\mu'=2g} \hat{B}_{\text{slot},\mu',v} \sin\left(\mu'(\theta'-\theta'_v)\frac{\pi}{\tau_N}\right) \tag{27}$$
$$\text{with} \quad g \in \mathbb{N}.$$

This function has the property that the radial field at the slot edges

$$\theta' = \theta'_v \pm \frac{\tau_N}{2} \tag{28}$$

equals zero. This is important, since the radial field is orthogonal to the boundary between an ideal magnetic conductor and air at this point. After the representation of the slot field is chosen, the amplitudes of the Fourier series can be

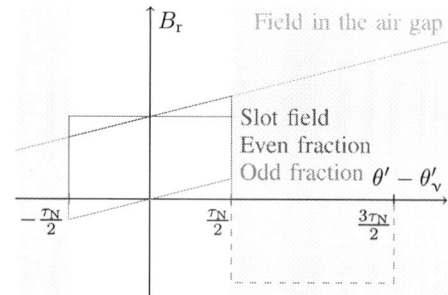

Fig. 4. At the slot opening, the radial field in the air gap is assumed to be equal to the field in the slot, which is separated into an even and an odd fraction.

calculated, using the Fourier transformation of the air-gap spatial harmonics (e.g., $\hat{B}_{n'_1,k'_1} \cos(n'_1\theta' - k'_1\delta' - \varphi'_1)$) as

$$
\begin{aligned}
\hat{B}_{\text{slot},\mu',\nu} =& \hat{B}_{n'_1,k'_1} \left(\text{si}\left(n'_1\frac{\tau'_N}{2} + \mu'\frac{\pi}{2}\right) + \right. \\
& \left. \text{si}\left(n'_1\frac{\tau'_N}{2} - \mu'\frac{\pi}{2}\right) \right) \cdot \\
& \cos(n'_1\theta'_\nu - k'_1\delta' - \varphi'_1)
\end{aligned}
\tag{29}
$$

for $\mu' = 1 + 2g$ (even fraction) and as

$$
\begin{aligned}
\hat{B}_{\text{slot},\mu',\nu} =& \hat{B}_{n'_1,k'_1} \left(\text{si}\left(n'_1\frac{\tau'_N}{2} + \mu'\frac{\pi}{2}\right) - \right. \\
& \left. \text{si}\left(n'_1\frac{\tau'_N}{2} - \mu'\frac{\pi}{2}\right) \right) \cdot \\
& \cos\left(n'_1\theta'_\nu - k'_1\delta' - \varphi'_1 - \frac{\pi}{2}\right)
\end{aligned}
\tag{30}
$$

for $\mu' = 2g$ (odd fraction), where $\text{si}(x) = \frac{\sin(x)}{x}$. Since the considered area can again be described in cylindrical coordinates, the results derived for the air gap can be used to calculate the circumferential field in the slot as

$$
\begin{aligned}
\hat{H}_{\text{slot},\mu',\nu} =& -\frac{1}{\mu_0} \frac{1}{\chi_{\mu'\frac{\pi}{\tau_N}}(r_{\text{bot}}, r_1, r_1)} \\
& \hat{B}_{\text{slot},\mu',\nu} \sin\left(\mu'\frac{2\pi}{2\tau_N}(\theta' - \theta'_\nu)\right)
\end{aligned}
\tag{31}
$$

for $\mu' = 1 + 2g$ and as

$$
\begin{aligned}
\hat{H}_{\text{slot},\mu',\nu} =& \frac{1}{\mu_0} \frac{1}{\chi_{\mu'\frac{\pi}{\tau_N}}(r_{\text{bot}}, r_1, r_1)} \\
& \hat{B}_{\text{slot},\mu',\nu} \sin\left(\mu'\frac{2\pi}{2\tau_N}(\theta' - \theta'_\nu) - \frac{\pi}{2}\right)
\end{aligned}
\tag{32}
$$

for $\mu' = 2g$. We can take advantage of the fact that the circumferential field is zero at the bottom of a slot (r_{bot}). Based on the circumferential field inside the slot, the influence on the circumferential field component in the air gap i.e., the spatial harmonic $\hat{H}_{n'_2,k'_2,\mu',\nu} \cos(n'_2\theta' - k'_2\delta' - \varphi'_2)$, can be calculated through a Fourier transformation as

$$
\begin{aligned}
\hat{H}_{n'_2,k'_2,\mu',\nu} =& \hat{H}_{\text{slot},\mu',\nu} \cdot \frac{\tau_N}{2\pi} \cdot \left(\text{si}\left(n'_2\frac{\tau'_N}{2} + \mu'\frac{\pi}{2}\right) - \right. \\
& \left. \text{si}\left(n'_2\frac{\tau'_N}{2} - \mu'\frac{\pi}{2}\right) \right) \cdot \\
& \cos\left(n'_2\theta'_\nu - k'_2\delta' - \varphi'_2 - \frac{\pi}{2}\right)
\end{aligned}
\tag{33}
$$

for $\mu' = 1 + 2g$ and as

$$
\begin{aligned}
\hat{H}_{n'_2,k'_2,\mu',\nu} =& \hat{H}_{\text{slot},\mu',\nu} \frac{\tau_N}{2\pi} \left(\text{si}\left(n'_2\frac{\tau'_N}{2} + \mu'\frac{\pi}{2}\right) + \right. \\
& \left. \text{si}\left(n'_2\frac{\tau'_N}{2} - \mu'\frac{\pi}{2}\right) \right) \cdot \\
& \cos(n'_2\theta'_\nu - k'_2\delta' - \varphi'_2)
\end{aligned}
\tag{34}
$$

for $\mu' = 2g$. It is assumed that the slots are evenly distributed at

$$
\theta'_\nu = \theta'_0 + \nu\frac{2\pi}{N}.
\tag{35}
$$

The influence of all slots can be calculated as the sum of the influence of each individual slot

$$
\hat{H}_{n'_2,k'_2,\mu'} = \sum_{\nu=1}^{N} \hat{H}_{n'_2,k'_2,\mu',\nu}.
\tag{36}
$$

The final result for $\mu' = 1 + 2g$ can be written as

$$
\begin{aligned}
& \hat{H}_{n'_2,k'_2,\mu'} \\
& = -\hat{B}_{n'_1,k'_1} \frac{\tau_N}{4\pi} \frac{1}{\mu_0} \frac{1}{\chi_{\mu'\frac{\pi}{\tau_N}}(r_{\text{bot}}, r_1, r_1)} \\
& \left(\text{si}\left(n'_1\frac{\tau'_N}{2} + \mu'\frac{\pi}{2}\right) + \text{si}\left(n'_1\frac{\tau'_N}{2} - \mu'\frac{\pi}{2}\right) \right) \cdot \\
& \left(\text{si}\left(n'_2\frac{\tau'_N}{2} + \mu'\frac{\pi}{2}\right) - \text{si}\left(n'_2\frac{\tau'_N}{2} - \mu'\frac{\pi}{2}\right) \right) \cdot \\
& \begin{cases} N; & (n'_1 - n'_2) = gN, \quad \text{with} \\ & g \in \mathbb{Z}, \quad k'_2 = k'_1, \\ & \varphi'_2 = \varphi'_1 - \frac{\pi}{2} - (n'_1 - n'_2)\theta'_0 \\ N; & (n'_1 + n'_2) = gN, \quad \text{with} \\ & g \in \mathbb{Z}, \quad k'_2 = -k'_1, \\ & \varphi'_2 = -\varphi'_1 - \frac{\pi}{2} + (n'_1 + n'_2)\theta'_0 \\ 0; & \text{otherwise} \end{cases} \\
& = \hat{B}_{n'_1,k'_1} \cdot \hat{\varsigma}_{\mu'}(n'_1, n'_2)
\end{aligned}
\tag{37}
$$

and for $\mu' = 2g$

$$
\begin{aligned}
& \hat{H}_{n'_2,k'_2,\mu'} \\
& = \hat{B}_{n'_1,k'_1} \frac{\tau_N}{4\pi} \frac{1}{\mu_0} \frac{1}{\chi_{\mu'\frac{\pi}{\tau_N}}(r_{\text{bot}}, r_1, r_1)} \\
& \left(\text{si}\left(n'_1\frac{\tau'_N}{2} + \mu'\frac{\pi}{2}\right) - \text{si}\left(n'_1\frac{\tau'_N}{2} - \mu'\frac{\pi}{2}\right) \right) \cdot \\
& \left(\text{si}\left(n'_2\frac{\tau'_N}{2} + \mu'\frac{\pi}{2}\right) + \text{si}\left(n'_2\frac{\tau'_N}{2} - \mu'\frac{\pi}{2}\right) \right) \cdot \\
& \begin{cases} N; & (n'_1 - n'_2) = gN, \quad \text{with} \\ & g \in \mathbb{Z}, \quad k'_2 = k'_1, \\ & \varphi'_2 = \varphi'_1 + \frac{\pi}{2} - (n'_1 - n'_2)\theta'_0 \\ N; & (n'_1 + n'_2) = gN, \quad \text{with} \\ & g \in \mathbb{Z}, \quad k'_2 = -k'_1, \\ & \varphi'_2 = -\varphi'_1 - \frac{\pi}{2} + (n'_1 + n'_2)\theta'_0 \\ 0; & \text{otherwise} \end{cases} \\
& = \hat{B}_{n'_1,k'_1} \cdot \hat{\varsigma}_{\mu'}(n'_1, n'_2).
\end{aligned}
\tag{38}
$$

Here, ς was introduced as a ratio between the circumferential magnetic field strength caused by a machine part and the radial flux density at its surface. As a vectorized representation, it can be written as

$$
\underline{\hat{H}}_{n'_2,k'_2,\mu'} = \underline{\underline{\varsigma}}_{\mu'}(n'_1, n'_2)\underline{\hat{B}}_{n'_1,k'_1,\mu'},
\tag{39}
$$

where $\underline{\underline{\varsigma}}_{\mu'}(n'_1, n'_2)$ is for $n'_1 - n'_2 = gN$ and $\mu' = 1 + 2g$

$$
\begin{aligned}
& \underline{\underline{\varsigma}}_{\mu'}(n'_1, n'_2) = \hat{\varsigma}_{\mu'}(n'_1, n'_2) \\
& \begin{bmatrix} -\sin((n'_1 - n'_2)\theta'_0) & \cos((n'_1 - n'_2)\theta'_0) \\ -\cos((n'_1 - n'_2)\theta'_0) & -\sin((n'_1 - n'_2)\theta'_0) \end{bmatrix},
\end{aligned}
\tag{40}
$$

for $n_1' - n_2' = gN$ and $\mu' = 2g$

$$\underline{\underline{\varsigma}}_{\mu'}(n_1', n_2') = \hat{\varsigma}_{\mu'}(n_1', n_2')$$
$$\begin{bmatrix} \sin((n_1'-n_2')\theta_0') & -\cos((n_1'-n_2')\theta_0') \\ \cos((n_1'-n_2')\theta_0') & \sin((n_1'-n_2')\theta_0') \end{bmatrix}, \quad (41)$$

for $n_1' + n_2' = gN$ and $\mu' = 1 + 2g$

$$\underline{\underline{\varsigma}}_{\mu'}(n_1', n_2') = \hat{\varsigma}_{\mu'}(n_1', n_2')$$
$$\begin{bmatrix} \sin((n_1'+n_2')\theta_0') & -\cos((n_1'+n_2')\theta_0') \\ -\cos((n_1'+n_2')\theta_0') & -\sin((n_1'+n_2')\theta_0') \end{bmatrix}, \quad (42)$$

and for $n_1' + n_2' = gN$ and $\mu' = 2g$

$$\underline{\underline{\varsigma}}_{\mu'}(n_1', n_2') = \hat{\varsigma}_{\mu'}(n_1', n_2')$$
$$\begin{bmatrix} \sin((n_1'+n_2')\theta_0') & -\cos((n_1'+n_2')\theta_0') \\ -\cos((n_1'+n_2')\theta_0') & -\sin((n_1'+n_2')\theta_0') \end{bmatrix}. \quad (43)$$

Equations (40)-(43) show that a symmetrically slotted stator, subjected to a radial spatial harmonic of the order n', only causes spatial harmonics with the order $n' + gN$ with $g \in \mathbb{Z}$. This complies with the results found in literature, e.g., [1]. Next, not only a single harmonic is considered but multiple harmonics. This is achieved by summation:

$$\underline{\hat{H}}_{n_2', k_2'} = \sum_{\mu'} \underline{\underline{\varsigma}}_{\mu'}(n_1', n_2') \underline{\hat{B}}_{n_1', k_1'} = \underline{\varsigma}(n_1', n_2') \underline{\hat{B}}_{n_1', k_1'}. \quad (44)$$

So far, only the coupling between a single radial spatial harmonic and a single circumferential spatial harmonic has been considered. Using the vector representation

$$\vec{\hat{H}}_{\theta'} = \vec{\vec{\varsigma}}\vec{\hat{B}}_{\mathrm{r}} \quad (45)$$

with

$$\vec{\hat{B}}_{\mathrm{r}} = \begin{bmatrix} \underline{\hat{B}}_{n',k,r}(r_i) \\ \underline{\hat{B}}_{n',k,r}(r_o) \\ \underline{\hat{B}}_{n'-N,k,r}(r_i) \\ \underline{\hat{B}}_{n'-N,k,r}(r_o) \\ \underline{\hat{B}}_{n'+N,k,r}(r_i) \\ \underline{\hat{B}}_{n'+N,k,r}(r_o) \end{bmatrix}, \quad (46)$$

$$\vec{\hat{H}}_{\theta'} = \begin{bmatrix} \underline{\hat{H}}_{n',k,\theta'}(r_i) \\ \underline{\hat{H}}_{n',k,\theta'}(r_o) \\ \underline{\hat{H}}_{n'-N,k,\theta'}(r_i) \\ \underline{\hat{H}}_{n'-N,k,\theta'}(r_o) \\ \underline{\hat{H}}_{n'+N,k,\theta'}(r_i) \\ \underline{\hat{H}}_{n'+N,k,\theta'}(r_o) \end{bmatrix}, \quad (47)$$

and

$$\vec{\vec{\varsigma}} =$$
$$\begin{bmatrix} \underline{\underline{0}} & \underline{\underline{0}} & \underline{\underline{0}} & \underline{\underline{0}} & \cdots \\ \underline{\underline{0}} & \underline{\varsigma}(n',n') & \underline{\underline{0}} & \underline{\varsigma}(n'-N,n') & \cdots \\ \underline{\underline{0}} & \underline{\underline{0}} & \underline{\underline{0}} & \underline{\underline{0}} & \cdots \\ \underline{\underline{0}} & \underline{\varsigma}(n',n'-N) & \underline{\underline{0}} & \underline{\varsigma}(n'-N,n'-N) & \cdots \\ \underline{\underline{0}} & \underline{\underline{0}} & \underline{\underline{0}} & \underline{\underline{0}} & \cdots \\ \underline{\underline{0}} & \underline{\varsigma}(n',n'+N) & \underline{\underline{0}} & \underline{\varsigma}(n'-N,n'+N) & \cdots \end{bmatrix}, \quad (48)$$

the coupling between multiple radial and circumferential spatial harmonics can be considered.

B. Impact of the Circumferential Spatial Harmonics Caused by Stator Slots on the Air-Gap Field

In this section, the air-gap field excited by the electric loading will be investigated where one side of the air gap is slotted. The radial spatial harmonics can be calculated as the sum of the radial spatial harmonics caused by the current and the radial spatial harmonics caused by the circumferential spatial harmonics of the slots

$$\vec{B}_{\mathrm{r}} = \mu_0 \vec{\vec{\chi}}\vec{J} + \mu_0 \vec{\vec{\chi}}\vec{\vec{\varsigma}}\vec{B}_{\mathrm{r}}$$
$$= \mu_0 \vec{\vec{\chi}}\vec{J} + \vec{\vec{\lambda}}^{\#}\vec{B}_{\mathrm{r}}. \quad (49)$$

The matrix $\vec{\vec{\lambda}}^{\#} = \mu_0 \vec{\vec{\chi}}\vec{\vec{\varsigma}}$ was introduced to represent the influence of the slotting in one matrix. Each element of the matrix describes a coupling between spatial harmonics: Diagonal elements describe the influence that a harmonic has on itself, whereas off-diagonal elements describe the coupling between different spatial harmonics. Finally, the resulting radial air-gap field caused by the electric loading can be calculated as

$$\vec{B}_{\mathrm{r}} = \left(\vec{\vec{I}} - \vec{\vec{\lambda}}^{\#} \right)^{-1} \mu_0 \vec{\vec{\chi}}\vec{J}. \quad (50)$$

The advantage of the approach is that the cause of each spatial harmonic of the vector $\vec{B}_{\mathrm{r}}$ is visible in (49). $\mu_0 \vec{\vec{\chi}}\vec{J}$ describes the flux density in the absence of slots. $\vec{\vec{\lambda}}^{\#}\vec{B}_{\mathrm{r}}$ describes the change in flux density due to the slots. It directly shows how much the fundamental spatial harmonic is reduced and which spatial harmonics are excited.

Since $\vec{B}_{\mathrm{r}}$ fully described the flux density inside of the air gap, the circumferential field can be calculated using (24) and the torque using (25).

IV. VALIDATION USING FEM

In order to demonstrate the validity of the analytic approach, an FEM-based validation is performed using the FEM tool FEMAG. An arbitrarily chosen four-pole machine with 24 slots was used for the comparison. A double-layer winding with a 5/6 pitch was modelled in FEM, while the corresponding electric loading on the stator side was derived for the analytic analysis. The field solution is shown in Fig. 5. For the highly conducting areas, a relative permeability of $\mu_{\mathrm{r}} = 10^8$ is chosen. The air gap is split into five layers.

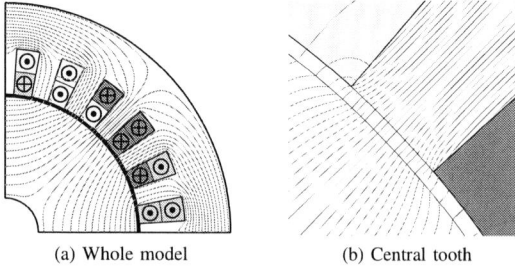

(a) Whole model   (b) Central tooth

Fig. 5. FEM field solution calculated and plotted with FEMAG.

The outer and the inner layer are modelled particularly thin in order to reduce the difference between the flux density at the geometric centre of the elements and the actual outer or inner radius. However, a difference between the radius, at which the field is evaluated, and the actual outer radius is unavoidable.

In the analytic calculation, orders up to the 480th circumferential order are considered. Fig. 6 and Fig. 7 show the comparison between the analytically calculated radial and circumferential flux density for the inner and the outer radius and the one determined with FEM. The comparison shows a good overall agreement. However, some differences can be seen. The flux density at the outer radius has additional, visible oscillations which are not visible in the FEM results. This can be partially contributed to the limited number of harmonics calculated and partially to the Gibbs phenomenon. Additionally, the amplitudes of the circumferential spatial harmonics at the outer radius are slightly smaller compared to the FEM results. One cause is the small difference in the evaluation radius. For the same reason, a small circumferential field component can be seen at the inner radius. Here, the same scaling is chosen as in Fig. 6 to demonstrate that the circumferential field at the inner radius with good approximation is zero. Since the rotor is smooth and the whole excitation is on the stator side, no torques can be analysed in this configuration. Concerning computation time, the whole analytical calculation took about 0.8 s on an Intel processor i3-5010U, while a numeric calculation of 90 rotary steps takes about 60 s.

V. DISCUSSION

In order to demonstrate the interpretability of the results achieved with the new approach, the analytical results are now discussed in more detail. The electric loading, which is located at the outer radius

$$
\vec{J} = \begin{bmatrix} \underline{\hat{J}}_{\text{p,p}}(r_{\text{i}}) \\ \underline{\hat{J}}_{\text{p,p}}(r_{\text{o}}) \\ \underline{\hat{J}}_{\text{p-N,p}}(r_{\text{i}}) \\ \underline{\hat{J}}_{\text{p-N,p}}(r_{\text{o}}) \\ \vdots \end{bmatrix} = \begin{bmatrix} 0 \\ 0 \\ 0 \\ -12 \\ 0 \\ 0 \\ 0 \\ 8 \\ \vdots \end{bmatrix} \frac{\text{kA}}{\text{m}} \tag{51}
$$

excites the whole magnetic field. The radial flux density

$$
\vec{B}_{\text{r}} = \begin{bmatrix} \underline{\hat{B}}_{\text{p,p,r}}(r_{\text{i}}) \\ \underline{\hat{B}}_{\text{p,p,r}}(r_{\text{o}}) \\ \underline{\hat{B}}_{\text{p-N,p,r}}(r_{\text{i}}) \\ \underline{\hat{B}}_{\text{p-N,p,r}}(r_{\text{o}}) \\ \vdots \end{bmatrix} = \begin{bmatrix} 315 \\ 0 \\ 309 \\ 0 \\ -119 \\ 0 \\ -125 \\ 0 \\ \vdots \end{bmatrix} \text{mT} \tag{52}
$$

is the result of the electric loading and the effects due to slotting. As an example, the cause of the fundamental spatial harmonic radial flux density at the outer radius of 309 mT

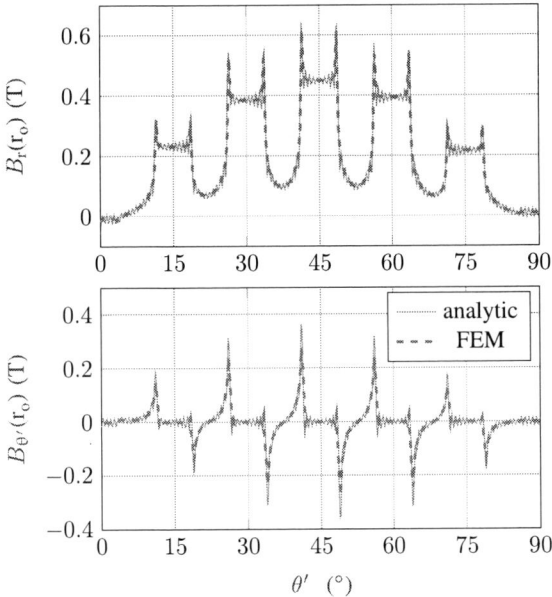

Fig. 6. Comparison of FEM and analytical results for the radial and the circumferential flux density components at the outer radius.

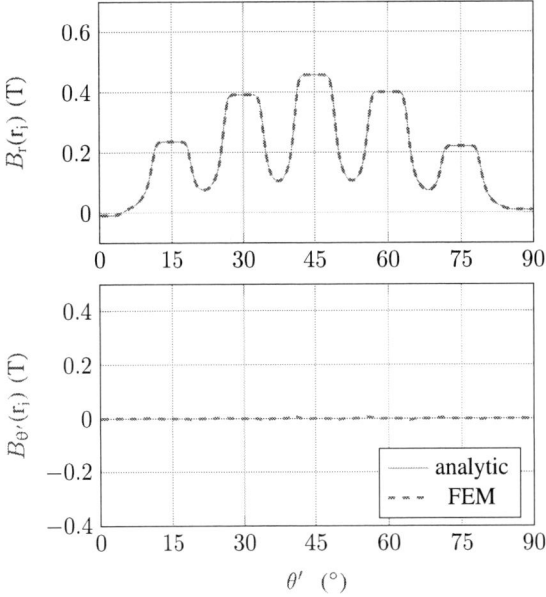

Fig. 7. Comparison of FEM and analytical results for the radial and the circumferential flux density components at the inner radius.

shall be analysed in more detail. Using the electric loading (51) and

$$
\vec{\chi} = \begin{bmatrix}
0 & 29.8 & 0 & -30.2 & \cdots \\
-29.8 & 0 & 30.2 & 0 & \\
0 & 29.2 & 0 & -29.8 & \\
-29.2 & 0 & 29.8 & 0 & \\
0 & 0 & 0 & 0 & \\
0 & 0 & 0 & 0 & \\
0 & 0 & 0 & 0 & \\
\vdots & & & & \ddots
\end{bmatrix}, \quad (53)
$$

the flux density amplitude would be $\mu_0 \cdot (-29.8) \cdot (-12)\,\mathrm{kA/m} = 443\,\mathrm{mT}$, if slotting was neglected. The slotting thus results in a reduction of $-134\,\mathrm{mT}$. Using

$$
\vec{\lambda}\# = \begin{bmatrix}
0 & 0 & -1.07 & 0 & \cdots \\
0 & 0 & 0 & -1.07 & \\
0 & 0 & -1.05 & 0 & \\
0 & 0 & 0 & -1.05 & \\
0 & 0 & -0.80 & 0 & \\
0 & 0 & 0 & 0.80 & \\
0 & 0 & -0.84 & 0 & \\
0 & 0 & 0 & 0.84 & \\
\vdots & & & & \ddots
\end{bmatrix}, \quad (54)
$$

the cause of the reduction can be analysed. The reduction of the fundamental spatial harmonic caused by the coupling with the slotting is $-1.05 \cdot 309\,\mathrm{mT} = -326\,\mathrm{mT}$. Since other spatial harmonics have a positive effect on the fundamental harmonic, considering all spatial harmonics results in a total reduction of only $-134\,\mathrm{mT}$. In a similar manner, any spatial harmonic could be analysed. For instance, (54) also shows the coupling between the fundamental spatial harmonic and the spatial harmonic of the circumferential order $p - N$ (0.80 and 0.84 for the inner and outer radius, respectively).

The approach to identify the cause of each spatial harmonic would be even more useful, if multiple effects were considered. Multiple effects similar to slotting could be considered through

$$
\vec{B}_r = \mu_0 \vec{\chi} \vec{J} + \sum_{l=1,2,3,\dots} \vec{\lambda}_l\# \vec{B}_r. \quad (55)
$$

Future work may also include permanent magnet excitation, consideration of saturation, calculation of iron losses and calculation of inductances.

VI. CONCLUSION

The approach introduced in this paper is a promising and time efficient tool for analysing the behaviour of electrical machines. For a machine with a simple slot geometry excited by an electric loading, the results of the new approach resemble FEM simulation results to a large extent. This is the case for radial and circumferential fields at the inner radius as well as at the outer radius of the air-gap, demonstrating that the new approach can calculate the required quantities.

The additional ability to deduce the source of each spatial harmonic is a unique feature that sets this approach apart from previously presented subdomain models as well as FEM. A typical application could be to identify the source of undesired harmonics. This knowledge could then be used to derive measures to reduce them.

REFERENCES

[1] J. F. Gieras, C. Wang, and J. C. Lai, *Noise of polyphase electric motors*, ser. Electrical and computer engineering. Boca Raton, FL, USA: CRC/Taylor & Francis, 2006, vol. 129.

[2] T. A. Lipo, *Introduction to AC machine design*, ser. IEEE Press series on power engineering. Hoboken, NJ, USA: Wiley IEEE Press, 2017, vol. 63.

[3] H. O. Seinsch, *Oberfelderscheinungen in Drehfeldmaschinen*. Stuttgart, Germany: Teubner, 1992.

[4] D. Zarko, D. Ban, and T. A. Lipo, "Analytical solution for electromagnetic torque in surface permanent-magnet motors using conformal mapping," *IEEE Trans. Magn.*, vol. 45, no. 7, pp. 2943–2954, July 2009.

[5] V. Ostvić, *Dynamics of Saturated Electric Machines*. Springer, New York, NY, USA, 1989.

[6] H. Tiegna, Y. Amara, and G. Barakat, "Overview of analytical models of permanent magnet electrical machines for analysis and design purposes," *Mathematics and Computers in Simulation*, vol. 90, pp. 162–177, 2013.

[7] E. Devillers, J. Le Besnerais, T. Lubin, M. Hecquet, and J.-P. Lecointe, "A review of subdomain modeling techniques in electrical machines: Performances and applications," in *Proc. Int. Conf. Elect. Mach. (ICEM), Lausanne, Switzerland*, 2016, pp. 86–92.

[8] Z. Q. Zhu, D. Howe, E. Bolte, and B. Ackermann, "Instantaneous magnetic field distribution in brushless permanent magnet dc motors. I. open-circuit field," *IEEE Trans. Magn.*, vol. 29, no. 1, pp. 124–135, 1993.

[9] T. Lubin, S. Mezani, and A. Rezzoug, "2-D exact analytical model for surface-mounted permanent-magnet motors with semi-closed slots," *IEEE Trans. Magn.*, vol. 47, pp. 479–492, 2011.

[10] ——, "Analytic calculation of eddy currents in the slots of electrical machines: Application to cage rotor induction motors," *IEEE Trans. Magn.*, vol. 47, pp. 4650–4659, 2011.

[11] E. Devillers, J. Le Besnerais, T. Lubin, M. Hecquet, and J.-P. Lecointe, "An improved 2-D subdomain model of squirrel-cage induction machine including winding and slotting harmonics at steady state," *IEEE Trans. Magn.*, vol. 54, no. 2, pp. 1–12, 2018.

[12] R. Sprangers, J. Paulides, B. Gysen, and E. Lomonova, "A fast semi-analytical model for the slotted structure of induction motors," *Mathematics and Computers in Simulation*, vol. 131, pp. 316 – 327, 2017.

[13] B. Hannon, P. Sergeant, and L. Dupre, "2D analytical torque study of slotted high-speed PMSMs considering pole pairs, slots per pole per phase and coil throw," in *Proc. Int. Conf. Elect. Mach. (ICEM), Berlin, Germany*, 2014, pp. 2524–2530.

[14] ——, "Time- and spatial-harmonic content in synchronous electrical machines," *IEEE Trans. Magn.*, vol. 53, no. 3, pp. 1–11, 2017.

[15] ——, "Computational-time reduction of fourier-based analytical models," *IEEE Trans. Energy Convers.*, vol. 33, no. 1, pp. 281–289, 2018.

[16] H. Ghoizad, M. Mirsalim, M. Mirzayee, and W. Cheng, "Coupled magnetic equivalent circuits and the analytical solution in the air-gap of squirrel cage induction machines," *International Journal of Applied Electromagnetics and Mechanics*, vol. 25, pp. 749–754, 2007.

[17] G. Madescu, I. Boldea, and T. J. E. Miller, "An analytical iterative model (AIM) for induction motor design," in *Conf. Rec. IEEE-IAS Annu. Meeting, San Diego, CA, USA*, vol. 1, Oct 1996, pp. 566–573 vol.1.

[18] B. Heller and V. Hamata, *Harmonic field effects in induction machines*. Amsterdam, Netherlands: Elsevier Scientific Publ. 1977.

[19] G. Müller and B. Ponick, *Theorie elektrischer Maschinen*, 6th ed., ser. Elektrische Maschinen. Weinheim, Germany: Wiley-VCH, 2009.

[20] D. Gerling, *Electrical Machines*. Berlin/Heidelberg, Germany: Springer Berlin Heidelberg, 2015, vol. 4.

[21] D. Braunisch, "Kombinierte analytisch-numerische berechnung der magnetgeräusche elektrischer maschinen," Ph.D. dissertation, Leibniz Univ. Hannover, Hannover, Germany, 2015.

[22] Z. Q. Zhu and D. Howe, "Instantaneous magnetic field distribution in brushless permanent magnet dc motors. III. effect of stator slotting," *IEEE Trans. Magn.*, vol. 29, no. 1, pp. 143–151, 1993.

978-1-5386-7688-2/19 $31.00 © 2019 IEEE

Early Detection of Turn-to-Turn Faults in Power Transformer Winding: An Experimental Study

Arash Moradzadeh
Department of Electrical Engineering,
Tabriz Branch, Islamic Azad University,
Tabriz, Iran
Stu.arash.moradzadeh@iaut.ac.ir

Kazem Pourhossein
Department of Electrical Engineering,
Tabriz Branch, Islamic Azad University,
Tabriz, Iran
k.pourhossein@iaut.ac.ir

Abstract— **Turn-to-turn insulations of transformer windings may degrade gradually because of mechanical forces, thermal stresses or chemical corrosion. Degradation decreases impedances of inter-turn insulations that finally may lead to a solid turn-to-turn short circuit. In this paper, early detection of turn-to-turn faults in transformers windings has been studied, in its high-impedance stage, using Artificial Neural Networks (ANN) based on its Frequency Response (FR). For this purpose, a model winding has been used as test object to approve capability of the proposed approach. A variety of low impedance and high impedance short circuit faults were tested on the model winding. Then the frequency response of winding in both intact and defected conditions is measured using Low Voltage Impulse (LVI) test. A mapping between frequency response and exact location of each fault was made using multi-layer perceptron (MLP) neural network. Extracted features from frequency responses are used to train and test the proposed MLP. The results show that this method is able to detect turn-to-turn faults in transformer winding even in their early stages.**

Keywords— *Turn-to-Turn fault, frequency response analysis, transformer winding, multi-layer perceptron*

I. INTRODUCTION

Due to increased electricity consumption, competition in the electricity market, the need for delivery with reliability and high quality is inevitable. In this regard, power transformers are considered as the most important and expensive equipment in terms of their key role in supplying power [1, 2]. These equipment can sometimes be damaged for environmental reasons. Creating a defect in transformers reduces the reliability and power quality of the network and causes fundamental problems in the power system. Therefore, early detection of transformer faults should be an integral part of the power grid monitoring system. The percentage of faults in different parts of the power transformers is presented in Fig. 1. According to two reports in [3] and [4] and other similar reports, OLTC (on-line tap changer) faults and winding faults have high percentage in comparison with other faults.

Windings of transformers are not easily accessible in comparison to OLTC thus detection and repair of winding faults are harder than OLTC faults [5-11]. If early detection of these mechanical defects is not performed, they may lead to a direct short circuit.

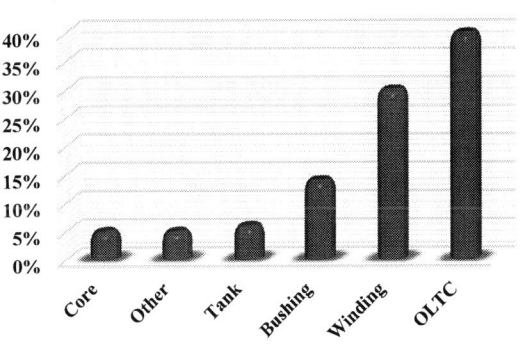

Fig. 1. Frequency of faults in different parts of power transformers [5]

Mechanical defects apply excessive force on winding insulation which start progressive deterioration of insulation and finally lead to a damaged insulation i.e. a low-impedance turn-to-turn short circuit (Fig. 2).

A turn-to-turn short circuit is a potential source of heat. Therefore, this type of fault must be detected in its early stages (high-impedance stage) to prevent subsequent sever faults e.g. low-impedance short circuit (direct short circuit) in adjacent turns, insulation destruction and so on [12, 13].

So far, many methods have been introduced to identify the mechanical defects of transformer windings. Winding impedance measurement [14], Vibro-acoustic analysis [11], Dissolved gas analysis (DGA) [15], Frequency response analysis (FRA) [9, 16, 17] are commonly used methods. In this regard, the research and results presented in several works have shown that the frequency response analysis method is the most important method among the above-mentioned methods [18, 19]. The basic idea behind FRA is effect of any physical change in transformer active part on its electrical circuit parameter and then on its frequency response.

Numerical indices [20, 21, 22], vector fitting for estimated transfer function [23], estimating parameters of transformer model [24], artificial neural networks [25, 26], support vector machine (SVM) for transfer function analysis [27], cross correlation [28], enhanced magnetic optimal algorithm [29] and used of finite element method [30] are well-known used methods for interpreting frequency responses of transformers.

978-1-5386-7688-2/19 $31.00 © 2019 IEEE

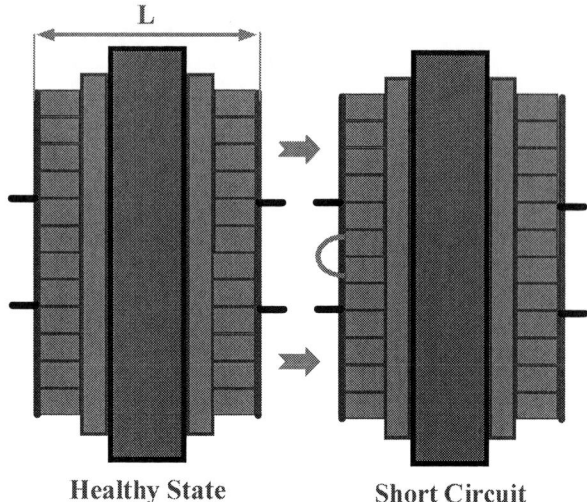

Fig. 2.   Short circuit fault in transformer winding

In the case of short circuit locating using FRA some valuable works have been done. Most of them used statistical or mathematical indices [31], index of similarity index to short circuit faults [32], three phase FRA comparison to detect short circuits [33], support vector machines to classify different mechanical faults of tested power transformer [27], Used of artificial neural networks to localize short circuits in transformer windings [25, 26], There is another paper reported neural network approach to short circuit location in time domain (without using FRA) [34].

Almost all of the above-mentioned methods that have been applied to transformer models or measured frequency responses suffer from two deficiencies. They have been used to diagnose only direct short circuits (low-impedance short circuit). It means detection of faults after their full occurrence, In some of the mentioned literature, detection exactly fault location is inaccuracy. In this paper uses a trained multi-layer perceptron (MLP) to accurate interpretation of frequency responses and locate winding short circuit in both low and high impedance states. Ability to detect high impedance short circuits guaranties early detection of turn-to-turn short circuits in transformer windings to prevent damage to the insulation system and transformer windings.

II. FREQUNCY RESPONSE ANALYSIS

Frequency Response Analysis (FRA) is a powerful diagnostic test method [17, 35]. This involves measuring the impedance of the transformer winding in different frequencies and comparing the results of these measurements with the reference state. Changes in frequency responses indicate defects to the transformer, which can be examined using other methods and techniques [17,18, 35].

FRA can be of two types: low voltage impulse method (LVI) or sweep frequency response analysis (SFRA) [17, 18]. In this paper, the LVI method was used. In this method, the surge voltage is applied to the transformer and its output signal (voltage or current) is measured (Fig. 3). Then, the calculated signals are transmitted using a Fourier transfer to the frequency

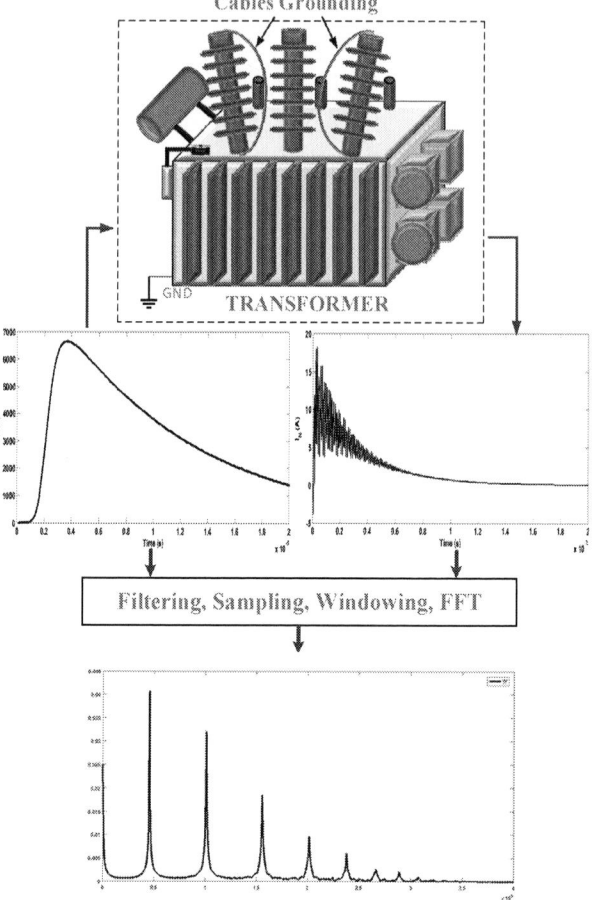

Fig. 3.   Frequency Response Measurement

space, finally, the frequency response can be calculated by dividing these signals in the frequency space [35].

III. MULTI-LAYER PERCEPTRON

Artificial neural networks (ANN) are suitable tools used in modeling and estimation purposes. ANNs learn patterns of data and then can show generalized behavior after learning process and classify new patterns of data [36, 37]. An ANN include some building blocks named neurons. The output vector (Y) in neural networks is calculated from the following equation [36]:

$$Y = f(b + \sum_{i=1}^{n} w_i x_i) \qquad (1)$$

where X_i, W_i and b are input vector, weight vector and bias, respectively. Multi-layer perceptron (MLP) is one of well-known ANNs that widely used to nonlinear mapping, regression and modeling [38, 39]. MLP has input, output and hidden layers in own structure (Fig. 4). The number of neurons in hidden layer depends on the complexity of the problem that the network wants to solve it. [40, 41]. Flowchart of Fig. 5 the presents design procedure of MLP.

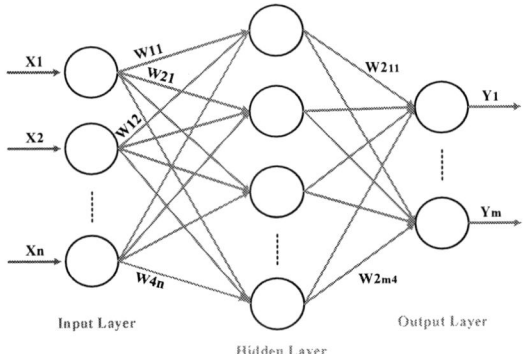

Fig. 4. A typical multi-layer perceptron

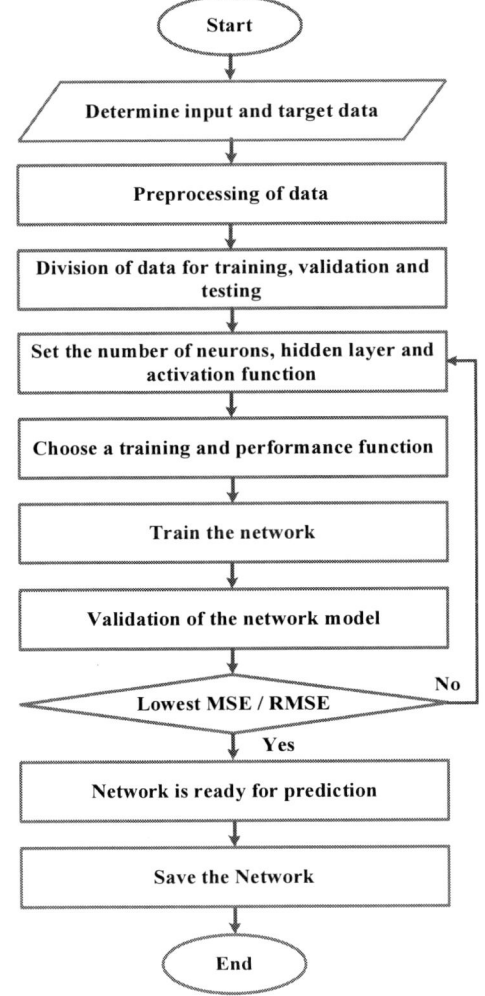

Fig. 5. Design procedure of MLP

IV. CASE STUDY

In this paper, an Experimental setup was used to perform the tests. This setup involves a 1-phase winding of a 3-phase transformer winding, which all defects were tested on a high voltage winding, and the low voltage section was open circuit.

In the design of this winding, wire with a diameter of 7 mm, 160 rounds has been wrapped around a pipe of 160 cm in length. Every 4 round of this wire was considered a disk. eventually a 40-disk winding with a length of 120 cm was prepared to perform the short circuit fault tests. Fig. 6 shows this setup. Surge voltage was applied as input to a high voltage winding and the Oscilloscope device GPS-1102B was utilized for measure and save the voltage and current at the end of the winding. the frequency response calculated for each fault from this voltage and current.

V. SIMULATION RESULTS

Frequency response of the winding has been defined as below [8, 10]:

$$FR = \left| \frac{I_o(f)}{V_i(f)} \right| \qquad (2)$$

$I_o(f)$ and $V_i(f)$ are earth current and input voltage, respectively, both in frequency space. To determine the defect in the winding, it is necessary to compare the frequency response of the healthy state and damaged. The Intact Frequency Response of winding given in Fig. 7. Condition of insulation between two adjacent turns of conductor defines short circuit resistance. To localize short circuit faults in both high and low impedance states using MLP, construction of a short circuit database is necessary. Short circuits in all sections of model winding have been considered by 0 Ω, 1 Ω, 2 Ω, 3 Ω, 4 Ω, 8 Ω, 10 Ω, 15 Ω, 22 Ω, 32 Ω and 47 Ω resistances.

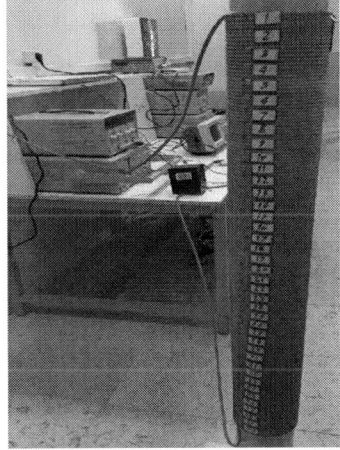

Fig. 6. Winding model for experimental study

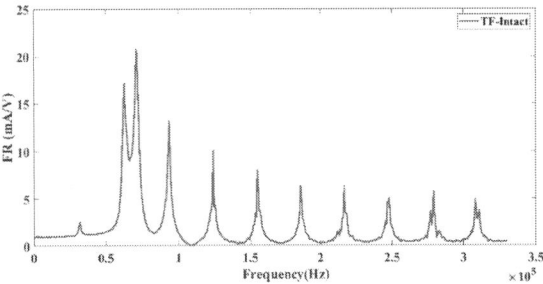

Fig. 7. Intact Frequency Response of winding

How to produce high-impedance and low-impedance short circuit faults are shown in Fig. 8-a and fig. 8-b respectively. Frequency responses for 30th unit of model winding depicted in Fig. 9. Frequency response of low impedance short circuits in all sections of the winding are presented in Fig. 10.

To locate short circuit faults correctly, a fit mapping between fault location and frequency response of winding must be developed. This mapping is constructed via multilayer perceptron (MLP). Simulated frequency responses used as data set for MLP design as input. Indexes should be introduced as the target, which fault locations (regardless of short-circuit impedance) are considered as target data. Data set divided into three parts: training, validation and testing dataset [36]. Dataset is pairs of input and target in vector or scalar format.

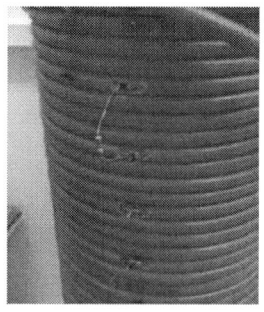 

Fig. 8-a   Fig. 8-b

Fig. 8. Fig. 8-a show Short circuit fault with Resistance (High Impedance Fault) and section 8-b illustrate Short circuit fault without Resistance (Low Impedance Fault)

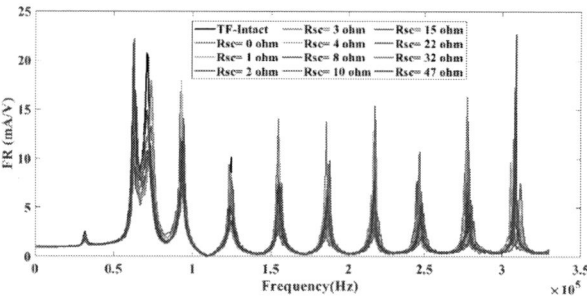

Fig. 9. Effect of short circuit impedance on frequency response (30th section is short circuited)

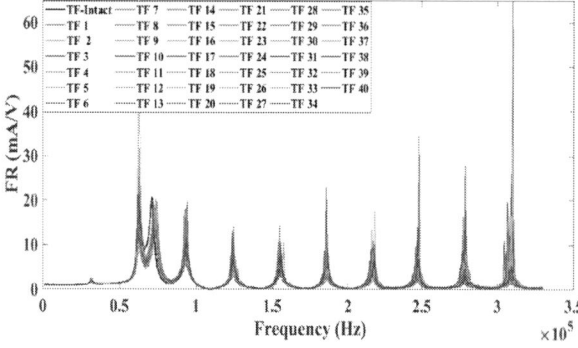

Fig. 10. Frequency response of low impedance short circuits in all sections of the model winding

Training dataset used to adjust the parameters of the network (weights of connections between neurons) via comparing network output and target. Then, the trained MLP is applied on the validation dataset. The validation dataset used to unbiased evaluation of the trained network and tuning number of the neurons in the hidden layer to avoid overfitting [36, 37, 41]. finally, the validated MLP is tested by test dataset to approve generalization characteristic of the network. The network was constructed with 15 hidden layers. 80% of the data was given as training data and 20% for the network test. The results of the training and testing of the network to identify the location of short-circuit faults and fit regression between fault location and frequency response of faults are presented in Fig. 11.

Fig. 11 shows a good overlap and excellent correlation between the target data and the output of the designed MLP. In this Fig., the acceptable correlation between target and output data for train, validation and testing operations is also presented Separately. Fig. 12 illustrate testing error in the forms of mean squared error (MSE), root mean square error (RMSE) and histogram. Fig. 13 depicted Train, validation and testing performance. For test the trained network and determine the exact location of new high-impedance faults that their location is unknown for MLP, Short circuit faults with 12, 15, 18 and 223 ohm resistances created in some sections of the winding and the frequency response for each fault is calculated. This frequency responses were used as input in the trained network. The network was able to detect each new fault using the previous training.

Turn-to-Turn fault by 223 ohms resistance is a high impedance fault and does not have a very damaging effect on the insulation system. High-impedance fault detection is the early detection of the fault in the winding that prevents winding and insulation system from damaging. Results of early detection of Turn-to-Turn fault in transformer windings by MLP are presented in Tables I.

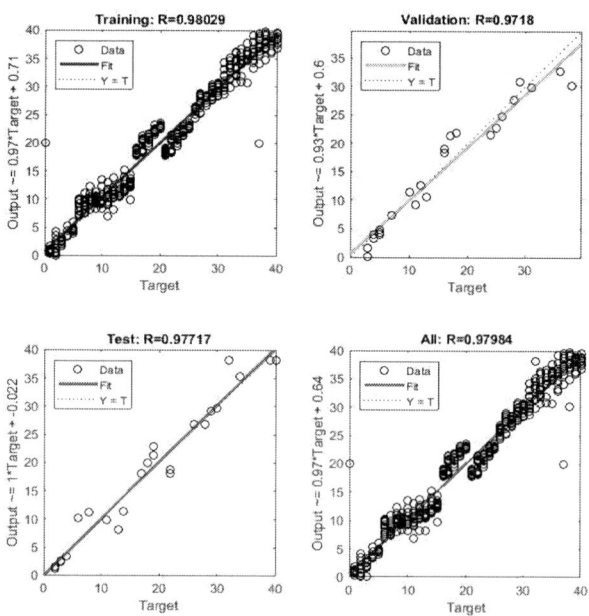

Fig. 11. Correlation between the target data and the output of the designed MLP

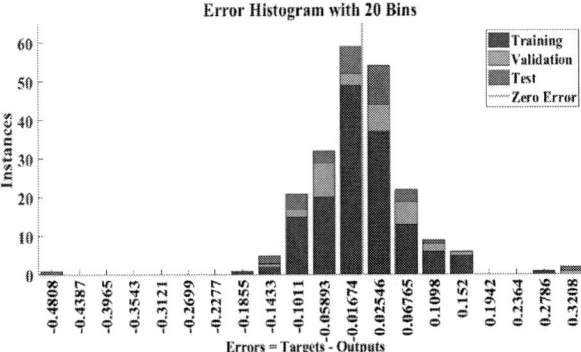

Fig. 12. Error Histogram for Testing Network

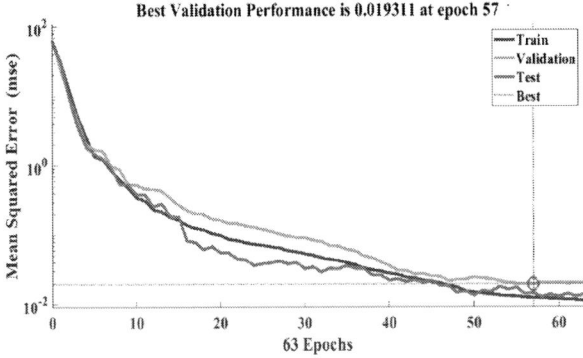

Fig. 13. Train, validation and testing performance

TABLE I.   DETECTION OF TURN-TO-TURN FAULTS BY MLP

Location	R (ohm)	MLP Prediction	Location	R (ohm)	MLP Prediction
1	12	1.0225≈1	19	12	19.0116≈19
	18	1.0095≈1		18	18.9851≈19
	15	0.9987≈1		55	18.8903≈19
	223	0.7972≈1		223	19.1160≈19
3	12	2.9990≈3	21	12	20.9960≈21
	18	2.9907≈3		18	21.0164≈21
	15	3.0007≈3		15	21.0507≈21
	223	2.7828≈3		223	21.3553≈21
6	12	5.9925≈6	26	12	26.0157≈26
	18	6.0342≈6		18	26.1325≈26
	15	6.0025≈6		15	25.9815≈26
	223	6.1801≈6		223	26.1007≈26
9	12	9.1458≈9	30	12	30.0005≈30
	18	9.0238≈9		18	29.9989≈30
	15	9.1487≈9		15	29.9083≈30
	223	9.2008≈9		223	30.2962≈30
12	12	11.9959≈12	34	12	33.9299≈34
	18	11.8925≈12		18	33.9005≈34
	15	12.0248≈12		15	34.1353≈34
	223	12.1985≈12		223	34.2715≈34
14	12	14.0020≈14	39	12	39.0165≈39
	18	13.9758≈14		18	39.0040≈39
	15	14.1056≈14		15	39.0005≈39
	223	14.2758≈14		223	39.1631≈39
18	12	17.9258≈18	40	12	40.0202≈40
	18	17.9699≈18		18	40.0030≈40
	15	17.9905≈18		15	39.9908≈40
	223	17.8940≈18		223	40.1043≈40

VI. CONCLUSION

Turn-to-turn faults of transformer winding may lead to a progressive major fault because of insulation degradation due to extra heat produced by short circuit. Therefore, detection of turn-to-turn faults in their early stages is very useful action. In this regard, this paper used a designed neural network to diagnose winding fault location in its high impedance stage. Here, low-impedance and high-impedance short circuit faults were applied to the model winding in various locations and their frequency responses were calculated. A Turn-to-Turn fault was observed using FRA in the winding. The results reveals precision of the method used to early detection of Turn-to-Turn faults in transformer winding.

REFERENCES

[1] Zh. Liang, A. Parlikad "A Markovian model for power transformer maintenance," *Electrical Power and Energy Systems,* Vol. 99, pp. 175–182, 2018.

[2] R. Murugan, R. Ramasamy, "Failure analysis of power transformer for effective maintenance planning in electric utilities," *Engineering Failure Analysis,* Vol. 55, pp. 182–192, 2015.

[3] M. Koch, M. Krüger, "A New Method for On-Line Monitoring of Bushings and Partial Discharges of Power Transformers," *2012 IEEE International Conference on Condition Monitoring and Diagnosis,* pp. 23-27, Bali, Indonesia, September 2012.

[4] I. Metwally, "Failures, Monitoring and New Trends of Power Transformers," IEEE Potentials · July 2011.

[5] G. Rigatos, P. Siano, "Power transformers' condition monitoring using neural modeling and the local statistical approach to fault diagnosis," Electrical Power and Energy Systems 80 (2016) 150–159.

[6] H. Tarimoradi, G.B. Gharehpetian, "A Novel Calculation Method of Indices to Improve Classification of Transformer Winding Fault Type, Location and Extent," *IEEE Transactions on Industrial Informatics,* Vol. 13, No. 4, pp. 1531-1540, 2017.

[7] E. Rahimpour, M. Jabbari, S. Tenbohlen, "Mathematical Comparison Methods to Assess Transfer Functions of Transformers to Detect Different Types of Mechanical Faults," IEEE TRANSACTIONS ON POWER DELIVERY, VOL. 25, NO. 4, OCTOBER 2010.

[8] K. Pourhossein, G. B. Gharehpetian, E. Rahimpour, N. Araabi, "A Vector-based Approach to Discriminate Radial Deformation and Axial Displacement of Transformer Winding and Determine Defect Extent," Electric Power Components and Systems, 40:597–612, 2012.

[9] E. Rahimpour, J. Christian, K. Feser, H. Mohseni, "Transfer Function Method to Diagnose Axial Displacement and Radial Deformation of Transformer Windings," IEEE TRANSACTIONS ON POWER DELIVERY, VOL. 18, NO. 2, APRIL 2003.

[10] A. R. Abbasi, M. R. Mahmoudi, Z. Avazzadeh, "Diagnosis and clustering of power transformer winding fault types by crosscorrelation and clustering analysis of FRA results," *IET Generation, Transmission & Distribution,* Vol. 12, no. 19, p.p. 4301-4309, 2018.

[11] H. Zhou, K. Hong, H. Huang, J. Zhou, "Transformer winding fault detection by vibration analysis methods," Applied Acoustics, Vol. 114, pp. 136–146, 2016.

[12] V. Behjat, A. Vahedi, "Numerical Modeling of transformers interturn faults and characterizing the faulty transformers behavior under various faults and oprating conditions," IET Electric Power Applications, Vol. 5, No. 5, pp. 415 – 431, June 2011.

[13] M.R Barzegaran, M. Mirzaie, A. Shayegani Akmal, "Investigating Short-circuit in Power Transformer Winding with Quasi-static Finite Element Analysis and Circuit-based Model," *Transmission and Distribution Conference and Exposition, IEEE PES,* 2010, pp. 1-8.

[14] T. Chiulan, B. Pantelimon, "A Practical Example of Power Transformer Unit Winding Condition Assessment by Means of Short-Circuit Impedance Measurement," *IEEE Bucharest Power Tech Conference,* Bucharest, Romania, June 28th-July 2nd 2009.

[15] Sherif S.M. Ghoneim, Ibrahim B.M. Taha, "A new approach of DGA interpretation technique for transformer fault diagnosis," Electrical Power and Energy Systems, Vol. 81, pp. 265–274, 2016.

[16] K. Pourhossein, G.B. Gharehpetian, E. Rahimpour, B.N. Araabi, "A probabilistic feature to determine type and extent of winding mechanical defects in power transformers," Electric Power Systems Research, Vol. 82, pp. 1– 10, 2012.

[17] R. Khalili Senobari, J. Sadeh, H. Borsi, "Frequency response analysis (FRA) of transformers as a tool for fault detection and location: A review," Electric Power Systems Research, Vol. 155, pp. 172–183, 2018.

[18] J.C. Gonzales, E.E. Mombello, "Fault Interpretation Algorithm Using Frequency Response Analysis of Power Transformers," *IEEE Transactions on Power Delivery*, Vol. 31, No. 3, pp. 1034-1042, Jun 2016.

[19] M, Florkowski, J. Furgał, "Detection of transformer winding deformations based on the transfer function—measurements and simulations," MEASUREMENT SCIENCE AND TECHNOLOGY, Vol. 14, pp. 1986–1992, 2003.

[20] K. Pourhossein, G.B. Gharehpetian, E. Rahimpour. "Buckling severity diagnosis in power transformer windings using Euclidean Distance classifier." In *Electrical Engineering (ICEE), 2011 19th Iranian Conference on IEEE*, pp. 1-4, 2012.

[21] K. Pourhossein, G. B. Gharehpetian, E. Rahimpour, "Discrimination of Axial Displacement and Radial Deformation in Power Transformer Windings Using Manhattan Distance Function," In *Proc. 25th International Power System Conference (PSC 2010)*, Nov 2010.

[22] K. Pourhossein, G. B. Gharehpetian, E. Rahimpour, "Axial Displacement Extent Determination in Power Transformer Windings Using an Adjustable Index," In *Proc. 25th International Power System Conference (PSC 2010)*, Nov 2010.

[23] P. Karimifard, G. B. Gharehpetian, S. Tenbohlen, "Localization of winding radial deformation and determination of deformation extent using vector fitting-based estimated transfer function," Euro. Trans. Electr. Power, Vol. 19, pp. 749–762, 2009.

[24] L. Satish, Subrat K. Sahoo, "Locating faults in a transformer winding: An experimental study," Electric Power Systems Research, Vol. 79, pp. 89–97, 2009.

[25] M. Faridi, M. Kharezi, E. Rahimpour, H. R. Mirzaei, A, Akbari, "Localization of Turn-to-Turn Fault in Transformers Using Artificial Neural Networks and Winding Transfer Function," *2010 International Conference on Solid Dielectrics*, Potsdam, Germany, July 4-9, 2010.

[26] H. Firoozi, M. Kharezi, H. Bakhshi, "Turn- to -Turn Fault Localization of Power Transformers Using Neural Network Techniques," *Proceedings of the 9th International Conference on Properties and Applications of Dielectric Materials*, July 19-23, Harbin, China, 2009.

[27] M. Bigdeli, M. Vakilian, E. Rahimpour, "Transformer winding faults classification based on transfer function analysis by support vector machine," IET Electr. Power Appl., Vol. 6, Iss. 5, pp. 268–276, 2012.

[28] A. R. Abbasi , M. R. Mahmoudi, Z. Avazzadeh, "Diagnosis and clustering of power transformer winding fault types by cross correlation and clustering analysis of FRA results," IET Gener. Transm. Distrib., Vol. 12 Iss. 19, pp. 4301-4309, 2018.

[29] M. S. Jahan, R. Keypour, H. R. Izadfar, M. T. Keshavarzi, "Locating power transformer fault based on sweep frequency response measurement by a novel multistage approach," IET Science, Measurement & Technology, Vol. 12, No. 8, pp. 949-957, 2018 May 29.

[30] J. Jiang, L. Zhou, Sh. Gao, W. Li, D. Wang, "Frequency response features of axial displacement winding faults in autotransformers with split windings," IEEE Transactions on Power Delivery, Vol. 33, No. 4, pp.1699-1706, Aug 2018.

[31] V, Behjat, A. Vahedi, A. Setayeshmehr, H. Borsi, E. Gockenbach, "Sweep frequency response analysis for diagnosis of low level short circuit faults on the windings of power transformers: An experimental study," Electrical Power and Energy Systems, Vol. 42, pp. 78–90, 2012.

[32] J. N. Ahour, S. Seyedtabaii , G. B. Gharehpetian, "Determination and localisation of turn-to-turn fault in transformer winding using frequency response analysis," IET Sci. Meas. Technol., Vol. 12 Iss. 3, pp. 291-300, 2018.

[33] A. A. Pandya, B.R. Parekh, "Interpretation of Sweep Frequency Response Analysis (SFRA) traces for the open circuit and short circuit winding fault damages of the power transformer," Electrical Power and Energy Systems, Vol. 62, pp. 890–896, 2014.

[34] M. Rahmatian, B. Vahidi , A.J. Ghanizadeh, G.B. Gharehpetian, H.A. Alehosseini, "Insulation failure detection in transformer winding using cross-correlation technique with ANN and k-NN regression method during impulse test," Electrical Power and Energy Systems, Vol. 53, pp. 209–218, 2013.

[35] S.A. Ryder, "Transformer Diagnosis Using Frequency Response Analysis: Results from Fault Simulations," Power Engineering Society Summer Meeting, 2002 IEEE, Vol. 1, pp. 399-404, Jul 2002.

[36] Simon Haykin, Neural Networks and Learning Machines, Third Edition, McMaster University Hamilton, Ontario, Canada.

[37] Ian H. Witten, E. Frank, M. A. Hall, "Data Mining: Practical machine learning tools and techniques, third edition" Morgan Kaufmann, 2011.

[38] M. W. Gardner, S. R. Dorling, "ARTIFICIAL NEURAL NETWORKS (THE MULTILAYER PERCEPTRON) A REVIEW OF APPLICATIONS IN THE ATMOSPHERIC SCIENCES," Atmospheric Environment Vol. 32, No. 14/15, pp. 2627-2636, 1998.

[39] Tin-Yau Kwok, Dit-Yan Yeung, "Constructive Algorithms for Structure Learning in Feedforward Neural Networks for Regression Problems," IEEE TRANSACTIONS ON NEURAL NETWORKS, VOL. 8, NO. 3, MAY 1997.

[40] Georg Thimm, Emile Fiesler, "High-Order and Multilayer Perceptron Initialization," IEEE TRANSACTIONS ON NEURAL NETWORKS, VOL. 8, NO. 2, MARCH 1997.

[41] Ramón Velo, Paz López, Francisco Maseda, "Wind speed estimation using multilayer perceptron," Energy Conversion and Management, Vol. 81, pp. 1–9, 2014.

Eccentricity Fault diagnosis in PMSM using Motor Current Signature Analysis

Zakaria GHERABI [1], Noureddine BENOUZZA [1], Djilali TOUMI [2], Azeddine BENDIABDELLAH [1]

[1] *Electrical Engineering Faculty, Diagnosis Group, University of Sciences and Technology of Oran (USTO-MB), Algeria*
[2] *Electrical Engineering Department, L2GEGI Laboratory, University of Ibn-Khaldoun of Tiaret, Tiaret, Algeria.*
Email : zakariagherabi@gmail.com, benouza@yahoo.com, toumi_dj@yahoo.fr, bendiazz@yahoo.fr

Abstract - **This work deals with the modeling and the diagnosis of Permanent Magnet Synchronous Motor (PMSM) with consideration of all space harmonics. To do this, a mathematical model based on the approach of the winding function has been developed in the case of a constant air gap, then in the other case where the machine may have an eccentricity fault. To highlight the interest of the model employed and the efficiency of the spectral analysis technique of the stator current, a series of simulations are performed in both cases of operation: healthy and in the presence of the eccentricity fault.**

Keywords – **PMSM; Modified Winding Function (MWF); Mixed Eccentricity; Motor Current Signature Analysis (MCSA).**

I. INTRODUCTION

Synchronous machines and more particularly PMSM have been used more and more in recent decades in different industrial applications. This growing interest is due to several parameters, including its high reliability (due to the absence of ring-brush contacts), its robust rotor construction and its high efficiency making this motor a serious competitor to DC motors and induction motor [1] - [3]. Nevertheless, in the presence of certain electrical, thermal, mechanical or environmental constraints, this motor can undergo certain faults, which mainly affect the stator or the rotor. These faults can be electrical, mechanical or magnetic [4], [5].

Eccentricity as a mechanical fault is a phenomenon that is the subject of numerous publications [6] - [9], because it is responsible for multiple damage on electrical machines [6], [7]. Generally, there are two types of eccentricities: Static Eccentricity (SE) where the rotor is moved from the center of the stator bore but always rotates around its axis. It is generally due either to an imperfection of the stator bore, or to a bad positioning of the rotor or stator during the construction phase. Dynamic Eccentricity (DE) or the rotor is positioned in the center of the stator bore but does not rotate around its axis. This type of fault can be caused by various factors such as bearing wear, mechanical resonance at critical speed and misalignment. Indeed, an inherent level of static eccentricity exists in electrical machines regardless of their manufacturing process. Its existence causes bending of the rotor shaft, wear or misalignment of the bearing, which consequently leads to the appearance of a few degrees of dynamic eccentricity. Therefore, in reality, if eccentricity exists, it can only be a mixed one (ME).

To diagnose this type of fault, many techniques have been proposed in the literature, analyzing phase currents and voltages, magnetic flux, electromagnetic torque and other types of signals [6] - [9]. Monitoring systems based on phase current analysis is the most common. Amplitudes of Side Band Components (ASBC) at well-defined frequencies, extracted from the current spectrum, are used to detect the eccentricity fault in a PMSM [6]. In addition, Ebrahimi et al. in [7] have

proposed a new indicator for the detection of static and dynamic eccentricities in PMSM. It has been defined as the linear combination of the coefficients of the discrete wavelet transform and the Auto-Regressive Model (ARM), extracted from the stator current. The static and dynamic eccentricity faults in a PMSM are detected by additional frequency components in the magnetic flux spectra of the air gap and vibratory signals [8]. In addition, and as presented in [9], the eccentricity fault produces Special Side Band Components (SSBCs) in the torque spectrum as well as special harmonics or in the Spectral Power Density (SPD) of the rotation speed.

In this paper, we begin by presenting the mathematical model of the PMSM without and in the presence of the eccentricity fault. Then we will use a detection method based on the signature analysis of the stator current to highlight the spectral content of the stator current of the machine in two previous cases. Finally, a simulation study is presented followed by a conclusion.

II. MODELING THE HEALTHY PMSM

The modeling approach is based on a semi-analytical method of the PMSM. This modeling method is rather generic in the sense that it relies on a description of the electromagnetic couplings within the machine based on the geometric and constitutive topology of the machine. This approach has already been proven for the modeling of squirrel cage induction machines [11] - [13]. It has also been adapted to the PMSM in [14], [15]. In this section, we will present more precisely the model where the taking into account of permanent magnets (PM) is carried out, to preserve the notion of magnetic coupling with the stator, by the use of the fictitious coils.

A. Modeling of permanent magnets

Magnet modeling involves considering the magnets on the rotor as equivalent electrical circuits, which will allow us to implement the magnetically coupled electrical circuit approach [14], [16]. The modeling method is based on the exploitation of an Ampere model which makes it possible to identify a permanent magnet whose magnetization $\vec{M}$ to a distribution of fictitious currents constituted by:

➢ A surface current density σ defined by:

$$\vec{\sigma} = \vec{M} \wedge \vec{ds} \tag{1}$$

➢ A volume current density ρ defined by

$$\vec{\rho} = rot(\vec{M}) \tag{2}$$

In our case, permanent magnet is characterized by a constant magnetization. Thanks to the equations (1) and (2) of the ampere model, we can deduce that there exists a fictitious surface current density, whereas the volume current density is zero because of the constant magnetization. In conclusion, we can represent the permanent magnet of the rotor by fictitious coils traversed by currents as illustrated in Fig. 1.

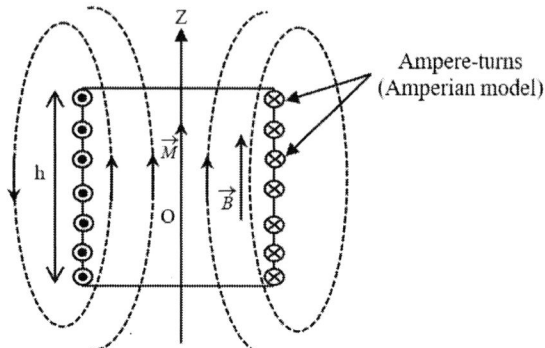

Fig. 1. Illustration of the ampere model on a cylindrical magnet [15].

A. Machine Modeling

The equation system of the PMSM can be represented in the following matrix form [14], [16]:

$$\left[V_{sabc}\right] = [R]\left[i_{sabc}\right] + \frac{d}{dt}\left[\phi_{sabc}\right] = [R]\left[i_{sabc}\right] + \frac{d}{dt}\left[\phi_i\right] + \frac{d}{dt}\left[\phi_m\right] \quad (3)$$

Such that:

$\left[V_{sabc}\right] = \left[V_{sa} \quad V_{sb} \quad V_{sc}\right]^T$: Stator voltages vector;

$\left[i_{sabc}\right] = \left[i_{sa} \quad i_{sb} \quad i_{sc}\right]^T$: Stator currents vector;

$[R] = R_s[I]$: Stator résistance matric;

$\left[\phi_i\right] = \left[\phi_{ia} \quad \phi_{ib} \quad \phi_{ic}\right]^T$: Stator fluxes vector;

$\left[\phi_m\right] = \left[\phi_{ma} \quad \phi_{mb} \quad \phi_{mc}\right]^T$: Mutual fluxes vector between stator and rotor;

$\left[\phi_{sabc}\right] = \left[\phi_{sa} \quad \phi_{sb} \quad \phi_{sc}\right]^T$: Total fluxes vector produced in the machine.

The matrix system (3) can be rewritten as follows:

$$\begin{bmatrix} V_{sa} \\ V_{sb} \\ V_{sc} \end{bmatrix} = \begin{bmatrix} R_s & 0 & 0 \\ 0 & R_s & 0 \\ 0 & 0 & R_s \end{bmatrix}\begin{bmatrix} i_{sa} \\ i_{sb} \\ i_{sc} \end{bmatrix} + \frac{d}{dt}\begin{bmatrix} \phi_{ia} \\ \phi_{ib} \\ \phi_{ic} \end{bmatrix} + \frac{d}{dt}\begin{bmatrix} \phi_{ma} \\ \phi_{mb} \\ \phi_{mc} \end{bmatrix} \quad (4)$$

Where:
$$\begin{bmatrix} \phi_{ia} \\ \phi_{ib} \\ \phi_{ic} \end{bmatrix} = \begin{bmatrix} L_{aa} & M_{ab} & M_{ac} \\ M_{ba} & L_{bb} & M_{bc} \\ M_{ca} & M_{cb} & L_{cc} \end{bmatrix}\begin{bmatrix} i_{sa} \\ i_{sb} \\ i_{sc} \end{bmatrix} \quad (5)$$

The mutual fluxes between the stator and the rotor of a machine consisting of 4 poles (corresponding to 4 fictitious coils) are given by the following matrix system:

$$\begin{bmatrix} \phi_{ma} \\ \phi_{mb} \\ \phi_{mc} \end{bmatrix} = \begin{bmatrix} M_{saf1} & M_{saf2} & M_{saf3} & M_{saf4} \\ M_{sbf1} & M_{sbf2} & M_{sbf3} & M_{sbf4} \\ M_{scf1} & M_{scf2} & M_{scf3} & M_{scf4} \end{bmatrix}\begin{bmatrix} i_{f1} \\ i_{f2} \\ i_{f3} \\ i_{f4} \end{bmatrix} \quad (6)$$

The modeling of the rotating part of the machine is based on the use of Newton's second law.

We will have:
$$J\frac{d}{dt}\Omega + F\Omega = T_e - T_r \quad (7)$$

By carrying out an energy balance, it is shown that the electromagnetic torque is equal to the partial derivative of the magnetic Co-energy with respect to the position of the rotor and consequently the torque is expressed by the following equation [14], [16]:

$$T_e = \frac{\partial W_{comag}}{\partial \theta_r} = \left[i_{sabc}\right]\frac{\partial\left[M_{ss}\right]}{\partial \theta_r}\left[i_{sabc}\right] + \left[i_{sabc}\right]\frac{\partial\left[M_f\right]}{\partial \theta_r}\left[i_f\right] \quad (8)$$

III. CALCULATION OF INDUCTANCES

It is very important to know that the accuracy of inductance calculation is the key point for a successful PMSM simulation. All inductances are calculated by using the winding function approach [11] - [13], which is based on the following formula:

$$L_{A.B} = \mu_o rl \int_o^{2\pi} N_A(\varphi, \theta_r) n_B(\varphi, \theta_r) g^{-1}(\varphi, \theta_r) d\varphi \quad (9)$$

Where, μ_o is the air permeability, r is the mean radius of the gap, l is the active length of the machine, $N_A(\varphi, \theta_r)$ is the winding function of the winding " a ", $n_B(\varphi, \theta_r)$ is the distribution function of the winding " b " and $g^{-1}(\varphi, \theta_r)$ is the inverse of the gap function.

➤ Magnetisation inductance of phase " a "

The magnetization inductance of the phase " a " of the stator is calculated thanks to the following relation:

$$L_{saa} = \mu_o rl \int_o^{2\pi} N_a(\varphi, \theta_r) n_a(\varphi, \theta_r) g^{-1}(\varphi, \theta_r) d\varphi \quad (10)$$

The use of the equation (10) for the parameters of the studied machine gives after all calculated fact, the magnetization inductance L_{saa} in the following form:

$$L_{saa} = \frac{\mu_o r.l.N_s^2}{g_o} 220.\alpha_s, \quad and \quad L_{saa} = L_{sbb} = L_{scc} \quad (11)$$

➤ Mutual inductances between stator phases

To calculate these inductances, it is sufficient to base on the following integral:

$$M_{sab} = \frac{\mu_o rl}{g_o} \int_o^{2\pi} N_a(\varphi, \theta_r) n_b(\varphi, \theta_r) d\varphi \quad (12)$$

The use of the equation (12) using the parameters of the machine studied gives us the following results:

$$M_{sab} = -\frac{\mu_o r.l.N_s^2}{g_o} 104.\alpha_s, \quad and \quad M_{sac} = M_{sbc} = M_{sab} \quad (13)$$

➤ Mutual inductances between the stator and the rotor

The mutual inductance between phase " a " and the fictitious coil " f_1 " is obtained using the following integral:

978-1-5386-7688-2/19 $31.00 © 2019 IEEE

$$M_{af1} = \frac{\mu_o r l}{g_o} \int_0^{2\pi} N_a(\varphi,\theta_r) n_{f1}(\varphi,\theta_r) d\varphi \qquad (14)$$

To calculate this integral, we vary the position of the fictitious coil "f_1" with respect to the stator, and for each interval the value of the analytically integral is calculated. The analytical integration gives us the results recorded in TABLE I. For each given interval, we have a value of the mutual inductance function of (θ_r).

TABLE I. MUTUAL INDUCTANCES BETWEEN THE FIRST STATORIC PHASE AND THE FIRST FICTIONAL COIL

Inductance M_{saf1} (H)	(θ_r) (rad)
$\frac{\mu_o.r.l.N_s.N_r}{g_o}(5.\theta_r + 3.\alpha_r - 12.\alpha_s)$	$0 \le \theta_r \le \alpha_s$
$\frac{\mu_o.r.l.N_s.N_r}{g_o}(4.\theta_r + 3.\alpha_r - 11.\alpha_s)$	$\alpha_s \le \theta_r \le 2\alpha_s$
$\frac{\mu_o.r.l.N_s.N_r}{g_o}(2.\theta_r + 3.\alpha_r - 7.\alpha_s)$	$2\alpha_s \le \theta_r \le 9\alpha_s - \alpha_r$
$\frac{\mu_o.r.l.N_s.N_r}{g_o}(\theta_r + 2.\alpha_r + 2.\alpha_s)$	$9\alpha_s - \alpha_r \le \theta_r \le 3\alpha_s$
$\frac{\mu_o.r.l.N_s.N_r}{g_o}(2.\alpha_r + 5.\alpha_s)$	$3\alpha_s \le \theta_r \le 10\alpha_s - \alpha_r$
$\frac{\mu_o.r.l.N_s.N_r}{g_o}(-\theta_r + \alpha_r + 15.\alpha_s)$	$10\alpha_s - \alpha_r \le \theta_r \le 4\alpha_s$
$\frac{\mu_o.r.l.N_s.N_r}{g_o}(-2.\theta_r + \alpha_r + 19.\alpha_s)$	$4\alpha_s \le \theta_r \le 11\alpha_s - \alpha_r$
$\frac{\mu_o.r.l.N_s.N_r}{g_o}(-4.\theta_r - \alpha_r + 41.\alpha_s)$	$11\alpha_s - \alpha_r \le \theta_r \le 12\alpha_s - \alpha_r$
$\frac{\mu_o.r.l.N_s.N_r}{g_o}(-5.\theta_r - 2.\alpha_r + 53.\alpha_s)$	$12\alpha_s - \alpha_r \le \theta_r \le 13\alpha_s - \alpha_r$
$\frac{\mu_o.r.l.N_s.N_r}{g_o}(-6.\theta_r - 3.\alpha_r + 66.\alpha_s)$	$13\alpha_s - \alpha_r \le \theta_r \le 9\alpha_s$

Where, $\alpha_s = \pi/18$ and "α_r" represent the opening of magnet.

The mutual inductances between phase "a" and the other fictitious coils "f_2, f_3, f_4" are calculated in the same way as that of "M_{af1}" but shifted respectively by an angle "$k.\pi/2$", with $k = 1,2,3$.

For the mutual inductances between the other stator phases "b, c" and the first fictitious coil "f_1", the same expressions are obtained but shifted respectively by $-2\pi/3$ and $-4\pi/3$.

A. Calculation of inductances in the case of ME

The calculation of the inductances is done in the same way as in the one of a uniform air gap, with the exception of the inverse function of the gap, which is replaced by the following equation [6], [7], [13]:

$$g^{-1}(\varphi,\theta_r) = \frac{1}{g_0}(1 + \delta_s \cos(\varphi) + \delta_d \cos(\varphi - \theta_r)) \qquad (15)$$

Where g_o is the average value of the air gap length, φ is the position of the stator, θ_r is the position of the rotor, δs is the degree of static eccentricity, δ_d is the degree of dynamic eccentricity.

➢ Magnetization inductance of phase "a"

Replacing the expression of the inverse function of the gap given by equation (15) in equation (10) gives us the magnetization inductance of phase "a" as follows:

$$L_{saa} = \frac{\mu_o.r.l.N_s^2}{g_o} \left(\begin{array}{l} -2.(\delta_s(\sin(\alpha_s) - \sin(0)) + \delta_d(\sin(\alpha_s - \theta_r) - \sin(0 - \theta_r))) + 220.\alpha_s \\ -2.(\delta_s(\sin(2.\alpha_s) - \sin(\alpha_s)) + \delta_d(\sin(2.\alpha_s - \theta_r) - \sin(\alpha_s - \theta_r))) \\ +4.(\delta_s(\sin(3.\alpha_s) - \sin(2.\alpha_s)) + \delta_d(\sin(3.\alpha_s - \theta_r) - \sin(2.\alpha_s - \theta_r))) \\ +10.(\delta_s(\sin(4.\alpha_s) - \sin(3.\alpha_s)) + \delta_d(\sin(4.\alpha_s - \theta_r) - \sin(3.\alpha_s - \theta_r))) \\ +18.(\delta_s(\sin(9.\alpha_s) - \sin(4.\alpha_s)) + \delta_d(\sin(9.\alpha_s - \theta_r) - \sin(4.\alpha_s - \theta_r))) \\ +10.(\delta_s(\sin(10.\alpha_s) - \sin(9.\alpha_s)) + \delta_d(\sin(10.\alpha_s - \theta_r) - \sin(9.\alpha_s - \theta_r))) \\ +4.(\delta_s(\sin(11.\alpha_s) - \sin(10.\alpha_s)) + \delta_d(\sin(11.\alpha_s - \theta_r) - \sin(10.\alpha_s - \theta_r))) \\ -2.(\delta_s(\sin(12.\alpha_s) - \sin(11.\alpha_s)) + \delta_d(\sin(12.\alpha_s - \theta_r) - \sin(11.\alpha_s - \theta_r))) \\ -2.(\delta_s(\sin(13.\alpha_s) - \sin(12.\alpha_s)) + \delta_d(\sin(13.\alpha_s - \theta_r) - \sin(12.\alpha_s - \theta_r))) \end{array} \right) \qquad (16)$$

➢ Mutual inductance between stator phases

The mutual inductance between the winding of phase "a" and phase "b" is given by equation (17).

$$L_{sab} = \frac{\mu_o.r.l.N_s^2}{g_o} \left(\begin{array}{l} -2.(\delta_s(\sin(\alpha_s) - \sin(0)) + \delta_d(\sin(\alpha_s - \theta_r) - \sin(0 - \theta_r))) - 104.\alpha_s \\ -6.(\delta_s(\sin(2.\alpha_s) - \sin(\alpha_s)) + \delta_d(\sin(2.\alpha_s - \theta_r) - \sin(\alpha_s - \theta_r))) \\ -12.(\delta_s(\sin(3.\alpha_s) - \sin(2.\alpha_s)) + \delta_d(\sin(3.\alpha_s - \theta_r) - \sin(2.\alpha_s - \theta_r))) \\ -15.(\delta_s(\sin(4.\alpha_s) - \sin(3.\alpha_s)) + \delta_d(\sin(4.\alpha_s - \theta_r) - \sin(3.\alpha_s - \theta_r))) \\ -18.(\delta_s(\sin(6.\alpha_s) - \sin(4.\alpha_s)) + \delta_d(\sin(6.\alpha_s - \theta_r) - \sin(4.\alpha_s - \theta_r))) \\ -12.(\delta_s(\sin(7.\alpha_s) - \sin(6.\alpha_s)) + \delta_d(\sin(7.\alpha_s - \theta_r) - \sin(6.\alpha_s - \theta_r))) \\ -6.(\delta_s(\sin(8.\alpha_s) - \sin(7.\alpha_s)) + \delta_d(\sin(8.\alpha_s - \theta_r) - \sin(7.\alpha_s - \theta_r))) \\ +6.(\delta_s(\sin(9.\alpha_s) - \sin(8.\alpha_s)) + \delta_d(\sin(9.\alpha_s - \theta_r) - \sin(8.\alpha_s - \theta_r))) \\ +10.(\delta_s(\sin(10.\alpha_s) - \sin(9.\alpha_s)) + \delta_d(\sin(10.\alpha_s - \theta_r) - \sin(9.\alpha_s - \theta_r))) \\ +12.(\delta_s(\sin(11.\alpha_s) - \sin(10.\alpha_s)) + \delta_d(\sin(11.\alpha_s - \theta_r) - \sin(10.\alpha_s - \theta_r))) \\ +6.(\delta_s(\sin(12.\alpha_s) - \sin(11.\alpha_s)) + \delta_d(\sin(12.\alpha_s - \theta_r) - \sin(11.\alpha_s - \theta_r))) \\ +3.(\delta_s(\sin(13.\alpha_s) - \sin(12.\alpha_s)) + \delta_d(\sin(13.\alpha_s - \theta_r) - \sin(12.\alpha_s - \theta_r))) \end{array} \right) \qquad (17)$$

➢ Mutual inductance between phase "a" and fictitious coils

The mutual inductance between phase "a" and the fictitious coil "f_1" for the first interval from 0 to α_s is given by:

$$M_{saf1} = \frac{\mu_o \cdot r.l.N_s.N_r}{g_o} \begin{pmatrix} -2.(\delta_s (\sin(\alpha_s) - \sin(\theta_r)) + \delta_d (\sin(\alpha_s - \theta_r) - \sin(0))) + 5.\theta_r - 12.\alpha_s \\ -1.(\delta_s (\sin(2.\alpha_s) - \sin(\alpha_s)) + \delta_d (\sin(2.\alpha_s - \theta_r) - \sin(\alpha_s - \theta_r))) + 3.\alpha_r \\ +1.(\delta_s (\sin(3.\alpha_s) - \sin(2.\alpha_s)) + \delta_d (\sin(3.\alpha_s - \theta_r) - \sin(2.\alpha_s - \theta_r))) \\ +2.(\delta_s (\sin(4.\alpha_s) - \sin(3.\alpha_s)) + \delta_d (\sin(4.\alpha_s - \theta_r) - \sin(3.\alpha_s - \theta_r))) \\ +3.(\delta_s (\sin(\theta_r + \alpha_r) - \sin(4.\alpha_s)) + \delta_d (\sin(\alpha_r) - \sin(4.\alpha_s - \theta_r))) \end{pmatrix} \qquad (18)$$

Proceeding in the same way we can obtain expressions for different intervals as well as for other inductances.

IV. SPECTRAL CONTENT OF STATOR CURRENT WITH AND WITHOUT FAULT

Because of the wealth of information contained in the stator current spectrum, several studies [4] - [7], [10] have shown that the analysis of this spectrum can detect and even identify all types of fault affecting the PMSM. This identification is made possible by a simple monitoring of the position and the frequency amplitude of certain harmonics.

A. Healthy case

In PMSM, the stator current contains in addition to the fundamental a series of space harmonics [3] - [5]. The frequencies of these harmonics are given by the following expression:

$$f_{sh} = [2.k + 1] f_s \qquad (19)$$

B. Case with mixed eccentricity faults

Several studies [6] - [8] have shown that the mixed eccentricity defect is manifested by the creation of a series of harmonics in the stator current spectrum, at the frequencies given by the following equation:

$$f_{ex} = \left[1 \pm \frac{k}{P} \right] f_s \qquad (20)$$

With: f_s the supply frequency, P the number of pole pairs and k a positive integer.

V. SIMULATION RESULTS

The simulation is performed on a 3.6 kW PMSM, 4 poles, 36 stator slots and 4 fictitious rotor coils. The arrangement of the stator and rotor winding of this motor and its parameters is indicated in the appendix. The mathematical model obtained and all the expressions of the inductances are implemented under MATLAB environment. In all the simulations considered in this work, the motor is fed by a perfectly symmetrical and sinusoidal three-phase voltage supply system and we started the no-load simulation and we applied from the instant 0.5 s a load torque equal to that of the nominal torque, which is of the order of 17.5 Nm.

A. Healthy Motor Operation Mode

Fig. 2. (a), illustrates the speed of rotation of the PMSM. The speed response has ripples of ± 10 rpm around the synchronism speed (2000 rpm). These ripples are due to the effect of space harmonics. Fig. 2. (b), depicts the waveforms of the stator current absorbed by the simulated motor. This current is not sinusoidal; it presents fluctuations, which are due to the effect of space harmonics. Fig. 2. (c), gives the spectrum of the stator current of phase " a ". We note that the spectrum contains in addition to the fundamental a series of harmonics which represent the space harmonics whose frequencies of

these harmonics can easily be verified using the expression of equation (19). This spectrum will be considered as the reference spectrum.

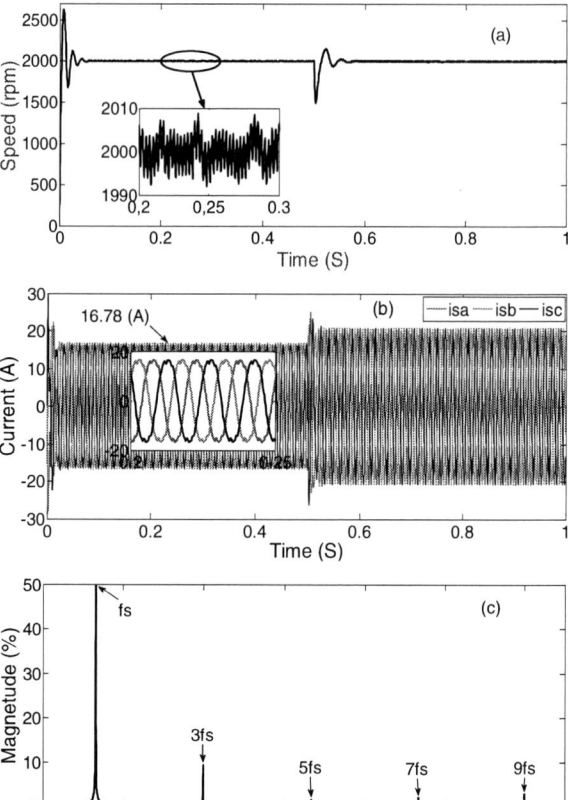

Fig. 2. Healthy operation mode. (a) - Speed of rotation, (b) - Currents of the three stator phases, (c) - Current spectrum of the phase " a ".

B. Motor Operation With Eccentricity Fault

Simulation of the operation of the machine with eccentricity fault is realized with simultaneous consideration of the static and dynamic eccentricity. To do this we simultaneously vary the degree of static eccentricity ($\delta_s \neq 0$) and the degree of dynamic eccentricity ($\delta_d \neq 0$).

Fig. 3. (a), shows the magnetization inductance of the phase " a " as a function of the rotor position (θ_r) for an eccentricity of 20 % ($\delta_s = 20\%$ and $\delta_d = 20\%$). The mutual inductance between the stator phases is also calculated and shown in Fig. 3. (b). We notice that these inductances become constant when the degree of eccentricity is zero. Figs. 4. (a) and 4. (b), respectively show the variations of the mutual inductances between the phase " a " and the fictitious coil " f_1 " and their derivative as a function of the rotor position for a mixed

978-1-5386-7688-2/19 $31.00 © 2019 IEEE

eccentricity of 20% and 40%. From these figures, it is possible to notice that the deformation undergone by the shape of these inductances evolves according to the degree of eccentricity. Fig. 5. (a), shows the increase of rpm fluctuations from ± 10 rpm for a zero eccentricity level up to ± 50 rpm for an eccentricity level of 20%. Fig. 5. (b), Illustrate the stator current for an eccentricity defect of 20%. Figs. 5. (c) and 5. (d), present the spectrum of the stator current for an eccentricity of 20% and 40%. TABLE II summarizes the variation of the amplitudes of the most significant harmonics as a function of the degree of eccentricity.

From the results obtained, we find that the increase of the fault severity is manifested by the increase of fluctuations in the temporal form of different electromechanical quantities, and by the increase of the amplitudes of the harmonics characteristic of the fault.

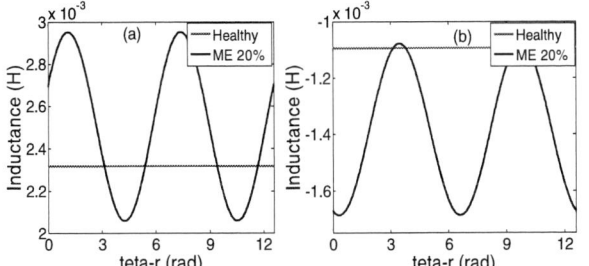

Fig. 3. (a) - Magnetization inductance of phase " a ", (b) - Mutual inductance between stator phases, for ME of 20%.

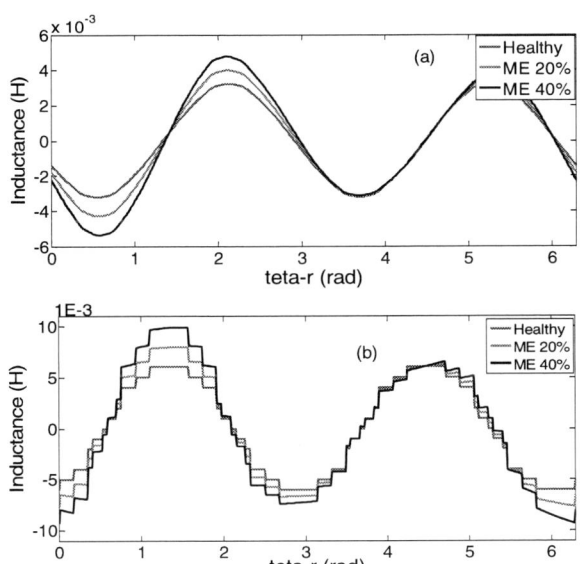

Fig. 4. (a) - Mutual inductance between phase " a " and the fictitious coil " f_1 ", for ME of 20% and 40% (b) - Their derivative.

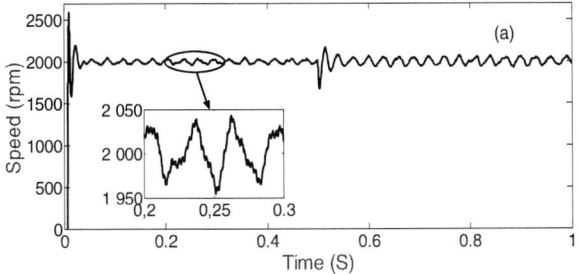

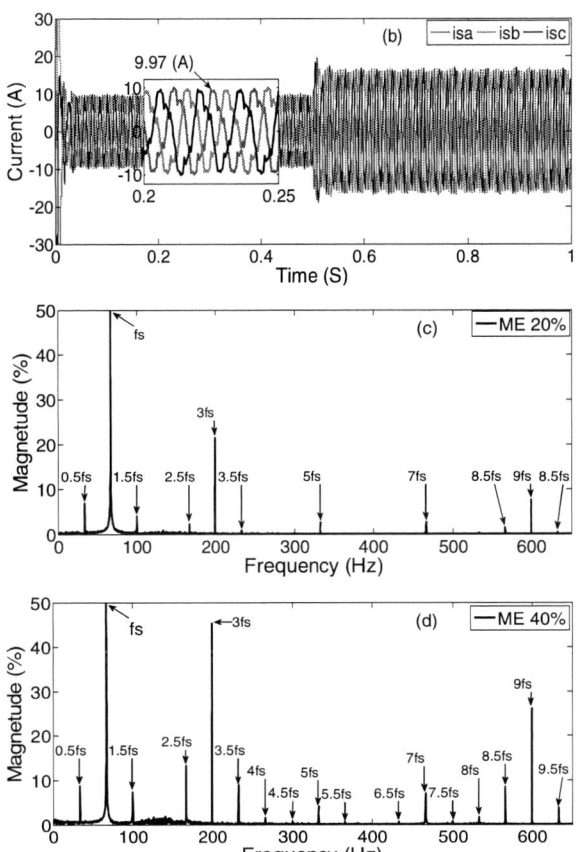

Fig. 5. Operation with ME fault (a) – Speed of rotation, (b) - Currents of three stator phases, (c) and (d) Current spectra of the phase "a".

TABLE II. AMPLITUDES OF FREQUENCE SPECTRUM COMPONENTS.

Eccentricity Severity (%)	0	10	20	40
Frequency (Hz)	Amplitude (%)	Amplitude (%)	Amplitude (%)	Amplitude (%)
$(1-1/P)f_s$	---	3.20	6.82	8.69
$(1+1/P)f_s$	---	1.64	3.93	7.28
$(1+3/P)f_s$	---	0.84	2.29	13.23
$(1+5/P)f_s$	---	0.16	0.78	8.88
$(1+7/P)f_s$	---	0.1	0.14	0.67
$(1+9/P)f_s$	---	0.09	0.14	0.52
$(1+11/P)f_s$	---	0.07	0.16	0.64
$(1+13/P)f_s$	---	0.14	0.17	0.61
$(1+15/P)f_s$	---	0.54	1.43	8.47
$(1+17/P)f_s$	---	0.18	0.39	3.64

VI. CONCLUSION

The work developed in this paper concerns the detection of eccentricity in PMSM. The proposed model based on the approach of the modified winding function of which we considered the permanent magnet as a fictitious coil traversed by an excitation current " i_f ". The simulation results showed that the lack of eccentricity is manifested by the creation of harmonics at fixed frequencies (constant parameters such as: the number of pairs of poles and the frequency of supply). Moreover, the severity of the eccentricity fault is manifested by the increase of the amplitude of the characteristic harmonics of the fault. In general, the simulation results show the interest of the model adopted and the efficiency of the spectral analysis technique of the stator current for the detection of the eccentricity fault.

APPENDIX

TABLE : PMSM Parameters.

Symbol	Description	Values	Units
V_n	Rated voltage	150	V
f_s	Rated frequency	66,7	Hz
I_n	Rated current	15.1	A
C_n	Rated torque	17.5	Nm
Ω_n	Rated speed	2000	rpm
R_s	Stator resistance	0.295	Ω
L_s	Synchronous inductance	3.5	mH
J	Moment of the inertia	3.10^{-4}	$Kg.m^2$
F	Viscous rubbing	0.017	$Nm/rad/s$
P	Number of pole pairs	2	-----
ε_s	Stator notch opening	0,06	rad
g_o	Nominal air gap	2	mm
r	Stator radius	64	mm
rr	Rotor radius	52	mm
l	Length of the machine	250	mm
N_s	Number of turns per Coil	6	-----
α_r	Opening of magnets	1,15	rad
h	Thickness of magnets	10	mm

Fig. 6.a, represents the $n_a(\varphi)$ distribution function of phase $''\,a\,''$ with a mean value $\langle n_a(\varphi) \rangle = 3Ns$. This value allows us to find the winding function of phase $''\,a\,''$ as shown in Fig. 6.b. Fig. 6.c, represents the distribution function of the fictitious coil $''f_1\,''$.

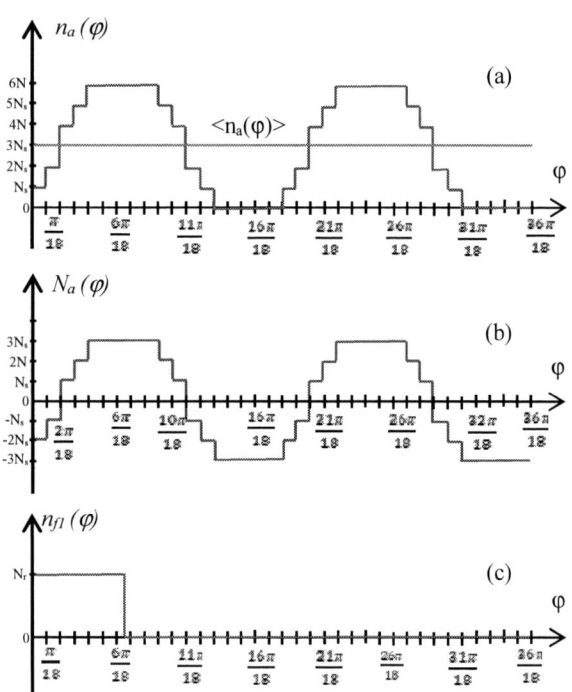

Fig. 6. (a) - Distribution function of phase $''\,a\,''$. (b) - Its winding function.
(c) - Fictitious coil distribution function $''f_1''$.

REFERENCES

[1] J. Faiz et M. R. Hassan-Zadeh, « Impacts of eccentricity fault on permanent magnet generators for distributed generation », in *International Conference on Optimization of Electrical and Electronic Equipment (OPTIM) Intl Aegean Conference on Electrical Machines and Power Electronics (ACEMP)*,p.434–441, 2017.

[2] L. M'hamed, G. Zakaria, et D. Khireddine, « A Robust Sensorless Control of PMSM Based on Sliding Mode Observer and Model Reference Adaptive System », *Int. J. Power Electron. Drive Syst. IJPEDS*, vol. 8, no 3, p. 1016–1025, 2017.

[3] H. Zhou, G. Liu, et W. Zhao, « Dynamic Performance Improvement of Five-Phase Permanent-Magnet Motor With Short-Circuit Fault », *IEEE Trans. Ind. Electron*, vol. 65, no 1, p. 145–155, 2018.

[4] J. Zheng, Z. Wang, D. Wang, et Y. Li, « Review of fault diagnosis of PMSM drive system in electric vehicles », *in Proceedings of the 36th Chinese Control Conference, China*, p. 7426–7432, July 26-28, 2017.

[5] J. Faiz et S. A. H. Exiri, « Short-circuit fault diagnosis in permanent magnet synchronous motors- an overview », *in Intl Aegean Conference on Electrical Machines Power Electronics (ACEMP), 2015 Intl Conference on Optimization of Electrical Electronic Equipment (OPTIM) Intl Symposium on Advanced Electromechanical Motion Systems (ELECTROMOTION)*, p. 18–27, 2015.

[6] B. M. Ebrahimi, J. Faiz, et M. R, « Static-, dynamic-, and mixed-eccentricity fault diagnoses in permanent-magnet synchronous motors », *IEEE Trans. Ind. Electron*, vol.56,no11,p.4727–4739, 2009.

[7] B. M. Ebrahimi, M. J. Roshtkhari, J. Faiz, et S. V. Khatami, « Advanced eccentricity fault recognition in permanent magnet synchronous motors using stator current signature analysis », *IEEE Trans. Ind. Electron.*, vol. 61, no 4, p. 2041–2052, 2014.

[8] B. M. Ebrahimi et J. Faiz, « Magnetic field and vibration monitoring in permanent magnet synchronous motors under eccentricity fault », *IET Electr. Power Appl.*, vol. 6, no 1, p. 35–45, 2012.

[9] W. S. Abu-Elhaija, J. Faiz, et B. M. Ebrahimi, « Analytical prediction of instantaneous torque and speed for induction motors with mixed-eccentricity fault using magnetic-field equations », *Electromagnetics,* vol. 30, no 6, p. 525–540, 2010.

[10] W. Liang et W. Fei, « An improved sideband current harmonic model of interior PMSM drive by considering magnetic saturation and cross-coupling effects », *IEEE Trans. Ind. Electron.*, vol. 63, no7,p.4097–4104, 2016.

[11] S. Nandi, H. A. Toliyat, et A. G. Parlos, « Performance analysis of a single phase induction motor under eccentric conditions », *in Thirty-Second Industry Applications Conference IAS,* vol 1,p.174–181, 1997.

[12] G. M. Joksimovic, M. D. Durovic, J. Penman, et N. Arthur, « Dynamic simulation of dynamic eccentricity in induction machines-winding function approach », *IEEE Trans. Energy Convers,* vol. 15, no 2, p. 143–148, juin 2000.

[13] S. Nandi, R. M. Bharadwaj, et H. A. Toliyat, « Mixed eccentricity in three phase induction machines: analysis, simulation and experiments », *in 37th IAS Annual Meeting. Conference Record of theIndustry Applications Conference*, vol. 3, p. 1525–1532, 2002.

[14] A. A. Abdallah, J. Regnier, J. Faucher, et B. Dagues, « Simulation of internal faults in permanent magnet synchronous machines », *in Power International Conference on Electronics and Drives Systems PEDS*, vol. 2, p. 1390–1395, 2005.

[15] Z. Liang, D. Liang, et S. Jia, « Inductance Calculation for the Symmetrical Non-Salient Dual Three-Phase PMSM Based on Winding Function Approach », *in 21st International Conference on Electrical Machines and Systems (ICEMS)*, p. 269–274, 2018.

[16] B. Aubert, « Détection des défaut de courts-circuits entre-spires dans les Générateurs Synchrones à Aimants Permanents: Méthodes basées modèles et filtre de Kalman étendu », *Thèse de doctorat,* université de toulouse, mars–2014.

978-1-5386-7688-2/19 $31.00 © 2019 IEEE

Highly efficient Multi-Junction Solar Cells Performance Improvement for AC Induction Motor Control Using the dsPIC30F Microcontroller

Abdelkader HADJ DIDA*[1,2]

[1]Department of Electrical Engineering, Laboratory of
Applied Power Electronics LAPE
University of Science and Technology of Oran Mohamed
Boudiaf USTO-MB, BP 1505, El M'naouer
31000 Oran, Algeria
E-mail: abdelkader.hadjdida@univ-usto.dz

[2]Department of Research and Space Instrumentation,
Satellites Development Centre, Algerian Space Agency
Po Box 4065, Ibn Rochd USTO, Bir El Djir,
31130 Oran, Algeria
E-mail: ahadjdida@cds.asal.dz

Mohamed BOURAHLA**,[1]

[1]Department of Electrical Engineering, Laboratory of
Applied Power Electronics LAPE
University of Science and Technology of Oran Mohamed
Boudiaf USTO-MB, BP 1505, El M'naouer
31000 Oran, Algeria
E-mail: Bourah3@yahoo.fr

H. Bülent ERTAN***,[3]

[3]Electrical Engineering Department, Middle East Technical
University METU, Ankara, Turkey
E-mail: ertan@metu.edu.tr

Abstract— Solar energy is an abundant renewable source which is expected to play an increasing role in terrestrial and space systems future infrastructure for distributed power generation. Improving of solar performance and efficiency is a key goal of research and the prominent factor in photovoltaic systems which make PV technologies cost-competitive with conventional sources of energy. Multi-junction solar cells are the most efficient technology for generation of electricity from solar irradiation. It's estimated that these cells have an incredible efficiency for more than 40%. They have the advantage of improved performance. Highly efficient multi-junction solar cells find their use in many applications. In regard to terrestrial industrial applications, AC electrics motors are being fed from the multi-junction photovoltaic systems when exposed to high light concentrations using a boost converter and three phase inverter for variable speed AC induction motor drives based on full digital control. Another application for multi-junction solar cells is the use in space vehicles and satellites under low light concentrations. The aim of this research work was modeled and simulated the multi-junction solar cells to improve their performance in terms of efficiency for further exploration in future photovoltaic power generation systems and implemented a variable speed control of an AC Induction Motor using the dsPIC30F Microcontroller for high performance industrial applications.

Keywords— Solar energy, Performance improvement, Generation of electricity, Multi-junction Solar cells, Three phase inverter; variable speed control, Induction motor.

I. INTRODUCTION

AC drives full digital control based on solar energy have reached the status of a maturing technology in a broad range of industrial applications compared to already existing power

generation technology. It's the efficiency of solar cells which growing this technology speedily. Low cost, high reliability, high robustness, high efficiency and direct connection to an ac power source have made the induction motors fed by multi-junction solar cell generators, the most attractive applications in industry and drawn the interest of many researchers in the field of renewable energy. The three phase representation of this machine provides an option for variable-speed applications. Multi-junction cells are used to power efficiently a boost converter to feed an induction motor through a three phase solar inverter. In this research work, we had modeled and simulated the multi-junction solar cells MJSC to improve their performance in terms of efficiency for further exploration in future photovoltaic power generation systems to implement a variable speed control of an AC Induction Motor using the dsPIC30F Microcontroller for industrial applications [1].

II. MODELLING AND SIMULATION OF MULTI-JUNCTION SOLAR CELLS

Electrical behaviour is one of the most important aspects that characterize a solar cell device and it's important to know what the fundamental equations are and how they are linked to the physics parameters to easily make simulations which permit the extraction of some important I-V and P-V curves.

A. Equivalent Electrical circuit for Multi-junction Solar Cells

Structure of these cells is demonstrated in fig. 1 [2]. MJSC are capable to convert a huge amount of sun's energy into electrical energy at much greater efficiency and hence differ from the traditional solar cells. The model for its equivalent electrical circuit is presented in fig. 2 [3].

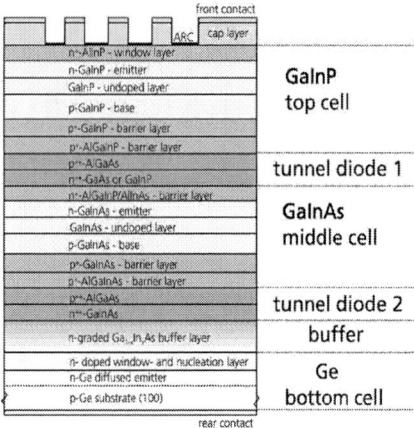

Fig. 1. The structure of MJSC

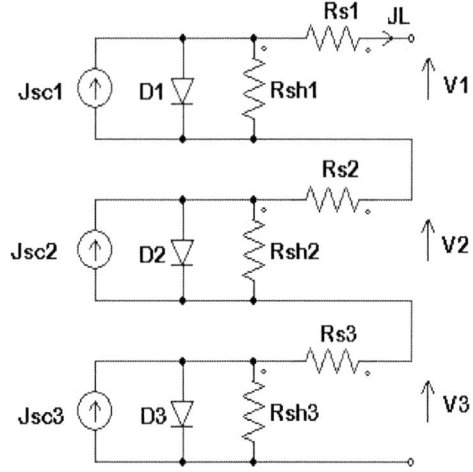

Fig. 2. Equivalent electrical circuit of MJSC

From equivalent electrical circuit, the total current density J_L is as follows:

$$J_L = J_{sc,i} - J_{o,i} \left(\frac{q\,(V_i + AR_{s,i}\,J_L)}{n_i\,K_B\,T} - 1 \right) - \frac{V_i + AR_{s,i}\,J_L}{AR_{sh,i}} \quad (1)$$

It's assumed that temperature of the cell is uniform. The diode's dark saturation current is calculated as follows [4]:

$$J_{o,i} = k_i \times T^{\frac{3+\gamma_i}{2}} \times e^{\frac{-q \times E_{gi}}{n_i \times K_B \times T}} \quad (2)$$

Where: γ_i is a constant and E_g is band gap energy in eV.

The relation between temperature and band gap energy is demonstrated as follows:

$$E_{gi} = E_g(0) - \frac{\alpha_i \times T^2}{T + \sigma_i} \quad (3)$$

Where: α, σ are constants dependent on materials used.

The band gap energy given in equation (4) is strongly affected by the mixture alloys of the Indium Gallium Arsenide "InGaAs" and Indium Gallium Phosphate "InGaP". However, Germanium "Ge" is referred as pure material.

$$E_g(A_{1-x}\,B_x) = (1-x)\,E_g(A) + x\,E_g(A) + x\,E_g(B) - x\,(1-x)\,P \quad (4)$$

Where: $A_{1-x}\,B_x$ is the composition of alloy material and P is an alloy dependent parameter in eV. Voltage at terminals of triple junction solar cell is given in equation (5) as following:

$$V = \sum_{i=1}^{3} V_i \quad (5)$$

Where:

$$V_i = \frac{n_i\,K_B\,T}{q} \ln\left(\frac{J_{sc,i} - J_L}{J_{o,i}} + 1 \right) - J_L\,AR_{s,i} \quad (6)$$

Open circuit voltage per cell is achieved by canceling the operating current density J_L and it's described by the equation below:

$$V_i = \frac{n_i\,K_B\,T}{q} \ln\left(\frac{J_{sc,i}}{J_{o,i}} + 1 \right) \quad (7)$$

B. Multi-Junction Solar Cells Simulations Results

To well understand the behavior of these cells in term of device efficiency, we had introduced programs under PSpice software to determine the electrical characteristics I-V and P-V curves which is the most efficient techniques for extracting their electrical parameters such as : short circuit current density J_{sci}, open circuit voltage V_{oci}, maximum power P_m, Fill Factor FF and the power conversion efficiency of solar cell η.

Fig. 3 shows simulation results of electrical characteristics curves of "GaInP/GaAs/Ge" multi-junction solar cells [5, 7]. The influences of irradiation and temperature on the electrical parameters of multi-junction solar cells were investigated

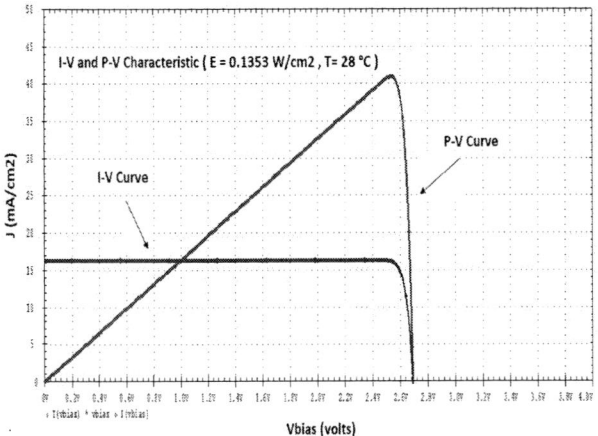

Fig. 3. Electrical characteristic curves of "GaInP/GaAs/Ge" solar cells.

C. MJSC performance under temperature effect

Simulation results of varying temperature effect on the energy band gap of MJSC at constant solar irradiation are shown in figure 4. We fixed the solar irradiation at constant value of 0.1353 W/cm² and we varied the temperature of solar cells between -80 to 90 °C. We observed that the maximum power increases with decrease in temperature. More the temperature decreases, the open circuit voltage V_{co} increases and the short circuit current J_{sc} decreases with a very small drop or the maximum power of solar cells increases.

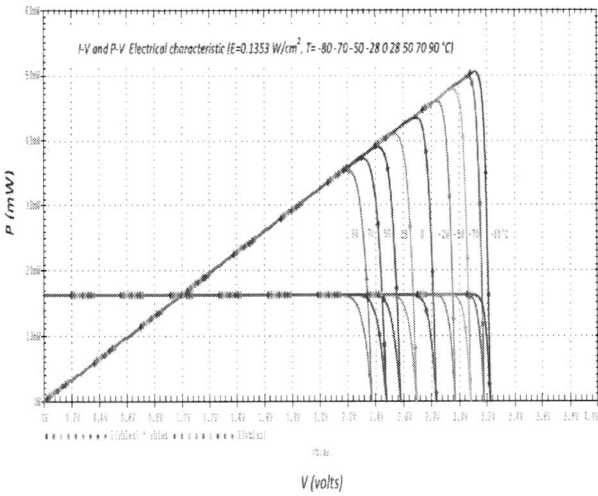

Fig. 4. The effect of varying temperature on the GaInP/GaAs/Ge multi-junction solar cells performance and P-V characteristics.

III. EXPERIMENTAL RESULTS

A. Experimental Prototype

The fig.5 presents the global block diagram of designed prototype and DSPic implemented program control

(Homme/Machine), it's formed of several blocks. To simplify the realization and conception of the prototype, every block was developed on PCB [8, 9, 10].

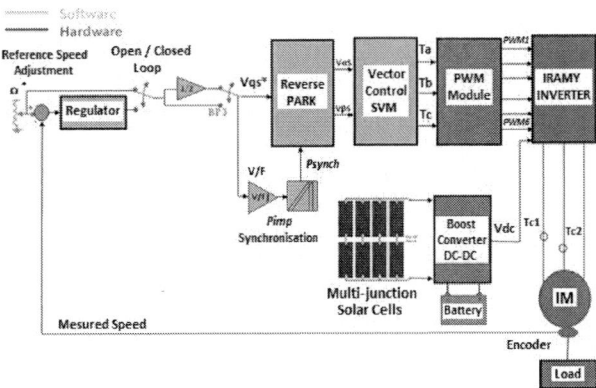

Fig. 5. Global block diagram of designed prototype

Picture on fig. 6 presents the built prototype at the laboratory of Applied and Power Electronics LAPE. It includes all PCB: sensors, power inverter module and DSPic microcontroller board [11, 12].

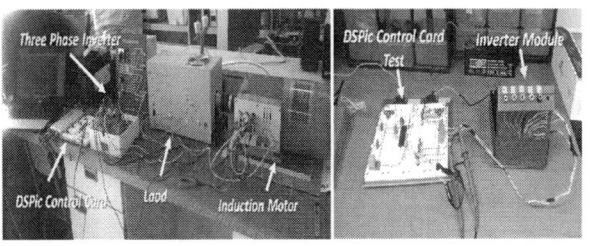

Fig. 6. Experimental platform

The parameters of the IM are presented in table.2.

IM Parameters	
Voltage	220V / 380V
Current	4.32A / 2.5A
Speed	1385 rpm
Rotor resistance Rr	3.553 Ω
Stator resistance Rs	7 Ω
Stator Ls	0.278 H
Rotor inductance Lr	0.278 H
Magnetizing inductance Lm	0.270 H
Inertia	0.0036
Friction	0.0017
Pole number	2

Table. 2. The IM motor parameters

B. Experimental Results Interpretations

The fig. 8 shows the output pulses control of DSPic of arm concerned of inverter. The fig. 9 presents the form of the output voltage and the line current of the three-phase inverter. We observed that the shape of the current is almost sinusoidal and the harmonics are due to the vector control technique.

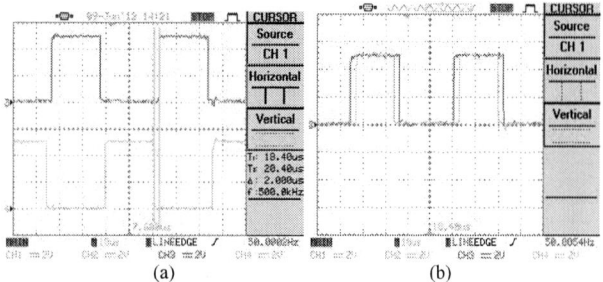

(a)   (b)

Fig. 8. The output pulse control of DSPic (a) PWM1H and PWM1L (b) PWM1H and PWM2H

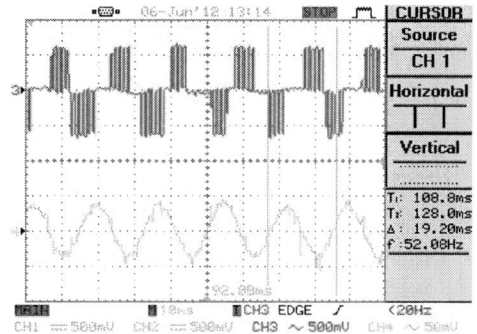

Fig. 9. The waveform of the phase-to-phase voltage and the shape of the line current of three-phase inverter output.

The fig. 10 presents the waveform of the output voltage of the generator in closed loop control. It's about 91.2 volts.

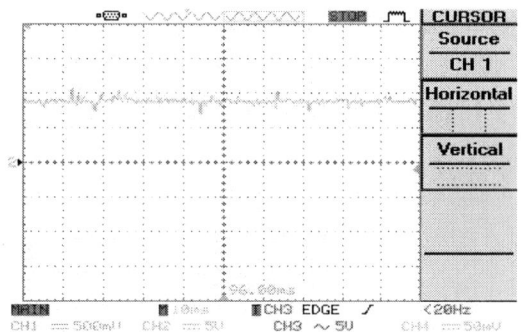

Fig. 10. The waveform of the output voltage of the generator.

Fig. 11 presents a typical scalar control profile of our application for different speed references values. In practice, this V/F profile can take any form. The control voltage can be increased at certain operating frequencies to provide more torque at these speeds. It has a low frequency boost zone to help the motor start from zero speed. In this supercharging region, the voltage is limited to a minimum value. To keep the ratio *V/F* fixed, the controller increase or decrease the voltage and frequency of $U\alpha$ and $U\beta$ to run the motor for variable speed applications. In scalar control, the IM supply voltage is

proportional to the frequency from which the maximum torque will be constant in a range of speed variation. For speeds above the nominal speed, the voltage is constant; it is characterized by a decreasing flux which causes a decrease in torque (overspeed regime).

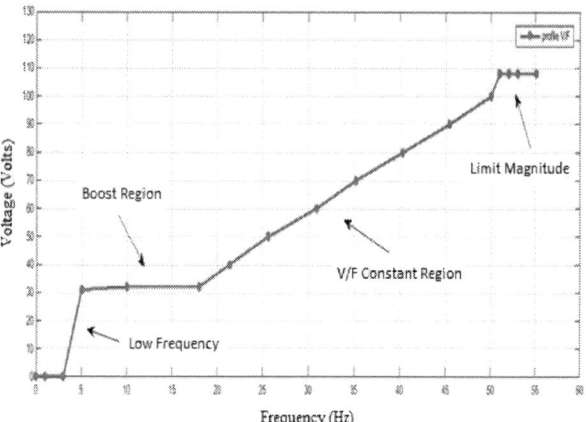

Fig. 11. The scalar control profile in closed loop

The results of the tests carried out on the hardware and software of our research project show the good performance of designed prototype. We notice that experimental results in open and closed loop are very satisfactory. But the most important problem remains when we use the motor for low speed.

IV. CONCLUSION

The paper attempted to contribute and implement a variable speed control of an AC Induction Motor using the dsPIC30F Microcontroller fed by highly efficient MJSC through a three phase inverter and boost converter for industrial control applications. The frequency and amplitude of the drive voltage must be varied to change the motor speed. MJSC had the advantage of improved performance; they had a good performance for a high radiation and low temperature. These experiments research shows a good understanding of how an ACIM responds to different voltages and frequencies in a variable speed applications. A successful result was obtained of the tests: Hardware/Sofware of V/f digital scalar control of an AC induction motor. Simulation and experimental results were very satisfactory for further exploration and contributions such as: highly efficient solar water pumping systems. Various other speed digital control techniques can be applied on the same and compared [13, 14].

REFERENCES

[1] Nauf Nissa, Sushma Gupta, Monish Gupta "Multi-junction fed field oriented control of induction motor" International Journal of Industrial Electronics and Electrical Engineering, ISSN: 2347-6982, Volume 4, Issue 8, Aug 2016.

[2] Jialin Liu, Raymond Hou "Solar Cell Simulation Model for Photovoltaic Power Generation System" International Journal of Renewable Energy Research IJRER, Vol.4, No.1, 2014.

[3] Md. Tofael Ahmed; Teresa Gonçalves; Mouhaydine Tlemcani "Single diode model parameters analysis of photovoltaic cell" IEEE International Conference on Renewable Energy Research and Applications (ICRERA 2016), Birmingham UK , 20-23 Nov. 2016.

[4] Vandana Khanna, Bijoy Kishore Das, Dinesh Bisht "Matlab/Simelectronics Models Based Study of Solar Cells" IJRER, Vol. 3, No. 1, 2013.

[5] Maithili Ganpati Kumbhare, P. Sathya "Design and Analysis of Indium Gallium Nitride based PIN solar cell" IJRER, Vol.6, No.3, 2016.

[6] Yuwei Sun; Xinping Yan ; Chengqing Yuan; Hu Luo; Qizhen Jiang "The I–V characteristics of solar cell under the marine environment: Experimental research" IEEE International Conference on Renewable Energy Research and Applications (ICRERA 2016), Palermo, Italy, 22-25 Nov. 2015.

[7] A. Hadj Dida; M. Bekhti "Study, modeling and simulation of the electrical characteristic of space satellite solar cells" Published in: IEEE International Conference on Renewable Energy Research and Applications (ICRERA 2017), San Diego, California, USA, 5-8 Nov. 2017.

[8] Nguyen Phung Quang, Jörg-Andreas Dittrich " Vector Control of Three-Phase AC Machines : System Development in the Practice" ISBN: 978-3-540-79028-0, Power Systems 2008 , ISSN: 1612-1287. Springer.com.

[9] Haitham Abu-Rub, Atif Iqbal, Jaroslaw Guzinski " High Performance Control of AC Drives with Matlab/Simulink models " Published on 2012 , John Wiley & Sons Ltd, The Atrium, Southern Gate, Chichester, West Sussex, PO19 8SQ, United Kingdom. ISBN: 978047097829.

[10] Amogh Jain B A, Smt. S. Poornima " DsPIC based Fixed Speed Induction Motor Drive" Volume I, Issue II, July 2014, IJRSI ISSN 2321 -2705.

[11] Sandip A. Waskar U.L. Bombale, Tanaji B.S. "dsPIC based SPWM controlled Three Phase Inverter Fed Induction Motor Drive" International Journal of Computer Applications (0975 –888) Vol. 47 - No. 16, 2012.

[12] A. Hadj Dida, M. Bourahla, H. B. Ertan, M. Benghanem " Design, development and implementation of sensorless digital control of an electric motorization " Power Electronics and Motion Control Conference and Exposition (PEMC), 2014 16th International.

[13] Gudimetla Ramesh*, Kari Vasavi**, S.Lakshmi Sirisha* "Photovoltaic Cell Fed 3-Phase Induction Motor Using MPPT Technique" International Journal of Power Electronics and Drive System (IJPEDS) Vol. 5, No. 2, October 2014, pp. 203~210 ISSN: 2088-8694.

[14] Neethu Raj P R, Vasanthi V "Performance Analysis of Photovoltaic Induction Motor Drive for Agriculture Purpose" International Journal of Power Electronics and Drive Systems 7(4) :1252. December 2016. DOI: 10.11591/ijpeds.v7.i4.pp1252-1260.

Improved Diagnosis of Induction Motor's Rotor Faults using the Papoulis Window

Mohamed Boudiaf KOURA, Ahmed Hamida BOUDINAR, Ameur Fethi AIMER

Laboratory of Electric Drives Development, Diagnosis Group
Department of Electrical Engineering, University of Sciences and Technology of Oran MB, Algeria
Email : mohamedkoura@outlook.fr, boud_ah@yahoo.fr , fethi.aimer@yahoo.fr

Abstract— This paper discusses the rotor faults diagnosis in induction motors through stator current analysis based on the estimation of the Power Spectral Density (PSD) using the Periodogram technique. This approach is the most used technique in industry thanks to its simplicity of programming and its relatively low computation time compared to high resolution signal processing methods. However, this method has some disadvantages such as a frequency resolution depending on acquisition time and an almost impossible identification of low harmonics (case of incipient faults). This second disadvantage is due to the side lobes effect related to the selected weighting window, such as Blackman, Hamming, Hanning or Kaiser. To reduce this effect, a new weighting window named Papoulis window is proposed in this paper. To highlight the performance of this new window in rotor faults diagnosis of induction motors, several simulation tests are achieved and discussed.

Keywords— *Induction motor, diagnosis, rotor fault, spectral analysis, Papoulis.*

I. INTRODUCTION

Unlike the stator design, the manufacture of induction motor's cage rotors has known few changes over the years. As a result, the rotor fault now accounts for about 5% to 10% of the total of induction motor failures [1]–[3]. This fault occurs when one or more bars of the squirrel cage rotor breaks, so that the flow of current through this bar is interrupted. The breaking of rotor bars is usually caused by [3]–[5]:

- Thermal stresses due to overload, sparks and temperature increase of the various constituents of the motor;

- Magnetic stresses due to electromagnetic forces, unbalanced magnetic traction, electromagnetic noise or vibrations;

- Mechanical stresses due to worn parts, bearing faults, etc.;

- Residual stresses due to manufacturing problems;

- Environmental stress caused for example by contamination and abrasion of the rotor material due to chemicals or moisture.

This fault usually starts at the junction point between the rotor bar and the short circuit ring. Bar breakage does not cause necessarily an induction motor to fail, but can cause other serious faults in the stator windings if it is not detected in time [6]. This will obviously lead to a degradation of the induction motor and reduce its lifetime.

In order to diagnose rotor faults, many works have been proposed, analyzing: the air gap flux [7], vibration [8], electromagnetic torque [9], instantanious power [10] and stator current [1], [11]–[14].

Note that the most used technique in recent years is based on the analysis of the stator current (also called MCSA for Motor Current Signature Analysis). Its particularity lies in the easy installation of the stator current sensors in a non-invasive way (which does not require access to the motor) and without disturbing the normal operation of the entire electric drive. In addition, the obtained stator current spectrum has indications on almost all types of the induction motor faults [12], [13], [15], [16].

Indeed, several studies [5], [17]–[19] have shown that the presence of a rotor fault is manifested by the appearance of new frequency components or the increase of the amplitude of some faults characteristic frequencies around the fundamental's frequency [12], [18]. To identify this frequency signature characterizing the rotor fault from the stator current analysis, the estimation of the Power Spectral Density (PSD) using the Periodogram is the most used method in industry because of its easy programming and its fast execution due to the use of the FFT algorithm (Fast Fourier Transform).

Unfortunately, this method can not detect the fault frequency signatures when they are very close to the fondamantal (case of a very low load for example). To solve this problem, various approaches of signal processing called high resolution spectral analysis methods [1], [15], [20], [21] have been applied to analyze this kind of case. For example, PRONY [15], MUSIC [1], or ESPRIT [20], [21]. These methods have a very good frequency resolution even with measurements made on a very short acquisition time; Unfortunately, they have a major drawback regarding the computation time. Indeed, they require a significant computation time because of the complexity of their algorithms.

A second drawback of the PSD estimation by Periodogram has been reported by most recent studies that an incipient rotor fault can not be detected due to the side lobe effect associated with the selected weighting window, such as Blackman, Hamming, Hanning or Kaiser. To overcome this problem, we propose in this paper a new weighting window called Papoulis window whose characteristics can improve the detection of the incipient faults.

Several simulations are performed to demonstrate the effectiveness of the use of papoulis window on the reliability of the diagnosis. The results obtained show the merits and the positive impact of the proposed approach.

978-1-5386-7688-2/19 $31.00 © 2019 IEEE

II. ROTOR FAULT FREQUENCY SIGNATURES

Several scientific studies [22]–[24], have shown that the existence of a rotor fault of an induction motor is manifested by two types of frequency signature.

The first signature appears as side components f_{bb} around the fundamental harmonic of the stator current spectrum:

$$f_{bb} = f_s(1 \pm 2ks) \qquad k = 1, 2, 3, \ldots \tag{1}$$

Where f_s is the supply frequency (fundamental), f_{bb} is the broken rotor bars frequency signature and s the motor slip.

Fig. 1, represent an ideal spectrum of the stator current around the fundamental in faulty rotor case, according to (1).

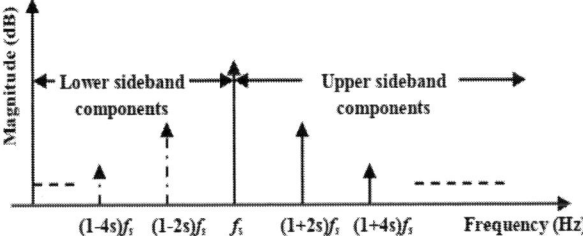

Fig. 1.   Rotor fault signature frequency around the fundamental.

The second signature proposed by [25], [26] is localized around space harmonics. It is also a decisive indicator for rotor faults diagnosis [16].

$$f_{bb} = \left[\frac{k}{p}(1-s) \pm v \right] f_s \tag{2}$$

Where p is the number of pole pairs, $k/p=1, 3, 5, \ldots$ and $v=\pm 1, \pm 3, \pm 5, \ldots$

III. POWER SPECTRAL DENSITY ESTIMATION USING THE PERIODOGRAM TECHNIQUE WITH THE PAPOULIS WINDOW

A. Principle of the PSD estimation by Periodogram

The PSD estimation by Periodogram is widely used in the electric motors diagnosis when the signal to be analyzed is acquired in steady state. This approach is a frequency analysis technique for stationary random signals. Under these conditions, the PSD by Periodogram of the stator current signal $i(n)$ can be defined directly in terms of the observed samples of $i(n)$ as follows [1], [27]:

$$PSD(f) = \frac{1}{N} |I(f)|^2 \tag{3}$$

Where N is the number of samples, $I(f)$ the Fourier transform of $i(n)$. $i(n)$ is the discrete version of the acquired stator current signal.

Moreover, and because of the discretization phase, the signal to be processed is limited in time. Mathematically, this amounts to performing the following operation:

$$i_T(n) = i(n).\Pi(n) \tag{4}$$

Where $i_T(n)$ is the truncated discrete signal, $\Pi(n)$ is a rectangular function of length N which represents the effect of time limitation and n = 0,1,2,... N-1. Frequency analysis of the truncated signal is given by [1]:

$$I_T(f) = I(f) \otimes N.\sin c(Nf) \tag{5}$$

$I_T(f)$ is the Fourier transform (FT) of signal $i_T(n)$ and the Cardinal sine function « $sinc(Nf)$ » is the FT of the rectangular function $\Pi(n)$. The operator $\otimes$ represents the convolution product.

Equation (5) shows that the truncation operation introduces negative effects on spectrum estimation, as:

- Inability to produce an accurate estimation of the power spectrum, particularly with signals obtained over short acquisition times. Indeed, for this case, this technique is unable to identify harmonics close to each other. We then define the frequency resolution (Δf) as [27]:

$$\Delta f = \frac{F_e}{N} = \frac{1}{T_{aqc}} \tag{6}$$

Where F_e is the sampling frequency and T_{aqc} is the acquisition time.

- Inability to identify very low harmonics due to the appearance of sidelobes due to $sinc(Nf)$. This affects the detection of incipient faults reducing thus, the fineness of the analysis.

To avoid these negative effects, we introduce the weighting windows $\omega(n)$. This means that instead of processing the truncated signal $i_T(n)$, the weighted signal is processed according to the following equation:

$$i_W(n) = i_T(n).\omega(n) \tag{7}$$

A weighting window is characterized by a given mathematical function, whose main parameters to take into account are:

✓ The width of the main lobe (L)at half magnitude at-*3dB*.

✓ The maximum magnitude of side lobes (A)

These parameters that characterize a weighting window are illustrated in Fig. 2.

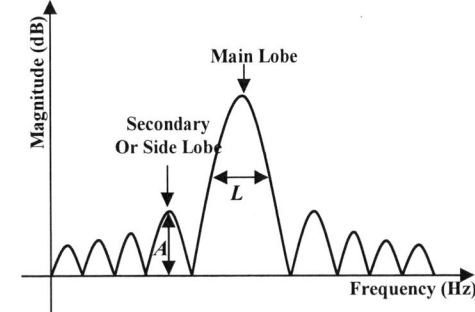

Fig. 2.   Representation of the parameters of a weighting window.

Depending on the used window, the function (*n*) allows:

- To reduce the width of the main lobes, allowing thus, the identification of frequency signatures close to each other;

- To reduce the magnitude of the side lobes, allowing thus, the identification of incipient faults.

Unfortunately, it is not possible to have both of these characteristics simultaneously.

However, in induction motor diagnosis, it is recommended to have main lobe as narrow as possible in order to identify frequency signatures close to each other and to side lobes.

B. Papoulis window

The Papoulis window $w_p(n)$ can be defined as the product of the triangular window with a single cycle of a cosine window of the same period [28]–[30].

$$w_p(n) = \begin{cases} \frac{1}{\pi}\sin\left(\frac{\pi|n|}{N/2}\right) + \left(1 - \frac{|n|}{N/2}\right)\cos\left(\frac{\pi n}{N/2}\right), & 0 \le |n| \le \frac{N}{2} \\ 0 & \text{elsewhere} \end{cases} \quad (8)$$

So, the Fourier transform of $w_p(n)$ gives us the following optimal spectral window [28]–[30]:

$$W_P(j\omega) = \begin{cases} 4\pi^2\tau\,\dfrac{(1+\cos\omega\tau)}{\left(\pi^2 - \omega^2\tau^2\right)^2} & 0 \le |n| \le \frac{N}{2} \\ 0 & \text{elsewhere} \end{cases} \quad (9)$$

In this context, the notion of "optimal" window is in the sense that the greatest energy content is concentrated in the main lobe of its Fourier transform (similar to the modified Kaiser function family), while the side lobes contain less energy. The width of the main lobe of this window is almost identical to that of the Blackman window while the level of the side lobe of this window is -46 dB.

TABLE I compares the frequency characteristics of the rectangular window with those of the Papoulis window.

TABLE I.   Characteristics of Weighting Windows

Window	Side Lobe Level	Main Lobe Width (at -3 dB)
Rectangular	-13 dB	$0.89\Delta f$
Papoulis	-46 dB	$1.69\Delta f$

According to TABLE I, the side lobe level of the Papoulis window (-46 dB) is lower than that of the rectangular window (-13 dB). This shows that the Papoulis window is recommended for the detection of incipient faults. On the other hand, the rectangular window has a main lobe of 0.89 Δf, which is narrower than that of Papoulis 1.69 Δf. For this reason, the rectangular window is recommended for the detection of harmonics close to each other.

IV. SIMULATION RESULTS

The stator current can be simulated in its simplified form as the sum of several sinusoids representing the fundamental harmonic and the harmonics of the rotor faults. The stator current model is written as follows:

$$i(t) = 5\sin(2\pi f_s t) + 0,066\sin(2\pi f_{c1} t) \\ + 0,068\sin(2\pi f_{c2} t) + b(t) \quad (10)$$

Where, $b(t)$ is a white noise simulating the random phenomena of the current. This noise is added to the signal according to a Signal Noise Ratio (SNR) defined as follows:

$$SNR_{dB} = 20\log_{10}\frac{P_s}{P_b} \quad (11)$$

Where P_s and P_b are, respectively, the signal and the noise powers.

In order to see the effect of the proposed window on the reliability of the incipient rotor faults diagnosis of induction motor. All the simulations are carried out with a sampling frequency equal to 1500 Hz and a number of samples N=4096, a frequency resolution of 0.366 Hz. The supply frequency (fs) is equal to 50 Hz.

A.   1st case: SNR = 50dB and s = 5%

For a moderately noisy signal, i.e. an SNR equal to 50 dB and for a motor slip of 5%.

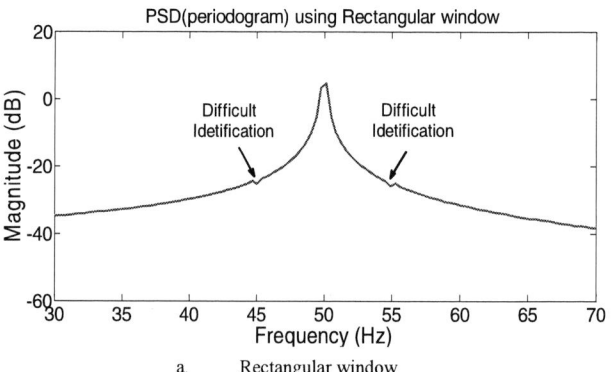

a.   Rectangular window

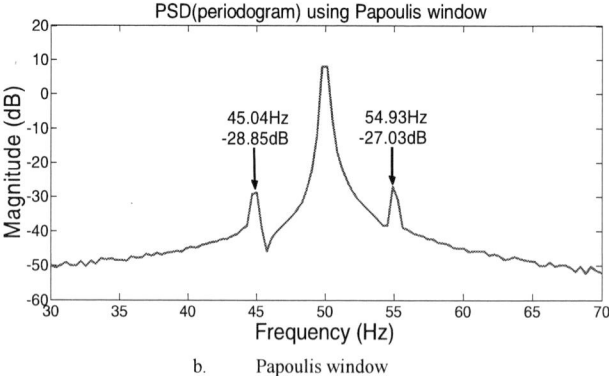

b.   Papoulis window

Fig. 3. PSD analysis of the stator current with both weighting windows. Case of SNR = 50dB and s = 5%.

The frequency signatures of the simulated fault according to eq. (1) should appear (theoretically for k = 1) at the frequencies: f_{c1} = 45Hz and f_{c2} = 55Hz. The spectral analysis of the stator current by the PSD estimation based on the algorithm of the Periodogram is given in Fig. 3.

The comparison of the obtained results shows that the use of the rectangular window in the PSD estimation can not locate the harmonics representing the simulated fault as shown by Fig. 3.a. On the other hand, Fig. 3.b, shows the effectiveness of the proposed window. Indeed, from the spectrum obtained using the proposed window, we can read the frequency signature of the fault, at 45.04 Hz and 54.93 Hz. This confirms the detection and localization ability of the proposed approach. It should be noted that this slight difference in values with respect to the theoretical frequencies is due to the calculation errors and the random effect of the programmed signal.

B. 2nd case: SNR = 15dB and s = 5%

To check the performance of the proposed weighting window in the location of frequencies contained in a highly noisy signal, considering a SNR of 15 dB for an unchanged motor slip of 5%. The obtained results once again show the clarity and readability of the proposed Papoulis window versus the rectangular window while identifying all the desired harmonics (of the rotor fault), as shown in Fig. 4.

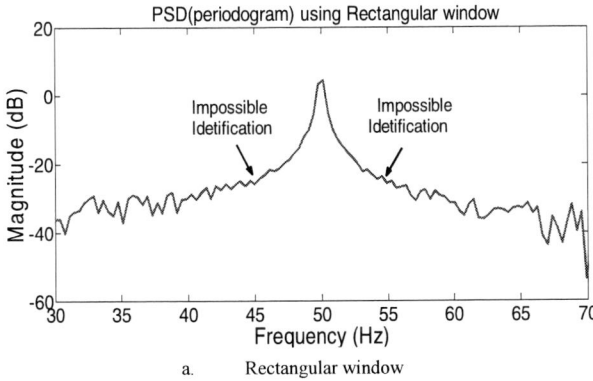

a.   Rectangular window

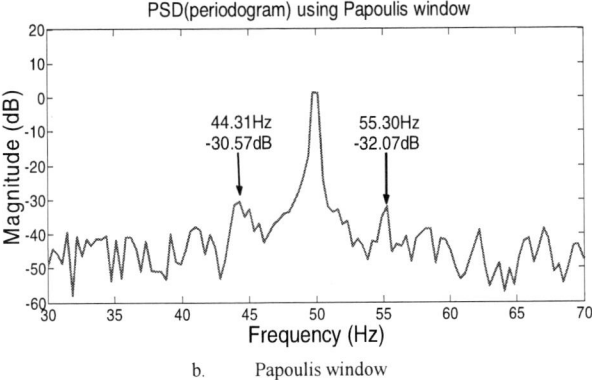

b.   Papoulis window

Fig. 4. PSD analysis of the stator current with both weighting windows. Case of SNR = 15dB and s = 5%.

C. 3rd case: SNR = 50dB and s = 2%

This simulation aims to show the localization ability of the frequency signatures of the incipient rotor fault of each window when these signatures are close to the fundamental.

Fig. 5. represents the PSD estimation by Periodogram of the stator current for a SNR of 50 dB and a motor slip of 2%. In these conditions and according to eq. (1) the frequencies representing the fault should appear, theoretically for k = 1, at the frequencies: f_{c1} = 48 Hz et f_{c2} = 52 Hz.

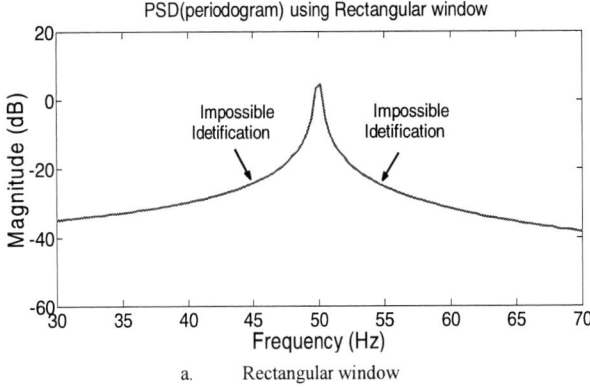

a.   Rectangular window

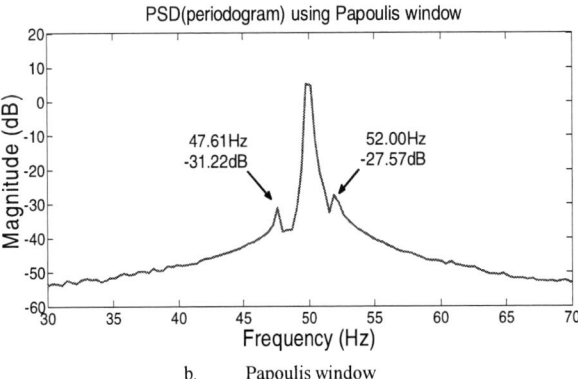

b.   Papoulis window

Fig. 5. PSD analysis of the stator current with both weighting windows. Case of SNR = 50dB and s = 2%.

From the obtained results, we find that the use of the rectangular window for PSD estimation of the simulated signal still does not allow the detection of the fault frequency signatures as shown in Fig.5.a. On the other hand, the use of the proposed window shows a better readability of the stator current spectrum as is illustrated in Fig.5.b.

Indeed, both searched frequencies (47.61 Hz and 52.00 Hz) around the fundamental frequency are easily visible, which allows an easier reading of their magnitude.

V. CONCLUSION

In this paper, a new diagnosis approach based on the use of the Papoulis window is proposed in order to improve the reliability of the diagnosis of incipient rotor faults in induction motors. To highlight the effectiveness of the proposed window and verify its performance in the PSD estimation by Periodogram of the stator current, several simulations were performed.

In the light of the obtained results, it has been shown the effectiveness of this approach and its ability to identify signatures of rotor faults compared to the use of the rectangular window and its robustness against noise. This shows that the proposed Papoulis window is recommended for fault diagnosis when frequency signatures have low magnitude and close to the fundamental.

978-1-5386-7688-2/19 $31.00 © 2019 IEEE

REFERENCE

[1] A. H. Boudinar, N. Benouzza, A. Bendiabdellah, et M. Khodja, « Induction Motor Bearing Fault Analysis Using a Root-MUSIC Method », *IEEE Transactions on Industry Applications*, vol. 52, no 5, p. 3851-3860, sept. 2016.

[2] A.H. Bonnett et all« Increased Efficiency Versus Increased Reliability », *IEEE Industry Applications Magazine*, vol.14, no 1, p. 29-36, janv. 2008.

[3] J. A. Antonino-Daviu, J. Pons-Llinares, et S. B. Lee, « Advanced Rotor Fault Diagnosis for Medium-Voltage Induction Motors Via Continuous Transforms », *IEEE Transactions on Industry Applications*, vol. 52, no 5, p. 4503-4509, sept. 2016.

[4] D. G. Jerkan, D. D. Reljić, et D. P. Marčetić, « Broken Rotor Bar *Fault Detection of IM Based on the Counter-Current Braking Method* », *IEEE Transactions on Energy Conversion*, vol.32, no4, p.1356-1366, 2017.

[5] H. Henao et al., « Trends in Fault Diagnosis for Electrical Machines: A Review of Diagnostic Techniques », *IEEE Industrial Electronics Magazine,* vol. 8, no 2, p. 31-42, juin 2014.

[6] M. Iorgulescu et R. Beloiu, « Faults diagnosis for electrical machines based on analysis of motor current », in *International Conference on Optimization of Electrical and Electronic Equipment,*p 291-297, 2014.

[7] P. A. Panagiotou, I. Arvanitakis, N. Lophitis, J. Antonino-Daviu, et K. N. Gyftakis, « A New Approach for Broken Rotor Bar Detection in Induction Motors Using Frequency Extraction in Stray Flux Signals », *IEEE Transactions on Industry Applications,* p. 1-1, 2019.

[8] M. Z. Ali et all « Machine Learning-Based Fault Diagnosis for Single- and Multi-Faults in Induction Motors Using Measured Stator Currents and Vibration Signals », *IEEE Transactions on Industry Applications*, vol. 55, no 3, p. 2378-2391, mai 2019.

[9] Praveen Kumar N et Isha T B, « Electromagnetic field analysis of 3-Phase induction motor drive under broken rotor bar fault condition using FEM », in *IEEE International Conference on Power Electronics, Drives and Energy Systems (PEDES)*, 2016, p. 1-6.

[10] M. Drif, H. Kim, J. Kim, S. B. Lee, et A. J. M. Cardoso, « Active and Reactive Power Spectra-Based Detection and Separation of Rotor Faults and Low-Frequency Load Torque Oscillations », *IEEE Transactions on Industry Applications*, vol. 53, no 3, p. 2702-2710, mai 2017.

[11] I. Ouachtouk, S. E. Hani, S. Guedira, K. Dahi, et H. Mediouni, « Broken rotor bar fault detection based on stator current envelopes analysis in squirrel cage induction machine », in *2017 IEEE International Electric Machines and Drives Conference (IEMDC)*, 2017, p. 1-6.

[12] M.-E.-A. Khodja, A. H. Boudinar, et A. Bendiabdellah, « Effect of Kaiser Window Shape Parameter for the Enhancement of Rotor Faults Diagnosis », *International Review of Automatic Control (IREACO)*, vol. 10, no 6, p. 461-467-467, nov. 2017.

[13] A. Bendiabdellah, A. H. Boudinar, N. Benouzza, et M. Khodja, « The enhancements of broken bar fault detection in induction motors », in 2015 Intl Aegean Conference on Electrical Machines Power Electronics (ACEMP), *Intl Conference on Optimization of Electrical Electronic Equipment (OPTIM) Intl Symposium on Advanced Electromechanical Motion Systems (ELECTROMOTION)*, 2015, p. 81-86.

[14] N. Benouzza, A. H. Boudinar, A. Bendiabdellah, et M. Khodja, « Slot harmonic frequency detection as a technique to improve stator current spectrum approach for broken rotor bars fault diagnosis », *in Intl Aegean Conference on Electrical Machines Power Electronics*

(ACEMP), *Intl Conference on Optimization of Electrical Electronic Equipment (OPTIM) Intl Symposium on Advanced Electromechanical Motion Systems (ELECTROMOTION)*, 2015, p. 118-122.

[15] M. Sahraoui, A. J. M. Cardoso, et A. Ghoggal, « The Use of a Modified Prony Method to Track the Broken Rotor Bar Characteristic Frequencies and Amplitudes in Three-Phase Induction Motors », *IEEE Trans. on Ind. Applicat.*, vol. 51, no 3, p. 2136-2147, mai 2015.

[16] J. Grande-Barreto et all « Half-broken bar detection using MCSA and statistical analysis », in *IEEE International Autumn Meeting on Power, Electronics and Computing (ROPEC)*, 2017, p. 1-5.

[17] A. Bellini et al., « On-field experience with online diagnosis of large induction motors cage failures using MCSA », *IEEE Transactions on Industry Applications*, vol. 38, no 4, p. 1045-1053, juill. 2002.

[18] W. T. Thomson et I. Culbert, « Motor Current Signature Analysis (MCSA) to Detect Cage Winding Defects », *in Current Signature Analysis for Condition Monitoring of Cage Induction Motors: Industrial Application and Case Histories*, IEEE, 2017.

[19] J. Zhuzhi, Z. Hongyu, L. Xuyang, et S. Hang, « Incipient Broken Rotor Bar Fault Diagnosis Based on Extended Prony Spectral Analysis Technique », *in 37th Chinese Control Conference*, 2018, p. 5705-5710.

[20] Y. Kim, Y. Youn, D. Hwang, J. Sun, et D. Kang, « High-Resolution Parameter Estimation Method to Identify Broken Rotor Bar Faults in Induction Motors », *IEEE Transactions on Industrial Electronics,* vol. 60, no 9, p. 4103-4117, sept. 2013.

[21] B. Xu, L. Sun, L. Xu, et G. Xu, « Improvement of the Hilbert Method via ESPRIT for Detecting Rotor Fault in Induction Motors at Low Slip », *IEEE Transactions on Energy Conversion*, vol.28, no1, p.225-233, 2013.

[22] S. Williamson et A. C. Smith, « Steady-state analysis of 3-phase cage motors with rotor-bar and end-ring faults », *IEE Proceedings B - Electric Power Applications*, vol. 129, no 3, p. 93-, mai 1982.

[23] G. B. Kliman et all « Noninvasive detection of broken rotor bars in operating induction motors », *IEEE Transactions on Energy Conversion,* vol. 3, no 4, p. 873-879, déc. 1988.

[24] F. Filippetti, G. Franceschini, C. Tassoni, et P. Vas, « AI techniques in induction machines diagnosis including the speed ripple effect », *IEEE Transactions on Industry Applications,* vol.34, no1, p. 98-108, 1998.

[25] H. Henao, G. A. Capolino, et H. Razik, « Analytical approach of the stator current frequency harmonics computation for detection of induction machine rotor faults », *in 4th IEEE International Symposium on Diagnostics for Electric Machines, Power Electronics and Drives, SDEMPED,* 2003, p. 259-264.

[26] G. B. Kliman et all « Methods of Motor Current Signature Analysis », *Electric Machines & Power Systems*, vol.20, no5, p. 463-474, 1992.

[27] A. F. Aimer, A. H. Boudinar, N. Benouzza, et A. Bendiabdellah, « Simulation and experimental study of induction motor broken rotor bars fault diagnosis using stator current spectrogram », *in 3rd International Conference on Control, Engineering Information Technology (CEIT)*, 2015, p. 1-7.

[28] A. Papoulis, « Minimum-bias windows for high-resolution spectral estimates », *IEEE Transactions on Information Theory*, vol. 19, p.9-12, 1973.

[29] F. J. Harris, « On the use of windows for harmonic analysis with the discrete Fourier transform », *Proceedings of the IEEE*, vol. 66, no 1, p. 51-83, 1978.

[30] K. M. M. Prabhu, Window Functions and Their Applications in Signal Processing. *CRC Press*, 2018.

Model Predictive Current and Capacitor Voltage Control of Post-Fault Three-Level NPC Inverter-Fed Synchronous Reluctance Motor Drives

Yong-Chao Liu[1], Salah Laghrouche[1], Abdoul N'Diaye[1], Maurizio Cirrincione[2]

[1]Energy Department, FEMTO-ST Institute (UMR 6174), National Center for Scientific Research (CNRS), UTBM,
Université Bourgogne Franche-Comté, Belfort, France
[2]School of Engineering and Physics, The University of the South Pacific, Laucala Campus, Suva, Fiji
cn.yong-chao.liu@ieee.org, salah.laghrouche@utbm.fr, abdoul-ousman.n-diaye@utbm.fr, maurizio.cirrincione@usp.ac.fj

Abstract—This paper proposes a finite-control-set model predictive current and capacitor voltage control strategy for the synchronous reluctance motor drive systems fed by the eight-switch three-phase inverter (ESTPI), which is a type of post-fault reconfigured three-level neutral-point-clamped inverter with open-circuit fault in a leg. The proposed control strategy uses the discrete-time models of the drive system to predict the behavior of the stator currents and the neutral point potential (NPP) error for zero and small voltage vectors provided by the ESTPI. The errors between stator currents and their references in the rotor reference frame as well as the NPP error are evaluated in a predefined cost function. The basic voltage vector, which minimizes this cost function, is selected. The proposed control strategy can achieve not only the reference stator current tracking but also the capacitor voltage offset suppression. Simulation results have shown the effectiveness of the proposed control scheme.

Keywords—*synchronous reluctance motor, three-level neutral-point-clamped inverter, eight-switch three-phase inverter, model predictive control, capacitor voltage balancing control*

I. INTRODUCTION

The synchronous reluctance motor (SynRM) has gained much attention owing to its robustness and low-cost rotor structure [1]-[10]. To obtain the better traction performance, the three-level neutral-point-clamped (3L-NPC) inverter has been proposed to replace the two-level (2L) inverter as the traction inverter for the SynRM drive system in recent years [8]-[10]. Compared to the 2L inverter, the 3L-NPC inverter has more power switching devices, which are one of the most vulnerable components in the inverter [11]. To improve the reliability of the 3L-NPC inverter, it can be converted to a fault-tolerant topology by adding three triacs to connect the dc-link neutral point to three-phase loads, respectively [12], [13]. Once the open-circuit fault occurs in a leg of the fault-tolerant 3L-NPC inverter, for example, leg *a*, based on the corresponding triac, it can be reconfigured to a post-fault inverter, which is equivalent to the eight-switch three-phase inverter (ESTPI) shown in Fig. 1, to maintain the continuous operation of the SynRM drive system.

The vector control (VC) strategy, which mainly contains the outer speed control loop, the inner current control loop and the pulse-width modulator, is widely used in the ESTPI-fed AC

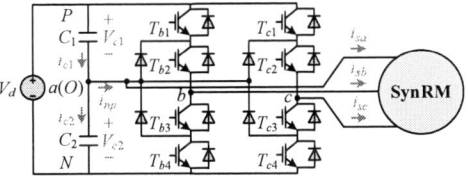

Fig. 1. ESTPI-fed SynRM drive system.

motor drive system [14]-[16]. In the conventional VC strategy, the speed and current control loops are based on the proportional-integral (PI) controllers, and the command signals for the power switching devices of the ESTPI are generated by the pulse-width modulator, where the space-vector pulse-width modulation (SVPWM) algorithm and the carrier-based PWM (CBPWM) algorithm are two most popular choices [17]-[19]. As the post-fault reconfigured topology of the 3L-NPC VSI-fed AC motor drive system, the capacitor voltages balancing control is a challenge for the ESTPI-fed AC motor drive system whose conventional VC strategy cannot maintain two balanced dc-link capacitor voltages over a wide speed range [13]. The imbalance between two dc-link capacitor voltages results in the imbalanced output voltages and currents of the ESTPI, which seriously affect the performance of the drive system. To maintain two balanced dc-link capacitor voltages in the vector-controlled ESTPI-fed AC motor drive system, few schemes have been proposed. In [20], a modified CBPWM algorithm based on varying the amplitude of the carrier waves was proposed for the ESTPI-fed induction motor drive system. However, the real-time Fourier transformation is required, which increase the complexity of the control system, and the operation frequency is a constant.

Finite-control-set model predictive current control (FCS-MPCC) strategy is a potential alternative of the conventional VC strategy for the AC drive system [21]-[23]. In the FCS-MPCC strategy, the pulse-width modulator is eliminated and the PI-controller-based current control loop is replaced by the model-based predictor. In this predictor, the discrete stator current model is adopted to predict the behavior of the stator currents for all basic voltage vectors provided by the inverter up to the prediction horizon. The switching state, which corresponds to

978-1-5386-7688-2/19 $31.00 © 2019 IEEE

the basic voltage vector minimizing the predefined cost function, is chosen to be applied to the inverter. Besides the current tracking, other control objectives, such as capacitor voltage balancing control and switching frequency reduction, can also be achieved by the FCS-MPCC strategy [24], [25]. All these control objectives can be integrated into a cost function. Thus, the FCS-MPCC strategy provides huge control flexibility.

In this paper, a FCS model predictive current and capacitor voltage control (FCS-MPC²VC) strategy is proposed for the ESTPI-fed SynRM drive system. The discrete stator current models considering the self- and cross-magnetic saturation effects in the rotor reference frame, namely the dq frame, as well as the discrete model of the neutral point potential (NPP) error are used to predict the behavior of the stator currents and the NPP error, respectively, for the zero and small voltage vectors provided by the ESTPI. The cost function is made up of the errors between stator current dq components and their references as well as the NPP error. The basic voltage vector minimizing this cost function is selected to be applied.

II. MODEL OF THE ESTPI-FED SynRM DRIVE SYSTEM

The ESTPI-fed SynRM drive system is presented in Fig. 1. i_{sa}, i_{sb} and i_{sc} are three-phase stator currents. T_{x1}-T_{x4} ($x = b$, c) are power switching devices in two normal legs. C_1 and C_2 are two dc-link capacitors, respectively. The capacitance values of C_1 and C_2 are identical and labelled as C. V_{c1} and V_{c2} are two dc-link capacitor voltages. i_{c1} and i_{c2} are two dc-link capacitor currents. i_{np} is the neutral point current. V_d is the constant dc-bus voltage.

A. Basic Voltage Vectors Provided by the ESTPI

The switching function labelled S_x, which is used to describe the switching states of each normal leg of the ESTPI, can be defined as.

$$S_x = \begin{cases} 1, & \text{if } T_{x1} \text{ \& } T_{x2} \text{ on}, T_{x3} \text{ \& } T_{x4} \text{ off} \\ 0, & \text{if } T_{x2} \text{ \& } T_{x3} \text{ on}, T_{x1} \text{ \& } T_{x4} \text{ off} \\ -1, & \text{if } T_{x3} \text{ \& } T_{x4} \text{ on}, T_{x1} \text{ \& } T_{x2} \text{ off} \end{cases} \quad (1)$$

According to Fig. 1 and (1), the three-phase pole voltages labelled V_{ao}, V_{bo} and V_{co}, respectively, can be expressed as

$$\begin{cases} V_{ao} = 0 \\ V_{bo} = S_b \left[(S_b +1)V_{c1} - (S_b -1)V_{c2} \right]/2 \\ V_{co} = S_c \left[(S_c +1)V_{c1} - (S_c -1)V_{c2} \right]/2 \end{cases} \quad (2)$$

The basic voltage vectors provided by the ESTPI can be obtained by the following equation.

$$V_s = \frac{2}{3} \left(V_{ao} + V_{bo} e^{j2\pi/3} + V_{co} e^{j4\pi/3} \right) \quad (3)$$

According to (1), the ESTPI can provide nine different switching states. Based on (2) and (3), nine different voltage vectors including a zero voltage vector **OO**, six small voltage vectors **NN**, **ON**, **PO**, **PP**, **OP**, **NO**, and two medium voltage vectors can be obtained, as shown in Table □. As presented in Fig. 1, one-phase current flows through V_{c1} and V_{c2}. Thus, the difference between V_{c1} and V_{c2} is nonzero. In fact, it consists of

TABLE I
BASIC VOLTAGE VECTORS PROVIDED BY THE ESTPI

(S_b, S_c)	Symbol	V_s
$(0, 0)$	**OO**	0
$(-1, -1)$	**NN**	$2V_{c2}/3$
$(0, -1)$	**ON**	$(1+j\sqrt{3})V_{c2}/3$
$(1, -1)$	**PN**	$(V_{c2}-V_{c1})/3+j\sqrt{3}(V_{c2}+V_{c1})/3$
$(1, 0)$	**PO**	$(-1+j\sqrt{3})V_{c1}/3$
$(1, 1)$	**PP**	$-2V_{c1}/3$
$(0, 1)$	**OP**	$-(1+j\sqrt{3})V_{c1}/3$
$(-1, 1)$	**NP**	$(V_{c2}-V_{c1})/3-j\sqrt{3}(V_{c2}+V_{c1})/3$
$(-1, 0)$	**NO**	$(1-j\sqrt{3})V_{c2}/3$

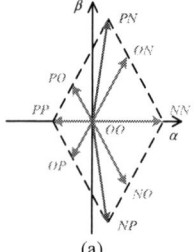 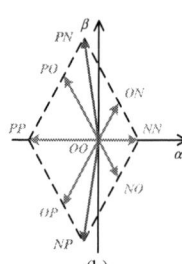

Fig. 2 Basic voltage vectors distribution. (a) $V_{c1}<V_{c2}$, and (c) $V_{c1}>V_{c2}$

the voltage fluctuation component and the voltage offset component [20]. From Table I, it can be observed that nonzero basic voltage vectors change with V_{c1} and V_{c2}. In the stator reference frame, namely the $\alpha\beta$ frame, the distribution of basic voltage vectors provided by the ESTPI in the cases of $V_{c1} < V_{c2}$ and $V_{c1} > V_{c2}$ are illustrated in Fig. 2(a) and (b), respectively. According to Fig. 2, the maximum effective gain of the ESTPI is affected by the difference between V_{c1} and V_{c2}. Because the voltage fluctuation component of the difference between V_{c1} and V_{c2} cannot be eliminated, the voltage offset component of the difference between V_{c1} and V_{c2} should be suppressed as much as possible.

B. Model of the SynRM

The stator voltage, stator flux and electromagnetic torque equations of the SynRM in the dq frame, can be expressed as

$$\begin{cases} v_{sd} = R_s i_{sd} + \dfrac{d\lambda_{sd}\left(i_{sd}, i_{sq}\right)}{dt} - \omega_e \lambda_{sq}\left(i_{sd}, i_{sq}\right) \\ v_{sq} = R_s i_{sq} + \dfrac{d\lambda_{sq}\left(i_{sd}, i_{sq}\right)}{dt} + \omega_e \lambda_{sd}\left(i_{sd}, i_{sq}\right) \end{cases} \quad (4)$$

$$\begin{cases} \lambda_{sd}\left(i_{sd}, i_{sq}\right) = L_d\left(i_{sd}, i_{sq}\right) i_{sd} \\ \lambda_{sq}\left(i_{sd}, i_{sq}\right) = L_q\left(i_{sd}, i_{sq}\right) i_{sq} \end{cases} \quad (5)$$

$$T = \frac{3}{2} n_p \left[\lambda_{sd}\left(i_{sd}, i_{sq}\right) i_{sq} - \lambda_{sq}\left(i_{sd}, i_{sq}\right) i_{sd} \right] \quad (6)$$

where v_{sd} and v_{sq} are the stator voltage dq components, i_{sd} and i_{sq} are the stator current dq components, λ_{sd} and λ_{sq} are the stator flux linkage dq components, L_d and L_q are the apparent inductances, R_s is the stator resistance, ω_e is the rotor electrical speed, T is the electromagnetic torque, and n_p is the pole pairs.

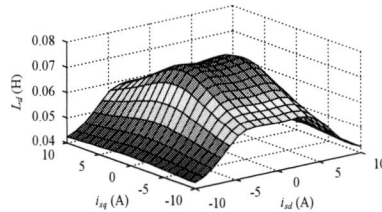

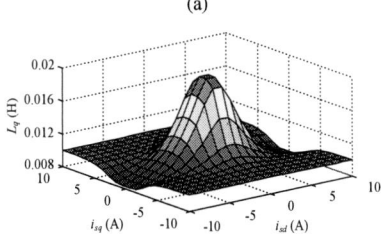

(b)

Fig. 3 Apparent inductance maps. (a) $L_d(i_{sd}, i_{sq})$, (b) $L_q(i_{sd}, i_{sq})$.

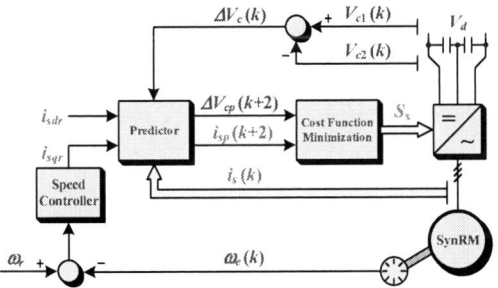

Fig. 4 Proposed FCS-MPC²VC strategy for the ESTPI-fed SynRM drive system.

As shown in (4) and (5), the effects of self- and cross-magnetic saturation on apparent inductances and stator flux linkages are taken into account. The apparent inductance maps of the SynRM adopted in this paper are shown in Fig. 3 [5].

The derivatives of λ_{sd} and λ_{sq} can be described as

$$
\begin{cases}
\dfrac{d\lambda_{sd}\left(i_{sd},i_{sq}\right)}{dt} = \dfrac{\partial \lambda_{sd}\left(i_{sd},i_{sq}\right)}{\partial i_{sd}}\dfrac{di_{sd}}{dt} + \dfrac{\partial \lambda_{sd}\left(i_{sd},i_{sq}\right)}{\partial i_{sq}}\dfrac{di_{sq}}{dt} \\
\dfrac{d\lambda_{sq}\left(i_{sd},i_{sq}\right)}{dt} = \dfrac{\partial \lambda_{sq}\left(i_{sd},i_{sq}\right)}{\partial i_{sd}}\dfrac{di_{sd}}{dt} + \dfrac{\partial \lambda_{sq}\left(i_{sd},i_{sq}\right)}{\partial i_{sq}}\dfrac{di_{sq}}{dt}
\end{cases}
\tag{7}
$$

The incremental inductances, which are labeled as L_{dd}, L_{dq}, L_{qd}, and L_{qq}, respectively, can be defined as

$$
\begin{cases}
L_{dd} = \dfrac{\partial \lambda_{sd}\left(i_{sd},i_{sq}\right)}{\partial i_{sd}}, \quad L_{dq} = \dfrac{\partial \lambda_{sd}\left(i_{sd},i_{sq}\right)}{\partial i_{sq}} \\
L_{qd} = \dfrac{\partial \lambda_{sq}\left(i_{sd},i_{sq}\right)}{\partial i_{sd}}, \quad L_{qq} = \dfrac{\partial \lambda_{sq}\left(i_{sd},i_{sq}\right)}{\partial i_{sq}}
\end{cases}
\tag{8}
$$

C. Model of the DC-Link Capacitors

The dc-link capacitor current equations can be expressed as

$$
i_{c1} = C\frac{dV_{c1}}{dt}, \qquad i_{c2} = C\frac{dV_{c2}}{dt}
\tag{9}
$$

Based on Fig. 1 and (9), applying the Kirchhoff's current law to the dc-link of the ESTPI, i_{np} can be expressed as

$$
i_{np} = i_{c1} - i_{c2} = -S_b^2 i_{bn} - S_c^2 i_{cn}
\tag{10}
$$

III. PROPOSED FCS-MPC²VC STRATEGY

The signal flow graph of the proposed FCS-MPC²VC strategy is illustrated in Fig. 4. The outer speed control loop is based on the PI controller, while the inner stator current and capacitor voltage controllers depend on the discrete-model-based predictor and the cost function. The reference d-axis stator current labelled i_{sdr} is constant, while the reference q-axis stator current labelled i_{sqr} is obtained from the outer speed control loop. The applied basic voltage vector is selected by the cost function. According to [16] and [26], the use of medium voltage vectors results in poor performance of the ESTPI-fed AC drive system. Thus, only the zero and small voltage vectors are considered by the cost function in this paper.

A. Stator Current and NPP Error Prediction

Based on (4), (5), (7) and (8), the dynamics of the stator current dq components can be expressed as

$$
\begin{cases}
\dfrac{di_{sd}}{dt} = \dfrac{L_{qq}}{L_{dd}L_{qq} - L_{dq}L_{qd}}\left(v_{sd} - R_s i_{sd} + \omega_e L_q i_{sq}\right) \\
\qquad - \dfrac{L_{dq}}{L_{dd}L_{qq} - L_{dq}L_{qd}}\left(v_{sq} - R_s i_{sq} - \omega_e L_d i_{sd}\right) \\
\dfrac{di_{sq}}{dt} = -\dfrac{L_{qd}}{L_{dd}L_{qq} - L_{dq}L_{qd}}\left(v_{sd} - R_s i_{sd} + \omega_e L_q i_{sq}\right) \\
\qquad + \dfrac{L_{dd}}{L_{dd}L_{qq} - L_{dq}L_{qd}}\left(v_{sq} - R_s i_{sq} - \omega_e L_d i_{sd}\right)
\end{cases}
\tag{11}
$$

For stator current prediction, using the first-order forward Euler method to discretize (11), the predicted stator current dq components at the $(k+1)$th sampling period labelled $i_{sdp}(k+1)$ and $i_{sqp}(k+1)$, respectively, can be calculated by

$$
\begin{cases}
i_{sdp}\left(k+1\right) = i_{sd}\left(k\right) \\
\qquad + \dfrac{L_{qq}T_s\left(v_{sd}\left(k\right) - R_s i_{sd}\left(k\right) + \omega_e\left(k\right)L_q i_{sq}\left(k\right)\right)}{L_{dd}L_{qq} - L_{dq}L_{qd}} \\
\qquad - \dfrac{L_{dq}T_s\left(v_{sq}\left(k\right) - R_s i_{sq}\left(k\right) - \omega_e\left(k\right)L_d i_{sd}\left(k\right)\right)}{L_{dd}L_{qq} - L_{dq}L_{qd}} \\
i_{sqp}\left(k+1\right) = i_{sq}\left(k\right) \\
\qquad - \dfrac{L_{qd}T_s\left(v_{sd}\left(k\right) - R_s i_{sd}\left(k\right) + \omega_e\left(k\right)L_q i_{sq}\left(k\right)\right)}{L_{dd}L_{qq} - L_{dq}L_{qd}} \\
\qquad + \dfrac{L_{dd}T_s\left(v_{sq}\left(k\right) - R_s i_{sq}\left(k\right) - \omega_e\left(k\right)L_d i_{sd}\left(k\right)\right)}{L_{dd}L_{qq} - L_{dq}L_{qd}}
\end{cases}
\tag{12}
$$

where T_s is the sampling period.

To describe the difference between V_{c1} and V_{c2}, the NPP error labelled ΔV_c can be defined as

$$\Delta V_c = V_{c1} - V_{c2} \qquad (13)$$

Based on (9), (10) and (13), the dynamics of the NPP error can be expressed as

$$\frac{d\Delta V_c}{dt} = \frac{1}{C} i_{np} = -\frac{1}{C}\left(S_b^2 i_{bn} + S_c^2 i_{cn}\right) \qquad (14)$$

For the prediction of the NPP error, similarly, using the first-order forward Euler method to discretize (14), the predicted NPP error at the $(k+1)$th sampling period labelled $\Delta V_{cp}(k+1)$ can be calculated by

$$\Delta V_{cp}(k+1) = \Delta V_c(k) - \frac{T_s}{C}\left[S_b^2 i_{bn}(k) + S_c^2 i_{cn}(k)\right] \quad (15)$$

B. Cost Function Optimization

To make stator current dq components track their references, the estimation of the reference stator current dq components at the $(k+1)$th sampling period is required, which will increase the computation burden. Fortunately, because i_{sdr} is a constant, $i_{sdr}(k+1)$ is equal to $i_{sdr}(k)$. Moreover, the frequency of the reference current is much smaller than the sampling frequency. Thus, $i_{sqr}(k+1)$ is approximately equal to $i_{sqr}(k)$.

After obtaining the predicted stator current dq components and their references as well as the predicted NPP error, a cost function labelled g is designed to evaluate the zero and small voltage vectors provided by the ESTPI, as expressed in the following equation.

$$g = \frac{\left|i_{sdr}(k) - i_{sdp}(k+1)\right|}{I_s} + \frac{\left|i_{sqr}(k) - i_{sqp}(k+1)\right|}{I_s} \\ + \lambda \frac{\left|\Delta V_{cp}(k+1)\right|}{V_d} \qquad (16)$$

where I_s is the rated value of the stator current, λ is the weighting factor.

As shown in (16), I_s and V_d are adopted to normalize the cost function terms. After that zero and small voltage vectors are evaluated by g, the one minimizing g is selected for the $(k+1)$th sampling period. However, the optimal voltage vector is obtained after the execution of the control strategy. Therefore, this voltage vector will be applied at the $(k+1)$th sampling period. It means that the implementation of the FCS-MPC²VC strategy suffers from the one-step delay. To address this problem, the predicted control variables at the $(k+2)$th sampling period rather than the $(k+1)$th sampling period should be adopted [21]. Thus, the cost function can be rewritten as

$$g = \frac{\left|i_{sdr}(k) - i_{sdp}(k+2)\right|}{I_s} + \frac{\left|i_{sqr}(k) - i_{sqp}(k+2)\right|}{I_s} \\ + \lambda \frac{\left|\Delta V_{cp}(k+2)\right|}{V_d} \qquad (17)$$

IV. SIMULATION RESULTS

The proposed FCS-MPC²VC strategy for the ESTPI-fed SynRM drive system is tested with the MATLAB/Simulink-based simulation. The system parameters are listed as follows: rated power is 1.1 kW, rated current is 6.3 A, dc-bus voltage is 400 V, rated torque is 4.8 N·m, rated rotor mechanical speed is 1500 rpm, the rotor inertia is 0.02 kg·m², R_s is 1.05 Ω, C is 2200 μF, T_s is 10 μs.

As shown in (17), λ is needed to be tuned in the proposed control strategy, which represents the relative importance of the capacitor voltage offset suppression. The higher the value of λ is, the smaller the capacitor voltage offset is. However, a very high value of λ will result in an unacceptable output performance of the drive system. Currently, there is no generic approach to tune the weighting factors in the cost function, and they are chosen by means of empirical procedures [21]. Thus, to achieve the control objectives, through a simulation-based trial-and-error approach [21], the value of λ is selected as 10 in this paper.

In the first test, the steady-state performance of the proposed FCS-MPC²VC strategy is tested. The reference value of d-axis component of stator current labelled i_{sdr} is kept at 5 A, the reference rotor mechanical speed is 1000 rpm and the load is set at 2.4 N·m. The simulation results are presented in Fig. 5. It can be seen that the stator current dq components and the rotor mechanical speed labelled ω_m track their references well at the steady state. Moreover, the capacitor voltage offset component in V_{c1} and V_{c2} is effectively suppressed at the steady state.

In the second test, the dynamic performance of the proposed FCS-MPC²VC strategy for a torque step is tested. i_{sdr} is kept at 5 A, the reference rotor mechanical speed is 1000 rpm. The load declines from 2.4 N·m to 1.2 N·m at 0.1 s and steps to 2.4 N·m at 0.2 s. The simulation results are presented in Fig. 6. It can be observed that the torque response is fast. The performance of the stator current tracking, the rotor mechanical speed tracking and the capacitor voltage offset suppression are acceptable.

In the third test, the steady-state and dynamic performance of the proposed FCS-MPC²VC strategy for the speed reversal maneuver from 1000 rpm to -1000 rpm is tested. i_{sdr} is kept at 5 A and the load is set at 2.4 N·m. The speed reversal signal is applied at 0.2s. The simulation results are presented in Fig. 7. It can be seen that the steady-state and dynamic performance of the proposed control strategy with the speed reversal maneuver is acceptable.

V. CONCLUSIONS

This paper proposes a novel FCS-MPC²VC strategy for the ESTPI-fed SynRM drive system considering the self- and cross-magnetic saturation effects. The basic voltage vectors provided by the ESTPI considering the imbalance between two dc-link capacitor voltages is given for the control of stator currents and dc-link capacitor voltages. The discrete stator current dq component models and the discrete NPP error model are used to perform the predictive control. The proposed control strategy has the capability of achieving the reference stator currents tracking and the capacitor voltage offset suppression. The simulation test results have demonstrated the excellent steady-state and dynamic performance of the proposed control strategy.

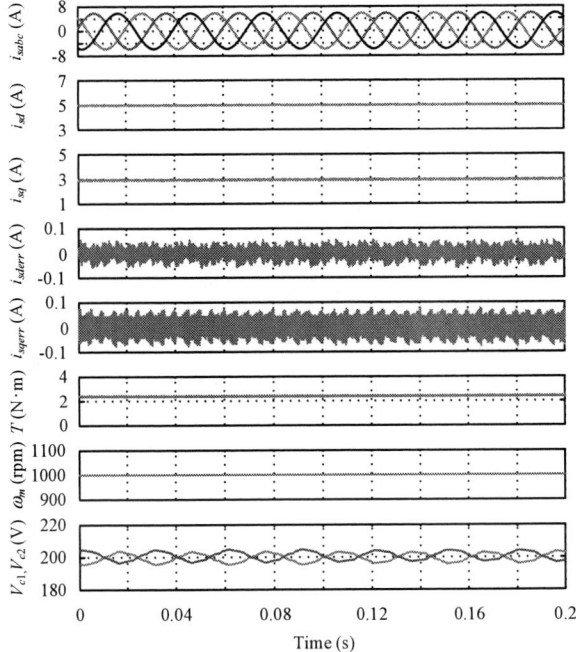

Fig. 5. Simulation results of the first test. From top to bottom: Three-phase stator currents, stator current d-axis component, stator current q-axis component, stator current d-axis component error, stator current q-axis component error, electromagnetic torque, rotor mechanical speed, two dc-link capacitor voltages.

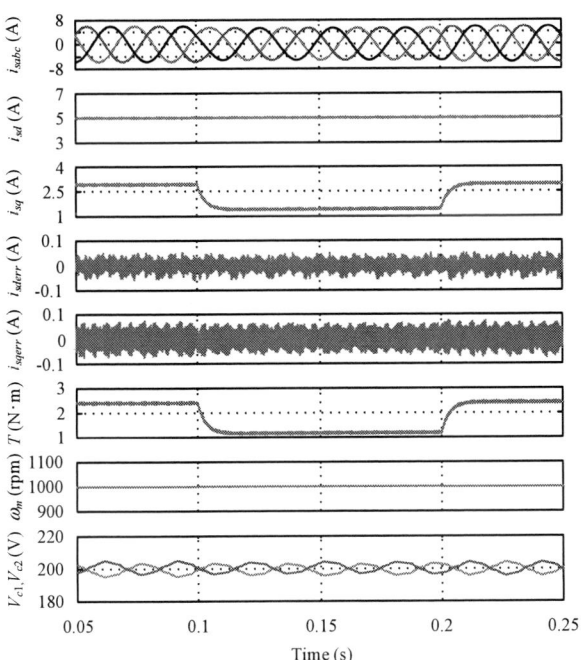

Fig. 6. Simulation results of the second test. From top to bottom: Three-phase stator currents, stator current d-axis component, stator current q-axis component, stator current d-axis component error, stator current q-axis component error, electromagnetic torque, rotor mechanical speed, two dc-link capacitor voltages.

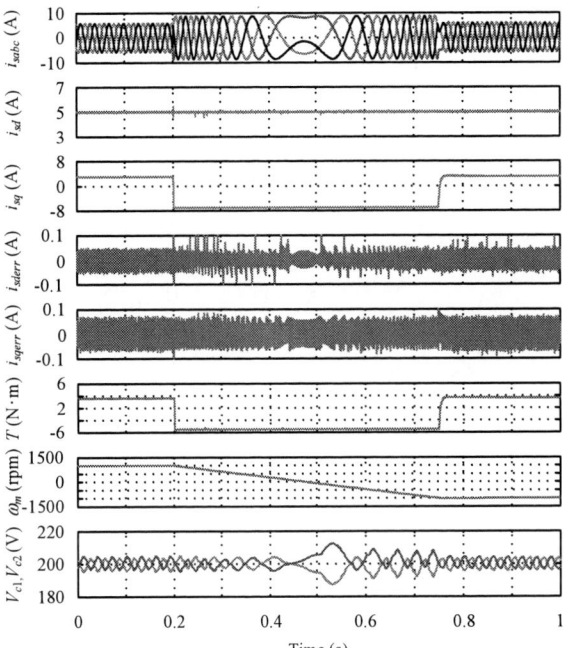

Fig. 7. Simulation results of the third test. From top to bottom: Three-phase stator currents, stator current d-axis component, stator current q-axis component, stator current d-axis component error, stator current q-axis component error, electromagnetic torque, rotor mechanical speed, two dc-link capacitor voltages.

ACKNOWLEDGMENT

This work was supported by the USP-SRT Research Project: Advanced Control of Synchronous Reluctance Motors for Electrical Vehicles (ACOSREV).

REFERENCES

[1] G. Pellegrino, T. M. Jahns, N. Bianchi, W. L. Soong, and F. Cupertino, The Rediscovery of Synchronous Reluctance and Ferrite Permanent Magnet Motors. Basel: Springer, 2016.

[2] G. Eason, B. Noble, and I. N. Sneddon, "Theoretical and experimental reevaluation of synchronous reluctance machine," *IEEE. Trans. Ind. Electron.*, vol. 57, no. 1, pp. 6-13, Jan. 2010.

[3] Y. -C. Liu, S. Laghrouche, A. N'Diaye, and M. Cirrincione, "Active-flux-based super-twisting sliding mode observer for sensorless vector control of synchronous reluctance motor drives," in *Proc. 7th IEEE Int. Conf. on Renew. Energy Res. Appl.*, Paris, 2018, pp. 402-406.

[4] Y. -C. Liu, S. Laghrouche, A. N'Diaye, S. Narayan, G. Cirrincione, and M. Cirrincione, "Sensorless control of synchronous reluctance motor drives based on the TLS EXIN neuron," in *Proc. 7th IEEE Int. Electr. Mach. Drives Conf.*, San Diego, 2019, pp. 1-5.

[5] S. Yamamoto, T. Ara, and K. Matsuse, "A method to calculate transient characteristics of synchronous reluctance motors considering iron loss and cross-magnetic saturation," *IEEE Trans. Ind. Appl.*, vol. 43, no. 1, pp. 47–56, Jan./Feb. 2007.

[6] S. Yamamoto, H. Hirahara, J. B. Adawey, T. Ara, and K. Matsuse, "Maximum efficiency drives of synchronous reluctance motors by a novel loss minimization controller with inductance estimator," *IEEE Trans. Ind. Appl.*, vol. 49, no. 6, pp. 2543–2551, Nov./Dec. 2013.

[7] N. Bianchi, S. Bolognani, E. Carraro, M. Castiello, and E. Fornasiero, "Electric vehicle traction based on synchronous reluctance motors," *IEEE Trans. Ind. Appl.*, vol. 52, no. 6, pp. 4762–4769, Nov./Dec. 2016.

[8] L. Masisi, P. Pillay, and S. S. Williamson, "A three-level neutral-point-clamped inverter synchronous reluctance machine drive," *IEEE Trans. Ind. Appl.*, vol. 51, no. 6, pp. 4531–4540, Nov./Dec. 2015.

[9] L. Masisi, P. Pillay, and S. S. Williamson, "A modulation strategy for a three-level inverter synchronous reluctance motor (SynRM) drive," *IEEE Trans. Ind. Appl.*, vol. 52, no. 2, pp. 1874–1881, Mar./Apr. 2016.

[10] L. Masisi, P. Pillay, and S. S. Williamson, "The effect of two- and three-level inverters on the core loss of a synchronous reluctance machine (SynRM)," *IEEE Trans. Ind. Appl.*, vol. 52, no. 5, pp. 3805–3813, Sep./Oct. 2016.

[11] W. Zhang, D. Xu, P. N. Enjeti, H. Li, J. T. Hawke, and H. S. Krishnamoorthy, "Survey on fault-tolerant techniques for power electronic converters," *IEEE Trans. Power Electron.*, vol. 29, no. 12, pp. 6319–6331, Dec. 2014.

[12] P. Lezana, J. Pou, T. A. Meynard, J. Rodriguez, S. Ceballos, and F. Richardeau, "Survey on fault operation on multilevel inverters," *IEEE Trans. Ind. Electron.*, vol. 57, no. 7, pp. 2207–2218, Jul. 2010.

[13] S. Ceballos, J. Pou, E. Robles, J. Zaragoza, and J. L. Martin, "Performance evaluation of fault-tolerant neutral-point-clamped converters," *IEEE Trans. Ind. Electron.*, vol. 57, no. 8, pp. 2709–2718, Aug. 2010.

[14] B. R. O. Baptista, M. B. Abadi, A. M. S. Mendes, and S. M. A. Cruz, "The performance of a three-phase induction motor fed by a three-level NPC converter with fault tolerant control strategies," in *Proc. IEEE 9th Int. Symp. Diag. Electr. Mach., Power Electron. Drives*, Valencia, 2013, pp. 497–504.

[15] L. M. A. Caseiro, A. M. S. Mendes, and S. M. A. Cruz, "Fault tolerance in back-to-back three-level neutral-point-clamped induction motor drives," in *Proc. IET 7th Power Electron., Mach. Drives*, Manchester, 2014, pp. 1–6.

[16] Q. Tang, X. Ge, and Y. -C. Liu, "Performance analysis of two different SVM-based field-oriented control schemes for eight-switch three-phase inverter-fed induction motor drives," in *Proc. IEEE 8th Int. Power Electron. Motion Contr. Conf.*, Hefei, 2016, pp. 3374–3378.

[17] Y. -C. Liu, X. -L. Ge, X. -Y. Feng, and R. -J. Ding, "Relationship between SVPWM and carrier-based PWM of eight-switch three-phase inverter," *Electron. Lett.*, vol. 51, no. 13, pp. 1018–1019, Jun. 2015.

[18] Y. -C. Liu, X. Ge, and Q. Tang, "Relationship between two different space-vector modulation methods of eight-switch three-phase inverters," in *Proc. IEEE 8th Int. Power Electron. Motion Contr. Conf.*, Hefei, 2016, pp. 3206–3210.

[19] Y. -C. Liu, X. Ge, Q. Tang, and B. Gou, "Two modified SVPWM algorithms for common-mode voltage reduction in eight-switch three-phase inverters," *Electron. Lett.*, vol. 53, no. 10, pp. 676–678, May 2017.

[20] M. Zhang, X. Zhang, Y. Guo, Z. Li, and J. Xia, "An improved CBPWM method to balance the output voltage of eight-switch three-phase inverter under neutral-point potential unbalance condition," in *Proc. 43rd Annu. Conf. IEEE Ind. Electron. Soc.*, Beijing, 2017, pp. 786–791.

[21] J. Rodriguez, and P. Cortes, Predictive control of power converters and electrical drives. New York: Wiley, 2012.

[22] M. Preindl, and E. Schaltz, "Sensorless model predictive direct current control using novel second-order PLL observer for PMSM drive systems," *IEEE Trans. Ind. Electron.*, vol. 58, no. 9, pp. 4087–4095, Sep. 2011.

[23] F. Wang, S. Li, X. Mei, W. Xie, J. Rodríguez, and R. M. Kennel, "Model based predictive direct control strategies for electrical drives: an experimental evaluation of PTC and PCC methods," *IEEE Trans. Ind. Informat.*, vol. 11, no. 3, pp. 671–681, Jun. 2015.

[24] R. Vargas, P. Cortés, U. Ammann, J. Rodríguez, and J. Pontt, "Predictive control of a three-phase neutral-point-clamped inverter," *IEEE Trans. Ind. Electron.*, vol. 54, no. 5, pp. 2697–2705, Oct. 2007.

[25] Y. -C. Liu, X. Ge, Q. Tang, Z. Deng, and B. Gou, "Model predictive current control for four-switch three-phase rectifiers in balanced grids," *Electron. Lett.*, vol. 53, no. 1, pp. 44–46, Jan. 2017.

[26] Q. Tang, X. Ge, Y. -C. Liu, and M. Hou, "Improved switching-table-based DTC strategy for the post-fault three-level NPC inverter-fed induction motor drive," *IET Electr. Power Appl.*, vol. 12, no. 1, pp. 71–80, Jan. 2018.

Modeling and Diagnosis of Stator Winding Faults in PMSM using Motor Current Signature Analysis

Zakaria GHERABI [1], Djilali TOUMI [2], Noureddine BENOUZZA [1], Noureddine HENINI [3]

[1] Electrical Engineering Faculty, Diagnosis Group, University of Sciences and Technology of Oran (USTO-MB), Oran, Algeria.
[2] Electrical Engineering Department, L2GEGI Laboratory, University of Ibn-Khaldoun of Tiaret, Tiaret, Algeria.
[3] Electrical Engineering Department, Yahia Fares University, Medea, Algeria.
Email : zakariagherabi@gmail.com, toumi_dj@yahoo.fr, benouza@yahoo.com, n.d.henini@gmail.com

Abstract — **Inter-Turn Short Circuit (ITSC) faults in stator winding of Permanent Magnet Synchronous Motors (PMSMs) is one of the major cause of motor failure. This paper deals with the diagnosis of ITSC faults in PMSM tacking into account of all space harmonics. To do this, it's necessary to build a mathematical model of the machine object of the diagnosis. This model must be reliable and able to describe the behavior of the machine in deferent operation modes. Winding Function Approach (WFA) is among the most significant modeling methods because it takes into account the real distribution of the windings in the stator slots. To highlight the interest of the model employed and the efficiency of the spectral analysis technique, a series of simulations are performed in both cases of operation: healthy and in the presence of ITSC faults.**

Keywords — *PMSM; WFA; ITSC faults; Spectral analysis.*

I. INTRODUCTION

Synchronous machines and more particularly PMSM have been used more and more in recent decades in different industrial applications. This growing interest is due to several parameters, including its high reliability (due to the absence of ring-brush contacts), its robust rotor construction and its high efficiency making this motor a serious competitor to DC motors and induction motor [1] – [3].

Nevertheless, in the presence of a certain electrical, thermal, mechanical or environmental constraints, this motor can undergo certain faults, which mainly affect the stator or the rotor. These faults can be electrical, mechanical or magnetic [4], [5]. In fact, several statistical studies [4] have shown that faults due to stator winding faults account between 25% and 35% of total break downs depending on the power of the motors. Winding faults are mainly due to the deterioration of the insulators of the conductors during the operation of the machine. In most cases, this degradation leads to an ITSC and which can degenerate into short circuits between coils, between phases and between phases and neutral, until complete shutdown of the machine [5]. It is then logical from an industrial, scientific and financial point of view to focus research on the detection and diagnosis of faults affecting PMSMs. This diagnosis allows to preserve the safety of personal, to avoid the stop of the whole chain of production which allows to increase the life of these engines and to minimize the financial losses.

In the literature, many techniques have been proposed for the detection of ITSC by analyzing: phase currents and voltages, magnetic flux and other types of signals [6] - [10]. In this context, most of the signal-based fault-detection methods proposed in the literature are based on harmonic monitoring in both the stator current and voltage in order to identify frequency components introduced by different faults [6]. In [7] an indicator of the ITSC in a stator phase, for a PMSM, Fresnel

diagrams of currents were extracted from the two healthy phases for healthy and faulty cases. In addition, the Short-Circuit (SC) fault was simulated in [8] by a finite element method in a PMSM. The work presented in [9] proposed fault indicators extracted from the main frequency spectrum of the magnetic flux in the air gap and leakage, measured by coil-type and Hall effect sensors. In addition, and according to [10], a short-circuit fault leads to the appearance of torque ripples with a frequency equal to twice the supply frequency, for asynchronous and PMSMs.

The paper is organized as follows: Section 2 presents the mathematical model of a healthy PMSM. Section 3 presents the PMSM model in the presence of ITSC fault. These models use the winding function approach to calculate the differential inductances of the machine by considering the real distribution of the windings in the stator slots. These models are then used to study the behavior of the machine and analyze the stator current to identify the frequency signatures characterizing the ITSC fault. Simulation results are presented in section 4 to demonstrate the effectiveness of the spectral analysis technique. The conclusions of the paper are summarized in Section 5.

II. MODELING THE HEALTHY PMSM

The modeling approach is based on a semi-analytical method of the PMSM. This modeling method is rather generic in the sense that it relies on a description of the electromagnetic couplings within the machine based on the geometric and constitutive topology of the machine. This approach has already been proven for the modeling of squirrel cage induction machines [11]. It has also been adapted to the PMSM in [12], [13]. In this section, we will present more precisely the model where the taking into account of permanent magnets is carried out, to preserve the notion of magnetic coupling with the stator, by the use of the fictitious coils.

A. Modeling of permanent magnets

The modeling method is based on the exploitation of an Ampere model which makes it possible to identify a permanent magnet whose magnetization $\vec{M}$ to a distribution of fictitious currents (ampere currents) constituted by:

➤ A surface current density σ defined by:

$$\vec{\sigma} = \vec{M} \wedge \vec{ds} \tag{1}$$

➤ A volume current density ρ defined by

$$\vec{\rho} = rot(\vec{M}) \tag{2}$$

In our case, the permanent magnet is characterized by a constant magnetization. Thanks to the equations (1) and (2) of the ampere model, we can deduce that there exists a fictitious surface current density, whereas the volume current density is zero because of the constant magnetization. In conclusion, we can represent the permanent magnet of the rotor by fictitious coils traversed by currents as illustrated in Fig. 1.

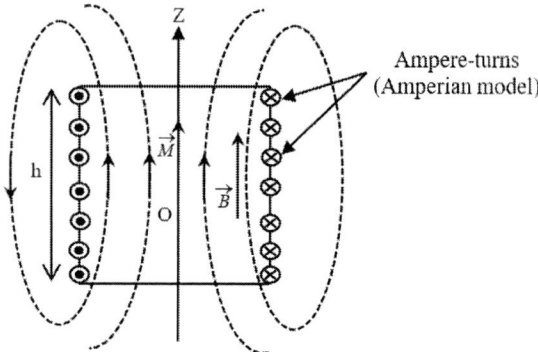

Fig. 1. Illustration of the ampere model on a cylindrical magnet [14].

B. Modeling of the machine

The equation system of the PMSM can be represented in the following matrix form [13], [14]:

$$\left[V_{sabc}\right]=[R]\left[i_{sabc}\right]+\frac{d}{dt}\left[\phi_{sabc}\right]=[R]\left[i_{sabc}\right]+\frac{d}{dt}\left[\phi_i\right]+\frac{d}{dt}\left[\phi_m\right] \quad (3)$$

Such that:

$\left[V_{sabc}\right]=\left[V_{sa}\ V_{sb}\ V_{sc}\right]^T$: Stator voltages vector;

$\left[i_{sabc}\right]=\left[i_{sa}\ i_{sb}\ i_{sc}\right]^T$: Stator currents vector;

$[R]=R_s[I]$: Stator résistance matric;

$\left[\phi_i\right]=\left[\phi_{ia}\ \phi_{ib}\ \phi_{ic}\right]^T$: Stator fluxes vector;

$\left[\phi_m\right]=\left[\phi_{ma}\ \phi_{mb}\ \phi_{mc}\right]^T$: Mutual fluxes vectors between stator and rotor;

$\left[\phi_{sabc}\right]=\left[\phi_{sa}\ \phi_{sb}\ \phi_{sc}\right]^T$: Total fluxes vector produced in the machine.

The matrix system (3) can be rewritten as follows:

$$\begin{bmatrix} V_{sa} \\ V_{sb} \\ V_{sc} \end{bmatrix} = \begin{bmatrix} R_s & 0 & 0 \\ 0 & R_s & 0 \\ 0 & 0 & R_s \end{bmatrix}\begin{bmatrix} i_{sa} \\ i_{sb} \\ i_{sc} \end{bmatrix} + \frac{d}{dt}\begin{bmatrix} \phi_{ia} \\ \phi_{ib} \\ \phi_{ic} \end{bmatrix} + \frac{d}{dt}\begin{bmatrix} \phi_{ma} \\ \phi_{mb} \\ \phi_{mc} \end{bmatrix} \quad (4)$$

Where:
$$\begin{bmatrix} \phi_{ia} \\ \phi_{ib} \\ \phi_{ic} \end{bmatrix} = \begin{bmatrix} L_{aa} & M_{ab} & M_{ac} \\ M_{ba} & L_{bb} & M_{bc} \\ M_{ca} & M_{cb} & L_{cc} \end{bmatrix}\begin{bmatrix} i_{sa} \\ i_{sb} \\ i_{sc} \end{bmatrix} \quad (5)$$

The mutual fluxes between the stator and the rotor of a machine consisting of 4 poles (corresponding to 4 fictitious coils) are given by the following matrix system:

$$\begin{bmatrix} \phi_{ma} \\ \phi_{mb} \\ \phi_{mc} \end{bmatrix} = \begin{bmatrix} M_{saf1} & M_{saf2} & M_{saf3} & M_{saf4} \\ M_{sbf1} & M_{sbf2} & M_{sbf3} & M_{sbf4} \\ M_{scf1} & M_{scf2} & M_{scf3} & M_{scf4} \end{bmatrix}\begin{bmatrix} i_{f1} \\ i_{f2} \\ i_{f3} \\ i_{f4} \end{bmatrix} \quad (6)$$

The modeling of the rotating part of the machine is based on the use of Newton's second law.

We will have:
$$J\frac{d}{dt}\Omega + F\Omega = T_e - T_r \quad (7)$$

By carrying out an energy balance, it is shown that the torque is expressed by the following equation [14]:

$$T_e = \frac{\partial W_{comag}}{\partial \theta_r} = \left[i_{sabc}\right]\frac{\partial \left[M_{ss}\right]}{\partial \theta_r}\left[i_{sabc}\right] + \left[i_{sabc}\right]\frac{\partial \left[M_f\right]}{\partial \theta_r}\left[i_f\right] \quad (8)$$

C. Calculation of inductances

It is very important to know that the accuracy of inductance calculation is the key point for a successful PMSM simulation. All inductances are calculated by using the winding function approach [11] - [13], which is based on the following formula:

$$L_{A.B} = \mu_o rl \int_o^{2\pi} N_A(\varphi,\theta_r)n_B(\varphi,\theta_r)g^{-1}(\varphi,\theta_r)d\varphi \quad (9)$$

Where, μ_o is the air permeability, r is the mean radius of the gap, l is the active length of the machine, $N_A(\varphi,\theta_r)$ is the winding function of the winding " a ", $n_B(\varphi,\theta_r)$ is the distribution function of the winding " b " and $g^{-1}(\varphi,\theta_r)$ is the inverse of the gap function.

➢ Magnetization inductance of phase " a "

The magnetization inductance of the phase " a " of the stator is calculated thanks to the following relation:

$$L_{saa} = \mu_o rl \int_o^{2\pi} N_a(\varphi,\theta_r)n_a(\varphi,\theta_r)g^{-1}(\varphi,\theta_r)d\varphi \quad (10)$$

The use of the equation (10) for the parameters of the studied machine gives after all calculated fact, the magnetization inductance L_{saa} in the following form:

$$L_{saa} = \frac{\mu_o r.l.N_s^2}{g_o}220.\alpha_s, \quad and \quad L_{saa} = L_{sbb} = L_{scc} \quad (11)$$

➢ Mutual inductances between stator phases

To calculate these inductances, it is sufficient to base on the following integral:

$$M_{sab} = \frac{\mu_o rl}{g_o}\int_o^{2\pi} N_a(\varphi,\theta_r)n_b(\varphi,\theta_r)d\varphi \quad (12)$$

The use of the equation (12) using the parameters of the machine studied gives us the following results:

$$M_{sab} = -\frac{\mu_o r.l.N_s^2}{g_o}104.\alpha_s, \quad and \quad M_{sac} = M_{sbc} = M_{sab} \quad (13)$$

➢ Mutual inductances between the stator and the rotor

The mutual inductance between phase " a " and the fictitious coil "f_1 " is obtained using the following integral:

$$M_{af1} = \frac{\mu_o r l}{g_o} \int_o^{2\pi} N_a(\varphi, \theta_r) n_{f1}(\varphi, \theta_r) d\varphi \qquad (14)$$

To calculate this integral, we vary the position of the fictitious coil " f_1 " with respect to the stator, and for each interval the value of the analytically integral is calculated. The analytical integration gives us the results recorded in TABLE I. For each given interval, we have a value of the mutual inductance function of (θ_r).

TABLE I. MUTUAL INDUCTANCES BETWEEN THE FIRST STATORIC PHASE AND THE FIRST FICTIONAL COIL

Inductance M_{saf1} [H]	(θ_r) [rad]
$\frac{\mu_o . r . l . N_s . N_r}{g_o}(5.\theta_r + 3.\alpha_r - 12.\alpha_s)$	$0 \leq \theta_r \leq \alpha_s$
$\frac{\mu_o . r . l . N_s . N_r}{g_o}(4.\theta_r + 3.\alpha_r - 11.\alpha_s)$	$\alpha_s \leq \theta_r \leq 2\alpha_s$
$\frac{\mu_o . r . l . N_s . N_r}{g_o}(2.\theta_r + 3.\alpha_r - 7.\alpha_s)$	$2\alpha_s \leq \theta_r \leq 9\alpha_s - \alpha_r$
$\frac{\mu_o . r . l . N_s . N_r}{g_o}(\theta_r + 2.\alpha_r + 2.\alpha_s)$	$9\alpha_s - \alpha_r \leq \theta_r \leq 3\alpha_s$
$\frac{\mu_0 . r . l . N_s . N_r}{g_o}(2.\alpha_r + 5.\alpha_s)$	$3\alpha_s \leq \theta_r \leq 10\alpha_s - \alpha_r$
$\frac{\mu_0 . r . l . N_s . N_r}{g_o}(-\theta_r + \alpha_r + 15.\alpha_s)$	$10\alpha_s - \alpha_r \leq \theta_r \leq 4\alpha_s$
$\frac{\mu_o . r . l . N_s . N_r}{g_o}(-2.\theta_r + \alpha_r + 19.\alpha_s)$	$4\alpha_s \leq \theta_r \leq 11\alpha_s - \alpha_r$
$\frac{\mu_o . r . l . N_s . N_r}{g_o}(-4.\theta_r - \alpha_r + 41.\alpha_s)$	$11\alpha_s - \alpha_r \leq \theta_r \leq 12\alpha_s - \alpha_r$
$\frac{\mu_o . r . l . N_s . N_r}{g_o}(-5.\theta_r - 2.\alpha_r + 53.\alpha_s)$	$12\alpha_s - \alpha_r \leq \theta_r \leq 13\alpha_s - \alpha_r$
$\frac{\mu_o . r . l . N_s . N_r}{g_o}(-6.\theta_r - 3.\alpha_r + 66.\alpha_s)$	$13\alpha_s - \alpha_r \leq \theta_r \leq 9\alpha_s$

Where, $\alpha_s = \pi/18$ and " α_r " represent the opening of magnet.

The mutual inductances between phase " a " and the other fictitious coils " f_2, f_3, f_4 " are calculated in the same way as that of " M_{af1} " but shifted respectively by an angle " $k.\pi/2$ ", with $k = 1, 2, 3$.

For the mutual inductances between the other stator phases " b, c " and the first fictitious coil " f_1 ", the same expressions are obtained but shifted respectively by $-2\pi/3$ and $-4\pi/3$.

III. MODELING OF THE PMSM WITH ITSC FAULT

The presence of the fault modifies the equation system of the PMSM, so when the Short circuit occurs between the turns of the phase " a ", the system of the equation (4) becomes:

$$\begin{bmatrix} V_{sa} \\ V_{sb} \\ V_{sc} \\ V_{sd} \end{bmatrix} = \begin{bmatrix} R'_{sa} & 0 & 0 & 0 \\ 0 & R_{sb} & 0 & 0 \\ 0 & 0 & R_{sc} & 0 \\ 0 & 0 & 0 & R_{sd} \end{bmatrix} \begin{bmatrix} i_{sa} \\ i_{sb} \\ i_{sc} \\ i_{sd} \end{bmatrix} + \frac{d}{dt}\begin{bmatrix} \phi_{la} \\ \phi_{lb} \\ \phi_{lc} \\ \phi_{ld} \end{bmatrix} + \frac{d}{dt}\begin{bmatrix} \phi_{ma} \\ \phi_{mb} \\ \phi_{mc} \\ \phi_{md} \end{bmatrix} \qquad (15)$$

Fig. 2, represents the equivalent stator circuit of the new system with two additional branches, " sd " relative to the short-circuited turns and " sf " relative to the SC branch.

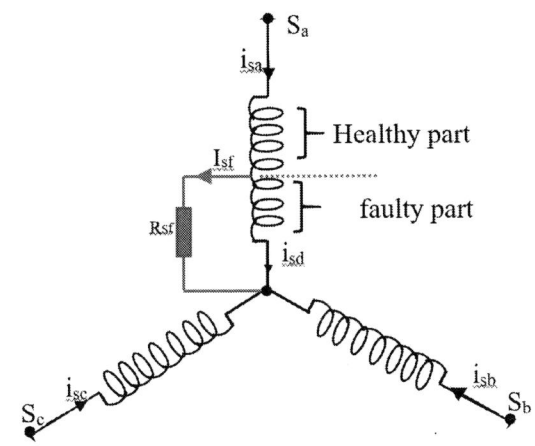

Fig. 2. Representation of the PMSM containing ITSC faults.

A. Calculation of inductances

During an ITSC, the winding function of the faulty phase changes as well as its own inductance, its resistance and its mutual inductances. After any calculation done, we get the expressions of given inductances as follows:

➢ Magnetization inductance of the faulty phase

The use of the equation (10) for the parameters of the studied machine gives after all calculated fact, the magnetization inductance in the following form:

$$L'_{saa} = \frac{\mu_o r.l.N_s^2.\alpha_s}{g_o}\left(220 - 36.cc + \frac{203.cc^2}{36}\right) \qquad (16)$$

➢ Magnetization inductance of the healthy phases

$$L_{sbb} = \frac{\mu_o r.l.N_s^2}{g_o}220.\alpha_s, \quad with \quad L_{scc} = L_{sbb} \qquad (17)$$

➢ Mutual inductances between stator phases

The use of the equation (12) using the parameters of the machine studied gives us the following results:

$$M_{sab} = \frac{\mu_o r.l.N_s^2.\alpha_s}{g_o}(-104 + 14.cc) \quad and \quad M_{sac} = M_{sab} \qquad (18)$$

$$M_{sbc} = -\frac{\mu_o r.l.N_s^2}{g_o}104.\alpha_s \qquad (19)$$

➢ Magnetization inductance of the short-circuit (SC) part

$$L_{sd} = \frac{\mu_o r.l.N_s^2.\alpha_s}{g_o}\frac{203.cc^2}{36} \qquad (20)$$

➢ Mutual inductance between phase " a " and SC part

$$M_{sad} = \frac{\mu_o r.l.N_s^2.\alpha_s}{g_o}\left(18.cc - \frac{203.cc^2}{36}\right) \qquad (21)$$

➢ Mutual inductances between phases " b, c " and short-circuit part

$$M_{sbd} = -\frac{\mu_o.r.l.N_s^{2}.\alpha_s}{g_O}14.cc \quad and \quad M_{scd} = M_{sbd} \qquad (22)$$

➤ Mutual inductance between the fictitious coil $"f_1"$ and the short-circuit part

The mutual inductances between the short-circuit part and the fictitious coils are obtained as a function of the relative position of the fictitious coils with respect to the faulty section $"s_d"$ which will be expressed by the following expression:

$$M_{df1} = \frac{\mu_o rl}{g_O}\int_{o}^{2\pi} N_d(\varphi,\theta_r)n_{f1}(\varphi,\theta_r)d\varphi \qquad (23)$$

To calculate this integral, we vary the position of the fictitious coil $"f_1"$ with respect to the short-circuit part, and for each interval the value of the analytically integral is calculated.

The analytical integration gives us the results recorded in TABLE II. For each given interval, we have a value of the mutual inductance function of (θ_r).

TABLE II.   MUTUAL INDUCTANCE BETWEEN THE SHORT CIRCUIT PART AND THE FICTITIOUS COIL.

Inductance M_{saf1} on (H)	(θ_r) on (rad)
$\frac{\mu_0.r.l.N_s.N_r}{g_o}\left(cc.\theta_r - 2cc.\alpha_s + \frac{29cc.\alpha_r}{36}\right)$	$0 \le \theta_r \le 2\alpha_s$
$\frac{\mu_o.r.l.N_s.N_r}{g_o}\frac{29cc.\alpha_r}{36}$	$2\alpha_s \le \theta_r \le 9\alpha_s - \alpha_r$
$\frac{\mu_o.r.l.N_s.N_r}{g_o}\left(9cc.\alpha_s - cc.\theta_r - \frac{7cc.\alpha_r}{36}\right)$	$9\alpha_s - \alpha_r \le \theta_r \le 9\alpha_s$
$-\frac{\mu_o.r.l.N_s.N_r}{g_o}\frac{7cc.\alpha_r}{36}$	$9\alpha_s \le \theta_r \le 38\alpha_s - \alpha_r$

with, cc represents the ratio between the shorted turns and the number of total turns.

➤ Mutual inductance between the phase $"a"$ and the fictitious coil $"f_1"$

The mutual inductance between phase $"a"$ and the fictitious coil $"f_1"$ for the range of θ_r varying from 0 to α_s is given by equation (24):

$$M_{saf1} = \frac{\mu_o.r.l.N_r}{g_o}\begin{pmatrix}\left(\frac{7}{36}cc-1\right)N_s.(\alpha_s)+\left(\frac{7}{36}cc-2\right)N_s.(\alpha_s-\theta_r)\\ \left(1-\frac{29}{36}cc\right)N_s.(\alpha_s)+\left(2-\frac{29}{36}cc\right)N_s.(\alpha_s)\\ \left(3-\frac{29}{36}cc\right)N_s.(\theta_r+\alpha_r-4.\alpha_s)\end{pmatrix} \qquad (24)$$

Proceeding in the same way, we can obtain the expressions for different intervals as well as for other mutual stator / rotor inductances.

IV. SIMULATION RESULTS

The simulation is performed on a 3.6 Kw PMSM, 4 poles, 36 stator slots and 4 fictitious rotor coils. The arrangement of the stator and rotor winding of this motor and its parameters are indicated in the appendix. The mathematical model obtained and all the expressions of the inductances are implemented under MATLAB environment. In all the simulations considered in this work, we started the no-load simulation and we applied from the instant 0.5 s a load torque equal to that of the nominal torque, which is of the order of 17.5 Nm.

A. Healthy motor operation

Fig. 3. (a), illustrate the speed of rotation of the PMSM. The speed response has ripples of ± 10 rpm around the synchronism speed (2000 rpm). These ripples are due to the effect of space harmonics. Fig. 3. (b), depicts the waveforms of the stator current absorbed by the simulated motor. This current is not sinusoidal; it presents fluctuations, which are due to the effect of space harmonics. Fig. 3. (c), gives the spectrum of the stator current of phase $"a"$. We note that the spectrum contains in addition to the fundamental a series of harmonics which represent the space harmonics whose frequencies of these harmonics can easily be verified using the following expression. This spectrum will be considered as the reference spectrum.

$$f_{sh} = [2.k+1]f_s \qquad (25)$$

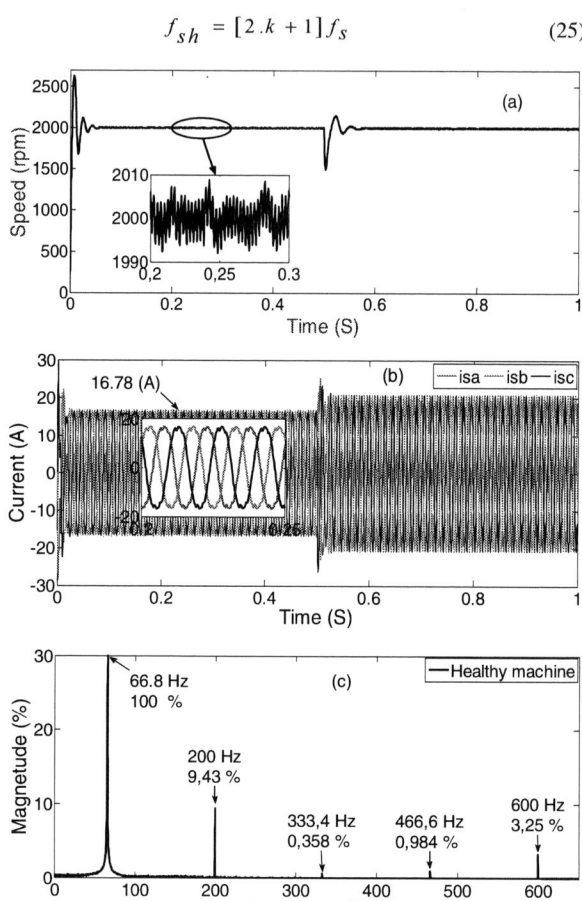

Fig. 3. Healthy Operation Mode. (a) - Speed of rotation, (b) - Currents of the three stator phases, (c) - Current spectrum of the phase $"a"$.

B. Motor operation with ITSC fault

In this part we consider the case of the short-circuit fault between 6 turns of the coil placed respectively in the slots 1 and 8 (the number of shorted turns is the equivalent of 8.33% of the turns of the phase $"a"$). Fig. 4. (a), presents the new mutual inductances between the stator phases and the fictitious coil $"f_1"$ as a function of the rotor position, On the other hand their derivatives are illustrated in Fig. 4. (b). Fig. 5. (a), shows the increase of the rpm fluctuations from ± 10 rpm for a zero ITSC level up to ± 100 rpm for an ITSC level of 8.33 %. while Fig. 5. (b), illustrate the three stator currents. Fig. 5. (c), present the spectrum of the stator current for an ITSC fault of 8.33 %.

TABLE III summarizes the variation of the amplitudes of the most significant harmonics as a function of the ITSC fault severity.

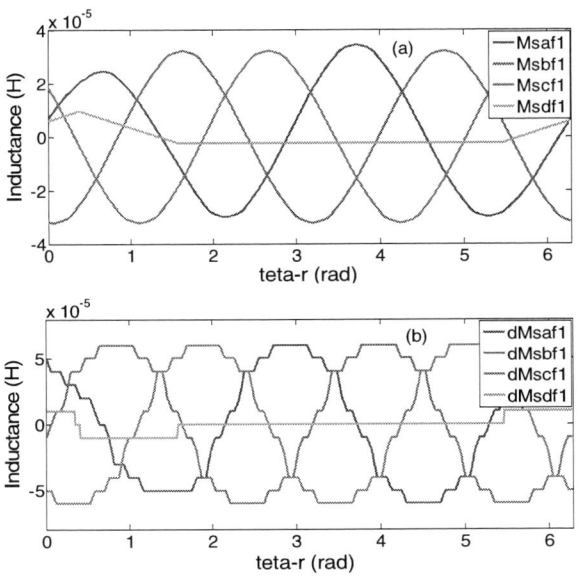

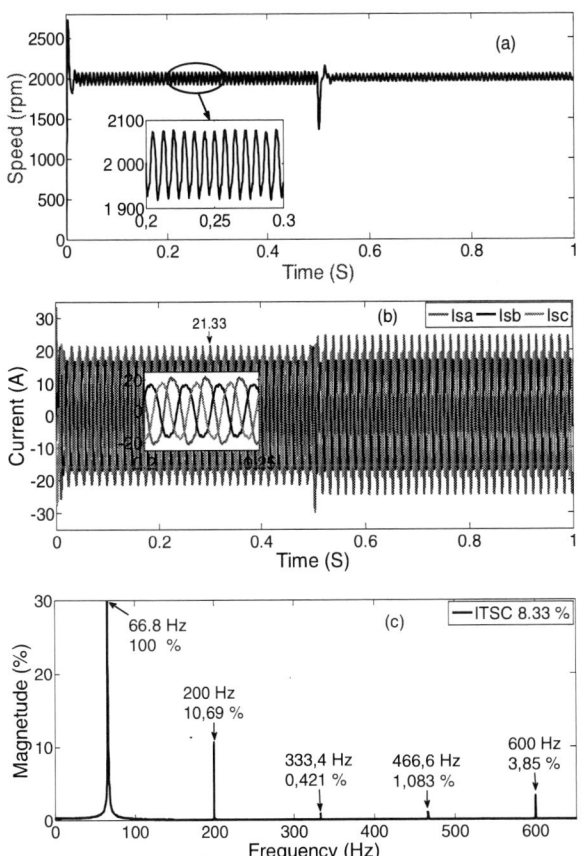

Fig. 4. (a) - Mutual inductances between the stator phases and the fictitious coil "f_1", (b) - their derivative, with an ITSC faults of 8.33 %.

Fig. 5. Operation with an ITSC fault (a) – Speed of rotation, (b) - Currents of the three stator phases, (c) Current spectrum of the phase "a".

TABLE III. AMPLITUDES OF THE HARMONIC CHARACTERISTIC OF THE ITSC FAULT.

ITSC severity (%)	0	1.38	4.164	8.33
Frequency (Hz)	Amplitude (%)	Amplitude (%)	Amplitude (%)	Amplitude (%)
$3f_s$	9.437	9.868	10.32	10.69

According to these results we find that the Inter-Turn Short Circuit fault is manifested by: the increase of the current circulating in the faulty phase. The increase of the currents in the other phases is due to the magnetic coupling between the three phases. Ripples in deferential electromechanical quantities appeared as well as the increase of the magnitude of third harmonic. We conclude the tracking of the magnitude of the third harmonic can be a decisive indicator to diagnose the Inter-Turn Short Circuit fault.

V. CONCLUSION

The work developed in this paper concerns the detection of the ITSC fault in Permanent Magnet Synchronous Motors. The proposed model based on the approach of the modified winding function of which we considered the permanent magnet as a fictitious coil traversed by an excitation current "i_f". The simulation results showed that the lack of ITSC fault is manifested by the increase of the current in the faulty phase and the ripples in deferential electromechanical quantities. Moreover, the severity of the ITSC fault is manifested by the increase of the magnitude of the third harmonic of the line current. In general, the simulation results show the interest of the model adopted and the efficiency of the spectral analysis technique of the stator current for the detection of the Inter-Turn Short Circuit fault.

ANNEXE

TABLE : PMSM Parameters.

Symbol	Description	Values	Units
V_n	Rated voltage	150	V
f_s	Rated frequency	66,7	Hz
I_n	Rated current	15.1	A
C_n	Rated torque	17.5	Nm
Ω_n	Rated speed	2000	rpm
R_s	Stator resistance	0.295	Ω
L_s	Synchronous inductance	3.5	mH
J	Moment of the inertia	3.10^{-4}	$Kg.m^2$
F	Viscous rubbing	0.017	$Nm/rad/s$
P	Number of pole pairs	2	-----
ε_s	Stator notch opening	0,06	rad
g_o	Nominal air gap	2	mm
r	Stator radius	64	mm
rr	Rotor radius	52	mm
l	Length of the machine	250	mm
N_s	Number of turns per Coil	6	-----
α_r	Opening magnets	1,15	rad
h	Thickness of magnets	10	mm

Fig. 6.a, represents the distribution function $n_a(\varphi)$ of phase "a" with a mean value $\langle n_a(\varphi) \rangle$ = 3Ns. This value allows us to find the winding function of phase "a" as shown in Fig. 6.b. Fig. 6.c, represents the distribution function of the fictitious coil "f_1".

978-1-5386-7688-2/19 $31.00 © 2019 IEEE   231

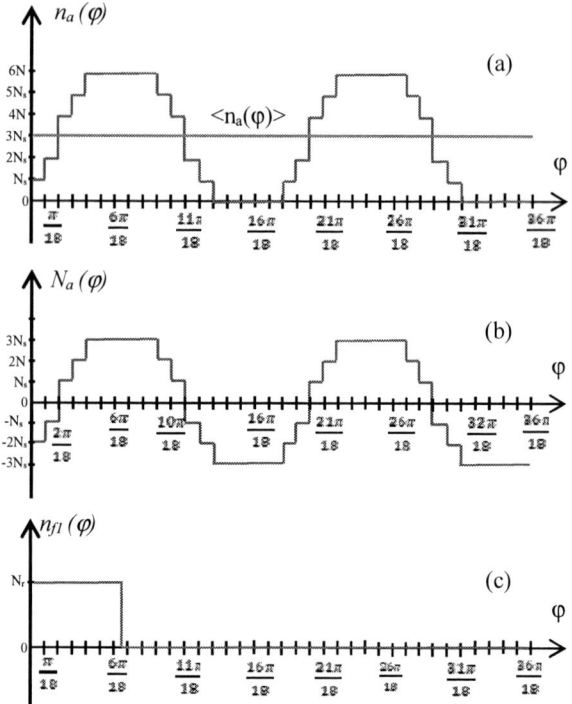

Fig. 6. (a) - Distribution function of phase " a ". (b) - Its winding function. (c) - Fictitious coil distribution function " f_1 ".

REFERENCES

[1] J. Faiz et M. R. Hassan-Zadeh, « Impacts of eccentricity fault on permanent magnet generators for distributed generation », in *International Conference on Optimization of Electrical and Electronic Equipment (OPTIM) 2017 Intl Aegean Conference on Electrical Machines and Power Electronics (ACEMP)*, p. 434–441, 2017.

[2] L. M'hamed, G. Zakaria, et D. Khireddine, « A Robust Sensorless Control of PMSM Based on Sliding Mode Observer and Model Reference Adaptive System », *Int. J. Power Electron. Drive Syst. IJPEDS*, vol. 8, no 3, p. 1016–1025, 2017.

[3] H. Zhou, G. Liu, et W. Zhao, « Dynamic Performance Improvement of Five-Phase Permanent-Magnet Motor With Short-Circuit Fault », *IEEE Trans. Ind. Electron*, vol. 65, no 1, p. 145–155, 2018.

[4] J. Zheng, Z. Wang, D. Wang, et Y. Li, « Review of fault diagnosis of PMSM drive system in electric vehicles », *in Proceedings of the 36th Chinese Control Conference, China*, p. 7426–7432, July 26-28, 2017.

[5] J. Faiz et S. A. H. Exiri, « Short-circuit fault diagnosis in permanent magnet synchronous motors- an overview », *in Intl Aegean Conference on Electrical Machines Power Electronics (ACEMP), 2015 Intl Conference on Optimization of Electrical Electronic Equipment (OPTIM) Intl Symposium on Advanced Electromechanical Motion Systems (ELECTROMOTION)*, p. 18–27, 2015.

[6] R. Z. Haddad, E. G. Strangas, "On the accuracy of fault detection and separation in permanent magnet synchronous machines using MCSA/MVSA and LDA", *IEEE Trans. Energy Convers.*, vol. 31, no. 3, pp. 924-934, Sep. 2016.

[7] W. Liang et W. Fei, « An improved sideband current harmonic model of interior PMSM drive by considering magnetic saturation and cross-coupling effects », *IEEE Trans. Ind. Electron.*, vol. 63, no 7, p. 4097–4104, 2016.

[8] G. Vinson, M. Combacau, T. Prado, et P. Ribot, « Permanent magnets synchronous machines faults detection and identification », in *38th Annual Conference on Industr, IEEE ial Electronics Society IECON*, p. 3925–3930, 2012.

[9] G. Mirzaeva et K. Saad, « Ac motor instrumentation and main air gap flux measurement for fault diagnostics », in *Power Engineering Conference (AUPEC), Australasian Universities*, p. 1–6, 2013

[10] J. Hang, J. Zhang, M. Cheng, and J. Huang, "Online inter-turn fault diagnosis of permanent magnet synchronous machine using zero sequence components," *IEEE Trans. Power Electron.*, vol. 30, no. 12, pp. 6731-6741, Dec. 2015

[11] G. M. Joksimovic, M. D. Durovic, J. Penman, et N. Arthur, « Dynamic simulation of dynamic eccentricity in induction machines-winding function approach », *IEEE Trans. Energy Convers*, vol. 15, no 2, p. 143–148, juin 2000.

[12] J. Gojko, M. W. Thomas, S. Goran, et Z. Ana, « Dynamic Model of Surface Mounted Permanent Magnet Synchronous Machine », *in Proceedings of 1st International Conference on Electrical, Electronic and Computing Engineering IcETRAN 2014, Vrnjačka Banja, Serbia*, ISBN 978-86-80509-70-9 June 2 – 5, 2014.

[13] Z. Liang, D. Liang, et S. Jia, « Inductance Calculation for the Symmetrical Non-Salient Dual Three-Phase PMSM Based on Winding Function Approach », *in 21st International Conference on Electrical Machines and Systems (ICEMS)*, p. 269–274, 2018.

[14] B. Aubert, « Détection des défaut de courts-circuits entre-spires dans les Générateurs Synchrones à Aimants Permanents: Méthodes basées modèles et filtre de Kalman étendu », *Thèse de doctorat*, université de toulouse, mars–2014.

Motor Efficiency Determination of SynRM and Measurement Uncertainty

Youn-Hwan Kim
Rotating Machinery Center
Korea Testing Certification
Gunpo, South Korea
younhwan@ktc.re.kr

Hee-Deuk Jun
Rotating Machinery Center
Korea Testing Certification
Gunpo, South Korea
jhdeuk@ktc.re.kr

Jae-Won Moon
Rotating Machinery Center
Korea Testing Certification
Gunpo, South Korea
moon@ktc.re.kr

Rae-Eun Kim
Intelligent Mechatronics Center
KETI
Bucheon, South Korea
kre2567@keti.re.kr

Se-Hyun Rhyu
Intelligent Mechatronics Center
KETI
Bucheon, South Korea
rhyush@keti.re.kr

Sang-Young Jung
Electronic and Electrical Engineering
Sungkyunkwan University
Suwon, South Korea
syjung@skku.edu

Abstract— **Electric motors are one of the largest electricity usage in the world. It is estimated that they consume between 43% and 46% of global electricity consumption. Recent studies show that efficiency values of Synchronous reluctance motors (SynRMs) reach IE5 Ultra-Premium Efficiency Class. The experimental determination of SynRM efficiency is getting more and more important, because of the need to place these motors in the right energy efficiency levels defined by international regulations. According to IEC 60034-2-3 standard, the efficiency of SynRM is determined by applying the direct or indirect efficiency similar to converter-fed AC induction motors. In this paper, a comparative analysis between direct and indirect efficiency determination for 75 kW SynRM, according to IEC 60034-2-3 standard is proposed. Measurement uncertainty analysis of the direct efficiency determination is carried out from the accuracy requirements for measurement equipment, as required by the GUM and IEC standard.**

Keywords— *Efficiency, Loss Measurement, Measurement Uncertainty, Motor Efficiency, SynRM.*

I. INTRODUCTION

Excessive greenhouse gas emissions, environmental problems such as abnormal temperatures and natural disasters are becoming serious. Due to environmental problems such as global warming, the world has recognized the importance of the environment, and governments around the world have strengthened environmental policies.

The eco-design directive is a European Commission initiative that establishes a framework to set mandatory ecological requirements for energy-using and energy-related products. Under this framework, the EC issued a regulation that set minimum energy performance standards (MEPS) for electric motors. MEPS for electric motors are based on the International Electro-technical Commission (IEC) standards for motor efficiency, which is divided into International Efficiency (IE) classes[1]. The major movements to expand the market of high efficiency motor are the specifications related to the efficiency test method, efficiency rating and marking, and the regulations according to the local situation.

Electric motors consume a major proportion of the world's electric energy[2]. The induction motors(IMs) are the most widely used in the industry due to its robustness and low cost. Motors of IE4 Super-Premium Efficiency Class are already available in the market, and a new IE5 Ultra-Premium Efficiency Class is being considered. Within the IE4 Super-Premium Class, line-start permanent-magnet motors (LSPMs) are a recent entrance in the industrial motor market [3]. Compared to IMs, the Synchronous reluctance motors(SynRMs) provides higher efficiency alternative without the penalty of increased cost/kW. However, SynRMs have disadvantages in terms of high torque ripple and low power factor[4]. The machine torque can be maximized by finding the best insulation ratios, while the torque ripple can be minimized by determining the best rotor slot pitch in the d-axis[5].

Recent studies have shown that SynRMs achieves IE5 energy efficiency class[6-10]. It is likely to be possible for the next generation of SynRMs, leading to an IE5-class general purpose motor, without rare-earth PMs[11]. There are also studies using PM-assisted SynRM technology when there is a size limitation[12]. Although its development is a rising industry trend, it is difficult for manufacturers and users to accept high-efficiency motors due to their high cost and price. In recent years, due to the rising costs of raw materials and labor, the industry faces tremendous pressure to control product costs and yields[13].

The current standard covers from IE1 to IE4 about 3-phase cage induction motors with a grid supply of 50 or 60 Hz and with 2, 4, 6 or 8 poles (IEC 60034-30-1 1st Edition 2014-3) [14]. The nominal limits for efficiency classes of IE5 have not yet been established. Motor technologies for IE5 are currently not well developed and not commercially available [15]. As the use of motors supplied by converters has increased, international and national committees have led to the definition of standardized procedures for evaluating the efficiency of the motors which are supplied by converters. IEC TS 60034-30-2 1st Edition 2016-12 [16] presents the nominal limits for efficiency class according to the rated speed and the rated output of motor from IE1 to IE5.

Efficiency is an important issue in various motor topologies. An analysis of efficiency may also require measurement of each loss. The loss measurements are a quite challenging step in the characterization of electrical machines. The existing standard procedures for measuring losses in electrical machines are adapted for standardized testing in the industry. The core losses obtained for such a test include several other sources of losses. As a result, the losses calculated by the use of traditional core loss models will most likely seem to be underestimated. Different methods are proposed to measure the core losses and mechanical losses or to segregate the mechanical loss components, the rotor of an

978-1-5386-7688-2/19 $31.00 © 2019 IEEE

interior PMSMs machine is replaced with a magnetless rotor with an identical geometry to measure the mechanical losses at different speeds[17].

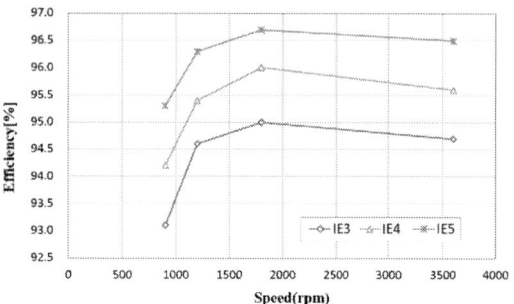

Fig. 1. Efficiency requirements of 75 kW variable speed AC motor

The efficiency requirements of the 75 kW variable speed AC motor are sketched in Fig.1 and the test results of the 75 kW 4poles three-phase induction motor by IEC standards are discussed in this paper.

The demand for analysis of measurement uncertainty(MU) for motor efficiency analysis is increasing. Measurement uncertainty is very important in metrology because in the real environment, it is almost impossible to measure a true value. A measurand always drifts from a true value and because of this, a user should determine the measurement uncertainty experimentally, theoretically and statistically considering the total system wherever possible[18]. Determination of the measurement uncertainty has been studied for several years in the direct and indirect efficiency measurement methods. There have been a number of studies on IM in [19] and [20] and recent analysis has also been made on PMSM in [18]. Evaluating the efficiency of the motor from the sum of the losses described in the detailed studies is a very plausible and reliable method for evaluating the loss of the motor. The objective of this study is to evaluate and verify the measurement uncertainty of efficiency measurements about various motor topologies.

This paper is organized into five sections. Section II outlines the background of measurement uncertainty determination according to the Guide to the Expression of Uncertainty in Measurement (GUM). The motor efficiency test method is described in Section III. Section IV gives an experimental result in the direct efficiency determination method of SynRM. The experimental results for SynRM in indirect efficiency determination are presented in Section V, and Section VI summarizes the results of this work and draws the conclusions.

II. General Understanding of Measurement Uncertainty in Motor Test

The measurement uncertainty is specified by the variance of the measurable value of the variable related to the measurement result. The measurement uncertainty is specified by the variance of the measurable value of the variable related to the measurement result. That is, the output variable Y is expressed by the estimated output y and the measurement uncertainty U as (1).

$$Y = y \pm U \qquad (1)$$

A. Classification of Uncertainty Components

The uncertainty component is classified into type A evaluation, which is a method of evaluation of uncertainty by the statistical analysis of series of observations, and type B evaluation, which is a method of evaluation of uncertainty by means other than the statistical analysis of series of observations. In order to reduce the uncertainty of type A evaluation, it is necessary to obtain enough raw measured data through repeated tests about torque, rotational speed, voltage, current, temperature and resistance. Type B evaluation is obtained using previous measurement data, experience with or related to the behavior and properties of relevant materials and instruments, manufacturer's specifications, data provided in calibration and other certificates.

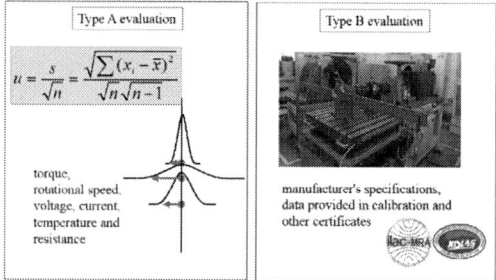

Fig. 2. Type A and Type B evaluation for Motor Efficiency Determination

B. Combined Standard Uncertainty by Propagation of Uncertainty

The combined standard uncertainty of the measurement results is denoted by $u_c(y)$ and represents the estimated standard deviation of the measured object. In general, the combination of standard uncertainties is obtained by propagation of uncertainty. The method of estimating the combined standard uncertainty is divided into the case where there is no correlation, as in (2) and the case where the input amounts are correlated with each other , as in (3):

$$u_c(y) = \sqrt{\sum_{i=1}^{n}\left(\frac{\partial f}{\partial x_i}\right)^2 u^2(x_i)} \qquad (2)$$

$$u_c(y) = \sqrt{\sum_{i=1}^{n}\sum_{j=1}^{n}\left(\frac{\partial f}{\partial x_i}\right)\left(\frac{\partial f}{\partial x_j}\right)u^2(x_i, x_j)} \qquad (3)$$

In the direct methods, there is no need for correlation. However, in the indirect methods, in the case of correlated quantities, as the iron losses and additional losses that are estimated by applying extrapolation methods.

III. Motor Efficiency Test Method

IEC 60034-2-1 is applicable for the measurement of efficiency of DC machines, AC synchronous and induction machines[21]. However, these methods may not be applied to

978-1-5386-7688-2/19 $31.00 © 2019 IEEE

other types of machines such as converter-fed AC induction motors because additional harmonic losses are not taken into account. This is the reason why this specification excludes machines for traction vehicles. IEC TS 60034-2-3 specifies test methods for determining losses and efficiencies of converter-fed AC induction motors[22]. Inverters for traction vehicles using DC power are tested by this specification.

Generally, when an electric motor is supplied from a inverter, the motor loss is higher than when operating in a sinusoidal voltage power system because of the additional harmonic losses. Additional harmonic losses are dependent on the spectrum of the output current or the output voltage of the inverter that is affected by the circuit and control method.

TABLE I. EFFICIENCY TEST METHODS

Test Method		Description	Required facility
Direct Method	Input-output	Torque measurement	Dynamometer for full-load
	Dual-supply back-to-back Test	Dual-supply, back-to-back Test	Two identical unit
Indirect Method	Summation of separate losses	P_{LL} determined from residual loss	Sinusoidal supply and test converter supply
	Equivalent circuit	P_{LL} from assigned value	If test equipment for other tests is not available
	Calorimetric method	Loss determination form coolant temperature rise	Calorimetric instrumentation: Flowmeters, Coolers, Thermal detectors

Direct method is test method by measuring input and output on a single machine. This involves the measurement of electrical or mechanical power input, and mechanical or electrical power out of a machine. In the direct methods, the efficiency is calculated by the input to output ratio.

Measurements of the separate losses in a machine under a particular condition are considered to be indirect. This is not usually the total loss but comprises certain loss components. The method may, however, be used to calculate the total loss or to calculate a loss component. Equivalent circuit and calorimetric method are also considered to be indirect.

In the indirect methods, the efficiency is determined by the summation of separate losses. The respective loss components are iron loss, windage and friction losses, stator and rotor copper losses, additional load losses. To determine the stator-winding losses, the stator winding resistance is measured and corrected to a reference coolant temperature of 25 °C. The rotor winding losses of IM are determined by subtracting the stator-winding and iron losses from the input and multiplying the value by the slip but it is ignored in non-IM. Temperature compensation is also included in determining the rotor winding losses. Friction and windage losses at approximately synchronous speed are determined by noload test from the 4 or more consecutive no-load loss points between approximately 60 % of voltage and 30 % of voltage in case of IM but it is measured by the rotation of dynamometer in non-IM.

Iron losses require the calculation of the iron loss voltage by using the power factor measured in the load test and the constants by using the ratio applied to the iron losses at no load. The residual losses are determined for each load point by subtracting from the input power: the output power, the uncorrected stator winding losses at the resistance of the test, the iron losses, the windage and friction losses, and the uncorrected rotor winding losses corresponding to the determined value of slip. After the smoothing of the residual loss data procedure for the slope constant establishment, a value of additional load losses for each load point are determined by using the product of the torque squared and the slope constant. The additional load losses are often referred as "not clearly know" losses that include, e.g., the extra losses caused by the current linkage harmonics created by stator windings, permeance harmonics produced by stator and rotor slots, saturation harmonics from core lamination, and the extra losses caused by the difference of ac and dc resistance[23].

The total losses are taken as the sum of the iron losses, the friction and windage losses, the winding losses and the additional load losses.

IV. DIRECT EFFICIENCY DETERMINATION METHOD OF SYNRM

A. Test Bench for SynRM of Direct Method

Fig.3 shows the test equipment layout for efficiency determination of SynRM. The torque and speed signals are generated from the torque transducer and a speed encoder. The generated signals are fed to a power analyzer, where the values of the torque and the speed are displayed. The measuring current transformer of 1 MHz bandwidth is used to measure the current signal that is fed from the inverter to the motor, whereas the voltage is directly measured by a three phase four-wire connection with power analyzer. The induced current measured from the current transformer(CT) is sent to the shut resistance and then to the power analyzer. The power analyzer is used to measure the currents, the voltages and the power factors as well. The LabVIEW graphical user interface (GUI) is used to automatically read the power analyzer measured data. The measured torque and rotational speed information are stored in the PC in synchronization with the electrical data. All data are measured at a sampling rate of 0.5 seconds and averaged over 30 seconds. About measuring equipment, the power meters shall have the same nominal accuracy for frequencies above 10 times the switching frequency of the invertor. The bandwidth of the current sensors and acquisition channels shall be at least 0 Hz to 100 kHz. Internal filters in digital power meters are turned off. The laboratory test setup for experimental test of SynRM is shown in Fig. 4

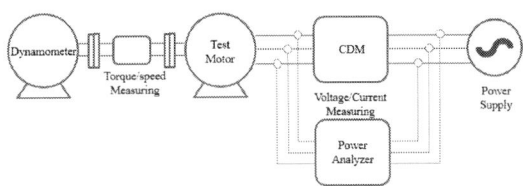

Fig. 3. Test equipment layout for direct efficiency determination of 75kW SynRM

978-1-5386-7688-2/19 $31.00 © 2019 IEEE

Fig. 4. Laboratory test setup for experimental test of 75kW SynRM.

The Complete drive module (CDM) applied voltage of SynRM is AC 380V and it is powered by AC power supply. The rated output of SynRM is 75 kW and 50 % load point of the rated output is 37.5 kW. The rated torque is 398 Nm and 600 Nm torque sensor range is used for the test.

B. Test Results and Measurement Uncertainty Analysis of SynRM

The test for MU of SynRM is conducted at the maximum output point, and the maximum speed point after temperature saturation. For the calculation of type A evaluation, each test was carried out 5 times at each operating points. IEC TS 60034-2-3 explains that the measuring instruments shall have the equivalent of an accuracy class of 0.2 in case of direct test and 0.5 in case of an indirect test. The measuring equipment shall reach an overall uncertainty of 0.2% of reading at power factor 1.0 and shall include all errors of instrument transducers. The test is carried out with equipment satisfying the standard.

TABLE II. CENTRAL VALUES AND STANDARD UNCERTAINTIES(DIRECT METHOD)

Measured quantities		Central Value	Standard Uncertainty (A-type)
100 % output point	Rotating speed	1798.9 rpm	0.2 rpm
	Torque	398.50 Nm	0.08 Nm
	Phase current	195.10 A	0.05 A
	Input power	77.09 kW	0.03 kW
50 % output point	Rotating speed	1799.4 rpm	0.3 rpm
	Torque	198.22 Nm	0.08 Nm
	Phase current	119.97 A	0.21 A
	Input power	38.44 kW	0.07 kW

TABLE III. INSTRUMENTS STANDARD UNCERTAINTIES (DIRECT METHOD)

Measuring Instruments	Range	Standard Uncertainty (B-type)
Voltmeter	600 V	0.4 V
Ammeter(CT)	500 A	0.3 A
Wattmeter	240 W	0.02 W
Torque meter	600 Nm	0.3 Nm
Tachometer	3600 rpm	0.3 rpm

A-type standard uncertainties are calculated by the standard deviation and the degree of freedom. A degree of freedom is a variable that can change freely under given conditions and in this case, it is 4 from the number of measurements. B-type standard uncertainties are determined from the measurement uncertainty of the instrument calibration report and the coverage factor $k = 2$.

The direct efficiency determination requires the direct measurement of motor output power P_2 and input power P_1

$$\eta_d = \frac{P_2}{P_1} \tag{4}$$

Whose uncertainty is

$$u_c(\eta_d) = \sqrt{\left(\frac{1}{P_1}\right)^2 \cdot u(P_2)^2 + \left(-\frac{P_2}{P_1^2}\right)^2 \cdot u(P_1)^2} \tag{5}$$

Motor output power P_2 is calculated by using the measured torque N and speed n values from torque meter and tachometer.

$$P_2 = \frac{2\pi \cdot n \cdot N}{60} \tag{6}$$

$$u_c(P_2) = \sqrt{\left(\frac{2\pi \cdot n}{60}\right)^2 \cdot u(N)^2 + \left(\frac{2\pi \cdot N}{60}\right)^2 \cdot u(n)^2} \tag{7}$$

$u(N)$ and $u(n)$ is standard uncertainties for torque and speed. They are derived using (2), which combines A-type and B-type evaluation.

The expanded uncertainties are obtained by using the coverage factor that transformed the combined standard uncertainty into a 95% confidence interval. The value of the coverage factor k is usually between 2 and 3. In this study, $k = 2$ is used in consideration of the effective degrees of freedom. . Equation (8) shows calculation of the effective degree of freedom from the combined standard uncertainty.

$$v_{eff} = \frac{u_c^4(y)}{\sum_{i=1}^{N} \frac{c_i^4 u^4(x_i)}{v_i}} \tag{8}$$

The overall measurement results for MU determination for the direct efficiency method of SynRM are summarized in Table IV. C_n is higher at the 100 % output point than at the 50% output point. The reason for the higher C_n is that a larger torque measurement result from the 100 % output point is applied in the calculation of the sensitivity coefficient. C_N has the same value because it is affected by the rotation speed.

C_{P1}, C_{P2} are 2 times larger than at the 50% output point and at the 100 % output point. C_{P1}, C_{P2} depend on the output power and input power, and the larger the output power and input power, the smaller C_{P1}, C_{P2} are.

$u_c(P_2)$ is determined under the influence of C_n, C_N. In this test it appears to be more dependent on C_N than C_n. $u_c(P_1)$ appears to be reflected by the oscillation of the input value at the 50% output point. As a result, $u_c(\eta_d)$ is calculated to be

0.375 times larger than at the 100 % output point to at the 50 % output point.

TABLE IV.   PARAMETERS FOR MEASUREMENT UNCERTAINTY CALCULATION OF SYNRM (DIRECT METHOD)

Quantities	100% output point	50% output point
C_n	41.7	20.8
C_N	188.4	188.4
C_{P1}	-0.0126	-0.0253
C_{P2}	0.0130	0.0260
$u_c(P_1)$	0.036	0.073
$u_c(P_2)$	0.060	0.059
$u_c(\eta_d)$	0.0009	0.0024
$u_e(\eta_d)$	0.0018	0.0048
η_d	$97.38\% \pm 0.18\%$	$97.17\% \pm 0.48\%$

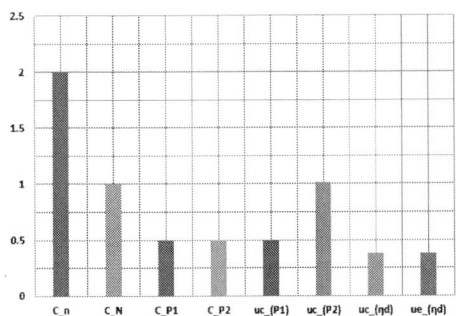

Fig. 5. Values of 100% output point /50% output point Ratio (Direct Method)

V.   INDIRECT EFFICIENCY DETERMINATION METHOD OF SYNRM

A. Test Bench for SynRM of Indirect Method

Fig.6 shows the test equipment layout for efficiency determination of SynRM of indirect method. It is similar to the test equipment layout for direct method except that ohmmeter and thermometer are added.

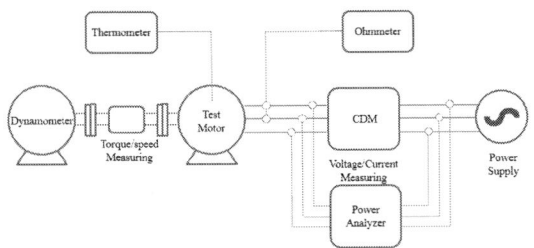

Fig. 6.   Test equipment layout for indirect efficiency determination of 75kW SynRM

B. Test Results Analysis of SynRM(Indirect Method)

The efficiency test of SynRM is conducted by the indirect efficiency method. The test is carried out with equipment satisfying the IEC TS 60034-2-3.

For the calculation of the separated losses, load test is carried at 125%, 115%, 100%, 75%, 50%, 25% of the rated load after the temperature saturation at 75 kW output power. No-load test is carried at 105%, 103%, 100%, 95% of the no-load- phase-voltage. The no-load-phase-voltage means the voltage determined by the control algorithm of the CDM when no load is applied. IEC 60034-2-1 recommends that it should be tested at voltages of 110%, 100%, 95% and 90%, but the test has been carried out with the voltage adjustable at the installed CDM. The current angle is kept constant while the voltage is being adjusted.

TABLE V.   CENTRAL VALUES AND STANDARD UNCERTAINTIES (INDIRECT METHOD)

Measured quantities		Central Value	Standard Uncertainty (A-type)
S1 Test	Stator resistance (before S1)	0.0137 Ω	0.0026 Ω
	Room temperature (before S1)	18.6 °C	0.7 °C
	Stator resistance (after S1)	0.01608 Ω	0.0021 Ω
	Room temperature (after S1)	22.7 °C	0.6 °C
No-load Test	Phase current	51.86 A	0.2 A
	Input power	1.05 kW	0.1 kW
	F/W loss	287 W	5 W
Load Test	Rotating speed	1798.9 rpm	0.2 rpm
	Torque	398.50 Nm	0.08 Nm
	Phase current	195.10 A	0.05 A
	Input power	77.09 kW	0.03 kW

TABLE VI.   INSTRUMENTS STANDARD UNCERTAINTIES(INDIRECT METHOD)

Measuring Instruments	Range	Standard Uncertainty (B-type)
Voltmeter	600 V	0.4 V
Ammeter(CT)	500 A	0.3 A
Wattmeter	240 W	0.02 W
Torque meter	600 Nm	0.3 Nm
Tachometer	3600 rpm	0.3 rpm
Ohmmeter	200 mΩ	0.005 mΩ
Thermometer	200 ˚C	0.5 ˚C

Friction and windage losses are measured at 1800rpm by the rotation of dynamometer. Except for friction and windage losses, tests are conducted according to IEC TS 60034-2-3-A Summation of losses method. As a result, the efficiency at 100% load by indirect method is calculated as 97.25%. The test results and the instrument information are summarized in Table V and Table VI.

Fig. 7 and Fig. 8 show the iron loss data and residual loss data. The polynomial fit and linear fit are used for the loss estimation. There also exist uncertainties in this process. The

978-1-5386-7688-2/19 $31.00 © 2019 IEEE

final separated loss results for each load are shown in Table VII.

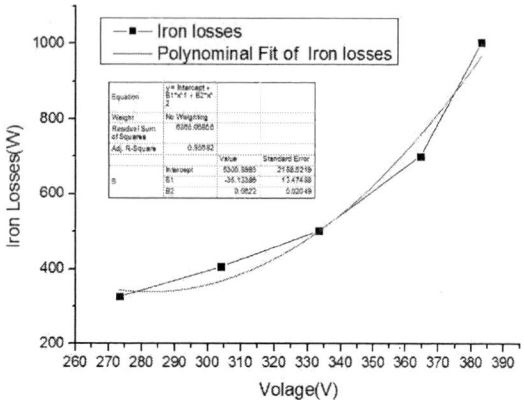

Fig. 7. Iron loss data and the polynominal fit

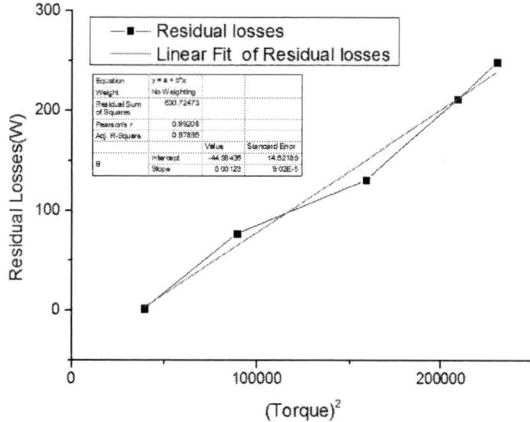

Fig. 8. Residual loss data and the linear fit

TABLE VII. SEPARATED LOSSES(W)

Losses	115% load	100% load	75% load	50% load
Friction and windage losses	286.65	286.65	286.65	286.65
Iron losses	694.06	710.39	576.79	464.63
Stator-winding losses	1176.81	891.73	574.11	337.25
Additional load losses	257.50	195.95	110.06	48.48
The total losses	2464.04	2123.56	1572.81	1151.62

VI. CONCLUSION

The efficiency test of SynRM is conducted by direct and indirect efficiency method. The test is carried out with equipment satisfying the IEC TS 60034-2-3. In the direct method, the efficiency measurement result is $97.38\% \pm 0.18\%$ and in the indirect method, the efficiency measurement result is 97.25%. Measurement uncertainty analysis of the direct efficiency determination is carried out from the accuracy requirements for measurement equipment, as required by the GUM and IEC standard. The result of the indirect method is within the uncertainty range of the result of the direct method.

In several countries, Premium/IE3 motors have been mandatory. A pass / fail decision is determined by a slight difference in the efficiency measurement result. Uncertainty about measurement is at the center of controversy. As the motor drive using CDM increases, a modified efficiency evaluation method is needed. An uncertainty assessment should be performed taking these factors into account. The uncertainty of the result of the indirect method is more complex and is not included in the content of this study. A study of uncertainty by the indirect method on SynRM is also needed for efficiency comparison with IM.

VII. ACKNOWLEDGMENT

This work was supported by the Ministry of Trade, Industry and Energy, Korea, supervised by the Korea Institute of Energy Technology Evaluation and Planning (KETEP) (No. 20172010105920)

REFERENCES

[1] E. C. Bortoni, J. V. Bernardes Jr., P.V.V. da Silva, V.A.D. Faria, and P. A.V. Vieira, "Evaluation of manufacturers strategies to obtain high-efficient induction motors," Sustainable Energy Technologies and Assessments 31, pp. 221–227, 2019.

[2] H. Karkkainen, L.Aarniovuori, M.Niemela, and J.Pyrhonen, "Converter-Fed Induction Motor Efficiency," IEEE Trans. Ind. Elect. Magazine, vol. 11, no. 2, pp. 45–57, Jun. 2017.

[3] A. de Almeida, F. Ferreira, and A. Quintino Duarte, "Technical and economical considerations on super high-efficiency three-phase motors," IEEE Trans. Ind. Appl., vol. 50, no. 2, pp. 1274–1285, Mar. 2014.

[4] M. A. Kabir, and I. Husain, "Application of a Multilayer AC Winding to Design Synchronous Reluctance Motors," IEEE Trans. Ind. Appl., vol. 54, no. 6, pp. 5941–5953, Nov/DEC. 2018.

[5] Reza–Rajabi Moghaddam and Freddy Gyllensten, "Novel High-Performance SynRM Design Method: An Easy Approach for A Complicated Rotor Topology," IEEE Trans. Ind. Elect., vol. 61, no. 9, pp. 5058 - 5065, SEP. 2014.

[6] V. A. Dmitrievskii, V. A. Prakht, and V. M. Kazakbaev, "Ultra Premium Efficiency (IE5 EnergyEfficiency Class) Synchronous Reluctance Motor with Fractional Slot Winding," 2018 International Conference on Electrical Machines (ICEM), pp. 1015-1020.

[7] V. Dmitrievskii, V. Prakht, V. Kazakbaev, S. Oshurbekov, and I. Sokolov, "Developing ultra premium efficiency (IE5 class) magnet-free synchronous reluctance motor," 2016 6th International Electric Drives Production Conference (EDPC), pp. 2-7.

[8] N. Safin, V. Kazakbaev, V. Prakht, and V. Dmitrievskii, "Calculation of the efficiency and power consumption of induction IE2 and synchronous reluctance IE5 electric drives in the pump application based on the passport specification according to the IEC 60034-30-2," 2018 25th International Workshop on Electric Drives: Optimization in Control of Electric Drives (IWED), pp. 1-5.

[9] S. Stipetic, D. Zarko, and N. Cavar, "Design Methodology for Series of IE4/IE5 Synchronous Reluctance Motors Based on Radial Scaling," 2018 International Conference on Electrical Machines (ICEM), pp. 146-151.

[10] V. Prakht, V. Dmitrievskii, and V. Kazakbaev, "Developing and experimental study a 1.5 kW synchronous reluctance motor of ultra premium efficiency (IE5 class)," 2017 IEEE 58th International Scientific Conference on Power and Electrical Engineering of Riga Technical University (RTUCON), pp. 1-5.

[11] A. T. de Almeida, F. J. T. E. Ferreira, G.e Baoming, "Beyond Induction Motors—Technology Trends to Move Up Efficiency," IEEE Trans. Ind. Appl., vol. 50, no. 3, pp. 2103–2114, May/Jun. 2014.

[12] V. Prakht, V. Dmitrievskii, and V. Kazakbaev, "Mathematical modeling ultra premium efficiency (IE5 class) PM assisted synchronous reluctance motor with ferrite magnets," 2018 25th International Workshop on Electric Drives: Optimization in Control of Electric Drives (IWED), pp. 1-5.

[13] S. M. Lu, "A review of high-efficiency motors: Specification, policy, and technology," Renewable and Sustainable Energy Reviews 59, pp. 1–12, 2016.

[14] D. G. Dorrell, "A Review of the Methods for Improving the Efficiency of Drive Motors to Meet IE4 Efficiency Standards," Journal of Power Electronics, vol. 14, no. 5, pp. 842-851, Sep. 2014.

[15] Rotating Electrical Machines - Part 30-1: Efficiency classes of line operated AC motors (IE code), IEC 60034-30-1, Mar. 2014.

[16] Rotating Electrical Machines - Part 30-2: Efficiency classes of variable speed AC motors (IE code), IEC TS 60034-30-2, Dec. 2016.

[17] A. Takbash, Maged Ibrahim, L. Masisi, and P. Pillay, "Core Loss Calculation in a Variable Flux Permanent Magnet Machine for Electrified Transportation," IEEE Trans. Transp. Electrific., vol. 4, no. 4, pp. 857–866, Dec. 2015.

[18] N. Yogal, C. Lehrmann, and M. Henke, "Determination of the Measurement Uncertainty of Direct and Indirect Efficiency Measurement Methods in Permanent Magnet Synchronous Machines," 2018 International Conference on Electrical Machines (ICEM). pp.

1149–1156.

[19] G. Bucci, F. Ciancetta, E. Fiorucci, and A. Ometto "Uncertainty Issues in Direct and Indirect Efficiency Determination for Three-Phase Induction Motors: Remarks About the IEC 60034-2-1 Standard," IEEE Trans. Instrum. Meas., vol. 65, no. 12, pp. 2701-2716, Dec. 2016.

[20] L. Aarniovuori, J. Kolehmainen, A. Kosonen, M. Niemelä, and J. Pyrhönen, "Uncertainty in Motor Efficiency Measurements," 2014 International Conference on Electrical Machines (ICEM). pp. 323–329.

[21] Rotating electrical machines – Part 2-1: Standard methods for determining losses and efficiency from tests (excluding machines for traction vehicles), IEC 60034-2-1, Jun. 2014.

[22] Rotating electrical machines – Part 2-3: Specific test methods for determining losses and efficiency of converter-fed AC induction motors, IEC TS 60034-2-3, Nov. 2013.

[23] Y. H. Kim, H. D. Jun, J. W. Moon, and S. Y. Jung, "Comparison on the IM efficiency measurement results according to IEC standards," 2018 International Conference on Electrical Machines (ICEM). pp. 1130–1135.

Number of Turns Influence on the Parameters of High Speed Switched Reluctance Motor

S. Kocan, P. Rafajdus, P. Makys
Faculty of Electrical Engineering and Information Technology,
University of Zilina
Department of Power Systems and Electric Drives
Zilina, Slovakia
stefan.kocan@fel.uniza.sk

R. Bastovansky
Faculty of Mechanical Engineering, University of Zilina
Department of Design and Machine Elements
Zilina, Slovakia
ronald.bastovansky@fstroj.uniza.sk

Abstract— In this paper, an analysis of the influence of the number of phase turns on the other parameters of high speed Switched Reluctance Motor (SRM) is presented. The paper starts with state of the art from the point of view of electrical machines high-speed utilization and a description of the Switched Reluctance Motor. The initial design of the high speed motor is described. In the main part of the paper, the influence of the number of turns is analyzed. This analysis is performed for different rotational angular speed. Finally, the results of torque ripple, average torque and losses are presented for different high speed.

Keywords—high speed, switched reluctance motor, analysis

I. INTRODUCTION

The design of high-speed electrical machines is more difficult than conventional high-speed electrical machines. It is necessary to pay attention to improving electrical and magnetic characteristics, mechanical properties, vibration damping, cooling and lubrication design. However, this complication of the machine design leads to certain advantages. The main advantage is improving the conversion of output power to machine volume. This means that at higher speed, the machine can be lighter and smaller with the same rated power. However, this fact also increases the conversion of electrical losses to volume, so it is necessary to use a special material designed for high speed machines that have low iron losses. Another advantage of high speed machines is increased reliability due to the elimination of the mechanical gear between the motor and the driven device [1].

High-speed machines are used in many applications, such as gas and air compressors, spindles, turbochargers, flywheel energy storage systems, micro-turbines or turbo-molecular pumps. The choice of machine type depends on its application. Currently, induction machines (IM), permanent magnet synchronous machines (PMSM) and reluctance machines (SRM) are most frequently used as high speed machines. Each of these machine types has its advantages and disadvantages. The advantage of induction motor is a simple and robust design and lower cost, due to the absence of permanent magnets. However, this type of machine has lower efficiency compared with PMSM [2]. The disadvantages of PMSM are higher costs, risk of demagnetization due to high temperatures and lower mechanical strength [3].

Another type of machine that is suitable for high-speed operation is switched reluctance motor. This motor has teeth on the stator and rotor. The concentrated winding is located on the stator only. The rotor is without winding or permanent magnets. This fact reduces costs and winding losses. The winding is formed by concentric coils wound on each tooth [4]. In addition to low cost and winding losses, this motor has a high starting torque, high fault tolerance and it can work in

high temperature environments. The disadvantages of this motor are higher torque ripple, noise, vibration and difficult control [5].

SRM is most frequently used in high-speed applications below 100 000 rpm. Only a few researches were conducted above this speed. Design of 4-phase SRM for 30 000 rpm and 500 W was presented in [6]. In [7] design of 2-phase SRM with asymmetric rotor structure for 45 000 rpm and 700 W was presented. Design for automotive traction of 60 kW and 50 000 rpm was presented in [5]. Another design for the automotive industry, but for turbochargers application was presented in [4] and [8]. Spindle application designs were presented in [9] and [10]. In the first case for 50 000 rpm and 1 kW and in the second for 750 000 rpm and 100 W. The ultra-high speed SRM design, 1,000,000 rpm was presented in article [11].

In this paper, the analysis of the influence of the number of turns on selected electromagnetic parameters of the high-speed switched reluctance motor is presented. These selected electromagnetic parameters include: average static torque, average dynamic torque, torque ripple, peak current, winding losses, iron losses, efficiency, input and output power. The results thus obtained are commented and shown graphically.

II. INITIAL DESIGN OF HIGH SPEED SRM

A. Design Goals

This motor was designed for the automotive industry, specifically for turbocharger application. The motor design was based on the existing high speed motor also used in the automotive industry. However, this motor was of a different type. The aim of the design was to achieve the closest electromagnetic properties while maintaining the same motor volume. Another goal was to achieve the highest efficiency and the lowest losses.

Other resources for this design were articles [4] and [8], where the SRM was designed with a rated speed 110 000 rpm and rated power 1.5 kW.

B. Designed motor parameters

The motor was designed for a speed of 100 000 rpm. It was selected in a three-phase topology, the number of stator poles is 6 and the number of rotor poles is 4. Stator pole arc and rotor pole arc are different, to achieve less torque ripple in the design process. The size of the air gap was chosen according to the articles mentioned above. A summary of the mechanical parameters of the designed motor is given in Table 1.

Since it is an automotive application, the supply voltage was limited to 48V. The number of turns was 2 and the supply (peak) current was limited to 200 A, due to the switching capacity of the inverter. Other parameters were obtained by analytical calculation and dynamic simulation. A summary of the electrical parameters of the designed motor are given in Table 2.

TABLE I.   SUMMARY OF MECHANICAL PARAMETERS

Parameter	Symbol	Value	Unit
Mechanical speed	n_{mech}	100 000	rpm
Number of phases	m	3	-
Number of stator poles	N_s	6	-
Number of rotor poles	N_r	4	-
Stator pole arc	β_s	30	°
Rotor pole arc	β_r	32	°
Air gap	δ	0.22	mm
Axial length	l_{Fe}	55	mm
Stator outer diameter	d_s	90	mm
Rotor outer diameter	d_r	36	mm

TABLE II.   SUMMARY OF ELECTRICAL PARAMETERS

Parameter	Symbol	Value	Unit
Supply votage	V_{DC}	48	V
Number of turns per phase	N	2	-
Supply current	I	200	A
Phase resistance	R_{ph}	0.424	mΩ
Average torque	T_{av}	0.46	Nm
Total losses	ΔP	2.04	kW
Rated power	P	2.74	kW
Efficiency	η	57.3	%

The material of the stator NO10 and the material of the rotor M235-35A was selected. The detailed design was listed in the article [12]. The cross-section of the machine is shown in Fig.1.

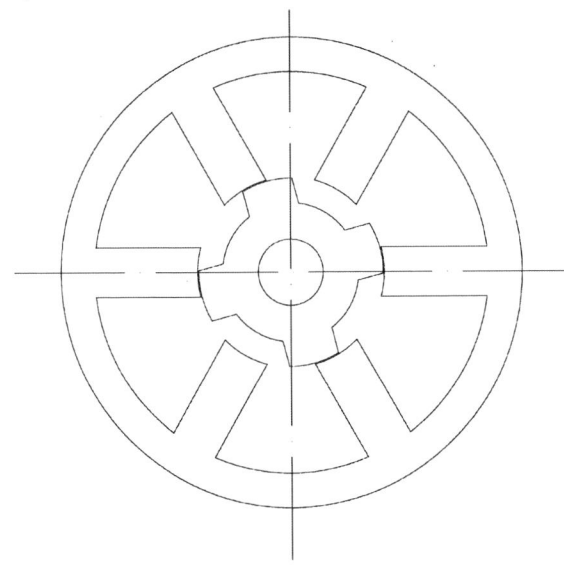

Fig. 1.   The cross-section of the designed motor

III.   ANALYSIS OF SRM PARAMETERS

This section describes the analysis procedure and the procedure for obtaining individual parameters.

A.   Simulation model

First, static parameters were calculated using the Finite Element Method (FEM) and then used in simulation model. This model is based on mathematical model of SRM. The mathematical model of SRM consists of three equations. Since only steady state was observed at different speeds, only 2 equations were used:

$$i = \int \frac{v - \left(R_{ph20} + \dfrac{dL(i,\vartheta)}{d\vartheta} \cdot \omega \right) \cdot i}{L(i,\vartheta)}, \tag{1}$$

where i is phase current, v is phase voltage, R_{ph20} is phase resistance at 20 °C, L is phase inductance as function of phase current and rotor position and ω is angular speed.

$$\vartheta = \int \omega \cdot dt , \tag{2}$$

where ϑ is rotor position. Based on these two equations, a 3-phase simulation model was developed in Matlab – Simulink. In the simulation model, the phase resistance obtained by the analytical calculation and the values of the switching angles were also entered. The values of these angles have been set to achieve maximum torque.

The simulation results are following waveforms: voltage, current, inductance, torque and flux linkage as a function of time. These dynamic parameters were used to calculate other parameters.

B.   SRM parameters calculation

Using dynamic parameters, following ones were calculated: average torque, torque ripple, winding losses, iron loss, mechanical loss, efficiency, input and output power. The average torque was calculated as:

$$T_{av} = \frac{1}{T} \int t_3 , \tag{3}$$

where T is period of magnetic flux and t_3 is sum of three phases torque. Torque ripple was calculated as:

$$\Delta T = \frac{T_{max} - T_{min}}{T_{av}}, \tag{4}$$

where T_{max} is maximum torque value in all three phases and T_{min} is minimum torque. Winding losses were calculated as [13]:

$$\Delta P_w = 3 \cdot R_{pf75} \cdot I_{rms}^2 , \tag{5}$$

where R_{ph75} is phase resistance at temperature 75 °C, I_{rms} is rms value of phase current. Higher harmonics of current were not included in this calculation. Iron losses were calculated as:

$$\Delta P_{Fe} = w \cdot k_t , \tag{6}$$

where w is machine weight and k_t is total loss coefficient, determined from the material datasheet. Mechanical losses were calculated as:

$$\Delta P_{mech} = K_T \cdot \left(\frac{n}{1000} \right)^2 \cdot \left(10 \cdot d_r \right)^3, \qquad (7)$$

where K_T is coefficient of mechanical losses (typically 2.9 − 5), n is rotating speed and d_r is rotor outer diameter. Input power was calculated as:

$$P_1 = T_{av} \cdot \omega = T_{av} \cdot \frac{2 \cdot \pi \cdot n}{60}, \qquad (8)$$

where ω is angular speed, n is rotating speed. Output power was calculated as:

$$P_2 = P_1 - \Delta P, \qquad (9)$$

where ΔP are total losses. Finally, efficiency was calculated as ratio of output power over input power.

IV. ANALYSIS RESULTS FOR DIFFERENT NUMBER OF TURNS

This analysis was performed for a number of turns from 2 up to 5 with constant air gap. A typical air gap value for high speed SRM is between 0.1 - 0.25 mm [11]. In this case, the air gap was selected to be 0.22 mm, based on [4].

Moreover, the analysis was performed for different rotating speeds. All parameters were obtained by calculating from dynamic parameters in addition to the average static torque.

A. Average static torque

First, the average static torque values for the different supply current values were calculated. The calculation results are shown in Fig. 2. This calculation was made to see how the average dynamic torque at given speed approached static torque.

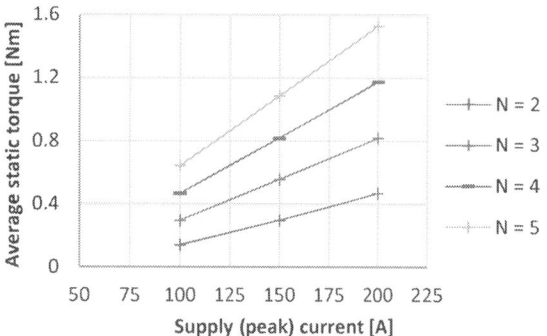

Fig. 2. Average static torque a function of supply current

The calculation was made for different supply currents because the 200 A supply current value was not always achieved in dynamic simulation. It can be seen from the figure that with increasing number of turns the average torque increases. However, this does not apply to dynamic parameters.

Fig. 3 and Fig. 4 show the waveform of static torque at the lowest number of turns $N = 2$ and at the highest number of turns $N = 5$.

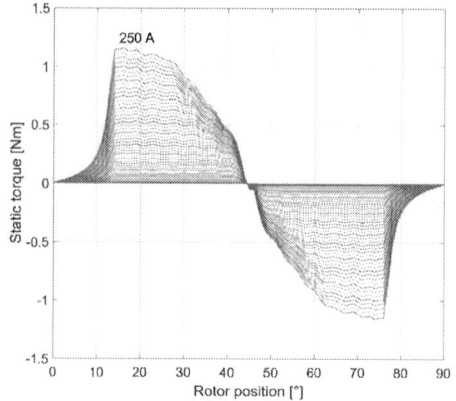

Fig. 3. The dependence of static torque $T = f(\vartheta, I)$, for $N = 2$

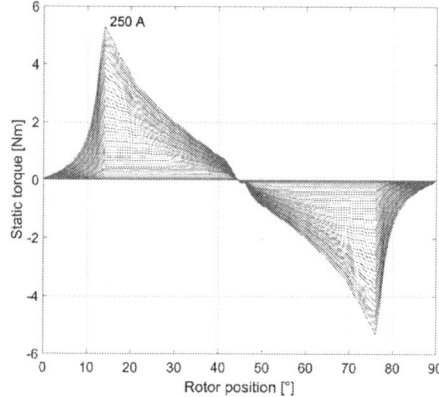

Fig. 4. The dependence of static torque $T = f(\vartheta, I)$, for $N = 5$

Static torque was calculated for all rotor positions with step 1 degree and for current values from 5 to 250 A, in step 5 A. The value 0 on the x-axis corresponds unaligned rotor position and value 45 corresponds aligned rotor position.

B. Supply current of designed SRM

Supply current for number of turns 2 and 3 did not have a problem to rise to a peak value of 200 A. At 4 turns, current increased to 200 A only at half rotating speed and at higher number of turns, the current did not reach 200 A. This is shown in Fig. 5.

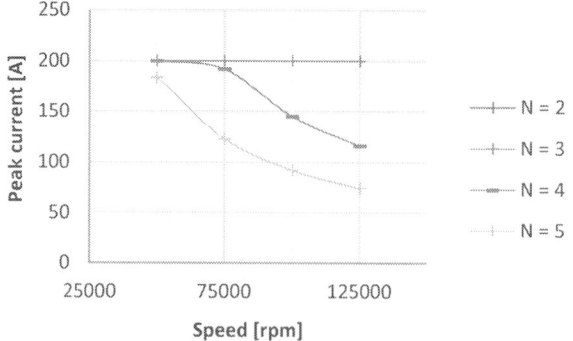

Fig. 5. Peak current a function of speed

This is probably due to high value of back electromotive force (emf) compared to the available supply voltage. In addition, Fig. 6 shows the dynamic current waveform for 2 and 5 turns, at 100 000 rpm. These waveforms confirm what is shown in Fig. 5.

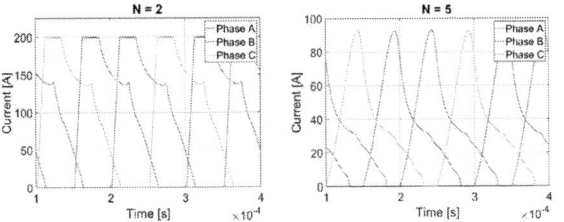

Fig. 6. Dynamic current as function of time for all 3 phases of the motor at 100 000 rpm

In this figure, you can see, that motor is able to commutate the current at speed 100 000 rpm.

C. Average torque of designed SRM

The average torque was calculated from the dynamic torque obtained from the simulation model, which was calculated using (3). The average torque calculation results are shown in Fig. 7.

According to the results of the static average torque, the highest torque at 5 turns should be achieved. However, because of high back electromotive force at high speed, this average torque is the lowest. At rotating speed of 100 000 rpm, the highest torque was achieved at 3 turns. At 2 and 4 turns, the average torque was lower and almost the same.

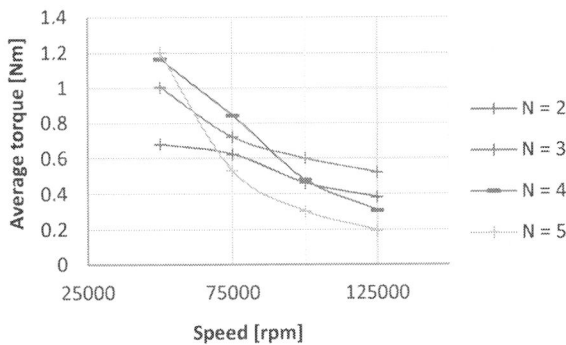

Fig. 7. Average torque a function of speed

The dynamic torque waveforms for 2 and 5 turns, at 100000 rpm, are shown in Fig. 8.

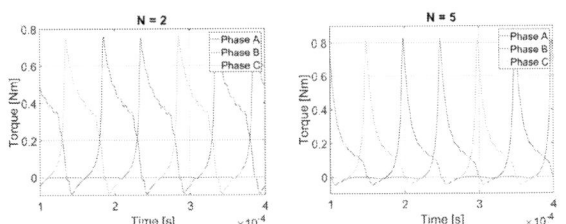

Fig. 8. Dynamice torque as function of time for all 3 phases of the motor at 100 000 rpm

D. Torque ripple

The torque ripple was calculated from the dynamic torque obtained from the simulation model using (4). The torque ripple calculation results are shown in Fig. 9.

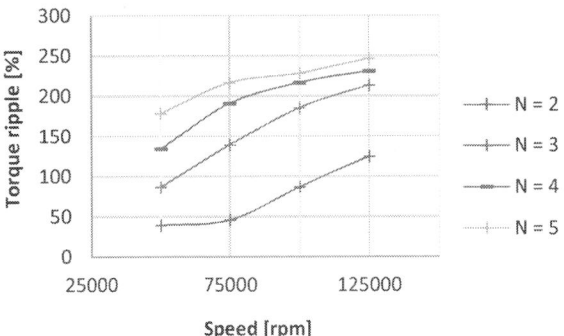

Fig. 9. Torque ripple as function of speed

It can be seen from Fig. 9 that, as the number of turns increase, the torque ripple also increases. The greatest increase in torque ripple is between 2 and 3 turns. This increasing is due to the large torque peak in the constant inductance interval near the unaligned position. Fig. 10 shows the sum of torque for all 3 phases of the motor for 2 and 3 turns.

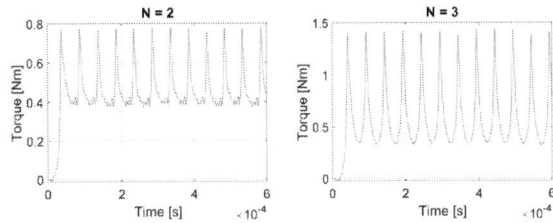

Fig. 10. The sum of torque for all 3 phases of the motor at 100 000 rpm

It can be seen from this Fig.10, that the peak torque at 3 turns reaches twice the peak torque at 2 turns. Although the average torque for 3 turns is higher, the minimum torque is almost identical in both cases. These facts increase torque ripple.

E. Calculated losses of designed SRM

This section shows the results for all types of losses calculated using (5), (6) and (7). The results of calculations are shown in Fig. 11, Fig. 12 and Fig. 13.

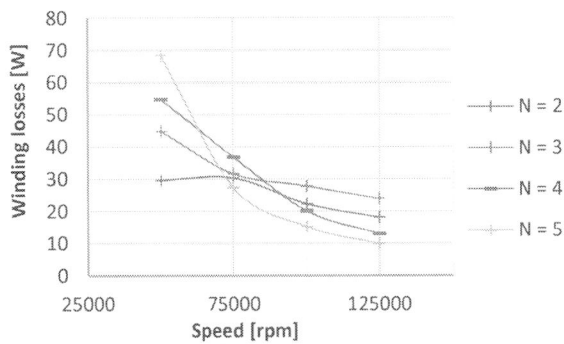

Fig. 11. Winding losses as function of speed

It can be seen from Fig. 11 that at lower speed, these losses are higher than at a higher number of turns. In this case, the current is approximately equal to current shown in Fig. 5 and the parameter that determines these losses is the phase resistance, which increases with the number of turns. However, at higher speed and higher turns than 3, the supply current is not constant and does not correspond to Fig. 5. As a result, with a lower number of turns these losses are higher.

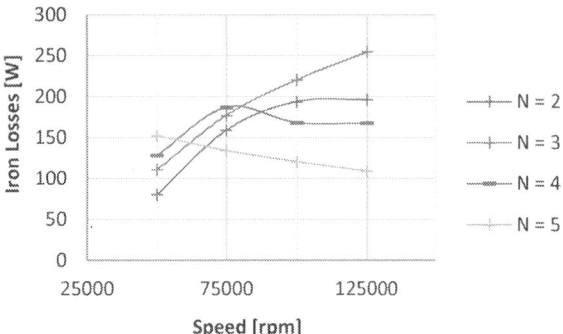

Fig. 12. Iron losses as function of speed

Mechanical losses do not depend on the number of turns, but depend on speed. Since this is only an analytical calculation, these losses are shown in Fig. 13 for 3 values of loss coefficient K_t. In other calculations, the losses at the loss coefficient $K_t = 3.95$ were considered.

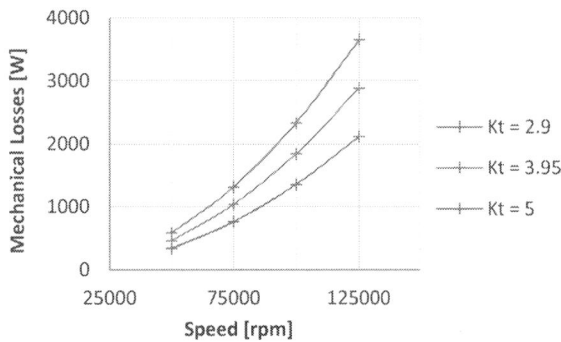

Fig. 13. Mechanical losses as function of speed

Mechanical losses consist of windage losses and bearing losses. Windage losses are not yet calculated, but Fig. 14 shown the frictional torque that corresponds to the bearing losses. Mechanical bearings 618/7 without sealing, lubricated with oil class VG 32 were considered in the calculation. The calculation was performed according to the bearing manufacturer's catalog for 4 load levels.

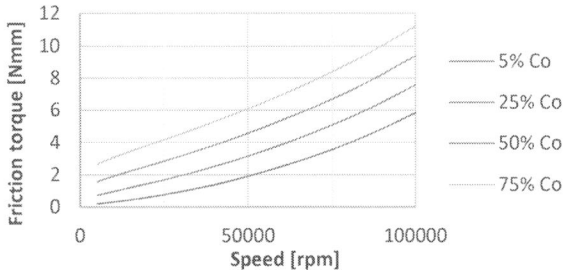

Fig. 14. Friction torque as function of speed

At high speeds, windage losses are very high. Especially for SRM due to the salient poles on the rotor [11]. The calculation of these losses will be presented in future works.

F. Input power, output power and efficiency of designed SRM

The input and output power were calculated using (8) and (9). The results are shown in Fig. 15 and Fig. 16.

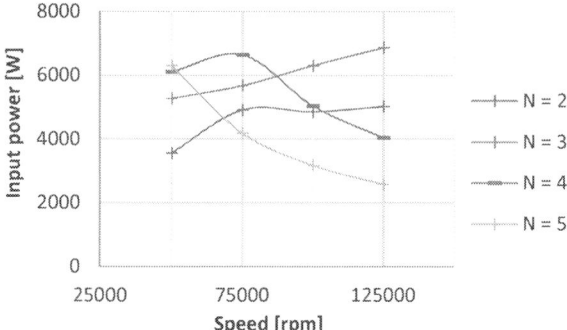

Fig. 15. Input power as function of speed

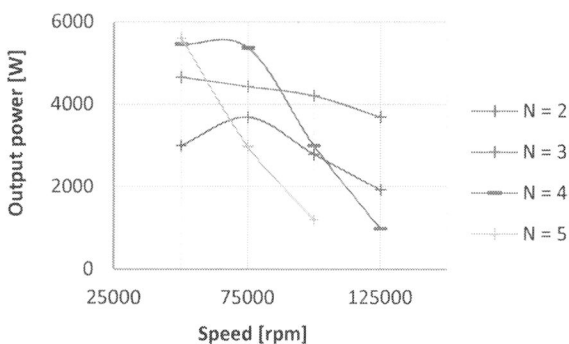

Fig. 16. Output power as function of speed

It can be seen from these figures that at higher speed above 85 000 rpm, the highest output power is achieved at 3 turns. Conversely, at lower speed, higher output power is achieved at higher number of turns.

The last investigated parameter was efficiency. The results are shown in Fig. 17.

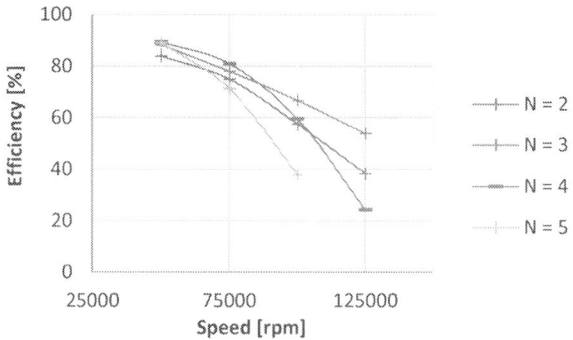

Fig. 17. Efficiency as function of speed

For a more detailed view, parameters at 100 000 rpm are summarized in Tab. 3.

978-1-5386-7688-2/19 $31.00 © 2019 IEEE

TABLE III. SUMMARY OF PARAMETERS AT 100 000 RPM

N [-]	T_{av} [Nm]	ΔT [%]	ΔP_w [W]	ΔP_{Fe} [W]	ΔP_{mech} [kW]	P_2 [kW]	η [%]
2	0.46	86.8	22.4	194		2.79	57.6
3	0.6	185	27.8	220	1.84	4.2	66.9
4	0.48	216	20.2	168		3.01	59.8
5	0.3	228	15.2	120		1.98	37.8

V. CONCLUSION

In this paper, an analysis of the influence of the number of turns on the properties of the high-speed switched reluctance motor was presented. The results of this analysis are waveforms average torque, torque ripple, losses, efficiency, input and output power as a function of speed for different number of turns. These waveforms have been discussed in this paper.

From the obtained waveforms it can be concluded that at 100 000 rpm the maximum average torque at 3 turns is achieved. If the number of turns increases from 2 to 3, the average torque increases, but it also increases torque ripple. Therefore, the control method needs to be adjusted to reduce torque ripple. At lower speed, a higher number of turns is preferable.

ACKNOWLEDGMENT

This work was supported by Slovak Scientific Grant Agency VEGA No. 1/0774/18, and by projects ITMS: 26220120046, cofounded from EU sources and European Regional Development Fund.

REFERENCES

[1] D. Gerada, A. Mebarki, N. L. Brown, C. Gerada, A. Cavagnino, and A. Boglietti, "High-Speed Electrical Machines: Technologies, Trends, and Developments", IEEE Transactions on Industrial Electronics, vol. 61, pp. 2946-2959, 2014.

[2] N. Uzhegov, J. Barta, J. Kurfürst, C. Ondrusek, and J. Pyrhönen, "Comparison of High-Speed Electrical Motors for a Turbo Circulator Application", IEEE Transactions on Industry Applications, vol. 53, pp. 4308 – 4317, 2017.

[3] Han-Wook Cho, Kyoung-Jin Ko, Jang-Young Choi, Hyun-Jae Shin, Seok-Myeong Jang, "Rotor Natural Frequency in High-Speed Permanent-Magnet Synchronous Motor for Turbo-Compressor Application", 9th International Conference on Electrical and Electronics Engineering (ELECO), vol. 47, pp. 4258 – 4261, 2011.

[4] N. Chayopitak, R. Pupadubsin, S. Karukanan, P. Champa, P. Somsiri, Y. Thinphowong, "Design of a 1.5 kW High Speed Switched Reluctance Motor for Electric Supercharger with Optimal Performance Assessment", 15th International Conference on Electrical Machines and Systems (ICEMS), pp. 1 – 5, 2012.

[5] M. Besharati, K. R. Pullen, J. D. Widmer, G. Atkinson, V. Pickert, "Investigation of the mechanical constraints on the design of a super-high-speed Switched Reluctance Motor for automotive traction", 15th International Conference on Electrical Machines and Systems (ICEMS), pp. 1 – 6, 2014.

[6] T. H. Pham, D. Lee, J. Ahn, "Design of novel high speed 2-phase 4/3 switched reluctance motor for air-blower application", IEEE International Conference on Advanced Intelligent Mechatronics (AIM), 2015

[7] P. Bogusz, M. Korkosz, A. Powrózek, J. Prokop, "A two-phase switched reluctance motor with asymmetrical rotor for a high-speed drive", International Conference on Electrical Drives and Power Electronics (EDPE), 2015

[8] Kozuka, S., Tanabe, N., Asama, J., Chiba, A., "Basic characteristics of 150,000r/min switched reluctance motor drive", IEEE Power and Energy Society General Meeting - Conversion and Delivery of Electrical Energy in the 21st Century, pp. 1 – 4, 2008.

[9] J. Kunz, S. Cheng, Y. Duan, J. R. Mayor, R. Harley, T. Habetler, "Design of a 750,000 rpm switched reluctance motor for micro machining", IEEE Energy Conversion Congress and Exposition, 2010

[10] J. Dang, S. Haghbin, Y. Du, C. Bednar, H. Liles, J. Restrepo, J. R. Mayor, R. Harley, T. Habetler, "Electromagnetic design considerations for a 50,000 rpm 1kW Switched Reluctance Machine using a flux bridge", International Electric Machines & Drives Conference, 2013

[11] C. Gong, T. Habetler, "Electromagnetic design of an ultra-high speed switched reluctance machine over 1 million rpm", IEEE Energy Conversion Congress and Exposition (ECCE), 2017

[12] S. Kocan, P. Rafajdus, P. Makys, R. Bastovansky, "Design of High Speed Switched Reluctance Motor", International Conference and Exposition on Electrical And Power Engineering (EPE), pp. 421 - 426, 2018.

[13] P. Rafajdus, V. Hrabovcova, P. Hudak, "Investigation of Losses and Efficiency in Switched Reluctance Motor", 12th International Power Electronics and Motion Control Conference, pp. 296 - 301, 2006.

NVH-Simulation of Salient-Pole Synchronous Machines for Traction Applications

Stephan-Akash Vip
*Institute for Drive Systems
and Power Electronics
Leibniz University Hannover*
Hannover, Germany
stephan.vip@ial.uni-hannover.de

Jan Hollmann
*Institute for Thermodynamics
Leibniz University Hannover*
Hannover, Germany
hollmann@ift.uni-hannover.de

Bernd Ponick
*Institute for Drive Systems
and Power Electronics
Leibniz University Hannover*
Hannover, Germany
ponick@ial.uni-hannover.de

Abstract—In comparison to electrical machines designed for industrial applications, the noise, vibration and harshness (NVH) behavior is more important in traction applications. Naturally, costumers demand a smooth running drive train over a wide operating range. In contradiction to that, electrical machines can emit high pitch discrete sounds which might be interpreted as a defect or failure of the drive train. In previous investigations, permanent magnet synchronous machines (PMSM) or induction machines (IM) have been investigated for traction applications. However, salient-pole synchronous machines (SPSM) are used for traction applications as well. Nonetheless, their NVH-behavior has not yet been analyzed in detail.

In this paper, a method to simulate the vibration at the stator surface of a SPSM is discussed. At first, the exciting electromagnetic forces are calculated analytically as well as numerically. Moreover, a 2D-FEM modal analysis is done. The results of this simulation are compared to experimental results. Furthermore, the method which combines 2D-FEM eletromagnetic simulations and analytical calculations to calculate the surface deflection is presented. In a final step, the calculated results are compared to measurement results of the surface vibration.

Index Terms—Traction motor, modal analysis, acoustic, NVH-simulation, noise, synchronous machine, vibration

Nomenclature

Symbols

A	Electric loading
B	Flux density
f	Frequency
l	Length in axial direction
L_p	Sound pressure level
p	Number of pole pairs
r	Ordinal number of spatial force harmonics
R	Radius
V	MMF curve
γ'	Circumferential angle
δ	Air-gap length
η	Harmonic response
λ	Air-gap permeance
μ_0	Permeability of vacuum
μ_r	Relative permeability
μ'	Ordinal number of spatial harmonics
ν'	Ordinal number of spatial harmonics
ξ	Radial deflection
σ	Force density
φ	Phase of spatial harmonic
Ψ	Modal matrix
ω_0	Eigenfrequency

I. Introduction

In order to keep up with stricter regulations regarding pollution, electric drive trains are used increasingly in the automotive sector. Three types of electrical machines are suitable for traction applications. IM or PMSM are commonly used in most electrified vehicles. But SPSM have become more attractive for traction applications, as well [1] [2] [3]. In contrast to PMSM, phenomena like demagnetization are irrelevant for SPSM [4]. Nonetheless, SPSM offer a similar power density as PMSM. As these machines are being used more and more in traction applications, a methodology to comprehend the vibrational behavior has been investigated.

In order to predict the acoustic noise emission of a SPSM, a multi-physics approach is necessary [5] [6]. Fig. 1 illustrates such a process. At first, the exciting forces of electromagnetic origin have to be obtained from an analytical calculation or an electromagnetic FEM simulation. The resulting time- and space-dependent force density waves σ_r excite the mechanical system leading to radial deflection ξ_{rad} of the motor surface. Coupling the vibration of the surface with an acoustic model allows the calculation of the radiated acoustic noise. Consequently, the acoustic noise emission of the motor can be quantified through the sound pressure level L_p.

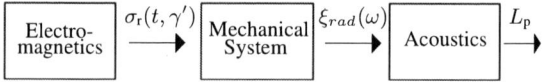

Fig. 1: Overview of calculation approach

In this paper, this entire process is presented. First of all, the electromagnetic forces are calculated using the FE program *FEMAG-DC* and the analytical software *ALFRED* [7]. The principles of *ALFRED* are explained in section II. In a next step, the modal characteristics of the motor are investigated. On the one hand, the results of a numerical modal analysis are compared to analytical equations to estimate the eigenfrequencies. On the other hand, a modal analysis is performed on a specimen to measure the eigenfrequency. Combing the results from the force calculation and the numerical modal analysis, a methodology to calculate the stator surface vibration is introduced and compared to measurements. Finally, the method is

978-1-5386-7688-2/19 $31.00 © 2019 IEEE

coupled to an acoustic model to calculate the sound pressure level L_p in various operating points.

II. Electromagnetic Forces in SPSM

Thanks to the spatial harmonic theory of electrical machines [6], the forces which excite the stator vibrations can be linked to the winding design and the geometry of the motor. The software *ALFRED* is based on this theory [7] and therefore, it is explained in brief. Starting with the current distribution in the slots that can be described as electric loading waves

$$A(\gamma', t) = -\sum_{\mu'} \hat{A}_{\mu'} \sin(\mu'\gamma' - 2\pi f_{\mu'} t + \varphi_{\mu'}) \quad (1)$$

is the first step to characterize the electromagnetic excitation. The integration of the electric loading leads to the MMF curve

$$V(\gamma', t) = \int A(\gamma', t)\, R\, \mathrm{d}\gamma' + c(t) \quad (2)$$
$$= \sum_{\mu'} \hat{V}_{\mu'} \cos(\mu'\gamma' - 2\pi f_{\mu'} t + \varphi_{\mu'}) \quad .$$

Further on, the geometry of the air-gap needs to be characterized by the air-gap permeance

$$\lambda_\delta(\gamma', t) = \frac{\mu_0}{\delta(\gamma', t)} \quad (3)$$
$$= \lambda_{\delta 0} + \sum_{\epsilon'} \hat{\lambda}_{\delta\epsilon'} \cos(\epsilon'\gamma' - 2\pi f_{\epsilon'} t + \varphi_{\epsilon'}) \,.$$

Cause	Type of wave	ν'	$f_{\nu'}$
$I_1(f_1)$	fundamental	p	f_1
	spatial winding harmonics	$p(1 + 2\frac{m_1}{q}g_1)$	f_1
	slotting	$p(1 + 2\frac{m_1}{q}g_1) + g_{N1}N_1 + g_{N2}2p$	$f_1(1 + g_{N2})$
	stator slotting fundamental	$p + g_{N1}N_1$	f_1
	rotor slotting fundamental	$p + g_{N2}2p$	$f_1(1 + g_{N2})$
I_{fd}	fundamental	p	f_1
	spatial rotor harmonics	$p(1 + 2g_{fd})$	$f_1(1 + 2g_{fd})$
	parametric rotor harmonics	$p(1 + 2g_{fd}) + g_{N1}N_1$	$f_1(1 + 2g_{fd})$

TABLE I: Important flux density spatial harmonics in SPSM

The air-gap permeance takes slotting effects and the change of the air-gap length through the distinct poles of the rotor into account. The product of permeance and MMF curve leads to the spatial harmonics of the flux density in the form of

$$B(\gamma', t) = V(\gamma', t) \cdot \lambda_\delta(\gamma', t) \quad (4)$$
$$= \sum_{\nu'} \hat{B}_{\nu'} \cos(\nu'\gamma' - 2\pi f_{\nu'} t + \varphi_{\nu'})$$

with

$$\hat{B}_{\nu'} = \frac{1}{2}\hat{V}_{\mu'}\hat{\lambda}_{\delta\epsilon'} \qquad \nu' = \mu' \pm \epsilon' \quad (5)$$
$$f_{\nu'} = f_{\mu'} \pm f_{\epsilon'} \qquad \varphi_{\nu'} = \varphi_{\mu'} \pm \varphi_{\epsilon'} \quad .$$

It can be seen that the spatial orders and the frequencies of the flux density harmonics are differences of the MMF curve orders and the air-gap permeance orders (Eq. 5). An alternative way to compute these spatial harmonics are reluctance circuits [8].

The most important flux density spatial harmonics of salient-pole synchronous machines are summarized in Table I with the number of slots per phase and pole q, the number of stator slots N_1, the number of pole pairs p and the variables g_1, g_{N1}, g_{N2} and g_{N2} being integer numbers. Consecutively, the radial force density

$$\sigma_{\mathrm{rad}} = \frac{B_{\mathrm{rad}}^2 - B_{\mathrm{tan}}^2}{2\mu_0} \quad (6)$$

and the tangential force density

$$\sigma_{\mathrm{tan}} = \frac{B_{\mathrm{rad}} \cdot B_{\mathrm{tan}}}{\mu_0} \quad (7)$$

can be derived from the Maxwell stress tensor [9]. For the analytical approach in this investigation, the tangential component is ignored. Thus, radial force density waves are simplified to

$$\sigma_r(\gamma', t) = \frac{B_{rad}^2(\gamma', t)}{2\mu_0} \quad (8)$$
$$= \sum_r \hat{\sigma}_r \cdot \cos(r\gamma' - 2\pi f_r t + \varphi_r) \,.$$

Here, the spatial force order r, the force frequency f_r and the amplitude of the force density wave $\hat{\sigma}_r$ characterize the exciting forces.

Specimen data			
No. of pole pairs	p	4	
Stator slots	N_1	36	
Inner stator diameter	$d_{i,1}$	180	mm
Outer stator diameter	$d_{a,1}$	266	mm
Outer housing diameter	$d_{f,a}$	300	mm
Stator core length	l_{axial}	100	mm
Stator yoke height	h_J	17.2	mm

TABLE II: Data of the SPSM specimen

The electrical machine in this study is a three-phase eight-pole SPSM with a two-layer fractional-slot winding. Table II summarizes the important machine data. The radial flux and

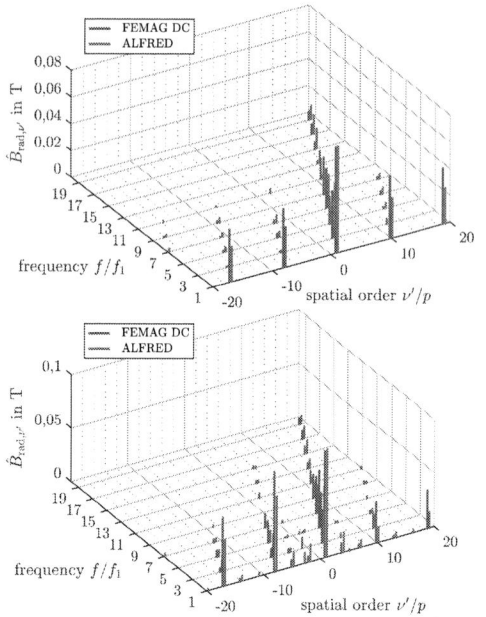

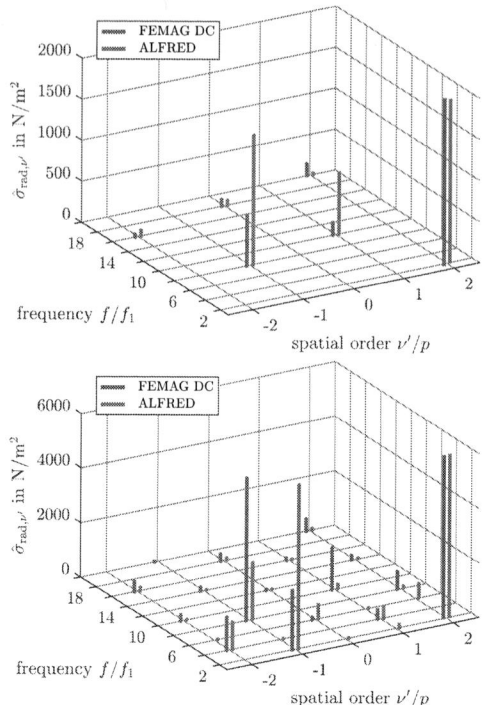

Fig. 2: Radial flux density spectrum of *OP1* (upper) and *OP2* (lower)

force density waves calculated by analytical means (*ALFRED*) and by a 2D-FFT performed on FEM simulation results (*FEMAG-DC*) are compared for two different operating points (Table III).

		OP1: no load-operating point	OP2: load-operating Point
d-axis current	I_d	0 A	150 A
q-axis current	I_q	0 A	50 A
Field current	I_fd	3 A	2 A

TABLE III: Reference operating points

In Fig. 2, the two-dimensional radial flux density spectrum of both operating points is visualized. For the reason of clarity, the axis has been limited to 8 mT cutting off the amplitude of the fundamental wave ($\nu' = p$, $f_{\nu'} = f_1$). For both operating points, it can be seen that the difference between the analytical calculation and FEM is up to 30%. A reason could be the simplified consideration of saturation effects in *ALFRED*.

Since the force density is proportional to the square of the flux density, the force dense amplitudes presented in Fig. 3 differ significantly. Nonetheless, the analytical calculation is capable of identifying all force excitation harmonics in less than a second. In contrast, the 2D-FEM simulation of 96 rotational steps took more than 20 min on the same computer.

III. MODAL ANALYSIS

The next step to calculate the surface vibration of the motor is to analyze the structural behavior of the machine. On that account, the eigenfrequencies have to be determined. On the

Fig. 3: Radial force density spectrum of *OP1* (upper) and *OP2* (lower)

one hand, analytical estimations like Nau's [10] equations, which are based on the work of Jordan [11] and Frohne [12], are used to predict the eigenfrequencies of the stator. Here, the mechanical system is approximated by a steel ring. The additional masses of the housing and the winding are considered by a mass factor. On the other hand, a numerical modal analysis is performed for the machine consisting of stator core, coils and housing. In this way, the complex geometry can be considered. From a structural point of view, the machine is a multi-mass system (Fig. 4) and can be described through

$$\mathbf{M}\frac{\mathrm{d}^2\vec{\xi}(t)}{\mathrm{d}t^2} + \mathbf{D}\frac{\mathrm{d}\vec{\xi}(t)}{\mathrm{d}t} + \mathbf{K}\vec{\xi}(t) = \vec{f}(t) \tag{9}$$

with the mass matrix $\mathbf{M}$, the stiffness matrix $\mathbf{K}$ and the damping matrix $\mathbf{D}$. Carrying out the modal decoupling of the system as explained in [13] leads to the harmonic force response vector

$$\underline{\vec{\eta}}_\mathrm{i} = \frac{\vec{\Psi}_0'^T \underline{\vec{f}}(\omega)}{\vec{\omega}_{0,\mathrm{i}}^2} \cdot \frac{1}{1 - \left(\frac{\omega}{\bar{\omega}_{0,\mathrm{i}}}\right)^2 + j \cdot 2 \cdot D_\mathrm{i} \cdot \frac{\omega}{\bar{\omega}_{0,\mathrm{i}}}} \tag{10}$$

in the frequency domain. The eigenfrequency vector $\vec{\omega}_{0,\mathrm{i}}$ and the modal matrix $\mathbf{\Psi}$ can be measured or extracted from a numerical simulation, whereas the damping D_i is assumed to be constant 2.8%. The harmonic force vector $\underline{\vec{f}}(\omega)$ inherits the shape of the force as well. Accordingly, this equation is valid for all force shapes.

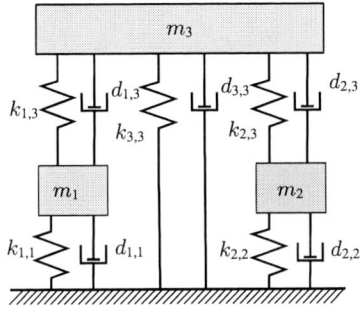

Fig. 4: Multi mass system

Performing the inverse modal transformation [13]

$$\vec{\xi}(\omega) = \mathbf{\Psi}'_0 \cdot \underline{\vec{\eta}}_i(\omega) \qquad (11)$$

on these two vectors finally leads to the deflection of each degree of freedom (DOF). For a numerical analysis, the number of DOFs is equal to two times the number of nodes (x and y direction in 2D).

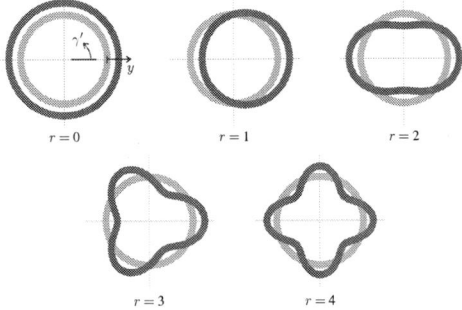

Fig. 5: First five spatial orders

The relevant circumferential eigenmodes are displayed in Fig. 5. As the deflection amplitude of a circular ring

$$\hat{\xi}_r \sim \frac{\hat{\sigma}_r}{(r^2 - 1)^2} \qquad (12)$$

resulting from an exciting force (mode shape $r \geq 2$) substantially decreases (Eq. 12) [6], higher spatial orders do not contribute significantly to acoustic noise emissions of a motor [14] [15]. For this reason, the modes $r \leq 4$ are considered to be of high importance to characterize the structural behavior.

Component	Density [kg/m³]	E-Module [Pa]	Length [mm]
Stator core	7305	2, 15e11	100.0
Coils	4200	1, 20e9	187.5
Housing	7850	2, 15e11	205.0

TABLE IV: Material parameters in ANSYS

Using the software *ANSYS Mechanical*, a 2D modal analysis was executed. The model is shown in Fig. 6 and the initially used material parameters are given in Table IV.

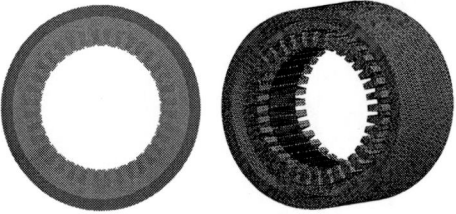

Fig. 6: 2D Model in ANSYS

To validate the approach, accelerometers (*Brüel&Kjaer 4508*) have been mounted on the motor as indicated by Fig. 7. The sensors were positioned in the middle of the stator core avoiding the water cooling ducts to have a direct metal connection to the stator. The inlet and the outlet of the cooling and the cabling did not allow to completely close the sensor ring. The machine was excited with a hammer and the reaction to the impact was evaluated. Analyzing the phase shift of each sensor and considering their positon on the machine allowed to determine the mode shape and the frequency.

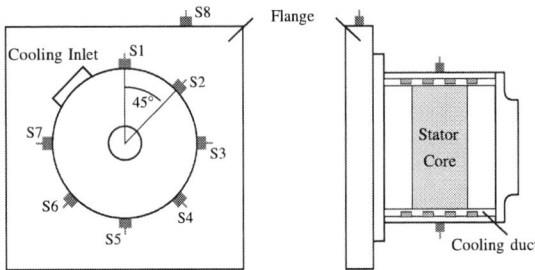

Fig. 7: Position of sensors

In Fig. 8, the identified modes are displayed. As only seven sensors were placed on the circumference, the mode shape $r = 4$ cannot be identified confidently. However, all seven sensors were placed on a quarter circle showing results which confirmed the assumption of the initial test. Table V shows the comparison between the measurement, Nau's [10] analytical approach, and the results of the numerical simulation.

	Eigenfrequency $f_{0,i}$ in Hz			
r	Measured	Analytical	ANSYS	ANSYS adjusted
0	5050	4728	4972	4883
2	1300	893	1014 ;1018	769 ; 772
3	2200	2392	2642 ; 2660	2002 ; 2010
4	3500	4284	4633 ; 4695	3491 ; 3515

TABLE V: Eigenfrequencies of the motor

It is apparent that the relative error increases with the spatial order in *ANSYS*. This is critical because the spatial order $r = 4$ is of high importance for this machine.

978-1-5386-7688-2/19 $31.00 © 2019 IEEE

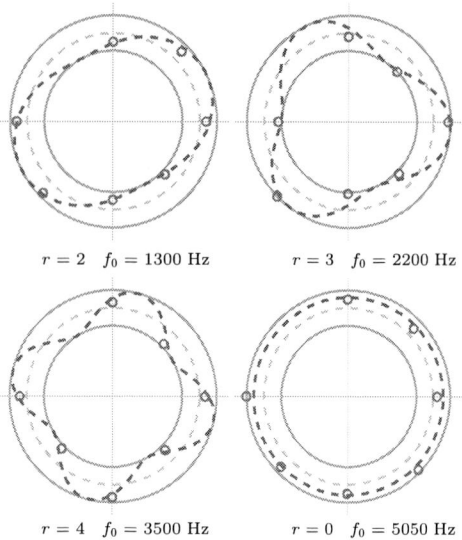

Fig. 8: Results of modal analysis measurement

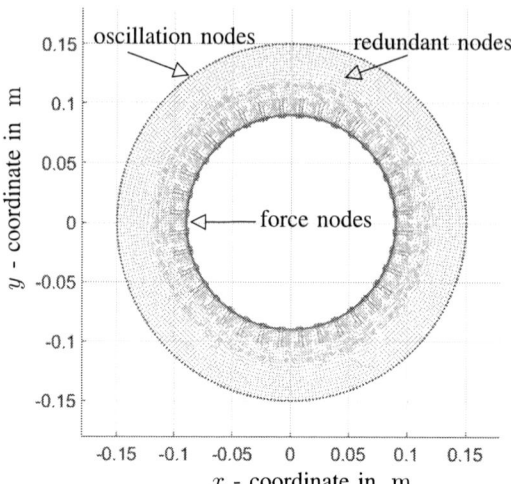

Fig. 9: Reduced model

In order to get more suitable results, the material properties have been adjusted. To consider mass and stiffness of the end shields, the density of the housing was changed to 9920 kg/m^3 and the E-Module was adjusted to $2.75 \cdot 10^{11}$ Pa. Furthermore, the E-Modul of the stator core was modified to $1.75 \cdot 10^{11}$ Pa which is in good accordance with van der Giet's observations [16]. In this way, it was achieved to improve the results except for the spatial order $r = 2$ which considers the force spectrum in Fig. 3 of lesser interest.

IV. VIBRATION CALCULATION

A. Model reduction

Each node of the structural numerical model in ANSYS has two vibration degrees of freedom (x- and y-direction). Therefore, two modal matrices with the dimension number of nodes times number of nodes have to be extracted. This leads to large matrices and high computational effort.

In contrast, the information needed is the transfer function of the nodes where the forces excite the structure to the deflection of the nodes at the outer surface. The deflection of nodes within the structure are acoustically irrelevant. For that reason, the model has to be reduced to be able to get accurate results in a reasonable amount of time. Consequently, the model is reduced according to Fig. 9. The vibration of the grey nodes are not of intrest and therefore ignored. The force and the oscillation nodes need to be saved. In addition, the nodes at the edges of each tooth need to be identified. On these nodes, the tangential forces are applied in the next section. For this model, the number of nodes was reduced from 10172 to 1214 resulting in a significantly smaller matrix.

B. Force decomposition

The next step in order to be able to compute the response vector (Eq. 10) is to calculate the force vector $\vec{\hat{f}}(\omega)$. After obtaining the forces through a 2D-FFT of the air-gap field, they are given in the form of force density waves (Eq. 8). The force waves have to be modified to suit the force decomposition. A spatial harmonic can be separated into two alternating harmonics:

$$
\begin{aligned}
\sigma_r(\gamma', t) &= \hat{\sigma}_\mathrm{r} \cdot \cos(r\gamma' - 2\pi f_r t + \varphi_r) \\
&= \hat{\sigma}_\mathrm{r} \cdot \cos(r\gamma' + \varphi_r) \cdot \cos(2\pi f_\mathrm{r} t) \\
&\quad + \hat{\sigma}_\mathrm{r} \cdot \sin(r\gamma' + \varphi_r) \cdot \sin(2\pi f_\mathrm{r} t) \quad .
\end{aligned} \tag{13}
$$

Hereby, two force components (cosine and sine time-dependent) are determined. Consequently, two force vectors and two modal response vectors need to be taken into account. The radial force of a node k for one force density wave

$$
\begin{aligned}
f_{k,\mathrm{cos,rad}} &= A_\mathrm{node} \cdot \hat{\sigma}_\mathrm{r} \cdot \cos(r\gamma'(k) + \varphi_r) \\
f_{k,\mathrm{sin,rad}} &= A_\mathrm{node} \cdot \hat{\sigma}_\mathrm{r} \cdot \sin(r\gamma'(k) + \varphi_r)
\end{aligned} \tag{14}
$$

can be calculated. Here, A_node is the stator tooth surface at the air-gap per node and $\gamma'(k)$ is the circumferential angular position of node k.

Furthermore, the tangential forces need to be mapped on the force nodes. The tangential force density waves $\sigma_t(\gamma', t)$ are given in the same form as in (13) and are separated in the same way. Thus, the tangential forces act on the stator tooth in circumferential direction and induce a torque in the yoke of the machine leading to additional radial vibration [17]. To consider this, the forces for the right node

978-1-5386-7688-2/19 $31.00 © 2019 IEEE

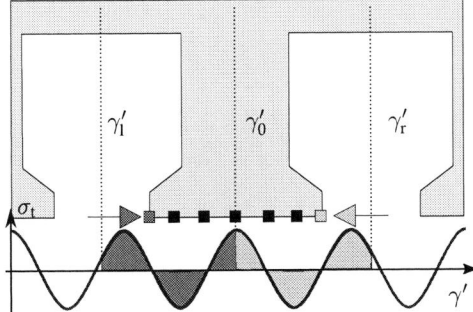

Fig. 10: Mapping of tangential forces

Fig. 11: Test bench setup

$$f_{k,\cos,\tan,r} = l_{axial} \cdot \int_{\gamma_0'}^{\gamma_r'} \hat{\sigma}_t \cdot \cos(r\gamma' + \varphi_r) \quad (15)$$

$$f_{k,\sin,\tan,r} = l_{axial} \cdot \int_{\gamma_0'}^{\gamma_r'} \hat{\sigma}_t \cdot \sin(r\gamma' + \varphi_r)$$

and for the left node

$$f_{k,\cos,\tan,l} = l_{axial} \cdot \int_{\gamma_l'}^{\gamma_0'} \hat{\sigma}_t \cdot \cos(r\gamma' + \varphi_r) \quad (16)$$

$$f_{k,\sin,\tan,l} = l_{axial} \cdot \int_{\gamma_l'}^{\gamma_0'} \hat{\sigma}_t \cdot \sin(r\gamma' + \varphi_r)$$

are integrated and mapped on each edge node of the stator teeth. Here, l_{axial} is the axial length of the stator. By this means, force vectors are established where the forces are calculated for each force node. The force for the oscillation nodes are set to zero.

C. Surface deflection

The final step to calculate the surface deflection is to use the two force vectors in (10) and evaluate the two modal response vectors $\vec{\underline{\eta}}_{i,\cos}(\omega)$ and $\vec{\underline{\eta}}_{i,\sin}(\omega)$. These vectors are 90° phase shifted in the time domain. Hence, they are combined into a complex modal response vector

$$\vec{\underline{\eta}}_i(\omega) = \vec{\underline{\eta}}_{i,\cos}(\omega) + j \cdot \vec{\underline{\eta}}_{i,\sin}(\omega) \quad . \quad (17)$$

This vector is then applied to (11). Consequently, two vectors (real and imaginary part) are used to compute the radial deflection $\vec{\xi}_{rad}(\omega)$. Analyzing the radial deflection using a FFT in circumferential direction and considering the time phase shift between real and imaginary part through sine and cosine provides the deflection in the form of alternating waves.

$$\xi_{real,rad}(\gamma', t) = \hat{\xi}_{real,rad} \cdot \cos(r\gamma' + \varphi_{real}) \cdot \cos(2\pi f_r t) \quad (18)$$

$$\xi_{img,rad}(\gamma', t) = \hat{\xi}_{img,rad} \cdot \sin(r\gamma' + \varphi_{img}) \cdot \sin(2\pi f_r t)$$

Again, the alternating waves can be described by two rotating waves of half the amplitude to compute the radial deflection. The rotating waves of the real

$$\xi_{real,rad}(\gamma', t) = \hat{\xi}_{real,rad} \cdot \cos(r\gamma' + \varphi_{real}) \cdot \cos(2\pi f_r t) \quad (19)$$

$$= \frac{\hat{\xi}_{real,rad}}{2} (\cos(r\gamma' + 2\pi f_r t + \varphi_{real}) + \cos(r\gamma' - 2\pi f_r t + \varphi_{real}))$$

and the imaginary component

$$\xi_{img,rad}(\gamma', t) = \hat{\xi}_{img,rad} \cdot \cos(r\gamma' + \varphi_{img}) \cdot \sin(2\pi f_r t) \quad (20)$$

$$= \frac{\hat{\xi}_{img,rad}}{2} (\cos(r\gamma' + 2\pi f_r t + \varphi_{img}) - \cos(r\gamma' - 2\pi f_r t + \varphi_{img}))$$

are therefore computed for one force density wave. This has to be repeated for all force waves. At the end of this loop, all rotating waves of the radial deflection have to be superimposed depending on their spatial order and frequency.

In order to validate the calculation, an operating deflection shape analysis (ODS) was done with the test bench setup displayed in Fig. 11. The visible accelerometers are marked with a red circle around it. To compare the measurement with the calculated radial surface velocity amplitude $\hat{v}_{rad}$, the significant force waves are investigated (Fig. 12). The first force wave $a)$ harmonic is caused by an interaction of rotor spatial harmonics and the stator slotting (Table I). $b)$ and $c)$ originate from the interaction of parametric rotor harmonics and the spatial fundamental. At last, $d)$ and $e)$ are combinations of the spatial winding harmonics and the spatial rotor harmonics.

First of all, looking at the comparison in Fig. 12, it can be seen that, without adjusting the *ANSYS* model, the resonances for b)-e) (3500 Hz) would not be predictable. No peaks would have been determined. For this reason, a future investigation is to extend the structural model to 3D in order to be able to get more accurate simulation results. Moreover, applying the tangential forces from FEMAG did not lead to a significant difference. In contrast, the forces taken from the analytical tool *ALFRED* are appropriate to predict the modes a) and b). For the other cases the forces calculated by *ALFRED* are too small. On closer examination of the cases d) and e), the methodology is not able to predict the radial vibration for frequencies far above the eigenfrequency. The reason for this might be missing

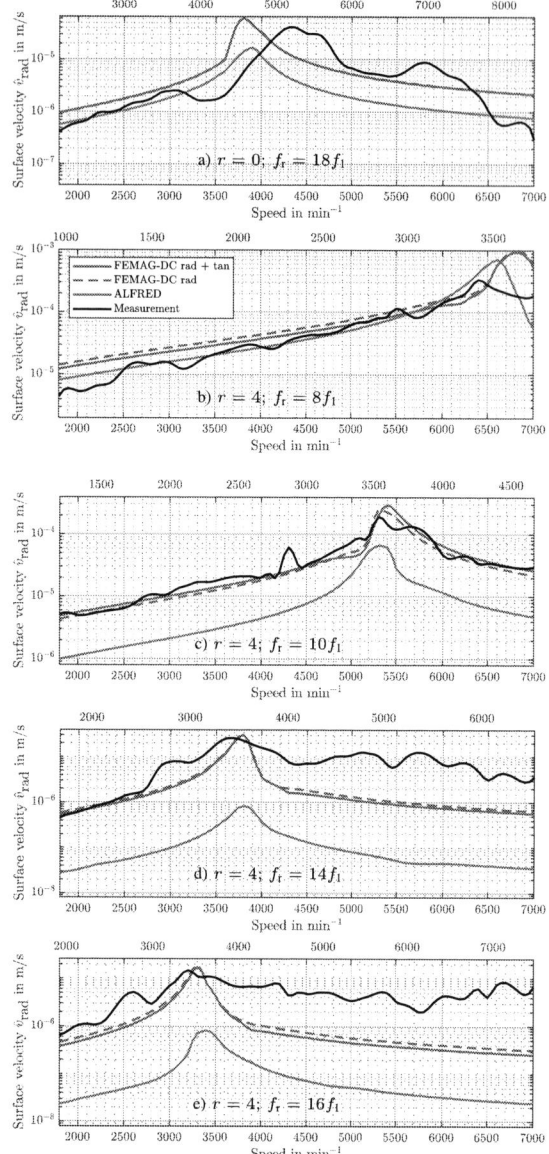

Fig. 12: Comparison of simulated to measured data for a run up

extended to the third dimension in the future in order to integrate the contribution of the end shields to the structural stiffness which promises great potential.

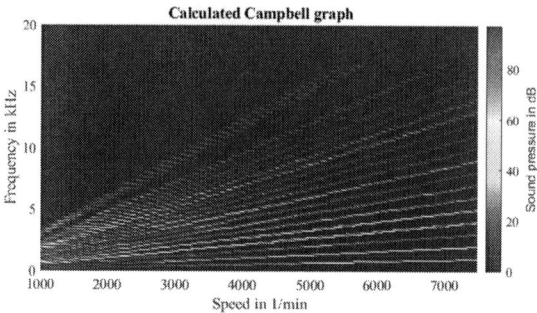

Fig. 13: Calculated Campbell diagram for operating point *OP2* using forces from *FEMAG*

The validated vibration calculation is the base for the acoustic noise calculation. Coupling the calculated vibration with an analytical model to predict the acoustic noise [13] [14], the sound pressure level can be calculated. The results for a run-up with the currents from operating point *OP2* is visualized in the Campbell diagram in Fig. 13. As it can be seen from the graph, frequencies of the noise caused by electromagnetic forces are proportional to the rotational speed.

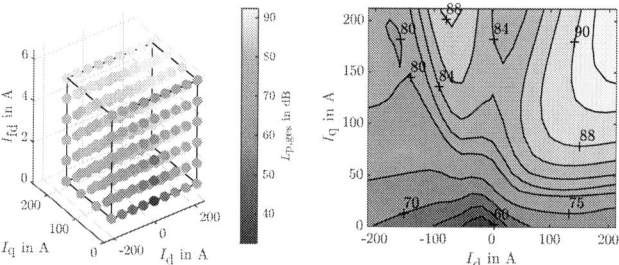

Fig. 14: Total sound pressure $L_{\mathrm{p,ges}}$ for various operating points at a speed of $n = 6600$ /min

Indeed, traction drives have to be examined for various operating points. As it was seen before, the forces are significantly dependent on the stator currents. For this purpose, the methodology is adjusted to be able to calculate an entire characteristic map of the machine. This is done using the *SPOK FAST* software [18] to calculate SPSM in various operating points. In Fig. 14, on the left, an identification for a characteristic map calculation is displayed. On the right, the plane for $I_{\mathrm{fd}} = 3$ A is shown. The given total sound pressure level $L_{\mathrm{p,ges}}$ is calculated for a speed of $n = 6600$ /min. For this machine, the sound pressure increases with I_{fd}. From the noise perspective, $I_{\mathrm{d}} < 0$ A is preferable as it is reducing the flux density and therefore the force.

CONCLUSION

In this paper, a promising method to calculate the deflection of the outer machine surface of a SPSM is proposed. This

modes of higher frequencies in the modal matrix extracted from *ANSYS*. In addition to that, higher spatial harmonics could have interfered in the measurement as they were not identifiable due to the limited amount of sensors. Besides, the mounting of the sensors using *Brüel&Kjaer* fastening rails could have an impact on the measurement of higher frequencies (> 5000 Hz) in the cases *d*) and *e*).
All in all, the tool chain using the forces from *FEMAG* was able to estimate the deflection in resonance adequately. To achieve more accurate results, numerically calculated forces should be used. Nonetheless, the structural model should be

978-1-5386-7688-2/19 $31.00 © 2019 IEEE

can be further used to estimate the acoustic noise emission of the motor. Forces from analytical and numerical origin are used to compute the deflection. With regard to a run-up deflection measurement, it is recommended to use numerically determined forces. The influence of tangential forces seems to be negligible for the investigated motor. Moreover, it is of utmost significance to know the eigenfrequencies of the structure. The 2D mechanical model neglected the end shields resulting in too high eigenfrequencies for the spatial mode $r = 4$. Without the theoretical adjustments of the material properties to include the additional mass of the end shields, resonances would not have been predictable in the run-up simulation. Concluding, the approach is valid to predict the vibration in resonance and over a wide speed range. However, an extended structural model including the end shields of the housing is desired for the future. Hereby, more vibration modes could be identified and included to be able to deliver good results for higher frequencies and therefore a wider operating range.

REFERENCES

[1] J. Jürgens, A. Brune, and B. Ponick, "Electromagnetic design and analysis of a salient-pole synchronous machine with tooth-coil windings for use as a wheel hub motor in an electric vehicle," in *International Conference on Electrical Machines (ICEM)*, 2014.

[2] J. Redlich, J. Jürgens, K. Brune, and B. Ponick, "Synchronous machines with very high torque density for automotive traction applications," in *IEEE International Electric Machines and Drives Conference (IEMDC)*, 2017.

[3] H. Krupp and A. Mertens, "Rotary Transformer Design for Brushless Electrically Excited Synchronous Machines," *IEEE Vehicle Power and Propulsion Conference (VPPC), Montreal, QC*, 2015.

[4] S. Hamidizadeh, N. Alatawneh, R. R. Chromik, and D. A. Lowther, "Comparison of different demagnetization models of permanent magnet in machines for electric vehicle application," *IEEE Transactions on Magnetics*, vol. 52, no. 5, 2016.

[5] J. Le Besnerais, A. Fasquelle, M. Hecquet, J. Pelle, V. Lanfranchi, S. Harmand, P. Brochet, and A. Randria, "Multiphysics modeling: electro-vibro-acoustics and heat transfer of induction machines," in *18th International Conference on Electrical Machines*, 2008.

[6] J. Gieras, C. Wang, and J. Lai, *Noise of polyphase electric motors*. CRC Press, 2005.

[7] B. Ponick, "Fehlerdiagnose bei Synchronmaschinen," Dissertation, Ph.D. dissertation, Univ. Hannover, Hannover, Germany, 1994.

[8] A. Ebrahimi, "Analytical Modeling of Permanent Magnetic Synchronous Motors Considering Spatial Harmonics," *Brazilian Power Electronic Conference (COBEP), Brazil*, 2017.

[9] L. Timár-P and P. Tímár, *Noise and vibration of electrical machines*, ser. Studies in electrical and electronic engineering. Elsevier, 1989.

[10] S. L. Nau, H. L. Bork, dos Santos, H. L. V., N. Sadowski, and R. Carlson, "The influence of the frame and windings on the natural frequencies of stator of induction motors," *Proceedings of the International Conference on Electrical Machines (ICEM)*, 2006.

[11] H. Jordan, *Geräuscharme Elektromotoren : Lärmbildung und Lärmbeseitigung bei Elektromotoren*. Essen: Girardet, 1950.

[12] C. Frohne, *Eigenschwingungen eingespannter Kreisringe am Beispiel der elektrischen Maschine*. Aachen: Shaker, 1999.

[13] D. Braunisch, B. Ponick, and G. Bramerdorfer, "Combined analytical-numerical noise calculation of electrical machines considering nonsinusoidal mode shapes," *IEEE Transactions on Magnetics*, 2013.

[14] Z. Q. Zhu and D. Howe., "Improved methods for prediction of electromagnetic noise radiated by electrical machines," *IEE Proceedings - Electric Power Applications 141, No. 2*, 1994.

[15] A. Hofmann, F. Qi, T. Lange, and R. W. D. Doncker, "The breathing mode-shape 0: Is it the main acoustic issue in the PMSMs of todays electric vehicles?" *17th International Conference on Electrical Machines and Systems (ICEMS)*, 2014.

[16] M. van der Giet, K. Kasper, R. W. D. Doncker, and K. Hameyer, "Material parameters for the structural dynamic simulation of electrical machines," *XXth International Conference on Electrical Machines*, 2012.

[17] H. Weh, "Zur elektromagnetischen Schwingungsanregung bei Asynchronmaschinen," *ETZ-A 85, pp.193-197*, 1964.

[18] O. Korolova, P. Dück, A. Brune, J. Jürgens, and B. Ponick, "Prediction of efficiency-optimized salient-pole synchronous machines' operating range using a coupled numerical-analytical method," in *International Conference on Electrical Machines (ICEM)*, 2014.

Outer Race Fault Diagnosis by Comparison between the Power Spectral Density and the Kurtogram

Mohammed-El-Amine Khodja, *Student member, IEEE*, Ahmed Hamida Boudinar, *member, IEEE*, Ameur Fethi Aimer, Azeddine Bendiabdellah, *member, IEEE*

Electrical Engineering Faculty, Diagnosis Group
University of Sciences and Technology of Oran
Oran, Algeria
E-Mails : khodjamea@gmail.com, boud_ah@yahoo.fr, fethi.aimer@yahoo.fr, bendiazz@yahoo.fr

Abstract—Statistically, the outer race of the rolling-element bearing is the most sensitive element used in induction motors. Among the known diagnosis methods, the analysis of the stator current envelope based on its demodulation, makes it possible to directly obtain the signature of the searched faults without the fundamental effect. This paper discusses one of these methods known as the Wavelet-Kurtogram. For this purpose, a comparison between the Wavelet-Kurtogram and the Power Spectral Density using the Periodogram is carried out for bearing faults diagnosis. The experimental results obtained show the superiority and the effectiveness of Wavelet-Kurtogram in the detection of the outer race fault.

Keywords—Induction Motor; Bearing Fault; Diagnosis; Spectral Analysis; Spectral Kurtosis; Kurtogram.

I. INTRODUCTION

The early diagnosis and localization of faults in induction motors is very important in the industrial sector, given their impact on the operational reliability of the various industrial processes. Among these faults, the bearing fault represents more than 60% of the faults affecting induction motors [1]. This high percentage is due to the function of this rolling-element that allows the transfer of mechanical energy [2-7].

A bearing fault is called localized bearing fault, if the fault is at one of the different elements constituting this bearing. In this case, the fault is characterized by well-defined frequencies [8], [9]. On the other hand, if the fault affecting the bearing is not characterized by particular frequencies, then its detection becomes very difficult or impossible in some cases. This fault is called a generalized roughness fault [9-11].

Motor Current Signature Analysis (MCSA) is a very promising technique given the richness of the stator current spectrum and the simplicity of sensor location compared to other techniques [12-14]. For this purpose, several signal processing methods have been developed to extract the signature of the searched faults from the stator current spectrum [15-21]. The estimation of the Power Spectral Density by Periodogram (PSDP) is the most widely used method in the industry because of its simplicity and speed due to the use of the Fast Fourier Transform (FFT) algorithm. However, this method has several disadvantages such as the detection

difficulty of faults frequency signatures of low amplitude and the fault detection for a non-stationary operation.

To overcome the limitation of PSDP estimation in the case of non-stationary signals, several envelope analysis methods are developed [20-25]. Among them, Spectral Kurtosis (SK) is an envelope analysis method defined by Dwyer [26]. Antoni [27] proposes a new definition of the SK from an estimator based on the Short Time Fourier Transform (STFT). The main disadvantage of this estimator is its sensitivity to the choice of the length of the window for the STFT calculation. For this, Antoni proposes an image representation of all the SKs of the processed signal with different lengths of windows in order to choose the best window. From this image, the Kurtosis maximum is located to determine the level of decomposition of the envelope to be analyzed. This approach is known as Kurtogram. However, this Kurtogram approach requires significant computation time because of the STFT-based estimator [28-29]. To improve this computation time, Sawalhi [30]; Lei [31] and Wang [32] propose an improved version of the Kurtogram based on filters obtained by the wavelet transform. This method is known as "Wavelet-Kurtogram". The effectiveness of the Wavelet-Kurtogram is demonstrated in several papers [30-32] dealing with the detection of bearing faults by vibration analysis. On the other hand, the analysis of the stator current does not give satisfactory results according to Leite [21]. This is explained by the use of Morlet's complex wavelet.

To solve this problem, this paper proposes the use of the Wavelet Packet Transform (WPT) for the Wavelet-Kurtogram computation in order to detect the bearing fault using the stator current analysis. To show the superiority of this new approach, a comparative study is carried out between the Wavelet-Kurtogram and the PSDP estimation. This comparative study is achieved on the basis of experimental tests to detect outer race fault of a rolling-element bearing.

II. BEARING FAULT FREQUENCY SIGNATURE

The rolling-element bearing is an important and sensitive element in electric motors especially in induction motors. This bearing acts as an electromechanical interface between the stator and the rotor. In addition, it maintains the axis of the motor to ensure good rotation of the rotor. The bearing consists of two races, one inside and another outside, balls and a cage that ensures an equidistance between the balls. Fig. 1 represents the geometry of a rolling-element bearing.

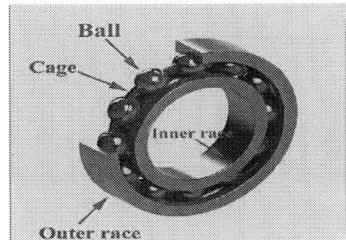

Fig. 1. Geometry of a rolling-element bearing.

The vibration characteristic frequency of the outer race fault is defined by [5]:

$$f_o = \frac{N_b}{2} f_r (1 - \frac{B_D}{C_D} \cos \beta) \qquad (1)$$

Where N_b is the number of balls, B_D and C_D are respectively the ball and cage diameters, β the contact angle and f_r is the mechanical rotor frequency.

We must note that, in the spectrum of the stator current, the frequency signature of the outer race fault is given by the following equation [6]:

$$f_{bear}(Hz) = |f_s \pm k.f_o| \quad ; \quad k = 1,2,3 \dots \quad (2)$$

Where f_s is the supply frequency.

III. PSDP ESTIMATION PRINCIPLE

The PSDP estimation can be defined directly in terms of the observed samples of the stator current $i_s(n)$ as [16]:

$$PSDP(f) = \frac{1}{N} |I_s(f)|^2 \qquad (3)$$

Where $I_s(f)$ is the Fourier Transform of $i_s(n)$. N is the number of samples.

In other words, the PSDP estimation is simply the square of the modulus of the discrete Fourier transform of the measured signal (the stator current in our case), as shown in (3).

IV. SPECTRAL KURTOSIS AND KURTOGRAM

A. Spectral Kurtosis

Spectral Kurtosis (SK) is the extension of scalar Kurtosis in the frequency domain to detect and locate non-stationary parts in a signal. According to Antoni [27], the SK can be defined as the standardized fourth-order cumulant:

$$SK(f) = \frac{\langle |H(n,f)|^4 \rangle}{\langle |H(n,f)|^2 \rangle^2} - 2 \qquad (4)$$

Where $H(n,f)$ represents the complex envelope of the digitized stator current $i(n)$ at frequency f. Constant 2 is used in this equation instead of constant 3 because $H(n,f)$ is a complex function. The symbol $\langle . \rangle$ represents the average value of the signal to be processed.

The main properties of this definition are [27], [28]:

➤ In the case of a stationary process, the SK is a constant function of the frequency.

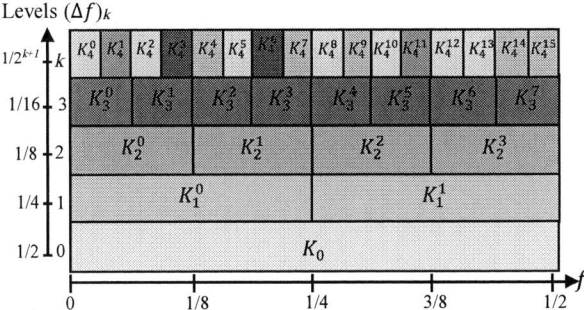

➤ In the case of a stationary Gaussian process, the SK is a function identically zero.

➤ In the case of a non-stationary process in the presence of an additive stationary noise, the $SK_2(f)$ become:

$$SK_2(f) = \frac{SK_1(f)}{[1+\rho(f)]^2} \qquad (5)$$

Where $\rho(f)$ is the signal-to-noise ratio dependent on the frequency, $SK_1(f)$ is the SK of the process without the presence of noise.

These three properties show the ability of the SK to identify the frequency signatures of the searched faults even in the case of load or speed variation.

B. Wavelet-Kurtogram

The Kurtogram is a four-dimensional spectral analysis tool that detects frequency signatures of non-stationary phenomena in a signal. This tool allows blind analysis of the signal without prior knowledge of the signal to be processed [28], [29]. It is clear that the exhaustive exploitation of all the decomposition levels of the Kurtogram takes an important computation time and is difficult to achieve in practice. As a solution to this drawback, a fast Kurtogram algorithm, known as Wavelet-Kurtogram, based on the wavelet transform is used instead [30-32].

The Wavelet-Kurtogram principle is based on the filtering of the signal to be processed by the WPT using a cascade of a low-pass filter (LPF) high-pass filter (HPF) with sub-sampling of 2 as shown in Fig. 2.

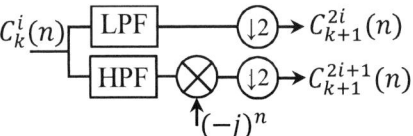

Fig. 2. Low pass and high pass filtering.

This decomposition sequence gives a tree structure as shown in Fig. 3. The symbol $C_k^i(n)$ indicates the decomposition

978-1-5386-7688-2/19 $31.00 © 2019 IEEE

of the signal obtained by the ith filter, $i = 0, \ldots, 2^k - 1$, while k is the decomposition level.

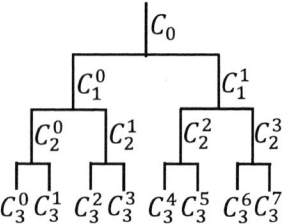

Fig. 3. Principle of the Wavelet-Kurtogram with a tree structure of filters benches.

The Kurtosis of each decomposition is calculated. All the Kurtosis K_k^i calculated are then plotted as an image, as shown in Fig. 4 with f is the normalized frequency.

Fig. 4. Wavelet-Kurtogram of the 4th level of decomposition.

From this Figure, the maximum Kurtosis, which is a scalar value, is localized. In this example, it is the position K_3^0. The position of the determined maximum Kurtosis allows us to plot the envelope of the decomposition $C_3^0(n)$ corresponding to this maximum value (K_3^0). The analysis of the resulting structure shown in Fig. 4, allows us to define a pass-band filter that can be located according to the frequency band of the decomposition to be studied. This pass-band filter is characterized by a center frequency f_c^i and a bandwidth B_k given by (6) and (7) respectively.

$$f_c^i = (i + 2^{-1})2^{-k-1} \qquad (6)$$

$$B_k = 2^{-k-1} \qquad (7)$$

As an example, for the decomposition C_3^0 and considering a sampling frequency of $3\ KHz$, the characteristics of the pass-band filter are: $f_c^0 = 93.75\ Hz$ and $B_3 = 187.5\ Hz$.

V. EXPERIMENTAL RESULTS

The induction motor used in these experimental tests is a three-phase squirrel cage coupled to a DC generator used as a load. The motor parameters are: $3\ kW$; $1410\ rpm$; $50\ Hz$; $4\ pôles$. The measuring chain consists of two Hall-effect current sensors, an Anti-aliasing filter and an acquisition card. The bench is connected to a computer for displaying and processing the measured signals as shown in Fig. 5. In addition, a tachometer is used to measure the real mechanical speed of the motor. All acquisitions were made in steady state. The duration of the data acquisition is 40 seconds with a sampling frequency of $3\ kHz$, which gives us a frequency resolution of $0.025\ Hz$.

Fig. 5. Photo of the test bench.

The studied bearing has a reference 6205-ZZ (coupling opposite side), whose geometrical parameters are: $B_D = 7.835\ mm$; $C_D = 38.5\ mm$; $N_b = 9$ and $\beta = 0$. In order to simulate an outer race fault, diameter holes of $3\ mm$ and $6\ mm$ are created artificially by drilling. Fig. 6 illustrates the created faults in bearings used in our tests.

Fig. 6. Rolling-element bearings without and with holes of $3\ mm$ and $6\ mm$.

Under these conditions, and according to (1) and (2), the theoretical frequency signature of the outer race fault must appear around $85.72\ Hz$ for the Wavelet-Kurtogram and around $35.72\ Hz$ for the PSDP estimation. This frequency is obtained where the rotational speed during our experimental tests is $1435\ rpm$, which corresponds to a rotation frequency of $23.9\ Hz$.

The different operation modes performed to validate the diagnosis procedure are:

➢ Motor operation with healthy bearings (without holes).

➢ Motor operation with outer race fault "3 mm hole".

➢ Motor operation with outer race fault "6 mm hole".

A. Motor Operation with Healthy Bearings

In this first case, the stator current is analyzed by the PSDP estimation (see Fig. 7) and by the Wavelet-Kurtogram method (see Fig. 8) where both bearings (coupling side and coupling opposite side) are healthy. Fig. 7 represents the PSDP estimation of the stator current on the frequency band of [20Hz, 80Hz]. We choose this particular band because the characteristic frequencies of the outer race bearing faults are likely to appear on this band. From the obtained results, it is clear that the PSDP method can detect: the supply frequency f_s ($49.8\ Hz$); the eccentricity harmonics f_{ecc} ($25.98\ Hz$ and $73.68\ Hz$) and the rotor cage imbalance harmonics f_{brb} ($45.55\ Hz$ and $54.03\ Hz$). Note that the harmonics of eccentricity and rotor cage imbalance can be expressed as a function of the fundamental by the following equations:

$$f_{ecc} = f_s \pm f_r \qquad (8)$$

$$f_{brb} = f_s \pm f_{br} \qquad (9)$$

Where f_{br} is the signature of the rotor cage imbalance.

Based on these two equations and the results of Fig. 7, we can deduce f_r and f_{br}: $f_r = 23.82\,Hz$ and $f_{br} = 4.25\,Hz$. Moreover, it should be noted that no outer race fault signature can be detected on the spectrum using this method.

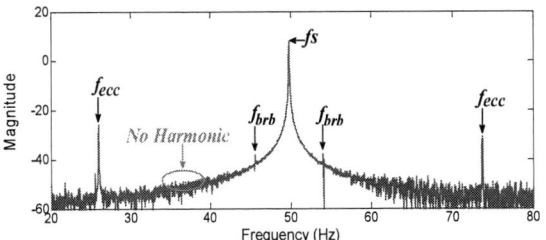

Fig. 7. Stator current PSDP estimation for healthy bearings.

From Fig. 8.a, the Wavelet-Kurtogram gives a maximum value of Kurtosis at decomposition C_4^0. This decomposition is characterized by a central frequency $f_c^0 = 46.875\,Hz$ and a frequency band $B_4 = 93.75\,Hz$. Note that the decomposition level is equal to 4 depending on (7) and given the sampling frequency used in these tests ($Fe = 3\,KHz$). Under these conditions, Fig. 8.b represents the analysis of the current envelope obtained by the Wavelet-Kurtogram for the obtained level of decomposition.

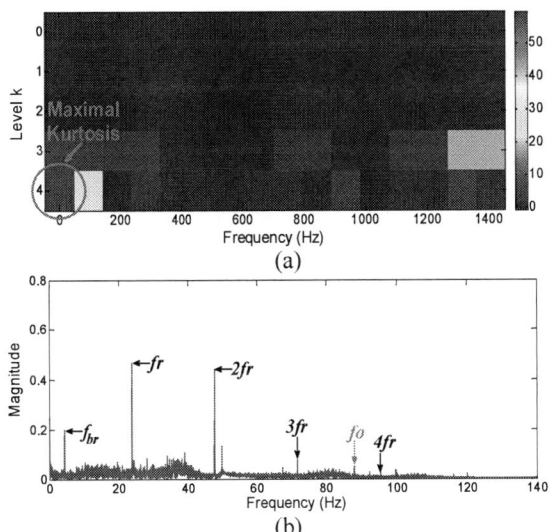

Fig. 8. a) Wavelet-Kurtogram and b) the spectral envelope obtained from the stator current with healthy bearings.

The analysis of Fig. 8.b shows that the Wavelet-Kurtogram detects the same frequency signatures obtained using the PSDP estimation without the effect of the fundamental frequency (i.e. the frequency of the rotor cage imbalance $f_{br} = 4.28\,Hz$ and the rotation frequency and its multiples $f_r = 23.85\,Hz$; $2f_r = 47.68\,Hz$ ...). Moreover, this method allows to detect a signature with very low magnitude $0.0497\,A/Hz$ at the

frequency $f_o = 87.93\,Hz$. Based on (1), we assume that this frequency represents the signature of the outer race fault. The slight difference between the theoretical and actual frequency is quite normal because:

➢ Measurement errors on rotation speed during tests.

➢ Frequency resolution used in our tests.

The appearance of this signature is probably due to the existence of scratches at the outer race of the bearing. The presence of these scratches can be explained by the repetitive process of mounting / dismounting of bearings during our tests (several dozens of experiments tests performed with the same bearing). Indeed, the outer race is the most exposed part of the bearing regarding its geometric position. These results show the reliability and the performances of the Wavelet-Kurtogram compared to the PSDP estimation approach in the bearing faults diagnosis. Moreover, this method allows to directly give the rotation frequency as well as that of the rotor cage imbalance without the supply frequency effect. This analysis will be considered as a reference for other tests.

B. Motor Operation with Outer Race Fault "3 mm hole"

In this second operation mode, the outer race of the used bearing, is perforated with a hole of 3 mm. During these tests, the rotation speed is kept the same as for the first operation mode, equal to 1435 tr/mn. Fig. 9 represents the stator current spectrum obtained by the PSDP estimation.

According to this figure, this method detects the same harmonics as the first operation mode ($f_s = 50.01\,Hz$; $f_{ecc} = 26,23\,Hz$; $f_{brb} = 45.03\,Hz$). Unfortunately, this method fails to detect the signature of outer race fault (3 mm hole) despite its presence. Moreover, based on both equations (8), (9) and the results of Fig. 9, we can deduce f_r and f_{br} as: $f_r = 23.78\,Hz$ and $f_{br} = 4.98\,Hz$.

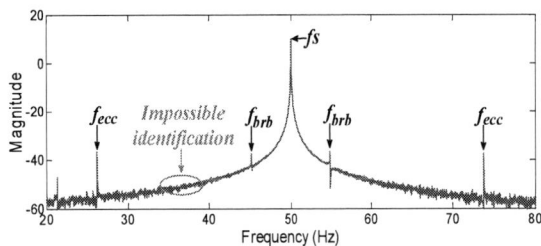

Fig. 9. Stator current PSDP estimation with an outer race fault "3 mm hole".

Fig. 10 represents the results obtained by the Wavelet-Kurtogram method for the same level of decomposition i.e. 4. From Fig. 10.a, the maximum Kurtosis is obtained at the same location as in the previous case at decomposition C_4^0. An in-depth analysis of Fig. 11.a shows that the color levels are not the same as for the healthy case. This shows that the stator current frequency signature is not the same.

The spectral analysis of the stator current envelope obtained from this decomposition (C_4^0) is illustrated in Fig. 10.b. According to this figure, the Wavelet-Kurtogram shows in addition to frequencies previously detected ($f_{br} =$

978-1-5386-7688-2/19 $31.00 © 2019 IEEE

$4.95\ Hz$; $f_r = 23.75\ Hz$; $2f_r = 47.53\ Hz \ldots$), the searched outer race fault signature. This signature is located at the frequency $f_o = 87.53\ Hz$ with a magnitude of $0.1163\ A/Hz$. This increase in magnitude compared to the first case $(0.0497\ A/Hz)$, shows that the outer race fault is more pronounced. These results prove the superiority of the Wavelet-Kurtogram method compared to the PSDP estimation in the detection of the outer race fault.

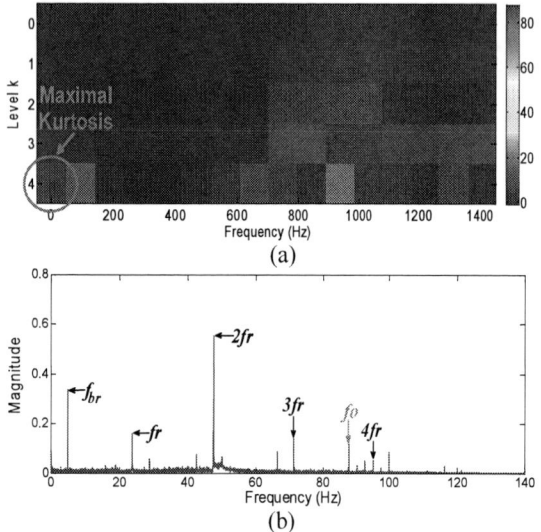

(a)

(b)

Fig. 10. a) Wavelet-Kurtogram and b) the spectral envelope obtained from the stator current with outer race fault "3 mm hole".

C. Motor Operation with Outer Race Fault "6 mm hole"

This last operation mode is achieved with an outer race bearing fault simulated by a $6\ mm$ hole diameter. The rotational speed of the motor is also chosen at $1435\ rpm$. Fig. 11 shows the stator current spectrum obtained by the PSDP estimation for this operation mode. According to this spectrum, this method can detect the same frequencies as in the previous cases but it is still unable to detect the outer race fault signature despite the hole diameter is larger ($6\ mm$). Finally, note that based on both equations (8), (9) and the results of Fig. 11, we can obtain $f_r = 23.87\ Hz$ and $f_{br} = 4.4\ Hz$.

Fig. 12 shows the results obtained by the Wavelet-Kurtogram method for the 4th level of decomposition. From Fig. 12.a the maximum Kurtosis is always localizable in the same location as in the two previous cases (C_4^0). Fig. 12.b represents the spectral analysis of the stator current envelope obtained from the decomposition (C_4^0). This figure shows that Wavelet-Kurtogram always gives good results. Indeed, this method is able to clearly and reliably detect the outer race fault signature ($87.53\ Hz$-$0.1620\ A/Hz$), in addition to the other characteristic frequencies ($f_{br} = 4.4\ Hz$; $f_r = 23.9\ Hz$; $2f_r = 47.78\ Hz \ldots$).

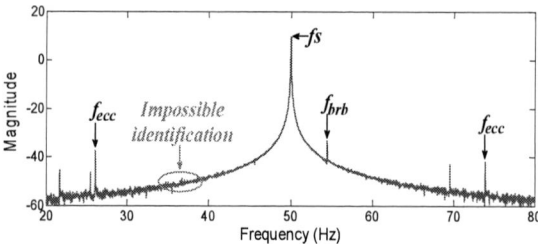

Fig. 11. Stator current PSDP estimation with an outer race fault "6 mm hole".

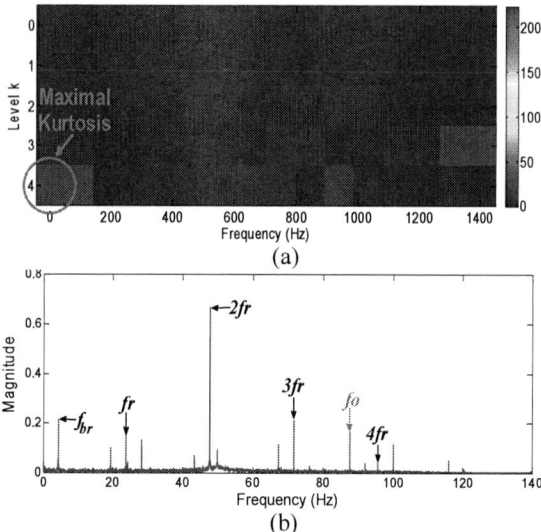

(a)

(b)

Fig. 12. a) Wavelet-Kurtogram and b) the spectral envelope obtained from the stator current with outer race fault "6 mm hole".

VI. CONCLUSION

The experimental results obtained in this paper, clearly show that the Wavelet-Kurtogram method has a better ability to detect faults affecting the outer race of a rolling-element bearing compared to the PSDP estimation. In addition, this method enabled us to have a better readability of the stator current spectrum allowing an easy analysis of the data. This readability also makes it possible to directly obtain the characteristic frequency signatures of the induction motor without the fundamental effect.

ACKNOWLEDGMENT

The authors wish to thank Prof. J. Antoni and Prof. D. Wang for their contributions and help in this work.

REFERENCES

[1] A. H. Bonnett, C. Yung, "Increased efficiency versus increased reliability," IEEE Ind. Appl. Mag, vol. 14, no. 1, pp. 29-36, Jan/Feb 2008.

[2] EPRI, "Improved motors for utility applications," final report publication, EPRI EL-2678 Project 1763-1, 2 (1982).

[3] F. Filippetti, A. Bellini and G. A. Capolino, "Condition monitoring and diagnosis of rotor faults in induction machines: State of art and future

perspectives," Electrical Machines Design Control and Diagnosis (WEMDCD), 2013 IEEE Workshop on, Paris, 2013, pp. 196-209.

[4] H. Henao, G. A. Capolino, M. Fernandez-Cabanas, F. Filippetti, C. Bruzzese, E. Strangas, R. Pucsa, M. Riera-Guasp, S. HedayatiKia, "Trends in Fault Diagnosis for Electrical Machines: A review of Diagnostic Techniques," IEEE Industrial Electronics Magazine, vol. 8, no.2, June 2014, pp. 31-42.

[5] E. Elbouchikhi, V. Choqueuse, Y. Amirat, M. E. H. Benbouzid and S. Turri, "An Efficient Hilbert–Huang Transform-Based Bearing Faults Detection in Induction Machines," in IEEE Transactions on Energy Conversion, vol. 32, no. 2, pp. 401-413, June 2017.

[6] A. Bellini, F. Filippetti, C. Tassoni and G. A. Capolino, "Advances in Diagnostic Techniques for Induction Machines," in IEEE Transactions on Industrial Electronics, vol. 55, no. 12, pp. 4109-4126, Dec. 2008.

[7] J. Antonio-Daviu, S. Aviyente, E. G. Strangas, and M. Riera-Guasp, "Scale invariant feature extraction algorithm for the automatic diagnosis of rotor asymmetries in induction motors," IEEE Trans. Ind. Informat., vol. 9, no. 1, pp. 100–108, Feb. 2013.

[8] W. Zhou, T. G. Habetler, and R. G. Harley, "Bearing fault detection via stator current noise cancellation and statistical control," IEEE Trans. Ind. Electron., vol. 55, no. 12, pp. 4260–4269, Dec. 2008.

[9] L. Frosini and E. Bassi, "Stator current and motor efficiency as indicators for different types of bearings faults in induction motors," IEEE Trans. Ind. Electron., vol. 57, no. 1, pp. 244–251, Jan. 2010.

[10] F. Immovilli, M. Cocconcelli, A. Bellini, and R. Rubini, "Detection of generalized-roughness bearing fault by spectral-kurtosis energy of vibration or current signals," IEEE Trans. Ind. Electron., vol. 56, no. 11, pp. 4710–4717, Nov. 2009.

[11] M. Blodt, P. Granjon, B. Raison, and G. Rostaing, "Models for bearing damage detection in induction motors using stator current monitoring," IEEE Trans. Ind. Electron., vol. 55, no. 4, pp. 1813–1822, Apr. 2008.

[12] E. H. El Bouchikhi, V. Choqueuse, M. Benbouzid and J. A. Antonino Daviu, "Stator current demodulation for induction machine rotor faults diagnosis," 2014 First International Conference on Green Energy ICGE 2014, Sfax, 2014, pp. 176-181.

[13] A. F. Aïmer, A. H. Boudinar, N. Benouzza, A. Bendiabdellah, "Simulation and Experimental Study of Induction Motor Broken Rotor Bars Fault Diagnosis using Stator Current Spectrogram", In Proc. of IEEE 3rd International Conference on Control, Engineering & Information Technology (CEIT), 25-27 May, 2015, Tlemcen, Algeria.

[14] H. A. Toliyat, S. Nandi, S. Choi, and H. Meshgin-Kelm, Electric Machines: Modeling, Condition Monitoring and Fault Diagnosis. New York, NY, USA: Taylor & Francis, 2012, pp. 1–23.

[15] Boqiang Xu, Liling Sun, Lie Xu, Guoyi Xu, "Improvement of the Hilbert Method via ESPRIT for Detecting Rotor Fault in Induction Motors at Low Slip," IEEE Trans on Energy Conversion, vol. 28, no. 1, pp. 225-233, March 2013.

[16] A. H. Boudinar, N. Benouzza, A. Bendiabdellah and M. E. A. Khodja, "Induction Motor Bearing Fault Analysis Using a Root-MUSIC Method," in IEEE Transactions on Industry Applications, vol. 52, no. 5, pp. 3851-3860, Sept.-Oct. 2016.

[17] A. Ghods and H. H. Lee, "A frequency-based approach to detect bearing faults in induction motors using discrete wavelet transform," 2014 IEEE International Conference on Industrial Technology (ICIT), Busan, 2014, pp. 121-125.

[18] A. Bouzida et al., "Fault diagnosis in industrial induction machines through discrete wavelet transform," IEEE Trans. Ind. Electron., vol. 58, no. 9, pp. 4385–4395, Sep. 2011.

[19] J. Pons-Llinares, J. A. Antonino-Daviu, M. Riera-Guasp, M. Pineda-Sanchez, and V. Climente-Alarcon, "Induction motor diagnosis based on a transient current analytic wavelet transform via frequency B-splines," IEEE Trans. Ind. Electron., vol. 58, no. 5, pp. 1530–1544, May 2011.

[20] R. Puche-Panadero, M. Pineda-Sanchez, M. Riera-Guasp, J. Roger-Folch, E. Hurtado-Perez and J. Perez-Cruz, "Improved Resolution of the MCSA Method for the Diagnosis of Rotor Asymmetries at Very Low Slip," in IEEE Transactions on Energy Conversion, vol. 24, no. 1, pp. 52-59, March 2009.

[21] V. C. M. N. Leite et al., "Detection of Localized Bearing Faults in Induction Machines by Spectral Kurtosis and Envelope Analysis of Stator Current," in IEEE Transactions on Industrial Electronics, vol. 62, no. 3, pp. 1855-1865, March 2015.

[22] R. B. Randall, Vibration-Based Condition Monitoring: Industrial, Aerospace and Automotive Applications. Chichester, U.K.: Wiley, 2011, pp. 24–62, pp. 167–227.

[23] F. Immovilli, A. Bellini, R. Rubini, and C. Tassoni, "Diagnosis of bearing faults in induction machines by vibration or current signals: A critical comparison," IEEE Trans. Ind. Appl., vol. 46, no. 4, pp. 1350–1359, Jul./Aug. 2010.

[24] W. Sui and D. Zhang, "Research on envelope analysis for bearings fault detection," in Proc. ICCSE, Hefei, China, 2010, pp. 973–976.

[25] M. Pineda-Sanchez et al., "Application of the Teager–Kaiser Energy Operator to the Fault Diagnosis of Induction Motors," in IEEE Transactions on Energy Conversion, vol. 28, no. 4, pp. 1036-1044, Dec. 2013.

[26] R. F. Dwyer, "Detection of non-Gaussian signals by frequency domain kurtosis estimation," in Proc. IEEE ICASSP, vol. 8, pp. 607-610, April 1983.

[27] J. Antoni, "The spectral kurtosis: a useful tool for characterising nonstationary signals," Mechanical Systems and Signal Processing, vol. 20, no. 2, pp. 282–307, 2006.

[28] J. Antoni and R. Randall, "The spectral kurtosis: application to the vibratory surveillance and diagnostics of rotating machines," Mechanical Systems and Signal Processing, vol. 20, no. 2, pp. 308–331, 2006.

[29] J. Antoni, "Fast computation of the kurtogram for the detection of transient faults," Mech. Syst. Signal Process., vol. 21, no. 1, pp. 108–124, Jan. 2007.

[30] N. Sawalhi, "Rolling element bearings: Diagnostic, prognostic and fault simulations," Ph.D. dissertation, Faculty Eng. Mech. Manuf. Eng., Univ. New South Wales, Sydney, Australia, 2007.

[31] Y. Lei, J. Lin, Z. He, Y. Zi, "Application of an improved kurtogram method for fault diagnosis of rolling element bearings," Mechanical Systems and Signal Processing, vol. 25, pp. 1738-1749, 2011.

[32] D. Wang, P. W. Tse, K. L. Tsui, "An enhanced Kurtogram method for fault diagnosis of rolling element bearings," Mechanical Systems and Signal Processing, vol. 35, pp. 176-199, 2013.

Sensorless Synchronous Reluctance Generator Control Based on q Axis Estimated Current

Liviu – Dănuț Vitan
Department of Electrical Engineering
Politehnica University of Timisoara
Timisoara, Romania
liviu.vitan@student.upt.ro

Lucian Tutelea
Department of Electrical Engineering
Politehnica University of Timisoara
Timisoara, Romania
lucian.tutelea@upt.ro

Nicolae Muntean
Department of Electrical Engineering
Politehnica University of Timisoara
Romanian Academy, Timisoara Branch
Timisoara, Romania
nicolae.muntean@upt.ro

Ion Boldea
Department of Electrical Engineering
Politehnica University of Timisoara
Romanian Academy, Timisoara Branch
Timisoara, Romania
ion.boldea@upt.ro

Abstract—This paper presents a sensorless Synchronous Reluctance Machine used as a generator with a different method for rotor position estimator. The estimator is based on the machine model and the error between measured q-axis current versus estimated current. The rotor position error is analyzed in various running conditions considering uncertain machine parameters. Stability of the rotor position estimator has been approached using model linearization by small signal deviations. Simulation and experimental results validate the theoretical assumptions.

Keywords—Synchronous reluctance machine, rotor position estimator, rotor position sensitivity, stability, sensorless control.

I. INTRODUCTION

The Synchronous Reluctance Machine (SynRM) has gained researchers attention in the last period due to its advantages as a simple and rugged structure, no rotor winding and no PMs [1]–[5]. Over the years, substantial research on SynRM design [2], [6]–[9], directly conected to the grid [10]–[12], and inverter fed with position sensor control [13]–[16] has been done. Also, sensorless methods regarding rotor position has been approached in [17]–[22]. For small power and low speed applications, sensorless control is essential to have a low-cost implementation. State-of-the-art estimators are based on measured voltages integration, usually implemented as low pass filters that introduce distortions at low speed, or on signal injection [23] which needs a DC voltage reserve.

This paper proposes a rotor position estimator for a SynRM based on machine model and q-axis current error (measured and estimated by model). In this method, the integration operations for the flux observer are naturally included in the closed loop of the machine model thus additional loops for offset and fluxes initial values compensation are not required. Also, the aperiodic flux component damping from the observer is similar with the electrical machine phenomena. An existing SynRM was controlled in generator mode with dq current references in order to verify the study.

II. SYSTEM LAYOUT

The proposed configuration of sensorless control used for controlling the Synchronous Reluctance Generator (SynRG) is shown in Fig.1. It consists of a Prime Mover, a SynRG and a power converter which delivers the energy in a DC grid.

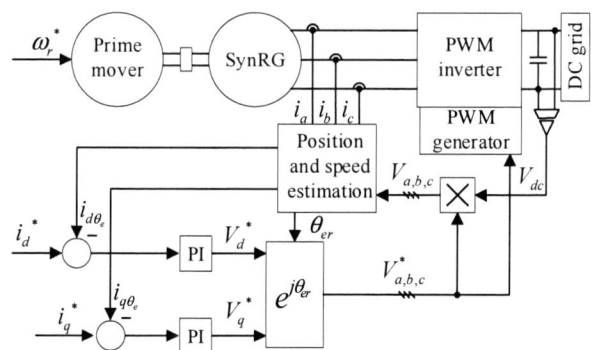

Fig. 1. Sensorless control diagram of SynRG

The Prime Mover can be a hydro turbine, a wind turbine or a variable speed drive (VSD) with speed control. The power reference of the SynRG is obtained by prescribing the d- and q-axes currents. The current errors are processed through a PI controller in order to produce the dq reference voltages. The dq reference voltages are transformed into polar coordinates to be prescribed to the PWM inverter. In order to calculate the rotor frame coordinates, the angular position is needed and it is estimated by measuring the phase currents and DC voltage.

A. SynRG Model

The SynRG is characterized by the following equations in rotating frame reference:

$$\frac{d\psi_d}{dt} = V_d - i_d \cdot R_s + p_1 \cdot \omega_r \cdot \psi_q, \tag{1}$$

$$\frac{d\psi_q}{dt} = V_q - i_q \cdot R_s - p_1 \cdot \omega_r \cdot \psi_d, \tag{2}$$

$$\frac{d\omega_r}{dt} = \frac{1}{J}\left(T_{em} - T_L\right), \tag{3}$$

$$\frac{d\theta_r}{dt} = \omega_r, \quad T_{em} = p_1\left(\psi_d \cdot i_q - \psi_q \cdot i_d\right), \tag{4}$$

$$\psi_d = L_d \cdot i_d, \quad \psi_q = L_q \cdot i_q. \tag{5}$$

978-1-5386-7688-2/19 $31.00 © 2019 IEEE

where ψ_d, ψ_q are the d- and q-axis linkage flux, V_d, V_q are the d- and q-axis voltages, R_s is the stator resistance, i_d, i_q are the d- and q-axis currents, p_1 is the number of pole pairs, L_d, L_q are the d- and q-axis inductances, ω_r is the rotor speed, θ_r is the rotor position, T_{em} is the electromagnetic torque, T_L is load torque and J is the entire system (prime mover and SynRG) inertia.

B. Rotor position-speed estimator

The overall performances of the system depends on the rotor position estimation, which is based on the same model as the machine, except for the mechanical equation which is replaced with a PI transfer function acting on the estimated and the machine q axis currents (with the machine q axis current in the estimated dq frame). Also for estimation of torque, the active flux concept is used [24]. Fig. 2 shows the implementation of position-speed estimator. The equations which describe the estimator subsystem are given below:

$$\frac{d\psi_{de}}{dt} = V_{d\theta_e} - i_{de} \cdot R_{se} + p_1 \cdot \omega_{re} \cdot \psi_{qe}, \tag{6}$$

$$\frac{d\psi_{qe}}{dt} = V_{q\theta_e} - i_{qe} \cdot R_{se} - p_1 \cdot \omega_{re} \cdot \psi_{de}, \tag{7}$$

$$\frac{d\omega_{re}}{dt} = k_p \cdot \frac{d\varepsilon_{\omega re}}{dt} + k_i \cdot \varepsilon_{\omega re}, \tag{8}$$

$$\frac{d\theta_{re}}{dt} = \omega_{re}, \tag{9}$$

$$T_{em_e} = \left(\psi_{de} - i_{d\theta_e} \cdot L_{qe}\right) \cdot p_1 \cdot i_{q\theta_e}, \tag{10}$$

$$i_{de} = \frac{\psi_{de}}{L_{de}}, \quad i_{qe} = \frac{\psi_{qe}}{L_{qe}}, \quad \varepsilon_{\omega re} = i_{qe} - i_{q\theta_e}. \tag{11}$$

where the terms with index ' e ' are the estimated values and they have the same signification as in SynRG model. The terms: $V_{d\theta_e}$, $V_{q\theta_e}$, $i_{d\theta_e}$, $i_{q\theta_e}$ are d- and q-axis voltages and currents in estimator coordinates (these were obtained from stator coordinates through inverse Park transform using estimated rotor position), k_p, k_i are the PI transfer function constants, $\varepsilon_{\omega re}$ is the error between estimated and measured q-axis currents.

III. POSITION ERORR ANALYSIS REGARDING PARAMETERS VARIATION

The estimated angle error, versus uncertain machine parameters is analyzed in this chapter. The machine and estimator subsystems both use the same phase voltages that are provided from V_d^*, V_q^*, in the estimated dq reference frame. This differs from the machine dq frame with error angle ε_θ.

$$V_d = V_d^* \cdot \cos\left(\varepsilon_\theta\right) + V_q^* \cdot \sin\left(\varepsilon_\theta\right) \tag{12}$$

$$V_q = -V_d^* \cdot \sin\left(\varepsilon_\theta\right) + V_q^* \cdot \cos\left(\varepsilon_\theta\right) \tag{13}$$

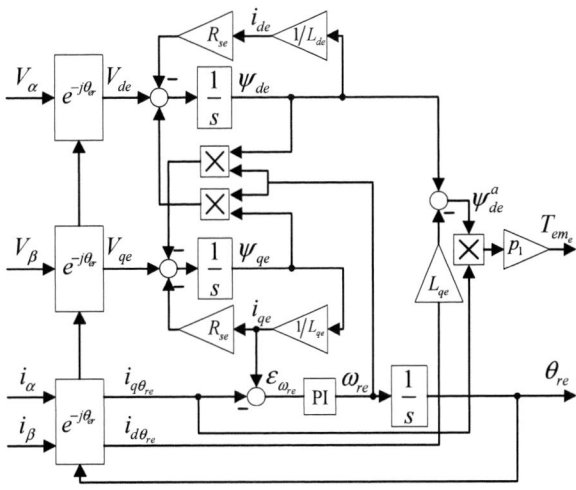

Fig. 2. Block diagram of the Position Estimator

$$i_{d\theta e} = \frac{\psi_d}{L_d} \cdot \cos\left(\varepsilon_\theta\right) - \frac{\psi_q}{L_q} \cdot \sin\left(\varepsilon_\theta\right) \tag{14}$$

$$i_{q\theta e} = \frac{\psi_q}{L_q} \cdot \cos\left(\varepsilon_\theta\right) + \frac{\psi_d}{L_d} \cdot \sin\left(\varepsilon_\theta\right) \tag{15}$$

$$\varepsilon_\theta = p_1\left(\theta_r - \theta_{re}\right) \tag{16}$$

Taking into consideration the machine voltages (12) and (13), the machine currents in the estimator reference (14) and (15), the machine equations and estimator equations, the entire system (machine and estimator) equations becomes:

$$\frac{d\psi_d}{dt} = V_d^* \cdot \cos\left(\varepsilon_\theta\right) + V_q^* \cdot \sin\left(\varepsilon_\theta\right) - \frac{R_s}{L_d} \cdot \psi_d + \\ + p_1 \cdot \omega_r \cdot \psi_q, \tag{17}$$

$$\frac{d\psi_q}{dt} = -V_d^* \cdot \sin\left(\varepsilon_\theta\right) + V_q^* \cdot \cos\left(\varepsilon_\theta\right) - \frac{R_s}{L_q} \cdot \psi_q \\ - p_1 \cdot \omega_r \cdot \psi_d, \tag{18}$$

$$\frac{d\omega_r}{dt} = \frac{1}{J}\left[p_1 \cdot \psi_d \cdot \psi_q \left(\frac{1}{L_q} - \frac{1}{L_d}\right) - T_L \right], \tag{19}$$

$$\frac{d\theta_r}{dt} = \omega_r, \quad \frac{d\theta_{re}}{dt} = \omega_{re}, \tag{20}$$

$$\frac{d\psi_{de}}{dt} = V_d^* - \frac{R_{se}}{L_{de}} \cdot \psi_{de} + p_1 \cdot \omega_{re} \cdot \psi_{qe}, \tag{21}$$

$$\frac{d\psi_{qe}}{dt} = V_q^* - \frac{R_{se}}{L_{qe}} \cdot \psi_{qe} - p_1 \cdot \omega_{re} \cdot \psi_{de}, \tag{22}$$

$$\frac{d\omega_{re}}{dt} = k_p \cdot \frac{d}{dt}\left(\frac{\psi_{qe}}{L_{qe}} - \frac{\psi_q}{L_q} \cdot \cos\left(\varepsilon_\theta\right) - \frac{\psi_d}{L_d} \cdot \sin\left(\varepsilon_\theta\right) \right) \\ + k_i \cdot \frac{\psi_{qe}}{L_{qe}} - \left(\frac{\psi_q}{L_q} \cdot \cos\left(\varepsilon_\theta\right) + \frac{\psi_d}{L_d} \cdot \sin\left(\varepsilon_\theta\right) \right), \tag{23}$$

A steady state solution of the system equations (17)-(23), result in constant state variables, so their derivatives are zero, except for (20) in this case. Solving the system equations, the following results are obtained:

$$\psi_{de} = \frac{L_{de} \cdot \left(R_{se} \cdot V_d^* + p_1 \cdot \omega_r \cdot L_{qe} \cdot V_q^* \right)}{\Delta_e}, \quad (24)$$

$$\psi_{qe} = \frac{L_{qe} \cdot \left(R_{se} \cdot V_q^* - p_1 \cdot \omega_r \cdot L_{de} \cdot V_d^* \right)}{\Delta_e}, \quad (25)$$

$$\psi_d = \frac{L_d}{\Delta} \cdot \Big[R_s \cdot \left(V_d^* \cdot \cos(\varepsilon_\theta) + V_q^* \cdot \sin(\varepsilon_\theta) \right) + \\ + X_q \cdot \left(-V_d^* \cdot \sin(\varepsilon_\theta) + V_q^* \cdot \cos(\varepsilon_\theta) \right) \Big], \quad (26)$$

$$\psi_q = \frac{L_q}{\Delta} \cdot \Big[R_s \cdot \left(-V_d^* \cdot \sin(\varepsilon_\theta) + V_q^* \cdot \cos(\varepsilon_\theta) \right) - \\ - X_d \cdot \left(V_d^* \cdot \cos(\varepsilon_\theta) + V_q^* \cdot \sin(\varepsilon_\theta) \right) \Big]. \quad (27)$$

where:

$$\Delta_e = R_{se}^2 + p_1^2 \cdot \omega_r^2 \cdot L_{de} \cdot L_{qe}, \quad X_d = p_1 \cdot \omega_r \cdot L_d, \\ \Delta = R_s^2 + L_d \cdot L_q \cdot \omega_r^2 \cdot p_1^2, \quad X_q = p_1 \cdot \omega_r \cdot L_q. \quad (28)$$

from (23) result:

$$\frac{\psi_{qe}}{L_{qe}} = \left(\frac{\psi_q}{L_q} \cdot \cos(\varepsilon_\theta) + \frac{\psi_d}{L_d} \cdot \sin(\varepsilon_\theta) \right). \quad (29)$$

The equation (30) is calculated by using (24) - (27) into (29).

$$a \cdot \cos(2\varepsilon_\theta) - b \cdot \sin(2\varepsilon_\theta) + e = 0 \quad (30)$$

where:

$$\begin{cases} a' = R_{se} \cdot V_q^* - X_{de} \cdot V_d^*, \quad a = -\frac{V_d^*}{2} \cdot X_d + \frac{V_d^*}{2} \cdot X_q \\ b = \frac{V_q^*}{2} \cdot \left(X_d - X_q \right) \\ f = R_s \cdot V_q^* - \frac{V_d^*}{2} \cdot \left(X_d + X_q \right), \quad e = f - a' \cdot k_\Delta \end{cases} \quad (31)$$

Solutions:

$$\begin{cases} \varepsilon_\theta = \pi \cdot k - \tan^{-1}\left(\frac{b - \sqrt{a^2 + b^2 - e^2}}{a - e} \right) \\ \varepsilon_\theta = \pi \cdot k - \tan^{-1}\left(\frac{b + \sqrt{a^2 + b^2 - e^2}}{a - e} \right) \end{cases} \quad (32)$$

Conditions:

$$\begin{cases} e^2 \le a^2 + b^2 \\ \frac{b \mp \sqrt{a^2 + b^2 - e^2}}{a - e} \in \mathbb{R} \end{cases} \quad (33)$$

The per unit stator resistance k_r, d-axis inductance k_d and q-axis inductances k_q, were introduced to consider the

uncertain parameters in the estimator. The estimator parameters related to machine parameters become:

$$R_{se} = k_r \cdot R_s, \quad L_{de} = k_d \cdot L_d, \quad L_{qe} = k_q \cdot L_q. \quad (34)$$

The algebraic equation system does not have real solutions for the high difference between estimator and machine parameters, which means there is no steady state regime. In Fig. 3 and Fig. 4 it is shown how ε_θ depends according to the variation of k_r, k_d and k_q (0.5 to 1.5 with 0.05 step) at several rotor speeds and different i_q/i_d ratio. The extensive simulations show that position error depends only on the i_q/i_d currents ratio, and their absolute values have no influence.

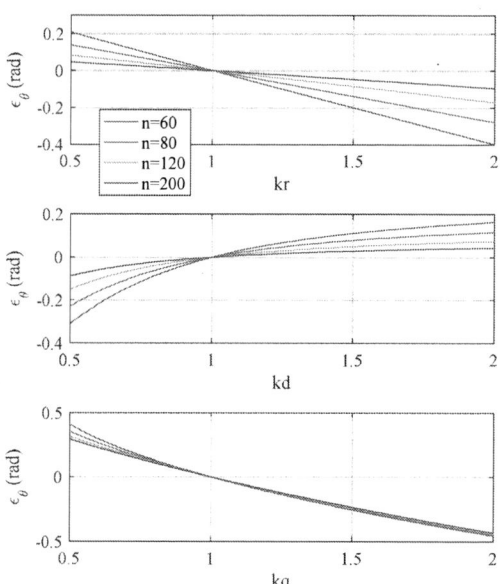

Fig. 3. Position error sensitivity at equal currents ($i_d = 5$, $i_q = -5$)

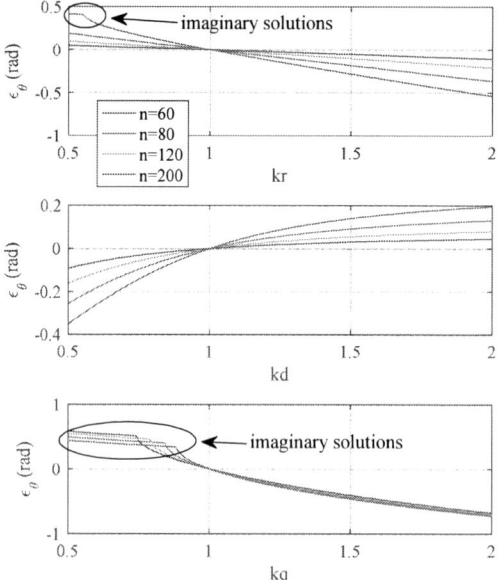

Fig. 4. Potion error sensitivity at current ratio $i_q/i_d = 2$ ($i_d = 5$, $i_q = -10$)

IV. STABILITY ANALYSIS

The steady state analysis from previous chapter shows the position error according to parameter variation, but it is not able to prove that a steady state solution is stable and how the PI transfer function constants of the estimator, k_p, k_i influence the stability, so a stability analysis is required. The analysis was performed by linearization, considering small variations in the state variables around a steady state point.

$$\psi_d = \psi_{d_0} + \tilde{\psi}_d \ , \ \psi_q = \psi_{q_0} + \tilde{\psi}_q \tag{35}$$

$$\psi_{de} = \psi_{de_0} + \tilde{\psi}_{de} \ , \ \psi_{qe} = \psi_{qe_0} + \tilde{\psi}_{qe} \tag{36}$$

$$\omega_{re} = \omega_{re_0} + \tilde{\omega}_{re} \ , \ \theta_r = \theta_{r_0} + \tilde{\theta}_r \ , \ \theta_{re} = \theta_{re_0} + \tilde{\theta}_{re} \tag{37}$$

$$\frac{d}{dt}\varepsilon_\theta = p_1 \cdot (\omega_r - \omega_{re}) \tag{38}$$

Considering system equations from (17)-(23) and equation (38) and substituting the state variables with their expressions from (35) to (37), the linearized system equation are obtained:

$$\frac{d}{dt}\tilde{\psi}_d = -\frac{R_s}{L_d}\cdot\tilde{\psi}_d + p_1 \cdot \omega_r \cdot \tilde{\psi}_q + \\ + \tilde{\varepsilon}_\theta \cdot \left[V_q^* \cdot \cos\left(\varepsilon_{\theta_0}\right) - V_d^* \cdot \sin\left(\varepsilon_{\theta_0}\right)\right], \tag{39}$$

$$\frac{d}{dt}\tilde{\psi}_q = -p_1 \cdot \omega_r \cdot \tilde{\psi}_q - \frac{R_s}{L_d}\cdot\tilde{\psi}_d - \\ - \tilde{\varepsilon}_\theta \cdot \left[V_d^* \cdot \cos\left(\varepsilon_{\theta_0}\right) - V_q^* \cdot \sin\left(\varepsilon_{\theta_0}\right)\right], \tag{40}$$

$$\frac{d}{dt}\tilde{\psi}_{de} = -\frac{R_{se}}{L_{de}}\cdot\tilde{\psi}_{de} + p_1 \cdot \omega_{re_0} \cdot \tilde{\psi}_{qe} + p_1 \cdot \psi_{qe_0} \cdot \tilde{\omega}_{re}, \tag{41}$$

$$\frac{d}{dt}\tilde{\psi}_{qe} = -p_1 \cdot \omega_{re_0} \cdot \tilde{\psi}_{de} - \frac{R_{se}}{L_{qe}}\cdot\tilde{\psi}_{qe} - p_1 \cdot \psi_{de_0} \cdot \tilde{\omega}_{re}, \tag{42}$$

$$\frac{d}{dt}\tilde{\omega}_{re} = \left(\begin{bmatrix} a_{11} & a_{12} \\ a_{21} & a_{22} \\ a_{31} & 0 \\ a_{41} & a_{42} \\ a_{51} & 0 \\ a_{61} & a_{62} \end{bmatrix}^T \cdot \begin{bmatrix} k_p \\ k_i \end{bmatrix}\right) \cdot \begin{bmatrix} \tilde{\psi}_d \\ \tilde{\psi}_q \\ \tilde{\psi}_{de} \\ \tilde{\psi}_{qe} \\ \tilde{\omega}_{re} \\ \tilde{\varepsilon}_\theta \end{bmatrix}, \tag{43}$$

where:

$$a_{11} = \frac{R_s \cdot \sin\left(\varepsilon_{\theta_0}\right)}{L_d^2} - p_1 \cdot \left(\omega_r - \omega_{re_0}\right)\cdot\frac{\cos\left(\varepsilon_{\theta_0}\right)}{L_d} + \\ + \frac{p_1 \cdot \omega_r \cdot \cos\left(\varepsilon_{\theta_0}\right)}{L_q}, \tag{44}$$

$$a_{12} = -\frac{\sin\left(\varepsilon_{\theta_0}\right)}{L_d^2} \ , \ a_{22} = \frac{\cos\left(\varepsilon_{\theta_0}\right)}{L_q^2}, \tag{45}$$

$$a_{21} = \frac{R_s \cdot \cos\left(\varepsilon_{\theta_0}\right)}{L_q^2} - p_1 \cdot \left(\omega_r - \omega_{re_0}\right)\cdot\frac{\sin\left(\varepsilon_{\theta_0}\right)}{L_q} - \\ - \frac{p_1 \cdot \omega_r \cdot \sin\left(\varepsilon_{\theta_0}\right)}{L_d}, \tag{46}$$

$$a_{31} = -\frac{p_1 \cdot \omega_{re_0}}{L_{qe}} \ , \ a_{41} = -\frac{R_{se}}{L_{qe}^2} \ , \ a_{42} = \frac{1}{L_{qe}}, \tag{47}$$

$$a_{51} = p_1 \cdot \frac{\psi_{d_0} \cdot \cos\left(\varepsilon_{\theta_0}\right)}{L_d} - \frac{\psi_{q_0} \cdot \sin\left(\varepsilon_{\theta_0}\right)}{L_q} - p_1 \cdot \frac{\psi_{de_0}}{L_{qe}}, \tag{48}$$

$$a_{61} = \left(\frac{1}{L_d} - \frac{1}{L_q}\right)\cdot\left\{V_d^* \cdot \cos\left(2\varepsilon_{\theta_0}\right) + V_q^* \cdot \sin\left(2\varepsilon_{\theta_0}\right) + \\ + p_1 \cdot \omega_r \cdot \left[\psi_{q_0} \cdot \cos\left(\varepsilon_{\theta_0}\right) - \psi_{d_0} \cdot \sin\left(\varepsilon_{\theta_0}\right)\right]\right\} - \\ - R_s \cdot \left[\frac{\psi_{d_0}}{L_d^2}\cdot\cos\left(\varepsilon_{\theta_0}\right) + \frac{\psi_{q_0}}{L_q^2}\cdot\sin\left(\varepsilon_{\theta_0}\right)\right] + \\ + p_1 \cdot \omega_{re_0} \cdot \left[\frac{\psi_{d_0}}{L_d}\cdot\sin\left(\varepsilon_{\theta_0}\right) + \frac{\psi_{q_0}}{L_q}\cdot\cos\left(\varepsilon_{\theta_0}\right)\right], \tag{49}$$

$$a_{62} = \frac{\psi_{q_0}}{L_q}\cdot\sin\left(\varepsilon_{\theta_0}\right) - \frac{\psi_{d_0}}{L_d}\cdot\cos\left(\varepsilon_{\theta_0}\right), \tag{50}$$

$$\frac{d}{dt}\tilde{\varepsilon}_\theta = -p_1 \cdot \omega_{re_0}. \tag{51}$$

The Hurwitz criterion was used to study the system stability. Based on the six order equation system (39)-(43) and (51), the system transfer function was calculated considering the machine parameters (TABLE I.), several values for rotor speed (60, 120 and 200 rpm) and several id, iq combinations. Although Hurwitz minors are available in analytical form, their expressions are too complex to be analyzed and aggregated for thousands of combinations (around 90000). Again, the numerical solution should be used by giving values to k_p and k_i. To explore a large domain (1 to one million) without impractical computation effort, an exponential distribution, with only 101 values per constant, (10201 total values) is used.

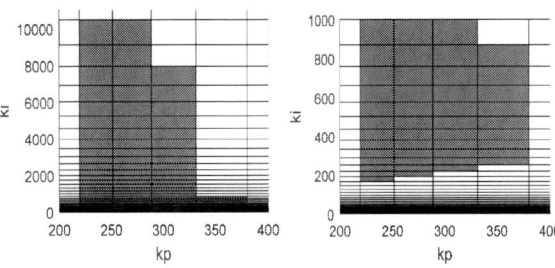

Fig. 5. Feasibility domain for k_p and k_i: all domain - left, close-up - right

TABLE I.   PARAMETERS OF THE SYNRG

Parameters	Value	Unit
Rated power P_N	1800	W
Line voltage V_L	400	V
Rated current I_N	10	A
Rated speed n_N	200	rpm
Pole pairs p_1	6	
d-axis inductance L_d	0,822	H
q-axis inductance L_q	0,289	H
Stator phase resistance R_s	6,17	Ω

A feasibility matrix, containing 1 for stable points and zero for unstable was built. The final feasibility matrix was computed as a set of intersection of all feasibility matrix. The feasibility domain (matrix) for k_p and k_i is shown in Fig. 5.

V. SIMULATION RESULTS

Stability analysis studied in previous chapter needs a validation through simulation of the entire system (Fig. 1) because the simulation provides access to the machine parameters. The digital simulation is based on the diagram block presented in Fig. 1, where SynRG and Prime Mover model are shown in Fig. 9, and estimator model from Fig. 2. In Fig. 9, the prime mover is a torque controller in order to drive the SynRG.

The results of the simulation were obtained for $k_p = 250$ and $k_i = 1500$ from Fig. 5 and the parameters for SynRG from TABLE I. The following cases are presented:

- Around rated power, n=200rpm: Fig. 6, Fig. 7;

- Rated torque, n=120rpm: Fig. 8, Fig. 10.

The saturation and variation of stator resistance with temperature were considered by modifying the machine parameters.

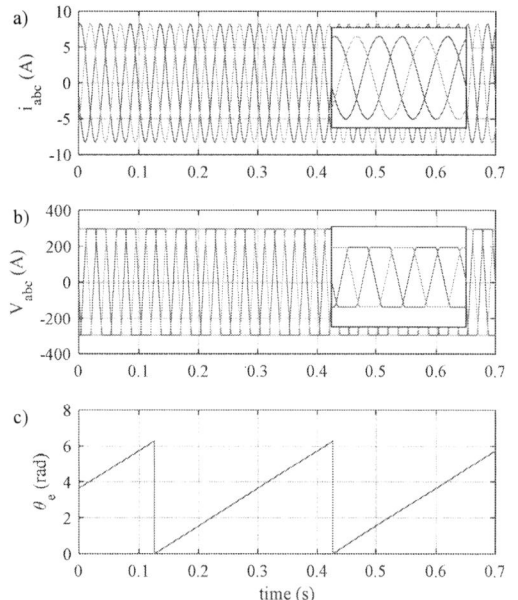

Fig. 6. High power results: $i_d = 5$; $i_q = -10$, $Rs = 6.7$; $Ld = 0.8$; Lq=0.254, a) stator currents, b) stator voltages, c) estimated rotor position.

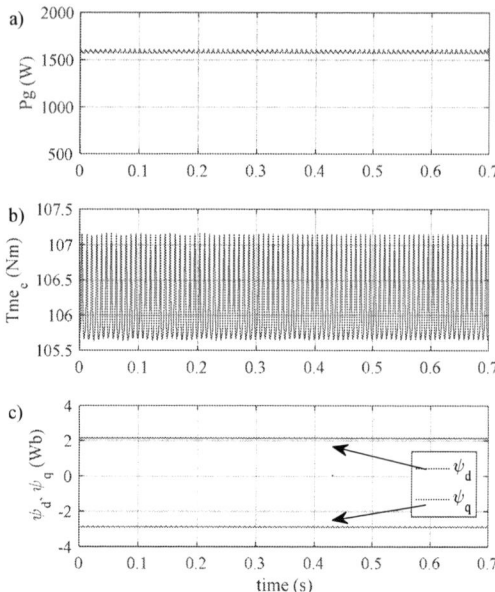

Fig. 7. High power results: $i_d = 5$; $i_q = -10$, $Rs = 6.7$; $Ld = 0.8$; $Lq = 0.254$, a) active output power, b) torque, c) dq flux linkages.

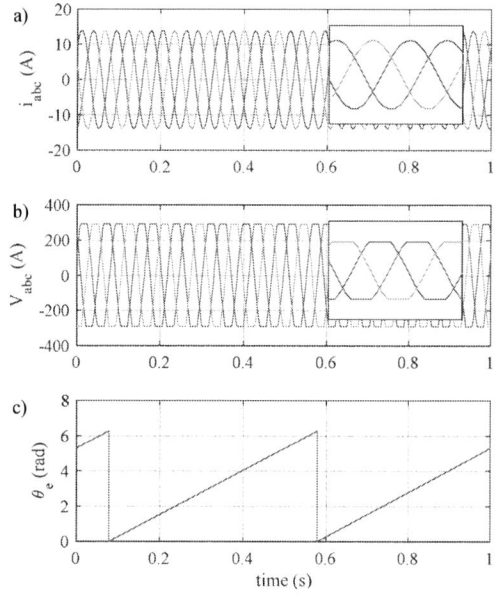

Fig. 8. Rated torque: $i_d = 7.5$; $i_q = -15.5$; $Rs = 6.7$; $Ld = 0.555$; $Lq = 0.255$, a) stator currents, b) stator voltages, c) estimated rotor position.

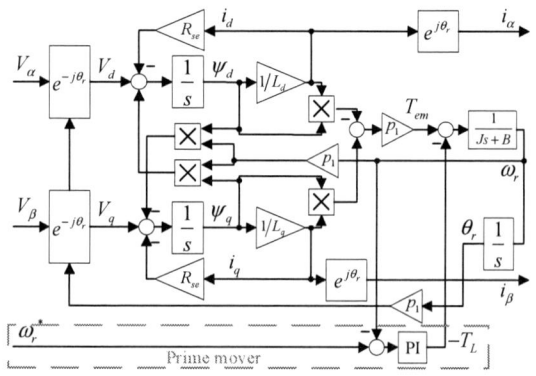

Fig. 9. SynRG with prime mover – block diagram

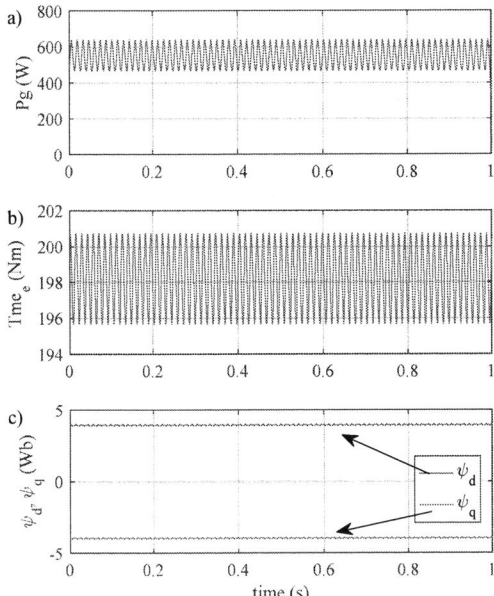

Fig. 10. Rated torque: i_d = 7.5; i_q = -15.5; Rs = 6.7; Ld = 0.555; Lq = 0.255, a) active output power, b) torque, c) dq flux linkages.

VI. EXPERIMENTAL RESULTS

The proposed control method was validated on a test bench as depicted in Fig. 11 and Fig. 12. The results for the corresponding setup are presented in Fig. 13 - Fig. 19. The SynRG has the parameters given in TABLE I. The SynRG is driven by a 11 kW three phase squirrel cage induction motor (IM), controlled by a back-to-back ABB (ACS 800) inverter. A 4 kVA Danfoss inverter feeds the SynRG. The control technique is implemented on a digital signal processor (dSpace DS1103). The Danfoss inverter control board was replaced with another interface card which provides full control over the inverter IGBT gate drivers.

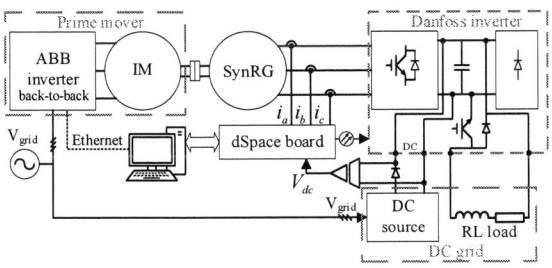

Fig. 11. Test setup diagram

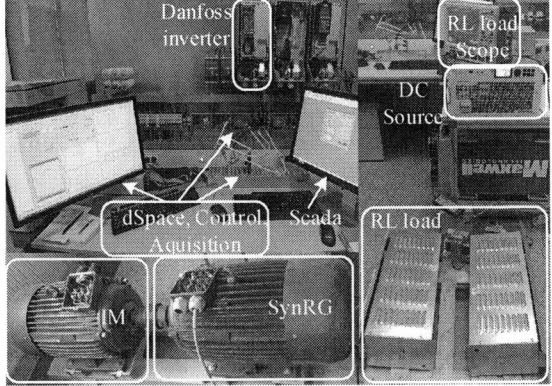

Fig. 12. Experimental platform

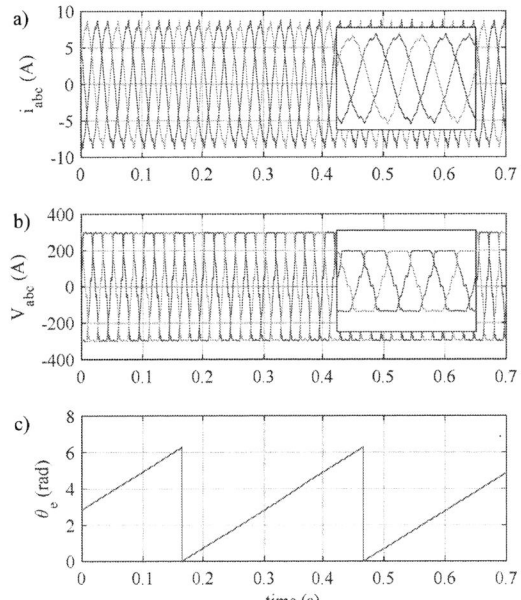

Fig. 13. Experimental results: $i_d^* = 5A$, $i_q^* = -10A$, $n = 200rpm$, a) stator currents, b) stator voltages, c) estimated rotor position.

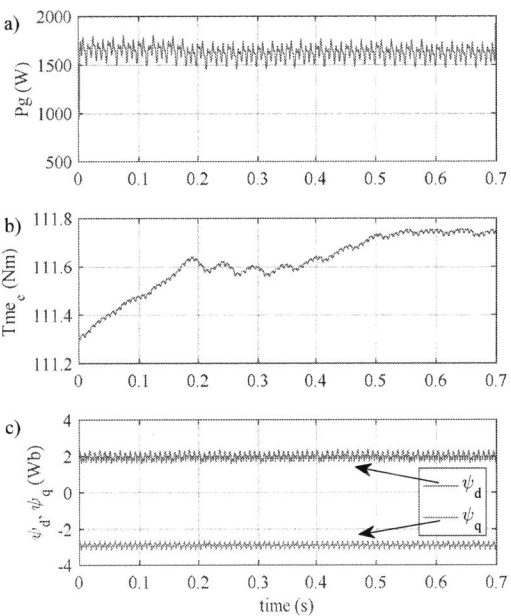

Fig. 14. Experimental results: $i_d^* = 5A$, $i_q^* = -10A$, $n = 200rpm$, a) active output power, b) torque, c) dq flux linkages.

The communication between dSpace and inverter is done through fiber optics for noise immunity and galvanic isolation. The phase currents and DC voltage are measured by the dSpace board. In order to emulate the DC grid, a DC source is used for keeping the voltage at 350V for inverter and measurement equipment feeding when the generator doesn't produce power. The load is represented by a chopper (inside the Danfoss inverter) with a RL load (R=60 Ω, L=9,36 mH). The chopper is controlled with a PI controller for maintaining the $V_{DC} = 580V$, implemented in dSpace. The reference DC voltage can be modified. The load voltage and current (V_{RL}, i_{RL}) are measured using an oscilloscope.

978-1-5386-7688-2/19 $31.00 © 2019 IEEE   265

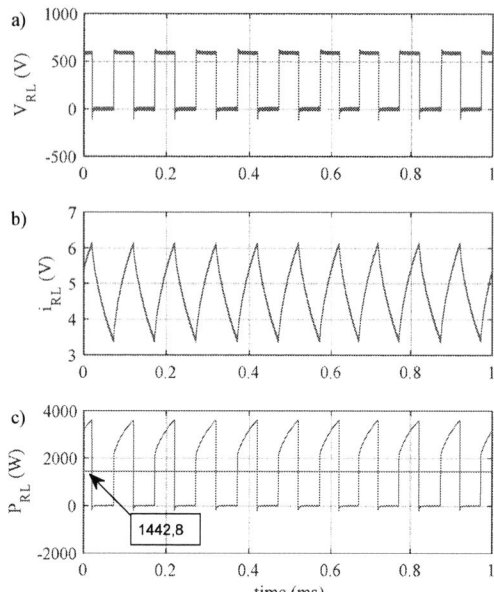

Fig. 15. Experimental results (RL load): $i_d^* = 5A$, $i_q^* = -10A$, $n = 200rpm$, a) load voltage, b) load current, c) load power.

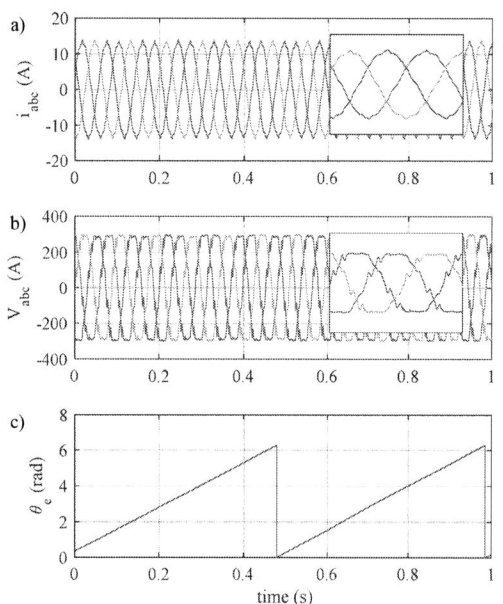

Fig. 16. Experimental results: $i_d^* = 7.5A$, $i_q^* = -15.5A$, $n = 120rpm$, a) stator currents, b) stator voltages, c) estimated rotor position..

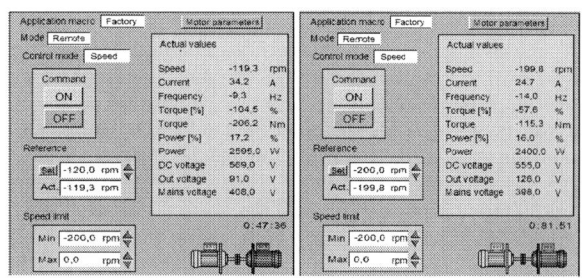

Fig. 17. SCADA actual values (left: $i_d^* = 5A$, $i_q^* = -10A$, $n = 200rpm$; right: $i_d^* = 7.5A$, $i_q^* = -15.5A$, $n = 120rpm$).

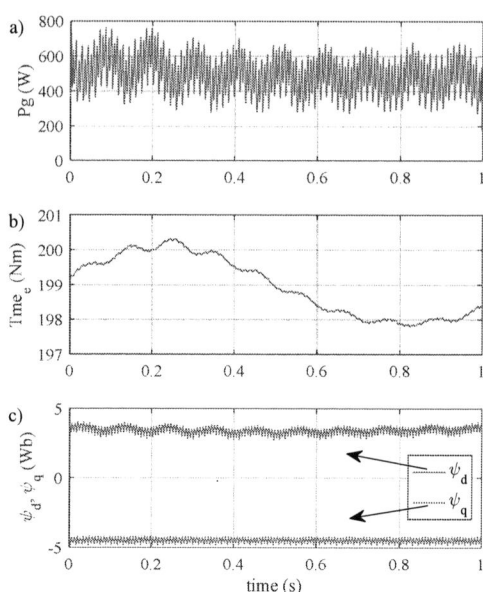

Fig. 18. Experimental results: $i_d^* = 7.5A$, $i_q^* = -15.5A$, $n = 120rpm$ a) active output power, b) torque, c) dq flux linkages.

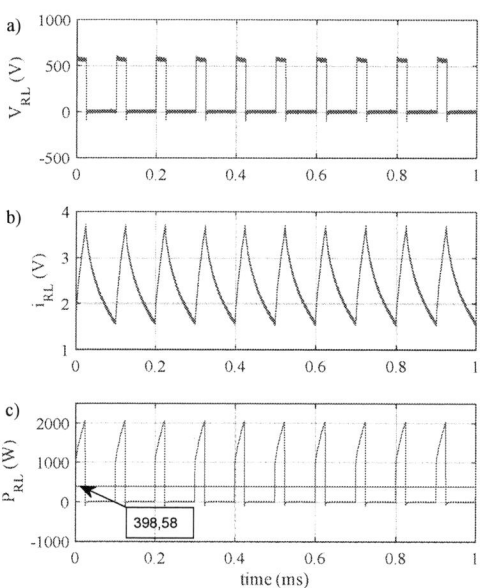

Fig. 19. Experimental results (RL load) $i_d^* = 7.5A$, $i_q^* = -15.5A$, $n = 120rpm$, a) load voltage, b) load current, c) load power.

Communication with ABB inverter is done through ethernet, protocol Modbus TCP/IP, using an adapter module (RETA-01). A SCADA system was developed (in Control Maestro software) for reference and actual values.

VII. CONCLUSIONS

Presented paper propose a different method for estimating the rotor position of a SynRG. The proposed method was successfully tested on a large speed respectively torque range, starting from 60 rpm up to 200 rpm respectively required torque for zero output power up to full load (around 200 Nm). The speed range covers requirements for wind and hydro generator applications. The experimental results show that torque and speed estimation by the proposed method are very close with the values given by the inverter of the driving machine Fig. 18. Lower speed in generator mode cannot be obtained due to the drop voltage on internal resistance of the machine. Extensive simulations and experimental results prove that the system is stable for the next ranges of the per unit estimator parameters: resistance variation between 0.71-1.35, d-axis inductance between 0.67-1.98, and q-axis inductance between 0.89- 1.18 from rated values.

This method could be an alternative for the existent sensorless control methods of the SynRG in order to avoid integrators offset, initial flux values compensation, and a DC voltage reserve (for the signal injection method).

The sensitivity and stability of position error method presented in this paper can be used also to analyze the performance of other rotor position estimators from literature (stator flux position angle [17], [21], [22] or active flux position angle [25]). The simulation and experimental results validated the proposed control method and with a larger Ld/Lq ratio (presented SynRG has only Ld/Lq=2.7 ratio), better energetical results can be obtained.

ACKNOWLEDGMENT

This work was supported by two grant of the Romanian Ministry of Research and Innovation, CCCDI – UEFISCDI, project number PN-III-P1-1.2-PCCDI-2017- 0391 / CIA_CLIM – *"Smart buildings adaptable to the climate change effects"*, within PNCDI III, and also, project number 10PFE/16.10.2018, PERFORM-TECH-UPT - *The increasing of the institutional performance of the Polytechnic University of Timișoara by strengthening the research, development and technological transfer capacity in the field of "Energy, Environment and Climate Change"*, Institutional Development Projects - Excellence Funding Projects in RDI, PNCDI III.

REFERENCES

[1] P. Roshanfekr, „Energy-efficient Generating System for HVDC Off-shore Wind Turbine", Chalmers University of Technology, Goteborg, Sweden, 2013.

[2] L. Castellini, M. D'Andrea, G. Fabri, D. Macera, și M. Villani, „Design of a Synchronous Reluctance Machine for a Flywheel-Based Energy Storage System", în *2018 XIII International Conference on Electrical Machines (ICEM)*, 2018, pp. 2099–2104.

[3] „Reluctance Electric Machines: Design and Control", *CRC Press*. [Online]. Disponibil la: https://www.crcpress.com/Reluctance-Electric-Machines-Design-and-Control/Boldea-Tutelea/p/book/9781498782333. [Data accesării: 25-nov-2018].

[4] P. Roshanfekr, S. T. Lundmark, T. Thiringer, și M. Alatalo, „Comparison of a 5MW permanent magnet assisted synchronous reluctance generator with an IPMSG for wind application", în *2014 International Conference on Electrical Machines (ICEM)*, 2014, pp. 711–715.

[5] P. Roshanfekr, S. Lundmark, T. Thiringer, și M. Alatalo, „A synchronous reluctance generator for a wind application-compared with an interior mounted permanent magnet synchronous generator", în *7th IET International Conference on Power Electronics, Machines and Drives (PEMD 2014)*, 2014, pp. 1–5.

[6] R. H. Moncada, B. J. Pavez, J. A. Tapia, și J. Pyrhönen, „Operation analysis of synchronous reluctance machine in electric power generation", în *2014 International Conference on Electrical Machines (ICEM)*, 2014, pp. 2734–2739.

[7] A. Vagati, G. Franceschini, I. Marongiu, și G. P. Troglia, „Design criteria of high performance synchronous reluctance motors", în *Conference Record of the 1992 IEEE Industry Applications Society Annual Meeting*, 1992, pp. 66–73 vol.1.

[8] T. J. E. Miller, A. Hutton, C. Cossar, și D. A. Staton, „Design of a synchronous reluctance motor drive", *IEEE Transactions on Industry Applications*, vol. 27, nr. 4, pp. 741–749, iul. 1991.

[9] T. Matsuo și T. A. Lipo, „Rotor design optimization of synchronous reluctance machine", *IEEE Transactions on Energy Conversion*, vol. 9, nr. 2, pp. 359–365, iun. 1994.

[10] S. Guha și N. C. Kar, „Saturation Modeling and Stability Analysis of Synchronous Reluctance Generator", *IEEE Transactions on Energy Conversion*, vol. 23, nr. 3, pp. 814–823, sep. 2008.

[11] S. Maroufian și P. Pillay, „Self-excitation criteria of the synchronous reluctance generator in stand-alone mode of operation", în *2016 IEEE International Conference on Power Electronics, Drives and Energy Systems (PEDES)*, 2016, pp. 1–5.

[12] R. Sharma și and B. Singh, „SyRG-PV-BES Based Standalone Microgrid Using Appoximate Multipliers Based Adaptive Control", în *2018 5th IEEE Uttar Pradesh Section International Conference on Electrical, Electronics and Computer Engineering (UPCON)*, 2018, pp. 1–6.

[13] R. H. Moncada, H. A. Young, B. J. Pavez-Lazo, și J. A. Tapia, „A commercial-off-the-shelf synchronous reluctance motor as a generator for wind power applications", în *2015 IEEE International Electric Machines Drives Conference (IEMDC)*, 2015, pp. 6–12.

[14] J. C. Mitchell, M. J. Kamper, și C. M. Hackl, „Small-scale reluctance synchronous generator variable speed wind turbine system with DC transmission linked inverters", în *2016 IEEE Energy Conversion Congress and Exposition (ECCE)*, 2016, pp. 1–8.

[15] S. Tokunaga și K. Kesamaru, „FEM simulation of novel small wind turbine generation system with synchronous reluctance generator", în *2011 International Conference on Electrical Machines and Systems*, 2011, pp. 1–6.

[16] M. Alnajjar și D. Gerling, „Medium-Speed Synchronous Reluctance Generator as Efficient, Reliabile and Low-Cost Solution for Power Generation in Modern Wind Turbines", în *2018 International Symposium on Power Electronics, Electrical Drives, Automation and Motion (SPEEDAM)*, 2018, pp. 1233–1238.

[17] D. V. M, B. Singh, și B. G, „Position Sensor-less Synchronous Reluctance Generator Based Grid-Tied Wind Energy Conversion System with Adaptive Observer Control", *IEEE Transactions on Sustainable Energy*, pp. 1–1, 2019.

[18] T. Mabuchi *et al.*, „Position sensorless control of synchronous reluctance motors at very low speeds region using high-frequency current control system", în *2017 20th International Conference on Electrical Machines and Systems (ICEMS)*, 2017, pp. 1–6.

[19] I. Boldea, Z. Fu, și S. A. Nasar, „Sensorless DC output control of a high performance reluctance generator system", în *Proceedings of 1994 IEEE Industry Applications Society Annual Meeting*, 1994, vol. 1, pp. 16–22 vol.1.

[20] I. Boldea și S. C. Agarlita, „The active flux concept for motion-sensorless unified AC drives: A review", în *International Aegean Conference on Electrical Machines and Power Electronics and Electromotion, Joint Conference*, 2011, pp. 1–16.

[21] X. Dianguo, J. Xinhai, și C. Wei, „Sensorless control of synchronous reluctance motors", în *2017 IEEE Transportation Electrification Conference and Expo, Asia-Pacific (ITEC Asia-Pacific)*, 2017, pp. 1–4.

[22] I. BOLDEA, Z. X. FU, și S. A. NASAR, „Torque Vector Control (tvc) of Axially-Laminated Anisotropic (ala) Rotor Reluctance Synchronous Motors", *Electric Machines & Power Systems*, vol. 19, nr. 4, pp. 533–554, iul. 1991.

[23] S. Agarlita, I. Boldea, și F. Blaabjerg, „High-Frequency-Injection-Assisted "Active-Flux"-Based Sensorless Vector Control of Reluctance Synchronous Motors, With Experiments From Zero Speed", *IEEE Transactions on Industry Applications*, vol. 48, nr. 6, pp. 1931–1939, nov. 2012.

[24] I. Boldea, M. C. Paicu, și G. Andreescu, „Active Flux Concept for Motion-Sensorless Unified AC Drives", *IEEE Transactions on Power Electronics*, vol. 23, nr. 5, pp. 2612–2618, sep. 2008.

[25] S. Agarliță, M. Fătu, L. N. Tutelea, F. Blaabjerg, și I. Boldea, „I-f starting and active flux based sensorless vector control of reluctance synchronous motors, with experiments", în *2010 12th International Conference on Optimization of Electrical and Electronic Equipment*, 2010, pp. 337–342.

Short Circuit Location in Transformer Winding Using Deep Learning of Its Frequency Responses

Arash Moradzadeh
Department of Electrical Engineering,
Tabriz Branch, Islamic Azad University,
Tabriz, Iran
Stu.arash.moradzadeh@iaut.ac.ir

Kazem Pourhossein
Department of Electrical Engineering,
Tabriz Branch, Islamic Azad University,
Tabriz, Iran
k.pourhossein@iaut.ac.ir

Abstract—**A turn-to-turn short circuit fault is One of the most important defects in transformer windings that is most difficult to diagnosis. Degradation decreases impedances of inter-turn insulations that finally may lead to a solid turn-to-turn short circuit. In this paper, early detection of turn-to-turn faults in transformers windings has been studied, in its high-impedance stage, using Convolutional Neural Network (CNN) based on extracting features from frequency response traces. For this purpose, a model winding has been used as test object to approve capability of the proposed approach. A variety of low impedance and high impedance short circuit faults were tested on the model winding. The results show that this method is able to detect turn-to-turn faults in transformer winding even in their early stages.**

Keywords— *Power transformer, frequency response analysis, Short Circuit Fault, Convolutional Neural Networks*

I. INTRODUCTION

Power transformers are expensive and critical components in power systems. The Sustainability of power system depends on health of the power transformers, but these components suffer from various electrical and mechanical defects over their lifetime [1]. The occurrence of any defect in transformers reduces the reliability and power quality of power systems [2]. The types of mechanical and electrical faults in power transformers with regard to the results presented in the researches done are presented in Figure 1 [3, 4].

According to the results presented, defects of OLTC and winding are the most important transformers faults. Winding faults have resonant nature and could be destructiveness, therefore they are important via early stage diagnosis viewpoint [5-11]. Occurrence of any mechanical defect in winding can lead to a short circuit (Fig. 2).

When a power transformer confronts with a short circuit fault, great short circuit current flows through the winding, and then a massive electromagnetic force is generated. This invasive flow increases heat generation in the winding and can cause serious damage to the insulation system [5, 6]. If this short circuit fault is detected at early stage, damage in insulation system and internal equipment of transformer can be prevented.

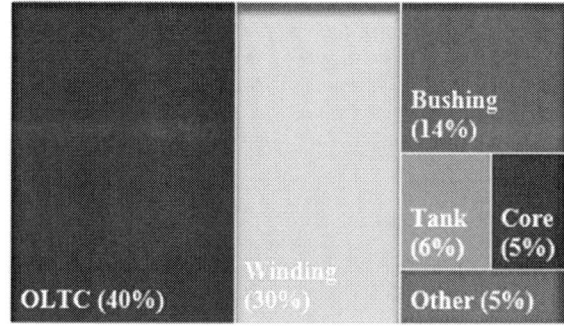

Fig. 1. Power Transformer faults percentage

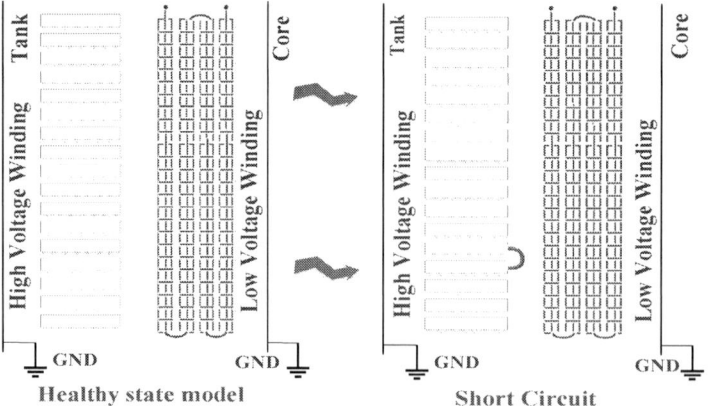

Fig. 2. Relevant mechanical defects in transformer windings

Monitoring the condition of power transformers for mechanical defects of winding can be done using Winding impedance measurement [12], dissolved gas analysis (DGA) [13], Vibro-acoustic analysis [7] and frequency response analysis (FRA) [14]. The results of several work in this regard have showed that among above-mention methods, FRA is best and powerful tool for identify mechanical faults in transformer windings [14, 15]. The basic idea behind FRA is effect of any physical change in transformer active part on its electrical circuit parameter and then on its frequency response.

Numerical indices [16-18], estimating parameters of transformer model [19], artificial neural networks [20, 21], vector fitting for estimated transfer function [22], support vector machine (SVM) for transfer function analysis [23], cross correlation [24, 25], enhanced magnetic optimal algorithm [26] and used of finite element method [27] are well-known used methods for interpreting frequency responses of transformers. Some of the valuable works that have been done about the short-circuit diagnosis via the FRA have used statistical or mathematical indices [28], three phase FRA comparison to detect short circuits [29], index of similarity index to short circuit faults [30], support vector machines to classify different mechanical faults of tested power transformer [23], Used of artificial neural networks to localize short circuits in transformer windings [20, 21], There is another paper reported neural network approach to short circuit location in time domain (without using FRA) [31]. The FRA method has problems in interpreting the results and is not able to detect the exact location of the fault. Each fault has its own unique frequency response. By extracting the features of each frequency response can be interpreted its results and be detected type and location of the fault.

In this paper, CNN method is used to interpret frequency response traces to locate SC fault exactly. To locate defect along transformer winding, SC is performed by low-impedance and high-impedance in forty locations in a model winding.

II. FREQUENCY RESPONSE ANALYSIS

Frequency response analysis is a comparative based diagnostic test. It is include measuring the impedance of the transformer windings over a vast range of frequencies and comparing the results with initial state [14, 15]. Frequency response analysis done by using low voltage impulse method (LVI). In this method, a low amplitude impulse voltage is applied to transformer and its output signal (voltage or current) is measured. Then, frequency response can be calculated using division of these signals in frequency space [14, 32].

III. CONVOLUTIONAL NEURAL NETWORK

Feature extraction facilitates the analysis, classification, visualization and communication of high-dimensional and complex data. A Convolutional Neural Network (CNN) is a powerful tool in feature learning, classification and high-dimensional data analysis [33, 34]. A CNN transforms the features of the sample in the original space into a new feature space by an intelligent learning mechanism.

This makes classification or diagnosis easier, and shows a powerful ability to learn the essential features of a dataset from a small number of samples. As shown in Figure 3, CNN composed of convolutional layers, pooling layers and fully connected layers [35]. A convolution layer is mainly used for feature extraction through a convolution operation. In the convolution layer, there are a number of convolution kernels, which are equal to a filter, which can extract the features of the input. Each convolution kernel can obtain a feature map, but the features extracted by different convolution kernels are not the same. The feature map of this layer will be used as the input of the next convolutional layer [36]. The Convolution Step using non-linear activation is as follows [37]:

$$C_r^m = ReLU\left(\sum_m v_{r-1}^n * w_r^m + b_r^m\right) \qquad (1)$$

Where C_r^m is the output of nth filter in convolutional layer r, v_{r-1}^n is the nth output of previous layer r−1; $*$ demonstrates the convolution and ω_r^m defines the mth filter kernel of the current layer r; b_r^m is the bias and Rectified Linear Unit (ReLU) function $f(x) = \max(0, x)$ define the nonlinear activation function [37, 38].

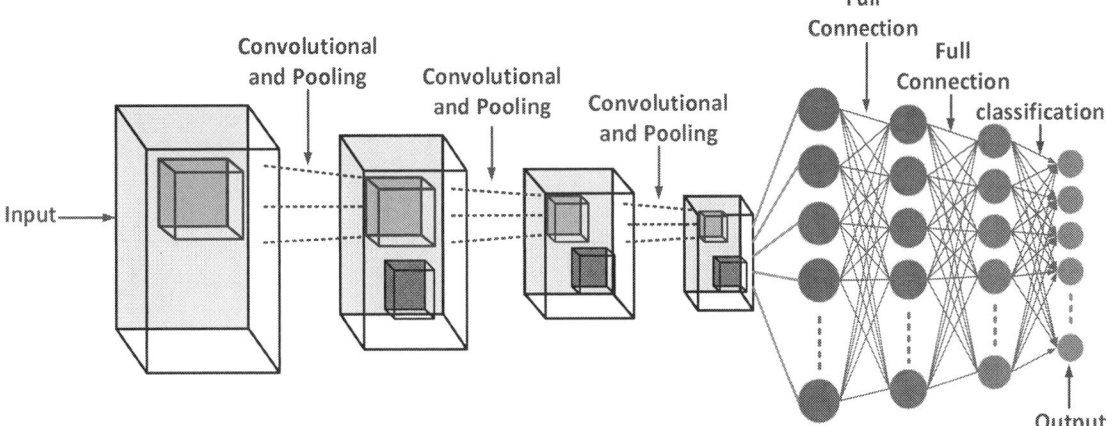

Fig. 3. Structure of CNN

After the formation of the feature spaces in each convolution layer by filters, the pooling layers to assemble the features and prevent overfitting appended to the convolutional layer [34, 36]. Max pooling is the most used type of pooling. This type of pooling takes the most important part of the input. The max pooling is described as [37]:

$$h_{r+1}^m = \max \; C_r^m(t) \; , (i-1)l+1 \le t \le il \tag{2}$$

Where C_r^m is the feature map, h_{r+1}^m is the outcome of pooling layer, and l is the length of a local region for pooling. With the features extracted of data by continuous the convolution layer and pooling layer, the feature map is stretched into a one-dimensional feature vector. A fully connected layer performed to classify and predict the feature vector in the final output of the network [36]. A softmax regression is used on the top classification layer. The output of softmax function is defined as [37, 39]:

$$O_j = \begin{bmatrix} P(y=1)|x;\theta \\ P(y=2)|x;\theta \\ ... \\ P(y=k)|x;\theta \end{bmatrix} = \frac{1}{\sum_{j=1}^{k} \exp(\theta^j x)} \begin{bmatrix} \exp(\theta^1 x) \\ \exp(\theta^1 x) \\ ... \\ \exp(\theta^k x) \end{bmatrix} \tag{3}$$

Where k is the number of samples and $\theta^j x$ is the parameters of the classification layer.

IV. CASE STUDY

Experimental setup involves a 1-phase winding of a 3-phase transformer winding. All defects were applied on HV winding while LV winding was open circuit. Every 4 turns of HV winding considered as a section and eventually a 40-section winding was prepared for SC fault tests to be performed (Fig. 4). A surge voltage applied as input to winding and the Oscilloscope device GPS-1102B utilized for measure and save the voltage and current at the end of the winding.

Fig. 4. Experimental study setup

V. SIMULATION RESULTS

Frequency response of the winding has been defined as below [9, 11]:

$$FR = \left| \frac{I_o(f)}{V_i(f)} \right| \tag{4}$$

$I_o(f)$ and $V_i(f)$ are earth current and input voltage, respectively, both in frequency space. To determine the defect in the winding, it is necessary to compare the frequency response of the healthy state and damaged. Before the fault occurred on winding, the frequency response of winding was calculated in healthy state (Fig. 5). To localize short circuit faults in both high-impedance and low-impedance states using CNN, construction of a short circuit database is necessary.

Condition of insulation between two adjacent turns of conductor defines short circuit resistance. In this study, 0 Ω, 1 Ω, 2 Ω, 3 Ω, 4 Ω, 8 Ω, 10 Ω, 15 Ω, 22 Ω, 32 Ω and 47 Ω resistances are considered to applied low-impedance and high-impedance short circuits in model winding. How to produce high-impedance and low-impedance short circuit faults are shown in Fig. 6-a and fig. 6-b respectively.

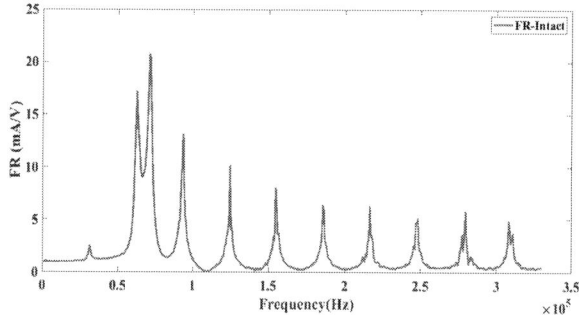

Fig. 5. Frequency Response of healthy state

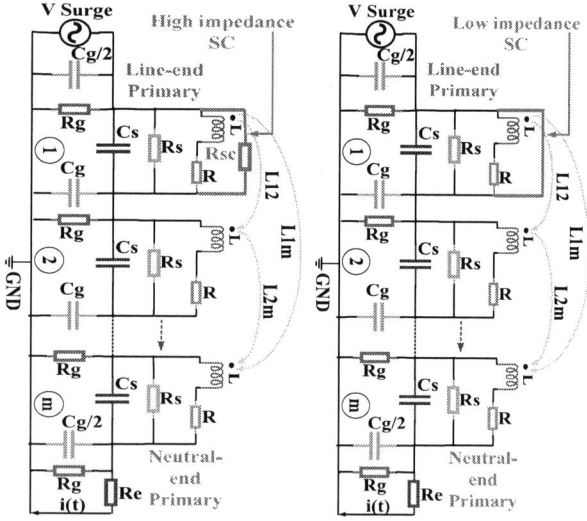

Fig. 6-a   Fig. 6-b

Fig. 6. Fig. 6-a show SC fault with Resistance (High Impedance Fault) and section 6-b illustrate SC fault without Resistance (Low Impedance Fault)

Effect of short circuit impedance on frequency responses in 5nd unit of model winding are depicted in Fig. 7. Frequency response of low impedance short circuits in all sections of the model winding are presented in Fig. 8. Changes in the frequency range and amplitude of each frequency response relative to the intact state indicate defect in the winding. In this paper, the CNN method has been used to determine the exact location of each fault via the extracted features in each frequency response.

All frequency responses are collected as a dataset and provided as input to the CNN network. Fault locations (irrespective of short circuit impedance) considered as target data. Data set divided into two parts: training, testing dataset. 80% of the dataset was considered as network training data and 20% of the data as test data. The convolutional and pooling layers are formed to identify and extract features. After extracting the features by this layers, the fully connected layers have been forming by 50 hidden layer in structure. Finally, the Softmax function was used to classify the faults. Fig. 9 and Fig. 10 are shows the location of faults that the network considered as training and testing data respectively with 99.74% and 97.98% accuracy.

This results show the ability of CNN method in extracting features from the frequency responses and determining the exact location of each fault. After the training CNN network by training data, a trained network is saved to form of a Black Box. This is used to detect new faults that their location is unknown.

For test the trained network, new short circuit faults with 18, 55, 99 and 223 ohm resistances created in some sections of the winding. Frequency response for each fault is calculated and used as input to the trained network. Results of new short circuit faults detection and testing the trained CNN are presented in Tables I.

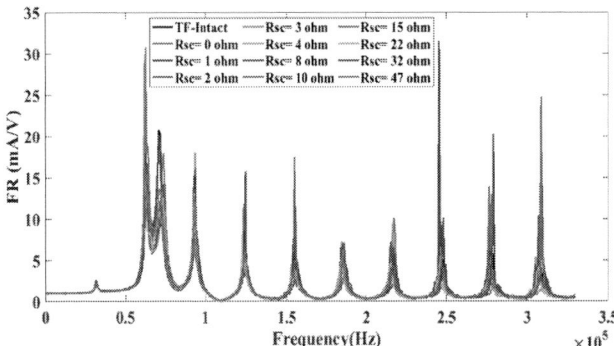

Fig. 7. Effect of short circuit impedance on frequency responses in 5nd unit of model winding

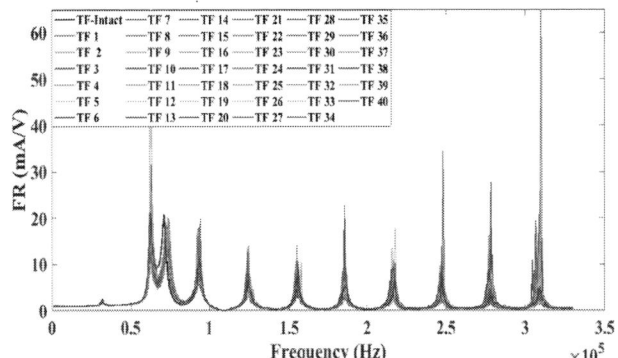

Fig. 8. Frequency response of low impedance short circuits in all sections of the model winding

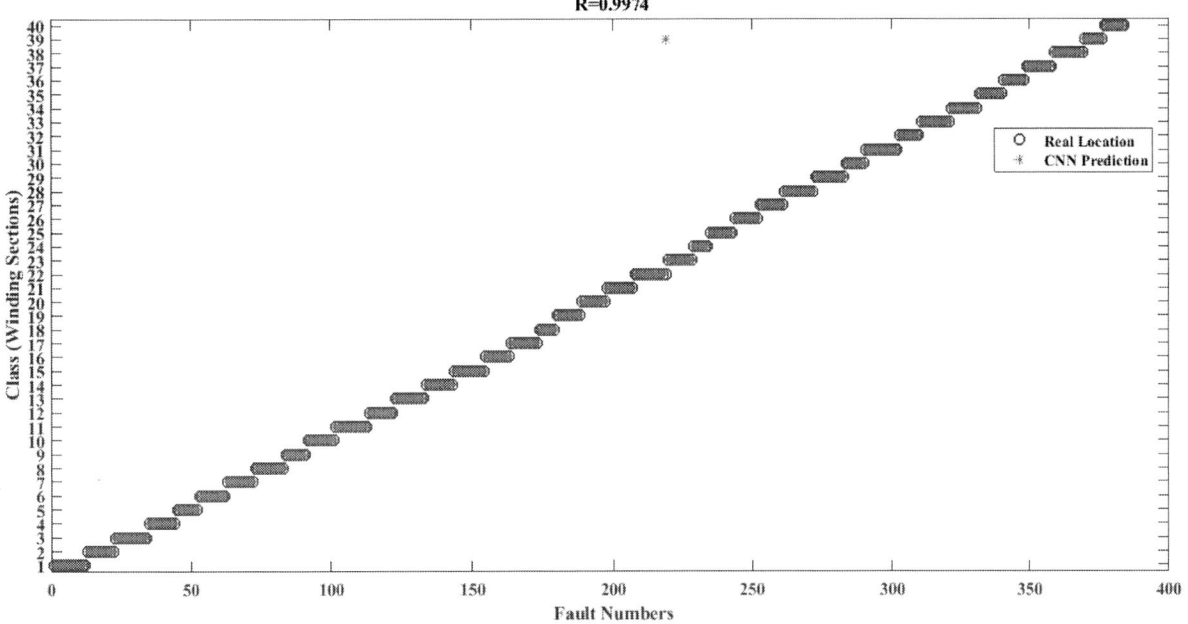

Fig. 9. Classification of SC faults by trained CNN using train data

978-1-5386-7688-2/19 $31.00 © 2019 IEEE

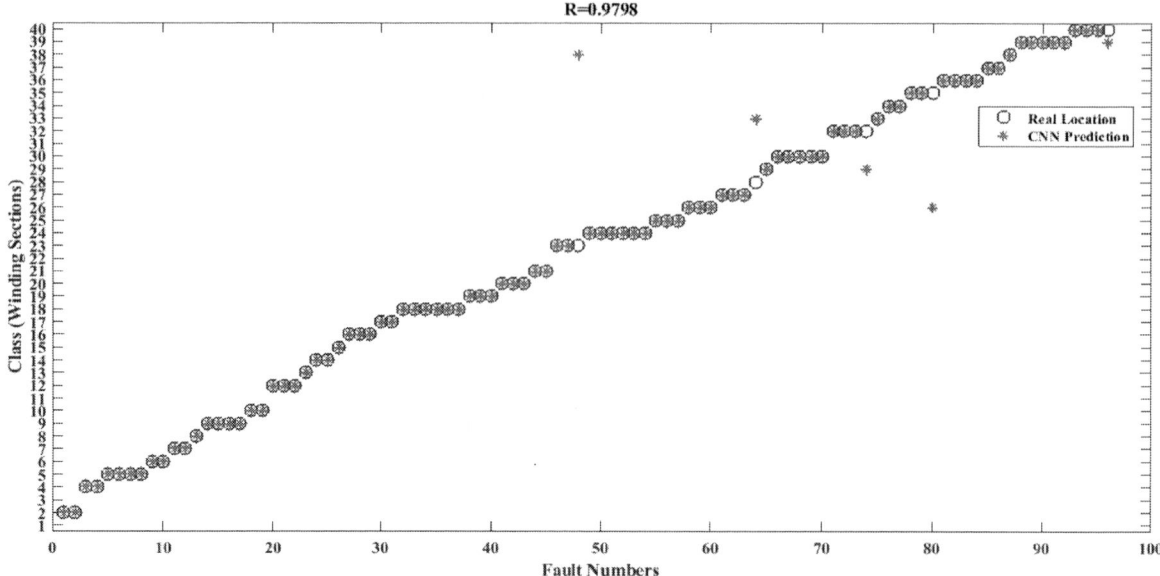

Fig. 10. Classification of SC faults by trained CNN using test data

TABLE I.   DETECTION OF TURN-TO-TURN FAULTS BY CNN

Location	R (ohm)	CNN Prediction	Location	R (ohm)	CNN Prediction
1	18	1	21	18	21
	55	1		55	21
	99	1		99	21
	223	1		223	21
3	18	3	23	18	23
	55	3		55	23
	99	3		99	23
	223	3		223	23
5	18	5	25	18	25
	55	5		55	25
	99	5		99	25
	223	5		223	25
7	18	7	27	18	27
	55	7		55	27
	99	7		99	27
	223	7		223	27
10	18	10	30	18	30
	55	10		55	30
	99	10		99	30
	223	10		223	30
14	18	14	33	18	33
	55	14		55	33
	99	14		99	33
	223	14		223	33
16	18	16	35	18	35
	55	16		55	35
	99	16		99	35
	223	16		223	35
18	18	18	37	18	37
	55	18		55	37
	99	18		99	37
	223	18		223	37
20	18	20	39	18	39
	55	20		55	39
	99	20		99	39
	223	20		223	39

VI. CONCLUSION

Turn-to-turn faults of transformer winding are potential source of heat and may lead to a progressive major fault because of insulation degradation. Therefore, early detection of turn-to-turn faults is very useful action. In this paper a convolutional neural network used to diagnose winding fault location in its high impedance stage. Here, low-impedance and high-impedance short circuit faults were applied to the model winding in various locations and their frequency responses were calculated. This network was used to locate faults based on frequency response pattern. The results shows precision of the method used to early detection of Turn-to-Turn faults (high impedance short circuit) in transformer winding.

REFERENCES

[1] Zh. Liang, A. Parlikad "A Markovian model for power transformer maintenance," *Electrical Power and Energy Systems,* Vol. 99, pp. 175–182, 2018.

[2] R. Murugan, R. Ramasamy, "Failure analysis of power transformer for effective maintenance planning in electric utilities," *Engineering Failure Analysis,* Vol. 55, pp. 182–192, 2015.

[3] M. Koch, M. Krüger, "A New Method for On-Line Monitoring of Bushings and Partial Discharges of Power Transformers," *2012 IEEE International Conference on Condition Monitoring and Diagnosis,* pp. 23-27, Bali, Indonesia, September 2012.

[4] I. Metwally, "Failures, Monitoring and New Trends of Power Transformers," IEEE Potentials · July 2011.

[5] A. Mejia-Barron, M. Valtierra-Rodriguez, D. Granados-Lieberman, J.C. Olivares-Galvan, R. Escarela-Perez, "Experimental data-based transient-stationary current model for inter-turn fault diagnostics in a transformer." *Electric Power Systems Research* 152 (2017): 306-315.

[6] M.R Barzegaran, M. Mirzaie, A. Shayegani Akmal, "Investigating Short-circuit in Power Transformer Winding with Quasi-static Finite Element Analysis and Circuit-based Model," *Transmission and Distribution Conference and Exposition, IEEE PES,* 2010, pp. 1-8.

[7] H. Zhou, K. Hong, H. Huang, J. Zhou, "Transformer winding fault detection by vibration analysis methods," Applied Acoustics, Vol. 114, pp. 136–146, 2016..

[8] H. Tarimoradi, G.B. Gharehpetian, "A Novel Calculation Method of Indices to Improve Classification of Transformer Winding Fault Type, Location and Extent," *IEEE Transactions on Industrial Informatics*, Vol. 13, No. 4, pp. 1531-1540, 2017.

[9] E. Rahimpour, M. Jabbari, S. Tenbohlen, "Mathematical Comparison Methods to Assess Transfer Functions of Transformers to Detect Different Types of Mechanical Faults," IEEE TRANSACTIONS ON POWER DELIVERY, VOL. 25, NO. 4, OCTOBER 2010.

[10] K. Pourhossein, G. B. Gharehpetian, E. Rahimpour, N. Araabi, "A Vector-based Approach to Discriminate Radial Deformation and Axial Displacement of Transformer Winding and Determine Defect Extent," Electric Power Components and Systems, 40:597–612, 2012.

[11] E. Rahimpour, J. Christian, K. Feser, H. Mohseni, "Transfer Function Method to Diagnose Axial Displacement and Radial Deformation of Transformer Windings," IEEE TRANSACTIONS ON POWER DELIVERY, VOL. 18, NO. 2, APRIL 2003.

[12] T. Chiulan, B. Pantelimon, "A Practical Example of Power Transformer Unit Winding Condition Assessment by Means of Short-Circuit Impedance Measurement,*" IEEE Bucharest Power Tech Conference*, Bucharest, Romania, June 28th-July 2nd 2009.

[13] Sh. Li, G. Wu, B. Gao, Ch. Hao, D. Xin, X. Yin, "Interpretation of DGA for Transformer Fault Diagnosis with Complementary SaE-ELM and Arctangent Transform," *IEEE Transactions on Dielectrics and Electrical Insulation*, Vol. 23, No. 1, pp. 586-895, Feb 2016.

[14] R. Khalili Senobari, J. Sadeh, H. Borsi, "Frequency response analysis (FRA) of transformers as a tool for fault detection and location: A review," Electric Power Systems Research, Vol. 155, pp. 172–183, 2018.

[15] K. Pourhossein, G.B. Gharehpetian, E. Rahimpour, B.N. Araabi, "A probabilistic feature to determine type and extent of winding mechanical defects in power transformers," Electric Power Systems Research, Vol. 82, pp. 1– 10, 2012.

[16] K. Pourhossein, G.B. Gharehpetian, E. Rahimpour. "Buckling severity diagnosis in power transformer windings using Euclidean Distance classifier." In *Electrical Engineering (ICEE), 2011 19th Iranian Conference on IEEE*, pp. 1-4, 2012.

[17] K. Pourhossein, G. B. Gharehpetian, E. Rahimpour, "Discrimination of Axial Displacement and Radial Deformation in Power Transformer Windings Using Manhattan Distance Function," In *Proc. 25th International Power System Conference (PSC 2010)*, Nov 2010.

[18] K. Pourhossein, G. B. Gharehpetian, E. Rahimpour, "Axial Displacement Extent Determination in Power Transformer Windings Using an Adjustable Index," In *Proc. 25th International Power System Conference (PSC 2010)*, Nov 2010.

[19] L. Satish, Subrat K. Sahoo, "Locating faults in a transformer winding: An experimental study," Electric Power Systems Research, Vol. 79, pp. 89–97, 2009.

[20] M. Faridi, M. Kharezi, E. Rahimpour, H. R. Mirzaei, A, Akbari, "Localization of Turn-to-Turn Fault in Transformers Using Artificial Neural Networks and Winding Transfer Function," *2010 International Conference on Solid Dielectrics*, Potsdam, Germany, July 4-9, 2010.

[21] H. Firoozi, M. Kharezi, H. Bakhshi, "Turn- to -Turn Fault Localization of Power Transformers Using Neural Network Techniques," *Proceedings of the 9th International Conference on Properties and Applications of Dielectric Materials*, July 19-23, Harbin, China, 2009.

[22] P. Karimifard, G. B. Gharehpetian, S. Tenbohlen, "Localization of winding radial deformation and determination of deformation extent using vector fitting-based estimated transfer function," Euro. Trans. Electr. Power, Vol. 19, pp. 749–762, 2009.

[23] M. Bigdeli, M. Vakilian, E. Rahimpour, "Transformer winding faults classification based on transfer function analysis by support vector machine," IET Electr. Power Appl., Vol. 6, Iss. 5, pp. 268–276, 2012.

[24] A. R. Abbasi , M. R. Mahmoudi, Z. Avazzadeh, "Diagnosis and clustering of power transformer winding fault types by cross correlation and clustering analysis of FRA results," IET Gener. Transm. Distrib., Vol. 12 Iss. 19, pp. 4301-4309, 2018.

[25] S. M. Saleh, S. H. EL-Hoshy, O. E. Gouda, "Proposed diagnostic methodology using the cross-correlation coefficient factor technique for power transformer fault identification," IET Electr. Power Appl., Vol. 11, Iss. 3, pp. 412–422, 2017.

[26] M. S. Jahan, R. Keypour, H. R. Izadfar, M. T. Keshavarzi, "Locating power transformer fault based on sweep frequency response measurement by a novel multistage approach," IET Science, Measurement & Technology, Vol. 12, No. 8, pp. 949-957, 2018 May 29.

[27] J. Jiang, L. Zhou, Sh. Gao, W. Li, D. Wang, "Frequency response features of axial displacement winding faults in autotransformers with split windings," IEEE Transactions on Power Delivery, Vol. 33, No. 4, pp.1699-1706, Aug 2018.

[28] V, Behjat, A. Vahedi, A. Setayeshmehr, H. Borsi, E. Gockenbach, "Sweep frequency response analysis for diagnosis of low level short circuit faults on the windings of power transformers: An experimental study," Electrical Power and Energy Systems, Vol. 42, pp. 78–90, 2012.

[29] A. A. Pandya, B.R. Parekh, "Interpretation of Sweep Frequency Response Analysis (SFRA) traces for the open circuit and short circuit winding fault damages of the power transformer," Electrical Power and Energy Systems, Vol. 62, pp. 890–896, 2014.

[30] J. N. Ahour, S. Seyedtabaii , G. B. Gharehpetian, "Determination and localisation of turn-to-turn fault in transformer winding using frequency response analysis," IET Sci. Meas. Technol., Vol. 12 Iss. 3, pp. 291-300, 2018.

[31] M. Rahmatian, B. Vahidi , A.J. Ghanizadeh, G.B. Gharehpetian, H.A. Alehosseini, "Insulation failure detection in transformer winding using cross-correlation technique with ANN and k-NN regression method during impulse test," Electrical Power and Energy Systems, Vol. 53, pp. 209–218, 2013.

[32] S.A. Ryder, "Transformer Diagnosis Using Frequency Response Analysis: Results from Fault Simulations," Power Engineering Society Summer Meeting, 2002 IEEE, Vol. 1, pp. 399-404, Jul 2002.

[33] L. Yann, Y. Bengio, G. Hinton, "Deep learning," nature, vol. 521, no. 7553, p.p. 436, 2015.

[34] X. Chen, X. LIN, "Big Data Deep Learning: Challenges and Perspectives," IEEE access, vol. 2, pp. 514-525, 2014.

[35] R. Huang, Y. Liao, SH. Zhang, W. LI, "Deep Decoupling Convolutional Neural Network for Intelligent Compound Fault Diagnosis," IEEE Access, vol. 7, pp. 1848-1858, 2019.

[36] L. Geng, J. Sun, Zh. Xiao, F. Zhang, J. Wu, "Combining CNN and MRF for road detection," Computers and Electrical Engineering, Vol. 70, pp. 895–903, 2018.

[37] T. Han, Ch. Liu, W. Yang, D. Jiang, "A novel adversarial learning framework in deep convolutional neural network for intelligent diagnosis of mechanical faults," Knowledge-Based Systems, vol. 165, pp. 474–487, 2019.

[38] J. Patterson, A. Gibson, "Deep learning: A practitioner's approach," O'Reilly Media, 2017 Jul 28.

[39] Ch. Dong, Ch. Change Loy, K. He, X. Tang, Fellow, "Image Super Resolution Using Deep Convolutional Networks," IEEE TRANSACTIONS ON PATTERN ANALYSIS AND MACHINE INTELLIGENCE, Vol. 38, No. 2, pp. 295-307, FEBRUARY 2016.

Temporal Envelope Estimation of Stator Current by Peaks Detection for IM Fault Diagnosis

Hamid Khelfi, and Samir Hamdani
Labortoire des Systèmes Electroniques et Industriels
Université des Sciences et de la Technologie Houari Boumedien
BP 32 El Ali 16111 Bab Ezzouar, Alger, Algérie
Emails : hkhelfi@usthb.dz and shamdani@usthb.dz

Youcef Chibani
Laboratoire d'Ingénierie des Systèmes Intelligents
Université des Sciences et de la Technologie Houari Boumedien
BP 32 El Ali 16111 Bab Ezzouar, Alger, Algérie
Email : ychibani@usthb.dz

Abstract—**This paper presents a simple and reliable method to extract the temporal envelope of the stator current for induction motor fault diagnosis. The proposed method is based on the detection of the prominent peaks of the stator current and the desired envelope is obtained by spline interpolation. This method is theoretically introduced and experimentally validated by testing three induction motors: under different load conditions. Experimental results performed under various operational situations show the effectiveness of the proposed method against the Hilbert Transform to show that the proposed method successfully for detecting the fault.**

Keywords—Broken bars, diagnosis, Induction machine, peaks detection.

I. INTRODUCTION

THE use of electrical machines in industrial systems has been accompanied by greater demand for their availability and reliability. Indeed, it is financially useless to design and integrate these machines in these systems, if they must regularly break down and present a danger for people and the environment [1]-[4].

Squirrel cage induction motor is often considered as the most reliable electric machine, the most robust of its generation and the least expensive to manufacture. With power ranging from a few kilowatts to hundreds of megawatts, the induction motor is often used in industrial applications, not only in low-risk locations (pumps, conveyors, machine tools and compressors) but also in dangerous and aggressive places (gas plants, petrochemicals). However, it is not excluded that this motor will eventually be subjected to various defects due to the operating constraints and location conditions.

According to their causes, induction motor faults can be internal or external. Internal defects are caused by the motor components: magnetic circuits, stator or rotor windings, airgap, and rotor cage. External faults can appear due to the supply system, the mechanical load and the operating environment [3], [5].

Increasing the availability of the induction motor can be achieved either by improving its reliability during its design and manufacture or by implementing a well-adapted diagnostic strategy, for the early faults detection before a total failure occurrence [1].

Many researchers have proposed and used several techniques for induction motor faults diagnostics. Among of them, signal-based techniques which are based on the

control and monitoring of motor current, voltage, speed, torque, flux or vibration, then used signal processing methods to extract fault indicators [5],[6].

Signal processing methods based on Motor Current Signal Analysis (MCSA) are widely used in several studies to detect an electrical and mechanical failure in induction motor [1], [4], [7-9]. Stator current monitoring is a simple, low cost and non-invasive diagnosis technique. It can be carried out in the time domain, in the frequency domain or in the mixed field [1],[6]. In the time domain, fault diagnosis can be done by calculating statistical parameters like average, standard deviation, skewness and kurtosis of the signal [1],[10]. In the frequency domain, it can be achieved by extracting components in the current spectrum which characterizes the fault [1].

MCSA method for induction motor faults diagnosis has been used in several works.-The researchers concluded that performing this method in an industrial environment gives good results when the machine is heavily loaded. Unfortunately, it has been found that this technique is not suitable when the load varies with time, when the supply is polluted or when the machine is low loaded. The current spectrum is influenced by static and dynamic load conditions, motor geometry, noise and fault conditions [1], [9]. In addition, when the supply network is polluted, the current spectrum becomes very rich in harmonics making it very difficult to identify and extract the characteristic components of defects. Also, when the motor is low loaded, the characteristic components are very close to the fundamental component, which makes their identification very difficult.

Stator current demodulation technique is one of several techniques found in the literature to overcome this problem. It has been seen that the failures of induction motors introduce frequency and/or amplitude modulation in the stator currents. Classical demodulation techniques can be classified into two types: mono-dimensional and multi-dimensional techniques. The first type includes the Synchronous Demodulation, the Hilbert Transform, Teager-Kaiser Energy Operator and other approaches. In the second type, we can find the Concordia Transform, the Principal Component Analysis (PCA) and Maximum Likelihood approach-based and other combination of Hilbert and Park Transform [7], [8],[11],[12].

Hilbert transform has been investigated in many papers to detect broken rotor bar [1] and [12], bearing faults [14] and eccentricities [7]. The spectrum of the envelope detected

978-1-5386-7688-2/19 $31.00 © 2019 IEEE

by Hilbert transform has been used in [1] for broken bar fault diagnosis. In this work, authors confirmed that this technique can be used even at low slip through the elimination of the DC component of the stator current envelope, before spectral analysis. In addition, characteristic components and their frequencies, in the presence of broken bar fault, have been clearly identified in [9]. These components have been introduced by [2],[15] as inputs of Neuron Network to quantify the number of broken bars in a faulty induction motor under different load conditions.

The drawback of the Hilbert transform method is the amount of calculation operations to extract the signal envelope. In the interest to simplify the envelope detection process, this paper presents a simple method allowing the extraction of the true envelope of the stator current. This method is based on the localization and the detection of the prominent peaks and computes the envelope through a spline interpolation. The spectrum of the obtained envelope will be performed to detect characteristic components, in the case of rotor bar failure in an induction motor. The obtained results are compared with the Hilbert Transform method to show the effectiveness of the proposed method.

The remaining of the paper is organized as follows: Section 2 reviews the stator current modulation. Section 3 describes the proposed method for peaks detection y means the envelope demodulation. Experimental results are presented in Section 4. Finally, a conclusion is presented in Section 5

II. STATOR CURRENT MODULATION

For a healthy motor, the stator current produces a forward magnetic field which rotates at the synchronous speed $\omega_s=2.\pi.f_s$. By crossing the rotor, this field creates EMFs which causes a current circulation in the rotor windings. However, the rotor rotates and reaches a speed ω_r less than the synchronous speed. Simultaneously, rotor currents with frequency $f_r=s.f_s$ introduce an additional forward magnetic field which rotates at the same synchronous related to the stator. When a rotor bar fault occurs, it appears a backward Magnetomotive Force MMF which rotates at the speed of $-s\omega_s$ with respect to the rotor and the speed of $(1-2.s).\omega_s$ with respect to the stator [13],[16]. Consequently, the stator current is amplitude and can be expressed as:

$$ I_f(t) = I_h(t) \,(1+m_a \cos(\omega_f t)) \qquad (1) $$

Where $I_h(t)$ is the phase current in healthy induction machine, it can be considered purely sinusoidal and written by the following equation [1],[2],[13],[16],[17]:

$$ I_h(t) = I_m \cos(2\pi f_s t) \qquad (2) $$

$\omega_f = 2.\pi f_f$ with f_f is the characteristic frequency of the faulty component, which for a machine with a broken bar, is given by $2.s.f_s$ [1],[16],[17]. The modulation index m_a can be expressed as:

$$ m_a = \frac{n_a}{N_b} \qquad (3) $$

Where N_b, and n_a are the number of rotor bars and the number of consecutive broken rotor bars respectively.

III. ENVELOPE DEMODULATION THROUGH PEAKS DETECTION

Usually, the proposed method is used in musical sound and heart sound analysis [18], [19]. The principle consists to detect peaks using an appropriate search algorithm and connection of these peaks through spline interpolation. The current signal envelope, it is commonly agreed that varies slowly between two peaks corresponding to the maximum and minimum amplitudes. In some empirical views, it should passes through the prominent data peaks smoothly [18]. In more details, the method implemented in this work to obtain the signal current envelope is performed in three steps as summarized by the flowchart illustrated in figure 1. As illustrated by the figure (3.a), in the first step, the negative part of the current signal is eliminated to get only the positive one noted I_a^+. After that, the main signal is divided into K sub-signals and an iterative search algorithm is used to find the local maxima for each sub-signal.

$$ I_a^+(t) = I_{a\,1}^+(t) + \cdots + I_{a\,i}^+(t) + \cdots + I_{a\,k}^+(t) \qquad (4) $$

The number of sub-signals can be calculated as follows:

$$ K = \frac{N}{N_p} \qquad (5) $$

Where N is number of samples in the main signal and N_p is number of samples per period. N_p can be calculated by:

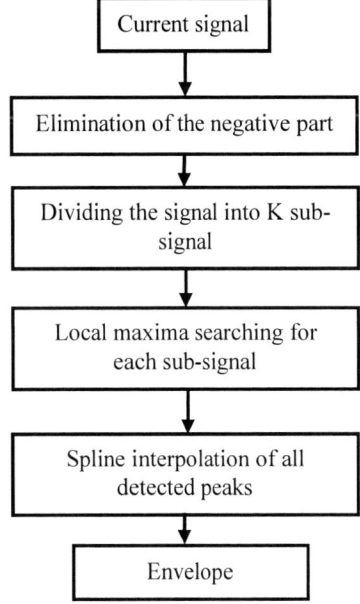

Fig. 1. Flowchart of the proposed method

$$N_\text{p} = \frac{f_e}{f_s} \qquad (6)$$

f_s and f_e are the main and the sampling frequencies respectively.

In the last step, the envelope of the current signal is obtained through a spline interpolation using all the detected peaks [20],[21]. Figure 3 illustrates an example of detecting peaks through different steps.

It is worth noting that the search algorithm can have two local maxima in some sub-signals caused by the power supply perturbations. So, the larger of the two maxima is taken as being the desired peak, while the other one is eliminated.

IV. EXPERIMENTAL RESULTS

The experimental setup displayed in Fig.2 consists of an electromechanical system which includes a 4kW industrial induction motor that drives a DC generator which acts as a load. A sensing card and a data acquisition card connected to a PC are used to sample motor current signal and store the data for off-line analysis.

Fig.2. Experimental banc setup

Different tests are performed using two different motors. The first motor is healthy, which serves as a reference for comparison. The two others have one broken bar and two broken bars, respectively.

The experimental tests of these motors have been performed under different load conditions, to evaluate the feasibility of the proposed method. Currently, all the tests are performed with the same sampling frequency (equal to 10 kHz). The signals are recorded for a duration of 10s, which gives a frequency resolution of 0.1 Hz.

First, the MCSA technique is used for a low load operating and results are illustrated in figure 4. In these conditions, the slip is relatively small, and the current spectrum contains lateral components very close to the fundamental which make it difficult to detect the fault, in particular for a motor with one broken bar. To overcome these difficulties, one of the solutions is the use of a high-resolution spectral analysis to separate the fault frequencies from the fundamental component.

Among the solutions proposed to overcome these difficulties is the use of a high-resolution spectral analysis or loading the motor at a heavy level, in order to separate the fault frequencies from the fundamental component. The first solution requires the increasing of data length and involves

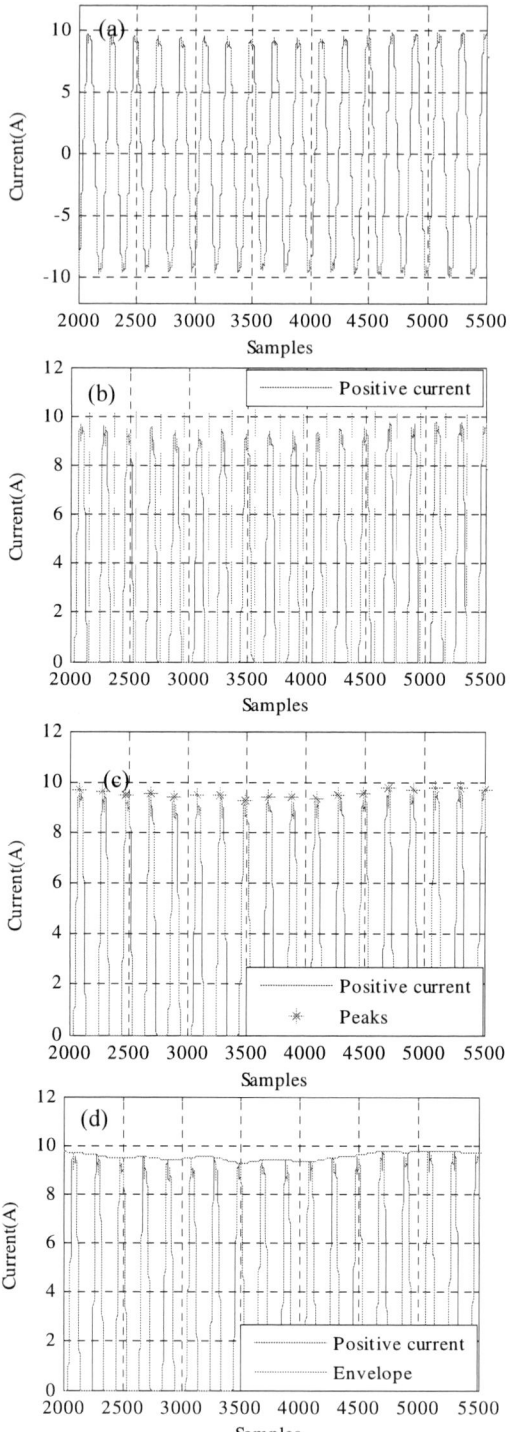

Fig. 3. Stator current modifications for envelope detection, (a): Input current, (b) positive part of current, (c) peaks detection, (d) obtained envelope

an important storage space. However, overloading a machine is undesirable since it reduces the machine's operating lifetime and is not generally under control of the operator.

Envelope analysis offers an alternative method to detect rotor fault by investigating stator current. In this section, results obtained by the proposed method are compared to those obtained by Hilbert Transform. (Fig. 5.) shows the shape of the envelope detected by the proposed method and Hilbert method. It is worth noting that the shape of the envelope obtained by the proposed method is very clear and passes through the maximum points of stator current, against the shape of the envelope obtained by Hilbert Transform presents some oscillations and does not follow current maximal points. Consequently, the proposed method has a high accuracy to extract current envelope signal and presents more advantages than the Hilbert envelope.

The second step after envelope detection is the spectrum

analysis of the obtained envelope. Figure 5 a comparison of the envelope spectrum using the Hilbert transform and the peaks detections method, when the three motors were tested under three load level conditions (low, medium and high load). Both methods can identify clearly the specific component with frequency *2sfₛ*. The magnitude of this component increases according to the number of broken bars and its location is extremely sensitive to the load. Whether under the conditions of high or low load, the proposed method allows, in a clear and precise manner, the identification of the specific component of the fault which overcome the drawback of the MCSA method

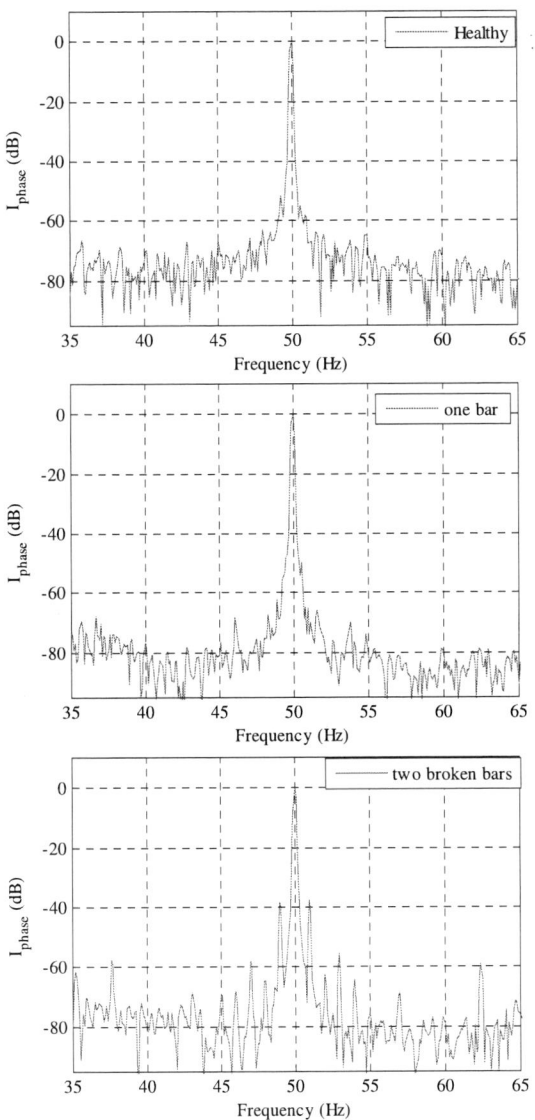

Fig. 4. The spectrum of the current stator at low load condition

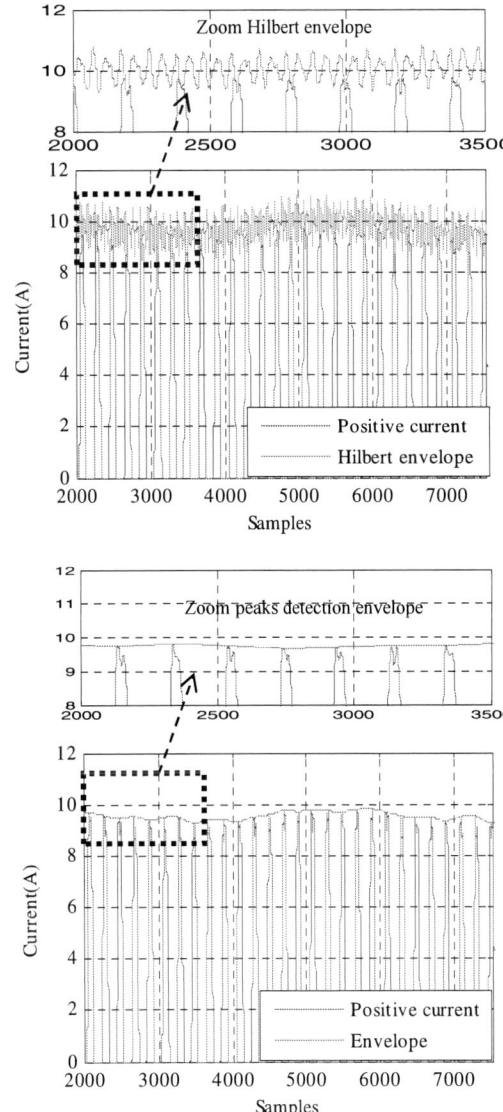

Fig. 5. Envelopes detection with its zoom

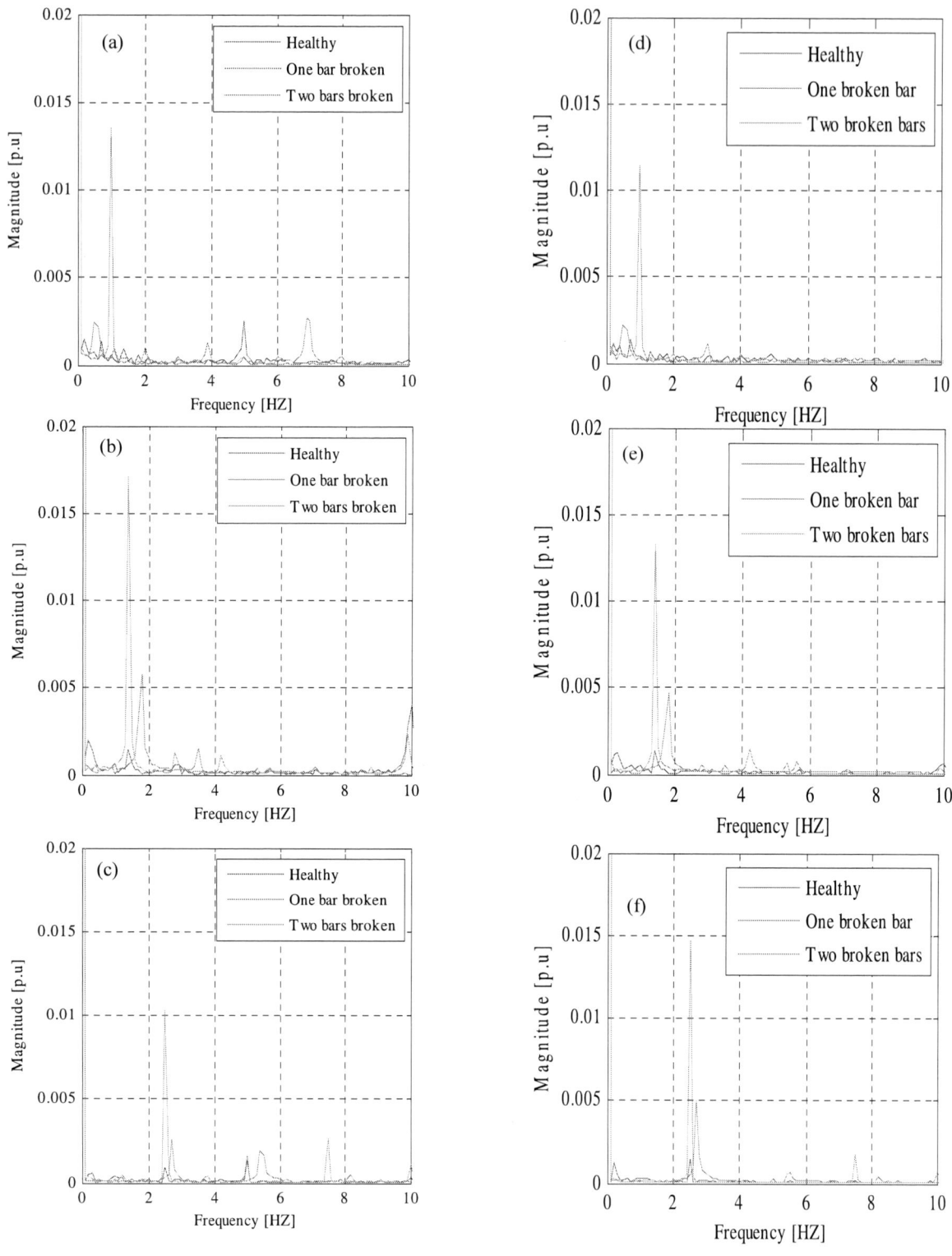

Fig. 5. Spectrum of the stator current envelope (a, b, c) peak detection method for low, medium, high load respectively, (d, e, f) Hilbert method for low, medium, high load respectively

V. CONCLUSION

In this paper, an accurate, simple and effective method for rotor bar fault diagnosis in induction motor under different load conditions is presented. This method is based on the detection of stator current peaks which can be used to form the desired envelope by spline interpolation. It has been shown that the estimated envelope can follow quite well the changing peaks of the stator current allowing it to have a good performance on the temporal envelope estimation compared to Hilbert Transform. The promising results obtained by this method testify the effectiveness of the method for fault detection even for low loads.

APPENDIX

Squirrel cage induction machine parameters

Pn = 4kW; Un =220/380 V; (•/Y); In=15.2/8.8 A; Nn=1435 tr/mn ; p=2; f=50Hz; cos =0.83; Number of rotor bars: 28; air-gap: 0.28mm; slots number: 36; number of turns in series per phase : 156; stator outer diameter: 140mm; axial stack length: 120mm

REFERENCES

[1] D. Matic, F. Kulic, M. Pineda-Sánchez, and I. Kamenko "Support vector machine classifier for diagnosis in electrical machines: Application to broken bar," in Journal of Expert Systems with Applications, Vol.39, pp.8681–8689,2012.

[2] W. Laala, S. Guedidi, S. Zouzou, ''Novel Approach for Diagnosis and Detection of Broken Bar in Induction Motor at Low Slip Using Fuzzy Logic,'' in IEEE Transactions on Industrial Electronics,pp. 511-516. 2011.

[3] Y. Trachi, El Houssin El Bouchikhi, and V. Choqueuse, M. Benbouzid, "Induction Machines Fault Detection Based on Subspace Spectral Estimation," in IEEE Transactions on Industrial Electronics Vol.63, No.9, Aug .2016.

[4] M. Benbouzid, G. Kliman, "What Stator Current Processing Based Technique to Use for Induction Motor Rotor Faults Diagnosis," in IEEE Transactions on Energy Conversion, Vol.18, No.2, pp.238-244.2003.

[5] P. S. Bhowmik, S. Pradhan, and M. Prakash, "Fault Diagnostic and Monitoring Method of Induction Motor: A REVIEW," IJACEEE International Journal of Applied Control, Electrical and Electronics Engineering Vol. 1, No. 1, May.2013.

[6] F. Tafinine, K. Mokrani, J. Antoni, A. Kabla, and Z. Asradj, «Introduction des SVM en MCSA,»SETIT 4th International Conference: Sciences of Electronic, Technologies of Information and Telecommunications March 25-29, 2007 – TUNISIA

[7] El Houssin El Bouchikhi, V. Choqueuse, M. Benbouzid, J. A. Antonino-Daviu, "Stator current demodulation for induction machine rotor faults diagnosis," ICGE First International Conference on Green Energy, Mar 2014, Sfax, Tunisia. pp.176 - 181, 2014.

[8] Elhoussin Elbouchikhi, V. Choqueuse and M. Benbouzid, "Condition Monitoring of Induction Motors based on Stator Currents Demodulation," IREE International Review of Electrical Engineering, Vol.10, No. 6. pp.704-715. 2015.

[9] R. Puche-Panadero, M. Pineda-Sanchez, J. Roger-Folch, E. Hurtado-Perez, and J. Perez-Cruz, "Improved Resolution of the MCSA Method Via Hilbert Transform, Enabling the Diagnosis of Rotor Asymmetries at Very Low Slip," in IEEE Trans. Energy Convers. Vol. 24, pp.52–59, No. 1, MARCH. 2009.

[10] K.M. Arunkumar, T.C. Manjunath, "A brief review/survey of vibration signal analysis in time domain," in SSRG International Journal of Electronics and Communication Engineering (SSRG-IJECE), Vol. 3,Issue. 3, March .2016.

[11] K. Bacha, S. Ben Salem, and A. Chaari "An improved combination of Hilbert and Park transforms for fault detection and identification in three-phase induction motors," in Journal of Electrical Power and Energy Systems, Vol. 43, pp. 1006–1016.2012.

[12] M. Pineda-Sanchez, R. Puche-Panadero, J. Pons-Llinares, V. Climente-Alarcon, and J. A. Antonino-Daviu, "Application of the Teager–Kaiser Energy Operator to the Fault Diagnosis of Induction Motors," in IEEE Transactions on Energy Conversion, Vol. 28, No. 4, pp. 1036-1044,DECEMBER. 2013.

[13] H. Khelfi, S. Hamdani, K. Nacereddine, and Y. Chibani "Stator Current Demodulation Using Hilbert Transform for Inverter-Fed Induction Motor at Low Load Conditions" 3th International Conference in Electrical Sciences and Technologies in Maghreb (CISTEM) Octobre 28-31th,2018.

[14] V. Choqueuse, M. Benbouzid, Y. Amirat, and S. Turri, "Diagnosis of three-phase electrical machines using multidimensional demodulation techniques," in IEEE Transactions on Industrial Electronics, vol. 59, no. 4, pp. 2014 –2023, april.2012.

[15] B. Bessam, A. Menacer, M. Boumehraz, H. Cherif, "Detection of broken rotor bar faults in induction motor at low load using neural network," in Journal of ISA Transactions, vol. 64, pp.241–246. 2016.

[16] H. Khelfi, S. Hamdani ",Stator Current Demodulation Using Square Roots Current Stator for Inverter-Fed Induction Motor at Low Load Conditions" International Conference on Communications and Electrical Engineering (ICCEE),Decembre,16-17th,2018.

[17] A. Sapena-Bãno, M.Pineda-Sanchez, R. Puche-Panadero, J. Martinez-Roman, and Z. Kanovic , "Low-Cost Diagnosis of Rotor Asymmetries in Induction Machines Working at a Very Low Slip Using the Reduced Envelope of the Stator Current,", in IEEE Transactions on Energy Conversion, Vol. 30, NO. 4, DECEMBER .2015.

[18] C. Jarne, "Simple Empirical Algorithm to Obtain Signal Envelope in Three Steps," March ,21. 2017.

[19] M. Qinglin, Y. Meng, Y. Zhenya, and F. Haihong, "An Empirical Envelope Estimation Algorithm", IEEE CISP 6th International Congress on Image and Signal Processing, pp.1132 - 1135.2013

[20] P. K. Mohanty, M. Reza, P. Kumar3, P. Kumar, "Implementation of Cubic Spline Interpolation on Parallel Skeleton using Pipeline Model on CPU-GPU Cluster," IEEE 6th International Conference on Advanced Computing, 2016, PP. 747-751

[21] N. Sun, T. Ayabe, K. Okumura "An Animation Engine with the Cubic Spline Interpolation," IEEE International Conference on Intelligent Information Hiding and Multimedia Signal Processing, 2008.PP, 109-112.

Torque Error Reduction of Interior Permanent Magnet Synchronous Motor Drives using a Stator Flux Linkage Observer

Sungmin Choi
Dept. of Electrical Engineering
Chonbuk National University
Jeonju, Republic of Korea
zpfzpfzpdls@gmail.com

Seung-Hwan Lee
Dept. of Electrical and Computer Engineering
University of Seoul
Seoul, Republic of Korea
seunghlee16@uos.ac.kr

Jae Suk Lee
Dept. of Electrical Engineering
Chonbuk National University
Jeonju, Republic of Korea
jaesuk@jbnu.ac.kr

Abstract— A torque error reduction algorithm targeting for an interior permanent magnet synchronous machine (IPMSM) is presented in this paper. In an IPMSM drive, Maximum Torque per Ampere (MTPA) lookup table (LUT) is typically applied for the minimization of copper loss. Estimated permanent magnet (PM) flux linkage is used for MTPA-LUT development, and torque error occurs due to change of PM flux linkage. In this paper, an IPMSM control algorithm is proposed to compensate torque error using a stator flux linkage observer. PM flux linkage estimated through the stator flux linkage observer is applied to the MTPA LUT to compensate for the torque error.

Keywords—Torque error, Gopinath flux observer, IPMSM drives

I. INTRODUCTION

In many industrial applications, PMSMs have been widely used due to its high efficiency and power density [1]. Comparing to IPMSMs, SPMSMs are simpler to manufacture and control. However, operation speed range of Surface mount PMSMs (SPMSMs) is narrower than IPMSMs. Therefore, an IPMSM is preferred for the applications desires a wide speed operation range, for example, automotive applications. Though PMSMs are attractive as described above, parameters are dependent to external conditions, for example, temperature and operating conditions. The parameter variation affects torque development and efficiency of PMSM drives [2]. An Maximum Torque Per Ampere (MTPA) approach is one of control algorithms for efficient operation of PMSMs. An MTPA trajectory is typically developed in a form of a look up table (LUT) and machine parameters are used for development of the MTPA-LUT. When parameters of PMSMs are varied by external conditions,

undesired current vector command is selected from the MTPA-LUT. It results in undesired torque development from PMSMs [3]. Among parameters, effect of PM flux linkage is significant for torque development of PMSMs. PM flux linkage decreases as magnet temperature of PMSMs increases due to demagnetization. Solution for torque error reduction with respect to magnet temperature variation have been proposed. In [4], online parameter identification method is applied to keep effect of the MTPA. However online parameter identification is typically complicated for implementation. An algorithm using a stator flux linkage observer based on a voltage model is proposed in [5]. However, performance degradation of the algorithm is expected at low speed because only the voltage model is used for flux observer development. In [6,7], a perturbation searching method is proposed and an MTPA is achieved regardless parameter variation. However, dynamic performance is limited due to long perturbation search time. In [8,9], a high frequency signal injection methods are introduced and MTPA operation of PMSMs is achieved regardless machine parameter variations However, the injected signal results in loss in a power converter and a motor and it results in undesired effect to operation of the PMSMs. In [10,11], algorithms for PM temperature estimation for PMSMs and variable flux PMSMs are proposed but torque error reduction for PMSMs has not been included in the algorithms. In [12], a torque error compensation algorithm using a stator flux linkage observer with respect to magnet temperature variation is proposed but the proposed algorithm is only limited to SPMSM drives.

978-1-5386-7688-2/19 $31.00 © 2019 IEEE

In this paper, PM flux linkage of PMSMs is estimated using a Gopinath style stator flux linkage observer and the estimated PM flux linkage of the PMSMs is used for torque error reduction for IPMSMs with magnet temperature variation in real time. In following sections of the paper, a characteristic of the Gopinath style stator flux linkage observer is described and analyzed. The torque error reduction algorithm for IPMSM drives is presented for improvement of torque development accuracy at various magnet temperature conditions. The proposed real time torque error reduction algorithm is implemented and verified through MATLAB/Simulink simulation results at various temperature conditions in this paper.

II. ANALYSIS OF GOPINATH STYLE STATOR FLUX LINKAGE OBSERVER

Gopinath style stator flux linkage observer has been applied for sensored and sensorless control of induction motor drives [13,14] and PMSM drives [15,16]. The stator flux linkage observer for PMSM drives shown in Fig. 1 is used for implementation of the torque error compensation algorithm.

Gopinath style stator flux linkage observer applied to PMSM drive is typically composed of two models, which are a current model and voltage model. During low speed operation, the voltage developed from an inverter is distorted due to external factors, for example, dead time and noise from a sensor or an interface circuit board. As a result, the voltage model generates distorted stator flux linkage estimation values. Therefore, the stator flux linkage of PMSMs is estimated using the current model based on (1) and (2) at low speed operation. In equations in the paper, subscripts d and q represent the d-axis and q-axis.

The superscripts r and s represent the rotor reference and stator reference frames. * is the command value, ^ is the estimation value.

$$\hat{\lambda}_{di}^r = \lambda_{pm} + L_d i_d^r \tag{1}$$

$$\hat{\lambda}_{qi}^r = L_q i_q^r \tag{2}$$

Though a current model based stator flux linkage observer is not affected by voltage distortion at low speed, it is influenced by changes in the parameters used in stator flux estimation because the current model estimates the stator flux linkage in the open loop as shown in (1) and (2). At high speed operation, external factors such as noise and deadtime can be ignored and the voltage drop of the stator can be also ignored due to the high back-emf voltage of IPMSM drives. The voltage equation in the stator reference frame is shown in (3) and (4). Equation (3) and (4) can be expressed by (5) and (6), and the voltage model is developed based on (5) and (6).

$$V_d^s = R_s i_d^s + \frac{d}{dt}\lambda_{dv}^s \tag{3}$$

$$V_q^s = R_s i_q^s + \frac{d}{dt}\lambda_{qv}^s \tag{4}$$

$$\hat{\lambda}_{dv}^s = (V_d^s - R_s i_d^s)\frac{1}{s} \tag{5}$$

$$\hat{\lambda}_{qv}^s = (V_q^s - R_s i_q^s)\frac{1}{s} \tag{6}$$

In Fig. 2, frequency response simulation results of a Gopinath style stator flux linkage observer over a wide operation range is shown. During simulation, IPMSM parameters are intentionally varied. The bandwidth of the stator flux linkage observer controller is tuned to be 0.1[pu] of the rated speed of the test

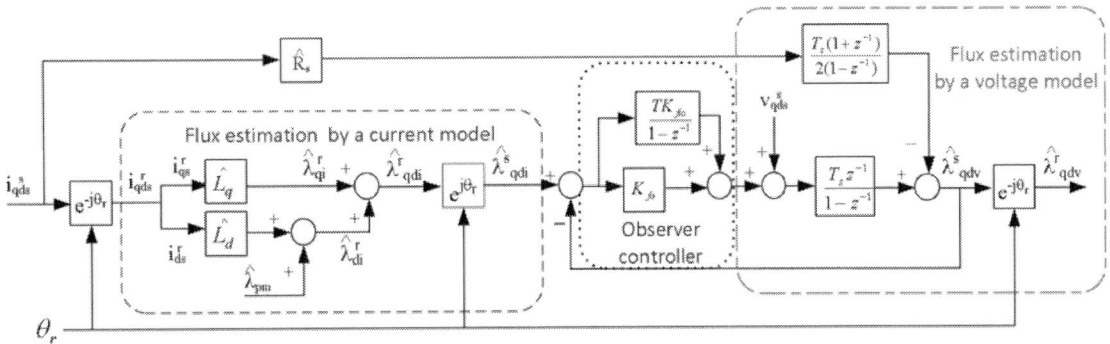

Fig. 1. A block diagram of a Gopinath style stator flux linkage observer for torque error compensation control system [12],

978-1-5386-7688-2/19 $31.00 © 2019 IEEE   281

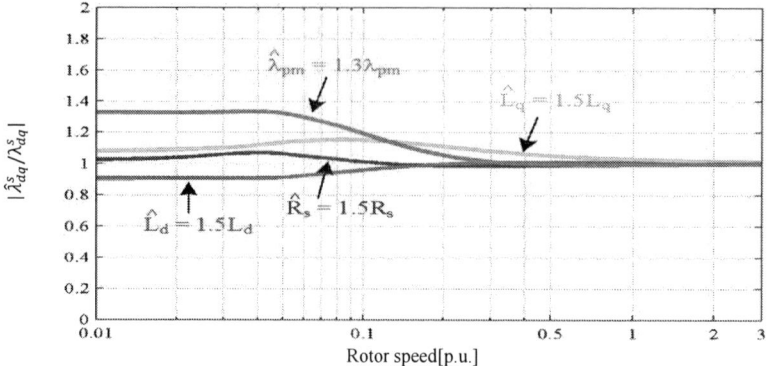

Fig. 2. Simulation results of a frequency response of the Gopinath style stator flux linkage observer with respect to motor parameter variations [12].

IPMSM. The bandwidth determines a crossover point between a current model and a voltage model. From the simulation result, accuracy of estimated stator flux linkage can be known over a wide operating range. From the characteristic of the stator flux linkage, inaccurate estimation of stator flux linkage is expected when the test IPMSM operates under the crossover frequency. Accuracy of stator flux linkage estimation becomes higher when operation speed of the test IPMSM is higher than the crossover frequency of the stator flux linkage observer. In Fig.2, it is verified that estimation accuracy of stator flux linkage is influenced by the parameter variation at low speeds. Accuracy of stator flux linkage estimation becomes higher regardless parameter variations of the test IPMSM beyond a crossover frequency of the stator flux linkage observer controller. Therefore, the estimated stator flux linkage at high speed can be used for development of the proposed torque error compensation algorithm.

III. PROPOSED TORQUE ERROR COMPENSATION ALGORITHM

Copper loss occurring in the electric machine is proportional to the square of the current. MTPA is a control technique that minimizes copper loss by generating a specific torque using minimum current. Torque equation is shown in (7) because the IPMSM has a characteristic of Lq > Ld in its construction. Also, torque equation can be expressed through the relationship between the stator flux linkage and current, as shown in (8).

$$T_{em} = \frac{3P}{4}[\lambda_{pm}i_q + (L_d - L_q)i_d i_q] \qquad (7)$$

$$T_{em} = \frac{3P}{4}(\lambda_d^r i_q^r - \lambda_q^r i_d^r) \qquad (8)$$

As shown in (7) and (8), PM flux linkage is an important parameter for torque generation, and the estimated PM flux linkage is used for constructing the

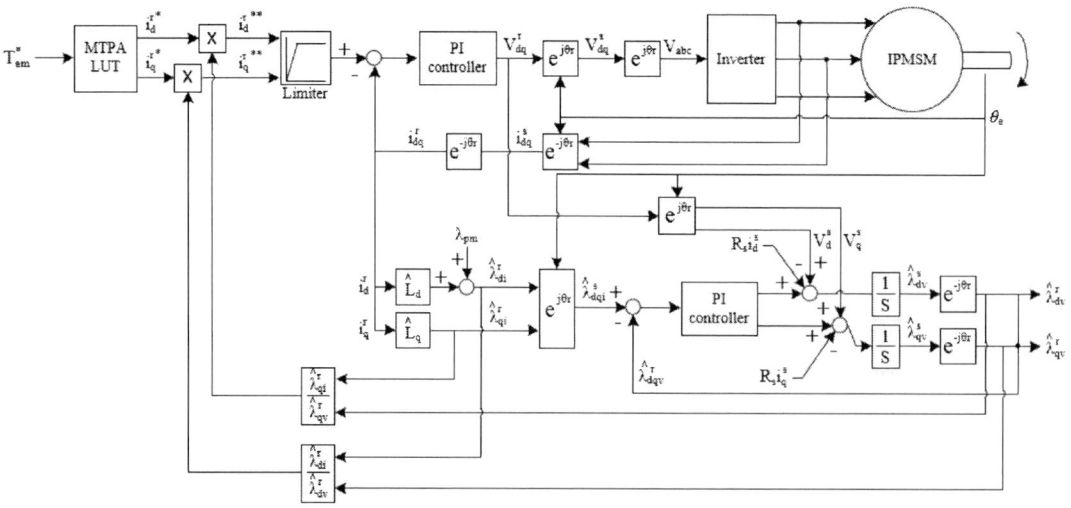

Fig. 3. A block diagram of the proposed torque error compensation control system for IPMSM drives.

MPTA LUT. However, PM flux linkage changes depending on the temperature of the PM, which causes an error between the torque command and the actually generated torque. In this paper, a method is proposed to compensate the torque error by estimated PM flux linkage through the Gopinath style stator flux linkage observer and reflecting the flux linkage change of the PM to the MTPA LUT.

$$K_{crd} = \frac{\hat{\lambda}_{qi}^r}{\hat{\lambda}_{qv}^r} \qquad (9)$$

$$K_{crq} = \frac{\hat{\lambda}_{di}^r}{\hat{\lambda}_{dv}^r} \qquad (10)$$

The torque compensation parameters Kcrd and Kcrq are shown in (9) and (10). As shown in (8), torque generation is related to the product of the estimated value of PM flux in d-axis and q-axis current, and the product of the PM flux in q-axis and d-axis current. Therefore, the flux of the q-axis estimated by the current and voltage model of the stator flux linkage observer is reflected in the d-axis MTPA LUT and estimated flux of d-axis is reflected in the q-axis MTPA LUT to compensate for the torque error.

IV. SIMULATION RESULT

Proposed torque error compensation algorithm is implemented and verified through simulation. Overall system block diagram of the torque error compensation algorithm is shown in Fig. 3. i_d^{r*} and i_q^{r*} are the initial current command and are calibrated using the torque error compensation parameters. The calibrated currents i_d^{r**} and i_q^{r**} are limited to not exceed the maximum current of the system. Fig. 4. shows a simulation model of torque error reduction algorithm for an IPMSM drives.

Table 1. shows the parameters and system characteristic used in the simulation. In order to implement the torque error compensation algorithm, PM flux linkage is set to 80% of the actual flux in the simulation model.

TABLE I. SIMULATION PARAMETERS AND SYSTEM CHARACTERISTIC

Parameters	Value
Stator resistance	1.5[Ω]
d axis stator inductance	5.5[mH]
q axis stator inductance	12.5[mH]
PM flux linkage	0.1[T]
Poles	4
PWM sampling time	10[kHz]

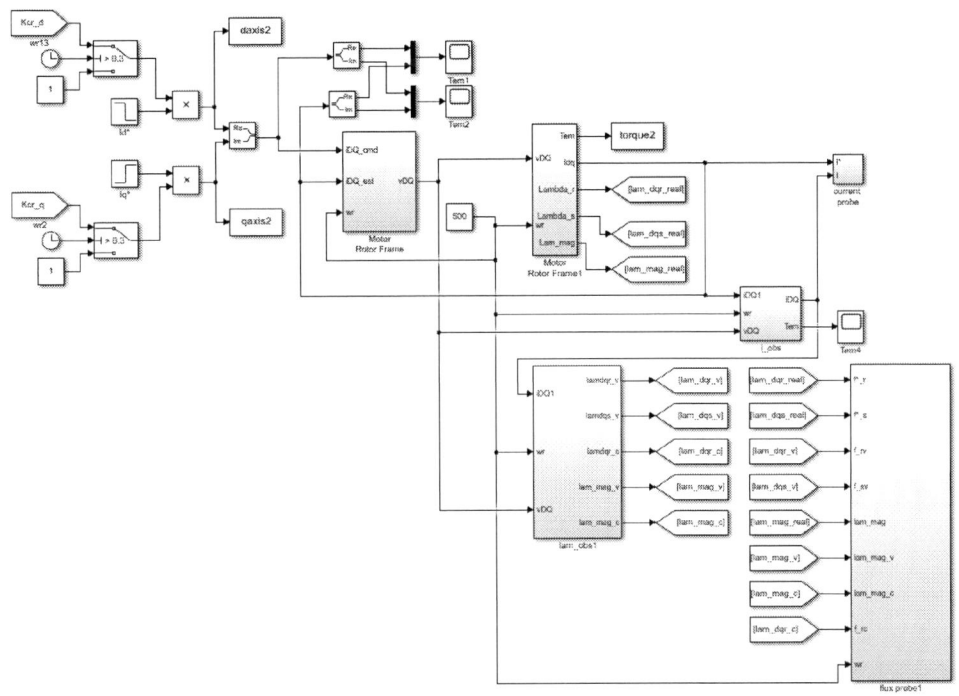

Fig. 4. A block diagram of the proposed torque error compensation control system for IPMSM drives.

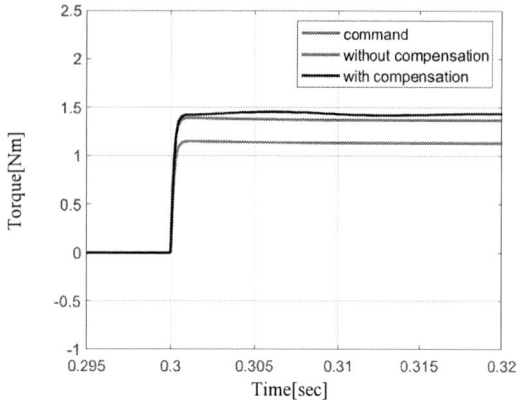

Fig.5 Simulation result of the torque response with and without compensation algoritum.

Fig. 5 shows the torque generation with and without the torque error compensation algorithm for the torque command. Without the algorithm, it can be confirmed that the change in PM flux linkage is not reflected in the MTPA LUT, and torque lower than the torque command is generated. With the algorithm, it can be confirmed that the change in PM flux linkage is reflected in MTPA LUT, and torque similar to the torque command is generated.

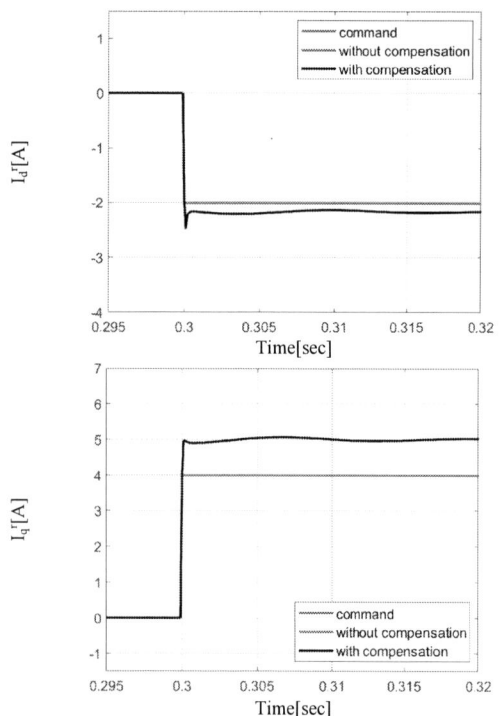

Fig. 6. Simulation reslut of the current command with and without compensation algorithm.

Fig. 6 shows the d-axis and q-axis current. Without algorithm, the same current output is commanded regardless parameter variation because the change in PM flux linkage is not reflected. With the algorithm, it can be confirmed that the current command is increased and the torque error is compensated for when the PM flux linkage decreases.

CONCLUSION

In this paper, an algorithm is proposed to compensate the torque error by estimated the PM flux linkage through the stator flux linkage observer and reflecting it to the MTPA LUT. Simulation results show that the proposed algorithm compensates the torque error when the permanent magnet flux decreases at high speed. Proposed algorithm will be verified through experiments.

REFERENCES

[1] T. Zhen, Z. Chengning and Z. Shuo, "Analytical Calculation of Magnetic Field Distribution and Stator Iron Losses for Surface-Mounted Permanent Magnet Synchronous Machines", *Energies*, vol.10, no.3, pp.320-332, Mar. 2017.

[2] S. Brigette, N. Donald W and L. Thomas A, "Field-weakening in buried permanent magnet AC motor drives", *IEEE Trans. on Ind. Appl*, vol.IA-21, no.2, pp.398-407, Mar. 1985.

[3] S. Morimoto, M. Sanada, Y. Takeda, "Wide-speed operation of interior permanent magnet synchronous motors with high performance current regulator", *IEEE Trans. on Ind. Appl*, vol.30, no.4, pp.920-926, Jul/Aug. 1994.

[4] K. Hyunbae, L. Robert D, "Using on-line parameter estimation to improve efficiency of IPM machine drives", *IEEE Power Electr. Specialist Conf*, pp.815-820, Nov. 2002.

[5] K. Yeon-Su, S. Seung-Ki, "Torque control strategy of an IPMSM considering the flux variation of the permanent magnets", *IEEE Ind. Appl. Society Annual Meeting*, pp.1301-1307, Oct. 2007.

[6] H. David, C. Yu, Y; Z. Ziqiang, "Online optimal flux weakening control of permanent magnet brushless drives", *IEEE Trans. on Ind. Appl*, vol.36, no.6, pp.1661-1668, Nov/Dec. 2000.

[7] D. Anton, K. Young-Kwan, L. Sang-Joon, L. Sang-Taek "Robust self-tuning MTPA algorithm for IPMSM drives", *34th Annual Conf of IEEE Ind. Electr. Society*, pp.1355-1360, Jan. 2008.

[8] B. Silverio, P. Roberto, P. Antonio, S. Luca, "Automatic tracking of MTPA trajectory in IPM motor drives based on AC current injection", *IEEE Trans. on Ind. Appl*, vol.47, pp.105-114, Nov. 2011.

[9] A. Riccardo, C. Matteo, Z. Mauro, "theory and implementation of an MTPA tracking controller for anisotropic PM motor drives", *38th Annual Conf of IEEE Ind. Electr. Society*, pp.2061-2066, Dec. 2012.

[10] G. Brent S, S. Kensuke, A. Apoorva, K. Takashi, L. Robert D, "Magnet Temperature Effects on the Useful Properties of Variable Flux PM Synchronous Machines and a Mitigating Method for Magnetization Changes", *IEEE Trans. on Ind. Appl*, vol.53, no.3, pp.2189-2199, Feb. 2017.

[11] R. David, F. Daniel, T. Tsutomu, K. Takashi, T, B. Rernando, "Comparative Analysis of BEMF and Pulsating High-

Frequency Current Injection Methods for PM Temperature Estimation in PMSMs", *IEEE Power Electr. Society,* Vol.32, no.5, pp. 3691-3699, Jul. 2017.

[12] C.Park and J.Lee, "Torque error compensation algorithm for surface mounted PMSMs with respect to magnet temperature variations", *Energies,* vol.10, no.9, pp. 1365-1379, Sep. 2017.

[13] P.Jansen and R.D.Lorenz, "Physically insightful approach to the design and accuracy assessment of flux observers field oriented induction machine drives", *IEEE Trans. on Ind. Appl,* vol.20, pp. 101-110, Aug. 1994.

[14] G. Shady M, G. Damian, F. John W, "Sensorless control of induction motor drives at very low and zero speeds using neural network flux observer", *IEEE Trans. on Ind. Elect,* vol.56, no.8, pp.3029-3039, Jun 2009.

[15] B. Ion, P. Mihaela Codruta, A. Gheorghe-Daniel, B. Frede, "Active flux DTFC-SVM sensorless control of IPMSM", *IEEE Trans. on Energy Converg,* vol.24, no.2, pp.314-322, May. 2009.

[16] A. Yoo and S.Sul, "Design of flux observer robust to interior permanent magnet synchronous motor flux variation", *IEEE Trans. on Ind. Appl,* vol.45, no.5, pp.314-322, Jul. 2009.

978-1-5386-7688-2/19 $31.00 © 2019 IEEE

Technical track on Power Electronics and Power Conversion

978-1-5386-7688-2/19 $31.00 © 2019 IEEE

A Bidirectional Hybrid Switched-Capacitor DC-DC Converter with a High Voltage Gain

Dan Hulea
Department of Electrical Engineering
Politehnica University of Timisoara
Timisoara, Romania
dan.hulea@student.upt.ro

Mihaita Gireada
Department of Electrical Engineering
Politehnica University of Timisoara
Timisoara, Romania
mihaita.gireada@student.upt.ro

Nicolae Muntean
Department of Electrical Engineering
Politehnica University of Timisoara
Timisoara, Romania
nicolae.muntean@upt.ro

Octavian Cornea
Department of Electrical Engineering
Politehnica University of Timisoara
Timisoara, Romania
octavian.cornea@upt.ro

Abstract—This paper presents a new non-isolated bidirectional hybrid switched-capacitor DC-DC converter (BHSC) which achieves a high voltage conversion ratio. The hybrid characteristic is due to the switched capacitor cell that is inserted into a conventional buck/boost bidirectional converter, which also helps achieve the high conversion ratio. Apart from the high conversion ratio, the switched cell also helps reduce all passive components and the stress on the active switches, all without interrupting the input to output ground path. The stability of the converter is also addressed by using the state space averaging method (SSA), prior to the actual construction of the converter, in order to design the passive components so that possible instabilities that might occur are eliminated. The simulation results are used to confirm the initial theoretical considerations.

Keywords— dc-dc power converters, bidirectional, high conversion ratio, non-isolated, stability.

I. INTRODUCTION

Bidirectional high ratio converters are beneficial in applications where a large voltage ratio between the input and output are present, such as in microgrid applications in storage applications [1]–[3] or as an interface converter [4], [5] or for electric vehicles in vehicle to grid configuration [6], [7], or supercapacitor storage [8], [9].

High ratio converters are especially beneficial in applications where supercapacitor storage is used, as the voltage of the supercapacitor varies proportionally to the square root of the stored energy, so a large variation is needed for a better utilization of the supercapacitor [10], [11].

The common ground is also beneficial in order to parallel multiple converters or to translate the topology into a multi-level one [12].

II. CONVERTER TOPOLOGY AND ANALYTICAL DESCRIPTION

A. Converter Topology

The proposed converter is a hybrid DC-DC converter as it uses switched capacitors cells, initially proposed in [13], and by doing so it achieves a higher conversion ratio, useful for a wider operating voltage. The topology is based on the unidirectional buck converter, enhanced with a switched capacitor cell, topology which was tested experimentally in the literature in [14].

The proposed converter, shown in Fig. 1, is a modified bi-directional buck/boost converter with the bidirectional

switched capacitor cell at the high voltage side. In comparison to other high ratio converter, such as, [15], [16], the proposed converter also has a common ground between the input and output signal which is a beneficial feature in many applications. In comparison to other topologies, [10], [17], with same conversion ratio, the proposed one uses an additional transistor, but as it is shown in the following sections, the total active switch stress is lower.

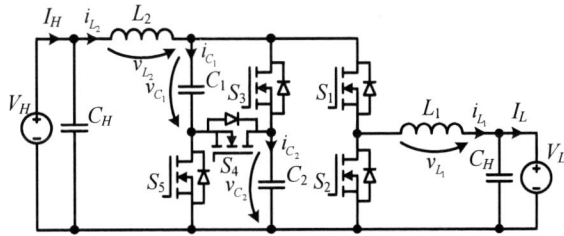

Fig. 1. The Bidirectional Hybrid Switched Capacitor (BHSC) converter topology.

B. BHSC Operation mode

The BHSC converter can be viewed as either a buck converter from the high voltage (V_H) perspective, or as a boost converter from the low voltage (V_L) perspective. The switched capacitive cell is therefore connected at the input of the buck side, or at the output of the boost side. In the following pages the buck mode is presented in diagrams, the boost mode being similar, with the only difference being in the sign of the currents.

The two switching states can be observed in Fig. 2 and Fig. 3, and are defined by the t_{on} and t_{off} periods. Only one driving signal is needed for the transistors, applied directly to S_1, S_3 and S_5 transistor, and then inverted to drive S_2 and S_4, resulting in a simpler control.

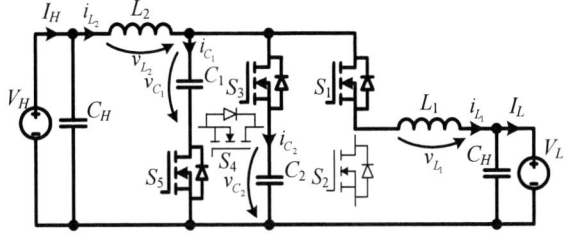

Fig. 2. BHSC equivalent schematic during t_{on} switching period.

978-1-5386-7688-2/19 $31.00 © 2019 IEEE

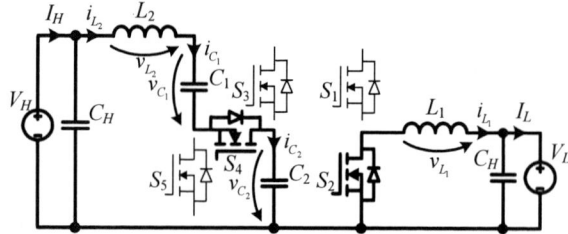

Fig. 3. BHSC equivalent schematic during t_{off} switching period.

In buck mode ($i_{L_1} > 0$) the two capacitors from the switching cells are discharged in parallel to the low voltage source, V_L, during t_{on}. The i_{L_2} current also charges the capacitors, but it is much smaller than the discharge current so it can be neglected. The capacitors are then charged in series from the V_H voltage during t_{off}.

C. Analytical description

Few presumptions are made in order to simplify the analysis: all components are ideal; the capacitors are considered large enough in order to obtain a negligible voltage ripple and the converter is operating in steady state.

The equations for the inductor voltages and for the switched capacitor currents in the two equivalent states are described in (1).

$$t_{on} : \begin{cases} v_{L_1} = V_{C_1} - V_L \\ v_{L_2} = V_H - V_{C_2} \\ i_{C_1} + i_{C_2} = i_{L_2} - i_{L_1} \end{cases} \quad t_{off} : \begin{cases} v_{L_1} = -V_L \\ v_{L_2} = V_H - V_{C_2} - V_{C_1} \\ i_{C_1} = i_{C_2} = i_{L_2} \end{cases} \quad (1)$$

In order to simplify the equations, the two capacitors and their currents are considered identical ($C_1 = C_2 = C_{sw}$) therefore (1) can be simplified as (2).

$$t_{on} : \begin{cases} v_{L_1} = V_C - V_L \\ v_{L_2} = V_H - V_C \\ 2 \cdot i_{C_{sw}} = i_{L_2} - i_{L_1} \end{cases} \quad t_{off} : \begin{cases} v_{L_1} = -V_L \\ v_{L_2} = V_H - 2 \cdot V_C \\ i_{C_{sw}} = i_{L_2} \end{cases} \quad (2)$$

In order to obtain a good comparison between the BHSC and the conventional buck converter, the duty cycle (D) is defined as in (3).

$$D = \frac{t_{on}}{T} = t_{on} \cdot f \quad (3)$$

By applying volt-second balance principle on the inductors, (2) is used to write (4).

$$\begin{cases} V_{L1} = D \cdot (V_C - V_L) + (1-D) \cdot (-V_L) = 0 \\ V_{L2} = D \cdot (V_H - V_C) + (1-D) \cdot (V_H - 2V_C) = 0 \end{cases} \quad (4)$$

Using (4), the switched capacitor voltage is determined in (5).

$$\begin{cases} V_{C_{sw}} = \frac{V_H}{2-D} \\ V_{C_{sw}} = \frac{V_L}{D} \end{cases} \quad (5)$$

Using (5) the relation between the high and the low voltage is determined in (6), and the duty cycle is determined in (7).

$$V_L = V_H \cdot \frac{D}{2-D} \quad (6)$$

$$D = \frac{2 \cdot V_L}{V_H + V_L} \quad (7)$$

The capacitor voltage is also determined in (8) and, as it can be observed, it has a constant voltage only dependent by the two input voltages, considering a steady state.

$$V_C = \frac{V_H + V_L}{2} \quad (8)$$

The theoretical waveforms presented in Fig. 4 are extracted from the two equivalent schematics (Fig. 2 and Fig. 3), and from the previous equations. Apart from the two inductor voltages and currents, the switched capacitor currents and voltage ripple ($\Delta V_{C_{sw}}$) are also presented. The input and output voltage ripples (ΔV_{C_H}, ΔV_{C_L}) are represented considering that ideal capacitors are also used at the input ports (not shown in schematics Fig. 1 - Fig. 3). The ripple voltage representation is made by considering a constant current load at the inputs.

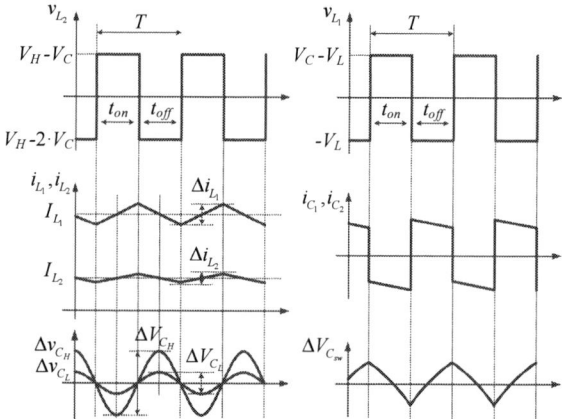

Fig. 4. Theoretical waveforms for BHSC (ΔV_C voltages are not to scale).

III. CONVERTER SIZING

An important aspect for any converter is its sizing and choosing its passive components, therefore the following two subsections will address this aspect.

A. Inductor Sizing

Considering the inductor voltages from (1), the inductor can be calculated considering t_{on} switching period and a specific value for the ripple current, therefore (9) can be written.

$$t_{on} : \begin{cases} L_1 \cdot \frac{di_{L_1}}{dt} = V_{C_{sw}} - V_L \Leftrightarrow L_1 \cdot \frac{\Delta i_{L1}}{t_{on}} = V_{C_{sw}} - V_L \\ L_2 \cdot \frac{di_{L_2}}{dt} = V_H - V_{C_{sw}} \Leftrightarrow L_2 \cdot \frac{\Delta i_{L2}}{t_{on}} = V_H - V_{C_{sw}} \end{cases} \quad (9)$$

978-1-5386-7688-2/19 $31.00 © 2019 IEEE

Based on (9), the inductor values can be rewritten as in (10).

$$\begin{cases} L_1 = \dfrac{D \cdot T \cdot (V_{C_{sw}} - V_L)}{\Delta i_{L1}} \\ L_2 = \dfrac{D \cdot T \cdot (V_H - V_{C_{sw}})}{\Delta i_{L2}} \end{cases} \tag{10}$$

In order to compare this topology to other topologies, the design of the passive components is done considering a percentage ripple (Δi_{Lp}) which is used to calculate the inductor ripple as in (11). Based on (5), (10) and (11) the inductor values can be calculated as in (12).

$$\begin{cases} \Delta i_{L1} = \Delta i_{Lp} I_{L1} \\ \Delta i_{L2} = \Delta i_{Lp} I_{L2} \end{cases} \tag{11}$$

$$\begin{cases} L_1 = \dfrac{V_L \cdot (V_H - V_L)}{\Delta i_{Lp} \cdot f \cdot I_L \cdot (V_H + V_L)} \\ L_2 = \dfrac{V_H \cdot (V_H - V_L)}{\Delta i_{Lp} \cdot f \cdot I_L \cdot (V_H + V_L)} \end{cases} \tag{12}$$

B. Capacitors Sizing

Capacitor sizing is done similar to the inductor sizing. First, (13) is used, to show the dependency between currents and voltages. The difference here is that the capacitor currents are not constant so the integral must be calculated in (14).

$$\begin{cases} C_{sw} \cdot \dfrac{dv_{C_{sw}}}{dt} = \dfrac{i_{L2} - i_{L1}}{2} \\ C_L \cdot \dfrac{dv_{CL}}{dt} = i_{L1} - I_{L1} \\ C_H \cdot \dfrac{dv_{CH}}{dt} = i_{L2} - I_{L2} \end{cases} \tag{13}$$

$$\begin{cases} C_{sw} = \dfrac{-1}{2 \cdot \Delta v_{C_{sw}}} \int_0^{t_{on}} (i_{L2} - i_{L1}) dt \\ C_L = \dfrac{1}{\Delta v_{C_L}} \int_{t_{on}/2}^{t_{on}+t_{off}/2} (i_{L1} - I_{L1}) dt \\ C_H = \dfrac{1}{\Delta v_{C_H}} \int_{t_{on}/2}^{t_{on}+t_{off}/2} (i_{L2} - I_{L2}) dt \end{cases} \tag{14}$$

As for the inductors, a ripple voltage percentage on capacitors is considered in (15).

$$\begin{cases} \Delta v_C = \Delta v_{Cp} V_{C_{sw}} \\ \Delta v_{C_L} = \Delta v_{Cp} V_{C_L} \\ \Delta v_{C_H} = \Delta v_{Cp} V_{C_H} \end{cases} \tag{15}$$

Calculating (13) by using (15) and the currents from Fig. 4, the result from (16) is obtained.

$$\begin{cases} C_{sw} = \dfrac{2 \cdot I_L \cdot V_L \cdot (V_H - V_L)}{\Delta v_{Cp} \cdot f \cdot V_H \cdot (V_H + V_L)^2} \\ C_L = \dfrac{\Delta i_{Lp} \cdot I_L}{8 \cdot \Delta v_{Cp} \cdot f \cdot V_L} \\ C_H = \dfrac{\Delta i_{Lp} \cdot I_L \cdot V_L}{8 \cdot \Delta v_{Cp} \cdot f \cdot V_H^2} \end{cases} \tag{16}$$

It is important to note that the values obtained here for the capacitor values will not necessarily be used in the final design because instabilities might occur, and this aspect will be addressed in the following sections. Most importantly the values for the capacitors will be used for comparison to other topologies.

C. Converter comparison

In order to compare the topology to other topologies, the total energy from inductors, total energy from capacitors, and total device switch stress is used as metrics. The total energy from the passive elements is more or less proportional to the cost and volume. The total switch stress provides information about the cost of the switches and the switch losses, therefore the efficiency.

The inductor energy calculated in (17). The total inductor energy is given in (19), considering the previous aspects.

$$\begin{cases} W_{L1} = \dfrac{L_1 \cdot I_{L1}^2}{2} = \dfrac{I_L \cdot V_L \cdot (\frac{V_H}{2} - \frac{V_L}{2})}{\Delta i_{Lp} \cdot f \cdot (V_H + V_L)} \\ W_{L2} = \dfrac{L_2 \cdot I_{L2}^2}{2} = \dfrac{I_L \cdot V_L^2 \cdot (V_H - V_L)}{2 \cdot \Delta i_{Lp} \cdot f \cdot V_H \cdot (V_H + V_L)} \end{cases} \tag{17}$$

$$W_{L_{Tot}} = W_{L1} + W_{L2} = \dfrac{I_L \cdot V_L \cdot (V_H - V_L)}{2 \cdot \Delta i_{Lp} \cdot f \cdot V_H} \tag{18}$$

The capacitor energies are calculated as in (20) and the result is given in (21). The capacitors from the two inputs (C_H and C_L) give the same calculated energy, because of similar parameters and sizing equations.

$$\begin{cases} W_{C_{sw}} = \dfrac{C_{sw} \cdot V_{C_{sw}}^2}{2} \\ W_{C_L} = \dfrac{C_L \cdot V_L^2}{2} \\ W_{C_H} = \dfrac{C_H \cdot V_H^2}{2} \end{cases} \tag{19}$$

$$\begin{cases} W_{C_{sw}} = \dfrac{I_L \cdot V_L \cdot (V_H - V_L)}{4 \cdot \Delta v_{Cp} \cdot f \cdot V_H} \\ W_{C_L} = W_{C_H} = \dfrac{\Delta i_{Lp} \cdot I_L \cdot V_L}{16 \cdot \Delta v_{Cp} \cdot f} \end{cases} \tag{20}$$

The total capacitor energy is calculated as in (22) and the result is given in (23).

$$W_{C_{Tot}} = 2 \cdot W_{C_{sw}} + W_{C_L} + W_{C_H} \tag{21}$$

$$W_{C_{Tot}} = \frac{I_L \cdot V_L \cdot (V_H \cdot (4 + \Delta i_{Lp}) - 4 \cdot V_L)}{4 \cdot \Delta v_{Cp} \cdot f \cdot V_H} \qquad (22)$$

The total switch stress is calculated as the product between the maximum voltage and maximum current on each switching device, and is used as metrics for comparison between different topologies, and is calculated as in (24).

$$S_{Total} = \sum_{j=1}^{5} V_{Sj} I_{Sj} \qquad (23)$$

To calculate the total switch stress, the maximum voltage and currents on the switching devices are expressed in (25). The result of the calculation are given in (26).

$$\begin{cases} V_{S1} = V_H + V_L \\ V_{S2} = V_{S3} = V_{S4} = \dfrac{V_H + V_L}{2} \end{cases} \begin{cases} I_{S1} = I_{S2} = I_L \\ I_{S3} = I_{S5} = \dfrac{I_H - I_L}{2} \\ I_{S4} = I_H \end{cases} \quad (24)$$

$$S_{Total} = \frac{I_L (V_H + V_L)^2}{V_H} \qquad (25)$$

In order to have a better understanding of the resulting calculations from (18), (22) and (25), and in order to compare the proposed topology to the quadratic converter from [17] and the conventional buck/boost, the relations are divided with the corresponding results from the conventional converter, and graphically represented in Fig. 5. The voltages were chosen to be V_H=400V, V_L ranging from 20V to 100V, and $\Delta i_{Lp} = 20\%$ which is common for inductors, or other optimal values can be used [18]. From these results it can be observed that the BHSC converter achieves a better conversion ratio than the conventional converter using the same passive components, and that it has a lower total stress on the active switches, even if it has a larger number of switches. Also, over a wide operating range the BHSC needs smaller passive and active components than the quadratic converter.

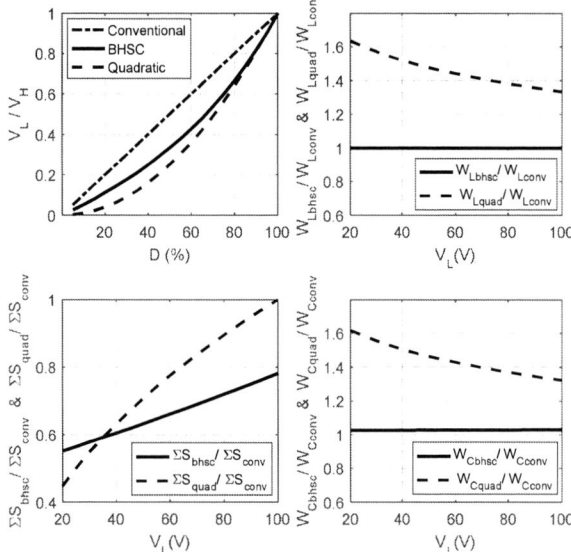

Fig. 5. BHSC comparison with the conventional bidirectional Buck/Boost converter [19] and a Quadratic converter [17] (V_H=400V).

The quadratic converter has the best conversion ratio, followed by the BHSC and then the conventional converter, but it needs larger inductors, capacitors and more expensive switches in order to achieve this characteristic.

IV. STABILITY ANALYSIS

Because this topology has a larger number of passive components, therefore is a higher order system, it is important to analyze the stability by mathematical and simulation means, prior to the final design stage. The capacitor values can be calculated as in (16), but as it is seen in [20] and [11], some converters might behave as a non-minimum-phase system or their control performance might be improved by appropriate hardware design.

In order to make the stability analysis the SSA method is used. To obtain the mathematical model, the schematic from Fig. 6, including the parasitic resistors and input capacitors (C_H, C_L), is used. The two equivalent switching schematics are presented in Fig. 7 and Fig. 8, for the t_{on} and t_{off} periods respectively. Considering the state (x) and the input (u) vectors as in (28), the state space representation for the two switching periods can be expressed as in (27) (where $i = 1$ for t_{on} or 2 for t_{off}). The A_i and B_i matrixes are defined in (29), and their elements from (30) to (46).

$$\dot{x} = A_i \cdot x + B_i \cdot u \qquad (26)$$

$$x^T = \begin{bmatrix} i_{L_1} & i_{L_2} & v_{C_{sw}} & v_{C_L} & v_{C_H} \end{bmatrix} \quad u = \begin{bmatrix} V_L \\ V_H \end{bmatrix} \quad (27)$$

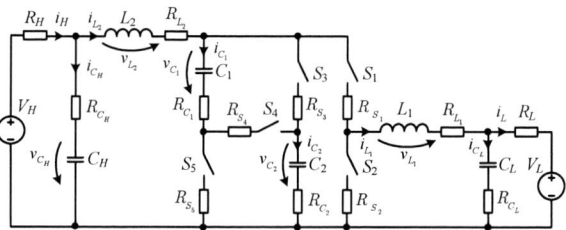

Fig. 6. BHSC schematic with parasitic components.

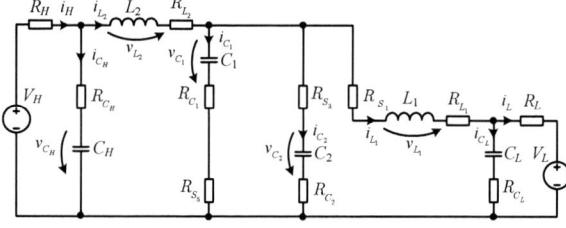

Fig. 7. BHSC schematic with parasitic components during t_{on}.

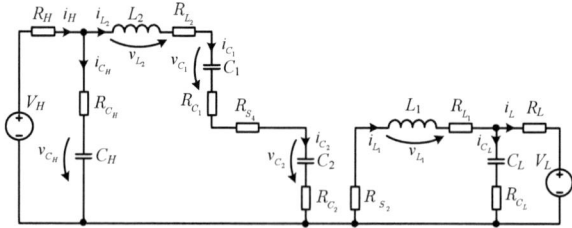

Fig. 8. BHSC schematic with parasitic components during t_{off}.

$$A_i = \begin{bmatrix} a_{11_i} & a_{12_i} & a_{13_i} & a_{14_i} & 0 \\ a_{21_i} & a_{22_i} & a_{23_i} & 0 & a_{25_i} \\ a_{31_i} & a_{32_i} & 0 & 0 & 0 \\ a_{41_i} & 0 & 0 & a_{44_i} & 0 \\ 0 & a_{52_i} & 0 & 0 & a_{55_i} \end{bmatrix} \quad B_i = \begin{bmatrix} b_{11i} & 0 \\ 0 & b_{22i} \\ 0 & 0 \\ b_{41i} & 0 \\ 0 & b_{52i} \end{bmatrix} \quad (28)$$

$$a_{11_1} = -\left(\frac{\frac{R_{C_{sw}}}{2} + R_{L_1} + R_{S_1} + \frac{R_{S_3}}{2} + \frac{R_{C_L} \cdot R_L}{R_{C_L} + R_L}}{L_1} \right) \quad (29)$$

$$a_{12_1} = \frac{R_{C_{sw}} + R_{S_3}}{2 \cdot L_1} ; \quad a_{13_1} = \frac{1}{L_1} ; \quad a_{14_1} = \frac{-R_L}{L_1 \cdot (R_{C_L} + R_L)} \quad (30)$$

$$a_{21_1} = \frac{R_{C_{sw}} + R_{S_3}}{2 \cdot L_2} ; \quad a_{23_1} = -\frac{1}{L_2} ; \quad a_{25_1} = \frac{R_H}{L_2 \cdot (R_{C_H} + R_H)} \quad (31)$$

$$a_{22_1} = -\left(\frac{\frac{R_{C_{sw}}}{2} + R_{L_2} + \frac{R_{S_3}}{2} + \frac{R_{C_H} \cdot R_H}{R_{C_H} + R_H}}{L_2} \right) \quad (32)$$

$$a_{31_1} = -\frac{1}{2 \cdot C_{sw}} ; \quad a_{32_1} = -a_{31_1} ; \quad a_{41_1} = \frac{R_L}{C_L \cdot (R_{C_L} + R_L)} \quad (33)$$

$$a_{44_1} = \frac{-1}{C_L \cdot (R_{C_L} + R_L)} ; \quad a_{52_1} = \frac{-R_H}{C_H \cdot (R_{C_H} + R_H)} \quad (34)$$

$$a_{55_1} = \frac{-1}{C_H \cdot (R_{C_H} + R_H)} \quad (35)$$

$$b_{11_1} = \frac{-R_{C_L}}{L_1 \cdot (R_{C_L} + R_L)} ; \quad b_{22_1} = \frac{R_{C_H}}{L_2 \cdot (R_{C_H} + R_H)} \quad (36)$$

$$b_{41_1} = \frac{1}{C_L \cdot (R_{C_L} + R_L)} ; \quad b_{52_1} = \frac{1}{C_H \cdot (R_{C_H} + R_H)} \quad (37)$$

$$a_{11_2} = -\left(\frac{R_{L_1} + R_{S_2} + \frac{R_{C_L} \cdot R_L}{R_{C_L} + R_L}}{L_1} \right) ; \quad a_{12_2} = 0 ; \quad a_{13_2} = 0 \quad (38)$$

$$a_{14_2} = \frac{-R_L}{L_1 \cdot (R_{C_L} + R_L)} ; \quad a_{21_2} = 0 ; \quad a_{23_2} = -\frac{2}{L_2} \quad (39)$$

$$a_{22_2} = -\left(\frac{2 \cdot R_{C_{sw}} + R_{L_2} + R_{S_4} + \frac{R_{C_H} \cdot R_H}{R_{C_H} + R_H}}{L_2} \right) \quad (40)$$

$$a_{25_2} = \frac{R_H}{L_2 \cdot (R_{C_H} + R_H)} ; \quad a_{31_2} = 0 ; \quad a_{32_2} = \frac{1}{C_{sw}} \quad (41)$$

$$a_{41_2} = \frac{R_L}{C_L \cdot (R_{C_L} + R_L)} ; \quad a_{44_2} = \frac{-1}{C_L \cdot (R_{C_L} + R_L)} \quad (42)$$

$$a_{52_2} = \frac{-R_H}{C_H \cdot (R_{C_H} + R_H)} ; \quad a_{55_2} = \frac{-1}{C_H \cdot (R_{C_H} + R_H)} \quad (43)$$

$$b_{11_2} = \frac{-R_{C_L}}{L_1 \cdot (R_{C_L} + R_L)} ; \quad b_{22_2} = \frac{R_{C_H}}{L_2 \cdot (R_{C_H} + R_H)} \quad (44)$$

$$b_{41_2} = \frac{1}{C_L \cdot (R_{C_L} + R_L)} ; \quad b_{52_2} = \frac{1}{C_H \cdot (R_{C_H} + R_H)} \quad (45)$$

Taking a variable duty cycle into account (d), the average model of the converter can be described as in (47).

$$\dot{x} = (A_1 \cdot d + A_2 \cdot (1-d)) \cdot x + (B_1 \cdot d + B_2 \cdot (1-d)) \cdot u \quad (46)$$

Linearizing the system around a steady state duty cycle D for a small signal variation $\tilde{d}$, the system from (48) is obtained, with the equivalent matrixes described in (49). The C_e matrix can be either [1 0 0 0 0] or [0 1 0 0 0], in order to express the desired transfer function for one of the two inductor currents in (50), which are needed to control the bidirectional converter.

$$\begin{cases} \dot{\tilde{x}} = A_e \cdot \tilde{x} + B_e \cdot \tilde{d} \\ \tilde{y} = C_e \cdot \tilde{x} \end{cases} \quad (47)$$

$$\begin{aligned} A_e &= (A_1 \cdot D + A_2 \cdot (1-D)) \\ B_e &= [(A_1 - A_2) \cdot X + (B_1 - B_2) \cdot u] \end{aligned} \quad (48)$$

$$H_1(s) = \frac{\tilde{i}_{L_1}(s)}{\tilde{d}(s)} ; \quad H_2(s) = \frac{\tilde{i}_{L_2}(s)}{\tilde{d}(s)} \quad (49)$$

The above mentioned transfer functions can be obtained by calculating (51).

$$\tilde{y} = C_e \cdot (s \cdot I - A_e)^{-1} \cdot B_e \cdot \tilde{d} \quad (50)$$

Initial design parameters and initial results are set in TABLE I.

TABLE I. INITIAL DESIGN PARAMETERS

Element	Design values		Element	Calculated values	
	Value	Unit		Value	Unit
V_H	400	V	L_1	75	μH
V_L	100	V	L_2	300	μH
I_L	50	A	C_{sw}	18.7	μF
Δi_{Lp}	20	%	C_L	7.8	μF
Δv_{Cp}	2	%	C_H	0.48	μF
f_{sw}	80	kHz	D	40	%

Unfortunately the stability of the converter is affected if the capacitors from the previous table are used because of the low ESR ($\approx 1..3$ mΩ) of the C_{sw}, which introduce right half-plane zero (around $1350.153 \pm 14782.163i$) in $H_1(s)$. This case is also studied in [20], and in order to avoid it, electrolytic capacitors are chosen for the design, with the final values from TABLE II.

TABLE II. SELECTED COMPONENTS (CASE A)

Part	Specifications					
	Value	Unit	ESR	ESR value	Unit	Component
L_1	100	μH	R_{L_1}	20	mΩ	2x DEHF-42/0,047/50
L_2	400	μH	R_{L_2}	53	mΩ	DEHF-42/0,47/16
C_{sw}	20	mF	$R_{C_{sw}}$	15.4	mΩ	ALC70(1)202EL3 50
C_L	10	mF	R_{C_L}	34	mΩ	ALC70103EH100
C_H	0.22	mF	R_{C_H}	328	mΩ	ALC70221BD400

Taking the values from the previous table, the transfer functions can be calculated, and the new poles and zeros of the system can be extracted in TABLE III. In order to further reduce the oscillations that might appear the introduction of an additional resistor is considered, so that $R_{C_{SW}} = 0.12\Omega$ (for Case B), and the new poles and zeros are presented in TABLE IV. The bode diagram of the two transfer functions, $H_1(s)$ and $H_2(s)$, for the two cases, are presented in Fig. 9 and Fig. 10 respectively. It can be observed from the new values of the poles that the converter will have smaller oscillations with an increased ESR, but this will also increase significantly the losses (approximately 1% more losses in the ESR of C_{sw}). Because of this the controller is designed with the capacitors from the case A, and a low-pass filter is used for the reference current (f_c=100 Hz), as implemented in [11]. The current through L_1 inductor is used for control, as it has higher current values and a higher dynamic.

The chosen current controller is a PI controller, and its transfer function is presented in (52). The frequency response of $C(s) \cdot H_1(s)$ from Fig. 11 shows good stability using this controller, obtaining a phase margin of 87.2 degrees.

$$C(s) = \frac{0.02754 \cdot (s + 4075)}{s} \qquad (51)$$

TABLE III. SYSTEM WITH COMPONENTS A.

Element	Pole and Zero values
	Pole and Zero values for set A of capacitors
$H_1(s)$ Poles	-12872.11578, -2240.59207, -194.02532 ± 579.94806i, -622.67995
$H_1(s)$ Zeros	-12872.10936, -2272.72727, -163.15708 ± 608.53028i
$H_2(s)$ Poles	-12872.11578, -2240.59207, -194.02532 ± 579.94806i, -622.67995
$H_2(s)$ Zeros	-12876.64177, -2241.63320, -407.04398 ± 511.63483i

TABLE IV. SYSTEM WITH COMPONENTS B.

Element	Pole and Zero values
	Pole and Zero values for 1st set of capacitors
$H_1(s)$ Poles	-12871.97709, -2235.19829, -390.80492 ± 457.04867i, -829.11094
$H_1(s)$ Zeros	-12871.94915, -2272.72727, -378.86668 ± 504.53682i
$H_2(s)$ Poles	-12871.97709, -2235.19829, -829.11094, -390.80492 ± 457.04867i
$H_2(s)$ Zeros	-12876.64177, -2230.04382, -628.40118 ± 167.60799i

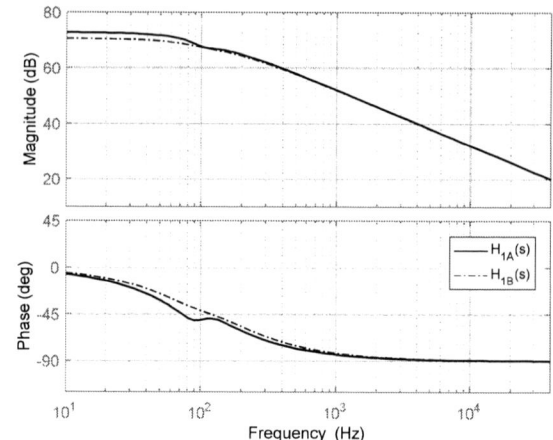

Fig. 9. Bode diagram for the transfer functions of $H_1(s)$ for the two capacitor sets: A and B.

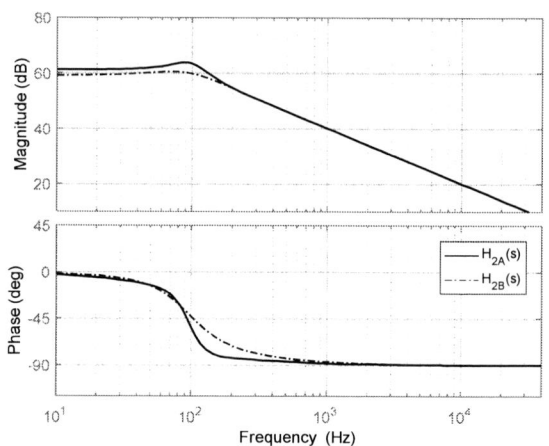

Fig. 10. Bode diagram for the transfer functions of $H_2(s)$ for the two capacitor sets: A and B.

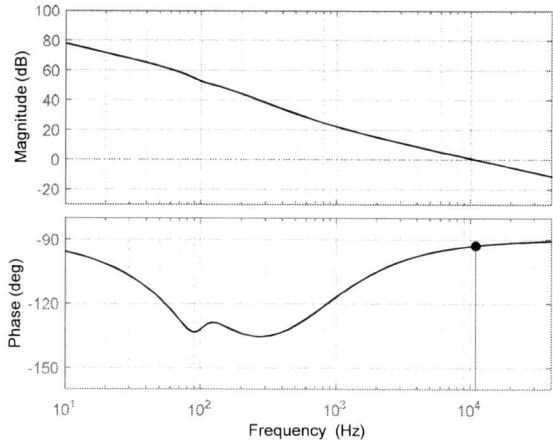

Fig. 11. Frequency response of the converter with the designed controller $C(s) \cdot H_1(s)$ (PM = 87.2 deg, f_t = 11 kHz).

V. SIMULATION RESULTS

The results, obtained from PSIM software simulations, show a good operation of the converter under various conditions.

For the first case, with the capacitors from A, the results are shown in Fig. 12 and Fig. 13. First, a current reference of ±50A is applied to test a large step variation. The I_{L_1} current has no overshoot and the I_{L_2} current has little oscillations.

Small oscillations are also present on the switched capacitors. On the second figure a ±10A reference is applied, and here it can be observed that the oscillations are largely reduced.

In the second case, with the additional ESR (B), the same conditions are tested. In Fig. 14 a reference of ±50A, is applied, and here it can be observed that the overshoot in the I_{L_2} current is much lower, and the settling time for the oscillations is shorter. The same are correct for the ±10A reference. The reduced oscillations are also present in this case. Another important aspect to mention is the increased voltage ripple on the capacitors due to the increased ESR.

A close-up on the steady state waveforms is also shown in Fig. 16, and it shows a good correspondence with the expected results.

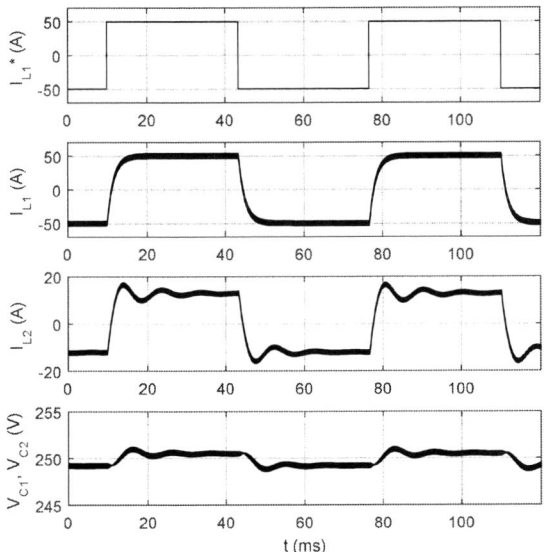

Fig. 12. Simulation results for $I_{L_1}* = \pm 50A$, V_H=400V, V_L=100V for Case A capacitors.

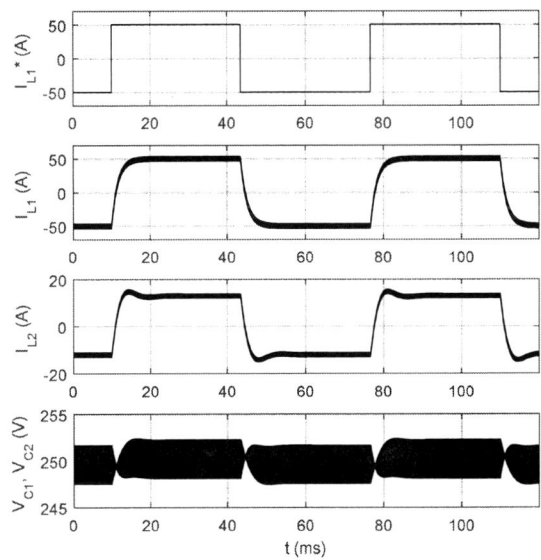

Fig. 14. Simulation results for $I_{L_1}* = \pm 50A$, V_H=400V, V_L=100V for Case B capacitors.

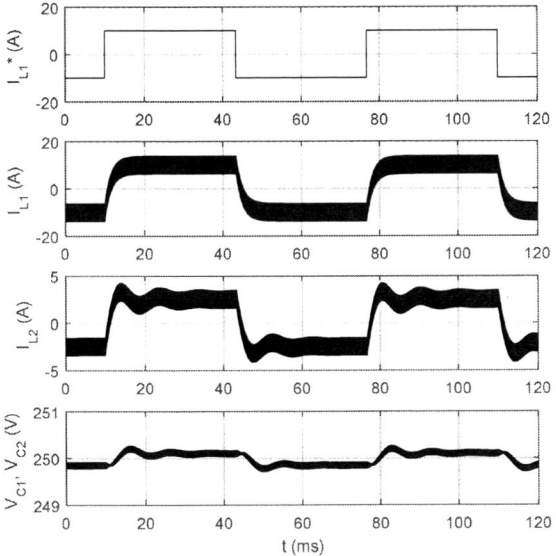

Fig. 13. Simulation results for $I_{L_1}* = \pm 10A$, V_H=400V, V_L=100V for Case A capacitors.

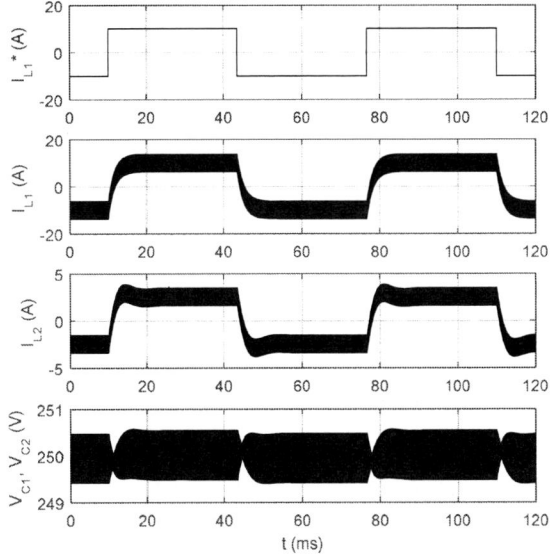

Fig. 15. Simulation results for $I_{L_1}* = \pm 10A$, V_H=400V, V_L=100V for Case B capacitors.

978-1-5386-7688-2/19 $31.00 © 2019 IEEE

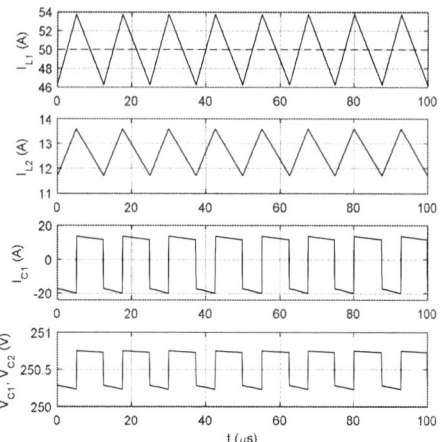

Fig. 16. Close up on the simulation waveforms, $I_{L_1}^* = 50A$, V_H=400V, V_L=100V for Case A capacitors.

VI. CONCLUSIONS

This paper presented a new bidirectional DC-DC converter topology, which can achieve a high conversion ratio with the help of a switched capacitor cell. Apart from the increased conversion ratio, the benefits of this topology also include smaller passive components, lower stress on the active devices, lower ripple current at the inputs and a common ground between the two ports. The stability of the converter was analyzed with emphasis on the design of the passive components, in order to achieve a stable design prior to the actual construction of the converter. The simulation results confirm the performances and the stability of the proposed converter.

ACKNOWLEDGMENT

This work was supported by a grant of the Romanian Ministry of Research and Innovation, CCCDI – UEFISCDI, project number PN-III-P1-1.2-PCCDI-2017- 0391 / CIA_CLIM – *"Smart buildings adaptable to the climate change effects"*, within PNCDI III.

This work was also supported by a grant of the Romanian Ministry of Research and Innovation, project number 10PFE/16.10.2018, PERFORM-TECH-UPT - *The increasing of the institutional performance of the Polytechnic University of Timişoara by strengthening the research, development and technological transfer capacity in the field of "Energy, Environment and Climate Change"*, within Program 1 - Development of the national system of Research and Development, Subprogram 1.2 - Institutional Performance - Institutional Development Projects - Excellence Funding Projects in RDI, PNCDI III.

REFERENCES

[1] Y. Karimi, H. Oraee, M. S. Golsorkhi, and J. M. Guerrero, "Decentralized Method for Load Sharing and Power Management in a PV/Battery Hybrid Source Islanded Microgrid," *IEEE Transactions on Power Electronics*, vol. 32, no. 5, pp. 3525–3535, May 2017.

[2] X. Zhao, X. Wu, Y. Li, and H. Tian, "Energy management strategy of multiple supercapacitors in an autonomous DC microgrid using adaptive virtual impedance," in *2016 IEEE 7th International Symposium on Power Electronics for Distributed Generation Systems (PEDG)*, 2016, pp. 1–8.

[3] Y. Zhang and Y. Li, "Energy management strategy for supercapacitor in autonomous DC microgrid using virtual impedance," in *2015 IEEE*

Applied Power Electronics Conference and Exposition (APEC), 2015, pp. 725–730.

[4] A. S. Morais and L. A. C. Lopes, "Interlink Converters in DC nanogrids and its effect in power sharing using distributed control," in *2016 IEEE 7th International Symposium on Power Electronics for Distributed Generation Systems (PEDG)*, 2016, pp. 1–7.

[5] J. Ma, M. Zhu, X. Cai, and Y. W. Li, "Configuration and operation of DC microgrid cluster linked through DC-DC converter," in *2016 IEEE 11th Conference on Industrial Electronics and Applications (ICIEA)*, 2016, pp. 2565–2570.

[6] B. Lee, J. Kim, S. Kim, and J. Lee, "An Isolated/Bidirectional PWM Resonant Converter for V2G(H) EV On-Board Charger," *IEEE Transactions on Vehicular Technology*, vol. 66, no. 9, pp. 7741–7750, Sep. 2017.

[7] P. He and A. Khaligh, "Comprehensive Analyses and Comparison of 1 kW Isolated DC–DC Converters for Bidirectional EV Charging Systems," *IEEE Transactions on Transportation Electrification*, vol. 3, no. 1, pp. 147–156, Mar. 2017.

[8] S. Dusmez, A. Hasanzadeh, and A. Khaligh, "Loss analysis of non-isolated bidirectional DC/DC converters for hybrid energy storage system in EVs," in *2014 IEEE 23rd International Symposium on Industrial Electronics (ISIE)*, 2014, pp. 543–549.

[9] Y. Zhang, Y. Gao, L. Zhou, and M. Sumner, "A Switched-Capacitor Bidirectional DC–DC Converter With Wide Voltage Gain Range for Electric Vehicles With Hybrid Energy Sources," *IEEE Transactions on Power Electronics*, vol. 33, no. 11, pp. 9459–9469, Nov. 2018.

[10] D. Hulea, B. Fahimi, N. Muntean, and O. Cornea, "High Ratio Bidirectional Hybrid Switched Inductor Converter Using Wide Bandgap Transistors," in *2018 20th European Conference on Power Electronics and Applications (EPE'18 ECCE Europe)*, 2018, p. P.1-P.10.

[11] O. Cornea, G. Andreescu, N. Muntean, and D. Hulea, "Bidirectional Power Flow Control in a DC Microgrid Through a Switched-Capacitor Cell Hybrid DC–DC Converter," *IEEE Transactions on Industrial Electronics*, vol. 64, no. 4, pp. 3012–3022, Apr. 2017.

[12] F. S. Garcia, J. A. Pomilio, and G. Spiazzi, "Comparison of non-insulated, high-gain, high-power, step-up DC-DC converters," in *2012 Twenty-Seventh Annual IEEE Applied Power Electronics Conference and Exposition (APEC)*, 2012, pp. 1343–1347.

[13] B. Axelrod, Y. Berkovich, and A. Ioinovici, "Switched-Capacitor/Switched-Inductor Structures for Getting Transformerless Hybrid DC–DC PWM Converters," *IEEE Transactions on Circuits and Systems I: Regular Papers*, vol. 55, no. 2, pp. 687–696, Mar. 2008.

[14] N. Muntean, O. Cornea, O. Pelan, and C. Lascu, "Comparative evaluation of buck and hybrid buck DC-DC converters for automotive applications," in *2012 15th International Power Electronics and Motion Control Conference (EPE/PEMC)*, 2012, p. DS2b.3-1-DS2b.3-6.

[15] S. Li, K. M. Smedley, D. R. Caldas, and Y. W. Martins, "A hybrid bidirectional DC-DC converter for dual-voltage automotive systems," in *2017 IEEE Applied Power Electronics Conference and Exposition (APEC)*, 2017, pp. 355–361.

[16] P. S. Tomar, A. K. Sharma, and K. Hada, "Energy storage in DC microgrid system using non-isolated bidirectional soft-switching DC/DC converter," in *2017 6th International Conference on Computer Applications In Electrical Engineering-Recent Advances (CERA)*, 2017, pp. 439–444.

[17] H. Ardi, A. Ajami, F. Kardan, and S. N. Avilagh, "Analysis and Implementation of a Nonisolated Bidirectional DC–DC Converter With High Voltage Gain," *IEEE Transactions on Industrial Electronics*, vol. 63, no. 8, pp. 4878–4888, Aug. 2016.

[18] He Li-gao and Wu Jian, "Selection of the current ripple ratio of converters and optimal design of output inductor," in *2010 5th IEEE Conference on Industrial Electronics and Applications*, 2010, pp. 1163–1167.

[19] K. Tytelmaier, O. Husev, O. Veligorskyi, and R. Yershov, "A review of non-isolated bidirectional dc-dc converters for energy storage systems," in *2016 II International Young Scientists Forum on Applied Physics and Engineering (YSF)*, 2016, pp. 22–28.

[20] Y. Zhang, J. Liu, and X. Ma, "Using RC type damping to eliminate right-half-plane zeros in high step-up DC-DC converter with diode-capacitor network," in *2013 IEEE ECCE Asia Downunder*, 2013, pp. 59–65.

978-1-5386-7688-2/19 $31.00 © 2019 IEEE

A Comparative Study of Capacitive and Inductive Pulsed Power Supply Topologies for Electromagnetic Launcher Applications

Doğa Ceylan
Middle East Technical University
Ankara, Turkey
doga.ceylan@metu.edu.tr

Siamak Pourkeivannour
Middle East Technical University
Ankara, Turkey
siamak.pourkeivannour@metu.edu.tr

Ozan Keysan
Middle East Technical University
Ankara, Turkey
keysan@metu.edu.tr

Abstract—Inductive and capacitive types are the most common pulsed power supply (PPS) topologies. In this paper, the comparison of inductive XRAM generator and capacitor-based (C-based) generator topologies is discussed for the excitation of an electromagnetic launcher (EML). In addition, the effect of capacitance or inductance of the storage element on the load current and laucher efficiency is investigated. The circuit simulation results of these PPS topologies are presented, each of which having 200 kJ PPS energy. The EML used in the study has 0.1 kg total mass of projectile, 3 m long rail. Although the energy density of the XRAM generators is larger than C-based PPSs, the design of an XRAM generator is more challenging than C-based PPS due to the large voltage drop of its opening switches. Moreover, the efficiency of the total system is highly dependent on the design of the storage element. For the XRAM generator, the efficiency is limited by the capability of the opening switches. In this study, using RC snubber circuit, the voltage stress on the GTO (gate turn-off) thyristor opening switches of the XRAM generator is decreased to 2 kV peak voltage, which is available in the market.

Index Terms—Pulsed power generation, XRAM generator, C-based pulsed power supply, electromagnetic launcher, opening switch

I. INTRODUCTION

Electromagnetic launcher (EML), also known as railgun, is an electromagnetic device that converts electrical energy into mechanical energy. It consists of two parallel conducting rails, a conducting armature and a non-conducting projectile as shown in Fig. 1.

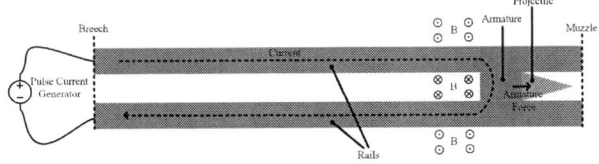

Fig. 1. Working principle of an electromagnetic launcher.

The authors are with the Department of Electrical and Electronics Engineering, Middle East Technical University, Ankara, Turkey (e-mail: doga.ceylan@metu.edu.tr; siamak.pourkeivannour@metu.edu.tr; keysan@metu.edu.tr).

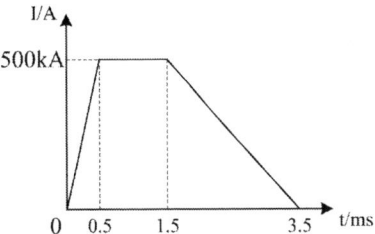

Fig. 2. Typical pulse shaped current waveform, [1].

Pulse shaped current is generated by pulsed power supply (PPS) to excite the rails. This current flows through the rails and armature. Electric current creates a magnetic field which causes an electromagnetic force on the armature due to Lorentz force, which accelerates the armature and projectile together. In order to increase the electromagnetic force acting on the armature, large amount of current generation is required in the PPS side. However, the excitation current should not be large when the armature leaves the barrel to reduce the arc due to high dI/dt in the muzzle side. Therefore, a pulse shaped current excitation is used in EMLs as in Fig. 2, [1]. The design of PPS units is critical to build an efficient EML. In the literature, there exist three ways to generate pulse shaped current in the electromagnetic launch systems: inductive pulsed power supplies, pulsed alternators, and capacitive pulsed power supplies.

Pulsed alternators use the stored inertial energy of an electrical machine to generate electrical pulse power as explained in [2]. In order to provide the desired mechanical stored energy, rotor of the alternator is accelerated by a motor which is attached to the alternator shaft. However, the required alternator size increases to store enough inertial energy in the rotor to get a large muzzle energy, and the magnitude of the field current of the pulsed alternators is limited by the excitation losses as explained in [3]. Therefore, the capacitor-based (C-based) and inductive PPS topologies are used more than pulsed alternators for the excitation of EMLs.

The most favorable way to generate pulsed shape current is the C-based PPS because of its ability to deliver the large amount of power pulse [4]. In C-based PPS, the power capacitors with initial voltage of several kVs discharges through pulse shaping inductors and the launcher. During the capacitor discharging period, the launcher current rises rapidly. After the capacitors are fully discharged, the charged pulse shaping inductors discharge through the free-wheeling diodes and the launcher which decreases the launcher current. In [5], a C-based pulsed power system is analyzed in terms of its efficiency. Moreover, Gou *et al.* work on the electric parameters of the C-based PPS circuit topology in [6]. Although it is possible to reach several MJs stored electrical energy using C-based PPS topology, this topology suffers from its small energy densities in comparison with inductive energy storage systems as explained in [7]. Since the electromagnetic launching is a military application, the size and weight of its PPS are significant. Inductive energy storage systems reach one order of magnitude higher energy densities in comparison with capacitive storage systems as explained in [8]. In addition, XRAM generator is one of the most favorable inductive PPS topologies as explained in [9]. In XRAM PPS topology, the inductors, connected in series, are charged by a DC source to store the electrical energy. Then, the stored energy in the inductors is injected to the launcher through a parallel configuration. In this paper, a comparative study of C-based and XRAM PPS topologies is addressed for electromagnetic launcher applications. These two topologies are investigated in the aspects of rail current waveform, muzzle velocity of the projectile, efficiency, and electrical characteristics of the circuit elements.

In the following sections, working principles of the pre-mentioned topologies are detailed. Then, the analytical modeling of the launcher to be used in the simulations is explained. Finally, simulation results of two PPS topologies are discussed.

II. PPS Topologies Used in the Study

In this section, C-based PPS and XRAM generator circuit topologies are explained in detail.

A. C-based PPS

The working principle of C-based PPS is very similar to a conventional buck converter. In Fig. 3, parallel connected n-stage C-based PPS topology is presented.

In each module, the capacitor is charged to an initial voltage through an external DC source to reach the required initial electrical energy. The total initial electrical energy of the PPS can be calculated using (1) where C is the capacitance of the capacitors, V_c is the initial voltage of the capacitors.

$$E_{initial} = n\frac{1}{2}CV_c^2 \tag{1}$$

After one of the switches is closed, the capacitor starts discharging through the pulse shaping inductor and the rails and the armature of the launcher. During the capacitor discharging process, the terminal current of the launcher increases rapidly. When the capacitor is fully discharged, free wheeling

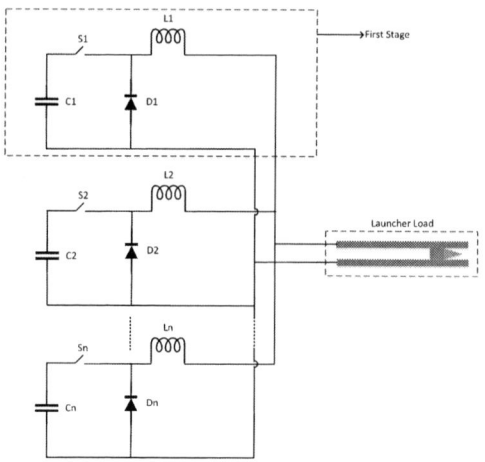

Fig. 3. N-stage C-based PPS topology.

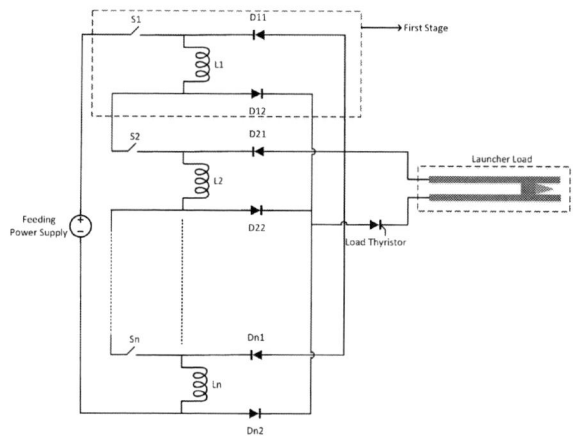

Fig. 4. N-stage XRAM generator topology.

diode starts conducting. Hence, the pulse shaping inductor starts to discharge through the launcher and diode. During the inductor discharging process, the terminal current of the launcher decreases. Each module of the whole PPS unit has its own switching instants to generate the desired rail current waveform as investigated in [10].

B. XRAM PPS

The concept of XRAM generator was proposed by W. Koch at Marx' High Voltage Institute at Brunswick Technical University [11]. The name of XRAM generator originates from the well-known Marx generator by reversing its letters. While the Marx generators are used to generate pulse shaped high voltages by charging and discharging the capacitors, the goal of the XRAM generators is to create pulse shaped current waveforms with large peak values as explained in [7]. In the

Fig. 5. Photograph of the designed 20-stage toroidal XRAM generator in ISL, [12].

XRAM generators, the initial electrical energy is stored in the power inductors, instead of capacitors as in C-based PPS. The circuit topology of an n-stage XRAM generator is given in Fig. 4. When the switches, in Fig. 4, are closed and the load thyristor is not triggered, feeding power supply charges the series connected inductors. The initial energy of the inductors can be found using (2) where L is the inductance of each inductor, I_L is the initial current of the inductors. When inductors are charged up to the required initial energy, switches are opened and the load thyristor triggered simultaneously. Then, diodes start conducting and inductors discharge through the launcher.

$$E_{initial} = n\frac{1}{2}LI_L^2 \qquad (2)$$

The most important advantage of the XRAM generators with respect to the C-based PPS topology is their small size and mass which provide to develop fieldable electromagnetic launch systems. One of the most compact XRAM generators with the diameter of 100 cm, and the height of 25 cm is developed in GermanFrench Research Institute of Saint Louis (ISL) and presented in [12]. The designed XRAM generator given in Fig. 5 has the energy storage capability of 0.5 MJ and energy storage density of 2.54 MJ/m³, while C-based PPS with the same amount of stored energy has 1 MJ/m³ energy storage density as presented in [13]. Moreover, according to Liebfried's review study in [14], the maximum inductively stored energy recorded in the literature using XRAM generator is 12.5 MJ. However, for that application, the load was an X-ray system instead of a railgun. For the electromagnetic launcher load, while the maximum stored energy in an XRAM generator is 200 kJ, the load current can reach 40 kA with using a four-stage toroidal XRAM generator. Therefore, it can be concluded that although XRAM generators are suitable only for small caliber launchers, it is possible to decrease the size and mass of the PPS of the launcher using XRAM generator topology. Moreover, the difficulty of using XRAM genetors

in large caliber railguns comes from the need of opening of switches in Fig. 4 for the commutation of high currents. When these swithces are opened to inject the stored energy to the launcher, commutation voltages in kilovolt range occur due to the wire and load impedance as explained in [12]. Although the opening switches are required to block that voltage spike only for a few hundert microseconds, commutation voltage may exceed the voltage tolerance of the opening switches for the railgun loads with relatively large impedances (i.e. large caliber railguns). In [15], using gate turn-off (GTO) thyristors with snubber circuits for the opening switches is recommended to decrease the peak of the voltage spike. Also, more effective and complex countercurrent-thyristor package called Inverse Current Commutation with Semiconductor (ICCOS) is proposed in [16]. ICCOS circuitry is able to commutate maximum current of 28 kA from the inductive storage coil into the load. Also, in [16], it is recommended to use fast-switching devices rather than ultrahigh-power-handling capability for the opening thyristors to decrease the switching losses. Specifically, ABB fast-switching thyristors (5STF09F1620) with a pulse-current capability of 16 kA during 10 ms, a reverse recovery time of 20 μs and 1.6 kV forward and reverse blocking voltages are used as opening switches. Therefore, increasing the initial current of the energy storage inductors more than a certain level is not possible due to the limitations of the opening switch technology. Increase the rail current may be possible with increasing the number of the stages of the XRAM generator. However, in this case another challenge appears about the load thyristor, since it is required to carry the whole rail current during the commutation. In order to overcome the problem of on-state current capability of the load thyristor, the topology of XRAM generator is selected as in Fig. 6 instead of the one presented in Fig. 4. Although the number of load thyristors is increased with this solution, the peak value of on-state current of the load thyristor is also decreased. Therefore, it is possible to increase the rail current by increasing the number of stages without any negative effect on neither opening swithces nor load thyristors.

In the next sections, the analytical modeling of the launcher and the designed PPS circuits is explained with the simulation results.

III. LAUNCHER MODELING

An electromagnetic launcher is used as the load of the PPS topologies explained in the previus section. The launcher can be modeled as series connected resistor and inductor as explained in [17]. The inductance and resistance of the launcher is dependent of armature velocity and position as in Fig. 7. Therefore, they change during the launch process.

Velocity and position dependent resistance of the launcher is given as (3) in [18]. In (3), v is the armature velocity, x is the armature position, R is the position dependent DC resistance of the launcher, L is the position dependent inductance of the launcher and R_{vc} is velocity skin effect resistance constant which is dependent of rail and armature material.

978-1-5386-7688-2/19 $31.00 © 2019 IEEE

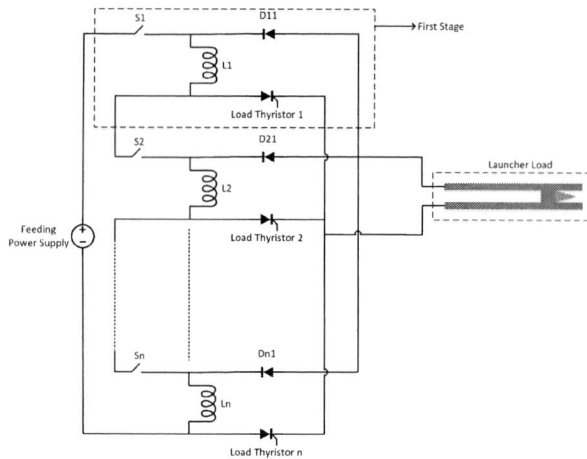

Fig. 6. N-stage XRAM generator topology used in the study.

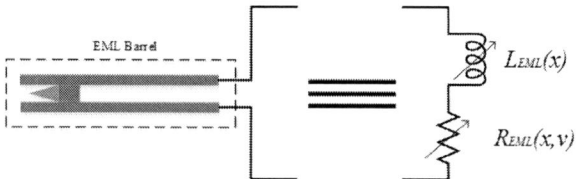

Fig. 7. Equivalent circuit model of launcher.

$$R_{EML} = \frac{\partial R}{\partial x}x + \frac{\partial L}{\partial x}v + R_{vc}v^{3/2} \qquad (3)$$

Inductance of the launcher can be calculated as in (4) where $\frac{\partial L}{\partial x}$ is the change of inductance with respect to the armature position also called inductance gradient which can be calculated using numerical methods [19] and x is the position of the armature.

$$L_{EML} = \frac{\partial L}{\partial x}x \qquad (4)$$

Inductance gradient of a launcher is only dependent of the launcher geometry. It influences the electromagnetic Lorentz force acting on the armature as in (5) where F_{arm} is the Lorentz force, I is the pulsed shaped rail current.

$$F_{arm} = \frac{1}{2}\frac{\partial L}{\partial x}I^2 \qquad (5)$$

Acceleration of the armature together with the projectile can be calculated using the electromagnetic force acting on the armature and the total mass of armature and projectile as in (6). Moreover, armature velocity and position can be calculated using (7) and (8) where t_{exit} is the required time for the armature to reach the muzzle.

$$a = \frac{F_{arm}}{m_{total}} \qquad (6)$$

$$v = \int_0^{t_{exit}} a\, dt \qquad (7)$$

$$x = \int_0^{t_{exit}} v\, dt \qquad (8)$$

The pre-mentioned PPS topologies are simulated with explained analytical model of the launcher using MATLAB Simulink software. In the next section, the results of the developed simulation model is discussed.

IV. SIMULATION RESULTS

The characteristics of the launcher, fed by C-based and XRAM generator PPSs, is given in Table I. In Table I, l, x_0, and m_{total} are determined according to the design of the launcher. Also, L' is calculated for the designed launcher geometry using the finite element method. Moreover, R_{vc} and R' are selected from [18] for a conventional launcher with aluminum armature and copper rails.

The circuit parameters of two C-based PPSs with different capacitor parameters but the same initial stored energies are given in Table II. In addition to the comparison of C-based PPS and XRAM generator, another goal of this study is to investigate the effects of the design or selection of the energy storage capacitors and inductors of these topologies with the same initial energy. Capacitance and initial voltage of the capacitors are determined to achieve 200 kJ total initial electrical energy for 10 number of modules as in Table II. Also, the resistance and inductance of the cables that connect the PPS to the launcher are neglected. In the simulation model, 10 number of parallel modules are fired at the same time.

TABLE I
THE PARAMETERS OF THE USED LAUNCHER

Launcher Length: l	3 m
Armature Initial Position: x_0	0.7 m
Total Mass of Armature and Projectile m_{total}	0.1 kg
Inductance Gradient of the Launcher: L'	0.45 μ H/m
Velocity Skin Effect Resistance Constant R_{vc}	3.1 10⁻⁹ (m/s)⁻³ᐟ²
DC Resistance Gradient: R'	50 $\mu\Omega$/m

TABLE II
THE CIRCUIT PARAMETERS OF THE C-BASED PPS 1 AND 2

	C-based PPS-1	C-based PPS-2
Capacitance: C	12 mF	48 mF
Capacitor Initial Voltage: V_c	1830 V	915 V
Pulse Shaping Inductance L_c	21.5 μH	21.5 μH
Number of Parallel Modules: n	10	10
Total Initial Energy: $E_{initial}$	200 kJ	200 kJ

The circuit parameters of two XRAM generators with different inductor parameters but the same initial energy (200 kJ) are given in Table III. Inductor charging process with an external DC voltage supply is also simulated in order to observe the opening switch concept. Inductance, initial current of the inductors and the number of stages are selected to achieve the same amount of stored energy as in the C-based PPSs

TABLE III
THE CIRCUIT PARAMETERS OF THE XRAM GENERATOR PPS

	XRAM-1	XRAM-2
Inductance: L	25 μH	100 μH
Inductor Initial Current: I_L	40 kA	20 kA
Charging Voltage (DC): V_{supply}	500 V	1000 V
Charging Time: $t_{charging}$	20 ms	20 ms
Number of Parallel Modules: n	10	10
Total Initial Energy: $E_{initial}$	200 kJ	200 kJ

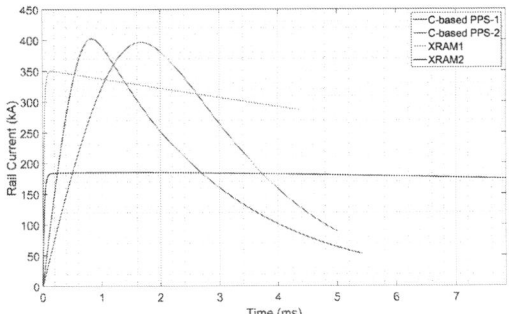

Fig. 8. Rail current for C-based and XRAM generator PPS topologies.

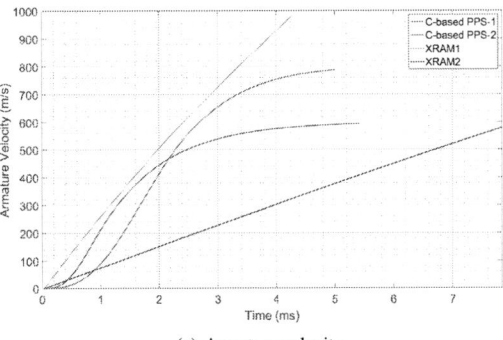

(a) Armature velocity.

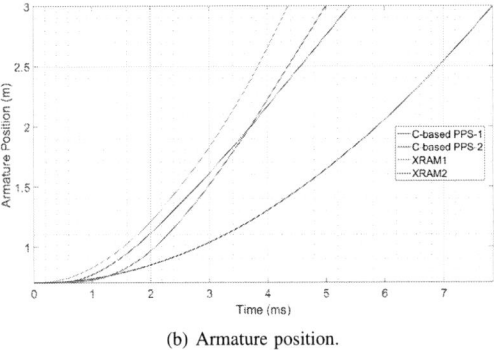

(b) Armature position.

Fig. 9. Armature velocity and position waveforms for C-based and XRAM generator PPS topologies.

using the equation given in (2). After the selection of these parameters, voltage of the DC supply and the charging time is selected using (9).

$$V_{supply} = n \, L \, \frac{I_L}{t_{charging}} \tag{9}$$

Four circuit models with C-based and XRAM generator PPS topologies with the same amount of initial electrical energy are simulated with the same analytic launcher model. In the simulation models, the resistance and inductance of the launcher are updated in each time step with respect to the value of the armature position and velocity. In addition, the simulations end when the armature position is equal to the length of the launcher. The simulation result for the rail current, armature velocitiy, position, and breech voltage waveform for both PPS topologies are given in Fig. 8, 9, and 10, respectively. Moreover, some critical outputs of these results are listed in Table IV where I_{peak}, I_{exit}, t_{rise}, t_{exit}, $V_{breech_{peak}}$, v_{exit}, and η are the peak value of the rail current, rail current when the armature exits the launcher, the amount of time until the rail current reaches its peak value, the peak

TABLE IV
SOME CRITICAL OUTPUTS OF THE SIMULATIONS.

	C-based PPS-1	C-based PPS-2	XRAM-1	XRAM-2
I_{peak}	403 kA	397 kA	416 kA	184 kA
I_{exit}	53 kA	90 kA	319 kA	174 kA
t_{rise}	0.85 ms	1.65 ms	0.13 ms	0.22 ms
t_{exit}	5.42 ms	5 ms	3.76 ms	7.84 ms
$V_{breech_{peak}}$	0.2 kV	0.1 kV	4 kV	2 kV
v_{exit}	594 m/s	787 m/s	995 m/s	578 m/s
η	8.8%	15.4%	24%	8.6%

value of the breech voltage, the armature velocity when it exits the launcher, and the efficieny of the total system including both PPS and launcher.

In addition, some significant conclusions are listed below taking the simulation results into consideration.

- During the rail current rising period, while C-based PPS has series connected RLC circuit, XRAM generator has only series connected RL circuit. The rising slope of the rail current is larger with XRAM generator than the one with C-based PPS due to the extra capacitance term in C-based generator for this period as in Fig. 8.
- Increasing the storage capacitance of the C-based PPS or the energy storage inductance of the XRAM generator increases the rail current rising time as in Fig. 8.
- While changing the capacitance of the C-based PPS does not have large influence on the peak of the rail current, increasing the inductance of the XRAM generator decreases the peak value of the rail current as in Fig. 8. Since the energy stored in an inductor is dependent of inductor current, increasing the inductance with the same stored energy decreases the rail current.
- During the rail current falling period, both of the topologies have series connected RL circuit. While for C-based PPS pulse shaping inductors are connected to the launcher in series, for XRAM generator, storage inductors are

978-1-5386-7688-2/19 $31.00 © 2019 IEEE

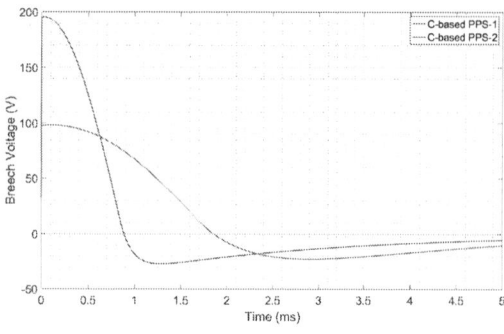

(a) Breech voltage for C-based PPSs.

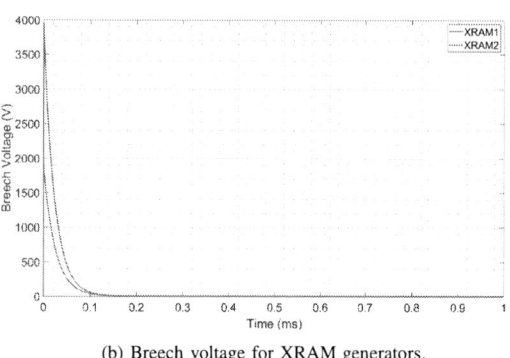

(b) Breech voltage for XRAM generators.

Fig. 10. Breech voltage waveforms for C-based and XRAM generator PPS topologies.

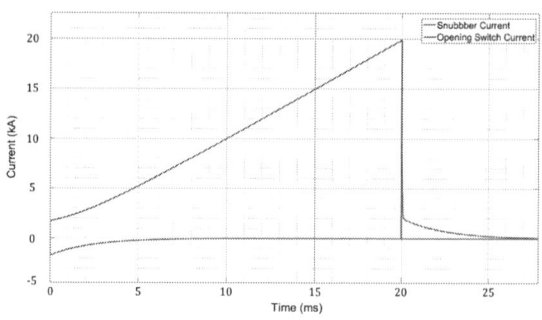

Fig. 11. Snubber and opening GTO thyristor current waveforms for XRAM-2 generator.

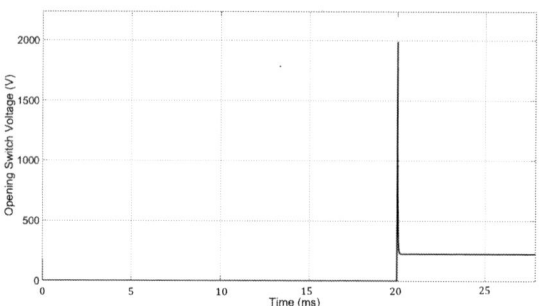

Fig. 12. Voltage stress on the opening GTO thyristor.

connected to the launcher in series. The falling slope of the rail current depends on the inductance of the pulse shaping inductor for C-based PPS and the inductance value of the energy storage inductor for the XRAM generator as in the Fig. 8. Since the inductance value of XRAM-2 generator is selected as 100 μH, the falling slope of the rail current is smaller than the other circuits.

- It is not possible to conclude that one of the investigated topologies is more efficient than the other. From Table IV, it can be observed that using an energy storage inductor with smaller inductance increases the efficiency of the XRAM generator. However, Fig, 10.(b) shows that this increases also the breech voltage. Moreover, using smaller inductor with the same amount of stored energy increases the initial current of the energy storage inductors which is a problem for the opening switches as discussed in the previous section.
- From Fig, 10.(b), it is observed that although for C-based PPS it takes approximately 0.5 ms to start moving for the armature after firing the PPS switches, for XRAM generator, armature start moving between the rails just after switching. If the armature spend much time at its initial position as in C-based PPS, during this period, the rail current causes melting on the contact surfaces between the rails and armature. This phenomena is known

as current melt-wave erosion which may cause armature transition as explained in [20]. Therefore it can be concluded that influence of current melt-wave erosion can be reduced by using XRAM generator topology as PPS for the electromagnetic launcher applications.

- From Fig. 10, it can be concluded that breech voltage becomes larger by using XRAM generator instead of C-based PPS due to large rail current rising slope of the XRAM generator. Therefore, the isolation of the cables, connecting the PPS to the launcher, becomes challenging for the XRAM generator. Also, the large amount of breech voltage causes voltage stress on the free-wheeling diode of the XRAM generator.
- During the rail current falling period, breech voltage of the launcher with C-based PPS becomes negative due to the large negative dI/dt on the inductance of the launcher.

In addition to the launcher parameters, it is required to investigate the electrical parameters of the PPS systems for the comparison of two topologies. One of the most critical parameters on the PPS side is the opening switches of the XRAM generator for the commutation of the high currents as discussed in the previous sections. GTO thyristors with RC snubber circuit is used in XRAM generators for the opening switches. In both XRAM-1 and XRAM-2, snubber resistance and capacitance are selected as 100 mΩ and 24 mF, respectively. Snubber and opening GTO thyristor current waveforms for XRAM-2 generator is given in Fig. 11. It can be observed that snubber circuit eleminates the spike current on the opening switch. However, due to the current spike on the snubber resistance, there exist a voltage stress on the opening switch as given in Fig. 12. The voltage stress on the opening switch is linearly dependent on the resistance of the snubber circuit. On one hand decreasing the snubber resistance decreases the voltage stress on the opening switch. On the other hand, decreasing the snubber resistance increases the power loss desipated in the snubber resistance.

V. CONCLUSION

In this study, XRAM generator and C-based PPS topologies are compared for the electromagnetic launcher application. The total mass of 0.1 kg launch package is accelerated using two C-based PPSs and two XRAM generators each of them having 200 kJ stored electrical energy. The electromagnetic launcher and the designed PPS topologies are simulated using the developed circuit model. The maximum efficiency is achived by XRAM generator with small storage inductance. However, for this design (XRAM-1) opening switches are suffered from the large voltage stress due to its small storage inductance. The simulation results also show that it is possible to achive 8.6% efficiency with XRAM-2 which provides a compact PPS system for the fieldable EML applications. In addition, it is observed that using large capacitance in C-based PPS system increases the efficiency and the size of the PPS. Moreover, using large inductance in XRAM generator decreases the efficiency and voltage stress on the opening switch.

REFERENCES

[1] L. Tang, J. He, L. Chen, S. Xia, and D. Feng, "Study of some influencing factors of armature current distribution at current ramp-up stage in railgun," IEEE Transactions on Plasma Science, vol. 43, no. 5, pp. 1585-1591, 2015.

[2] J. E. King, R. M. Kobuck, and J. R. Repp, "High speed water-cooled permanent magnet motor for pulse alternator-based pulse power systems," 2008 14th Symposium on Electromagnetic Launch Technology, EML, Proceedings, pp. 475-480, 2008.

[3] S. Cui, S. Wang, S. Wu, O. S. Yuryevich, and I. M. Milyaev, "Research of a modular pulsed alternator power system," IEEE Transactions on Plasma Science, vol. 45, no. 7, pp. 1406-1413, 2017.

[4] J. S. Bernardes, M. F. Stumborg, and T. E. Jean, "Analysis of a capacitor based pulsed-power system for driving long-range electromagnetic guns," IEEE Transactions on Magnetics, vol. 39, no. 1, pp. 486-490, 2003.

[5] P. Liu, J. Li, Y. Gui, S. Li, Q. Zhang, and N. Su, "Analysis of energy conversion efficiency of a capacitor-based pulsed-power system for railgun experiments," IEEE Transactions on Plasma Science, vol. 39, no. 1 PART 1, pp. 300-303, 2011.

[6] X. Guo, L. Dai, Q. Zhang, F. Lin, Q. Huang, and T. Zhao, "Influences of electric parameters of pulsed power supply on electromagnetic railgun system," IEEE Transactions on Plasma Science, vol. 43, no. 9, pp. 3260-3267, 2015.

[7] R. D. Ford, R. D. Hudson, and R. T. Klug, "Novel hybrid XRAM current multiplier," IEEE Transactions on Magnetics, vol. 29, no. 1, pp. 949-953, 1993.

[8] P. Dedi, V. Brommer, and S. Scharnholz, "Experimental realization of an eight-stage XRAM generator based on ICCOS semiconductor opening switches, fed by a magnetodynamic storage system," IEEE Transactions on Magnetics, vol. 45, no. 1, pp. 266-271, 2009.

[9] O. Libfried, V. Brommer, and S. Scharnholz, "Development of XRAM generators as inductive power sources for very high current pulses," Pulsed Power Conference, pp. 2-7, 2013.

[10] C. Gong, X. Yu, and X. Liu, "Study on the system efficiency of the capacitive pulsed-power supply," IEEE Transactions on Plasma Science, vol. 43, no. 5, pp. 1441-1447, 2015.

[11] W. Koch, Ein induktiver Stostromgenerator fr Hochstromexperimente, German Federal Ministry Sci. Res. [Bundesministerium fr wissenschaftliche Forschung], Bonn, Germany, Tech. Rep. BMwF-FB K 67-35, 1967.

[12] P. Dedi, V. Brommer, and S. Scharnholz, "Twenty-stage toroidal XRAM generator switched by countercurrent thyristors," IEEE Transactions on Plasma Science, vol. 39, no. 1 PART 1, pp. 263-267, 2011.

[13] Z. Li, Y. Zhang, J. Wu, Y. Jin, B. Li, "Development and properties of a 500kJ pulsed power supply," IEEE Pulsed Power Conference, pp. 1-4, 2015.

[14] O. Liebfried, "Review of inductive pulsed power generators for railguns," IEEE Transactions on Plasma Science, vol. 45, no. 7, pp. 1108-1114, 2017.

[15] M. Kanter, S. Singer, R. Cerny, and Z. Kaplan, "Multikilojoule inductive modulator with solid-state opening switches," IEEE Transactions on Power Electronics, vol. 7, no. 2, pp. 402-424, 1992.

[16] P. Dedi, V. Brommer, and S. Scharnholz, "ICCOS countercurrent-thyristor high-power opening switch for currents up to 28 kA," IEEE Transactions on Magnetics, vol. 45, no. 1, pp. 536-539, 2009.

[17] T. G. Engel, J. M. Neri, and M. J. Veracka, "The maximum theoretical efficiency of constant inductance gradient electromagnetic launchers," IEEE Transactions on Plasma Science, vol. 37, no. 4 PART 2, pp. 608-614, 2009.

[18] T. G. Engel, J. M. Neri, and M. J. Veracka, "Characterization of the velocity skin effect in the surface layer of a railgun sliding contact," IEEE Transactions on Magnetics, vol. 44, no. 7, pp. 18371844, 2008.

[19] Sanghyuk An, Byungha Lee, Youngseok Bae, Young-Hyun Lee, and Seong-Ho Kim, "Numerical Analysis of the Transient Inductance Gradient of Electromagnetic Launcher Using 2-D and 3-D Finite-Element Methods," IEEE Transactions on Plasma Science, vol. 45, no. 7, pp. 16351638, 2017.

[20] F. Stefani, and R. Merrill, "Experiments to measure melt-wave erosion in railgun armatures," IEEE Transactions on Magnetics, vol. 39, no. 1, pp. 188192, 2003.

A family of quadratic DC/DC converters with one low-side switch and a tapped inductor at the output side

Felix A. Himmelstoss
University of Applied Science
Technikum Wien
Vienna, Austria
felix.himmelstoss@technikum-wien.at

Helmut L. Votzi
University of Applied Science
Technikum Wien
Vienna, Austria
votzi@technikum-wien.at

Abstract — **Four new quadratic step-up-down converters with one low side switch and coupled inductors are presented. One is treated in detail, for the other converters results are shown. The voltage transformation rate and the connection of the inductor currents in dependence of the load current are calculated. An idealized mathematical description of the converter is derived and the large signal and the small signal model are determined. Furthermore the transfer functions between the output voltage and the duty cycle and the input voltage are calculated. A feed-forward control to cancel the influence of the changes of the input voltage is also determined.**

Keywords — quadratic, DC/DC converter, tapped inductor.

I. Introduction

A rich literature exists about quadratic converters (e.g. [1-9]. The here treated new converters (Fig. 1 – Fig. 4) consist of an input stage comprising a low-side active switch S, two passive switches (D_1, D_2), an inductor L, and a capacitor C_1 and an output stage comprising two magnetically coupled windings (N_{21}, N_{22}), a diode D_3 and an output capacitor C_2. The convertor shown in Fig. 1 will be treated in detail, the different results for the other converters will be shown too. First the converters are analyzed in the steady state and in the continuous mode, and for large capacitors, so that the voltage across them is nearly constant during the switching period. When the active switch is on, also the diode D_1 is conducting (mode M1). When the switch is turned off, also D_1 blocks and the other diodes D_2, D_3 are now conducting (mode M2).

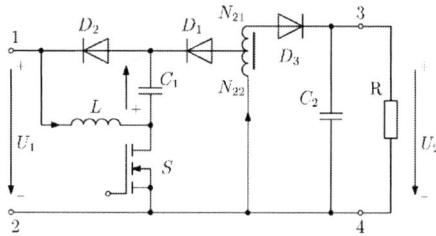

Fig. 1.   Converter A.

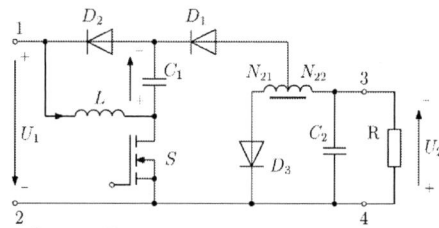

Fig. 2.   Converter B.

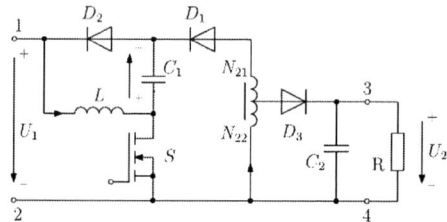

Fig. 3.   Converter C.

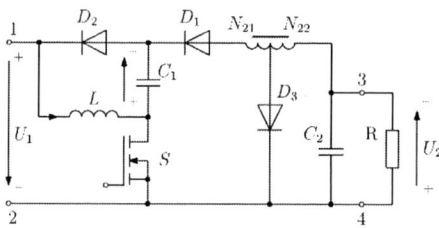

Fig. 4.   Converter D.

II. Voltage Transformation Ratio

The voltage across the inductor is zero in the mean. With the duty cycle as the ratio between the on-time of the active switch and the switching period the positive voltage-time area must be of the same size as the negative voltage-time area, one can write for the inductor

$$U_1 d = U_{C1}(1-d) . \tag{1}$$

This equation is valid also for the other modifications and leads to the value of the voltage across the capacitor according to

$$U_{C1} = \frac{d}{1-d} U_1 . \tag{2}$$

For the voltage across winding N_{22} (using the voltage transformation rate of coupled inductors) on must write

$$U_{C1} d = U_2 \frac{N_{22}}{N_{21}+N_{22}}(1-d) . \tag{3}$$

The output voltage of the converter in dependence of the duty cycle d and the input voltage U_1 can be written according to

$$U_2 = \frac{d^2}{(1-d)^2} \frac{N_{21}+N_{22}}{N_{22}} U_1 . \tag{4}$$

978-1-5386-7688-2/19 $31.00 © 2019 IEEE

For the modifications one gets

B: $\qquad U_2 = \dfrac{d^2}{(1-d)^2} \dfrac{N_{22}}{N_{21}+N_{22}} U_1$ (5)

C: $\qquad U_2 = \dfrac{d^2}{(1-d)} \dfrac{N_{21}+N_{22}}{N_{21}d+N_{22}} U_1$ (6)

D: $\qquad U_2 = \dfrac{d^2}{(1-d)} \dfrac{N_{22}}{N_{21}(1-d)+N_{22}} U_1$ (7)

III. CONNECTIONS BETWEEN THE CURRENTS

To get relations between the currents one has to inspect the currents through the capacitors. In steady-state the current through the capacitors must be zero in the mean. We start with the load current which is constant.

With $\bar{I}_{N22,2}$ as the mean value of the current through winding N_{22} referred to the duration of mode M2, the current-time balance for C_2 is therefore

$$I_{Load} d = \left(\bar{I}_{N22,2} - I_{Load} \right)(1-d)$$ (8)

which leads to

$$\bar{I}_{N22,2} = \frac{1}{1-d} I_{Load} \ .$$ (9)

During mode M2 both windings produce the current linkage (ampere turns), in Mode M1 only winding N_{22} is working. The current through N_{22} must therefore be higher. The mean value in reference to the duration of mode M1 must therefore be written according to

$$\bar{I}_{N22,1} = \frac{N_{21}+N_{22}}{N_{22}} \bar{I}_{N22,2} \ .$$ (10)

Inspecting the current through C_1 one gets for the charge balance

$$\bar{I}_{N22,1} d = \bar{I}_L (1-d) \ .$$ (11)

The mean value of the current through the inductor is therefore

$$\bar{I}_L = \frac{d}{(1-d)^2} \frac{N_{21}+N_{22}}{N_{22}} I_{Load} \ .$$ (12)

The mean value of the input current results in

$$\bar{I}_1 = \bar{I}_L d = \frac{d^2}{(1-d)^2} \frac{N_{21}+N_{22}}{N_{22}} I_{Load} \ .$$ (13)

For the modifications we get
Converter B

$\bar{I}_{N22,2}$ is equal to type A, but $\bar{I}_{N22,1}$ changes, because of the different connection of the coupled windings leading to

$$\bar{I}_{N22,1} = \frac{N_{22}}{N_{21}+N_{22}} \bar{I}_{N22,2} \ .$$ (14)

The mean value of the inductor current and the input current are therefore

$$\bar{I}_L = \frac{d}{(1-d)^2} \frac{N_{22}}{N_{21}+N_{22}} I_{Load}$$ (15)

$$\bar{I}_1 = \bar{I}_L d = \frac{d^2}{(1-d)^2} \frac{N_{22}}{N_{21}+N_{22}} I_{Load} \ .$$ (16)

Converter C

$$\bar{I}_{N22,2} = \frac{N_{22}}{N_{21}+N_{22}} I_{Load}$$ (17)

$$\bar{I}_{N22,1} = \frac{N_{21}+N_{22}}{dN_{21}+N_{22}} I_{Load}$$ (18)

$$\bar{I}_L = \frac{d}{(1-d)} \frac{N_{21}+N_{22}}{dN_{21}+N_{22}} I_{Load}$$ (19)

$$\bar{I}_1 = \frac{d^2}{(1-d)} \frac{N_{21}+N_{22}}{dN_{21}+N_{22}} I_{Load}$$ (20)

Converter D

$$\bar{I}_{N22,2} = \frac{N_{21}+N_{22}}{N_{21}(1-d)+N_{22}} I_{Load}$$ (21)

$$\bar{I}_{N22,1} = \frac{N_{22}}{N_{21}(1-d)+N_{22}} I_{Load}$$ (22)

$$\bar{I}_L = \frac{d}{1-d} \frac{N_{22}}{N_{21}(1-d)+N_{22}} I_{Load}$$ (23)

$$\bar{I}_1 = \frac{d^2}{1-d} \frac{N_{22}}{N_{21}(1-d)+N_{22}} I_{Load} \ .$$ (24)

IV. IDEALIZED MODEL

The state vector consists of the variables: current through the inductor L, flux per winding of the coupled windings, and the voltages across the capacitors.
Mode 1: S and D_1 are conducting
One gets the state equations

$$\frac{di_L}{dt} = \frac{u_1}{L}$$ (25)

$$\frac{d\psi_2}{dt} = N_{22} \frac{d\phi_2}{dt} = u_{C1}$$ (26)

The flux is produced by the current. The proportional factor between current and flux is the inductance, which can be calculated by a material factor AL multiplied with the square of the number of turns. Therefore one can write for the flux linkage

$$N_{22}\phi_2 = \psi_2 = A_L N_{22}^2 \cdot i \ .$$ (27)

Now one can write for the derivative of the voltage across C_1

$$\frac{du_{C1}}{dt} = -\frac{\phi_2}{A_L N_{22} C_1} \ .$$ (28)

The current through the output capacitor is the negative load current. The fourth state equation is therefore

$$\frac{du_{C2}}{dt} = -\frac{u_{C2}}{C_2 R} \ . \tag{29}$$

Summarizing the state equation in matrix form leads to the state-space description according to

$$\frac{d}{dt}\begin{pmatrix} i_L \\ \phi_2 \\ u_{C1} \\ u_{C2} \end{pmatrix} = \begin{bmatrix} 0 & 0 & 0 & 0 \\ 0 & 0 & \dfrac{1}{N_{22}} & 0 \\ 0 & -\dfrac{1}{A_L N_{22} C_1} & 0 & 0 \\ 0 & 0 & 0 & -\dfrac{1}{C_2 R} \end{bmatrix} \begin{pmatrix} i_L \\ \phi_2 \\ u_{C1} \\ u_{C2} \end{pmatrix} + \begin{bmatrix} \dfrac{1}{L} \\ 0 \\ 0 \\ 0 \end{bmatrix}(u_1) \tag{30}$$

Mode 2: D_2 and D_3 are conducting
The state equations can be written according to

$$\frac{di_L}{dt} = -\frac{u_{C1}}{L} \tag{31}$$

$$\left(N_{21} + N_{22}\right)\frac{d\phi_2}{dt} = \frac{d\psi_2}{dt} = -u_{C2} \tag{32}$$

$$\frac{du_{C1}}{dt} = \frac{i_L}{C_1} \tag{33}$$

$$\frac{du_{C2}}{dt} = \frac{\phi_2/A_L\left(N_{21}+N_{22}\right) - u_{C2}/R}{C_2} \ . \tag{34}$$

In the last equation the flux linkage

$$\left(N_{21}+N_{22}\right)\phi_2 = \psi_2 = A_L\left(N_{21}+N_{22}\right)^2 \cdot i \tag{35}$$

was used.

In matrix form now one can write

$$\frac{d}{dt}\begin{pmatrix} i_L \\ \phi_2 \\ u_{C1} \\ u_{C2} \end{pmatrix} = \begin{bmatrix} 0 & 0 & -\dfrac{1}{L} & 0 \\ 0 & 0 & 0 & -\dfrac{1}{N_{21}+N_{22}} \\ \dfrac{1}{C_1} & 0 & 0 & 0 \\ 0 & \dfrac{1}{A_L\left(N_{21}+N_{22}\right)C_2} & 0 & -\dfrac{1}{C_2 R} \end{bmatrix} \begin{pmatrix} i_L \\ \phi_2 \\ u_{C1} \\ u_{C2} \end{pmatrix} + \begin{bmatrix} 0 \\ 0 \\ 0 \\ 0 \end{bmatrix}(u_1) \tag{36}$$

V. LARGE SIGNAL MODEL

To get a compact model both state-space descriptions are weighted by d for M1 and by 1-d for M2 and added (state-space averaging). The large signal model can be obtained according to

$$\frac{d}{dt}\begin{pmatrix} i_L \\ \phi_2 \\ u_{C1} \\ u_{C2} \end{pmatrix} = \begin{bmatrix} 0 & 0 & \dfrac{d-1}{L} & 0 \\ 0 & 0 & \dfrac{d}{N_{22}} & \dfrac{d-1}{N_{21}+N_{22}} \\ \dfrac{1-d}{C_1} & -\dfrac{d}{A_L N_{22} C_1} & 0 & 0 \\ 0 & \dfrac{1-d}{A_L\left(N_{21}+N_{22}\right)C_2} & 0 & -\dfrac{1}{C_2 R} \end{bmatrix} \begin{pmatrix} i_L \\ \phi_2 \\ u_{C1} \\ u_{C2} \end{pmatrix} + \begin{bmatrix} \dfrac{d}{L} \\ 0 \\ 0 \\ 0 \end{bmatrix}(u_1) \tag{37}$$

This model is a non-linear one. To get a linear one, one has to linearize it around the operating point.

Converter B

$$\frac{d}{dt}\begin{pmatrix} i_L \\ \phi_2 \\ u_{C1} \\ u_{C2} \end{pmatrix} = \begin{bmatrix} 0 & 0 & \dfrac{d-1}{L} & 0 \\ 0 & 0 & \dfrac{d}{N_{21}+N_{22}} & \dfrac{d-1}{N_{22}} \\ \dfrac{1-d}{C_1} & -\dfrac{d}{A_L\left(N_{21}+N_{22}\right)C_1} & 0 & 0 \\ 0 & \dfrac{1-d}{A_L N_{22} C_2} & 0 & -\dfrac{1}{C_2 R} \end{bmatrix} \begin{pmatrix} i_L \\ \phi_2 \\ u_{C1} \\ u_{C2} \end{pmatrix} + \begin{bmatrix} \dfrac{d}{L} \\ 0 \\ 0 \\ 0 \end{bmatrix}(u_1) \tag{38}$$

Converter C

$$\frac{d}{dt}\begin{pmatrix} i_L \\ \phi_2 \\ u_{C1} \\ u_{C2} \end{pmatrix} = \begin{bmatrix} 0 & 0 & \dfrac{d-1}{L} & 0 \\ 0 & 0 & \dfrac{d}{N_{22}} & -\left(\dfrac{d}{N_{22}}+\dfrac{1-d}{N_{21}+N_{22}}\right) \\ \dfrac{1-d}{C_1} & -\dfrac{d}{A_L N_{22} C_1} & 0 & 0 \\ 0 & \dfrac{1-d}{A_L\left(N_{21}+N_{22}\right)C_2} & 0 & -\dfrac{1}{C_2 R} \end{bmatrix} \begin{pmatrix} i_L \\ \phi_2 \\ u_{C1} \\ u_{C2} \end{pmatrix} + \begin{bmatrix} \dfrac{d}{L} \\ 0 \\ 0 \\ 0 \end{bmatrix}(u_1) \tag{39}$$

Converter D

$$\frac{d}{dt}\begin{pmatrix} i_L \\ \phi_2 \\ u_{C1} \\ u_{C2} \end{pmatrix} = \begin{bmatrix} 0 & 0 & \dfrac{d-1}{L} & 0 \\ 0 & 0 & \dfrac{d}{N_{21}+N_{22}} & -\left(\dfrac{d}{N_{21}+N_{22}}+\dfrac{1-d}{N_{22}}\right) \\ \dfrac{1-d}{C_1} & -\dfrac{d}{A_L\left(N_{21}+N_{22}\right)C_1} & 0 & 0 \\ 0 & \dfrac{1-d}{A_L N_{22} C_2} & 0 & -\dfrac{1}{C_2 R} \end{bmatrix} \begin{pmatrix} i_L \\ \phi_2 \\ u_{C1} \\ u_{C2} \end{pmatrix} + \begin{bmatrix} \dfrac{d}{L} \\ 0 \\ 0 \\ 0 \end{bmatrix}(u_1) \tag{40}$$

VI. LINEARIZATION

All variables are written as the sum of the working point value marked by a capital letter with a 0 in the index and a perturbation of this variable, marked by a small letter with a roof on top. This results in a small signal model around the operating point

Written in a more formal form

$$\frac{d}{dt}\begin{pmatrix} \hat{i}_L \\ \hat{\phi}_2 \\ \hat{u}_{C1} \\ \hat{u}_{C2} \end{pmatrix} = \begin{bmatrix} 0 & 0 & A_{13} & 0 \\ 0 & 0 & A_{23} & A_{24} \\ A_{31} & A_{32} & 0 & 0 \\ 0 & A_{42} & 0 & A_{44} \end{bmatrix} \begin{pmatrix} \hat{i}_L \\ \hat{\phi}_2 \\ \hat{u}_{C1} \\ \hat{u}_{C2} \end{pmatrix} + \begin{bmatrix} B_{11} & B_{12} \\ 0 & B_{22} \\ 0 & B_{32} \\ 0 & B_{42} \end{bmatrix} \begin{pmatrix} \hat{u}_1 \\ \hat{d} \end{pmatrix} \tag{41}$$

978-1-5386-7688-2/19 $31.00 © 2019 IEEE

one gets for converter A

$$\frac{d}{dt}\begin{pmatrix} \hat{i}_L \\ \hat{\phi}_2 \\ \hat{u}_{C1} \\ \hat{u}_{C2} \end{pmatrix} = \begin{bmatrix} 0 & 0 & \dfrac{D_0-1}{L} & 0 \\ 0 & 0 & \dfrac{D_0}{N_{22}} & \dfrac{D_0-1}{N_{21}+N_{22}} \\ \dfrac{1-D_0}{C_1} & -\dfrac{D_0}{A_L N_{22} C_1} & 0 & 0 \\ 0 & \dfrac{1-D_0}{A_L(N_{21}+N_{22})C_2} & 0 & -\dfrac{1}{C_2 R} \end{bmatrix} \begin{pmatrix} \hat{i}_L \\ \hat{\phi}_2 \\ \hat{u}_{C1} \\ \hat{u}_{C2} \end{pmatrix} + \begin{bmatrix} \dfrac{D_0}{L} & \dfrac{U_{C10}+U_{10}}{L} \\ 0 & \dfrac{U_{C10}}{N_{22}}+\dfrac{U_{C20}}{N_{21}+N_{22}} \\ 0 & -\left(\dfrac{I_{L0}}{C_1}+\dfrac{\Phi_{20}}{A_L N_{22} C_1}\right) \\ 0 & -\dfrac{\Phi_{20}}{A_L(N_{21}+N_{22})C_2} \end{bmatrix} \begin{pmatrix} \hat{u}_1 \\ \hat{d} \end{pmatrix} \tag{42}$$

and for Converter B

$$\frac{d}{dt}\begin{pmatrix} \hat{i}_L \\ \hat{\phi}_2 \\ \hat{u}_{C1} \\ \hat{u}_{C2} \end{pmatrix} = \begin{bmatrix} 0 & 0 & \dfrac{D_0-1}{L} & 0 \\ 0 & 0 & \dfrac{D_0}{N_{21}+N_{22}} & \dfrac{D_0-1}{N_{22}} \\ \dfrac{1-D_0}{C_1} & -\dfrac{D_0}{A_L(N_{21}+N_{22})C_1} & 0 & 0 \\ 0 & \dfrac{1-D_0}{A_L C_2 N_{22}} & 0 & -\dfrac{1}{C_2 R} \end{bmatrix} \begin{pmatrix} \hat{i}_L \\ \hat{\phi}_2 \\ \hat{u}_{C1} \\ \hat{u}_{C2} \end{pmatrix} + \begin{bmatrix} \dfrac{D_0}{L} & \dfrac{U_{C10}+U_{10}}{L} \\ 0 & \dfrac{U_{C10}}{N_{21}+N_{22}}+\dfrac{U_{C20}}{N_{22}} \\ 0 & -\left(\dfrac{I_{L0}}{C_1}+\dfrac{\Phi_{20}}{A_L(N_{21}+N_{22})C_1}\right) \\ 0 & -\dfrac{\Phi_{20}}{A_L N_{22} C_2} \end{bmatrix} \begin{pmatrix} \hat{u}_1 \\ \hat{d} \end{pmatrix}$$

$$\tag{43}$$

for Converter C

$$\frac{d}{dt}\begin{pmatrix} \hat{i}_L \\ \hat{\phi}_2 \\ \hat{u}_{C1} \\ \hat{u}_{C2} \end{pmatrix} = \begin{bmatrix} 0 & 0 & \dfrac{D_0-1}{L} & 0 \\ 0 & 0 & \dfrac{D_0}{N_{22}} & -\left(\dfrac{D_0}{N_{22}}+\dfrac{1-D_0}{N_{21}+N_{22}}\right) \\ \dfrac{1-D_0}{C_1} & -\dfrac{D_0}{A_L N_{22} C_1} & 0 & 0 \\ 0 & \dfrac{1-D_0}{A_L(N_{21}+N_{22})C_2} & 0 & -\dfrac{1}{C_2 R} \end{bmatrix} \begin{pmatrix} \hat{i}_L \\ \hat{\phi}_2 \\ \hat{u}_{C1} \\ \hat{u}_{C2} \end{pmatrix} + \begin{bmatrix} \dfrac{D_0}{L} & \dfrac{U_{C10}+U_{10}}{L} \\ 0 & \left(\dfrac{U_{C10}}{N_{22}}+\dfrac{U_{C20}}{N_{21}+N_{22}}-\dfrac{U_{C20}}{N_{22}}\right) \\ 0 & -\left(\dfrac{I_{L0}}{C_1}+\dfrac{\Phi_{20}}{A_L N_{22} C_1}\right) \\ 0 & -\dfrac{\Phi_{20}}{A_L(N_{21}+N_{22})C_2} \end{bmatrix} \begin{pmatrix} \hat{u}_1 \\ \hat{d} \end{pmatrix}$$

$$\tag{44}$$

for Converter D

$$\frac{d}{dt}\begin{pmatrix} \hat{i}_L \\ \hat{\phi}_2 \\ \hat{u}_{C1} \\ \hat{u}_{C2} \end{pmatrix} = \begin{bmatrix} 0 & 0 & \dfrac{D_0-1}{L} & 0 \\ 0 & 0 & \dfrac{D_0}{N_{21}+N_{22}} & -\left(\dfrac{D_0}{N_{21}+N_{22}}+\dfrac{1-D_0}{N_{22}}\right) \\ \dfrac{1-D_0}{C_1} & -\dfrac{D_0}{A_L(N_{21}+N_{22})C_1} & 0 & 0 \\ 0 & \dfrac{1-D_0}{A_L C_2 N_{22}} & 0 & -\dfrac{1}{C_2 R} \end{bmatrix} \begin{pmatrix} \hat{i}_L \\ \hat{\phi}_2 \\ \hat{u}_{C1} \\ \hat{u}_{C2} \end{pmatrix} + \begin{bmatrix} \dfrac{D_0}{L} & \dfrac{U_{C10}+U_{10}}{L} \\ 0 & \dfrac{U_{C10}}{N_{21}+N_{22}}-\dfrac{N_{21}}{(N_{21}+N_{22})N_{22}}U_{C20} \\ 0 & -\left(\dfrac{I_{L0}}{C_1}+\dfrac{\Phi_{20}}{A_L(N_{21}+N_{22})C_1}\right) \\ 0 & -\dfrac{\Phi_{20}}{A_L N_{22} C_2} \end{bmatrix} \begin{pmatrix} \hat{u}_1 \\ \hat{d} \end{pmatrix}$$

$$\tag{45}$$

It is also possible to get the connection of the working point values

$$(1-D_0)U_{C10} = D_0 U_{10} \tag{46}$$

$$\frac{D_0}{N_{22}}U_{C10} = \frac{(1-D_0)}{N_{21}+N_{22}}U_{C20} \tag{47}$$

$$(1-D_0)I_{L0} = \frac{D_0}{A_L N_{22}}\Phi_{20} \tag{48}$$

$$\frac{1-D_0}{A_L(N_{21}+N_{22})}\Phi_{20} = \frac{U_{C20}}{R} = I_{LOAD} \tag{49}$$

$$\frac{D_0}{N_{21}+N_{22}}U_{C10} = \frac{(1-D_0)}{N_{22}}U_{C20} \tag{51}$$

$$(1-D_0)I_{L0} = \frac{D_0}{A_L(N_{21}+N_{22})}\Phi_{20} \tag{52}$$

$$\frac{1-D_0}{A_L N_{22}}\Phi_{20} = \frac{U_{C20}}{R} = I_{LOAD} \tag{53}$$

Converter C

$$(1-D_0)U_{C10} = D_0 U_{10} \tag{54}$$

Converter B

$$(1-D_0)U_{C10} = D_0 U_{10} \tag{50}$$

$$\frac{D_0}{N_{22}}U_{C10} = \frac{D_0 N_{21}+N_{22}}{(N_{21}+N_{22})N_{22}}U_{C20} \tag{55}$$

$$(1 - D_0)I_{L0} = \frac{D_0}{A_L N_{22}} \Phi_{20} \quad (56)$$

$$\frac{1 - D_0}{A_L(N_{21} + N_{22})} \Phi_{20} = \frac{U_{C20}}{R} = I_{LOAD} \quad (57)$$

Converter D

$$(1 - D_0)U_{C10} = D_0 U_{10} \quad (58)$$

$$\frac{D_0}{N_{21} + N_{22}} U_{C10} = \frac{(1 - D_0)N_{21} + N_{22}}{N_{22}(N_{21} + N_{22})} U_{C20} \quad (59)$$

$$(1 - D_0)I_{L0} = \frac{D_0}{A_L(N_{21} + N_{22})} \Phi_{20} \quad (60)$$

$$\frac{1 - D_0}{A_L N_{22}} \Phi_{20} = \frac{U_{C20}}{R} = I_{LOAD} \quad (61)$$

VII. TRANSFER FUNCTIONS

The linear model can be used to calculate the transfer functions. These are useful for designing the controller and for drawing Bode plots.
Starting from the common form

$$\frac{d}{dt} \begin{pmatrix} \hat{i}_L \\ \hat{\phi}_2 \\ \hat{u}_{C1} \\ \hat{u}_{C2} \end{pmatrix} = \begin{bmatrix} 0 & 0 & A_{13} & 0 \\ 0 & 0 & A_{23} & A_{24} \\ A_{31} & A_{32} & 0 & 0 \\ 0 & A_{42} & 0 & A_{44} \end{bmatrix} \begin{pmatrix} \hat{i}_L \\ \hat{\phi}_2 \\ \hat{u}_{C1} \\ \hat{u}_{C2} \end{pmatrix} + \begin{bmatrix} B_{11} & B_{12} \\ 0 & B_{22} \\ 0 & B_{32} \\ 0 & B_{42} \end{bmatrix} \begin{pmatrix} \hat{u}_1 \\ \hat{d} \end{pmatrix} . \quad (62)$$

the Laplace transformation leads to

$$\begin{bmatrix} s & 0 & -A_{13} & 0 \\ 0 & s & -A_{23} & -A_{24} \\ -A_{31} & -A_{32} & s & 0 \\ 0 & -A_{42} & 0 & s - A_{44} \end{bmatrix} \begin{pmatrix} I_L(s) \\ \Phi_2(s) \\ U_{C1}(s) \\ U_{C2}(s) \end{pmatrix} = \begin{bmatrix} B_{11} & B_{12} \\ 0 & B_{22} \\ 0 & B_{32} \\ 0 & B_{42} \end{bmatrix} \begin{pmatrix} U_1(s) \\ D(s) \end{pmatrix} \quad (63)$$

For the denominator one gets with Crammer's law

$$N = s^4 - A_{44} \cdot s^3 - (A_{13}A_{31} + A_{23}A_{32} + A_{24}A_{42}) \cdot s^2 +$$
$$+ (A_{13}A_{31}A_{44} + A_{23}A_{32}A_{44}) \cdot s + A_{13}A_{24}A_{31}A_{42} = \quad (64)$$
$$= s^4 + a_3 s^3 + a_2 s^2 + a_1 s + a_0 .$$

The numerator of the transfer function U_{C2}/D for the control signal can be calculated from (63) according to

$$Z_D = B_{42}s^3 + A_{42}B_{22} \cdot s^2 +$$
$$+ [A_{23}A_{42}B_{32} - (A_{13}A_{31} + A_{23}A_{32})B_{42}] \cdot s +$$
$$+ (A_{23}A_{31}B_{12} - A_{13}A_{31}B_{22})A_{42} = b_3 s^3 + b_2 s^2 + b_1 s + b_0 \quad (65)$$

and the numerator for the transfer function U_{C2}/U_1 is given by

$$Z_{U1} = A_{23}A_{31}A_{42}B_{21} = c_0 . \quad (66)$$

The transfer function between the output voltage $U_2(s)$ and the control signal $D(s)$ can be written according to

$$G_{U2D}(s) = \frac{U_2(s)}{D(s)} = \frac{Z_D}{N} = \frac{b_3 s^3 + b_2 s^2 + b_1 s + b_0}{s^4 + a_3 s^3 + a_2 s^2 + a_1 s + a_0} . \quad (67)$$

For the influence of the disturbance (changes of the input voltage) one gets the transfer function

$$G_{U2U1}(s) = \frac{U_2(s)}{U_1(s)} = \frac{Z_{U1}}{N} = \frac{c_0}{s^4 + a_3 s^3 + a_2 s^2 + a_1 s + a_0} \quad (68)$$

VIII. DISTURBANCE FEED-FORWARD

Fig. 5 shows the signal flow graph of the disturbance feed-forward control. The input voltage $U_1(s)$ represents the disturbance and $U_2^*(s)$ is the reference value of the output signal $U_2(s)$. The feed-back controller is marked by R(s).

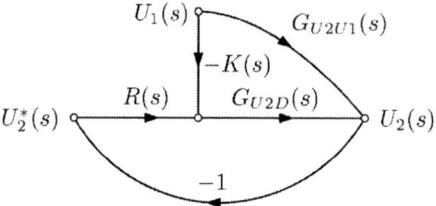

Fig. 5: Signal flow graph representing the converters A to D with disturbance feed-forward K, disturbance transfer function G_{U2U1} and plant controlled system G_{U2D}.

The output voltage can be calculated with the help of the principle of superposition according to

$$U_2(s) = \frac{G_{U2U1} - KG_{U2D}}{1 + RG_{U2D}} U_1 + \frac{RG_{U2D}}{1 + RG_{U2D}} U_2^* . \quad (69)$$

To cancel the influence of the input voltage to the output voltage, K must be

$$K(s) = \frac{G_{U2U1}(s)}{G_{U2D}(s)} . \quad (70)$$

This leads to

$$K(s) = \frac{c_0}{b_3 s^3 + b_2 s^2 + b_1 s + b_0} . \quad (71)$$

Compensation of the input voltage changes by this forward controller reduces the work of the feed-back controller.

IX. VOLTAGE STRESS ACROSS THE SEMICONDUCTOR DEVICES

The voltage stresses of the semiconductors are

$$U_S = U_{C1} + U_1 = \left(1 + \frac{d}{1 - d}\right)U_1 = \frac{1}{1 - d}U_1 \quad (72)$$

$$U_{D1} = \frac{N_{22}}{N_{21} + N_{22}} U_2 - U_1 \quad (73)$$

$$U_{D2} = -(U_{C1} + U_1) = \frac{1}{1 - d}U_1 \quad (74)$$

$$U_{D3} = -\frac{N_{21} + N_{22}}{N_{22}} U_{C1} - U_2 \quad (75)$$

The voltage stress can be easily calculated using Kirchhoff's voltage law. The value of the duty cycle can be determined from the voltage transformation rate of the converter. The voltage across D_1 must be negative when the active is switch is turned off. Therefore the duty cycle must not exceed 0.5!

X. DIMENSIONING

We start with the output capacitor C_2. During the on-time of the active switch S diode D_3 is blocked. The load must be supplied by C_2. Therefore the voltage across it decreases with the load current according to

$$\Delta u_{C2} = \frac{1}{C_2} \int_0^{dT} I_{Load} \, dt = \frac{1}{C_2} I_{Load} \frac{d}{f} \quad . \tag{76}$$

The capacitor has to be designed for the maximum load current

$$C_2 = \frac{I_{Load,\max} d}{\Delta u_{C2} f} \quad . \tag{77}$$

When the active switch is turned off, the current of the inductor L commutates into C1 and charges it according to

$$\Delta u_{C1} = \frac{1}{C_2} \int_{dT}^{T} \bar{I}_L \, dt = \frac{1}{C_1} \bar{I}_L (1-d) T \tag{78}$$

leading to

$$C_2 = \frac{1}{\Delta u_{C1}} \int_{dT}^{T} \bar{I}_L \, dt = \frac{1}{\Delta u_{C1}} \bar{I}_L \frac{1-d}{f} \quad . \tag{79}$$

When choosing the capacitors the equivalent series resistor has to be taken into account, which causes an additional voltage drop. With this in mind the value of the capacitor will be two or three times higher.

During the on-time of the active switch, the input voltage is across the coil L. The current increases during this time interval with ΔIL

$$U_1 = L \frac{\Delta \bar{I}_L}{dT} \quad . \tag{80}$$

With a chosen current ripple (e.g. of 20%) one can calculate the inductor according to

$$L = \frac{U_1 d}{\Delta \bar{I}_L f} \quad . \tag{81}$$

For the coupled coils we design first the winding N_{22} which is always under current. One chooses first the ripple of the current during the on-time of the switch S. For the mode M1 one gets

$$U_{C1} = L_{22} \frac{\Delta I_{22}}{dT} = \frac{d}{1-d} U_1 \tag{82}$$

$$L_{22} = \frac{d^2}{(1-d) f \Delta I_{22}} U_1 \quad . \tag{83}$$

The value of d can be found by using the voltage transformation rate. Including it into the equations leads to confusing expressions, therefore it is omitted. There is a huge potential of optimization relating to size, cost and volume. The tolerance of the input voltage and the task of the converter (has it a fixed output voltage or shall it be changeable) have to be kept in mind.

XI. SIMULATION

To better understand the functionality of the converter type A a simulation is shown (Fig. 6). Up to down one has input current, current through the inductive elements and the load, output voltage, and control signal of the active switch.

CONCLUSION

Four single low side-switch converters with a tapped inductor were treated and the voltage transformation rate, the relationship between the currents through the inductors and the load current were calculated. The large signal and the small signal model and a feed-forward controller to cancel the influence of input voltage disturbances were derived. Besides of the quadratic term in the voltage transformation rate (4, 5, 6, 7) which flattens the curve, and an additional degree of freedom due to the winding ratio of the coupled coils, a better adaption of the input and output voltages is given for special applications. The converters can be used for lighting applications with dimming feature, as battery charger and for driving small permanent magnet DC motors or actuators.

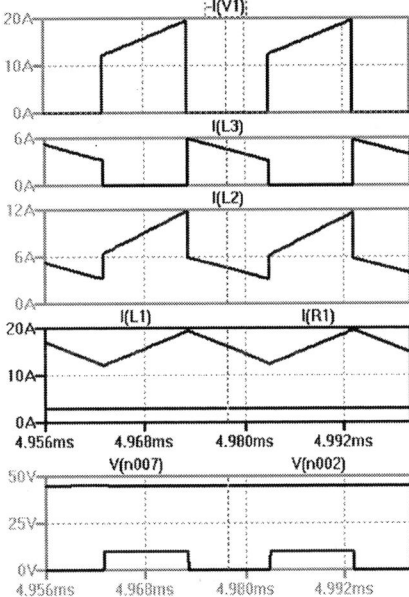

Fig. 6. Simulation results (up to down: Input current, current through N_{21}, current through N_{22}, current through the inductor L, and through the load, output voltage and control signal).

REFERENCES

[1] D. Maksimovic, and S. Cuk, "Switching converters with wide DC conversion range," *IEEE Transactions on Power Electronics*, vol. 6, pp. 151 – 157, January 1991.

[2] A. Ayachit, and M. Kazimierczuk, "Open-loop small-signal transfer functions of the quadratic buck PWM DC-DC converter in CCM," *IECON 2014*, pp. 1643 – 1649, Nov. 2014.

[3] A. Ayachit, and M. Kazimierczuk. "Power losses and efficiency analysis of the quadratic buck converter in CCM," *Midwestern Symp on Circuits and Systems*, pp. 463 – 466, August 2014.

[4] V.M. Pacheco, V.J. Farias, J.B. Vieira, J.B., A.J. Nascimento, and L.C. Freitas, "A quadratic buck converter with lossless commutation," *Applied Power Electronics Conference APEC '96*, pp. 311 – 317, March 1996.

[5] E. Carbajal-Gutierrez, J. Morales-Saldana, and J. Leyva-Ramos, "Modeling of a single-switch quadratic buck converter," *IEEE Tran. on Aerospace and Electronic Syst.*, vol. 41, pp. 1450 – 1456, Oct. 2005.

[6] H. Cheng, K. Ma Smedley, and A. Abramovitz, "A Wide-Input–Wide-Output (WIWO) DC–DC Converter," *IEEE Transactions on Power Electronics*, vol. 25, pp. 280 – 289, February 2010.

[7] R. Loera-Palomo, and Morales-Saldana, "Analysis of Quadratic Step-Down DC-DC Converters Based on Noncascading Structures," *CERMA 2012*, pp. 305 – 310, November 2012.

[8] D. S Wijeratne, and G. Moschopoulos, "Quadratic Power Conversion for Power Electronics: Principles and Circuits," *IEEE Transactions on Circuits and Systems I*, vol. 59, pp. 426 – 438, Feb. 2012.

[9] K. Pirghaibi and F. A. Himmelstoss, "A step-down converter with high down-scale ratio," 2017 International Conference on Optimization of Electrical and Electronic Equipment (OPTIM) & 2017 Intl Aegean Conference on Electrical Machines and Power Electronics (ACEMP), Brasov, 2017, pp. 567-572.

978-1-5386-7688-2/19 $31.00 © 2019 IEEE

A Framework for Fast Simulation of Power Electronic Circuits*

Hadhiq Khan[1], Mohammad Abid Bazaz[1] and Shahkar Ahmad Nahvi[2]

Abstract— An adaptive multi-resolution framework based on model order reduction to accelerate the simulations of high-fidelity power electronic circuits (PECs) is presented in this paper. The wide span of eigenvalues makes the simulation of PECs prohibitively slow. The proposed framework tackles this problem. We use singular perturbation approximation to extract reduced-order models where the transient component of non-dominant eigenvalues is ignored. Each reduced-order model corresponds to a particular level of accuracy or resolution. The simulation engine is set such that it adaptively switches across different resolutions. This approach is advantageous in the sense that instead of simulating the original system over the entire time-range, a combination of the full and reduced models is simulated. This accelerates the simulation considerably and at the same time ensures that the complete response of the system is obtained. An error bound for the approximation method is derived. The method is illustrated using a DC-DC Buck-boost converter.

I. INTRODUCTION

To minimize repetitive breadboarding and design alterations of power electronic circuits, accurate simulation is necessary. The goal is to simulate the performance of power electronic circuits (PECs) using software for precise prediction of complex phenomena such as parasitic effects, electromagnetic interference, thermal and mechanical effects etc. The demand for accurate simulation is further escalated during the design optimization process where multiple iterations are required to be performed so that the design constraints are satisfied and the target performance is met [1].

PECs are known to have time-constants spanning multiple orders of magnitude [2]. The wide span of time-constants requires that an extremely small time-step be chosen to capture the fast evolving transients. To capture the dynamics associated with the large time-constants, the simulation must continue till steady state, which is reached after a few hundred or more conversion periods. To effectively capture the dynamics of the circuit, the simulation must continue for a long time, albeit with sufficiently small time-steps. The simulation of the complete response of PECs by conventional simulation tools turns out to be computationally prohibitive and inefficient even with use of variable step-size based solvers [3], [4].

*This work was supported by the Central Power Research Institute (CPRI), Ministry of Power, Government of India under the Research Scheme on Power grant RSOP/2018/GD/04.

[1]H. Khan and M. A. Bazaz are with the Department of Electrical Engineering, National Institute of Technology Srinagar, J&K, India. emails: hadhiq@nitrsi.ac.in, abid@nitsri.ac.in

[2]S. A. Nahvi is with the Department of Electrical Engineering Islamic University of Science and Technology, Awantipora J&K, India. email: shahkar.nahvi@islamicuniversity.edu.in

To accelerate the simulation speed, Model Order Reduction (MOR) methods can be used [5]. MOR aims at approximating a given system of order n with a reduced system of order r such that the input-output relation of the system is preserved. An MOR framework for accelerating the simulation speeds of PECs has been presented by the authors in [6], [7]. Even though the straightforward implementation of the framework accelerates the simulation and captures the long-term response, the fast-evolving transients are not captured. The approximation error may sometimes be too large to be neglected.

In this work, we present an adaptive multi-resolution simulation (AMRS) approach which captures the fast transients, while accelerating the speed of the simulation. MOR is used to extract models of different orders or resolutions using singular perturbation approximation (SPA). In this method, the dominant eigenvalues and only the steady-state components of the non-dominant eigenvalues are retained. Each reduced-order model corresponds to a particular resolution (level of accuracy or abstraction) [8]. The simulation engine is set such that it adaptively changes the resolution of the simulation. The engine begins simulating the original system first. The engine tracks the evolution of the states of the system and as the transients die down, the reduced-order model takes over and is simulated till the transients occur again. This approach is advantageous in the sense that instead of simulating the original system over the entire time-range, a combination of the full and reduced models are simulated. The high-fidelity model is simulated for only a fraction of the simulation cycle. We will show in the following sections that considerable savings in terms of simulation time are achieved while ensuring that the complete response of the system is obtained with negligible approximation error.

II. MULTI-RESOLUTION FRAMEWORK

A. Modeling

Since PECs are switched in nature, and assuming that the circuit elements are piecewise linear-time-invariant (LTI), a state-space model of the form

$$\dot{\mathbf{x}}(t) = \mathbf{A}\mathbf{x}(t) + \mathbf{B}\mathbf{u}(t)$$
$$\mathbf{y}(t) = \mathbf{C}\mathbf{x}(t) \tag{1}$$

can be derived for each switching mode. Here $\mathbf{A} \in \mathbb{R}^{n \times n}$, $\mathbf{B} \in \mathbb{R}^{n \times m}$ and $\mathbf{C} \in \mathbb{R}^{p \times n}$ are the system matrices and $\mathbf{u}$, $\mathbf{x}$ and $\mathbf{y}$ are the input, state and output vectors respectively. n is the order of the system, m is the number of inputs and p is the number of outputs. A PEC can be represented as a switched system [9]

$$\dot{\mathbf{x}}(t) = \mathbf{A}^{\sigma(t)}\mathbf{x}(t) + \mathbf{B}^{\sigma(t)}\mathbf{u}(t)$$
$$\mathbf{y}(t) = \mathbf{C}^{\sigma(t)}\mathbf{x}(t) \qquad (2)$$

where $\sigma(t) : [0, \infty) \to \Sigma, \Sigma = \{1, 2, \ldots, N\}$ is the mode selector function. N is the total number of switching modes of the circuit. When the ith mode is active, the representation is given by

$$\dot{\mathbf{x}}(t) = \mathbf{A}^i \mathbf{x}(t) + \mathbf{B}^i \mathbf{u}(t)$$
$$\mathbf{y}(t) = \mathbf{C}^i \mathbf{x}(t) \qquad (3)$$

B. Model Reduction

A number of natural and artificial processes and systems exist where the system order is quite large and typically ranges from several hundred to hundreds of thousands. Such systems are known as large-scale systems (LSS). Working with these systems is by no means an easy task. Such LSS are approximated by simplified, easy-to-handle reduced-order models using MOR. The aim is to obtain a reduced-order system which approximates the dynamics of the original system (1)

$$\dot{\hat{\mathbf{x}}}(t) = \hat{\mathbf{A}}\hat{\mathbf{x}}(t) + \hat{\mathbf{B}}\mathbf{u}(t)$$
$$\hat{\mathbf{y}}(t) = \hat{\mathbf{C}}\hat{\mathbf{x}}(t) \qquad (4)$$

such that for the same input function $\mathbf{u}$, we obtain $\mathbf{y}(t) \approx \hat{\mathbf{y}}(t)$ or $\mathbf{Y}(s) \approx \hat{\mathbf{Y}}(s)$ in the frequency domain. Here $\hat{\mathbf{A}} \in \mathbb{R}^{r \times r}$, $\hat{\mathbf{B}} \in \mathbb{R}^{r \times m}$, $\hat{\mathbf{C}} \in \mathbb{R}^{p \times r}$ are the reduced-order system matrices. Commonly used model reduction methods include modal truncation/singular perturbation approximation (SPA), balanced truncation (BT) and Krylov subspace based methods [5].

Since the ordinary differential equations (ODEs) describing PECs have eigenvalues spanning several orders of magnitude, the simulation is extremely slow. Modal truncation, where the eigenvalues closest to the imaginary axis (dominant eigenvalues) are retained discarding the remaining ones, can be used to overcome this problem. A drawback of the this approach is the possibility of a large steady-state error. Zero steady-state error can be obtained by using the singular perturbation approximation (SPA) where only the transient part of the non-dominant eigenvalues is neglected [10]. The steady-state contribution of all the eigenvalues is taken into account.

C. Problem Formulation

In conventional model reduction techniques, one always has to choose the order of the reduced model *a priori*. Model reduction may have to be performed repeatedly before an acceptable solution is obtained. In most cases, a single reduced order model may not give sufficiently accurate results [11] and yield computational savings at the same time. Usually, a trade-off has to be made between the speed and the accuracy of the reduced-order model. These drawbacks are taken care of using the proposed method. Here, models of different orders, each corresponding to a certain accuracy

or resolution are derived on the fly at no additional computational cost. The engine then adaptively switches across resolutions during the course of the simulation.

We define similarity transformation,

$$\mathbf{x}(t) = \mathbf{P}^i \mathbf{z}(t), \quad \mathbf{P}^i \in \mathbb{R}^{n \times n} \qquad (5)$$

where $\mathbf{P}^i$ is the eigenvector matrix that diagonalizes the system in (3) into the form

$$\dot{\mathbf{z}}(t) = \mathbf{\Lambda}^i \mathbf{z}(t) + \boldsymbol{\beta}^i \mathbf{u}(t)$$
$$\mathbf{y}(t) = \mathbf{\Gamma}^i \mathbf{z}(t) \qquad (6)$$

Here $\mathbf{\Lambda}^i = (\mathbf{P}^i)^{-1}\mathbf{A}^i\mathbf{P}^i$, $\boldsymbol{\beta}^i = (\mathbf{P}^i)^{-1}\mathbf{B}^i$ and $\mathbf{\Gamma}^i = \mathbf{C}^i\mathbf{P}^i$. Since (5) applies to each switching mode, the time index "t" in the following equations is dropped for brevity. Consider a partition of $\mathbf{\Lambda}^i$

$$\Lambda^i = \begin{bmatrix} \Lambda_k^i & 0 \\ 0 & \Lambda_{n-k}^i \end{bmatrix} \qquad (7)$$

with eigenvalues arranged in descending order of dominance (with eigenvalues closest to the imaginary axis being considered dominant). The subscript $k \leq n$ is the index of the eigenvalues being retained. Here,

$$\mathbf{\Lambda}_k^i = \mathrm{diag}(\lambda_1^i, \lambda_2^i, \ldots, \lambda_k^i),$$
$$\mathbf{\Lambda}_{n-k}^i = \mathrm{diag}(\lambda_{k+1}^i, \lambda_{k+2}^i, \ldots, \lambda_n^i),$$
$$|\mathrm{Re}(\lambda_1^i)| < |\mathrm{Re}(\lambda_2^i)| < \ldots < |\mathrm{Re}(\lambda_k^i)| < \ldots < |\mathrm{Re}(\lambda_n^i)|$$

Then (6) can be written as

$$\begin{bmatrix} \dot{\mathbf{z}}_k \\ \dot{\mathbf{z}}_{n-k} \end{bmatrix} = \begin{bmatrix} \mathbf{\Lambda}_k^i & 0 \\ 0 & \mathbf{\Lambda}_{n-k}^i \end{bmatrix} \begin{bmatrix} \mathbf{z}_k \\ \mathbf{z}_{n-k} \end{bmatrix} + \begin{bmatrix} \boldsymbol{\beta}_k^i \\ \boldsymbol{\beta}_{n-k}^i \end{bmatrix} \mathbf{u}$$
$$\mathbf{y} = \begin{bmatrix} \mathbf{\Gamma}_k^i & \mathbf{\Gamma}_{n-k}^i \end{bmatrix} \begin{bmatrix} \mathbf{z}_k \\ \mathbf{z}_{n-k} \end{bmatrix} \qquad (8)$$

where

$$\mathbf{z}_k = \begin{bmatrix} z_1 & z_2 & \ldots & z_k \end{bmatrix}^T$$
$$\mathbf{z}_{n-k} = \begin{bmatrix} z_{k+1} & z_{k+2} & \ldots & z_n \end{bmatrix}^T$$

and matrices $\boldsymbol{\beta}^i, \mathbf{\Gamma}^i$ have been appropriately partitioned such that $\boldsymbol{\beta}_k^i \in \mathbb{R}^{k \times m}$, $\boldsymbol{\beta}_{n-k}^i \in \mathbb{R}^{n-k \times m}$, $\mathbf{\Gamma}_k^i \in \mathbb{R}^{p \times k}, \mathbf{\Gamma}_{n-k}^i \in \mathbb{R}^{p \times (n-k)}$.

In the singular perturbation approach, the transient component of the non-dominant states is ignored. Setting $\dot{\mathbf{z}}_{n-k} = 0$ in (8), we obtain a differential algebraic equation (DAE) system

$$\begin{bmatrix} \dot{\mathbf{z}}_k \\ 0 \end{bmatrix} = \begin{bmatrix} \mathbf{\Lambda}_k^i & 0 \\ 0 & \mathbf{I} \end{bmatrix} \begin{bmatrix} \mathbf{z}_k \\ \mathbf{z}_{n-k} \end{bmatrix} + \begin{bmatrix} \boldsymbol{\beta}_k^i \\ (\mathbf{\Lambda}_{n-k}^i)^{-1}\boldsymbol{\beta}_{n-k}^i \end{bmatrix} \mathbf{u}$$
$$\mathbf{y} = \begin{bmatrix} \mathbf{\Gamma}_k^i & \mathbf{\Gamma}_{n-k}^i \end{bmatrix} \begin{bmatrix} \mathbf{z}_k \\ \mathbf{z}_{n-k} \end{bmatrix} \qquad (9)$$

which can be re-written in the standard descriptor form as

$$\mathbf{E}_k^i \dot{\mathbf{z}} = \mathbf{A}_k^i \mathbf{z} + \mathbf{B}_k^i \mathbf{u}$$
$$\mathbf{y} = \mathbf{\Gamma}^i \mathbf{z} \qquad (10)$$

Equation (10) is the DAE model of the ith mode where dynamics of eigenvalues whose magnitude is greater than λ_k^i are ignored.

$$\mathbf{E}_k^i = \begin{bmatrix} \mathbf{I}_{k\times k} & 0 \\ 0 & 0 \end{bmatrix}, \quad \mathbf{A}_k^i = \begin{bmatrix} \mathbf{\Lambda}_k^i & 0 \\ 0 & \mathbf{I}_{(n-k)\times(n-k)} \end{bmatrix}$$

$$\mathbf{B}_k^i = \begin{bmatrix} \boldsymbol{\beta}_k^i \\ (\mathbf{\Lambda}_{n-k}^i)^{-1}\,\boldsymbol{\beta}_{n-k}^i \end{bmatrix} \tag{11}$$

When a single eigenvalue (or a conjugate eigenvalue pair) is retained, it is said to be the lowest resolution model. The simulation engine keeps track of the dynamics of the non-dominant states and as soon as the steady-state corresponding to a particular non-dominant state is reached, a lower resolution model takes over. Instead of simulating the original full-order model over the entire time range, a combination of full and reduced-order models is simulated.

D. Error Analysis

For the system described in (6), the transfer function matrix (TFM) is of the form

$$\mathbf{G}^i(s) = \mathbf{\Gamma}^i (s\mathbf{I} - \mathbf{\Lambda}^i)^{-1} \boldsymbol{\beta}_n^i \tag{12}$$

where $\mathbf{I} \in \mathbb{R}^{n\times n}$ and $s \in \mathbb{C}$. The TFM of the reduced model (10), tfwhere k eigenvalues are retained, is

$$\hat{\mathbf{G}}^i(s) = \mathbf{\Gamma}_k^i (s\mathbf{I} - \mathbf{\Lambda}_k^i)^{-1} \boldsymbol{\beta}_k^i - \mathbf{\Gamma}_k^i (\mathbf{\Lambda}_k^i)^{-1} \boldsymbol{\beta}_k^i \tag{13}$$

Subtracting (12) & (13),

$$\mathbf{G}^i(s) - \hat{\mathbf{G}}^i(s) = \mathbf{\Gamma}_{n-k}^i (s\mathbf{I} - \mathbf{\Lambda}_{n-k}^i)^{-1} \boldsymbol{\beta}_{n-k}^i \\ + \mathbf{\Gamma}_{n-k}^i (\mathbf{\Lambda}_{n-k}^i)^{-1} \boldsymbol{\beta}_{n-k}^i \tag{14}$$

For stable systems, it follows from [10], [5] that

$$\left\| \mathbf{G}^i - \hat{\mathbf{G}}^i \right\|_\infty = \max_{\omega\in\mathbb{R}} \left\| \mathbf{\Gamma}_{n-k}^i (s\mathbf{I} - \mathbf{\Lambda}_{n-k}^i)^{-1} \boldsymbol{\beta}_{n-k}^i \right. \\ \left. + \mathbf{\Gamma}_{n-k}^i (\mathbf{\Lambda}_{n-k}^i)^{-1} \boldsymbol{\beta}_{n-k}^i \right\|_2 \\ \leq \left\| \mathbf{\Gamma}_{n-k}^i \right\|_2 \left\| \boldsymbol{\beta}_{n-k}^i \right\|_2 \max_{\omega\in\mathbb{R}} \left\| (\jmath\omega\mathbf{I} \right. \\ \left. - \mathbf{\Lambda}_{n-k}^i)^{-1} + (\mathbf{\Lambda}_{n-k}^i)^{-1} \right\|_2 \tag{15}$$

Now,

$$(\jmath\omega\mathbf{I} - \mathbf{\Lambda}_{n-k}^i)^{-1} + (\mathbf{\Lambda}_{n-k}^i)^{-1} \\ = \mathrm{diag}\left(\frac{1}{\phi_{k+1}^i + \jmath\psi_{k+1}^i}, \ldots, \frac{1}{\phi_n^i + \jmath\psi_n^i} \right) \tag{16}$$

where $\phi_k^i = \left(a_k^i - \dfrac{2a_k^i b_k^i}{\omega} \right)$,
$\psi_k^i = \left(\dfrac{(a_k^i)^2 - (b_k^i)^2}{\omega} + b_k^i \right)$, and (a, b) are the real and imaginary parts of the eigenvalue λ. The maximum value is attained at

$$\omega = \frac{(a_{k+1}^i)^2 + (b_{k+1}^i)^2}{b_{k+1}^i}$$

Equation (15) simplifies to

$$\left\| \mathbf{G}^i - \hat{\mathbf{G}}^i \right\|_\infty \leq \left\| \mathbf{\Gamma}_{n-k}^i \right\|_2 \left\| \boldsymbol{\beta}_{n-k}^i \right\|_2 \max_{\lambda\in\{\lambda_{k+1},\ldots,\lambda_n\}} \frac{1}{|a_{k+1}^i|} \tag{17}$$

Given that out of the N topologies, some topologies may be redundant and not practically possible, the error bound for the AMRS framework is

$$\epsilon = \max\left\{ \left\| \mathbf{G}^i - \hat{\mathbf{G}}_{min}^i \right\|_\infty \right\}, \forall\ i \in \Sigma' \subseteq \Sigma \tag{18}$$

where $|\Sigma'| = M$, the number of admissible topologies. Here $\hat{\mathbf{G}}_{min}^i$ is the TFM corresponding to the lowest resolution model of the ith switching mode.

III. Case Study

We demonstrate the efficiency of the proposed framework on a DC-DC Buck-boost converter. The equivalent circuit and high-fidelity models are shown in Figs. 1, 2. The simulations are carried out using MATLAB 2018a on an Intel Xeon W-2145 Workstation with a clock speed of 3.70 GHz. We perform three sets of simulations: We first simulate the original full-order model. Then, we extract a reduced order model (ROM) of appropriate order in each switching topology of the circuit and simulate it. The frequency of the converter is taken as a measure of selecting the order of the reduced-order model [7]. Finally, we perform simulations using the AMRS framework. To assess the accuracy, we compare the states obtained by solving the original model with that of the reduced model and the AMRS framework respectively. Relative error and root-mean-squared error (RMSE) are used as error indices. The state-space model of the converter is given in the Appendix. The order of the original system is 7. For $V_i = 10$ V and frequency, $f_s = 30$ kHz of the MOSFET with duty cycle of 49%, simulation of the original (full-resolution), reduced (low-resolution) and the AMRS framework is performed. Waveforms of the diode voltage are shown in Fig. 3.

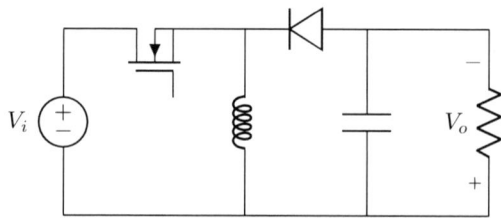

Fig. 1. DC-DC Buck-Boost Converter: Equivalent Circuit

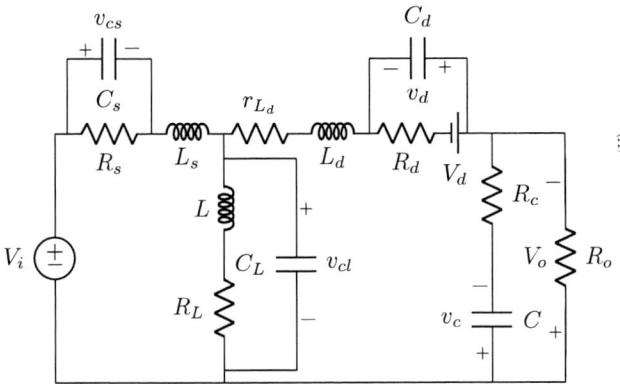

Fig. 2. DC-DC Buck-Boost Converter: High-fidelity Model

TABLE I
COMPARISON OF CPU SIMULATION TIMES & SOLUTION MATRIX
SIZES: BUCK-BOOST CONVERTER

	CPU Time (s)	Size of Solution Matrix
Original Simulation	416.52	3828882 × 7
ROM Simulation	11.89	35861 × 7
MRS	142.54	584422 × 7

TABLE II
COMPARISON OF ERROR INDICES: BUCK-BOOST CONVERTER

States	Error Indices			
	Rel. Error		RMSE	
	ROM	MRS	ROM	MRS
i_L	0.0003	0.0001	0.0003	0.0001
v_c	0.0003	0.0000	0.0025	0.0000
i_{L_s}	0.5940	0.0314	0.9333	0.0670
v_{c_s}	0.2958	0.0190	2.5654	0.1659
i_{L_d}	0.5412	0.0282	0.4860	0.0421
v_{c_d}	0.2717	0.0121	2.0093	0.1764
v_{c_l}	1.1215	0.0241	9.1813	0.2295

IV. CONCLUSION

We have presented an adaptive multi-resolution framework based on model order reduction for accelerating simulation speeds of PECs. MOR is applied to each switching mode of the circuit independently to obtain a reduced-order model. Each reduced-order model corresponds to a particular resolution. The simulation engine is set up in such a way that it automatically switches across different switching resolutions. This yields a powerful tool for faster simulation while ensuring that the approximation error is within limits. The

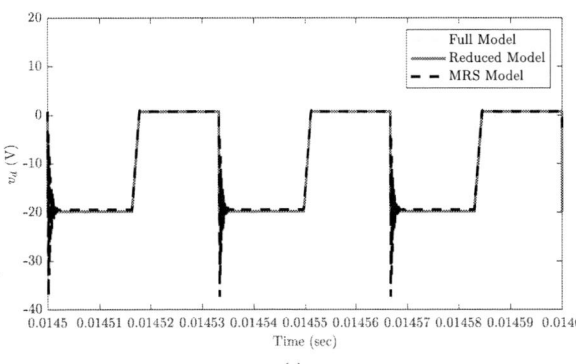

(a)

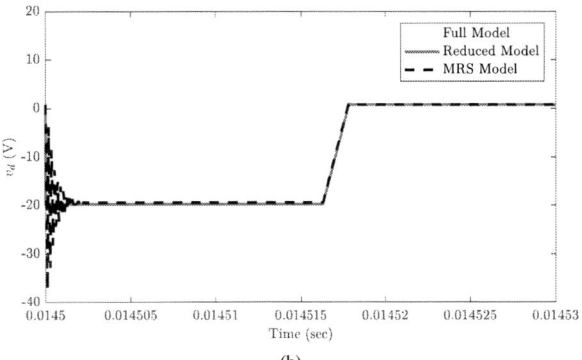

(b)

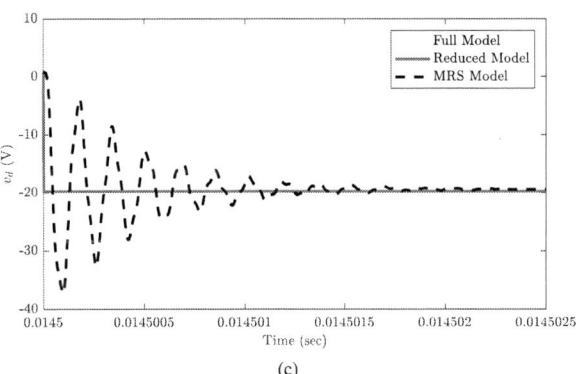

(c)

Fig. 3. Diode Voltage v_d (a) Multiple Cycles (b) Single cycle (c) Zoomed-in Waveform

method overcomes the drawbacks of variable step-size methods in tackling PEC simulation. Considerable improvements in simulation speeds and reduction in sizes of the solution matrices is achieved without any significant loss in accuracy.

APPENDIX

The state and input vector are:

$$\mathbf{x} = \begin{bmatrix} i_l & v_c & i_{Ls} & v_{cs} & i_{ld} & v_d & v_{cl} \end{bmatrix}^\top \quad \mathbf{u} = \begin{bmatrix} V_i \\ V_d \end{bmatrix}$$

$$\mathbf{A}^i = \begin{bmatrix} -\dfrac{R_L}{L} & 0 & 0 & 0 & 0 & 0 & \dfrac{1}{L} \\[2mm] 0 & -\dfrac{1}{C(R_o+R_c)} & 0 & 0 & \dfrac{R_o}{C(R_o+R_c)} & 0 & 0 \\[2mm] 0 & 0 & 0 & -\dfrac{1}{L_s} & 0 & 0 & -\dfrac{1}{L_s} \\[2mm] 0 & 0 & \dfrac{1}{C_s} & -\dfrac{1}{C_s R_s} & 0 & 0 & 0 \\[2mm] 0 & -\dfrac{R_o}{L_d(R_o+R_c)} & 0 & 0 & -\dfrac{R_d}{L_d}-\dfrac{R_c R_o}{L_d(R_o+R_c)} & -\dfrac{1}{L_d} & -\dfrac{1}{L_d} \\[2mm] 0 & 0 & 0 & 0 & \dfrac{1}{C_d} & -\dfrac{1}{C_d R_d} & 0 \\[2mm] -\dfrac{1}{C_L} & 0 & \dfrac{1}{C_L} & 0 & \dfrac{1}{C_L} & 0 & 0 \end{bmatrix}$$

$$\mathbf{B}^i = \begin{bmatrix} 0 & 0 \\ 0 & 0 \\ \dfrac{1}{L_s} & 0 \\ 0 & 0 \\ 0 & 0 \\ 0 & \dfrac{1}{R_d C_d} \\ 0 & 0 \end{bmatrix} \qquad \mathbf{C}^i = \begin{bmatrix} I \end{bmatrix}_{7\times 7}$$

TABLE III
CIRCUIT PARAMETERS: BUCK-BOOST CONVERTER

V_i	10 V	f	30 kHz	L	4.54 mH
C_s	10 nF	L_s	20 nH	$R_{s,on}$	20 mΩ
$R_{s,off}$	20 MΩ	C_d	5 nF	$R_{d,on}$	220 mΩ
$R_{d,off}$	10 MΩ	$V_{d,on}$	0.7 V	$V_{d,off}$	0 V
R_o	64.5 Ω	L_d	50 nH	r_L	20 mΩ
C	96 μF	L_c	1 nH	R_c	345 mΩ

TABLE IV
EIGENVALUES

Switch Modes	sw_m : On sw_d : On	sw_m : On sw_d : Off	sw_m : Off sw_d : Off	sw_m : Off sw_d : On
λ_1	-8.80	-8.81	-1.12e1 + j9.38e4	-1.44e2 + j1.50e3
λ_2	-1.78e4	-1.60e2	-1.12e1 - j9.38e4	-1.44e2 - j1.50e3
λ_3	-8.34e6	-1.52e6 + j1.20e8	-1.60e2	-9.66e5 + j1.33e8
λ_4	-1.96e6 + j1.18e8	-1.52e6 - j1.20e8	-7.21e5 + j1.35e8	-9.66e5 - j1.33e8
λ_5	-1.96e6 - j1.18e8	-2.41e6 + j3.69e7	-7.21e5 - j1.35e8	-4.68e6 + j3.31e7
λ_6	-3.02e9	-2.41e6 - j3.69e7	-2.72e6 + j5.22e7	-4.68e6 - j3.31e7
λ_7	-4.99e9	-4.99e9	-2.72e6 - j5.22e7	-3.02e9

[8] P. L. Chapman, "Multi-resolution switched system modeling," in *2004 IEEE Workshop on Computers in Power Electronics*, pp. 167–172, August 2004.

[9] C. Pedicini, L. Iannelli, F. Vasca, and U. Jönsson, "Averaging for power converters," in *Dynamics and Control of Switched Electronic Systems: Advanced Perspectives for Modeling, Simulation and Control of Power Converters* (F. Vasca and L. Iannelli, eds.), ch. 5, pp. 163–188, Springer London, 2012.

[10] P. Benner, "Numerical linear algebra for model reduction in control and simulation," *GAMM-Mitteilungen*, vol. 29, no. 2, pp. 275–296, 2006.

[11] L. Feng, J. G. Korvink, and P. Benner, "A fully adaptive scheme for model order reduction based on moment matching," *IEEE Transactions on Components, Packaging and Manufacturing Technology*, vol. 5, pp. 1872–1884, Dec 2015.

REFERENCES

[1] P. L. Evans, A. Castellazzi, and C. M. Johnson, "Design tools for rapid multidomain virtual prototyping of power electronic systems," *IEEE Transactions on Power Electronics*, vol. 31, pp. 2443–2455, March 2016.

[2] C. Deml and P. Turkes, "Fast simulation technique for power electronic circuits with widely different time constants," *IEEE Trans. Ind. Appl.*, vol. 35, pp. 657–662, May 1999.

[3] L. F. Shampine and M. W. Reichelt, "The MATLAB ODE suite," *SIAM J. Sci. Comput.*, vol. 18, no. 1, pp. 1–22, 1997.

[4] J. Schönberger, "An overview of simulation tools," in *Dynamics and Control of Switched Electronic Systems: Advanced Perspectives for Modeling, Simulation and Control of Power Converters* (F. Vasca and L. Iannelli, eds.), ch. 13, pp. 391–416, Springer London, 2012.

[5] A. Antoulas, *Approximation of Large-Scale Dynamical Systems*. Society for Industrial and Applied Mathematics, 2005.

[6] H. Khan, M. A. Bazaz, and S. A. Nahvi, "Simulation acceleration of high-fidelity nonlinear power electronic circuits using model order reduction," *IFAC-PapersOnLine*, vol. 51, no. 1, pp. 273 – 278, 2018.

[7] H. Khan, M. A. Bazaz, and S. A. Nahvi, "Singular perturbation based model reduction of power electronic circuits," *IET Circuits Devices and Systems*, 2019. to appear in.

A Full Soft Switched Bridgeless Power Factor Corrected AC-DC Converter

Sevilay Cetin
Technology Faculty
Pamukkale University
Denizli, Turkey
scetin@pau.edu.tr

Veli Yenil
Cardak OSB MYO
Pamukkale University
Denizli, Turkey
veliyenil@pau.edu.tr

Abstract-In this study, a soft swithed and bridgeless power factor corrected (BPFC-FSS) boost converter with an active snubber cell is presented. The converter is operated with pulse with modulation (PWM) and the average current mode control is used to generate PWM signals. The soft switching operation of all semiconductors is achieved by a snubber circuit in introduced converter. The snubber circuit allows zero voltage transition (ZVT) turn on and zero voltage switching (ZVS) turn off for the boost switch. In addition, zero current switching (ZCS) turn on and ZVS turn off of the snubber switch are provided. The boost diode and the other snubber diodes work with soft switching. Moreover, the current stress of the snubber switch is descended by the soft switching energy delivery to the output. Thes soft switching operation of all semiconductors is accomplished for different load case. Thus, the conduction and switching losses are reduced and the efficiency is increased. The theoretical analysis of the BPFC-FSS is presented and validated with a simulation work operating at 100 kHz, with 1 kW output power and 400 V output voltage.

I. INTRODUCTION

In recent years, the nonlinear electric appliances which create harmonic currents cause decrease of power quality of the system. Therefore, power factor correction (PFC) circuits have become important to improve the power factor of system. A variety of circuit topologies have been developed for PFC applications. The PFC topologies can be employed in switching mode power supplies (SMPS), battery chargers, electronic ballasts and the other industrial applications fed from AC line.

Boost converters have been used widely in different industrial areas, due to high power density, fast transition response, the simple structure and easy to implement. The boost converter following a diode bridge rectifier is the most commonly used in the PFC applications. In conventional boost PFC converter, the current flows through three semiconductor devices, two diodes are at the rectifier stage and one is at the boost stage. These diodes exhibit a forward voltage that leads to conduction losses. Therefore, researchers start to develop new alternatives known as bridgeless PFC to reduce conduction losses. In the BPFC converter, current flows through only two semiconductor devices. Unlike the traditional boost PFC converter, BPFC converter improves efficiency by removing two rectifier diodes.

PFC converters can operate in discontinuous conduction mode (DCM), boundary conduction mode (BCM), and continuous conduction mode (CCM). The CCM operation of the boost converter is generally preferred at high power levels. However, reverse recovery power loss of boost diode leads to reduced efficiency and this power loss worsens when the switching frequency is increased. Electromagnetic interference (EMI) is the important issue at high frequency applications as well. Therefore, soft switching (SS) techniques should be used to overcome problems mentioned above [1]-[4].

In [1], the conventional zero voltage transition (ZVT) PWM converter is proposed. The ZVT turn-on for the boost switch and zero current switching (ZCS) turn off for the boost diode are accomplished very well. However, the snubber switch hardy turn off and it has high current stress. To overcome these problems, different methods have been reported [5]-[18]. In [5], [9] and [15], the boost switch has an extra current stress and the boost diode has extra voltage stress in [6]. The snubber diode has an extra voltage stress in [5], [9]. In [16], extra voltage stress is occurred across the snubber switch. In [11], soft switched turn on process of the main switch worsens at partial load conditions. In [14], the bridgeless SEPIC converter with positive output voltage is introduced. The converter has one switch but three semiconductor components are in power flow path. The main switch works with hard switching condition as well. In [17], [18], The ZVT and zero current transition (ZCT) techniques are adopted for bridgeless PFC converter. Thanks to auxiliary circuit, there are no extra voltage and current stress on the main switches.

In the soft switching techniques addressed above, the current stress of the snubber switch is high because of the discharge of the capacitor parallel connected to the boost switch. The soft switching techniques presented in [16], [19]-[22], reduce current stress of the snubber switch in the boost converter. In these studies, soft switched turn on for the boost switch is accomplished and low current stress of the snubber switch as well. The low current stress is accomplished by a transformer used in a snubber circuit. The presented snubber circuit in [16] and [21], works a transformer and this transformer requires high magnetizing inductance. Besides, the energy of the magnetizing inductance is absorbed by passive components. This magnetizing energy results in extra voltage stress across the snubber switch as well. The introduced snubber cell in [20] needs a center tapped transformer to provide low current stress across the snubber

978-1-5386-7688-2/19 $31.00 © 2019 IEEE

switch. The snubber switch operates with soft switching as well. However, SS turn off for the snubber switch deteriorates at light load conditions.

In this study, a bridgeless power factor corrected full soft switched (BPFC-FSS) converter which overcomes many of problems discussed before is constructed. In the converter, the boost switch work with soft switching; it turns on with ZVT and turns off with almost ZVS. The snubber switch turns on with almost ZCS and turns off with almost ZVS. All diodes including boost and the other snubber diodes work with soft switching and they have no an extra voltage stress. The boost switch and diode have no extra current stress. In the snubber cell of the BPFC-FSS converter, since the most of the SS energy is delivered to the output, the current stress of the snubber switch is reduced. In addition, the BPFC architecture reduces the conduction losses and used SS technique work well for different load cases. The operation modes of the BPFC-FSS converter is analyzed in detail, and simulation results are given at 100 kHz with 1 kW output power and 400 V output voltage.

II. Operation Principles of The Proposed Converter

The circuit diagram of the proposed BPFC-FSS converter in Fig. 1 consist of two parts; the boost and the snubber circuits. In the boost circuit, v_{ac} is the input voltage and rectified from the input voltage, i_{ac} is input current, V_o is the output voltage, L_B is the boost inductor, T_{B1} and T_{B2} are the boost switches, driven with same PWM signals, D_{B1} and D_{B2} are the boost diodes, D_{TB1} and D_{TB2} are the body diodes of the boost switch. In the snubber circuit, T_S is snubber switch, L_{S1} and L_{S2} are the snubber inductors, C_{S1}, C_{S2} and C_{S3} are the snubber capacitors and D_{S1}-D_{S5} are used as the snubber diodes.

In a half line cycle, the boost inductor of the BPFC-FSS is energized with the conduction of T_{B1} and D_{TB2} then stored energy in the boost inductor is transferred to the output by D_{B1}. In the second half line cycle, another boost operation is occurred by the conduction of T_{B2}, D_{TB1} and D_{B2}.

In the analysis of the BPFC-FSS, the operation of the first half line cycle is took into consideration. All of used semiconductor components are assumed as ideal except D_{B1}, D_{B2}. The current of L_B inductance and the voltage of C_o are accepted as constant in one switching cycle. Based on these assumptions, the converter operation in a switching cycle can be divided into eleven operations. The waveforms for the operation of the BPFC-FSS converter is illustrated in Fig. 2.

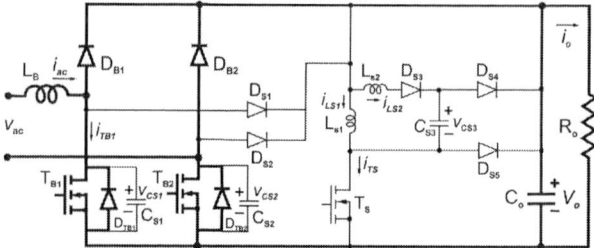

Fig. 1 The circuit scheme of the proposed BPFC-FSS converter.

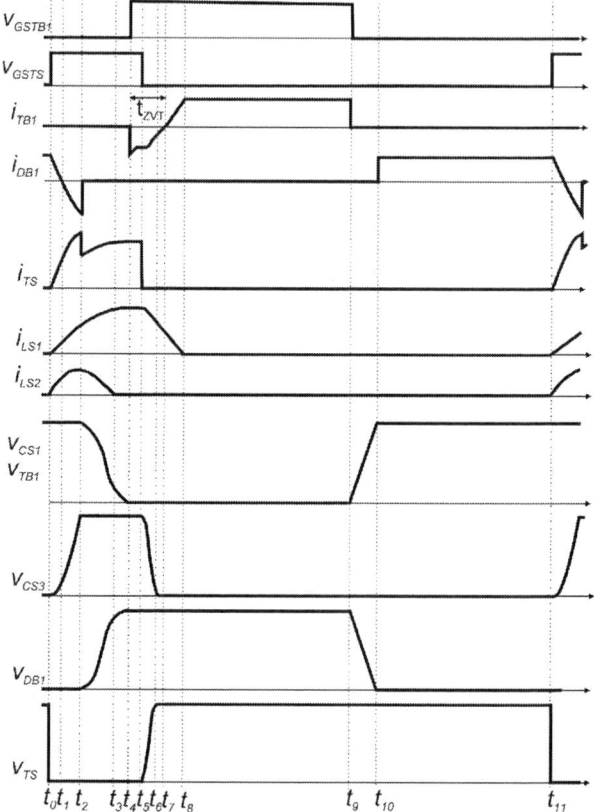

Fig. 2 The waveforms of proposed BPFC-FSS converter.

At $t=t_o$, it is assumed that the input current flows through the D_{B1} diode. When the PWM signal is applied to T_S, it turns on with almost ZCS and i_{LS2} current increases with the resonance occurred between L_{S2} and C_{S3}. At the same time, C_{S3} capacitor charges and the current of L_{S1} increases. The equation for this mode can be defined as follows:

$$L_{S1} \frac{di_{L1}}{dt} = V_o \tag{1}$$

$$L_{S2} \frac{di_{LS2}}{dt} = V_o - v_{CS3} \tag{2}$$

$$C_{S3} \frac{dv_{s3}}{dt} = I_i - i_{LS1} \tag{3}$$

At $t=t_1$, i_{DB1} drop to zero and the i_{LS1} reaches to the input current. The reverse recovery current begins to flow in D_{B1}.

At $t=t_2$, v_{CS3} equals to V_o, D_{S4} turns on with ZVS and D_{B1} turns off with ZCS. When the boost diode turns off, another resonance starts between L_{S1}, L_{S2} and C_{S1}. In this resonant mode, C_{S1} capacitor discharges, L_{S1} current increases and L_{S2} current decreases. At the same time, D_{S4} is still on state and the current of L_{S2} is delivered to the output. The equations of this operation can be extracted as below:

$$L_{S1}\frac{di_{LS1}}{dt} + L_{S2}\frac{di_{LS2}}{dt} = V_o \tag{4}$$

$$C_{S1}\frac{dv_{CS1}}{dt} = i_{LS1} \tag{5}$$

$$i_{TS} = i_{LS1} \tag{6}$$

At $t=t_3$, the current flowing through L_{S2} current descends to zero, D_{S4} is turned off. Another resonance occurs between L_{S1} and C_{S1}. This resonance maintains the decrease of the voltage of C_{S1} and the increase of the current of L_{S1}. The equations representing the behavior of the operation can be given as follow:

$$L_{S1}\frac{di_{LS1}}{dt} = v_{CS1} \tag{7}$$

$$C_{S1}\frac{dv_{CS1}}{dt} = i_{LS1}. \tag{8}$$

Then, C_{S1}'s voltage drops to zero and the body diode of the T_{B1} begins to carry the current. Thus, during the conduction of D_{TB1}, T_{B1} switch can be turned on with ZVT. At time $t=t_4$, C_{S1} voltage drops to zero.

At $t=t_5$, the snubber switch turns off and the current of L_{S1} discharges the C_{S3} capacitor by the conduction of D_{S4}. Because of the resonance happened between L_{S1} and C_{S3}, T_S turns off with almost ZVS.

At $t=t_6$, when the voltage of C_{S3} capacitor drops to zero and the snubber switch turns off, D_{S5} diode is turned on with ZVS.

At $t=t_7$, when the current of L_{S1} drops to input current I_i, T_{B1} switch turns on with ZVT.

At $t=t_8$, D_{S5} diode turns off when the current of L_{S1} drops to zero and the current of T_B reaches I_i. As a result, the on stage of traditional boost converter starts to work.

At $t=t_9$. T_{B1} and T_{B2} are turned off with ZVS. The current of L_B charges the C_{S1} capacitor by turn off of T_{B1} and T_{B2}.

At $t=t_{10}$, D_{B1} diode turns on under the ZVS condition when C_{S1} capacitor charges to V_o.

At $t=t_{11}$, the one switching cycle is completed.

III. DESIGN PROCEDURE

A. SNUBBER CIRCUIT DESIGN

To achieve soft switching conditions of the snubber switch and the boost diodes, following equations can be used.

$$L_{S1} \geq \frac{V_o}{I_i} t_{rTs} \tag{9}$$

$$L_{S1} \geq \frac{V_o}{I_i} 3t_{rr}. \tag{10}$$

Here, t_{rTs} is the rising time of T_S and t_{rr} is the reverse recovery time of D_{B1}. The ZCS turn on for snubber switch and turn off for boost diode is provided by the snubber inductance.

To achieve ZVS turn off for T_{B1} and T_{B2} switch, the voltage of switches must reach V_o in the falling time, t_{fTB}. Thus, C_{S1} and C_{S2} can be calculated as follows:

$$C_{S1} \geq \frac{I_i}{V_o} t_{fTB1} \tag{11}$$

$$C_{S2} \geq \frac{I_i}{V_o} t_{fTB2}. \tag{12}$$

Above, t_{fTB1} and t_{fTB2} represent the falling time of T_{B1} and T_{B2}.

The snubber capacitor C_{S3} provides the ZVS turn off for the snubber switch, it can be calculated as follows:

$$C_{S3} \geq \frac{I_i}{V_o} t_{fTs} \tag{13}$$

Here, t_{fTs} is the falling time of the snubber switch.

B. CONTROL CIRCUIT DESIGN

In the control method of BPFC-SS converter, average current mode control is used to generate PWM signals both boost and snubber switch. T_{B1} and T_{B2} can be driven with same signal. The PWM signal of T_s should be applied just before the control signal of T_{B1} or T_{B2} and ends after turn on of T_{B1} or T_{B2}. It is also assumed that the converter is operated with CCM which means that the current of L_B never falls to zero.

The average current mode control used to obtain sinusoidal input current for the proposed BPFC-SS converter, consists of two parts which are the current control loop design and the voltage control loop design. The block diagram of average current mode control is shown in Fig. 3.

In the voltage control loop design, the sensed output voltage $v_{o\text{-}sensed}$ is compared to the reference output voltage v_{ref} and an error is produced. This error is multiplied with sensed sinusoidal reference current obtained from the rectified line voltage $v_{i\text{-}sensed}$. Thus, a reference signal i_{ref} is obtained then compared to the measured inductor current $i_{LB\text{-}sensed}$. The boost switches T_{B1} and T_{B2} are switched according to produced error to provide high power factor.

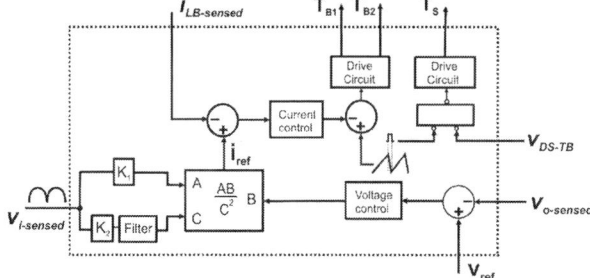

Fig. 3 The block diagram of average current mode control.

C. POWER CIRCUIT DESIGN

In the power stage design, L_B inductor is determined to

provide PFC and operate in CCM. Thus, L_B can be defined based on following equation for universal line voltage range.

$$L_B = \frac{V_{ac(min)} \cdot D}{f_s \cdot \Delta I} \quad (14)$$

Above, ΔI represents the inductor ripple current flowing in L_B and D is the maximum duty ratio at low line voltage $V_{ac(min)}$. The acceptable ripple current can be selected as 20% to provide PFC.

The current stress of power semicondcutors can be defined based on the peak line current. The peak line current, $I_{ac(pk)}$, can be defined as follows at low line

$$I_{ac(pk)} = \frac{2 \cdot P_o}{\sqrt{2} \cdot V_{ac(min)}} \quad (15)$$

The filter capacitor C_o is determined took into consideration of the output voltage ripple, output power and hold-up time requirement.

The voltage stress of power semiconductors are limited by the output voltage.

IV. SIMULATION RESULTS

A simulation study is performed to validate the proposed operation of the BPFC-SS converter. The simulation work is operated for 400 V output voltage and 1 kW output power. The switching frequency is selected as 100 kHz and 220 V_{ac} input voltage is applied to the converter. PSIM program is used to validate the operation of BPFC-SS converter. The simulated circuit schematic of proposed converter is shown in Fig. 4.

According to design procedure given in previous section, the circuit parameters are determined as, L_B=200 µH, C_o=960µF, L_{S1}=12 µH, L_{S2}=6 µH, C_{S1}, C_{S2}=2 µF, C_{S3}=1 µF. The power semiconductors are selected according to their voltage and current stress defined in previous section. The used components and their performance are summarized in Table I.

TABLE I. THE POWER SEMICONDUCTORS USED IN THE SIMULATION OF BPFC-SS CONVERTER.

Semiconductors	Part Name / Specifications
T_{B1}, T_{B2}	IXFK36N60 / 600 V – 36 A
D_{B1}, D_{B2}	DSEI19-06AS / 600 V – 20 A
T_S	IXFH20N60Q / 600 V – 20 A
D_{S1}-D_{S4}	DSEI19-06AS / 600 V – 20 A

The implemented control signals for the boost and snubber switches are shown in Fig. 5.

The boost switch turns on with ZVT and turns off with ZCS as shown in Fig. 6. ZVT turn on is achieved with the conduction of D_{TB1}. At the turn off process, ZVS turn off of T_B is achieved by the C_{S2}'s charge. The voltage stress of the switch is reduced.

The soft switching operation of the snubber switch is given in Fig. 7. The snubber switch turns on with ZCS and turns off with ZVS. The reduced current stress of the snubber switch achieved and additional voltage stress across the snubber switch is not occurred.

The waveforms for the boost diode is given in Fig. 8. The voltage stress of the diode is restricted by the output voltage. It turns on with ZVS and turns off with ZCS.

As it can be seen in Fig 9, the waveforms of the input voltage and the current are almost in same phase. The power factor (PF) is obtained as 0.998, very close to unity, at full power and total harmonic distortion of the line current (THD$_i$) is obtained as %4, at full power. Fig. 10 gives the input voltage and current waveform at with PF measurement, at 50% load condition. The PF is measured as 0.989 at half power. The PSIM simulation have function providing PF and THDi measurement. The obtained PF and THD values are extracted directly from transient analysis in the simulation. In the simulation work, the efficiency of the BPFC-FSS is measured as %97.4 at full load condition.

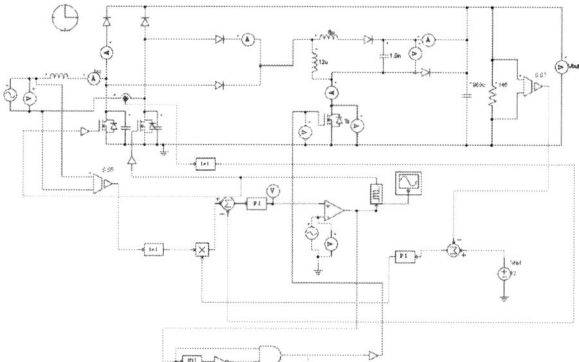

Fig. 4 The simulation schematic of the proposed BPFC-FSS converter.

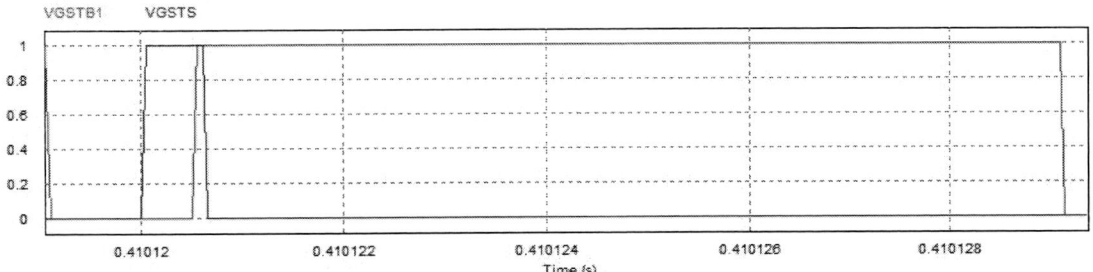

Fig. 5. The control signals of the boost switch, V_{GS-TB1}, and the snubber switch, V_{GS-TS}.

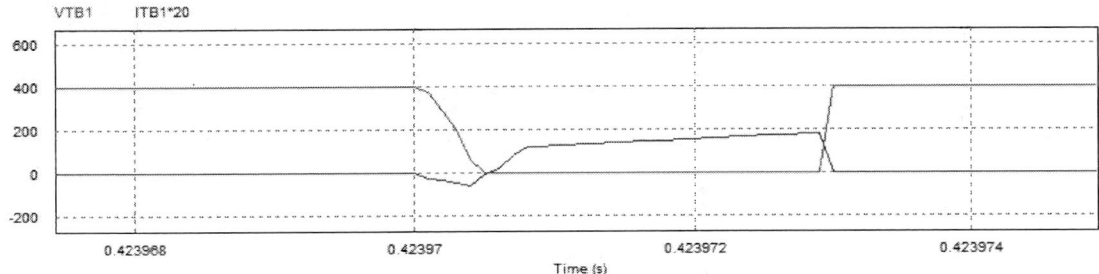

Fig. 6 The current and voltage waveforms of the boost switch, i_{TB1} and v_{TB1}.

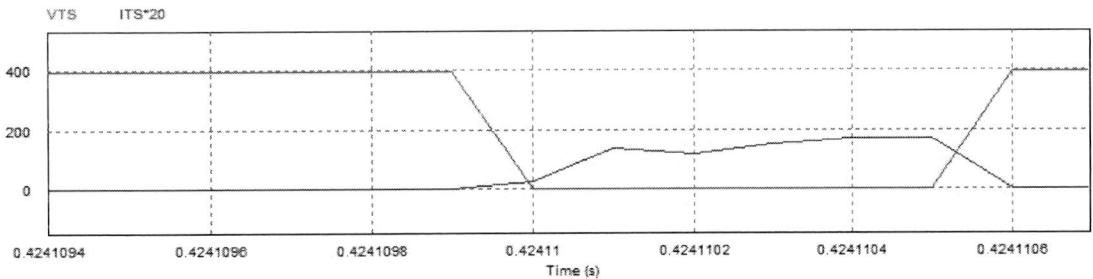

Fig. 7 The current and voltage waveforms of the snubber switch, i_{TS} and v_{TS}.

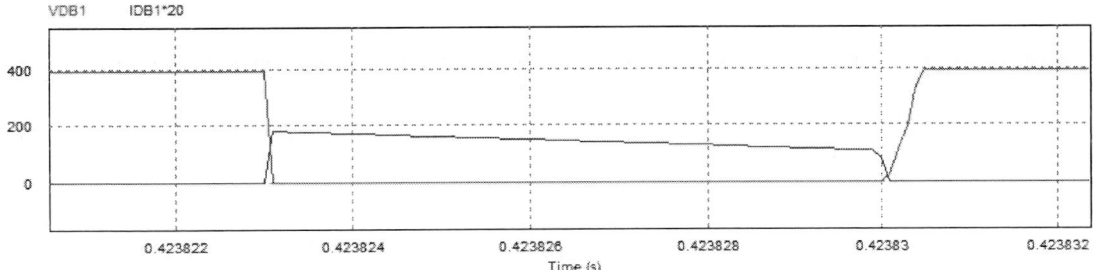

Fig. 8 The current and the voltage waveforms of the boost diode, i_{DB1} and v_{DB1}.

978-1-5386-7688-2/19 $31.00 © 2019 IEEE

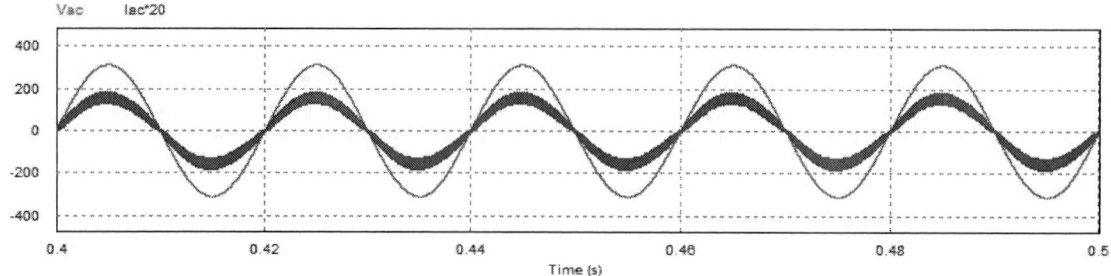

Fig. 9 The line voltage (v_{ac}) and input line current (i_{ac}) waveforms of the BPFC-SS converter at full power.

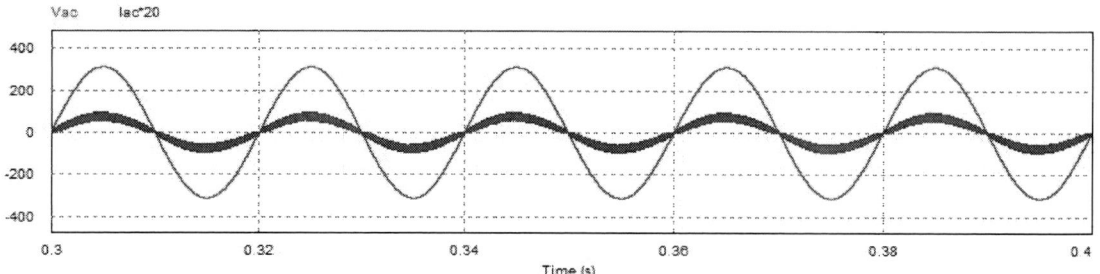

Fig. 10 The line voltage (v_{ac}) and input line current (i_{ac}) waveforms of the BPFC-SS converter at 50% load.

V. CONCLUSIONS

In this work, a full soft switched bridgeless power factor corrected AC-DC converter is presented. In the presented converter, all semiconductor devices are soft switched and their voltage stress are suppressed by the output voltage. The presented snubber circuit accomplishes low current stress for the snubber switch by the transfer of soft switching energy to the output. In order to verify the system performance, the converter performed by a simulation study which operates with 1 kW output power and 400 V output voltage, at 100 kHz operation frequency. The simulation results give coherent results compared to the presented theoretical analysis.

ACKNOWLEDGMENT

This work is supported by Pamukkale University Scientific Research Coordination Unit, under grant number 2019KKP081.

REFERENCES

[1] G. Hua, C.S. Leu, Y. Jiang, and F.C. Lee, "Novel Zero-Voltage-Transition PWM Converters", *IEEE Trans. on Power Electron.*, vol. 9, pp. 213-219, March 1994.

[2] G. Hua, E.X. Yang, Y. Jiang, and F.C. Lee, "Novel Zero-Current-Transition PWM Converters", *IEEE Trans. on Power Electron.*, vol. 9, no.6, pp. 601-606, November 1994.

[3] H. Mao, F.C. Lee, X. Zhou, H. Dai, M. Cosan, and D. Boroyevich, "Improved Zero-Current-Transition Converters for High-Power Applications", *IEEE Trans. on Ind. Applicat.*, vol.33, no.5, pp. 1220-1232, September/October 1997.

[4] J.G. Cho, J.W. Baek, G.H. Rim, and I. Kang, "Novel Zero-Voltage-Transition PWM Multiphase Converters", *IEEE Trans. on Power Electron.*, vol.13, no.1, pp. 152-159, January 1998.

[5] C.J. Tseng, and C.L. Chen, "Novel ZVT-PWM Converters with Active Snubbers", *IEEE Trans. on Power Electron.*, vol.13, no.5, pp. 861-869, September 1998.

[6] V. Grigore, and J. Kyyra, "A New Zero-Voltage-Transition PWM Buck Converter", *9th Mediterranean Electrotechnical Conf. (MELECON'98)*, Tel Aviv, Israel, pp. 1241-1245, April 1998.

[7] K.M. Smith, and K.M. Smedley, "Properties and Synthesis of Passive Lossless Soft-Switching PWM Converters", *IEEE Trans. on Power Electron.*, Vol.14, No.5, pp. 890-899, September 1999.

[8] D.Y. Lee, B.K. Lee, S.B. Yoo, and D.S. Hyun, "An Improved Full-Bridge Zero-Voltage-Transition PWM DC/DC Converter with Zero-Voltage / Zero-Current Switching of the Auxiliary Switches", *IEEE Trans. on Ind. Applicat.*, vol.36, no.2, pp. 558-566, March/April 2000.

[9] C.M.O. Stein, C.M.O., H. L. Hey, "A True ZCZVT Commutation Cell for PWM Converters", *IEEE Trans. on Power Electron.*, vol.15, no.1, pp. 185-193, January 2000.

[10] T.W. Kim, H. S. Kim, and H. W. Ahn, "An Improved ZVT PWM Boost Converter", *31th Power Electron. Spec. Conf. (PESC'00)*, Galway, Ireland, pp. 615-619, June 2000.

[11] H. Bodur H., and A.F. Bakan, "A New ZVT-PWM DC-DC Converter", *IEEE Trans. on Power Electron.*, Vol.17, No.1, pp. 40-47, January 2002.

[12] H. Yu, B.M. Song, and J.S. Lai, "Design of a Novel ZVT Soft-Switching Chopper", *IEEE Trans. on Power Electron.*, Vol.17, No.1, pp. 101-108, January 2002.

[13] D.Y. Lee, M.K. Lee, D.S. Hyun, and I. Choy, "New Zero-Current-Transition PWM DC/DC Converters Without Current Stress", *IEEE Transactions on Power Electronics*, vol.18, no.1, pp. 95-104, January 2003.

[14] H.-T. Yang, H.-W. Chiang, and C.-Y. Chen, "Implementation of bridgeless cuk power factor corrector with positive output voltage," *IEEE Trans. Ind. Appl.*, vol. 51, no. 4, pp. 3325–3333, Jul. 2015.

[15] A.F. Bakan, H. Bodur, and I. Aksoy, "A Novel ZVT-ZCT PWM DC-DC Converter", *11th Europen Conference on Power Electronics and Applications (EPE2005)*, Dresden, Germany, pp. 1-8, Sept. 2005.

[16] Y. Jang, M.M. Jovanovic, K.H. Fang, and Y.M. Chang, "High-Power-Factor Soft-Switched Boost Converter", *IEEE Trans. on Power Electron.*, vol. 21, no.1, pp. 98-104, January 2006.

[17] M. Mahdavi andH. Farzanehfard, "Zero-current-transition bridgeless PFC without extra voltage and current stress," *IEEE Trans. Ind. Electron.*, vol. 56, no. 7, pp. 2540–2547, Jul. 2009.

[18] M. Mahdavi and H. FarzanehFard, "Zero-voltage transition bridgeless single-ended primary inductance converter power factor correction rectifier," *IET Power Electron*, vol. 7, no. 4, pp. 895–902, Apr. 2014.

978-1-5386-7688-2/19 $31.00 © 2019 IEEE

[19] H. Bodur, S. Cetin, and G. Yanik, "A new zero-voltage transition pulse width modulated boost converter", *IET Power Electronics*, vol. 4, no.4, pp. 827-834, August 2011.

[20] H. Bodur and S. Cetin, "An improved high-power factor AC-DC soft-switched AC-DC converter", *International Review of Electrical Engineering*, vol. 7, no. 5, pp. 5299-5309, October 2012.

[21] H. Y. Tsai, T. H. Hsia, and D. Chen, "A family of zero voltage-transition bridgeless power-factor-correction circuits with a zero-current-switching auxiliary switch", *IEEE Transactions on Industrial Electronics*, vol. 58, no.5, pp. 1848-1855, May 2011.

[22] S. Cetin, "Power Factor Corrected and Fully Soft Switched PWM Boost Converter, IEEE Transactions on Industry Applications, Volume: 54, Issue: 4, pp. 3508-3517, July-August 2018.

A Method for Accelerating FPGA Based Digital Control of Switched Mode Power Supplies

Tudor Gherman
Applied Electronics Department
Technical University of Cluj-Napoca
Cluj-Napoca, Romania
Tudor.Gherman@ael.utcluj.ro

Dorin Petreus
Applied Electronics Department
Technical University of Cluj-Napoca
Cluj-Napoca, Romania
Dorin.Petreus@ael.utcluj.ro

Remus Teodorescu
Department of Energy Technology
Aalborg University
Aalborg, Denmark
ret@et.aau.dk

Abstract— **As Field Programmable Gate Arrays (FPGAs) became available in small packages and at low prices, these devices have been increasingly used to replace traditional Digital Signal Processors (DSPs) in digital control applications. Besides the considerable advantages such as parallelism, high frequency operation and great flexibility, FPGAs also come with some drawbacks. While code development time can be reduced by the use of high level synthesis tools, synthesis and implementation times drastically reduce productivity. This paper presents a novel digital platform meant to accelerate the design process of FPGA based control systems by developing a real time configurable hardware system that allows modifying the topology and the controller's parameters at run time. The design procedure proposed will be exemplified on a custom Zynq based board featuring a high performance analog interface used to control a synchronous buck converter.**
Keywords—digital control, FPGA, high level synthesis

I. INTRODUCTION

The increased flexibility, the greater complexity of the control algorithms that can be implemented, the ease of interfacing with other systems are some of the advantages that made digital control the preferred choice in many industrial control applications, designers having now two main types of digital hardware platforms available: Field Programmable Gate Arrays (FPGAs) and Digital Signal Processors (DSPs) [1]. DSPs are generally regarded as a cheaper solution, but as electronic systems require more complex algorithms to meet the expected efficiency and performance and as more and more functionalities need to be added to differentiate the product and make it more profitable on market, using an FPGA may result in a lower cost per implemented function [1]. The parallel architecture, fast on chip static RAM memories (BRAM) and the presence of specialized powerful digital signal processing blocks (DSP Slices) differentiate the FPGAs from conventional DSPs and make them more suitable in systems where high performance and integration are required. Several comparisons between FPGAs and DSPs in applications such as synchronous motor control [2], Static Compensator (STATCOM) control [3], artificial intelligence [4][5] or signal processing applications [6] prove that drastically reduced execution times, increased controller bandwidth and increased throughput are the major advantages of using FPGA devices. Surprisingly, also a superior performance/Watt in the favor of FPGAs was reported [4][6]. However, DSPs benefit of the advantage of being widely used, of the large amount of code available and a reduced compilation time [5] leading to a reduced time to market of the developed product. Another important aspect is related to the programming languages, DSPs being associated with more popular programming languages than FPGAs. To mitigate these shortcomings, the concept of high level synthesis (HLS) was introduced. A HLS tool takes as

input code designed in C or other high level descriptions languages and transforms it in a low level cycle accurate register level transfer (RTL) specification [7]. HLS products have been available for almost two decades [8] and, while many of them were discontinued or have limited use, some have reached maturity and are successfully used in industry. One of the most used HLS tool for rapid prototyping is the MATLAB's HDL Coder which accepts as input MATLAB script files or Simulink designs [9]. DSP Builder and Xilinx System Generator also use Simulink for describing the input specifications, but have the major drawback of being limited by the available component libraries [10]. MATLAB script files and Simulink provide a very convenient mean of describing the input specifications since they are also used to design and simulate digital control systems. However, this paper will focus on Xilinx's Vivado HLS because it comes at no extra cost with the Vivado design suite and offers users a high degree of flexibility.

Vivado HLS is part of the Vivado HLx Edition and it allows users to describe a module's behavior in C/C++ transforming it into a register transfer level implementation that can be synthetized for a Xilinx FPGA [11]. The output of the Vivado HLS design flow is a reusable functional block, known as an IP core, that can be added to user repositories and can be used in Vivado block design projects. This approach is very well suited for designing IP cores required to implement computational demanding arithmetical operations such as those specific to digital control of power converters since describing arithmetical operations in C/C++ is often straight forward. However convenient this may be for the controller design, the high speed ADC and DAC controllers required for the analog interface and other timing sensitive blocks, such as the PWM generator and the clocking architecture, still need to be developed in a low level hardware description language if they are not already available.

This paper presents a digital development platform that enables fast testing and tuning of FPGA based control systems. The goal of the developed system is to offer the possibility of performing several measurements (the power converter transfer function bode plot, controller step response or open loop transfer function bode plot) that help characterizing the controller and the power system. Debugging errors is made possible before closing the loop, thus reducing the risk of hardware damage. Verifying that the frequency characteristic of the system is the expected one (while operating in closed loop) helps confirming that the mathematical models used are correct. The digital system architecture (Fig. 1), the perturbation injection method, the loop transfer function measurement technique and the enhanced PWM generator are innovative elements that confer novelty to the paper. Besides the ability of modifying the architecture and parameters at run time, one

978-1-5386-7688-2/19 $31.00 © 2019 IEEE

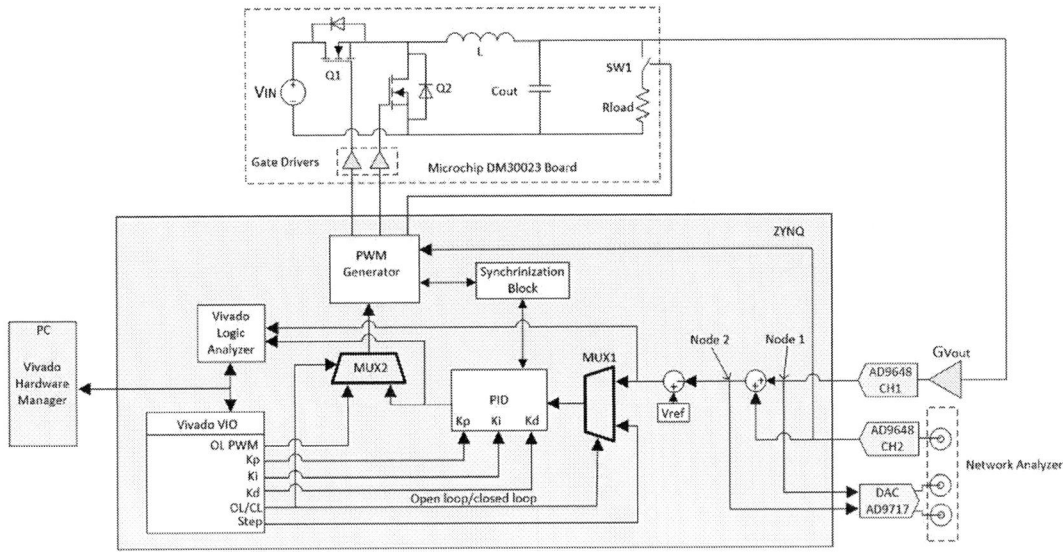

Fig. 1: Control System Architecture.

of the major advantages of the presented system is that it eliminates the need of soldering or adding extra components (such as an extra resistor in the feedback and an injection transformer [12]) to the power converter (which may be an already developed board) for perturbation injection.

The reminder of this paper is organized as follows. In section II the system architecture and the design procedure are described. Section III will present the controller design and implementation. Section IV will present the experimental results, while conclusions will be drawn in Section V.

II. SYSTEM DESCRIPTION

A. Control System Architecture

The implemented system's architecture is presented in Fig. 1. The power converter used to test the control algorithm will be one of the converters of a DM300023 Microchip demo board while the digital control is implemented on a ZYNQ SoC from Xilinx (XC7Z020). The ZYNQ is a hybrid device containing a dual-core ARM processing system (PS) and programmable logic (PL) [13]. The PL part is composed of the typical Xilinx FPGA hardware structures and is the only part used for this paper. The PL will be further referred to as FPGA to outline the compatibility of the presented system with any 7 Series Xilinx FPGAs. A 2 channel external ADC (AD9648) is used to monitor the converter's output voltage and to read the disturbance output of a network analyzer. An external DAC (AD9717) is also used for the system's open loop transfer function measurement. The voltage reference, the error amplifier, the compensator and the pulse width modulator (PWM) generator will be developed in the programable logic of the ZYNQ device. The interaction between the hardware system and the user is made possible through two IP cores provided by Xilinx: the Integrated Logic Analyzer (ILA) and the Virtual Input Output (VIO). Both feature a JTAG interface that, over a USB-JATG bridge, assure the connection with the Vivado Hardware Manager that runs on the PC. The ILA is used to capture any design internal signal and provides an advanced trigger mechanism that facilitates capturing any desired event. The VIO core will be used to

TABLE 1: SYSTEM PARAMETERS

Parameter	Symbol	Value
Buck converter filter inductor	L	39uH
Buck converter filter capacitor	C_{out}	660μF
Buck converter load resistor	Rload	0.94Ω
Buck converter inductor DCR	R_L	0.1
Buck converter filter Capacitor ESR	R_C	20mΩ
Buck converter nominal output voltage	V_O	1.8V
Buck converter nominal input voltage	V_{IN}	9V
Output voltage sensor gain	G_{Vout}	0.16
ADC delay	t_{ADC}	170ns
ADC full scale voltage	V_{FS}	1V
Compensator delay	t_{cntrl}	40ns
ADC and DAC sampling rate	ADC_{SR}	100MSPS
ADC and DAC resolution	n_{ADC}	14
System clock frequency	SYS_CLK	100MHz

dynamically configure the control system architecture and parameters at run time. The entire design, including the VIO, ILA and the controller IP cores, runs in a single 100MHz clock domain thus eliminating any clock domain crossing synchronization issues.

As already mentioned in Section II, the main advantage of the system is the possibility of dynamically reconfiguring the digital circuit topology and update the controller's parameters. To take advantage of these features, the following design procedure is proposed. Once the FPGA is programmed, the VIO core is accessible from the Vivado Hardware manager interface and is able to configure MUX1 and MUX2 multiplexors (Fig. 1) for open loop functioning. In this mode of operation, the controller will be isolated from the rest of the loop and will have its inputs driven by the VIO IP core and its output connected to one of the ILA channels. The digital PWM (DPWM) modulator will also have its input driven by the VIO core and will generate a constant PWM signal. A network analyzer can be used to inject a disturbance signal through ADC channel2 while monitoring the converter's output, enabling the power stage transfer function measurement. The controller's response to a step input (generated by the user through the VIO core) can be captured with the ILA core and viewed in the Vivado Hardware Manager window on the PC. As explained above, the controller is totally isolated from the rest of the system in the open loop operation mode and can be tested

978-1-5386-7688-2/19 $31.00 © 2019 IEEE

independently. Having the power converter's transfer function measured, the controller's coefficients can be computed, updated and its step response measured. Once the desired response is obtained, the loop can be closed with minimum risk of malfunctions or hardware damage. To verify the closed loop behavior, MUX1 and MUX2 can be configured to operate the system in the closed loop mode, which represents the normal mode of operation. Since the closed loop stability is usually assessed by studying the closed loop transfer function, obtaining its bode plot by experimental means is also useful. As described in [12], it is possible to make this measurement with the circuit functioning in closed loop. However, instead of injecting the disturbance signal through an injection transformer, ADC channel2 will be used for this purpose. The ADC channel1 output will be looped back to the DAC's channel1 (Node1 in Fig. 1), while Node2 which represents the superposition of the output voltage and the disturbance signal is sent to the DAC's channel2. The network analyzer can then measure the complex gain between Node1 and Node2 which represents the open loop gain of the system [12]. The error signal, also monitored by the ILA, will also indicate if the system is stable and if limit cycling is present. It is important to note that all these tests and the controller tuning is done without reconfiguring the FPGA. In Section IV a more detailed description of the configuration options for each measurement is presented.

Regarding the system's implementation options, it is a matter of choice which functional blocks are developed in Vivado HLS and which are developed in a different environment. However, the clocking architecture, the reset signal generation and the sampling period generation (the Synchronization Block) need to be implemented as separate blocks using the preferred hardware description language (VHDL was the choice for the project described). The discrete PWM generator block was also developed in VHDL since it requires the instantiation of specific IO FPGA primitives to maximize the output resolution.

B. Controller Design Process

The development process proposed is summarized in the following steps:
1. Design the top level function, which will describe the IP core behavior. C was the choice for the programming language used.

2. Generate the stimuli for the designed top level function and obtain the expected results (golden data) through simulation means. The simulation tool used for this purpose was Matlab. An example of how to use Matlab with the Vivado HLS design flow can be found in [14].

3. Design the Test Bench that reads the stimuli for the top level function's inputs, calls the top level function and compares the outputs against the golden data to verify that the input code functionality is correct before moving to the next steps in the design flow. This step, called C simulation, helps reducing development time.

4. Once the desired functionality is obtained, proceed to synthesizing the design. Vivado HLS generates a synthesis report that provides detailed information about the estimated performance of the IP core.

5. Run the C/RTL co-simulation to validate the resulting RTL. Vivado HLS will automatically create the simulation files using the stimuli provided for the Test Bench. The RTL outputs are compared against the golden data and Vivado HLS will prompt the result of this comparison.

6. Once the C/RTL simulation has passed, the RTL output files will be packed to be imported in a Vivado IP library.

7. Instantiate the control IP core developed in the block design described in Section A, and test the controller behavior on hardware.

III. CONTROLLER DESIGN AND IMPLEMENTATION

A. Discrete Time Small Signal Modeling of the Synchronous Buck Converter

For designing a stable switched mode power supply (SMPS) it is essential to have an accurate small signal model for the power converter. The averaged model is widely used in the case of analog control implementations. However, it does not take into account the effects of time and amplitude quantization that digital control techniques introduce. This paper will use the discrete time small signal modeling technique described in [15, pp. 121-165] that determines the accurate small signal response of SMPSs. The converter's relevant circuit parameters are summarized in Table 1.

The first step of the method described in [15] is the determination of the State Space representation matrices A_c, B_c, C_c, E_c where index c represents the switching interval that the matrices describe. U will denote the input vector, X the state vector and Y the output vector. The switching period will be labeled T_s, while the duty factor D. During interval 0, Q1 is in the on state and Q2 in the off state, while in interval 1 Q1 is in the off state and Q2 in the on state. Equations (1), (2), (3) and (4) describe the synchronous buck converter state space representation.

$$A_1 = A_0 = \begin{pmatrix} -\dfrac{R_C + R_L}{L} & -\dfrac{1}{L} \\ \dfrac{1}{C_{out}} & 0 \end{pmatrix} = A \qquad (1)$$

$$B_1 = \begin{pmatrix} \dfrac{1}{L} & \dfrac{R_C}{L} \\ 0 & -\dfrac{1}{C_{out}} \end{pmatrix} \qquad (2)$$

$$B_0 = \begin{pmatrix} 0 & \dfrac{R_C}{L} \\ 0 & -\dfrac{1}{C_{out}} \end{pmatrix} \qquad (3)$$

$$C_0 = C_1 = \begin{pmatrix} 0 & 0 \\ R_C & 1 \end{pmatrix} = C \qquad (4)$$

The delay the control loop introduced is caused by the ADC acquisition and conversion time (t_{ADC}), the computational delay of the compensator (t_{cntrl}) but also by the modulation delay (t_{DPWM}). t_{DPWM} represents the time interval between the latching of the control command and the generation of the modulated edge [15, p. 104].

$$t_d = t_{ADC} + t_{cntrl} + t_{DPWM} \qquad (5)$$

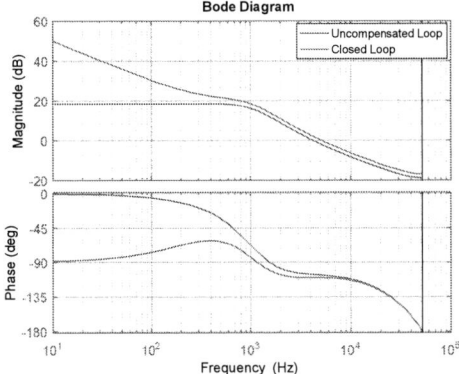

Fig. 2: Uncompensated loop and closed loop transfer functions.

The state equations describing the discrete time small signal model becomes:

$$\hat{x}(z) = (zI - \phi)^{-1} \gamma \hat{u}(z) \qquad (6)$$

$$\hat{y}(z) = \delta \hat{x}(z) \qquad (7)$$

The control to output transfer matrix is defined as:

$$W(z) = \begin{pmatrix} \dfrac{\hat{i}_L(z)}{\hat{u}(z)} \\[2mm] \dfrac{\hat{v}_0(z)}{\hat{u}(z)} \end{pmatrix} = \delta(zI - \phi)^{-1}\gamma \qquad (8)$$

Where γ, δ and ϕ are defined below:

$$\phi = e^{A_0(T_S - t_d)} e^{A_1 DT_S} e^{A_0(t_d - DT_S)} \qquad (9)$$

$$\gamma = \frac{T_S}{Nr} e^{A_0(T_S - t_d)} F \qquad (10)$$

$$\delta = C_0 \qquad (11)$$

$$F = (A_1 X + B_1 V) - (A_0 X + B_0 V) \qquad (12)$$

$$X = (I - e^{A_1 DT_S} e^{A_0(1-D)T_S})^{-1} *$$
$$* [-e^{A_1 DT_S} A_0^{-1}(I - e^{A_0(1-D)T_S})B_0 -$$
$$- A_1^{-1}(I - e^{A_1 DT_S})B_1] V \qquad (13)$$

The uncompensated loop transfer function (T_u) and the open loop transfer function are computed with the aid of Matlab based on the state space representation. The ideal bode plots are presented in Fig. 2.

B. Compensator Design

Based on the results obtained in Section III A, one can note that a PI structure is suitable for obtaining a 70 degree phase margin (φ_m) and a crossover frequency (ω_c) equal to 1/10 of the switching frequency (F_S), so the derivative coefficient of the PID controller will be set to 0. The digital controller design also follows the procedure described in [15, pp. 165-217]. A detailed description of this procedure is outside the scope of this paper, however the steps followed and the relations used are briefly described. The design starts by selecting a digital controller structure (with a corresponding transfer function expressed in the z domain) suitable to compensate the power converter's discretized transfer function. The controller's transfer function is then mapped in the continuous time domain where the design is easier to be performed and, finally, the design is back mapped to z-domain. The z domain transfer function of the PI compensator is presented in (14).

$$G_{PI}(z) = K_P + \frac{K_I}{1 - z^{-1}} \qquad (14)$$

By applying the bilinear transform to the digital model of the PID compensator, the equivalent s-domain transfer function is obtained (15). To obtain accurate results, all the z-domain frequency specifications are prewarped. The uncompensated loop transfer function becomes T_u' in the s domain while the crossover frequency becomes ω_c'.

$$G_{PI}(s) = G_{PI\infty}(1 + \frac{\omega_{PI}}{s}) \qquad (15)$$

By imposing the phase margin and crossover frequency constraints mentioned above, the equivalent continuous time domain coefficients can be determined:

$$\omega_{PI} = \frac{\omega_C'}{\tan(-\frac{\pi}{2} + \varphi_m - \angle(Tu'(j\omega_C')))} \qquad (16)$$

$$G_{PI\infty} = \frac{\omega_C'}{|Tu'(j\omega_C')|\sqrt{\omega_{PI}^2 + (\omega_C')^2}} \qquad (17)$$

Finally, the z-domain proportional and integrative coefficients of the PI controller are obtained:

$$K_P = G_{PI\infty}(1 - \frac{\pi\omega_{PI}}{\omega_S}) \qquad (18)$$

$$K_I = \frac{2\pi}{\omega_S} G_{PI\infty}\omega_{PI} \qquad (19)$$

Where ω_s is the switching angular frequency. To avoid limit cycling, the quantization effects of the discrete system need to be considered [15, pp. 217-248]. The ADC full scale voltage and resolution as well as the DPWM resolution are summarized in Table 3. The DPWM generator is assumed to use a digital counter counting up to Nr. The following conditions need to be imposed:

$$q_{vo(AD)} > q_{vo(DPWM)} \qquad (20)$$

$$q_{vo(AD)} > q_{vo(K_I)} \qquad (21)$$

Where:

$$q_{vo(AD)} = \frac{V_{FS}}{G_{Vout} 2^{n_{ADC}}} \qquad (22)$$

$$q_{vo(DPWM)} = \frac{V_{IN}}{2^{n_{DPWM}}} \qquad (23)$$

$$q_{vo(K_I)} = \frac{V_{IN} K_I q_{vo(AD)}}{Nr} \qquad (24)$$

Based on (20) and (21), the minimum number of DPWM resolution bits and the minimum integrative coefficient can be determined:

$$n_{DPWM} > \frac{V_{IN} G_{Vout} 2^{n_{AD}}}{V_{FS}} \qquad (25)$$

$$K_I < \frac{Nr}{V_{IN} G_{Vout}} \qquad (26)$$

Equations up to now have assumed a unitary gain for the ADC and, also a unitary value for the DPWM Nr. Thus, a scaling operation is required. The scaling coefficient is computed below:

$$\lambda = q_{vo(AD)} G_{Vout} Nr \qquad (27)$$

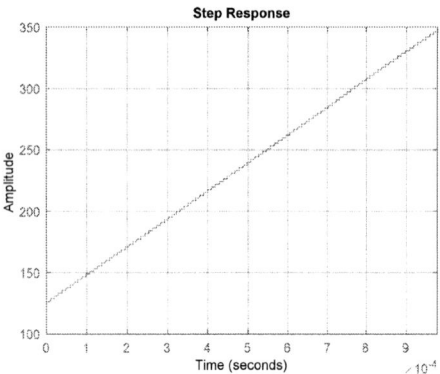

Fig. 3: PID controller step response.

Imposing a loop gain error of less than 1% at the crossover frequency and a loop gain error of less than 10% at low frequency, a <18,6> fixed point format was selected for Kp, Ki and Kp. The coefficients' computed values are summarized in Table 3. MATLAB can easily plot the step response (Fig. 3). The input excitation and the output data can be formatted in two distinct files (referred to as Input file and Golden Data file in the rest of this paper) to be later read from the Vivado HLS Test Bench.

C. Designing the Top Level Function and the Test Bench

The compensator will need to provide an input for the sensed output voltage and one output. To enable the real time modification of the reference, the proportional, integrative and derivative coefficients are also considered as inputs. The error signal computation will also be part of the HLS IP core. An anti-windup mechanism and output saturation were also added. The conceptual method of implementing the described PID controller is presented in [15]. Implementing the required operations in C is straight forward.

The Test Bench will read the stimuli from the Input file, call the top level function for each input sample and compare the top level function output against the samples read from the Golden Data file created by the Matlab environment. If the mean error is less than a predefined value, the test is considered passed. A graphical representation of the Golden Data file is presented in Fig. 3.

D. RTL Synthesis and Packiging

The Synthesis options can control clocking and reset behavior, interface properties, throughput optimization, latency and area, all configurable through a wide range of directives and control settings. Considering the low complexity of the PID design and the excellent performance obtained with the default environment options no effort to configure directives or control settings was required. The default synthesis settings [16] were also used for the IO protocol options. A clock port, a reset port, handshake signals and the top level function ports are automatically created. The Synchronization Block (Fig. 1) will manage the handshake signals. The "ap_start" handshake signal [16] will effectively define the sampling rate of the PID compensator, which will be equal to the switching frequency. The PID IP core will be clocked at 100MHz.

After running the C synthesis, the Synthesis Report is available. The utilization estimates generated for the designed compensator are presented in Fig. 4. By opening

Utilization Estimates

⊟ Summary

Name	BRAM_18K	DSP48E	FF	LUT
DSP	-	-	-	-
Expression	-	-	0	330
FIFO	-	-	-	-
Instance	-	12	498	147
Memory	-	-	-	-
Multiplexer	-	-	-	44
Register	-	-	272	-
Total	0	12	770	521
Available	280	220	106400	53200
Utilization (%)	0	5	~0	~0

Fig. 4: PID controller utilization Estimates.

	Operation\Control Step	C0	C1	C2	C3	C4
1	Vin V read(read)					
2	Vref V read(read)					
3	p Val2 2 (-)					
4	node 20(write)					
5	p Val2 4 (+)					
6	Uc i prev V load(read)					
7	p Val2 7 (+)					
8	tmp 3 (icmp)					
9	p Val2 3 (*)					
10	tmp 4 (select)					
11	tmp 5 (icmp)					
12	p Val2 8 (select)					
13	p Val2 5 (read)					
14	tmp2 V (-)					
15	p Val2 6 (*)					
16	node 58(write)					
17	node 59(write)					
18	tmp1(+)					
19	tmp s(+)					
20	Uc k A V(+)					
21	Uc k A V cast(+)					
22	tmp 7(icmp)					
23	Uc k A V 1(select)					
24	tmp 10(icmp)					
25	Uc k A V 2(select)					
26	node 60(write)					

Fig. 5: PID controller latency.

the Analysis Perspective, the synthesis results can be examined in more detail. In the Performance window (Fig. 5) the way Vivado HLS schedules the operations described in the top level function and their latency can be observed. A total of 4 clock cycles are necessary to compute the output of the compensator. The first clock cycle is necessary for reading the inputs and computing the error signal. The 2nd clock cycle is required for computing the proportional, integrative and derivative components. The integral anti-windup related operations are also executed in the 2nd cycle. The 3rd clock cycles is necessary to add the proportional, integrative and derivative components, while the last clock cycle is required for writing the outputs to the output port.

Once the synthesis results are considered satisfactory, the C/RTL cosimulation can be run. The input stimuli for the synthetized design simulation are the same used for the Test Bench, thus the output of the step response will be tested. The simulation results can be viewed using the Vivado Wave Viewer. Since the C/RTL simulation results are identical to the load step response captured with the ILA core on hardware and both match the Matlab simulation results, only the experimental results will be displayed (Fig. 8).

E. Discrete PWM Generator Implementation

The usual approach of implementing the discrete PWM generator relies on comparing the control input (u[k]) with the output of a digital ramp clocked by a clock signal with a period of T_{clk}[15]. Supposing the digital ramp's maximum value is a power of 2, the number of resolution bits of the DPWM generator would be given by (28).

978-1-5386-7688-2/19 $31.00 © 2019 IEEE

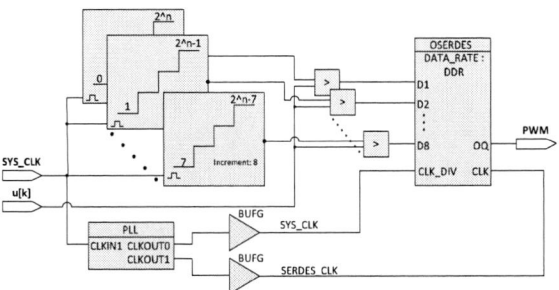

Fig. 6: Discrete PWM Generator.

TABLE 2: XC7Z020 CLOCKING TREE AND IO SPEED LIMITATIONS

FPGA Primitive	XC7Z020 Speed Grade			Units
	-3	-2	-1	
BUFG, BUFH	628	628	464	MHz
BUFIO	680	680	600	MHz
BUFR	420	375	315	MHz
OSERDES DDR Transmitter	1250	1250	950	Mb/s

$$n_{DPWM} = \log_2 \left(\frac{T_{clk}}{T_{sw}} \right) \qquad (28)$$

So, the DPWM resolution is limited by the frequency at which the digital ramp can run. Xilinx 7 Series FPGAs are divided into regions and logic in each region can have its clock sourced through one of the dedicated buffers presented in Table 2. Running the digital ramp at the maximum speed allowed by the clocking infrastructure, besides the frequency limitation and the increased power consumption, would also add an extra clocking domain to the design. As a consequence, the design implementation may fail with timing errors. A method of overcoming this problem is presented in Fig. 6. Instead of one counter counting from 0 to $2^{n_{DPWM}}-1$, 8 counters will be used. The reset value of each counter will correspond to its index (i.e. counter1 will start counting from 1, counter 2 will start counting from 2, etc.), each counter will have an increment of 8 and will be implemented on n_{DPWM} +3 bits. The output of each counter will be compared against the control input. Each comparator output will be connected to an input port of the parallel interface of an OSERDES parallel to serial converter. The OSERDES primitive features a variable width parallel input interface and is capable of outputting serial data in a single data rate mode (SDR) or a double data rate mode (DDR). The input data width and the operation mode (SDR or DDR) determine the frequency ratio between the input clock (divided clock) and the output clock (high speed clock). The implementation presented uses the DDR mode. So, for the 8 bit wide parallel interface, a 4:1 ratio between the high speed clock and the divided clock is required. For a 100MHz parallel clock, a 400MHz high speed clock would be required, increasing the DPWM resolution by 3 bits. The clock domain crossing is managed by the primitive itself, so no extra effort is required to synchronize the clock domains. Another advantage of this approach is the maximized output data rate.

IV. EXPERIMENTAL RESULTS

The design parameters used to capture the experimental results are summarized in Table 3. The Buck converter's circuit parameters are listed in Table 1.

Before testing the PID controller, the open loop transfer function of the buck converter is determined by experimental means. The open loop/closed loop VIO output (OL/CL) configures MUX2 to select the OL PWM signal as an input for the PWM generator. The disturbance signal generated by an Analog Discovery's waveform generator is read on the second channel of the ADC and added to OL PWM reference generated by the VIO. The gain measured with the Analog Discovery's Network Analyzer represents the convertor's gain multiplied by the ADC's conditioning circuit, ADC and DPWM gains (-30.1 dB are added to the converter's gain). The measurement result is presented in Fig. 7, representing the continuous domain transfer function of the converter. The controller responds to the sampled output value, so the discrete transfer function would be more of interest. Because the output voltage is well filtered (the case of small aliasing approximation [15, p. 116]) and the control loop delay is small, the behavior of the continuous and discretized transfer functions near the crossover frequency should be close enough for this measurement to be useful.

The second proposed measurement is the step response of the PID controller. The OL/CL VIO output configures MUX1 to select the Step VIO output (which is toggled from 0 to 1) as an input to the PID controller. The PWM outputs are disabled during this test. The Vivado Logic Analyzer is used to capture the controller's response to this excitation (Fig. 8). It can be observed that the PID response is identical with the MATLAB simulation output (Fig. 3).

After the controller is validated, the loop can be closed and the open loop transfer function can be measured (the principle is described in Section II). The OL/CL VIO output configures MUX1 to select the ADC's channel1 output as the controller's input and MUX2 to select the controller's output as the DPWM generator input. The ADC's channel2 will read the Network Analyzer's disturbance signal. Node 1 (before the injection point) and Node 2 (after the injection point) will each be converted into analog voltages. The open loop transfer function gain and phase margin of the system can be obtained by plotting the ratio between Node 2 and Node 1 (Fig. 9). Finally, a load step is applied by closing SW1. The load step response can be observed in Fig. 10.

TABLE 3: EXPERIMENTAL SETUP PARAMETERS

Parameter	Value
ADC bits used	9
Switching frequency	100KHz
DPWM resolution	13
Scaled PID proportional coefficient (Kp)	124.26
Scaled PID Integrative coefficient (Ki)	2.22
Scaled PID Derivative coefficient	0

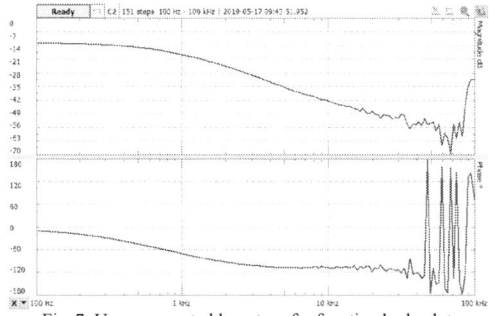

Fig. 7: Uncompensated loop transfer function bode plot.

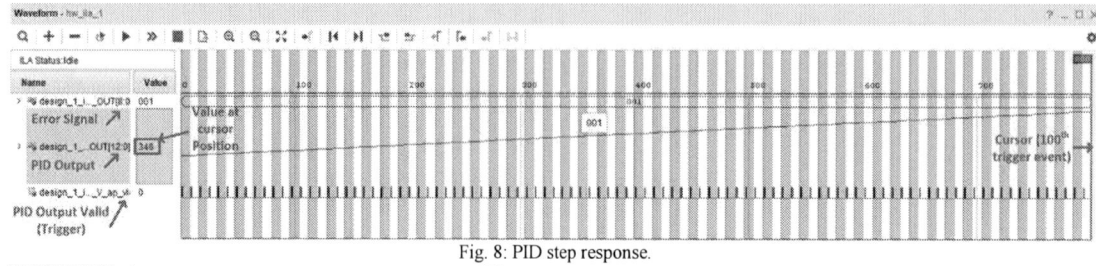

Fig. 8: PID step response.

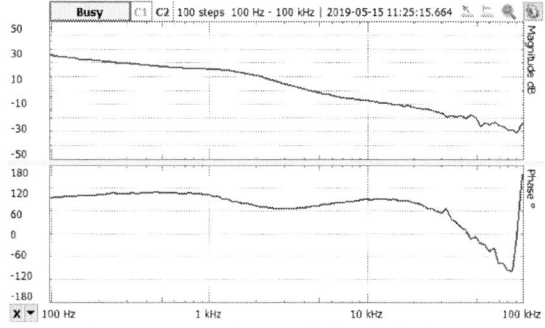

Fig. 9: System open loop transfer function bode plot.

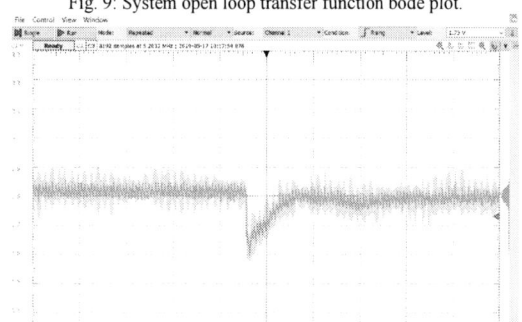

Fig. 10: Converter load step response. (horizontal scale: 150us/div, vertical scale: 100mV/div,).

V. CONCLUSIONS

In this paper a FPGA based digital hardware platform capable of accelerating the design process of digital control was presented. The disadvantage of long synthesis and implementation time associated with FPGA design was mitigated by allowing system parameters and topology to be changed at run time. The powerful debug capabilities and the validation methods enabled by the proposed architecture contribute to reducing development time as well. The FPGA technology used allows controlling a high performance analog interface featuring two 100MHz double date rate buses, computing the controller's output and processing the disturbance signal for enabling transfer function measurements. All these tasks are executed in parallel and with extremely low latencies, performance that can't be obtained using conventional DSPs. The HLS tools that the design process is based on represent a quick and friendly way of designing the controller. An extremely low latency PID controller that uses minimum FPGA resources was obtained by using Vivado HLS, tool that proves to be extremely reliable and that allows a high degree of customization.

VI. REFERENCES

[1] E. Monmasson *et al*, "FPGAs in Industrial Control Applications," in IEEE Transactions on Industrial Informatics, vol. 7, no. 2, pp. 224-243, May 2011.

[2] L. Idkhajine, E. Monmasson and A. Maalouf, "Extended Kalman Filter for AC drive Sensorless Speed Controller - FPGA-based solution or DSP-based solution,"2010 IEEE International Symposium on Industrial Electronics, Bari, 2010, pp. 2759-2764.

[3] C. A. Sepúlveda, J. A. Muñoz, J. R. Espinoza, M. E. Figueroa and C. R. Baier, "FPGA v/s DSP Performance Comparison for a VSC-Based STATCOM Control Application," in IEEE Transactions on Industrial Informatics, vol. 9, no. 3, pp. 1351-1360, Aug. 2013.

[4] E. Nurvitadhi *et al*, "Accelerating recurrent neural networks in analytics servers: Comparison of FPGA, CPU, GPU, and ASIC,"2016 26th International Conference on Field Programmable Logic and Applications (FPL), Lausanne, 2016, pp. 1-4.

[5] M. Kaminski and T. Orlowska-Kowalska, "Comparison of DSP and FPGA realization of neural speed estimator for 2-mass system,"2011 IEEE International Symposium on Industrial Electronics, Gdansk, 2011, pp. 1543-1548.

[6] Y. Bai *et al.*, "FPGA vs DSP: A throughput and power efficiency comparison for Hierarchical Enumerative Coding,"2013 IFIP/IEEE 21st International Conference on Very Large Scale Integration (VLSI-SoC), Istanbul, 2013, pp. 318-321.

[7] J. Cong *et al*, "High-Level Synthesis for FPGAs: From Prototyping to Deployment," IEEE Transactions on Computer-Aided Design of Integrated Circuits and Systems, vol. 30, no. 4, pp. 473-491, Apr 2011.

[8] R. Nane *et al.*, "A Survey and Evaluation of FPGA High-Level Synthesis Tools," in IEEE Transactions on Computer-Aided Design of Integrated Circuits and Systems, vol. 35, no. 10, pp. 1591-1604, Oct. 2016.

[9] Y. P. Siwakoti and G. E. Town, "Design of FPGA-controlled power electronics and drives using MATLAB Simulink," 2013 IEEE ECCE Asia Downunder, Melbourne, VIC, 2013, pp. 571-577.

[10] D. Navarro, Ó. Lucı́a, L. A. Barragán, I. Urriza and Ó. Jiménez, "High-Level Synthesis for Accelerating the FPGA Implementation of Computationally Demanding Control Algorithms for Power Converters," in IEEE Transactions on Industrial Informatics, vol. 9, no. 3, pp. 1371-1379, Aug. 2013.

[11] K. Georgopoulos *et al.*, "An Evaluation of Vivado HLS for Efficient System Design,"2016 International Symposium *ELMAR*, Zadar, 2016, pp. 195-199.

[12] S. Synkule, L. Heinzle, and F. Hämmerle, "DC/DC Converter Stability Measurement," 2018. [Online], Available: https://www.omicron-lab.com/fileadmin/assets/Bode_100/ApplicationNotes/DC_DC_Stability/App_Note_DC_DC_Stability_V3_3.pdf.

[13] Xilinx, "Zynq-7000 All Programmable SoC Tehnical Reference Manual," vol. 585, 2016.

[14] E. Monmasson and M. Ricco, "Advanced Design of FPGA-based Controllers for Power Electronic and Drive Applications," PhD Course, held at Aalborg University, Denmark, April 2018.

[15] L. Corradini, D. Maksimovic, P. Mattavelli and R. Zane, Digital Control of High-Frequency Switched Mode Power Converters, 1st ed. Wiley-IEEE Press, 2015.

[16] Xilinx, "High-Level Synthesis Vivado Design Suite User Guide," June 2018.

A Real Time Simulator of a PEV's On Board Battery Charger

Tudor Gherman
Applied Electronics Department
Technical University of Cluj-Napoca
Cluj-Napoca, Romania
Tudor.Gherman@ael.utcluj.ro

Dorin Petreus
Applied Electronics Department
Technical University of Cluj-Napoca
Cluj-Napoca, Romania
Dorin.Petreus@ael.utcluj.ro

Remus Teodorescu
Department of Energy Technology
Aalborg University
Aalborg, Denmark
ret@et.aau.dk

Abstract—**This paper presents an FPGA based real time (RT) simulator for a Plug in Electrical Vehicle's (PEV) on board battery charger. Common values for the switching frequencies of the AC-DC and DC-DC stages range between tens of kilohertz to hundreds of kilohertz, so the design effort aimed obtaining a very small time step in order to account for the switching effects. By taking advantage of the parallel structure of the FPGA and using Xilinx's Vivado High Level Synthesis tool, a time step of 40ns was obtained. The RT simulator was validated through simulation means and experimentally on a custom ZYNQ based board. The purpose of the developed intellectual property (IP) is to reduce the development cost and time of the charger's control system with the aid of hardware in the loop (HIL) simulations.**

Keywords—Real-time simulation, Hardaware in the Loop, FPGA, High Level Synthesis, EV charger

I. Introduction (*Heading 1*)

Real time simulations have gained an increased interest in automotive, aviation or industrial applications. HIL testing has become part of the development process helping to reduce the system software development time and cost [1]. Failures of the control system while testing at limit conditions or at fault conditions can result in expensive hardware damage or other sorts of hazards, risk that can be mitigated by using real time models of the plant [2] [3]. HIL simulations can be classified as signal level, power level or mechanical level simulations [2]. For signal level simulations, all other system components except de controller are simulated. Power level HIL simulations require that at least part of the power electronic system is implemented. The power converter's outputs will be part of the interface between the system under test and the RT simulator. In the case of mechanical level HIL simulations, it is just the mechanical component that is simulated.

A typical PEV on-board battery charger is designed with two stages: an AC-DC stage and a DC-DC stage [4]. The AC-DC stage is required to control the input harmonic content in accordance with the IEC 61000-3-2 standard, having a power factor correction (PFC) functionality. The requirements for the DC-DC stage are to provide galvanic insulation and to implement the charging method (continuous current – continuous voltage in the presented example) of the battery. For obtaining the high efficiency required for such applications, the two charger's stages are based on switched mode power supplies (SMPS) with soft switching techniques applied if possible. The behavior of SMPSs can be described by a system of state space equations for each switching interval. The task of solving the circuit equations reduces to solving a first order ordinary differential equations (ODE) system. ODE solvers can be classified as explicit or implicit,

depending on whether they use only present and past states of the system to predict the response of the circuit or they require future states as well. Another classification divides ODE solvers in single step methods and multi step methods. Single step methods rely exclusively on the values of the state variables at time step t_n to predict the solution for the following time step t_{n+1}. Multi step methods require the values of the state variables at multiple past or future time steps to make this prediction. A hybrid method used for real time simulations of power converters is the corrector-predictor method which uses an explicit method (predictor) to predict the values of the state variables at t_{n+1} and use this result in an implicit fashion to refine the prediction [5].

There are two approaches that can be adopted for SMPS RT simulations: using an averaged model, which drastically simplifies the simulator's design and relaxes the time step requirements or using a model that accounts for the switching effects as well. While using averaged models can give some insight for the power system's stability analysis, there are some significant disadvantages with this approach: the lack of accuracy [6], the fact that the discrete PWM (DPWM) generator will not be tested (an averaged model must be used instead) and the effect of the sampling time will not be predicted, aspect particularly important for current mode control where the high output current ripple can cause significant errors [7]. Another important item related to RT simulations of SMPSs is the ability to model limit cycling. Limit cycling is an effect caused by the time and amplitude quantization associated with digital control, one of the factors involved being the discrete PWM (DPWM) generator resolution. An averaged model will not be able to model this behavior. The state of the switches, which is determined based on the DPWM inputs, is evaluated by the simulator on each time step. It is therefore essential that the time step is small enough not to reduce the effective DPWM resolution "seen" by the converter's model. A false limit cycling condition can result otherwise. In order to achieve this objective, an explicit single step method (Numerical Integrator Substitution [8]) is proposed. As explained in [9], a circuit partitioning method is also essential for modeling discrete circuit elements such as inductors, shunt capacitors or switches.

The chosen hardware platform for implementation is a ZYNQ based custom board featuring a high performance (100 Msps) Digital to Analog Converter (DAC) capable of reproducing the outputs of the RT simulator which need to be updated on each time step. The unit under test (the charger's controller) can be either placed on a different platform or embedded on the same ZYNQ device as the RT simulator. For simplicity, the proposed solution will consider the controller and the RT charger model developed on the same

978-1-5386-7688-2/19 $31.00 © 2019 IEEE

device. The analog to digital converter (ADC) involved in the digital control system will be modeled in the FPGA fabric as well. Several high level software tools like System Generator or MATLAB's HDL Coder are suitable for control or real time simulation using FPGA technology. This paper will however use Xilinx's Vivado HLS which may imply a more elaborate development process, but it provides the possibility of controlling the IO interface behavior, latency and resource usage thorough a wide set of directives [10].

RT simulators for SMPS or for the PEV battery charging system have been discussed in other papers as well. [2] and [11] present an averaged RT model of power supplies implemented on a DSP system with an unspecified time step. In [12] a battery charger RT simulator is presented, but little details are offered regarding the SMPSs model design and performance. A very brief presentation of a similar modeling approach to the one used in this paper was presented in [13]. The novelty of this paper consists in applying RT simulation techniques that allow accounting for switching effects on a battery charger's power converters, in the detailed presentation of the converters' modeling and in the reduced time step obtained by using Vivado HLS.

The rest of this paper is organized as follows. Section II will discuss the RTE modelling of the battery charger's AC-DC and DC-DC stages. Section III will present the implemented HIL system, while in Section IV the simulation and experimental results will be discussed. Conclusions will be finally drawn in Section V.

II. BATTERY CHARGER REAL TIME SIMULATOR

A. AC-DC Stage

In the proposed example, the AC-DC stage of the charger is based on a boost converter. The continuous and discretized models are presented in Fig. 1. The PWM signal will divide the behavior of the converter in two intervals: Ton, when Q1 is on and Q2 is off, and Toff, when Q1 is off and Q2 is on. The switch network can be modeled as follows:

$$\begin{pmatrix} v_{sw\_o} \\ i_{sw\_o} \end{pmatrix} = \begin{pmatrix} S & 0 \\ 0 & S \end{pmatrix} \begin{pmatrix} v_{sw\_in} \\ i_{sw\_in} \end{pmatrix} \quad (1)$$

In the above equation, S will be 0 during time interval Ton and 1 during time interval Toff. In the continuous time domain, the inductor current can be expressed as follows:

$$i_L(t) = i_L(t - T_s) + \frac{1}{L_{PFC}} \int_{t-T_s}^{t} (v_{I\_PFC}(\tau) - v_{sw\_in}(\tau)) d\tau \quad (2)$$

$$v_C(t) = v_C(t - T_s) + \frac{1}{C_{PFC}} \int_{t-T_s}^{t} (i_{sw\_o}(\tau) - i_o(\tau)) d\tau \quad (3)$$

Where T_s represents the RT simulation time step and i_o represents the output current and it is deduced below:

$$i_o(t) = \frac{v_o(t)}{R_{Load}} + i_{Load}(t) \quad (4)$$

Equations (2) and (3) can be discretized using the Forward Euler method:

$$i_L(n+1) = i_L(n) + \frac{T_s}{L_{PFC}} (v_{I\_PFC}(n) - v_{sw\_in}(n)) \quad (5)$$

$$v_C(n+1) = v_C(n) + \frac{T_s}{C_{PFC}} (i_{sw\_o}(n) - i_o(n)) \quad (6)$$

B. DC-DC Stage

The proposed topology for the DC-DC stage is the Phase Shifted (PS) converter shown in Fig. 2. Zero Voltage Switching (ZVS), which is essential for obtaining a high efficiency, relies on the resonant circuit formed by the drain to source capacitance of the primary FETs and the parasitic inductance of the transformer. Typical values for the resonant circuit range between tens of MHz and hundreds of MHz, so this phenomenon is difficult to be model accurately. However, from the controller perspective, ZVS is irrelevant as long as the necessary dead time is provided. The PS converter's switch network comprised of Q1, Q2, Q3, Q4, D1 and D2 which is controlled by 4 PWM signals can be simplified, an equivalent switch network controlled by only 2 PWM signals being obtained. The resulting two control signals (PWM1 and PWM2) necessary to emulate the voltage across the two inductors are obtained using combinational logic from two of the gate drive signals (as shown in Fig. 3).

Fig. 1: Synchronous Boost converter a) continuous model b) Discretized model.

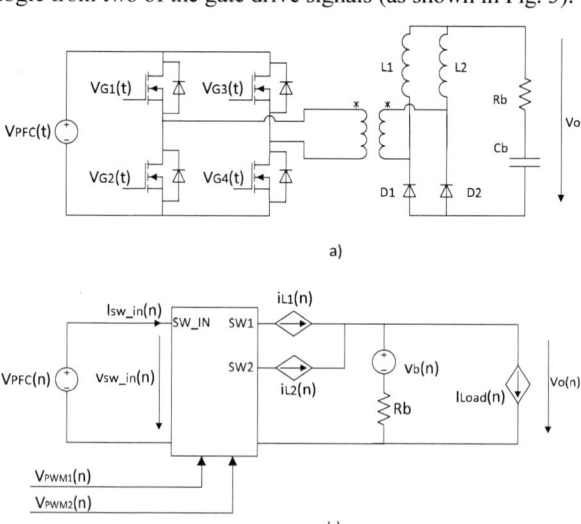

Fig. 2: Phase Shifted converter a) continuous model b) Discretized model.

978-1-5386-7688-2/19 $31.00 © 2019 IEEE

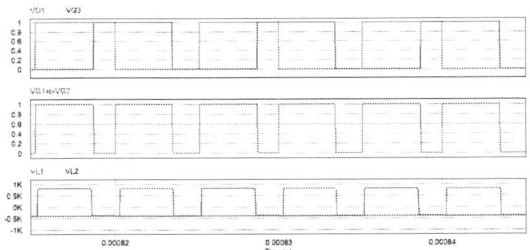

Fig. 3: Phase Shifted converter's inductor voltage as a function of the PWM inputs.

These relations are described in (6) and (7). As a result, the phase shifted PWM generator can be tested with the RT model developed.

$$PWM1 = (VG1 \text{xor} VG3) \text{and} VG1 \quad (7)$$

$$PWM1 = (VG1 \text{xor} VG3) \text{and} VG3 \quad (8)$$

The switch network can be modeled as follows:

$$
\begin{pmatrix} v_{sw1} \\ i_{sw1} \\ v_{sw2} \\ i_{sw2} \end{pmatrix} =
\begin{pmatrix} n*PWM1 & 0 \\ 0 & \dfrac{1}{n}*PWM1 \\ n*PWM2 & 0 \\ 0 & \dfrac{1}{n}*PWM2 \end{pmatrix}
\begin{pmatrix} v_{sw\_in} \\ i_{sw\_in} \end{pmatrix} \quad (9)
$$

Where n is the transfer ratio of the transformer. The voltage on the secondary inductors is expressed in (10).

$$v_{Lx}(t) = v_{SWx}(t) - v_{bat}(t) \quad (10)$$

In the previous equation, x represents the inductors' index. In the continuous time domain, the inductor current is expressed by relation (11), while the output capacitor (battery) voltage by relation (12).

$$i_{Lx}(t) = i_{Lx}(t-T_s) + \frac{1}{Lx}\int_{t-T_s}^{t}(v_{SWx}(\tau) - v_{bat}(\tau))d\tau \quad (11)$$

$$v_o(t) = v_b(t-T_s) + \frac{1}{C_b}\int_{t-T_s}^{t}(i_{Cb}(\tau))d\tau + R_b i_{Cb}(t) \quad (12)$$

The battery charging current (i_{Cb}) is the sum of the inductor currents. However, an external current source (i_{Load}) is added to the model so that the stability of the control loop can be easily assessed by generating a load step.

$$i_{Cb}(t) = i_{L1}(t) + i_{L2}(t) - i_{Load}(t) \quad (13)$$

In a similar manner as for the synchronous Boost converter, the state space equations can be discretized using the Forward Euler method.

$$i_{Lx}(n+1) = i_{Lx}(n) + \frac{T_s}{Lx}(v_{SWx}(n) - v_{bat}(n)) \quad (14)$$

$$v_o(n+1) = v_b(n) + \frac{T_s}{C_b}i_{cb}(n) + R_b i_{cb}(n) \quad (15)$$

$$i_{Cb}(n+1) = i_{L1}(n) + i_{L2}(n) - i_{Load}(n) \quad (16)$$

III. SIMULATION AND IMPLEMENTATION

A. RT Simulation Model Implementation

The parallelism provided by the FPGA fabric structures is extremely suitable for implementing a first order differential equation system solver. Look up tables (LUTs) are used to implement combinatorial logic such as additions and

subtractions. The switch network behavior can be implemented using multiplexors (in the FPGA fabric multiplexers are also implemented in LUTs), while multiplication operations are best handled by DSP slices. Divisions are costly in terms of hardware usage and latency, therefore are better to be avoided wherever possible. A conceptual hardware implementation of the synchronous Boost converter and of the PS converter are presented in Fig. 4 and Fig. 6.

To avoid the burden of describing all the required operations in a low level hardware description language, the model was described in C and Vivado HLS was used to translate it to a register level transfer implementation. The Vivado HLS Analysis Performance window can be used to visualize the manner in which the tool divides and schedules operations. Fig. 5 which is interpreted below, presents the results obtained for the Boost converter, while Fig. 7 presents the results for the PS converter.

The inductor voltage at time step t_n is computed based on the output capacitor voltage and the input voltage at the same time instant. In a similar manner, the capacitor current at time step t_n is computed based on the load current and the inductor current. These operations are completed in the first two cycles along with the input variables read operations and the switch assessment, being identified as operations 1-9 in the Vivado HLS Analysis Performance window (Fig. 5). The multiplexor functionality described in Fig. 4 can be described by an if statement in the Vivado HLS source file (operation 7 and 8). These combinatorial operations take one cycle to be completed. The inductor voltage and capacitor current are

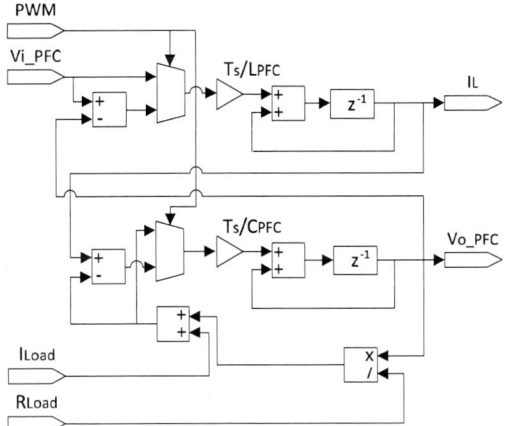

Fig. 4: Synchronous Boost RT simulation model.

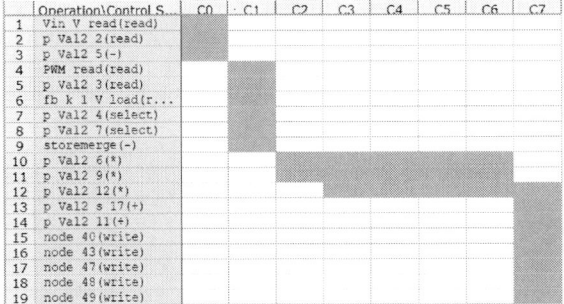

	Operation\Control S.	C0	C1	C2	C3	C4	C5	C6	C7
1	Vin V read(read)								
2	p Val2 2(read)								
3	p Val2 5(-)								
4	PWM read(read)								
5	p Val2 3(read)								
6	fb k 1 V load(r...								
7	p Val2 4(select)								
8	p Val2 7(select)								
9	storemerge(-)								
10	p Val2 6(*)								
11	p Val2 9(*)								
12	p Val2 12(*)								
13	p Val2 s 17(+)								
14	p Val2 11(+)								
15	node 40(write)								
16	node 43(write)								
17	node 47(write)								
18	node 48(write)								
19	node 49(write)								

Fig. 5: Vivado HLS operation scheduling for the Boost RT simulation model.

978-1-5386-7688-2/19 $31.00 © 2019 IEEE

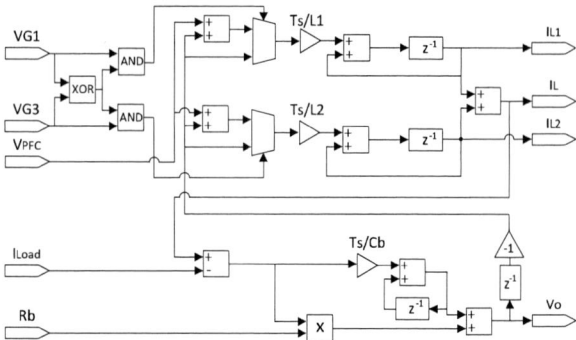

Fig. 6: Phase Shifted converter RT simulation model.

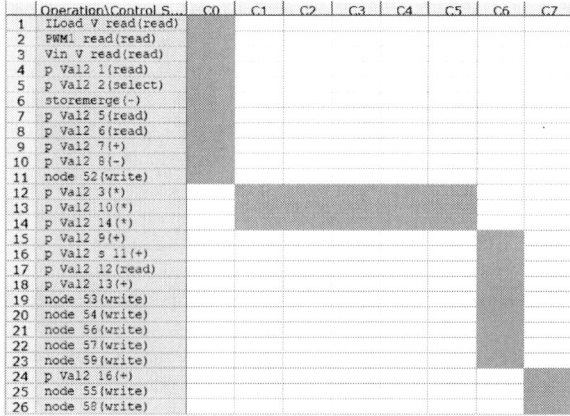

Fig. 7: Vivado HLS operation scheduling for the Phase Shifted converter RT simulation model.

further multiplied by integrative constant, operations that takes 5 clock cycles (operations 10-12). Finally, the multiplication output is added to the state's variable previous state value (operations 13 and 14), and the IP outputs will be updated during cycle C7. With a 200MHz clock, a time step of 40ns was obtained. The scheduling of the PS converter model (Fig. 7) can be interpreted in a similar manner.

B. Test Platform Implementation

The hardware platform used to test the developed RT simulation models of the charger's two stages is described in Fig. 9. The entire system is developed on the ZYNQ device, except for the digital to analog converter (DAC).

The AC-DC stage signal level HIL simulation system is composed of a sinusoidal input generator, the synchronous Boost RT simulation model, a sensor and the ADC model for the input current, input and output voltages and, finally, the control block. A DDS Compiler LogiCore IP from Xilinx's Vivado Repository was used to generate the input sinusoidal voltage. The absolute value will be further computed to model the input full wave rectifier. The ADC and the conditioning amplifier circuit are modeled by a gain block having the gain equal to the intended ADC gain multiplied by the voltage/current sensor gain. The output of the gain block is delayed so that the acquisition and conversion time are also modeled. The same model is used for all ADC required in the system.

For both the synchronous Boost converter and the PS converter the inductor currents and the output voltages are sent to an AD9717 two channel DAC. The DAC and the output

TABLE 1: AC-DC STAGE TEST PLATFORM PARAMETERS

Parameter	Symbol	Value
Filter inductor	L_{PFC}	500µH
Filter Capacitor	C_{PFC}	220µF
Nominal Output voltage	$V_{o\ PFC}$	400V
Input voltage amplitude	$V_{i\ PFC}$	255V
Current sensor and ADC gain	$G_{cs\ PFC}$	7.75
Input voltage sensor & ADC gain	$K_{Vin\ PFC}$	0.2
Output voltage sensor & ADC gain	$K_{Vout\ PFC}$	0.2
ADC ENOB	n_{ADC}	8
ADC Delay	t_{ADC}	170ns
DAC sampling rate	DAC_{SR}	100MSPS
DAC and output filter gain	G_{DAC}	1.327e-3V
PWM generator resolution bits	$n_{DPWM\ PFC}$	10
Switching frequency	$F_{sw\ PFC}$	100kHz
Current loop PI proportional coefficient	$KP_{I\ PFC}$	4.87
Current loop PI integrative coefficient	$KI_{I\ PFC}$	0.73
Voltage loop PI proportional coefficient	$KP_{V\ PFC}$	0.0014
Voltage loop PI integrative coefficient	$KI_{V\ PFC}$	2.46e-5

TABLE 2: DC-DC STAGE TEST PLATFORM PARAMETERS

Parameter	Symbol	Value
Filter inductor	L_1, L_2	660uH
Nominal Output voltage	V_o	400V
Inductor current sensor & ADC gain	$G_{cs\ PS}$	7.75
Output voltage sensor & ADC gain	$G_{Vout\ PS}$	0.388
PWM generator resolution bits	$n_{DPWM\ PS}$	10
Battery equivalent capacitance	C_b	660uF
Battery equivalent resistance	R_b	0.2Ω
Switching frequency	$F_{sw\ PS}$	100kHz
Current loop PI proportional coefficient	$KP_{I\ PS}$	8.62
Current loop PI integrative coefficient	$KI_{I\ PS}$	0.56
Voltage loop PI proportional coefficient	$KP_{V\ PS}$	98.23
Voltage loop PI integrative coefficient	$KI_{V\ PS}$	0.43
Voltage loop PI integrative coefficient	$KD_{V\ PFC}$	120.3

filter gain (G_{DAC}) is listed in Table 1. The output voltage and current are represented in a <32;20> fixed point format which need to be converted to a 14 bit format so that the DAC can process them. Depending on the output variable values, higher order or lower order bits are passed to the DAC, effectively multiplying the output variable by a power of 2. For example, if bits [25:22] are selected, an effective conversion to a <14;7> format is performed. The highest order bit selected (25 in the previous example) is labeled as MSB_SEL in (17). The DAC channel output voltage becomes:

$$V_{DAC} = Output\_variable * G_{DAC} * 2^{2+31-MSB\_SEL} \quad (17)$$

The control block is designed to emulate a resistive load on the input side of the converter and to regulate the DC voltage at the output (the loss free resistor model [7]). A current loop compensated by a PI controller is therefore required to keep the input current sinusoidal and in phase with the input voltage and a voltage loop, also compensated by a PI controller, will keep the output voltage constant. The sampling rate of the emulated ADCs will be equal to switching frequency, the PWM generator block being responsible for controlling the sample rate. The input parameters used for the HIL simulation are listed in Table 1. The DC-DC stage test platform (Fig. 8 b)) is composed of the PS converter RT simulation model, the sensor and ADC models for the output current and output voltage measurement, a DPWM generator and the control block. A constant current – constant voltage (CC-CV) charging method is designed by using two control loops, one for regulating the output current and one for regulating the output voltage.

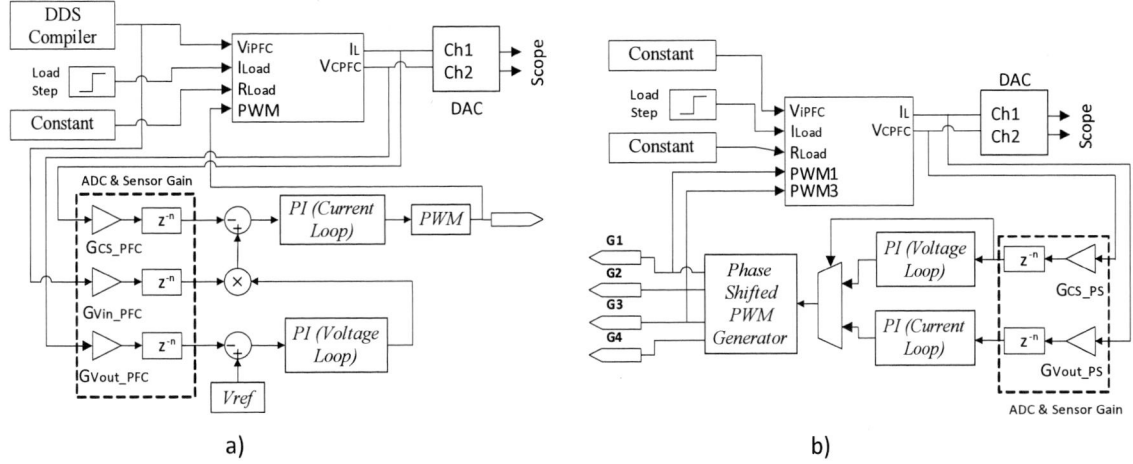

Fig. 8: Charger RT Model test platform for the a) AC-DC stage b) DC-DC stage.

Thus, the control block will implement a PI controller for the current loop and a PID controller for the voltage loop. The small signal model of the PS converter is identical to that of a Buck converter. A detailed description of the Buck converter controller design can be found in [7]. A simple RC model was used for the battery.

IV. SIMULATION AND EXPERIMENTAL RESULTS

A two step validation strategy is proposed. First, the RT models are developed in Plecs and compared against the synchronous Boost and PS topologies simulated using circuit elements. Once the models are validated through simulation, they will be implemented in FPGA and the digital outputs, converted to analog format, will be monitored with a scope and compared against the simulation results. The circuit schematic used for the Plecs simulation validation step applied to the synchronous Boost converter is illustrated in Fig. 9 a), while the RT simulation model Plecs implementation is presented in Fig. 9 b). The results for the synchronous Boost converter continuous and discrete models are presented in Fig. 10. The inductor current and the output voltage have been selected for the comparison. A 200V constant input voltage and a 0.5 duty factor for the PWM signal were used. For the discrete model, the internal variables have been represented in a <32;20> fixed point format. The DC-DC stage RT simulation model has been validated using the same approach, therefore the simulation circuit diagram is not displayed in this paper. As proved for the synchronous Boost converter, implementing the PS converter's discretized model (Fig. 9 b)) in Plecs is straightforward. The same <32;20> fixed point format was used to represent the internal variables for the PS converter. A 400V constant input voltage and a 0.666 duty cycle were used. The comparison between the continuous model and the discretized one is illustrated once again on the inductor current and output voltage (Fig. 11). A DC error can be observed in the case of the PS converter's output voltage. This can be explained by the fact that the discrete model samples the continuous PWM input with a sampling period equal to T_s, thus effectively transforming it into a discrete PWM signal. The resulting PWM resolution causes the DC error. The same effect is not visible for the synchronous

Boost converter because the on and off times of the PWM input are multiples of T_s.

For the second step of the validation process, the Plecs simulation parameters are recreated as closely as possible in the signal level HIL system. As described in Section III, the output variables are first converted in a 14 bit format and then converted by a DAC into an analog format. The relation between the output variables' values and the DAC output voltage is given by (17) and summarized in Table 3 for the performed measurements. The MSB_SEL column

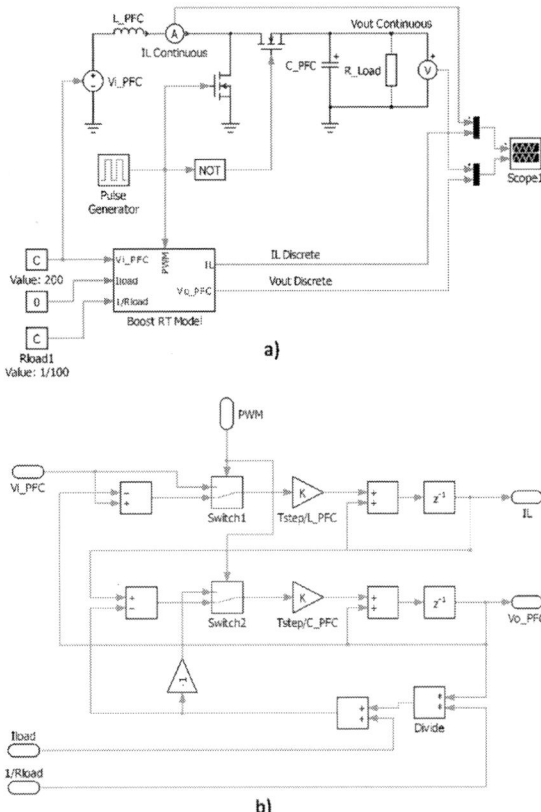

Fig. 9 Synchronous Boost RT model validation through simulation means (Plecs circuit schematic diagram).

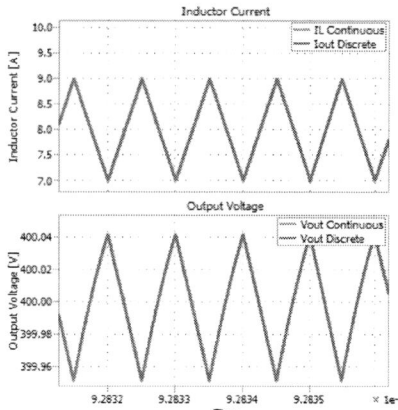

Fig. 10: Comparison between the continuous and the discretized models for the synchronous Boost converter.

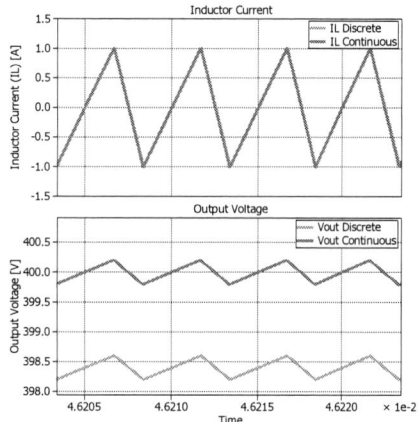

Fig. 11: Comparison between the continuous and the discretized models for the Phase Shifted converter

indicates which 14 bits out of the variable's 32 bits are sent to the DAC. An Analog Discovery USB oscilloscope was used to capture the waveforms. Channel 1 (orange) measures the RT model's output voltage, while channel 2 measures the RT model's output current. AC coupling was used on both channels for the output ripple measurements (Fig. 12 and Fig. 14). For the AC-DC stage, the synchronous Boost converter RT model was first run with a constant load and duty factor. The same parameters as for the Plecs simulation were used. The results for the closed loop test of the AC-DC stage are presented in Fig. 13.

The PS converter model is tested exclusively with the loop closed, the CC-CV voltage operation being illustrated in Fig. 15. At power up, the converter enters the constant current mode. A soft start technique is used to avoid overcurrent conditions while the charging current reaches its 4A nominal value. After the output voltage reaches its nominal value, the converter enters the constant voltage mode. The load current in this mode of operation should have a small mean value to compensate for battery self discharging. The output ripple waveforms are captured while the model operates in the constant voltage mode (Fig. 14). The measurements performed prove that the RT models' outputs closely match the expected values. For the output of channel1 an offset voltage of approximately 100mV is added by the DAC's output analog filter. The numeric values are however correct.

TABLE 3: Expected Measurements Results

Output variable	Numeric Value	MSB_ SEL	DAC output voltage [V]
Boost current ripple (peak to peak)	1	24	0.68
Boost voltage ripple (peak to peak)	0.09	19	1.977
PFC mean output voltage	400	32	2.12
PFC input current amplitude	12.55	25	4.26
PS converter current ripple (peak to peak)	2.02	23	2.74
PS converter voltage ripple (peak to peak)	0.4	21	2.21
PS converter nominal charging current	4	23	5.43
PS converter nominal output voltage	400	32	2.12

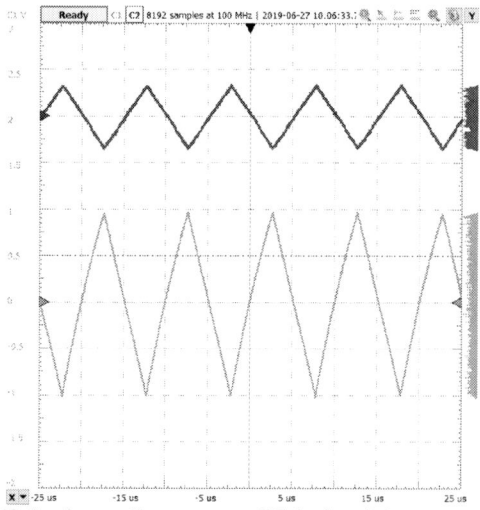

Fig. 12: Synchronous Boost converter RT simulator inductor current and output voltage ripple (horizontal scale: 5μs/div, Channel1 vertical scale: 0.5V/div, Channel2 vertical scale: 0.5V/div, Channel1 offset: -0.5V, Channel2 offset: 1.5V).

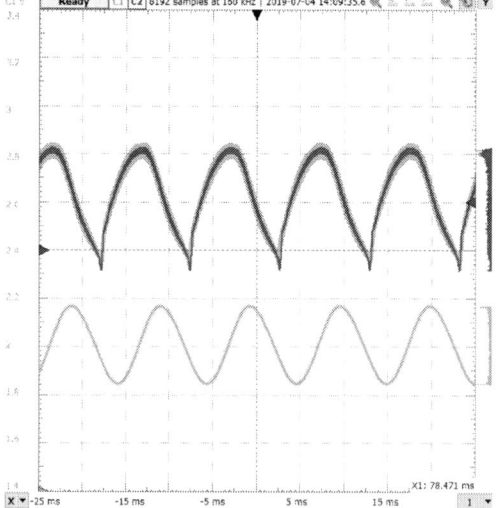

Fig. 13: PFC inductor current and output voltage (horizontal scale: 5ms/div, Channel1 vertical scale: 0.2V/div, Channel2 vertical scale: 2V/div, Channel1 offset: -2.4V, Channel2 offset: 0V).

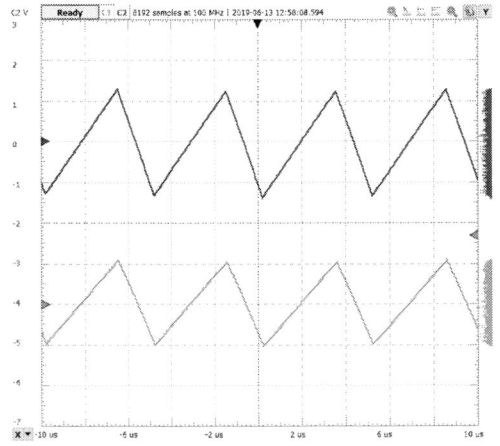

Fig. 14: Phase Shifted converter RT simulator inductor current and output voltage ripple (horizontal scale: 2µs/div, Channel1 vertical scale: 1V/div, Channel2 vertical scale: 1V/div, Channel1 offset: -2V, Channel2 offset: 2V).

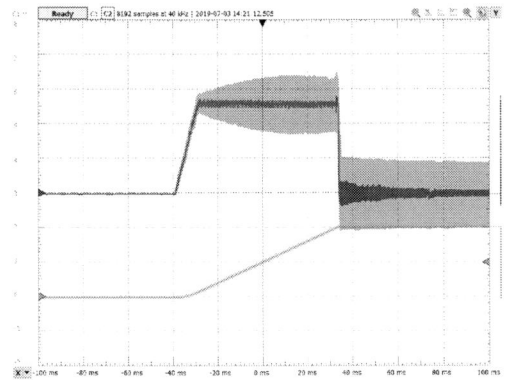

Fig. 15: Closed Loop behavior for the AC-DC stage based on the synchronous Boost RT model (horizontal scale: 20ms/div, Channel1 vertical scale: 1V/div, Channel2 vertical scale: 2V/div, Channel1 offset: -3V, Channel2 offset: 0V).

V. CONCLUSIONS

In this paper an RT simulation model for an on board battery charger was presented. The 40ns time step obtained will result in a maximum effective DPWM resolution of 9 bits, accurately accounting for switching effects up to this limit. External controllers or FPGA based controllers implemented on the same device can be tested with the developed model. Errors related to sampling, ADC amplitude quantization and ADC delay can also be modeled in the FPGA. Because of the low resource utilization, the charger's RT model can be embedded on the same low cost FPGA as the digital control system, thus making it ideal for reducing the time and cost of the digital control system development but also for applications such as system observation or fault detection. The model behavior was compared against the circuit behavior through simulations means. By converting the model's outputs through a DAC, experimental validation was also performed.

REFERENCES

[1] C. Washington and S. Delgado, "Improve design efficiency and test capabilities with HIL Simulation," 2008 IEEE AUTOTESTCON, Salt Lake City, UT, 2008, pp. 593-594.

[2] A. Ben Hadid and K. Ben Saad, "HIL simulation of a DC-DC converter controller on a Zynq," 2015 7th International Conference on Modelling, Identification and Control (ICMIC), Sousse, 2015, pp. 1-5.

[3] L. Xin, Z. Zaimin and X. Yun, "A HIL test bench for FCHV control units," 2009 IEEE Vehicle Power and Propulsion Conference, Dearborn, MI, 2009, pp. 1783-1787.

[4] M. Yilmaz and P. T. Krein, "Review of charging power levels and infrastructure for plug-in electric and hybrid vehicles," 2012 IEEE International Electric Vehicle Conference, Greenville, SC, 2012, pp. 1-8.

[5] K. Popovici and P. J. Mosterman, Real Time Simulation Technologies: Principles, Methodologies, Applications, 1st ed CRC Press, 2017

[6] M. Dagbagi, A. Hemdani, L. Idkhajine, M. W. Naouar, E. Monmasson and I. Slama-Belkhodja, "ADC-Based Embedded Real-Time Simulator of a Power Converter Implemented in a Low-Cost FPGA: Application to a Fault-Tolerant Control of a Grid-Connected Voltage-Source Rectifier," in IEEE Transactions on Industrial Electronics, vol. 63, no. 2, pp. 1179-1190, Feb. 2016.

[7] L. Corradini, D. Maksimovic, P. Mattavelli and R. Zane, Digital Control of High-Frequency Switched Mode Power Converters, 1st ed. Wiley-IEEE Press, 2015.

[8] N. Watson and J. Arrillaga, Power Sysyems Electromagnetic Transients Simulation, 1st ed, The Institution of Engineering and Technology, London, 2007.

[9] C. Liu, R. Ma, H. Bai, F. Gechter, and F. Gao, "A new approach for FPGA-based real-time simulation of power electronic system with no simulation latency in subsystem partitioning" Electr. Power Energy Syst., vol. 99, pp. 650–658, July 2018.

[10] K. Georgopoulos et al., "An evaluation of vivado HLS for efficient system design," 2016 International Symposium ELMAR, Zadar, 2016, pp. 195-199.

[11] Y. Yonezawa, H. Nakao and Y. Nakashima, "Novel Hardware-in-the-Loop Simulation (HILS) technology for virtual testing of a power supply," 2018 IEEE Applied Power Electronics Conference and Exposition (APEC), San Antonio, TX, 2018, pp. 2947-2951.

[12] L. Bao, L. Fan and Z. Miao, "Real-Time Simulation of Electric Vehicle Battery Charging Systems," 2018 North American Power Symposium (NAPS), Fargo, ND, 2018, pp. 1-6.

[13] T. Gherman, M. Ricco, J. Meng, R. Teodorescu and D. Petreus, "Smart Integrated Charger with Wireless BMS for EVs," IECON 2018 - 44th Annual Conference of the IEEE Industrial Electronics Society, Washington, DC, 2018, pp. 2151-2156."

978-1-5386-7688-2/19 $31.00 © 2019 IEEE

A Single-Switch ZCS Boost Converter with Low Conducted EMI

Mohammad Rouhollah Yazdani
dept. of electrical and computer engineering
Isfahan (Khorasgan) Branch, Islamic Azad University
Isfahan, Iran
m.yazdani@khuisf.ac.ir

Mohammad Pahalvandust
dept. of electrical and computer engineering
Isfahan (Khorasgan) Branch, Islamic Azad University
Isfahan, Iran

Abstract— **Soft switching DC-DC converters with high power density and simple topology are needed for compact and high-reliability applications. In this paper, a new resonant boost converter with zero current switching (ZCS) is introduced. The proposed converter has only one switch and a simple auxiliary circuit including an inductor and a capacitor. Due to the low number of components and small inductors, the power density of converter can be increased. There is no extra voltage stress on the switch and the diode. Theoretical analysis of the proposed converter is presented and verified by the Simulation results. Also, conducted common-mode electromagnetic interference (EMI) of the proposed converter is predicted.**

Keywords— DC-DC Converter, EMI, Resonant Converter, Zero current switching (ZCS)

I. INTRODUCTION

Switching power converters with reduced loss, size and weight are needed for portable electronics systems. Soft switching techniques are introduced to reduce switching losses and electromagnetic interference (EMI). Soft switching PWM converters need at least one extra switch or a considerable number of passive components to achieve soft switching conditions [1] - [2].

Quasi-resonant converters (QRCs) are a family of soft switching converter that derived by adding an LC tank to basic PWM converters [3]-[6]. Although QRCs have many advantages such as soft switching conditions and high efficiency but need a bulky inductor. Resonant converters such as series resonant converters (SRC) and parallel resonant converters (PRC) employ either more than two switches or a large number of components. Non-isolated PRCs are can be employed in step-up applications with more than one switch and a large filter inductor [7]. A switched resonator converter is introduced in [8], which has some advantages such as high efficiency with employing one inductor but needs at least two or three unidirectional switches. The resonant boost converter proposed in [9] needs two unidirectional switches with floating grounds, leading to complex control and drive circuit. The resonant converter introduced in [10] has characteristics similar to SRC and its voltage gain can reach up to unity at discontinuous conduction mode (DCM). However, it needs two switches and four diodes.

In this paper, a new resonant boost converter is presented in which the switch and the diode are turned on and off under ZCS. The proposed converter has only one switch, leading to a simple control circuit. The auxiliary circuit is quite simple and consists of an inductor and a capacitor. In addition, the main inductor of proposed converter becomes a small resonant inductor, so the size and weight of the converter are considerably reduced in comparison with QRC, PWM and other resonant converters. The proposed converter can operate in CCM and DCM conditions without extra voltage stress on the switch and main diode. Besides providing soft switching, EMI is an important issue that should be taken into account by power converter designers. Thus, the conducted EMI of the proposed converter is predicted.

This paper is organized as follows. In section II, the operation principles of the proposed converter are presented. The design procedure of the converter is described in section III. Section IV presents the simulation results of the proposed converter. The conducted EMI prediction results are shown in section V and conclusions are given in section VI.

II. OPERATION PRINCIPLES OF THE PROPOSED CONVERTER

Fig. 1 shows the proposed ZCS converter. The auxiliary circuit consists of a resonant inductor (L_r), a resonant capacitor (C_r) and small resonant inductor which acts as the main inductor (L_1) of the boost converter. D_s is placed to prevent the power returning from to the input at DCM operation and for CCM operation it is omitted because the input current is not negative. To simplify the analysis, it is assumed that all circuit components are ideal and the output capacitor is considered as a voltage source. The important quantities of the converter are defined below.

$$\alpha = \frac{L_r}{L_1} \tag{1}$$

$$\omega_1 = \frac{1}{\sqrt{L_r C_r}} \quad , \quad f_r = \frac{1}{T_r} = \frac{\omega_1}{2\pi} \tag{2}$$

$$Z_1 = \sqrt{\frac{L_r}{C_r}} \tag{3}$$

$$\omega_2 = \frac{1}{\sqrt{(L_1 + L_r)C_r}} \tag{4}$$

$$Z_2 = \sqrt{\frac{L_1 + L_r}{C_r}} \tag{5}$$

$$r = \frac{R}{Z_1} \tag{6}$$

$$f_n = \frac{f_s}{f_r} \tag{7}$$

978-1-5386-7688-2/19 $31.00 © 2019 IEEE

$$M = \frac{V_o}{V_{in}} \tag{8}$$

Depending on the current waveform of the L_1, the proposed converter can work in Deep DCM, DCM or CCM condition.

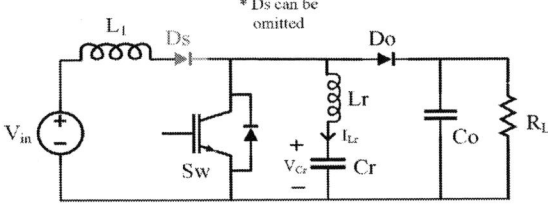

Fig. 1. Circuit configuration of the proposed boost converter.

A. Deep-DCM

Deep-DCM waveforms and equivalent circuit in each operating interval are shown in Fig. 2 and 3. If I_{L1} becomes zero before I_{Lr} reaches zero, the converter will operate in Deep-DCM. Prior to the first interval, it is assumed that all semiconductor devices are off and the output capacitor feeds the load. L_1 and L_r currents are zero and C_r voltage is V_{C0}.

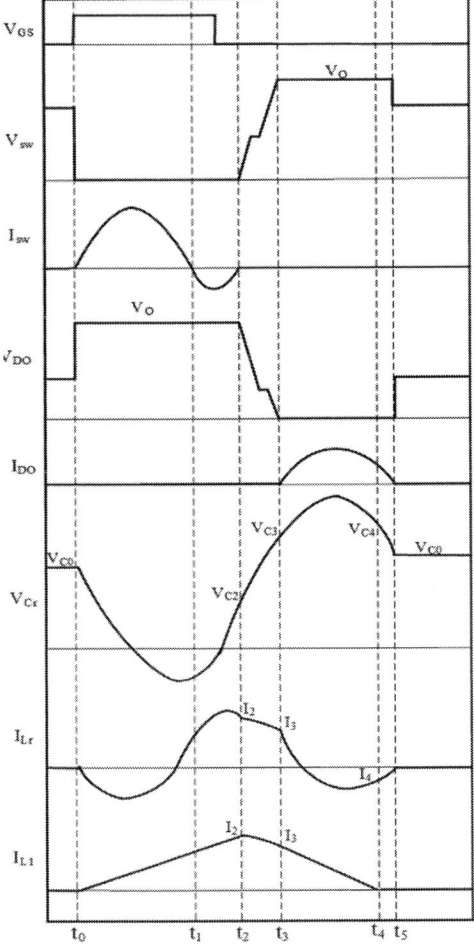

Fig. 2. Key waveforms of the proposed converter at Deep-DCM.

Mode I (t_0-t_1): At the beginning of this mode, the switch turns on under ZCS condition. The current of L_1 tends to rise linearly and L_r current changes in a sinusoidal fashion. At the end of this mode, I_{Lr} and I_{L1} become equal and I_{sw} reaches zero so the switch can turn off under ZCS. Important voltage and current equations of this mode are as follows.

$$I_{L1} = \frac{V_{in}}{L_1}(t - t_0) \tag{9}$$

$$V_{Cr}(t) = V_{C0} \cos(\omega_1(t - t_0)) \tag{10}$$

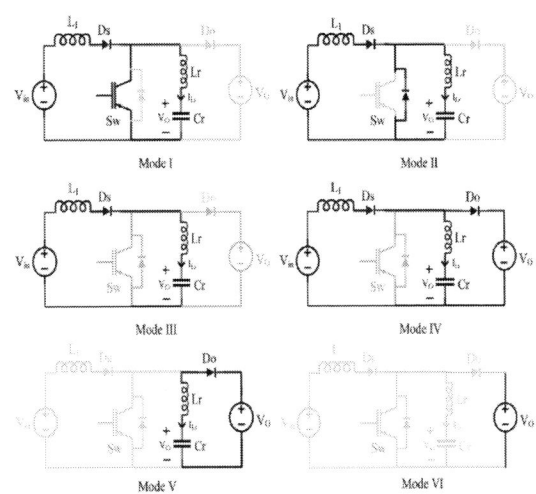

Fig. 3. Equivalent circuits in each Operating Mode (Deep- DCM).

Mode II (t_1-t_2): At t_1, the body diode of the switch turns on and carries the difference between I_{Lr} and I_{L1}. At the end of this mode, I_{Lr} and I_{L1} become equal and the body diode of the switch turns off. Voltage and current of this mode can be calculated by (9) - (10). According to Fig. 2, I_{L1} intersects I_{Lr} between $3/4T_r$ and T_r. Accordingly, I_2, V_{C0}, and V_{C2} are calculated as follows.

$$I_2 = \frac{7}{4}\pi\frac{V_{in}}{L_1\omega_1} = \frac{7}{4}\pi\frac{\alpha V_{in}}{Z_1} \tag{11}$$

$$V_{C0} = 2.5\pi\alpha V_{in} \tag{12}$$

$$V_{C2} = \frac{7}{4}\pi\alpha V_{in} \tag{13}$$

Mode III (t_2-t_3): In this mode, the current of L_1 flows through L_r and C_r and V_{Cr} increases until it approximately reaches V_O, which prepares ZVS condition for D_O to be turned on. The current of L_1 can be written by

$$I_{L1}(t) = I_{Lr}(t) = I_2\cos(\omega_2(t - t_2)) + \frac{V_{in} - V_{C2}}{Z_2}\sin(\omega_2(t - t_2)) \tag{14}$$

Mode IV (t_3-t_4): At t_3, D_O turns on. I_{L1} decreases linearly and I_{Lr} changes in a sinusoidal fashion. At the end of this interval, I_{L1} becomes zero and cannot be negative due to the existence of D_s. So, the energy could not return to the input source.

$$I_{L1}(t) = I_3 + \frac{V_{in} - V_O}{L_1}(t - t_3) \tag{15}$$

$$I_{Lr}(t) = I_3\cos(\omega_1(t - t_3)) \tag{16}$$

978-1-5386-7688-2/19 $31.00 © 2019 IEEE

Mode V (t_4-t_5): In this mode, I_{Lr} flows through D_O until it reaches zero and D_O turns off under ZCS condition. The duration of IV-V modes is calculated as below.

$$t_5 - t_3 = \frac{3\pi}{2\omega_1} = \frac{3}{4}T_r$$

(17)

$$I_3 = \frac{V_{in}}{Z_1}(M - 2.5\pi\alpha)$$

(18)

Mode VI (t_5-t_6): In this mode, both switch and diodes are off and the load is supplied by the output capacitor. The duration of this mode is determined by the control circuit to achieve the desired output voltage (dead time control).

B. DCM

The operation of the converter in DCM condition is quite similar to Deep-DCM, except in fourth and fifth intervals. At the end of the fourth interval, the current of the D_O becomes zero, so D_O turns off under ZCS. In the fifth interval, I_{L1} flows through L_r and C_r until it reaches zero. All current and voltage equations of Deep-DCM are valid for DCM, except in the fifth interval.

C. CCM

In CCM operation, I_{L1} is always continuous and does not reach zero. The key waveforms of the proposed converter in CCM operation are similar to DCM except that I_{L1} in CCM operation is always continuous.

III. DESIGN PROCEDURE

This section addresses important equations and relations that are necessary for designing the proposed converter. The value of the input voltage and its variation, the output voltage, and the output power are given. In this paper, the following equations are derived for the Deep-DCM state. However, DCM and CCM equations can be extracted in the same manner.

A. Voltage Gain (M)

The voltage gain of the converter can be calculated by using energy conversion principle in one switching cycle at the steady-state condition as below.

$$\int_{T_S} V_{in} I_{L1} dt = \int_{T_S} \frac{V_O^2}{R} dt$$

(19)

M can be computed by substituting (9), (15) and (16) into (19).

$$M = \frac{1}{2} + \frac{7}{8}\sqrt{r.\pi.\alpha.f_n + \frac{16}{49}}$$

(20)

B. Inductors Ratio (α)

The Deep-DCM operation can be achieved if the discharge time of L_1 takes less than $3/4T_r$. Therefore, the boundary condition between Deep-DCM and DCM is calculated as follows.

$$t_4 - t_3 = \frac{3}{4}T_r$$

(21)

$$\frac{L_1 I_3}{V_o - V_{in}} = \frac{3\pi}{2\omega_1}$$

(22)

Substituting (1), (2), and (18), into (22) yields the value of α at the boundary of Deep-DCM and DCM:

$$\alpha_B = \frac{2M}{\pi(3M + 2)}$$

(23)

By selecting $\alpha > \alpha_B$ converter operates at Deep-DCM. Choosing $\alpha < \alpha_B$ leads to DCM and further lowering of α takes the converter to the CCM state.

In order to guarantee the soft switching condition for the switch, I_{L1} should intersect I_{Lr} at the end of the first interval. In the worst case, I_{L1} should become equal to I_{Lr} at $3/4\ T_r$.

$$\frac{V_{C0}}{Z_1} \geq \frac{3\pi V_{in}}{2L_1\omega_1}$$

(24)

Considering (1), (2) and (24) yields:

$$\alpha \leq \frac{2V_{C0}}{3\pi V_{in}}$$

(25)

The maximum value of α, which ensures soft switching condition for the switch, obtained by (25). α_{max} and α_B are plotted in Fig. 4. The boundary between Deep-DCM and CCM is shown by a blue dashed line. Maximum allowable α with respect to V_{C0}/V_{in}, which defines the soft and hard switching regions, is also illustrated by the red line.

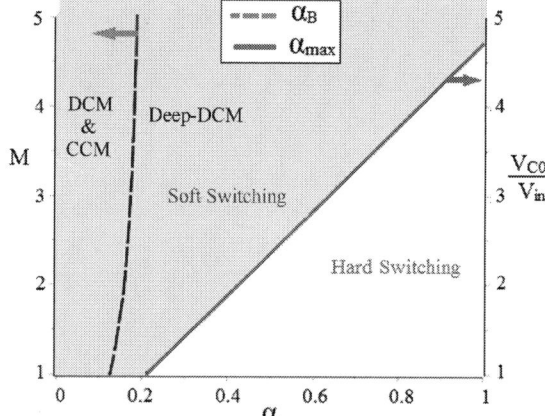

Fig. 4. α_{max} versus V_{C0}/V_{in} (red line) and α_B against M (blue dashed line).

C. Normalized Switching Frequency (fn)

In the absence of dead time, the converter operates in maximum power capability. In this condition, the voltage gain and the switching frequency are also maximized. Considering the duration of all intervals yields:

$$T_{s,min} = \frac{1}{f_{s,max}} = \frac{7}{8}T_r + \frac{3}{4}T_r + \frac{17\pi\alpha - 4M_{max}}{8\pi\alpha(M_{max} - 1)}T_r$$

(26)

Therefore, the switching frequency normalized to the resonant frequency is obtained as follows.

$$f_{n,max} = \frac{f_{s,max}}{f_r} = \frac{8\pi\alpha(M_{max} - 1)}{13\pi\alpha M_{max} + 4\pi\alpha - 4M_{max}}$$

(27)

f_s should be lower than $f_{s,max}$ to prevent the soft switching condition.

D. Resonant Elements

The switching frequency (f_s) can be chosen by the designer. Once the switching frequency is determined, the resonant frequency (f_r) is obtained by (7). Since r and R are calculated so far, Z_1 can be found using (6). By solving (2) and (3), L_r and C_r values are determined. Finally, L_1 is calculated by (1).

IV. SIMULATION RESULTS

In order to examine the theoretical analysis, the proposed converter is designed and simulated in Deep-DCM using OrCAD software. The key parameters of the converter are shown in Table I. Fig. 5 shows the simulated switch voltage and current waveforms. It can be seen that the switch turns on with ZCS and turns off with ZCS and almost ZVS conditions. Also, there is no extra voltage stress on the switch. Fig. 6 illustrates the simulated voltage and current waveforms of the diode D_O. It can be observed that the diode turns on with ZCS and almost ZVS and turns off with ZCS. There is no extra voltage stress on the output diode of the proposed converter. The voltage of the resonant capacitor (C_r) and resonant inductor current (L_r) are shown in Fig. 7, which established the theoretical analysis.

TABLE I. KEY PARAMETERS OF SIMULATED CONVERTER

Parameter	Value
Vin	50V ±10%
Vo	100 V
Po	200 W
f_{sw}	200 kHz
L_1	7 µH
L_r	1.5 µH
C_r	62 nF
C_o	10 µf
Switch	IRG4BC20UD
Diode	MUR815

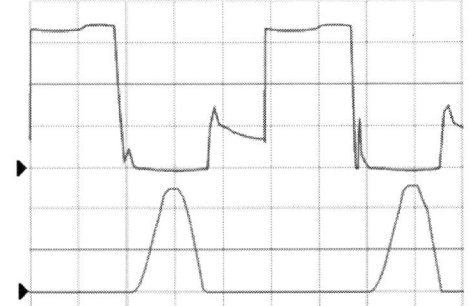

Fig. 5. Switch voltage (top) and current (bottom) waveforms. (voltage: 30V/div, current: 5A/div, time: 1µs/div).

Fig. 6. Output diode voltage (top) and current (bottom) waveforms. (voltage: 30V/div, current: 5A/div, time: 1µs/div)

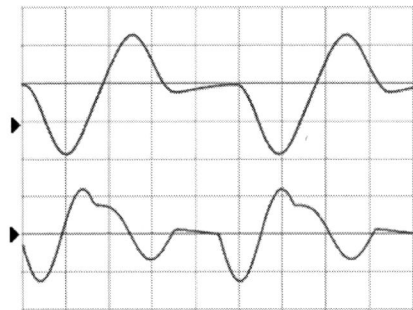

Fig. 7. Resonant capacitor (Cr) voltage (top) and resonant inductor (Lr) current waveforms. (voltage: 60V/div, current: 5A/div, time: 1µs/div)

V. CONDUCTED COMMON MODE EMI PREDICTION

This section provides the prediction results of the conducted common-mode (CM) EMI of the proposed converter. The high di/dt and dv/dt of the switch are reduced by soft switching method in the proposed converter and consequently, it is expected to have proper EMC performance. To simulate conducted electromagnetic emissions, a line impedance stabilization network (LISN) according to CISPR 11 standard is used in the simulation software [11]. In the simulation model of the converter, the drain-earth parasitic capacitor of the switch which is a major pass of the CM noise current is assumed 15 pF. If the heat-sink size of the switch is increased for the thermal issues, this stray capacitor becomes larger leading to more CM- EMI.

To separate differential mode (DM) and common-mode conducted EMI, a simple DM rejection network (DMRN) introduced in [12] is connected to the input port of the spectrum analyzer model for CM emissions simulation. The simulation model of LISNs and DMRN are shown in Fig. 8.

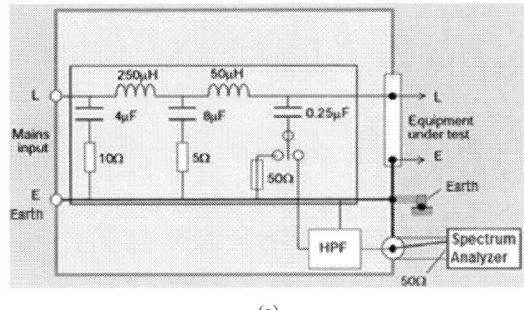

(a)

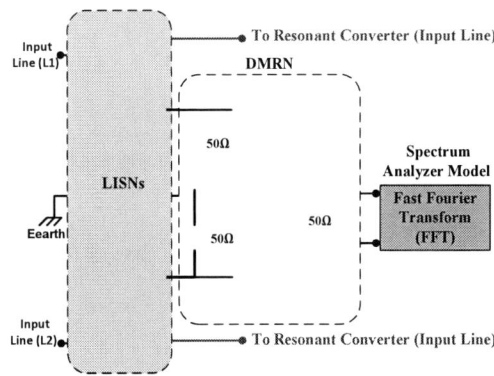

Fig. 8. a) CISPR 11 LISN b) DM noise rejection network (DMRN)

The spectrum analyzer port is modeled by a 50Ω resistor as illustrated in Fig. 8. As shown in Fig. 8, the FFT of the simulated voltage across a 50Ω resistor yields the predicted EMI spectrum. Fig. 9 shows the conducted CM-EMI spectrums of the designed converter in 150 kHz- 20 MHz frequency band using FFT in OrCAD software. As it can be observed, the EMI peak of the proposed converter is around 5.6 mV which is a low noise value.

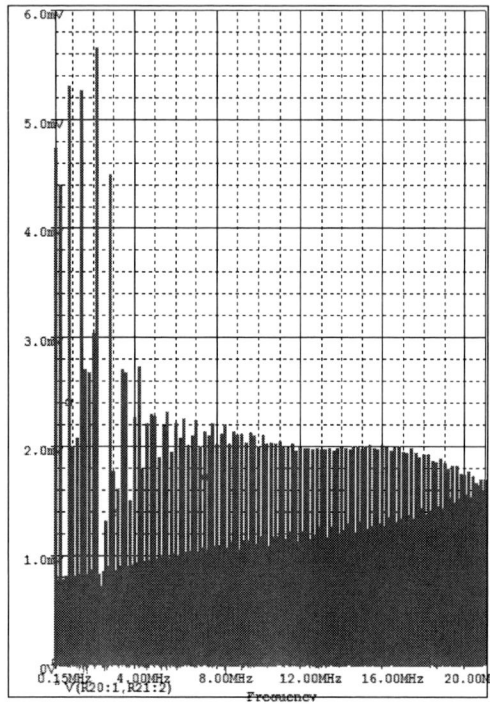

Fig. 9. Predicted CM conducted electromagnetic emissions (Vertical scale:0-6mV, Horizontal scale: 0.15-20MHz)

VI. CONCLUSION

In this paper, a resonant ZCS boost converter is introduced. The zero current switching conditions are provided for the switch and the diode at turn-on and turn-off instants. Theoretical analysis and operation intervals of the proposed converter are presented. To verify the theoretical analysis, the simulation results of Deep-DCM operation are shown using OrCAD simulation software. Moreover, the CM conducted electromagnetic emissions is predicted that shows the proposed converter has low and the main peak is lower than 5.7 mV. Simple auxiliary circuit, small inductors and low conducted EMI are benefits of the proposed converter.

REFERENCES

[1] M. Mohammadi, E. Adib and M. R. Yazdani, "Family of soft-switching single-switch PWM converters with lossless passive snubber," *IEEE Trans. Ind. Electron.*, vol. 62, no. 6, pp. 3473-3481, 2015.

[2] M. R. Yazdani, M. Pahlavandust, and P. Hemmati, "A New Family of Non-Isolated Zero Current Transition PWM Converters," *Journal of power Electronics,* vol. 16, no. 5, pp. 1669-1677, 2016.

[3] B.T. Lin, Y.S. Lee, "A Unified Approach to Modeling Synthesizing and Analyzing Quasi-Resonant Converters," *IEEE Trans. Power Electron*, vol. 12, no. 6, pp. 983-992, 1997.

[4] B. Kang, K. S. Low, J. J. Soon, Q. V. Tran, "Single-Switch Quasi-Resonant DC–DC Converter for a Pulsed Plasma Thruster of

Satellites," *IEEE Trans. Power Electron.* , vol. 32, no. 6, pp. 4503–4513, 2017.

[5] S. Sharifi, M. Jabbari, "Family of single-switch quasi-resonant converters with reduced inductor size," *IET Power Electron.*, vol. 7, no. 10, pp. 2544 - 2554, 2014.

[6] S. Salehi Dobakhshari, J. Milimonfared, M. Taheri, H. Moradisizkoohi, "A Quasi-Resonant Current-Fed Converter With Minimum Switching Losses," *IEEE Trans. Power Electron.*, vol. 32, no. 1, pp. 353–362, 2017.

[7] S. Li, Y. Zheng, B. Wu, K. M. Smedley, "A family of resonant two-switch boosting switched-capacitor converter with ZVS operation and a wide line regulation range," *IEEE Trans. Power Electron.*, vol. 33, no.1, pp. 448-459, 2018.

[8] M. Jabbari, "Unified analysis of switched-resonator converters," *IEEE Trans. Power Electron.*, vol. 26, no. 5, pp. 1364-1376, 2011.

[9] M. Jabbari, F. Barati, "New resonant LCL boost converter," *IET Power Electron.*, vol. 7, no. 7, pp. 1770 - 1776, 2014.

[10] M. Jabbari, and H. Farzanefard, "Resonant inverting-buck converter," *IET Power Electron.*, vol.3, no.43, pp.571-577, 2010.

[11] IEC International Special Committee on Radio Interference-C.I.S.P.R., Industrial, scientific and medical equipment – Radio-frequency disturbance characteristics – Limits and methods of measurement, *CISPR 11*, 2015.

[12] H.W. Ott , *Electromagnetic Compatibility Engineering*, John Wiley & Sons, 2009.

A TLBO Algorithm for Design Optimization of DVRs in an Interline DVR (IDVR)

Mahdi Jabbari
Roozbeh Institute of Higher
Education, Faculty of Electrical
Engineering Department
Zanjan, Iran
Email: jabbari.mahdi1361@gmail.com

Majid Moradlou
Department of Electrical
Engineering, Zanjan Branch, Islamic
Azad University
Zanjan, Iran
Email: majid.moradloo@iauz.ac.ir

Mehdi Bigdeli
Department of Electrical
Engineering, Zanjan Branch, Islamic
Azad University
Zanjan, Iran
Email: mehdi.bigdeli@iauz.ac.ir

Abstract—Because of the limitation of active power exchange between dynamic voltage restorers (DVRs) in an interline DVR (IDVR), selection of the appropriate voltage for each DVR in an IDVR should be carried out according to two conditions. First, voltage sags in each of the feeders are compensated, and second, IDVR rated kVA and total cost are reduced. To serve this purpose, a convenient strategy based on teaching-learning based optimization algorithm is proposed which can minimize sum of rated kVAs of DVRs. Comparing the results of this new method with the well-known published work in the literature reveals the importance and efficiency of the proposed method.

Keywords— Power Quality; Dynamic Voltage Restorer (DVR); Interline DVR (IDVR); Teaching-Learning Based Optimization (TLBO).

I. INTRODUCTION

Voltage Sag is one of the most important occurrences in power quality area which causes a lot of damage to industrial and commercial producers [1, 2]. To counteract these destructive phenomena some common approaches have been proposed which could apply to network or industrial loads and can reduce, to some extent, damaging effects of this occurrence. By the growing development of power electronic converters, application of devices named Custom Power as a proper and flexible method, with the purpose of power quality enhancement has been proposed [3]. One of these devices is Dynamic Voltage Restorer (DVR).

DVR is a series compensator positioned between a load and a utility bus which mainly aims to protect loads against voltage sags, swells and imbalance. The mentioned purpose is carried out by injection of series voltage with an appropriate magnitude and angle in order to restore load-side voltages to acceptable and desirable amplitude in the presence of voltage disturbance source. A DVR can accomplish the compensation using the injection of active power or without it. The compensation range of DVRs with the capacity of active power injection is wider than the others. They, however, demand a converter or an additional energy storage device. Hence, applying them would be more expensive [3, 8]. In some environments such as industrial parks, due to reliability enhancement and some network

considerations, loads feed is carried out through two or more feeders connected to different substations. These feeders are electrically independent, although they are location-wise near each other in the mentioned industrial environments. If in each independent feeder a DVR is installed for voltage disturbance compensation, these DVRs can exchange active power via a common dc bus. Therefore, when a voltage disturbance occurs in one of the feeders, the active power demanded to operate the DVR installed in that feeder can be obtained from the other healthy feeder or feeders. This structure is named Interline Dynamic Voltage Restorer (IDVR) [9, 10]. The ability of active power exchange among DVRs in an IDVR not only eliminates the need for energy storage devices and decreases expenses but also expands the compensation range of DVRs.

Considering the fact that DVR voltage rating determines its apparent power rating and consequently the price and size of it, presentation of an appropriate strategy for selecting DVR voltage rating in steady state seems necessary. Each of the DVRs of an IDVR operates in two moods which are compensation and energy injection moods. When voltage disturbance occurs in the feeder of installed DVR, the mood is compensation and in the case of disturbance occurrence in the adjacent feeder, the mood is called energy injection. Moreover regarding the feeders load (in terms of apparent power and power factor) and also DVRs voltage rating, there is limitation in transferable active power from a healthy feeder to a faulty one [11]. Therefore design strategy should be in a way that firstly, makes it possible to compensate a variety of voltage disturbance characteristics in each feeder using IDVR structure and secondly, chooses the minimum possible amount for DVRs voltage rating in IDVR structure considering their both operation moods. In [12] considering the sum of apparent power ratings of DVRs as an objective function (which needs to be minimized), the optimization problem has been defined and its constraints and related equations have been presented. This problem has

---

978-1-5386-7688-2/19 $31.00 © 2019 IEEE

been solved using the genetic algorithm (GA). The GA, however, has low convergence speed and today intelligent algorithms of great efficiency compared to the GA, have been introduced.

Therefore, proposing more effective algorithm (i.e., with higher accuracy, faster convergence, smaller number of parameters, easier use, etc.) for design optimization of DVRs in an IDVR is practically important.

Among the intelligent optimization algorithms, teaching-learning based optimization (TLBO) is an efficient method and its ability to solve different optimization problems can be proven. Hence, in this paper, TLBO method is used to solve optimization problem. The final results attest that the TLBO algorithm is able to reach the optimal solution with excellent accuracy and speed.

II. PROBLEM DESCRIPTION

The steady state modeling and expression of the relations in an IDVR have been stated in [12].

The size and cost of the DVR are determined by the rated kVA of a structure based on it. Therefore, minimization of the rated kVA for converter-based structures can always be an important aim in the design process. In an IDVR, minimizing the sum of rated kVAs of individual DVRs can be an acceptable objective. For this purpose, a design strategy to select DVRs optimal values of rated voltage employed in an IDVR is presented in order to compensate for balanced voltage sags [12]. In [12], objective function has been proposed as the sum of kVA ratings of DVRs in IDVRs i.e. S_{total}):

$$S_{total}^{p.u} = S_{Lrated,1}^{p.u} \times V_{DVR,1,i}^{p.u} + S_{Lrated,2}^{p.u} \times V_{DVR,2,i}^{p.u} \quad (1)$$

The S_{total} needs to be minimized. Where, $V_{DVR,i}$ is a function of $P_{ex,i,k}$, $P_{ex,j,k}$ and $V_{L,h,i,k}$ ($k = 1,2, ... T$). For example, if two intervals exist for load curve (i.e. $T = 2$), there will be eight variables, and each of the $V_{DVR,1}$ and $V_{DVR,2}$ will be a function of six variables. It can be described as follows:

$$V_{DVR,i} = \text{Max} \begin{Bmatrix} V_{inj,f,i,1}(P_{ex,i,1}), V_{inj,h,i,1}(P_{ex,j,1}, V_{L,h,i,1}), ..., \\ V_{inj,f,i,T}(P_{ex,i,T}), V_{inj,h,i,T}(P_{ex,j,T}, V_{L,h,i,T}) \end{Bmatrix} \quad (2)$$

$$i = 1, 2, \qquad j = 1, 2, \qquad i \neq j$$

In the kth interval, $V_{inj,f,i,k}$ states DVR injected voltage in feeder (i) to compensate for the voltage sag with VF_i amplitude in this feeder. $V_{L,h,j,k}$ is the load voltage of feeder (j) when this feeder is assumed as a healthy feeder (in the kth interval) and can be considered in the range of $V_{L_{min,j}} \leq V_{L,h,j,k} \leq 1pu$. Other variables are defined as follows [12]:

$$V_{inj,f,i,k}(P_{ex,i,k}) = \sqrt{A_{i,k} + B_{i,k} \times P_{ex,i,k} + C_{i,k} \times} \sqrt{D_{i,k} + E_{i,k} \times P_{ex,i,k} + F_{i,k} \times P_{ex,i,k}^2} \quad (3)$$

$$V_{inj,h,i,k}(P_{ex,i,k}, V_{L,h,j,k})$$
$$= \sqrt{1 + V_{L,h,j,k}^2(G_{j,k} - E_{j,k} \times P_{ex,i,k}) +}$$
$$H_{j,k}\sqrt{\frac{1}{V_{L,h,j,k}^2} + M_{j,k} - E_{j,k} \times P_{ex,i,k} + F_{j,k} \times P_{ex,i,k}^2} \quad (4)$$

$$i = 1, 2, \qquad j = 1, 2, \qquad i \neq j$$

$$P_{ex-min,i,k} = S_{L,i,k}\left[\frac{1}{V_{L_{min,i}}} - \cos(\varphi_{i,k})\right] \quad i = 1, 2 \quad (5)$$

$$P_{min,i,k} = S_{L,i,k}\left[\cos(\varphi_{i,k}) - \frac{VF_i}{V_{L_{min,i}}}\right] \quad i = 1, 2 \quad (6)$$

Where, coefficients A to M are constant parameters in each interval and are defined as follows:

$$A_{i,k} = VF_i^2 - V_{L_{min,i}}^2[2\cos^2(\varphi_{i,k}) - 1]$$
$$= VF_i^2 - V_{L_{min,i}}^2\cos(2\varphi_{i,k})$$

$$B_{i,k} = 2\cos(\varphi_{i,k})\frac{V_{L_{min,i}}^2}{S_{L,i,k}}$$

$$C_{i,k} = -2V_{L_{min,i}}^2|\sin(\varphi_{i,k})|$$

$$D_{i,k} = \left(\frac{VF_i}{V_{L_{min,i}}}\right)^2 - \cos^2(\varphi_{i,k})$$

$$E_{i,k} = 2\frac{\cos(\varphi_{i,k})}{S_{L,i,k}}$$

$$F_{i,k} = -\frac{1}{S_{L,i,k}^2}$$

$$G_{i,k} = -\cos(2\varphi_{i,k})$$

$$H_{i,k} = -2|\sin(\varphi_{i,k})|$$

$$M_{i,k} = -\cos^2(\varphi_{i,k})$$

$S_{Lrated,1}$ and $S_{Lrated,2}$ indicate the rated load kVA (pu) for feeders 1 and 2, respectively.

$V_{Lmin,1}$ and $V_{Lmin,2}$ are the minimum permissible voltage of loads (pu) for feeders 1 and 2, respectively.

$S_{L,1,k}$ and $S_{L,2,k}$ represent the load kVA (pu) in the kth interval for feeders 1 and 2, respectively.

$\cos(\varphi_{1,k})$ and $\cos(\varphi_{2,k})$ indicate the load power factor in the kth interval for feeders 1 and 2, respectively.

It was shown that S_{total} is a function of $P_{ex,1,1}$, $P_{ex,2,1}$, $V_{L,h,1,1}$, $V_{L,h,2,1}$,..., $P_{ex,1,T}$, $P_{ex,2,T}$, $V_{L,h,1,T}$, and $V_{L,h,2,T}$ (The number of variables equals 4T). Hence, by searching in the range of these variables, $V_{DVR,1}$ and $V_{DVR,2}$ are selected so that S_{total} is minimized. The values of $V_{DVR,1}$ and $V_{DVR,2}$ which correspond to minimum S_{total} are final rated values of voltage of DVRs, observed as the variables of $V_{DVR, rated,1}$ and $V_{DVR, rated,2}$.

III. THE PROPOSED DESIGN PROCEDURE

The optimization problem described in the previous section is such that actual minimum S_{total} is obtained by searching the entire solution space [11]. However, considering the great number of time intervals for load curve in feeders, the solution space will be extremely large. For instance, if there are 8 intervals in every curve in a day (T=8), S_{total} will be a function of 32 variables. Hence, it will be impossible to solve the problem by searching the whole solution space and applying intelligent optimization methods is necessary for solving the problem. The TLBO is one of the best intelligent optimization methods that its ability in solving various optimization problems has been proven. For this purpose, the TLBO is used to optimize the objective function. Therefore, in this section the TLBO algorithm is briefly introduced.

A. Teaching-Learning-Based Optimization Algorithm

The TLBO is one of the algorithms based on population which has been introduced by Rao and his colleagues in 2011 [13] and simulate the process of training and learning in a class. This algorithm is an intelligent optimization method, established according to the effect of a teacher on the students in order to improve the scientific level of the class. This method is fundamentally based on a principal that a teacher is trying to near the class level to his level of knowledge so that the students besides exploiting the teacher's knowledge, could improve their level by interaction with the other students and using their achievements. Since the teacher cannot lead the level of each and every student to himself, he makes an effort to improve the average level and would be able to evaluate it according to the examinations and the students' scores.

In this algorithm a group of students are considered as a population. The variable courses suggested for the students are considered as different decision-making variables of the optimization problem. Therefore, the students' scores in different courses are compared to objective amounts. The best response among the whole population is selected as a teacher. Decision-making variables actually are parameters that optimization objective functions have been defined according to them and the best result is the best amount of the objective function. This algorithm is divided into two phases: teacher phase and student phase [13].

Teacher Phase

In this phase the teacher tries to make the average of the class reach himself. But since this task is very difficult, he makes an effort to increase the average of the class from M_i to M_new . Every set of problem variables are updated based on the difference of these values. This difference could be saved in the Diff_Mean$_i$ parameter as follows.

$$\text{Diff\_Mean}_i = r_i(\text{M\_new} - T_f M_i) \tag{7}$$

T_F is the training factor or teacher parameter which decides how much the average value changes and r_i is a random number in the range of [0,1]. T_F could be randomly 1 or 2 which is an exploratory stage and randomly with the same probability is equal to:

$T_F = \text{round}[1 + \text{rand}(0,1)\{2 - 1\}]$

It should be noted that the difference value of the Eq. 7 would correct the answer as Eq. 8.

$$X_{new,i} = X_{old,i} + \text{Diff\_Mean}_i \tag{8}$$

Student Phase

The students use each other's knowledge besides the teacher's information. The mathematical expression of this phase is in a way that in each iteration, each variable set (the students) select one of the others by chance. For example the student i chooses the student j. If the student j has a higher level of knowledge compared to i, the student i will be updated according to Eq. 9.

$$X_{new,i} = X_{old,i} + r_i(X_i - X_j) \tag{9}$$

Else, the student status is varied as follows:

$$X_{new,i} = X_{old,i} + r_i(X_j - X_i) \tag{10}$$

After all the students change their status, their level is evaluated using the objective function. In this situation the best student is compared to the teacher of previous step and if the student was better, he would be substituted for the teacher of previous iteration. This process continues until the convergence conditions would be acquired.

B. Implementation of TLBO to the problem

The steps of solution for IDVR optimal design problem using TLBO algorithms are as follows:

Step 1: initialization of the variables
DVR active power injected in each feeder and DVR injected voltage in the faulty feeder are selected as decision-making variables (according to Eq. 11).

$$X_{in} = \begin{bmatrix} P_{ex,1,1} \cdots P_{ex,1,T}, P_{ex,2,1} \cdots P_{ex,2,T}, \\ V_{L,h,1,1} \cdots V_{L,h,1,T}, V_{L,h,2,1} \cdots V_{L,h,2,T} \end{bmatrix} \tag{11}$$

The initial value of these variables is determined at random. In addition, the $V_{L,h}$ ranges between 0.5 to 1, and the range of the P_{ex} is obtained by the equations 5, 6.

Step 2: objective function evaluation
The objective function should be minimum while all the constraints should be satisfied. For this purpose, according to Eq. 12 the penalty coefficients are applied in order to satisfy the constraints.

$$\begin{aligned} S_{total,i} = {} & S_{Lrated,1}V_{DVR1,i} + S_{Lrated,2}V_{DVR2,i} \\ & + \beta_1(V_{DVR1} - V_{max1}) \\ & + \beta_2(V_{DVR2} - V_{max2}) \end{aligned} \tag{12}$$

Where, β_1 and β_2 are penalty factors and their values are assumed to be 100000 in this research. These are determined by trial and error.

Step 3: the teacher is selected according to each set of variables.

Step 4: the variables are updated based on the phases of teacher and student.

Step 5: considering the stop criterion

The algorithm will be stopped if it reaches the determined number of repetitions; otherwise it returns to Step 2.

According to the mentioned steps, the flowchart of the TLBO algorithm to solve the proposed problem is given in Fig 1.

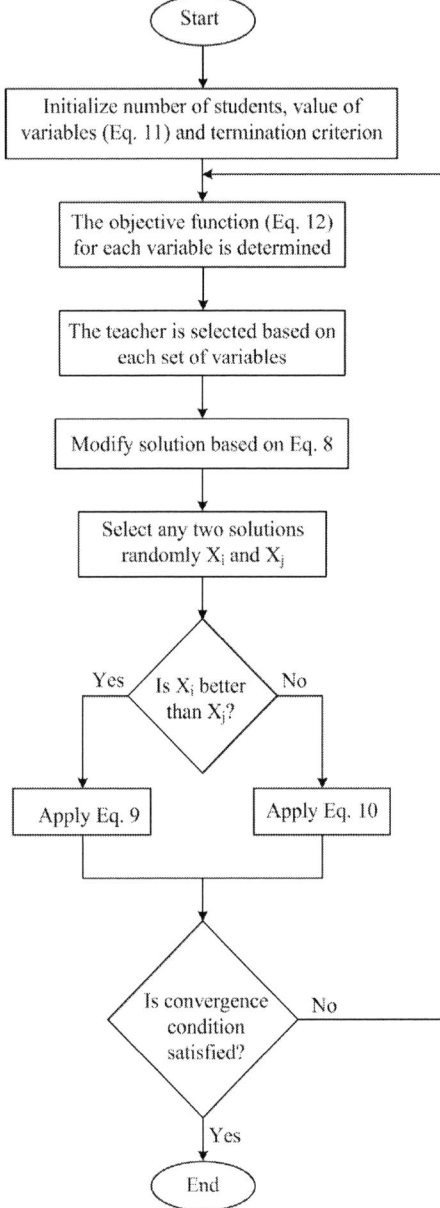

Fig. 1. Flowchart of the TLBO algorithm to solve the proposed problem

IV. RESULTS AND DISCUSSIONS

In this section different examples are presented to show the efficiency of the TLBO algorithm in the discussed optimization problem. Moreover, the results from applying the proposed algorithm compared to the results from the GA

[12] in order to prove the superiority of the TLBO algorithm in this research. In all cases IDVR includes two DVRs in two independent feeders with $S_{Lreated,1}=S_{Lrated,2}=1$p.u and $V_{Lmin}=0.9$ p.u in each feeder. In addition it is assumed that the purpose is to compensate for the voltage sag with the amplitude of $V_{F1}=0.65$ in the first and $V_{F2}=0.65$ in the second feeder.

A. First example

In the first example, it is assumed that the load curve for each feeder has two time intervals (T=2) which their apparent power and power factor data are shown in table I. The number of the students is equal to 200 and the number of iterations equals 50.

TABLE I.   IDVR OPTIMIZATION RESULTS PERTAINING TO EXAMPLE 1

Load data Interval	Interval 1	Interval 2
Feeder 1	$S_{L,1,1}=0.80^{p.u}$ $\cos\varphi_{1,1}=0.85$	$S_{L,1,2}=0.70^{p.u}$ $\cos\varphi_{1,2}=0.75$
Feeder 2	$S_{L,2,1}=1^{p.u}$ $\cos\varphi_{2,1}=0.90$	$S_{L,2,2}=0.90^{p.u}$ $\cos\varphi_{2,2}=0.99$

Table II shows the obtained results of optimization according to different methods. The first row of the table presents the actual optimal solution through searching the whole solution space. The second and third rows are devoted to the GA and TLBO algorithms respectively. Table II makes it clear that the results of intelligent algorithms exactly equal the optimum results obtained by the search of the whole solution space. Hence, it can be claimed that all the methods have reached the optimum solution with an excellent accuracy.

TABLE II.   IDVR OPTIMIZATION RESULTS PERTAINING TO EXAMPLE 1

Optimization method	Rating of individual DVRs (p.u)	Total IDVR rating (S_{total}) (p.u)	Running time (second)
Search the entire solution space [11]	$V_{DVR,rated,1}=0.5569$ $V_{DVR,rated,2}=0.2771$	0.8340	14.4
GA [12]	$V_{DVR,rated,1}=0.5569$ $V_{DVR,rated,2}=0.2771$	0.8340	13.86
TLBO	$V_{DVR,rated,1}=0.5569$ $V_{DVR,rated,2}=0.2771$	0.8340	1.02

To evaluate the speed of the algorithms in achieving an optimal solution, in the last column of table II the running time of the program is given. In addition, Fig. 2 shows the graph of the objective function in terms of the iterations number for the intelligent algorithms. As shown in table II, the optimized response in the TLBO algorithm has been reached faster than in the GA. Nevertheless, considering the small number of time intervals and the fact that solution space is not too large, applying smart algorithms is not necessary.

B. Second example

In the second example, it is assumed that the load curve of each feeder includes three time intervals (T=3). Table III demonstrates their apparent power and power factor data. The number of students and the number of iterations are assumed 200 and 100, respectively. Table IV shows the results of optimization using different methods and Fig. 3 represents the curve of the objective function in terms of the number of iterations. In this case the responses are virtually the same. Therefore, all methods have achieved an optimal response with a good accuracy. Nonetheless, the TLBO algorithm is much faster than the classic method and even the GA. Although using smart algorithms in this case does not seem necessary, it could considerably lower the time of achieving an optimal response (as seen in table IV).

reached the optimal solution in less iterations. It should be noted that by increasing the number of iterations to 5000 no change occurred in the final solution. Hence, it can be claimed that the TLBO method has reached the optimal response with high precision and low duration.

TABLE III. LOAD DATA IN FEEDERS TO EXAMPLE 2

Interval Load data	Interval 1	Interval 2	Interval 3
Feeder 1	$S_{L,1,1}=0.98^{p.u}$ $\cos\varphi_{1,1}=0.85$	$S_{L,1,2}=0.94^{p.u}$ $\cos\varphi_{1,2}=0.953$	$S_{L,1,3}=1^{p.u}$ $\cos\varphi_{1,3}=0.95$
Feeder 2	$S_{L,2,1}=0.70^{p.u}$ $\cos\varphi_{2,1}=0.90$	$S_{L,2,2}=0.70^{p.u}$ $\cos\varphi_{2,2}=0.75$	$S_{L,2,3}=0.80^{p.u}$ $\cos\varphi_{2,3}=0.80$

TABLE IV. LOAD DATA IN FEEDERS TO EXAMPLE 2

Optimization method	Rating of individual DVRs (p.u)	Total IDVR rating (S_{total}) (p.u)	Running time (second)
Search the entire solution space [11]	$V_{DVR,rated,1} = 0.3403$ $V_{DVR,rated,2} = 0.4129$	0.7533	10800
GA [12]	$V_{DVR,rated1}=0.3395$ $V_{DVR,rated2}=0.4134$	0.7525	24.75
TLBO	$V_{DVR,rated,1} = 0.3398$ $V_{DVR,rated,2} = 0.4127$	0.7528	2.31

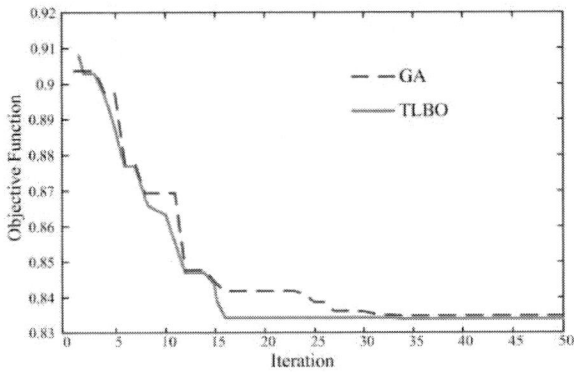

Fig. 2. Convergence curve of the algorithms in the Example 1

A. Third example

In this example it is assumed that the load curve of each feeder has six time intervals and their apparent power and power factor data are shown in table V. In this case the number of students and iterations is respectively set at 500 and 200. Table VI shows the results of the optimization using implementation of different methods. As it is seen in the first row of the table no results are achieved by the first approach. The running time of the program in this case is too long because of searching the entire solution space. Within 30 hours of running the program, the response had not been optimal. However, it sounds clear that the optimal response can be obtained in a very small period of time using the smart algorithms. Therefore, the importance of applying intelligent algorithms is revealed in this case. In addition, as seen in Fig. 4 the convergence point of the TLBO and GA is the same. However, the TLBO algorithm is faster and has

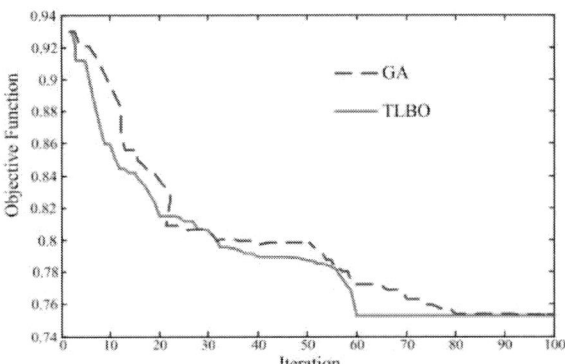

Fig. 3. Convergence curve of the algorithms in the Example 2

TABLE V. LOAD DATA IN FEEDERS TO EXAMPLE 3

Time interval Load data	Interval 1	Interval 2	Interval 3	Interval 4	Interval 5	Interval 6
Feeder 1	$S_{L,1,1}=0.9$ pu $\cos\varphi_{1,1}=0.75$	$S_{L,1,2}=1$ pu $\cos\varphi_{1,2}=0.8$	$S_{L,1,3}=1$ pu $\cos\varphi_{1,3}=0.9$	$S_{L,1,4}=0.7$ pu $\cos\varphi_{1,4}=0.8$	$S_{L,1,5}=1$ pu $\cos\varphi_{1,5}=0.95$	$S_{L,1,6}=0.7$ pu $\cos\varphi_{1,6}=0.85$
Feeder 2	$S_{L,2,1}=1$pu $\cos\varphi_{2,1}=0.9$	$S_{L,2,2}=1$pu $\cos\varphi_{2,2}=0.9$	$S_{L,2,2}=0.9$pu $\cos\varphi_{2,3}=0.9$	$S_{L,2,4}=0.9$pu $\cos\varphi_{2,4}=0.95$	$S_{L,2,5}=0.9$pu $\cos\varphi_{2,5}=0.85$	$S_{L,2,6}=0.8$pu $\cos\varphi_{2,6}=0.95$

V. CONCLUSIONS

Due to the importance of considering the load variations in the feeders for solving optimization problem related to optimal selection of DVRs voltages in an IDVR structure

and the necessity of using smart algorithms, applying the TLBO method was proposed. The ability of the proposed algorithm was compared to the GA (as an intelligent algorithm) and a classic method (which searches the entire solution space) in three presented examples.

The results of implementing smart algorithms show that both TLBO and GA methods lead to the optimal solution with good accuracy. TLBO accuracy, however, is higher than GA.

TABLE VI.   IDVR OPTIMIZATION RESULTS PERTAINING TO EXAMPLE 3

Optimization method	Rating of individual DVRs (p.u)	Total IDVR rating (S_{total}) (p.u)	Running time (second)
Search the entire solution space [11]	$V_{DVR,rated,1}$ = ---- $V_{DVR,rated,2}$ = ----	----	In 30 hours, no results achieve
GA [12]	$V_{DVR,rated1}$ =0.4957 $V_{DVR,rated2}$=0.4221	0.9173	106.58
TLBO	$V_{DVR,rated,1}$ = 0.4945 $V_{DVR,rated,2}$ = 0.4222	0.9162	15.56

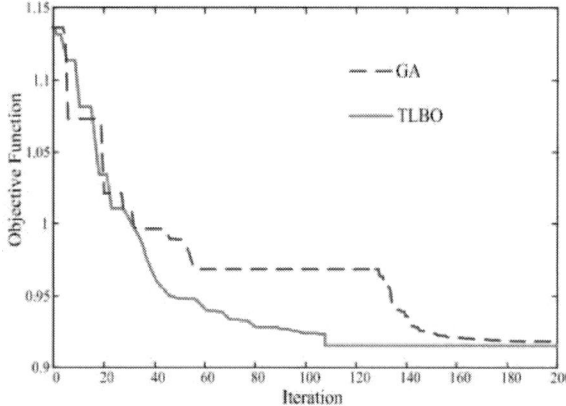

Fig. 4.   Convergence curve of the algorithms in the Example 3

REFERENCES

[1] R. C., Dugan, and M. F., Mc Granghan, "Electrical power system quality", 2th edition, Mc Graw-Hill, New York, 2004.

[2] M. H. J., Bollen, "Understanding power quality problems: voltage sag and interruptions", John Wiley, New York, 1999.

[3] A., Gosh, and G., Ledwich, "Power quality enhancement using custom power devices", Kluwer Academic Publishers, New York, 2002.

[4] G. J., Li, X. P., Zhang, S. S., Choi, T. T., Lie, and Y. Z., Sun, "Control strategy for dynamic voltage restorers to achieve minimum power injection without introducing sudden phase shift", IET Generation, Transmissions and Distribution, vol. 1, pp. 847-853, 2007.

[5] S. Priyavarthini, A. Ch. Kathiresan, Ch. Nagamani, S. I. Ganesan, "PV-fed DVR for simultaneous real power injection and sag/swell mitigation in a wind farm" IET Power Electronics, Vol.11, pp. 2385-2395, 2018.

[6] M. M., Norouzi, J., Olamaei, and G. B., Gharehpetian, "Compensation of voltage sag and flicker during thermal power plant turbo-expander operation by dynamic voltage restorer", International Transactions on Electrical Energy Systems, vol. 26, pp. 16-31, 2016.

[7] S., Taghizadeh, N. M., Lin Tan, and M., Karimi-Ghartemani, "Study of fuzzy based control algorithm for dynamic voltage restorer",

International Transactions on Electrical Energy Systems, vol. 25, pp. 3600-3617, 2015.

[8] F. H., Pai, "A new dynamic voltage restorer with separating active and reactive power circuit design", International Journal of Electronics, vol. 102, pp. 822-838, 2015.

[9] H. M., Vijekoon, D. M., Vilathgamuwa, and S. S., Choi, "Interline dynamic voltage restorer: an economical way to improve interline power quality", IEE Proceeding- Generation, Transmissions and Distribution, vol. 150, pp. 513-520, 2003.

[10] D. M., Vilathgamuwa, H. M., Vijekoon, and S. S., Choi, "Interline Dynamic Voltage Restorer: A Novel and Economical Approach for Multi-Line Power Quality Compensation", IEEE Transactions on Industry Applications, vol. 40, pp. 1678-1685, 2004.

[11] M., Moradlou, and H. R., Karshenas, "Design strategy for optimum rating selection of interline DVR", IEEE Transactions on Power Delivery, vol. 26, pp. 242-249, 2011.

[12] M., Moradlou, M. Bigdeli, P., Siano, and M. Jamadi, "Minimization of interline dynamic voltage restorers rated apparent power in an industrial area consisting of two independent feeders considering daily load variations", Electric Power System Research, vol. 149, pp. 65-75, 2017.

[13] R. V., Rao, D. P., Vakharia, V. J., Savsani, "Teaching–learning-based optimization: a novel method for constrained mechanical design optimization problems", Computer Aided Design, vol. 43, pp. 303–315, 2011.

An Approach for Space Vector PWM to Reduce Harmonics in Low Switching Frequency Applications

Ali BAKBAK
Department of EEE
Ege University
Izmir, Turkey
alibakbak@gmail.com

Erkan MEŞE
Department of EEE
Ege University
Izmir, Turkey
erkan.mese@ege.edu.tr

Abstract— The advantages of space vector PWM have made it preferred technique for power converters. Number of switchings and switching frequency have important effects on thermal operating conditions and losses of power switches. This paper presents a novel space vector PWM approach with reduced number of switchings and relatively low total harmonic distortion factor at low switching frequency. To verify the proposed approach, comparative simulation studies have been carried out using MATLAB/Simulink and the results have been presented and compared.

Keywords—space vector pulse width modulation, discontinuous pulse width modulation, low switching frequency, switching loss, harmonics, high-power converter.

I. INTRODUCTION

Space vector pulse width modulation (SVPWM) technique is widely used to control power electronic converters. Compared to the conventional PWM, SVPWM can be easily implemented and maximum output voltage can be increased by 15% [1]. Furthermore, SVPWM can obtain a better voltage total harmonic distortion factor.

Nowadays, the application range of power electronics is growing to Multi-Megawatt power level with the rapid development of semiconductor devices. In high-power applications, the converter switching frequency must be kept at minimum level to avoid switching losses and electromagnetic interference problems [2]. Also, in the application of power range of Megawatts, for example flexible alternating current transmission systems, the switching frequency of semiconductor is limited to a few hundred Hertz up to 1kHz [3]. Due to these constraints, the switching frequency and operating losses are of great concern for converters of high-power level [4]. On the other hand, as the switching frequency decreases, the SVPWM tends to produce more low-order harmonics [5].

In this study, the basic principle of SVPWM is presented. A novel SVPWM approach is proposed, in which space vector periods are recalculated every half of sampling period. The new approach is compared with two popular SVPWM schemes. Advantages and disadvantages of the proposed method are discussed in detail. The validity of new approach has been verified by simulation. Based upon this study, compared with the SVPWM schemes having the same number of switchings, proposed method results in harmonic reduction at low switching frequency.

II. PRINCIPLES OF SVPWM

SVPWM is based on the fact that there are only eight possible switching states for a three-phase inverter. A basic three phase voltage source inverter (VSI) circuit is shown in Fig. 1.

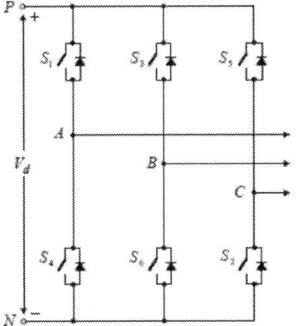

Fig. 1. Circuit topology of VSI

The task of SVPWM is to use the combinations of switching states to produce the desired voltage vector. The eight possible switching states of the inverter are represented as 2 zero vectors and 6 active vectors. The operating states and corresponding vectors are listed in Table I.

TABLE I.   SWITCHING STATES OF TWO LEVEL INVERTER

Space vector		Switching state	'On' switches S_i
Zero vector	V_7	[PPP]	1,3,5
	V_0	[OOO]	4,6,2
Active vector	V_1	[POO]	1,6,2
	V_2	[PPO]	1,3,2
	V_3	[OPO]	4,3,2
	V_4	[OPP]	4,3,5
	V_5	[OOP]	4,6,5
	V_6	[POP]	1,6,5

These vectors (V_0 -V_7) can be used to frame the vector plane which is illustrated in Fig. 2. The voltage vector V_{ref} in Fig. 2 represent the reference voltage vector to be synthesized.

978-1-5386-7688-2/19 $31.00 © 2019 IEEE

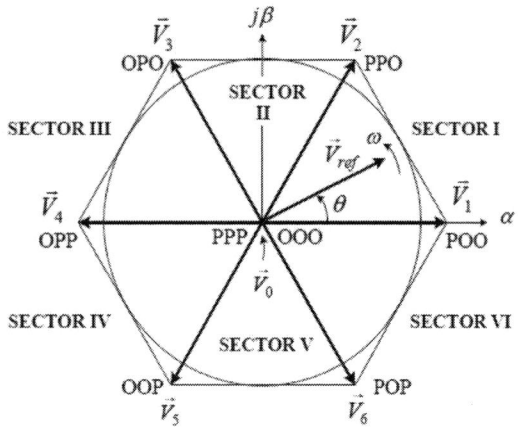

Fig. 2. Vector plane frame [6]

Vref is sampled at equal intervals of time. (Ts referred to as sampling period.) The sampled value of the *Vref* is produced by applying the two nearest active vectors and the zero voltage vectors over different time durations within the sampling period. The location and magnitude of reference vector can be expressed as:

$$\begin{bmatrix} V_\alpha(t) \\ V_\beta(t) \end{bmatrix} = \frac{2}{3} \begin{bmatrix} \cos 0 & \cos \frac{2\pi}{3} & \cos \frac{4\pi}{3} \\ \sin 0 & \sin \frac{2\pi}{3} & \sin \frac{4\pi}{3} \end{bmatrix} \begin{bmatrix} V_{AO} \\ V_{BO} \\ V_{CO} \end{bmatrix} \quad (1)$$

$$V_{ref}(t) = V_\alpha(t) + jV_\beta(t) \quad (2)$$

Application duration of vectors must satisfy the volt second balance. The generalized equations of volt second balance are given in (3) and (4) for sector 'k'.

$$V_{ref}Ts = V_k T_k + V_{k+1}T_{k+1} + V_0 T_0 + V_7 T_7 \quad (3)$$

$$T_S = T_k + T_{k+1} + T_0 \quad (4)$$

The active and zero voltage vectors' application duration can be calculated for sector '*k*' as in (5), (6) and (7).

$$T_k = \frac{\sqrt{3}T_s V_{ref}}{V_d} \sin\left(\frac{\pi}{3} - \theta\right) \quad (5)$$

$$T_{k+1} = \frac{\sqrt{3}T_s V_{ref}}{V_d} \sin(\theta) \quad (6)$$

$$T_0 = T_s - T_k - T_{k+1} \quad (7)$$

Where Vd is DC bus voltage, θ is the angle of between V_{ref} and V_k.

III. CONVENTIONAL AND PROPOSED SVPWM SCHEMES

A popular SVPWM technique is to alternate the zero voltage vectors in a sampling time period and to reverse sequence after each zero vector. This method called as symmetric seven-segment technique. Switching sequences of seven-segment technique for sector 1 are shown in Fig 3.

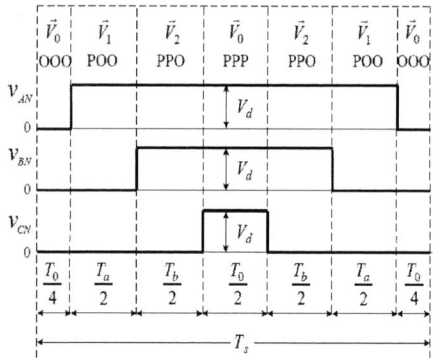

Fig. 3. Conventional 7-segment switching sequences of sector I

Another method with reduced number of switchings is 120° discontinuous PWM. Details can be found in [7]. Fig. 4. shows the switching sequences of 120° discontinuous PWM for sector 1.

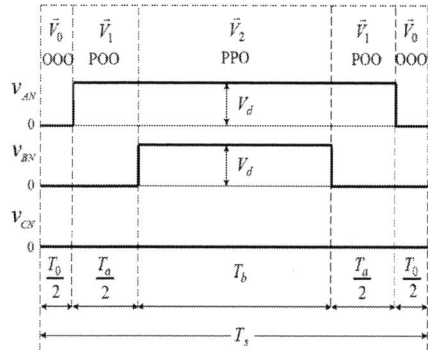

Fig. 4. 120° discontinuous PWM switching sequences of sector I

In the proposed method, sequence is the same as discontinuous PWM, but dwell times of vectors are recalculated every half of a period which means sampling period is reduced by half. Recalculating dwell times of vectors at half period allows to acquire the location of reference vector with higher fidelity and this improves resolution of the scheme. The symmetry of pattern would probably disappear due to the recalculation. Fig 5. shows the switching sequence of the proposed method.

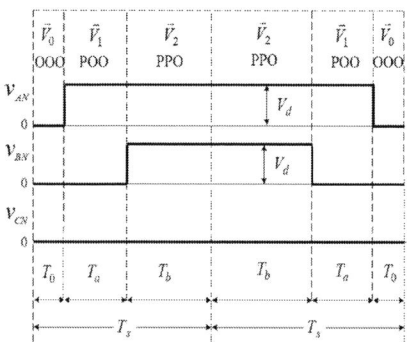

Fig. 5. Proposed method switching sequences of sector I

IV. SIMULATION RESULTS

The proposed and the other two techniques are simulated in MATLAB/Simulink. To validate the proposed approach, all techniques have been carried out under same conditions.

978-1-5386-7688-2/19 $31.00 © 2019 IEEE

For the simulation study switching frequency *fsw* is taken as 500 Hz, Vdc is taken as 780V, magnitude and frequency of reference voltage vector are taken as 720V and 50 Hz, respectively. In addition, a comparison was made in order to demonstrate the effect of switching frequency increasing on total harmonic distortion factor for SVPWM techniques. Furthermore, switching losses of each method are calculated based on a high-power IGBT module and results are given in Table II.

The applied line voltage V_{AB} and the phase current of A for a R-L load in the proposed method are shown in Fig. 6. Resistance and inductance values are taken as 1 ohm and 1 mH respectively.

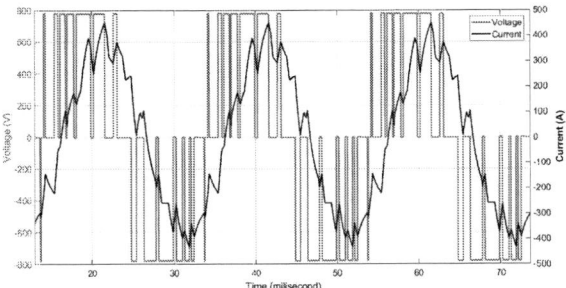

Fig. 6. The line voltage V_{AB} and the phase current of A in the proposed method

Amplitude spectrum of the output line voltages of all three algorithms are shown in Fig 7. to Fig. 9.

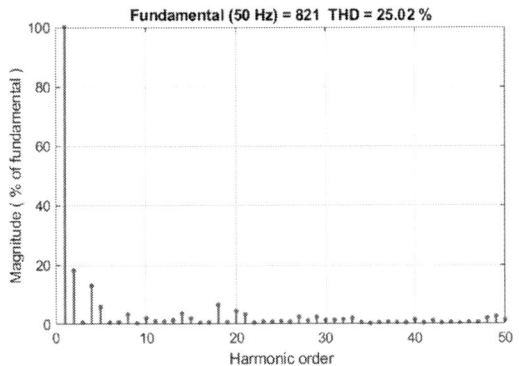

Fig. 7. Amplitude spectrum of proposed method

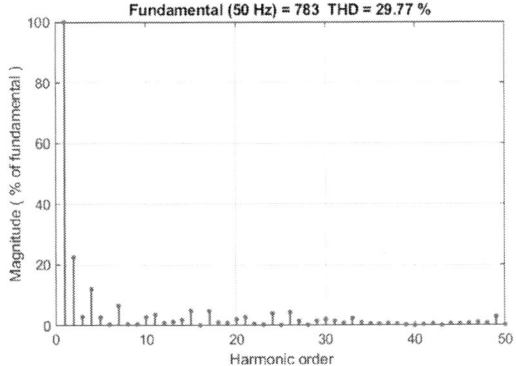

Fig. 8. Amplitude spectrum of discontinuous PWM

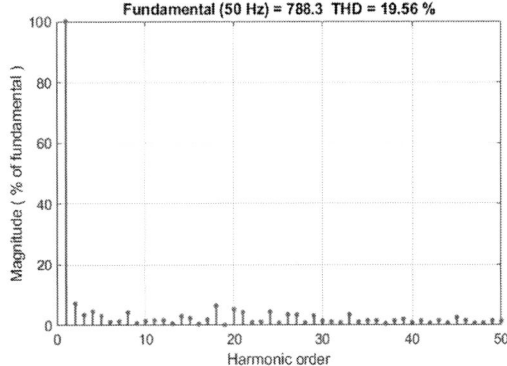

Fig. 9. Amplitude spectrum of 7-segment PWM

The number of switchings and total harmonic distortion factor is preferred to be low in high-power level applications for the reasons stated in the introduction. Seven segment method has 6 and the other two methods have 4 number of switchings in one sampling period. Thus, proposed and discontinuous PWM method are more preferable than seven segment method in terms of switching losses. As shown in the figures, seven segment method has the lowest, discontinuous PWM has the highest THD value. The proposed method performs better than discontinuous PWM in terms of THD. However, its THD value is higher than the seven-segment method.

As the switching frequency increases, the THD difference between the proposed and discontinuous PWM methods decreases. The comparison results of total harmonic distortion for seven segment, discontinuous PWM and proposed method in different switching frequencies are shown in Fig 10.

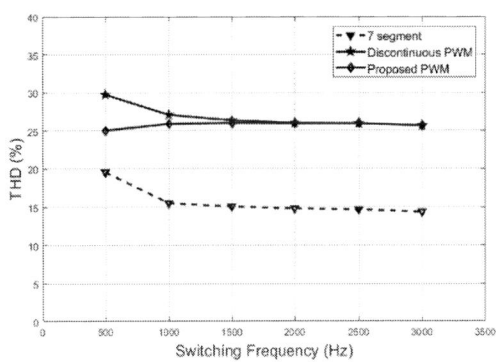

Fig. 10. Comparison of THD values

Average switching losses of each method are calculated in simulation based on SKM200GB12T4 IGBT module of SEMITRANS which has 1200V V_{CES} and 200A I_{Cnom} ratings. The average switching losses in nominal load are listed in Table II.

Seven segment method has higher number of switchings than the other two methods, consequently it has the highest average switching losses. Discontinuous PWM and proposed method have the same number of switchings, but proposed method has slightly more switching losses. The resolution improvement resulting from recalculation led to an increase in switching losses.

978-1-5386-7688-2/19 $31.00 © 2019 IEEE

TABLE II. AVERAGE SWITCHING LOSSES OF METHODS

Method	Average Switching Losses
7 segment	122.95 W
Discontinuous PWM	77.80 W
Proposed PWM	86.23 W

V. CONCLUSIONS

The speed of today's processors is at adequate level to calculate SVPWM algorithm every half of sampling period in low switching frequencies. To turn this feature into an advantage, a new SVPWM approach is presented. By calculating dwell times of vectors in every half of sampling period, a reduction in THD value was achieved. Simulation result have verified the validity of proposed SVPWM technique. For very low switching frequency applications, zero sequence component added carried based PWM schemes may give comparable results in terms of switching losses. However, the proposed method is anticipated to have less THD compared to carrier based schemes. Since the number of switchings (switching losses) and THD value is significant in high-power applications, the proposed method may offer more optimized solution by exploiting well known advantages of space vector modulation. In addition to comparison made between SVPWM and carrier based PWM, the proposed method presents certain advantages among other SVPWM schemes.

REFERENCES

[1] Ogasawara, Satoshi, Hirofumi Akagi, and Akira Nabae. "A novel PWM scheme of voltage source inverters based on space vector theoryEin neues Modulationskonzept für Wechselrichter mit eingeprägter Spannung." Archiv für Elektrotechnik 74.1, 1990, pp. 33-41.

[2] Lai, Jih-Sheng, and Fang Zheng Peng. "Multilevel converters-a new breed of power converters." Industry Applications Conference, 1995. Thirtieth IAS Annual Meeting, IAS'95., Conference Record of the 1995 IEEE. Vol. 3. IEEE, 1995.

[3] Hingorani, Narain G., Laszlo Gyugyi, and Mohamed El-Hawary. Understanding FACTS: concepts and technology of flexible AC transmission systems. Vol. 1. New York: IEEE press, 2000.

[4] Zhang, Hui, et al. "Selective harmonic controlling for three-level high power active front end converter with low switching frequency." Power Electronics and Motion Control Conference, 2006. IPEMC 2006. CES/IEEE 5th International. Vol. 1. IEEE, 2006.

[5] Li, Yun Wei, et al. "Space vector sequence investigation and synchronization methods for active front-end rectifiers in high-power current-source drives." IEEE Transactions on industrial Electronics 55.3 (2008): 1022-1034.

[6] Wu, Bin, and Mehdi Narimani. High-power converters and AC drives. Vol. 59. John Wiley & Sons, 2017.

[7] Holmes, D. Grahame, and Thomas A. Lipo. Pulse width modulation for power converters: principles and practice. Vol. 18. John Wiley & Sons, 2003.

Analysis of Current-Feedback PWM Procedures Based on Hysteresis and Current-Carrier-Wave Control for VSI-Fed Induction Motor Drive

Csaba Szabo
Dept. of Electrical Machines &Drives
Technical University of Cluj-Napoca
Cluj-Napoca, Romania
csaba.szabo@emd.utcluj.ro

Eniko Szoke
Dept. of Electrical Machines &Drives
Technical University of Cluj-Napoca
Cluj-Napoca, Romania
eniko.szoke@emd.utcluj.ro

Norbert Szekely
Dept. of Electrical Machines &Drives
Technical University of Cluj-Napoca
Cluj-Napoca, Romania
norbert.szekely@emd.utcluj.ro

Vlad Zacharias
Dept. of Electrical Machines &Drives
Technical University of Cluj-Napoca
Cluj-Napoca, Romania
vlad.zacharias@edr.utcluj.ro

Maria Imecs
Dept. of Electrical Machines &Drives
Technical University of Cluj-Napoca
Cluj-Napoca, Romania
maria.imecs@emd.utcluj.ro

Abstract— For voltage-source inverter-fed AC motor drive applications that need fast response to load transients, current-feedback PWM procedures offer better dynamic response than the voltage PWM techniques. Hysteresis control offers a simple solution but leads to variable switching frequency. Replacing the hysteresis band with a carrier-wave signal added to the reference phase current signals constant switching frequency operation can be achieved. Comparative analysis by numerical simulation of variable frequency hysteresis current control and constant switching frequency current-carrier-wave techniques applied for a rotor-field oriented control of a cage-induction motor is presented in this paper.

Keywords—pulse-width-modulated voltage-source inverter, hysteresis current control, carrier-wave-based current control, constant switching frequency, induction motor drive, field-oriented control.

I. INTRODUCTION

Nomenclature

PWM – Pulse Width Modulation
VSI – Voltage Source Inverter
CCW – current carrier–wave
RFO – rotor field orientation
d-q – stator-fixed stationary reference frame
$d\lambda_r - q\lambda_r$ – rotor-field oriented rotating reference frame

PhT – Phase transformation (direct: a,b,c to d-q; reverse d-q to a,b,c)

CooT – Coordinate transformation (direct: d-q to $d\lambda_r - q\lambda_r$; reverse: $d\lambda_r - q\lambda_r$ to d-q)
Ψ_s – stator flux; Ψ_r – rotor flux
λ_r – rotor-flux position related to the stationary reference frame
m_e – electromagnetic torque

For low and medium power variable speed, induction motor drives Field-Oriented Control (FOC) is widely used. Most of the dedicated digital controllers are designed with embedded PWM modules using Sinusoidal PWM (SPWM) or Space-Vector Modulation (SM) for the control of the Voltage-Source Inverter (VSI). Three-phase VSI can work also as active current filter on the line-side. In electrical drive applications the electric motor represents the actuator for the mechanical load that is attached to its shaft, while the power electronic converter is the actuator for the motor. By controlling the converter output parameters (the voltage amplitude and frequency) the energy transfer to the load via the motor can be optimized to achieve the required torque and speed and also to ensure the magnetization of the motor. The cage-induction machine is fed via the stator that means that the absorbed stator current is used both for magnetization (reactive component) and torque production. The optimal distribution of the reactive (magnetizing) and active (torque producing) component of the stator current may performed by vector control based on field-orientation principle that ensure the independent control of the machine torque and flux on two separate control loops. A fast and accurate control of the motor currents is an important requirement and different solutions lead to different control structures. One of the criteria that will determine how and where the current control is performed in a control structure is given by the inverter control procedure. Vector control structures with voltage controlled VSI include perform the current control using linear (usually PI) controllers in order to generate the reference voltage needed for inverter control, while in case of current-feedback based modulation techniques the current control is performed using non-linear (hysteresis or on-off) controllers. The most suitable vector control method for the induction motor is based on rotor-field orientation that combined with voltage-controlled VSI leads to a complex and motor parameter dependent control system configuration regarding the rotor-flux identification and also the reference stator-voltage generation which represents the control variable of the inverter [1], [2], [3]. The flux and speed controllers are cascaded on both control loops by the two current controllers working in synchronous rotor-field oriented (RFO) reference frame, meaning that there are four controllers that require an accurate tuning process, which affects the system stability over a wide operating range [4]. Moreover, the computation of the reference stator voltage performed in RFO-ed reference frame is highly motor-parameter dependent. The increased computational time and the time delays due to the stator and rotor time constants considered in the computation process of the converter control variable introduce time delays, decreasing the performances of the control system [2], [5]. However, if current-feedback control is performed on the VSI, two (flux and speed) PI controllers are required to generate the rotor-field-oriented $i_{sd\lambda r}$ reactive (direct) and $i_{sq\lambda r}$ active (quadrature) components of the stator-current that represents the inverter control variable. It means that in comparison with

978-1-5386-7688-2/19 $31.00 © 2019 IEEE

the proper voltage modulation based scheme the two linear current controllers and the voltage computation block are no longer needed. It gives a great advantage over the voltage controlled structure since The current control of the converter is performed using hysteresis controllers that ensure better dynamic response than the PI type current controllers used in voltage-controlled structures, are working without tracking errors and are robust to load parameter variations [4], [6]. It also ensures a better current waveform compared to the voltage-control based procedures. The inverter switching signal generation does not depend on the load parameters. The main disadvantage of the current-feedback control with hysteresis regulators is represented by the variable switching frequency [4]. The current (and torque) ripples can be adjusted by the hysteresis bandwidth. Narrow hysteresis width ensure a better current waveform, but it leads to very high switching frequency, while a large band will introduce distortions and increased current ripples. There are several techniques based on current-feedback control of VSI as an alternative solution to the classical hysteresis control, developed in the past decades, focus either on achieving constant, or nearly constant switching frequency, or on reducing the number of commutations. For large power applications, where high switching frequencies lead to high commutation losses, a relatively simple technique where the output signal of the simple-on-off controllers (based on the current error) is sampled by a constant frequency clock signal where its state is maintained until the next sampling instant [3], [6]. It does not ensure constant switching frequency, but the commutation process can be further optimized if the sampling instants are delayed with one third of the sampling period on each phase [3], [6]. Hysteresis control may be improved also by frequency limiting procedures may be applied in order to ensure a minimum switching frequency to ensure stable operation [14]. Nearly constant switching frequency may also be achieved by online adjustment of the hysteresis dead-band based on the phase current errors and the inverter output voltage fundamental value [7], [8], [15]. Variable hysteresis band computation may also be combined with variable sample time in order to achieve constant switching frequency [18]. Space-vector-based solutions also ensure optimal switching process by forcing the current error vector between boundaries of specific shapes [4], [6], [15], [17], [19].

This paper focuses on a solution that ensures constant switching frequency by adding a fixed frequency carrier-wave to the reference phase-currents that will replace the hysteresis band. The resulted reference current signal will have a low frequency sinusoidal component and a high frequency triangular-shape carrier signal superposed to the low frequency. The carrier signal that replaces the hysteresis band has fixed frequency, while its amplitude is comparable with the hysteresis bandwidth. The hysteresis controllers are replaced by simple *"on-off"* switches [9]. The commutation instants are generated at the intersections of the combined reference current signal with the actual phase current. The current ripples are limited by setting the amplitude of the triangular carrier-wave signal. To avoid multiple unwanted commutations, a ramp condition is applied for the carrier wave signal, as its slope must be greater than the current variation slope [10]. Validation was performed by numerical simulation, the AC induction motor drive being developed using physical models of DC-link Static Frequency Converter (SFC) consisting of a diode rectifier, a braking chopper and a VSI, and the squirrel-cage motor, modeled in two-phase

stationary coordinates. Rotor-field-oriented vector control is applied for motor while the VSI works with current-carrier-wave (CCW) modulation at a constant switching frequency.

II. CURRENT-FEEDBACK CONTROL OF THREE-PHASE INVERTERS

A. Hysteresis Band Current-Feedback Control

When a proper VSI is controlled in current, it will act as a current source inverter. The reference value of the stator current generated by the vector control structure will be strictly maintained, while the inverter output voltage will be modified in order to meet the power demand of the AC drive, acting as load for the VSI. The three-phase inverter currents are controlled independently using hysteresis regulators (non-linear controllers). The actual phase current $i_s(t)$ will have to follow the reference current $i_s^{Ref}(t)$ within a hysteresis dead-band. This modulation procedure presented in Fig.1 is known in the literature also as "*Bang-bang current control*" [6]. The switching instants will occur when the actual current value reaches the lower or upper limits of the hysteresis band placed around the reference signal.

The switching logic will be [2]:

$$\begin{cases} sw_{log} = 0 \ for \ i(t) \geq i_s^{Ref}(t) + (\Delta i)^{Ref}/2; \\ sw_{log} = 1 \ for \ i(t) \leq i_s^{Ref}(t) - (\Delta i)^{Ref}/2, \end{cases} \quad (1)$$

where the Δi hysteresis dead-band width will determine the current ripples, but also influence the switching frequency. Even though the motor phase currents are independently controlled, the mutual coupling between the motor windings leads to current slope variation, resulting random switching instants and consequently variable switching frequency. Very high switching frequencies mean high switching losses since all the power is flowing through the inverter and may cause overheating. These situations can alternate with relatively large time periods when no switching demand occurs. For a three-phase IGBT inverter where the stator voltage space phasor, which has intermittent motion, there are 8 possible switching states corresponding to six nonzero voltage vectors (forming a hexagon) and two zero voltage vectors. Optimal switching process is realized when at a certain moment only one commutation takes place, meaning the voltage space phasor can only move to an adjacent position, corresponding to the direction of the rotating magnetic field, or it can switch to a zero voltage vector (000 or 111, depending on which one can be performed by switching a single inverter leg) [4], [6], [11].

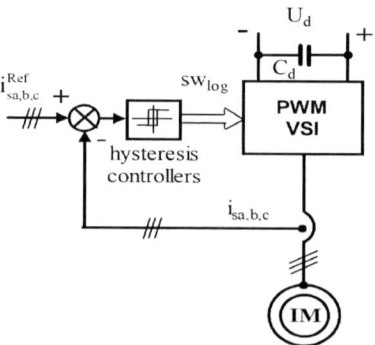

Fig. 1. Block diagram of the curent-feedbeck controlled VSI with hysteresis controllers.

With hysteresis control, simultaneous commutations cannot be avoided, which will force the voltage space vector to randomly jump to any position leading to a non-optimal switching process.

B. Carrier-Wave Current Feedback Control

The carrier-wave-based modulation technique is derived from the classical hysteresis one but the switching states will be determined by the intersection of the actual phase current $i_s(t)$ with a constant frequency triangular carrier-wave signal i_{cr} added to the reference phase current $i_s^{Ref}(t)$, as it is presented in Fig. 2 [9], [12]. Thus, the switching frequency will be equal to the constant carrier frequency. The inverter gate signals will be generated according to the following switching logic sw_{log} [2]:

$$\begin{cases} sw_{log} = 0 \ for \ i_s(t) \geq i_s^{Ref}(t) + i_{cr}; \\ sw_{log} = 1 \ for \ i_s(t) \leq i_s^{Ref}(t) + i_{cr}. \end{cases} \quad (2)$$

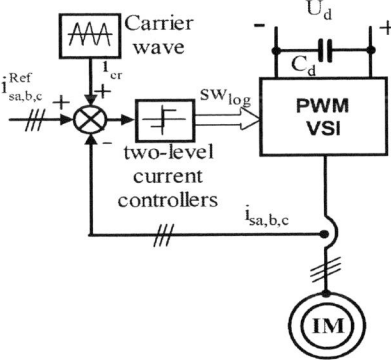

Fig. 2. Block diagram of the current-carrier-wave controlled VSI.

The triangular carrier expression for a full period T is:

$$i_{cr} = \begin{cases} 4A_{cr}f_{cr}t \quad for \ t \in 0 + kT, T/4 + kT; \\ -4A_{cr}f_{cr}t \quad for \ t \in (T/4 + kT, T + kT); \ (3) \\ 0 \quad for \ t \in \{0 + kT, T + kT\}. \end{cases}$$

where A_{cr} and f_{cr} represents the amplitude and frequency of the carrier. The reference current signal can be expressed as [2], [12]:

$$i_s^{Ref}(t) = I_s^{Ref} sin(2\pi f^{Ref}t + \varphi), \quad (4)$$

In order to maintain a constant switching frequency (related to the carrier-wave frequency) no multiple unwanted commutations are allowed during a period. This can be fulfilled if at any moment the m^{Ref} slope of the combined reference signal is less than the m^{i_s} slope of the actual load current. The slope of the reference signal is [2]:

$$m^{Ref} = \begin{cases} I_s^{Ref}2\pi f_s^{Ref} cos(2\pi f_s^{Ref}t + \varphi) \pm 4A_{cr}f_{cr}, \\ \quad for \ cos(2\pi f_s^{Ref}t + \varphi) \in [0,1]; \\ -I_s^{Ref}2\pi f_s^{Ref} cos(2\pi f_s^{Ref}t + \varphi) \pm 4A_{cr}f_{cr}, \\ \quad for \ cos(2\pi f_s^{Ref}t + \varphi) \in (0,-1]; \end{cases} \quad (5)$$

where f_s^{Ref} is the reference current frequency. The first term of the expressions (cosine function) represents the derivative of the reference current while the second term is the carrier-wave slope. During a period of the phase, current presents a variation between a minimum and a maximum value.

- If $cos(2\pi f_s^{Ref}t + \varphi) = 0$

At this point the reference current is at its peak value, consequently the combined reference current slope will be represented only by the carrier, with $4A_{cr}f_{cr}$ for rising slope and $-4A_{cr}f_{cr}$ for falling slope.

- If $cos(2\pi f_s^{Ref}t + \varphi) = \{1,-1\}$

These situations occur at the zero crossings of the current when its slope is at its maximum value [10], [2], [9]:

$$|m_{max}^{Ref}| = \pm I_s^{Ref}2\pi f_s^{Ref} \pm 4A_{cr}f_{cr} \quad (6)$$

However, the one case that will be taken into account for choosing the carrier-wave slope is when the combined reference is at its minimum [10], [2], [9]:

$$|m_{min}^{Ref}| = \pm I_s^{Ref}2\pi f_s^{Ref} \mp 4A_{cr}f_{cr} \quad (7)$$

The actual load-current slopes $m^{i_s} = di_s/dt$ are obtained by derivation of the motor phase currents [4]:

$$\begin{cases} di_{sa}/dt = 1/(3L_s)[2(u_{sa} - e_{sa}) - (u_{sb} - e_{sb}) - \\ \quad -(u_{sc} - e_{sc})] - (R_s/L_s)i_{sa} \\ di_{sb}/dt = 1/(3L_s)[-(u_{sa} - e_{sa}) + 2(u_{sb} - e_{sb}) - \\ \quad -(u_{sc} - e_{sc})] - (R_s/L_s)i_{sa} \\ di_{sc}/dt = 1/(3L_s)[-(u_{sa} - e_{sa}) - (u_{sb} - e_{sb}) + \\ \quad +2(u_{sc} - e_{sc})] - (R_s/L_s)i_{sc} \end{cases} \quad (8)$$

where u_s represents the voltages and e_s the e.m.f.-s on each motor phase. Based on above assumptions, the slope condition that needs to be fulfilled for an optimal number of switching instants during a carrier period is [10], [2], [12]:

$$|m_{min}^{Ref}| \geq |m^{i_s}| \quad (9)$$

For a fixed switching frequency the A_{cr} carrier has to be adjusted in order to fulfill the condition [2]:

$$A_{cr} > \frac{|m^{i_s}| + I_s^{Ref}2\pi f_s^{Ref}}{4f_{cr}} \quad (10)$$

III. THE VECTOR CONTROL STRUCTURE

In a vector control structure of an induction machine, based on field-orientation, the independent control of the machine torque and flux (based on the analogy with the DC machine) can be performed on two separate control loops that will indirectly determine the f_s frequency and U_s amplitude of the supply voltage [1]. In Fig. 3 there is presented the space phasor diagram of the squirrel-cage induction motor running with constant rotor flux, which represents the principle of the rotor-field-oriented vector control strategy. In Fig. 4 there is presented the vector control structure of a squirrel-cage induction machine fed by a three-phase IGBT-VSI controlled by means of carrier-wave current-feedback modulation that ensures constant switching frequency. On the electromagnetic control loop, the Ψ_r rotor-field amplitude is controlled, which will influence the amplitude of the supply voltage. On the electromechanical loop the rotor speed is controlled, which determine the frequency.

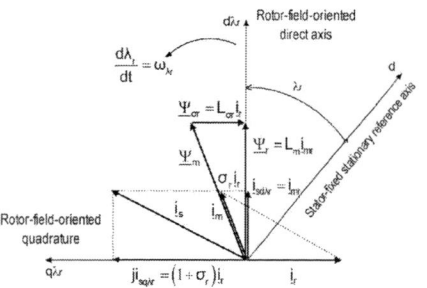

Fig. 3. VSI.Space-phasor diagram of the cage-induction motor running with constant rotor flux.

For RFO the control is performed in a rotating coordinate frame $(d\lambda_r - q\lambda_r)$ with the direct axis pointing in the direction of the $\underline{\Psi}_r$ rotor-flux space phasor, where λ_r represents its angular position related to the stator-fixed (stationary) reference frame. The flux components are [1], [4]:

$$\Psi_{rd\lambda r} = \Psi_r = |\underline{\Psi}_r| \quad \text{and} \quad \Psi_{rq\lambda r} = 0. \qquad (11)$$

The flux controller on the reactive loop generates the $i_{sd\lambda r}^{Ref}$ reactive stator current component that is equal to the rotor-flux-based magnetizing current i_{mr} [1], [4]:

$$i_{sd\lambda r} = i_{mr} = \Psi_r/L_m \qquad (12)$$

On the active loop, the speed controller generate the m_e^{Ref} torque reference, while the $i_{sq\lambda r}^{Ref}$ active (torque producing) of current component is obtained by:

where $k_{Mr} = 3/2\ z_p$ is the torque constant, σ_r the rotor leakage coefficient and i_r the rotor-current.

The RFO-ed current control variables are transformed into two-phase stator-fixed reference frame by a coordinate transformation (CooT) with matrix operator $[D(\lambda_r)]^{-1}$ then into three-phase components by a reverse phase transformation (PhT with a matrix operator $[A]^{-1}$) obtaining the $i_{sa,b,c}^{Ref}$ reference currents for the VSI control. Applying the switching logic (1) for hysteresis control or (2) for CCW control, the gate signals are generated for the three-phase IGBT inverter. The rotor flux is identified in two steps based on the following procedure [1]:

Initially, the stator-flux identification is performed by integration of the back e.m.f. in stator-fixed coordinates, obtaining the two components of the flux space phasor:

$$\Psi_{sd} = \int (u_{sd} - R_s i_{sd})\,dt \quad \text{and} \quad \Psi_{sq} = \int (u_{sq} - R_s i_{sq})\,dt. \quad (14)$$

The rotor flux is then computed by [1]:

$$\begin{cases} \Psi_{rd} = (1 + \sigma_r)\Psi_{sd} - (1 - \sigma)^{-1}\sigma L_m i_{sd} \\ \Psi_{rq} = (1 + \sigma_r)\Psi_{sq} - (1 - \sigma)^{-1}\sigma L_m i_{sq} \end{cases}; \qquad (15)$$

where $\sigma_r = L_{\sigma r}/L_m$ is the rotor leakage coefficient and σ is the total leakage coefficient. The vector analyzer (VA) block computes the $|\Psi_r|$ rotor-flux module used as feedback for the flux controller and also identifies the rotor flux space-phasor position. This identification method of the rotor flux leads to direct field orientation based control structure.

$$i_{sq\lambda r}^{Ref} = m_e^{Ref}/k_{Mr}\Psi_r = -(1+\sigma_r)\,i_r, \qquad (13)$$

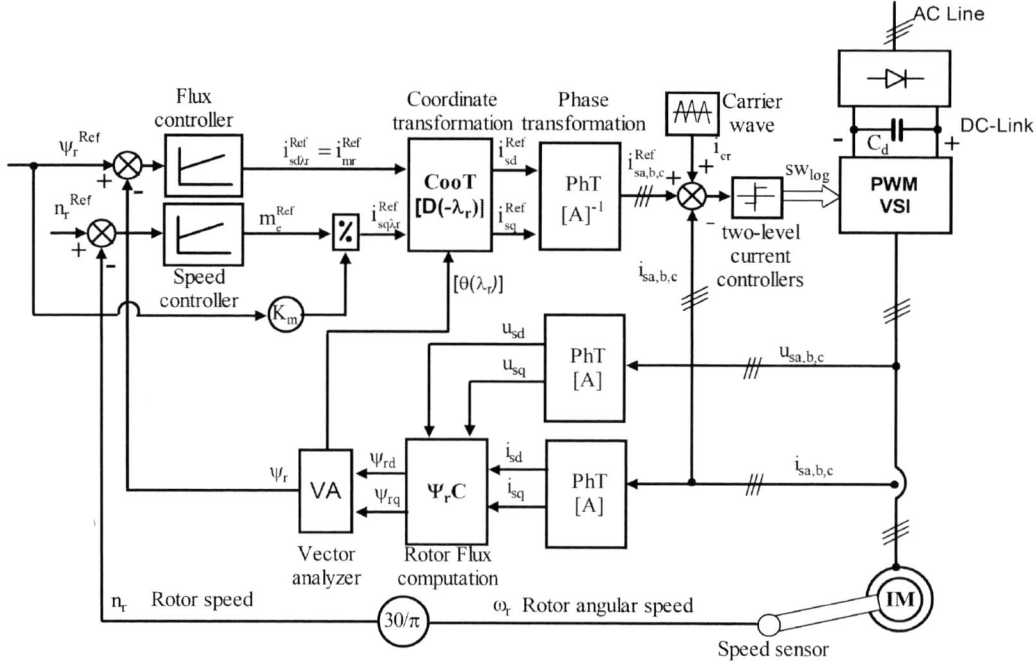

Fig. 4. Rotor-field-orientation-based vector control structure of a squirrel-cage induction motor fed by a VSI controlled with carrier-wave-based current-feedback modulation.

Numerical simulations were performed in MATLAB-Simulink environment for a 2.2 kW, 230V_{rms}, 50 Hz, 4.7 A_{rms} squirrel-cage induction motor. Other rated data: rotor speed n_N = 1420 rpm, electromagnetic torque m_{eN} = 15.5 Nm, rotor flux Ψ_{rN} = 0.9 Wb. The motor model is realized with d-q stator-fixed reference frame interfaced with the three-phase VSI by PhT[A]. The simulation structure also includes physical models of a DC-link SFC consisting of a diode-rectifier and a three-phase IGBT-VSI with options for both hysteresis and CCW modulation. The DC-link voltage is U_d = 560 V. A braking chopper is also included in the DC-link. The models are discretized, and a sampling time of T_s = 1e-6 s was used in simulations.

IV. SIMULATION RESULTS

Simulation results are presented both for hysteresis control and CCW for the following operating conditions: after a short time period when the DC-link voltage is stabilized, the motor is started with a load torque m_L= 5 Nm at with a reference speed of n^{Ref} = 1420 rpm (rising slope 5000 rpm/s) and reference rotor flux Ψ_r^{Ref}= 0.9Wb. At t = 0.75s a sudden load change is applied at m_L= 15 Nm, followed at t = 1.2s by a speed reversal command of n^{Ref} = -1420 rpm with the load torque unchanged. For hysteresis control, a dead-band of Δi = 0.6 A was set. For the CCW procedure, preliminary simulation tests resulted in a maximum value of the current derivative of 27000 A/s that occur at the motor starting, with a current peak of 15A. At steady state operation, its maximum value decreases to 19000 A/s. Based on (10) a minimum value of the carrier amplitudeA_{Cr} = 0.528 A resulted for a carrier frequency of f_{Cr}= 15 kHz. Chosen value is A_{Cr} = 0.6 A.

Fig. 5 shows the stator-voltage space-phasor trajectory with intermittent motion having possible non-zero states (voltage vectors V$_1$ to V$_6$) and two zero voltage states (V$_7$ and V$_8$). In Fig. 5a for hysteresis control, the space-phasor jumps randomly to any other possible position, meaning that simultaneous commutations occur in more than one inverter leg, leading to increased commutation losses. For CCW modulation (Fig. 5b) from an actual position, the space-phasor jumps only to an adjacent position or to a zero-voltage vector, in both cases only one of the three inverter legs is commutated.

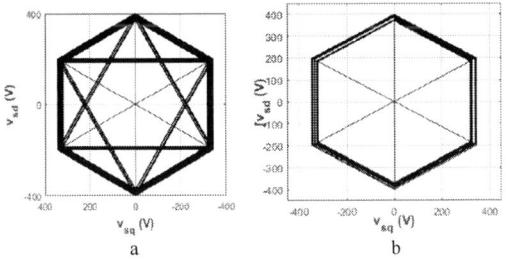

Fig. 5. The stator-voltage space-phasor diagram for: a) hysteresis control; b) Current carrier-wave modulation.

The voltage vector states are presented in Fig. 6 (for hysteresis control) and in Fig. 7 (for CCW). Random commutation states can be observed for hysteresis control, while for CCW an optimal switching process takes place in a similar manner as it happens when the VSI is controlled by voltage PWM procedures like sinusoidal PWM or SVM, as the actual voltage space phasor is generated using by switching only between two adjacent voltage vectors and the corresponding zero vectors. On the other hand, it can be

noticed that CCW modulation uses more frequently the V$_7$ and V$_8$ zero voltage vectors.

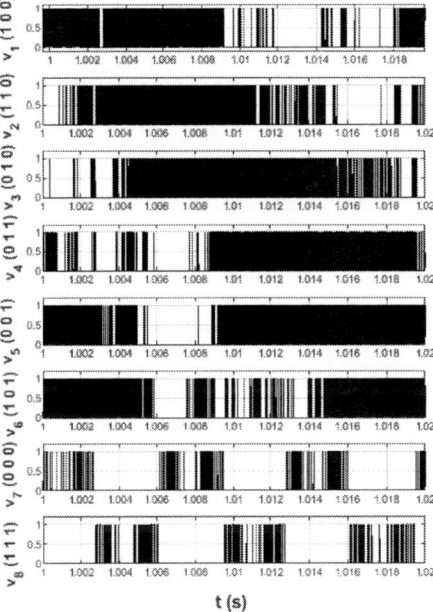

Fig. 6. V$_1$ to V$_8$ voltage vector states for hysteresis dead-band control.

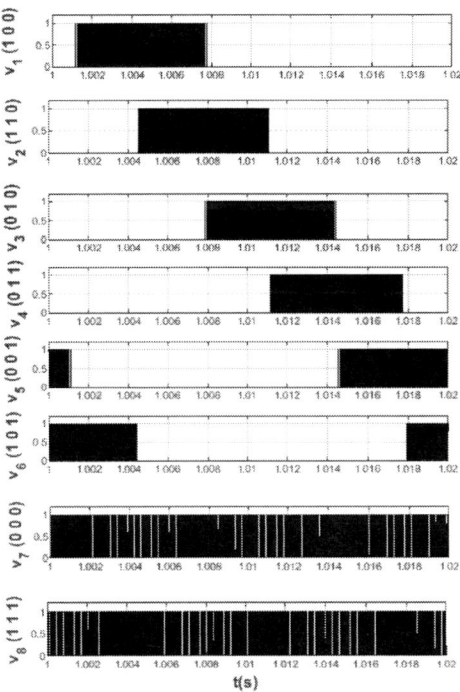

Fig. 7. V$_1$ to V$_8$ voltage vector states for current-carrier wave modulation.

The stator-current space phasor trajectories presented in Fig. 8 indicates that for CCW modulation the motor phase current follows the reference current in a very narrow band, resulting in a better current waveform.

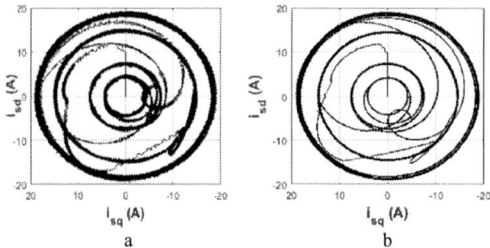

Fig. 8. The stator-current space-phasor diagram for: a) hysteresis control; b) Current carrier-wave modulation.

The phase *a* actual and reference current signals together with the generated switching logic signals (sw$_a$, sw$_b$ and sw$_c$) are presented in Fig. 9 for hysteresis control and Fig. 10 for CCW. In both cases, the current slope is influenced by the commutations that take place on the other two inverter legs, due to the mutual coupling between the three phases as it can be observed in (3).

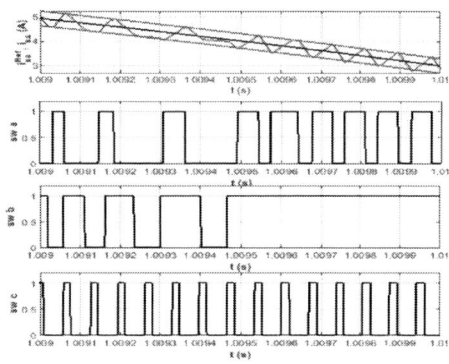

Fig. 9. Phase *a* stator current i_a^{Ref} reference and i_{sa} actual values; Inverter switching logic signals swa, swb and swc for hysteresis control.

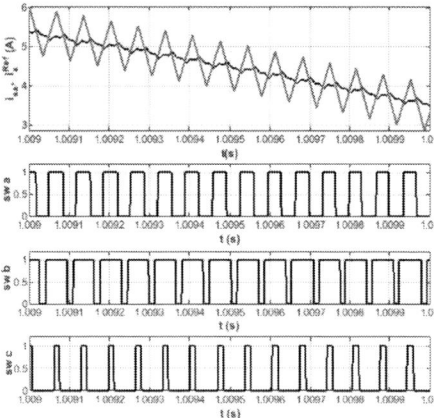

Fig. 10. Phase *a* stator current i_a^{Ref} combined reference and i_{sa} actual values; Inverter switching logic signals swa, swb and swc for current carrier-wave modulation.

Fig. 11 and Fig 2. show the evolution of the motor speed, torque and rotor flux and the DC-link voltage during the whole simulation cycle for the two analyzed modulation techniques.

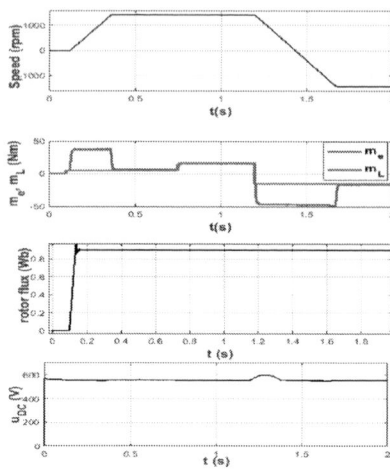

Fig. 11. The motor speed n, motor electromagnetic torque me, rotor flux module Ψ_r and the DC-link voltage for hysteresis control.

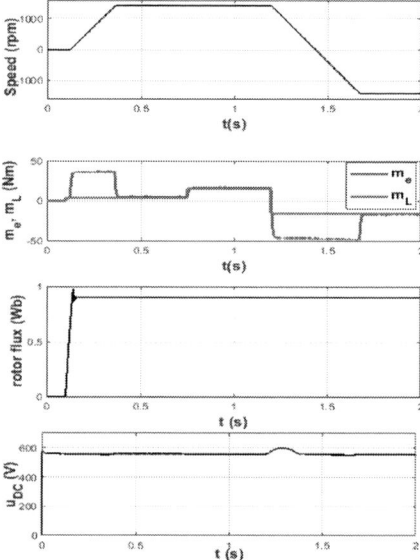

Fig. 12. The motor speed n, motor electromagnetic torque m$_e$ and load torque m$_L$, rotor flux module Ψ_r and the DC-link voltage for current carrier-wave modulation.

There are no noticeable differences regarding the dynamic response, only the torque ripples are more pronounced for hysteresis control. By analyzing the motor and load torque evolution it can be observed that during speed transients the dynamic torque is constant resulting in constant acceleration and deceleration values. The stator phase current evolution in Fig. 13 follows the torque demand – as the rotor flux is constant – and doesn't present overshoot or variations during transient operation. Such accuracy regarding current is difficult to achieve with voltage control-based PWM procedures. The DC-link voltage present variations only at the beginning of the speed reversal procedure at t=1.2 s, when the braking chopper is activated in order to adjust the voltage level.

978-1-5386-7688-2/19 $31.00 © 2019 IEEE   356

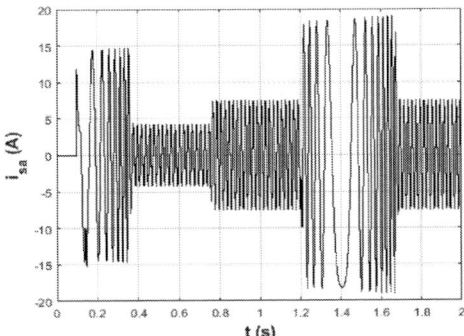

Fig. 13. Stator current on phase *a*.

Fig. 14 and Fig 15 present the harmonic spectrums of the phase *a* currents waveform based on 6 cycles when the motor operates at rated speed with full load of 15 Nm.

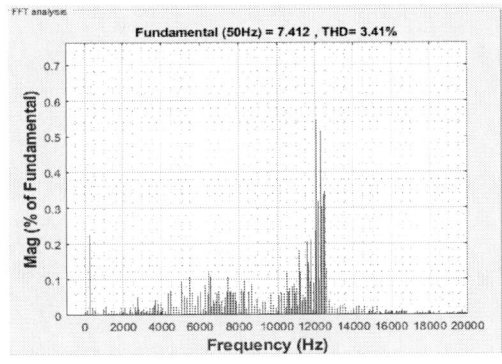

Fig. 14. THD for hysteresis controlled current-feedback procedure.

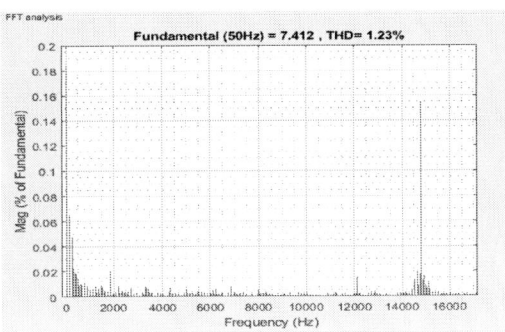

Fig. 15. THD for CCW modulated VSI inverter.

With hysteresis control the total harmonic distortion (THD) factor is 3.41%, while for CCW modulation is reduced to 1.23% in same operating conditions. Moreover, for CCW at higher order harmonics the spectrum is concentrated around the value of the imposed 15 kHz switching frequency, while for the hysteresis control there is a much wider harmonic spectrum. Simulation results show that especially at the motor starting process it results much higher values for the current derivative than in case of steady-state operation or even at speed reversal. To meet the slope condition (9) the A_{cr} value needs to be set high enough, but this may lead to higher current ripples.

V. CONCLUSIONS

Current-feedback control of VSI offers high dynamic response and robustness on load parameters in FOC-based AC drives, also ensuring sinusoidal current waveform. Current carrier-wave-based control procedure of VSI ensures both constant switching frequency and optimal commutation process, offering a viable alternative for the classical hysteresis control. As for hysteresis, the current presents a wide harmonic spectrum. This is significantly reduced by the CCW procedure, similar to the voltage PWM techniques. Higher order harmonics will have smaller magnitudes as in case of voltage-PWM due to the improved current waveforms.

Optimum switching process is achieved as the transition between two consecutive switching states is realized by commutation that significantly reduces the switching losses.

Due to the current-feedback control of VSI, there is a very fast response to load perturbations. Moreover, due to the current control transient operations that imply speed variation (starting or speed reversal) take place at constant current (torque) while the stator supply voltage will present variations in order to meet the actual power demand of the drive.

The possibility of online adjustment of the carrier-wave amplitude should also be considered.

ACKNOWLEDGMENT

"The results presented in this paper were obtained in the framework of the GNaC 2018 ARUT grant "Optimization of Current-Feedback PWM Procedures for Three-Phase Power Electronic Converters", research Contract no. 3046/2019, with the financial support of the Technical University of Cluj-Napoca".

REFERENCES

[1] Kelemen, A.; Imecs, Maria, "*Vector Control of AC Drives. Volume 1: Vector Control of induction Machine Drives.* Budapest: Omikk Publisher, 1991, ISBN 963-593-140-9.

[2] Pop, A. V., Incze, I. I.,Imecs, Maria; Negrea, C. A., Szabo, C., "Current-controlled PWM strategy with carrier wave for constant switching frequency," in *Proceedings of the 2012 IEEE International Conference on Automation Quality and Testing Robotics (AQTR)*, 2012, ISBN 978-1-4673-0701-7, pp. 347-351.

[3] Ionescu (b. Pop) A. V., Imecs Maria; Incze I. I., Szabó Cs., "Performance Analysis of Three Current-Controlled PWM-Procedures Applied in Vector Control of Induction Machine Drives", *Proceedings of the 4th International Conference in Recent Achievements in Mechatronics, Automation, Computer Science and Robotics MACRo2013*, Tg. Mures, Sapientia University, 2013, ISSN 2247 – 0948, ISSN-L 2247 – 0948, pp. 143-153

[4] Mohseni, M.; Islam, S. M., "A New Vector-Based Hysteresis Current Control Scheme for Three-Phase PWM Voltage-Source Inverters," *IEEE Transactions on Power Electronics*, vol. 25, pp. 2299-2309, 2010, ISSN 0885-8993, DOI 10.1109/TPEL.2010.2047270.

[5] Eniko, Szoke, Cs. Szabo, Maria Imecs, I.I. Incze, "Double Field Oriented Sensorless Control of Cage Induction Motor, CINTI 2014, Budapest, 29-21 Nov. 2014, pp. 403-408.

[6] Kazmierkowski, M.P. Malesani, L. "Current control techniques for three-phase voltage-source PWM converters: a survey" *Industrial Electronics, IEEE Transactions on*, Volume 45, Oct 1998, pp. 691-703.

[7] Malesani, L.; Tomasin, P., "PWM current control techniques of voltage source converters - A survey," in *Proceedings of the International Conference on Industrial Electronics, Control, and Instrumentation, IECON '93*, Maui, HI, 1993, ISBN 0-7803-0891-3, DOI 10.1109/IECON.1993.339000, pp. 670-675 vol.2.

[8] B. K. Bose, "An adaptive hysteresis-band current control technique of a voltage-fed PWM inverted for machine drive system," IEEE Trans. Ind. Electron., vol. 37, pp. 402–408, Oct. 1990

978-1-5386-7688-2/19 $31.00 © 2019 IEEE

[9] P. Thepsatorn, C. Sodaban, V. Tipsuwanporn, P. Jitnaknan, "Improvement signal output of DM–PWM inverter for driving high-efficient electrical load", *Proc. IEEE Power Electron. Motion Control Conf.*, pp. 913-918, 2006.

[10] V. Tipsuwanporn, C. Sodaban, P. Thepsatorn, "Adaptive Switching Frequency in Delta Modulation Inverter for Single Phase Induction Motor Drives," IEEE International Conference on Industrial Technology, pp. 599-603, Dec 2005.

[11] K. P. M. Shafi, R. Lakshmi, J. Peter and R. Ramchand, "Improved switching frequency variation control in hysteresis controlled VSI-fed induction motor drives for reduced line current ripple," *IECON 2017 - 43rd Annual Conference of the IEEE Industrial Electronics Society*, Beijing, 2017, pp. 6634-6639.

[12] Choochuan, C.; Kinnares, V.; Pirajnanchai, V., "A Simple and Low Cost of Delta Modulator for Static PWM Inverte", *The 4th International Power Electronics and Motion Control Conference, 2004. IPEMC 2004.*, Volume: 2, Page(s): 708 - 713 Vol.2.

[13] Jianwei Zhang, Li Li, Lei Zhang, David G. Dorrell "Hysteresis band current controller based field-oriented control for an induction motor driven by a direct matrix converter" *IECON 2017 - 43rd Annual Conference of the IEEE Industrial Electronics Society*, 29 Oct.-1 Nov. 2017, pp. 4633-4638, ISBN: 978-1-5386-1128-9.

[14] Youn-Ho Choi, Nae-Soo Cho, Woo-Hyen Kwon and Dong-Ha Lee, "Design of Switching Frequency Limiter for Hysteresis Current Controlled PWM VSI", *13th Int. Conf. on Control, Automation and Systems* (ICCAS 2013), pp. 431-433, Print ISSN: 2093-7121.

[15] K. Srikar, J. Peter, R. Ramchand, "Comparative Analysis of Hysteresis Current Control Strategies to Achieve Nearly Constant Switching Frequency for a Two-Level Inverter fed IM Drive", *IECON 2018 - 44th Annual Conference of the IEEE Industrial Electronics Society*, pp. 433-438, Electronic ISSN: 2577-1647.

[16] C.D. Tran, P. Brandstetter, B.H. Dinh, S.D.Ho, M.C.H. Nguyen, "Current-Sensorless Method for Speed Control of Induction Motor Based on Hysteresis Pulse Width Modulation Technique" *Journal of Advanced Engineering and Computation*, Vol2, Issue 4, Dec 2018, pp. 271-280, ISSN (Print): 1859-2244.

[17] Joseph Peter, Mohammed Shafi KP, Lakshmi R, Rijil Ramchand, "Nearly Constant Switching Space Vector Based Hysteresis Controller for VSI Fed IM Drive", IEEE Trans. on Industry Applications, Vol. 54, No. 4, July/August 2018, pp. 3366-3371, Electronic ISSN: 1939-9367.

[18] M. Kumar, R. Gupta, "Sampled-Time-Domain Analysis of a Digitally Implemented Current Controlled Inverter", *IEEE Transactions On Industrial Electronics*, Vol. 64, No. 1, January 2017, pp. 217-227.

[19] A. Fereidouni, M. A.S. Masoum, K.M. Smedley, "Supervisory Nearly Constant Frequency Hysteresis Current Control for Active Power Filter Applications in Stationary Reference Frame", IEEE Power and Energy Technology Systems Journal, Volume 3, No. 1, March 2016, pp. 1-12, ISSN 2332-7707.

Capacitor Voltage Balance on NPC Multilevel Converter

1st Juan Diego Nieto Cardona
Universidad Loyola Andalucía
Seville, Spain
jdnieto@uloyola.es

2nd Fabio Gómez-Estern Aguilar
Universidad Loyola Andalucía
Seville, Spain
fgestern@uloyola.es

3rd Francisco Gordillo
Universidad de Sevilla
Seville, Spain
gordillo@us.es

Abstract—In this paper, a modulation strategy is formulated to be integrated in the model of a 5-level AC-DC converter. The resulting formulation is able to deals with the unbalance voltage problem generated in the dc-link. Such modulation strategy can be calculated offline and the solution can be stored as a multidimensional array for its use during the online operation. The effectiveness and good performance of the converter under the proposed control approach is validated by simulation results.

Index Terms—Neutral-point clamped (NPC), pulse-width modulation (PWM), dc-link.

I. INTRODUCTION

The use of multilevel converters in power conversion applications is becoming more and more prevalent, because the increasing complexity of the devices that interact within the electrical networks requires that the architectures to be robust and scalable for power electronics tools. For this architecture, there is a diversity of topolgies [1] of which the neutral-point clamped converter, also known as NPC, has been the object of several studies [1]–[3].

The multilevel converters have in general better performance characteristics [4]. Several advantage can be highlighted:

- Pre-charging of capacitors is done in groups.
- The requirement of capacitance is reduced because a common DC bus is used for all three phases.
- Back-to-back configuration have great efficient in high power connections.
- Achieve higher power in the dc-link.
- Low cost.
- Lesser number of components.
- Reducing the harmonic distortion.

The control design in two level converters typically end at the power control stage. In multilevel NPC converters a problem appears even when the phase currents and the dc-link voltage have reached their desired behavior: the voltage difference among the capacitors may grow indefinitely yielding a poor performance of the converter.

A method that can be used to balance the dc link capacitor voltages uses additional power electronics to monitoring and balance the capacitor [5], however, the cost, volume and power losses are increased. Modulation techniques for multilevel converters have been studied for long, and many modulation approaches have been proposed [6], [7] and [8]. Space-Vector

(SV) based algorithms and Voltage level based algorithms are the main categories to address the modulation which have been proved in implementations as [9]. Also, voltage balancing algorithms for back-to-back connected 5-level NPC converters have been developed using SV Modulation, but this techniques treat the dc-link as a two level converter [10], [11]. In an effort to reduce the complexity of the above approach, the modulation technique proposed here deal with the unbalance problem of the capacitor voltage as simple as effective way. The value of the discrete inputs is chosen at every sampling instant incorporating in the model the duty cycle variables which allows the direct calculation of this. The main advantages of the proposed concept is the simplicity on account of taking advantage of the a degree of freedom given by the formulated average model. Simulation results using a detailed model tests the performance of the validity of the proposed approach.

II. MULTILEVEL CONVERTER TOPOLOGY

The converter considered in this paper is a five-level diode-clamped NPC operating in rectifier mode. The topology of this converter is shown in Fig. 1. An inductive filter is used to connect the multilevel converter to a grid with phase voltage v_{sa}, v_{sb} and v_{sc}. The dc-link is composed by four capacitor with equal capacitance C and their voltage are denoted by v_{c_1}, v_{c_2}, v_{c_3} and v_{c_4}. The dc terminals are connected to a resistive load R_{DC}.

Note that the total dc-link voltage is defined by $v_{dc} = v_{c_1} + v_{c_2} + v_{c_3} + v_{c_4}$. The voltage differences between the capacitor are defined as the state variables in the model of the controller. These variables are introduced to measure the capacitor voltage deviation. With this, the capacitor voltage differences are

$$
\begin{aligned}
v_{d_1} &= v_{c_4} - v_{c_1} \\
v_{d_2} &= v_{c_3} - v_{c_2} \\
v_{d_3} &= v_{c_2} - v_{c_1}
\end{aligned} \tag{1}
$$

These functions are derived from the model proposed by [12]. This model uses a polynomial approximation to describe the linear combination of the capacitor voltage. The expressions of the capacitor voltage difference dynamics are given by

$$v_{d_i}(kT) = v_{d_i}(k-1) + \Delta v_{d_i}(k), \quad \text{for } i = 1, 2, 3 \quad (2)$$

where

$$\Delta v_{d_i}(kT) = \int_{kT}^{(k+1)T} g_{i3}(t)\delta_\gamma^3(t)\,dt + \int_{kT}^{(k+1)T} g_{i2}(t)\delta_\gamma^2(t)\,dt$$
$$+ \int_{kT}^{(k+1)T} \left(g_{i1}(t)\delta_\gamma(t) + g_{i0}(t)\right)dt \quad (3)$$

being $g_{ij}(t)$ nonlinear functions which depend on $\delta_\alpha(t)$ and $\delta_\beta(t)$.

III. CONTROL STRATEGY

In order to control a NPC converter, a current controller, a total dc voltage controller and a controller to avoid differences between the capacitor voltage should be taken into account. The objective of this paper is the balance of the capacitor voltages, therefore it has been assumed that the variables in relation to the power control are roughly in their respective references. Multiple techniques are found in the literature concerning the current and power control in converters [13], [14]. The control law to deal with the capacitor voltage unbalances will be built using the averaged model of [12].

A. Capacitor voltage unbalance control

The objective of bringing the vector $v_d(t)$ as close as possible to zero will be achieved by making the signs of the components of the vector Δv_d (from Eq. (3)) opposite to the signs of v_d at the end of each sampling interval. This in turn is achieved by using the degree of freedom δ_γ remaining after the current controller has been implemented (through $\delta_\alpha, \delta_\beta$). Moreover, instead of keeping $\delta_\gamma(t)$ constant for the whole sampling interval $[kT, (k+1)T]$, this strategy can make it switch at some intermediate time instants between two functions of time that have been designed, each, for a different control goal.

$$\delta_\gamma(t) = \begin{cases} \delta_{\gamma_1}(t), & \text{if } g_{31}(t) \leq \psi \\ \delta_{\gamma_2}(t), & \text{if } g_{31}(t) > \psi \end{cases} \quad (4)$$

where

$$\psi = (\eta_{31} - \mu_{31} \cdot \cos(\pi\varepsilon)) \quad (5)$$

The inequality in (4) comes from the observation of the non-autonomous dynamics of the $v_d(t)$, which suggests that at some parts of the period, some coefficients of the control input are prevalent, while at other times, different coefficients are more important (please refer to [15] for further details). These periods of time are set by the parameter ε. This strategy takes advantage of the fact that the frequency of the g_{ij} is known. The parameters μ_{31} and η_{31} in (5) are the amplitude and the offset, respectively, of the sinusoidal wave g_{31}.

In [15], the values of δ_{γ_1} and δ_{γ_2} are fixed during the period corresponding to $3f$, and δ_γ switches between them. The choice of δ_{γ_1} and δ_{γ_2} will determine the sign of the elements

of the increment vector Δv_d. That approach has the limitation that, for some nominal values, not all sign combinations of Δv_d could be achieved by choosing δ_{γ_1} and δ_{γ_2}, and the framework resulted in a limited range of application.

In search for better ways of achieving all desired signs combinations of Δv_d this work proposes to pick of $\delta_{\gamma_1}(t)$ and $\delta_{\gamma_2}(t)$ between a choice of time-varying functions. These functions are defined by the (also time-varying) saturation levels of δ_γ, and because o this, the range of action of the control input is better exploited. The action of control in (4) is parametrized by ε. The parameter ε set when $\delta_{\gamma_1}(t)$ and $\delta_{\gamma_2}(t)$ is applied, that means that the components of v_d are controller in different periods of time. As a consequence, more combinations of signs of Δv_d can be reached for the same circuit values.

The strategy proposed here uses time-varying maximum and minimum functions for the control input δ_γ. An example of the waveform of the control inputs $\delta_\alpha(t)$, $\delta_\beta(t)$ and $\delta_\gamma(t)$ during a period is shown in Fig. 2. The waveform of $\delta_{\gamma max}$ and $\delta_{\gamma min}$ can be seen in the third plot of Fig. 2. These waveform are repeated three times in the period. That means a frequency of $3f$ with $f = 50$Hz, which matches with the frequency of functions $g_{31}(t)$. In view of this, the controller operates to a high sampling frequency, but $\delta_\gamma(t)$ is updated to $3f$. Hence, this frequency of updating is equal to the frequency of the g_{ij} dynamics.

B. Modulation strategy

The control inputs concerning to the current control and total dc-link are determined by $\delta_\alpha(t)$ and $\delta_\beta(t)$. For these fast variables, it have been assumed that they have reached their respective references. It means that the fast variables have achieved the steady state, hence the duty ratios in $\alpha\beta$-coordinates can be defined as

$$\delta_\alpha(t)|_{ss} = \mu \sin(\omega t + \theta) \quad (6)$$
$$\delta_\beta(t)|_{ss} = \mu \cos(\omega t + \theta) \quad (7)$$

In view of this, the calculation of the duty ratio in abc-coordinates can be performed using the invariant power Clarke-Transformation. Because of saturation, the duty ratios d_a, d_b and d_c must be in the range defined by

$$\begin{bmatrix} -1 \\ -1 \\ -1 \end{bmatrix} \leq \sqrt{\frac{2}{3}} \begin{bmatrix} 1 & 0 & \frac{1}{\sqrt{2}} \\ -\frac{1}{2} & \frac{\sqrt{3}}{2} & \frac{1}{\sqrt{2}} \\ -\frac{1}{2} & -\frac{\sqrt{3}}{2} & \frac{1}{\sqrt{2}} \end{bmatrix} \begin{bmatrix} \delta_\alpha \\ \delta_\beta \\ \delta_\gamma \end{bmatrix} \leq \begin{bmatrix} 1 \\ 1 \\ 1 \end{bmatrix} \quad (8)$$

Thus, taking into account Eq. (8), the following variables can be introduced

$$\tau_a(t) = \sqrt{\frac{2}{3}}\delta_\alpha(t) \quad (9)$$

$$\tau_b(t) = -\frac{1}{\sqrt{6}}\delta_\alpha(t) + \frac{1}{\sqrt{2}}\delta_\beta(t) \quad (10)$$

$$\tau_c(t) = -\frac{1}{\sqrt{6}}\delta_\alpha(t) - \frac{1}{\sqrt{2}}\delta_\beta(t) \quad (11)$$

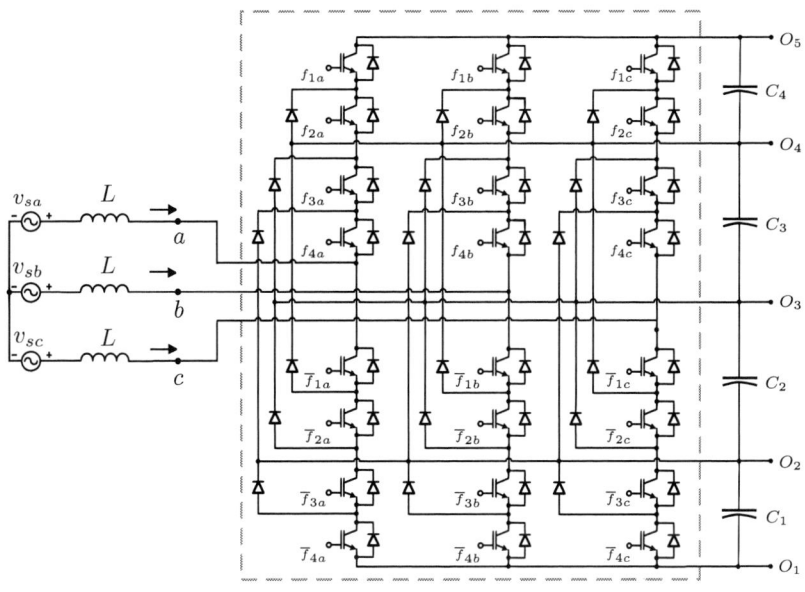

Fig. 1. Schematic circuit of 3-Phase 5-level diode-clamped converter operating as rectifier.

this yields the following expression of the duty ratios in abc-coordinates

$$d_a = \tau_a + \frac{1}{\sqrt{3}}\delta_\gamma \qquad (12)$$

$$d_b = \tau_b + \frac{1}{\sqrt{3}}\delta_\gamma \qquad (13)$$

$$d_c = \tau_c + \frac{1}{\sqrt{3}}\delta_\gamma \qquad (14)$$

From Eqs. (12)-(14) can be seen that all the duty ratio in abc-coordinates are affected in the same way for the control input $\delta_\gamma(t)$. Therefore, knowing that $d_i \in [-1, 1]$ for $i = a, b, c$, the waveform of maximum and minimum value of δ_γ can be calculated as follows

$$\delta_{\gamma_{min}}(t) = \sqrt{3}(-1 - \max(\tau_a(t), \tau_b(t), \tau_c(t))) \qquad (15)$$

$$\delta_{\gamma_{max}}(t) = \sqrt{3}(1 - \max(\tau_a(t), \tau_b(t), \tau_c(t))) \qquad (16)$$

Since variable $\delta_\gamma(t)$ is directly related to the balance of the capacitor voltages, then, the value of $\delta_\gamma(t)$ will chosen in each sampling instant between the three values ($\delta_{\gamma_{min}}(t), \delta_{\gamma_{min}}(t)$ and zero) according to (4). The waveform of $\delta_\gamma(t)$ corresponding to Eq. (15) and Eq. (16) can be seen in Fig. 2. With the control inputs in $\alpha\beta\gamma$-coordinates and using the transformation matrix presented in Eq. 8, the duty in abc-coordinates can be calculated. Then, the duty ratio in abc-coordinates is compared with a carried triangular waveform. In this manner, via PWM, the switching state of the converter are defined. Therefore, the variables d_a, d_b and d_c are related to the values of the generated voltage with the following constraint

$$d_{i1} + d_{i2} + d_{i3} - d_{i4} - d_{i5} = d_i \qquad (17)$$

for $i = a, b, c$. The definition in (17) allows that the modulation can be performed between the nearest levels which avoid strong jumps among levels. This methodology is based on [3] where the the duty ratio are break down. A smooth power control is accomplished with this modulation.

C. Implementation

Figure 3 shows a schematic diagram of the proposed control. The controller has three inputs, one of these is the measurement of the capacitor voltage difference $v_d(t)$ and the others are $\delta_\alpha(t)$ and $\delta_\beta(t)$ from the current controller. The block *Case Selector* makes reference to the selection of δ_{γ_1}, δ_{γ_2} and ε. The selection of the signals associated to the balance is made from a multidimensional array where every dimension is regard to Table I, whereby each case can generate a different combination of signs for the vector Δv_d. Each table is the result of a particular value of the parameter $\varepsilon \in [0, 1]$. The vector Δv_d is represented by the sign of the increment of $v_d(t)$ produced to apply an specific case ($\delta_{\gamma_1}, \delta_{\gamma_2}$). The sign of Δv_d can change depending on the case and the value of ε. Then, the control action is made by applying a case so that the sign of the vector $v_d = [v_{d_1}, v_{d_2}, v_{d_3}]$ can be modified. For example, if the signs of the vector $v_d = [+, +, +]$, *Case Selector* will select the case with opposite sign combination to conduct the capacitor voltage difference toward zero.

The component of Δv_d with the small increment at the end of the period $T_g = 1/3f$ will condition the chosen value of ε. The same combination of the sign of Δv_d is evaluated for the different values of ε. Thus, if any of the new vector evaluated has a greater than the before vector, so the new case will be selected.

Prior calculations should be made to know the combination of signs generated by each one of the cases presented in Table

978-1-5386-7688-2/19 $31.00 © 2019 IEEE

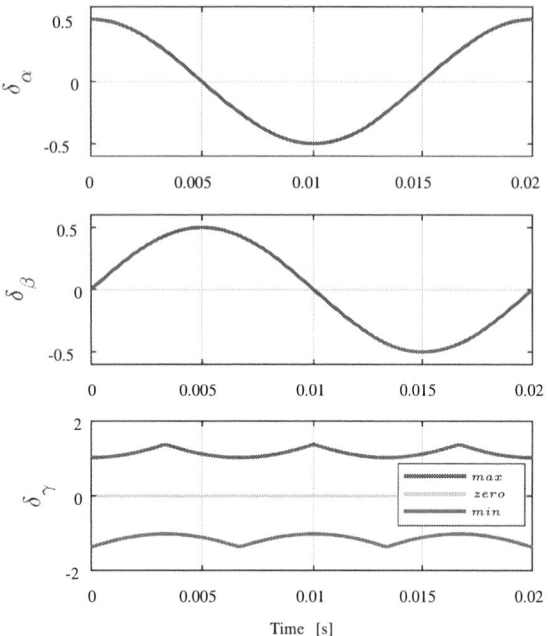

Fig. 2. Control input signals during a period of the fundamental wave. The signals of δ_α and δ_α have a frequency $f = 50Hz$ and $\delta_{\gamma_{max}}$ has a frequency of $3f$.

I. The signs of Δv_d produced by the action of each case is obtained offline and it is store together with the case table. An example of this is presented in Table II.

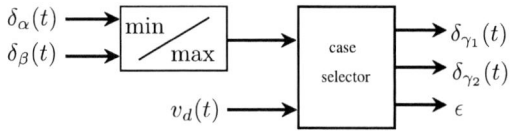

Fig. 3. Proposed control scheme.

TABLE I
TABLE CASES OF CONTROL

Case	δ_{γ_1}	δ_{γ_2}
1	$\delta_{\gamma_{max}}(t)$	$\delta_{\gamma_{max}}(t)$
2	$\delta_{\gamma_{max}}(t)$	$\delta_{\gamma_{min}}(t)$
3	$\delta_{\gamma_{max}}(t)$	0
4	$\delta_{\gamma_{min}}(t)$	$\delta_{\gamma_{max}}(t)$
5	$\delta_{\gamma_{min}}(t)$	$\delta_{\gamma_{min}}(t)$
6	$\delta_{\gamma_{min}}(t)$	0
7	0	$\delta_{\gamma_{max}}(t)$
8	0	$\delta_{\gamma_{min}}(t)$
9	0	0

D. Generation of the look-up table

A multidimensional array will store the look-up tables for different values of $\varepsilon \in [0, 1]$. A look-up table corresponds to a

specific value of parameter ε. Each row in these look-up tables is composed by the sign of the variables v_{d_1}, v_{d_2} and v_{d_3} for each case presented in Table I. An off-line computation to know the sign of v_d at the end of a sampling period is made. The process is executed for all cases with a fixed value of ε. After that, the same procedure is repeated for different values of ε. It is necessary to store the value of each component of the increment vector Δv_d, since this will be used to select the appropriate ε.

TABLE II
SIGN GENERATED WITH EACH CASES OF CONTROL FOR $\varepsilon = 0.15$

Case	Δv_{d_1}	Δv_{d_2}	Δv_{d_3}
1	+	+	−
2	−	−	−
3	+	+	+
4	+	+	−
5	−	−	−
6	−	+	+
7	+	+	+
8	−	−	−
9	+	+	−

IV. SIMULATION RESULTS

The simulation performance of the proposed controller is evaluated using MATLAB-Simulink. The circuit model of the NPC operating as rectifier with near-unity power factor implements ideal IGBTs and diodes in parallel with RC snubber circuit. The gating signal is updated with a frequency of 10kHz. In the dc side, the terminals of the dc-link are connected to a resistive load. The rest of the parameter values of the considered system are summarized in Table III.

It is worth stressing that the switching states have been generated by means of a PWM strategy from the duty ratios d_{ij}, for $i = a, b, c$ and $j = 1, 2, 3, 4, 5$. In this section, different operation conditions have been evaluated, for which the stationary state of the system is the starting point.

TABLE III
PARAMETERS OF THE 5-LEVEL NPC

Frequency	50 Hz
Sampling frequency	10 kHz
ac Voltage	110 V
dc Voltage	700 V
Capacitance	3300 μF
Inductance	2 mH

A. Operation under unbalanced condition

For the first test, a disturbance is introduced in the error signal measured by the controller and removed at 0.07 s. This disturbance is applied to generate unbalanced condition. Fig. 4 shows the behaviour of the four capacitor that constitute the dc-link. It can observed that all v_{c_i} for $i = 1, 2, 3, 4$, return to the their reference. The ripple around the reference is lower than 7 V.

Concerning the capacitor voltage differences, Fig. 5 shows the evolution during and after the disturbance. It can be seen that large initial values of v_d have been introduced in this test. The voltage differences between capacitors are kept under 10V which means that the ripple $|v_c| < 5\%$ with respect to its reference. Therefore, the controllers regulate properly these three variables.

Fig. 6 presents the instantaneous active and reactive power for a load $R_{DC} = 98\,\Omega$. Both powers are affected by the disturbance but the strategy of modulation brings back around their desired values. After a few control sampling times, all variables have converged to the steady state profile as result of the modulation. Fig. 7 shows the signal δ_γ and their transitions used in the modulation. From Fig. 7 is easy to identify the cases selected by the controller and also visualize the time duration of δ_{γ_1} and δ_{γ_2} defined in (4). In Fig. 8 can be seen the switching state concerning to the phase a. These resultant states are obtained by comparing the duty d_a with a triangular carrier signal. The jumps between levels are, most of the time, made in a stepwise manner.

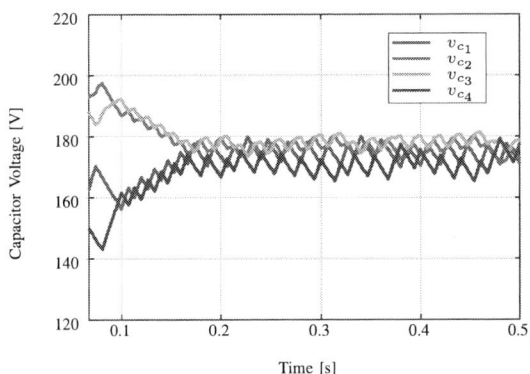

Fig. 4. Capacitor voltages v_{c1}, v_{c2}, v_{c3} and v_{c4} during a disturbance in the error signal.

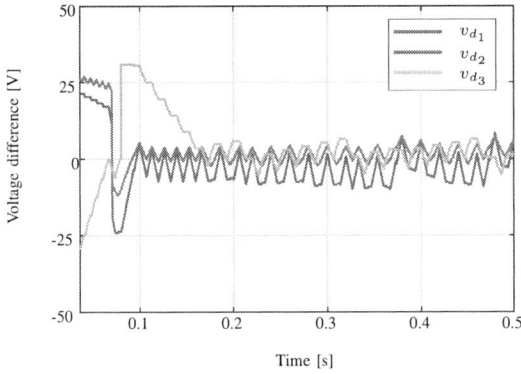

Fig. 5. Capacitor voltage differences during a disturbance in the error signal.

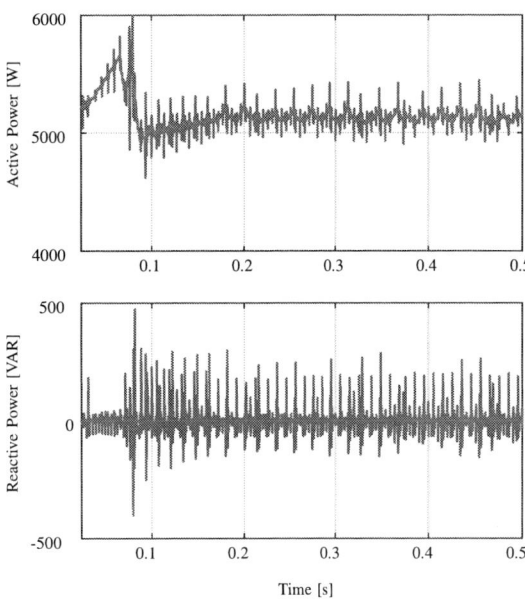

Fig. 6. Instantaneous active and reactive power with respective reference (red line) at ac side during a disturbance in the error signal.

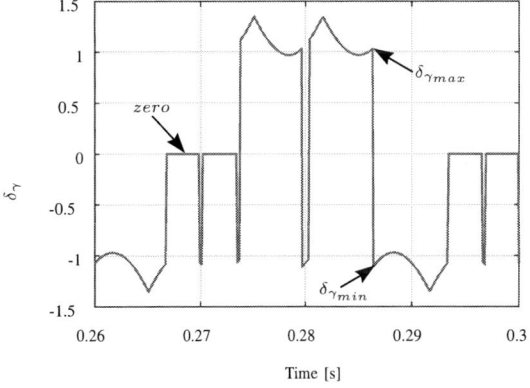

Fig. 7. Control input δ_γ used in the modulation.

B. Operation under load changes

Fig. 9 exhibits the effect of changing the load on the system. The initial active power is $5\,kW$, a stepwise increase to $8\,kW$ at 0.1s and a stepwise decrease to $2\,kW$ at 0.2s have been performed. The behavior of the active power tracks very quickly their respective references.

Fig. 10 shows that the voltage differences among the capacitors increase when the active power is larger. Despite this, v_{d_1}, v_{d_1} and v_{d_1} are in the acceptance range which does not exceed 10V. It is important to note that the proposed strategy brings v_d to zero before a few sampling time. Also, the capacitor voltage differences remain stable in spite of rough load changes.

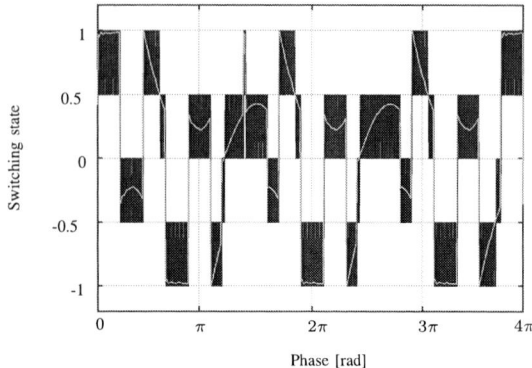

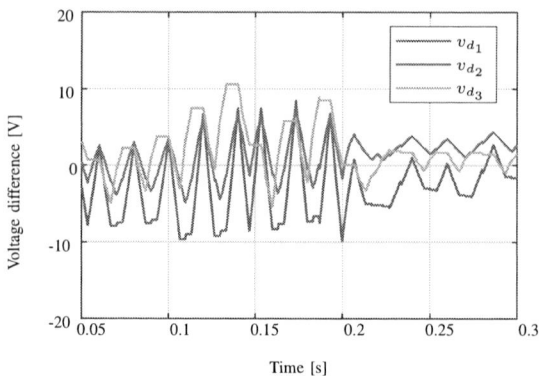

Fig. 8. Switching states of the NPC converter derived from the comparison of d_a (yellow line) with a carried.

Fig. 10. Capacitor voltage differences under load step-up and step-down changes.

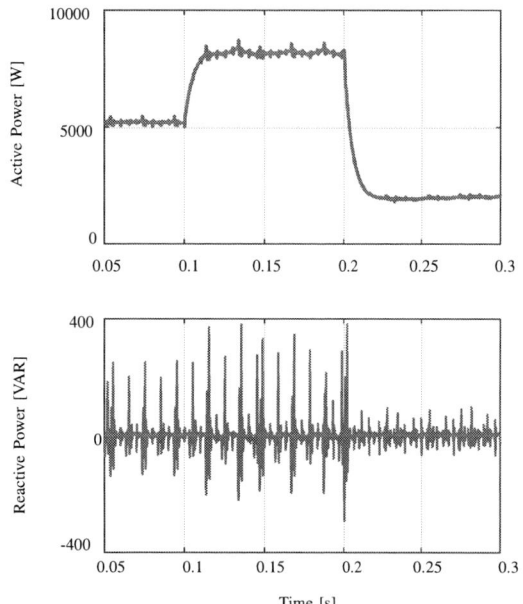

Fig. 9. Instantaneous active and reactive power with their respective reference (red line) during step changes in the load.

V. CONCLUSIONS

This paper presents a new modulation technique for a NPC multilevel converter. With the modulation proposed the capacitor balance is accomplished in a satisfactory way. The proposed method shows robustness for strong load changing conditions and for large voltage differences between capacitors.

An important element to emphasize is the fact that the design of the controller is simple and computed offline. It is also important to note that the proposed approach does not require the utilization of auxiliary converters or any other

kind of supplementary electronic device which represents a significant feature.

The modulation strategy carried out for the converter operating as rectifier, can be also used in inverter mode with a few modifications. One of the future directions of research of this work will be the extension of the modulation approach to the back-to-back five-level DCC topology.

ACKNOWLEDGMENT

This work has been funded under grant MINECO-FEDER DPI2016-75294-C2-1-R and DPI2016-75294-C2-2-R.

REFERENCES

[1] J. Rodriguez, J.-S. Lai, and F. Z. Peng, "Multilevel inverters: a survey of topologies, controls, and applications," *IEEE Transactions on Industrial Electronics*, vol. 49, no. 4, pp. 724–738, Aug 2002.

[2] J.-S. Lai and F. Z. Peng, "Multilevel converters-a new breed of power converters," in *IAS'95. Conference Record of the 1995 IEEE Industry Applications Conference Thirtieth IAS Annual Meeting*, vol. 3. IEEE, 1995, pp. 2348–2356.

[3] S. Busquets-Monge, J. Bordonau, D. Boroyevich, and S. Somavilla, "The nearest three virtual space vector pwm - a modulation for the comprehensive neutral-point balancing in the three-level npc inverter," *IEEE Power Electronics Letters*, vol. 2, no. 1, pp. 11–15, March 2004.

[4] R. Teodorescu, F. Blaabjerg, J. K. Pedersen, E. Cengelci, and P. N. Enjeti, "Multilevel inverter by cascading industrial vsi," *IEEE Transactions on Industrial Electronics*, vol. 49, no. 4, pp. 832–838, 2002.

[5] Z. Shu, H. Zhu, X. He, N. Ding, and Y. Jing, "One-inductor-based auxiliary circuit for dc-link capacitor voltage equalisation of diode-clamped multilevel converter," *IET Power Electronics*, vol. 6, no. 7, pp. 1339–1349, August 2013.

[6] F. Gordillo, F. Gómez-Estern, and F. Salas, "An optimization approach for modulation in multilevel converters," in *IECON 2016-42nd Annual Conference of the IEEE Industrial Electronics Society*. IEEE, 2016, pp. 5033–5038.

[7] J. Pou, J. Zaragoza, S. Ceballos, M. Saeedifard, and D. Boroyevich, "A carrier-based pwm strategy with zero-sequence voltage injection for a three-level neutral-point-clamped converter," *IEEE Transactions on Power Electronics*, vol. 27, no. 2, pp. 642–651, Feb 2012.

[8] K. Ma and F. Blaabjerg, "Modulation methods for three-level neutral-point-clamped inverter achieving stress redistribution under moderate modulation index," *IEEE Transactions on Power Electronics*, vol. 31, no. 1, pp. 5–10, Jan 2016.

978-1-5386-7688-2/19 $31.00 © 2019 IEEE

[9] S. Madishetti, B. Singh, and G. Bhuvaneswari, "Three-level npc-inverter-based svm-vcimd with feedforward active pfc rectifier for enhanced ac mains power quality," *IEEE Transactions on Industry Applications*, vol. 52, no. 2, pp. 1865–1873, March 2016.

[10] G. Mademlis and Y. Liu, "Dc link voltage balancing technique utilizing space vector control in sic-based five-level back-to-back-connected npc converters," in *2018 IEEE Energy Conversion Congress and Exposition (ECCE)*, Sep. 2018, pp. 3032–3037.

[11] A. Dekka and M. Narimani, "Capacitor voltage balancing and current control of a five-level nested neutral-point-clamped converter," *IEEE Transactions on Power Electronics*, vol. 33, no. 12, pp. 10 169–10 177, Dec 2018.

[12] R. Portillo, J. M. Carrasco, J. I. Leon, E. Galvan, and M. M. Prats, "Modeling of Five-Level Converter Used in a Synchronous Rectifier Application," *Proceedings of the 36th Annual IEEE Power Electronics Specialists Conference - PESC'05*, pp. 1396–1401, 2005.

[13] T. Noguchi, H. Tomiki, S. Kondo, and I. Takahashi, "Direct power control of pwm converter without power source voltage sensors," in *IAS'96. Conference Record of the 1996 IEEE Industry Applications Conference Thirty-First IAS Annual Meeting*, vol. 2. IEEE, 1996, pp. 941–946.

[14] S. Kouro, M. Malinowski, K. Gopakumar, J. Pou, L. G. Franquelo, B. Wu, J. Rodriguez, M. A. Perez, and J. I. Leon, "Recent advances and industrial applications of multilevel converters," *IEEE Transactions on Industrial Electronics*, vol. 57, no. 8, pp. 2553–2580, Aug 2010.

[15] F. Umbría, F. Gómez-Estern, F. Gordillo, and F. Salas, "Voltage balancing in five-level diode-clamped power converters," *IFAC Proceedings Volumes (IFAC-PapersOnline)*, vol. 9, no. PART 1, pp. 365–370, 2013.

978-1-5386-7688-2/19 $31.00 © 2019 IEEE

Cascaded Fuzzy Controller for Electric Vehicle Traction System Battery Energy Management

Ahmed Sayed Abdelaal Abdelaziz
Department of Electrical Engineering
American University of Sharjah
Sharjah, United Arab Emirates
b00060280@alumni.aus.edu

Habib-ur Rehman
Department of Electrical Engineering
American University of Sharjah
Sharjah, United Arab Emirates
rhabib@aus.edu

Shayok Mukhopadhyay
Department of Electrical Engineering
American University of Sharjah
Sharjah, United Arab Emirates
smukhopadhyay@aus.edu

Abstract— **This paper presents a battery energy management (BEM) technique for an indirect field oriented (IFO) induction motor driven electric vehicle (EV) traction system. The proposed technique is designed using two cascaded fuzzy logic controllers. The first fuzzy logic controller (FLC) generates the desired torque command current, Iq, to regulate the motor speed while the second fuzzy controller limits Iq based on the battery SOC. Computer simulations are developed for the entire EV traction system which is powered by a 450 V Li-ion battery bank. The simulation results even for a short period of 30 secs show that proposed BEM technique results in a reduction in the battery energy consumption and a lower degradation in the battery state-of-health (SOH) with little bit compromise on the motor acceleration.**

Keywords— *Electric Vehicle, Fuzzy logic controller, Indirect Field oriented Control, Battery energy management, induction motor.*

I. INTRODUCTION

Environment pollution and global warming are direct outcomes of the substantial fossil fuel consumption in the transportation industry [1]. As a result, the transportation industry has made efforts to advance towards a clean and efficient transportation system [2],[3]. One possible solution is through the use of battery electric vehicles (BEVs). Benefits provided by BEVs such as a lower operating and maintenance costs, little to no air pollution, and reduced dependency on fuel are all reasons to why BEVs are enticing [3]. However, a major drawback of BEVs is that the energy density of today's batteries is still significantly lower than the energy density of fossil fuel and therefore, this work will focus on an energy management strategy to optimize battery's energy utilization which in return leads to lower energy consumption as well as an overall longer battery life [1],[2].

Fuzzy Logic Controller (FLC) has been extensively applied in the management of multiple sources such as fuel cells, combustion engines, ultracapacitors and batteries where the other source shares the energy demand when the battery current exceeds the maximum current [4],[5]. A predictive protective algorithm was created in [6] in which a FLC monitored the road information and the SOH of the battery and produced a signal for the optimal battery charging or discharging. In addition, V.Galdi et al. claims to use an FLC that takes into account the SOC of the battery, the drive cycle and the driver's acceleration and produces a new output drive cycle with minimal degradation in motor performance; however, the results show a substantial steady state error compared to the actual drive cycle [7].

The work in literature involved estimated SOC equations and obtaining SOC using linearized equations instead of using the actual battery model. In addition, the previous work mostly relied on the road information such as traffic conditions and slopes which are not readily available for all roads. Furthermore, the drive system dynamics was never considered in [8], [7], [9] and the FLC was directly controlling the reference speed and/or acceleration of the vehicle that was modeled by newton's equations. Therefore, this paper develop the simulation model of complete EV traction system and proposes a BEM technique on a overall EV traction system. This technique aims to maximize the SOH of the battery by altering the dynamics of the drive system using the battery SOC which is estimated by a widely accepted Chen-Mora battery model [10].

The remainder of the paper is organized as follows: section II describes induction motor model for an indirect field-oriented drive system and Chen & Mora model of Li-ion battery which will be used in this work. Section III presents our proposed BEM technique for reducing the SOC consumption and battery SOH degradation. Section IV shares the simulation results using fuzzy logic alone for the drive system speed regulation and fuzzy speed regulator cascaded by another fuzzy controller for implementing the proposed BEM technique. Finally, section V makes the concluding remarks highlighting the contributions that are made by this work.

II. INDUCTION MOTOR AND LI-ION BATTERY LI-ION BATTERY MODELLING

The AC drive system can run on a switched reluctance motor (SC), permanent magnet synchronous (PMS) motor or an induction motor [11], [12].The SC motor has problems with torque ripple and vibration while the PMS is relatively more expensive and the magnet is sensitive to temperature variations therefore it is subject to demagnetization. On the contrary, the induction motor is well known for its ruggedness and its low cost; hence, this work focuses on an induction motor drive system for the EV traction system [13].

A. Induction Motor Modeling

The dynamic model of the induction motor in synchronously rotating frame of reference can be describe as [14].

$$V_{sd} = R_s i_{sd} + \frac{d\lambda_{sd}}{dt} - \omega_d \lambda_{sq} \tag{1}$$

$$V_{sq} = R_s i_{sq} + \frac{d\lambda_{sq}}{dt} - \omega_d \lambda_{sd} \tag{2}$$

978-1-5386-7688-2/19 $31.00 © 2019 IEEE

$$V_{rd} = R_r i_{rd} + \frac{d\lambda_{rd}}{dt} - \omega_{dA}\lambda_{rq} \qquad (3)$$

$$V_{rq} = R_r i_{rq} + \frac{d\lambda_{rq}}{dt} + \omega_{dA}\lambda_{rd} \qquad (4)$$

$$Tem = \frac{3p}{2}\frac{Lm}{Lr}(\lambda_{rd}i_{sq} - \lambda_{rq}i_{sd}) \qquad (5)$$

$$\frac{d\omega_m}{dt} = \frac{1}{J}(Tem - TL - B\omega) \qquad (6)$$

Where V, i, λ are the dq-components for the stator and rotor voltages, currents and flux respectively. The R_s and R_r are the stator and rotor resistances respectively and J, B, Tem, TL are the motor inertia, coefficient of friction, electromagnetic torque and load torque while , $\omega_m, \omega_d, \omega_{dA}$ represent the mechanical speed, d axis rotation speed and rotor axis rotational speed. ω_d is selected to be equal to the synchronous speed $\omega_{sync} = 2\pi f$ rads. Equations (1)-(5) represent the electrical equations and equation (6) shows the mechanical dynamics of the induction machine.

An indirect field-oriented controller (IFOC) is popular amongst the induction motor control schemes due to the decoupling effect it has on the torque and flux. Applying the principle of IFOC by setting $\lambda_{rq} = 0$, the torque equation (5) can be simplified to equation (7)

$$Tem = \frac{3p}{2}\frac{Lm}{Lr}(\lambda_{rd}i_{sq}) \qquad (7)$$

Therefore, a relation governing the speed ωm and the current isq can be derived by combining equation (6) and (7). The state space representation of the IFOC drive system is given by equations (8) and (9)

$$\begin{bmatrix}\ddot{\omega}_m \\ \dot{\omega}_m\end{bmatrix} = \begin{bmatrix} -\frac{B}{J} & 0 \\ 1 & 0 \end{bmatrix}\begin{bmatrix}\dot{\omega}_m \\ \omega_m\end{bmatrix} + \begin{bmatrix}\frac{3p}{2}\frac{Lm}{Lr}\lambda_{rd} \\ 0 \end{bmatrix}\frac{di_{sq}(t)}{dt} \qquad (8)$$

$$\omega_m = \begin{bmatrix} 0 & 1 \end{bmatrix}\begin{bmatrix}\dot{\omega}_m \\ \omega_m\end{bmatrix} \qquad (9)$$

B. Li-ion Battery Modeling

The battery is simulated using the Chen and Mora's equivalent circuit model shown in [10]. The state space equations for Chen and Mora's model as obtained in [10] are:

$$\dot{x}_1(t) = -\frac{1}{C_C}i(t) \quad C_C = 3600 C f_1 f_2 f_3 \qquad (10)$$

$$\dot{x}_2(t) = -\frac{x_2(t)}{R_{ts}(x_1)C_{ts}(x_1)} + \frac{i(t)}{C_{ts}(x_1)} \qquad (11)$$

$$\dot{x}_3(t) = -\frac{x_3(t)}{R_{tl}(x_1)C_{tl}(x_1)} + \frac{i(t)}{C_{tl}(x_1)} \qquad (12)$$

$$y = E_o(x_1) - x_2(t) - x_3(t) - i(t)R_s(x_1) \qquad (13)$$

x_1 represents the SOC and is between [0, 1]. The states x_2 and x_3 are non-negative real numbers when the current is flowing out of the battery. The variable C in equation (10) represents the capacity (A.h) of the battery. The factors $f1$, $f2$,

$f3 \in [0, 1]$ in equation (10) account for the effects of temperature, number of charge–discharge cycles, and self-discharge respectively and they are set to 1 for simplicity. The battery terminal voltage is given in equation (13) and it is dependent on the states x_2, x_3, the battery current i(t), the series resistance Rs and the open circuit voltage. E_o, R_{ts}, R_{tl}, C_{ts}, C_{tl} and R_s are given in equations (14) - (19) [15].

$$E_o(x_1) = -a_1 e^{-a_2 x_1} + a_3 + a_4 x_1 - a_5 x_1^2 + a_6 x_1^3 \qquad (14)$$

$$R_{ts}(x_1) = a_7 e^{-a_8 x_1} + a_9 \qquad (15)$$

$$R_{tl}(x_1) = a_{10}e^{-a_{11}x_1} + a_{12} \qquad (16)$$

$$C_{ts}(x_1) = -a_{13}e^{-a_{14}x_1} + a_{15} \qquad (17)$$

$$C_{tl}(x_1) = -a_{16}e^{-a_{17}x_1} + a_{18} \qquad (18)$$

$$R_s(x_1) = a_{19}e^{-a_{20}x_1} + a_{21} \qquad (19)$$

Where the parameters $a_1 \ldots a_{21}$ are obtained by the techniques described in [16]. The battery state-of-charge SOC can be obtained by coulomb counting method defined by equation (20). The SOH can be obtained by the equations in [17].

$$SOC(t) = SOC(t_0) + \frac{1}{C_c}\int_0^t i(t)\,dt \qquad (20)$$

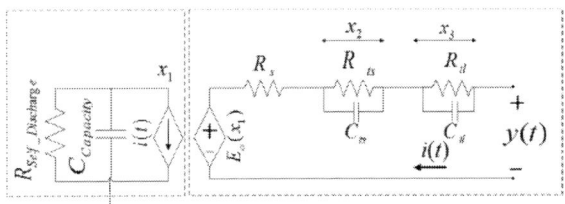

Fig. 1. Chen and Mora's equivalent circuit [10].

III. THE PROPOSED BATTERY ENERGY MANAGEMENT TECHNIQUE

The overall EV traction system with an indirect field-oriented induction motor drive system designed and simulated in this work is shown Figure (2).

Figure 2 shows the proposed BEM technique is based on two cascaded fuzzy logic controllers. The first fuzzy controller regulates the motor speed while the second one optimizes the torque command current based on the motor SOC. Since the proposed technique involves a fuzzy logic controller, then in subsection A we give a brief overview on FLC which is followed by the description of the proposed BEM technique in subsection B.

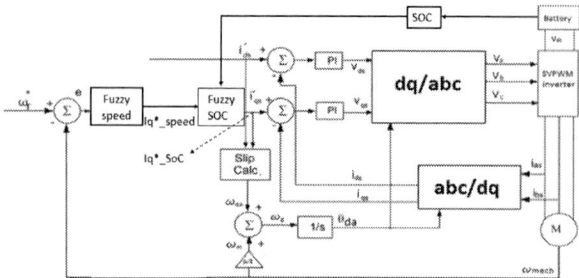

Fig. 2. Overall cascaded fuzzy drive system.

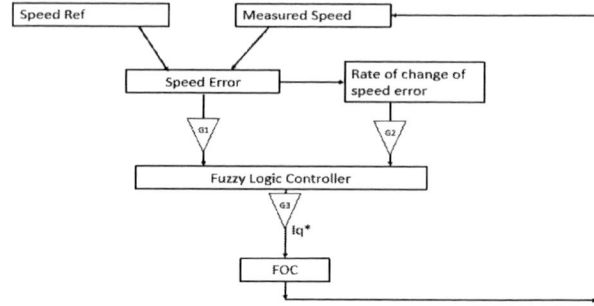

Fig. 4. Simple FLC based speed regulator.

A. Fuzzy Logic Controller (FLC)

A fuzzy logic controller is a controller that emulates the human thinking and incorporates the expert's deductive reasoning into the system [18]. Figure 3 shows the 3 stages that the signal passes through before a command is issued by the controller. The first stage is the fuzzification stage and it converts the signal into linguistic variables with the aid of membership functions. The second stage is the fuzzy inference engine which takes the linguistic variables and compares them with the rule base set by the expert then generates a linguistic output. The final stage is the defuzzification stage and it converts the linguistic outputs to the command signal [18].

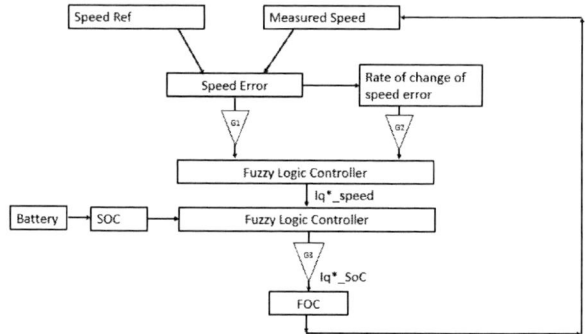

Fig. 5. BEM Technique 1: Cascaded Fuzzy drive system.

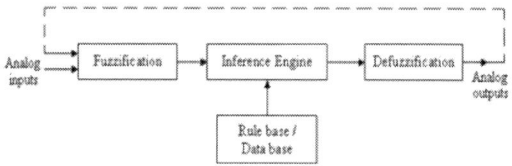

Fig. 3. Fuzzy Logic Controller Architecture.

In the next section, we describe the proposed technique for the BEM.

B. Cascaded FLC based BEM Technique

Figures 4 and 5 show the conventional FLC based drive system and the proposed cascaded FLC for BEM respectively. A fuzzy logic controller is used in figure 4 as a speed regulator that does not take into account the SOC of the battery and therefore, does not optimize the SOH. While in our proposed architecture in figure 5; a second FLC that monitors the SOC of the battery acts as a current restrictor by scaling the reference signal Iq such that it minimizes the SOH degradation. This interaction introduces a delay in motor rise time, in other words, reduces the motor acceleration. Figures 6 and 7 describe the surface that generates the control signal Iq for the drive system. The FLCs used in this work follow the center of gravity method of defuzzification. The cascaded fuzzy architecture excels due its ability to incorporate the expert's knowledge into reducing the discharge current.

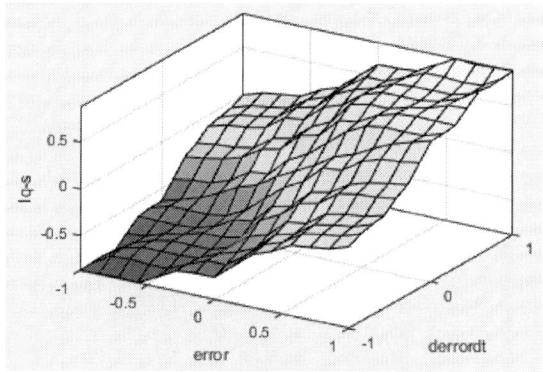

Fig. 6. Figure 1. FLC speed control.

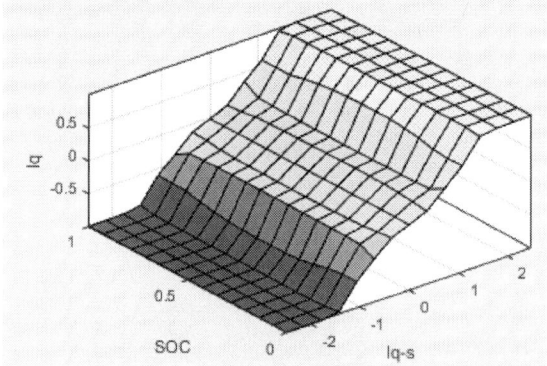

Fig. 7. Current restrictor.

IV. SIMULATION RESULTS

The computer simulations were conducted on Simulink on a 415V 50 Hz 4 pole induction motor and a 450 V Li-ion battery bank which is composed of 20 batteries each with a 22.5V, 6.6 Ah capacity. The sampling time is T_s=0.0001 and the simulations were performed on MALTAB/Simulink. The system is simulated for a drive cycle containing several step references as shown in figure 8.

The CS FLC represents the cascaded fuzzy architecture. Figures 8 and 9 show the speed response for the fuzzy and cascaded fuzzy architectures. It is clear that the cascaded fuzzy slows the acceleration of the motor in favor of conserving more SOC as shown in figures 13-15. The settling time of the cascaded fuzzy taken at the 1500 rpm step responses is 0.31 seconds while the fuzzy speed regulator when working alone without BEM has a settling time of 0.2 seconds. This step has the largest settling time delay of 0.11 seconds.

Figure 10 shows that the control signal I_{sq} which is used to command the motor speed. It is clear that the speed regulator demands a high current during speed transients which imposes a higher stress on the battery. This comes at the cost of the battery's SOC as shown in figures 13-15.

Figure 11 and Figure 12 show the reference and measured motor currents and further imply that the motor does not really consume as much current from the battery when regulated with a cascaded fuzzy architecture. However, the current drawn is much higher when it comes to the fuzzy speed regulator working alone and this negatively affects the SOC and SOH of the battery.

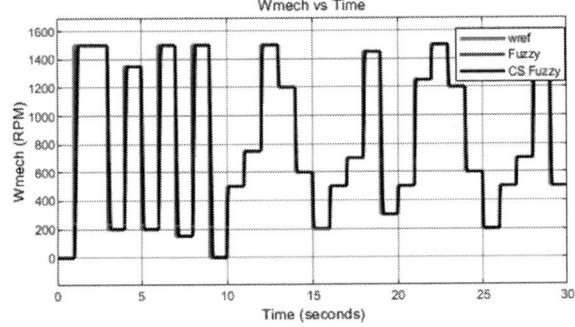

Fig. 8. Speed response for the different architectures.

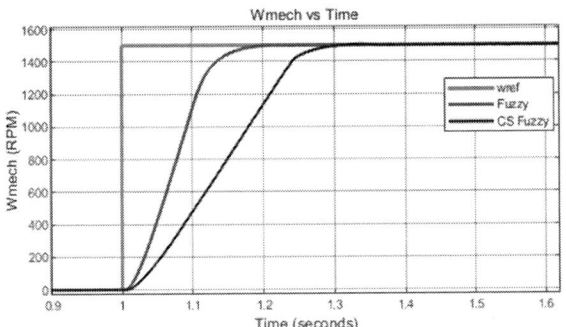

Fig. 9. Zoomed speed response.

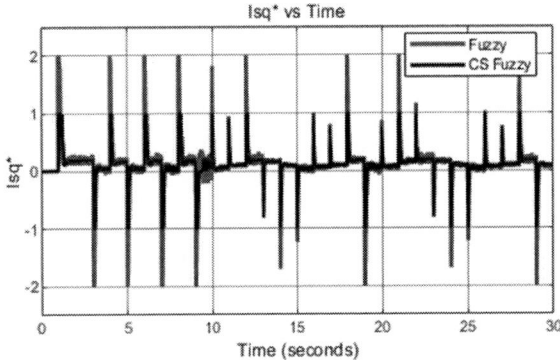

Fig. 10. Reference current vs time.

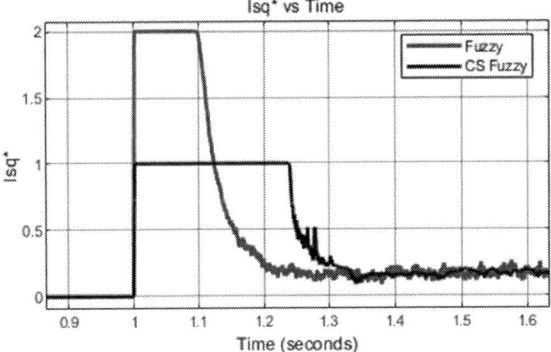

Fig. 11. Reference current zoomed at t=1 sec.

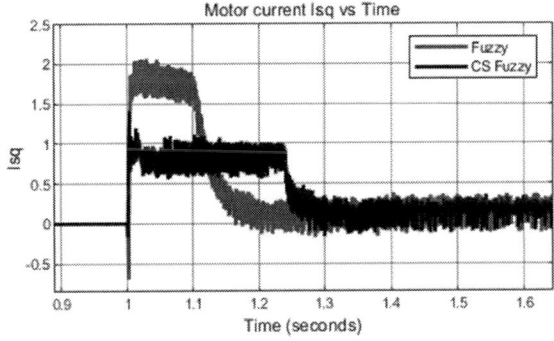

Fig. 12. Measured motor current vs time.

Figure 13-15 show that the cascaded fuzzy resists the abrupt changes in SOC when a reference signal is applied which means they can take a corrective action when a situation, such as an unplanned road slope stresses the battery by drawing a high current. Furthermore, the cascaded fuzzy architecture has effectively reduced the SOC consumption by the end of the driving cycle. Table 1 summarizes the consumed and conserved SOCs during the driving cycle and the average time delay required to reach the reference speed. It can be observed that for a given speed command, the cascaded fuzzy consumed less SOC with some compensation in the motor rise time. Finally, Table 2 summarizes the SOH degradation obtained from the simulation.

V. CONCLUSION

This paper proposed a BEM architecture that achieves the same reference speed whilst conserving more SOC. The BEM architecture is a cascaded fuzzy logic controller which optimizes the SOC consumption while maintaining an acceptable motor performance. The results show that using the proposed cascaded fuzzy architecture, the SOC consumption was reduced by 1.71% and the SOH of the battery improved by 2.02% in exchange for a delay of 0.2 seconds in motor rise time. This means that in exchange for a slower acceleration speed, the vehicle can run for longer distances, due to the reduction in SOC consumption, and the battery replacement will be less frequent, due to a slower degradation in SOH. Moreover, no prior information about the road was required and the optimization algorithm can run in real-time. It is worth mentioning that the improvement shown by this work is only for a duration of 30 Sec. This effect will be much more significant when the EV operates for a longer period of time.

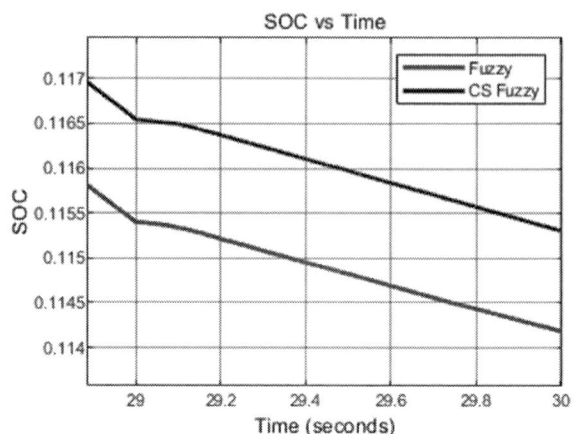

Fig. 15. SOC zoomed at t=29 sec.

Table 1. SOC consumed.

Architecture	Consumed SOC	Conserved SOC	Max Time Delay (s)
Fuzzy	0.0658130	-	-
Cascaded Fuzzy	0.0646905	1.71%	0.11

Table 2. SOH decay.

Architecture	SOH decay	Relative Improvement
Fuzzy	3.19824822e-5	-
Cascaded Fuzzy	3.13368858e-5	2.02 %

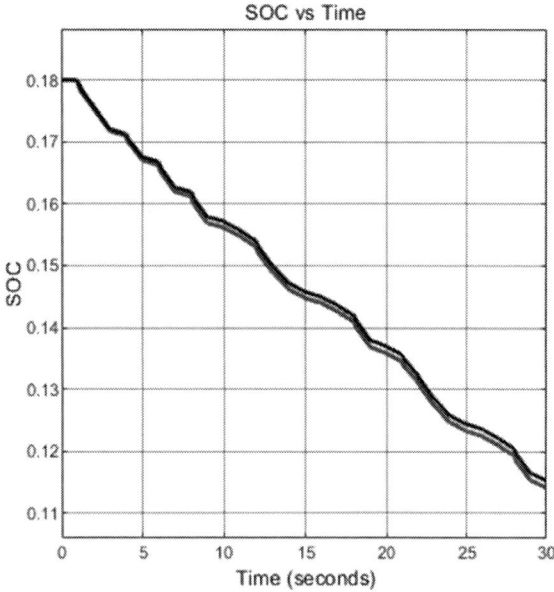

Fig. 13. SOC vs time

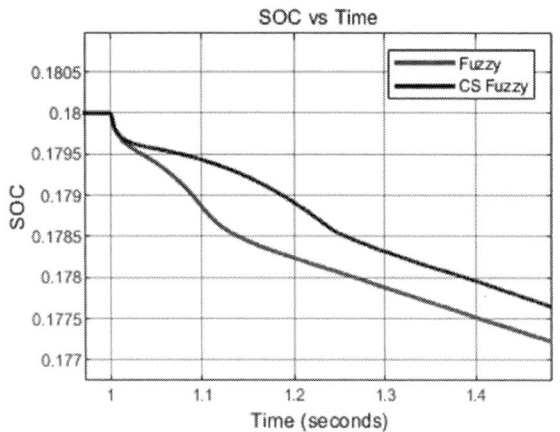

Fig. 14. SOC zoomed at t=1sec.

REFERENCES

[1] Salah and O. Wasseem, "Energy Management of a Multi-Source Power System," Master of Science in Engineering Systems Management (MSESM), Department of Industrial Engineering, American University of Sharjah, 2018.

[2] B. Frieske, M. Kloetzke, and F. Mauser, "Trends in vehicle concept and key technology development for hybrid and battery electric vehicles," in *2013 World Electric Vehicle Symposium and Exhibition (EVS27)*, 17-20 Nov. 2013 2013, pp. 1-12, doi: 10.1109/EVS.2013.6914783.

[3] Z. Liu, Y. Wang, and J. Du, "Fuzzy model predictive control of a permanent magnet synchronous motor in electric vehicles," in *2013 10th IEEE International Conference on Control and Automation (ICCA)*, 12-14 June 2013 2013, pp. 604-608, doi: 10.1109/ICCA.2013.6565084.

[4] N. Jinrui, S. Fengchun, and R. Qinglian, "A Study of Energy Management System of Electric Vehicles," in *2006 IEEE Vehicle Power and Propulsion Conference*, 6-8 Sept. 2006 2006, pp. 1-6, doi: 10.1109/VPPC.2006.364301.

[5] Q. Li, W. Chen, Y. Li, S. Liu, and J. Huang, "Energy management strategy for fuel cell/battery/ultracapacitor hybrid vehicle based on fuzzy logic," *International Journal of Electrical Power & Energy Systems,* vol. 43, no. 1, pp. 514-525, 2012, doi: https://doi.org/10.1016/j.ijepes.2012.06.026.

[6] M. H. Hajimiri and F. R. Salmasi, "A Fuzzy Energy Management Strategy for Series Hybrid Electric Vehicle with Predictive Control and Durability Extension of the Battery," in *2006 IEEE Conference on Electric and Hybrid Vehicles,* 18-20 Dec. 2006, pp. 1-5, doi: 10.1109/ICEHV.2006.352279.

[7] V. Galdi, A. Piccolo, and P. Siano, "A Fuzzy Based Safe Power Management Algorithm for Energy Storage Systems in Electric Vehicles," in *2006 IEEE Vehicle Power and Propulsion Conference,* 6-8 Sept. 2006 2006, pp. 1-6, doi: 10.1109/VPPC.2006.364267.

[8] C. Weng, X. Zhang, and J. Sun, "Adaptive model predictive control for hybrid electric vehicles power management," in *Proceedings of the 32nd Chinese Control Conference,* 26-28 July 2013 2013, pp. 7756-7761.

[9] S. Ebbesen, P. Elbert, and L. Guzzella, "Battery State-of-Health Perceptive Energy Management for Hybrid Electric Vehicles," *IEEE Transactions on Vehicular Technology,* vol. 61, no. 7, pp. 2893-2900, 2012, doi: 10.1109/TVT.2012.2203836.

[10] M. Chen and G. A. Rinc´on-Mora, *Accurate Electrical Battery Model Capable of Predicting Runtime and I–V Performance.* 2006, pp. 504-511.

[11] O. Hegazy, R. Barrero, J. V. Mierlo, M. E. Baghdad, P. Lataire, and T. Coosemans, "Control, analysis and comparison of different control strategies of electric motor for battery electric vehicles applications," in *2013 15th European Conference on Power Electronics and Applications (EPE),* 2-6 Sept. 2013 2013, pp. 1-13, doi: 10.1109/EPE.2013.6631906.

[12] A. Khurram, H.-U. Rehman, S. Mukhopadhyay, and D. Ali, *Comparative analysis of integer-order and fractional-order proportional integral speed controllers for induction motor drive systems.* 2018, pp. 723-735.

[13] H. Rehman and A. Khurram, "Fuzzy Logic Enhanced Sensorless Alternative Energy Vehicular Drive System," in *2017 IEEE Vehicle Power and Propulsion Conference (VPPC),* 11-14 Dec. 2017 2017, pp. 1-5, doi: 10.1109/VPPC.2017.8330885.

[14] O. Hegazy, R. Barrero, J. V. Mierlo, M. E. Baghdad, P. Lataire, and T. Coosemans, "Control, analysis and comparison of different control strategies of electric motor for battery electric vehicles applications," in *2013 15th European Conference on Power Electronics and Applications (EPE),* 2-6 Sept. 2013, pp. 1-13, doi:10.1109/EPE.2013.6631906.

[15] D. Ali, S. Mukhopadhyay, H. Rehman, and A. Khurram, "UAS based Li-ion battery model parameters estimation," *Control Engineering Practice,* vol. 66, pp. 126-145, 2017, doi: https://doi.org/10.1016/j.conengprac.2017.06.012.

[16] H. M. Usman, S. Mukhopadhyay, and H. Rehman, "Universal Adaptive Stabilizer Based Optimization for Li-Ion Battery Model Parameters Estimation: An Experimental Study," *IEEE Access,* vol. 6, pp. 49546-49562, 2018, doi:10.1109/ACCESS.2018.2867560.

[17] C. Depature *et al.*, "IEEE VTS Motor Vehicles Challenge 2017 - Energy Management of a Fuel Cell/Battery Vehicle," in *2016 IEEE Vehicle Power and Propulsion Conference (VPPC),* 17-20 Oct. 2016, pp. 1-6, doi: 10.1109/VPPC.2016.7791701.

[18] P. Ananto, F. Syabani, W. D. Indra, O. Wahyunggoro, and A. I. Cahyadi, "The state of health of Li-Po batteries based on the battery's parameters and a fuzzy logic system," in *2013 Joint International Conference on Rural Information & Communication Technology and Electric-Vehicle Technology (rICT & ICeV-T),* 26-28 Nov. 2013, pp. 1-4, doi:10.1109/rICTICeVT.2013.6741508

Control Strategy for Flywheel Energy Storage Systems on a Three-Level Three-Phase Back-To-Back Converter

M. di Benedetto, A.Lidozzi,
Roma Tre University
C-PED, Center for Power Electronics and Drives
Rome, Italy
marco.dibenedetto@uniroma3.it

D. M.Kumar, H. K.Mudaliar, M. Cirrincione
School of Engineering and Physics,
University of the South Pacific, Suva, Fiji

Abstract—**This paper studies the control structure for a flywheel energy storage system (FESS) used in the grid-connected applications. The power conversion structure uses a double conversion AC/AC through a three-phase thee level Neutral Point Clamp (NPC) inverter. The control structure allows a seamless connection of the FESS to the load and a simultaneous reduction of the current harmonic content to the load. A Fractional PI (FOPI) has been used to regulate the DC-link. Simulation results show excellent dynamic performance of the proposed control system, reducing both the DC-link voltage fluctuations and total harmonic distortion (THD) of the output currents.**

Keywords— energy storage system, multilevel converter, power semiconductor, fractional PI.

I. INTRODUCTION

The energy conversion in a Flywheel Energy Storage Systems (FESS) is generally accomplished by using an electrical machine and a bi-directional power converter. Particularly, the power electronic converter topologies that can be used for FESS applications are DC-AC, AC-AC, AC-DC-AC, or a combination of these. One of the most widely used configuration of power converters in FESS is the back-to-back configuration, connected to a DC-bus capacitor [1],[2].

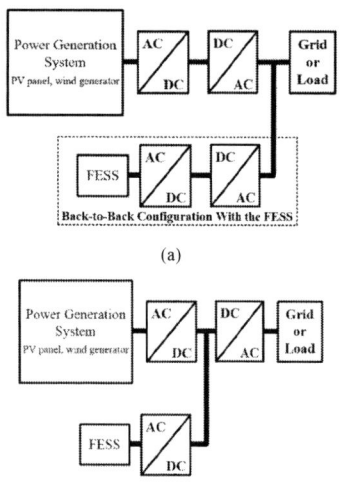

(a)

(b)

Fig. 1. Block scheme of the FEES connection to a renewable generation: a) ac mode, b) dc mode.

For instance, in renewable energy source (RES) applications the flywheel is usually installed in the AC-AC (back-to-back) power conversion with a common DC-bus [3],[4], and one converter keeps the DC-bus voltage constant to ensure a fast response, low cost of the conversion system and better energy management [5]. As shown in Fig. 1, there are two different modes to store electric energy by using the FESS, a) AC mode and b) DC mode. In wind energy source applications, the FESS can be connected to the DC-bus using the AC/DC converter; however, in photovoltaic (PV) applications the DC-bus voltage is a function of the operating point of the panels. Thus, an extra DC/DC converter should be used in order to avoid a DC-bus voltage variation, as shown in Fig. 2. The FESS can be coupled to an electric motor or generator that produces electrical energy. As well known, the flywheel can be classified as high-speed flywheel or low-speed flywheel, according to the different operating speeds [6], [7].

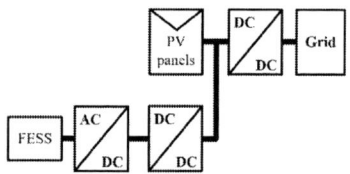

Fig. 2. Block scheme of the photovoltaic applications with FEES.

Low operating speed is used in RES generation and uninterruptible power supply (UPS), while high-speed flywheel are mostly employed in aerospace and military applications [8]-[11]. Furthermore, the FESS is well adapted due to its high efficiency, long lifetime, low cost, environmentally friendly, and high dynamic speed [10], [11]. Using high-speed generators to directly couple an energy conversion unit results in reduction of the system volume, maintenance, costs and increase of the efficiency.

The multi-level topology of converters can be considered as an important topology to be used for high-speed generation [12]-[14]. Indeed, due to the high fundamental frequency of the high-speed generation unit, the synchronous inductance value can be reduced with resulting decrease of the reactive voltage drop. However, it is preferable to increase the number of levels of the converter in order to have an acceptable current ripple value [12]-[16]. Additionally, multilevel converters show different advantages as compared to two-level converters, like reduced voltage stress on the power

978-1-5386-7688-2/19 $31.00 © 2019 IEEE

devices, lower harmonics contents, and lower instantaneous rate of voltage change (dv/dt) [12]-[16].

The back-to-back converter based on three-level NPC topology is a suitable conversion system in high power applications thanks to its excellent dynamic response compared to the conventional two-level three-phase converter. Many control techniques have been proposed in literature to control back-to-back converter [17]-[21]. In [17] different control techniques, namely Sequence Domain Control (SDC) and Direct Power Control (DPC), have been compared to identify the best control technique suitable for low-voltage ride-through applications of doubly-fed induction generator (DFIG)-based wind system converters. A dynamic model of back-to-back voltage source converter under synchronous rotating d-q coordinate system has been developed in [18]. In this case, a nonlinear decoupled controller for active and reactive power exchange between the converters has been presented. A properly modified Space-Vector Pulse-Width-Modulation (SVPWM) algorithm has been implemented in [19] to improve the DC-link voltage balancing of the back-to-back NPC converter in an electric drive application. Moreover, a direct model predictive current control scheme for both the grid and generator side power converter control of a 3L-NPC back-to-back power converter PMSG wind turbine system has been presented in [20].

This paper proposes the use of a three-phase three level (3Φ3L) back-to-back converter with the FESS, as shown in Fig. 3. In this topology, during the power generation the first converter, which is an NPC topology, works as the rectifier and the second converter, also an NPC topology, operates as the inverter. On the other hand, when the flywheel is charged, the first NPC converter works as the inverter and the second NPC converter operates as a rectifier. The proposed control structure allows to improve both the dynamic response of the DC-link voltage variations and THD of the output voltage.

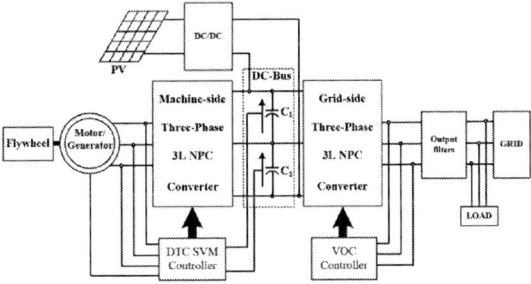

Fig. 3. Block scheme of the FESS conversion system.

The new control method for 3Φ3L NPC converter makes use of the Direct Torque Control with Space Vector Modulation (DTC SVM) and a Voltage Oriented Control (VOC). The DTC SVM is applied to the first NPC converter, the VOC controller is applied to the second converter. The paper is organized as follows. The FEES conversion system is analysed in Section II. The proposed control structure is explained in Section III. The simulation results are shown in Section IV, and conclusions are drawn in section V.

II. SYSTEM ARCHITECTURE WITH

The proposed conversion system is shown in Fig. 4. The configuration is made up of a flywheel directly coupled to a permanent magnetic synchronous motor/generator and two 3L NPC converter.

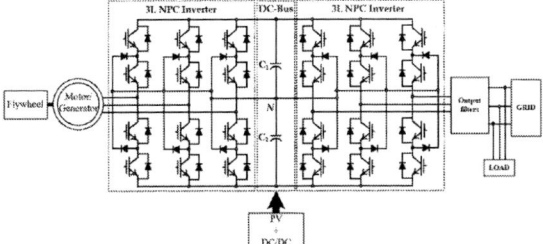

Fig. 4. Three-phase 3L NPC converter.

A Flywheel Energy Storage System (FESS) mainly consists of a rotating mass appropriately coupled with an electrical drive, which has to operate as either a motor or a generator. As a result, the FESS is able to store a certain amount of electrical energy, converting it into kinetic energy of the flywheel and delivering it back in accordance with specific application needs. The energy E that can be stored in the flywheel is given in (1), where J is the inertia of the flywheel and ω is the rotational speed.

$$E = \frac{1}{2}J\omega^2 \qquad (1)$$

The mechanical equation on the shaft of the motor connected to the flywheel can be written in (2), where T_{em} is the torque produced by the electrical motor, M_B is the torque produced by the mechanical bearings supporting the flywheel and T_{air} is the torque due to the air friction.

$$J\frac{d\omega}{dt} = T_{em} - M_B - T_{air} \qquad (2)$$

When such a system is applied, a DC voltage can be rectified from an AC variable frequency waveform, and an AC waveform can be created from a DC one with the required frequency and voltage level. During the discharge of the FESS, the electric machine works as a generator extracting mechanical energy from the flywheel and as a result the speed of the flywheel slows down changing the frequency of the AC output. In this paper, a flywheel with rated torque of 2.3 Nm has been selected to rotate at 10000 rpm. The flywheel is connected to the permanent magnetic generator having a rated power of 4kW and 2 poles.

III. CONTROL STRUCTURE OF THE FESS

The control scheme of the machine side converter is shown in Fig.5. The DTC-SVM control strategy has been used to control the inverter. The proposed control strategy works in the flux weakening region and is controlled in torque to maintain the DC link voltage constant. Starting from the line current of the inverter machine-side, the d and q-axis stator currents, are obtained, and then, the stator flux is estimated. In order to regulate the DC-Link voltage, the fractional order PI (FOPI) control has been used as a voltage controller, for its good dynamical performance [20].

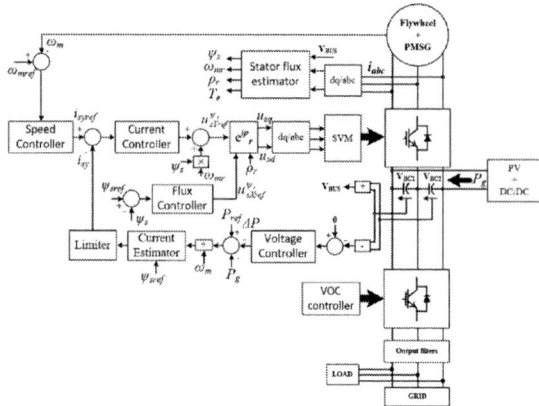

Fig. 5. Proposed control scheme for machine-side three-level NPC converter.

The transfer function (TF) of the FOPI controller is given in (3), where λ is the positive real parameter included between 0 and 1 and K_p, K_i are the proportional and integral gain, respectively.

$$C(s) = K_p + K_i s^{-\lambda} \qquad (3)$$

The parameters K_p and K_i are tuned according to a certain settling time. Furthermore, the real number λ has been selected that gives the best performance in terms of overshoot and other dynamical performance specifications. If P_{DC} represents the power exchanged with the DC link to be regulated (considered positive if it flows from the DC link towards the load), P_g is the power supplied by the PV array, and finally P_{req} is the power required by the grid-side converter, the power that the FESS has to supply to the load is given in (4).

$$P_{ref} = P_{req} - P_g - P_{DC} \qquad (4)$$

From the above equation the reference torque that is needed for the Field Oriented Control to drive the machine is given by the equation (5), where ω_m is the FW speed.

$$T_{ref} = \frac{P_{ref}}{\omega_m} + M_b + T_{air} \qquad (5)$$

It is can be noted that the value of the P_{ref} must not exceed the maximum power of the Power Magnet Synchronous Generator (PMSG), this action would result in the PMSG surpassing it torque capability. Thus, it is necessary to operate the PMSG close to its rated power or in the field weakening operation if the PMSG works above its rated speed.

The control scheme of the VOC controller is illustrated in Fig. 6. VOC [23], which stands for Voltage Oriented Control, is inspired to the FOC, since it realizes a control action on the direct (d) and quadrature (q) components of the current injected by the grid-side converter in the grid-voltage space vector reference frame.

In this reference frame, the dq components of the voltage space vector at the output of the converter is given by

$$\begin{cases} u_{gd} = u_{sd} + Ri_{sd} + L\frac{di_{sd}}{dt} - \omega_s Li_{sq} \\ u_{gq} = u_{sq} + Ri_{sq} + L\frac{di_{sq}}{dt} - \omega_s Li_{sd} \end{cases} \qquad (6)$$

where u_{gd} and u_{gq} are the direct and quadrature components in the dq reference frame of the grid voltage, u_{sd} and u_{sq} are the direct and quadrature components in the dq reference frame of the inverter voltage, i_{sd} and i_{sq} are the direct and quadrature components in the dq reference frame of the inverter current, L and R are respectively the inductance and resistance of the filter, ω_s is the pulsation of the electrical grid or the rotation speed of the grid voltage reference frame, and ρ_s is its angle w.r.t the stationary reference frame.

PI controllers can be used to control the inverter dq current components. However, as expected and like FOC, there is some cross-coupling between the 2 axes terms, to be compensated by suitable feedforward components, as shown in Fig. 6 and in [23]. The quadrature current reference is generally set to zero to keep a null reactive power injected to the grid. The direct current reference controls the injected active power.

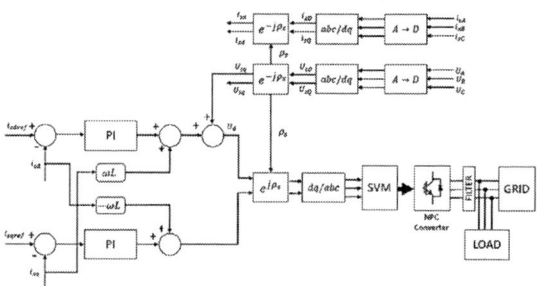

Fig. 6. Control scheme for load-side three-level NPC converter.

IV. SIMULATION RESULTS

The model of the proposed system has been implemented in Matlab/Simulink environment. The main parameters of the system are shown in Table I. Fig. 7 shows the speed response of the FESS in a back to back three-level converter application.

Table I. Parameters of the system.	
PMSG Parameters	
Rated motor power (kW)	2.2
Rated motor voltage (V)	220
Rated frequency machine (Hz)	50
Number of pole-pairs	2
Stator winding resistance (Ω)	3.88
Stator winding inductance (μH)	500
Rotor bar resistance (Ω)	1.87
Rotor bar inductance (mH)	252
magnetizing inductance (mH)	236
Inertia of motor J (kg·m2)	0.0089
Converter Parameters	
DC-link voltage (V)	500
DC-link capacitor (μF)	560
Switching frequency (kHz)	20
Nominal Power (kVA)	20
Flywheel Parameters	
Energy need (MJ)	216
Rotation speed (rpm)	10000
Material selection	Mild Steel
Working stress (MPA)	486
Mild Steel Density, ρ (kg/m^3)	7830
Poisson's Ratio, v	0.29
Rotor Disc Radius, r (m)	0.6
Rotor Disc Width, h (m)	0.25

The FESS starting curve to the reference speed of 260 rad/s is shown from t=0 second to t=27 seconds. Once the reference speed is attained, the dynamical response is shown for power variations of the RES generation units.

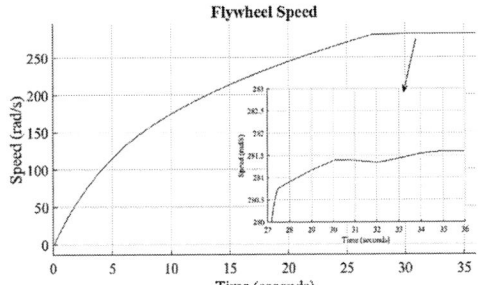

Fig. 7. Speed response of induction motor coupled with FESS.

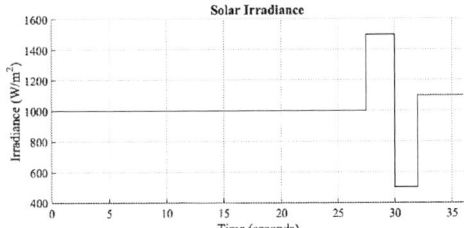

Fig. 8. Solar irradiance.

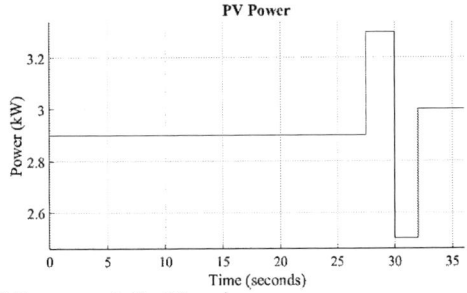

Fig. 9. Power generated by PV panels.

Fig. 8 gives the solar irradiance reference given to the PV system. It can be seen that the irradiance is equal to 1000 W/m² for t<27.5 seconds. The irradiance at t=27.5 seconds is about equal to 1500 W/m², while at t=30 seconds the irradiance is about equal to 500 W/m². After t>32 seconds, the irradiance is equal to 1100 W/m².

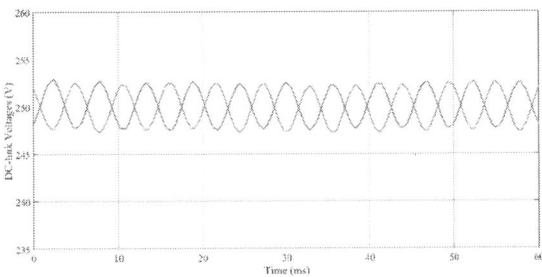

Fig. 10. DC-link voltages.

Consequently, the electrical power generated by the PV system for the given irradiance references is shown in Fig. 9. The DC link voltages are shown in Fig. 10, where the black

line is the voltage across the C_1 and the blue line is voltage across C_2. It is can be seen that the control algorithm allows voltages across the capacitors C_1, C_2 to be unbalanced. The line-to-line switching voltage, V_{AB}, of the machine-side converter when the motor speed is set at 1500 rpm is illustrated in Fig. 11.

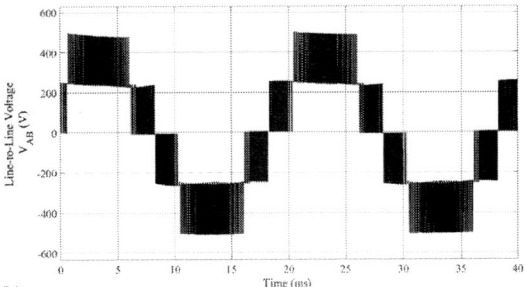

Fig. 11. Line-to-line switching voltage V_{AB} of the machine-side converter.

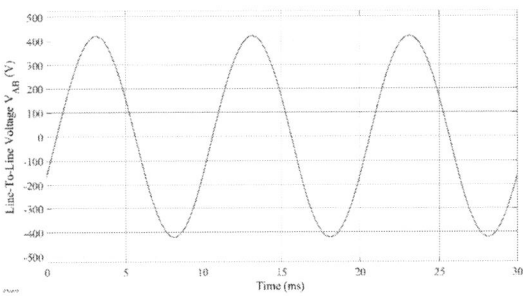

Fig. 12. Line-to-line voltage V_{AB} waveform of the machine-side converter.

As it can be noticed, the line-to-line voltage is composed by five voltage levels. In this condition, the power flows from the AC-side to DC-side. In the same conditions, the waveform of the line-to-line voltage V_{AB} is shown in Fig. 12. The waveforms of the phase currents are illustrated in Fig. 13.

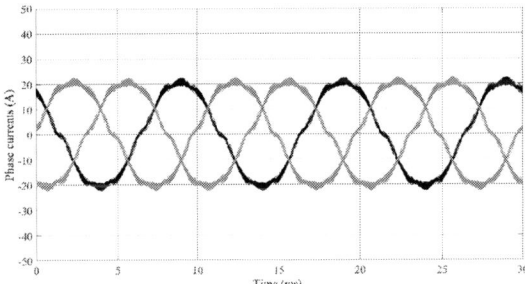

Fig. 13. Phase currents of the machine-side converter.

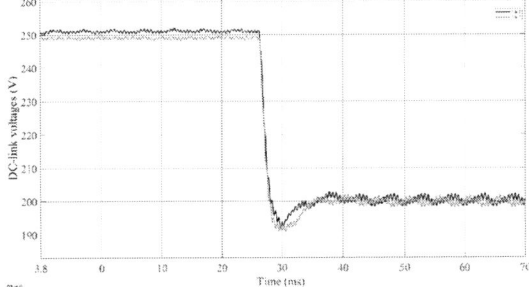

Fig. 14. Step load of the DC-link voltage.

In order to evaluate the DC-link voltage dynamic performance a step load from 500V to 400V is shown in Fig. 14. The blue line is the voltage across C_1 and the black line is the voltage across C_2. It can be seen that the proposed control exhibits a fast-dynamic performance. Finally, Fig. 15 shows the three-phase current after the output filters under resistive load condition. The resulting THD is about 4%.

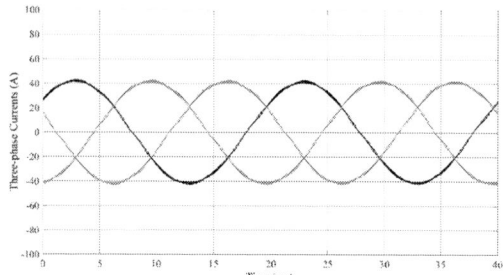

Fig. 15. Three-phase current after the output filters under resistive load condition.

V. CONCLUSION

The control structure for a flywheel energy storage system (FESS) used in the grid-connected applications has been presented in this paper. The topology converter makes use of three-phase 3 level NPC converter in back-to-back configuration. This configuration has been controlled using the direct torque control with space vector modulation and a voltage-oriented control (VOC). A FOPI controller has been used to regulate the DC link. The proposed control algorithm has been tested in Matlab/Simulink environment. Simulation results show the good dynamic response of the proposed system.

VI. ACKNOLEDGMENT

The manuscript has been realized within the framework of the project REFEPICS2 (Implementation of a Renewable energy source system with a Flywheel Energy storage system for supplying energy in Pacific Island CountrieS with weak grid, part II) funded by the Economic, Social and Cultural Cooperation Fund for the Pacific or "Pacific Fund" (PF) financed by the Ministry of Foreign Affairs and international development (MAEDI).

REFERENCES

[1] Itaru Ando, Junji Shibata, Hitoshi Haga and Kiyoshi Ohishi, "Long life ups based on active filter and flywheel without electrolytic capacitor," 2007 European Conference on Power Electronics and Applications, Aalborg, 2007, pp. 1-10.

[2] H. Haga, K. Ohishi and I. Ando, "Development of economical UPS system having active filter across DC-bus," 2005 European Conference on Power Electronics and Applications, Dresden, 2005, pp. 10 pp.-P.10.

[3] H. Zhang and Y. Li, "Constant voltage control on DC bus of PV system with flywheel energy storage source (FESS)," 2011 International Conference on Advanced Power System Automation and Protection, Beijing, 2011, pp. 1723-1727.

[4] X. Li, N. Erd and A. Binder, "Evaluation of flywheel energy storage systems for residential photovoltaic installations," 2016 International Symposium on Power Electronics, Electrical Drives, Automation and Motion (SPEEDAM), Anacapri, 2016, pp. 255-260.

[5] J. Wu, J. Wen and H. Sun, "A new energy storage system based on flywheel," 2009 IEEE Power & Energy Society General Meeting, Calgary, AB, 2009, pp. 1-6.

[6] H. Lee, B. Y. Shin, S. Han, S. Jung, B. Park and G. Jang, "Compensation for the Power Fluctuation of the Large Scale Wind

Farm Using Hybrid Energy Storage Applications," in IEEE Transactions on Applied Superconductivity, vol. 22, no. 3, pp. 5701904-5701904, June 2012, Art no. 5701904.

[7] G. Cimuca, S. Breban, M. M. Radulescu, C. Saudemont and B. Robyns, "Design and Control Strategies of an Induction-Machine-Based Flywheel Energy Storage System Associated to a Variable-Speed Wind Generator," in IEEE Transactions on Energy Conversion, vol. 25, no. 2, pp. 526-534, June 2010.

[8] J. Lee, S. Jeong, Y. H. Han and B. J. Park, "Concept of Cold Energy Storage for Superconducting Flywheel Energy Storage System," in IEEE Transactions on Applied Superconductivity, vol. 21, no. 3, pp. 2221-2224, June 2011.

[9] D. W. Swett and J. G. Blanche, "Flywheel charging module for energy storage used in electromagnetic aircraft launch system," in IEEE Transactions on Magnetics, vol. 41, no. 1, pp. 525-528, Jan. 2005.

[10] D. W. Swett and J. G. Bjanche, "Flywheel charging module for energy storage used in high-voltage pulsed and multi-mode mobility hybrid electric military vehicle power system," Conference Record of the Twenty-Sixth International Power Modulator Symposium, 2004 and 2004 High-Voltage Workshop., San Francisco, CA, 2004, pp. 153-156.

[11] S. Samineni, B. K. Johnson, H. L. Hess and J. D. Law, "Modeling and analysis of a flywheel energy storage system for Voltage sag correction," in IEEE Transaction on Industry Applications, vol. 42, no. 1, pp. 42-52, Jan.-Feb. 2006.

[12] M. Di Benedetto, A. Lidozzi, L. Solero, F. Crescimbini and P. J. Grbovic, "Five-level back to back E-Type converter for high speed gen-set applications," IECON 2016 - 42nd Annual Conference of the IEEE Industrial Electronics Society, Florence, 2016, pp. 3409-3414.

[13] M. di Benedetto, A. Lidozzi, L. Solero, F. Crescimbini and P. J. Grbović, "Five-Level E-Type Inverter for Grid-Connected Applications," in IEEE Transactions on Industry Applications, vol. 54, no. 5, pp. 5536-5548, Sept.-Oct. 2018.

[14] M. Di Benedetto, A. Lidozzi, L. Solero, F. Crescimbini and P. J. Grbovic, "Performance assessment of the 5-level 3-phase back to back E-type converter," 2017 IEEE Energy Conversion Congress and Exposition (ECCE), Cincinnati, OH, 2017, pp. 2106-2113.

[15] A. C. N. Maia, C. B. Jacobina and G. A. de Almeida Carlos, "A New Three-Phase AC–DC–AC Multilevel Converter Based on Cascaded Three-Leg Converters," in IEEE Transactions on Industry Applications, vol. 53, no. 3, pp. 2210-2221, May-June 2017.

[16] A. Dekka, B. Wu and N. R. Zargari, "Start-Up Operation of a Modular Multilevel Converter With Flying Capacitor Submodules," in IEEE Transactions on Power Electronics, vol. 32, no. 8, pp. 5873-5877, Aug. 2017.

[17] M. M. Baggu, B. H. Chowdhury and J. W. Kimball, "Comparison of Advanced Control Techniques for Grid Side Converter of Doubly-Fed Induction Generator Back-to-Back Converters to Improve Power Quality Performance During Unbalanced Voltage Dips," in IEEE Journal of Emerging and Selected Topics in Power Electronics, vol. 3, no. 2, pp. 516-524, June 2015.

[18] Yan Gan-gui et al., "Dynamic modeling and nonlinear decoupled control of back-to-back voltage source converter," 2008 International Conference on Electrical Machines and Systems, Wuhan, 2008, pp. 3862-3866.

[19] G. Mademlis and Y. Liu, "DC Link Voltage Balancing Technique Utilizing Space Vector Control in SiC-based Five-Level Back-to-Back-Connected NPC Converters," 2018 IEEE Energy Conversion Congress and Exposition (ECCE), Portland, OR, 2018, pp. 3032-3037.

[20] Z. Zhang, X. Cai, R. Kennel and F. Wang, "Model predictive current control of three-level NPC back-to-back power converter PMSG wind turbine systems," 2016 IEEE 8th International Power Electronics and Motion Control Conference (IPEMC-ECCE Asia), Hefei, 2016, pp. 1462-1467.

[21] B. Fan, K. Wang, P. Wheeler, C. Gu and Y. Li, "An Optimal Full Frequency Control Strategy for the Modular Multilevel Matrix Converter Based on Predictive Control," in IEEE Transactions on Power Electronics, vol. 33, no. 8, pp. 6608-6621, Aug. 2018

[22] D. M. Kumar, H. K. Mudaliar, M. Cirrincione, U. V. Mehta, and M. Pucci, "Design of a Fractional Order PI (FOPI) for the speed control of a high-performance electrical drive with induction motor," in IEEE 21st International Conference on Electrical Machines and System (ICEMS), October 7-10, 2018, Jeju , South Korea.

[23] D. Mukherjee and D. Kastha, "Voltage Sensorless Control of VIENNA Rectifier in the Input Current Oriented Reference Frame," in IEEE Transactions on Power Electronics, vol. 34, no. 8, pp. 8079-8091, Aug. 2019.

Digital Hybrid Current Mode Control with Asymmetric Slope Compensation for Three-Level Flying Capacitor Buck Converter

Abdulkerim Ugur[1,2], Murat Yilmaz[2]
[1]Electro-Optic and Laser Systems, TUBITAK BILGEM, Kocaeli, Turkey
[2]Electrical Engineering Department, Istanbul Technical University, Istanbul, Turkey
ugurab@itu.edu.tr, myilmaz@itu.edu.tr

Abstract— Due to high effective frequency and reduced voltage stress on the switching devices, the three level flying capacitor (3LFC) buck converter has potential to provide better efficiency and higher power density as compared to the traditional buck converter. However, the control implementation is challenging as a result of flying capacitor instability issue. In this paper, the digital hybrid current mode (HCM) control method which combines the average and peak current mode control techniques is modified and implemented to 3LFC buck converter. Flying capacitor voltage balancing is achieved by adjusting the slope compensation of the two switching pairs asymmetrically. The proposed asymmetric slope compensation technique for digital HCM control method is verified in simulations and experimentally with an ARM-based mixed-signal microcontroller on a 40 V input 13 V / 6 A output prototype.

Keywords—Digital hybrid current mode control, three-level flying capacitor converter, digital control

I. INTRODUCTION

The three level flying capacitor (3LFC) converters which are initially proposed for high power DC-AC and AC-DC applications [1], have recently gained an interest as alternative to the traditional buck converters for low power DC-DC practices. With 3LFC topology, it is possible to cut the voltage swing on the inductor into halves and double the frequency, thus the size of magnetics can be reduced significantly. The voltage stress on the switching devices are decreased as well, therefore it is possible to use components with lower voltage ratings. In addition, due to doubled effective frequency the output capacitor size can be reduced.

Several control methods are studied for output regulation of 3LFC converter [2]-[8] in literature. Although the current mode control methods are favorable due the desirable gain characteristics, the flying capacitor voltage balancing, which is necessary to have a symmetric inductor current waveform in each switching halves and keep the voltage stress equal across the switching devices, might be challenging. The Peak current mode (PCM) control, which is one of the most used control method in switch mode power supplies due to fast response and inherent cycle-by-cycle current protection, is recently shown as inherently unstable for 3LFC buck converter unless a large inductor current ripple exist in the converter [9]. Therefore for PCM control method a control mechanism to keep the flying capacitor voltage balanced is essential. Conventionally for multilevel converters with voltage mode control, capacitor balancing is achieved via a

feed forward loop in which the controller output is added to one of the switching pair duty cycle and subtracted from the other switching ones [10]. As a result, capacitor charge is programmed as desired without changing the effective duty cycle for a switching time. However, this method is not applicable if the duty cycle is not pre-determined as in PCM control. Therefore, in [6] a peak offsetting mechanism is introduced, in which the two switching pair current peaks are controlled separately. Nevertheless in this method the effect of slope compensation is not considered which can be effective on the flying capacitor voltage balancing.

On the other hand, recently a digital hybrid current mode (HCM) control is proposed for DC-DC converter which combines the Average Current Mode (ACM) and PCM control methods to provide accurate current control with fast response and inherent over-current protection [11]. Similar to the analog counterpart [12] in the digital HCM method, the inner loop is the PCM loop which is implemented via a continuous-time comparator built in microcontroller, therefore the peak current limiting is achieved cycle-by-cycle. The outer loop is the ACM loop which is implemented in discrete-time to achieve accurate current control. Since the PCM control enhances the overall controller bandwidth, the need for computation power and fast sampling in the digital ACM loop is reduced significantly.

In this paper, digital HCM control technique is modified and implemented to 3LFC buck converter. The inductor current sense voltage is provided directly to the analog comparator and via a low-pass filter to the Analog-to-Digital Converter (ADC) for digital loop. In digital HCM control technique, due to the inner current loop that is similar to PCM control, an appropriate flying capacitor balancing scheme is necessary. In this study, the flying capacitor balancing is achieved by programming the slope compensation of the two switching pairs separately which is named as Asymmetric Slope Compensation (ASC). For this purpose, a feed forward loop in which the flying capacitor is compared with the half of the input voltage is implemented. The slope of the two switching pairs are programmed with the output of this feed forward loop controller separately, so that the capacitor is charged or discharged cycle-by-cycle as required. By modifying the slope of the compensators, the proposed ASC method aims to counteract the inductor current slope deviation which is a result of flying capacitor voltage unbalance. In Section II, the modified digital HCM control technique is reviewed briefly and the proposed

978-1-5386-7688-2/19 $31.00 © 2019 IEEE

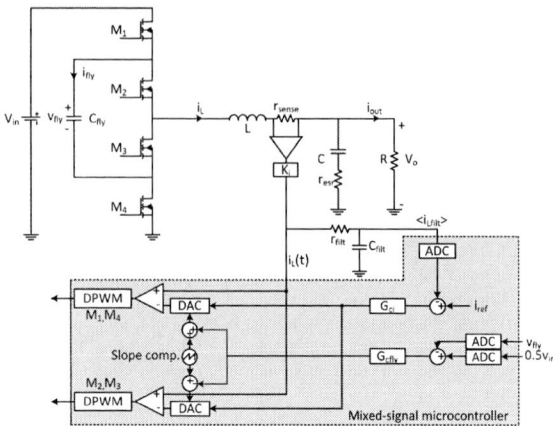

Fig. 1. Architecture of a mixed-signal microcontroller based digital HCM control with asymmetrical duty cycle compensation for a 3LFC buck converter

asymmetric slope compensation technique for flying capacitor voltage balancing is introduced. In Section III, the modified digital HCM control with asymmetrical slope compensation method is verified with simulations and experimental results on a 40 V input 13 V / 6 A output prototype.

II. DIGITAL HCM CONTROL AND CAPACITOR BALANCING WITH ASC

Basically, the digital HCM control method combines an analog PCM control loop with a digital ACM one. The inner PCM loop improves the response time of the system to the load or control signal changes. In addition to this, the PCM control loop is implemented in analog with the built-in analog comparator of the microcontroller, therefore high ADC sampling or geometric calculation based methods are eliminated. The on-time of the PWM is terminated with a compare-match event in the microcontroller. Furthermore, the peak current level is determined by the outer ACM control loop. The outer ACM control loop is implemented in digital; the average current information is gathered with a low pass filter implemented before the ADC. The output of the ACM controller is converted to the analog domain via the internal DAC of the microcontroller which is used the implement the slope compensation as well. Lastly, in the feed forward flying capacitor balancing control loop, the input and flying capacitor voltages are sampled, the error between the flying capacitor voltage and half of the input voltage is processed with the digital controller. The output of the feed forward loop Δs_e is added to the slope s_e for M_2/M_3 switches whereas it is subtracted from s_e for M_1/M_4. The overall control structure of the digital HCM control method modified for a 3LFC buck converter is shown in Fig. 1.

The ASC method for flying capacitor balancing is investigated under the following assumptions:

1. The ACM loop is stable and fast, as a result it regulates the output even with an unbalanced flying capacitor voltage. Therefore it is assumed that $I_0(t) = I_0(t + T_s)$ for a switching cycle where T_s is a single switching time.

2. The flying capacitor does not interfere the dynamics of the buck topology, i.e. the inductor current rise linearly during on-time of M_1 and M_2.

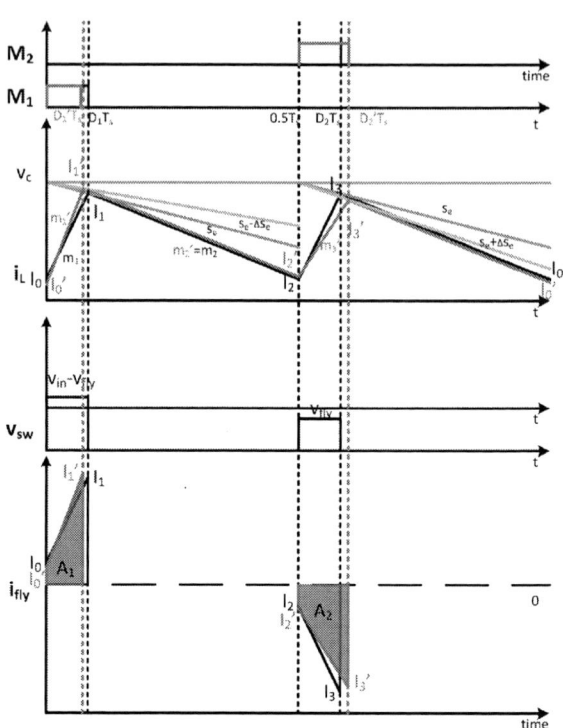

Fig. 2. Waveforms of digital HCM control with ASC for a 3LFC buck converter

3. The worst case scenario for flying capacitor voltage balancing is that under the small perturbation the average capacitor charge is constant which means that the total area under the flying capacitor current waveform for a switching cycle is zero, i.e. $A_1 = A_2$ (Fig. 2). To prevent flying capacitor voltage run-away phenomena [9], for a perturbation $\widehat{v_{fly}} < 0$, it is necessary to have $A_1 > A_2$, so that the capacitor is charged cycle-by-cycle.

Based on these assumptions, the operating point of the 3LFC buck converter with small perturbation at steady-state can be found analytically. From the inductor current geometry given in Fig. 2

$$m_1'D_1'T_s = I_1' - I_0' \tag{1}$$

$$m_3'D_2'T_s = I_3' - I_2' \tag{2}$$

$$m_2(0.5 - D_1')T_s = I_1' - I_2' \tag{3}$$

$$m_2(0.5 - D_2')T_s = I_3' - I_0' \tag{4}$$

where $m_1' = \frac{V_{in}/2 - (V_{fly} + \widehat{v_{fly}}) - V_{out}}{L}$, $m_2' = \frac{V_{out}}{L}$, $m_3' = \frac{(V_{fly} + \widehat{v_{fly}}) - V_{out}}{L}$ and $V_{fly} = \frac{V_{in}}{L}$ for D<0.5. Without considering the parasitics, the average inductor current is equal to the output current, i.e. $I_{avg} = I_{out}$, therefore

$$(I_0' + I_1')D_1' + (I_1' + I_2')(0.5 - D_1') + (I_2' + I_3')D_2' \\ + (I_3' + I_0')(0.5 - D_2') = 2I_{out} \tag{5}$$

The charge balance condition in one switching cycle can be calculated by using the area under the flying capacitor

978-1-5386-7688-2/19 $31.00 © 2019 IEEE   378

current waveform with $k=0$, where k is the additional charge ratio in the second half of a single switching cycle

$$(I_0' + I_1')D_1'T_s = (1 + k)(I_2' + I_3')D_2'T_s \qquad (6)$$

Then by using (1)-(6), the duty cycles for M_1/M_4 and M_2/M_3 switch pairs can be found for capacitor charge balance case as

$$D_1' = D_2' = D' = \frac{m_2'}{m_1' + 2m_2' + m_3'} \qquad (7)$$

For M_1, M_4 complementary switches the new slope is defined as $s_e - \Delta s_e$ whereas for M_2, M_3 it is $s_e + \Delta s_e$. Therefore I_1' and I_2' can be calculated via

$$I_1' = v_c - (s_e - \Delta s_e)D'T_s \qquad (8)$$

$$I_3' = v_c - (s_e + \Delta s_e)D'T_s \qquad (9)$$

where v_c is the peak current control voltage. By using (5), (8) and (9), v_c can be found as

$$v_c = 2I_{out} + s_e D'T_s - (I_0' + I_2')/2 \qquad (10)$$

With asymmetric slopes, the valley inductor currents are calculated by

$$I_0' = v_c - (m_1' + (s_e - \Delta s_e))D'T_s \qquad (11)$$

$$I_2' = v_c - (m_3' + (s_e + \Delta s_e))D'T_s \qquad (12)$$

And finally by using charge balance condition (6), with (11)-(12), one can found the required Δs_e for the flying capacitor perturbation as

$$\Delta s_e = \widehat{v_{fly}}/2L \qquad (13)$$

Equation (13) is the boundary condition to maintain the charge balance of the flying capacitor. For stabilization of the flying capacitor voltage, the area under the flying capacitor current is required to be positive which means that a net charge ΔQ is added to the flying capacitor to increase the voltage cycle by cycle to $V_{in}/2$ until the negative perturbation $\widehat{v_{fly}}$ is eliminated. In other words to eliminate a voltage error in the flying capacitor, $\Delta s_e > \widehat{v_{fly}}/2L$ is required. The analysis can be expanded using (1)-(13) for $k>0$ which is $A_1 > A_2$ by using mathematical tools such as

TABLE I.   CIRCUIT PARAMETERS

Parameter	Value	Parameter	Value
V_{in} (input voltage)	40 V	C_{out} (output capacitor)	5.7 μF
I_o (output current)	6 A	C_{fly} (flying capacitor)	20 μF
R_{out} (load resistance)	2.15 Ω	T_s (switching time)	4 μs
L (inductor)	2.2 μH		

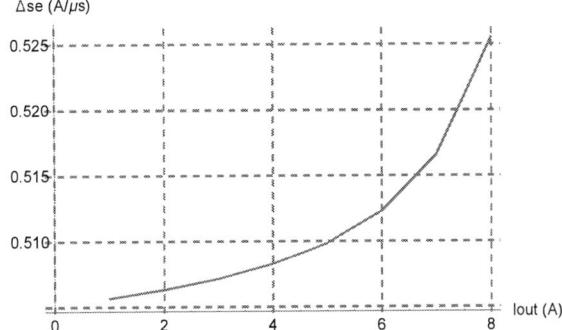

Fig. 3.   Output current I_o vs. Δs_e for $\widehat{v_{fly}} = -10\% V_{fly}$ and k = 1%

MATLAB or Mathcad.

In Fig. 3, Δs_e vs. output current for $\widehat{v_{fly}} = 10\% V_{fly}$ and $k = 1\%$ is shown for a stable, constant current load. The circuit parameters are given in Table I. In this case, the duty cycles D_1 and D_2 is not equal as a result (7)-(13) changes accordingly. A full expression is not given due to the complexity. In addition, the analysis above is conducted for $V_{out}/V_{in} < 0.5$, but can be extended with the same procedure for $V_{out}/V_{in} > 0.5$. As can be seen from the figure, by providing the necessary Δs_e with opposite signs to the switching pairs, it is possible to maintain the flying capacitor voltage balance for PCM and digital HCM control. It is important to note that the minimum slope should be provided in any case to prevent sub-harmonic oscillations [9].

III.   SIMULATIONS AND EXPERIMENTAL VERIFICATION

To support the mathematical analysis given in Section II, simulation studies are conducted by using PSIM. The circuit parameters are same as in Table I except that the output capacitor is high enough with the necessary initial charge to have a stable output and eliminate the effect of the control for the output regulation.

In Fig. 4, the response of the flying capacitor voltage to the negative perturbation ($\widehat{v_{fly}} = -10\% V_{fly}$) for different constant Δs_e values is shown. The loads are implemented as constant 3.2 A and 6 A current sink in Fig. 4a and Fig. 4b, respectively. For both simulations, the output capacitor is high enough to maintain the initial charge and to keep the output voltage at $I_{out} \times R_{out}$. In these simulations the digital HCM control method is applied although PCM control can be implemented similarly. The ACM loop of the HCM has a dummy controller which is not effective for output regulation due to the stable output but necessary to generate the required v_c control signal. As can be seen from the figures, for $\Delta s_e = 0.8$ A/μs, flying capacitor voltage rises as oppose to the negative perturbation until it reaches to V_{in}. Below this value, the flying capacitor voltage decreases, since the capacitor discharges i.e. positive feedback occurs.

By considering the mathematical analysis and the simulation results, one can conclude that the asymmetrical slope compensation can be used to eliminate flying capacitor voltage perturbation with the flying capacitor balancing feedback loop. One important point is that, the

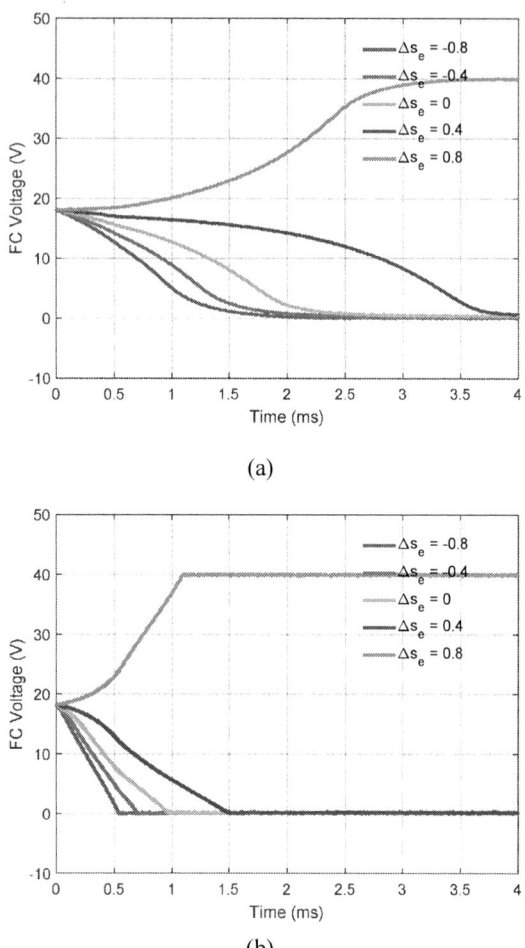

(a)

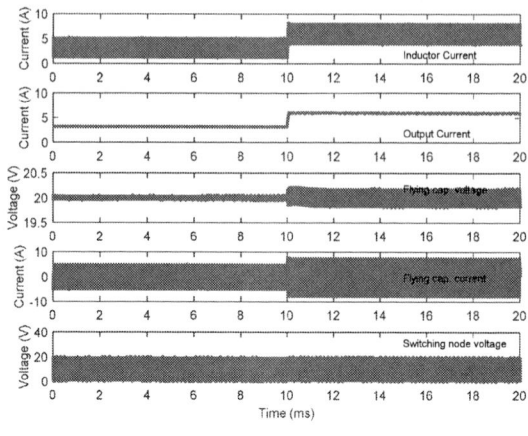

(a)

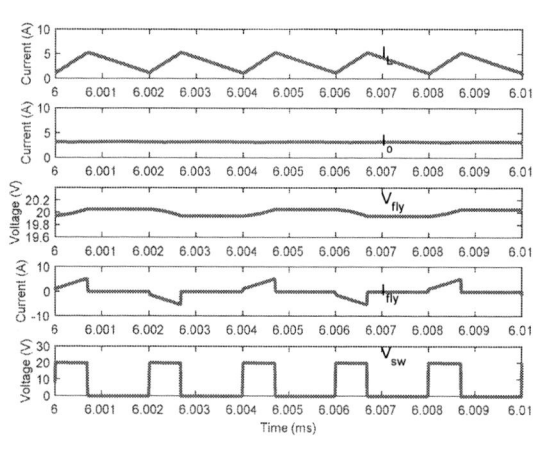

(b)

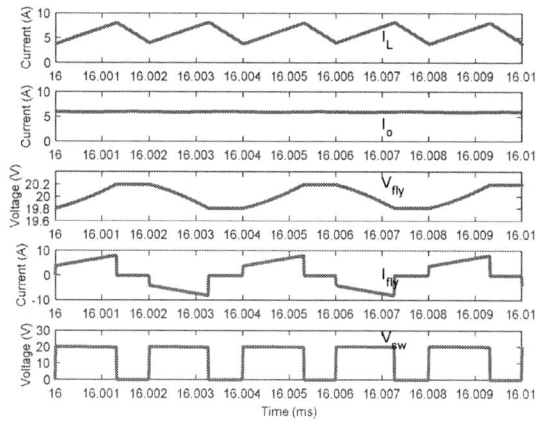

(c)

Fig. 4. Flying capacitor voltage change over time for $-0.8 < \Delta s_e (A/\mu s) < 0.8$ with $\widehat{v_{fly}} = -10\%V_{fly}$ a) $I_{out} = 3.2\,A$ b) $I_{out} = 6\,A$

controller output of the flying capacitor balancing feedback loop should be capable of providing the minimum Δs_e to prevent positive feedback which could result in voltage runaway.

Digital HCM control and the proposed ASC technique is evaluated with a simulation model of the 3LFC buck converter as well. The simulation results are given in Fig. 5 and Fig. 6. For the ACM control loop, a type-II PI compensator is implemented in the simulation model based on the calculation method given [11], [13]. For flying capacitor feedback loop a PI compensator exists. The response of the model and 3LFC buck converter to the control signal changes and to the flying capacitor perturbation are analyzed separately.

As it is seen in Fig. 5, both at 3.2 A and 6 A output the ASC method maintains the flying capacitor voltage balance at $V_{in}/2$. In addition, the digital HCM control acts fast and accurately as response to control command and the inductor current reaches to the steady-state within a couple of switching cycles. In Fig. 6, the simulation results of the control method for flying capacitor perturbation $\widehat{v_{fly}} = 2\,V$ are given for 6 A load operation. At the middle of the

Fig. 5. Simulation results of a digital HCM controlled 3LFC buck converter a) control command changes from 3.2 to 6 A, b) zoom in 3.2 A operation c) zoom in 6 A operation

simulation, the perturbation source is removed and the response is demonstrated. As expected, the flying capacitor perturbation is eliminated and the capacitor voltage has reached to $V_{in}/2$ by the asymmetric slope compensation.

978-1-5386-7688-2/19 $31.00 © 2019 IEEE

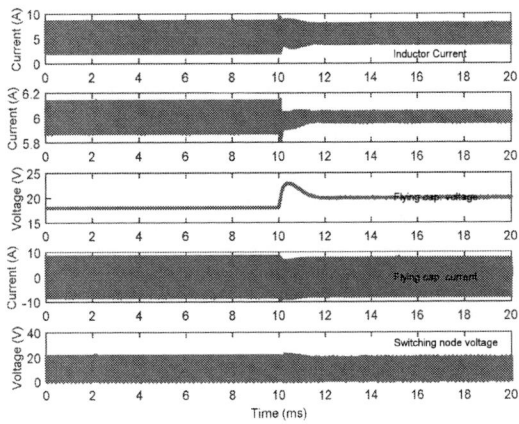

(a)

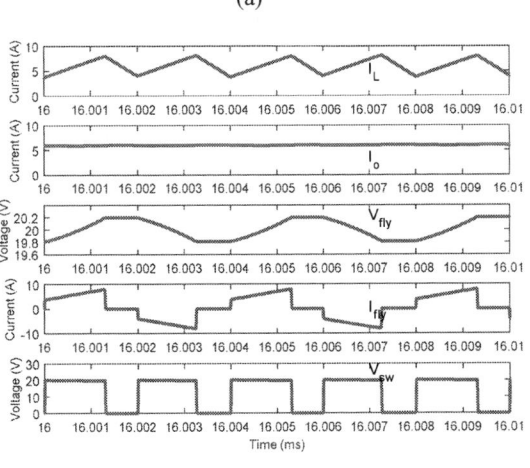

(b)

Fig. 6. Simulation results of a digital HCM controlled 3LFC buck converter a) with flying capacitor perturbation $\widehat{v_{fly}} = -10\% V_{fly}$ and 6 A current command b) Zoom-in to (a) at steady-state operation

In Fig. 7, the picture of the designed prototype for experimental study is given. In this study, an Infineon XMC4200 ARM based mixed-signal microcontroller is placed on the board. For current sensing Analog Devices LT1999-10 current sense amplifier is used for its wide bandwidth and high input common voltage range although Texas Instruments INA169 is an alternative. For flying capacitor voltage sensing Analog Devices LT1990 difference amplifier is employed on the board. It is important to note that the digital control loops run for every two switching cycles in this application to release the MCU for other actions such as communication or data acquisition. Clearly this reduces the controller response time, however the analog peak current loop helps to improve the bandwidth as it compares the inductor current with the control signal in each switching cycle.

In Fig. 8a, it is shown that the rise time of a 3.2 A to 6 A load transition is 470 μs, and the inductor current does not overshoot. In addition, the flying capacitor voltage balance is achieved by the asymmetric slope compensation successfully. In Fig. 8b and Fig. 8c, the steady-state operation of the digital HCM control with asymmetric slope

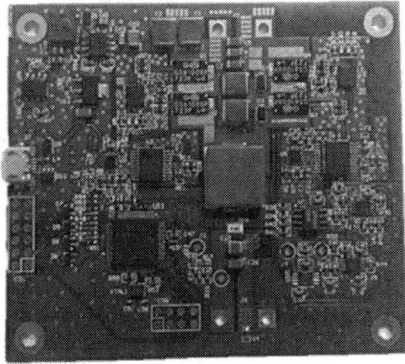

Fig. 7. Prototype of the 3LFC buck converter with on-board microcontroller

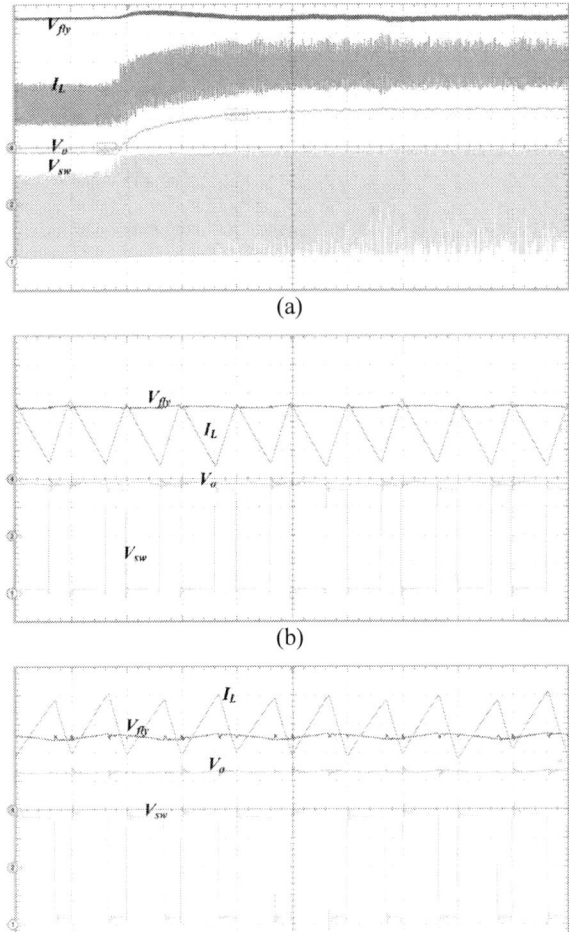

(a)

(b)

(c)

Fig. 8. Experimental results of a digital HCM controlled 3LFC buck converter (yellow: V_{sw} voltage (5 V/div.), blue: V_o voltage (4 V/div.), purple: V_{fly} voltage (3 V/div., offset: -2 div.), green: I_L current (2 A/div.)) a) control command changes from 3.2 A to 6 A (time scale: 2 μs/div.), b) zoom in 3.2 A operation (time scale: 2 μs/div.) c) zoom in 6 A operation (time scale: 250 μs/div.)

compensation for 3.2 A and 6 A load conditions in the 3LFC buck prototype is shown respectively. The flying

978-1-5386-7688-2/19 $31.00 © 2019 IEEE

capacitor voltage is maintained at $V_{in}/2$, therefore switching point voltage V_{sw} swings between $V_{in}/2$ and 0 V as expected.

IV. CONCLUSION

In this paper, the digital Hybrid Current Mode (HCM) control technique which combines the Average Current (ACM) and Peak Current Mode (PCM) control methods is modified and implemented to the three-Level Flying Capacitor (3LFC) buck converter. Similar to the PCM, digital HCM control requires a flying capacitor voltage balancing mechanism. Therefore in this study a method, which is named as Asymmetric Slope Compensation (ASC) is introduced. The ASC method is investigated using the inductor current geometry and confirmed in simulations. In addition, the digital HCM control method with ASC is verified for 3LFC buck converter in simulations as well and experimentally on a 40 V input 13 V / 6 A output prototype using an on-board mixed signal microcontroller. The proposed ASC method is not only useful for digital HCM control but also applicable to the PCM control as well. The proposed control method of the 3LFC buck converter is especially useful for LED and Laser diode drive applications where regulated output current is necessary.

ACKNOWLEDGMENT

This work was supported by Istanbul Technical University (ITU) Research Funding grant no. ITU-BAP-40236.

REFERENCES

[1] T. A. Meynard and H. Foch, "Multi-level conversion: high voltage choppers and voltage-source inverters," in Power *Electronics Specialists Conference, 1992*, PESC'92 Record., 23rd Annual IEEE, June 1992, pp. 397-403.

[2] A. Stillwell, E. Candan and R. C. N. Pilawa-Podgurski, "Constant effective duty cycle control for flying capacitor balancing in flying capacitor multi-level converters," *2018 IEEE 19th Workshop on Control and Modeling for Power Electronics (COMPEL)*, Padua, 2018, pp. 1-8.

[3] J. S. Rentmeister and J. T. Stauth, "A 48V:2V flying capacitor multilevel converter using current-limit control for flying capacitor balance," *Applied Power Electronics Conference and Exposition (APEC)*, pp. 367-372, 2017.

[4] J. S. Rentmeister, C. Schaef, B. X. Foo, J. T. Stauth, "A flying capacitor multilevel converter with sampled valley-current detection for multi-mode operation and capacitor voltage balancing," 2016 IEEE Energy Conversion Congress and Exposition (ECCE), pp. 1-8, Sept. 2016.

[5] N. Vukadinovic, A. Prodic, B. A. Miwa, C. B. Arnold, M. W. Baker, "Extended Wide-Load Range Model for Multi-Level DC-DC Converters and a Practical Dual-Mode Digital Controller" in *2016 IEEE Applied Power Electronics Conference and Exposition (APEC)*, March 2016, pp. 1597-1602

[6] L. Lu, S.M. Ahsanuzzaman, A. Prodic, G. Calabrese, G. Frattini and M. Granato, "Peak offsetting based cpm controller for multi-level flying capacitor converters," in *2018 IEEE Applied Power Electronics Conference and Exposition (APEC)*, March 2018, pp. 3102-3107.

[7] S. da Silva Carvalho, M. Halamíček, N. Vukadinović and A. Prodić, "Digital PWM for Multi-Level Flying Capacitor Converters with Improved Output Resolution and Flying Capacitor Voltage Controller Stability," *2018 IEEE 19th Workshop on Control and Modeling for Power Electronics (COMPEL)*, Padua, 2018.

[8] L. Lu, Y. Zhang, S.M. Ahsanuzzaman, A. Prodic, G. Calabrese, G. Frattini and M. Granato, "Digital Average Current Programmed Mode Control for Multi-level Flying Capacitor Converters," *2018 IEEE 19th Workshop on Control and Modeling for Power Electronics (COMPEL)*, Padua, 2018.

[9] E. Abdelhamid, G. Bonanno, L. Corradini, P. Mattavelli and M. Agostinelli, "Stability Properties of the 3-Level Flying Capacitor Buck Converter Under Peak or Valley Current-Programmed-Control," *IEEE Trans. Power Electron.*, 2018 (in press).

[10] X. Ruan, B. Li, Q. Chen, S. Tan and C. K. Tse, "Fundamental Considerations of Three-Level DC–DC Converters: Topologies, Analyses, and Control," in *IEEE Transactions on Circuits and Systems I: Regular Papers*, vol. 55, no. 11, pp. 3733-3743, Dec. 2008.

[11] A. Ugur and M. Yilmaz, "Digital hybrid current mode control for DC–DC converters," in *IET Power Electron.*, vol. 12, no. 4, pp. 891-898, April 2019.

[12] Y. Yan, F.C. Lee, P. Mattavelli, P. H. Liu, " I2 average current mode control for switching converters," *IEEE Trans. Power Electron.*, vol. 29, no. 4, pp. 2027-2036, 2014.

[13] M. Hallworth and S.A. Shirsavar, "Microcontroller-based peak current mode control using digital slope compensation," *IEEE Trans. Power Electron.*, vol. 27, no. 7, pp. 3340-3351, 2012.

Dual-Mode Operation of 3-Level 4-Leg AT-NPC Inverter for Microgrids

Emre Avci
Dept. of Electrical and Electronics Engineering
Duzce University
Duzce, Turkey
emreavci@duzce.edu.tr

Mehmet Ucar
Dept. of Electrical and Electronics Engineering
Duzce University
Duzce, Turkey
mehmetucar@duzce.edu.tr

Abstract—This study presents dual-mode operation and seamless transfer of 3-level 4-leg Advanced T-type Neutral Point Clamped (AT-NPC) inverter for low voltage microgrids. The proposed control technique is designed with Proportional Resonant (PR) based current controlled in grid-connected operation and with Proportional Multi Resonant (PMR) based voltage controlled in islanded operation. In order to ensure smooth transition between these operating modes, combined seamless transfer method is proposed. The proposed 3-phase 4-leg inverter system can supply both 1-phase and unbalanced 3-phase load groups in islanded mode. A laboratory prototype is installed for the experimental testing of the 3-phase 4-leg AT-NPC 3-level inverter system. The proposed system is simulated via PSIM software and the model-based embedded codes are generated with PSIM/SimCoder software. Real-time control algorithm of the inverter system is carried out with TMS320F28335 digital signal controller board. Finally, the operation performance of the inverter system is investigated both in the islanded/grid-connected mode and the seamless transfer process between these two operating modes.

Keywords—multi-level inverter, AT-NPC inverter, RB-IGBT, islanding, grid connection, seamless transfer, microgrids.

I. INTRODUCTION

Renewable distributed energy generation systems must ensure a sustainable and reliable operation in line with the electricity grid [1]. Distributed generators (DGs) are usually connected to the grid via an inverter. Therefore, controlling the inverters in terms of quality and continuity of the power generation is one of the most important issues for DG system. In many studies, inverters are classified according to their input source type, employed control technique, grid interactivity, output voltage level and so forth. Among the output voltage level categorized inverters, Multilevel Inverters (MLIs) can generate a sinusoidal waveform that has lower harmonic distortion level according to 2-level inverters [2]. Neutral Point Clamped (NPC), cascade H-bridge and flying capacitor inverter are well-known MLI topologies in the literature. Especially in low voltage applications, 3-level T-type NPC (T-NPC) inverter topology is one of the most favored topology among the MLIs [3]-[4]. Additionally, T-NPC inverter topology offers low Total Harmonic Distortion (THD) and has simple operational principles, which are advantages of conventional NPC topology and 2-level topology, respectively [3]. Different from the T-NPC inverter, in the Advanced T-NPC (AT-NPC) inverter, Reverse Blocking IGBTs (RB-IGBTs) are substituted for the conventional IGBTs at the midpoint of the structure. Whit this way, the T-NPC inverter structure has been transformed into a more efficient AT-NPC structure as the switching and conduction losses are reduced [5]-[6].

Conventionally, DG inverters are operated as a current source and as a voltage source in grid-connected mode and islanded mode, respectively. For these two operating modes, two different control methods have been developed for output voltage control in islanded mode and current control in grid-connected mode [7]-[8]. But, since two different control methods are used for two operating modes, it is necessary to change the control technique from grid-connected mode to islanded mode or vice versa when the grid status is changed. In the case of a grid failure, grid-connected inverters must be able to detect the grid interruption and to disconnect the distributed generation system quickly from the grid [9]-[10]. Therefore, the load voltage quality can be distorted during this transition process as the control system is changed when the islanding occurs. In the islanded mode operation, inverters must provide sinusoidal output voltage to 3-phase/1-phase, linear/non-linear and balanced/unbalanced load groups. 4-leg power converters are an innovative solution for supplying voltage to the unbalanced 3-phase load [11]-[14].

In the literature, there are studies regarding to seamless transition methods from grid-connected mode to islanded mode operation or vice versa for DG inverters [15]-[17]. In order to effectively prevent voltage or current increases during this transition period, islanded and grid-connected mode control techniques should be combined. And the only single control method should be developed. For this reason, various seamless transfer control strategies have been developed in which the inverter system is controlled with a single control method for grid-connected and islanded mode operations [18]-[21]. In these control methods, when the operating mode is changed, the inverter control technique does not need to be changed and so the quality of the load voltage and the grid currents are improved by the transition period.

In this study, a combined seamless transfer control method is proposed to ensure the smooth transition between the grid-connected and the islanded modes of operation for the 3-phase 4-leg AT-NPC inverter. In the case of non-linear local loads, the proposed control strategy is capable of providing sinusoidal grid current in the grid-connected operation and the sinusoidal load voltage in islanded operation of the inverter. Fig. 1 shows the general block diagram of the proposed inverter system in island mode and grid-connected mode.

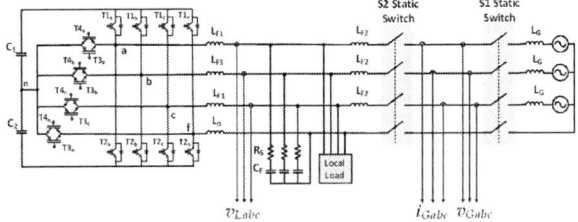

Fig. 1. General block diagram of the inverter system in dual mode operation.

978-1-5386-7688-2/19 $31.00 © 2019 IEEE

II. COMBINED SEAMLESS TRANSFER CONTROL TECHNIQUE

The proposed combined seamless transfer control method is developed in the *abc* plane consisting of an outer grid current control loop with Proportional Resonant (PR) controller and an inner load voltage control loop with Proportional Multi Resonant (PMR) controller. The PR controller is used to follow the reference inverter current with a low THD value. The transfer function of a damped PR controller is given in (1). As can be shown in the equation, the PR controller generates resonance for only the fundamental frequency (w_1). The proportional coefficient (K_p) generally adjusts the dynamics of the controller and the stability of the system, while the integrator coefficient (K_i) affects the resonance peak amplitude and the bandwidth around the corresponding frequency [22]. When the inverter feeds non-linear loads (typically have 3rd, 5th, 7th and 9th harmonics), the output voltage THD is affected by the load current. With the PMR controller, voltage harmonics are suppressed at their frequencies (w_h) [23]. The transfer function of the PMR controller is equal to (2). The dynamic performance of the fundamental frequency controller in the PMR controller is almost affected by the proportional part (K_p). Therefore, a good design is important for transient performance. In this study, passive damping method is used for the instability caused by the inverter output filter.

$$G_{PR}(s) = K_p + \frac{K_i w_{c1} s}{s^2 + 2w_c s + w_1^2} \tag{1}$$

$$G_{PMR}(s) = K_p + \sum_{h=1,3,5,7,9} \frac{K_{ih} w_{ch} s}{s^2 + 2w_{ch} s + w_h^2} \tag{2}$$

After obtaining the reference signals at the control algorithm output, appropriate switching signals must be established for these reference signals. In this study, Carrier Based PWM (CBPWM) method [24] is used to produce the switching signals. The proposed combined seamless transfer control technique is shown in Fig. 2. The flow diagram of the combined seamless transfer control technique for the proposed system is shown in Fig. 3.

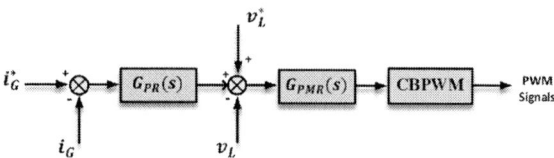

Fig. 2. Proposed combined seamless transfer control technique.

The proposed 3-phase 4-leg AT-NPC inverter system has three main operation modes: islanding, grid synchronization and grid-connected modes. As shown in Fig. 3, the state of the grid voltage is firstly investigated in order to decide a suitable mode operation for the inverter. The effective value of the grid voltage (V_G) calculated by (3) is monitored to decide mode operation.

$$V_G = \sqrt{v_{G\alpha}^2 + v_{G\beta}^2} \tag{3}$$

Fig. 4 shows the block diagram that detetecs the grid voltage and frequency parameters. In addition to the effective value of the grid voltage, the grid frequency is also monitored using a Second-Order Generalized Integrator (SOGI) based Phase-Locked Loop PLL [25]. According to the IEEE Std.

1547-2003, if the amplitude and frequency of the grid voltage is in range stated as in (4), the inverter can operate in grid-connected mode and transfers power to the grid. Otherwise, the inverter is disconnected from the grid and is operated in islanded mode.

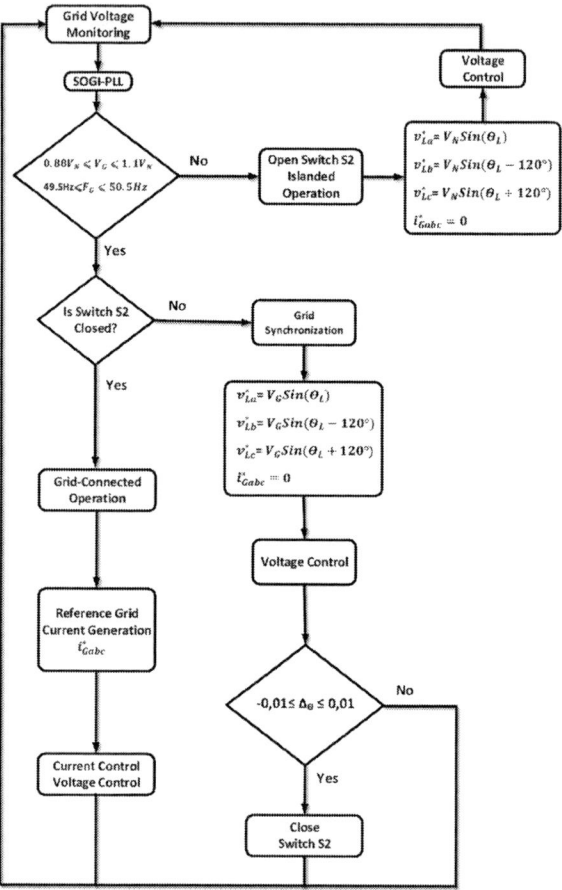

Fig. 3. Combined seamless transfer flow diagram.

$$0.88V_N \leq V_G \leq 1.1V_N$$
$$49.5\ \text{Hz} \leq f_G \leq 50.5\ \text{Hz} \tag{4}$$

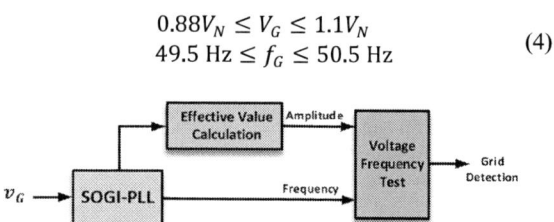

Fig. 4. Detection of grid voltage and frequency parameters.

In islanded mode operation, the inverter is controlled to ensure sinusoidal output voltage with nominal load voltage and frequency. Therefore, PMR based voltage control technique is used to obtain high voltage quality in the inverter output. In this mode, reference load voltages (5) are generated in the control loop. In (5), θ_L is the phase angle of the load voltage (V_L) at the 50 Hz output frequency.

$$v_{La}^* = V_{Nm} \sin(\theta_L)$$
$$v_{Lb}^* = V_{Nm} \sin\left(\theta_L - \frac{2\pi}{3}\right) \tag{5}$$
$$v_{Lc}^* = V_{Nm} \sin\left(\theta_L + \frac{2\pi}{3}\right)$$

Fig. 5 shows the phase angle generation block diagram for islanded mode operation. In this figure, θ_C is the phase angle of the grid voltage before the PLL detects a grid failure. The phase angle generation of the load voltages in the islanded mode is also obtained after operating in grid-connected mode. This prevents the load current interruption during the transition from grid-connected mode to islanded mode.

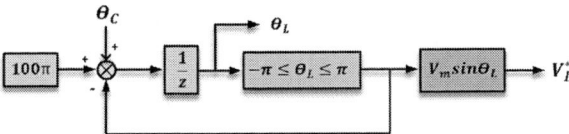

Fig. 5. Phase angle generation in islanded mode.

In grid synchronization mode, the SOGI-PLL generates the grid frequency and phase as shown in Fig. 6 when the grid voltages appear by turning-on the static switch S1. Even if the amplitude and frequency of the voltages for the incoming grid is equal to that of the voltage range in (4), the inverter is not transferred from islanded mode to grid-connected mode, as there is a difference between the grid voltage and the load voltage in terms of the amplitude and phase. Therefore, when transferring the mode to grid-connected operation, the inverter is operated by synchronization mode. In this process the voltage amplitude and frequency of the load must be same of the grid voltage. For this reason, the reference load voltages are generated with (6). In this equation V_G is the amplitude value of the grid voltage determined by (3). During the synchronization process, the load voltages are controlled by means of the PMR voltage control loop. Fig. 6 shows the synchronization block diagram.

$$
\begin{aligned}
v_{La}^* &= V_{Gm} \sin(\theta_L) \\
v_{Lb}^* &= V_{Gm} \sin\left(\theta_L - \frac{2\pi}{3}\right) \\
v_{Lc}^* &= V_{Gm} \sin\left(\theta_L + \frac{2\pi}{3}\right)
\end{aligned}
\tag{6}
$$

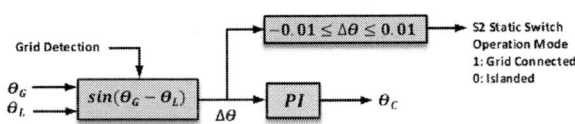

Fig. 6. Synchronization block diagram.

In Fig. 6, Δ_θ is the phase angle error between the load and the grid voltage. The PI controller is used to compensate the fault between the grid and load phase angle. When the phase angle error is between the range of +/- 0.01 radians, it is assumed that the grid and load voltages are synchronized. After that the S2 static switch is turned-on with the generated synchronization signal. Therefore, the inverter is automatically transferred from the islanded mode to the grid-connected mode without interruption. Otherwise, the operation mode of the inverter is not changed. This leads the controller to execute continuously the synchronization process.

In islanded mode operation, the phase angle of reference voltages is still being produced from the grid-connected operation mode so that the load current is not interrupted during the transition to the incoming grid-connected mode. In grid-connected mode operation, the main purpose of the inverter is to transfer the reference current to the grid with maximum current quality. In this study, the grid current is

regulated by PR current control. In the control technique, the reference grid currents are generated in accordance with the phase of the grid voltages by means of the SOGI-PLL structure. Fig. 7 shows the block diagram of the reference grid currents generation.

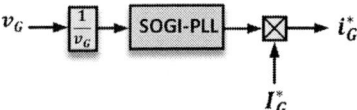

Fig. 7. Reference grid currents generation.

During the transition from grid-connected mode to islanded mode, these are assumed that the inverter is operated in current controlled mode and the local loads are connected in parallel to the grid. It means that the local loads are always parallel connected to the inverter output terminals. As a being grid fault, the state of the grid disconnection is simulated by turning-off the static switch S1. When the grid disconnection occurs with the turning-off the static switch S1, there will be an increase in the load voltage since the inverter is still in current controlled mode. When this increasing exceeds a certain threshold voltage level or the frequency of the grid is out of range of 49.5-50.5 Hz, the system automatically switches to islanded mode. At the same time, the static switch S2 is turned-on from to turned-off state. Once the phase angles of the voltages in the islanded mode and in the grid connected mode overlaps, the transition between the operating modes is ensured without any current interruption or without any the sudden current rise.

III. SIMULATION RESULTS

The proposed system is simulated with PSIM software. Table 1 shows the system parameters for the islanded and grid-connected mode operation of the 3-level 4-leg AT-NPC inverter.

TABLE I.   SYSTEM PARAMETERS.

Parameters	Values
Nominal Voltage and Frequency	110 V, 50 Hz
DC Bus Voltage	350 V
Filter Inductances (L_{F1}, L_{F2})	3 mH, 2 mH
Filter Capacitor	27 μF
Damping Resistor	10 Ω
Switching Frequency	20 kHz
Unbalanced RL Load (phase-a, b) and (phase-c)	50 Ω, 25 mH and 25 Ω, 25 mH

Fig. 8 shows the simulation results of the islanding and grid-connected mode operations. In this simulation study, the grid connection is achieved at the t=0.1s and it disconnects again at the t=0.2 s. According to the simulation results given in Fig. 8(a) and Fig. 8(b), the inverter is operated in islanded mode before the 0.1 s. The proposed inverter provides balanced and sinusoidal voltage to the load despite unbalanced load currents. At the t=0.1 s, the grid connection is suddenly occurring. And the turning-on signal is generated for static switch S2 when the θ_G and θ_L are set in the same phase with the synchronization algorithm. In this case, the inverter performs a seamless transfer from the islanded mode to the grid-connected mode, where the load is not affected by this transition. Also, as can be seen in the figure, the grid currents increase slowly.

In grid-connected mode operation, the given reference currents are successfully transferred to the grid. The turning-off signal for the static switch S2 is produced quickly when the grid is disconnected at the t= 0.2 s. A seamless transition to the islanding mode operation of the inverter is achieved with no effect on the local load. Then the inverter continues to be operated as in islanded mode.

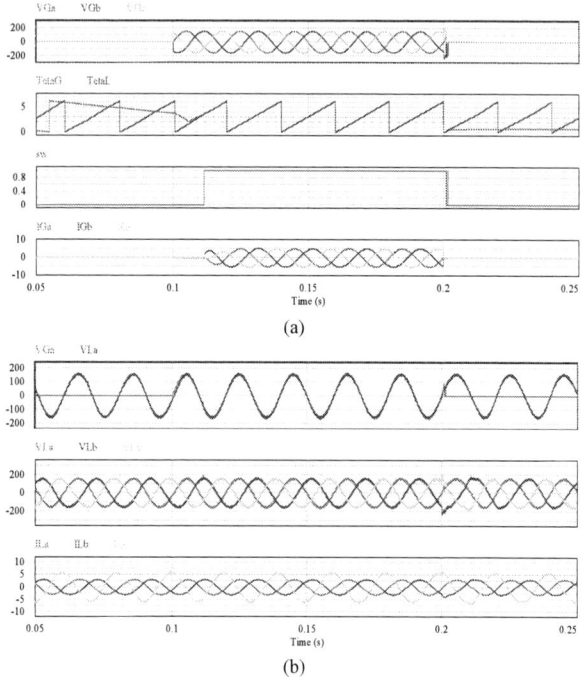

(a)

(b)

Fig. 8. Simulation results in islanded and grid-connected operation with (a) grid voltages V_{Gabc}, synchronization signal sw, grid phase angle θ_G and load phase angle θ_L, grid currents I_{Gabc}, and (b) a-phase source voltage V_{Ga} and load voltage V_{La}, load voltages V_{Labc}, load currents I_{Labc}.

IV. EXPERIMENTAL RESULTS

In this study, a TMS320F28335 digital signal controller based real-time control algorithm is implemented for the 3-level 4-leg AT-NPC inverter system. The model based embedded codes of the controller and modulation techniques are produced by PSIM/SimCoder. In the experimental studies, 110 V phase-neutral effective voltage value is used. The 4MBI300VG-120R-50 module is employed on the power stage of the 3-phase 4-leg AT-NPC inverter system. For the 3-level inverter, 20 kHz switching signals are generated with CBPWM technique. The experimental results are measured with Tektronix MDO3024 oscilloscope and analyzed with Tektronix MDO3PWR power analysis application module. Photograph of the experimental setup is shown in Fig. 9.

It is important that the proposed system, which is operated in islanding mode, can provide the automatic grid connection to the 4-leg AT-NPC in the event of the grid connection. And similarly, the proposed system, which is operated in grid connected mode, can provide automatic grid disconnection to the 4-leg AT-NPC inverter in the event of the grid disconnection. Therefore, by disabling of the control algorithm, only the PLL and the synchronization structure were used to test the performance of these structures for the islanding mode detection and the grid synchronization. To realize this, the 3-phase grid voltages were connected and disconnected to the system so that test the grid connection and disconnection signals which is produced by PLL structure for the static switch S2.

Fig. 9. Photograph of the experimental setup.

After the 3-phase grid is connected with the static switch S1, the synchronization is achieved with PLL and synchronization algorithm that generates switching signal for the static switch S2. This switching signal generation occurs in 1.5-2 period (this may vary depending on the grid initial angle). On the other hand, the disconnection prosses is completed after the grid is disconnected from the system with turning-off the static switch S1, the synchronization algorithm generates the turning-off signal for the static switch S2. In this way, the inverter disconnects quickly from the grid.

The performance of the seamless control algorithm in islanded operation is investigated under unbalanced RL-type local load condition. Fig. 10 shows the experimental results of islanded operation of the system. In the figure, 3-phase load voltage waveforms and S2 switch signal are given in islanded mode operation. Under unbalanced RL type local load, the inverter supplies balanced 110 V, 50 Hz nominal voltage with 2.87% THD value to the load unit.

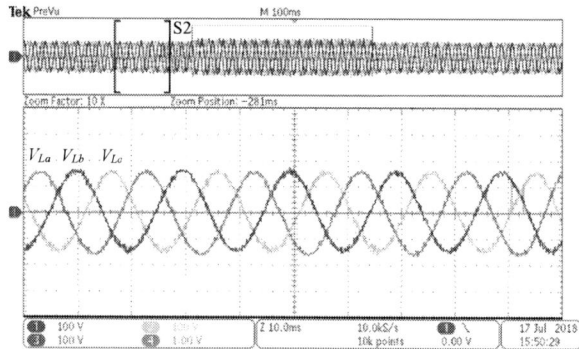

Fig. 10. Test results in islanded operation: 3-phase load voltage waveform V_{Labc}.

When the inverter is running in islanded mode, it is necessary to switch to grid-connected mode via PLL and synchronization algorithms when the grid is ON. In order for this transition to be smooth and uninterrupted, synchronization and PLL algorithms must be switched to inverter current control by generating S2 static switch signal. Fig. 11 shows the transition results from islanded mode to the

978-1-5386-7688-2/19 $31.00 © 2019 IEEE

grid-connected mode. In order to examine the transition performance of the system, as in Fig. 11, the grid voltage is existed with the S1 static switching ON signal. Then switching ON signal for transfer switch S2 is generated by synchronizing with the operation of PLL and synchronization algorithms. As shown in Fig. 11, after the grid is recovered, synchronization is performed quickly and the inverter is switched to grid-connected mode. A smooth transition is achieved during the transfer without interrupting the voltage at the load terminals.

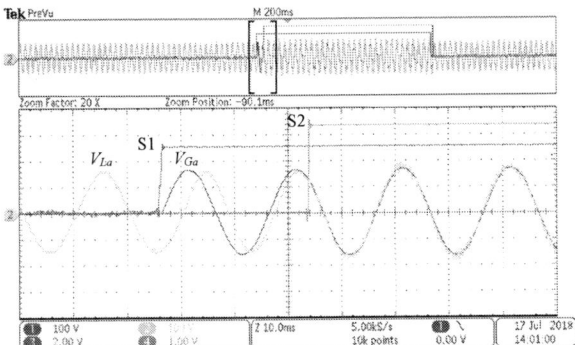

Fig. 11. Islanded mode to grid-connected mode transfer test results: grid voltage V_{Ga}, load voltage V_{La}, S1 and S2 switch signal.

After a seamless transition from the islanded mode to the grid-connected mode, the inverter starts current controlled operation and should follow the reference currents synchronized with the grid voltage. Fig. 12 shows the test results in grid-connected mode. In order to examine the performance of the presented algorithm in grid-connected mode, voltage waveforms of the grid and the load terminals are given in Fig. 12. As seen from the figure, the grid voltage and the load voltage are approximately the same although there is a filter inductance between the load and the grid.

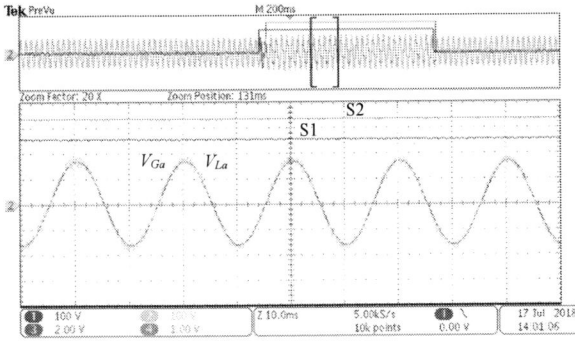

Fig. 12. Test results in grid-connected operation: grid voltage V_{Ga}, load voltage V_{La}, S1 and S2 switch signal.

Fig. 13 shows the transition process results of the system from grid-connected mode to islanded mode. The grid voltage is interrupted by the static switch S1. Then the S2 switch signal is interrupted with the synchronization and SOGI-PLL algorithm. This transfers the inverter to islanded mode operation. The figure shows that a smooth transition is provided during the transfer without any interruption in the voltage at the local load terminals. As a result, the local load is supplied during the transition from grid-connected mode to island mode. Moreover, there is not any sudden increase in the load and grid currents that will damage the inverter during both transitions.

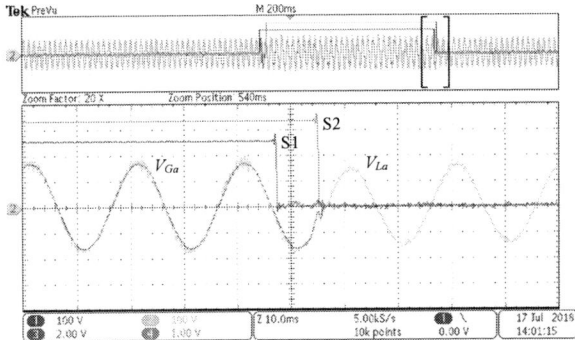

Fig. 13. Grid-connected mode to islanded mode transition test results: grid voltage V_{Ga}, load voltage V_{La}, S1 and S2 switch signal.

V. CONCLUSION

In DG for the microgrids, inverters are operated as a current source in grid-connected mode and as a voltage source in islanded mode. During the mode transition period, it is necessary to make a seamless transfer of the inverter between the grid-connected and islanded modes. In this study, dual-mode operation and seamless transfer of the 3-level 4-leg AT-NPC inverter system is achieved. In order to ensure the smooth transition between the operating modes, a combined seamless transfer control technique is proposed. The method has enabled the inverter to supply the local loads with nominal voltage and low THD value in islanded operation. In the transferring process from islanded mode to grid-connected mode, the grid synchronization is completed quickly which ensures the smooth and uninterrupted transition without distorting the voltage waveform at the local load terminals. In grid-connected mode operation, the inverter can feed both unbalanced local load unit and the grid with the current that is in phase with grid voltage. Finally, the dual-mode operation of the inverter is validated by the simulation and experimental results.

ACKNOWLEDGMENT

This study was supported by The Scientific and Technological Research Council of Turkey, TUBITAK (Grant No. 215E357).

REFERENCES

[1] T. Adefarati and R. C. Bansal, "Integration of renewable distributed generators into the distribution system: a review," in IET Renewable Power Generation, vol. 10, no. 7, pp. 873-884, 2016.

[2] J. Rodriguez, Jih-Sheng Lai and Fang Zheng Peng, "Multilevel inverters: a survey of topologies, controls, and applications," in IEEE Transactions on Industrial Electronics, vol. 49, no. 4, pp. 724-738, Aug. 2002.

[3] M. Schweizer and J. W. Kolar, "Design and Implementation of a Highly Efficient Three-Level T-Type Converter for Low-Voltage Applications," in IEEE Transactions on Power Electronics, vol. 28, no. 2, pp. 899-907, Feb. 2013.

[4] I. Staudt, "AN-11001: 3L NPC and TNPC topology," Semikron, Nuremberg, Germany, 2012.

[5] H. Uemura, F. Krismer and J. W. Kolar, "Comparative evaluation of T-type topologies comprising standard and reverse-blocking IGBTs," 2013 IEEE Energy Conversion Congress and Exposition, Denver, CO, 2013, pp. 1288-1295.

[6] L. Zhang, K. Sun, L. Huang and S. Igarashi, "Comparison of RB-IGBT and normal IGBT in T-type three-level inverter," 2013 15th European Conference on Power Electronics and Applications (EPE), Lille, 2013, pp. 1-7.

[7] R. Teodorescu and F. Blaabjerg, "Flexible control of small wind turbines with grid failure detection operating in stand-alone and grid-

978-1-5386-7688-2/19 $31.00 © 2019 IEEE

connected mode," in IEEE Transactions on Power Electronics, vol. 19, no. 5, pp. 1323-1332, Sept. 2004.

[8] T. Tran, T. Chun, H. Lee, H. Kim and E. Nho, "PLL-Based Seamless Transfer Control Between Grid-Connected and Islanding Modes in Grid-Connected Inverters," in IEEE Transactions on Power Electronics, vol. 29, no. 10, pp. 5218-5228, Oct. 2014.

[9] A. M. Massoud, K. H. Ahmed, S. J. Finney and B. W. Williams, "Harmonic distortion-based island detection technique for inverter-based distributed generation," in IET Renewable Power Generation, vol. 3, no. 4, pp. 493-507, December 2009.

[10] R. Teodorescu, M. Liserre, P. Rodriguez, "Grid Converters for Photovoltaic and Wind Power Systems", John Wiley & Sons, Ltd., 2011.

[11] M. Sedlak, S. Stynski, M. P. Kazmierkowski and M. Malinowski, "Operation of four-leg three-level flying capacitor grid-connected converter for RES," IECON 2013 - 39th Annual Conference of the IEEE Industrial Electronics Society, Vienna, 2013, pp. 1100-1105.

[12] S. Chee, S. Sul, Y. H. Roh and J. Lee, "Loss comparison of the 3 level topologies for four-leg voltage converters," 2014 IEEE International Conference on Industrial Technology (ICIT), Busan, 2014, pp. 324-329.

[13] D. Vyawahare and M. Chandorkar, "Distributed generation system with hybrid inverter interfaces for unbalanced loads," 2015 IEEE 6th International Symposium on Power Electronics for Distributed Generation Systems (PEDG), Aachen, 2015, pp. 1-7.

[14] M. R. Miveh, M. F. Rahmat, A. A. Ghadimi, M. W. Mustafa, "Control techniques for three-phase four-leg voltage source inverters in autonomous microgrids: A review," Renew. Sustain. Energy Rev., vol. 54, pp. 1592-1610, Feb. 2016.

[15] M. N. Arafat, S. Palle, Y. Sozer and I. Husain, "Transition Control Strategy Between Standalone and Grid-Connected Operations of Voltage-Source Inverters," in IEEE Transactions on Industry Applications, vol. 48, no. 5, pp. 1516-1525, Sept.-Oct. 2012.

[16] M. Fatu, F. Blaabjerg and I. Boldea, "Grid to Standalone Transition Motion-Sensorless Dual-Inverter Control of PMSG With Asymmetrical Grid Voltage Sags and Harmonics Filtering," in IEEE Transactions on Power Electronics, vol. 29, no. 7, pp. 3463-3472, July 2014.

[17] D. S. Ochs, B. Mirafzal and P. Sotoodeh, "A Method of Seamless Transitions Between Grid-Tied and Stand-Alone Modes of Operation for Utility-Interactive Three-Phase Inverters," in IEEE Transactions on Industry Applications, vol. 50, no. 3, pp. 1934-1941, May-June 2014.

[18] Z. Yao, L. Xiao and Y. Yan, "Seamless Transfer of Single-Phase Grid-Interactive Inverters Between Grid-Connected and Stand-Alone Modes," in IEEE Transactions on Power Electronics, vol. 25, no. 6, pp. 1597-1603, June 2010.

[19] R. Wai, C. Lin, Y. Huang and Y. Chang, "Design of High-Performance Stand-Alone and Grid-Connected Inverter for Distributed Generation Applications," in IEEE Transactions on Industrial Electronics, vol. 60, no. 4, pp. 1542-1555, April 2013.

[20] S. Yoon, H. Oh and S. Choi, "Controller design and implementation of indirect current control based utility-interactive inverter system," 2011 IEEE Energy Conversion Congress and Exposition, Phoenix, AZ, 2011, pp. 955-960.

[21] Z. Liu, J. Liu and Y. Zhao, "A Unified Control Strategy for Three-Phase Inverter in Distributed Generation," in IEEE Transactions on Power Electronics, vol. 29, no. 3, pp. 1176-1191, March 2014.

[22] K. Lim and J. Choi, "PR based indirect current control for seamless transfer of grid-connected inverter," 2016 IEEE 8th International Power Electronics and Motion Control Conference (IPEMC-ECCE Asia), Hefei, 2016, pp. 3749-3755.

[23] K. Lim, J. Choi, J. Jang, J. Lee and J. Kim, "P+ multiple resonant control for output voltage regulation of microgrid with unbalanced and nonlinear loads," 2014 International Power Electronics Conference (IPEC-Hiroshima 2014 - ECCE ASIA), Hiroshima, 2014, pp. 2656-2662.

[24] J. Kim, S. Sul and P. N. Enjeti, "A Carrier-Based PWM Method With Optimal Switching Sequence for a Multilevel Four-Leg Voltage-Source Inverter," in IEEE Transactions on Industry Applications, vol. 44, no. 4, pp. 1239-1248, July-aug. 2008.

[25] S. Golestan, J. M. Guerrero and J. C. Vasquez, "Three-Phase PLLs: A Review of Recent Advances," in IEEE Transactions on Power Electronics, vol. 32, no. 3, pp. 1894-1907, March 2017.

Improving the Modular Layer Method to Represent the Capacitive Effects of Overlapping Layers in Planar Transformers

İsmail Onur LORAZ
Aselsan AS
Ankara, TURKEY
ioloraz@aselsan.com.tr

M. Timur AYDEMİR
Gazi University, Engineering Faculty, Electrical and Electronics Engineering
Ankara, TURKEY
aydemirmt@gazi.edu.tr

Abstract— The demand for high power density and low-cost power electronic circuits increases day by day. Recently, planar transformers that can help meet this demand have been drawing more attention. Good thermal characteristics, compact profiles and high efficiency are the advantages of planar transformers. Despite these advantages, high common mode noise values are observed due to high capacitive effects between overlapped layers and turns of planar transformers. The Modular Layer Method (MLM) is used to model planar transformers by using ideal electrical circuit elements. However, it is not sufficient to represent the capacitive effects due to the overlapping layers. In this paper, two different approaches have been proposed to improve the MLM to include these capacitive effects. The results obtained from the experimental set-up show that the proposed models work very accurately. Also, a new approach to planar transformer structure is proposed that leads to very low common mode noise.

Keywords—Planar transformer, transformer modeling, simulation, M2Spice, common mode noise, modular layer method, capacitive effect, interwinding capacitance

I. INTRODUCTION

Nowadays, planar transformers are preferred to wire wound transformers due to their low profile, high power density and good thermal conduction capability. Repeatability and suitability for mass production are other advantages of planar transformers from industrial perspective [1-2]. Nevertheless, despite relatively short production duration and less workforce for mass production, design process of planar transformers is not short or straight forward compared with wire wound transformers. Because modification over the manufactured planar transformers is not possible, designers should design several prototypes with different options to obtain the best solution before finalizing the product and produce each prototype as identical as possible. This requires accurate modeling and simulation of planar transformers to reduce the design duration and cost.

Different modelling and simulation techniques have been proposed in the literature. Finite element (FE) models of planar transformers were discussed in detail and planar transformer characteristics were predicted from numerical data exported from other planar transformers produced or modelled in FEM simulations [3-8]. In addition to that, the Modular Layer Method (MLM) was also developed to model electromagnetic characteristics of planar transformer by ideal circuit elements such as resistor, ideal inductor and ideal transformer. The MLM is compatible with several circuit simulators such as Spice. These simulators simulate the circuit created by the MLM and all other components used around the planar transformer [9].

Inter-winding and interlayer capacitances of PCBs are critical issues for planar transformers. Routes, i.e. windings of planar transformers, are placed on layers of PCBs as close as possible in horizontal and vertical directions to get low profile by remarking the isolation requirement. However, this leads to an increase in capacitive effects between windings. For the whole frequency range, high capacitive effect of PCB layers and windings leads to high common mode noise. Different studies have been conducted to reduce the common mode noise problem of planar transformers [10-13]. Different interlayer capacitive effect models to derive a method to calculate common mode noise were also proposed [10,12-13].

In this paper, important techniques to suppress common mode noise in planar transformers are summarized. In addition, a new condition for low common mode noise based on the solution suggested in [10] is derived. The MLM has been modified to include the capacitive effects by using the models presented in [10] and [13], and the results are discussed. It is assumed that there is only one turn in each layer of PCB since the models proposed in [10] and [13] are valid only for this case. Therefore, inter-winding capacitance and interlayer capacitance have the same meaning for this study.

The remaining of this paper has been structured as follows: The models proposed in [10] and [13] are briefly explained in Section II. Section III explains how the proposed methods can be combined with the MLM and discusses the common mode noise problem. Simulation and experimental results are given in Section IV followed by the conclusion.

II. TWO-CAPACITANCE AND SIX-CAPACITANCE INTERLAYER CAPACITANCE MODELS FOR PLANAR TRANSFORMERS

In this paper, two different inter-winding capacitance modeling methods are chosen for discussion since they are applicable on the MLM. One of them is the "Two-capacitance method" derived in [13]. The other one is the "Six-capacitance method derived in [10] which is based on the method developed for wire wound transformers in [14]. These methods are briefly presented in this section.

A. Two-Capacitance Method

Planar transformer interlayer capacitances were modeled by using only two different capacitances in [12]. This method requires that some experimental measurements are conducted, and the other semiconductor circuit elements connected to the planar transformer are modeled as independent and dependent current or voltage sources. Since experimental work is necessary to calculate the capacitances, it is not possible to use this method before the production of the planar transformer. As the purpose of using the MLM is to perform simulations before the production, this method is not applicable to the MLM. A method to calculate the capacitances by mathematical expressions instead of experimental results was proposed in [13]. This method was chosen to implement in this paper.

Six different models for inter-winding capacitances were presented in [13]. These models and the formulas to calculate the capacitance values are shown in Fig. 1. The mathematical calculations of capacitances are based on λ_1 and λ_2 parameters that are calculated by using (1) and (2).

$$\sum C_{P_m S_n} = \lambda_1 \tag{1}$$

$$\sum C_{P_m S_n} \frac{(m - n)}{N_p} = \lambda_2 \tag{2}$$

where m and n denote the mth turn of the primary and the nth turn of the secondary windings. N_p, N_s are the turn numbers of the primary and secondary windings, respectively. $C_{P_m S_n} = C_0$ for each overlaying winding if $m \neq n$, and $C_{P_m S_n} = 0$ is for each overlaying winding if $m = n$. C_0 is the inter-winding capacitance between two overlaying windings and is defined as

$$C_0 = \epsilon_0 \epsilon_r \frac{W \times L}{d} \tag{3}$$

where W is the width and L is the length of the windings, d is the distance between two overlaying layers, and ϵ_r is the relative dielectric constant of the PCB material.

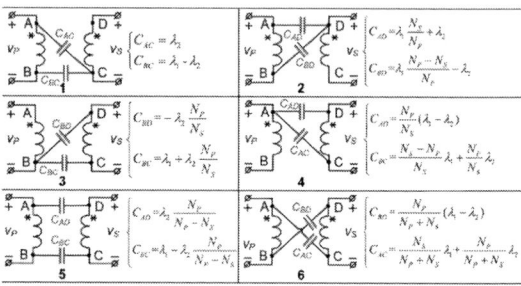

Fig 1: Two-capacitance models of the planar transformers and the formulas defining their values [13]

B. Six-Capacitance Model

Capacitive effects between two overlaying layers can be modelled with six different capacitances as shown in Fig. 2 for wire wound transformers [14]. The formulas given in the figure to calculate these capacitances are given in [10] for planar transformers. By using the terminal voltages of the overlaying turns, the energy stored in the isolation layer is calculated. Then, the energy stored in these capacitances are calculated. Since the two energy values should be equal, the capacitance values are calculated by using this equality. The values of these six capacitances can be written in terms of C_0 calculated in (3) and are shown in Figure 2 [10]. These values change depending on whether the currents passing through the overlaying turns are in the same direction or not. This is due to changing voltages at the terminals of the layers with respect to the direction of the currents.

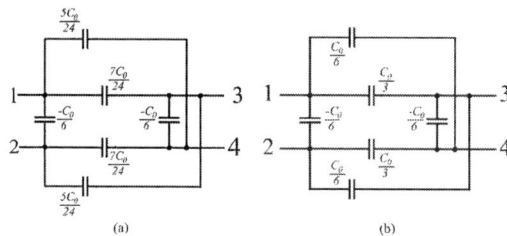

Fig. 2: Six-capacitive model of the capacitive effect between each overlaying turn of planar transformers for currents passing through overlaying turns in (a) the same direction and (b) the reverse direction

III. COMBINING THE PROPOSED METHODS WITH THE MLM AND DISCUSSION ON THE COMMON MODE NOISE

The MLM provides flexibility to connect terminals of planar transformer turns to each other. Preferred layer connection can be done by connecting them manually. In

978-1-5386-7688-2/19 $31.00 © 2019 IEEE

addition, interlayer capacitances can be added by connecting capacitances between these terminals as mentioned in [9]. Six-capacitance method is applied to the MLM that the same way. Two-capacitance method suggests that capacitances are placed between the terminals of the planar transformer windings, not between the layers. Because of that, these two capacitances connected between related terminals of planar transformer windings.

Five 3:6 planar transformers with different turn arrangements were modeled by using the MLM to prove the proposed method. Then, these transformers were manufactured and tested, and the measurements were compared with the simulation results. In parallel with this work a new condition for nearly zero common mode noise generation was developed as discussed next.

A condition for nearly zero common mode noise generated by planar transformers was proposed in [10]. This concept is based on the fact that the sum of the currents i_{14}, i_{13}, i_{24} and i_{23} given in Fig. 3 should be zero. The capacitance values yielding zero common mode noise were calculated by assuming that dv_2/dt and dv_4/dt are zero [10]. This means that the same numbered turns of the primary and the secondary windings (such as the third turn of the primary and the secondary) should be overlaid. These overlaid turns are called as "paired layers." If the number of turns are not equal, in this case the excess turns should not be overlaid with any turn of the other winding. It should also be noted that the numbering is started from the negative terminals of each winding for this assumption to work.

In addition to the condition given in [10], another condition for zero common mode noise can be obtained by setting that dv_1/dt and dv_3/dt to zero. This assumption yields the equations given in (4) and (5).

$$-C_{14}\frac{dv_4}{dt} + C_{23}\frac{dv_2}{dt} + C_{24}\left(\frac{dv_2}{dt} - \frac{dv_4}{dt}\right) = 0 \quad (4)$$

$$\frac{N_s}{N_p}(C_{14} + C_{24}) = (C_{23} + C_{24}) \quad (5)$$

C_{14}, C_{24} and C_{23} values are calculated in terms of the six capacitances given in [10]. These formulas are given in Table I. The a, b, c and d values used in Table I are given in Table II.

By combining (5) and the values given in Table I and II, equations (6-8) can be written.

$$\frac{N_s}{N_p}\frac{dv_2}{dt}(24 - 12c - 12d) = \frac{dv_2}{dt}(24 - 12a - 12b) \quad (6)$$

$$\frac{N_s}{N_p}\left(\frac{2n-1}{N_s} - 2\right) = \left(\frac{2m-1}{N_p} - 2\right) \quad (7)$$

$$N_s - n = N_p - m \quad (8)$$

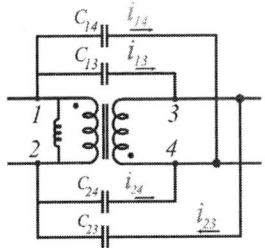

Fig. 3: Parasitic capacitances of transformers causing common mode noise [10]

Table I. Capacitance values given in fig.3 in terms of six capacitances between overlapping turns current passing through different directions [10]

Capacitance	Value
C_{13}	$\frac{C_0}{24}(5ad + 7bd + 7ac + 5bc)$
C_{14}	$\frac{C_0}{24}(12a + 12b - 5ad - 7bd - 7ac - 5bc)$
C_{23}	$\frac{C_0}{24}(12c + 12d - 5ad - 7bd - 7ac - 5bc)$
C_{24}	$\frac{C_0}{24}(24 + 5ad + 7bd + 7ac + 5bc - 12a - 12b - 12c - 12d)$

Table II. a, b, c and d expressions in terms of m, n, N_p and N_s

Expression	Value
a	$(m-1)\frac{1}{N_p}$
b	$m\frac{1}{N_p}$
c	$(n-1)\frac{1}{N_s}$
d	$n\frac{1}{N_s}$

Equation 8 means that, although the condition set in [10] is correct, if the number of turns are not equal, the same numbered turns do not have to be overlaid, as long as the overlaying is started either from the highest numbered turns or the lowest numbered turns of the primary and secondary; and the excess turns should not be overlaid with any turn of the other winding. Some examples are given below in the next section.

IV.   SIMULATION AND EXPERIMENTAL RESULTS

A. Adopting the Two-capacitor Method and the Six-capacitor Method to the MLM

In order to compare the two models representing the capacitive effects in planar transformers explained above, five different winding arrangements given in Table III were compared by using the simulation and experimental results. For each planar transformer type, the design parameters are given in Table IV. In the table,

Transformer A is designed without any concern for parasitic capacitances, transformer B and C are designed to verify the nearly zero common mode noise condition described in the previous section, and transformer D and E are designed to observe common mode noise when there is no overlapping.

In Table III, P and S letters are used to indicate the primary and secondary winding. The first number following the letter indicates the turn number and the number after the dash indicates the parallel turns. For instance, S6-1 indicates the 6th turn of one of the secondary windings in parallel, and S6-2 indicates the 6th turn of the other secondary winding connected in parallel. Numbering the turns begins from the negative terminal; i.e. the 1st turn indicates the lower terminal of the winding.

Table III. Planar transformer winding arrangements used in the simulations and experiments

	A	B	C	D	E
Layer 1	S6-1	S6-1	S1-1	P3-1	S6-1
Layer 2	P3-1	P3-1	S2-1	P3-2	S6-2
Layer 3	S5-1	P3-2	S3-1	P2-1	S5-1
Layer 4	S5-2	S6-2	S6-1	P2-2	S5-2
Layer 5	P3-2	S5-1	P3-1	P1-1	S4-1
Layer 6	S4-1	P2-1	P3-2	P1-2	S4-2
Layer 7	S4-2	P2-2	S6-2	S6-1	P3-1
Layer 8	P2-1	S5-2	S5-1	S6-2	P3-2
Layer 9	S3-1	S4-1	P2-1	S5-1	P2-1
Layer 10	S3-2	P1-1	P2-2	S5-2	P2-2
Layer 11	P2-2	P1-2	S5-2	S4-1	P1-1
Layer 12	S2-1	S4-2	S4-1	S4-2	P1-2
Layer 13	S2-2	S3-1	P1-1	S3-1	S3-1
Layer 14	P1-1	S3-2	P1-2	S3-2	S3-2
Layer 15	S1-1	S2-1	S4-2	S2-1	S2-1
Layer 16	S1-2	S2-2	S3-2	S2-2	S2-2
Layer 17	P1-2	S1-1	S2-2	S1-1	S1-1
Layer 18	S6-2	S1-2	S1-2	S1-2	S1-2

In order to test the concepts proposed in the paper an experimental set-up was built. The five different planar transformer configurations that were designed, were tested in a high-side active clamp forward converter. The topology is shown in Fig. 4. In this circuit, the switching frequency is 100 kHz and the low side MOSFET is driven by 25% duty cycle while the high side MOSFET is driven by 75% duty cycle with a short dead time.

The common mode noise is measured by the voltage across the 250Ω resistance which is a low impedance path for common mode noise [10]. As the load, a 45Ω non-inductive resistor was used and the measurements were taken by an isolation probe to isolate parasitic effects of the oscilloscope. In the supply side, an LISN was not used because no unexpected common mode noise is observed due to the supply used. But it may be needed with a different power supply used in the measurements.

Table IV: Planar transformer parameters used in the simulations and measurements

Parameter	Value
Relative permeability of the core	810
Layer thickness	70um for all layers
Isolation thickness between layers	100um between all layers
Number of turns in each layer	1 for all layers
Airgap of the core	0
Effective core area	260 mm^2
Turn length	91 mm
Turn width	71 mm
Thickness of top and bottom core	3.18 mm

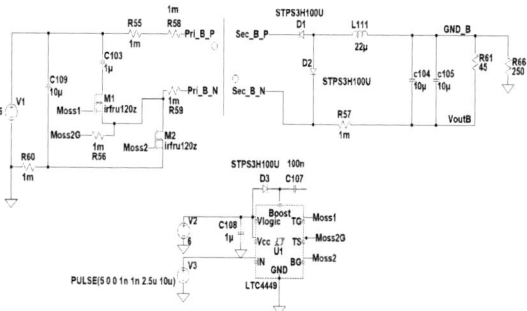

Fig. 4: The circuit used to test planar transformers

The simulations are conducted by using the method described in [10] and [13].

Table V lists the capacitance values obtained with the "Two-capacitance method" for the five planar transformers labeled A to E. The value of C_0 was found as 240 pF. The capacitance values for the "Six-capacitor method" are not given here but they can be calculated by using this C_0 value and the formulas given in Figure 2.

Results of the simulation and measurements are given in Fig. 5 to 9 for both models. As seen in the figures, both models proposed in [10] and [13] give nearly the same results, and are very close to the experimental results. Both simulations were conducted for the same PC and by using the same Spice tool. The simulation for the method proposed in [13] is about ten times faster because of its simplicity.

Table V. C$_{BD}$ and C$_{AC}$ values for transformer A to E

	A	B	C	D	E
C$_{BD}$	1440 pF	960 pF	960 pF	80 pF	240 pF
C$_{AC}$	1440 pF	480 pF	480 pF	160 pF	240 pF

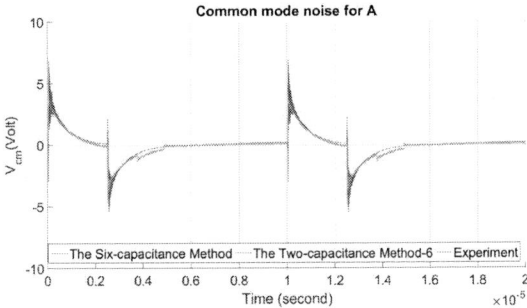

Fig. 5: Simulations and experimental results for planar transformer A

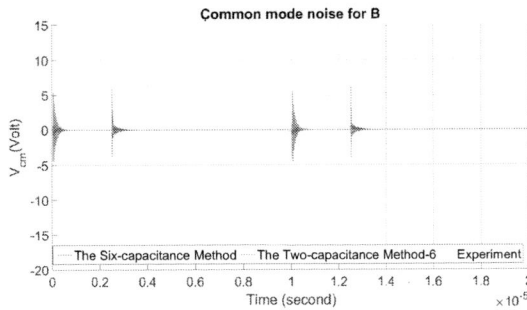

Fig. 6: Simulations and experimental results for planar transformer B

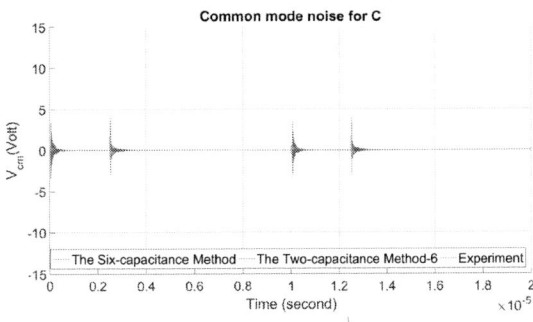

Fig. 7: Simulations and experimental results for planar transformer C

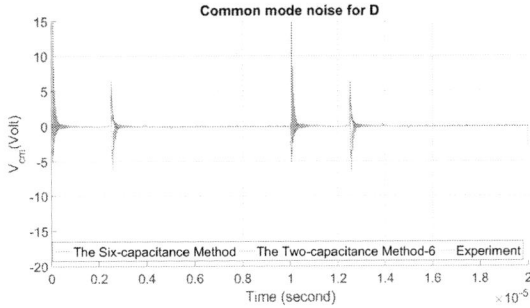

Fig. 8: Simulations and experimental results for planar transformer D

The results show that, planar transformers B and C generate very low common mode noise compared to the others as expected. In addition, less overlapping between the primary and the secondary turns reduces the common

mode noise significantly as seen in the cases of D and E compared to A which has the same turn numbers. Lastly, planar transformer A which was designed without any concern for parasitic capacitances generates the highest common mode noise compared with the others.

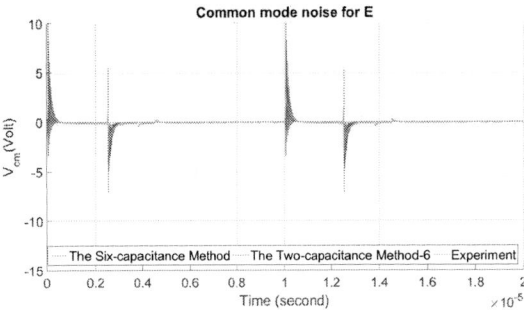

Fig. 9: Simulations and experimental results for planar transformer E

B. Comparing Performances of Different Two-capacitor Models

After observing that 6th model of the two-capacitor method gives nearly same results with the experimental results, the simulations were repeated with the 3rd, 4th and 5th models given in Figure 1 for the planar transformers A and B. Calculated capacitance values for these cases are given in Table VI. The simulations results are given in Figures 10 and 11.

Table VI. Capacitance values for the two-capacitor method for models 3, 4 and 5.

Models	Capacitances	A	B
3	C_{BD}	720 pF	720 pF
	C_{BC}	2160 pF	720 pF
4	C_{AD}	2160 pF	1440 pF
	C_{AC}	720 pF	0
5	C_{AD}	1440 pF	1440 pF
	C_{BC}	1440 pF	0

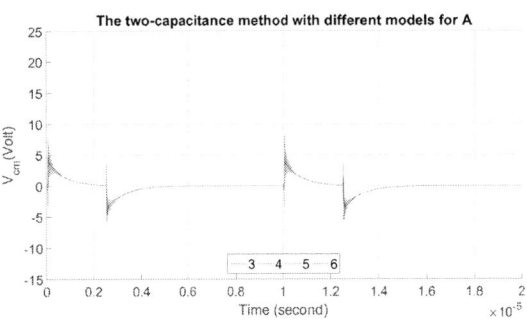

Fig. 10: Common mode noise obtained by the two-capacitance method with different models for planar transformer A

978-1-5386-7688-2/19 $31.00 © 2019 IEEE

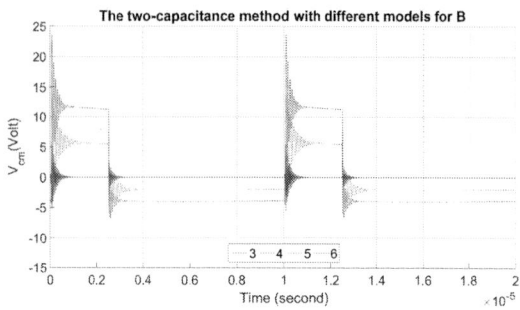

Fig. 11: Common mode noise obtained by the two-capacitance method with different models for planar transformer B

As seen clearly, two-capacitance method gives results in agreement with the experiments only if the values of both capacitances used in models are nonzero.

V. CONCLUSION

In this paper, two different approaches for modelling common mode noise of planar transformers are discussed and combined with the MLM. The results show the method proposed in [10] and [13] give nearly the same results with the experiments under the condition of non-zero capacitance because the reason for this is that a zero-capacitance term cancels a coefficient that causes unbalance in current in the circuit during simulations. Moreover, simulations used in the two-capacitance method yield results faster than the simulations of the method using six-capacitances because of its simplicity.

In this paper, a new approach to get nearly zero common mode noise is also introduced. This approach shows that nearly zero common mode noise can be achieved if paired layers are overlapped beginning from the high or low numbered terminals. This approach is verified by simulation and experimental results.

VI. ACKNOWLEDGEMENTS

This research is supported by Aselsan A.S. We would like to express our gratitude to Aselsan A.S. for the technical support.

REFERENCES

[1] M. T. Quirke, J. J. Barrett and M. Hayes, "Planar magnetic component technology-a review," *in IEEE Transactions on Components, Hybrids, and Manufacturing Technology*, vol. 15, no. 5, pp. 884-892, Oct. 1992.

[2] Z. Ouyang and M. A. E. Andersen, "Overview of Planar Magnetic Technology—Fundamental Properties," *in IEEE Transactions on Power Electronics*, vol. 29, no. 9, pp. 4888-4900, Sept. 2014.

[3] Y. Veli, A. M. Morega, M. Morega, and L. Pîslaru-Dǎnescu, "Numerical modeling of a planar transformer for micro-power controllers," 2017 10th Int. Symp. Adv. Top. Electr. Eng. ATEE 2017, pp. 206–210, 2017.

[4] A. Ammouri, H. Belloumi, T. Ben Salah, and F. Kourda, "High-frequency investigation of planar transformers," 2014 Int. Conf. Electr. Sci. Technol. Maghreb, Cist. 2014, pp. 1–5, 2014.

[5] J. Qin, Z. Yu, K. Sun, P. Wei, and Z. Lan, "Analysis and simulation of parasitic parameters for PCB planar transformer," Proc. - 2012 Int. Conf. Control Eng. Commun. Technol. ICCECT 2012, pp. 241–244, 2012.

[6] A. Ammouri, T. Ben Salah, and F. Kourda, "Modeling and simulation of a high- frequency planar power transformers," 2015 4th Int. Conf. Electr. Eng. ICEE 2015, pp. 1–6, 2016.

[7] S. R. Cove, M. Ordonez, and J. E. Quaicoe, "Modeling of planar transformer parasitics using design of experiment methodology," Can. Conf. Electr. Comput. Eng., no. Ccd, pp. 1–5, 2010.

[8] Y. Guan, N. Qi, Y. Wang, X. Zhang, D. Xu, and W. Wang, "Analysis and design of planar inductor and transformer for resonant converter," ECCE 2016 - IEEE Energy Convers. Congr. Expo. Proc., 2016.

[9] M. Chen, M. Araghchini, K. K. Afridi, J. H. Lang, C. R. Sullivan, and D. J. Perreault"A Systematic Approach to Modeling Impedances and Current Distribution in Planar Magnetics," IEEE Trans. Power Electron., vol. 31, no. 1, pp. 560–580, 2016.

[10] M. A. Saket, M. Ordonez, and N. Shafiei, "Planar Transformers with Near-Zero Common-Mode Noise for Flyback and Forward Converters," IEEE Trans. Power Electron., vol. 33, no. 2, pp. 1554–1571, 2018.

[11] M. Pahlevaninezhad, D. Hamza, and P. K. Jain, "An improved layout strategy for common-mode EMI suppression applicable to high-frequency planar transformers in high-power dc/dc converters used for electric vehicles," IEEE Trans. Power Electron., vol. 29, no. 3, pp. 1211–1228, 2014.

[12] H. Zhang, S. Wang, Y. Li, Q. Wang, and D. Fu, "Two-Capacitor Transformer Winding Capacitance Models for Common-Mode EMI Noise Analysis in Isolated DC-DC Converters," IEEE Trans. Power Electron., vol. 32, no. 11, pp. 8458–8469, 2017.

[13] Z. Zhang, B. He, D. D. Hu, X. Ren, and Q. Chen, "Common Mode Noise Modeling and Reduction for 1-MHz eGaN Multi-Output DC-DC Converters," IEEE Trans. Power Electron., vol. PP, no. 51722702, p. 1, 2018.

[14] M. B. Eric laveuve, Jean-pierre Keradec, "00178054.pdf," in Conference Record of the 1991 IEEE Industry Applications Society Annual Meeting, 1991.

Investigation of the Effects of Switching Technique on the Performance of Four Switch Buck-Boost Bidirectional DC/DC Converters

İbrahim Koçak
MGEO *Business Sector*
ASELSAN *Inc.*
Ankara, Turkey
Email: ikocak@aselsan.com.tr

Prof. Dr. Hulusi Bülent Ertan
Dept. of Electrical and Electronics Engineering
Middle East Technical University
Ankara, Turkey
Email: ertan@metu.edu.tr

Abstract - In general, a DC/DC converter provides energy transfer from a source to a load in one direction. Today, bidirectional power transfer is needed in many applications that include battery based energy storage. Serving a compact solution for this problem is crucial in order to reduce the total size of the system where the physical size is a critical subject. Various types of bidirectional DC/DC converters are implemented for this purpose in literature. Highlighting the reasons, one of them is selected, proposed and simulated for a special double battery suited system which has a rated power of 1 kW. In order to compare efficiency performance of the switching strategy design for this topology; proposed in [12], with conventioal one throughout the whole operating points, simulations are realized at various operating conditions. Since proposed method is a software based solution, there will be no need for extra circuit components. With the results of this study, it is seen that the selected topology and applied modulation technique are suitable for bidirectional power flow and the power density can be increased with a software-only solution for this kind of applications.

1. Introduction:

With the increasing concern about environmental pollution and energy crisis in the world, using renewable energy resources have become more important [1]. Since these sources are dependent on weather conditions, energy storage elements such as batteries or ultracapacitors should be fitted in these systems. They store the excessive energy when the demand is lower than the source (charging) or they provide the energy in the case of an extra demand, failure or source fluctuation (discharging). Therefore, the power should be transferred in both directions. The interface between the supply and storage elements can be achieved with the use of bidirectional DC/DC converters. These converters are used in solar power plant systems [1], wind power plants [2], uninterruptible power supplies [3] and power trains of hybrid or electrical vehicles (HEVs, EVs) [4].

Bidirectional DC/DC converters can be classified in isolated versus non-isolated ones with regard to needs of the system. Then, they should be categorized as buck type, boost type or buck-boost type according to placement of the energy storage device. Also, having voltage source or current source structure can be a classification criterion [5]. By considering these properties, several topologies have been proposed in the literature for bidirectional power flow such as single phase buck or boost [6], non-isolated buck-boost [7], dual active bridge [8], dual half bridge [9] and half bridge-push pull [10]. Among these topologies, a non-isolated one in which a buck and boost converter are cascaded (Fig.1) is selected and

analyzed in detail in this study. In short, this topology is referred as 4S-BBBC (Four Switch Buck Boost Bidirectional Converter).

Although switch count is much more than the other non-isolated ones, this topology brings the advantage of both buck and boost operation at each side which is also useful when input and output voltage levels are close to each other. Furthermore, this topology is flexible to implement zero voltage switching (ZVS) without any extra components unlike the resonant or active clamped converters, which require additional LLC circuits or active switches [11] for zero voltage switching operation. Also, the disadvantages of resonant converters, which are limited operating voltage and load conditions, higher ratings of switches [10], complicated EMI filter design due to variable switching frequency [12], are eliminated with this topology. ZVS can be achieved just by controlled phase shift with the control of gating signals of each switching device in a wide range and with a decrease in components' stress.

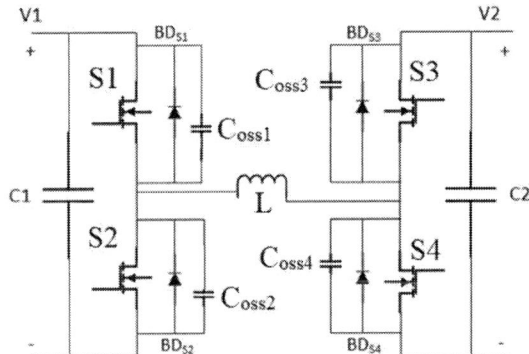

Figure1 Four Switch Buck-Boost Bidirectional DC/DC Converter (4S-BBBC)

There are two different approaches for operation of this topology, namely **conventional** and **phase shifted** ones. The main difference of them is the switching pattern of switching devices. They are explained in detail in the next section. In this paper, these two approaches are compared in an analytical way with respect to their efficiency under different input and output conditions while input varies from 12V to 28V and output varies between 24V to 56V or vice versa. Design specification of the converter is assumed to be as shown in Table1 in order to identify the feasibility of the topology, advantages and disadvantages of each operation scheme. The application on which the evaluation is made is a double battery suited vehicle in which, the main battery supplies the traditional loads such as lightning, entertainment,

978-1-5386-7688-2/19 $31.00 © 2019 IEEE 395

safety and security while the additional one is responsible for 8 kW special load which is operated only for a limited time (maximum for 1 hour) throughout a day. The main battery is charged by the alternator of the vehicle, while the additional one is charged from the main one with the help of the converter fitted for this purpose when aforementioned special loads go out of use. A rough calculation indicates that, 8 kWh energy is required for 1 hour operation of "special load" and 8 hours are enough for a 1 kW charger to recharge the additional battery. Moreover, when the engine is off, the energy in the additional battery can be used to charge the main battery, which is still responsible to supply the emergency loads. Therefore the charger between these two should be bidirectional which is the main scope of this study.

Table1 – Design Specifications

First Battery Voltage	24V-56V
Second Battery Voltage	12V-28V
Rated Power	1 kW
Switching Frequency	100 kHz

2.   Switching Techniques Comparison of 4S-BBBC

Two different gating patterns for switcing devices are applied for this topology. Only two switches are actively switched in conventional control while all four switches are gated simultaneously in phase shifted operation. These operations are clarified in this section.

A.   Conventional Operation

In conventional operation, only respective half bridge switches are gated at a switching frequency, while fixed control signals are applied to others. For example, in buck mode (Fig.2 (a)), during power flow from V1 to V2, S1 and S2 are switched complementarily at a determined switching frequency while S4 is off and S3 is on during entire period. Operation principle is similar to basic buck converter operation. That is to say, when S1 is switched on, inductor current begins to rise storing the energy and the power is transferred to load. Since the body diode of S2 is reversed biased, it plays no part in this time period. When S1 is switched off, the energy stored in the magnetic field around the inductor is released back to the output using the path formed by the body diode of S2 at the beginning and later than by S2 itself. Unfortunately, the high side switch S1 is hard switched, while the low side one (S2) is operated as synchronous rectifier. On the other hand, in boost mode S3 and S4 are modulated in a similar manner while S1 is on and S2 is off constantly as shown in Fig.2 (b).

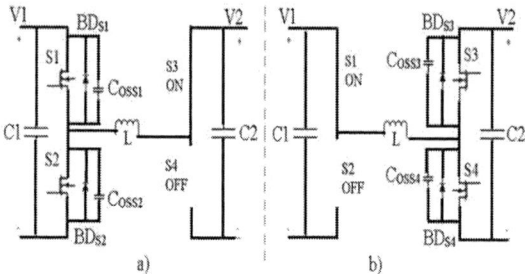

Figure2 Equivalent circuits in buck (a) and boost mode (b) from V1 to V2 (conventional operation)

With this conventional method, desired voltage levels can be reached by adjusting the duty cycles of switched devices. However, it is important to note that there should be a dead time between the same leg switches (i.e. complementary switches) in order to avoid shoot through operation. Moreover, zero voltage switching can be achieved with such a switching pattern if eq.1 conditions are satisfied resulting in a high efficiency converter. In eq.1, I_L, E_L, E_{coss} and V_{DS} show the inductor current, inductor energy, output capacitance energy of switching devices and the voltage across the switches respectively [13]. To give an example, in [13], 94% efficiency is reported with this topology and operation principle.

$$E_L \geq E_{Coss1} + E_{Coss2}$$

$$\frac{1}{2}LI_L^2 \geq \frac{1}{2}C_{oss1}V_{DS1}^2 + \frac{1}{2}C_{oss2}V_{DS2}^2 \tag{1}$$

B.   Phase Shifted Operation (Proposed)

In this operation, all four switching devices at both legs (S1 to S4 in Fig.1) are switched on and off during a period. Phase shift means that with such a gating pattern there will be a phase difference, θ, between bridge leg outputs as shown in Fig.3. Input-output voltage conversion ratio and direction of power flow is determined by the duty ratio of corresponding switches and the phase shift angle between them. For instance, the leading gating scheme of S1 over S3 makes power flow from V1 to V2. Moreover, it can be inferred from the eq.s 2 to 6 that, conversion ratio (V2/ V1) will be equal to ratio of duty cycles of S1 ($D_{S1} = t_2/T$) and S3 ($D_{S3} = (t_3-t_1)/T$). In eq.s 2 to 6, ΔI's are the variations of inductor current in respective time intervals.

$$V_L(t) = L * \frac{d}{dt} i_L(t) \tag{2}$$

$$V1 = L * \frac{\Delta I_1}{\Delta t} \qquad ; t_0 \leq t < t_1 \tag{3}$$

$$V1 - V2 = L * \frac{\Delta I_2}{\Delta t} \qquad ; t_1 \leq t < t_2 \tag{4}$$

$$-V2 = L * \frac{\Delta I_3}{\Delta t} \qquad ; t_2 \leq t < t_3 \tag{5}$$

$$\Delta I_1 + \Delta I_2 = \Delta I_3 \tag{6}$$

It is important to note that S1 and S2 are switched in a complementary way, likewise S3 and S4 are switched in the same manner where S refers to switches.

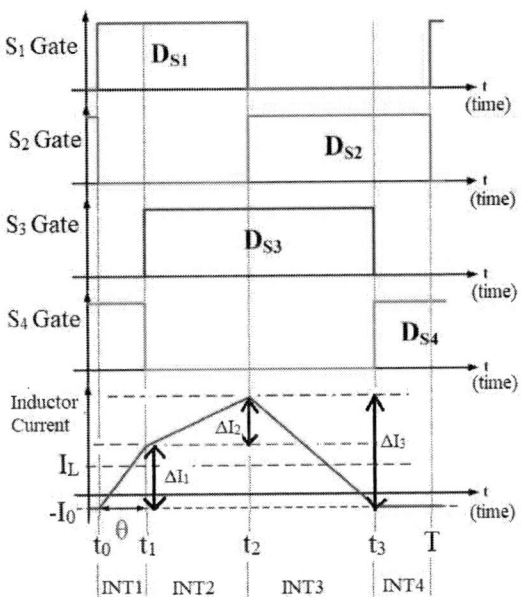

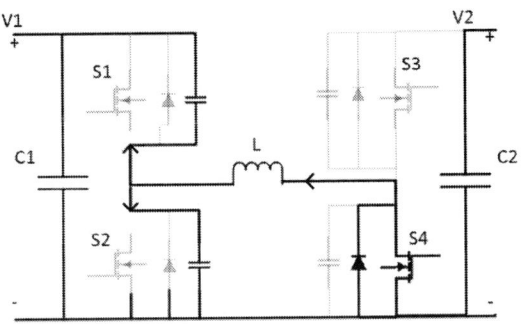

Figure3 Required gating signals and inductor current waveform during Buck Mode energy flow from V1 to V2 (Adopted from [14]) where INTs are respective time intervals

B.1. Buck Mode of Operation

In buck mode of operation, one switching period can be divided into four time intervals as shown in Fig.3. At t=t_0, switches S2 and S4 are conducting, therefore S2 can be turned off under ZVS due to voltage rise delay effect of parasitic output capacitances (C_{oss}) of mosfets. With the turn off of S2, inductor current begins to charge C_{oss2} and discharge C_{oss1} (Fig.4).

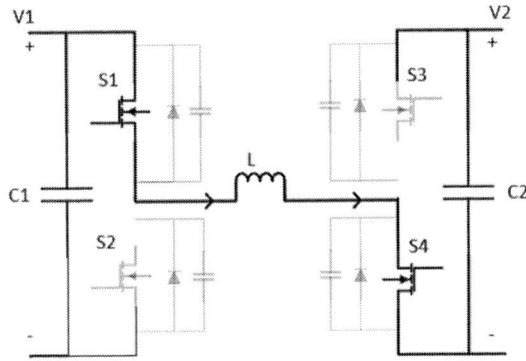

Figure4 Equivalent circuit of buck mode during power flow from V1 to V2 at t=t_0

When the voltage of C_{oss1} becomes zero, body diode of S1 begins to conduct and S1 can be switched with ZVS. At time interval $t_0 < t < t_1$ (INT1), switches S1 and S4 conduct (equivalent circuit and current path can be seen in Fig.5) resulting a linear increase in inductor current (Fig.3). During this time interval, voltage applied to inductor is $V_L(t) = V1$ (Eq. 3) and load power is supplied only by C2.

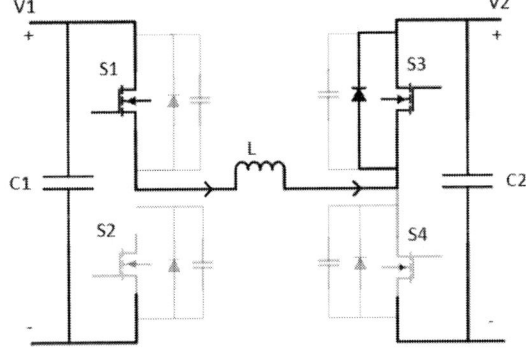

Figure5 Equivalent circuit of buck mode during power flow from V1 to V2 at INT1

At the end of this period; at t=t_1, S4 is switched off with ZVS since its output capacitance is discharged due to conduction. After that, since the inductor current cannot change suddenly, its energy will charge C_{oss4} and discharge C_{oss3}. When the voltage across C_{oss3} becomes zero, body diode of it takes over the current and S3 is ready to turn on with ZVS. Therefore, at time $t_1 < t < t_2$, S1 and S3 begin to conduct simultaneously (Fig.6), resulting an increase of inductor current with a smaller slope (due to voltage applied to inductor $V_L(t) = V1 - V2$ as seen in Fig.3 and Eq. 4) besides supplying the load.

Figure6 Equivalent circuit of buck mode during power flow from V1 to V2 at INT2

At t=t_2, S1 turns off and the inductor current charges C_{oss1} and discharges C_{oss2}. Body diode of S2 will begin to conduct when the voltage across C_{oss2} is zero and S2 turns on with ZVS. At time interval $t_2 < t < t_3$, S2 and S3 creates a path for inductor current to the load resulting a voltage across the inductor $V_L(t) = -V2$ (Eq. 5), which makes the inductor current decrease linearly (Fig.3). Equivalent circuit and inductor current path can be seen in Fig.7.

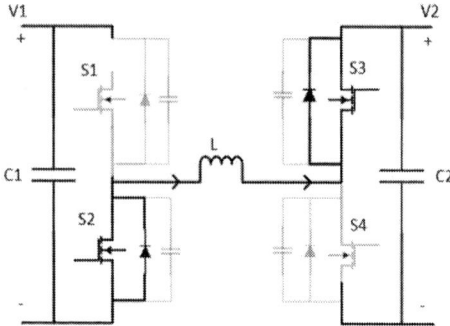

Figure7 Equivalent circuit of buck mode during power flow from V1 to V2 at INT3

At the end of this period (at t=t₃), S3 is turned off and C_{oss3} will be charged while C_{oss4} will be discharged as shown in Fig.8. It should be noted that, reversed inductor current means that the energy is transferred from output capacitances to the inductor during this period. Therefore, the energy required for other periods is recovered at this time interval that results lower switching losses [12].

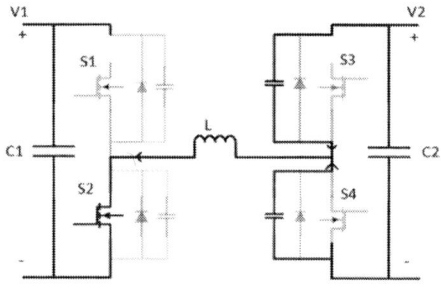

Figure8 Equivalent circuit of buck mode during power flow from V1 to V2 at t=t₃

When the voltage across C_{oss4} becomes zero, it turns on under ZVS making a zero voltage across inductor and constant current. During this period, load is supplied only by C2. Equivalent circuit, current path and shape of inductor can be seen in Fig.9 and Fig.3. This situation continues until the end of the switching period in order to provide a negative inductor offset current to realize ZVS at the beginning of next period. Moreover, since the switching frequency of converter is kept constant, EMI filter design concern is reduced.

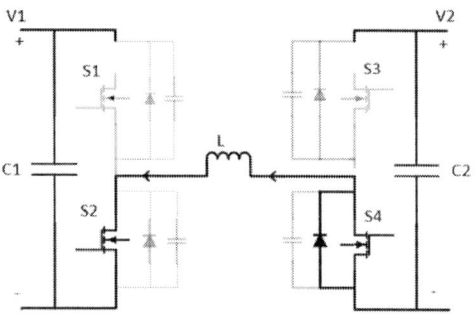

Figure9 Equivalent circuit of buck mode during power flow from V1 to V2 at INT4

B.2. Boost Mode of Operation

Similar to the buck mode operation, one complete switching cycle can be divided into four time intervals during boost operation as can be seen from the inductor current waveform in Fig.10. Conducting switches are the same in corresponding time intervals as in the case of buck mode. The only difference of this operation is that; since the output voltage level will be higher than the input voltage, at time interval $t_1 < t < t_2$, inductor current has a negative slope (Fig.10). Equivalent circuit configurations for buck mode shown in Fig.4 to 9 are valid for boost mode of operation too.

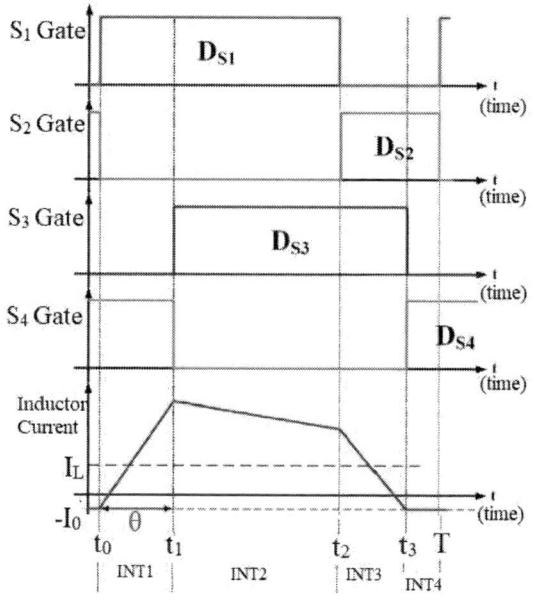

Figure10 Required gating signals and inductor current waveforms during Boost Mode while power flow from V1 to V2 (Adopted from [14]) where INTs are respective time intervals

3. Calculation of Switching Times of Proposed Operation:

From the operation principle and circuit analysis of the proposed topology stated in the preceding section, it can be observed that the turn on/off times of switching devices are important to improve efficiency and reliability of the converter. Furthermore, since the transferred power is directly proportional to inductor current, time function of it should be obtained in order to calculate the whole transferred power.

In [12], input and output power are calculated in terms of switching times t_1, t_2, t_3, inductance value and desired input and output voltages. An optimization process (explained in detail in [12]) is applied while determining the times to achieve maximum power transfer. Finally, eq.s (7) to (9) are obtained and used for different input, output and load conditions.

$$t_{1(for\ maximum\ power\ transfer)} =$$

$$\frac{V2^2*t_3+ V1*I_0*L}{V1^2+ V1*V2+ V2^2} \qquad (7)$$

$$t_{2(for\ maximum\ power\ transfer)} =$$

$$\frac{(V2^2 + V1 * V2)t_3 - V2 * I_0 * L}{V1^2 + V1 * V2 + V2^2} \qquad (8)$$

$$P_{transferred\ maximum} =$$

$$\frac{V1*V2(I_0^2*L^2-2I_0*L(V1+\ )t_3 + V1*V2*t_3^2)}{2L*t_s*(V1^2+ V1*V2+ V2^2)} \qquad (9)$$

Moreover, it is important to note that the mechanism for zero voltage turn on can be realized as; output capacitance of a switch should be completely discharged before gating command is applied to it. In other words, body diode or anti-paralleled diodes of that switch begin to conduct before it is triggered resulting nearly zero voltage across its output capacitance. The energy of inductor should be enough to charge or discharge the output capacitances of switch. Therefore, eq.1 (see section 2.A.) is also desired for this operation.

4. Comparison of operation techniques via simulation

In order to asses the performance of the phase shifted method with the conventional one; with respect to efficiency with different input and output conditions and under load range described in section 1, simulations are done via LTspice XVII.

➤ **Selection of circuit and simulation parameters**

Operation techniques and analytical calculations expressed in the preceeding sections are analyzed in detail and reported in this section. Eq.s 7 to 9 show that optimized switching times t_1, t_2 and t_3 depend on the variables V1, V2, output power, inductance value and the switching frequency. Selection of these parameters are realized to satisfy the requirements of the converter given in Table1. Required maximum inductance value (in order to keep output voltage ripple at minimum) is found as 1.58uH in the case of boost mode, while V2 is equal to 28V and V1 is equal to 56V at 1 kW rated power using eq.9. 100 kHz is selected as the switching frequency, which is compatible with rise and fall times of switching devices at these voltage levels. It should also be noted that minimum dead time between complementary switches is taken as 100 ns. After the decision of switching frequency and inductance value, it is checked whether zero voltage switching operation can be achieved throughout the whole power range. Using the determined parameters and eq.1, maximum allowable C_{oss} value for switching devices is found as 6.33nF for low voltage side (28V) and 1.58nF for high voltage side (56V). Therefore, the model of IRFP4568 (Infineon Tech.) mosfets are used as switching devices (S1, S2, S3, S4) having a maximum Coss value of 977pF (< 1.58nF).

In order to decrease simulation time, switching times (t_1, t_2 and t_3) for the proposed method are calculated from eq.s 7 to 9 for each operating condition and used at their respective conditions.

➤ **Evaluation of simulation results**

Gating signal commands and inductor current waveform; under buck mode while 1 kW power flowing from V1(=56V) to V2(=28V), shown in Fig.11 are similar to the theoretical waveform shown in Fig.3.

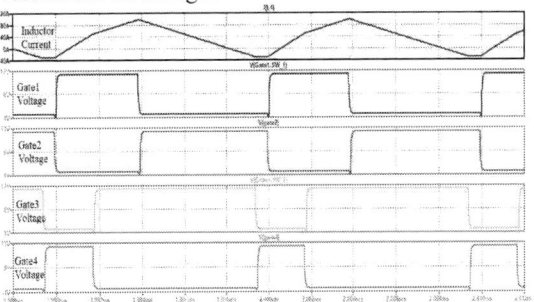

Figure11 Inductor current and gatings of switching devices over one period during buck mode (V1=56V, V2=28V at P=1 kW)

Fig.12 shows that zero voltage switching of S1 and S3 are achieved which supports the accuracy of the calculation of switching times.

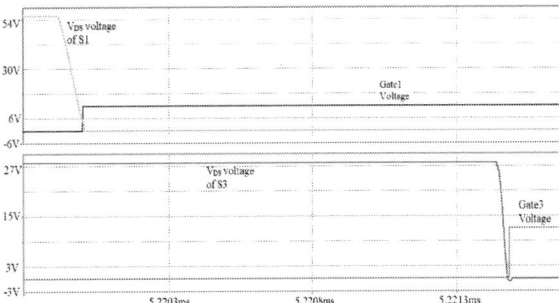

Figure12 Vds voltages and gatings of switching devices S1 and S3 at turn on during buck mode (V1=56V, V2=28V at P=1 kW)

Efficiency comparison of proposed and conventional methods are realized at various operating points while one of the variables is held costant and the others are changeable. Simulations are done by changing V1 from 24V to 56V and V2 from 12V to 28V while load is variable from 25W to 1 kW. It can be observed from the Fig.13 that the efficiency with phase shifting method is better at whole operating points and under full range of load conditions. Although both methods have close efficiency values under rated power levels, the proposed method overrides the conventional one under light load. To give an example for buck mode, 95% efficiency is obtained for V1=56V, V2=28V at 100w load with phase shifting control while it is 84% with conventional gating (Fig.13). It is also important to note that this improvement is independent of the operating mode and power flow direction, i.e. both buck and boost modes have higher efficiencies. In the case of boost mode while energy is transferred from V2(=28V) to V1(=56V), efficiency is observed as 93% at 50w with the phase shifting control method while it is 85% with the conventional one (Fig.14).

978-1-5386-7688-2/19 $31.00 © 2019 IEEE

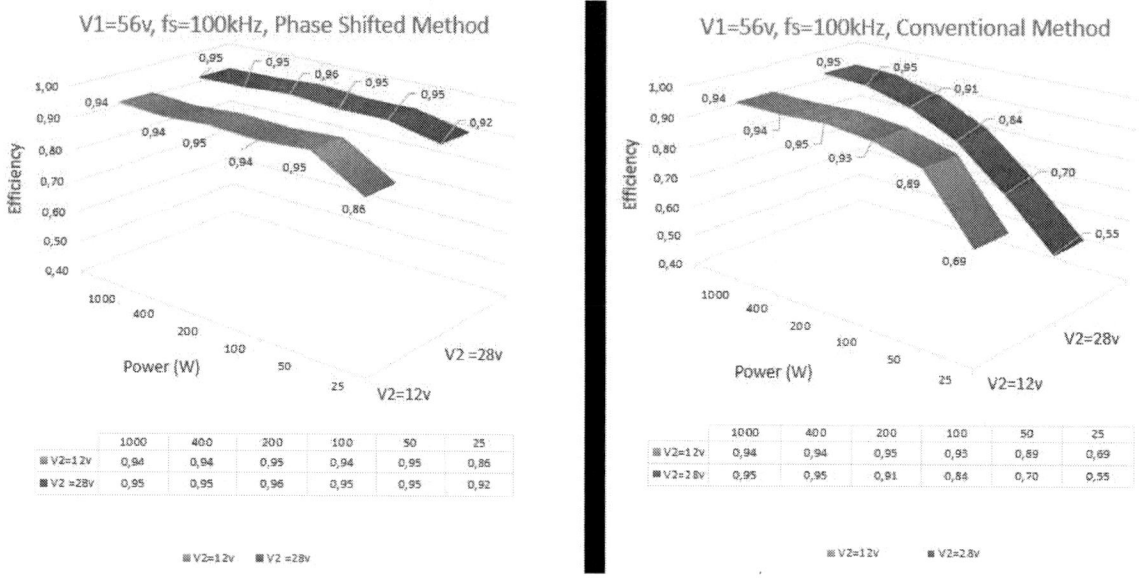

Figure13 Efficiency comparison of proposed and conventional methods while energy is transferred from V1 to V2 at different output voltages and load conditions

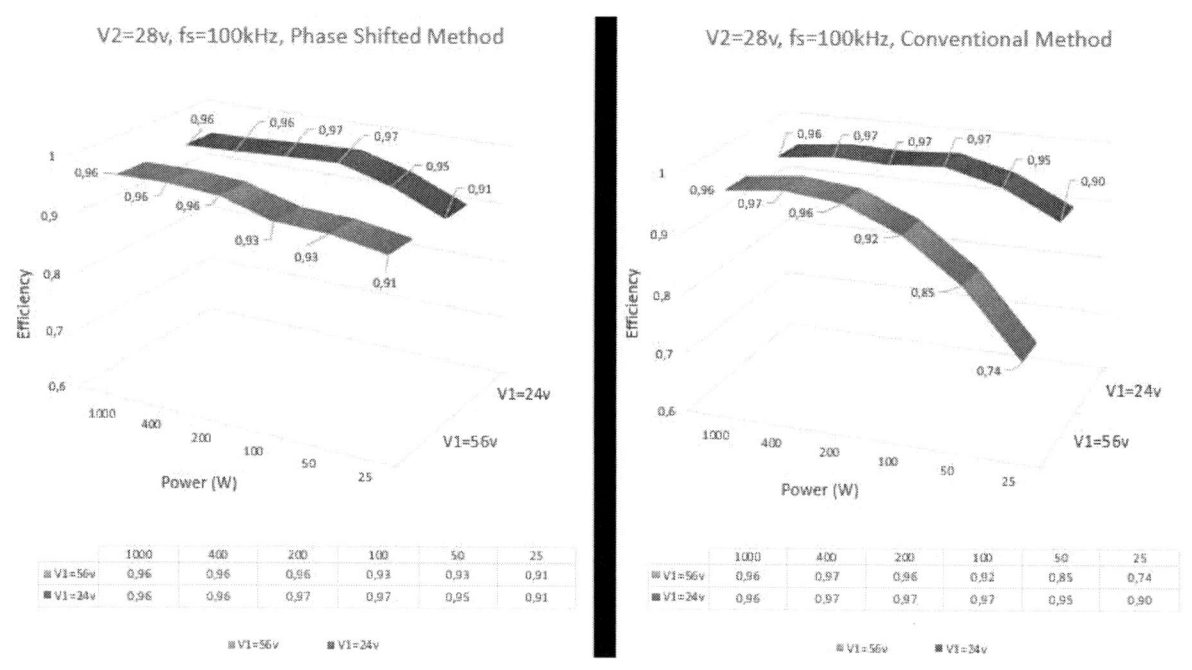

Figure14 Efficiency comparison of proposed and conventional methods while energy is transferred from V2 to V1 at different output voltages and load conditions

978-1-5386-7688-2/19 $31.00 © 2019 IEEE

5. Conclusion

In this paper, it is aimed to identify the advantages and disadvantages of these two control strategies mentioned above. Also limitation of the strategies regarding their energy transfer capability for different input/output voltage ratios and the variation of efficiency of the circuits are investigated. One of them which is the proposed method in [12] is discussed and analyzed in detail with description of operating principle. It provides a gate timing scheme and a negative offset current in order to achieve higher efficiency throughout the entire operating points. With this method, efficiency is improved at even very light loads without any extra component increasing power density which is very important for space limited applications. Simulations are done in order to support the theory and analytical calculations. According to simulation results, great efficiency improvement is achived at even very light loads while it is still high at full load operation by using the proposed gating scheme. On the other hand, when the voltage levels are decreased to levels of zero voltage switching requirements of it (see the results in Fig.14 at operating condition while V2=28 and V1=24V) or operating at near rated load conditions, conventional one gives more efficient results, too. Hence, if the operating voltage levels become such levels or while operating at rated load levels, modulation method can be changed to conventional one in order to reduce the complexity of switching patterns. Since the gating signals are controlled in a digital manner, they can be easily interchanged on the instant.

6. References

[1]Hamid R. Karshenas; Hamid Daneshpajooh; Alireza Safaee; Praveen Jain; Alireza Bakhshai, "Bidirectional DC - DC Converters for Energy Storage Systems, Energy Storage in the Emerging Era of Smart Grids", Prof. Rosario Carbone (Ed.), ISBN: 978-953-307-269-2, InTech, 2011

[2]Shafquat Ullah Khan; Ali I.Maswood; Hossein Dehghani Tafti; Muhammad M.Roomi; Mohd Tariq, "Control of Bidirectional Converter for Back to Back NPC-based Wind Turbine System under Grid Faults", in 4th ICDRET, 2016

[3]Muhammad Aamir; Kafeel Ahmed Kalwar; Saad Mekhilef, "Review: Uninterruptible Power Supply system", in Renewable and Sustainable Energy Reviews 58 1395–1410, 2016

[4]Di Han; Jukkrit Noppakunkajorn; Bulent Sarlioglu, "Comprehensive Efficiency, Weight, and Volume Comparison of SiC- and Si-Based Bidirectional DC–DC Converters for Hybrid Electric Vehicles" in IEEE transactions on vehicular technology, vol. 63, no.7, Sep. 2014

[5] Jih-Sheng Lai; Douglas J.Nelson, "Energy Management Power Converters in Hybrid Electric and Fuel Cell Vehicles", in Proceedings of the IEEE Vol. 95, No. 4, April 2007

[6]Kong Zhiguo; Zhu Chunbo; Yang Shiyan; Cheng Shukang, "Study of Bidirectional DC-DC Converter for Power Management in Electric Bus with Supercapacitors", in Vehicle Power and Propulsion Conference, in 2006

[7]F.Caricchi; F.Crescimbirii; A.Di Napoli, "20 kW Water-cooled Prototype of a BBB DC-DC Converter Topology for Electrical Vehicle Motor Drives", in APEC, 1995

[8]Giuseppe Guidi; Atsuo Kawamura; Yuji Sasaki and Tomofumi Imakubo, "Dual Active Bridge Modulation with Complete Zero Voltage Switching Taking Resonant Transitions into Account", Proceedings of the EPE, 2011

[9] Fang Z. Peng; Hui Li; Gui-Jia Su and Jack S. Lawler, "A New ZVS Bidirectional DC–DC Converter for Fuel Cell and Battery Application", IEEE Transactions on power electronics (Vol. 19, No. 1), Jan. 2004

[10]Manu Jain, "A Bidirectional DC–DC Converter Topology for Low Power Application", IEEE transactions on power electronics, (Vol. 15, No. 4), July 2000

[11]Pritam Das; Brian Laan; Seyed Ahmad Mousavi and Gerry Moschopoulos, "A Nonisolated Bidirectional ZVS-PWM Active Clamped DC–DC Converter", in IEEE Transactions on Power Electronics (Vol.24, No.2), Feb. 2009

[12]Stefan Waffler; Johann W. Kolar, "A Novel Low-Loss Modulation Strategy for High-Power Bidirectional Buck+Boost Converters", IEEE transactions on power electronics, (Vol. 24, No. 6), June 2009

[13]Kristian Kruse, "GaN-based High Efficiency Bidirectional DC-DC Converter with 10 MHz Switching Frequency", in APEC, 2017

[14]Kou-Bin Liu; Chen-Yao Liu; Yi-Hua Liu; Yuan-Chen Chien; Bao-Sheng Wang; Yong-Seng Wong, "Analysis and Controller Design of A Universal Bidirectional DC-DC Converter", in NSRRC, June 2016

Lifetime Estimation and Reliability of PV Inverter With Multi-Timescale Thermal Stress Analysis

Sara Bouguerra[1], Kamel Agroui[2], Oussama Gassab[3], Ariya Sangwongwanich[4], and Frede Blaabjerg[4]

[1] Signals and Systems Research Laboratory, Department of Electronic, Institute of Electrical and Electronics Engineering (IGEE), University M'Hamed Bougara of Boumerdes, Boumerdes, 35000, Algeria.
[2] Semiconductors Technology for Energetic Research Center (CRTSE) BP 140 Alger-7 Merveilles Algiers, Algeria.
[3] The Center for Microwave and RF Technologies (CMRFT), Key Lab of Ministry of Education for the Design and Electromagnetic Compatibility (EMC) of High-Speed Electronic Systems, Shanghai Jiao Tong University, Shanghai 200240, China.
[4] Department of Energy Technology, Aalborg University, 9220 Aalborg, Denmark.
alkharif2013@live.fr, kagroui@yahoo.fr, gassab@sjtu.edu.cn, ars@et.aau.dk, and fbl@et.aau.dk

Abstract— The reliability of the PV inverter is a critical issue because it is the less reliable component of the PV system. In order to lower the risks of failure and maintenance in PV systems, the factors that influence the PV inverter lifetime should be analyzed. Thermal stress is the main causes of IGBT failure in a PV inverter, which includes the fast cycling stress due to loss variations in an IGBT, and slow cycling due to mission profile fluctuations. In this paper, the design for reliability (DFR) approach based on mission profile analysis is used and demonstrated on a single phase, single stage, grid connected PV inverter installed in Algeria, where the lifetime is estimated by taking the effect of low frequency thermal cycling under long-term operation into consideration, in addition to the line frequency power cycling due to loss variations. The results reveal that low frequency thermal cycling have a significant impact on lifetime degradation and reliability of the PV inverter after many years of operation.

Keywords—Insulated-gate bipolar transistor, design for reliability, mission profile, PV inverters, thermal loading, Monte Carlo methods, reliability.

I. INTRODUCTION

Solar photovoltaic systems (PV) have achieved a grid parity in many countries, experts are optimistic of achieving 100% renewable energy systems by 2050, especially in industrialized countries. Moreover, many of them expect that the cost of renewables will likely undercut the fossil fuels in the next 10 years [1]. Solar energy is one of the most promising solutions for future energy supply. To achieve that goal, the energy harvesting from PV systems should be optimized, and this is done by increasing the efficiency of power conversion systems, which are based on power electronic technology. Therefore, the reliability of power electronics and lifetime estimation of power devices, used in PV converters, is an essential step to ensure high reliable operation of PV systems [2]. Solar PV systems can experience failures during the operation, which decrease their reliability and availability. Power electronic components are responsible for 37% of unscheduled maintenance in the PV system [3]. Hence, analyzing the lifetime and applying physics of failure approach is mandatory to improve the reliability of PV system [4].

Most power converters including DC-DC converters and DC-AC converters use IGBTs as switching devices. In fact, IGBTs are among the weakest components in a PV inverter, which usually fail after a certain period [5]. The heat generated by the IGBT chip during operation causes the power module's

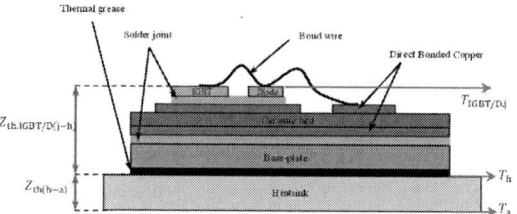

Fig. 1. Schematic cross section of a power module where the most non-reliable parts are the bond wire and the solder joint.

temperature to vary rapidly, referred to as thermal cycling [6]. Thermal cycling occurs at the different IGBT layers, which causes fatigue at the bonding wire and the solder layer [7]. Moreover, the thermal expansion difference between bond wire and chip causes bond wire liftoff due to stress [3]-[8]. The varying thermal expansion coefficient between the different layers of the IGBT module, see Fig. 1, induces fatigue in the device or even failure. In [10] it was investigated that semiconductor devices are the least reliable in the power converter. Hence, the reliability the power converter depends on the reliability of power electronic components inside. The reliability of power electronic devices is usually affected by external conditions, mainly solar irradiance and ambient temperature, they are referred to as mission profile data [11]. During the operation, the thermal stress is present in two modes. The first mode is the power cycling, or line frequency power cycling, it is caused by load variations which results in continuous power losses during operation, and this represents the high frequency power cycling. The second mode is thermal cycling due to variations of the surrounding thermal environment (mission profiles effect), this is known as low frequency thermal cycling [12]. In [13], a statistical approach is used to predict the bond wire fatigue in fourth generation 1200/50A IGBTs, used in three-phase grid-connected PV inverters. In this research, the accumulated damage corresponding to low frequency thermal cycling due to mission profiles and the line frequency (50Hz) thermal cycling are summed up to obtain the bond wire accumulated damage, and the effect of mission profiles is included in evaluating long term performance of this kind of IGBT. In [14], the design for reliability approach is applied for a 6 kW single-phase single-stage PV inverter installed in Denmark, where the line frequency thermal cycling (50 Hz) is considered, and the effect of low frequency thermal cycling is neglected when extracting the thermal loading static parameters of the IGBT device.

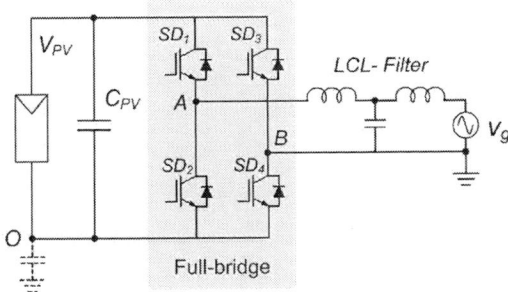

Fig. 2. Single-phase transformerless inverter with LCL filter.

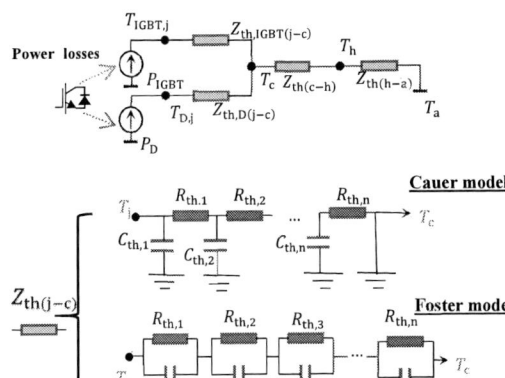

Fig. 4. Cauer and Foster thermal models for the IGBT and the diode.

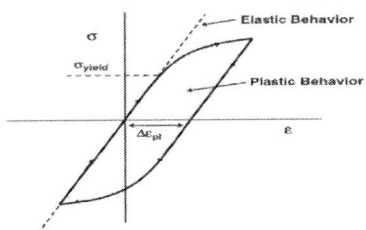

Fig. 3. Internal structure of a semiconductor device with thermal representation.

Fig. 5. Stress-strain curve (adapted from [19]).

In this paper, the IGBTs and inverter system lifetime for long-term operation is predicted using design for reliability approach. The effect of low frequency thermal cycling due to mission profile (irradiance and ambient temperature) is demonstrated by predicting the lifetime when only the line frequency is considered, and when both effects are taken into account (low frequency thermal cycling and line frequency power cycling). The paper is organized as follows: Section II explains the design for reliability approach, including thermal modelling for IGBTs, lifetime model development and reliability assessment using Monte Carlo simulation. In Section III the procedure is applied to a case study on 8.6 kW single phase, single stage PV inverter, installed in Algeria. The reliability assessment based on Monte Carlo Simulation is carried out with comparing the effects of line frequency and low frequency thermal cycling for a long-term operation. Finally, conclusions are drawn in Section IV.

II. DESIGN FOR RELIABILITY APPROACH FOR PV INVERTERS

The power inverter is an important electronic device since it converts direct current (DC) coming from the PV solar panels into an alternative current (AC) for the consumer grid. However, the PV inverter is the most non-reliable part, mainly due to the fact that the IGBTs are highly stressed by the

thermal cycling as it was mentioned earlier. To predict the reliability of the PV inverter, an accurate thermal model for the IGBTs should be developed.

A. Single-Phase Transformerless Inverters

The single-phase, single stage transformerless inverter topology is shown in Fig. 2. This inverter is made of two half bridge IGBT power modules, each module consists of an IGBT ship and freewheeling diode (FWD).

B. Thermal Model for the IGBT

The failure can happen in the IGBT or diode due to the thermomechanical breakdown that occurs in bond wire liftoff or solder cracking [15 -16]. The power losses in an IGBT or diode include the switching losses (on and off) and the conduction losses. The junction temperature T_j of the power electronic device (diode or IGBT) is related to the power loss and the thermal impedance of the power device as follows:

$$T_j = T_h + P_{loss}(Z_{th\ j-c} + Z_{th\ c-h}) \qquad (1)$$

where T_h is the heat sink temperature, $Z_{th\ j-c}$ is the thermal impedance from junction to case, $Z_{th\ c-h}$ is the thermal impedance from case to heat sink. The IGBT layers with thermal impedance representation are given in Fig. 3.

There are two common equivalent thermal circuits, Cauer network [17] and Foster network [18]. These networks describe the temperature distribution between the actual physical layers (die, solder...etc.). The R and C elements represent the thermal resistance and capacitance for each physical layer in the power electronic component. The topology for Foster and Cauer networks are given in Fig. 4.

978-1-5386-7688-2/19 $31.00 © 2019 IEEE   403

C. Temperature Cycling Effect and Lifetime Models

Temperature cycling is the main cause of IGBT module failure. In order to explain this effect, Fig. 5 shows stress-strain curve. In the elastic region, no damage is supported to occur during cycling. However, in the plastic region, each stress cycle will drive a certain plastic deformation ε to the material, and the degradation will induce material failure. In this case, during temperature cycling, stress is applied to different IGBT layers, more details are given in [19].

Since the power electronic device is the less reliable component of the PV inverter, several researches are done [20-21] in order to predict the lifetime (cycles to failure) of the power device as a function of the thermal cycling data. The most commonly used lifetime models are given below:

Coffin-Mason Model:

$$N_f = \alpha \cdot \Delta T_j^{-n} \qquad (2)$$

This is the simplest lifetime model, ΔT_j is the fluctuation of the junction temperature, α and n are constants that can be obtained experimentally.

Coffin-Mason Model-Arrhenius Model:

$$N_f = \alpha \cdot \Delta T_j^{-n} \exp\left(\frac{E_a}{K_b T_{jm}}\right) \qquad (3)$$

In this model the mean junction temperature T_{jm} , the Boltzmann constant K_b , and the activation energy E_a are introduced.

Bayerer Model:

$$N_f = K \cdot \Delta T_j^{-\beta_1} \exp\left(\frac{\beta_2}{T_{jm}+273}\right) t_{on}^{\beta_3} I^{\beta_4} V^{\beta_5} D^{\beta_6} \qquad (4)$$

This model was developed by Bayerer et al in 2008, where many influencing factors were included [21]. In this lifetime model, the effect of power-on time period t_{on}, chip thickness, bonding technology, diameter of bonding wire D, current per wire bond I, blocking voltage of the Chip V, and package type were all included, the parameters of this model are acquired by an experimental- accelerated test [22].

U. Scheuermann et al model:

$$N_f = A \cdot \Delta T_j^{\alpha} \cdot \exp\left(\frac{E_a}{K_b \cdot T_{jm}}\right) \cdot ar^{\beta_1 \cdot \Delta T_j + \beta_0} \cdot \left(\frac{C + t_{on}^{\gamma}}{C+1}\right) \cdot f_D \qquad (5)$$

This is the latest lifetime model, where a phenomenological approach is used to generalize the old model and add other factors into consideration. The parameters $A, \alpha, \beta_1, \beta_0, C \dots$, are given in TABLE I. [23]

TABLE I
PARAMETERS OF THE LIFETIME MODEL OF AN IGBT MODULE [23].

Parameters	Value	Experimental condition
A	3.4368×10¹⁴	
α	-4.923	64 K ≤ ΔT_j ≤ 113 K
β₁	-9.012×10⁻³	
β₀	1.942	0.149 ≤ ar ≤ 0.42
C	1.434	
γ	-1.208	0.07 s ≤ t_on ≤ 63 s
f_D	0.6204 for diode	
	1 for IGBT	32.5 °C ≤ T_j ≤ 122 °C
E_a	0.06606 eV	
k_b	8.6173324×10⁻⁵ eV/K	

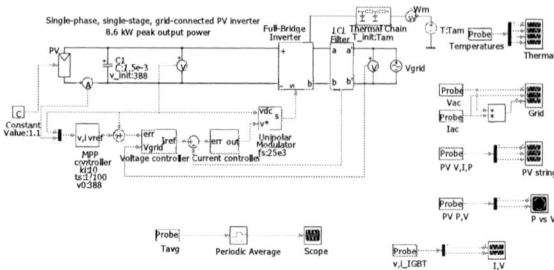

Fig. 6. Single phase - single stage grid connected PV inverter in PLECS software.

In order to estimate the lifetime of the PV inverter, mission profile data should be translated into thermal loading of the PV inverter (junction temperature T_j). Afterward, the cycle counting algorithm (e.g., the rainflow algorithm) is used [24]. This algorithm converts the irregular loading profile of the junction temperature T_j into regular loading with certain parameters $\Delta T_{ji}, n_i, T_{jmi}$ which denote the mean junction temperature, the number of cycles, the mean junction temperature, respectively for each regular loading i . Then, the lifetime model is applied and the lifetime period can be estimated [14].

The lifetime consumption LC is computed using Miner's rule given by:

$$LC = \sum_i \frac{n_i}{N_{fi}} \qquad (6)$$

where n_i is the number of cycles, obtained by Rainflow analysis, N_{fi} is number of the cycles to failure, calculated by the lifetime model. Finally, LC indicates how much time the device consumed during operation. If LC is unity ($LC = 1$), then the device reaches its end of life [25].

D. Monte Carlo Analysis and Reliability Assessement

The application of the lifetime model and minor's rule gives the life consumption of the IGBT module, but this result can be way far from reality, because the IGBT parameter variations and the statistical properties of the lifetime model are not taken into account. In order to overcome parameter uncertainties due to manufacturing process and application of the lifetime model, Monte Carlo Simulation is performed. Physical parameters and stress variations are considered, the parametric quantities: A, β_0, β_1 and stress parameters: ΔT_j and T_{jm} are modeled with normal distribution. After that, a large set of population is taken randomly from each distribution and then it is applied to the lifetime model to evaluate the damage accumulation [13-14]. The system level reliability assessment is performed considering the reliability bloc diagram [26], in the case of full bridge inverter topology (4 IGBT module components), see Fig. 2, the failure of one component would induce the whole design inverter system failure, so the unreliability function can be given by:

$$F_{tot} = 1 - \prod_4(1 - F_n(x)) \qquad (7)$$

III. CASE STUDY

In this section, a case study on 8.6 kW single phase, single stage grid connected PV inverter will be presented for lifetime estimation and reliability analysis of the PV inverter. The design for reliability approach discussed in [14] will be used. The PV system is configured according to Fig. 2, and the single phase single stage PV system given in [27] is used. The

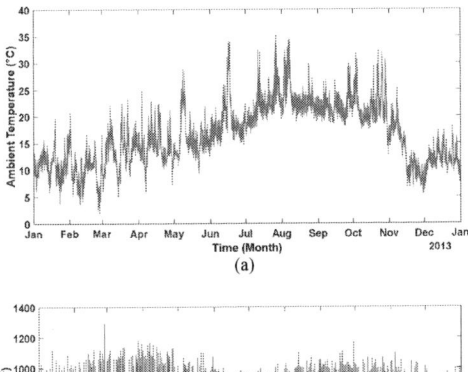

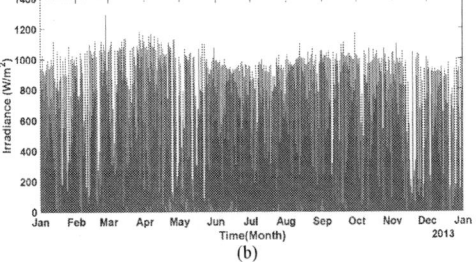

Fig. 7. Yearly mission profile data (ie, ambient temperature and irradiance with sampling rate of 5 minutes per sample) in Algeria: (a) Ambient temperature and (b) Solar irradiance.

PV panel BP 365/65W is chosen to represent a 8.6 kW PV system, where 6 strings are connected in parallel and each string consists of 22 modules. The model in [27] is modified by including the thermal model of the IGBTs. The simulation of the case study and thermal modeling are realized in PLECS software, as shown in Fig. 6. The IGBTs and diodes used are IKW50N60H3 (600V/50A) [28] and the thermal model of IGBTs and diodes are built according to the manufacturer datasheet.

A. Mission Profiles of the Case Study

The lifetime of single IGBT and PV inverter is determined according to mission profile data. The yearly mission profile: solar irradiances and ambient temperatures, are recorded from the installation site of Algiers energy center (CDER), Algeria. The sampling frequency is 5 minutes per sample and the maximum and minimum temperatures are given by $35.37°C$ in July and $1.85°C$ in February, respectively. The maximum solar irradiance throughout the year is $1293\ W/m^2$ in February. The mission profile data are shown in Fig. 7.

B. Translation of Thermal Loading

The mission profile data are translated into thermal loading of the PV inverter. The device junction temperature profile T_j is obtained by PLECS software after inserting the thermal model of the IGBTs and Diodes from the manufacturer datasheet. In order to get the thermal loading parameters: mean junction temperature T_{jm} and cycle amplitude of the junction temperature variation ΔT_j for a yearly mission profile, a lookup table is generated by performing the simulation several times using 4 ambient temperatures $(-5, 15, 35, 55)°C$ and 16 solar irradiance levels $(0, 100, \dots 1500)$ W/m^2. For each simulation, T_{jm} and ΔT_j values are extracted.

The lookup table is constructed in Simulink/MATLAB and the junction temperature parameters: T_j and ΔT_j are

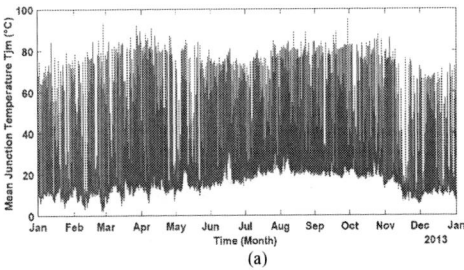

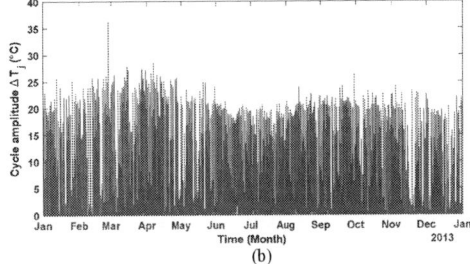

Fig. 8. Thermal loading of a single IGBT device in the PV inverter for yearly mission profile data: (a) Mean junction temperature T_{jm} and (b) Cycle amplitude of the junction temperature variation ΔT_j.

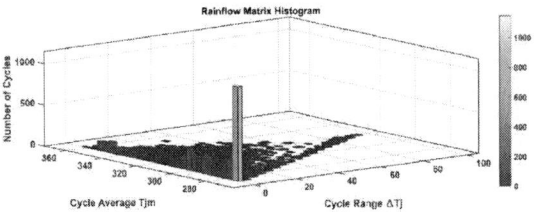

Fig. 9. Rainflow counting damage histogram of a single IGBT device.

obtained for the yearly mission profile. The obtained thermal loading data of a single IGBT device is shown in Fig. 8.

The mean junction temperature reaches its maximum value $T_{jm(Max)} = 95.33°C$ at the end of September and the maximum cycle amplitude is $\Delta T_{j(Max)} = 36.16°C$ at the end of February.

C. Rainflow Cycle Counting Algorithm and Lifetime Evaluation

The application of Rainflow cycle counting algorithm results in the extraction of new parameters: $T_{jmi}, \Delta T_{ji}, n_i$ for a cycle period t_{on} and regular loading i. For mission profile fluctuations, t_{on} is set to be 60 s, for line frequency power cycling it is chosen to be 0.01s. After that, the lifetime model is applied to find of number of cycles to failure for each regular loading i. The lifetime consumption of a single IGBT device under the high frequency power cycling and low frequency thermal cycling (due to mission profiles) is found to be 104 years. In order to obtain the life consumption due to mission profiles only, Rainflow cycle counting algorithm is used, in Fig. 9 it can be noticed that the damage is almost concentrated where cycle amplitude ΔT_j is very small.

The heating time for high frequency power cycling is below $60s$ ($t_{on} < 60s$), in order investigate the effect of this high frequency thermal cycling, the same lifetime model in (5)

TABLE II
DAMAGE ACCUMULATION RESULTS

Loading	Heating period t_{on}	Accumulated damage LC in one year	Lifetime
Line frequency	0.01 s	0.0084	119.23 years
Mission profiles (Irr & Temp)	60 s	0.0012	815.10 years
Total effects		0.0096	104.012 years

TABLE III
EQUIVALENT STATIC VALUES OF THE STRESS PARAMETERS

Parameters	Line frequency	Irr & Temp
Mean junction temperature $T_{jm(static)}$	18.030°C	18.030°C
Cycle amplitude $\Delta T_{j(static)}$	13.167°C	38.155°C
Heating period $t_{on(static)}$	0.01 s	60 s
Number of cycles per year n_i	1.577×10^9	8758
Number of cycles to failure N_f	1.8771×10^{11}	7.139×10^6
Damage accumulation per year LC	0.0084	0.0012

is used to evaluate the number of cycles to failure $N_{f(line\ freq)}$, but without using Rainflow analysis, the recorded thermal loading values T_{jm} and ΔT_j for each 5 minutes are used, and the number of cycles in 5 minutes is fixed to $n'_{i(line\ freq)}$ where $n'_{i(line\ freq)} = 50 \times 60 \times 5 = 1.5 \times 10^4$ cycles in 5 minutes. Then the damage accumulation $LC_{(line\ freq)}$ is obtained.

$$LC_{(line\ freq)} = n'_{i(line\ freq)} \sum_i \frac{1}{N_{f(line\ freq)}} \quad (8)$$

where $n'_{i(line\ freq)}$ is constant and $N_{f(line\ freq)}$ is a function of ΔT_{ji} and T_{jmi} and t_{on}=0.01s, for each loading i. The damage accumulation due to line frequency power cycling is found to be 0.84% ,which results in lifetime of 119 years, the results are summarized in TABLE II.

In order to evaluate the reliability of the IGBT and the PV inverter, Monte Carlo simulation is performed. The static values should be taken from stress parameters T_{jm}, ΔT_j, t_{on} and the lifetime model parameters : A, β_0, β_1 , the method of extracting these parameters is explained in [14] and the list of static parameters is given in TABLE III. After obtaining the static values, they are modelled as normal distribution with 5% parameter variation, and the simulation is repeated 10.000 times, where a random value from these static parameters is taken each time and the lifetime model is applied to obtain the damage accumulation for each simulation. Finally, a histogram of end of life for a large population of IGBTs is evaluated, as shown is Fig. 10. It is worth noticing that the best fit to lifetime distribution for this histogram is the Weibull probability density function (PDF), it is given as:

$$f(x) = \frac{\beta}{\eta^\beta} x^{\beta-1} \exp\left[-\left(\frac{x}{\eta}\right)^\beta\right] \quad (9)$$

where β and η are the shape and scale parameters respectively. The value of η corresponds to the time when 63.2 % of population is failed. The lifetime consumption of the device is evaluated for two cases. The first case only considers the line frequency power cycling, while the second case takes also low frequency thermal cycling effect into consideration. The results are shown Fig. 11 where the unreliability function (cumulative distribution function) of the Weibull distribution is shown for one single IGBT

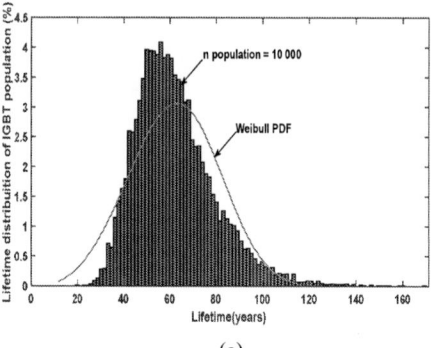

(a)

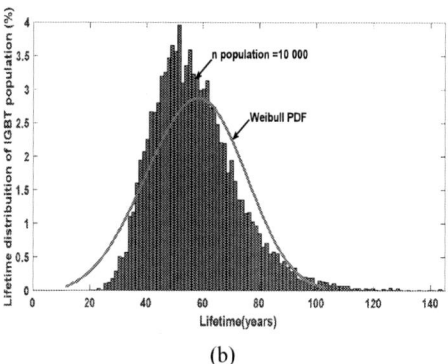

(b)

Fig. 10. Lifetime distribution of a single IGBT under the effect of: (a) line frequency thermal cycling (b) line frequency & mission profiles thermal cycling.

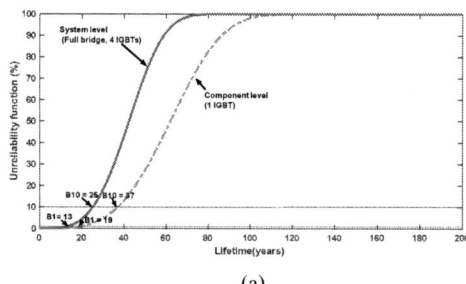

(a)

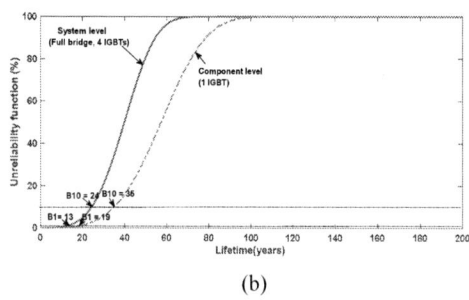

(b)

Fig. 11. Unreliability function (Weibull CDF function) of component and system level under the effects: (a) line frequency thermal cycling (b) line frequency & mission profiles thermal cycling.

(component level). In addition, the formula in (7) is used to evaluate the system level unreliability function. The B_x lifetime , the time when x% of the population of power

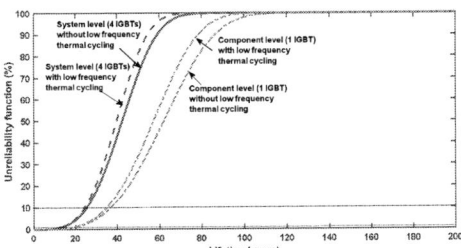

Fig. 12. Impact of low frequency thermal cycling on the unreliability function (Weibull CDF function) of component and system.

TABLE IV
LIFETIME EVALUATION RESULTS BASED ON MONTE CARLO SIMULATIONS

Operating Conditions	Component Level			System level		
	B_{50}	B_{10}	B_1	B_{50}	B_{10}	B_1
Line Frequency thermal cycling	62	37	19	42	25	13
Line frequency & mission profiles	57	35	19	40	24	13

devices is failed, can be obtained from the unreliability function, the results for different cases are shown in TABLE IV. The results reveal that when only the line frequency power cycling is considered, 1% of IGBT population will fail after 19 years, 10% will fail after 37 years, and 50% of IGBT population will fail after 62 years, whereas, when the low frequency thermal cycling is added, the device fails 2 years earlier for 10% of IGBT population, and 5 years earlier for 50 % of population. Fig. 12 shows the impact of low frequency thermal cycling due to mission profile, on the unreliability function of component level (single IGBT) and system level (i.e., 4 IGBTs). The comparison of the two unreliability graphs, when the line frequency is taken into account, and when it is neglected in lifetime prediction, reveals that the low frequency thermal cycling decrease the lifetime significantly after many years of operation, mainly after about 40 years of operation, and it is apparent that the effect of low frequency cycling is more dominant in component level, single IGBT, rather than system level.

IV. CONCLUSION

In this paper, the design for reliability (DFR) approach is applied to a case study on grid-connected PV systems. The mission profile based lifetime estimation is employed, and the effect of mission profile fluctuations in lifetime of PV inverter for long term operation is demonstrated through reliability assessment. It has been shown that mission profile low frequency thermal loading is a critical issue in lifetime degradation after many years of operation. It is worth to mention that the future view in reliability assessment of power electronics is to analyze the different factors influencing the thermal loading of power electronic devices used in inverter systems, and take these impacts into consideration in the design phase, in order to increase robustness and reduce the cost accordingly.

ACKNOWLEDGMENT

We would like to give special thanks to Mr Omar Boumia, Dr Amar Hadj Arab and Dr Mohamed Rédha Yaiche from CDER (Centre de Développement des Energies Renouvelables), Algiers, Algeria for their help in providing the needed mission profile data.

REFERENCES

[1] REN21,Paris,France,"Renewables 2017: Global Futures Report (GFR)," 2017.[Online]. Available: http://www.ren21.net/.

[2] F. Blaabjerg, K. Ma and Y. Yang, "Power electronics - The key technology for Renewable Energy Systems," *2014 Ninth International Conference on Ecological Vehicles and Renewable Energies (EVER),* Monte-Carlo, 2014, pp. 1-11.

[3] A.Albarbar, and C.Batunlu, *Thermal Analysis of Power Electronic Devices Used in Renewable Energy Systems,* Springer International Publishing, 2018.

[4] S. Yang, A. Bryant, P. Mawby, D. Xiang, L. Ran and P. Tavner, "An industry-based survey of reliability in power electronic converters," IEEE Trans. Ind. Appl., vol. 47, no. 3, pp. 1441-1451, May-June 2011.

[5] H. Wang et al., "Transitioning to physics-of-failure as a reliability driver in power electronics," *IEEE J. Emerg. Sel. Topics Power Electron.,* vol. 2, no. 1, pp. 97-114, March 2014.

[6] T.Y. Hung, C.J. Huang, C.C. Lee, C.C. Wang, K.C. Lu and K.C.Chiang, " Investigation of solder crack behavior and fatigue life of the power module on different thermal cycling period," *in Microelectronic Engineering,* 107, 125-129.

[7] M. Musallam, C.M. Johnson, C. Yin, H. Lu, C. Bailey, "In-service life consumption estimation in power modules, " *in Power Electronics and Motion Control Conference, 2008. EPE-PEMC 2008.* 13th, 2008, pp. 76–83.

[8] M.Ciappa, "Selected failure mechanisms of modern power modules," *in Microelectronics reliability* 42.4-5 (2002): 653-667.

[9] L.Zhou, J.Wu and P.SU, " Junction temperature management of IGBT module in power electronic converters," *in Microelectronics Reliability,* 2014, vol. 54, no 12, p. 2788-2795.

[10] S. Yang, A. Bryant, P. Mawby, D. Xiang, L. Ran and P. Tavner, "An industry-based survey of reliability in power electronic converters," *IEEE Trans. Ind. Appl.,* vol. 47, no. 3, pp. 1441-1451, May-June 2011.

[11] Y.Yang, H. Wang, A. Sangwongwanich, and F. Blaabjerg. "Design for reliability of power electronic systems." *In Power Electronics Handbook,* pp. 1423-1440. Butterworth-Heinemann, 2018.

[12] V. Smet, F. Forest, J.-J. Huselstein, F. Richardeau, Z. Khatir, S. Lefebvre, and M. Berkani, "Ageing and failure modes of IGBT modules in high temperature power cycling," *IEEE Trans. Ind. Electron.,* vol. 58, no. 10,pp. 4931–4941, Oct. 2011.

[13] P. D. Reigosa, H. Wang, Y. Yang and F. Blaabjerg, "Prediction of bond wire fatigue of IGBTs in a PV inverter under a long-term operation," *IEEE Trans. Power Electron.,* vol. 31, no. 10, pp. 7171-7182, Oct. 2016.

[14] Y. Yang, A. Sangwongwanich and F. Blaabjerg, "Design for reliability of power electronics for grid-connected photovoltaic systems," *CPSS Trans Power Electron. Appl.,* vol. 1, no. 1, pp. 92-103, Dec. 2016

[15] M. Musallam, C. Yin, C. Bailey and M. Johnson, "Mission Profile-Based Reliability Design and Real-Time Life Consumption Estimation in Power Electronics," *IEEE Trans. Power Electron.,* vol. 30, no. 5, pp. 2601-2613, May 2015.

[16] H. Huang and P. A. Mawby, "A Lifetime Estimation Technique for Voltage Source Inverters," *IEEE Trans. Power Electron.,* vol. 28, no. 8, pp. 4113-4119, Aug. 2013.

[17] W. Cauer, "Die Verwirklichung der Wechselstromwiderstände vorgeschriebener Frequenzabhängigkeit", *Electr. Eng.,* vol. 17, no 4, pp. 355–388, 1926.

[18] R.M. Foster, "Academic and theoretical aspects of circuit theory," *Proc. IRE.,* vol. 50, no. 5, pp. 866–871, May 1962.

[19] J. W. McPherson, *Reliability Physics and Engineering: Time-to Failure Modeling,* Springer, 2010.

[20] M. Held, P. Jacob, G. Nicoletti, P. Scacco and M. -. Poech, "Fast power cycling test of IGBT modules in traction application," *Proceedings of Second International Conference on Power Electronics and Drive Systems,* Singapore, 1997, pp. 425-430 vol.1.

[21] R. Bayerer, T. Herrmann, T. Licht, J. Lutz and M. Feller, "Model for power cycling lifetime of IGBT modules - various factors influencing lifetime," *5th International Conference on Integrated Power Electronics Systems,* Nuremberg, Germany, 2008, pp. 1-6.

[22] F. Blaabjerg, K. Ma and D. Zhou, "Power electronics and reliability in renewable energy systems," *2012 IEEE International Symposium on Industrial Electronics*, Hangzhou, 2012, pp. 19-30.

[23] U. Scheuermann, R. Schmidt and P. Newman, "Power cycling testing with different load pulse durations," *7th IET International Conference on Power Electronics, Machines and Drives* (PEMD 2014), Manchester, 2014, pp. 1-6.

[24] C. Lalanne, *Fatigue Damage* 3rd ed. John Wiley & Sons, Inc. 2014.

[25] H. Huang and P. A. Mawby, "A Lifetime Estimation Technique for Voltage Source Inverters," *IEEE Trans. Power Electron.*, vol. 28, no. 8, pp. 4113-4119, Aug. 2013.

[26] P .O'Connor and A. Kleyner, *Practical Reliability Engineering*, John Wiley & Sons, 2012.

[27] https://www.plexim.com/support/application-examples/solar.

[28] Infineon. IGBT module—IKW50N60H3. [Online]. Available: www.infineon.com.

Multiloop PR+P Controller of Integrated BESS-DVR for Power Quality Improvement

Abdul Muiz Sufianto
PT Perusahaan Listrik Negara Persero
Jakarta, Indonesia
abdul.muiz@pln.co.id

Jaeho Choi
School of Electrical Engineering
Chungbuk National University
Cheongju, Republic of Korea
choi@chungbuk.ac.kr

Nanang Hariyanto, Arwindra Rizqiawan
School of Electrical Engineering and
Informatics
Institute of Technology Bandung
Bandung, Indonesia
nanang.hariyanto@stei.itb.ac.id
windra@stei.itb.ac.id

Abstract— The power quality problems such as voltage sag are a great concern to consumers, especially for industrial consumers. Most modern industrial equipment is quite sensitive to instantaneous voltage sags, so it is necessary to keep the power quality to guarantee the reliability of the production process. The Dynamic Voltage Restorer (DVR) with Battery Energy Storage System (BESS) has been applied as one of the representative devices for compensating the voltage sags. This paper proposes a multiloop PR control based an integrated BESS-DVR, which is organized in the stationary reference frame, to mitigate the power quality problems such as the unbalanced voltage sag, deep voltage sag, and phase jump. The advantages of the proposed multiloop controller provide the rapid response to the disturbance on the grid to track the sinusoidal reference, while the configuration is simpler than the PI controller. In this paper, the design of the proposed controller is explained in detail, and the simulation results of comparative studies with the PI controller in the synchronous reference frame are described. The proposed model of an integrated BESS-DVR system and the control algorithm are validated by PSiM simulation under various voltage disturbances such as unbalanced and balanced voltage sag.

Keywords— *DVR, multiloop controller, power quality, PR controller*

Nomenclature

V_{ph_sag}	Phase sag voltage
I_l	Load current
$Cos\phi$	Load power factor
L_{dc}	Inductance of the bidirectional DC-DC converter
V_{bat}	Battery voltage
$V_{dc\text{-}dc}$	Output voltage of the bidirectional DC-DC converter
D_{boost}	Duty cycle of the boost converter
D_{buck}	Duty cycle of the buck converter
ΔI_l	Inductor ripple current
f_s	Switching frequency of the bidirectional DC-DC converter
C_{dc}	DC-link capacitance
k	Dynamic energy factor
a	Overloading factor
I_{ph}	Phase current
V_{ph}	Phase voltage
t	Minimum time required for achieving steady-state
V_{dcmin}	Minimum DC-link voltage
V_{dc}	Average DC-link voltage
V_{ll}	Line to line grid voltage
C_{fmax}	Maximum filter capacitance
P_n	Rated active power
ω_0	Fundamental frequency
L_{fmin}	Minimum filter inductance
f_{sw}	Switching frequency of the inverter
ΔI_{imax}	Maximum ripple of the inverter current

m	PWM modulation index
R_d	Damping resistance
ζ	Damping factor
L_f	Inductance of the filter
C_f	Capacitance of the filter
$G(s)$	Transfer function of the LC filter
I_{cf}	Capacitor filter current
V_{inv}	Inverter voltage
K_c	Gain of the inner current controller
K_{pwm}	Gain of the PWM
V_l	Load voltage
V_{lref}	Load reference voltage
ω_c	Cut off frequency
K_p	Gain of the proportional controller
K_r	Gain of the resonant controller
$G_{i(s)}$	Transfer function of the converter model for average current
$G_{v(s)}$	Transfer function of the converter model for average voltage
R_{dc}	Resistance of the bidirectional DC-DC converter
D	Duty cycle of the bidirectional DC-DC converter
$C_{i(s)}$	Compensator of the current controller
$C_{v(s)}$	Compensator of the voltage controller

Abbreviations

PR	Proportional Resonant
P	Proportional
BESS	Battery Energy Storage System
DVR	Dynamic Voltage Restorer
PI	Proportional Integrator
FESS	Flywheel Energy Storage System
UCAP	Ultracapacitor
SMES	Superconducting Magnetic Energy Storage
Li-on	Lithium-Ion
PCC	Point of Common Coupling
VSC	Voltage Source Converter
DoD	Depth of Discharge
PWM	Pulse Width Modulation

I. INTRODUCTION

In recent years, the power quality problems in the electrical distribution system are greatly concerned by consumers, especially for industrial consumers. Among the power quality issues, it is focused on the voltage sag defined as the decrease in rms voltage from the rated value by 10% to 90% with a duration of 0.5 cycles to 1 minute [1]. Most modern industrial equipment is quite sensitive to the instantaneous voltage sags, so it is necessary to keep the power quality to guarantee the reliability of the production process. Voltage sag that occurs just a split of a second can shut down the operation of sensitive equipment at industrial consumers. It results in the restarting of an industrial process or even damages the equipment, so it leading to an increase in the

978-1-5386-7688-2/19 $31.00 © 2019 IEEE

production cost. Voltage sags are caused by the fault on the transmission or distribution line, the fault in the consumer's installation, the switch-on of the heavy load, and the start-up of the large motor. It is known that the ratio of the phase number associated with the voltage drop below 85% and the duration in 60 seconds is composed of 68% for single-phase, 21% for two phases, and 11% for three phases [2]. Therefore, most of the voltage sags are unbalanced and the balanced voltage sags are only 11%. Furthermore, the voltage sags always accompanied by the phase angle jump which is depending on the line and fault impedances.

The Dynamic Voltage Restorer (DVR) has been used as one of the effective custom power devices for compensating the voltage sags. The DVR alone can compensate for the low voltage drop, but it requires an energy storage device or a separate energy source to compensate for the deep voltage sags[3]. The various types of rechargeable energy storage such as Flywheel Energy Storage System (FESS), Battery Energy Storage System (BESS), Ultracapacitor (UCAP) Energy Storage System, and Superconducting Magnetic Energy Storage (SMES) integrated with DVR are developed [4]–[8]. Among them, based on the technological developments and the cost reduction estimation for battery energy storage systems, the Lithium-Ion (Li-on) batteries are recommended to be suitable for integrating with DVR [9]. The estimated costs for providing BESS have declined from year to year, so the cost of the topology with energy storage for DVR is no longer a problem.

Generally, there are several algorithms to control the DVR. A multiloop control which is a combination of the feedback path and the feedforward path with using the proportional controller was presented in [10]. Furthermore, it is developed by replacing the Proportional controller with the Proportional Integral (PI) controller [11]. The result of this controller has a better performance in terms of steady-state and transient state for regulating various load voltage conditions rather than the open-loop controller which produces the poorly damped response. However, it has some disadvantages to using the PI controller, such as the voltage tracking errors and the limited disturbance rejection capability. Especially, to compensate for the unbalanced voltage sags, it needs the separately positive and negative sequence compensators in the synchronous reference frame. On the other hand, the Proportional Resonant (PR) controller of the stationary reference frame is simpler than the PI controller, because is equivalent separately to the positive and negative sequence of the PI controller in the synchronous reference frame [12].

This paper proposes the multiloop PR control based an integrated BESS-DVR to mitigate the power quality problems such as the unbalanced voltage sag, deep voltage sag, and phase jump. A multiloop controller with combining the PR controller for the voltage controller and the P controller for the current controller is built in the stationary reference frame. During the unbalanced and distorted grid condition, the synchronization signal often defected. To eliminate this effect, the method to extract the positive-sequence component in a stationary reference frame is used further to generate the voltage reference [13]. The proposed method has a simple structure and can mitigate the unbalanced voltage sags and phase jump.

The system configuration and the design of the proposed controller are described in detail. The performance of this

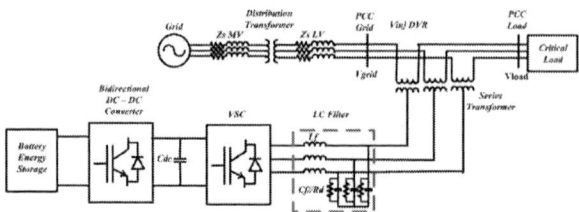

Fig. 1. Configuration of integrated BESS-DVR system.

method is compared with that of the PI controller in the synchronous reference frame to compensate for various voltage disturbance such as unbalanced and balanced voltage sag. The proposed model of BESS-DVR system and the proposed controller are validated by the PSiM simulation.

II. CONFIGURATION OF INTEGRATED BESS-DVR SYSTEM

The configuration of an integrated BESS-DVR system is shown in Fig. 1. It composes of a series transformer installed between the Point of Common Coupling (PCC) of the load and the grid, a LC filter with resistor damping that serves to reduce the harmonics generated by the switching frequency at the inverter output, a Voltage Source Converter (VSC), a DC-link capacitor, and a bidirectional DC-DC converter connected to the BESS. The DVR is connected in series between the PCC of the load and the grid. The DVR operates when the voltage sag occurred in the grid, so the DVR injects the missing voltage to keep the load voltage same as the normal value. The required power during the period of compensation is supplied from the BESS.

A. Capacity of BESS

The capacity of the BESS for DVR depends on the length of the compensation period and the depth of voltage sag. Further, the BESS must sufficiently to provide the power when it compensates the voltage sag. The capacity of the BESS is sized according to the power and duration required for compensating voltage sag, the depth of discharge and the efficiency of the battery. The minimum capacity of BESS can be sized by (1):

$$\text{Cap\_min(Wh)} = \frac{\text{Power(W)} \times \text{Duration(h)}}{\text{DoD(\%)} \times \text{Efficiency(\%)}} \quad (1)$$

where the required power, Power(W), during compensating the voltage sag is:

$$\text{Power(W)} = 3 \times V_{\text{ph\_sag}} \times I_l \times \text{Cos}\phi \quad (2)$$

where $V_{\text{ph\_sag}}$ is the phase sag voltage, I_l is the load current, and $\text{Cos}\phi$ is the load power factor. In this paper, the BESS-DVR is designed for the deep voltage sags with a duration of 1 minute and a load of 100kVA, 400V, and 0,9 lagging. The Lithium-ion battery has the Depth of Discharge (DOD) varying between 80% to 100% and these technologies have an efficiency approximate with a range between 92% to 96% [9].

B. The Bidirectional DC-DC Converter

Fig. 2 shows the block diagram of the bidirectional DC-DC converter. It has two operation mode depending on the power flow direction: Boost and buck mode. It operates in a boost mode when the BESS discharges energy to the grid, and it does in a buck mode when the BESS absorbs energy from

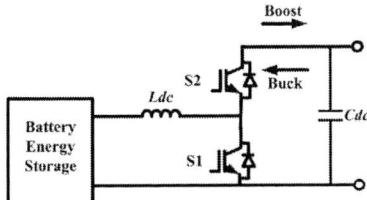

Fig. 2. Block diagram of a bidirectional DC-DC converter for BESS.

the grid or other sources. The inductor parameter of the bidirectional DC-DC converter is determined by (3) and (4) during boost mode and buck mode, respectively [14].

$$L_{dc} = \frac{V_{bat} \times D_{boost}}{2 \times \Delta I_l \times f_s} \qquad (3)$$

$$L_{dc} = \frac{(V_{dc\text{-}dc} - V_{bat}) \times D_{buck}}{2 \times \Delta I_l \times f_s} \qquad (4)$$

where V_{bat} is the battery voltage, $V_{dc\text{-}dc}$ is the DC-DC converter output voltage to be regulated at 700V, ΔI_l is the inductor ripple current of which typical value is 10% to 20% of full load value, f_s is the switching frequency. Since the output ripple current from the DC-DC converter is affected by the inductor parameter value, so the largest calculated inductor value between (3) and (4) is selected.

The value of the DC-link capacitor can be described as in (5) [15]:

$$C_{dc} = \frac{3 \times k \times a \times V_{ph} \times I_{ph} \times t}{0.5 \times (V_{dc}^{2} - V_{dcmin}^{2})} \qquad (5)$$

where k is the dynamic energy factor of which typical value is 10%, a is the overloading factor chosen as 1.2, I_{ph} and V_{ph} is a phase current and voltage, t is the minimum time required for achieving steady-state chosen as 1 cycle period of 20ms, V_{dcmin} and V_{dc} is the minimum and the average DC-link voltage. For a depth of modulation selected as 1, V_{dcmin} can be obtained as in (6) [15]:

$$V_{dcmin} = \frac{2 \times \sqrt{2} \times V_{ll}}{\sqrt{3}} \qquad (6)$$

where V_{ll} is the line to line grid voltage.

C. Parameter of LC Filter

The value of the filter component depends on the output ripple current, the filter resonance frequency, the drop voltage from the inductance filter, and the power factor of the system. Furthermore, the typical value of power factor ratio from the system is not to exceed 5% of the decrease in power factor, so the maximum filter capacitance can be determined as in (7) [16]:

$$C_{fmax} = \frac{5\% \times P_n}{3 \times \omega_0 \times V_{ll}^{2}} \qquad (7)$$

where P_n is the rated active power, and ω_0 is the fundamental frequency of the system. The large size inductor is required to eliminate the lower order harmonics of the output current, but it is caused more voltage drop. Therefore, the ripple current of inverter affects the inductor value, so the value of the inductor is given by (8) [16]:

$$L_{fmin} = \frac{2 \times V_{dc}}{3 \times f_{sw} \times \Delta I_{imax}} \times (1 - m) \times m \qquad (8)$$

where f_{sw} is the switching frequency of the inverter, ΔI_{imax} is the maximum ripple current given as 10% of the maximum inverter rated current, and m is the PWM modulation index which is typically given by 0.5 for sinusoidal PWM.

The drawback of the LC filter is that it produces a resonance effect that can result in harmonics generated by the switching frequency of the inverter and low order harmonics generated by over-amplification. Therefore, the value of resonance frequency should be chosen as between ten times of the fundamental frequency to not more than a half from the switching frequency. For damping of the resonance effect from the LC filter, the parallel resistor is added at the capacitor. The damping resistor can be obtained as in (9):

$$R_d = \frac{1}{2 \times \zeta} \times \sqrt{\frac{L_f}{C_f}} \qquad (9)$$

where ζ is the damping factor chosen with the natural damping factor value of 0.707.

III. CONTROLLER OF DVR AND BESS

An integrated BESS-DVR system has two controllers; a multiloop PR+P controller which is built in the stationary reference frame to mitigate the voltage sag thus the load voltage can be maintained constantly during the disturbance period and a controller of the bidirectional DC-DC converter for regulating the DC-link voltage to be constant.

A. Multiloop PR+P Controller

In general, the multiloop controller for the inverter consists of the outer loop voltage controller which has the feedback of the load voltage or the output DVR voltage and the inner loop current controller which has the feedback of the inductor current or the capacitor current. In this paper, the multiloop controller has an additional feedforward path to provide the instantaneous response under the transient change of the grid voltage. Even though with an additional feedforward path, the system stability is not affected because it doesn't change the characteristics equation of controller [11]. Further, the outer loop voltage controller considers the load voltage as the feedback to regulate the load voltage. While the inner loop current controller considers the capacitor current as of the feedback to improve the dynamic voltage controller, because the current distortion is less to affect the capacitor current and the distortion of output voltage filter can be compensated instantaneously.

The proposed multiloop controller is shown in Fig. 3. All controllers are based on a stationary reference frame which includes the inner current controller with a proportional controller and the outer voltage controller with a PR controller. If the voltage sag occurs in the grid, the actual load voltage is compared with the load reference voltage, and then the error between both is applied to the outer voltage PR controller. The output of the outer loop voltage controller

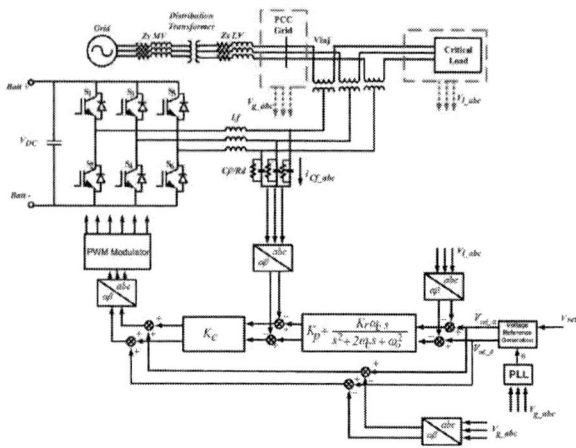

Fig. 3. Configuration of DVR and multiloop control block diagram using PR+P controller.

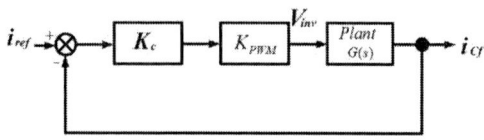

Fig. 4. Block diagram of an inner current controller.

generates the virtual current reference and it is compared with the actual capacitor current. The difference between both is applied to the inner current proportional controller. The resulting of the inner loop current controller is forward fed to generate the PWM signal for the inverter. The output voltage of the inverter is injected to the grid to maintain the load voltage constantly.

1) Design of Inner Current Controller

The inner current controller with a proportional controller is shown in Fig. 4. I_{cf} is the capacitor current as of the feedback, I_{ref} is the virtual current reference generated from the outer loop voltage controller, V_{inv} is the inverter voltage, and the plant represents the LC filter with passive damping resistor when the output current of the load as a disturbance is ignored. The transfer function of a plant can be derived as in (10):

$$G(s) = \frac{I_{cf}}{V_{inv}} = \frac{sC_fR_d + 1}{s^2 L_fC_fR_d + sL_f + R_d} \quad (10)$$

When the feedforward path is ignored, the closed-loop transfer function of the inner current controller can be derived as in (11):

$$\frac{I_{cf}}{I_{ref}} = \frac{K_cK_{pwm}(sC_fR_d + 1)}{s^2 L_fC_fR_d + s(L_f + K_cK_{pwm}C_fR_d) + R_d + K_cK_{pwm}} \quad (11)$$

Fig. 5 shows the corresponding root locus of the inner current controller for the closed-loop transfer function with various values of K_c and the value K_{pwm} is assumed of 1. Using the SISO tool in MATLAB software to analyze the stability of the system for the selected crossover frequency of 1kHz which

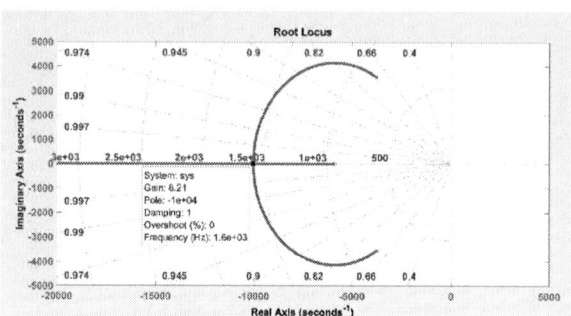

Fig. 5. Root locus of the inner current controller.

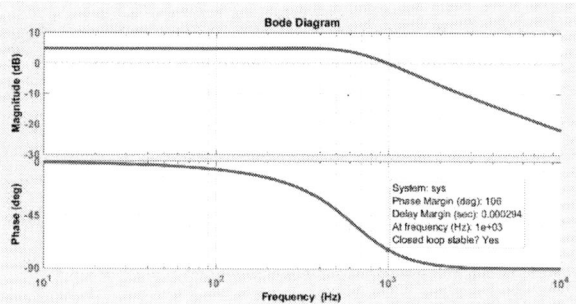

Fig. 6. Bode plot of the inner current controller with the selected value of K_c=3.26.

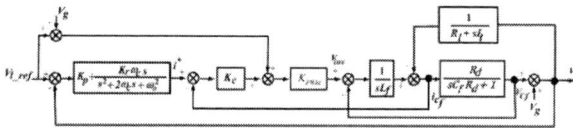

Fig. 7. Block diagram of outer voltage controller.

is one-tenth of the switching frequency, the gain value of the inner current controller determined of 3.26. Fig. 6 shows a bode plot of the inner current controller for the open-loop transfer function with the selected value of $K_c = 3.26$. It shows that the system is stable with a phase margin of 106° at frequency 1kHz.

2) Design of Outer Voltage Controller

Fig. 7 shows the block diagram of the outer voltage control loop with a PR controller. In the block of PR controller, ω_0 is the grid fundamental frequency, and ω_c is the cut off frequency of the fundamental component. In practically, the value of ω_c is in the range 5 to 15rad/s to provide good performance for transient response [17]. The closed-loop transfer function of the outer voltage controller shown in Fig. 7 can be derived as in (12):

$$\frac{V_l}{V_{lref}} = \frac{n_3s^3 + n_2s^2 + n_1s + n_0}{d_5s^5 + d_4s^4 + d_3s^3 + d_2s^2 + d_1s + d_0} \quad (12)$$

Fig. 8 shows the corresponding root locus of the closed-loop transfer function of (12) for the value of K_p=1 and the various value of K_r=10, 100, and 1000. For the values of K_r=10, the system is unstable because there is a pole of the closed-loop transfer function located on the right half-plane of the 's' field. As increasing the value of K_r, it shows that the pole is shifted to the left half-plane of the 's' field, and it indicates that the system becomes stable. The stability of the

978-1-5386-7688-2/19 $31.00 © 2019 IEEE   412

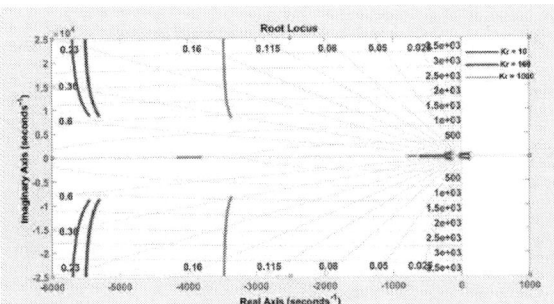

Fig. 8. Root locus of outer voltage controller for K_p = 1 and various values K_r = 10, 100, 1000.

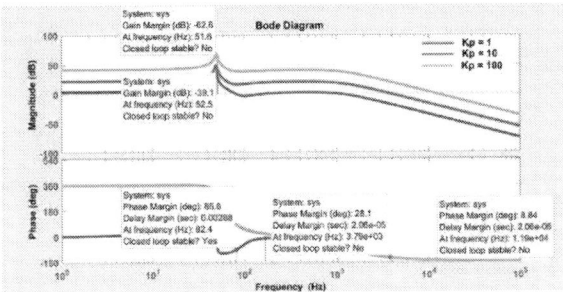

Fig. 9. Root locus of outer voltage controller for various value K_p = 1, 10, 100 and K_r =100.

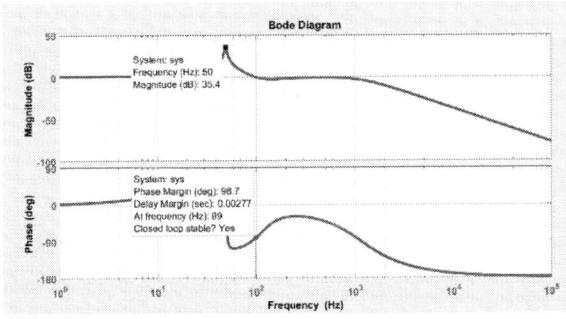

Fig. 10. Bode plot of outer voltage controller for K_p = 0.6 and K_r = 110.

system also has been studied by using the bode plot of the open-loop transfer function. The system may be stable if both the gain and phase margin in the open-loop system is positive. The simulation has been done with various value of K_p=1, 10, and 100 while the K_r value is kept constant at 100 and as shown in Fig. 9 With increasing the value of K_p, the gain is more negative, and the phase margin is decreasing.

For selected crossover frequency of 100 Hz which is one-tenth of the crossover frequency of the inner current controller, the gain K_p and K_r of the outer voltage controller is determined 0.6 and 110, respectively. Fig. 10 shows the bode plot of the open-loop transfer function for outer voltage controller with values of K_p, K_r, and ω_c is 0.6, 110, and 6 rad/s, respectively. The system is stable with a phase margin of 98.7° at frequency 99Hz and at the fundamental frequency have the gain 35.4dB.

B. Bidirectional DC-DC Converter

Fig. 11 shows the controller of the bidirectional DC-DC converter which has the current and voltage controllers. The function of the bidirectional DC-DC converter is to maintain the DC-link voltage to be constant during the compensation

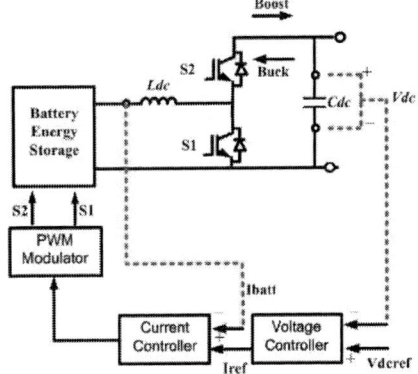

Fig. 11. Configuration of bidirectional DC-DC converter with the controller.

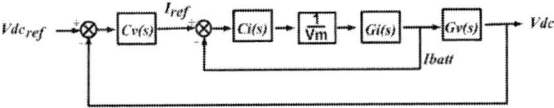

Fig. 12. Control block diagram of a bidirectional DC-DC converter.

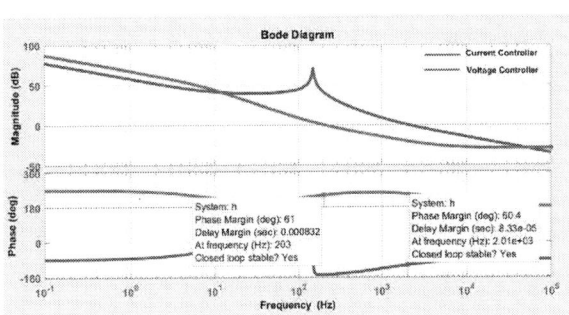

Fig. 13. Bode plot of a bidirectional DC-DC converter.

period.

The actual voltage of the DC-link capacitor is compared with the reference voltage set at 700V, and then the error is multiplied by the voltage controller and generates the reference current. Further, the reference current is compared with the actual inductor current and the result fed to the current controller to generate the signal for the PWM generator. The transfer function of the converter model for average current and voltage mode control can be derived as in (13) and (14) [18]:

$$G_{i(s)} = \frac{V_{dc\text{-}dc}(sC_{dc} + \frac{2}{R_{dc}})}{s^2 L_{dc}C_{dc} + s\frac{L_{dc}}{R_{dc}} + (1-D)^2} \tag{13}$$

$$G_{v(s)} = \frac{(1-D)(1 - \frac{sL_{dc}}{R_{dc}(1-D)^2})}{sC_{dc} + \frac{2}{R_{dc}}} \tag{14}$$

The control block diagram of the current and voltage controllers for the bidirectional DC-DC converter which is based on the PI controller is shown in Fig. 12 [18]. For tuning the gain parameters of the PI controller, it is used the SISO tool in MATLAB software. The crossover frequency for the current controller is chosen as 2kHz and 200 Hz for the voltage controller, respectively, and then the compensator of both controllers is given by:

$$C_{i(s)} = 1.879 \cdot 10^{-3} + \frac{13.422}{s} \qquad (15)$$

$$C_{v(s)} = 9.41 + \frac{6767.1}{s} \qquad (16)$$

The Bode plot of the open-loop transfer function for both controllers with the compensator (15) and (16) is shown in Fig. 13. It can be observed that the current controller has a phase margin of 60.4°. while the voltage controller has a phase margin of 60.9°.

IV. SIMULATION AND RESULT

To prove the validity of the proposed multiloop PR+P controller in the stationary reference frame, the dynamic performance of the proposed BESS+DVR is analyzed through the PSiM simulation. The simulation results of the proposed controller for unbalanced and balanced voltage sags are compared with those of the PI controller in the synchronously rotating reference frame. All system parameters for the simulation are shown in Table 1.

TABLE I. PARAMETERS FOR SIMULATION

Parameter	Value
Grid	400 V, 50 Hz
Load	100 kVA, 400 V, 50 Hz, 0.9 PF
Filter Capacitance	89.5 µF
Filter Inductance	0.65 mH
DC-Link Voltage	700 V
DC-DC Converter Inductance	0.12 mH
BESS Rating	624 V, 2.5 Ah
Inner Current Controller	$K_c = 3.26$
PR Controller	$K_p = 0.6, K_r = 110$
Switching Frequency	10 kHz and 20 kHz
Fundamental Frequency	$\omega_0 = 314$ rad/s
Cut-off frequency	$\omega_c = 6$ rad/s
PI Voltage Controller	$K_p = 1, K_i = 100$
DC-DC Current Controller	$K_p = 0.001879, K_i = 13.422$
DC-DC Voltage Controller	$K_p = 9.41, K_i = 6767.1$

A. Unbalanced Voltage Sag

The single line ground fault which causes the unbalanced voltage sag is the most popular among the power system faults. As a case study for the unbalanced voltage sag, phase C occurs the single line ground fault from 0.2s to 0.3s for 0.1s. Fig. 14 shows the simulation results of the compensation performance of the proposed integrated BESS-DVR system in the case of the unbalanced voltage sag. Fig. 14 illustrated the waveforms of the grid voltage, the injected DVR voltage, the load voltage, and the DC-link voltage. It shows that the phase C of the grid voltage is reduced to 60% from 320V to 195V, while phase A and B have sagged to 85% from 320V to 272V. It indicates that the healthy phase A and B are sensed the voltage sag due to the configuration of the transformer connection. It also shows that the load voltage is maintained constant with balanced voltage during the compensation

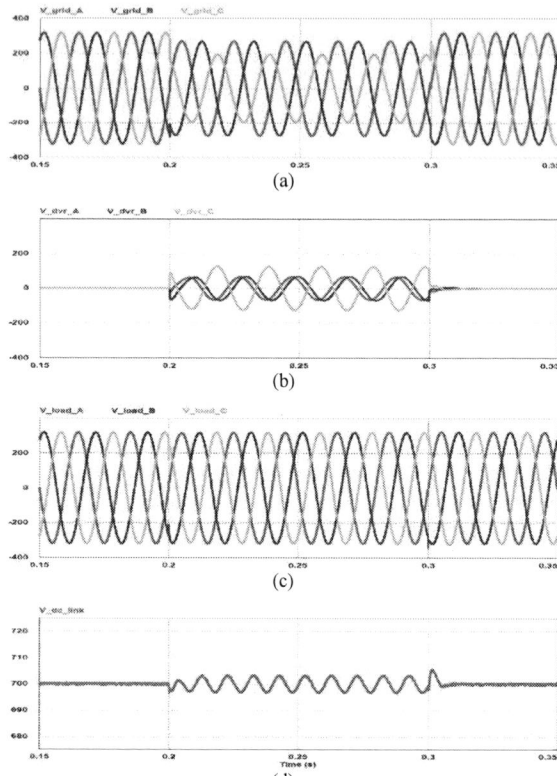

Fig. 14. Simulation waveforms of unbalanced voltage sag compensation using proposed integrated BESS-DVR: (a) Grid voltage, (b) injected DVR voltage, (c) load voltage, (d) DC-link voltage.

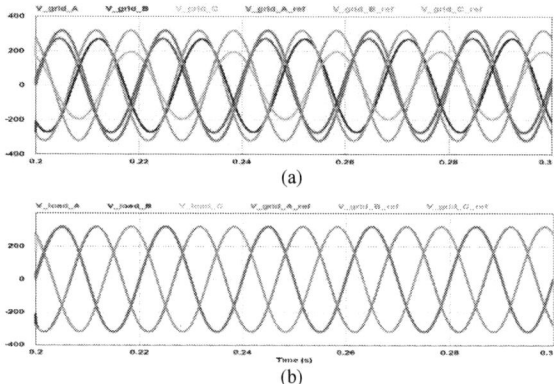

Fig. 15. Mitigation performance of phase jump during unbalanced voltage sag compensation by proposed integrated BESS-DVR: (a) Grid voltage and grid reference voltage waveforms, (b) load voltage and grid reference voltage waveforms.

period by injecting a missing voltage from DVR. Fig. 14 also illustrates the performance of the DC-DC converter for regulating the DC-link capacitor voltage. Generally, it indicates that the DC-link voltage is regulated constantly at the reference setting voltage of 700V during the voltage sag compensation. It shows that the peak to peak ripple from the DC-link voltage is 0.5% during the compensation period.

Fig. 15 illustrates the grid and load voltage waveforms with the reference grid voltage waveform during the unbalanced voltage sag event. The reference grid voltage is under the condition without any disturbance. In Fig. 15(a), the unbalanced voltage sag causes the phase jump at the healthy

978-1-5386-7688-2/19 $31.00 © 2019 IEEE

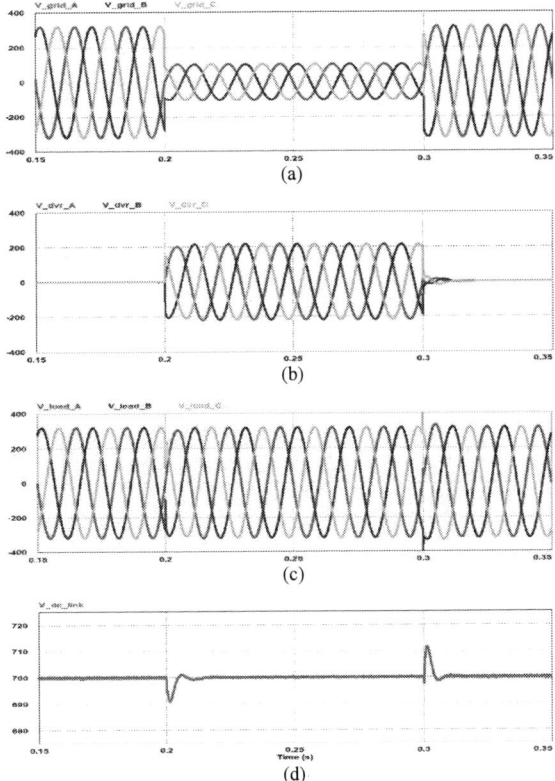

Fig. 16. Simulation waveforms of balanced voltage sag compensation using proposed integrated BESS-DVR with proposed controller: (a) Grid voltage, (b) injected DVR voltage, (c) load voltage, (d) DC-link voltage.

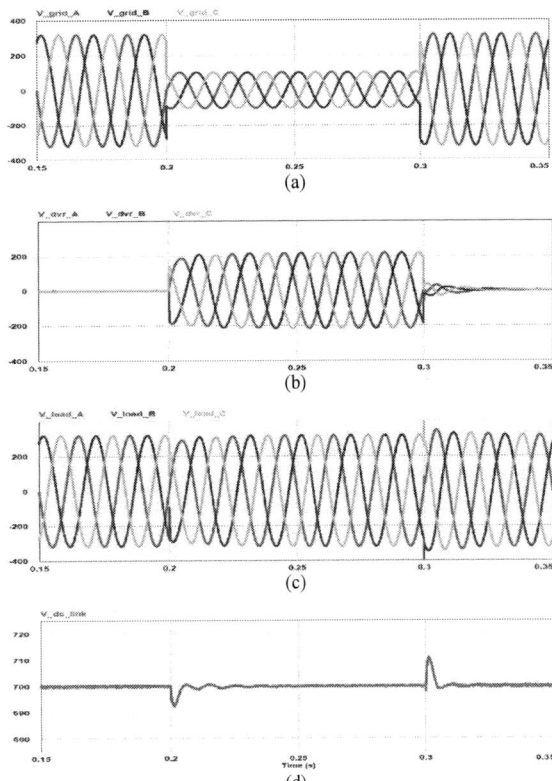

Fig. 17. Simulation waveforms of balanced voltage sag compensation using proposed integrated BESS-DVR with PI controller: (a) Grid voltage, (b) injected DVR voltage, (c) load voltage, (d) DC-link voltage.

phases, phase A and B around 8.5° and 9.5°, respectively. However, there is no phase jump at the fault phase, phase C. In Fig. 15(b), it shows that the proposed multiloop controller can mitigate the phase jump, so the load voltage waveform is in phase with the waveform of grid reference voltage during the occurrence of disturbance.

B. Balanced Voltage Sag

For the case study of the deep balanced voltage sag, a three-phase ground fault is considered in the distribution line. The voltage magnitudes of all phases are decreased from the nominal voltage from 0.2s to 0.3s for 0.1s. Fig. 16 shows the simulation results of the deep balanced voltage sag with the proposed multiloop controller. It shows that the phases A, B, and C of the grid voltage are sagged to around 32% from 320V to 102.4V, and then the proposed DVR injects the missing voltage to maintain the load voltage to be constant.

Fig. 16 also shows the performance of the DC-DC converter to regulate the DC-link capacitor voltage. It indicates that the DC-link voltage is well regulated with the peak to peak ripple of 0.065% from the DC-link voltage at the transition time of the compensation period.

As the comparative study with the proposed multiloop controller, the performance of the existing PI controller for the deep balanced voltage sag case is shown in Fig. 17. From the waveforms of the grid voltage, the injected DVR voltage, the load voltage, and the DC-link voltage, it is verified that the transient response of the proposed multiloop controller is a little bit better than that of the existing PI controller.

V. CONCLUSION

In this paper, a multiloop PR+P controller of an integrated BESS-DVR is proposed to mitigate the power quality problems such as the unbalanced voltage sag, the deep voltage sag, and the phase jump. The BESS system integrated with DVR provides power during compensation. A multiloop controller using the PR+P controller in a stationary reference frame has the load voltage and the capacitor current controller as a feedback path, furthermore a feedforward path is added. As the advantage of a multiloop controller, it provides the fast response to the disturbance that occurs on the grid, the good tracking performance to follow the sinusoidal reference while the configuration is simpler than that of the PI controller.

ACKNOWLEDGMENT

This paper was supported by PT Perusahaan Listrik Negara (persero) Indonesia, Human Capital Management, No.1773.K/SDM.00.03/DIR/2017.

APPENDIX

$n3 = K_{pwm}R_dL_l(K_pK_c+1)$

$n2 = K_{pwm}R_d(L_l\omega_c(K_c(2K_p+K_r)+2)+R_l(K_pK_c+1))$

$n1 = K_{pwm}R_d(L_l\omega_0^2(K_pK_c+1)+R_l\omega_c(K_c(2K_p+K_r)+2))$

$n0 = K_{pwm}R_dR_l\omega_0^2(K_pK_c+1)$

$d5 = L_fC_fR_d(L_l+L_t)$

$d4 = L_fC_fR_d(2\omega_c(L_l+L_t)+(R_l+R_t))+L_f(L_l+L_t)+K_{pwm}K_cC_fR_d(L_l+L_t)$

978-1-5386-7688-2/19 $31.00 © 2019 IEEE   415

$d3 = L_f C_f R_d(\omega_0^2(L_l+L_t)+2\omega_c(R_l+R_t))+L_f(2\omega_c(L_l+L_t)+(R_d+R_l+R_t))+R_d(L_l+L_t)+K_{pwm}K_c(K_p R_d L_l+C_f R_d(2\omega_c(L_l+L_t)+(R_l+R_t))+(L_l+L_t))$

$d2 = L_f\omega_0^2(C_f R_d(R_l+R_t)+(L_l+L_t))+2L_f\omega_c(R_d+R_l+R_t)+2\omega_c R_d(L_l+L_t)+R_d(R_l+R_t)+K_{pwm}K_c(R_d(K_p R_l+L_l\omega_c(2K_p+K_r))+C_f R_d(\omega_0^2(L_l+L_t)+2\omega_c(R_l+R_t))+2\omega_c(L_l+L_t)+(R_l+R_t))$

$d1 = L_f\omega_0^2(R_d+R_l+R_t)+R_d\omega_0^2(L_l+L_t)+2\omega_c R_d(R_l+R_t)+K_{pwm}K_c(R_d(R_l\omega_c(2K_p+K_r)+L_l K_p\omega_0^2)+\omega_0^2((L_l+L_t)+C_f R_d(R_l+R_t))+2\omega_c(R_l+R_t))$

$d0 = R_d\omega_0^2(R_l+R_t)+K_{pwm}K_c(\omega_0^2(K_p R_d R_l+(R_l+R_t)))$

REFERENCES

[1] IEEE, *IEEE Recommended Practice for Monitoring Electric Power Quality*, June. 2009.

[2] V. Sag and I. Analysis, "Distribution System Power Quality Assessment: Phase II—Voltage Sag and Interruption Analysis," *IEEE IAS 52nd Annual Petroleum and Chemical Industry Conference*, 2005.

[3] J. G. Nielsen and F. Blaabjerg, "A detailed comparison of system topologies for dynamic voltage restorers," *IEEE Trans. Ind. Appl.*, vol. 41, no. 5, pp. 1272–1280, 2005.

[4] R. S. Weissbach, G. G. Karady, and R. G. Farmer, "A combined uninterruptible power supply and dynamic voltage compensator using a flywheel energy storage system," *IEEE Trans. Power Deliv.*, vol. 16, no. 2, pp. 265–270, 2001.

[5] P. Jayaprakash, B. Singh, D. P. Kothari, A. Chandra, and K. Al-Haddad, "Control of reduced-rating dynamic voltage restorer with a battery energy storage system," *IEEE Trans. Ind. Appl.*, vol. 50, no. 2, pp. 1295–1303, 2014.

[6] D. Somayajula and M. L. Crow, "An Integrated Dynamic Voltage Restorer-Ultracapacitor Design for Improving Power Quality of the Distribution Grid," *IEEE Trans. Sustain. Energy*, vol. 6, no. 2, pp. 616–624, 2015.

[7] Jing Shi, Yuejin Tang, Kai Yang, Lei Chen, Li Ren, Jingdong Li, and Shijie Cheng, "SMES Based Dynamic Voltage Restorer for Voltage Fluctuations Compensation," *IEEE Trans. Applied Supercond.*, vol. 20, no. 3, pp. 1360–1364, 2010.

[8] A. M. Gee, F. Robinson, and W. Yuan, "A Superconducting Magnetic

[9] IRENA, *Electricity storage and renewables: Costs and markets to 2030*, October 2017.

Energy Storage-Emulator/Battery Supported Dynamic Voltage Restorer," *IEEE Trans. Energy Convers.*, vol. 32, no. 1, pp. 55–64, 2017.

[10] M. Vilathgamuwa, A. A. D. R. Perera, and S. S. Choi, "Performance improvement of the dynamic voltage restorer with closed-loop load voltage and current-mode control," *IEEE Trans. Power Electron.*, vol. 17, no. 5, pp. 824–834, 2002.

[11] A. Karthikeyan, D. G. Abhilash Krishna, S. Kumar, and C. Nagamani, "Design and Analysis of Multi-Loop Feed Forward Control Schemes for DVR under Distorted Grid Conditions," *2017 14th IEEE India Counc. Int. Conf. INDICON 2017*, 2018.

[12] L. Herman, I. Papic, and B. Blazic, "A proportional-resonant current controller for selective harmonic compensation in a hybrid active power filter," *IEEE Trans. Power Deliv.*, vol. 29, no. 5, pp. 2055–2065, 2014.

[13] T. V. Tran, T. W. Chun, H. H. Lee, H. G. Kim, and E. C. Nho, "PLL-based seamless transfer control between grid-connected and islanding modes in grid-connected inverters," *IEEE Trans. Power Electron.*, vol. 29, no. 10, pp. 5218–5228, 2014.

[14] R. W. Erickson and D. Maksimovic, *Fundamentals of Power Electronics, 2nd Edition*, Kluwer Academic, New York, 2004.

[15] S. Devassy and B. Singh, "Analysis and Design of Solar PV Integrated UPQC," *Natl. Power Electron. Conf. 2015*, vol. 54, no. 1, pp. 73–81, 2015.

[16] M. Büyük, A. Tan, M. Tümay, and K. Ç. Bayindir, "Topologies, generalized designs, passive and active damping methods of switching ripple filters for voltage source inverter: A comprehensive review," *Renew. Sustain. Energy Rev.*, vol. 62, pp. 46–69, 2016.

[17] N. F. Roslan, J. A. Suul, A. Luna, I. Candela, and P. Rodriguez, "A Simulation Study of Proportional Resonant Controller Based on the Implementation of Frequency- Adaptive Virtual Flux Estimation with the LCL Filter," *41st Annu. Conf. IEEE Ind. Electron. Soc. (IECON 2015)*, pp. 1934–1941, 2015.

[18] D. Somayajula, S. Member, M. L. Crow, and A. P. Stage, "An Ultracapacitor Integrated Power Conditioner for Intermittency Smoothing and Improving Power Quality of Distribution Grid," *IEEE Trans. Sustain. Energy*, vol. 5, no. 4, pp. 1145–1155, 2014.

Optimal Low-Pass Butterworth Filter Design by an Enhanced ACO Algorithm

Bachir Benhala

Laboratory of Electronics, Automatics and Biotechnology, LEAB,
University of Moulay Ismail, Faculty of Sciences, B.P. 11201, Zitoune, Meknes, Morocco
b.benhala@fs-umi.ac.ma

Abstract— In this paper, we present an optimization algorithm based on a BAcktracking Ant Colony Optimization (BA-ACO) technique for dealing with the active analog filter design. The BA-ACO algorithm is applied to the Low-Pass Butterworth filter design, realized with components (Resistors and capacitors) selected from different manufactured series, to satisfy the filter design criteria. PSPICE simulations are used to validate the obtained result/performances. A comparison with published works is highlighted

Keywords—Metaheuristics; Ant Colony Optimization; Backtracking algorithm; Low-Pass Butterworth Filter.

I. INTRODUCTION

Integrated circuit design is a complicated and delicate process activity due to the number of variables involved and to the set of technological constraints to respect. Generally, the circuit sizing is carried out thanks to the experiment and the intuition of the designer or according to the approaches based on fixed topologies and/or statistical techniques [1]. However, these techniques are time consuming and do not guarantee reaching the global optimum solution.

To efficiently resolve circuits sizing optimization problems, some (meta-)heuristics and algorithms were proposed in the literature and are used by the designers, such as Tabu Search [2,3], Simulated Annealing [4], Genetic Algorithms (GA) [4], [5], local search (LS) [6]. However, the metaheuristics that gave the best results are those that are nature inspired. They are inventive, resourceful, efficient, easy to use and known as SI: 'Swarm Intelligence Techniques' [7]. The SI techniques focus on animal conduct in order to develop some meta-heuristics which can mimic their problem resolution abilities, namely Particle Swarm Optimization (PSO) [8], Artificial Bee Colony (ABC) [9,10] and Ant Colony Optimization (ACO) [11,12].

In our previous works, we have presented, successfully, the ACO technique to deal with analog circuits design and sizing [11-13]. This optimization technique leads to the best optimum qualities, but the ACO requires a significant execution time compared to other metaheuristics [14,15].

To enhance the quality of the solution, several modifications to the original ACO were introduced [16-21]. Despite these changes, which have improved the performance of the ACO algorithm, they have not tackled the problem of excessive accumulation of pheromone which entraps ACO in local optima.

The BA-ACO presents a way to overcome this problem. By using the principle of the Backtracking algorithm [22] to the ACO, in order to reduce the search space, to solve the problem of excessive accumulation of pheromones which increases the speed of convergence and improves the overall research capacity. The Backtracking technique is a method that optimizes the search by returning back to new selection if it fails to achieve the objectives.

In this work, we focus on the use of the BA-ACO method to solve a typical analog circuit sizing problem: a low-pass Butterworth filter design mainly composed of discrete elements (capacitors and resistors) that are available in the market in the form of approximate E12, E24, E48, E96, and E192 series. They are produced in approximate logarithmic multiples of a defined number of constant values. PSPICE software was used for performing simulations in order to check reached performances. The rest of the paper is structured as follows: The second part presents an overview of the BA-ACO technique. The third part deals with the application of the proposed algorithm to the optimal design of the Low-Pass Butterworth filter. Simulation results and a comparison with the basic ACO are provided in the fourth section to show the validity of the proposed algorithm. Finally, section five provides some concluding remarks.

II. THE BACKTRACKING ANT COLONY OPTIMIZATION TECHNIQUE: AN OVERVIEW

A. The basic ACO algorithm

The ACO is an evolutionary stochastic computational discipline well adaptable for solving hard combinatorial optimization problems. Inspired from the natural behavior of ants in finding the shortest distance between their nests and food sources, it's based on indirect communication within a colony of simple agents, called (artificial) ants which exchange information about good routes through a chemical substance called pheromone that accumulates for short routes and evaporate for long routes [23, 24].

ACO was initially used to solve graph related problems, such as the traveling salesman problem (TSP) [25], vehicle routing problem [26]... For solving such problems, ants randomly select the vertex to be visited. When ant k is in vertex i, the probability of going to vertex j is given by expression (1) [23, 24].

978-1-5386-7688-2/19 $31.00 © 2019 IEEE

$$p_{ij}^k = \begin{cases} \dfrac{(\tau_{ij})^\alpha \cdot (\eta_{ij})^\beta}{\sum_{l \in J_i^k} (\tau_{il})^\alpha \cdot (\eta_{il})^\beta} & if \quad j \in J_i^k \\ \\ 0 & if \quad j \notin J_i^k \end{cases} \tag{1}$$

where J_i^k is the set of neighbors of vertex i of the kth ant, τ_{ij}^k is the amount of pheromone trail on edge *(i,j)*, α and β are weightings that control the pheromone trail and the visibility value, i.e. η_{ij}, which expression is given by (2) :

$$\eta_{ij} = \frac{1}{d_{ij}} \tag{2}$$

d_{ij} is the distance between vertices *i* and *j*.

The pheromone values are updated each iteration by all the m ants that have built a solution in the iteration itself. The pheromone τ_{ij}, which is associated with the edge joining vertices *i* and *j*, is updated as follows:

$$\tau_{ij} = (1 - \rho)\, \tau_{ij} + \sum_{k=1}^{m} \Delta \tau_{ij}^k \tag{3}$$

where ρ is the pheromone evaporation rate, *m* is the number of ants, and $\Delta \tau_{ij}^k(t)$ is the quantity of pheromone laid on edge *(i, j)* by ant *k*:

$$\Delta \tau_{ij}^k = \begin{cases} \dfrac{Q}{L^k} & \text{if ant } k \text{ used edge } (i, j) \text{ in its tour,} \\ \\ 0 & \text{otherwise} \end{cases} \tag{4}$$

Q is a constant and L^k is the length of the tour constructed by ant *k*.

Each ant k will randomly choose a path according to the probability given by expression (1), and form a directed graph while randomly generating a rate of pheromone at the formed graph edges. At each iteration, the path giving the minimum value of the objective function (OF) sees its rate increase, in contrast with the other paths which pheromone rates are partially evaporated with respect to expression (3).

The ACO algorithm conducts research intelligently to perform global optimizations, it is characterized by good strength, positive feedback, distributed calculation, and it can easily combine with other algorithms. Therefore, ACO provides a powerful tool for the optimization of many fields [27]. This algorithm did not cease to be developed and improved continuously, but there are still some gaps of consuming time, easy to stagnation and easily fall into local optima [28].

In the ACO algorithm, pheromones are the means for indirect communication between ants; they are the bracket passage of information which directly affects the convergence

and the efficient resolution of the ACO algorithm [29].

B. The BA-ACO algorithm

In the ACO algorithm, when a trail is preferred it automatically has continuous accumulation of pheromones as iterations go on. Actually, this easily leads the algorithm to be trapped in a local optimum.

In order to overcome this drawback, the backtracking technique which is an algorithm that is back slightly on decisions to get out of a blocking [22], [30]: Backtrack the pheromone to the initial value each time the algorithm is trapped in a local optimum. The update has to be performed once it is noted that the current 'optimal' value does not change for a certain number of iterations.

The improved algorithm operates according to the ACO technique by including the following detailed points:

- When the optimal value does not change for N-time, the pheromone are updated on the optimal path, in each backtracking period

$$\tau_{ij}(t + \Delta t) = \tau_{ij}(t) - \left(\frac{NQ}{L_L} \right) \tag{5}$$

where L_L is the length of current local optimal tour.

- When the optimal value does not change for M-time, it gets back to the backtracking point, and it re-initializes the pheromone value.
- To improve the convergence speed, pheromones are updated with respect to updating rules, in the proposed algorithm, using local and global updating rules, as given by expressions (3) and (6), respectively.

$$\tau_{ij}(t + \Delta t) = \tau_{ij}(t) + \left(\frac{Q}{L_G} \right) \tag{6}$$

where L_G is the length of the current global optimal tour.

The rule of updating pheromones reduces the solution search space, so it can lessen the number of 'bad' solutions reached so far and thus can improve the quality of solutions and enhance the algorithm's performances. More details on the BA-ACO principle are presented in [31].

III. APPLICATION TO THE OPTIMAL DESIGN OF THE BUTTERWORTH LOW PASS FILTER

Analog active filters are extensively used in the separation and demodulation of signals, frequency selection decoding, estimation of a signal from noise; they are the key components in mixed-signal circuit design [32].

The considered circuit is a fourth order low pass Butterworth filter formed by two operational amplifiers, four resistors and four capacitors. The schematic of this filter is given in Figure 1.

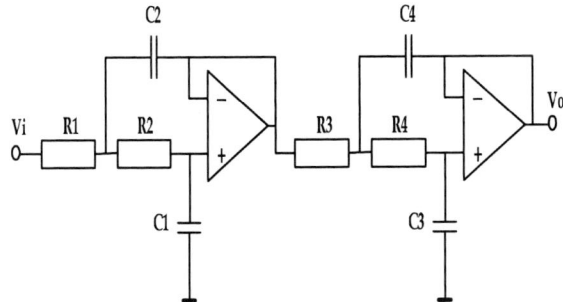

Fig. 1. Butterworth fourth order low-pass filter

The transfer function of this filter can be obtained as follows:

$$H(s) = \frac{\omega_{c1}^2}{s^2 + \dfrac{\omega_{c1}}{Q_1}s + \omega_{c1}^2} \times \frac{\omega_{c2}^2}{s^2 + \dfrac{\omega_{c2}}{Q_2}s + \omega_{c2}^2} \quad (5)$$

$$H(s) = \frac{1}{s^2 R_1 R_2 C_1 C_2 + s(R_1 C_1 + R_2 C_1) + 1}$$
$$\times \frac{1}{s^2 R_3 R_4 C_3 C_4 + s(R_3 C_3 + R_4 C_3) + 1} \quad (6)$$

The cutoff frequency $(\omega_{c1}, \omega_{c2})$ and the selectivity factor (Q_1, Q_2) of filter, which depend only on the values of the passives components, are given as follows:

$$\omega_{c1} = \frac{1}{\sqrt{R_1 R_2 C_1 C_2}} \quad (7)$$

$$\omega_{c2} = \frac{1}{\sqrt{R_3 R_4 C_3 C_4}} \quad (8)$$

$$Q_1 = \frac{\sqrt{R_1 R_2 C_1 C_2}}{R_1 C_1 + R_2 C_1} \quad (9)$$

$$Q_2 = \frac{\sqrt{R_3 R_4 C_3 C_4}}{R_3 C_3 + R_4 C_3} \quad (10)$$

For comparison reasons, the specification chosen here is [33]:

- $\omega_{c1} = \omega_{c2} = 10\ 000$ rad/s (1591.55 Hz)
- $Q_1 = 0.7654$
- $Q_2 = 1.8478$

The values of the resistors and capacitors to choose must be able to generate $\omega_{c1}, \omega_{c2}, Q_1$ and Q_2 approaching the specified values. For this, we define the Total Error (TE) [20] which expresses the offset values, of the cut-off frequency and the selectivity factor, compared to the desired values, by:

$$TE = 0.5 \Delta\omega + 0.5 \Delta Q \quad (11)$$

were

$$\Delta\omega = \frac{|\omega_{c1} - \omega| + |\omega_{c2} - \omega|}{\omega} \quad (12)$$

and

$$\Delta Q = \left| Q_1 - \frac{1}{0.7654} \right| + \left| Q_2 - \frac{1}{1.8478} \right| \quad (13)$$

The objective function considered is the Total Error (14). The decision variables are the resistors and capacitors forming the circuit. Each component must have a value of the standard series (E12, E24, E48, E96, E192). The resistors have values in the range of 10^3 to $10^6 \Omega$., similarly, each capacitor must have a value in the range of 10^{-9} to 10^{-6}F.

$$TE = \left(\begin{array}{c} 0.5 \dfrac{\left| \dfrac{1}{\sqrt{R_1 R_2 C_1 C_2}} - \omega \right| + \left| \dfrac{1}{\sqrt{R_3 R_4 C_3 C_4}} - \omega \right|}{\omega} \\ + 0.5 \left(\left| \dfrac{\sqrt{R_1 R_2 C_1 C_2}}{R_1 C_1 + R_2 C_1} - \dfrac{1}{0.7654} \right| + \left| \dfrac{\sqrt{R_3 R_4 C_3 C_4}}{R_3 C_3 + R_4 C_3} - \dfrac{1}{1.8478} \right| \right) \end{array} \right) \quad (14)$$

IV. SIMULATIONS AND COMPARISON RESULTS

In this section we applied BA-ACO algorithm to perform optimization of a low-pass Butterworth fourth order filter. The studied algorithm parameters are given in Table 1 with a generation algorithm of 1000. The optimization techniques work on MATLAB codes and are able to link PSPICE to measure performances.

TABLE I. THE ALGORITHMS' PARAMETERS

Number of Ants	100
Evaporation rate (ρ)	0.1
Quantity of deposit pheromone (Q)	0.2
Pheromone Factor (α)	1
Heuristics Factor (β)	1
N-time	3% of NCmax *
M-time	10% of NCmax *

* NC is the number of iterations

The optimal values of resistors and capacitors forming the considered filter and the performance associated with these values for the different series are shown in Table 2.

TABLE II. VALUES OF COMPONENTS AND RELATED FILTER PERFORMANCES

	E12	E24	E48	E96	E192
R1 (KΩ)	4.70	4.70	4.87	4.87	4.42
R2 (KΩ)	4.70	5.10	2.74	2.74	4.22
C1 (nF)	8.20	8.20	10.0	10.0	8.76
C2 (nF)	56.0	56.0	75.0	75.0	60.4
R3 (KΩ)	1.00	1.00	3.83	3.74	2.26
R4 (KΩ)	2.20	2.40	1.15	1.15	1.30
C3 (nF)	56.0	56.0	38.3	37.4	52.3
C4 (nF)	82.0	75.0	61.9	61.9	64.9
$\Delta\omega$	0.0122	0.0310	0.0218	0.0025	0.0018
ΔQ	0.0198	0.0268	0.0135	0.0125	0.0070
TE	**0.0160**	**0.0289**	**0.0176**	**0.0075**	**0.0044**

Notice that the selection of the optimal own parameters is very crucial on minimizing total design error value thus effecting filter performance. Indeed, the values of the E192 series are the smallest total error compared to other series.

Comparing the Total Errors given by the basic ACO and the BA-ACO (presented in Table 3), we note that the BA-ACO method has improved the designs and offers solutions that are more precise.

TABLE III. COMPARISONS BETWEEN THE BA-ACO AND ACO FOR THE TOTAL ERROR

	E12	E24	E48	E96	E192
BA-ACO	0.0160	0.0289	0.0176	0.0075	0.0044
ACO [35]	0.0160	0.0328	0.0251	0.0118	0.0049
Improvement	0 %	12 %	30 %	36 %	10 %

In order to check the validity of the results, the following figure shows the PSPICE simulation in the filter gain for the optimal values of the E192 series. The practical cut off frequency are equal to 1592 Hz.

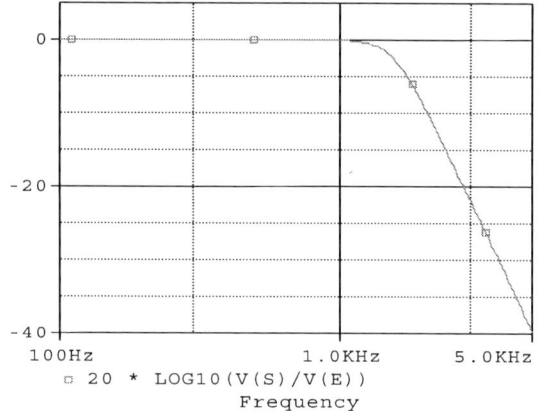

Fig. 2. Frequency responses of low-pass filter

The following table shows the comparison between the theoretical values and those practices for the error on the cut-off frequency for different series.

TABLE IV. COMPARISONS BETWEEN THE OPTIMIZATION AND SIMULATION RESULTS FOR THE ERROR ON THE CUT-OFF FREQUENCY

	$\Delta\omega$ (BA-ACO)	$\Delta\omega$ (PSPICE Simulation)
E12	0.0122	0.0182
E24	0.0310	0.0417
E48	0.0218	0.0286
E96	0.0025	0.0064
E192	0.0018	0.0029

From the results presented in Table 4, we notice that simulation results are in good agreement with those obtained using the BA-ACO algorithm. The slight difference between the two values is mainly due to imperfections of the Op-Amp which are considered perfect in the theoretical calculations.

Exact values of discrete components, deviations, and Total Error of ABC, PSO, ACO and the BA-ACO method are tabulated in Table 5. A comparison between these four techniques shows that the BA-ACO algorithm achieved a smaller design error.

TABLE V. COMPONENT VALUES AND PERFORMANCE OF GA, ABC, PSO AND ACO TECHNIQUES FOR BUTTERWORTH FILTER DESIGN

	ABC [33]	PSO [34]	ACO [35]	BA-ACO
R1 (KΩ)	4.70	4.58	4.70	4.42
R2 (KΩ)	4.70	4.70	4.99	4.22
C1 (nF)	8.20	8.20	7.87	8.76
C2 (nF)	56.0	56.0	54.2	60.4
R3 (KΩ)	1.00	1.10	1.00	2.26
R4 (KΩ)	39.0	1.00	2.34	1.30
C3 (nF)	4.70	87.6	54.9	52.3
C4 (nF)	56.0	102.2	77.7	64.9
$\Delta\omega$	0.0201	0.0135	0.0011	0.0018
ΔQ	0.0024	0.0018	0.0087	0.0070
TE	0.0113	0.0076	0.0049	**0.0044**

In order to check the convergence rate of the proposed algorithm, a robustness test was performed. *i.e.* the algorithm is applied a hundred times for optimizing the *TE* objective. In Table 6 we present the percentage of convergence to the same optimal value, for the BA-ACO and the ACO.

TABLE VI. CONVERGENCE TO THE SAME OPTIMAL VALUE (%)

	E12	E24	E48	E96	E192
BA-ACO	80	66	59	51	46
ACO	57	52	44	39	35

Results in Table 6 confirms that the robustness of the BA-ACO algorithm is better than the robustness of the basic ACO algorithm.

To complete the comparison, we check the computing time (average running times for 100 runs of the BA-ACO algorithm, using an Intel Core i5 3437U CPU - 2.40 GHz). Table 7 corresponds to computing time of the BA-ACO algorithm, compared to the ACO technique, where we clearly notice that BA-ACO is faster than the classical ACO.

TABLE VII.   COMPARISON OF THE COMPUTING TIME

BA-ACO	25.3	seconds
ACO	61.7	seconds

V.   CONCLUSION

The presented work proposes an application of the Backtracking Ant Colony Optimization technique for dealing with the optimal sizing of analog circuit. The BA-ACO algorithm was used for optimal design of a fourth order Butterworth low pass analog filter, for that, we investigated for the selection of passive components from different manufactured series. The design of the analog filter with high accuracy and short execution time is successfully realized using the BA-ACO method. Performances were compared to the ones obtained by using the ABC, PSO and the classical ACO algorithm, and then checked via PSPICE simulations. The optimization results show that the BA-ACO algorithm offers better results in terms of qualities of solutions, robustness and computing time than the classical ACO technique. We can argue that the BA-ACO is a priori an adequate technique to use in the field of analog circuit sizing.

REFERENCES

[1] F. Medeiro, R. R. MacíasFernández, F.V.R. DomínguezAstro, J.L. Huertas and A. R. Vázquez, "Global design of analogcells using statistical optimization techniques," Analog integrated circuits and signal processing, vol. 6, No. 3, pp. 179–195, 1994.

[2] F. Glover, "Tabu search-part I," ORSA Journal on computing, vol. 1, No. 3, pp. 190–206, 1989.

[3] F. Glover, "Tabu search-part II," ORSA Journal on computing, vol. 2, No. 1, pp. 4–32, 1990.

[4] B. Benhala and O. Bouattane, "GA and ACO techniques for the analog circuits design optimization", Journal of Theoretical and Applied Information Technology (JATIT), Vol. 64, No 2, pp. 413–419, 2014.

[5] J. B. Grimbleby, "Automatic analogue circuit synthesis using genetic algorithms," IEE Proceedings-Circuits, Devices and Systems, vol. 147, No. 6, pp. 319–323, 2000.

[6] E. Aarts and K. Lenstra, "Local search in combinatorial optimization," Princeton: Princeton University Press, 2003.

[7] B. Benhala, P. Pereira, A. Sallem, (Editors), "Focus on swarm intelligence research and applications," NOVA Science Publishers, 2017.

[8] M. Fakhfakh, Y. Cooren, A. Sallem, M. Loulou and P. Siarry, "Analog Circuit Design Optimization through theParticle Swarm Optimization Technique," Springer,Journal of Analog Integrated Circuits & Signal Processing, vol. 63, No. 1, pp. 71–82, April, 2010.

[9] H. Bouyghf, B. Benhala and A. Raihani, "Optimization of 60-GHZ down-converting CMOS dual-gate mixer using artificial bee colony algorithm", Journal of Theoretical and Applied Information Technology (JATIT), Vol. 95. No 4, pp. 890–902, 2017.

[10] H. Bouyghf, B. Benhala and A. Raihani, "Optimal design of RF CMOS circuits by means of an artificial bee colony technique", Chapter 11, Book: Focus on swarm intelligence research and applications, Eds., B. Benhala, P. Pereira and A. Sallem, NOVA Science Publishers, pp. 221–246, 2017.

[11] B. Benhala, "Sizing of an inverted current conveyors by an enhanced ant colony optimization technique", The International Conference on Design of Circuits and Integrated Systems, (DCIS 2016), November, 23-25, 2016, Granada, Spain.

[12] L. Kritele, B. Benhala, and I. Zorkani, "Ant Colony Optimization for Optimal Analog Filter Sizing", Chapter 10, Book: Focus on swarm intelligence research and applications, Eds., B. Benhala, P. Pereira and A. Sallem, NOVA Science Publishers, pp. 193–220, 2017.

[13] B. Benhala, "An improved ACO algorithm for the analog circuits design optimization", International Journal of Circuits, Systems and Signal Processing, ISSN: 1998-4464, Vol. 10, pp.128-133, 2016.

[14] M. Kotti, B.Benhala, M.Fakhfakh, A.Ahaitouf, B.Benlahbib, M.Loulou and A. Mecheqrane, "Comparison between PSO and ACO techniques for analog circuit performance optimization," the International Conference on Microelectronics, IEEE TN CEDA's, ENG-OPTIM'Contest "Engineering Applications of Optimization Techniques'', 19–22 December, Hammamat, Tunisia, 2011.

[15] A. Sallem, B. Benhala, M. Kotti, M. Fakhfakh, A. Ahaitouf and M. Loulou, "Application of Swarm Intelligence Techniques to the Design of Analog Circuits: Evaluation and Comparison," Analog Integrated Circuits and Signal Processing. Springer, vol. 75, No. 3, pp. 499–516, March 2013.

[16] M. Dorigo, V. Maniezzo and A. Colorni, "The ant system: Optimization by a colony of cooperating agents," IEEE Transactions on Systems, Man and Cybernetics, Part B, vol. 26, pp. 29–42, 1996.

[17] M. Dorigo and L. Gambardella, "Ant colony system: A cooperative learning approach to the traveling salesman problem," IEEE Transactions on Evolutionary Computation, vol. 1,pp. 53–66, 1997.

[18] T. Stützle and H. Hoos, "Max-Min ant system," Future Generation Computer Systems vol. 18, pp. 889–914, 2000.

[19] Z. Song, H. Di-Bo and Z. Ze-Kui, "Ant colony algorithm with dynamic transition probability," Control and Decision, vol.23, No.2, pp.225–228, 2008.

[20] H.M. Naimi and N. Taherinejad, "New robust and efficient ant colony algorithms: using new interpretation of local updating process," Expert Systems with Applications, No.36, pp. 481–488, 2009.

[21] B. Yingzhou, Z. Peng, Z. Zhi and D. Lixin, "A novel pheromone update with important solution components," 6th International Conference on Natural Computation (ICNC), vol. 5, pp.2551–2555, 2010.

[22] W. Li and L. Dong, "Special factor backtracking algorithm for optimizing," International Conference on Intelligent Computing and Integrated Systems (ICISS), pp.27–30, 2010.

[23] M. Dorigo, G. DiCaro and L. M. Gambardella, "Ant algorithms for discrete optimization," Artificial Life Journal, vol. 5, pp. 137–172, 1999.

[24] M. Dorigo and S, Krzysztof, "An Introduction to Ant Colony Optimization," a chapter in Approximation Algorithms and Metaheuristics, a book edited by T. F. Gonzalez, 2006.

[25] Y. Jinhui, S. Xiaohu, M. Maurizio and L.Yanchun, "An ant colony optimization method for generalized TSP problem", Progress in Natural Science, Vol. 18, pp 1417–1422, 2008.

[26] B. Yu, Z. Yang and B. Yao, "An improved ant colony optimization for vehicle routing problem", European Journal of Operational Research Vol. 196, pp. 171–176, 2009.

[27] A. Colorni, M. Dorigo and V. Maniezzo, "Distributed optimization by ant colonies," Proceedings of the 1st European Conference on Artificial Life, pp.134–142, 1991.

[28] S. Guo and Y. Meng, "An improved entropy-based ant colony optimization algorithm," International Conference on Computer Application and System Modeling, (ICCASM), vol.15, pp.1548–1550, 2010.

[29] M. Daniel and M. Martin, "Ant colony optimization with global pheromone evaluation for scheduling a single machine," Applied Intelligence, vol.18, pp.105–111, 2003.

[30] Z. Liu, T. Liu and X. Gao, "An Improved Ant Colony Optimization Algorithm Based on Pheromone Backtracking," IEEE 14th International Conference on Computational Science and Engineering (CSE), pp. 658–661, 2011.

[31] B. Benhala, M. Kotti, A. Ahaitouf and M. Fakhfakh, " Backtracking ACO for RF-Circuit Design Optimization," book: "Performance Optimization Techniques in Analog, Mixed-Signal, and Radio-Frequency Circuit Design", Edt. M. Fakhfakh, E. Tlelo-Cuautle, M.H. Fino, IGI-Global, Chapter 7, pp. 158-179, 2015.

[32] L. D. Paarman, *Design and Analysis of Analog Filters*, Norwell, MA: Kluwer, 2007.

[33] R.A. Vural, T. Yildirim, T. Kadioglu and A. Basargan, Performance Evaluation of Evolutionary Algorithms for Optimal Filter Design, *IEEE transactions on evolutionary computation* Vol. 16, No. 1, 2012, pp. 135–147.

[34] R. A. Vural and T. Yildirim, Component value selection for analog active filter using particle swarm optimization, *in Proc. 2nd ICCAE*, Vol. 1. 2010, pp. 25–28.

[35] B. Benhala, " Ant Colony Optimization for Optimal Low-Pass Butterworth Filter Design," WSEAS Transactions on Circuits and Systems, Vol. 13, pp. 313-318, June 2014.

Optimization of Efficiency and Harmonics for Gbit Flash Memory based PWM Inverters

Dorin O. Neacșu

Department of Applied Electronics and Intelligent Systems, Technical University of Iași, Romania
Email: dorin.neacsu@ieee.org

Abstract – **The Gbit-size Flash Memory based Pulse Width Modulation (PWM) algorithms for three-phase inverters differ from the conventional counter-based implementation since they follow a pre-programmed optimal PWM pattern. At a fixed sampling interval, each position within the complex plane may correspond to a different sequence of states or pulse structure. This paper extracts a set of rules for such an ideal PWM aiming at optimization of both current harmonics / flux error and power loss / efficiency. Patterns are optimized with MATLAB for harmonics and PSIM for thermo-electric modeling. Conclusions depict clearly the promising future of this novel technology.**

Keywords – thermal analysis, flash memory, PWM algorithms.

I. INTRODUCTION

Three-phase six-switch inverters are at the heart of any AC drive or three-phase grid interface platforms. Their control involves PWM [1,2], that is mostly implemented within microcontroller systems. The technology has been improved over the last 40 years and seems to be a closed topic for research. However, all PWM algorithms reported so far are based on a repetitive approach and this probably constitutes a limitation for further research. The newly reported flash-memory based approach to implementation of PWM algorithms allows the designer to go away from this stereotype [3,4]. Very complex PWM methods can be implemented without maintaining the repetitive scheme of preceding approaches.

For instance, the previously reported Frequency Modulation [5,6] has considered a sinusoidal modulation of the carrier frequency and relied on a variable carrier period. In a modern perspective, the PWM can have a pattern near the active vectors (for instance, Center-aligned PWM) and a different pattern near each sector bisector (for instance, Left-aligned PWM), with the general effect of pulse frequency modulation. The previous work in flash-memory based PWM tried to illustrate this powerful approach with a complete hardware implementation of a set of methods already reported in mid-90's [3,4].

The next evolutionary step is to take advantage of the flash-memory based infrastructure and to design the PWM without considering emulation of previous methods. The only consideration here should be exactly the unlimited number of possible patterns which can occur during a fundamental cycle. This paper uses this infrastructure, and extracts a set of rules for an ideal PWM aiming at optimization of both current harmonics / flux error and power loss / efficiency.

This paper develops a computer-based design procedure for a complex PWM algorithm able to optimize efficiency, junction temperature, and current harmonics, while maintaining the same sampling interval for the control system. The flux trajectory is considered as tool for qualifying PWM algorithms.

Performance is assessed with the content in harmonics of the output current. In this respect, we consider the Harmonic Current Factor (HCF) as figure of merit, similar to [1-9]:

$$HCF(\%) = \frac{100}{V_{(1)}} \cdot \sqrt{\sum_{n=5}^{\infty} \left[\frac{V_{(n)}}{n}\right]^2} \qquad (1)$$

where $V(n)$ are the voltage harmonics in the phase voltage. Results are reported in respect to the modulation index m that is defined herein to report Root-Mean-Square (RMS) at fundamental frequency to fundamental of square-wave operation, that is $(2/\pi) \cdot V_{dc}$, where V_{dc} is the DC bus voltage.

$$m = \frac{V}{\frac{2}{\pi} V_{dc}} \qquad (2)$$

Switching the focus from considering solely the current harmonics [3,4,7,8] towards considering the reduction of power loss at any operation point demonstrates the opportunity to reduce the switching frequency at low current, or low modulation index [7-12]. The complex design and analysis are performed with models in MATLAB and PSIM.

An analysis of Center-Aligned PWM using flux trajectory is developed in this paper. A coordinate system based on radial and axial (tangential) directions is leading to two orthogonal components for the flux error. The position in complex plane of the tip of the voltage vector influences the two components. Different approaches are demonstrated as being useful for reduction of each of the two error components. They can be applied separately or simultaneously. After a set of recommendations is developed, an example is herein fully developed for illustration, following the hardware architecture proposed in [3] and replicated in Fig. 1. Since newly proposed solution considers several states during the sampling interval, the resolution in pulse definition has been increased to 10-bit.

II. INVERTER POWER LOSS

Recent applications are more sensitive to efficiency requirements than the quality of current waveform. The generic efficiency's dependency on power level P_2 is shown with Fig. 2. For a fixed input voltage and load, P_2 changes with modulation index as in Fig. 3. This shape is due to stand-by loss for an inverter being almost the same for all output power levels, so the efficiency at lower output power levels is affected more. For more about the efficiency curve, [13] proposed to define three components $P_{I,0}, P_{I,I}, P_{I,II}$ of the inverter power loss (through constants k_0, k_1 and k_2), as suggested by next equation.

$$\eta = 1 - \frac{P_{loss}}{P_2} = 1 - \frac{P_{I,0} + P_{I,I} + P_{I,II}}{P_2} = 1 - \frac{k_0 + k_1 \cdot P_2 + k_2 \cdot P_2^2}{P_2}$$
$$= 1 - \frac{k_0}{P_2} - k_1 - k_2 \cdot P_2 \qquad (3)$$

978-1-5386-7688-2/19 $31.00 © 2019 IEEE

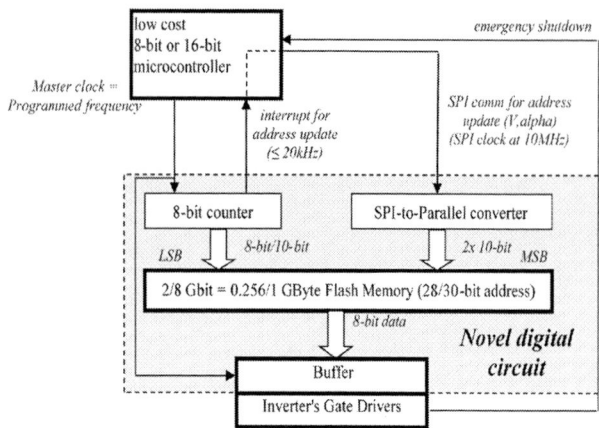

Fig. 1 Principle of PWM generator implemented with flash memory ICs, upgraded from 2 Gbit to 8 Gbit for increased resolution.

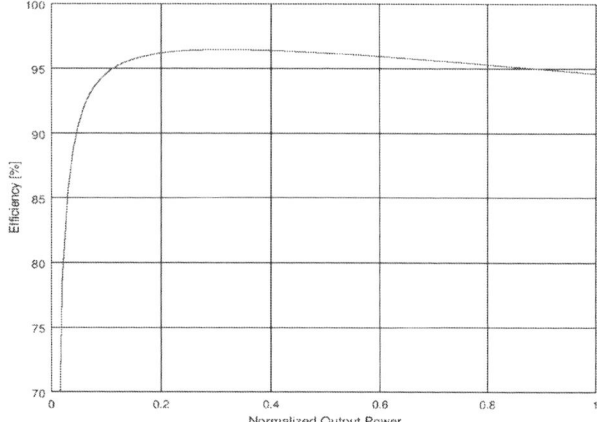

Fig.2 Inverter efficiency with output power (example generated in MATLAB)

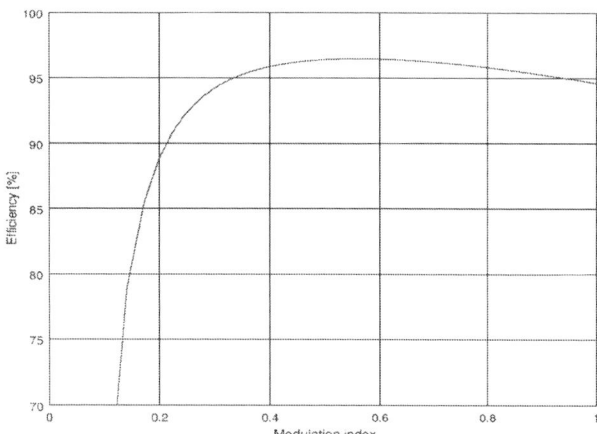

Fig. 3 Efficiency with modulation index, for constant DC bus and load (example generated in MATLAB)

First term ($P_{L,0}$) in (3) represents the nearly constant stand-by power loss due to auxiliary equipment and energy loss within parasitic capacitive components due to high frequency pulsations of voltage. While the auxiliary loss contribution (control, sensors) is constant, with minimal variation due to operation point, the capacitive loss contribution (called switching loss) increases with switching frequency, for a given current. Hence, the latter can be reduced with a decrease in switching frequency. The capacitive loss contribution is almost constant with load current or power level.

This recently justified several industrial-borne papers to reduce the switching frequency at low current or low modulation index. The inverter loss, Insulated Gate Bipolar Transistor (IGBT) junction temperatures, and Mean Time To Failure (MTTF) are analyzed in [10] when switching frequency changes. It is verified that the overload capability and MTTF of the inverter can be increased significantly with a lower switching frequency at motor drive's low speed. However, this solution is not very used since it is impractical to change sampling/switching frequency for the entire control system as it may introduce some control issues during operation.

By contrary, the solution proposed in this paper is changing the switching frequency at the same sampling interval, that means the control system is not influenced by the PWM pattern.

The second term ($P_{L,I}$) in (3) comes mostly from voltage drop on diodes and IGBT type devices and relates to the conduction loss. Power loss is slightly proportional to the current, and produces a nearly constant contribution to efficiency.

The third term ($P_{L,II}$) in (3) is mainly due to calorimetric loss in all resistive parasitic. Any reduction of the Root-Mean-Square (RMS) of the current would improve this efficiency term slightly. Thus, a reduction of harmonics and improvement of Harmonic Current Factor (HCF) is desirable at high currents where this term yields important. Hence the proposal of *Frequency Modulation* methods [5,6,10]. The curvature introduced by this term is often minor and not always be visible:

- especially for air-cooled systems where the limitation comes from junction temperature earlier than maximum current, maybe at around 33% of module's rated current;
- when reporting loss in power semiconductor module without busbar and connections.

Otherwise, choosing a large-area power semiconductor would introduce a large R_{dson} and a steeper decreased from this term. This third term does not depend on switching frequency (under assumption that resistive parasitic does not change with switching) and some efficiency improvement can be achieved with a better quality of the current.

The peak efficiency is achieved when the constant term equals the quadratic term (~ 35% in the graphical example from Fig. 4). Thence, many papers report efficiency after this point and disregard the low or high current regions. To counteract this trend, recent standards for motor drives and solar renewable energy ask for efficiency reported in multiple points or for a weighed efficiency, from multiple operation points [14].

III. ANALYSIS BASED ON FLUX TRAJECTORY

The flux vector (λ) is defined with dependency to the voltage vector (V_s):

$$\underline{\lambda} = \int \underline{V_s} dt \qquad (4)$$

The flux error represents difference between the ideal circular trajectory and the actual polygonal trajectory. This error can be decomposed on *Real* and *Imaginary* axes (err_{Real}, err_{Imag}, Fig. 5).

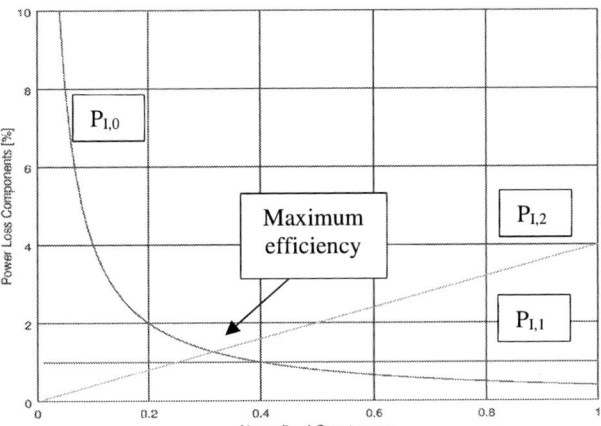

Fig. 4 Power loss components as defined in (3), for an example in MATLAB.

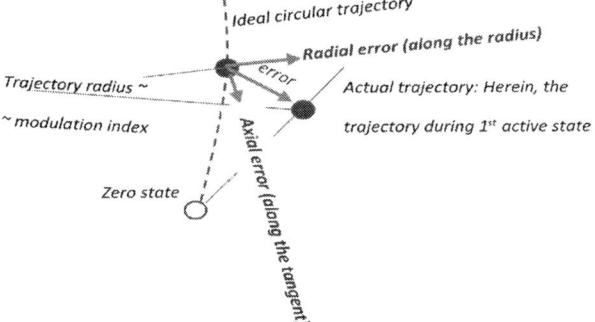

Fig. 5 Error decomposition in axial and radial errors.

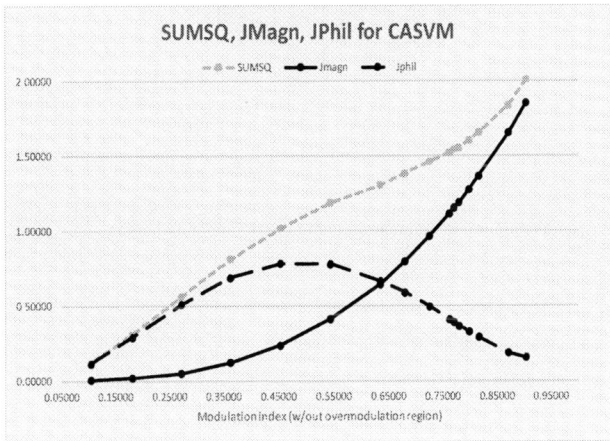

Fig. 6 Proposed performance indices calculated for a Center-Aligned Space Vector Modulation with a frequency ratio of 256 (for instance, $f_1 = 19.073$ Hz, and $f_{samp} = 4.88$ kHz). Pulses generated with a 10-bit resolution.

A cumulative effect yields as sum of squared components:

$$\begin{cases} err_{Real} = circ_{Real} - poly_{Real} \\ err_{Imag} = circ_{Imag} - poly_{Imag} \\ sq_{err} = err_{Real} \cdot err_{Real} + err_{Imag} \cdot err_{Imag} \end{cases} \quad (5)$$

An alternative to the orthogonal coordinate system, provided by the *Real* and *Imaginary* axes, can be used as an orthogonal system rotating in complex plane, and having the axes along the radius' direction and perpendicular to the radius of the desired

circular trajectory, that also means along a tangent to the circular trajectory. The transform yields:

$$\begin{cases} err_{Radial} = err_{Real} \cdot cos(\alpha) + err_{Imag} \cdot sin(\alpha) \\ err_{Axial} = -err_{Real} \cdot sin(\alpha) + err_{Imag} \cdot cos(\alpha) \\ sq_{Pol} = err_{Radial} \cdot err_{Radial} + err_{Axial} \cdot err_{Axial} \end{cases} \quad (6)$$

Therefore, the flux trajectory error can be decomposed into two terms: an angular (axial, tangential) error and a radial error (along the radius, Fig. 5). These components vary during a fundamental cycle, depending on the actual position within the complex plane, magnitude, or angular coordinate.

While the actual time-domain waveform for error components is also considered, a set of performance indices (SUMSQ, J_{Magn}, J_{Phil}) are considered for further analysis:

$$\begin{cases} SUMSQ = \sum sq_{Pol} = \sum sq_{err} \\ J_{Magn} = \sum err_{Radial}^2 \\ J_{Phil} = \sum err_{Axial}^2 \end{cases} \quad (7)$$

where the sum Σ is calculated from samples, over a full fundamental cycle. Based on this, the conventional Space Vector Modulation (SVM) can work up to m=0.907.

A series of conclusions can be drawn from this analysis of Center-aligned SVM method:

- Axial and radial error components are seeing their peak values at 90^0 phase-shift of each other;
- Axial error is seeing its maximum value in vicinity of the active vectors (0^0 or 60^0);
- Radial error is seeing its maximum value in vicinity of the sector bisector (30^0);
- At high modulation index, most of the flux error is coming from radial (magnitude) error;
- At low modulation index, most of the flux error is coming from axial (angular) error;
- Doubling the PWM sampling frequency (pulse frequency) reduces by roughly half the error, and both of its components, radial and axial.
- Analogous to the ripple of the phase current, the cumulative SUMSQ (5) has a lower value at low modulation index.

The latter allows a relaxation of the switching frequency (a lower switching frequency) at low modulation index [10-13]. This is reversed from the evolution of HCF with modulation index since HCF is calculated by division to the magnitude of the fundamental component.

IV. REDUCTION OF RADIAL ERROR

Furthermore, measurements in [3] showed an improvement of the current harmonics around 40% when using the memory-based *Merged-Pulse PWM* instead of the conventional center-aligned PWM (Fig. 7). Sample results are shown in Fig. 8. This allows the use of three split states for the active or zero state within the largest time interval. It also helps the digital implementation since the largest interval is easier to be split with minimal resolution loss. This means using three intervals of zero vector whenever the time interval associated to zero state is the largest in Fig. 9(a), or using a certain active vector whenever the time associated to that vector yields the shortest in Fig. 9(b).

978-1-5386-7688-2/19 $31.00 © 2019 IEEE

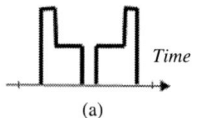 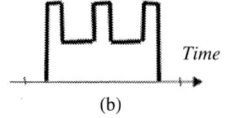

(a)   (b)

Fig. 7 Output phase voltages for one sampling interval: (a) Conventional Center-Aligned Space Vector Modulation; (b) Memory-Based Merged-Pulse PWM. Observe the same count of distinct switching states (7).

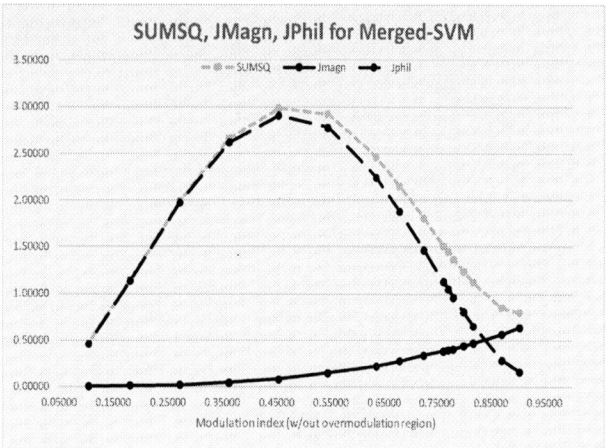

Fig. 8 Merged-Pulse SVM at any modulation index, for frequency ratio of 256: Direct comparison with results in Fig. 6 shows that at high modulation indices, the radial error (expressed with Jmagn) is reduced from 1.8525 to 0.63539.

In conclusion, using *Merged-pulse PWM* allows a radial error reduction at high modulation indices, roughly above m = 0.7619.

A *Sinusoidal Frequency Modulation* allows more PWM cycles in the middle of the sector so that trajectory bouncing due to the two active vectors is minimized. It is worthwhile to mention that implementation variants for the *Frequency Modulation* consider the actual variation of either pulse frequency or pulse period. These two variants are not the same, yet they target the same idea of crowding pulses near the sector bisector and spreading pulses thin near active vectors. Fig. 10 illustrates reduction of ripple at high modulation index, above m=0.76. When compared to Fig. 6, the improvement in radial error (J_{magn} reduced from 0.4126 to 0.0831) is slightly influencing the axial error (J_{phil} from 0.0539 to 0.0884), yet it is dominant in reduction of total error.

For implementation, the cosine can be approximated with a step waveform (Fig. 11) in order to keep the sampling interval constant for compatibility with the control systems. The pulse period is changed as integer multiples of the pulse period at active vectors, that is to generate four identical pulses near the sector bisector, for T/2 = (2T)/4. The middle region stays compatible with Center-Aligned PWM, with two equivalent pulses per period, for T = (2T)/2. The average value of this waveform should equal the sampling interval T. The *Frequency Modulation* can be associated with the *Merged-Pulse PWM* approach, with patterns for a sampling interval as shown in Fig. 12. Furthermore, the benefits of *Discontinuous PWM* can also be used herein [15-17].

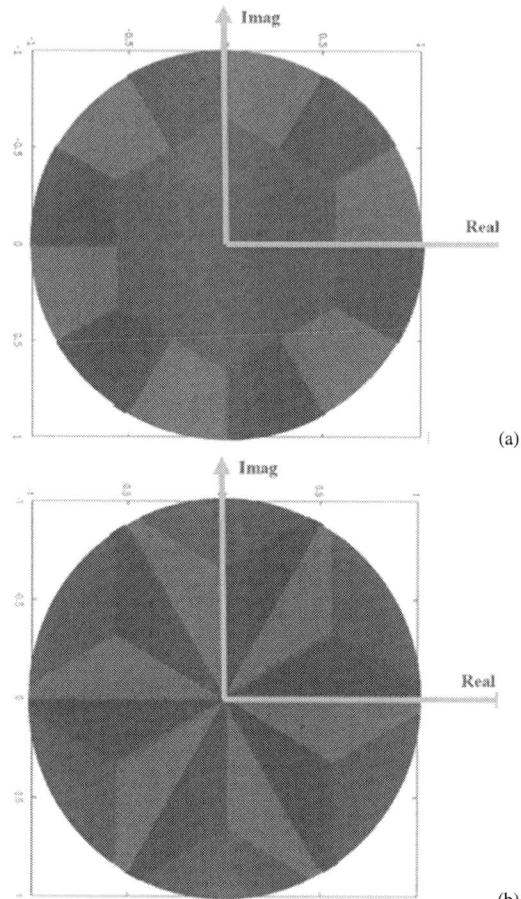

(a)

(b)

Fig. 9 (a) Definition of the **largest time interval** from SVM equations: blue = zero state, red = 1st active vector, black = 2nd active vector within sector. (b) Definition of the **shortest time interval** from SVM equations: blue = zero state, red = 1st active vector, black = 2nd active vector.

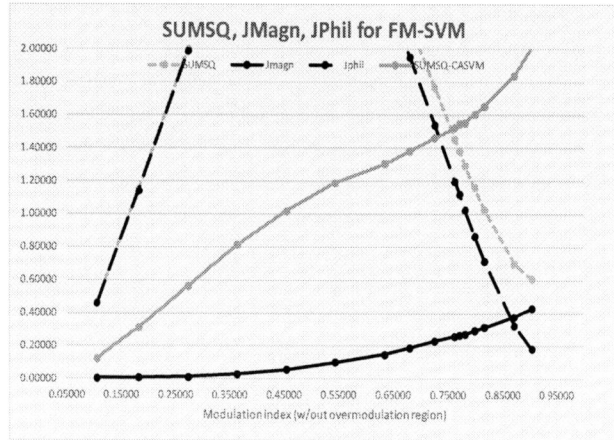

Fig. 10 Additional pulses near the 30⁰ bi-sector, as seen within the Frequency Modulated PWM method. To be compared with Fig. 6 for Center-aligned PWM at same frequency ratio of 256 (denoted here SUMSQ-CASVM). Pulses generated with a 10-bit resolution.

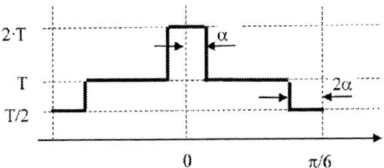

Fig. 11 Option for frequency modulation at constant sampling frequency, with variation of the pulse period in multiple increments.

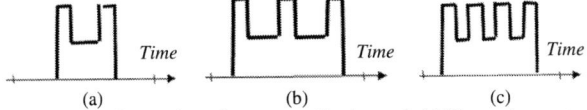

Fig. 12 Output phase voltage for one sampling interval: (a) Near active vectors, over an angular interval (0, stpa); (b) in the middle of the sector, for an angular interval (stpa, pi/6-stpa); (c) near the sector bisector, for an angular interval (pi/6-stpa, pi/6). "stpa" is the optimization parameter at 4.6⁰.

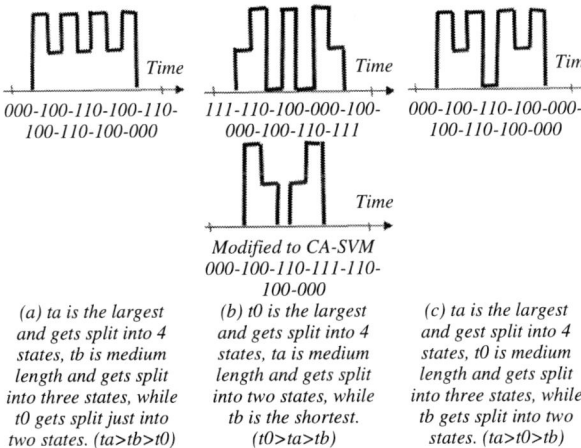

(a) ta is the largest and gets split into 4 states, tb is medium length and gets split into three states, while t0 gets split just into two states. (ta>tb>t0)

(b) t0 is the largest and gets split into 4 states, ta is medium length and gets split into two states, while tb is the shortest. (t0>ta>tb)

(c) ta is the largest and gets split into 4 states, t0 is medium length and gets split into three states, while tb gets split into two states. (ta>t0>tb)

Fig. 13 Output phase voltage patterns based on minimum and maximum pulse widths for ta, tb, t0, described for the first 30⁰ from the sector (when ta>tb).

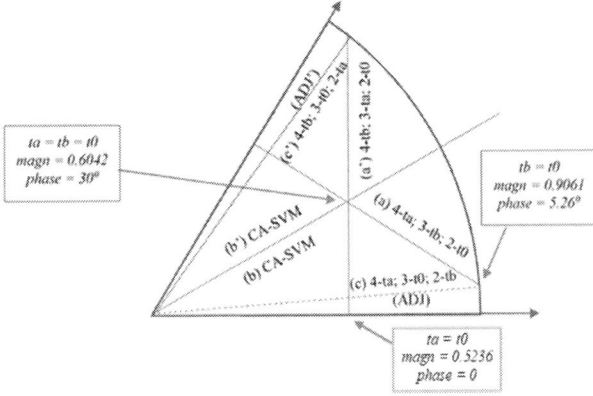

Fig. 14 Operation modes depicted from Figs. 11 and used to define patterns shown in Fig. 13 (herein, for a generic 60⁰ sector, characteristic for SVM).

V. REDUCTION OF AXIAL ERROR

The angular (or axial) error corresponds to the displacement along the tangent to the trajectory. This component is more important for positions near the active vectors and less important near the bisector of the 60⁰ sectors since the time intervals

allocated to the zero-state are also larger near 0⁰ and 60⁰ positions. When the trajectory sits on a zero-vector, there is no displacement and hence the angular error is larger. Fig. 6 shows that the axial flux error is more important than the radial error at medium values for modulation index. It features however an almost constant peak-to-peak waveform for the axial flux error, not depending on the actual position.

A common improvement for reduction of axial error consists of the increase in PWM frequency. However, our analysis suggests to include more zero states instead of the conventional solution of multiplying the PWM frequency. In conclusion:

- Zero states can break the longest active vector states since during such intervals the flux trajectory flies away more from the circular path, while the active state with a shorter time interval produces less flux variation.
- A certain ratio between control sampling period and pulse period can be used to allow several states to optimize with.

VI. CONSTRUCTION OF A NEW OPTIMAL PATTERN

The *Space Vector Modulation* approach proposes to generate a pulse structure with an average between adjacent vector states and zero state. This average concept resorts to the calculation of three intervals and it is the designer choice for the actual state sequence and sharing of such time intervals between similar states. Observing Figs. 9 allows a new concept:

- The longest time interval is split into four intervals, and the shortest interval is split into two intervals, while the interval with a medium length is split into three intervals. Individual patterns shown with Fig. 13.
- Various regions within the complex plane where different operation modes are used are shown in Fig. 14 and these can produce the results shown in Fig. 15.

This combination of PWM patterns produces a higher count of switching processes at high modulation since more states are involved over a sampling interval than the *Center-Aligned PWM*, and a relaxation of the switching frequency at lower modulation index. This result can be reversed: the nominal switching frequency can be considered as being the one from high modulation index region, while a relaxation is produced at lower modulation index. Overall, the same sampling frequency is involved, that is no change for the control sampling interval.

VII. DESIGN PROCEDURE

The design procedure for an inverter using the proposed memory-based PWM is explained with an example.

1) Consider conventional *Center-Aligned Space Vector Modulation*. Determine what is the highest switching frequency in full load for the given power stage. One can consider maximal junction temperature or the lifetime of the equipment for this determination.

2) Use a graph like Fig. 15(c) and observe the detected operation point. For instance, CA-SVM with a frequency ratio of 684. For 60 Hz, the maximal sampling frequency yields 41.04 kHz. Adopt this or go for an even lower value. This produces 522 switching state changes per fundamental, within the entire inverter.

978-1-5386-7688-2/19 $31.00 © 2019 IEEE

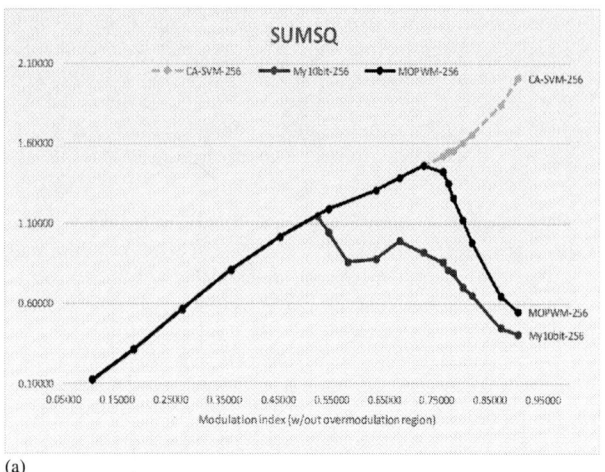

(a)

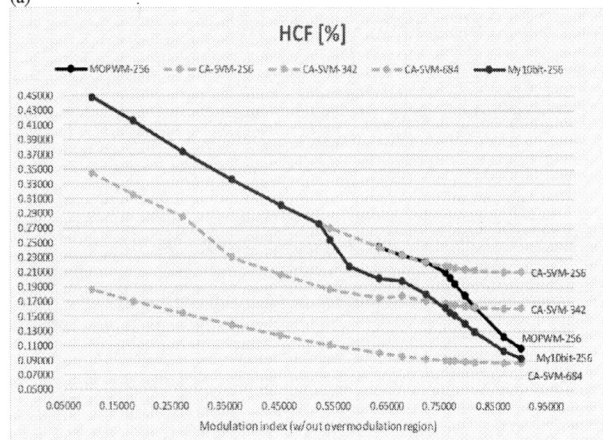

(b)

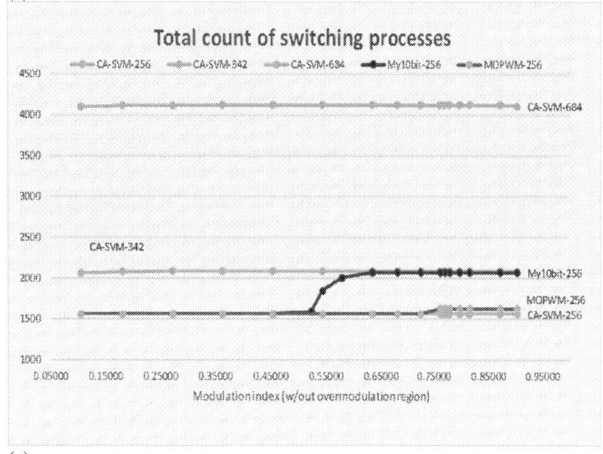

(c)

Fig. 15. Comparative Results for *Center-Aligned PWM* (CA-SVM, followed by the frequency ratio), the combination pattern in Fig. 14 digitized on 10 bit ("My10bit-256"), and the pattern for previous Multi-Optimal PWM ("MOPWM-256"): (a) SUMSQ performance index, as a measure of phase current ripple; (b) HCF performance index, as a measure of total harmonics in phase current normalized to fundamental; (c) Total count of state-switching processes per entire inverter during a fundamental cycle.

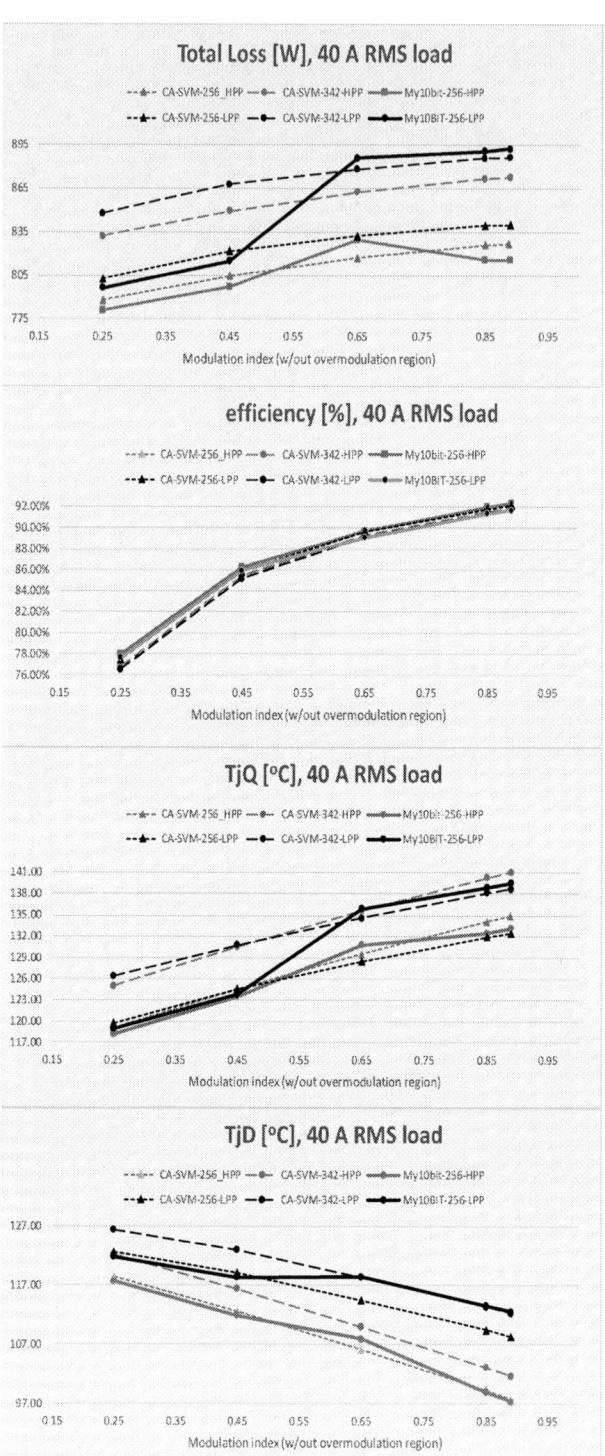

Fig. 16 Power loss and junction temperature for various PWM methods applied to Mitsubishi **CM100TU-12H** module: maintain the same 40 A quasi-sinusoidal current load, for any modulation index, for either unity power factor (HPP) or low power factor of 0.707 (LPP), with pulse frequencies of 4.096 kHz (ratio of 256) and 5.472 kHz (ratio of 342).

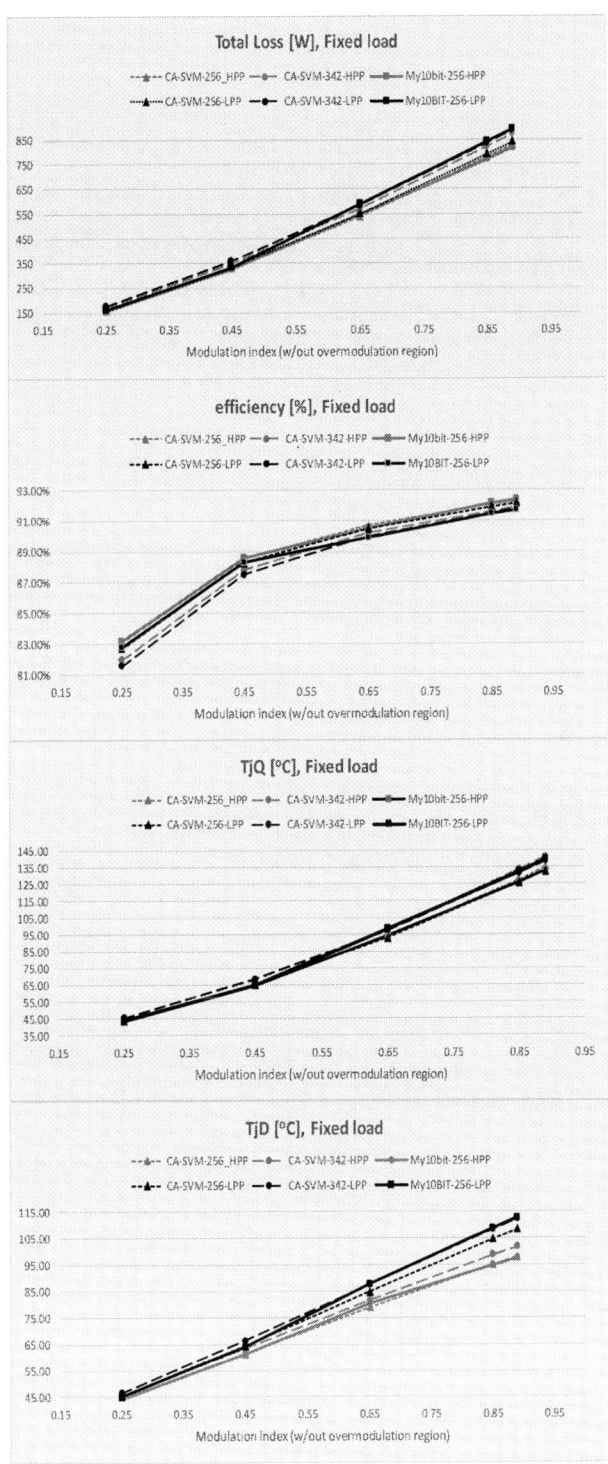

Fig.17 Power loss and junction temperature for various PWM methods applied to Mitsubishi **CM100TU-12H** module: maintain the same physical load at any modulation index, different for either unity power factor (HPP) or low power factor of 0.707 (LPP), with pulse frequencies of 4.096 kHz (ratio of 256) and 5.472 kHz (ratio of 342).

3) Identify the optimal pattern which leads to almost the same count of switching processes, at highest modulation index. In example from Fig. 15(c) that means the flash-memory pattern with a frequency ratio of 256 and 518 changes of the inverter switching states. The equivalent switching frequency yields also 41 kHz, while the control sampling frequency drops to 15.36 kHz and it is maintained throughout the operation range.

4) Use the memory look-up table with this pattern and the equivalence to the conventional SVM with a frequency ratio of 256 and a modulation index of m=0.5236.

The analyzed system is based on a CM100TU-12H power module (6 IGBT + Diode switches, rated 100 A, 600 V), that is supplied from a 300 Vdc bus. The load has been made up with resistive and inductive components, appropriate for each test.

The thermal model for semiconductors is coming from PSIM library and it had been extensively validated in various applications. Thermal model uses calculated power loss from electrical circuit, and follows power module datasheet and heatsink air-cooling. The model does not include busbars or wiring influence on calorimetric loss. Hence, efficiency results are for power semiconductors only. Results in Figs. 16-17 are provided as an example to illustrate the sense of dependencies and not as an ultimate numerical value. Other remarks:

- For all tests, peak efficiency is reached at limit for junction temperature (graph increasing with current);
- If a better heatsink or liquid-cooling would be used, the advantages of the novel PWM would be even more important since the calorimetric loss would prevail;
- Using proposed PWM at high modulation index and unity power factor decreases loss and improves efficiency;
- Power loss and efficiency are damaged when operating for long time (unlikely) with low power factor;
- As expected, changing the switching frequency for low modulation index (or current) does not alter overall performance at any power factor load;
- As expected, changing the switching frequency for low modulation index (or current) reduces power loss and improves efficiency at that operating point.

VIII. CONCLUSION

This paper develops an improvement for the three-phase inverter with a Gbit flash-memory based PWM generator, previously reported in [3,4].

This paper extracts a set of design rules for an ideal PWM aiming at optimization of both current harmonics / flux error and power loss / efficiency.

As a novelty, this paper:

- ... uses the theory of decomposing power loss into three components from [13], and translates this into requirements for the PWM generator.
- ... uses a comprehensive analysis of the flux error (radial and axial) components for depicting the optimal PWM setup for reduction of flux ripple and current harmonics.

978-1-5386-7688-2/19 $31.00 © 2019 IEEE

- ... demonstrates the relationship between either of Merged PWM, Frequency Modulation, Split-Zero Vector PWM, and flux error and current harmonics.
- ... embeds requirements derived from both power loss and current harmonics analyses into the previously reported hardware of Gbit flash memory IC based PWM.
- ... uses for the first time with flash memory based PWM, a set of patterns with different switching count at different modulation indices with benefits in reliability, similar to [10].
- ... defines a novel optimal setup based on regions in the complex plane where different patterns are used while preserving a constant sampling interval.
- ... provides a design procedure for the flash-memory based PWM with respect to loss and current harmonics.
- ... demonstrates efficiency improvement for operation of flash memory based PWM with high power factor when compared to conventional Space Vector Modulation.
- ... presents a comprehensive analysis of the current harmonics, power loss, efficiency, junction temperature for the inverter operated with flash memory based PWM.

These are promising results for the technology.

Acknowledgment

The Author would like to thank Professor Brad Lehman and his team for the encouragements and advice during the initial phases of this project, at Northeastern University.

References

[1] D. G. Holmes, T. A. Lipo, "Pulse Width Modulation for Power Converters: Principles and Practice", Ed. Wiley, USA, Oct. 3, 2003.

[2] A. Trzynadlowski, "Introduction to Modern Power Electronics", Ed. Wiley, third edition, New York, NY, USA, Nov. 2015.

[3] D.O. Neacşu, Jonathan Kim, Brad Lehman, A Three-Phase Multi-Optimal PWM Implemented on 2Gbit Flash Memory Integrated Circuits, IEEE Trans. on Power Electronics, vol.32, iss.7, pp. 5813 - 5826, July 2017.

[4] D.O. Neacşu, Yue Zheng, Brad Lehman, A SD Card Flash-Memory Based Implementation of a Multi-Optimal Three-Phase PWM Generator, IEEE Trans. on Power Electronics, vol. 31, no. 1, pp. 39-51, January 2016.

[5] Y. Iwaji and S. Fukuda, "A pulse frequency modulated PWM inverter for induction motor drive", IEEE Transactions on Power Electronics, vol. 7, no. 2, pp. 404–410, April 1992.

[6] Y. Murai, Y. Gohshi, K. Matsui, and I. Hosono, "High-frequency split zero-vector PWM with harmonic reduction for induction motor drive", IEEE Trans. on IA, vol. 28, no. 1, pp. 105–112, Jan./Feb. 1992.

[7] M. J. Meco-Gutiérrez, F. Pérez-Hidalgo, F. Vargas-Merino, J. R. Heredia-Larrubia, "A New PWM Technique Frequency Regulated Carrier for Induction Motors Supply", IEEE Transactions on Industrial Electronics, vol. 53, no. 5, pp.1750- 1753, October 2006.

[8] K. Yodpradit, A. Pichetjamroen, N. Teerakawanich, "An Inverse-Sinusoidal PWM Technique to Improve Thermal Performance of IGBT Module", 2018 IEEE Int'l Transportation Electrification Conference & EXPO Asia-Pacific, Bangkok, Thailand, pp.1-7, June 6-9, 2018.

[9] J.W. Kolar, H. Ertl, F.C. Zach, "Influence of the modulation method on the conduction and switching losses of a PWM converter system", IEEE Transactions on Industry Applications, vol.27, iss.6, pp.1063-1075, Nov/Dec 1991.

[10] L. Wei, J. McGuire, J. Hu, "Study of PWM Frequency and Its Impact to Adjustable Speed Drive Reliability", The 2017 IEEE Energy Conversion Congress and Exposition (ECCE), Cincinnati, OH, USA, pp.3844-3850, October 1-5, 2017.

[11] D. Finn, G. Walker, P. Sernia, "Method of extracting Switching Loss from High efficiency MOSFET Based Half Bridge Converter", Internet source, 2018.

[12] G. Ramírez, M. Aníbal-Valenzuela, M.D. Weaver, R.D. Lorenz, "The impact of switching frequency on PWM AC drive efficiency", The 2016 IEEE Pulp, Paper & Forest Industries Conference (PPFIC), pp. 153-163, June 19-23, 2016.

[13] J.W.Kolar, "X-treme Efficiency Power Electronics", Tutorial Presented at Fifteenth IEEE Workshop on Control and Modeling for Power Electronics, Santander, Spain, June 22 -25, 2014.

[14] K.Mertens, K.F.Hanser, "Photovoltaics: Fundamentals, Technology and Practice", Chapter 7, Section 7.2.4 Efficiency of Inverters, pp. 177-181, Ed. Wiley, 2 edition, July 23, 2018.

[15] A. M. Trzynadlowski, S. Legowski, "Minimum-loss vector PWM strategy for three-phase inverters," IEEE Transactions on Power Electron., vol. 9, no. 1, pp. 26–34, Jan. 1994.

[16] L. Dalessandro, S. D. Round, U. Drofenik, and J.W.Kolar, "Discontinuous space-vector modulation for three-level PWM rectifiers," IEEE Trans. on Power Electron., vol. 23, no. 2, pp. 530–542, Mar. 2008.

[17] C. Charumit, V. Kinnares, "Discontinuous SVPWM techniques of three-leg VSI-fed balanced two-phase loads for reduced switching losses and current ripple," IEEE Transactions on Power Electron., vol. 30, no. 4, pp. 2191–2204, Apr. 2015.

[18] Anon, PSIM User Manual, 2018.

Power Loss Analysis in Modular Multilevel Converters

Ahmed Eshwiage
Mechatronics Research Group
Faculty of Engineering and Physical
University of Southampton
Email: ane1r14@soton.ac.uk

Suleiman M. Sharkh
Mechatronics Research Group
Faculty of Engineering and Physical
University of Southampton
Email: S.M.Sharkh@soton.ac.uk

Sara Bouguerra
Signals and Systems Research Laboratory Group
Faculty of Electronic, Institute of Electrical
and Electronics Engineering(IGEE)
University of MHamed Bougara of Boumerdes
Email: alkharif2013@live.fr

Abstract—A mathematical method of averaging MMC's arm current(AMAC) method has been developed in this study for loss analysis and calculation in MMC. This paper analyses and compares losses in a Modular Multilevel Converter(MMC) using different sub-module converter topologies.

All the expressions have been derived and applied for loss calculation. The sub-module topologies described here include 2-level half-bridge converters (HBC), a 3-level Neutral-Point Clamped Converter(NPC) and 2-level interleaved (inte). In addition, the paper estimates the impact of the MMC arm inductor on overall losses. The devices involving the sub-module converters were selected from manufacturing data-sheet to allow for appropriate voltage and current ratings.

All the expressions mathematically have been derived and applied for loss calculation in MATLAB where all the results are obtained. The MMC using interleaved as sub-modules was found to have the lowest power losses when compared to those in MMC-HBC and MMC-NPC.

I. INTRODUCTION

Modular Multilevel Converters (MMC)were initially introduced in 2003 [1]. The key advantages of the MMC topology is its modularity and scalability, as it can be adapted to medium and high voltage levels. From a construction viewpoint, an MMC has an advantage in terms of its capacity to reduce the size of the entire converter by obviating the need for high-voltage dc-link capacitors or a series connected insulated gate bipolar transistor (IGBT) [2]. In addition, an MMC has the ability to synthesise a very high-quality sinusoidal waveform; thereby, minimizing losses [3]. It is for these reasons that MMC is being thought of as increasingly vital for long distance HVDC power transmission systems [4].

During the design stage of an MMC, loss estimation assumes importance, since it can allow engineers to optimise overall system performance, which is crucial when selecting heat sinking and system cooling equipment. Therefore, the calculation of losses is the first stage in development and optimisation of the MMC design. The majority of publications on MMC loses focus on MMC by using half-bridge converter sub-modules. In addition, the effect of an unsymmetrical cycling time for the arm current on switching and conduction loss analysis in MMC was not considered in of these studies.

Examples of papers on loss analysis can be found in [5] and [6].

Any evaluation of switching and conduction losses in MMC is based on different sub-modules topologies that can be found in [7] and [8], and a clear review of the subject of MMC losses was presented in[7]. Conduction and switching losses are estimated using both a thermal model and a mathematical model, based on second-order equations, which have been used to describe and calculate instantaneous conduction losses. The comparison between losses estimation, and unipolar and bipolar sub-modules topologies in MMC was reported. Reference in [9] reported the simplified analytical calculation model for average power loss in an MMC based on the root mean square (RMS) and arm current. The advantage of this method is that it can be used to investigate the relationship between the loss, voltage balancing methods, and main variables of an MMC. However, the main limitation of this technique is that "the analytical expressions of the conduction loss and the switching loss are derived in terms of the whole of MMC rather than a single sub-module. It avoids the effect of the complexity of its switching pattern on the calculation accuracy of the model". Thus, although the results have been verified, it is still real consequences of avoiding avoids the effect of the complexity of its switching pattern, such as the non-symmetrical losses for each switching device of the single sub-module as a result of the non-symmetrical sine wave duty cycle of the arm current. Furthermore, a study presented by reference in [10] proposed a method for calculation losses in MMCs based on a comparison between the IGBT and IGCT switching device efficiency. The author analysed the circuit of the one sub-module based on Kirchhoff's second law, which is based on the assumption that when the switch is on noted as 1 and if off is noted as 0, after which these equations are integrated based on a phase angle from 0π to 2π. However, this affects the accuracy of the final MMC loss results, since there is no clarity regarding the switching and conduction period of each switching device. Extinctive A method for calculating switching and conduction loss in isolated DC-DC MMC with AC-link was proposed by [11]. This method is based on an assumption that the harmonic in the arm current is neglected. Also, author in [12]presented a method for

978-1-5386-7688-2/19 $31.00 © 2019 IEEE

calculating loss in MMCs based on a proposed virtual arm mathematical model (VAMM). The author claimed that this method is suitable for fast loss evaluation in high voltage applications. However, loss calculation based on the virtual arm can negatively affect the accuracy of the results, since switching devices are not provided based on manufacturing data-sheet. The author in [13] conducted a very brief analysis of power losses in MMCs on both the DC side and AC side. However, this did not include details about the switching pattern and characteristics of the switching devices.

This paper develops an analytical method for losses calculation in an MMC. The method of averaging the MMC's arm current (AMAC)is developed as described by [6]. As shown in Fig. 1, the MMC's sub-modules topologies of 2-level HBC, 2-level interleaved and 3-level NPC sub-modules are investigated. Furthermore, this method can also be used to calculate losses in remaining topologies. From a construction viewpoint, the use of 3L NPC as sub-modules in MMC, with high voltage applications, such as HVDC system is advantageous in reducing the number of modules to half, when compared to 2-level HBC. Therefore, this topology can reduce the dimensions of an entire converter and overall losses[14]. However, a 3-level NPC converter requires a controller for the purpose of capacitor voltage balancing in each of the sub-modules, contributing additional complexity and cost.

The significant contribution of this paper is the use of an interleaved converter topology as sub-modules in MMC, which have not been reported in other articles. The interleaved converter topology entails advantages for low conduction loss, because the arm current flow into the switches will be reduced by half [15]. Hence, this topology offers several advantages: reducing the RMS current in the input sub-modules capacitors, thus, enabling the use of smaller capacitors, and reducing the size of semiconductor devices with a half current rate, which then prompts a reduction in heat-sink requirements. The MMC uses an interleaved converter, as sub-module topology can be beneficial for high power applications demanding high current levels, such as HVDC transmission systems.

II. POWER LOSSES ANALYSIS

Reference in [9] reported the simplified analytical calculation model for average power loss in an MMC based on the root mean square (RMS) and arm current. The advantage of this method is that it can be used to investigate the relationship between the loss, voltage balancing methods, and main variables of an MMC. However, the main limitation of this technique is that "the analytical expressions of the conduction loss and the switching loss are derived in terms of the whole of MMC rather than a single sub-module. It avoids the effect of the complexity of its switching pattern on the calculation accuracy of the model". Thus, although the results have been verified, it is still real consequences of avoiding avoids the effect of the complexity of its switching pattern, such as the non-symmetrical loss for each switching device of the single

sub-module as a result of the non-symmetrical sine wave duty cycle of the arm current.

Furthermore, a study presented by reference in [10] proposed a method for calculation losses in MMCs based on a comparison between the IGBT and IGCT switching device efficiency. The author analysed the circuit of the one sub-module based on Kirchhoff's second law, which is based on the assumption that when the switch is on noted as 1 and if off is noted as 0, after which these equations are integrated based on a phase angle from 0π to 2π. However, this affects the accuracy of the final MMC loss results, since there is no clarity regarding the switching and conduction period of each switching device.

Extinctive A method for calculating switching and conduction loss in isolated DC-DC MMC with AC-link was proposed by [11]. This method is based on an assumption that the harmonic in the arm current is neglected. Also, author in [12]presented a method for calculating loss in MMCs based on a proposed virtual arm mathematical model (VAMM). The author claimed that this method is suitable for fast loss evaluation in high voltage applications. However, loss calculation based on the virtual arm can negatively affect the accuracy of the results, since switching devices are not provided based on manufacturing data-sheet.

The author in [13] conducted a very brief analysis of power losses in MMCs on both the DC side and AC side. However, this did not include details about the switching pattern and characteristics of the switching devices.

A method for calculating power losses in 2-level and 3-level converters can be found in [6]; from which the expressions used for calculating conduction and switching losses are derived. In this method, the characteristics of the IGBTs and diodes are obtained from the manufacturer's data sheet. The curve of turn-on and off switches energetically is approximated as linear. The switching losses in each sub-module device can be considered in Equations(1) as the relationship between switching energy and the arm current; thus, the switching losses can be given as:

$$P_{sw} = f_{sw}\, \frac{V_{sw}}{V_{base}}\, \frac{1}{2\pi} \int_{\theta_1}^{\theta_2} \left(a_{c/d} \cdot I_{arm} + b_{c/d} \right) \mathrm{d}\theta \qquad (1)$$

Where (a_c, b_c) are the constants values for the IGBTs and (a_d, b_d) are the constants values of the Diodes. These constants can be calculated using the manufacturer's data-sheet identifying the characteristics of IGBTs and Diodes. (V_{base}) is the inverter typical switching voltage, and (f_{sw}) is the switching frequency of the switching devices IGBTs-Diodes.

The developed method AMAC can also be used to calculate any conduction losses in an MMC switching device during its duty cycle. The respective integral for conduction loss is given below;

$$P_{con} = \frac{1}{2\pi} \int_{\theta_1}^{\theta_2} \delta\left(\theta + \phi\right) I_{arm} \left(R_{c/d}\, I_{arm} + V_{0,c/d} \right) \mathrm{d}\theta$$

$$(2)$$

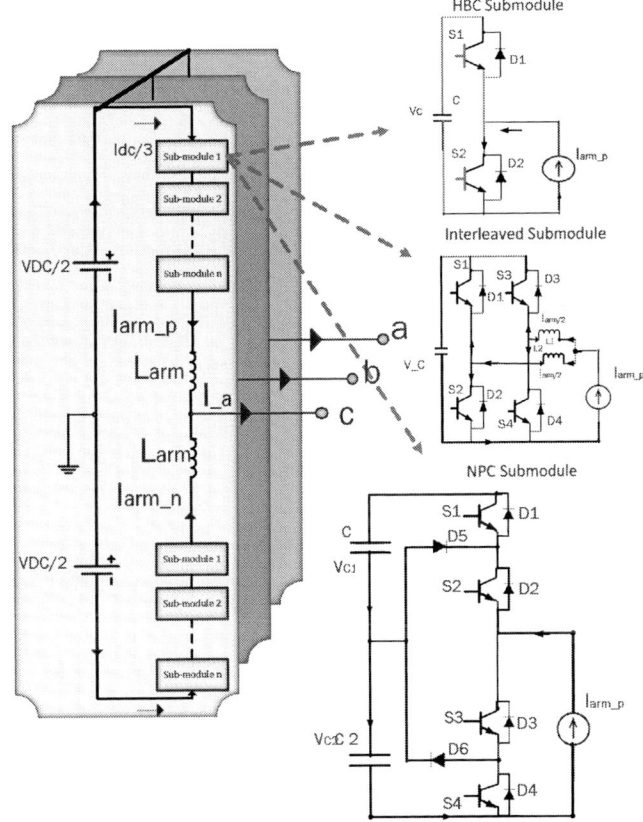

Fig. 1: Modular Multi level converter with different sub-modules topologies

TABLE I			
Switches sign[1]	Switches sign[0]	Conduction Switches	Arm Current and voltage conditions
S1	S2	D1	$I_{arm} \geq 0, /V_C \downarrow capacitor\ discharging$
S2	S1	S2	$I_{arm} \geq 0, /V_C = 0$
S1	S2	S2	$I_{arm} \leq 0, V_C \uparrow Capacitor\ charging$
S2	S1	D2	$I_{arm} \leq 0, /V_C = 0$

TABLE I: MMC-HBC Switching states

TABLE II			
Switches sign[1]	Switches sign[0]	Conduction Switches	Arm Current and voltage conditions
S1/S3	S2/S4	D1/D2	$I_{arm}/2 \geq 0, /V_C \downarrow Capacitor\ discharging$
S2/S4	S1/S3	S2/S4	$I_{arm}/2 \geq 0, /V_C = 0$
S1/S3	S2/S4	S1/S3	$I_{arm}/2 \leq 0, V_c \uparrow Capacitor\ charging$
S2/S4	S1/S3	D2/D4	$I_{arm}/2 \leq 0, /V_C = 0$

TABLE II: MMC-int Switching states

TABLE III			
Switches sign[1]	Switches sign[0]	Conduction Switches	Arm Current and voltage conditions
S1/S2	S3/S4/D5/D6	D1/D2	$I_{arm} \geq 0, /V_C1 + V_C2 \downarrow Capacitor\ discharging$
S3/D6	S1/S2/S4/D5	S3/D6	$I_{arm} \geq 0, /V_C2 \downarrow Capacitor\ discharging$
S3/S4	S1/S2/D5/D6	S3/S4	$I_{arm} \geq 0, V_C1 + V_C2 = 0$
S3/S4	S1/S2/D5/D6	D1/D2	$I_{arm} \leq 0, V_C1 + V_C2 = 0$
D5/S2	S1/S3/S4/D6	S2/D5	$I_{arm} \leq 0, /V_C1 \uparrow Capacitor\ charging$
S1/S2	S3/S4/D5/6	S1/S2	$I_{arm} \leq 0, /V_C1 + V_C2 \uparrow Capacitor\ charging$

TABLE III: MMC-NPC Switching states

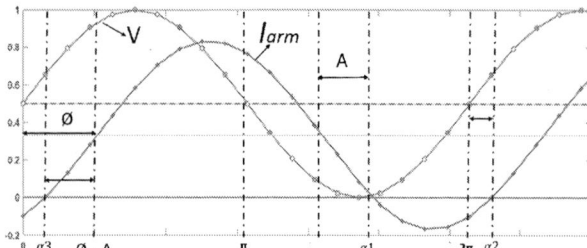

Fig. 2: The MMC upper arm current

where (I_{arm}) is the instantaneous arm current amplitude, $(R_{c/d})$ and $(V_{0,c/d})$ are the parameters of the IGBTs/diode determined from data sheet, and $([\theta_1, \theta_2])$ is the conduction interval during a fundamental period, and $(\delta(\theta + \phi))$ is the conduction duty cycle as a function of the voltage angle.

Based on the theory of MMC in [1], the devices in the upper and lower arms of each phase leg are symmetrical. Therefore, the methods of loss calculation can be applied to one arm only in MMC to ascertain overall loss. The MMC's arm current model (I_{arm}) is given in [5], as expressed in Equations (3) and (4). However, as shown in Fig. 2 the arm current is simply shifted up with the value of A, as given in Equations (3). Thus the average for the arm current is no longer equal to zero.

The arm current is $(I_{arm} > 0)$ between intervals of $(\alpha 3)$ to $(\alpha 1)$, and $(I_{arm} < 0)$ between intervals of $(\alpha 1)$ to $(\alpha 2)$. This phenomenon is very significant and must be taken into consideration during the loss calculation stage in MMC, because the levels of loss in the MMCs semiconductor devices will be unsymmetrical as a result.
$\alpha 1 = (\pi + \phi + A)$, $\alpha 2 = (2\pi + \phi - A)$ and $\alpha 3 = (\phi - A)$. The angle (A) is the same as:

$$A = \arcsin \frac{M \cos \phi}{2} \qquad (3)$$

The upper arm current $(I_{arm} - p)$ can be expressed thus:

$$I_{arm} - p = \frac{I_L}{2} \left(\frac{M \cos \phi}{2} + \sin(\theta - \phi) \right) \qquad (4)$$

The MMC lower arm current $(I_{arm} - n)$ is given as:

$$I_{arm} - n = \frac{I_L}{2} \left(\frac{M \cos \phi}{2} - \sin(\theta - \phi) \right) \qquad (5)$$

, where (I_L) is the peak value of the MMC load current, (ϕ) is the phase of the phase-current delay for the phase voltage, and (M) is the modulation index of the switching devices in the MMC.

In this study, the PWM switching technique is used to control the IGBTs and diodes of the MMC-HBC and MMC-inte and MMC-NPC. The switching stages and operations are illustrated in Tables I II and III, respectively.

In the MMC-HBC and MMC-inte sub-modul, the upper and lower switches are controlled by the PWM switching algorithm, as shown below:

$$SM_{upper}(\theta) = \frac{1}{2}[1 - M\sin\theta],$$

$$SM_{bottom}(\theta) = \frac{1}{2}[1 + M\sin\theta],$$

Under MMC-NPC sub-module, the upper and lower switches are controlled by PWM switching algorithm depicted as:

$$SM_{upper}(\theta) = [1 - M\sin\theta],$$

$$SM_{bottom}(\theta) = [1 + M\sin\theta],$$

In order to simplify the complexity of the losses analysis and calculations, several assumptions as applied in this study are as follows:

1) The additional switching loss in the MMC are neglected in this study because it is assumed that the voltage of the capacitors in the MMCs sub-modules is balanced. In view of this,

2) the dead time delay of these switches will not be taken into consideration in this study.

The total conduction loss of the arm conductor in MMC is given in [5] as:

$$P_{arm} = R_{arm\ l} \left(\frac{3M \cos^2(\phi) + 6}{8} \right)$$

The expressions derived for calculating conduction loss in MMC-HBC, MMC-int and MMC-NPC are given in Equations (6) ,(7) and (8) ,respectively. The expressions derived for calculating the switching loss in MMC-HBC, MMC-int and MMC-NPC are provided in Equations (9) ,(10) and (11), respectively. All the K coefficients are included in the appendix.

III. RESULT ANALYSIS AND DISCUSSION

The parameters of IGBTs and diodes are obtained from manufacturer's data sheets, as shown in TableIV. The suitable IGBTs and diodes for different sub-module topologies have been divided into two groups. Module group A $FF75R12RT4(1200V - 75A)$ is used for the 2-level sub-module HBC and the 3-level sub-module NPC; module B $FP25R12KE3(1200V - 40A)$ is used in the 2-level interleaved converter. The nominal load peak current $(I_L = 70A)$, the MMC module DC-link voltage $(VDC = 3KV)$ and the number of sub-modules $(N = 6)$ in MMC-HBC and MMC-int and in MMC-NPC $(N = 3)$ [4]. The switching frequency of the semiconductor devices (250 HZ).

TABLE IV: The Parameters of IGBTs and Diodes

Parameter	IGBT/diode Module A	IGBT/diode Module B	Unit
V_{base}	0.6	0.6	kV
$V_{0,c}$	0.7	0.7	V
R_c	20	52	mΩ
$V_{0,d}$	0.75	0.75	V
R_d	11	20.4	mΩ
a_c	0.3	0.05	mJA^{-1}
b_c	1	0	mJ
a_d	0.35	0.028	mJA^{-1}
b_d	0.75	0.5	mJ

$$P_{con}(MMC-HBC) = NI_L\frac{3}{4\pi}\Big[(2K1+K2+2K3+K4)\Big](Vc+Vd) + NI_L^2\frac{3}{32\pi}\Big[(2K5+2K6+K7+K8+K9+$$
$$K10) + 4\pi(M^2+2)(\cos(\phi))^2)\Big](Rc+Rd) \tag{6}$$

$$P_{con}(MMC-inte) = NI_L\frac{3}{8\pi}\Big[(2K1+K2+2K3+K4)\Big](Vc+Vd) + NI_L^2\frac{3}{128\pi}\Big[(2K5+2K6+K7+K8+K9+$$
$$K10) + 4\pi M^2((\cos(\phi))^2) + 8\pi(\cos(\phi))^2)\Big](Rc+Rd) \tag{7}$$

$$P_{con}(MMC-NPC) = NI_L\frac{6}{\pi}\Big[(4K1+2K2+4K3+2K4)\Big](Vc+Vd) + NI_L^2\frac{3}{4\pi}\Big[(4K5+4K6+2K8+2K9+2K7$$
$$+2K10) + 18\pi M^2((\cos(\phi))^2) + 16AM^2(\cos(\phi))^2 + 16A(\cos(\phi))^2 + 20\pi((\cos(\phi))^2)\Big](Rc+Rd) \tag{8}$$

$$P_{sw}(MMC-HBC) = F_{sw}\frac{VDC}{Vbase}6\left[\frac{I_L}{K}(a_c+a_d) - 2(b_c+b_d)\right] \tag{9}$$

$$P_{sw}(MMC-int) = F_{sw}\frac{VDC}{Vbase}6\left[\frac{I_L}{K}(a_c+a_d) - 4(b_c+b_d)\right] \tag{10}$$

$$P_{sw}(MMC-NPC) = F_{sw}\frac{VDC}{Vbase}6\left[\frac{2I_L}{K}(a_c+a_d) - 2(b_c+b_d)\right] \tag{11}$$

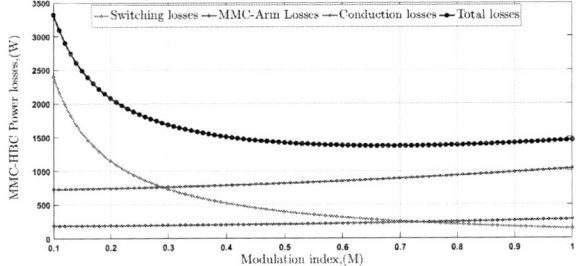

Fig. 3: Power loss in MMC-HBC against modulation index, (M)

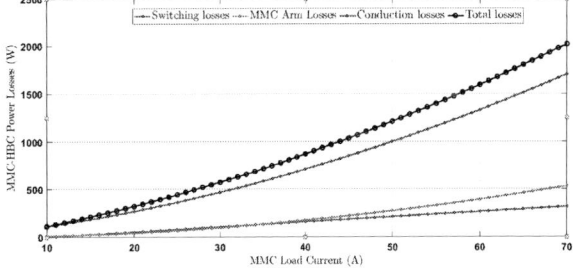

Fig. 4: Power loss in MMC-HBC against load current, (I_L)

The average of the conduction loss in MMC-NPC, NNC-inte and, MMC-HBC in Fig. 3, Fig. 6 and Fig. 9 are nearly constant with the increase in modulation index M. On the other hand, the switching loss reduce with increases in the modulation index (M) because with a high modulation index duty cycle the demand for the actioned number of turns ON and Off, in the switching devices will be reduced. Similarly, the average of the conduction losses in the MMC-arm has been found to be constant with the increases in the modulation index. However, it has a high impact on the load current as it was found that it

significantly increased with the greater load current.

Fig. 4, Fig. 7 and Fig. 10 shows that increases in load current will lead to an increase in the average for conduction loss. However, it is also shown that this will have a low impact on the switching loss in MMC.

The maximum discrepancy between the averages for switching losses and conduction losses are found with a phase angle of (π), as shown in Fig. 5, Fig. 11 and Fig. 14. In contrast, there is a lower discrepancy in the average of the losses with phase angles of (0) and (2π).

The comparative results between total losses in MMC-HBC,

978-1-5386-7688-2/19 $31.00 © 2019 IEEE

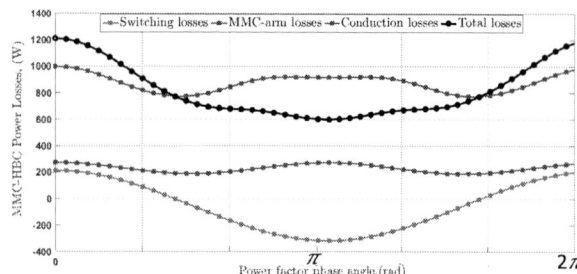

Fig. 5: Power loss in MMC-HBC against phase angle, (ϕ)

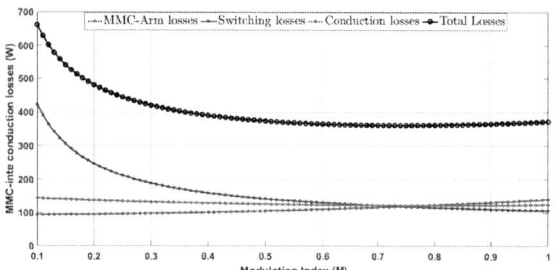

Fig. 6: Power loss in MMC-inte against modulation index, (M)

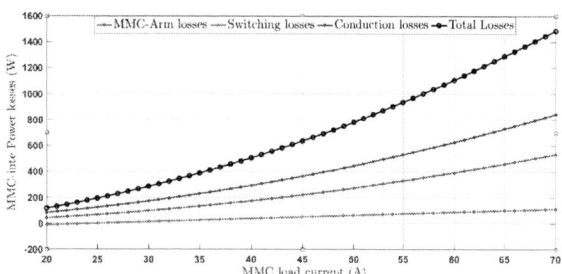

Fig. 7: Power loss in MMC-inte against load current, (I_L)

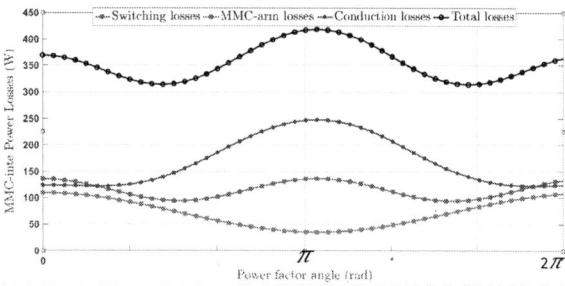

Fig. 8: Power loss in MMC-inte against phase angle, (ϕ)

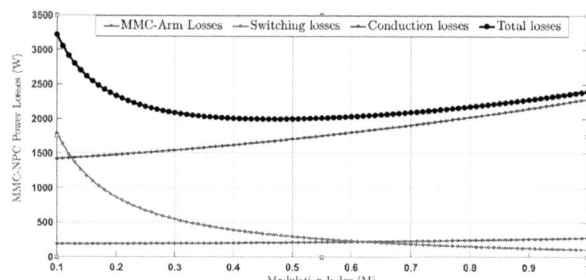

Fig. 9: Power loss in MMC-NPC against modulation index, (M)

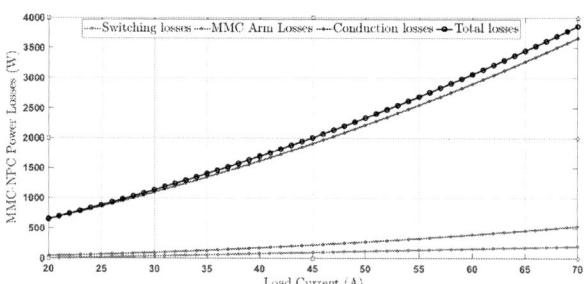

Fig. 10: Power loss in MMC-NPC against load current, (I_L)

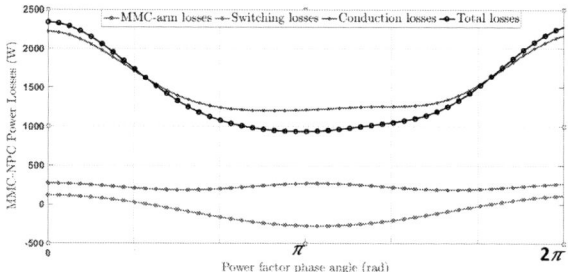

Fig. 11: Power loss in MMC-NPC against phase angle, (ϕ)

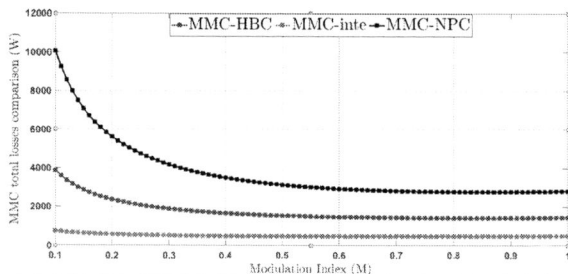

Fig. 12: Power loss comparison in MMC against modulation index, (M)

MMC-inte and MMC-NPC are shown in Fig. 12, Fig. 13 and Fig. 14. These results illustrate that in MMC-NPC, half the number of submodules compared to MMC-HBC will have an equal average losses level, increasing the modulation index (M). However, the lowest average loss was found in MMC-inte, when compared to those in MMC-NPC and MMC-HBC.

The effect of the phase angle (ϕ) on total losses in MMC-NPC, MMC-inte and MMC-HBC is shown in Fig. 14. The results indicate that MMC-NPC suffers lower losses at a phase angle of π, but it will experience higher losses with phase

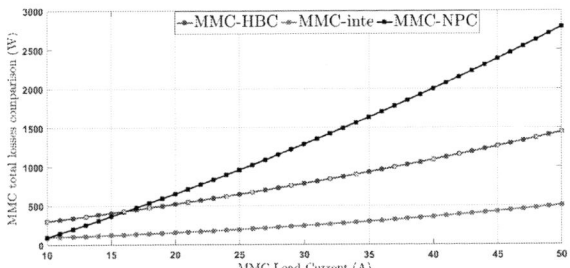

Fig. 13: Power loss comparison in MMC against load current, (I_L)

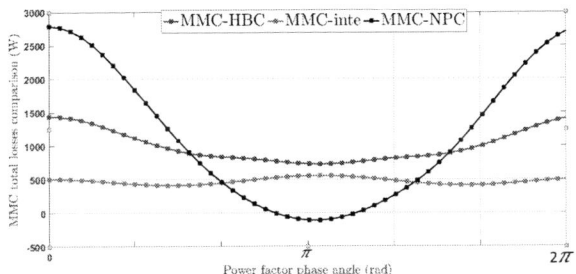

Fig. 14: Power loss comparison in MMC against phase angle, (ϕ)

angles of (0) and (2π). The effect of the phase angle (ϕ) on the total losses in MMC-HBC and MMC-inte is low when compared to MMC-NPC. However, the maximum discrepancy of losses was found in a phased angle of π.

IV. CONCLUSION

Averaging MMC's arm current(AMAC) method has been developed in this study for loss analysis and calculation in MMC. The study examined loss in MMC based the topologies of MMC-HBC's, MMC-int and MMC-NPC, where conduction loss, switching loss and arm loss are investigated.

All the expressions have been derived and applied for loss calculation. The 2-level interleaved converter has been first time introduced as sub-module for MMC (MMC-inte), where no publications are found in the literature presents this idea. This topology has shown an advantage of the lowest loss level compared to those in MMC-HBC, MMC-NPC. Although MMC-NPC has been applied with a half number of sub-module, it was found suffering from the highest loss compared to those in MMC-HBC, MMC-inte.

The conduction loss is decreasing, whereas switching loss are increasing with the increases of the Modulation index (M). The conduction loss will rise with an increase in load current. However, these will have a low impact on the switching loss which is affected directly by the switching frequency. The highest effect of the phase angle (ϕ) on the loss in MMC, as can be found from the phase angle ($\phi = \pi$) in MMC-NPC. The effect of the arm inductor on total losses in MMC-HBC, MMC-

inte and MMC-NPC investigated, and it has been approximated to be about (10%) from the average of the total loss.

REFERENCES

[1] A. Lesnicar and R. Marquardt, "An innovative modular multilevel converter topology suitable for a wide power range," *2003 IEEE Bologna PowerTech - Conference Proceedings*, vol. 3, pp. 272–277, 2003.

[2] I. J. of Power Electronics and D. S. (IJPEDS). (2015) 2015 ieee power energy. Accessed 27-March-2017. [Online]. Available: http://www.pes-gm.org/2015/

[3] C. Oates and C. Davidson, "A comparison of two methods of estimating losses in the modular multi-level converter," in *Power Electronics and Applications (EPE 2011), Proceedings of the 2011-14th European Conference on.* IEEE, 2011, pp. 1–10.

[4] S. Madichetty and A. Dasgupta, "Modular Multilevel Converters Part-I : A Review on Topologies , Modulation , Modeling and Control Schemes," *International Journal of Power Electronics and Drive System (IJPEDS)*, vol. 4, no. 1, pp. 36–50, 2014.

[5] M. Zygmanowski, "Analytical and Numerical Power Loss Analysis in Modular Multilevel Converter," no. 2, pp. 465–470, 2013.

[6] G. Orfanoudakis, "Loss comparison of two and three-level inverter topologies," *Power Electronics*, pp. 143–143, 2010.

[7] G. Konstantinou, J. Zhang, S. Ceballos, J. Pou, and V. G. Agelidis, "Comparison and evaluation of sub-module configurations in modular multilevel converters," *2015 IEEE 11th International Conference on Power Electronics and Drive Systems*, pp. 958–963, June 2015.

[8] A. Nami, J. Liang, F. Dijkhuizen, and G. D. Demetriades, "Modular multilevel converters for hvdc applications: Review on converter cells and functionalities," *IEEE Transactions on Power Electronics*, vol. 30, no. 1, pp. 18–36, Jan 2015.

[9] L. Yang, Y. Li, Z. Li, P. Wang, S. Xu, and R. Gou, "A simplified analytical calculation model of average power loss for modular multilevel converter," *IEEE Transactions on Industrial Electronics*, vol. 66, no. 3, pp. 2313–2322, March 2019.

[10] B. Zhao, R. Zeng, J. Li, T. Wei, Z. Chen, Q. Song, and Z. Yu, "Practical analytical model and comprehensive comparison of power loss performance for various mmcs based on igct in hvdc application," *IEEE Journal of Emerging and Selected Topics in Power Electronics*, vol. 7, no. 2, pp. 1071–1083, June 2019.

[11] S. Zhao, Y. Chen, and L. Peng, "Semiconductor loss calculation of dcdc modular multilevel converter for hvdc interconnections," *High Voltage*, vol. 3, no. 4, pp. 263–271, 2018.

[12] Y. Shiyuan, W. Yue, Y. Taiyuan, N. Cheng, D. Guozhao, and W. Zhang, "Instantaneous power loss calculation for mmc based on virtual arm mathematical model," pp. 2625–2629, May 2018.

978-1-5386-7688-2/19 $31.00 © 2019 IEEE

[13] S. Sajedi, M. Basu, and M. Farrell, "Power loss comparison of dc side and ac side cascaded modular multilevel inverters," pp. 1–5, Sep. 2018.

[14] E. Solas, G. Abad, J. a. Barrena, S. Aurtenetxea, a. Carcar, and L. Zajac, "Modular multilevel converter with different submodule concepts - Part II: Experimental validation and comparison for HVDC application," *Industrial Electronics, IEEE Transactions on*, vol. 60, no. 10, pp. 4536–4545, 2013.

[15] M. A. Abusara, J. M. Guerrero, and S. M. Sharkh, "Line-interactive UPS for microgrids," *IEEE Transactions on Industrial Electronics*, vol. 61, no. 3, pp. 1292–1300, 2014.

Appendix

$$K = \frac{2}{M\,\cos(\phi)}$$

$$K1 = M^2\cos(A)(\cos(\phi))^2 + M(\sin(\phi))^2\cos(\phi)\sin(A)\cos(A)$$
$$- 2M\cos(A)\cos(\phi)\sin(A)(\sin(\phi))^2$$

$$K2 = -M(\cos(\phi))^3\sin(A)\cos(A) + \pi M\cos(\phi)$$
$$+ 2AM\cos(\phi) + 2\cos(A))$$

$$K3 = -M^2\cos(A)(\cos(\phi))^2 - M\cos(A)(\sin(\phi))^2\cos(\phi)\sin(A)$$
$$+ 2M\cos(A)\cos(\phi)\sin(A)(\sin(\phi))^2$$
$$+ M\sin(A)\cos(A)(\cos(\phi))^3 + 2\cos(A)$$

$$K4 = -M(\cos(\phi))^3\sin(A)\cos(A) - \pi M\cos(\phi)$$
$$+ 2AM\cos(\phi) + 2\cos(A))$$

$$K5 = -(\sin(\phi))^2\sin(2(\phi - A)) + (\sin(\phi))^2$$
$$\sin(2(\phi + \pi + A))$$

$$K6 = -(\sin(\phi))^2\sin(2(\phi + \pi + A)) + (\sin(\phi))^2$$
$$\sin(2(\phi + 2\pi - A))$$

$$K7 = (2\pi - 4A)(\sin(\phi))^2 + (1 - M^2)(\cos(\phi))^2\sin(2(\phi + \pi + A))$$
$$+ (M^2 - 1)(\cos(\phi))^2\sin(2(\phi + 2\pi - A))$$
$$+ (M^3 - 4M)(\cos(\phi))^2\cos(\phi - A)$$

$$K8 = (2\pi + 4A)(\sin(\phi))^2 + (M^2 + 1)(\cos(\phi))^2$$
$$\sin(2(\phi - A)) - (M^2 + 1)(\cos(\phi))^2$$
$$\sin(2(\phi + \pi + A)) + (M^3 + 4M)(\cos(\phi))^2\cos(\phi - A)$$

$$K9 = (2\pi + 4A)(\sin(\phi))^2 + (1 - M^2)(\cos(\phi))^2$$
$$\sin(2(\phi - A)) + (M^2 - 1)(\cos(\phi))^2\sin(2(\phi + \pi + A)$$
$$+ (M^3 - M4)(\cos(\phi))^2\cos(\phi - A)$$

$$K10 = (2\pi - 4A)(\sin(\phi))^2 + (M^2 + 1)(\cos(\phi))^2\sin(2(\phi + \pi + A)$$
$$- (M^2 + 1)(\cos(\phi))^2\sin(2(\phi + 2\pi - A))$$
$$- (M^3 + 4M)(\cos(\phi))^2\cos(\phi - A)$$

Three-phase Modified Z-source Three-level T-Type Inverters with Continuous Source Current

Anh-Vu Ho [1], Anh-Tuan Huynh [2], Tae-Won Chun [2]

[1] School of Engineering, Eastern International University, Vietnam
[2] Department of Electrical Engineering, University of Ulsan, Korea
E-mail: vu.ho@eiu.edu.vn, huynhtuanddt@gmail.com, twchun@mail.ulsan.ac.kr

Abstract—This paper proposed a three-phase modified Z-source three-level T-type inverter (MZS-3LTI) topology implemented by integrating a modified Z-source impedance network with embedded dc source to the traditional three-level T-type inverter. The proposed MZS-3LTI provides the enhanced boost ability and ensures the continuous dc source current by adopting the embedded concept without any extra capacitor or filter. The modulation technique is proposed for controlling effectively the upper and lower shoot-through states with a simple logic circuit and reducing the number of switching. The performances of the proposed topology and the modulation technique are demonstrated through simulation results.

Keywords— Boost capability, embedded dc source, Z-source impedance network, modulation technique, three-level T-type inverter (3LTI).

I. INTRODUCTION

Recently, multilevel inverters have been an effective solution in the high-power applications. They can provide a lower total harmonic distortion (THD) due to a stepped output waveform more than two voltage levels and can reduce the voltage stress of the inverter switching devices. The multilevel inverters can be applied to various industrial applications such as the ac motor control and the renewal energy generation systems [1], [2]. Among various multilevel topologies, the neutral-point-clamped (NPC), cascaded H-bride (CHB), and flying capacitors (FC) inverters have been widely used.

As an alternative three-level inverter topology, the three-level T-type inverter (3LTI) was introduced, which is implemented by connecting a bidirectional switch between the midpoint of the dc-link voltage and the inverter outputs of the two-level inverter [3]-[5]. Compared to the three-phase three-level NPC inverter, it can provide a lower switching loss and reduce two diodes per bridge leg while maintaining the merits of the three-level inverter such as the enhanced output voltage waveform. In [6], the control scheme for minimizing a dc-link capacitor current at the transient situation in the T-type AC/DC/AC PWM converter is proposed, in order to reduce the value of dc-link capacitor.

However, the 3LTI can only step down its output voltage, because it can only offer the buck operation. Therefore, it has some troubles to apply some areas to require a higher voltage gain to obtain a desired ac output voltage for a renewal energy sources with low voltage. In order to design the 3LTI with the enhanced voltage boost capability, the concepts of the (quasi-)Z-source impedance network are employed to the 3LTI. The Z-source inverter (ZSI) or quasi-Z-source inverter (qZSI) can raise the dc-link voltage by shorting any phase leg of inverter [7]-[9]. A buck-boost operation with a single-stage power conversion can be achieved. The reliability is improved because the dead-time is not required.

Several topologies combing the (quasi-)Z-source impedance network with the 3LTI have been proposed, in order to obtain the advantages of (q-)ZSI as well as those of the 3LTI [10]-[14]. The topology and modulation method for a Z-source three- or five-level T-type inverter are described in [10], [11]. The space vector modulation scheme of the qZSI-3LTI topology realized by combining the two symmetrical quasi-Z-source networks and three-phase 3LTI is proposed for reducing both the slew-rate and the magnitude of the common-mode voltage [12]. The behavior for the normal and fault-tolerant operation modes of the qZSI-3LTI is analyzed in [13]. The topologies integrated the (q-)Z-source impedance and a T-type inverter have a boost capability, and achieve a neutral voltage balancing. However, their boost factor is not high enough in many industrial applications, which require the high voltage gain. The operation analysis and PWM control method of the three-level quasi-switched boost T-type inverter (3L-qSBT²I), which combines two symmetrical quasi-switched boost networks and 3LTI, is presented in [14]. The 3L-qSBT²I can provide a high voltage gain and suppress the inductor current ripples. However, it requires the two additional switching devices and a complex modulation technique.

This paper introduces a three-phase modified Z-source three-level T-type inverter (MZS-3LTI) implemented by integrating a modified Z-source impedance network [15] with embedded concept to the traditional 3LTI. The proposed MZS-3LTI provides the enhanced boost ability and ensures the continuous dc source current without any extra capacitor or filter by employing the embedded concept. The detailed operation for the shoot-through state and non-shoot-through state of proposed MZS-3LTI topology is analyzed. In order to control effectively the upper- and lower-shoot through states, the modulation technique based on an alternative phase opposition disposition (APOD) for the MZS-3LTI topology is proposed. The performances of the proposed topology and modulation technique are demonstrated through simulation results.

978-1-5386-7688-2/19 $31.00 © 2019 IEEE

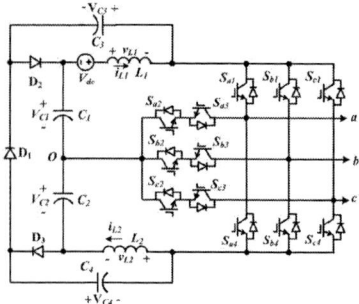

Fig. 1. Proposed three-phase embedded modified Z-source three-level T-source inverter (MZS-3LTI).

II. STEADY-STATE ANALYSIS OF MZS-3LTI

The steady-state analysis for the operation of proposed MZS-3LTI is performed, assuming that all of switches, diodes, inductors and capacitors used in the proposed topology are ideal for simplification. Fig. 1 shows the structure of the proposed three-phase MZS-3LTI, which is established by combining the modified Z-source impedance network [15] and the 3LTI. The modified Z-source impedance network consists of two inductors, four capacitors, and three diodes. The embedded concept is applied at the MZS. A single dc source is placed to be in serial with the inductor L_1 of the upper cell in the impedance network. The continuous dc source current can be achieved without any additional capacitor and filter. Three-phase bidirectional switches are connected between the mid-point of two serial capacitors C_1 and C_2 and the three-phase inverter outputs.

The operating state of the proposed topology can be divided into the shoot-through state and the non-shoot-through (NST) state like traditional ZSI. The equivalent circuits for both operating states are described, in order to perform a steady-state analysis of the operations of the proposed topology.

A. Shoot-through state

There are three kinds of the shoot-through state available at the three-phase MZS-3LTI such as an upper shoot-through (UST) state and a lower shoot-through (LST) state and a full shoot-through (FST) state. In FST state, a full dc-link of the inverter is shorted by switching on all of switches in any phase leg. As the output voltage is zero during FST state, the output voltage may be distorted. Both the upper and lower shoot-through states can be made by conducing the upper or lower switch and connecting the mid-point O to the output bridge. Because the output voltage can be generated during both the UST and LST states, the quality of the output voltage waveform can be improved.

Fig. 2 shows the equivalent circuits for the three-phase embedded MSI-3LTI in the shoot-through state and the non-shoot-through state. In the UST state as shown in Fig. 2(a), the switches S_{x1}, S_{x2}, and S_{x3} ($x = a$, b, or c) are switched on, and diodes D_1 and D_3 are in the conducting state, whereas diode D_2

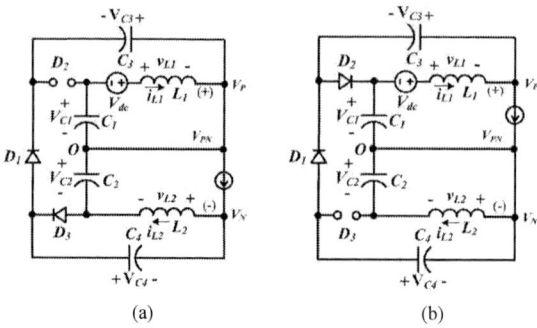

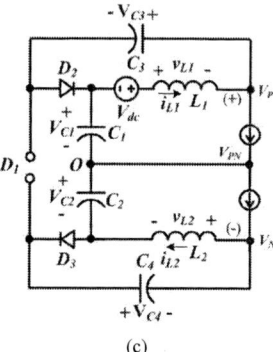

Fig. 2. Equivalent circuits for the three-phase embedded MSI-3LTI: (a) upper-shoot-through state, (b) lower shoot-through state, (c) non-shoot-through state.

is in the blocking state. The upper impedance cell is shorted, and the inductor L_1 charges the energy from the dc input source and the capacitor C_1. Both the inductor L_2 and capacitor C_2 delivery energy to the load side. The two inductor voltages and dc-link voltage can be written by

$$v_{L1} = V_{dc} + V_{C1} \ , \ v_{L2} = -V_{C4} \tag{1}$$

$$V_P = 0 \ , \ V_N = -(V_{C2} + V_{C4}) \tag{2}$$

$$V_{PN} = V_{C2} + V_{C4} . \tag{3}$$

In the LST state as shown in Fig. 2(b), the switches S_{x2} and S_{x3}, and S_{x4} ($x = a$, b, or c) are switched on, and diodes D_1 and D_2 are in the conducting state, whereas diode D_3 is in the blocking state. The lower impedance cell is shorted, and the inductor L_2 charges the energy from the capacitor C_2 at the lower impedance cell. Two capacitors C_1 and C_3 delivery energy to the ac load side. The two inductor voltages and dc-link voltage can be written by

$$v_{L1} = V_{dc} - V_{C3} \ , \ v_{L2} = V_{C2} \tag{4}$$

$$V_P = V_{C1} + V_{C3} \ , \ V_N = 0 \tag{5}$$

$$V_{PN} = V_{C1} + V_{C3} . \tag{6}$$

B. Non-shoot-through state

Fig. 2(c) shows the equivalent circuit of the NST, which consists of an active state and a zero state. The MZS-3LTI operates as the conventional 3LTI. The diodes D_2 and D_3 are

978-1-5386-7688-2/19 $31.00 © 2019 IEEE   440

TABLE I
Switching States of Three-Phase MZS-3LTI ($x = a$, b, or c)

State Type	ON switches	ON diodes	V_{Px}	V_{Nx}
UST	S_{1x}, S_{2x}, S_{3x}	D_1, D_3	0	$-(V_{C2}+V_{C4})$
LST	S_{2x}, S_{3x}, S_{4x}	D_1, D_2	$V_{C1}+V_{C3}$	0
NST	(S_{1x}, S_{2x}) or (S_{3x}, S_{4x})	D_2, D_3	$V_{C1}+V_{C3}$	$-(V_{C2}+V_{C4})$
NST	S_{2x}, S_{3x}	D_2, D_3	0	0

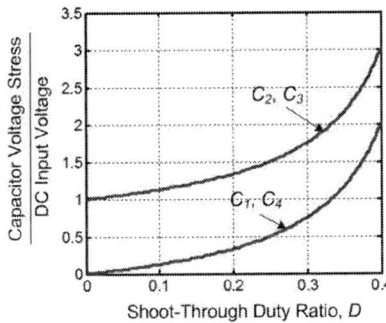

Fig. 3. Capacitor voltage stress with a variation of shoot-through duty ratio.

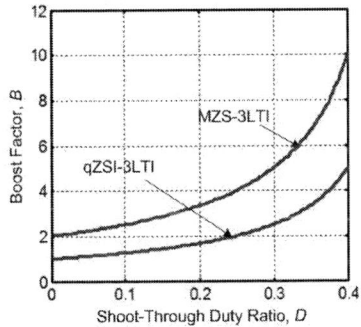

Fig. 4. Boost factors with a variation of shoot-through duty ratio.

in the conducting state, whereas diode D_1 is in the blocking state. When x-phase reference signal is positive in the active state, S_{x1} and S_{x2} are switched on. When x-phase reference signal is negative in the active state, S_{x3} and S_{x4} are switched on. The two inductor voltages and dc-link voltage can be written by

$$v_{L1} = V_{dc} - V_{C3}, \; v_{L2} = -V_{C4} \tag{7}$$

$$V_P = V_{C1} + V_{C3}, \; V_N = -(V_{C2} + V_{C4}) \tag{8}$$

$$V_{PN} = V_{C1} + V_{C2} + V_{C3} + V_{C4} = \hat{V}_{PN}. \tag{9}$$

It can be noted that the dc-link voltage in the active state can be established by adding four capacitor voltages, and it is a peak dc-link voltage. When S_{x2} and S_{x3} are turned on in the

zero state, the output voltage becomes zero as connecting the mid-point O to the inverter output.

Table I summaries the output phase terminal voltages V_{Px} and V_{Nx} ($x = a$, b, or c) and the turn-on switching devices and diodes according to the operating state of the MZS-3LTI.

C. Boost factor

The average values of UST and LST periods are the same over one period of the inverter frequency due to symmetrical operation of the modulation technique. Therefore, the period of UST and LST is represented as T_{ST}. By utilizing the voltage-second balance principle to two inductors L_1 and L_2 from (1), (4), and (7) over one switching period T, and by assuming $V_{C1} = V_{C4}$ and $V_{C2} = V_{C3}$, the four capacitor voltages are expressed as a function of the shoot-through duty ratio D, which is defined as $D = T_{ST}/T$.

$$V_{C1} = V_{C4} = \frac{D}{1-2D} V_{dc} \tag{10}$$

$$V_{C2} = V_{C3} = \frac{1-D}{1-2D} V_{dc} \tag{11}$$

Fig. 3 shows the plots of the four capacitor voltages to the dc input voltage with a variation of D. The voltage stress of two capacitors C_2 and C_3 is higher than that of two capacitors C_1 and C_4.

By substituting (10) and (11) into (9), the boost factor B, defined as the ratio of the peak dc-link voltage to the dc input voltage is given by

$$B = \frac{\hat{V}_{PN}}{V_{dc}} = \frac{2}{1-2D}. \tag{12}$$

Fig. 4 shows the plots of the boost factors of the proposed MZS-3LTI topology and the qZSI-3LTI introduced in [12], [13] with a variation of D. It can be noted that the boost factor of the proposed topology is twice higher than the qZSI-3LTI over all range of D.

III. Modulation Technique for a MZS-3LTI

Fig. 5 explains the modulation technique for a-phase of the proposed MZS-3LTI based on an alternative phase opposition disposition (APOD) [16] in order to produce the boosted ac output voltage. As shown in Fig. 5(a), the proposed modulation technique has two carrier signals V_{Tr1} and V_{Tr2} with 180° phase shift and the shoot-through envelope signal V_P. By using three-phase sinusoidal reference signals V_{ref_x} ($x = a$, b, or c), the three-phase positive and negative modulation signals V_{px} and V_{nx} are calculated as

$$V_{px} = \frac{1}{2}\left(\left|V_{ref\_x}\right| + V_{ref\_x}\right) \tag{13}$$

$$V_{nx} = \frac{1}{2}\left(\left|V_{ref\_x}\right| - V_{ref\_x}\right). \tag{14}$$

The PWM signals S_{1x}/S_{3x}, which are switched complementary, are generated by comparing the positive modulation signal V_{px} with a carrier signal V_{Tr1}. When the

978-1-5386-7688-2/19 $31.00 © 2019 IEEE

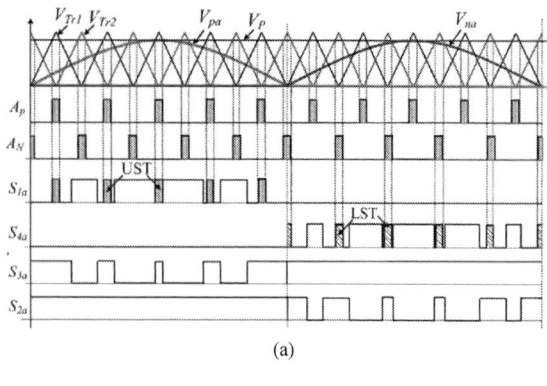

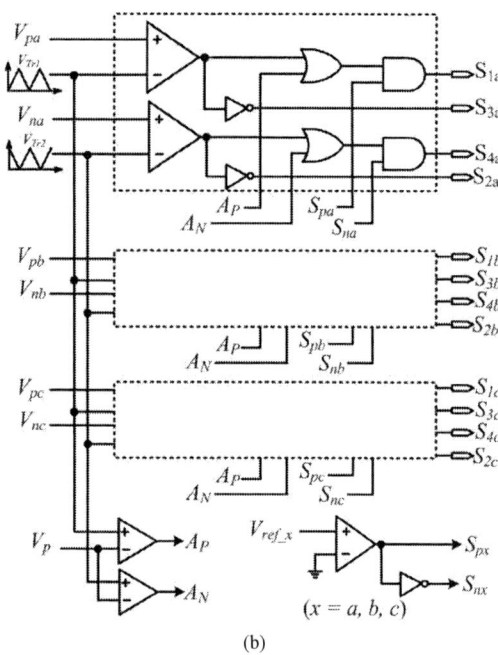

(a)

(b)

Fig. 5. Modulation technique of a MZS-3LTI: (a) switching pattern for proposed modulation technique, (b) logic circuit.

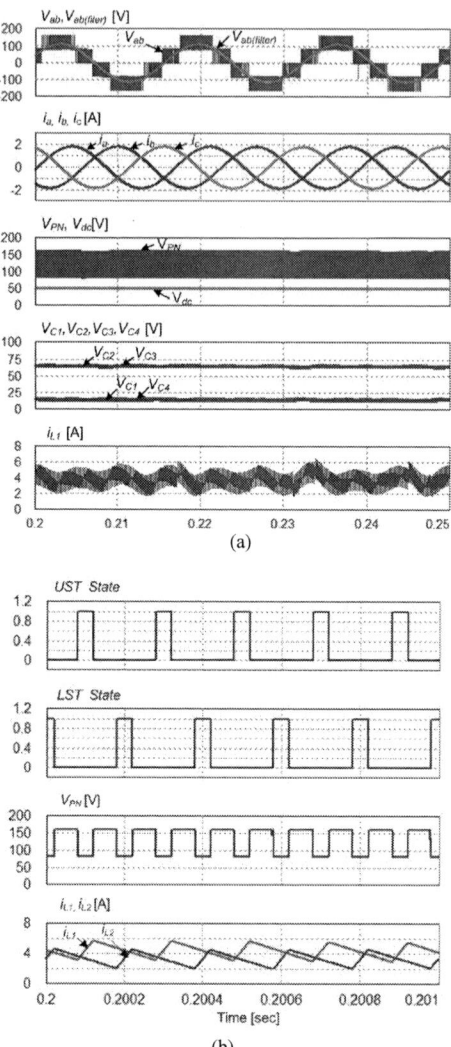

(a)

(b)

Fig. 6. Simulation results when $M = 0.8$ and $D = 0.2$: (a) ac output voltage and current, dc-link and dc input voltages, capacitor voltages, and dc input current, (b) UST and LST states, dc-link voltage and two inductor currents

signal V_{px} is higher than V_{Tr1}, the switch S_{1x} is switched on and the switch S_{3x} is switched off. The PWM signals S_{2x}/S_{4x}, which are switched complementary, are generated by comparing the negative modulation signal V_{nx} with a carrier signal V_{Tr2}. When the signal V_{nx} is higher than V_{Tr2}, the switch S_{4x} is switched on and the switch S_{2x} is switched off. The UST signal A_P and LST signal A_N can be generated by comparing the shoot-through envelope signal V_P with V_{tr1} and V_{tr2}, respectively. The UST is inserted in the PWM signal of S_{1a}, and the LST is inserted in the PWM signal of S_{4a}. The UST and LST are applied when the reference signal is positive and negative, respectively.

Fig. 5(b) shows a logic circuit for generating the three-phase 12 PWM signals from the three-phase modulation signals V_{px} and V_{nx} calculated by (13) and (14). The signals S_{px} and S_{nx} (x = a, b, or c) are utilized to insert the UST or LST in

the PWM signal when the reference signal of each phase is positive or negative, respectively. The modulation technique can be easily implemented by a simple logic circuit.

IV. SIMULATION RESULTS

The simulation study is carried out by using PSIM program in order to verify the performances of the proposed topology. The dc input voltage V_{dc} is 50 V, and the resistive-inductive load is used. The switching frequency of the inverter is 5 kHz. The system parameters used at the simulation are as follows:

- Inductor $L_1 = L_2 = 1$ mH
- Capacitor $C_1 = C_2 = C_3 = C_4 = 1000$ μF
- Three-phase LC filter: $L = 0.6$ mH, $C = 100$ μF

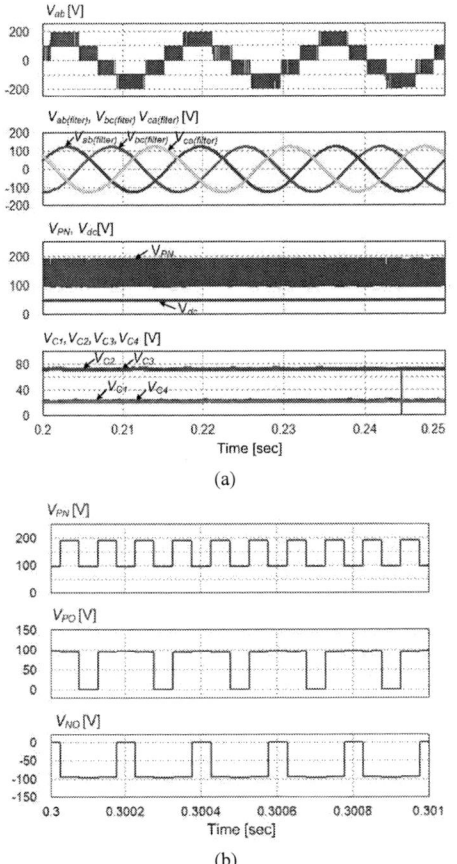

(a)

(b)

Fig. 7. Simulation results when $M = 0.75$ and $D = 0.25$: (a) three-phase ac output voltage, dc-link and dc input voltages, capacitor voltages, (b) dc link voltage and positive and negative pole voltages.

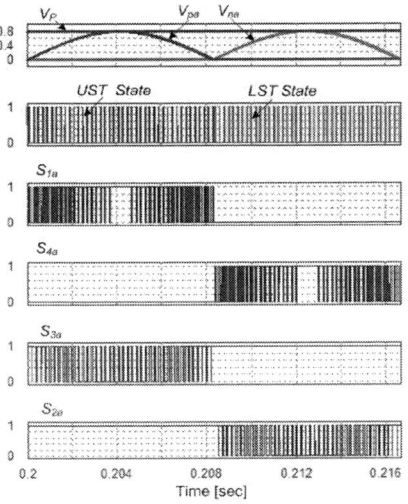

Fig. 8. Simulation results for a-phase gating signals.

• Three-phase RL load: $R = 50\ \Omega$, $L = 1.2$ mH

Fig. 6 shows simulation results of the proposed topology when the modulation index of $M = 0.8$ and $D = 0.2$. From 6(a), the generated ac output line-to-line voltage has five voltage steps, and the RMS value of the ac line-to-line voltage filtered by three-phase LC filter is about 79 V. The dc-link voltage is raised to 162 V, which is 3.1 times to the dc input voltage of 50 V. The average value of two capacitor voltages V_{C2} and V_{C3} is 66 V, and the average capacitor voltage of two capacitors C_1 and C_4 is 16 V, that is considerably lower than dc input voltage. It can be seen that the dc input current, which is the same as the inductor current i_{L1}, is continuous. As shown in Fig. 6(b), the inductor currents i_{L1} and i_{L2} increase during UST and LST states, respectively, in order to store up the energy in the inductors. The dc-link voltage during both shoot-through states is a half of the peak dc-link voltage.

Fig. 7 shows the simulation results when the shoot-through duty ratio of D increases from 0.2 to 0.25 and M decreases to 0.75. The RMS value of the filtered three-phase line-to-line voltage is 86 V. From Fig. 7(a), the dc-link voltage is boosted to 192 V, which is 3.85 times to V_{dc}. The average values of V_{C2} (V_{C3}) and V_{C1} (V_{C4}) are 73 V and 23 V, respectively. In spites of the decrease of M, the increase of D by 0.05 results in increasing dc-link and ac output voltages. Fig. 7(b) shows the dc-link voltage, the positive and negative pole voltages. The positive and negative pole voltages V_{PO} and V_{NO} have zero voltage during UST and LST states, respectively. It can be noted that the values of four capacitor voltages and dc-link voltage are nearly the same as those of theoretical analysis.

Fig. 8 describes the a-phase gating signals with the positive and negative modulation signals and shoot-through envelope signal of the proposed MZS-3LTI topology. The UST state is inserted in the PWM signal S_{1a} when a reference signal is positive and the LST state is inserted in the PWM signal S_{4a} when a reference signal is negative. Because the two switches among four switches are only modulated the pulse width, the switching losses of inverter switches can be reduced due to the reduction of the number of switching.

V. CONCLUSION

This paper proposed a three-phase modified Z-source three-level T-type inverter (MZS-3LTI) topology designed by integrating a modified Z-source impedance network to the traditional 3LTI. The proposed MZS-3LTI provides the enhanced boost ability. The continuous dc source current can be obtained without any extra capacitor or filter, by embedding a single dc source at the modified Z-source impedance network. A modified modulation technique based on APOD for adjusting effectively the upper and lower shoot-through states can be implemented by a simple logic circuit and it can reduce the switching losses of the inverter switches. Through simulation results, the dc-link voltage can be boosted to 3.85 times to dc input voltage, and the ac output line-to-line voltage of 86 V_{RMS} can be provided when $D = 0.25$.

REFERENCES

[1] H. Abu-Rub, J. Holtz, J. Rodriguez, and B. Ge, "Medium-voltage multilevel converters-state of art, challenges, and requirements in industrial applications," *IEEE Trans. Ind. Electron.*, vol. 57, no. 8, pp. 2581-2596, Aug. 2010.

[2] Rodriguez, S. Bernet, P. K. Steimer, and I. E. Lizama, "A survey on neutral-point-clamper inverters," *IEEE Trans. Ind. Electron.*, vol. 57, no. 7, pp. 2219-2230, Jul. 2010.

[3] S. M. Shin, J. H. Anh, and B. K. Lee, "Maximum efficiency operation of three-level T-type inverter for low-voltage and low-power home appliances", *Journal of Electrical Engineering and Technology*, vol. 10, no. 2, pp. 586-594, Mar. 2015.

[4] I. Won, K. B. Lee, and Y. Cho," An optimized switching scheme for DC-link current ripple reduction in three-level T-type inverter", in *Proc. IEEE-APEC*, 2017, pp. 3415-3419.

[5] M. Schweizer and J. W. Kolar, "Design and implementation of a highly efficiency three-level T-type converter for low-voltage applications," *IEEE Trans. Power Electron.*, vol. 28, no. 2, pp. 899-907, Feb. 2013.

[6] P. Alemi, Y. C. Jeung, and D. C. Lee, "DC-link capacitance minimization in T-type three-level AC/DC/AC PWM converters, *IEEE Trans. Ind. Electron.*, vol. 62, no. 3, pp. 1382- 1391, Mar. 2015.

[7] F. Z. Peng, "Z-source inverter," *IEEE Trans. Ind. Appl.*, vol. 39, no. 2, pp. 504-510, Mar. 2003

[8] J. Anderson and F. Z. Peng, "A class of quasi-Z-source inverters," *in Conf. Rec. IEEE-IAS Annu. Meeting*, Oct. 2008, pp.1-7.

[9] O. Husev, F. Blaabjerg, C. Roncero-Clemente, E. Romero-Cadaval, D. Vinnikov, Y. P. Siwakoti, and R. Strzelecki, "Comparison of impedance-source networks for two and multilevel buck-boost inverter applications," *IEEE Trans. Power Electron.*, vol. 31, no. 11, pp. 7564-7579, Nov. 2016.

[10] X. Xing, C. Zhang, A. Chen, J. He, W. Wang, C. Du, "Space-vector-modulated method for boosting and neutral voltage balancing in Z-source three-level T-type inverter," *IEEE Trans. Ind. Appl.*, vol. 52, no. 2, pp. 1621-1631, Mar./Apr. 2016.

[11] H. T. Luong, M. K. Nguyen, and T. T. Tran, "Single-phase five-level Z-source T-type Inverter," *IET Power Electron.*, vol. 11, no. 14, pp. 2367-2376, 2018.

[12] C. Qin, C. Zhang, A. Chen, X. Xing, and G. Zhang, "A space vector modulation scheme of quasi-Z-source three-level T-type inverter for common-mode voltage reduction," *IEEE Trans. Ind. Electron.*, vol. 65, no. 10, pp. 8340-8350, Oct. 2018.

[13] V. F. Pires, A. Cordeiro, D. Foito, and J. F. Martins, "Quasi-Z-source with a T-type converter in normal and fault mode," *IEEE Trans. Power Electron.*, vol. 31, no. 11, pp. 7462-7470, Nov. 2016.

[14] D. T. Do and M. K. Nguyen, "Three-level quasi-switched boost T-type inverter analysis, PWM control, and verification," *IEEE Trans. Ind. Electron.*, vol. 65, no. 10, pp. 8320-8329, Oct. 2018.

[15] A. V. Ho and T. W. Chun, "Single-phase modified quasi-Z-source cascaded hybrid five-level inverter," *IEEE Trans. Ind. Electron.*, vol. 65, no. 6, pp. 5125-5134, Jun. 2018.

[16] J. Chavarria, D. Biel, F. Guinjoan, C. Meza, and J. J. Negroni, "Energy-balance control of PV cascaded multilevel grid-connected inverters under level-shifted and phase-shifted PWMs," *IEEE Trans. Ind. Electron.*, vol. 60, no. 1, pp. 98-111, Jan. 2013.

Technical track on Renewable Electric Energy Conversion, Processing and Storage

978-1-5386-7688-2/19 $31.00 © 2019 IEEE

A Superimposed Frequency Method with an Adaptive Droop Characteristic for DC Microgrids

Mohammad Jafari Matehkolaei
The Center of Excellence in Power System
Management & Control
Department of electrical engineering
Sharif University of technology
Tehran, Iran
m.jfr.m@ee.sharif.edu

Hossein Mokhtari
The Center of Excellence in Power System
Management & Control
Department of electrical engineering
Sharif University of Technology
Tehran, Iran
mokhtari@sharif.edu

Abstract—This paper proposes an adaptive droop characteristic to enhance the superimposed droop frequency scheme for the control of DC microgrids. Conventional superimposed droop frequency scheme solves the poor current sharing and undesirable voltage regulation issues in DC microgrid. However, this method suffers from two main problems which are (i) instability in terms of load variation due to the limitation in transferrable reactive power, and (ii) poor voltage quality caused by the injected AC voltages in the DC system. In this paper, an adaptive droop characteristic is proposed to decrease the transferred reactive power and mitigate system overall voltage quality, and its performance is verified through simulation.

Keywords— *Microgrid, DC microgrid, Droop characteristic, Current-sharing, Voltage regulation, voltage quality, maximum load of the system, superimposed droop frequency method.*

I. Introduction

Nowadays, the conventional centralized power system is moving toward a more distributed structure [1]- [2]. Compared to a centralized generation, a distributed generation (DG) improves the overall system efficiency [3], [4], reliability [5]- [7] and stability [8], [9]. Large-scale application of DGs requires a base structure [2]. Microgrids, as small parts of the distribution system, are the base structure to accomplish the concept of DG, which integrate different kinds of generation, consumption and storage [10]- [12].

Generally, microgrids are categorized into two different types of DC and AC microgrids. Today, by the development of electronic devices and power electronic convertors most of the loads have a DC nature [13]. Energy storage systems (ESS) and most DGs such as photovoltaics, fuel cells and wind turbines are either inherently DC or require AC/DC conversion in their structure. Reactive power does not exist in DC systems, either. DC systems are more efficient and reliable than its counterpart i.e. AC systems [13]. That is why trends to use the DC microgrids have been widely increasing [14], [15].

Conventional control systems of DC microgrids are mostly based on droop characteristics [13]. In droop-based methods, the output power/current of each source is determined by its output voltage. In fact, the changes of the output voltage are employed as a signal indicating the power change among the loads and DGs. This method is called DC bus signaling (DBS) in the literature [16]- [18].

Conventional DBS methods have some major disadvantages. The local parameter of the voltage signal is used as an indicator of power changes which results in inaccurate control of the system. Also, transmission line resistances deteriorate the power sharing and voltage regulation in the system [13]. Using high droop gains decrease the droop characteristic sensitivity to line resistances, however it results in an undesirable voltage drop at high load levels which can be regarded as a third problem [13]. In order to solve these issues, several approaches are presented in the literature as follows. To mitigate the voltage drops, a secondary controller is proposed in [19] and [20] which automatically increases the voltage references. However, this scheme needs a communication infrastructure, which affects the overall system reliability and stability [13].

In [21], an adaptive droop-based control is proposed to increase droop gains as the load increases. But the load sharing is not accurate in this method. A master-slave based control algorithm is presented in [22]. However, because of neglecting line impedances in this method results in undesirable bus voltage.

A new control strategy based on a superimposed AC signal is proposed in [23] and [24]. In this scheme, an AC voltage with a constant amplitude is injected to the system. The frequency of the injected AC voltage is regulated using a frequency/current droop characteristic. Also, DC voltages are controlled using a voltage-reactive power droop characteristic [23], [24]. Unlike voltage, frequency is a global parameter. Therefore, the power sharing among sources is so accurate with the superimposed droop frequency method [23]. Also, line resistances do not cause any inaccuracy in the power sharing and voltage regulation [23]. However, limitation in transferred reactive power, limits the system loading [23], and the overall stability of the grid is questionable in terms of load variations [25]. Furthermore, higher load levels entail higher amplitude of the injected AC voltage [23]. But, it deteriorates overall voltage quality of the system.

In this paper, an adaptive droop characteristic is proposed to improve the performance of the conventional superimposed droop frequency method. The system stability is improved by adjusting the voltage droop gains, and the system load maximum level is increased by increasing the droop gains at high loads. Using the proposed method, it is possible to supply higher loads by injecting AC voltages with smaller amplitude, which improves the system power quality.

This paper is structured as follows. After explaining different types of droop characteristics in the superimposed frequency scheme, the effect of voltage droop gains on the

978-1-5386-7688-2/19 $31.00 © 2019 IEEE

transferred reactive power is analyzed in section II. Section III presents the proposed adaptive droop characteristic. In section IV, the dynamic behavior of the system is analyzed using small signal stability analysis. In section V, the proposed droop characteristic is verified using the simulation results and finally, section VI concludes the paper.

II. Droop Characteristics in The Superimposed Frequency Scheme

For simplicity, a test system of Fig. 1 with the parameters given in Table. I is considered. The test system contains two sources and a load connected to the point of common coupling (PCC).

Two types of droop characteristics are used in the superimposed droop frequency method, the frequency-current and the voltage-reactive power droop characteristics.

A. Frequency- current droop

At steady state operation of the system, the frequency of sources injected AC voltage is equal [23]. According to the frequency/current droop characteristic, the output currents of sources are as [23]:

$$\frac{I_1}{I_2} = \frac{k_{f2}}{k_{f1}}, \qquad (1)$$

where I_1 and I_2 are the DC currents of source 1 and 2, and kd_{f1} and k_{f2} are the frequency droop gains. The system frequency is as:

$$f = f^* - k_{fi}I_i, i = 1,2,...,N \qquad (2)$$

where N is the number of sources in the system and k_{fi} is the frequency droop gain of the ith convertor.

B. Voltage-reactive power droop

In the superimposed frequency method, the voltage of each convertor is controlled using the voltage/reactive power droop characteristic as:

$$V_i = V_{ref} - k_{vi}Q_i; i = 1,2 \qquad (3)$$

where k_{vi} is the voltage droop gain of the ith convertor, and V_{ref} is the reference DC voltage of the grid.

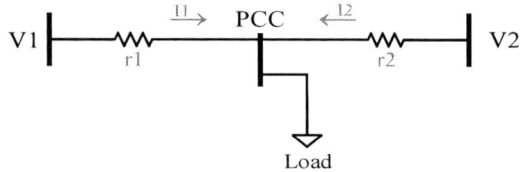

Fig. 1. Typical DC Microgrid (V_1 and V_2 are the power sources)

TABLE I. Parameters Of The Study Microgrid [23]

Quantity	Value
(frequency droop gains) k_{f1} , k_{f2}	0.15 , 0.15
r_1 , r_2	2 , 4 Ω
Voltage reference	700 V
Amplitude of the injected ac voltage	10 V
Reference Frequency	50 Hz
Filter Cutoff Frequency (w_c)	35 Hz

For the microgrid in Fig. 1, the load is modeled as a resistance. Therefore, the output voltage of the ith source can be calculated as:

$$V_i = V_{PCC} + r_i I_i; i = 1,2 \qquad (4)$$

$$V_{PCC} = R_{load}(I_1 + I_2) \qquad (5)$$

Noting the stability of the reactive power and using (1), (3) and (4), the transferred reactive power among the sources is as [23]:

$$Q = \frac{V_{ref}(k_{f1}r_2 - k_{f2}r_1)}{R_{load}(k_{v1} + k_{v2})(k_{f1} + k_{f2}) + r_1 k_{v2}k_{f2} + r_2 k_{v1}k_{f1}}. \qquad (6)$$

For the system in Fig. 1, the load does not consume any reactive power. Therefore, the sources injected reactive power and the maximum transferred reactive power are as:

$$\begin{cases} Q_1 = -\frac{A^2}{2(r_1+r_2)}\sin(\delta) \\ Q_2 = \frac{A^2}{2(r_1+r_2)}\sin(\delta) \end{cases} \qquad (7)$$

$$Q_{max} = \frac{A^2}{2(r_1+r_2)} \qquad (8)$$

A is the amplitude of the injected AC voltage, $\delta = \delta_1 - \delta_2$ where δ_1 and δ_2 are the angle of the buses AC voltages. According to (8), for high values of line resistances, the maximum transferrable reactive power is dramatically low. If the amplitude of the injected AC voltage is increased, the maximum transferrable reactive power will also increase. But, it deteriorates the overall voltage quality of the system.

According to (6), by increasing the load level i.e. decreasing R_{load}, the transferred reactive power increases. For the system in Fig. 1 with the parameters given in Table. I, the transferred reactive power in terms of load variations is depicted in Fig. 2. It should be noted that due to the limitation in the maximum transferred reactive power, the maximum loading is limited. The changes of the transferred reactive power in terms of voltage droop gains variations at a constant load is depicted in Fig. 3. As shown in Fig. 3, increasing voltage droop gains results in a decrease in the transferred reactive power.

As shown in Fig. 2, high droop gains increase the maximum loading of the system, but this may result in instability at low load levels. The instability is due to the location of the dominant poles of the system which is further discussed in section IV. For low values of the voltage droop gains, maximum loading of the system is limited but the system is stable at low load levels.

Therefore, it is desired to have low values of voltage droop gains for low load levels, and high values of droop gains for high loads. As a conclusion, for stable operation of the system at any load level, an adaptive droop characteristic is required.

III. Proposed Adaptve Droop Characteristic

As mentioned in previous section, voltage droop gains should change according to the load level of the system. Therefore, an indicator is required to evaluate the overall load of the system. In the superimposed frequency method both the frequency and output current of each convertor can be set as indicators to determine the system load level.

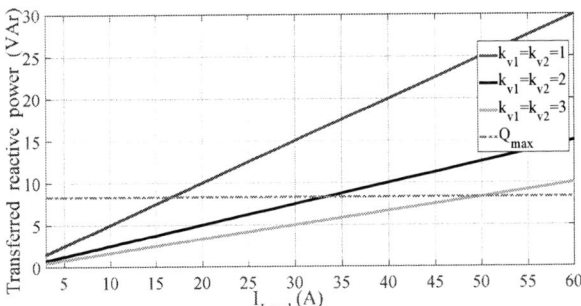

Fig. 2. Variation of the transferred reactive power in terms of load variations for different voltage droop gains

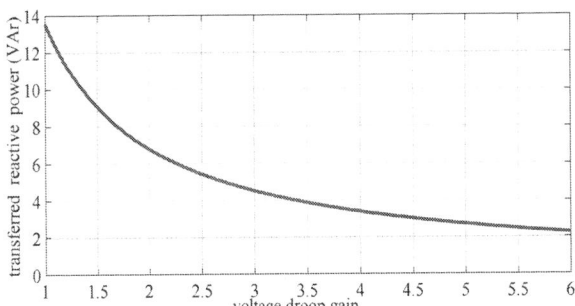

Fig. 3. Effect of voltage droop gains on the transferred reactive power in constant load level ($1<k_v<6$, $R_{load}=50\Omega$)

The proposed adaptive droop characteristic and its change according to the output DC current or the grid frequency is shown in Fig. 4. As shown in Fig. 4, the voltage droop gain is chosen according to the system loading. Low droop gains are adopted at low loads. As the load increases, the system frequency decreases, and according to Fig. 4, high values of droop gain are adopted. Voltage droop gains are calculated in real time according to:

$$k_{vi} = k_{v,min} + \alpha_k (f_{ref} - f) \quad (9)$$

The output current of the sources can also be used to determine the corresponding droop gain. The frequency and output current of the sources are related according to (1) and (2). Substituting (2) into (9), voltage droop gains can also be adopted according to:

$$k_{vi} = k_{v,min} + \alpha_d k_{fi} I_i \quad (10)$$

$$\alpha_k = \frac{k_{v,max} - k_{v,min}}{f_{ref} - f_{min}} = \frac{k_{v,max} - k_{v,min}}{k_{fi} I_{i,max}}. \quad (11)$$

$k_{v,min}$ is the voltage droop gain at no-load condition, and $k_{v,max}$ is the highest droop gain of the grid which is implemented at the highest loading of the system. α_k is the change ratio of the droop gain according to the frequency in Fig. 4. The control block diagram of the proposed method is depicted in Fig. 5 [23]. As shown in Fig. 5, the frequency of the injected AC voltage is proportional to the source output current, and the DC voltage is regulated using a voltage/reactive power droop characteristic [23], [24].

According to Fig. 2 and (6), increasing the grid load leads

to an increase in the transferred reactive power. Therefore, according to (8), for stable operation of the system, higher amplitude of an AC voltage must be injected to the system which deteriorates the overall voltage quality. With the proposed method, it is possible to supply higher load levels with smaller amplitude of the injected AC voltages which enhances the overall voltage quality of the system.

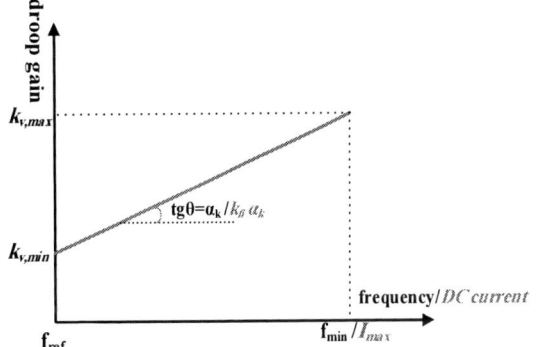

Fig. 4. Variation of the droop gain in terms of load variations

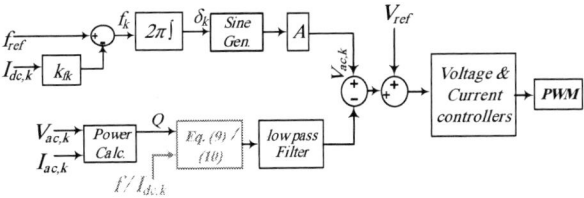

Fig. 5. Control block diagram of the proposed control method for the kth source

IV. SMALL SIGNAL STABILITY ANALYSIS

In this section, the dynamic behavior of the system is investigated by employing the small signal stability analysis. Linear models of (6) and (3) are as:

$$\begin{cases} \Delta Q_1 = -k_\delta \Delta \delta \\ \Delta Q_2 = k_\delta \Delta \delta \end{cases} \quad (12)$$

$$k_\delta = \frac{A^2}{2(r_1 + r_2)} \cos \delta_0 \quad (13)$$

$$\begin{cases} \Delta V_1 = -k_{v10} G_{(s)} \Delta Q_1 - Q_{10} G_{(s)} \Delta k_{v1} \\ \Delta V_2 = -k_{v20} G_{(s)} \Delta Q_2 - Q_{20} G_{(s)} \Delta k_{v2} \end{cases} \quad (14)$$

k_{v10} and k_{v20} are the voltage droop gains, and Q_{10} and Q_{20} are the injected reactive powers of source 1 and 2 at the operating point of the system. According to [25], δ is calculated as:

$$\delta = \frac{2\pi}{S} (k_{f2} I_2 - k_{f1} I_1). \quad (15)$$

Using (12), (14) and the linear form of (4), the small signal model of I_1 and I_2 are calculated. Placing ΔI_1 and ΔI_2 in the linear form of (15) yields the characteristic equation of the system as:

$$s^2 + w_c s + \frac{\beta}{h} = 0 \qquad (16)$$

where,

$$h = r_1 r_2 + R_{load}(r_1 + r_2) \qquad (17)$$

and

$$\beta = 2\pi w_c k_\delta R_{load} \left[\begin{array}{l} kv_{10}\left(k_{f2} + k_{f1}\dfrac{(r_2 + R_{load} - Q_{10}\alpha_k k_{f2})}{R_{load}}\right) \\ + k_{v20}\left(k_{f1} + k_{f2}\dfrac{(r_1 + R_{load} + Q_{10}\alpha_k k_{f1})}{R_{load}}\right) \end{array} \right] \qquad (18)$$

w_c is the cutoff frequency of the applied low pass filter.

For the system in Fig. 1 with the parameters given in table I, the variation of the system dominant pole location in terms of load change is shown in Fig. 6b. the system is unstable at high loads because the poles of the system reach the imaginary axis. For low load levels, the imaginary part of the system dominant poles are too high, and the damping ratio of the grid is too small which leads to instability. The impact of increasing the voltage droop gains is depicted in Fig. 6a. As shown in Fig. 6, increasing the voltage droop gains alleviates the impact of the load increase. This keeps the dominant pole of the system in an acceptable area.

For the proposed method, dominant pole placement for the system in Fig. 1 with the parameters in Table I is depicted in Fig. 7 for different load levels. Using the proposed method, system dominant poles are in an acceptable area in any level of the system load and system has a stable operation.

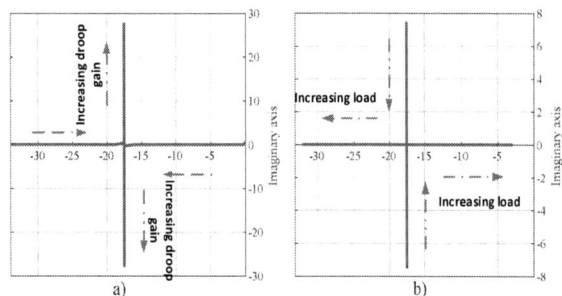

a)   b)

Fig. 6. System dominant pole placement for the convetional droop characteristic (a) Effect of increasing voltage droop gain($1<k_v<6$, R_{load}=80Ω)   (b) Effect of increasing load (k_v=2 , $20<R_{load}<200$ Ω)

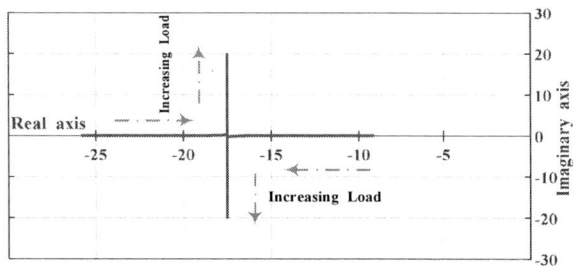

Fig. 7. System dominat pole placement for the proposed method ($k_{v,min}$=1 , α_k=1.15 , $15<R_{load}<200$ Ω)

V. SIMULATIONS

In this section, the superimposed frequency method is simulated for three different cases as:
Case1: low droop gain and high load,
Case2: high droop gain and low load, and
Case3: the proposed adaptive droop characteristic.

A. Case1

In this case, the performance of the system using low droop gains is analyzed for high load levels. As shown in Fig. 8, the system is unstable at high levels of loads.

B. Case2

In this case, the operation of the grid is analyzed using high droop gains. As Fig. 9 shows, unlike case1, the system performance is not desirable at low load levels.

C. Case3

The performance of the system with the proposed method is investigated in this section. In order to investigate the performance of the proposed method, the grid load is varied from low levels to high levels. The load current profile is depicted in Fig. 10. The output DC current of the sources are depicted in Fig. 11. Due to equality of frequency droop gain of the sources, the load is equally supplied by the sources at any level of the system loading.

The injected frequency of the two sources is shown in Fig. 12. As shown in Fig. 12, the injected frequency of the two sources converges to a constant value. Also, the stable frequency response at different load levels shows the desirable performance of the proposed adaptive droop characteristic.

The imposed voltage droop gains of the two sources are shown in Fig. 13 ($k_{v,min}$=1, α_k=1.15). As this figure shows, the voltage droop gains increase/decrease as the load increases/decreases.

The DC voltage of the convertors and the average DC voltage of the buses are shown in Fig. 14. The average DC voltage of the buses is equal to the grid reference voltage at any load level of the system.

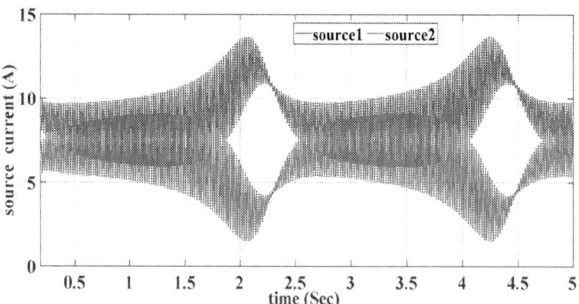

Fig. 8. Instability of grid with low voltage droop gains at high loads (k_v=1 , R_{load}=45Ω)

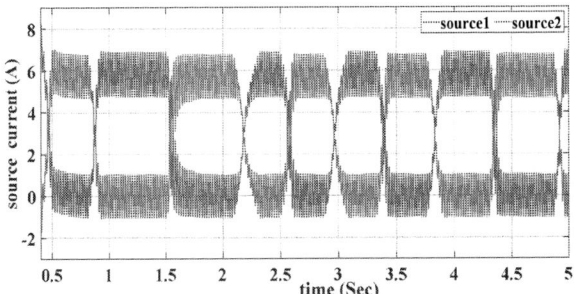

Fig. 9. Instability of grid with high voltage droop gains at low loads (k_v=3, R_{load}=120Ω)

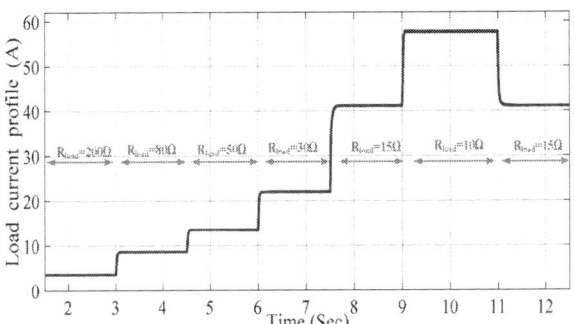

Fig. 10. Load current profile

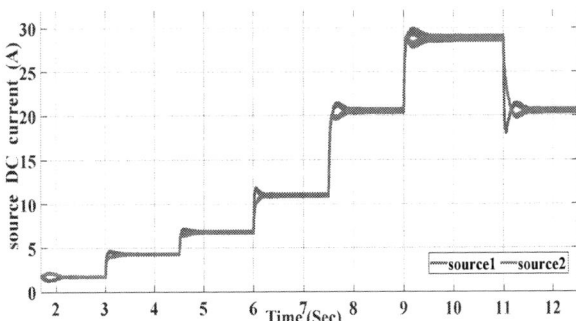

Fig. 11. Source DC currents ($k_{v,min}$=1, α_k=1.15)

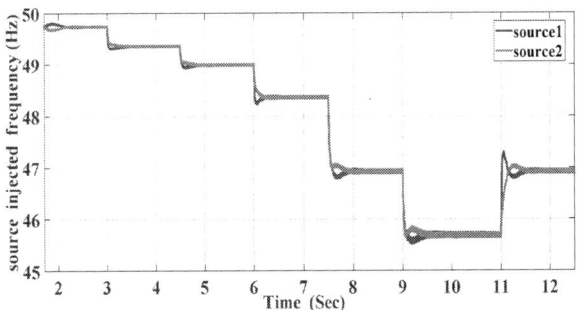

Fig. 12. Sources injected frequency

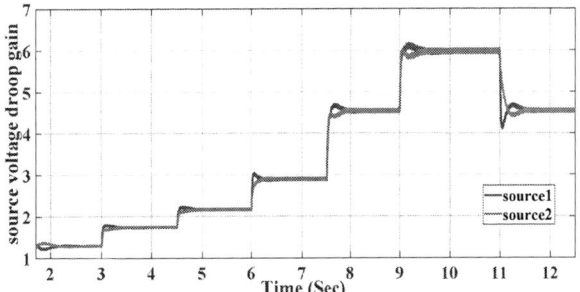

Fig. 13. Source voltage droop gains

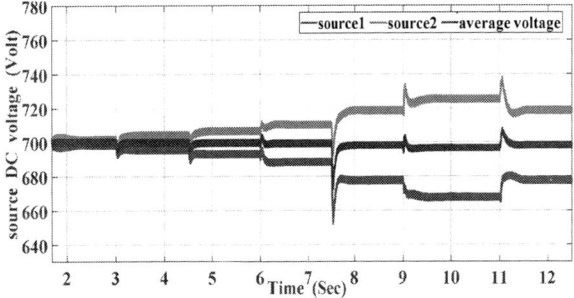

Fig. 14. Microgrid bus DC voltages and average bus voltage

VI. CONCLUSION

This paper proposes an adaptive droop characteristic in order to enhance the stability of the superimposed droop frequency method in DC microgrids. Due to the limitation in the transferred reactive power, the conventional droop characteristic leads to limited load variations. In this paper, the proposed method determines the voltage droop gains proportional to the grid overall load level. Using the proposed adaptive droop characteristic, the system stability is guaranteed for a wide range of load variations and the maximum loading of the system is increased accordingly. The proposed method automatically adjusts the sources droop gains, and does not require any communication infrastructure. The performance of the proposed method is investigated by different simulation studies in MATLAB/SIMULINK environment, and compared with that of the existing method. The results indicate that with the proposed technique, stable operation the system is achievable at all load levels.

REFERENCES

[1] J.M. Guerrero, J. C. Vasquez, J. Matas, L. G. de Vincu˘na, and M. Castilla, "Hierarchical control of droop-controlledAC andDCmicrogrids—A general approach toward standardization," IEEE Trans. Ind. Electron., vol. 58, pp. 158–172, Jan. 2011.

[2] H. Kakigano, Y. Miura, and T. Ise, "Distribution voltage control for DC microgrids using fuzzy control and gain-scheduling technique," IEEE Trans. Power Electron., vol. 28, no. 5, pp. 2246–2258, May 2013.

[3] H. Kakigano, Y. Miura, and T. Ise, "Low-voltage bipolar-type DC microgrid for super high quality distribution," IEEE Trans. Power Electron., vol. 25, no. 12, pp. 3066–3075, Dec. 2010.

[4] Y. K. Chen, Y. C.Wu, C. C. Song, and Y. S. Chen, "Design and implementation of energy management system with fuzzy control for

DC microgrid systems," IEEE Trans. Power Electron., vol. 28, no. 4, pp. 1563–1570, Apr. 2013.

[5] A. Kwasinski, "Quantitative evaluation of DC microgrids availability: Effects of system architecture and converter topology design choices," IEEE Trans. Power Electron., vol. 26, no. 3, pp. 835–851, Mar. 2011.

[6] Y. Gu, X. Xiang, W. Li, and X. He, "Mode-adaptive decentralized control for renewable DC microgrid with enhanced reliability and flexibility," IEEE Trans. Power Electron., vol. 29, no. 9, pp. 5072–5080, Sep. 2014.

[7] Y. C. Chang and C. M. Liaw, "Establishment of a switched-reluctance generator-based common DC microgrid system," IEEE Trans. Power Electron., vol. 26, no. 9, pp. 2512–2527, Sep. 2011.

[8] N. Bottrell, M. Prodanovic, and T. C. Green, "Dynamic stability of a microgrid with an active load," IEEE Trans. Power Electron., vol. 28, no. 11, pp. 5107–5119, Nov. 2013.

[9] S. R. Huddy and J. D. Skufca, "Amplitude death solutions for stabilization of DC microgrids with instantaneous constant-power loads," IEEE Trans. Power Electron., vol. 28, no. 1, pp. 247–253, Jan. 2013.

[10] F. Li, Z. Lin, Z. Qian and J. Wu, "Active DC bus signaling control method for coordinating multiple energy storage devices in DC microgrid," 2017 IEEE Second International Conference on DC Microgrids (ICDCM), Nuremburg, 2017, pp. 221-226, doi: 10.1109/ICDCM.2017.8001048 .

[11] D. Chen and L. Xu, "Autonomous DC Voltage Control of a DC Microgrid With Multiple Slack Terminals," in IEEE Transactions on Power Systems, vol. 27, no. 4, pp. 1897-1905, Nov. 2012. doi: 10.1109/TPWRS.2012.2189441 .

[12] J. M. Guerrero, P. C. Loh, T. Lee and M. Chandorkar, "Advanced Control Architectures for Intelligent Microgrids—Part II: Power Quality, Energy Storage, and AC/DC Microgrids," in IEEE Transactions on Industrial Electronics, vol. 60, no. 4, pp. 1263-1270, April 2013, doi: 10.1109/TIE.2012.2196889.

[13] S. Peyghami, H. Mokhtari, and F. Blaabjerg, "Hierarchical Power Sharing Control in DC Microgrids," in Microgrid: Advanced Control Methods and Renewable Energy System Integration, First Edition, Magdi S Mahmoud, Elsevier Science & Technology, 2017, pp. 63–100.

[14] Seyed Fariborz Zarei, Mohammad Amin Ghasemi, Hossein Mokhtari, Frede Blaabjerg, (2019). "Performance Improvement of AC-DC Power Converters under Unbalanced Conditions". Scientia Iranica, early access, doi: 10.24200/sci.2019.53337.3254.

[15] R. S. Balog, W. W. Weaver and P. T. Krein, "The Load as an Energy Asset in a Distributed DC SmartGrid Architecture," in IEEE Transactions on Smart Grid, vol. 3, no. 1, pp. 253-260, March 2012, doi: 10.1109/TSG.2011.2167722 .

[16] S K. Sun, L. Zhang, Y. Xing and J. M. Guerrero, "A Distributed Control Strategy Based on DC Bus Signaling for Modular Photovoltaic Generation Systems With Battery Energy Storage," in IEEE Transactions on Power Electronics, vol. 26, no. 10, pp. 3032-3045, Oct. 2011, doi: 10.1109/TPEL.2011.2127488 .

[17] L. Xu and D. Chen, "Control and Operation of a DC Microgrid With Variable Generation and Energy Storage," in IEEE Transactions on Power Delivery, vol. 26, no. 4, pp. 2513-2522, Oct. 2011, doi: 10.1109/TPWRD.2011.2158456.

[18] J. Schonbergerschonberger, R. Duke and S. D. Round, "DC-Bus Signaling: A Distributed Control Strategy for a Hybrid Renewable Nanogrid," in IEEE Transactions on Industrial Electronics, vol. 53, no. 5, pp. 1453-1460, Oct. 2006, doi: 10.1109/TIE.2006.882012.

[19] J. M. Guerrero, J. C. Vasquez, J. Matas, L. G. de Vicuna and M. Castilla, "Hierarchical Control of Droop-Controlled AC and DC Microgrids—A General Approach Toward Standardization," in IEEE Transactions on Industrial Electronics, vol. 58, no. 1, pp. 158-172, Jan. 2011, doi: 10.1109/TIE.2010.2066534.

[20] X. Lu, J. M. Guerrero, K. Sun and J. C. Vasquez, "An Improved Droop Control Method for DC Microgrids Based on Low Bandwidth Communication With DC Bus Voltage Restoration and Enhanced Current Sharing Accuracy," in IEEE Transactions on Power Electronics, vol. 29, no. 4, pp. 1800-1812, April 2014, doi: 10.1109/TPEL.2013.2266419.

[21] Amir Khorsandi, Mojtaba Ashourloo, Hossein Mokhtari, Reza Iravani, "Automatic droop control for a low voltage DC microgrid," IET Gener. Transm. Distrib., vol. 10, Iss. 1, pp. 41–47, 2016, doi: 10.1049/iet-gtd.2014.1228.

[22] B. Luis and E. Zubieta, "Are Microgrids the Future of Energy ?," IEEE Electrif. Mag., vol. 4, no. 2, pp. 37-44, 2016.

[23] S. Peyghami, H. Mokhtari, P. C. Loh, P. Davari and F. Blaabjerg, "Distributed Primary and Secondary Power Sharing in a Droop-Controlled LVDC Microgrid with Merged AC and DC Characteristics," IEEE Trans. Smart Grid, vol. 9, no. 3, pp. 2284-2294, May 2018, doi: 10.1109/TSG.2016.2609853.

[24] S. Peyghami, H. Mokhtari, and F. Blaabjerg, " Autonomous Power Management in LVDC Microgrids based on a Superimposed Frequency Droop," IEEE Trans. Power Electron., vol. 33, no. 6, pp. 5341-5350, June 2018, doi: 10.1109/TPEL.2017.2731785.

[25] S. Peyghami, H. Mokhtari and F. Blaabjerg, "Decentralized Load Sharing in a Low-Voltage Direct Current Microgrid With an Adaptive Droop Approach Based on a Superimposed Frequency," IEEE Trans. Power Electron., vol. 5, no. 3, pp. 1205-1215, Sept. 2017, doi: 10.1109/JESTPE.2017.2674300.

978-1-5386-7688-2/19 $31.00 © 2019 IEEE

Analysis of MMC HVDC System Using Symmetric Coordinate Method

Chan-Ki Kim*, Soo-Yeon Sim*, Sang-Min Kim**, Kyeon Hur**

KEPRI*, Yonsei University

chankikim@kepco.co.kr

Abstract- **This paper deals with the new analysis method of MMC HVDC system, which is based on the classical symmetric coordinate method. As the AC/DC system which is AC system combined with MMC HVDC converter, has the different characteristics from a conventional AC system, in order to analyze the faults of the system, the new analysis technique is needed. In this paper, the modified symmetric coordinate method is proposed for analyzing MMC HVDC system. Through the mathematical model and simulations, the validation of the proposed method is confirmed.**

I. INTRODUCTION

Until now, the fault analysis of AC system has been done by using 3-phase symmetric coordinate method which separates the 3-phases (A-phase, B-phase and C-phase) into positive sequence, negative sequence and zero sequence. However, 3-phase symmetric coordinate method can be used only in a 3-phase AC system, and is not useful in an AC/DC system which AC system combined with DC converter system.

In the AC/DC system, since the 3-phase symmetric coordinate method for analyzing faults of AC/DC system has the limitations, the fault analysis of AC/DC system has been analyzed through transient program such as PSCAD, EMTP and Matlab. Strictly speaking, the voltage and current of the DC converter are made of the voltage and current of AC system, conversely because the voltage and current of the AC system are also generated by the DC converter, to analyze the AC/DC system using 3-phase symmetric coordinate method could be possible and considered to be very useful.

The purpose of this paper is analyzing the 3-phase AC/DC system by the symmetric coordinate method and quantifying by mathematic model. As the targeted AC/DC system is the AC system connected to VSC HVDC based on MMC topology, when a ground faults or short circuit faults occur on the DC side or the AC side, to interpret fault currents and voltages should be possible. This method has several advantages, the first is that AC network and HVDC system are completely separated and interpreted, and the second is to find the suitable control algorithms of HVDC system in order to decouple AC system and HVDC system.

An MMC-based VSC HVDC system has different characteristics with 2-level converter, that is, the intrinsic characteristics of MMC topology such as the circulating current and instability due to control algorithm or the negative sequence of AC system. These problems have to be removed because the MMC system causes resonance or power unbalance of the converter. This paper represents the mathematical model of AC/DC system which could analyze the fault condition by using symmetric coordinate method.

II. ANALYSIS OF MMC SYSTEM

As Fig. 1 shows an 3-phase MMC system, this system is composed of six arms in which each arm on turn contains N x SMs(N is the Number of SMs), one inductor L_0, resistor R_0 respectively. The upper and lower arm within a single leg of MMC system comprises a phase unit; a half bridge SM circuit is shown on the right side of Fig. 1 a).

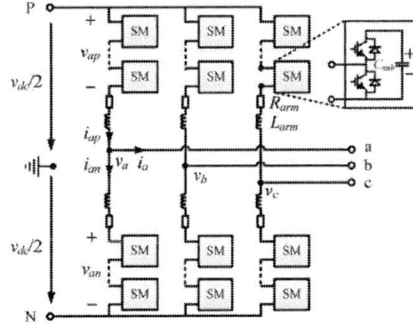

(a) Basic Structure of MMC

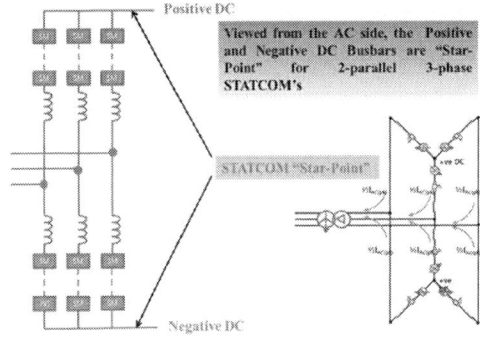

(b) MMC HVDC Concept

Fig. 1. MMC HVDC Topology

978-1-5386-7688-2/19 $31.00 © 2019 IEEE

The SM output voltage v_{SM} is determined based on the switching states of the upper and lower IGBTs. SM voltage, v_{SM}, becomes the capacitor voltage, v_c. On the other hand, if the upper and lower IGBTs are turn-off and turn-on respectively, the SM voltage become zero. This means a SM has two normal operation states: inserted ($v_{SM} = v_c$) or bypassed ($v_{SM} = 0$). The total DC bus voltage of the MMC system and the converter output voltage of phase j are represented by U_{dc} and u_{js} ($j = a, b, c$) respectively. The line current of the each phase are denoted by i_{js} ($j = a, b, c$), where j represents the three phases of the MMC system. The arm voltages generated by the cascaded SMs express as u_{ju} and u_{jl} where the subscripts u and l denote the upper and lower arms, respectively. The arm currents i_{ju} and i_{jl} in Fig.1 can be expressed by (1) and (2).

$$i_{ju} = i_{jo} + \frac{i_{js}}{2} \qquad (1)$$

$$i_{jl} = i_{jo} - \frac{i_{js}}{2} \qquad (2)$$

Where, i_{jo} is the inner difference current of phase j, which flow through both the upper and lower arms and is given by:

$$i_{jo} = \frac{i_{ju} + i_{jl}}{2} \qquad (3)$$

The MMC can be characterized by the following equations:

$$u_{js} = e_j - \frac{R_0}{2} i_{js} - \frac{L_0}{2} \cdot \frac{di_{js}}{dt} \qquad (4)$$

$$L_0 \frac{di_{jo}}{dt} + R_0 i_{jo} = \frac{U_{dc}}{2} - \frac{u_{ju} + u_{jl}}{2} \qquad (5)$$

Where, e_j in (4) is the inner EMF generated in phase j and is expressed as:

$$e_j = \frac{u_{jl} - u_{ju}}{2} \qquad (6)$$

From the relation in (4) and (5), considering u_{js} as the AC grid network voltage, the line current i_{js} can be regulated by adjusting the control variable e_j. The inner dynamic performance of the MMC which characterize the relation between inner unbalance voltage by and leg current is given by (5) and it is redefined as:

$$u_{jo} = L_0 \frac{di_{jo}}{dt} + R_0 i_{jo} = \frac{1}{2} [U_{dc} - (u_{ju} + u_{jl})] \qquad (7)$$

Where, u_{jo} is the inner voltage of phase j.

The voltage u_{jo} can be adjusted in order to regulate the difference current i_{p} as per (7). The upper and lower arm current reference can be derived by taking (6) and (7) in to consideration and are given in (8) and (9).

$$u_{ju\_ref} = \frac{U_{dc}}{2} - e_j^* - u_{jo}^* \qquad (8)$$

$$u_{jl\_ref} = \frac{U_{dc}}{2} + e_j^* - u_{jo}^* \qquad (9)$$

The AC grid dynamics are not affected by subtract the same voltage from upper and lower arm reference according to (4) ~ (6). As previously described, the e_j and u_{jo} is generated by the current controller and circulating current suppressing controller of the MMC system, respectively. As Fig.1 b) shows the concept diagram of the MMC system, in which a 3-phase power source of MMC converter is Y connected and the midpoint of Y connected source is called "star point" and the DC current flow through star point.

Assuming that as the previous states, a 3-phase power sources are completely balanced, Fig. (1) can be represented as Fig. (2).

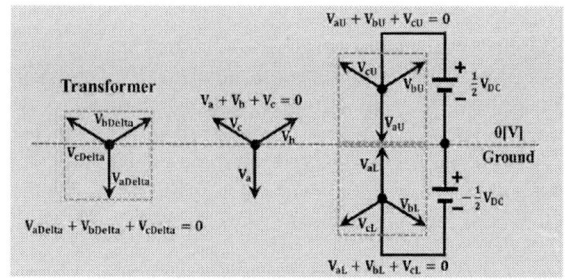

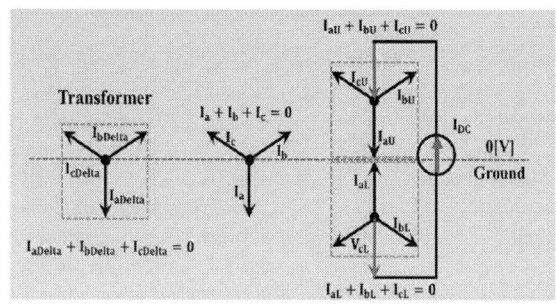

(a) Voltage Model

(b) Current Method

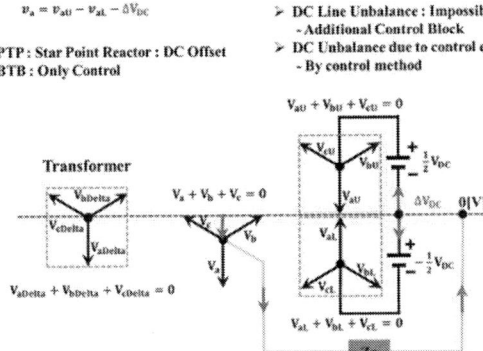

$$v_a = v_{aU} - v_{aL} + \frac{1}{2}V_{DC} - \Delta\frac{1}{2}V_{DC} - (\frac{1}{2}V_{DC} + \Delta\frac{1}{2}V_{DC})$$

$$v_a = v_{aU} - v_{aL} - \Delta V_{DC}$$

PTP : Star Point Reactor : DC Offset
BTB : Only Control

➢ DC Line Unbalance : Impossible
   - Additional Control Block
➢ DC Unbalance due to control error
   - By control method

(c) Zero Sequence Path through Star Point Reactor

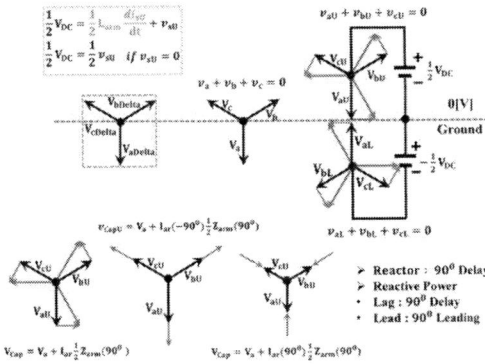

$$\frac{1}{2}V_{DC} = \frac{1}{2}L_{arm}\frac{di_{arm}}{dt} + v_{aU}$$
$$\frac{1}{2}V_{DC} = \frac{1}{2}v_{aU} \; if \; v_{aU} = 0$$

$v_{capU} = v_a + I_{ar}(-90^0)\frac{1}{2}Z_{arm}(90^0)$

➢ Reactor : 90^0 Delay
➢ Reactive Power
• Lag : 90^0 Delay
• Lead : 90^0 Leading

$V_{Cap} = v_a + I_{ar}\frac{1}{2}Z_{arm}(90^0)$ $V_{Cap} = v_a + I_{ar}(90^0)\frac{1}{2}Z_{arm}(90^0)$

(d) Voltage Model including Control signal

Fig. 2. Voltage Source and Current Source Model of MMC system

As discussed in the previous chapter, the since structure of the MMC system is a 3-phase power supply with a Y-connection of the converter symmetrically connected to an upper positive source and a lower negative source, a DC voltage is biased at the star point of DC line.

Fig. 2 shows the modeling of AC system including MMC system. From the view point of the 3-phase symmetric coordinate method, Fig. 2 a) and b) show the voltage model and the current model, respectively.

III. CONTROL CHARACTERISTICS ANALYSIS TO DECOUPLE AC GRID AND DC CONVERTER

A. AC voltage modelling

The converter arm voltages v_{pnx}, which are adjusted by a series connection of converter modules, are determined by the converter DC voltage v_{DC} and the grid voltages v_{sx}. The relationship between these voltages can be obtained from (10).

$$\begin{pmatrix} v_{px} \\ v_{nx} \end{pmatrix} = \begin{bmatrix} -1 & \frac{1}{2} \\ 1 & \frac{1}{2} \end{bmatrix} \begin{pmatrix} v_{sx} \\ v_{DC} \end{pmatrix} \quad (10)$$

An unbalanced 3-phase grid voltage can be split into a positive sequence voltage and a negative sequence voltage. Depending on the grid angle, these voltages can be dq-transformed in a positive reference frame with components v_{sd}^+ and v_{sq}^+ as well as in a negative reference frame with components v_{sd}^- and v_{sq}^-.

$$\begin{pmatrix} v_{sa} \\ v_{sb} \\ v_{sc} \end{pmatrix} = \begin{bmatrix} T_{ad}^+ & T_{aq}^+ & T_{ad}^- & T_{aq}^- \\ T_{bd}^+ & T_{bq}^+ & T_{bd}^- & T_{bq}^- \\ T_{cd}^+ & T_{cq}^+ & T_{cd}^- & T_{cq}^- \end{bmatrix} \begin{pmatrix} v_{sd}^+ \\ v_{sq}^+ \\ v_{sd}^- \\ v_{sq}^- \end{pmatrix} \quad (11)$$

Where, $T_{ad}^+ = \cos(\omega t + \delta)$, $T_{bd}^+ = \cos(\omega t + \delta - \frac{2\pi}{3})$,
$T_{cd}^+ = \cos(\omega t + \delta + \frac{2\pi}{3})$, $T_{aq}^+ = -\sin(\omega t + \delta)$,
$T_{bq}^+ = -\sin(\omega t + \delta - \frac{2\pi}{3})$, $T_{cq}^+ = -\sin(\omega t + \delta + \frac{2\pi}{3})$,
$T_{ad}^- = \cos(-\omega t - \delta)$, $T_{bd}^- = \cos(-\omega t - \delta + \frac{2\pi}{3})$,
$T_{cd}^- = \cos(-\omega t - \delta - \frac{2\pi}{3})$, $T_{aq}^- = -\sin(-\omega t - \delta)$,
$T_{bq}^- = -\sin(-\omega t - \delta + \frac{2\pi}{3})$, $T_{cq}^- = -\sin(-\omega t - \delta - \frac{2\pi}{3})$

B. AC current and Control current modelling

It is an established concept to model the phase arm currents i_{pnx} as a superposition of grid currents i_{sx}, the DC-current i_{DC} and circulating currents i_{Cx} that neither effect the DC-side nor the AC-side. This superposition can be mathematically described as shown in (12).

$$\begin{pmatrix} i_{px} \\ i_{nx} \end{pmatrix} = \begin{bmatrix} \frac{1}{2} & 1 & \frac{1}{3} \\ -\frac{1}{2} & 1 & \frac{1}{3} \end{bmatrix} \begin{pmatrix} i_{sx} \\ i_{Cx} \\ i_{DC} \end{pmatrix} \quad (12)$$

$$\begin{pmatrix} i_{sa} \\ i_{sb} \\ i_{sc} \end{pmatrix} = \begin{bmatrix} T_{ad}^+ & T_{aq}^+ & T_{ad}^- & T_{aq}^- \\ T_{bd}^+ & T_{bq}^+ & T_{bd}^- & T_{bq}^- \\ T_{cd}^+ & T_{cq}^+ & T_{cd}^- & T_{cq}^- \end{bmatrix} \begin{pmatrix} i_{sd}^+ \\ i_{sq}^+ \\ i_{sd}^- \\ i_{sq}^- \end{pmatrix} \quad (13)$$

Where, $T_{ad}^0 = \cos(2\omega t + \delta_{\bar{t}})$, $T_{bd}^0 = \cos\left(2\omega t + \delta_{\bar{t}} + \frac{2\pi}{3}\right)$
$T_{cd}^0 = \cos(2\omega t + \delta_{\bar{t}} - \frac{2\pi}{3})$, $T_{aq}^0 = -\sin(2\omega t + \delta_{\bar{t}})$,
$T_{bq}^0 = -\sin\left(2\omega t + \delta_{\bar{t}} - \frac{2\pi}{3}\right)$, $T_{cq}^0 = -\sin(2\omega t + \delta_{\bar{t}} + \frac{2\pi}{3})$

In order to calculate the voltage deviation of SM, (14) and (15) are introduced. Also, since the main variation of SM voltage is due to the number of capacitors, the number of capacitors actually can be represented as N-n with sinusoidal waveform. Therefore, the module voltages related to (12) can be expressed as (14).

$$\frac{V_C}{V_{DC}} \cdot \sin(\omega t) = \frac{n}{N}, N - n = N \cdot (1 - \frac{V_C}{V_{DC}} \cdot \sin(\omega t)) \quad (14)$$

Equation (14) is finally calculated to (15), which shows the reason that why the voltage of MMC module is distorted. Reasons of the distortion are categorized by five terms, each terms from (15) are following as:

1. A DC term that should cancel with the opposite arm.
2. A fundamental magnitude variation that will be catered for by the converter control.
3. A fundamental phase shift that will be catered for by the converter control.
4. A variation with time that should cancel with the opposite arm.
5. A second harmonic term.

Therefore, from (12) and (15), the circulating currents I_{Cx} generally consist of DC-parts $I_{C\alpha}$ and $I_{C\alpha}$ as well as AC-parts which are composed of the components i_{Cd}^+, i_{Cq}^+, i_{Cd}^-, i_{Cq}^- of an unbalanced 3-phase current and an 2nd harmonic currents with a zero phase angle $\delta_{\bar{t}}$ represented by the components i_{Cd}^0 and i_{Cq}^0. Again, a compact and complete mathematical description can be obtained from a matrix notation in (16) under consideration of the abbreviations from (13) and additional abbreviations from (14).

$$\Delta v_{CL1} = \frac{N \cdot I_{AC}}{2 \cdot \omega_0 \cdot C} \cdot \left[-\cos(\omega t + \beta) + \left(\frac{V_C}{4V_{DC}} - 1 \right) \right.$$

$$\cdot \sin(\beta) + \left(\frac{V_{AC}PF}{2V_{DC}} - \frac{V_C}{2V_{DC}} \cdot \cos(\beta) \right) \cdot \left(\omega t - \frac{\pi}{2} \right) + \quad (15)$$

$$+ \frac{V_C}{2V_{DC}} \cdot \left(-\frac{1}{2} \cdot \sin(2\omega t + \beta) + \frac{V_{AC} \cdot V_C \cdot PF}{2V_{DC}^2} \cdot \cos(\omega t) \right]$$

$$\begin{pmatrix} i_{Ca} \\ i_{Cb} \\ i_{Cc} \end{pmatrix} = \begin{pmatrix} 1 & 0 & T_{ad}^+ & T_{aq}^+ & T_{ad}^- & T_{aq}^- & T_{ad}^0 & T_{aq}^0 \\ \frac{-1}{2} & \frac{\sqrt{3}}{2} & T_{bd}^+ & T_{bq}^+ & T_{bd}^- & T_{bq}^- & T_{bd}^0 & T_{bq}^0 \\ \frac{-1}{2} & \frac{\sqrt{3}}{2} & T_{cd}^+ & T_{cq}^+ & T_{cd}^- & T_{cq}^- & T_{cd}^0 & T_{cq}^0 \end{pmatrix} \begin{pmatrix} I_{C\alpha} \\ I_{C\beta} \\ i_{Cd}^+ \\ i_{Cq}^+ \\ i_{Cd}^- \\ i_{Cq}^- \\ i_{Cd}^0 \\ i_{Cq}^0 \end{pmatrix} \quad (16)$$

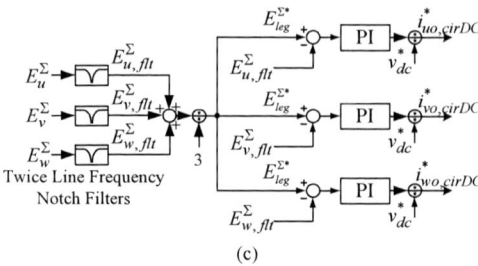

(c)

Fig.3. Controller Structure for Energy Balancing

B. Balancing Control modelling

When modeling the energy balance inside the MMC, the instantaneous power p_{sx} in each phase arm has to be taken into account. p_{xy} can easily be calculated as depicted in (17) and is transformed to the phase arm power into six components, corresponding to the overall converter energy by one component $p\Sigma0$, to the horizontal energy imbalance by two components $p\Sigma\alpha$ and $p\Sigma\beta$ and to the vertical energy imbalance by three components $p\Delta\alpha$, $p\Delta\beta$ and $p\Delta0$ as stated in (17) in order to get a suitable description of the energy distribution.

	Direct Modulation		Indirect Modulation	
Modulation Index Calculation	$n_{xu} = \frac{v_{xu}^*}{V_{dc\_rated}}$	$n_{xl} = \frac{v_{xl}^*}{V_{dc\_rated}}$	$n_{xu} = \frac{v_{xu}^*}{\sum_{i=1}^{N} v_{xu,i}}$	$n_{xl} = \frac{v_{xl}^*}{\sum_{i=1}^{N} v_{xl,i}}$
Number of Inserted Cells	$N_{xu} + N_{xl} = N_{arm}$		$N_{xu} + N_{xl} \neq N_{arm}$	
Synthesized Voltage	$v_{xu} \neq v_{xu}^*, v_{xl} \neq v_{xl}^*$		$v_{xu} = v_{xu}^*, v_{xl} = v_{xl}^*$	

(a)

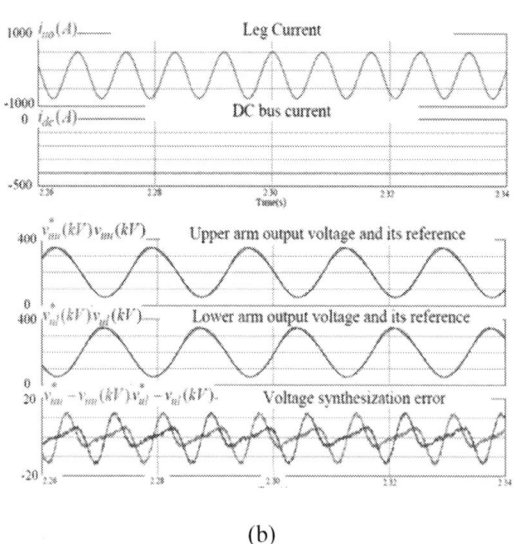

(b)

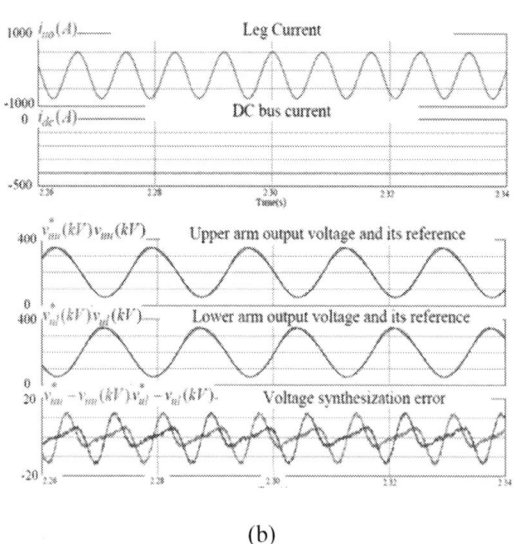

(a)

(b)

978-1-5386-7688-2/19 $31.00 © 2019 IEEE   456

2. Positive sequence, negative sequence, and zero sequence should be modeled during dq modeling.
3. Controller modeling to control the circulating current inside the controller should be added.

Considering these conditions, Fig. 3 is introduced. In Fig. 3, the energy balancing algorithm is introduced from (15) and (17).

Also, because the MMC system uses the phase lock loop, the 2nd harmonic is generated due to the negative sequence and the positive sequence of AC system. The 2nd harmonic induces a new voltage distortion of the control loop which can lead to the instability. As shown in (18)~(20), the negative sequence and the positive sequence can lead to the system instability of the controller using energy balancing algorithm.

Fig.4 shows the simulation results according to the control methods. Fig. 4(a) is the case for the direct control method and Fig. 4(b) is for the indirect control method.

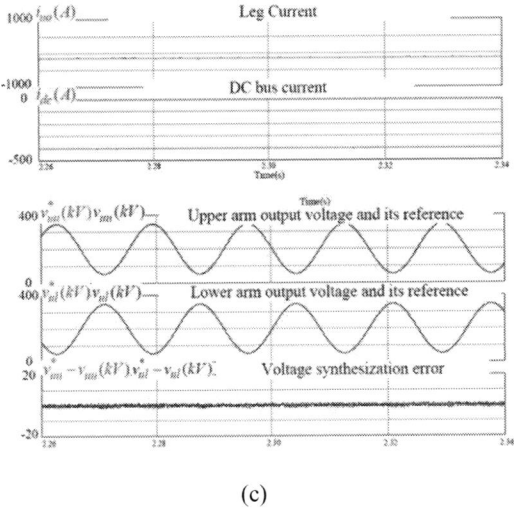

(c)

Fig.4. Control characteristics according to control structure

$$p_{xy} = v_{xy} \cdot i_{xy}$$

$$\begin{pmatrix} p_{\Sigma\alpha} \\ p_{\Sigma\beta} \\ p_{\Sigma 0} \\ p_{\Delta\alpha} \\ p_{\Delta\beta} \\ p_{\Delta 0} \end{pmatrix} = \begin{bmatrix} 2/3 & 2/3 & -1/3 & -1/3 & -1/3 & -1/3 \\ 0 & 0 & 1/\sqrt{3} & 1/\sqrt{3} & -1/\sqrt{3} & -1/\sqrt{3} \\ 1/3 & -1/3 & 1/3 & 1/3 & 1/3 & 1/3 \\ 2/3 & -2/3 & -1/3 & 1/3 & -1/3 & 1/3 \\ 0 & 0 & 1/\sqrt{3} & -1/\sqrt{3} & 1/\sqrt{3} & -1/\sqrt{3} \\ 1/3 & -1/3 & 1/3 & -1/3 & 1/3 & -1/3 \end{bmatrix} \begin{pmatrix} p_{pa} \\ p_{na} \\ p_{pb} \\ p_{nb} \\ p_{pc} \\ p_{nc} \end{pmatrix}$$

$$\begin{pmatrix} p_{\Sigma\alpha} \\ p_{\Sigma\beta} \\ p_{\Sigma 0} \\ p_{\Delta\alpha} \\ p_{\Delta\beta} \\ p_{\Delta 0} \end{pmatrix} = \begin{bmatrix} v_{DC} & 0 & 0 & 0 & 0 & 0 & 0 & 0 & 0 & -\frac{1}{2}v_{sd}^- & -\frac{1}{2}v_{sq}^- & -\frac{1}{2}v_{sd}^+ & \frac{1}{2}v_{sq}^+ \\ 0 & v_{DC} & 0 & 0 & 0 & 0 & 0 & 0 & 0 & -\frac{1}{2}v_{sq}^- & \frac{1}{2}v_{sd}^- & \frac{1}{2}v_{sq}^+ & \frac{1}{2}v_{sd}^+ \\ 0 & 0 & v_{DC} & 0 & 0 & 0 & 0 & 0 & 0 & -\frac{3}{2}v_{sd}^- & -\frac{3}{2}v_{sq}^+ & -\frac{3}{2}v_{sd}^- & \frac{3}{2}v_{sq}^- \\ 0 & 0 & 0 & -v_{sd}^- & -v_{sq}^- & -v_{sd}^+ & v_{sq}^+ & -v_s^0 & 0 & -\frac{3}{2}v_{sd}^- & -\frac{3}{2}v_{sq}^+ & -\frac{3}{2}v_{sd}^- & \frac{3}{2}v_{sq}^- \\ 0 & 0 & 0 & -v_{sq}^- & v_{sd}^- & v_{sq}^+ & v_{sd}^+ & 0 & -v_s^0 & 0 & 0 & 0 & 0 \\ 0 & 0 & 0 & -v_{sd}^+ & -v_{sq}^+ & -v_{sd}^- & v_{sq}^- & 0 & 0 & 0 & 0 & 0 & 0 \end{bmatrix} \begin{pmatrix} I_{C\alpha} \\ I_{C\beta} \\ I_d \\ i_{Cd}^+ \\ i_{Cq}^+ \\ i_{Cd}^- \\ i_{Cq}^- \\ i_{Cd}^0 \\ i_{Cq}^0 \\ i_{sd}^+ \\ i_{sq}^+ \\ i_{sd}^- \\ i_{sq}^- \end{pmatrix} \quad (17)$$

In above section, an extended symmetric coordinate system of AC/DC system was introduced. The limitation of these equation is the fact that the decoupling between AC voltage and DC current. Taking this limit into consideration, an extended symmetric coordinate system including the controller is introduced as follows.

1. When converting DC voltage and current with AC voltage and current, dq transformation algorithm should be used for decoupling.

Briefly speaking, a direct control method is as shown in Fig. 4(a), DC voltage is fixed or referenced to calculate the control signal. On the other hand, an indirect control method is to measure the DC voltage

$$\begin{bmatrix} P \\ Q \\ P_0 \end{bmatrix} = \begin{bmatrix} \bar{P} + P_{c2} \cos (2\omega t) + P_{s2} \sin (2\omega t) \\ \bar{Q} + Q_{c2} \cos (2\omega t) + Q_{s2} \sin (2\omega t) \\ \bar{P}_0 + P_{0c2} \cos (2\omega t) \end{bmatrix} \quad (18)$$

$$\begin{bmatrix} \bar{p} \\ P_{c2} \\ P_{s2} \\ \bar{Q} \\ Q_{c2} \\ Q_{s2} \end{bmatrix} = \frac{3}{2} \begin{bmatrix} V_{sd}^+ & V_{sq}^+ & V_{sd}^- & V_{sq}^- \\ V_{sd}^- & V_{sq}^- & V_{sd}^+ & V_{sq}^+ \\ V_{sq}^- & -V_{sd}^- & -V_{sq}^+ & V_{sd}^+ \\ V_{sq}^+ & -V_{sd}^+ & V_{sq}^- & -V_{sd}^- \\ V_{sq}^- & -V_{sd}^- & V_{sq}^+ & -V_{sd}^+ \\ -V_{sd}^- & -V_{sq}^- & V_{sd}^+ & V_{sq}^+ \end{bmatrix} \begin{bmatrix} i_d^+ \\ i_q^+ \\ i_d^- \\ i_q^- \end{bmatrix} \qquad (19)$$

$$i_d^{+\,ref} = \frac{2v_{sd}^+ P_s^{ref}}{3\left[\left(v_{sd}^+\right)^2 - \left(v_{sd}^-\right)^2\right]}$$

$$i_q^{+\,ref} = -\frac{2v_{sd}^+ Q_s^{ref}}{3\left[\left(v_{sd}^+\right)^2 + \left(v_{sd}^-\right)^2\right]}$$

$$i_d^{-\,ref} = \frac{2v_{sd}^- P_s^{ref}}{3\left[\left(v_{sd}^+\right)^2 - \left(v_{sd}^-\right)^2\right]} \qquad (20)$$

$$i_q^{-\,ref} = -\frac{2v_{sd}^- Q_s^{ref}}{3\left[\left(v_{sd}^+\right)^2 + \left(v_{sd}^-\right)^2\right]}$$

IV. CONTROL MODELLING AND AC VOLTAGE MODELLING USING SYMMETRIC COORDINATE METHOD

Fig.5 shows the results of the faults analysis of the MMC system as the results of Fig. (2). Fig.5 shows the AC voltage in case of the line to ground faults at DC line. From Fig.5 (a), it is shown that DC bias is applied to the AC voltage. In addition, As Fig.5 (b) shows the vector diagram of AC voltage when a single-phase ground fault occurs on the AC side, the phase of the AC voltage is shifted to 30 degrees and the magnitude of AC voltage becomes $\sqrt{3}$. And through the ground point, AC fault current flow to DC lines in both directions. Fig. 5 c) shows the inner control modeling of the MMC HVDC system considering the circulating signal inside the MMC system.

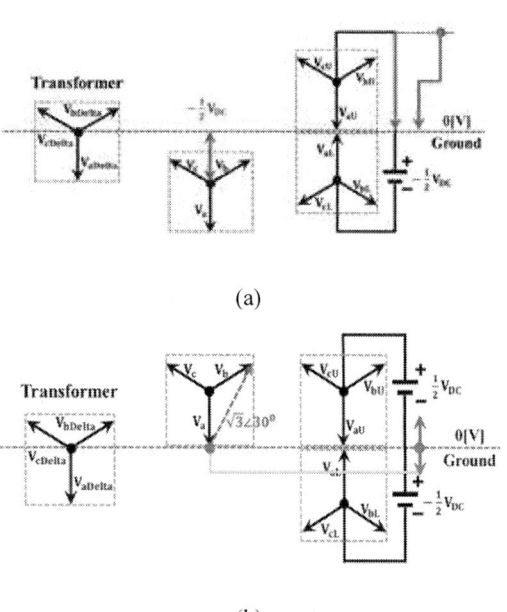

(a)

(b)

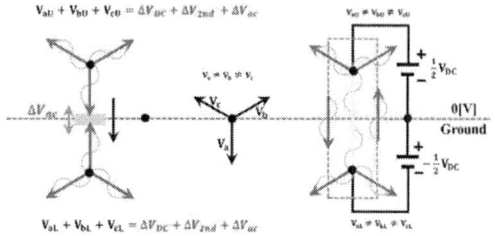

(c) MMC Inner Control Model

Fig.5 Faults of MMC HVDC
by the modified symmetric coordinate method

III. CONCLUSION

This paper presents the method to analyze MMC HVDC system through symmetric coordinate system and d/q transformation algorithm. The advantages of this method are that firstly, the AC system, DC system, and the inside of the controller are completely separated and analyzed under the assumption that the controller operates perfectly, secondly, DC fault and AC fault are analyzed similar to AC system. Additionally, in order to decouple AC system and DC converter, optimal control structure (energy balancing) is introduced.

REFERENCES

[1] S. Shah, R. Hassan, and S. Jian, "HVDC transmission system architectures and control - A review," in Control and Modeling for Power Electronics, 2013 IEEE 14th Workshop on, 2013, pp. 1-8.
[2] N. M. Kirby, "HVDC system solutions," in Transmission and Distribution Conference and Exposition T&D, IEEE PES, 2012, pp. 1-3.
[3] M. H. Okba, M. H. Saied, M. Z. Mostafa, and T. M. Abdel-Moneim, "High voltage direct current transmission - A review, part I," in Energytech, 2012 IEEE, 2012, pp. 1-7.
[4] R. Radzuan, M. A. A. Raop, M. K. M. Salleh, M. K. Hamzah, and R. A. Zawawi, "The designs of low power AC-DC converter for power electronics system applications," in Computer Applications and Industrial Electronics, IEEE Symposium on, 2012, pp. 113-117.
[5] N. Flourentzou, V. G. Agelidis, and G. D. Demetriades, "VSC-Based HVDC Power Transmission Systems: An Overview," Power Electronics, IEEE Transactions on, vol. 24, pp. 592-602, 2009.
[6] M. P. Bahrman, "HVDC transmission overview," in Transmission and Distribution Conference and Exposition, 2008; IEEE/PES, 2008, pp.1-7.
[7] H. K. Müller, S. S. Torbaghan, M. Gibescu, M. M. Roggenkamp, and M. A. M. M. van der Meijden, "The need for a common standard for voltage levels of HVDC VSC technology," Energy Policy, vol. 63, pp. 244-251, 12// 2013.
[8] L. de Andrade and T. P. de Leao, "A brief history of direct current in electrical power systems," in HISTory of ELectro-technology CONference (HISTELCON), 2012 Third IEEE, 2012, pp. 1-6.
[9] S. M. Yousuf and M. S. Subramaniyan, "HVDC and Facts in Power System," International Journal of Science and Research, vol. 2, 2013.
[10] B. K. Bose, "Evaluation of modern power semiconductor devices and future trends of converters," Industry Applications, IEEE Transactions on, vol. 28, pp. 403-413, 1992.
[11] S. Tamai, "High power converter technologies for saving and sustaining energy," in Power Semiconductor Devices & IC's (ISPSD), 2014 IEEE 26th International Symposium on, 2014, pp. 12-18.
[12] E. Kontos, R. T. Pinto, S. Rodrigues, and P. Bauer, "Impact of HVDC Transmission System Topology on Multiterminal DC Network Faults," Power Delivery, IEEE Transactions on, vol. 30, pp. 844-852, 2015.
[13] C. Kim, S. Sim, C. Park, J. Kim, " Virtual Group Control Algorithm for Modular Multilevel Converter," in 2018 icSmartGrid, Dec. 2018.

978-1-5386-7688-2/19 $31.00 © 2019 IEEE

Control of Multi-Sources Energy PV/Fuel Cell and Battery Based Multi-Level Inverter for AC Load

Mostefa Koulali
Laboratory of L2GEGI
Department of Electrical Engineering
University Tahar Moulay of Saida
Saida, Algeria
mostefa.koulali@univ-tiaret.dz

Siamak Pourkeivannour.
Electrical and Electronics Engineering
Department
Middle East Technical University
Ankara, Turkey
siamak.pourkeivannour@metu.edu.tr

Bachir Boumediene
Laboratory of L2GEGI
Department of Electrical Engineering
University Ibn Khaldoun of Tiaret
Tiaret, Algeria
bachir.boumediene@univ-tiaret.dz

Mohamed Mankour
Laboratory of Electrical Engineering
Department of Electrical Engineering
University Tahar Moulay of Saida
Saida, Algeria
mankourmohamed312@yahoo.fr

Karim Negadi
Laboratory of L2GEGI
Department of Electrical Engineering
University Ibn Khaldoun of Tiaret
Tiaret, Algeria
karim.negadi@univ-tiaret.dz

Attalah Smaili
Laboratory of L2GEGI
Department of Electrical Engineering
University Ibn Khaldoun of Tiaret
Tiaret, Algeria
smaili_at@yahoo.fr

Abstract—**The considered system was implemented in the Matlab/Simulink, the results show the effectiveness of the proposed method and can be realized with experimented setup. This work deals with the study of an electrical energy production system made up of three sources of energy: photovoltaic energy, a fuel cell and a battery. The optimization of this hybrid production system is ensured by the control of each part. The hybrid system chain also contains a multilevel inverter that improves the quality of energy injected into the alternating load and consequently reduces the harmonic rate. In order to optimize the power flow in the different parts of the production line, an energy management algorithm is developed in order to mitigate the fluctuations of the load. To analyze our approach, a prototype is modeled, simulated with Matlab / simulink and can be realized in an experimental test bench**

Keywords—*Fuel Cell, Photovoltaic PV, Battery, Hybrid System, MPPT tracking, Three Level Inverter, Fuzzy Logic Control (FLC).*

I. INTRODUCTION

Renewable energy sources have received greater attention during the past few decades and considerable efforts have been made to develop efficient renewable energy conversion system. The major objectives of these approaches are to have reduced environmental damage, conservation of energy, exhaustible sources and increased safety. The renewable energy systems can be used to supply power either directly to a utility grid or to an isolated load. The stand-alone systems find wider applications in isolated areas which are far away from the utility grid.

The integration of Energy storage systems (ESSs) have some significant applications in operations like grid stabilization, stable power quality, load shifting, grid operational support and smooth power injection to grid. Several power smoothing methods have been introduced in literature [2], battery energy storage system (BESS) is selected as energy storage and incorporated into fuel cell, PV to maintain power and energy balance as well as to improve power quality.

In our research, the proposed Energy Management of multi-sources Power System PV/Fuel Cell and Battery Based Three Level Inverter consists of a fuel cell energy conversion system FC/PV generators as the primary energy source and a battery energy storage system (as short and medium, time storage devices) [1].

They are all connected to a DC voltage link, and a three phase three level Neutral Point Clamping (NPC) converter is used to connect the whole system to the load. The DC coupled structure makes the overall system more flexible since the number and the types of energy sources can be freely chosen and it requires fewer controllers as no synchronization is needed to integrate the energy sources.

Regarding the control of the three-level inverter connected to the load, we used fuzzy logic control (FLC) to ensure the RMS value of the load voltage.

The structure of the presented work is organized as follow: The description of the proposed multi-sources system approach and the physical modeling different part of our system with their equations model is set in section 2. The control of different component is set in section 3. The simulations results of the studied are presented in section 4. Section 5 summarizes the work done in the conclusion.

II. DESCRIPTION OF THE PROPOSED APPROACH

The figure 1 shows the main elements of the studied multi-source system connected to an AC load.

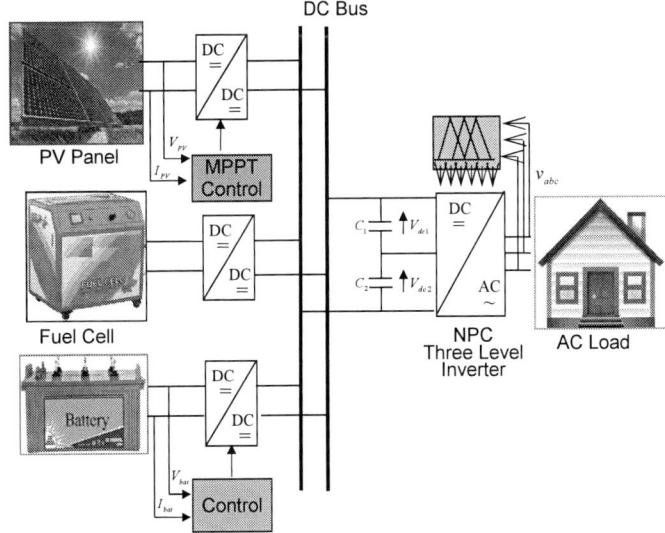

Fig. 1. Schematic of the proposed PV/FC and Battery multi-sources system

978-1-5386-7688-2/19 $31.00 © 2019 IEEE

A. Modeling of solar PV

PV system is based on solar energy, where PV cell is the most basic generation part in PV. As figure 2 shows, the PV cell is formed from a diode and a current source was connected anti parallel with a series resistance [2].

The relation of the current and voltage in the single-diode cell can be written as follows:

$$I_{PV} = I_{ph} - I_0 \left(\exp\left(\frac{q(V_{PV} + R_{s\,\text{mod}}I_{PV})}{AKT} \right) - 1 \right) \qquad (1)$$

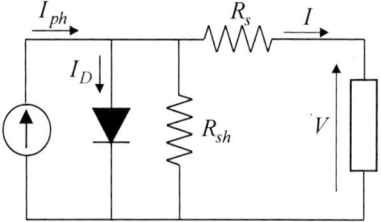

Fig. 2. Equivalent circuit of PV cell

B. DC/DC boost converter and Maximum Power Point Tracking (MPPT)

A boost converter is a step up DC/DC converter which increases the solar voltage to desired output voltage as required by load. The configuration is shown in figure 3, which consists of a DC input voltage V_{in}, inductor L, switch S, diode D1, capacitor C for filter, and load resistance R.

When the switch S is ON the boost inductor stores the energy fed from the input voltage source and during this time the load current is maintain by the charged capacitor so that the load current should be continuous. When the switch S is OFF the input voltage and the stored inductor voltage will appear across the load hence the load voltage is increased. Hence, the load voltage is depends upon weather switch S in ON or OFF and this is depends upon the duty ratio D.

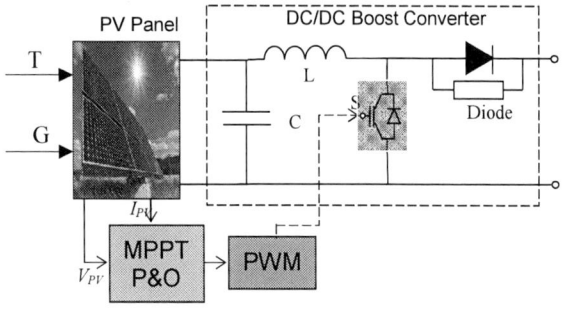

Fig. 3. Boost converter with MPPT control

The solar panel efficiency is increased by the use MPPT technique. The MPPT is the application of maximum power transfer theorem which says that the load will receive maximum power when the source impedance is equal to load impedance. The MPPT is a device that extracts maximum power from the solar cell and changes the duty ratio of DC/DC converter in order to match the load impedance to the source.

C. FC System Modeling

A dynamic model of the PEMFC is based on the relationship between the output voltage and potential pressure of hydrogen, oxygen and water. The overall output voltage of the fuel cell stack can be obtained as [3, 4].

$$V_{cell} = E_{nerst} - V_{act} - V_{ohmic} - V_{con} \qquad (2)$$

Where E_{nerst} is the Nernst voltage, which is the thermodynamics voltage of the cells and depends on the temperatures and partial pressures of reactants and products inside the stack, E_0 is the standard reversible cell potential (V), N_0 is number of cells in stack, R is the universal gas constant (8.3145 J·mol^{-1}·K^{-1}), T is the stack temperature (K), F is the Faraday's constant (96485 A·C·mol^{-1}), $P_{H_2}, P_{O_2}, P_{H_2O}$, are the partial pressures of hydrogen, oxygen and water (atm) respectively.

$$\begin{cases} E_{nerst} = N_0[E_0 + \dfrac{RT}{2F}\log(\dfrac{P_{H2}p_{O2}^{0.5}}{P_{H2O}})] \\ V_{ohmic} = R_m I \end{cases} \qquad (3)$$

Where $K_{O_2}, K_{H_2}, K_{H_2O}$ are the valve molar constant for oxygen, hydrogen and water in (Kmol·s^{-1}·atm^{-1}) respectively.

$$\begin{cases} P_{O2} = \dfrac{1/k_{O2}}{1+\tau_{O2}s}(q_{O2}^{in} - 2k_r I) \\ p_{H2} = \dfrac{1/K_{H2}}{1+\tau_{H2}s}(q_{H2}^{in} - 2k_r I) \\ P_{H2O} = \dfrac{1/K_{H2O}}{1+\tau_{H2O}s}(2K_r I_{fc}) \\ q_{H2}^{in} = \dfrac{1}{1+T_f s}[\dfrac{2k_r}{U_{opt}}I_{fc}) \\ q_{O2}^{in} = \dfrac{1}{rHO}q_{H2}^{in} \end{cases} \qquad (4)$$

$q_{H_2}^{in}, q_{O_2}^{in}$ and are the hydrogen and oxygen input flow (kMol/s), I is the stack current (A), $K_r = \dfrac{N}{4F}$ is the modeling constant, with N being the number of the series – wound fuel cells in the stack. $\tau_{H_2}, \tau_{O_2}, \tau_{H_2O}$ are the time constants for hydrogen, oxygen and water in (sec), U_{opt} is the optimum fuel utilization, T_f is the fuel time constant (sec), rHO is ratio of hydrogen to oxygen [5, 6],

$V_{act}, V_{ohmic}, V_{con}$ are the activation, Ohmic and concentration polarizations losses respectively.

$$V_{act} = [\xi_1 + \xi_2 T + \xi_{3T} \times \ln(C_{O2}) + \xi_4 T \times \ln(I)] \quad (5)$$

Where $\xi_i \left(i = 1, 2, 3, 4 \right)$ are the parametric coefficients defined based on the kinetic, thermodynamic and electrochemical phenomena. Co_2 is the concentration of oxygen dissolved in a water film interface in the catalytic of the cathode in (mol/m³). It is expressed as follows [5]

$$Co_2 = \frac{P_{O2}}{5.08 \times 10^6 e^{-\frac{498}{T}}} \quad (6)$$

The Ohmic polarization loss is given as:

$$V_{ohmic} = IR_m \quad (7)$$

R_m is ohmic resistance calculate in the paper [6]. A concentration polarization is expressed as:

$$V_{con} = -B \times \ln(1 - \frac{I}{I_{lim}}) \quad (10)$$

With I_{lim} being the current density where fuel is used in a same rate as the maximum input rate (A/cm²).

To size the fuel cell, the amount of electric energy extracted from the FC should be calculated. Therefore, it is necessary to estimate the amount of energy generated from the FC per 1 Kg of hydrogen which can be obtained as follows [7]:

$$E_g^{FC} = H_2^{used} \xi_{fc} \frac{H_2 \, heating \, value}{H_2 \, density} \quad (8)$$

Where H_2^{used} represents the quantity of hydrogen input to the FC in Kg, ζ_{fc} is the FC efficiency, H_2 heating value is equal to 3.4 kWh/m³ in the standard condition and H_2 density is 0.09 Kg/m³.

III. CONTROL OF SYSTEM COMPONENTS

The multi-sources system shown in figure 1 contains control strategies that will be developed in this section.

A. Battery Modeling

Several authors have proposed models for the battery and the results of experiments carried out on lead/acid batteries deduce a model named "CIEMAT model" representing the battery operation during the charge, discharge and overcharge processes. In our case study, from the carried out experiments, a validated model is proposed with respect to

the battery capacity for any size and type of lead/acid battery [8].

This model represented by an equivalent circuit model contains a voltage source which is the open circuit voltage V, in series with an internal resistance R. Thus, the output voltage of the battery is:

$$V_{bat} = V - RI_{bat} \quad (9)$$

where the both V_{bat} and I_{bat} depend on the battery state of charge (SOC), temperature and internal resistance variations

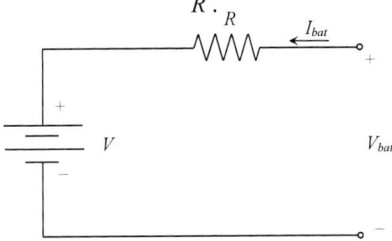

Fig. 4. The battery control

In this study, this simple model based on the CIEMAT model for the battery is considered as enough accurate to assess power management objectives and to compare performance of several strategies. During the charging and discharging process, the state of charge (SOC) in terms of time (t) can be described by [9]

$$SOC(t) = \begin{cases} SOC(t - \Delta t) + P_{bat} \cdot \dfrac{\eta_{ch}}{C_n \cdot V_{dc}} \cdot \Delta t \\[2mm] SOC(t - \Delta t) + P_{bat} \cdot \dfrac{1}{\eta_{dis} \cdot C_n \cdot V_{dc}} \cdot \Delta t \end{cases} \quad (10)$$

where Δt is the time step, P_{bat} represents the battery power, C_n is the nominal capacity of the battery, η_{ch} and η_{dis} are respectively the battery efficiencies during charging and discharging phase. V_{dc} denotes the nominal DC bus voltage. At any time step Δt, the SOC must comply with the following constraints

$$SOC_{min} \leq SOC(t) \leq SOC_{max} \quad (11)$$

where SOC_{min} and SOC_{max} are maximum and minimum allowable storage capacities, respectively.

B. Battery control

The objective of the control system is to regulate the battery current in order to obtain the required power. Charging and discharging current limits and maximum SOC limitations are also included in the model. The BESS is connected to the DC grid via a bi-directional Buck-Boost DC/DC converter, as shown in figure 5.

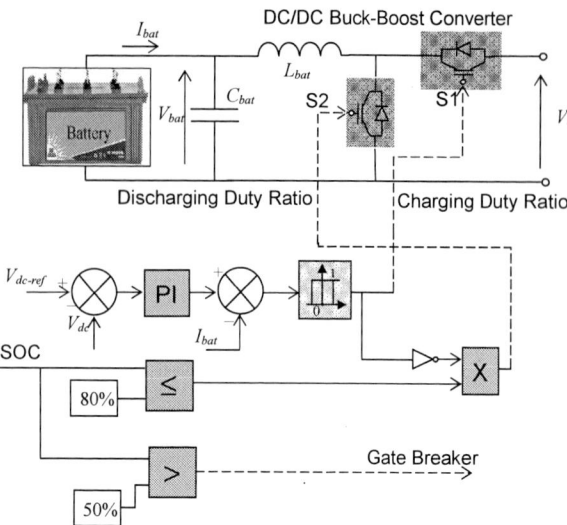

Fig. 5. The battery control

The BESS will operate in charging, discharging or floating modes depending on the energy requirements and these modes are managed according to the DC bus voltage at the BESS point of coupling. Consequently, the BESS is required to provide necessary DC voltage level under different operating modes of the microgrid or AC load as in our case. When charging, switch S2 is activated and the converter works as a boost circuit; otherwise, when discharging, switch S1 is activated and the converter works as a buck circuit. When the voltage at the DC link is lower than the voltage reference, switch S1 is activated. Alternatively, when the voltage at the DC link is higher than the voltage reference, switch S2 is activated. The PV-battery system response to transient variations is characterized by an inherent time constant. In such cases, capacitors along the DC grid can act as virtual inertia to supply the shortfall or absorb the surplus of energy [10, 11, 12; 13, 14].

The DC-link power balance can be expressed by the following differential equation:

$$V_{dc}i_{dc} = P_{PV} + P_{FC} + P_{bat} - P_{load} \qquad (12)$$

Neglecting the losses in the power converters, battery, filtering inductors and transformer and also the harmonics due to switching actions, the power balance of the integrated hybrid distributed generation system (DGS) with energy storage is governed by:

$$V_{dc}i_{dc} = CV_{dc}\frac{dV_{dc}}{dt} = P_{PV} + P_{FC} + P_{bat} - P_{load} \qquad (13)$$

The objective of the battery converter is to maintain constant voltage at the DC link, so the ripple in the capacitor voltage is much lower than the steady-state voltage.

If the powers injected by the two back-to-back voltage source converters (VSC) are assumed constant at any particular instant, the power from the battery is responsible for adjusting the capacitor voltage [15, 16, 17].

IV. SIMULATION RESULTS AND DISCUSSIONS

The simulations carried out to check the validity of the proposed scheme used Fuzzy Logic Control (FLC) to control the DC voltage of a three level (NPC) inverter connected to the multi-sources (PV-FC-battery) system integrated with an AC load.

It is evaluated using MATLAB/Simulink software under variable load conditions. The proposed system parameters are listed in Table 1, 2, 3 and 4. The amount of power generated/supplied by the PV array depends on the variable solar irradiation G = [600, 1000, 800] W/m2 at the time t=[0, 2, 3.5] s respectively and the temperature T = 293 K. The proposed system is operated in three possible operating modes depending on the variable load. The system performance for this situation is shown in Figs. 6 to 16.

In figures 6, 14 and 15, the active power, current, and voltage on the load with both a PI controller and a FLC are shown. The increase of the loads can be seen clearly. First a 7-kW, then a 14-kW, then 7-kW, then 10-kW and finally a 7-kW load were added to the system at [0, 3s], [3, 4s], [4, 5s], [5, 6s] and [6, 7s] respectively. As a result, a total load of 14- kW is supplied by the system.

As it can be seen in Figure 12, with changing power, the DC link voltage V_{dc} is well kept constant at the specified value (640V) which constitutes an important advantage and proves the effectiveness of the proposed schema. It has allowed us to equalize the different input DC link voltages of the multilevel inverter. Then, the input voltages are practically equal by pairs as shown in figure 12.

In figure 8, we present the global (total) state of charge SOCG of storage devices. From 0 s to 2 s the SOC equals 60%, and the system operates with full charge. All the storage devices are switched off because the total fuel cell and PV generated power is higher than the load demand as shown in figure 6.

When a SOCG decreases, and the system enters the critical mode. The supervisory controller reacts properly and switches off the load of the lowest priority in order to save the equilibrium of the overall system. If we have lack of power, we have to request for the utility of delesting process, and hence, all storage devices would be disconnected. Else, we authorize the charge mode of the storage system. These results show the efficiency of the management and the controls used for this hybrid system.

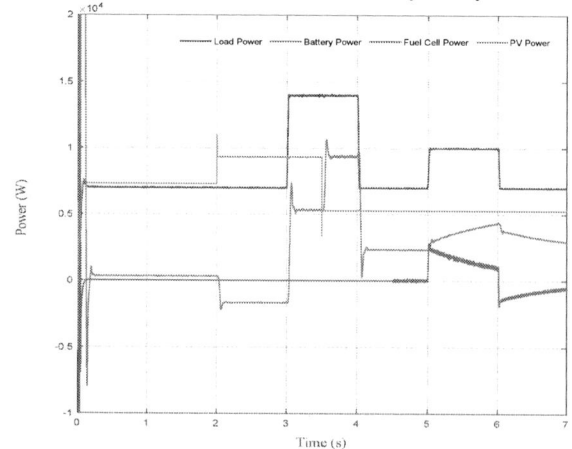

Fig. 6. Power generation of the hybrid system under varying AC load

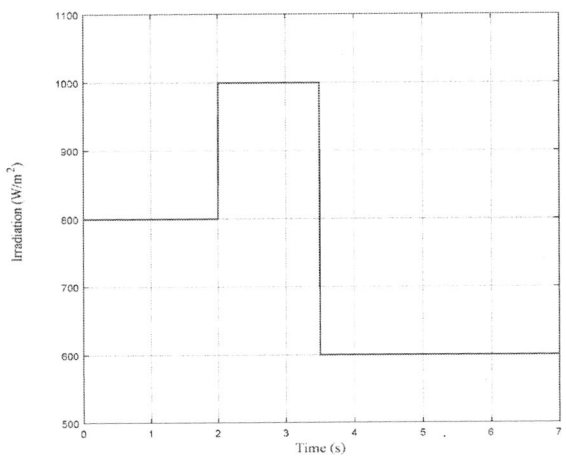

Fig. 7.   Example of a figure caption. (figure caption)

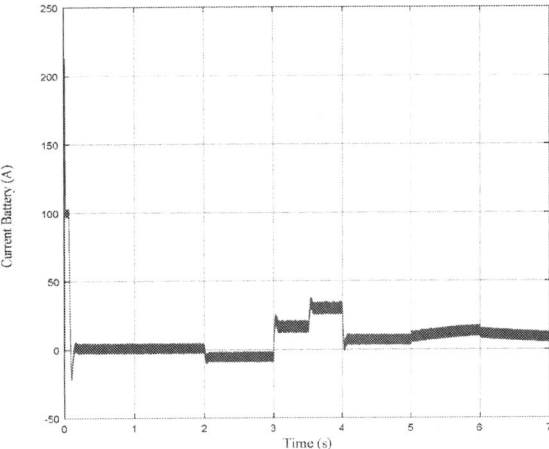

Fig. 10. Current battery

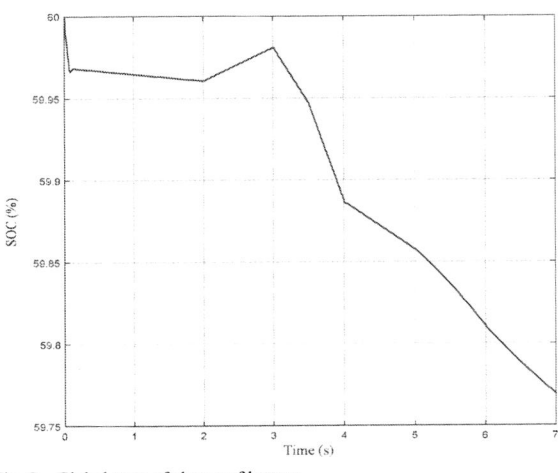

Fig. 8.   Global state of charge of battery

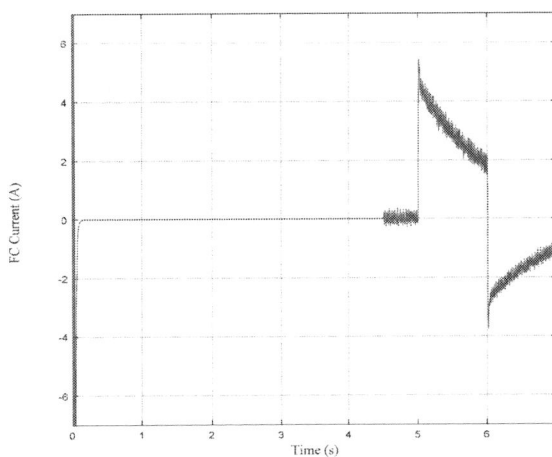

Fig. 11. Fuel Cell current

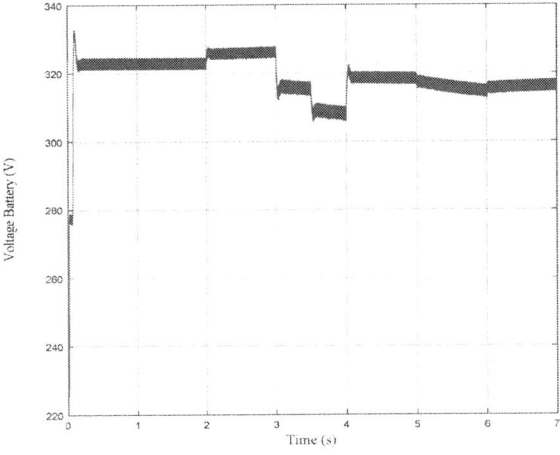

Fig. 9.   Voltage battery

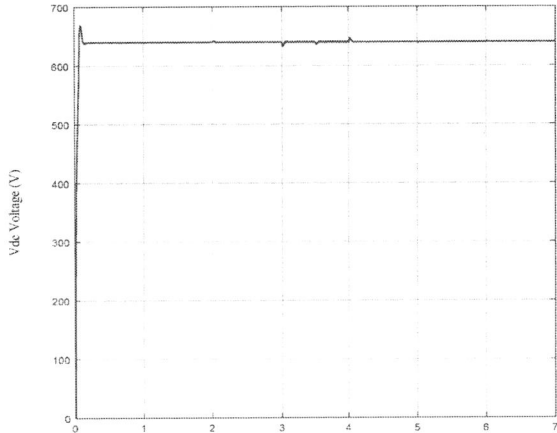

Fig. 12. DC link voltage

978-1-5386-7688-2/19 $31.00 © 2019 IEEE

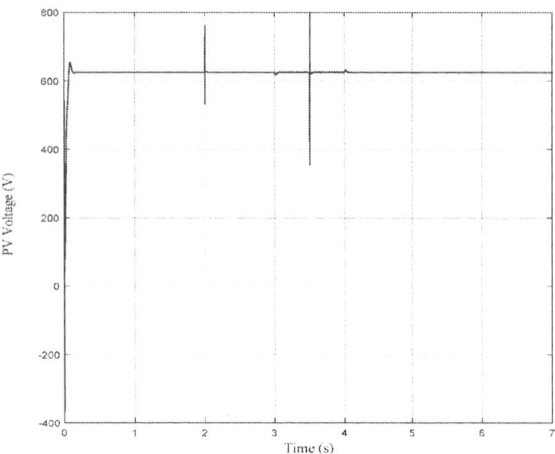

Fig. 13. Photovoltaic Model Output Voltage

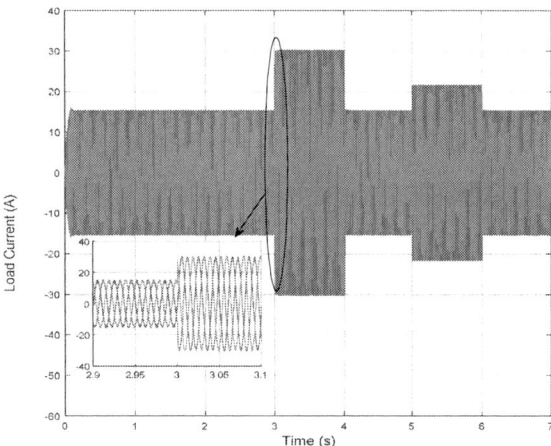

Fig. 14. AC load current

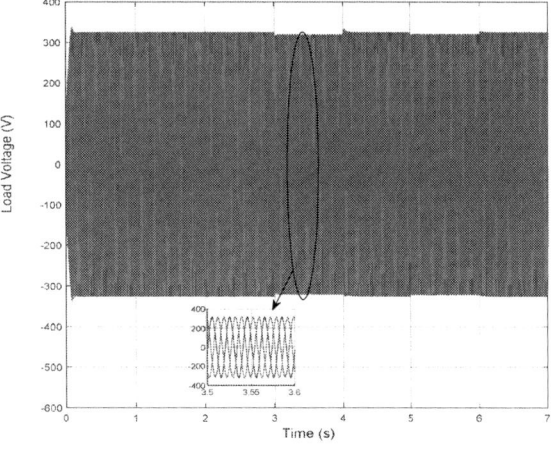

Fig. 15. AC load voltage

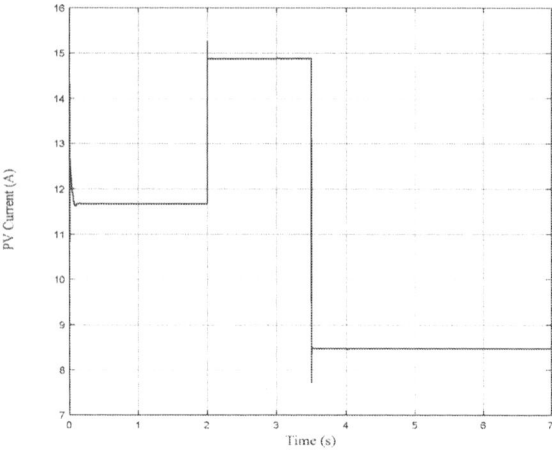

Fig. 16. Photovoltaic Model Output Current

V. CONCLUSIONS

In this work a multi-sources energy system fuel cell/PV and battery have been presented. Dynamic modeling and simulations of the hybrid system is proposed using MATLAB/SIMULINK. A hybrid energy system and its supervisory-control of battery voltage was developed and tested. Load demand is met from the combination of PV array, FC and the battery. A three level inverter is used to convert output from solar and FC systems into AC power output. Circuit Breaker is used to connect and disconnect an additional load in the given time. This multi-sources system is controlled to give maximum output power under all operating conditions to meet the load. Either FC or solar system is supported by the battery to meet the load. Also, simultaneous operation of FC and the PV system is supported by battery for the same load.

These results show the efficiency of the management and the controls used for this hybrid system and can be implemented easily with DSP or Dspace platform.

TABLE I.   FUELL CELL PARAMATERS

Absolute temperature (K)	1273
Initial current (A)	50
Faraday's constant (C/kmol)	96.487e6
Universal gas constant J/(kmol K)	8314
Ideal standard potential (V)	1.18
Number of cells in series	450
Maximum, minimal and optimal fuel utilization	[8.43e-4 2.81e-4 2.52e-3]
Response time for hydrogen, water and oxygen flow (s)	[26.1 78.3 2.91]
Ohmic loss per cell (ohms)	3.2813e-004
Electrical response time (s)	2
Fuel processor response time (s)	5
Ratio of hydrogen to oxygen	0.145

978-1-5386-7688-2/19 $31.00 © 2019 IEEE

TABLE II. PV PARAMETERS

Components	Rating values
Peak Power	200 W
Peak Voltage	660 V
Peak current	7.52 A
Open Circuit Voltage	33.2 V
Short Circuit Current	8.36 A

TABLE III. BATTERY AND AC LOAD PARAMETERS

Components	Rating values
Load 1, 2, 3 (R)	7 kW, 7 kW, 3 kW
Battery type	Nickel Metal hydride
Nominal voltage	300 V
Capacity rating	6.5 AH

TABLE IV. DC/DC BI-CONVERTER AND THREE LEVEL NPC INVERTER

Components	Rating values
$C_1=C_2$	2.2 mF
Dc link Voltage	640 V
Frequency	50 Hz
Converter Inductor	5 mH
DC link Voltage	640 V
Converter Capacitor	2.2 mF

REFERENCES

[1] Nabil Karami, NazihMoubayed, RachidOutbib, "Energy management for a PEMFC–PV hybrid system,"http://dx.doi.org/10.1016/j.enconman. 2014.02 070, Energy Conversion and Management 82 (2014) 154–168.

[2] V. Boscaino, R. Miceli, G. Capponi, G. RiccoGalluzzo, "A review of fuel cell based hybrid power supply architectures and algorithms for household appliances," 0360-3199/$ e see front matter Copyright ª 2013, Hydrogen Energy Publications, LLC. Published by Elsevier Ltd, http://dx.doi.org/10.1016/j.ijhydene.2013.10.165.

[3] Th.F. El-Shatter, M.N. Eskandar, M.T. El-Hagry, "Hybrid PV/Fuel Cell System Design and Simulation," 0960-1481/02/$ - see front matter 2002 Elsevier Science Ltd, PII: S09 60 -1481(01)00062-3, Renewable Energy 27 (2002) 479–485.

[4] PhatiphatThounthong, ArkhomLuksanasakul, PoolsakKoseeyaporn, " Intelligent Model-Based Control of a Standalone Photovoltaic/Fuel Cell Power Plant With Supercapacitor Energy Storage," IEEE Transactions on Sustainable Energy, VOL. 4, NO. 1, January 2013, 1949-3029/$31.00 © 2012 IEEE.

[5] Akbar Maleki, AlirezaAskarzadeh, "Optimum Configuration of Fuel Cell-B PV/Wind Hybrid System using a Hybrid Metaheuristic Technique," International Journal of Engineering & Applied Sciences (IJEAS) Vol.5, Issue 4 (2014)1-12.

[6] M. Caldero´n, A.J. Caldero´n, A. Ramiro, J.F. Gonza´lez, "Automatic management of energy flows of a stand-alone renewable energy supply with hydrogen support," International Journal of hydrogen energy 3 5 (2 0 1 0) 2 2 2 6 – 2 2 3 5.

[7] ErkanDursun, Osman Kilic, "Comparative evaluation of different power management strategies of a stand-alone PV/Wind/PEMFC hybrid power system," 0142-0615/$ - see front matter _ 2011 Elsevier Ltd. doi:10.1016/j.ijepes.2011.08.025, , "Electrical Power and Energy Systems 34 (2012) 81–89.

[8] A. Berkani, K. Negadi, T. Allaoui, F. Marignetti, "Fuzzy Direct Torque Control for Induction Motor Sensorless Drive Powered by Five Level Inverter with Reduction Rule Base," Przegląd Elektrotechniczny, doi:10.15199/48.2019.07.14, No/Vol: 07/2019 Page no. 66.

[9] JuanP. Torreglosa, Pablo García-Triviño, LuisM. Fernández-Ramirez, Francisco Jurado, Control based on techno-economic optimization of renewable hybrid energy system fors tand-alone applications," http://dx.doi.org/10.1016/j.eswa.2015.12.038 0957-4174/© 2016 Elsevier Ltd, Expert SystemsWith Applications 51 (2016) 59–75.

[10] Koulali M., Mankour M., Negadi K., Mezouar A. (2019). Energy management of hybrid power system PV Wind and battery based three level converter, TECNICA ITALIANA-Italian Journal of Engineering Science, Vol. 63, No. 2-4, pp. 297-304. https://doi.org/10.18280/ti-ijes.632-426.

[11] Jingang Han, Jean-Frederic Charpentier and Tianhao Tang, "An Energy Management System of a Fuel Cell/Battery Hybrid Boat," Energies 2014, 7, 2799-2820; doi:10.3390/en7052799.

[12] Omar Hazem Mohammed, YassineAmirat, Mohamed Benbouzid, Adel Elbast. Optimal Design of a PV/Fuel Cell Hybrid Power System for the City of Brest in France, IEEE ICGE 2014, Mar 2014, Sfax, Tunisia.pp.119-123. hal-01023490.

[13] Pablo García, Juan P. Torreglosa, Luis M. Fernández , Francisco Jurado, "Improving long-term operation of power sources in off-grid hybrid systems based on renewable energy, hydrogen and battery," Journal of Power Sources 265 (2014) 149e159

[14] A. Elbaset, "Design, Modeling and Control Strategy of PV/FC Hybrid Power System," Journal of Electrical System 7.2 (2011): 270-286.

[15] Pablo Garcı´a, Juan P. Torreglosa, Luis M. Ferna´ndez , Francisco Jurado, "Optimal energy management system for standalone wind turbine/photovoltaic/hydrogen/battery hybrid system with supervisory control based on fuzzy logic," http://dx.doi.org/10.1016/j.ijhydene.2013.08.106, International Journal of Hydrogen Energy 38 (2013) 14146-14158.

[16] A. Berkani, K. Negadi, T. Allaoui, F. Marignetti, "Sliding mode control of wind energy conversion system using dual star synchronous machine and three level converter," TECNICA ITALIANA-Italian Journal of Engineering Science, Vol. 63, No. 2-4, pp. 243-250. https://doi.org/10.18280/ti-ijes.632-418, 2019.

[17] Power management system for off-grid hydrogen production based on uncertainty," International Journal of Hydrogen Energy 40 (2015) 7260 -7272.

Control Strategy for Optimizing Energy Management in Microgrid System Using Adaptive Control

R Dimas Dityagraha[1,2,3], Jaeho Choi[2], Nanang Hariyanto[3]
[1]PT PLN (Persero), Jakarta, Indonesia
[2]School of Electrical Engineering, Chungbuk National University, Cheongju, Republic of Korea
[3]School of Electrical Engineering and Informatics, Bandung Institute of Technology, Bandung, Indonesia
dimas.dityagraha@pln.co.id; choi@chungbuk.ac.kr; nanang.hariyanto@stei.itb.ac.id

Abstract — Recently, the development of microgrid system is growing rapidly. Energy such as sun, wind, ocean, wave currents, and geothermal are alternative energy in developing Renewable Energy Sources (RES) in the microgrid system. Most commonly hybrid RES system in the microgrid is a combination between Photovoltaic (PV), ESS and diesel generation. To implement the hybrid RES, the combination strategy control is needed. One of the main motivations using energy management in the microgrid is that they are capable of managing and coordinating diesel generators, storages, RES and loads to optimize the RES for cost efficiency. In this paper, the optimized energy management for the hybrid RES system combined by PV, Li-ion battery ESS, and diesel generator is proposed. To accommodate the diesel generation condition, the developed control strategy is the economic dispatch for fuel cost optimization in diesel generation using combination of the Lagrange multiplier and lambda iteration method and the optimization in energy management operation using an Adaptive Model Predictive Control (AMPC) with the lifetime constraint of batteries. Finally, the proposed control algorithm is verified through the PSiM simulation results.

Keywords — *Adaptive MPC, Control Strategy, EMS, Fuel Cost Optimization, Microgrid*

I. INTRODUCTION

Recently, the development of microgrid system is growing rapidly. Energy such as sun, wind, ocean and wave currents, and geothermal is alternative energy in developing Renewable Energy Sources (RES) in the microgrid system. But these energy sources are intermittent and cannot be controlled so that a combination of strategy control is needed for implementation. Most commonly hybrid system in the microgrid is a combination between Photovoltaic (PV), ESS and diesel generation. Many types of research were developed in recent technologies of equipment for this hybrid system [1-2]. In fact, the control strategy for the hybrid energy system considering SOC and fuel consumption of diesel generator based on the BSFC was applied in [3]. In Indonesia, the operation of most of diesel generators cannot be controlled remotely, so they are manually operated as their own local control strategy by the operator. Most of diesel generators were installed in 1980s, so there is no recent technology applied, for example, any SCADA equipment for data monitoring and remote control. Such a situation makes it impossible to accept new technologies that are being developed recently.

One of the main motivations using energy management in the microgrid is that they are capable of managing and coordinating diesel generators, storages, RES and loads to optimize the RES for cost efficiency. The optimal energy management problem for energy storage with wind power generation was investigated with the effect of battery lifetime characteristic on the system [4]. The adaptive control is suitable for implementing energy management to accommodate the optimal dispatching of power system. Some works can be found in the literature that addressed the model predictive control for the optimal dispatch of power system [5]. From other studies, in [6] there is an Adaptive Model Predictive Control (AMPC) that could be a new solution for the energy management system.

In this paper, the optimized energy management for the hybrid RES combined by PV, Li-ion battery ESS, and diesel generator is proposed. To accommodate the diesel generation condition in Indonesia, the developed control strategy is the economic dispatch for fuel cost optimization in diesel generation using combination of the Lagrange multiplier and lambda iteration method and the optimization in energy management operation using an AMPC with the lifetime constraint of batteries. The purpose of using AMPC is to maintain the diesel generators to operate in a certain value as a base load operation. Then, the economic dispatch can be easily applied with more optimal generation control operation than the diesel generation operation without energy management system.

Rest of this paper is organized as follows: The configuration of microgrid system is introduced in section II. Then the control strategy of microgrid system is proposed to meet the fuel cost optimization of a hybrid RES system with multi generators under local control and the AMPC for the control of battery power considering the battery lifetime in section III. After that simulation and result of the microgrid control strategy is applied in section IV. Finally, the proposed algorithm is verified through the PSiM simulation in section V.

II. CONFIGURATION OF MICROGRID SYSTEM

The typical structure of microgrid usually consists of multiple diesel generators, ESS, customer loads and a central energy management unit which is responsible for the power dispatch [7]. Fig. 1 shows the configuration of the hybrid RES system considered in this paper. Four diesel generators operate as the main source to supply the power to the load and maintain the voltage and frequency in the microgrid system. They cannot be controlled remotely, so their power outputs are maintained as a base load level in the system. The PV system operates depending on the irradiance in that coverage area. Then it delivers the generating power to the

This paper is supported by PT. PLN (Persero)

978-1-5386-7688-2/19 $31.00 © 2019 IEEE

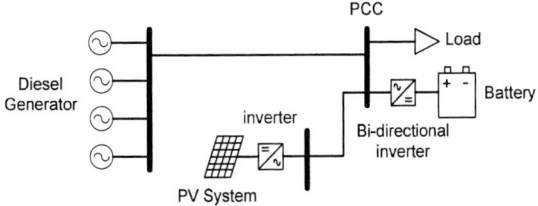

Fig. 1. Configuration of microgrid system.

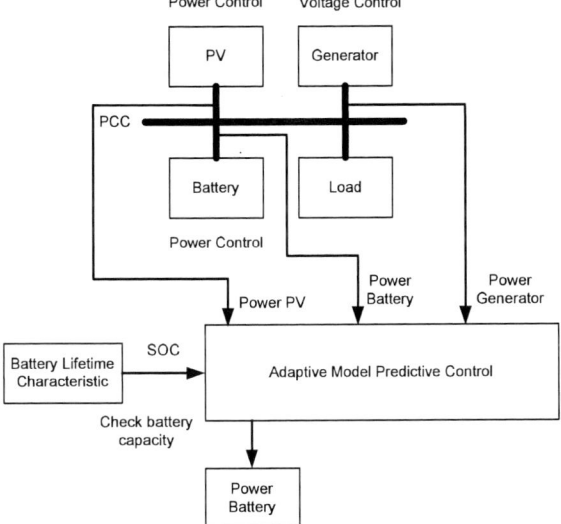

Fig. 2. Proposed control strategy in microgird system.

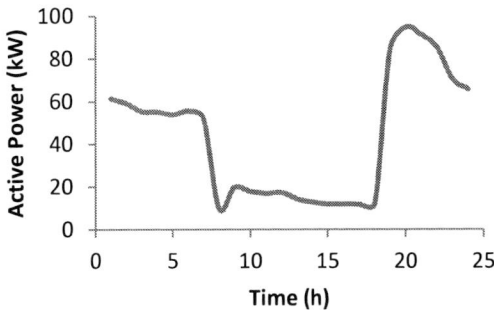

Fig. 3. Electricity load profile in Nule Island for 24 hours.

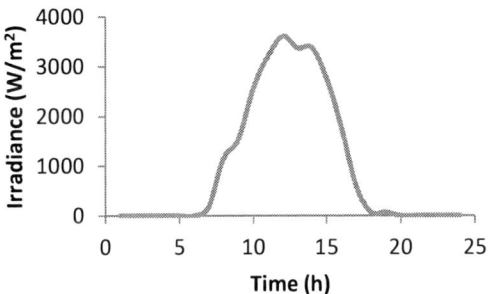

Fig. 4. PV irradiance data on August 1st, 2018.

PCC bus without control. Only the battery operates in the power control to meet the power balance between the supplied power from the diesel generators and PV and the load demand power.

The system modelling is done using the PSiM software for simulation. For simplicity, the diesel generator is modeled by AC source because its characteristic in the system is the same even though the response of diesel generators is slower. Commonly, the load in the microgrid system is radial configuration. Therefore, the battery as an ESS should be connected to the Point of Common Coupling (PCC) bus that the load is also connected directly.

III. CONTROL STRATEGY FOR MICROGRID

The control strategy implemented in this paper is shown in Fig. 2. In real condition, the frequency change is affected by the change of load in the system. In the simulation, the change of active power in diesel generator can be represented as the change of frequency. Therefore, the changes of power transfer from diesel generators will be the references by the AMPC to have the optimized power supply output for the battery. Power transfer from PV and SOC of the battery will be the constraint for AMPC to optimized the battery lifetime in the system.

To maintain the frequency stability, the diesel generator can be followed the change of load. Because there is no SCADA applied, all active power measurements are carried out on the PCC bus. The AMPC will calculate the active power in the battery (charge/ discharge) to optimize the PV power supply in the system and maintain the diesel generation power supply to the load in a certain value as a

base load. The Nule Island electricity data system in Nusa Tenggara, Indonesia is used for simulation. Data consist of diesel generators, load profile, and PV irradiance data. The electricity load profile of Nule Island for 24 hours is described in Fig. 3. The peak load of Nule Island is ±98kW at night time and ±20kW at day time. The total capacities of diesel generators are 188kW, which consists of two diesel generators 1 and 2 with 20kW each, a diesel generator 3 with 48kW, and a diesel generator 4 with 100kW. The PV irradiance data are described in Fig. 4 which were measured on 1st August 2018.

Before the AMPC control strategy is carried out, it is necessary to do the frequency stability analysis with the PV penetration active power in the system. The aim of this calculation is to determine the stability of the system if there is a fault or power loss in PV. In this paper, the simulation of system stability is done using Digsilent software. The grid code in Indonesia is 47.5~52Hz. Normal operation system is 49.5~50.5Hz. Fig. 5 describes that the system stability is simulated when the power loss is happened suddenly from PV about 20~60% of peak load. Until 50% of power loss, the frequency system is still within the range of grid code. But in 60% of PV power loss, the system frequency drops to 47.2 Hz, and it is out of the grid code range. Therefore, it can be seen that there is a stability awareness of the system when the PV penetration is higher than 60%. Besides that, there are three PV penetration criteria that have to be concern in the hybrid power system [8]: the low level PV penetration (PV intermittence does not cause interference or outage to the system), the middle level PV penetration (PV intermittence causes interference or outage to the system and ESS installation is still optional), the high level PV penetration (PV intermittence causes outage to the system and ESS installation is essential). Where the middle PV penetration

978-1-5386-7688-2/19 $31.00 © 2019 IEEE

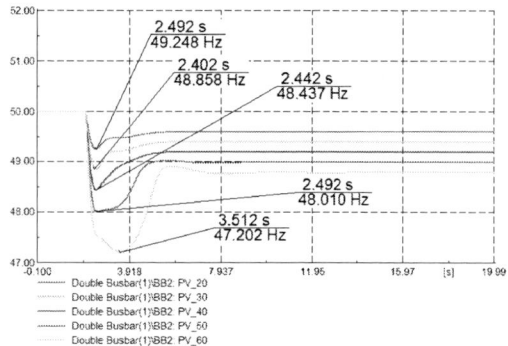

Fig. 5. Simulation of system reliability when power loss happened suddenly in PV penetration respectively from 20-60% of system.

level is 40% of system. Based on that, the simulation is applied at diesel generators supplying 60% of peak load demand. Then the expected PV system with battery will cover 40% of peak load demand at day and night time operation.

The proposed control strategies for the microgrid system are fuel cost optimization for multi diesel generators and AMPC in maximizing PV power supply in the system with battery SOC consideration for optimizing battery lifetime. It is done separately to accommodate the optimization of diesel generations operation.

A. Fuel Cost Optimization

As mentioned above, the control diesel generator is a local control operation and cannot automatically control remotely. Then, one of the control strategies which is suitable to optimize these multi diesel generators are economic dispatch. The economic dispatch problem is how to minimize the total generation cost of power system with various constraints including power balance and generation power limits of each unit [9]. From [9-11], the optimizing generation cost is described in (1) to (6):

Minimize:

$$F_T = \sum_{i=1}^{n} F_i(P_i) \tag{1}$$

$$F_i(P_i) = (a_i P_i^2 + b_i P_i + c) \tag{2}$$

Subject to:

$$\sum_{i=1}^{n} P_i = P_{load} \tag{3}$$

$$P_{min,i} \leq P_i \leq P_{max,i} \tag{4}$$

The economic dispatch is optimized by finding the minimum value of fuel cost described in (1) and (2) with constraints described in (3) and (4) where F is the fuel cost and P is the active power in a diesel generator. The fuel cost equation in (2) has the coefficients, a, b, and c, which can be obtained either through measurement or specification from each generator. The Lagrange multiplier method is a strategy to find the optimal value of the active power diesel generator with the Lagrange multiplier, λ. The Lagrange formulation can be rewritten in (5).

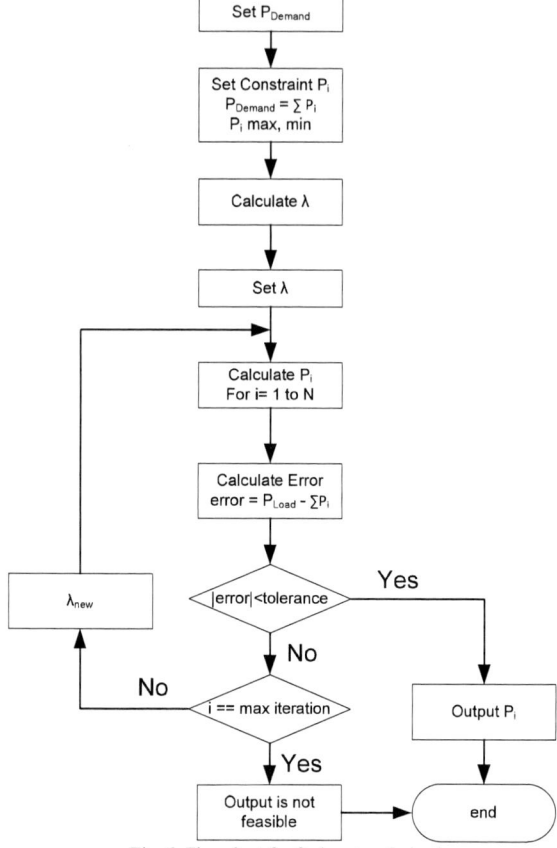

Fig. 6. Flowchart for fuel cost optimization.

$$L = F_T + \lambda\emptyset = \sum_{i=1}^{n} F_i(P_i) + \lambda\left(P_{load} - \sum_{i=1}^{n} P_i\right) \tag{5}$$

By differentiation (5), the Lagrange multiplier, λ is obtained in (6).

$$\frac{\partial F_i}{\partial P_i} + \lambda(0 - 1) = 0 \rightarrow \frac{\partial F_i}{\partial P_i} = \lambda \tag{6}$$

After then, the optimum active power, P, of each generator is acquired to serve the system load. But the result is not accurate enough for economic dispatching multi diesel generators if the coefficients a, b, and c, are not obtained accurately for a single iteration, and then another iteration is necessary to have the optimum value. It takes a long time to do the next iteration until the result is accurate enough. In order to make the calculation more accurate and faster, the Lagrange multiplier and the lambda iteration method are combined. The Lagrange multiplier method is used to estimate the lambda value. After then, for more accurate optimization operation, the lambda iteration method is used. Fig. 6 shows the flowchart of the fuel cost optimization method. As the result of this calculation, the optimization value of the active power output for each diesel generator is determined. Where, the Lagrange multiplier method is used to have the value of the lambda value. With lambda value result, the lambda iteration method is used to have the optimized value of power transfer from each generator.

978-1-5386-7688-2/19 $31.00 © 2019 IEEE 468

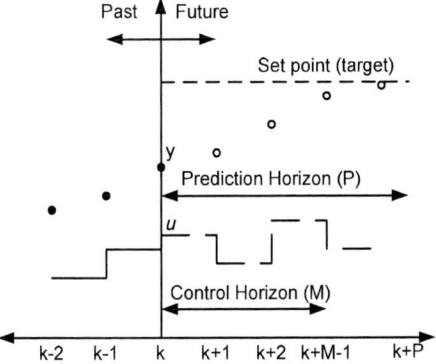

Fig. 7. Basic concept for model predictive control [11].

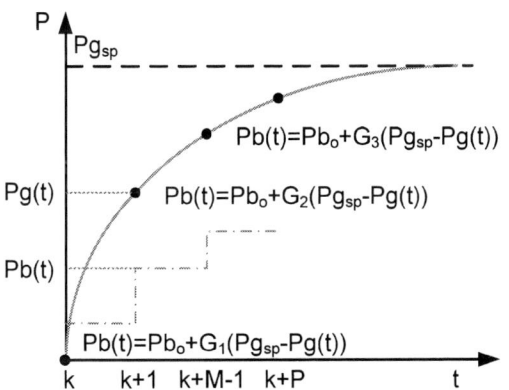

Fig. 8. Basic concept for gain scheduling method.

B. Adaptive Model Predictive Control

The MPC method is an advanced control technique for multivariable control problems [12]. The basic concept of MPC is predicting the current and future values of output with an accurate dynamic model. Predictions calculated in MPC are set-point calculation and control calculation. Fig. 7 shows the basic concept of MPC scheme. It uses the control horizon and the prediction horizon to perform the setpoint calculation and the control calculations at every sampling time. Even though a sequence of control is calculated at every sampling time in the control horizon, only the first step of the control strategy is implemented. The prediction horizon and the control horizon must be more than 10 sample prediction to have the accurate MPC results. Hence, the MPC can be used in a linear system.

For a nonlinear system, there has been an AMPC method. True adaptive nonlinear MPC algorithm must address the robustness of model uncertainty during the evolution of estimates [7]. Since the system is not linear, the control horizon and the prediction horizon of the AMPC are smaller than those of the MPC. Fig. 8 shows the basic concept of the gain scheduling method which is one of the representative AMPC methods. The gain scheduling method is practical and powerful for the control of a nonlinear system. The result of the gain scheduled controller is linear whose parameter (gain) is adjusted at each operating point. The gain is obtained by the MPC optimization using (7) for certain control horizon in several prediction horizons.

$$Cost\ Function\ \min(J) = \sum_{i=1}^{p} w_e e_{k+i}^2 + \sum_{i=0}^{p-1} w_{\Delta u} \Delta u_{k+i}^2 \quad (7)$$

Where e is an error obtained from setpoint and current output. Δu is the difference of control output value after and before the control calculation. The weighting factor, w, is determined from the AMPC priority control. The main priority value is bigger than the others.

Fig. 9 shows the proposed AMPC algorithm. First, define the prediction horizon, the control horizon, and the weighting factor which are important factors for this algorithm. Then, define the setpoint as the power of diesel generators to maintain in a certain value as a base load. The error tolerance for this algorithm is $\pm5\%$. After then, the

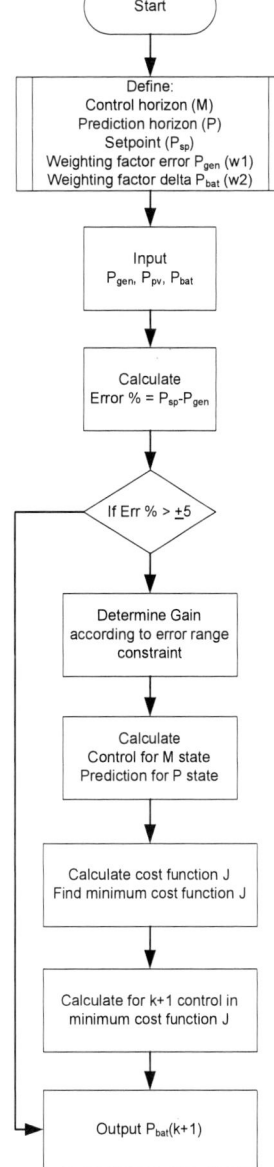

Fig. 9. Algorithm for adaptive model predictive control.

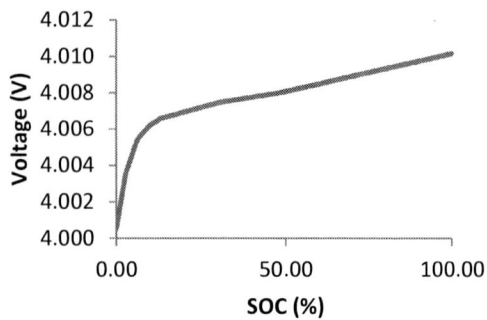

Fig. 10. OCV table for SOC.

MPC algorithm is applied to find the optimal gain for a certain error. At last, the first move calculation is implemented.

From basic control theory, typically a simpler mathematical model is chosen in order to simplify the calculation. Otherwise, the true system dynamics can be so complicated that a complete model is impossible. Therefore, the used model in this paper is expressed in (8) referring the model in [12].

$$Pb(t) = Pb_0 + G\big(Pg_{sp} - Pg(t)\big) \qquad (8)$$

Where $Pb(t)$ and Pb_0 are the output variable of battery active power at k+1 and k, respectively. G is the gain parameter and Pg_{sp} and $Pg(t)$ is the setpoint value and current value of the active power of diesel generator, respectively. With this simple model, the references are only the changes of diesel generators power transfer to have the optimize power output from battery. Therefore, it does not need data from any other power transfer to have the optimization.

C. Battery lifetime consideration

It is necessary to consider the battery lifetime in the control strategy for energy management optimization because the usage of battery is very important with this strategy. The battery lifetime is reduced when it is not properly maintained. Fig. 10 shows the characteristics of battery between the Open Circuit Voltage (OCV) and the State of Charge (SOC) of the battery. As can be seen from this figure, the battery has a limited SOC usage area to meet the necessary condition of DC link voltage. Furthermore, considering the battery lifetime, it was suggested to use the range of SOC within the range of 20 to 60%.

In other references [13], an interval with SOC set-points between 40 and 80% is suggested for lifetime dependency on SOC levels of the battery. Fig. 11 describes the control strategies that can be applied refer to [13]. If the battery SOC reaches 40%, then the battery is automatically charging constantly certain value until it reaches at 75 or 80% of the battery SOC. But if the PV power is supplying to the system, then that algorithm cannot be applied because the diesel generator response is too slow for PV power supply to maintain the stabilization of the system.

IV. SIMULATION AND RESULT

Simulation for the optimization of energy management can be summarized as follows: First, as mentioned above, the

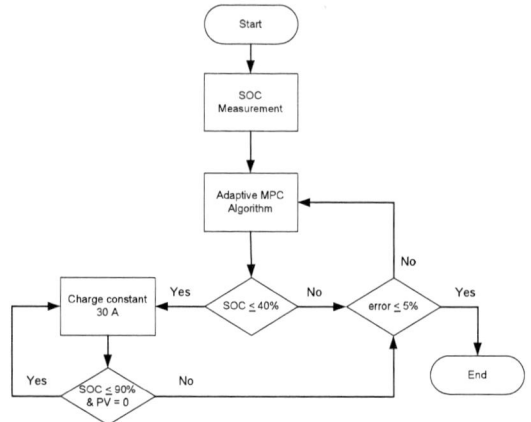

Fig. 11. Control strategies for battery.

```
----------------------------------------------------------
 i    Lambda    Ptot     P1      P2      P3     error
----------------------------------------------------------
 1    2.0306    61.2    20.0    20.0    21.2     1.2
 2    1.8275    57.6    18.8    18.8    20.0     2.4
 3    1.9643    60.4    20.0    20.0    20.4     0.4
 4    1.9451    60.2    20.0    20.0    20.2     0.2
 5    1.7506    55.8    17.9    17.9    20.0     4.2
 6    1.9376    60.1    20.0    20.0    20.1     0.1
----------------------------------------------------------
error = 0.0808769%
```

Fig. 12. Lambda iteration result for night time peak load.

PV and battery in this paper will cover 40% of the peak load in day time and night time. In day time, the peak load is 20kW, so the PV and battery are covering $\pm$8kW and the diesel generator must supply $\pm$12kW. In night time, the peak load increases to 98kW, then the PV and battery are covering +39.2kW and the diesel generation must supply $\pm$58.8kW. For simplification, the simulation carried out at 400V and 50Hz system. After then, the fuel cost optimization is simulated using the MATLAB software. From [14], the fuel cost coefficients are 0.0453, 0.2106, 0.45 where a, b, and c are intercept coefficient, slope coefficient, and fuel price coefficient. Constraint for each diesel generator can be seen in TABLE 1. The Lambda calculation for night time peak load is 1.9376. The diesel generator 4 is not operated or under the standby mode. In day time peak load, only one diesel generator operates with 12kW power supply as a base load and the others are not operated.

TABLE 1.
CONSTRAINT FOR MULTI DIESEL GENERATORS

Diesel Generators	P_{min}	P_{max}
Diesel Generator 1	8	20
Diesel Generator 2	8	20
Diesel Generator 3	20	48

The Lambda iteration results for night time power supply are shown in Fig. 12, where P1 (diesel generator 1) is in 20kW, P2 (diesel generator 2) is in 20kW, and P3 (diesel generator 3) is in 20.1kW while diesel generator 4 is in standby mode. Then, the AMPC can be applied considering the battery lifetime to maintain the operation of diesel generators at the base load. With this optimization, the total fuel cost per hour can be reduced from 247.49$/h to 68.55$/h.

The PSiM simulation model for AMPC is shown in Fig. 13, where AC source is represented as the active power supply from multi diesel generators. The PV system model includes the MPPT, a boost DC-DC converter, and a DC-AC inverter connected to the PCC. Fig. 14 shows the PSiM model of battery with AMPC. The battery model includes a bidirectional DC-DC converter and a DC-AC inverter connected to the PCC, which is done after referring the model in [15]. The AMPC inputs are the setpoints of active power supply for multi diesel generators, PV system, and battery, respectively, and the battery SOC. Weighting factors are 3 and 1 for w_e and $w_{\Delta u}$, respectively. In this simulation, there

are 5 different gains and 5 future prediction outputs that are optimized at each point of control horizon. The AMPC simulation starts at 1s.

The AMPC simulation results are shown in Fig. 15. Where the active power of multi diesel generator (P_gen) is maintained at 60% of the active power load demand as a base load in day time and night time operation. The changes in active power of battery (P_inv_bat) is following the changes of the active power of load (P_Load) and the active power of PV (P_inv_pv). The error range of simulation results is shown in Fig. 16. The error range for this AMPC is maintained in the range of ± 5%. It shows some transients occur at 8s and 20s caused by the operation change from night time to day time or vice versa.

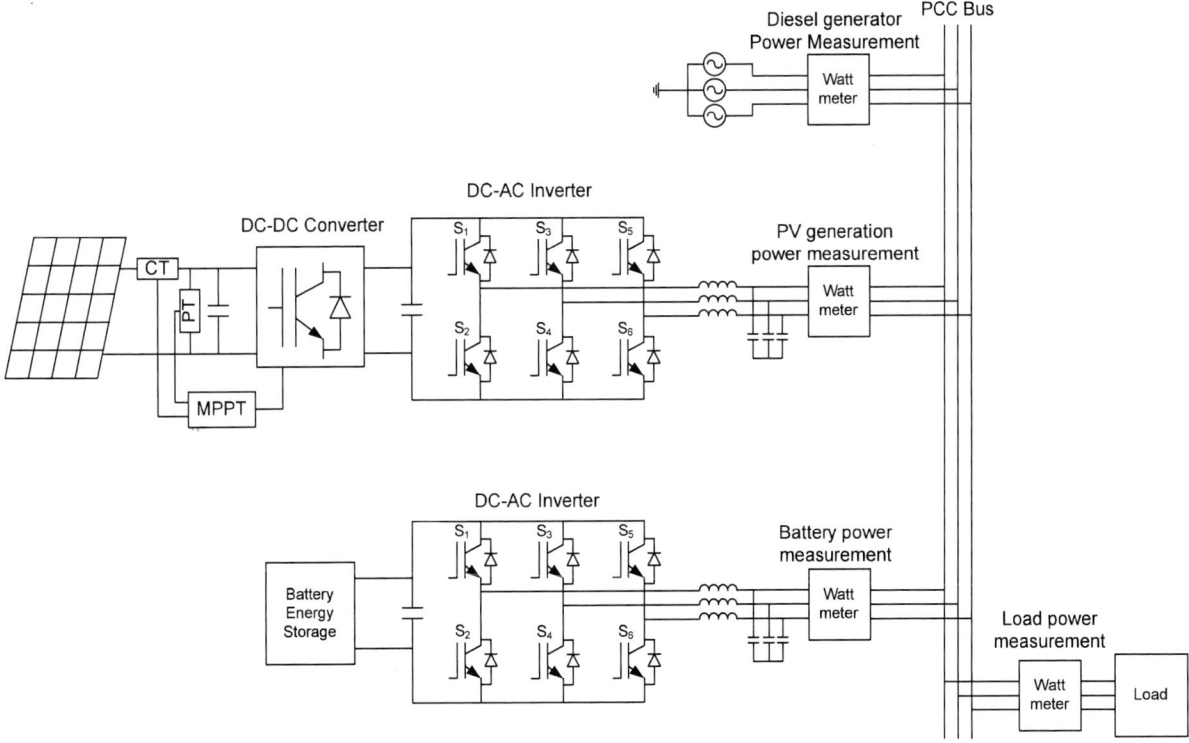

Fig. 13. PSiM simulation for microgrid system.

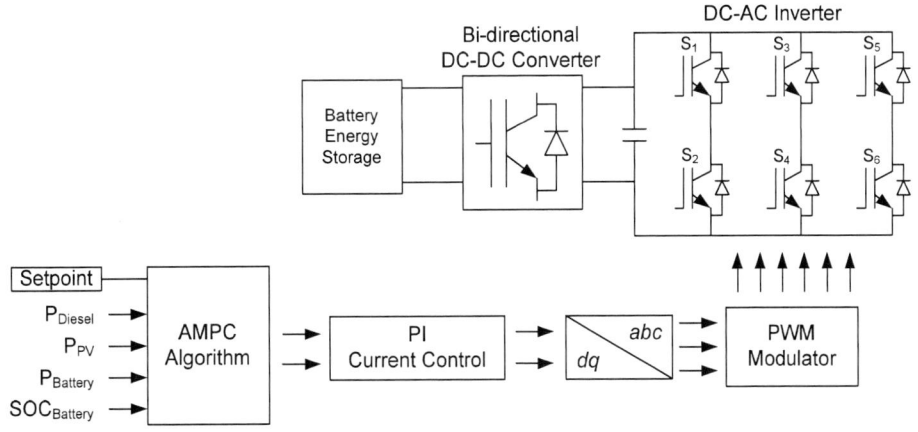

Fig. 14. PSiM simulation for AMPC based battery.

978-1-5386-7688-2/19 $31.00 © 2019 IEEE

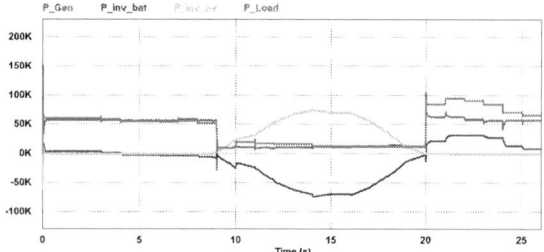

Fig. 15. Simulation result of AMPC.

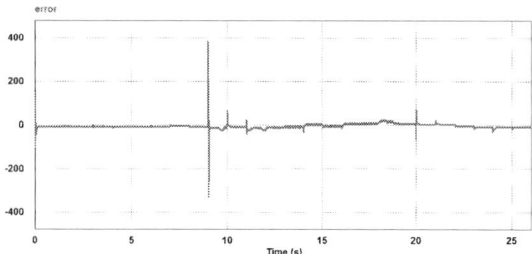

Fig. 16. Simulation result of error calculation.

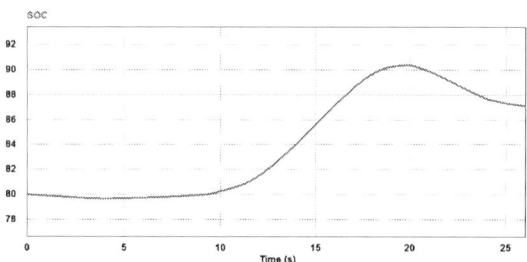

Fig. 17. Simulation result of battery SOC.

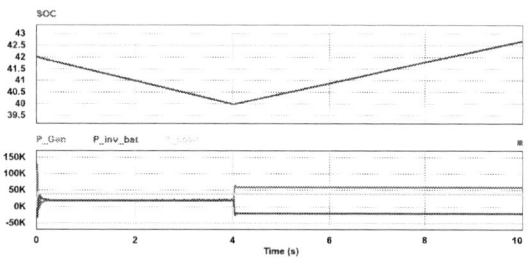

Fig. 18. Simulation result at battery mode transition.

The battery simulation results are shown in Figs. 17 and 18. Fig. 17 shows the battery SOC simulation result. During the night time operation, the battery is operated under the discharge mode while it is operated under the charging mode during the day time operation. The battery constraint simulation also has done and Fig. 18 shows the results. When the battery SOC is arrived at 40%, the battery automatically changes the mode from the discharging mode to the charging mode with constant current. Then the diesel generator will compensate the lack of power in the system.

V. CONCLUSION

Simulation results indicate that the proposed AMPC can be applied in the control strategy of microgrid energy management system where the diesel generators are only operated in local control. Economic dispatch is suitable for optimizing multi diesel generators in this microgrid system. The battery lifetime consideration is needed for this optimization because the role of battery in this strategy is very important. It is suggested to do the experimental verification for future research to have more accurate result.

ACKNOWLEDGMENT

This paper was supported by PT. PLN (Persero),Human Capital Management, 1765.K/SDM.00.03/DIR/2017, Indonesia.

REFERENCES

[1] Pawel Malysz, Shahin Siroupour, and Ali Emadi, "An Optimal Energy Storage Control Strategy for Grid-connected Microgrids," *IEEE Transactions on Smart Grid,* vol. 5, no. 4, pp. 1785-1796, 2014.

[2] Li Jing, Shen Yanxia, Wu Dinghui, and ZhaoZhipu, "A control strategy for Islanded DC Microgrid with Battery/Ultra-capacitor Hybrid Energy Storage System," *28th Chinese Control and Decision Conference (CCDC'2016),* 2016.

[3] Kyungkyu Lee, Boguen An, Muhammad Wardi Hadi, Jaeho Choi, and Yujin Soang, "Optimized Control Strategy for Hybrid Energy System," *9th International Conference on Power Electronics (ICPE'2015)-ECCE Asia,* 2015.

[4] Avijit Das, Zhen Ni, and Xiangnan Zhong, "Near Optimal Control for Microgrid Energy System Considering Battery Lifetime Characteristics," *IEEE Symposium Series on Computational Intelligence (SSCI'2016),* 2016.

[5] Alessandra Parisio, Evangelos Rikos, and Luigi Glielmo, "A Model Predictive Control Approach to Microgrid Operation Optimization," *IEEE Transactions on Control Systems Technology,* vol. 22, no. 5, pp. 1813-1827, 2014.

[6] Veronica Adetola, Darryl DeHaan, and Martin Guay, "Adaptive Model Predictive Control for Constraint Nonlinear Systems," *Systems & Control Letters,* Vol. 58, no. 5, pp. 320-326, 2009.

[7] Jinyong Lei, Xuzhu Dong, Yuyong Jiang, Zhiyong Huang, Xiaoyun Huang, Feijin Peng, and Zhan Shen, "Control Strategy and Its Dynamic Simulation of Energy Storage System in Microgrid," *China International Conference on Electricity Distribution CICED'2014),* 2014.

[8] Alyssa Diva Mustika, Rizki Rahmani, Nanang Hariyanto, and Muhammad Nurdin, "Optimized Operation Scheme of On-Grid PV Farm to Grid Case Lombok Island," *International Conference on High Voltage Engineering and Power System Indonesia (ICHVEPS'2017),* 2017.

[9] Adelhard Beni Rehiara Sabar Setiawidayat, and Elias Kondorura Bawan, "Optimal operation scheme for diesel power plant units of PT. PLN-Manokwari Branch using Langrange Multiplier Method," *3rd International Conference on Sustainable Future for Human Security (SUSTAIN'2012),* pp. 557-565, 2012.

[10] Allen J. Wood, Bruce F.Woolenberg, and Gerald B. Sheble, *Power Generation, Operation, and Control (3rd Edition,* John Wiley & Sons Inc, 2013.

[11] Zivic Djurivic, M., Milacic, A., and Krsulja, M., "A Simplified Model of Quadratic Cost Function for Thermal Generator," *Annals of DAAAM for 2012 & Proceedings of 23rd International DAAAM Symposium,* vol. 23, no.1, pp. 25-28, 2012.

[12] Dale E. Seborg, Thomas F. Edgar, Duncan A. Mellichamp, and Francis J. Doyle, *Process Dynamics and Control, 4th Edition,* John Wiley & Sons Inc, 2016.

[13] Egill Thorbergsson, Vaclav Knap, Maciej Swierczynski, Daniel Stroe, and Remus Teodorescu, "Primary Frequency Regulation with Li-ion Battery Energy Storage System– Evaluation and Comparison of Different Control Strategies," *Proceeding of the 35th International Telecommunications Energy Conference (IEEE INTELEC'2013),* pp. 178-183, 2013.

[14] Muhammad Said, and et. al., *Kajian Kelayakan Penerapan Manajemen Energi Pada Sistem Pembangkit Listrik Hybrid,* Jakarta, PLN Puslitbang, 2015.

[15] Desmon Petrus Sim atupang and Jaeho Choi, "PV Source Inverter with Voltage Compensation for Weak Grid Based on UPQC Configuration," *IEEE 18th International Power Electronics and Motion Control Conference (IEEE PEMC'2018),* 2018.

978-1-5386-7688-2/19 $31.00 © 2019 IEEE

Hybrid Storage System Associated with a Grid-Connected Wind Generator

Karima Boulaam and Akkila Boukhelifa

Instrumentation Laboratory, Faculty of Electronics and Computers
University of Sciences and Technology Houari Boumediene, Algiers, Algeria
kboulaam@yahoo.fr, aboukhelifadz@yahoo.fr

Abstract- **One of the solutions to increase the integration of wind generators to the grid is their association with a battery storage system. This later which offers the possibility of filtering the fluctuations of the generated wind power, remedies also its unpredictability. However, batteries have a limited charge/discharge cycles number, and the rapid fluctuations of wind power will reduce their lifespan. This paper proposes a hybrid energy storage system (HESS) combining batteries and supercapacitors in order to alleviate constraints on batteries. So, the whole system will be capable to satisfy at best grid power demand in terms of both quality and quantity. An appropriate supervisory control algorithm is developed to manage the hybrid storage system with the aim of adapting the generated wind power to the grid power requirement. The studied wind energy conversion system (WECS) is based on a doubly fed induction generator (DFIG), and three-level converters are used to connect the rotor to the grid. Modeling and control of the global system are developed and the control performances are analyzed through simulations on Simulink/Matlab software.**

I. INTRODUCTION

Wind power is known by its fluctuation and unpredictability; so large wind penetration causes problems of stability and electrical energy availability [1-3].

In the aim of increasing wind generators integration to the grid, the wind-caused power variations have to be erased in order to deliver to the grid a smooth power.

The energy storage systems (ESS) are considered to be an effective solution to balance the generation and demand, supporting the wind energy deficit when necessary, and storing the excess when possible, according to a given grid power demand [3].

Various types of ESS could be used to perform this function, such as batteries, flywheels, supercapacitors, fuel cells, compressed air storage and superconducting magnetic energy storage [3-5]. Long-term storage systems are suitable to overcome the uncertainty of energy availability, since they can store much energy for long-term operation; but cannot provide fast varying power due to their low dynamic. When, short-term energy storage devices can overcome the fast fluctuations of wind power, as they can provide fast and high power variations.

Batteries are the "trade-off" product which is usually used to provide medium performances in both energy density and power density, and their use for wind turbines power smoothing has been largely found in literature [6-11].

However, the principle drawback of batteries is their limited charge/discharge cycles number, and the fast variations of wind power decrease their lifespan. Moreover, high wind power variations can lead to an over-sizing of the storage system.

Different methods have been proposed by researchers to extend batteries lifespan for wind energy applications. In [6], a rule based control strategy has been proposed considering the SOC (state of charge) and the depth of discharge limitations. In [9], they proposed a control strategy based on multi-branch battery-bank configuration using switches to connect just the necessary number of branches according to the required power from the control system.

With the objective of meeting rapid and high wind power variations, while maximizing batteries lifespan, we propose to combine them with supercapacitors (SCs). These later are the best candidates as fast dynamic energy storage devices, particularly for smoothing fluctuant energy production, like wind energy generators. Compared to batteries, SCs are capable of very fast charges and discharges, and can achieve a very large number of cycles without degradation [12-14].

The use of a supercapacitor storage system (SCSS) in combination with the battery storage system (BSS) should reduce considerably constraints on batteries as well as their size. It has been already proposed by researchers in different energy supply systems [15-18].

The considered wind energy generator associated with a battery/SC-based hybrid energy storage system (HESS) is described in section 2, followed by the presentation of the adopted models of the main system components, namely the generator and the two types of storage systems. In section 3, the proposed control structure of the whole system is presented; where control schemes of both stator powers and storage systems currents are detailed. In section 4, a supervisory control algorithm is developed to determine the power references must be generated by/stored in the energy storage systems. Finally, simulation tests are carrying out under Matlab/Simulink software to analyze the performances of the whole system towards the proposed control and supervisory algorithms to achieve the desired objectives.

II. WIND - BATTERY/SC ENERGY CONVERSION SYSTEM DESCRIPTION AND MODELING

The whole system structure is represented in Fig. 1. It consists of a wind generator based on a doubly fed induction

978-1-5386-7688-2/19 $31.00 © 2019 IEEE

generator (DFIG). Two three-level PWM bidirectional power converters are used to connect rotor winding to the grid. These converters are linked by a two capacitors DC-bus. The hybrid energy storage system (HESS) which combines batteries and supercapacitors (SCs) is joined to the wind generator via the DC-bus by two DC/DC bidirectional power converters (choppers). Each chopper controls the power flow of the storage unit to which it is linked according to an appropriate control algorithm.

A. Modeling of the wind turbine

The turbine model is based on the aerodynamic power it develops, given by the following expression [19-22]:

$$P_{aer} = \frac{1}{2} \cdot C_p(\lambda, \beta) \cdot \rho \cdot \pi \cdot R^2 \cdot v^3 \tag{1}$$

$$\lambda = \frac{R \cdot \Omega_{turb}}{v} \tag{2}$$

where ρ is the air density; v is the wind speed; C_p is the power coefficient; λ is the tip-speed ratio; β is the blade pitch angle; R is rotor radius and Ω_{turb} is the turbine rotational speed.

Various expressions of power coefficient C_p have been found in literature. In this paper, it is expressed by [22]

$$C_p = (0.5 - 0.00167(\beta - 2)).\sin\left(\frac{\pi(\lambda + 0.1)}{18.5 - 0.3(\beta - 2)}\right) - 0.00184(\lambda - 3)(\beta - 2). \tag{3}$$

B. Generator and its Converters

The wind generator is based on a doubly fed induction machine (DFIG) for which a simplified dynamic model is used. This model adopts the oriented-flux strategy defined in the synchronous reference frame (d-q) fixed to the stator flux [19]

$$\begin{cases} V_{sd} = 0 \\ V_{sq} = V_s = \omega_s \Phi_s \end{cases} \tag{4}$$

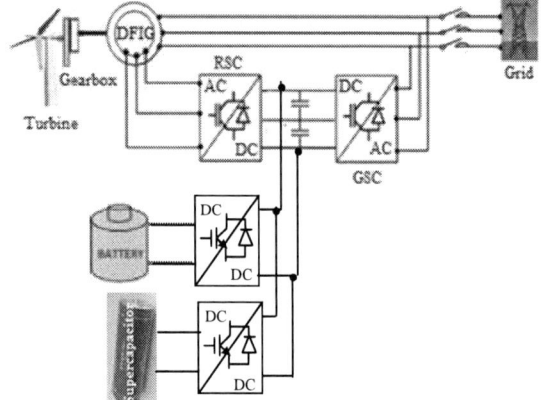

Fig. 1. Battery/SC-based HESS associated with a DFIG-based wind generator.

$$\begin{cases} V_{rd} = R_r I_{rd} + \sigma L_r \frac{d}{dt} I_{rd} - s\omega_s \sigma L_r I_{rq} \\ V_{rq} = R_r I_{rq} + \sigma L_r \frac{d}{dt} I_{rq} + s\omega_s \sigma L_r I_{rd} + s\frac{M}{L_s} V_s \end{cases} \tag{5}$$

$$\begin{cases} P_s = -V_s \frac{M}{L_s} I_{rq} \\ Q_s = -V_s \frac{M}{L_s} I_{rd} + \frac{V_s^2}{\omega_s L_s} \end{cases} \tag{6}$$

where s (r) is stator (rotor) index, d (q) is synchronous reference frame index, V (I) is voltage (current), Φ is a flux, $P(Q)$ is active (reactive) power, R is a resistance, L (M) is an inductance (mutual inductance), ω (ω_s) is angular speed (synchronous speed), s is the slip and σ is the leakage coefficient ($\sigma = 1 - M^2/L_sL_r$).

The rotor winding is fed by two bidirectional PWM three-level converters, commonly named rotor side converter (RSC) and grid side converter (GSC) (as shown in Fig. 1).

The use of three-level converters reduces the voltage stress on the switching devices and delivers a less harmonic distortion output voltage.

The structure of a three-level NPC (Neutral Point Clamped) inverter is presented in [23] and [24].

C. Batteries Modeling

Lead-acid batteries have been considered. Comparing with other types, this type of batteries is the most used since it presents low self discharge, cost effectiveness, high specific power and good temperature performance [4], [11].

Among different models found in literature [7], [25], we have adopted the model represented in Fig. 2, where I_b, I_{MR}, and I_G are respectively the battery current, the reaction main current and the gasification current; T_b, C_b, V_b and SOC are respectively the temperature, the capacity, the voltage and the state of charge of the battery. Each block of this battery model is defined by an equation system developed in [25].

D. Supercapacitors modeling

We have considered the "two RC branches" model [16]. The equivalent electric circuit is presented in Fig. 3.

The main branch (R_1C_1) determines the immediate behavior of the supercapacitor during rapid charge and discharge cycles in a few seconds, when the slow branch (R_2C_2) completes the first branch in longtime range in the order of a few minutes, and describes the internal energy distribution at the end of the charge (or discharge). The equivalent parallel resistance R_f represents the leakage current and can be neglected during fast charge/discharge of the supercapacitor.

E. Storage systems converters

Two-quadrant DC/DC PWM converters are used to connect storage devices to the wind generator. These bidirectional choppers will ensure the power flow of batteries and supercapacitors while adapting their voltage to that of the DC-bus. Fig. 4 illustrates the electrical scheme of a two-quadrant

DC/DC converter, where V_{cho} and I_{cho} are the modulated voltage and current of the chopper, V and I are respectively the storage device voltage and current, and V_{dc} is the DC-bus voltage.

III. CONTROL SCHEME OF THE CONVERSION SYSTEM

The proposed control scheme of the wind generator associated to battery/supercapacitor hybrid storage system is represented in Fig. 5. The rotor-side converter (RSC) controls stator powers via rotor voltages; when grid-side converter (GSC) controls DC-bus voltage and line currents (flowing out between the converter and the grid). DC/DC converters control the power flows of storage units through the DC-bus so that they supply or absorb their reference powers. The latter are delivered by the supervisory control system. The aim is to smooth the fluctuating wind power in order to satisfy the grid power requirement.

A. Generator control

To be able to easily control the wind power generation, an independent control of stator active and reactive powers has been realized using the generator model presented in section II.B. Equation (6) shows that stator active and reactive powers P_s and Q_s can be controlled separately by rotor current d, q components I_{rq} and I_{rd} respectively.

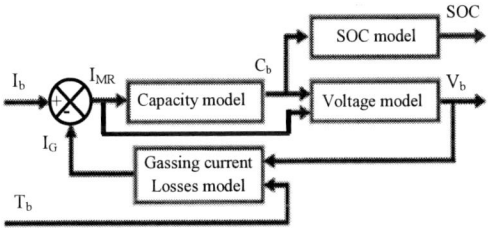

Fig. 2. General structure of the battery model.

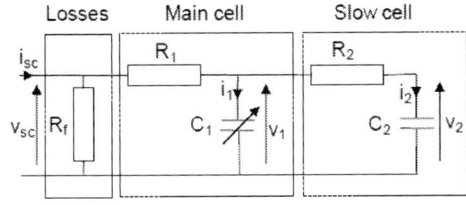

Fig. 3. Equivalent circuit of supercapacitor model.

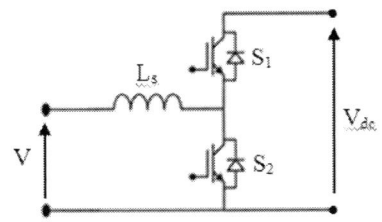

Fig. 4. Two-quadrant DC/DC converter

This control is made via rotor voltage components V_{rq} and V_{rd} according to (5).

The block diagram of stator powers control is shown in Fig. 6, where powers' references are chosen so that to ensure global energy efficiency. So, desired active power, noted $P_{s\text{-}ref}$, corresponds to maximum power point (MPPT) for values below rated power; otherwise it is set to the latter. Concerning reactive power reference $Q_{s\text{-}ref}$, it is set equal to zero in order to operate at unitary power factor on stator side.

B. Grid side converter control

The GSC control serves to regulate the DC-bus voltage to a constant value regardless of rotor power flow direction.

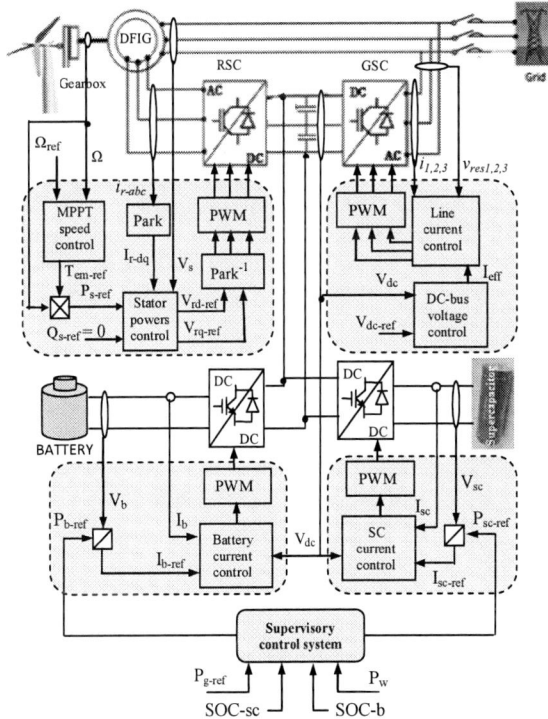

Fig. 5. Wind-battery/SC-based HESS control scheme.

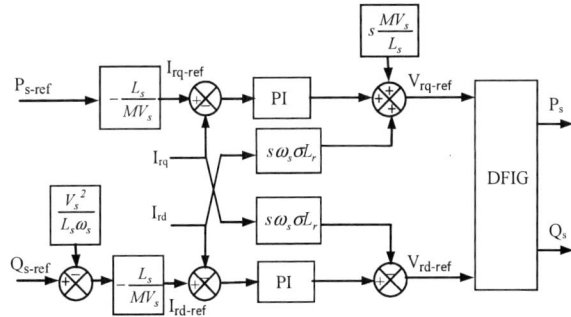

Fig. 6. Generator's stator powers control scheme.

978-1-5386-7688-2/19 $31.00 © 2019 IEEE

The control algorithm contains three inner loops regulating line currents so that to operate at unitary power factor on rotor side and an outer loop regulating the DC-bus voltage [24].

C. Storage systems control algorithm

The storage systems BSS and SCSS are both controlled each by its shopper. This control allows storage units to adjust their powers to the reference values (P_{b-ref} and P_{sc-ref}) determined by the supervisory control system.

The same control scheme is adopted for the two storage systems. The block diagram is represented in Fig. 7.

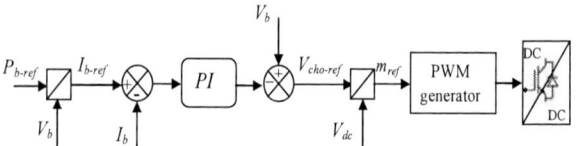

Fig. 7. Control scheme of the battery and its converter.

IV. SUPERVISORY CONTROL SYSTEM

The supervisory control system allows determining reference powers of storage units. Three stages have been considered:

1) First, the difference ΔP between power grid requirement P_{g-ref} and available wind power P_w is estimated.

2) Second, for a good management of HESS, each storage device has to be requested according to its storage characteristics, i.e. SC-bank will be called to compensate rapid variations as well as high values of wind power, since SC is short-term storage device characterized by high dynamics. While the battery-bank will be solicited for slow variations of wind power, as battery is long-term storage device characterized by slow dynamics.

To do this, power difference ΔP is decomposed into two signals, namely a high frequency signal noted ΔP_{hf} and a signal at low frequencies noted ΔP_{lf}, by means of a low-pass filter (LPF).

3) Third, according to their states of charge (SOC), reference powers of BSS and SCSS (P_{b-ref} and P_{sc-ref} respectively) are set to ΔP_{bf} and ΔP_{hf} (respectively) or to zero.

For efficiency and security reasons, storage levels are limited between usually allowed values. So, the battery-bank SOC is limited from 30% to 70% of the total capacity, and the SC-bank SOC from 50% to 100% of its voltage terminal.

The flow diagram of the proposed supervisory control system is depicted in Fig. 8.

In order to achieve a suitable operation of wind- battery/SC conversion system, the following scenarios have been considered:

- If the SOC of one storage unit is within its thresholds, it is in normal operating mode, i.e. it absorbs the power difference component that has been assigned to it if this amount is positive (charging mode) or generates it if it is negative (discharging mode).

- If the SOC of one storage unit reaches its maximum limit, it is only allowed to discharge, and if it reaches its minimum limit, only the charge process is allowed.

- In the case where it is the SCSS that reaches its maximum or minimum storage level, the BSS will ensure the corresponding process of charge or discharge with the totality of power difference as long as its storage level allows it.

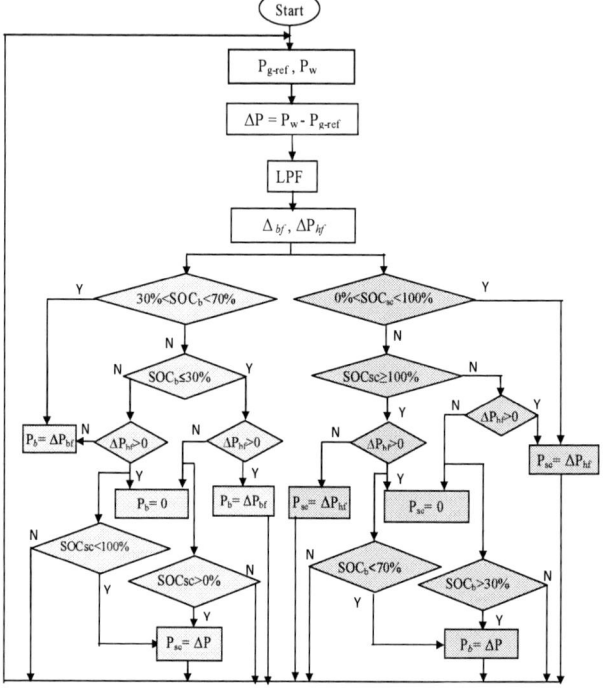

Fig. 8. Flow diagram of the supervisory control system.

- If battery SOC is too high to absorb the power difference, we propose to enslave the wind generator power to the grid power requirement.

The power delivered to the grid is calculated then as follow:

$$P_g = P_w + P_b + P_{sc} \qquad (7)$$

with:

$$\begin{cases} P_w = P_s + P_r \\ P_r = -s \cdot P_s \end{cases} \qquad (8)$$

where P_w is the wind generator power, P_s and P_r are generator's stator active and reactive powers respectively, and s is the generator slip.

V. SIMULATION AND RESULTS

To validate the proposed supervisory control algorithm and evaluate the behavior of conversion chain, simulation tests using Matlab/Simulink software have been applied on a 1.5MW rated power DFIG-based wind generator. Conversion system parameters are given in Appendix.

Wind profile used in simulation is depicted in Fig. 9. Simulation results showing performances of the whole wind conversion system towards the proposed control algorithms are presented from Fig. 10 to Fig. 21.

Fig. 10 illustrates: stator power and its reference, rotor power and the produced wind power. The latter, as we can see, is fluctuating and does not meet grid needs (Fig. 11).

In Fig. 12, we have reported the power produced by the whole system, i.e. the wind generator in combination with the hybrid storage system. So, we can clearly observe that this power is smooth and satisfies at best grid demand.

Energy storage systems performances, namely battery-bank and SC-bank, are respectively presented from Fig. 13 to Fig. 16 and from Fig. 17 to Fig. 20. These figures show the evolution of different quantities characterizing storage devices. From Fig. 13 and Fig. 17, we note that both batteries power and SCs power follow perfectly their references delivered by supervisory control system. In fact, these powers reflect images of storage currents represented by Fig. 14 and Fig. 18 which explains the good behavior of their controllers. In addition, ESSs powers evolution shows that battery-bank has supported the slow variations of the power to be compensated by the HESS, while SC-bank has been responsible for rapid variations of this power. So, batteries have presented a continuous charge until 70s, where their voltage has increased by 1V (Fig. 15) and their SOC has undergo an increase of 0.8 % (Fig. 16). At the same time, SCs have marked several phases of charge and discharge, reflected by their voltage (Fig. 19) which has presented increasing and decreasing variations between 380V and 530V, as well as their SOC (Fig. 20) between 10 % and 50 %. Then, between the moments 70s and 73s, where the wind speed has passed from 10m/s to 7m/s, wind power has decreased from 1.2MW to 250kW. During this time, SC-bank which was only 18% charged, has been completely discharged. Consequently, battery-bank has compensated the power peak by a fast discharge indicated by a 0.1% decrease in its SOC and 5V decrease in its voltage. At 112s, when wind power has reached 2MW, batteries have arrived at their maximum allowable power (750W), so they have continued their charge with constant current until 120s. During this time, SCs have compensated power difference between grid set power and power transmitted by batteries resulting in a storage level of 80% of their total capacity. As this fact does not happen much in practice, it cannot affect good operating of HESS.

Finally, DC-bus voltage is shown in Fig. 21. It is noted that it oscillates around its reference which is equal to 1400V, throughout simulation duration.

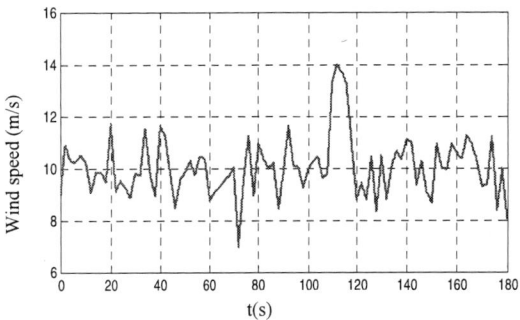

Fig. 9. Wind profile.

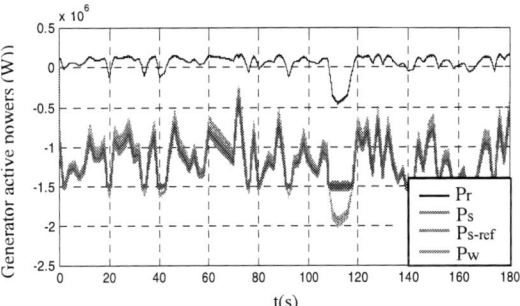

Fig. 10. Actual and reference stator power, rotor power and wind power.

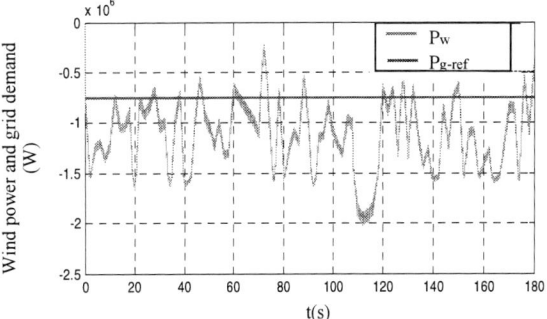

Fig. 11. Wind generator power and grid demand.

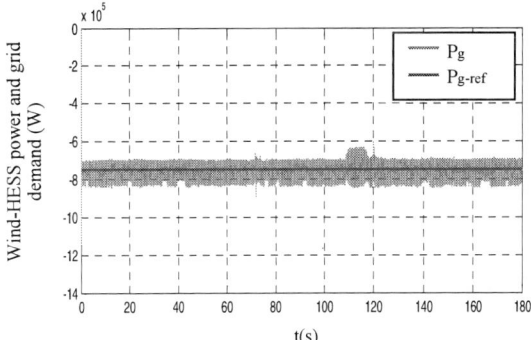

Fig. 12. Wind-Battery/SC HESS power and grid demand.

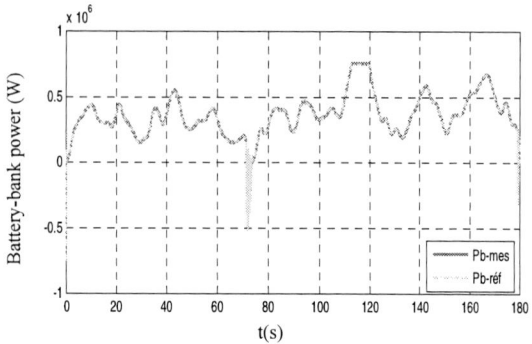

Fig. 13. Battery-bank actual and reference powers.

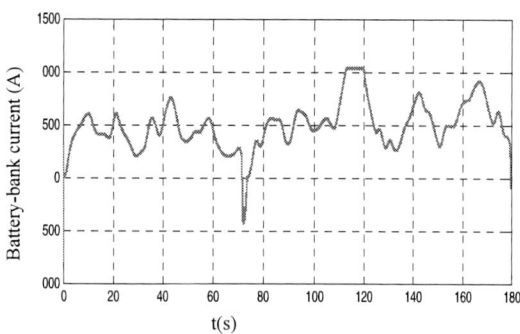

Fig. 14. Battery-bank current.

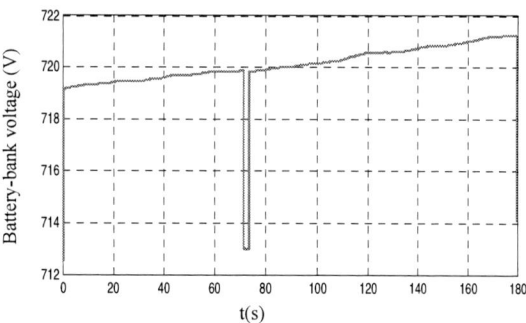

Fig. 15. Battery-bank voltage evolution.

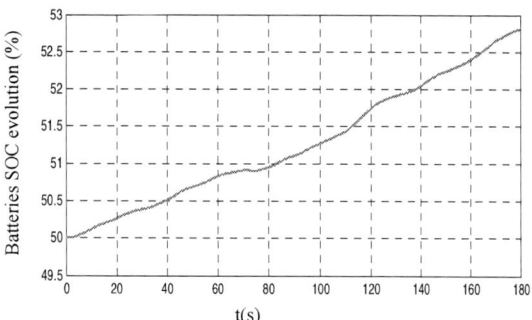

Fig.16. Batteries SOC evolution.

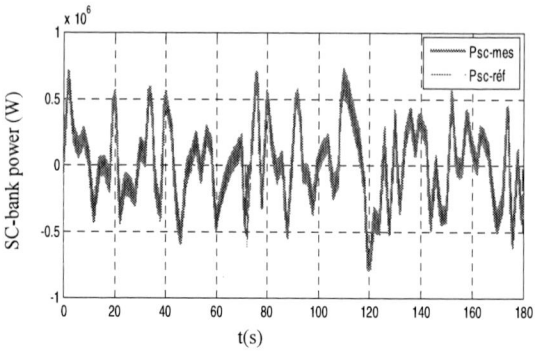

Fig. 17. SC-bank actual and reference powers.

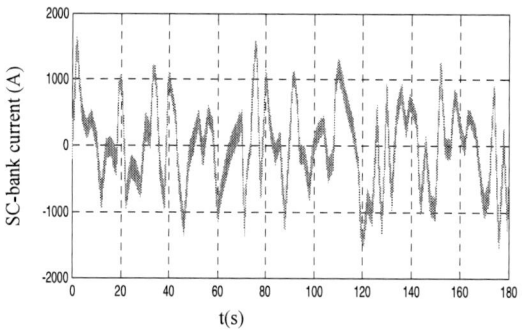

Fig. 18. SC-bank current.

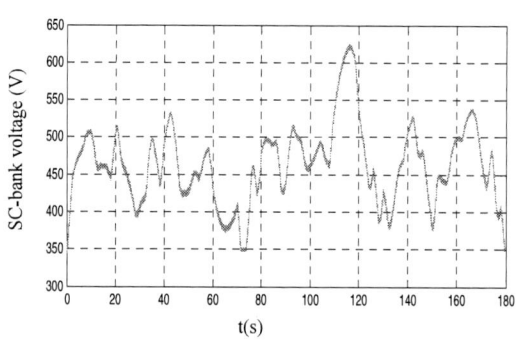

Fig. 19. SC-bank voltage evolution.

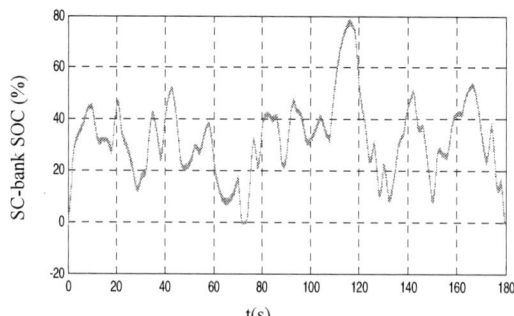

Fig. 20. SC-bank SOC evolution.

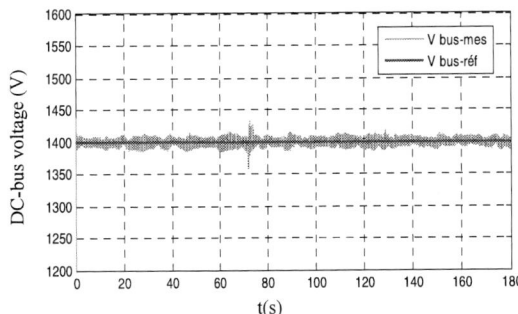

Fig. 21. DC-bus voltage evolution.

VI. CONCLUSION

In this paper, a wind energy conversion system associated with a battery/SC-based hybrid storage system has been studied. The objective was to solve the wind power fluctuation problem and ensure good energy availability while reducing constraints on batteries by means of supercapacitors. The proposed configuration uses a doubly fed induction generator and three level converters to benefit from their advantages. In order to optimize wind power, active and reactive powers of wind generator have been independently controlled using the stator oriented field control. The main part in this work is the supervisory control system necessary for hybrid energy storage management. It has allowed both batteries and supercapacitors to operate in the appropriate mode.

Simulation results have shown satisfactory behavior of the whole system, so while battery-bank has been solicited for slow variations in wind power, SC-bank has been called to compensate rapid variations caused by wind power as well as high power values. As a result, wind-battery/SC HESS system has been able to satisfy grid power requirement by smoothing the wind power in a manner which increases batteries lifespan.

APPENDIX
WIND GENERATOR AND STORAGE SYSTEMS PARAMETERS

TABLE I
WIND TURBINE PARAMETERS

Rated power (MW)	ρ (kg/m³)	R (m)	G	J_t (kg.m²)	f_t (N.m/s)
1.5	1.225	30.5	90	1000	0.0024

TABLE II
GENERATOR PARAMETERS

Rated power (MW)	p	V_s (V)	R_r (Ω)	L_s (H)	L_r (H)
1.5	2	690	0.021	0.0137	0.0136

TABLE III
BATTERY PARAMETERS

Rated capacity (A.h)	Rated voltage (V)
118	12

TABLE IV
SUPERCAPACITORS MODEL PARAMETERS

Rated capacitance (F)	Rated voltage (V)	R1(mΩ)	C0 (F)	Cv (F/V)	R2(Ω)	C2 (F)
3000	2.7	0.36	2100	623	1.92	172

REFERENCES

[1] P. S. Georgilakis, "Technical challenges associated with the integration of wind power into power systems," *Renewable and Sustainable Energy Reviews*," vol. 12, iss. 3, pp. 852–863, April 2008.

[2] R. D. Fernández, R. J. Mantz, and P. E. Battaiotto, "Impact of wind farms on a power system. An eigenvalue analysis approach," *Renewable Energy*, vol. 32, iss. 10, pp. 1676–1688, August 2007.

[3] A. M. Howlader, N. Urasaki, A. Yona, T. Senjyu, and A.Y. Saber , "A review of output power smoothing methods for wind energy conversion systems," *Renewable and Sustainable Energy Reviews*, vol. 26, pp. 135–146, Oct. 2013.

[4] F. Díaz-González, A. Sumper, O. Gomis-Bellmunt, and R. Villafáfila-Robles, "A review of energy storage technologies for wind power applications," *Renewable and Sustainable Energy Reviews*, vol. 16, iss. 4, pp. 2154–2171, May 2012.

[5] A. Rabiee, H. Khorramdel, and J. Aghaei, "A review of energy storage systems in microgrids with wind turbines," *Renewable and Sustainable Energy Reviews*, vol. 18, pp. 316–326, Feb 2013.

[6] S. Teleke, M.E. Baran, S. Bhattacharya, and A. Huang, "Validation of battery energy storage control for wind farm dispatching," *IEEE Power and Energy Society General Meeting*, Providence, RI, USA, 25–29 July, 2010.

[7] A. Shaltout and H. Gamal, "Power coordination of grid-connected wind turbine doubly fed induction generator augmented with battery storage," *IEEE International Conference on Smart Energy Grid Engineering* (SEGE'13), Oshawa, ON, Canada, 28–30 August, 2013.

[8] R. Sarrias, L. M. Fernández, C. García, and F. Jurado, "Supervisory control system for DFIG wind turbine with energy storage system based on battery," *2011 International Conference on Power Engineering, Energy and Electrical Drives*, Málaga, Spain, 11–13 May, 2011.

[9] H. Babazadeh, W. Gao, and K. Duncan, "A new control scheme in a battery energy storage system for wind turbine generators," *2012 IEEE Power and Energy Society General Meeting*, San Diego, CA, USA, 22–26 July, 2012.

[10] H. Borhan, M.A. Rotea, and D. Viassolo, "Optimization-based power management of a wind farm with battery storage," *Wind Energy*, vol. 16, iss. 8, pp. 1197–1211, November 2013.

[11] A. S. Subburaj, P. Kondur, S. B. Bayne, M. G. Giesselmann, and M. A. Harral, "Analysis and review of grid connected battery in wind applications," *2014 Sixth Annual IEEE Green Technologies Conference*, Corpus Christi, TX, USA, 3–4 April, 2014.

[12] T. Zhou and B. François, "Energy management and power control of a hybrid active wind generator for distributed power generation and grid integration," *IEEE Transactions on Industrial Electronics*, vol. 58, pp. 95-104, 2011.

[13] L. Qu and W. Qiao, "Constant power control of DFIG wind turbines with supercapacitor energy storage," *IEEE Transactions on Industry Applications*, vol. 47, pp. 359-367, 2011.

[14] C. Abbey and G. Joos, "Supercapacitor energy storage for wind energy applications," *IEEE Transactions on Industry Applications*, vol. 43, pp. 769-776, 2007.

[15] A. Tani, M. B. Camara, and B. Dakyo, "Energy Management in the Decentralized Generation Systems Based on Renewable Energy-Ultracapacitors and Battery to Compensate the Wind/Load Power Fluctuations," *IEEE Transactions on Industry Applications*, vol. 51, pp. 1817-1827, 2015.

[16] A. Lahyani, P. Venet, A. Guermazi, and A. Troudi, "Battery/supercapacitors combination in uninterruptible power supply (UPS)," IEEE transactions on power electronics, vol. 28, pp. 1509-1522, 2013.

[17] H. Babazadeh, W. Gao, J. Lin, and L. Cheng, "Sizing of battery and supercapacitor in a hybrid energy storage system for wind turbines," in *Transmission and Distribution Conference and Exposition (T&D), 2012 IEEE PES*, 2012, pp. 1-7.

[18] N. Mendis, K. M. Muttaqi, and S. Perera, "Active power management of a super capacitor-battery hybrid energy storage system for standalone operation of DFIG based wind turbines," in *Industry Applications Society Annual Meeting (IAS), 2012 IEEE*, 2012, pp. 1-8.

[19] K. Boulâam and A. Boukhelifa, "Output power control of a wind energy conversion system based on a doubly fed induction generator," *2013 International Renewable and Sustainable Energy Conference (IRSEC)*, Ouarzazate, Morocco, 7–9 March, 2013.

[20] M. Boutoubat, L. Mokrani, and M. Machmoum, "Control of a wind energy conversion system equipped by a DFIG for active power generation and power quality improvement," *Renewable Energy,* vol. 50, pp. 378–386, February 2013.

[21] S. Taraft, D. Rekioua, D. Aouzellag, and S. Bacha, "A proposed strategy for power optimization of a wind energy conversion system connected to the grid," *Energy Conversion and Management*, vol. 101, pp. 489–502, September 2015.

[22] S. El Aimani, " Modélisation de différentes technologies d'éoliennes intégrées dans un réseau de moyenne tension," *Thèse de Doctorat*, Lille 2004.

[23] M. Abbes, J. Belhadj, and A. Ben Abdelghani Bennani, "Design and control of a direct drive wind turbine equipped with multilevel converters," *Renewable Energy*, vol. 35, iss. 5, pp. 936–945, May 2010.

[24] R. Chibani, E. Berkouk, and M. Boucherit, "Input DC voltages of three-level neutral point clamped voltage source inverter balancing using a new kind of clamping bridge," *International Journal of Computer and Electrical Engineering*, vol. 2, no. 5, pp. 879–886, October 2010.

[25] B. Wichert, "Control of photovoltaic-diesel hybrid energy systems," *PhD Thesis*, Curtin University of Technology, April 2000.

978-1-5386-7688-2/19 $31.00 © 2019 IEEE

Identifying Internal Defects of Photovoltaic Panels Using Sweep Frequency Response Analysis

Kazem Pourhossein
Department of Electrical Engineering,
Tabriz Branch,
Islamic Azad University,
Tabriz, Iran
Email: k.pourhossein@iaut.ac.ir

Meysam Asadi
Department of Electrical Engineering,
Tabriz Branch,
Islamic Azad University,
Tabriz, Iran
Email: stu.Meysam.asadi@iaut.ac.ir

Abstract—**Solar irradiation is the main energy source which can sustainably provide all of energy needs on the earth. Because of fast progress in direct conversion of solar energy to electricity, photovoltaic (PV) converters are going to become major devices to supply world energy requirements. Similar to other electricity generators, PV panels may be defected in production stage or in the field. Thus, it is necessary to detect these defects as soon as possible. This paper proposes a frequency domain approach to diagnose defects of PV panels. The proposed method investigates effect of various defects of PV panels on their frequency domain characteristics.**

Keywords— Photovoltaic Panel, Defect, Frequency Domain Characteristic, Ac Equivalent Circuit

I. INTRODUCTION

Over the last decade, more attention has been paid to photovoltaic (PV) energy systems because they are silent, modular, easy to install and less pollutant [1]. PV systems are reliable energy conversion systems and can be installed in energy farms or in buildings.

PV panels are heart of PV energy systems therefore occurrence of any fault in them greatly impedes the performance, reliability and safety of the system [2-4].

As any other power conversion system, PV panels may encounter various anomalies. To detect these anomalies, there are two main classes of diagnostic methods [5-6]: Visual and thermal methods and Electric methods. The first class is used to detect anomalies like color changes, hot spot and other similar ones.

The second class is used to detect faulty PV panels, ground fault, etc [7,8]. Electric methods like calculation of solar panels' DC equivalent circuit parameters [9-11], estimating PV series resistance and impedance spectroscopy [12,13] are recent methods in PV anomaly detection.

This paper aims to detect internal defects of PV panels via sweep frequency response analysis. AC equivalent circuit of PV panels is used to illustrate diagnostic capability of the proposed approach.

II. EQUIVALENT CIRCUIT OF A PV MODULE

Figure 1 represents DC and AC equivalent circuits of a PV module. It is seen that a PV module is modeled as a current source, a diode, two resistors, a capacitor and an inductor. The shunt resistor represents the leakage current and the series resistor shows all series resistance in current pass [7]. Capacitor and inductor model parasitic capacitance and magnetic flux in the module.

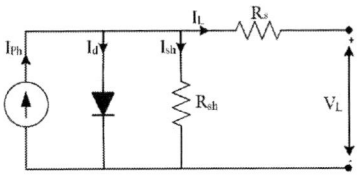

(a) DC Equivalent circuit

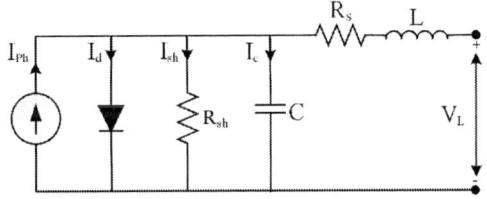

(b) AC Equivalent circuit

Fig. 1. Equivalent circuit of a PV module

III. PROPOSED METHOD

Any defect in a PV module changes one or more elements in equivalent circuit (Table I), thus it can be unfolded by frequency sweep of the PV module [14-15]. In PV frequency sweep, a constant and low amplitude sinusoidal voltage with swept frequency is injected to PV panel and then input current is measured. Division of input current amplitude by injected voltage (in frequency domain) in used here in diagnostic process as the frequency response of PV panel. Intact frequency response of a PV panel is considered as its fingerprint i.e. any change in its frequency response trace can be interpreted as an anomaly.

TABLE I.   RELATIONS OF EQUIVALENT CIRCUIT AND ANOMALIES [14-15]

Element in equivalent circuit	Anomaly
Increased series resistance (R_s)	Encapsulation Delamination Discoloration Back sheet adhesion loss
Decreased shunt resistance (R_{sh})	Hot spot
Increased series resistance (R_s) & Decreased shunt resistance (R_{sh})	Corrosion Partial shading Dust/bird dropping/leaves
Bypass diode (D)	Open-circuit bypass diode short-circuit bypass diode
Variation of Capacitor(C)	Variation of module thickness or p-n junction
Variation of inductance (L)	Relative displacement of PV modules

978-1-5386-7688-2/19 $31.00 © 2019 IEEE

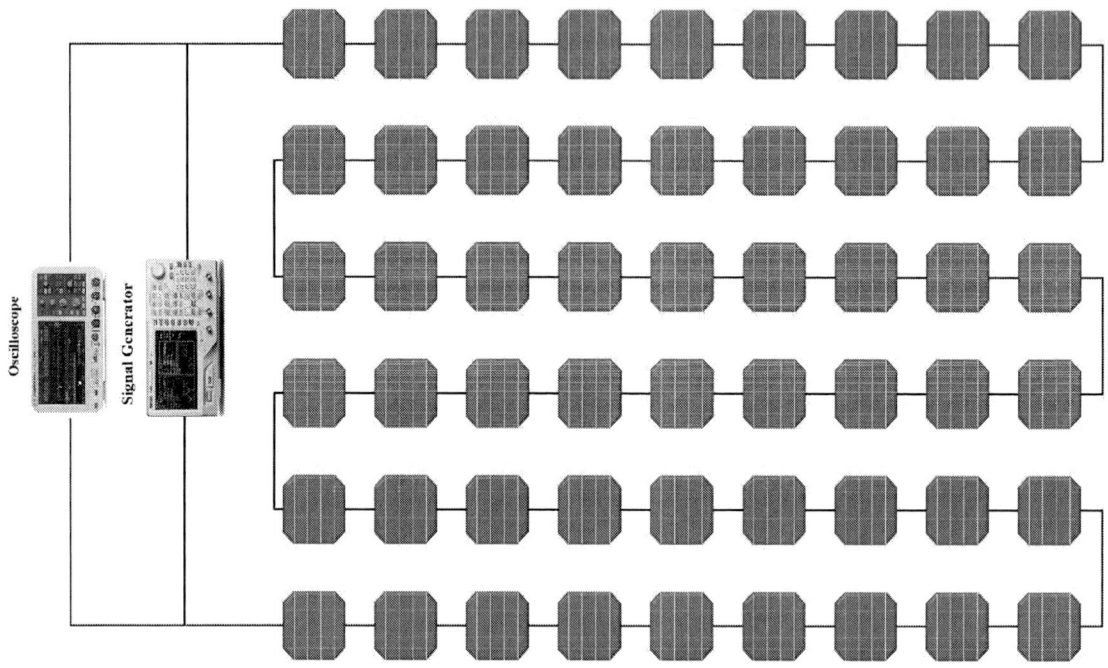

Fig. 2. Simulation diagram

IV. CASE STUDY

To show capability of the proposed method, Kyocera-KC200GT PV panel has been simulated in electromagnetic transient program (EMTP). The simulation diagram has been presented in Figure 2.

KC200GT panel is composed of 54 photovoltaic modules in series. This panel can generate current up to 8.21A and voltage up to 32.9V. Specifications of Kyocera-KC200GT PV panel are given in Table II [16-19]. As mentioned earlier, any physical change in modules of the PV panel can be reflected in its frequency response traces. To show this fact, frequency response of KC200GT PV panel is simulated in the following conditions:

- Increased series resistance

- Decreased shunt resistance

- Increased series resistance and decreased shunt resistance

- Diode fault

- Capacitance change

The base capacitance is considered 1nF. Inductance matrix of the panel has been presented in appendix.

Simulation results are depicted in Figure 3. Due to Table I it can be seen that, variations in any element of the equivalent circuit changes frequency response trace of the panel thus easily can be detected visually.

TABLE II. PARAMETERS OF KYOCERA-KC200GT PV MODULE

Parameters	Values
Maximum Power (P_{Max})	200.143 W
Maximum Power Voltage (V_{MP})	26.3 V
Maximum Power Current (I_{MP})	7.61 A
Open Circuit Voltage (V_{OC})	32.9 V
Short Circuit Current (I_{SC})	8.21 A
Temperature coefficient of open circuit voltage (K_V)	- 0.1230 V/K
Temperature coefficient of short circuit current (K_I)	0.0032 A/K
Number of cells per module (n_s)	54
Series Resistance (R_S)	0.221Ω
Parallel Resistance (R_{Sh})	415.405 Ω
Diode ideality factor (a)	1.3

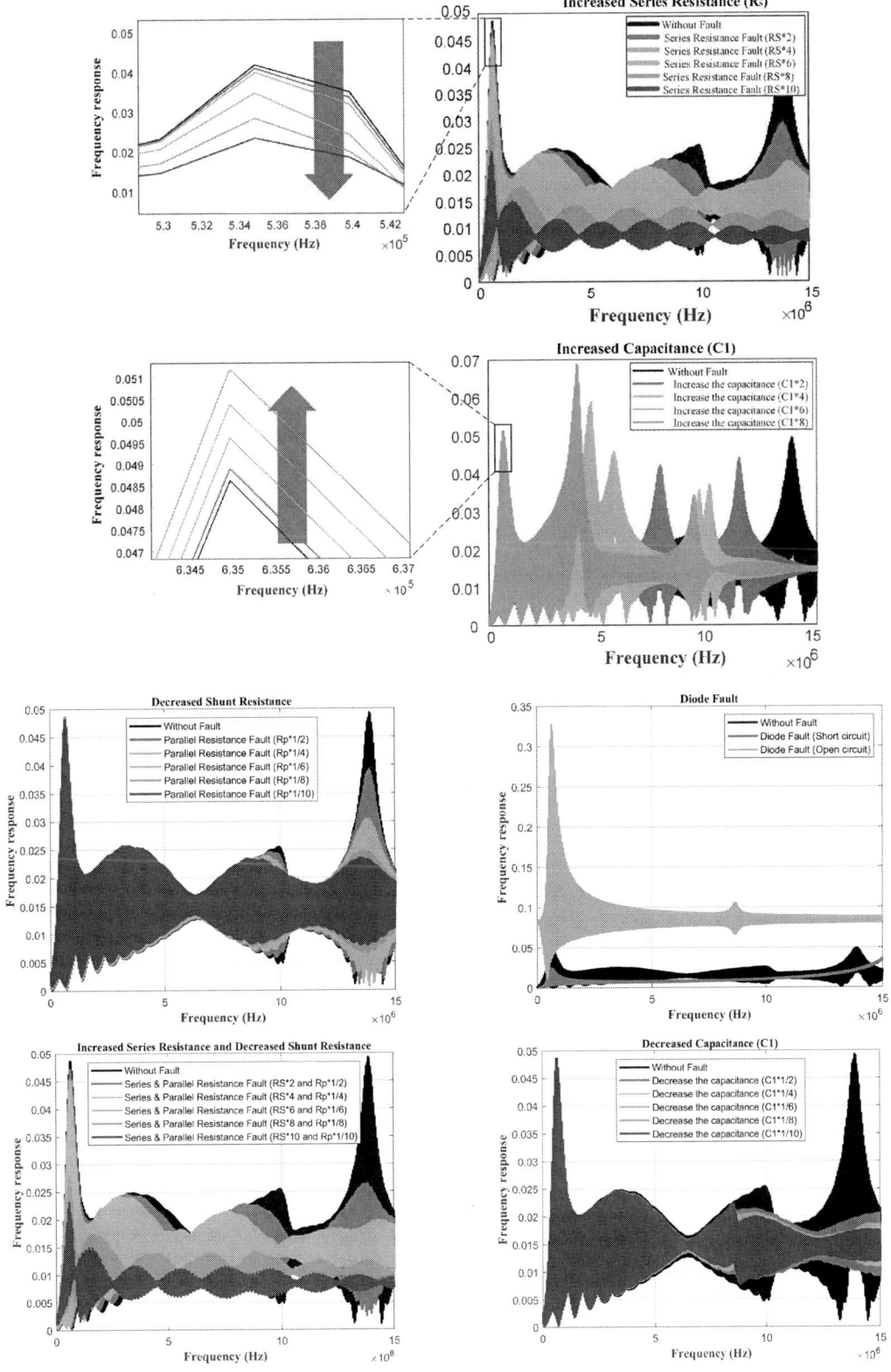

Fig. 3.   Simulation results with variations in elements of the PV panel

V. CONCLUSIONS

Utilization of sweep frequency response of photovoltaic panels has been investigated in this paper to detect internal anomalies of them. Simulations show that any defect in photovoltaic panels can be diagnosed using this approach if it affects the equivalent circuit elements. Case study of Kyocera-KC200GT PV module presents capability of the proposed method to reveal internal changes in PV panels. As shown in Figure 3, any change in a PV panel has a regular effect on its frequency response trace. Comparing any newly measured trace by the intact one (fingerprint of PV panel), leads to a decision about existence of anomalies in the tested PV panel.

REFERENCES

[1] Mellit A, Tina G, Kalogirou S. Fault detection and diagnosis methods for photovoltaic systems: areview. Renew Sustain Energy Rev 2018; 91:117

[2] Pillai DS, Rajasekar N. A comprehensive review on protection challenges and fault diagnosis inPV systems. Renew Sustain Energy Rev 2018; 91:18–40.

[3] Karim, I. A. (2015, May). Fault analysis and detection techniques of solar cells and PV modules. In *2015 International Conference on Electrical Engineering and Information Communication Technology (ICEEICT)* (pp. 1-4). IEEE.

[4] Phinikarides, A., Kindyni, N., Makrides, G., & Georghiou, G. E. (2014). Review of photovoltaic degradation rate methodologies. *Renewable and Sustainable Energy Reviews*, *40*, 143-152

[5] Harrou F, et al. Reliable fault detection and diagnosis of photovoltaic systems based on statisticalmonitoring approaches. Renew Energy 2018; 116:22–37.

[6] Madeti SR, Singh SN. A comprehensive study on different types of faults and detection techniquesfor solar photovoltaic system. Solar Energy. 2017 Dec 1; 158:161-85

[7] Yu, K., Qu, B., Yue, C., Ge, S., Chen, X., & Liang, J. (2019). A performance-guided JAYA algorithm for parameters identification of photovoltaic cell and module. *Applied Energy*, *237*, 241-257.

[8] Ndiaye, A., Charki, A., Kobi, A., Kébé, C. M., Ndiaye, P. A., & Sambou, V. (2013). Degradations of silicon photovoltaic modules: A literature review. Solar Energy, 96, 140-151.

[9] Khezzar, R., Zereg, M., & Khezzar, A. (2009, November). Comparative study of mathematical methods for parameters calculation of current-voltage characteristic of photovoltaic module. In 2009 International Conference on Electrical and Electronics Engineering-ELECO 2009 (pp. I-24). IEEE.

[10] Şentürk, A. (2018). New method for computing single diode model parameters of photovoltaic modules. Renewable energy, 128, 30-36.

[11] Ayang, A., Wamkeue, R., Ouhrouche, M., & Saad, M. (2018, October). Faults Diagnosis and Monitoring of a Single Diode Photovoltaic Module Based on Estimated Parameters. In 2018 IEEE Electrical Power and Energy Conference (EPEC) (pp. 1-6). IEEE.

[12] Bastidas-Rodriguez, J. D., Franco, E., Petrone, G., Ramos-Paja, C. A., & Spagnuolo, G. (2015). Model-based degradation analysis of photovoltaic modules through series resistance estimation. IEEE Transactions on Industrial Electronics, 62(11), 7256-7265.

[13] Bora, B., Sastry, O. S., Singh, R., Bangar, M., Rai, S., Singh, Y. K., ... & Kuber, R. (2016). Series resistance measurement of solar PV modules using mesh in real outdoor condition. Energy Procedia, 90, 503-508.

[14] Zhao, Y. (2010). Fault analysis in solar photovoltaic arrays.

[15] SabbaghpurArani M, Hejazi MA. The comprehensive study of electrical faults in PV arrays. JElectrComputEng 2016.

[16] Aloui, F., &Dincer, I. (Eds.). (2018). *Exergy for A Better Environment and Improved Sustainability 2: Applications.* Springer.

[17] Azizi, A., Logerais, P. O., Omeiri, A., Amiar, A., Charki, A., Riou, O.&Durastanti, J. F. (2018). Impact of the aging of a photovoltaic module on the performance of a grid-connected system. *SolarEnergy*, *174*, 445-454.

[18] Huang, C., & Wang, L. (2018). Simulation study on the degradation process of photovoltaic modules. *Energy conversion and management, 165*, 236-243.

[19] (2010) The KC200GT module datasheet. [Online]. Available:

APPENDIX

TABLE III. INDUCTANCE MATRIX OF KYOCERA-KC200GT PV PANEL ($\times 10^{-4} \mu H$) (SELF AND MUTUAL INDUCTANCES)

	M_1	M_2	M_3	M_4	M_5	M_6	M_7	M_8	M_9	M_{10}	M_{11}	M_{12}	M_{13}	M_{14}	M_{15}	M_{16}	M_{17}	M_{18}	M_{19}	M_{20}	M_{21}	M_{22}	M_{23}	M_{24}	M_{25}	M_{26}	M_{27}
M_1	927	218	83	54	40	32	26	23	20	20	22	26	31	39	51	72	113	149	77	70	56	44	36	29	25	22	19
M_2	218	927	218	83	54	40	32	26	23	22	26	31	39	51	72	113	149	113	70	77	70	56	44	36	29	25	22
M_3	83	218	927	218	83	54	40	32	26	26	31	39	51	72	113	149	113	72	56	70	77	70	56	44	36	29	25
M_4	54	83	218	927	218	83	54	40	32	31	39	51	72	113	149	113	72	51	44	56	70	77	70	56	44	36	29
M_5	40	54	83	218	927	218	83	54	40	39	51	72	113	149	113	72	51	39	36	44	56	70	77	70	56	44	36
M_6	32	40	54	83	218	927	218	83	54	51	72	113	149	113	72	51	39	31	29	36	44	56	70	77	70	56	44
M_7	26	32	40	54	83	218	927	218	83	72	113	149	113	72	51	39	31	26	25	29	36	44	56	70	77	70	56
M_8	23	26	32	40	54	83	218	927	218	113	149	113	72	51	39	31	26	22	22	25	29	36	44	56	70	77	70
M_9	20	23	26	32	40	54	83	218	927	149	113	72	51	39	31	26	22	20	19	22	25	29	36	44	56	70	77
M_{10}	20	22	26	31	39	51	72	113	149	927	218	83	54	40	32	26	23	20	20	22	26	31	39	51	72	113	149
M_{11}	22	26	31	39	51	72	113	149	113	218	927	218	83	54	40	32	26	23	22	26	31	39	51	72	113	149	113
M_{12}	26	31	39	51	72	113	149	113	72	83	218	927	218	83	54	40	32	26	26	31	39	51	72	113	149	113	72
M_{13}	31	39	51	72	113	149	113	72	51	54	83	218	927	218	83	54	40	32	31	39	51	72	113	149	113	72	51
M_{14}	39	51	72	113	149	113	72	51	39	40	54	83	218	927	218	83	54	40	39	51	72	113	149	113	72	51	39
M_{15}	51	72	113	149	113	72	51	39	31	32	40	54	83	218	927	218	83	54	51	72	113	149	113	72	51	39	31
M_{16}	72	113	149	113	72	51	39	31	26	26	32	40	54	83	218	927	218	83	72	113	149	113	72	51	39	31	26
M_{17}	113	149	113	72	51	39	31	26	22	23	26	32	40	54	83	218	927	218	113	149	113	72	51	39	31	26	22
M_{18}	149	113	72	51	39	31	26	22	20	20	23	26	32	40	54	83	218	927	149	113	72	51	39	31	26	22	20
M_{19}	77	70	56	44	36	29	25	22	19	20	22	26	31	39	51	72	113	149	927	218	83	54	40	32	26	23	20
M_{20}	70	77	70	56	44	36	29	25	22	22	26	31	39	51	72	113	149	113	218	927	218	83	54	40	32	26	23
M_{21}	56	70	77	70	56	44	36	29	25	26	31	39	51	72	113	149	113	72	83	218	927	218	83	54	40	32	26
M_{22}	44	56	70	77	70	56	44	36	29	31	39	51	72	113	149	113	72	51	54	83	218	927	218	83	54	40	32
M_{23}	36	44	56	70	77	70	56	44	36	39	51	72	113	149	113	72	51	39	40	54	83	218	927	218	83	54	40
M_{24}	29	36	44	56	70	77	70	56	44	51	72	113	149	113	72	51	39	31	32	40	54	83	218	927	218	83	54
M_{25}	25	29	36	44	56	70	77	70	56	72	113	149	113	72	51	39	31	26	26	32	40	54	83	218	927	218	83
M_{26}	22	25	29	36	44	56	70	77	70	113	149	113	72	51	39	31	26	22	23	26	32	40	54	83	218	927	218
M_{27}	19	22	25	29	36	44	56	70	77	149	113	72	51	39	31	26	22	20	20	23	26	32	40	54	83	218	927
M_{28}	19	21	24	27	32	37	44	50	52	77	70	56	44	36	29	25	22	19	20	22	26	31	39	51	72	113	149
M_{29}	21	24	27	32	37	44	50	52	50	70	77	70	56	44	36	29	25	22	22	26	31	39	51	72	113	149	113
M_{30}	24	27	32	37	44	50	52	50	44	56	70	77	70	56	44	36	29	25	26	31	39	51	72	113	149	113	72
M_{31}	27	32	37	44	50	52	50	44	37	44	56	70	77	70	56	44	36	29	31	39	51	72	113	149	113	72	51
M_{32}	32	37	44	50	52	50	44	37	32	36	44	56	70	77	70	56	44	36	39	51	72	113	149	113	72	51	39
M_{33}	37	44	50	52	50	44	37	32	27	29	36	44	56	70	77	70	56	44	51	72	113	149	113	72	51	39	31
M_{34}	44	50	52	50	44	37	32	27	24	25	29	36	44	56	70	77	70	56	72	113	149	113	72	51	39	31	26
M_{35}	50	52	50	44	37	32	27	24	21	22	25	29	36	44	56	70	77	70	113	149	113	72	51	39	31	26	22
M_{36}	52	50	44	37	32	27	24	21	19	19	22	25	29	36	44	56	70	77	149	113	72	51	39	31	26	22	20
M_{37}	39	38	35	32	28	25	22	20	18	19	21	24	27	32	37	44	50	52	77	70	56	44	36	29	25	22	19
M_{38}	38	39	38	35	32	28	25	22	20	21	24	27	32	37	44	50	52	50	70	77	70	56	44	36	29	25	22
M_{39}	35	38	39	38	35	32	28	25	22	24	27	32	37	44	50	52	50	44	56	70	77	70	56	44	36	29	25
M_{40}	32	35	38	39	38	35	32	28	25	27	32	37	44	50	52	50	44	37	44	56	70	77	70	56	44	36	29
M_{41}	28	32	35	38	39	38	35	32	28	32	37	44	50	52	50	44	37	32	36	44	56	70	77	70	56	44	36
M_{42}	25	28	32	35	38	39	38	35	32	37	44	50	52	50	44	37	32	27	29	36	44	56	70	77	70	56	44
M_{43}	22	25	28	32	35	38	39	38	35	44	50	52	50	44	37	32	27	24	25	29	36	44	56	70	77	70	56
M_{44}	20	22	25	28	32	35	38	39	38	50	52	50	44	37	32	27	24	21	22	25	29	36	44	56	70	77	70
M_{45}	18	20	22	25	28	32	35	38	39	52	50	44	37	32	27	24	21	19	19	22	25	29	36	44	56	70	77
M_{46}	17	18	20	22	25	27	29	31	31	39	38	35	32	28	25	22	20	18	19	21	24	27	32	37	44	50	52
M_{47}	18	20	22	25	27	29	31	31	31	38	39	38	35	32	28	25	22	20	21	24	27	32	37	44	50	52	50
M_{48}	20	22	25	27	29	31	31	31	29	35	38	39	38	35	32	28	25	22	24	27	32	37	44	50	52	50	44
M_{49}	22	25	27	29	31	31	31	29	27	32	35	38	39	38	35	32	28	25	27	32	37	44	50	52	50	44	37
M_{50}	25	27	29	31	31	31	29	27	25	28	32	35	38	39	38	35	32	28	32	37	44	50	52	50	44	37	32
M_{51}	27	29	31	31	31	29	27	25	22	25	28	32	35	38	39	38	35	32	37	44	50	52	50	44	37	32	27
M_{52}	29	31	31	31	29	27	25	22	20	22	25	28	32	35	38	39	38	35	44	50	52	50	44	37	32	27	24
M_{53}	31	31	31	29	27	25	22	20	18	20	22	25	28	32	35	38	39	38	50	52	50	44	37	32	27	24	21
M_{54}	31	31	29	27	25	22	20	18	17	18	20	22	25	28	32	35	38	39	52	50	44	37	32	27	24	21	19

	M_{28}	M_{29}	M_{30}	M_{31}	M_{32}	M_{33}	M_{34}	M_{35}	M_{36}	M_{37}	M_{38}	M_{39}	M_{40}	M_{41}	M_{42}	M_{43}	M_{44}	M_{45}	M_{46}	M_{47}	M_{48}	M_{49}	M_{50}	M_{51}	M_{52}	M_{53}	M_{54}
M_1	19	21	24	27	32	37	44	50	52	39	38	35	32	28	25	22	20	18	17	18	20	22	25	27	29	31	31
M_2	21	24	27	32	37	44	50	52	50	38	39	38	35	32	28	25	22	20	18	20	22	25	27	29	31	31	31
M_3	24	27	32	37	44	50	52	50	44	35	38	39	38	35	32	28	25	22	20	22	25	27	29	31	31	31	29
M_4	27	32	37	44	50	52	50	44	37	32	35	38	39	38	35	32	28	25	22	25	27	29	31	31	31	29	27
M_5	32	37	44	50	52	50	44	37	32	28	32	35	38	39	38	35	32	28	25	27	29	31	31	31	29	27	25
M_6	37	44	50	52	50	44	37	32	27	25	28	32	35	38	39	38	35	32	27	29	31	31	31	29	27	25	22
M_7	44	50	52	50	44	37	32	27	24	22	25	28	32	35	38	39	38	35	29	31	31	31	29	27	25	22	20
M_8	50	52	50	44	37	32	27	24	21	20	22	25	28	32	35	38	39	38	31	31	31	29	27	25	22	20	18
M_9	52	50	44	37	32	27	24	21	19	18	20	22	25	28	32	35	38	39	31	31	29	27	25	22	20	18	17
M_{10}	77	70	56	44	36	29	25	22	19	19	21	24	27	32	37	44	50	52	39	38	35	32	28	25	22	20	18
M_{11}	70	77	70	56	44	36	29	25	22	21	24	27	32	37	44	50	52	50	38	39	38	35	32	28	25	22	20
M_{12}	56	70	77	70	56	44	36	29	25	24	27	32	37	44	50	52	50	44	35	38	39	38	35	32	28	25	22
M_{13}	44	56	70	77	70	56	44	36	29	27	32	37	44	50	52	50	44	37	32	35	38	39	38	35	32	28	25
M_{14}	36	44	56	70	77	70	56	44	36	32	37	44	50	52	50	44	37	32	28	32	35	38	39	38	35	32	28
M_{15}	29	36	44	56	70	77	70	56	44	37	44	50	52	50	44	37	32	27	25	28	32	35	38	39	38	35	32
M_{16}	25	29	36	44	56	70	77	70	56	44	50	52	50	44	37	32	27	24	22	25	28	32	35	38	39	38	35
M_{17}	22	25	29	36	44	56	70	77	70	50	52	50	44	37	32	27	24	21	20	22	25	28	32	35	38	39	38
M_{18}	19	22	25	29	36	44	56	70	77	52	50	44	37	32	27	24	21	19	18	20	22	25	28	32	35	38	39
M_{19}	20	22	26	31	39	51	72	113	149	77	70	56	44	36	29	25	22	19	19	21	24	27	32	37	44	50	52
M_{20}	22	26	31	39	51	72	113	149	113	70	77	70	56	44	36	29	25	22	21	24	27	32	37	44	50	52	50
M_{21}	26	31	39	51	72	113	149	113	72	56	70	77	70	56	44	36	29	25	24	27	32	37	44	50	52	50	44
M_{22}	31	39	51	72	113	149	113	72	51	44	56	70	77	70	56	44	36	29	27	32	37	44	50	52	50	44	37
M_{23}	39	51	72	113	149	113	72	51	39	36	44	56	70	77	70	56	44	36	32	37	44	50	52	50	44	37	32
M_{24}	51	72	113	149	113	72	51	39	31	29	36	44	56	70	77	70	56	44	37	44	50	52	50	44	37	32	27
M_{25}	72	113	149	113	72	51	39	31	26	25	29	36	44	56	70	77	70	56	44	50	52	50	44	37	32	27	24
M_{26}	113	149	113	72	51	39	31	26	22	22	25	29	36	44	56	70	77	70	50	52	50	44	37	32	27	24	21
M_{27}	149	113	72	51	39	31	26	22	20	19	22	25	29	36	44	56	70	77	52	50	44	37	32	27	24	21	19
M_{28}	927	218	83	54	40	32	26	23	20	20	22	26	31	39	51	72	113	149	77	70	56	44	36	29	25	22	19
M_{29}	218	927	218	83	54	40	32	26	23	22	26	31	39	51	72	113	149	113	70	77	70	56	44	36	29	25	22
M_{30}	83	218	927	218	83	54	40	32	26	26	31	39	51	72	113	149	113	72	56	70	77	70	56	44	36	29	25
M_{31}	54	83	218	927	218	83	54	40	32	31	39	51	72	113	149	113	72	51	44	56	70	77	70	56	44	36	29
M_{32}	40	54	83	218	927	218	83	54	40	39	51	72	113	149	113	72	51	39	36	44	56	70	77	70	56	44	36
M_{33}	32	40	54	83	218	927	218	83	54	51	72	113	149	113	72	51	39	31	29	36	44	56	70	77	70	56	44
M_{34}	26	32	40	54	83	218	927	218	83	72	113	149	113	72	51	39	31	26	25	29	36	44	56	70	77	70	56
M_{35}	23	26	32	40	54	83	218	927	218	113	149	113	72	51	39	31	26	22	22	25	29	36	44	56	70	77	70
M_{36}	20	23	26	32	40	54	83	218	927	149	113	72	51	39	31	26	22	20	19	22	25	29	36	44	56	70	77
M_{37}	20	22	26	31	39	51	72	113	149	927	218	83	54	40	32	26	23	20	20	22	26	31	39	51	72	113	149
M_{38}	22	26	31	39	51	72	113	149	113	218	927	218	83	54	40	32	26	23	22	26	31	39	51	72	113	149	113
M_{39}	26	31	39	51	72	113	149	113	72	83	218	927	218	83	54	40	32	26	26	31	39	51	72	113	149	113	72
M_{40}	31	39	51	72	113	149	113	72	51	54	83	218	927	218	83	54	40	32	31	39	51	72	113	149	113	72	51
M_{41}	39	51	72	113	149	113	72	51	39	40	54	83	218	927	218	83	54	40	39	51	72	113	149	113	72	51	39
M_{42}	51	72	113	149	113	72	51	39	31	32	40	54	83	218	927	218	83	54	51	72	113	149	113	72	51	39	31
M_{43}	72	113	149	113	72	51	39	31	26	26	32	40	54	83	218	927	218	83	72	113	149	113	72	51	39	31	26
M_{44}	113	149	113	72	51	39	31	26	22	23	26	32	40	54	83	218	927	218	113	149	113	72	51	39	31	26	22
M_{45}	149	113	72	51	39	31	26	22	20	20	23	26	32	40	54	83	218	927	149	113	72	51	39	31	26	22	20
M_{46}	77	70	56	44	36	29	25	22	19	20	22	26	31	39	51	72	113	149	927	218	83	54	40	32	26	23	20
M_{47}	70	77	70	56	44	36	29	25	22	22	26	31	39	51	72	113	149	113	218	927	218	83	54	40	32	26	23
M_{48}	56	70	77	70	56	44	36	29	25	26	31	39	51	72	113	149	113	72	83	218	927	218	83	54	40	32	26
M_{49}	44	56	70	77	70	56	44	36	29	31	39	51	72	113	149	113	72	51	54	83	218	927	218	83	54	40	32
M_{50}	36	44	56	70	77	70	56	44	36	39	51	72	113	149	113	72	51	39	40	54	83	218	927	218	83	54	40
M_{51}	29	36	44	56	70	77	70	56	44	51	72	113	149	113	72	51	39	31	32	40	54	83	218	927	218	83	54
M_{52}	25	29	36	44	56	70	77	70	56	72	113	149	113	72	51	39	31	26	26	32	40	54	83	218	927	218	83
M_{53}	22	25	29	36	44	56	70	77	70	113	149	113	72	51	39	31	26	22	23	26	32	40	54	83	218	927	218
M_{54}	19	22	25	29	36	44	56	70	77	149	113	72	51	39	31	26	22	20	20	23	26	32	40	54	83	218	927

Integration of Offshore Wind Farm Plants to the Power Grid using an HVDC line Transmission

Abderrahmane Berkani
Laboratory of L2GEGI
Deptement of Electrical Engineering
University Ibn Khaldoun of Tiaret
Tiaret, Algeria
abderrahmane.berkani@univ-tiaret.dz

Bachir Boumediene
Laboratory of L2GEGI
Deptement of Electrical Engineering
University Ibn Khaldoun of Tiaret
Tiaret, Algeria
bachir.boumediene@univ-tiaret.dz

Siamak Pourkeivannour.
Electrical and Electronics Engineering
Department
Middle East Technical University
Ankara, Turkey
siamak.pourkeivannour@metu.edu.tr

Tayeb Allaoui
Laboratory of L2GEGI
Deptement of Electrical Engineering
University Ibn Khaldoun of Tiaret
Tiaret, Algeria
tayeb.allaoui@univ-tiaret.dz

Karim Negadi
Laboratory of L2GEGI
Deptement of Electrical Engineering
University Ibn Khaldoun of Tiaret
Tiaret, Algeria
karim.negadi@univ-tiaret.dz

H. Bülent Ertan
Middle East Technical University
Atilim University
Ankara, Turkey
ertan@metu.edu.tr

Abstract—**This paper investigates an integration of Offshore Wind Farm Plants with Power Grid Based on an HVDC line Interconnection. Large offshore wind farms are installed in the North Sea area using modern multi-megawatt wind turbines. The Voltage source converter - high voltage direct current VSC-HVDC is a suitable means of integrating such large and distant offshore Wind Power Plants (WPP) which need long submarine cable transmission to the onshore grid. The offshore network then becomes very different from the conventional grid, in that it is only connected to electronic power converters. A wind farm model with VSC-HVDC connection is developed. This work presents the modeling and simulation of such a system. The dynamic study of system performance under the fluctuations of wind energy and wind speed was studied to demonstrate the effectiveness of the control strategy. The validity of the proposed control technique is verified by Matlab/Simulink. Simulation results presented in this paper confirm the validity and feasibility of the proposed control approach, and can be tested on experimental setup.**

Keywords—*Wind farm, Wind Power Plants, Permanent Magnet Synchronous Generator (PMSG), High Voltage Direct Current (HVDC), Multilevel inverter.*

I. INTRODUCTION

Renewable energy sources such as solar, wind, hydro, geothermal have emerged as a new paradigm for meeting global energy needs.

Their benefits lie in the fact that they are clean, abundant, naturally replenished renewable energy sources, available over large geographical areas and have little or no impact on the environment. These renewable energy sources are mainly used for electricity generation, heating, fuel transportation and rural energy supply.

The wind energy industries have experienced strong growth, which has led to major improvements in offshore applications [1]. This development will be a huge step forward in the planning of the electrical system. Offshore wind farms have seen an upward trend compared to onshore wind farms due to the unavailability of onshore sites and the increased and persistent power of wind at offshore sites.

The connections between offshore wind farm and power grid can be high voltage ac (HVAC) or high voltage dc (HVDC) with classical HVDC LCC (line commutated converter) using thyristor bridge, or HVDC VSC (voltage source converter) using IGBTs. HVDC VSC has better

dynamic behaviors than HVDC CSC, but they are expensive and have larger power loss. The three integration methods are compared in [2] and [3]. HVDC transmissions are more favorable for large power capacities and long transmission distances,. Considering the cost and efficiency, HVDC using thyristor bridges can be used for extremely large wind farm, 600 MW for example, which is connected with strong ac grid. HVDC using IGBTs can be integrated with relatively weak grid for its capability to provide dynamic reactive power.

Several control strategies for VSC-based back-to-back high-voltage DC (HVDC) conversion system applications are mentioned in the literature [4, 5]. Among these control technique that prove their efficiency, decoupled control of the active and reactive power on the dq frame as well as the setting of the DC bus voltage.

This work contains five main parts :

First, modeling of wind turbine based PMSG with its HVDC delivery system are presented in section 2. The second part talks about control strategies for both wind farm side converter and grid side converter are considered in section 3. The third part studies the simulation results analysis in Matlab / Simulink of the studied system. Finally, in a last part, we present the conclusions of this study.

II. PHYSICAL MODELING OF THE PROPOSED SYSTEM

The considered wind farm connected to the grid using HVDC line transmission is shown in Figure 1.

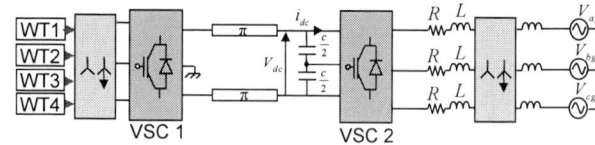

Fig. 1. Description wind farms plants connected to the grid with HVDC

A. Wind turbine

The wind generator consists of a wing which captures the kinetic energy of the wind coupled directly with a synchronous generator which delivers on a DC bus via a diode rectifier, this is the retained structure for this modeling and simulation work (figure 2).

978-1-5386-7688-2/19 $31.00 © 2019 IEEE

The air action applied to the turbine results in a mechanical torque that is proportional to the wind speed, however the turbine can only extract a part of the kinetic energy of the wind, its relation is expressed as follows [6 , 7]:

$$P_m = P_{mt} C_p = \frac{1}{2} C_p(\lambda) \rho S v^3 \qquad (1)$$

where:

ρ is the volume density of air,

S and the surface swept by the turbine blades,

v the wind speed,

C_p is the power coefficient which represents the percentage of power which can be extracted from the turbine, it is dependent on the calibration angle β and the specific velocity λ, this coefficient is limited by the value 0.59 which is called the BETZ limit. The coefficient C_p is often derived from practical measurement and is estimated by the following equation:

$$C_p(\lambda, \beta) = (0,44 - 0,01167\beta) \sin(\pi \frac{\lambda - 3}{15 - 0,3\beta}) \qquad (2)$$
$$-0,00184(\lambda - 3)\beta$$

The ratio of turbine velocity to wind velocity is expressed as (3), where Ω_m is the rotational speed of the turbine, R_t is the radius of the blades [8].

$$\lambda = \frac{(\Omega_m R_t)}{v} \qquad (3)$$

The mechanical torque obtained from the mechanical power is given by (4):

$$T_m = \frac{P_t}{\Omega_m} \qquad (4)$$

The mechanical equation of the system is expressed by (5), where J_t and J_m are respectively the moment of inertia of the turbine and that of the generator, f_v is the coefficient of viscous friction of the generator, Ω_m is the rotational speed of the generator [9]:

$$\begin{cases} J = J_t + J_m \\ J \frac{d\Omega_m}{dt} = T_m - T_{em} - f_v \Omega_m \end{cases} \qquad (5)$$

B. PMSG Model

In order to linearize the system, we use the transformation of Park, whose passage matrix is given by [10]:

$$[P(\theta)] = K_t \begin{vmatrix} \cos\theta & -\sin\theta & \frac{1}{\sqrt{2}} \\ \cos(\theta - \frac{2\pi}{3}) & -\sin(\theta - \frac{2\pi}{3}) & \frac{1}{\sqrt{2}} \\ \cos(\theta - \frac{4\pi}{3}) & -\sin(\theta - \frac{4\pi}{3}) & \frac{1}{\sqrt{2}} \end{vmatrix} \qquad (6)$$

Where:

K_t is a positive constant.

The dynamic model of the PMSG in the reference dq is presented by (7), where R_s is the resistance of the stator, L_d and L_q are the cyclic inductances in the reference dq,

i_{sd} and i_{sq} are the stator currents in the reference dq, p is the number of pairs of poles and is the residual flow of the PMSG [11].

$$\begin{bmatrix} \frac{di_{sd}}{dt} \\ \frac{di_{sq}}{dt} \end{bmatrix} = \begin{bmatrix} -\frac{R_s}{L_d} & \frac{\omega_r L_q}{L_d} \\ -\frac{\omega_r L_d}{L_q} & -\frac{R_s}{L_q} \end{bmatrix} \begin{bmatrix} i_{sd} \\ i_{sq} \end{bmatrix}$$
$$+ \begin{bmatrix} \frac{1}{L_d} & 0 & 0 \\ 0 & \frac{1}{L_q} & -\frac{\omega_r}{L_q} \end{bmatrix} \begin{bmatrix} V_{sd} \\ V_{sq} \\ \psi_f \end{bmatrix} \qquad (6)$$

We adopted a PMSG with smooth pole which implies: $(L_d = L_q = L_s)$.

The expressions of the voltages V_{sd} and V_{sq} and of the fluxes in the PMSG are given by (8) and (9) respectively:

$$\begin{cases} V_{sd} = R_s i_{sd} + \frac{d\psi_{sd}}{dt} - \omega_r \psi_{sq} \\ V_{sq} = R_s i_{sq} + \frac{d\psi_{sq}}{dt} + \omega_r \psi_{sd} \end{cases} \qquad (7)$$

$$\begin{cases} \psi_{sd} = L_d i_{sd} + \psi_f \\ \psi_{sq} = L_q i_{sq} \end{cases} \qquad (8)$$

Hence the matrix representation of the tensions is given by [5]:

$$\begin{bmatrix} V_{sd} \\ V_{sq} \end{bmatrix} = \begin{bmatrix} R_s + SL_d & -\omega_r L_q & 0 \\ \omega_r L_d & R_s + SL_q & \omega_r \end{bmatrix} \begin{bmatrix} I_{sd} \\ I_{sq} \\ \psi_f \end{bmatrix} \qquad (9)$$

The electromagnetic torque is given by the following relation:

$$T_{em} = p.(\psi_{sd}.i_d - \psi_{sq}.i_q) \qquad (10)$$

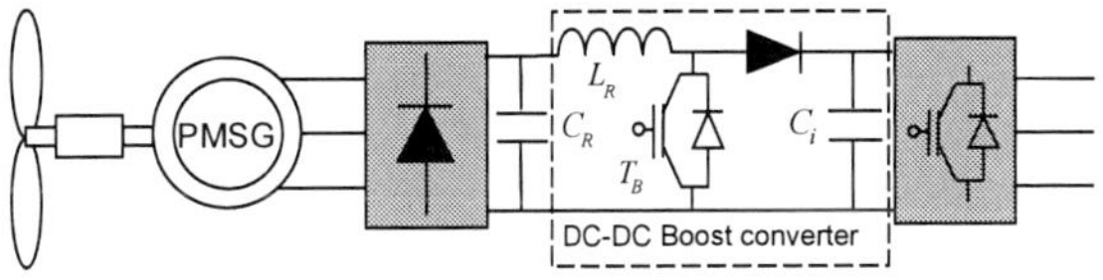

Fig. 2. Description of wind turbine

C. Structure of HVDC System

The HVDC conversion system is composed of two back-to-back connected VSC systems. Both VSC systems employ the three-level NPC as their power converters, labeled as NPC1 and NPC2. Each three-level NPC has a DC-side capacitive voltage divider with two nominally identical capacitors (Fig. 1)

When VSC-HVDC is used for PMSG-based wind farms, the wind farm side VSC collects the wind energy from the wind farm, while the grid side VSC transmits the collected energy to onshore grid. As an indication of power balance, the dc-voltage is kept constant through wind farm side VSC control [12, 13].

The mathematic model of VSC-HVDC in three phases:

$$\begin{cases} v_a = L\dfrac{di_a(t)}{dt} + Ri_a(t) + v_{a\_conv}(t) \\[2mm] v_b = L\dfrac{di_b(t)}{dt} + Ri_b(t) + v_{b\_conv}(t) \\[2mm] v_c = L\dfrac{di_c(t)}{dt} + Ri_c(t) + v_{c\_conv}(t) \end{cases} \qquad (12)$$

$$\begin{cases} i_{dc}(t) = C\dfrac{du_{dc}(t)}{dt} + i_L \end{cases} \qquad (13)$$

Where:

$v_a(s)$, $v_b(s)$ and $v_c(s)$: are component of the AC voltage of transmission network,

$i_a(s)$, $i_b(s)$ and $i_c(s)$ are phase currents.

L is the inductance of the phase reactor,

R is the resistance of the phase reactor,

$v_a(t)_{conv}$, $v_b(t)_{conv}$ and $v_c(t)_{conv}$ are phases voltage at the side of the converter.

The *dq* frame transformation of equation (13), the *dq* components of the voltage become:

$$\begin{cases} v_d(t) = L\dfrac{di_d(t)}{dt} + Ri_d(t) - \omega L i_q(t) + v_{dg}(t) \\[2mm] v_q(t) = L\dfrac{di_q(t)}{dt} + Ri_q(t) - \omega L i_d(t) + v_{qg}(t) \end{cases} \qquad (14)$$

The equations of the active and reactive power in the *dq* frame as:

$$\begin{cases} P_{ac} = \dfrac{3}{2}\left(v_d i_d + v_d i_d\right) \\[2mm] Q_{ac} = \dfrac{3}{2}\left(v_q i_d - v_d i_q\right) \end{cases} \qquad (15)$$

Whit $v_q = 0$ thus the active and reactive power are represented as [13]:

$$\begin{cases} P_{ac} = +\dfrac{3}{2}v_d i_d \\[2mm] Q_{ac} = -\dfrac{3}{2}v_d i_q \end{cases} \qquad (16)$$

The active power balanced relationship between the AC input and DC output are given as [14, 15]:

$$P_{ac} = P_{dc} \Leftrightarrow \dfrac{3}{2}v_d i_d = v_{dc} i_{dc} \qquad (17)$$

D. Back-to-back Converter Model

A three-level inverter differs from a conventional two-level inverter in that it is capable of producing three different levels of output phase voltage. The structure of a three-level neutral point clamped inverter is shown in figure 3. When switches 1 and 2 are on the output is connected to the positive supply rail. When switches 3 and 4 are on, the output is connected to the negative supply rail. When switches 2 and 3 are on, the output is connected to the supply neutral point via one of the two clamping diodes [16, 17].

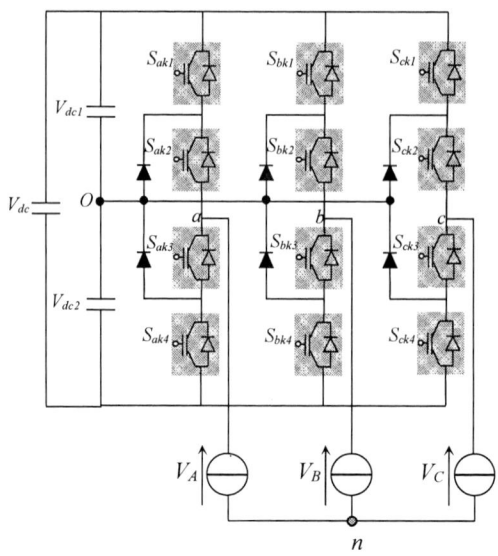

Fig. 3. Three level inverter

The functions F_{km}^b of connection are given by:

$$\begin{cases} F_{k1}^b = F_{k1}\,F_{k2} \\[2mm] F_{k0}^b = F_{k3}\,F_{k4} \end{cases} \qquad (18)$$

where: $m = 1$: the upper half arm and $m = 0$: the lower half arm.

The phase voltage V_{AO}, V_{BO}, V_{CO} can be written as:

$$\begin{cases} V_{AO} = F_{11}^b V_{c1} - F_{10}^b V_{c2} \\[2mm] V_{BO} = F_{21}^b V_{c1} - F_{20}^b V_{c2} \\[2mm] V_{CO} = F_{31}^b V_{c1} - F_{30}^b V_{c2} \end{cases} \qquad (19)$$

Simple output voltages are written as:

$$\begin{bmatrix} V_A \\ V_B \\ V_C \end{bmatrix} = \frac{1}{3}\begin{bmatrix} 2 & -1 & -1 \\ -1 & 2 & -1 \\ -1 & -1 & 2 \end{bmatrix}\left\{\begin{bmatrix} F_{11}^b \\ F_{21}^b \\ F_{31}^b \end{bmatrix}V_{c1} - \begin{bmatrix} F_{10}^b \\ F_{20}^b \\ F_{30}^b \end{bmatrix}V_{c2}\right\} \qquad (20)$$

III. CONTROL STRATEGY

This part describes the basic control scheme for the PMSG-based wind farm and HVDC.

A. Control of wind turbine

PMSGs are predominantly used in full-variable wind speed energy conversion systems (WECS), and are the most popular choice for high-power WT manufacturers.

The power conversion illustrated in figure 4 is composed of a unset rectifier, DC-DC boost converter to increase the DC voltage and an inverter linked by DC-link capacitors. The control schemes employed for the power converters play an important role in maximizing energy capture and in complying with HVDC connexion requirements.

To obtain high dynamic performance and decoupled torque and flux control similar to the separately excited DC motor, vector control strategies are adopted for PMSGs (figure 4).

978-1-5386-7688-2/19 $31.00 © 2019 IEEE

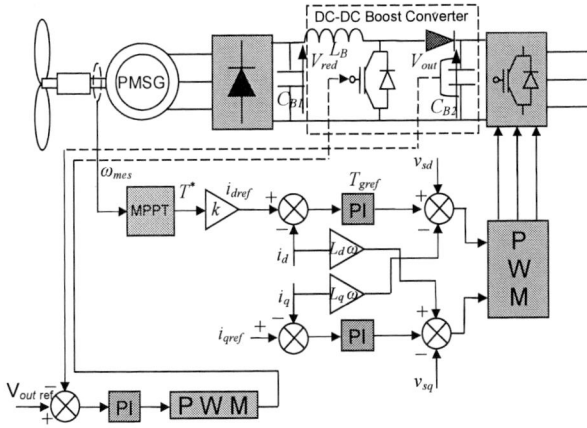

Fig. 4. Control of the wind turbine

B. Control of back-to-back converter

The back-to-back HVDC converter system show in figure 1 can be considered as the composition of two constant-frequency VSC systems: the left-hand side VSC system (wind farmas plants side) is a real-/reactive-power controller and the right-hand-side VSC system (grid) is a controlled DC-voltage and the reactive-power controller as illustrated in figure 5.

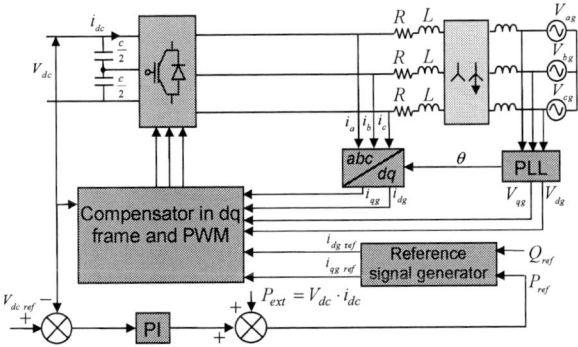

Fig. 5. Control of the HVDC

IV. SIMULATION RESULTS AND DISCUSSION

This section is based on simulation studies of the performance of an offshore wind farm connected to VSC based HVDC for the control of wind turbine generator, offshore converter, onshore converter, and to ensure quality output of power to the grid from the offshore wind farm. The simulation software was used for designing and simulation of the model. In the study, converter level models in which the converters are modeled in the semiconductor switching-level were used.

The figure 6, 8, 10, 12 shows the dynamic behavior of the wind turbine WT1, WT2, WT3 and WT4. We can well notice that the speed of the generator follows its reference with a good response time and without exceeding. The electromagnetic torque depends on the variation of the mechanical torque.

In the figures 7, 9, 11, 13 It can be seen that the boost converter produces a fixed voltage 20 kV to power the inverter.

From figures 14 and 15, we note that the active and reactive powers follow their references with a transient regime very fast and without overtaking.

It can be seen that the current present variations of the power variations which shows that the current is the image of the powers.

The DC voltage illustrated in figure 16 follows its reference with small fluctuations in the moments of the variations of the powers.

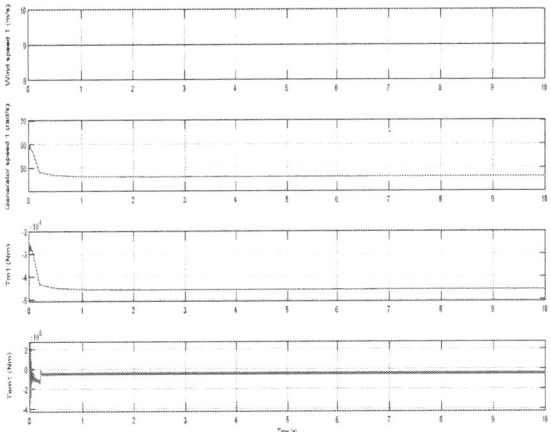

Fig. 6. Speed wind, generator speed, mechanical torque and electromagnetic torque reponses of WT1

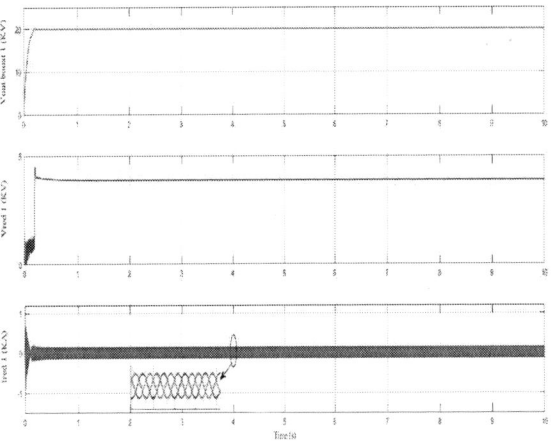

Fig. 7. Voltage output boost V_{out1}, input voltage inverter V_{red1} and output current of inveter I_{red1}

978-1-5386-7688-2/19 $31.00 © 2019 IEEE

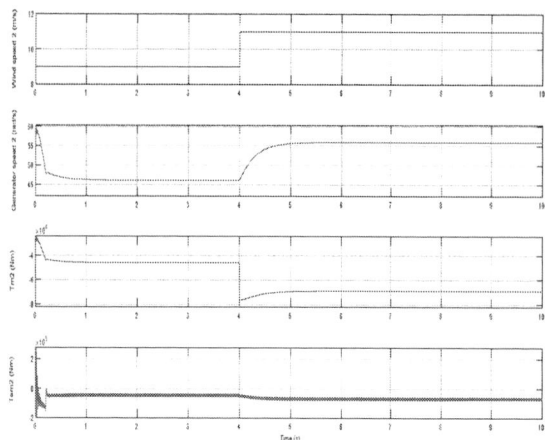

Fig. 8. Wind speed, generator speed, mechanical torque and electromagnetic torque reponses of WT2

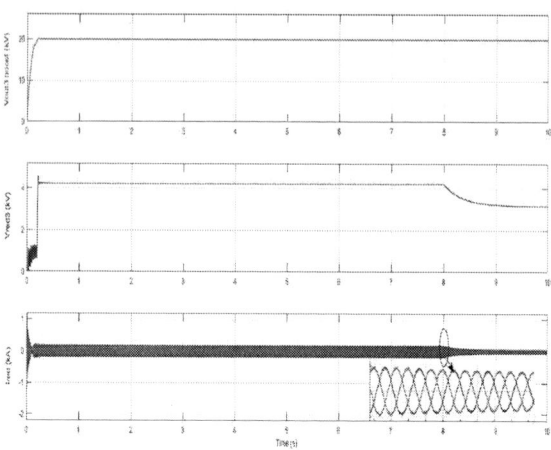

Fig. 11. Voltage output boost V_{out3}, input voltage inverter V_{red3} and output current of inveter I_{red3}

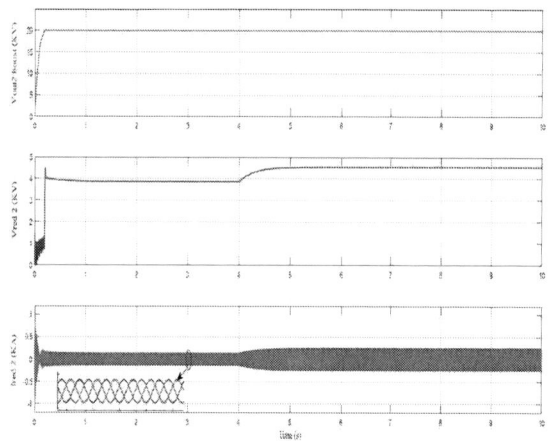

Fig. 9. Voltage output boost V_{out2}, input voltage inverter V_{red2} and output current of inveter I_{red2}

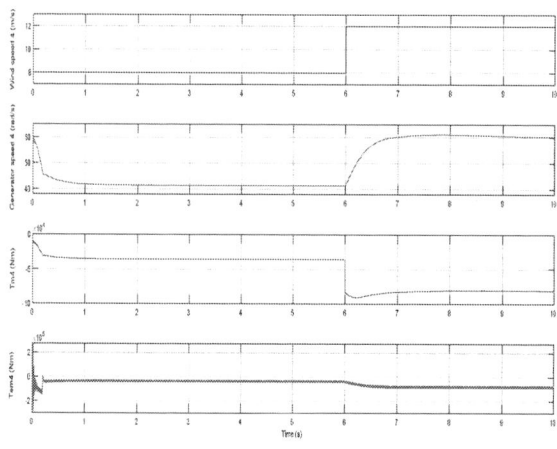

Fig. 12. Wind speed, generator speed, mechanical torque and electromagnetic torque reponses of WT4

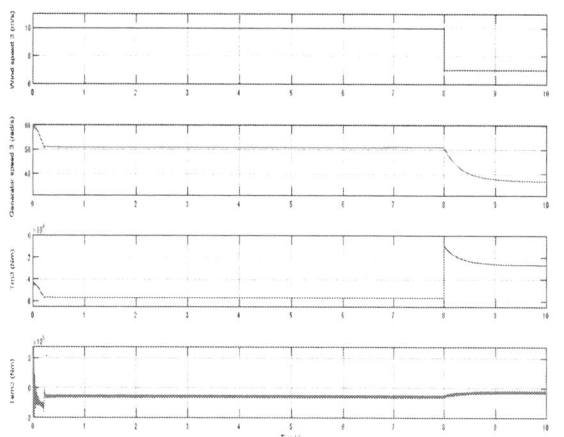

Fig. 10. Wind speed, generator speed, mechanical torque and electromagnetic torque reponses of WT3

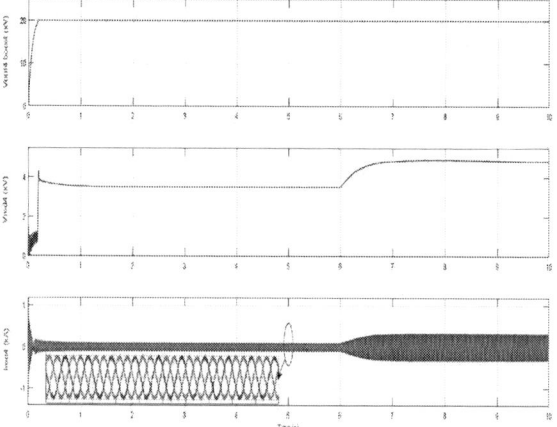

Fig. 13. Example of a figure caption. Voltage output boost V_{out4}, input voltage inverter V_{red4} and output current of inveter I_{red4}

978-1-5386-7688-2/19 $31.00 © 2019 IEEE   490

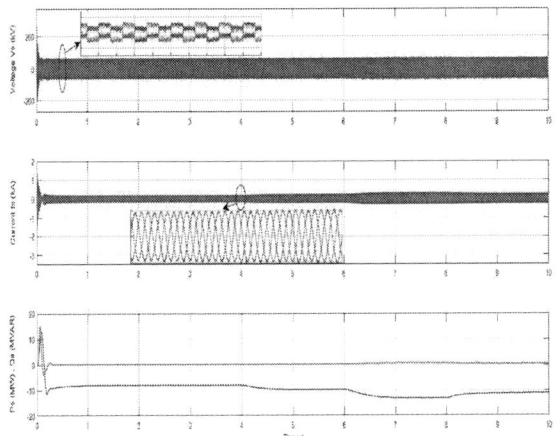

Fig. 14. Wind farm voltage and current, active and reactive power during the whole time

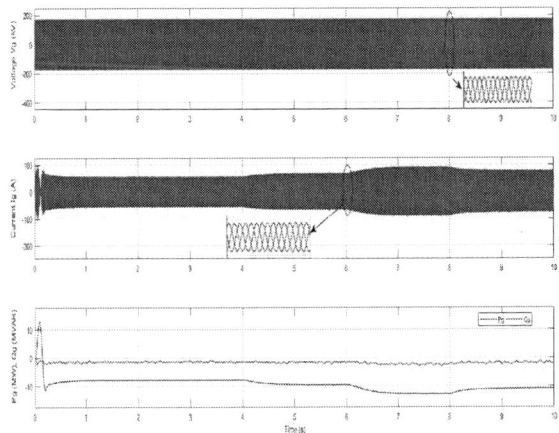

Fig. 15. Grid voltage and current, active and reactive power during the whole time

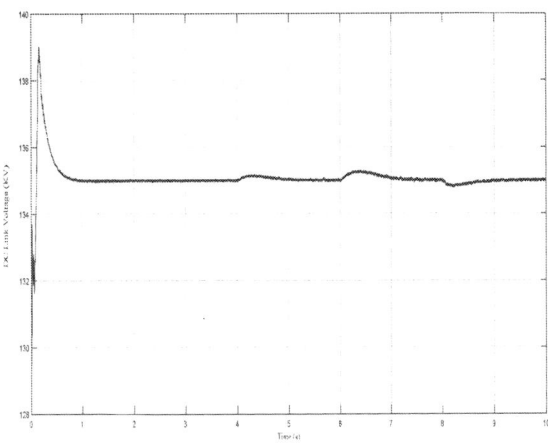

Fig. 16. DC Link voltage

V. CONCLUSION

This paper has proposed an Integration of Offshore Wind Farm Plants to the Power Grid using an HVDC line Transmission. The offshore wind farm is designed to deliver power to an AC main grid using HVDC by an undersea cable

transmission. A PMSG is used for driving the wind turbines due to its variable speed operation. The transmission of power is achieved by the use of rectifier at offshore side for converting the AC voltage supplied by wind farm plant into DC and then transmitted through undersea cables. This transmitted power is then converted by inverter back to AC to feed the grid. The control of this system is achieved through the control of wind turbine generators and wind speed by ensuring that high quality power is delivered to the grid.

The results of the simulation obtained by the control applied to this system, clearly show the satisfactory performances.

VI. APPENDIX

TABLE I. PMSG PARAMATER

Components	Rating values
PMSG rated power	6 MW
RMS Voltage in stator	9795 V
Inductance stator and rotor	1.5 mH
Resistance in stator	12 mΩ
Inductance in stator	12 mH
Number of pair of poles	6
PMSG inertia moment	17 Kg.m^2
Induced magnetic flux	9 Wb

TABLE II. HVDC TRANSMISSION PARAMATERS

Components	Rating values
DC Voltage	135 kV
Simple AC system at offshore station	120 kV/50 Hz
DC line inductance per kilometer	0.9337 mH/km
DC line resistance per kilometer	0.05 Ω/Km
DC line capatitor per kilometer	1e-9 F/km
Length	40 km

TABLE III. DC-DC BOOST CONVERTER

Components	Rating values
Capacitor C$_{B1}$	7 mF
Capacitor C$_{B2}$	4.1 mF
Inductance L$_B$	5 mH

REFERENCES

[1] V. Yaramasu Bin Wu, "Model Predictive Control Of Wind Energy Conversion Systems," Copyright © 2017 by The Institute of Electrical and Electronics Engineers, Inc, 2017.

[2] A. Yazdani, R. Iravani, "Voltage-Sourced Converters In Power Systems, Modeling, Control and Applications Amirnaser Yazdani, Copyright © 2010 by John Wiley & Sons, Inc.

[3] Y. ZHOU, P. Bauer, J. Pierik, J. A. Ferreira, "Integration of Large Offshore Wind Farm - Doubly Fed Induction Generators With

Classical HVDC," Acta Electrotechnica et Informatica Vol. 9, No. 2, 8–14, 2009.

[4] M. Raza, O. Gomis-Bellmunt, "Dynamic Modelling and Implementation of VSC-HVDC System The Grid Connected Large Offshore Wind Power Plant Application," SMARTGREENS 2014 - 3rd International Conference on Smart Grids and Green IT Systems.

[5] J. Bartomeu Pons Perelló, "Modeling and Simulation of an HVDC Network for Offshore Wind Farms," Barcelona, 2015.

[6] S. K. Chaudhary, R. Teodorescu, P. Rodriguez, P. C. Kjaer and P. W. Christensen, "Modelling and Simulation of VSC-HVDC Connection for Offshore Wind Power Plants," PhD Seminar on Detailed Modelling and Validation of Electrical Components and Systems 2010 in Fredericia, Denmark, February 8th, 2010.

[7] M. Suwan, "Modeling and Control of VSC-HVDC Connected Offshore Wind Farms," Duisburg-Essen University.

[8] E. Miguel. D. J. Montilla, S. Arnaltes, E. D. Castronuovo and D. Santos-Martin, "Optimal Power Transmission of Offshore Wind Power Using a VSC-HVdc Interconnection," Energies 2017, 10, 1046; doi:10.3390/en10071046.

[9] A. Mete Vural , I. Auwalu İbrahim, "Modeling and Control of an Offshore Wind Farm connected to Main Grid with High Voltage Direct Current Transmission," Electrica 2018; 18(2): 198-209, 2018

[10] Y. Haiping, "Modeling and Control of Wind Generation and Its HVDC Delivery System," University of South Florida Scholar Commons, 2011.

[11] A. Bidadfar; S. Romano, O. Altin, Müfit; Göksu, Ömer; Cutululis, N. Antonio; Sørensen, P. Ejnar, "Offshore Wind Farms and HVDC Grids Modeling as a Feedback Control System for Stability Analysis," 16th International Workshop on Large-Scale Integration of Wind Power into Power Systems as well as on Transmission Networks for Offshore Wind Power Plants, Berlin, Germany, 2017.

[12] Rios, Bardo; Garcia-Valle, Rodrigo, "Dynamic modelling of VSC-HVDC for connection of offshore wind farms," 9th IET International Conference on AC and DC Power Transmission (pp. 1-4), 2011.

[13] M. Pinaki, Z. Lidong, "Real-Time Simulation of a Wind Connected HVDC Grid," IET ACDC 2012 conference in Birmingham, UK, December, 4-5, 2012.

[14] A. Berkani, K. Negadi, T. Allaoui, F. Marignetti, "Fuzzy Direct Torque Control for Induction Motor Sensorless Drive Powered by Five Level Inverter with Reduction Rule Base," Przegląd Elektrotechniczny, doi:10.15199/48.2019.07.14, No/Vol: 07/2019 Page no. 66.

[15] J. Rodríguez D´Derlée, "Control strategies for offshore wind farms based on PMSG wind turbines and HVDC connection with uncontrolled rectifier," Polytechnic University of Valencia, 2013.

[16] K. Liao, Zheng-youhe, and S. Bin, "Small Signal Model for VSC-HVDC Connected DFIG-Based Offshore Wind Farms," Journal of Applied Mathematics, Volume 2014, Article ID 725209.

[17] A. Berkani, K. Negadi, T. Allaoui, F. Marignetti, "Sliding mode control of wind energy conversion system using dual star synchronous machine and three level converter," TECNICA ITALIANA-Italian Journal of Engineering Science, Vol. 63, No. 2-4, pp. 243-250. https://doi.org/10.18280/ti-ijes.632-418, 2019.

MATLAB/SIMULINK Model For HVDC Fault Calculations

Ahmad Mustapha Usman
Electrical Engineering
Yaşar University
İzmir, Turkey
ahmadusmanmustapha@gmail.com

Mahir Kutay
School of Applied Sciences
Yaşar University
İzmir, Turkey
mahir.kutay@yasar.edu.t

Tuncay Ercan
School of Applied Sciences
Yaşar University
İzmir, Turkey
tuncay.ercan@yasar.edu.tr

Abstract—**Power transmission has faced many challenges as the days goes by. Power demand has increased and its becoming increasingly difficult to acquire right way for new lines, due to growth, population, urbanizations and environmental issues. This research aims to provide methods for the rationale selections of HVDC system, essential HVDC equations and their control. A MATLAB/SIMULINK model is developed for a HVDC link using IGBT/DIODE converters between a 500KV, 50Hz system to a 330KV, 60Hz system over the distance of 450km. Analysis by using the model on this system identifies any abnormal behavior on the AC and DC sides and calculates the time elapsed for recovery to its reliable state conditions.**

Keywords—matlab/simulink, converters, reliable state conditions, abnormal analysis on AC and DC systems, IGBT, thyristors.

I. INTRODUCTION

Electrical energy is generated in the form alternating current (AC). After the production process, the electrical energy sent out as AC. Power transmitted to various locations and distributed to the consumers as AC.

In some situations, it is more advantageous to use direct current (DC) scheme for supplying electrical energy. This is true because in some cases, it may be the most effective technique to transport the energy. AC systems have some specific restrictions more especially when the network will not be able to synchronize due to the frequency difference. In HVDC, energy is created mostly in the form of an AC power, transported as DC and retransformed back as an AC energy again at the ending end [1].

II. BACKGROUND OF STUDY

The world has been developing at a very fast rate over the past few years. In order to sustain the development, power systems have had to expand. This has led to the interconnection of all kinds of power systems worldwide [2]. The escalating rate of industrialization worldwide creates a huge demand for the utilization of electrical energy. Demand for electrical energy has led to the search of more efficient means of an electrical energy transportation at increasing voltage and power levels. High voltage alternating current (HVAC) used all over the world tends to be problematic over longer distances and it creates some environment problems [2]. Therefore, HVDC use is been suggested.

III. LITERATURE REVIEW

Literature review here deals with the classifications of HVDC configurations and the components that make up the entire HVDC system described below

A. Types of HVDC Systems

The types of HVDC systems are given in detail below.

1) Mono-polar HVDC system: This system mostly it comprises of more than one units of six-pulse converters, in which they are either arranged in the series or parallel manner via the ending paths. It has only one conductor in it and the return is either through the earth or ocean [2]. Mono-polar line with return path as ground shown in Fig. 1.

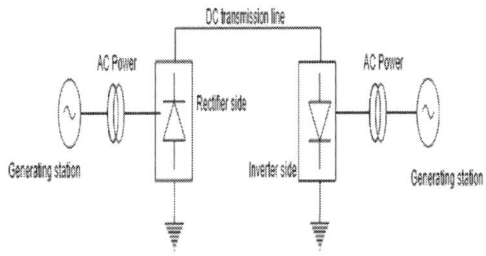

Fig. 1. Mono-polar line with return path as ground.

2) Bi-polar HVDC system: Bi-polar dc lines configuration system as the name implies it is comprising of two different kinds of polarity or conductors in the system. These polarities that are existing in the system are the positive and negative terminals. Mostly this conductor that are available in the circuit are the positive terminal and the negative terminal. These two conductors are of the same rated voltage and are been configured in a series arrangement at the end of the dc lines [3]. Bi-polar DC line system is shown in Fig. 2.

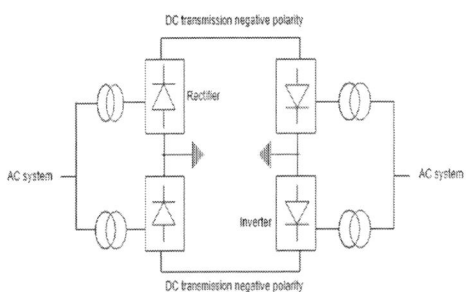

Fig. 2. Bi-polar DC line system.

3) Homo-polar HVDC system: Homo-polar as the can express also, it consists of more than one or two conductors that are linked together having the same polarity which can either be the negative or positive electrodes and they also function in a parallel arrangement. This connection between

978-1-5386-7688-2/19 $31.00 © 2019 IEEE

the rectifier and the inverter of the system is done with the use of DC bus system [4]. Homo-polar DC line system is shown in Fig. 3.

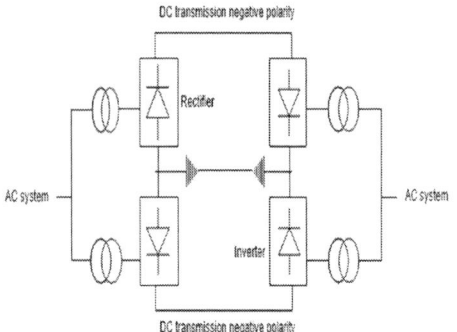

Fig. 3. Homo-polar DC line system.

4) Back to back HVDC system: In this system configuration, these two converters that's the rectifier and the inverter has no any separation of distances between them. This is mainly used to make an interconnection between two AC links with different frequencies [5].

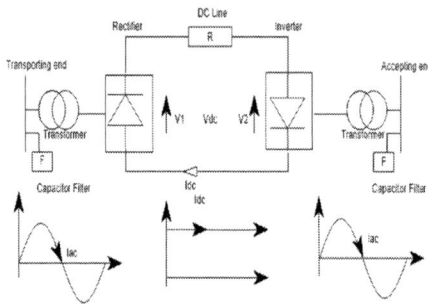

Fig. 4. Back to back DC line system.

B. Components of HVDC

HVDC as a network system, it comprises of many different sections of units or components that are associating with each other in the entire systems, so as to function or operated efficiently. A simple representation of the entire electrical system of the HVDC will is shown below in Fig. 5.

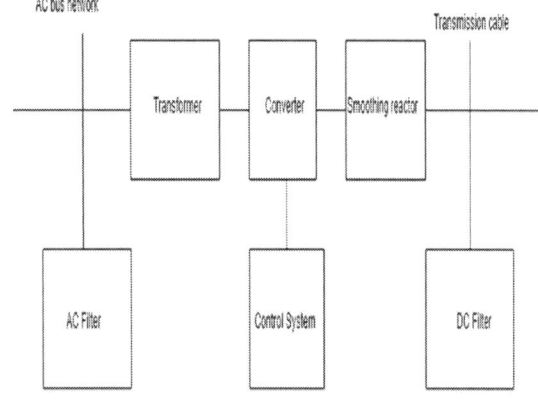

Fig. 5. HVDC network systems.

1) AC harmonic filter system: Harmonics is said to be any presence of an unwanted signals in the systems that's causes

any interruption or changes in the waveform. Converters are said to creates harmonics that are associating to the AC bus network in the system as can also be viewed from the above figure. The functioning of the HVDC converter systems are where the sources of the unwanted signals or harmonics creation of the AC harmonic currents comes from in the system [6]. The HVDC being one of the nonlinear-loads such as the power electronic converters, it creates harmonics that are been taken up by the AC filters which in reverse it gives it to the power reactive in the systems In any converter locations, there is the production of unwanted harmonics due to the following factors been consider, shunt join together to the switchable AC system filters that are connected to the ac bus network. AC harmonics filters are being easily open to be in on or off state automatically with the help of the circuit breaker of the AC network in the systems [7]. On the other hand, state that on the side of this AC system in the converter it has two main purpose to execute which are to be listed below.

- To take up the harmonic currents created by the HVDC network

- To give or produce the reactive power to the system

2) Transformer converter: This converter transformer (TR) is the connection between two devices i.e. the thyristors or IGBT and the AC networks. Typically, this transformer converter system is connected as "an earthed grounded star system, twisted and fluctuating-star and secondary delta rotation point" [8]. The converter system serves as a connection in between these two systems that is the converter and the AC system, in other to provides various purposes which includes the following below:

- It gives separations between the systems.

- It gives the appropriate or needed amount of voltage to the HVDC converter in the system.

This HVDC converter transformer changes the voltage point level of this AC network bus to the desired level of the voltage control entry to the system [9].

3) HVDC smoothing reactor systems: HVDC network for the transportation of power it needed a HVDC system smoothing reactor [10]. This equipment gives a certain purpose as to which can be listed below:

- Restriction of specific fault current in the DC movement

- Reduction of the current harmonic, consisting of the limited communication devices interaction such as the telephone interactions

- Reduction of the unwanted ripples that are present in the DC lines systems.

A representation of the entire electrical system of the HVDC will is shown below in Fig.6.

4) HVDC protection network and control system: Similarly, like the functioning of the AC systems for the DC abnormalities, they are entirely being affected by the inefficient functioning of the controller's system and others. The interruption of the power system transmission, protection

978-1-5386-7688-2/19 $31.00 © 2019 IEEE 494

systems and control systems, are entirely view by the use of switching method and the equipment controls such as the

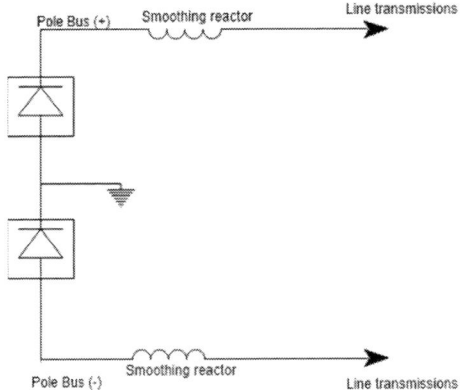

Fig. 6. HVDC smoothing reactor system

5) HVDC converter systems: Generally, the purpose of HVDC rectifier and inverter which are called the converter system in any electrical system is a need to replace any electrical power component. These components can be either be the current, frequency and voltage. The converter system is the place where the transformation or interchange do takes place such as the changes from the AC-DC and DC-AC. HVDC, now uses modern systems which is the thyristor system based converter. With the use of commutation techniques which can be define as the natural interruption of any current in a given circuit or system, it consists of more than one classification of the power electronic converter [11].

6) DC transducer networks: The DC linking transducers can be classified into two different classes namely:

a) DC voltage measurement of the system: This measurement is carried out by the use of the visual division of voltage or the resistive voltage dc division. The resistive division of the voltage, it consists of a series connection between the resistors in the system and are therefore, the measuring of the voltage can be extracted over the lower end of the voltage resistor. The visual transducer voltage senses the power and durability of the electric field that are close to the network bus bar [12].

b) DC current measurement of the system: This measurement is carried out all on the protection and control system that needed the action of a computerized systems. This computation can be made by the production or creation of a magnetic field within the computation head that is enough or adequate to neutralize the magnetic field that is close to the network bus bar via the computation head system.

IV. MULTI-TERMINAL DC LINES SYSTEMS

A multi-terminal dc lines (MTDC) system, express that this kind of configurations is consist of more than one or two converter stationary. However, some of these converters can functioning as the rectifier for the transformation of AC-DC, while the others can be functioning as an inverter for the conversion of DC-AC. This mode of arrangement techniques can be interchangeable by the switching processes. The easiest path of creating an MTDC configuration from a two

point terminal system in other to bring in the idea of tappings. Parallel working of the converters and the bi-polar named as MT-operations.

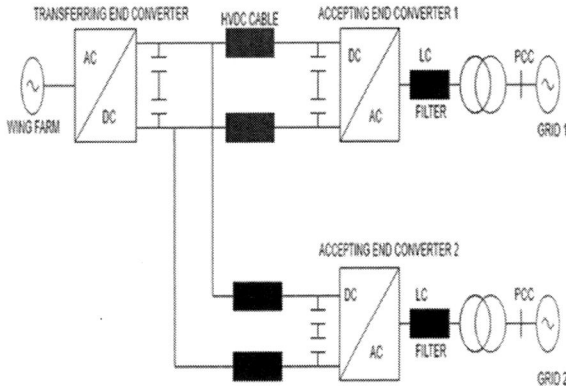

Fig. 7. MTDC Lines system

As in the Fig.7. shown above, it can be seen how the wind farm (WF) is joined together with the rectifier side for the transformation of the AC-DC, the power of the DC is move through the line cable to the other side of the converter which are known as the "Accepting End Converter (AEC)". DC is retransformed again back to the AC and connected to the AC grid passing through the filtering point and the transformer side. [13].

7) Classification of the multi-terminal DC lines systems: The multi-terminal dc lines are divided into two (2) different classes which can be classified and explained below with their various circuit diagrams and explanations.

a) Series mode connections: This mode of connection as the name points out, they are being joined together in a series arrangement to each other in the systems. However, a simple representation of this kind of connections will be seen below in Fig. 8. with a three-point terminal systems for the MTDC [14].

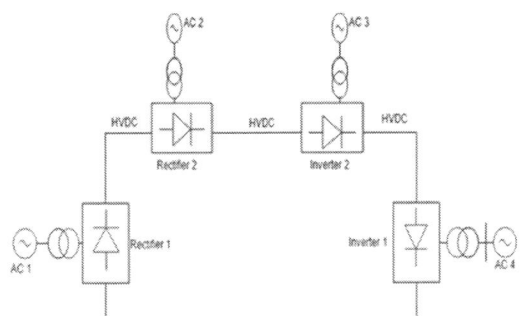

Fig.8. A series mode connection MTDC network

8) Parallel mode connections: This mode of connection, as the name suggest they are being connected adjacent or opposite to each other in the circuit systems. In this adjacent or opposite connection, the dissociation of one or other single parts of the sending sections can or will cause a break or interference of the power within the power converters that are presents inside the circuit system as can be shown in the Fig.

978-1-5386-7688-2/19 $31.00 © 2019 IEEE

9. below. This opposite connection in the network can also work without imploring the use of the HVDC circuit breaker.

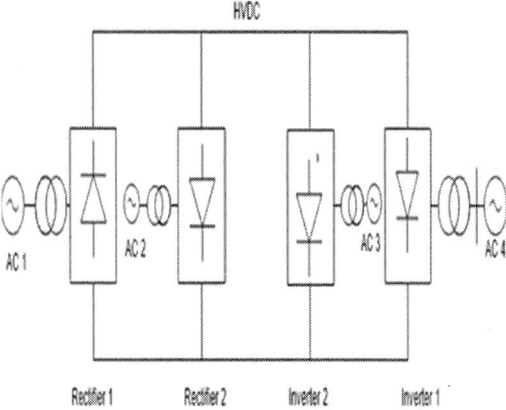

Fig. 9. A parallel mode connection MTDC network.

The most common challenges for the control of this adjacent inverter system is that this high voltage dc inverter system is working or function in the most productive or well organized modes [15].

V. ACTORS FOR CHOOSING THE HVDC TRANSMISSION

The current movement in the DC moves only on an onward direction. While in the AC cases, the electric current is not stable it changes its direction of flowing frequently. Not only in the current movement but also in the voltage situations because its reverses its movement due to the changes that occurs in the flowing of the current and also presented some points that needs to be put into considerations which can be detailed below:

- Cost of the transportation of the power.

- The efficiency in the system.

- Performance of the system.

The extensive on most of the electricity power transportation uses the 3-phase of the alternating current. The logic after this idea of the HVDC over the AC to transport the electric power in a particular situation are usually countless. The idea that tends to support the HVDC usage is the "Investment Cost".

HVDC transportation line value is smaller than that of the AC line for the same purpose of transporting the amount of electric power. Alternating current is commonly used for short ranges that is they are commonly used for household and industrial purpose. The use of the AC for transportation purposes in this area it can cost less in its procedures and its frequency can simply be controlled unlike when trying to apply the use of AC for a very big project its frequency tends to be very complex to control and also as it can be viewed that it has some specific restrictions. Direct Current does not have any specific clampdown attached to it and require less investment cost in it. DC transportation does not require too much use of conductor like the AC systems. The use of both AC and DC for transmitting purpose have been in use recently due to how we can both used in them in transporting an electrical power from one far away location to the other by the means of converters such as the rectifier and inverter. Below the investment cost of both AC and DC transmissions shown in Fig. 10.

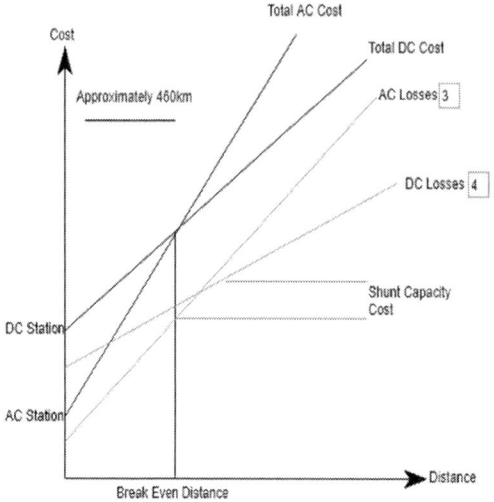

Fig. 10. Variation of cost with length line for AC and DC transmissions.

VI. METHODOLOGY

A) Rectifier Mode of Operation

Rectification can be defined as the transformation of AC into a DC system by the use of a constant dc voltage value. The "on and off" device is the "diode" in the network circuit. The valve system basically works in a single direction to which it is moving from the positive (+) terminal points of the system to the negative (-) terminal points in the circuit [16]. If the thyristor based three-phase rectifier is operating with a zero (0) angle delay its median DC voltage output is calculated by using equation (1).

$$V_o = 3 * \sqrt{3} * \sqrt{2} * V_{rms} \div \pi \qquad (1)$$

If the rectifier is fired with a delay angle α for voltage regulation purposes output DC voltage is calaculated by using equation (2).

$$V_{dc} = V_o * cos\alpha \qquad (2)$$

B) Control of HVDC Converter System

The movement of the current in the dc line transportation is defined by the total differences in the dc voltage among the two system of the converters that is between sending end voltage A and receiving end voltage B as can be seen in the below expression. Noting that all the terminals in the system which are the positive and negative poles are working under the same state.

$$I_d = \frac{V_{dA} - V_{dB}}{R_{dc}} \qquad (3)$$

And the power transportation to the voltage B will be written as

$$P_d = V_{dB} * \frac{V_{dA} - V_{dB}}{R_{dC}} \qquad (4)$$

where, R_{dc} is our dc resistance for the positive transportation line conductor. Experimentally, the R_{dc} is low and its I_d becomes as a consequence to the low difference among the two big voltages in the above equation (3). Therefore, one side of the converter is put on to monitor and control the transportation line voltages and also monitor the I_d. Since we know that the inverter is functioning at a fixed extinction angle, it is preferably to be selected to the inverter to monitor the V_d. Then Id and therefore the level of the power to be monitored by the rectifier. In Fig. 11, rectifier and the inverter control characteristics in the Vd-Id plane is shown [17].

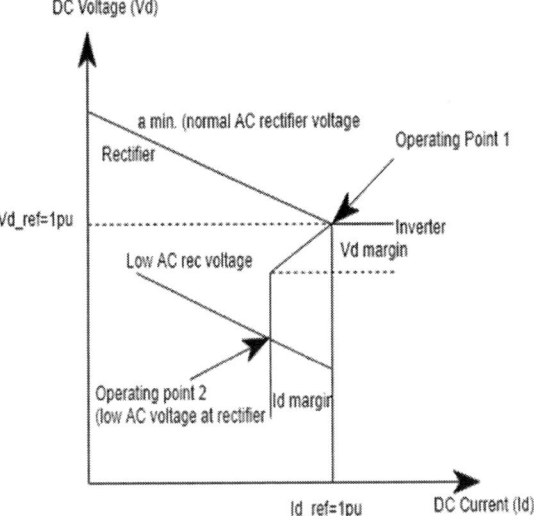

Fig. 11. Rectifier and the inverter control characteristics in the Vd-Id plane.

VII. HVDC MATLAB/SIMULINK MODEL

The MATLAB/SIMULINK design is a DC connection that is usually operating for the transportation of an electrical power from 450Kv, 60Hz system to 345Kv, 50Hz system. The converters in the system are the rectifier and the inverter which are created based upon the "12-pulse converters" that are implore to be working with the combination of a "6-pulse bridge IGBT/DIODE system" that are being join together in a sequence or series arrangement. This converter is being linked through the use of a long line transmission modelling of about 450km that is being disseminated for the line specifications or parameters, and also there is a presence of two (2) smoothing system reactors consisting of a value of 0.75H. The tap interchanger transformer is not run by the system and the tap constant is presumed. Inside the transformers on both the inverter and the converter rectifier blocks subsystem a fixed number of value is applied for the factor value on the TR voltage corresponding to the primary location has a value of 0.90 on the rectifier system side and also the inverter side is having a value of about 0.96 corresponding to it also. The needed power reactive which is always needed by this system that is the rectifier and the inverter is accessible or provided by the pair group of the capacitors banks of 11th, 13th and the higher pass filtering having a number of 150Mvar on the four different filters that where available in the systems. Circuit breakers are also implored in which they are been use to add faults on the AC side of the inverter and also on the DC side of the system rectifier. The DC protection functions are created inside the rectifier and the inverter side of the networks. On the DC fault rectifier protection system, it is used for sensing and to

pressure the angle delay onto the location of the inverter side in other to put out the abnormal current in it. The function of the controlling master block is to commence the taking off operation and ending operation of the converter as well as the ramping up and down of the reference current. As we can notice, also the system is given at a discrete time of about 50 microseconds which can be shown in Fig. 12.

VIII. FAULT SIMULATIONS

We observe the behaviors of the HVDC network under some various working states examples are the AC network fault or any kind of faults. Different fault conditions were applied at both the transferring point to the accepting point of the network and observe their performances and the effects of this applied AC faults and the DC line faults in the system. In order for us to classify the fault condition characteristics in the system simulations were been carried out by the use of MATLAB/SIMULINK package. This simulation is carried out with an AC line fault at the rectifier and DC line fault at the inverter.

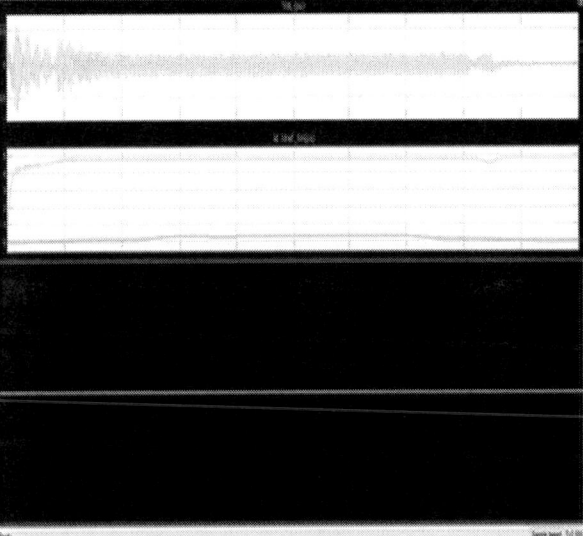

Figure 13. Reliable state condition at rectifier condition.

A) DC line Fault at the rectifier side

The DC line abnormal condition is one of the most often type of fault that usually occurs all the time in a system, which alters the voltage and the current situations. This abnormal fault condition is caused by the use of the power electronic devices that is used in the conversion process and also caused by the instability of the voltage. When this kind of fault problem happens we can observe that the voltage at the transferring point is not stable due to the occurrence of so many fluctuations from the starting point at t=0-0.25seconds as can been seen from the graph with a little stabilization starting from 0.25-0.69 seconds before it finally results to becoming zero at 0.7-1.4seconds and also we can notice the current is also decremented exactly to 3.3pu at t= 0.7seconds-1seconds before it finally goes to zero fully at 1-1.4 seconds. as can be shown in the Fig. 14.

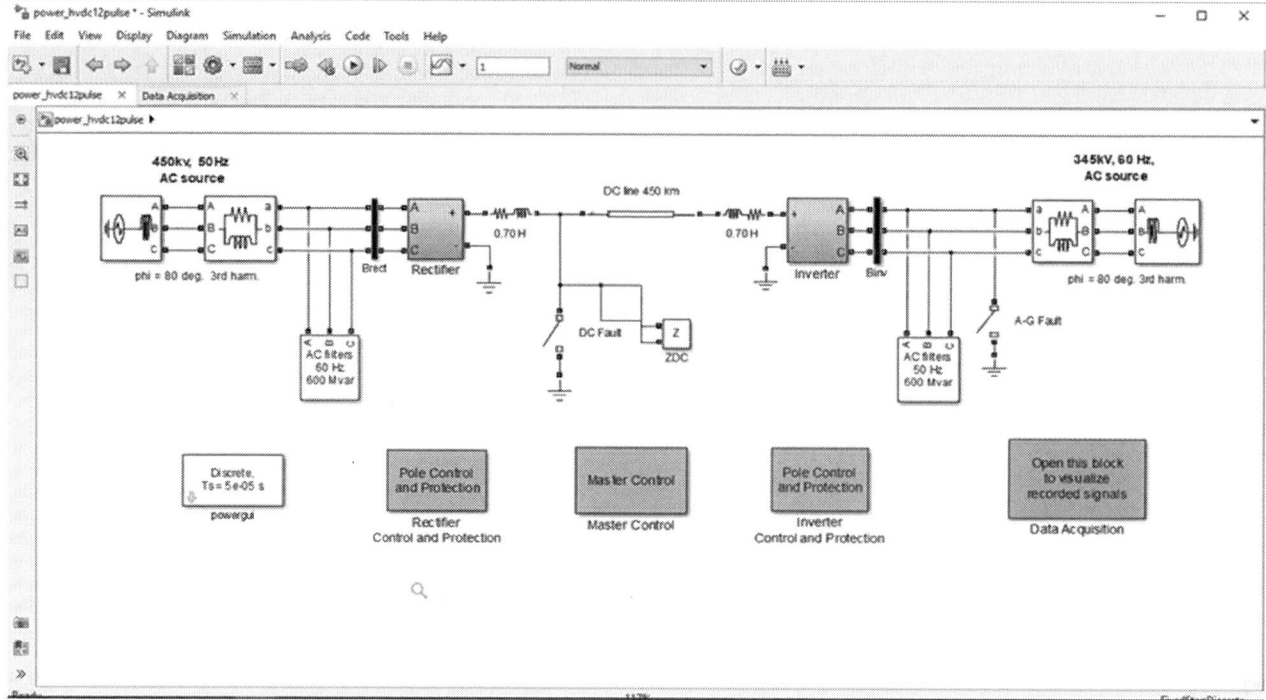

Fig. 12. Matlab/Simulink HVDC IGBT/DIODE based transmission model.

Figure 14. Rectifier at DC line abnormal conditions.

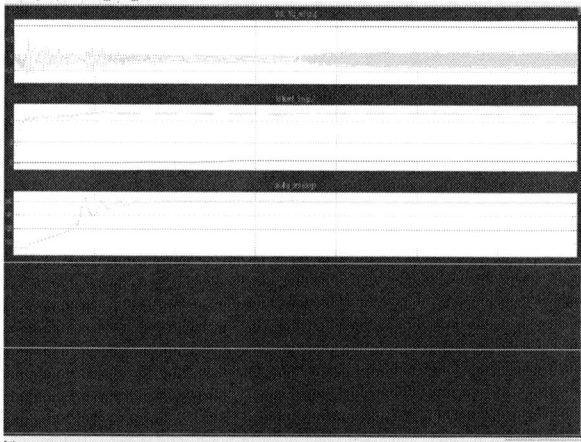

Figure 15. Inverter at DC line abnormal condition

Now, after the clearance out of the faulty condition the voltage finally results to be in constant. While the reference current can be seen, that it maintains or retains its accuracy after the occurrence of the dc line fault conditions. We can also notice that after the recovery of the faults at time 0.8-0.95seconds the voltage results to becoming too damping which may result or cause the lowering performance of the entire transmission process and it may lead to serious effects stronger than the AC systems. It can be seen in Fig. 15.

A) AC Line to Ground Fault condition

During the AC abnormal conditions on the DC transportation we can observe that the instability of the voltage with value of Vdc=0.6pu in which the abnormal behaviors results to dropping of the voltage to Vdc=-0.6pu and it begins to recover slowly at a t=0.9-1second. While the Id_ref also drops to 4pu.

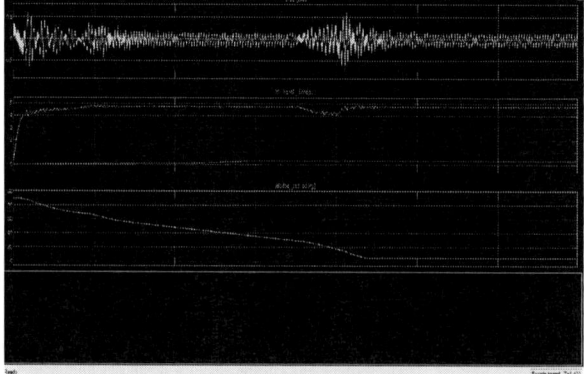

Figure 16. Single line to ground fault at the rectifier.

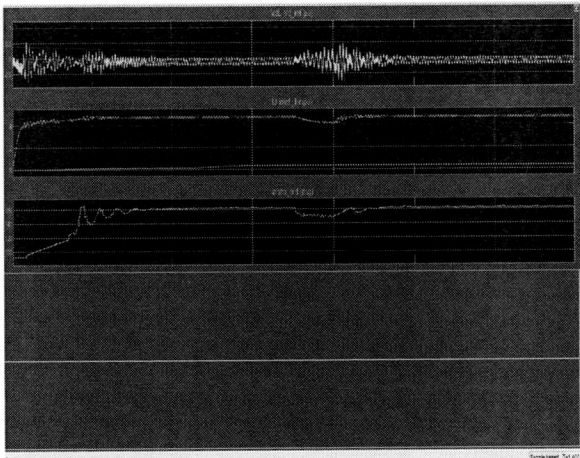

Figure 17. Single phase to ground fault at the inverter.

As can be seen the instability is reducing and are steadier at the inverter side having a Vdc=0.49pu when the abnormal occurs while the Id ref maintains its stability.

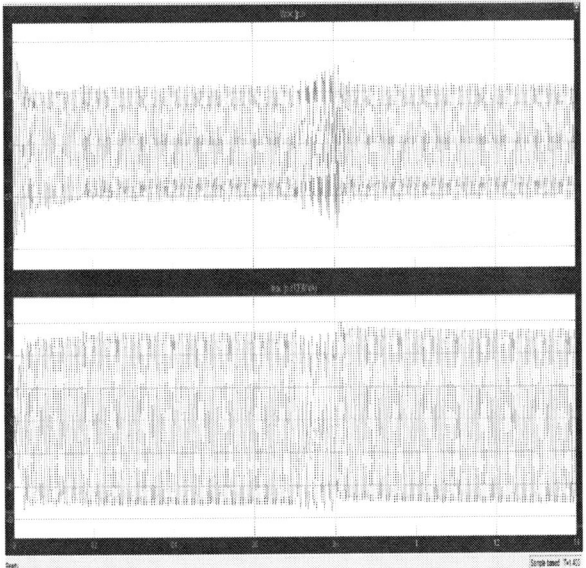

Figure 18. Variations of voltage and current on the 50Hz part for an AC line fault at the inverter section.

We can observe that the phases that are having the problem during the occurrences of the AC line to ground fault are phase A and phase C while phase becomes zero during the AC line faults.

IX. CONCLUSIONS

This paper has analyzed the nature behavior of the entire HVDC network process when they are subjected to some various abnormal conditions. The power electronic device that is responsible for this transformation/conversion process is the IGBT. It has duly been observed that whenever the fault conditions occur in the process such as the DC line transportation fault and single line fault, it can be noted that there is a huge instability or fluctuations at the voltage levels resulting to poor transmission of the process due to the problems that it causes during the processes. IGBT performance is poor in this situations because whenever the faults occurs it causes some parts of the voltage levels to be damping which in turns results to slowing the performance of the overall system. Voltage and current as regards to this they react firmly on the type of the abnormal conditions that occurs in the system. IGBT has very low performance in back-to-back HVDC configurations. And also it has zero recovery to any kind of fault that occurs in a system that is using the IGBT/DIODE as a means of conversion.

X. REFERENCES

[1] Alstom HVDC for Beginners and Beyond, Journal sept 2010, pp. 22-25, 31-36.

[2] M. Desta, "Design and Evaluation of HVDC System for the Grand Renaissance Hydroelectric Power Plant in Ethiopia", December 2012.

[3] A. M. Eltamaly, Y. Sayed, Amer Nasr A. Elghaffar, "A survey: HVDC System Operation and Fault Analysis", International Journal of Engineering Tome (IJET) November, 2017.

[4] K Sri Lakshmi, G. Sravanthi, L. Ramadevi, "A review paper on technical data of present HVDC links in India", International journal of Recent and Innovations trends in Computing and Communication, vol. 3, April 2015.

[5] S. C. Rajpoot, P. S. Rajpoot, K. Gupta, D. Sharma, "Analysis of Power Transmission Line with Unique Power Control Room", International Journal of Science and Research (IJSR) ISSN (Online): 2319-7064, January 2017.

[6] S. R. Durdhavale, Dyaneshwar D. Ahire "A rewiev of Harmonic Detection and Measurement in Power system", International Journal of Computer Applications (0975-8887), vol. 143, no.10, June 2016.

[7] Siemens, 2005, "High Voltage Direct Current Transmission" Retrieved January 1, 2019, from:https://www.energy.siemens.com/us/pool/hq/power-transmission/HVDC/ HVDC_Proven_Technology.pdfB.

[8] B. Kumar, "Design of Harmonic Filters for Renewable Energy Applications", Journal of Electrical & Electronic Systems, Sweden 2011.

[9] Larruskain, I. Zamora1, A.J. Mazón, O. Abarrategui1, J. Monasterio, "Transmission and Distribution Networks: AC versus DC", Department of Electrical Engineering University of the Basque Country, Bilbao (Spain).

[10] K. R. Padiyar, Smoothing Reactor in HVDC Power Transmission Systems Technology and System Interactions, New Delhi: New age international (P) Ltd, Publishers, p.14., 2005.

[11] M. Giraneza, High Voltage Direct Current in Applications for Distributed Independent Power Provider, Thesis, Cape Peninsula University of Technology, December 2013.

[12] R. Rudervvall, J .P. Charpentier, R. Sharma "High Voltage Direct Current (HVDC) Transmission Systems Technology Review Paper", Presented at Energy Week, Washington, D.C, USA, March 2000.

[13] J. Yu, G. G. Karady, L. Gu, "Applications of Embedded HVDC in Power System Transmission". Conference Paper, September 2012.

[14] Ishtaiwi, "An Overview of HVDC Application: A Study on Medium Voltage Distribution Networks", Power Engineering and Automation Conference, December 2017.

[15] C. L. Wadhwa, Scilab Textbook Companion for Electrical Power Systems, Sixth Edition New Age International Publishers New Delhi, 2010.

[16] H. Lasisi, S. O. Olayemi, "Power Improvement of Transmission Line Using High Voltage Direct Current (HVDC) Transmission System" American Journal of Engineering Research (AJER), vol.3, no. 4, pp. 66-75, 2014.

[17] N. Mohan, T. M. Undeland, W. P. Robbins, Power Electronics: Converters, Applications and Design, Third Edition, John Wiley and Sons, 2003.

Model Comparison and Parameter Estimation of Polymer Exchange Membrane (PEM) Fuel Cell Based on Nonlinear Least Squares Method

Krishnil R Ram
School of Engineering and Phsyics
University of the South Pacific
Suva, Fiji
ram_k@usp.ac.fj

Karteek Naidu
School of Engineering and Phsyics
University of the South Pacific
Suva, Fiji
s11131679@student..usp.ac.fj

Ravinesh Kumar
School of Engineering and Phsyics
University of the South Pacific
Suva, Fiji
s11134281@student.usp.ac.fj

Maurizio Cirrincione
School of Engineering and Phsyics
University of the South Pacific
Suva, Fiji
maurizio.cirrincione@usp.ac.fj

Ali Mohammadi
School of Engineering and Phsyics
University of the South Pacific
Suva, Fiji
ali.mohammadi@usp.ac.fj

Abstract— The comparison of different single cell PEMFC models and the estimation of model parameters are described in this article. Three different models are selected from literature and reviewed. For each PEMFC model the polarization curves were developed and then compared. Nonlinear Least Squares Method have been implemented to find the parameters of each model. The results show that some of the parameters estimated are in agreement with the experimental and simulation model polarization curves while there are more errors in the estimation of the parameters describing the activation losses.

Keywords—Least Square, PEMFC, Modeling, Estimation Parameters

I. INTRODUCTION

A Fuel Cell is an electrochemical device that produces electricity from the chemical reaction between hydrogen and oxygen. The by-products are heat and water – making Fuel Cells (FC) an obvious choice in modern sustainable transportation and stationary power applications. Several electromechanical auxiliary components add up to a FC to form the Fuel Cell System (FCS) [1] . There are, actually, different types of FCS [2]. This paper focusses on the Proton Exchange Membrane Fuel Cell (PEMFC) within these systems. As well known, in the case of a PEMFC a thin permeable polymer electrolyte is used in the cell [3] which also leads to it being called a polymer electrolyte membrane fuel cell. PEMFC have gained popularity due to the possibility to be used in low temperature mobile applications like electric vehicles[4], [5] . One of the key areas of development in FCS technology is their design, simulation and control. However, the simulation or control of FCS is not possible unless a detailed mathematical model of the FCS exists. Accurate system identification and modelling of the FCS, therefore, becomes an essential prerequisite for efficient use of FCS. While mathematical models exist in literature for FCS, there are certain parameters in these models that are influenced by the geometry, physical and chemical properties of elements inside the cell and are generally considered as constants. In addition, there are coefficients in the FCS model, which need to be established empirically to completely model the system. Pukrushpan [6] gives a very detailed model of the PEMCF for control purposes after applying nonlinear regression to determine unknown parameters. A lot of PEMFC modelling relies on the works of J C Amphlett [7] on a Ballard Mark IV

PEM cell. The single cell PEMFC model forms the basis of FC research and this paper looks at some static models and their parameter estimation using least squares methods. Limited work can be found in the area of parameter estimation for PEMFC and this still remains an area under improvement. Askarzadeh and Rezazadeh [8] proposed an Innovative Global Harmony Search (IGHS) Algorithm for the parameter estimation in PEM fuel cells. Forrai et al [9] presented parameter estimation results using a current interrupt test and a system identification approach for diagnostics purposes. In another study, Priya et al [10] have utilized a simple Genetic Algorithm Optimization to estimate the parameters of the Fuel Cell. Genetic algorithms however, may suffer from premature convergence and weak exploitation capabilities [11] [12]. Particle Swarm Optimization (PSO) is another artificial intelligence algorithm which has been successfully used for parameter estimation of a PEMFC [13]. Meiying Ye et al [14] report on a PSO for a PEMFC parameter estimation using experimental and simulated data. They claim that the PSO method outperforms the GA and other optimization methods for parameter estimation. It is also a known fact that the PEMFC model parameters are influenced by the changes in operating conditions [15] such as moisture content, temperature , pressures , flow rates, humidity, reactant purity, load current, membrane electrode assembly properties etc. While most artificial intelligence and complex algorithms do provide reasonable accuracy, their computation time is large and they require several runs to achieve the results. A faster method would be very useful and may lend itself to use in real-time operation to monitor any change in parameters with regard to operating conditions. Wu and Chen [16] apply a moving –horizon least squares estimator in order to estimate FC parameters in the presence of time varying disturbances. They approximate the static nonlinearity parts by a series of piecewise linear functions. Moreira and Silva [17] also presented a simpler method of parameter estimation and modelling using a linear least squares method. The least squares method simply tries to fit the simulated and experimental values to minimize the sum of squared residuals. The method of least squares is also used for parameter estimation of several other systems of interest with ease [18]. The experimental data (polarization curve) of the fuel cell is normally used to approximate a close fit curve using the estimated parameters. Several researchers have proposed PEMFC models, which have very slight differences. Horng

978-1-5386-7688-2/19 $31.00 © 2019 IEEE

Wen Wu [19] presents a review of transport processes and performance modeling of FC systems. In the review, [19] highlighted the need for a more detailed model of the FC system. The improvements in FC technology is directly related to accurate modeling of the systems and estimation of parameters. In a recent review of parameter estimation techniques [15], stated that there is no literature which serves as a reference point for parameter estimation in FC technology and the authors in [15] do present a comprehensive discussion on different parameter estimation techniques. Parameter estimation needs to be based on a sound model. This paper attempts to focus on the differences and similarities between the models used by different researchers and the effect of parameter estimation on them. The current study is part of a larger study focused on comparing different models and parameter estimation techniques. This paper compares the three wide-spread PEMFC models for FCS [6], [20], [21] in same conditions and experimentally by using a nonlinear least-squares method. Following the model discussion, the parameters identified in each model will be estimated and their respective errors compared.

II. MATHEMATICAL MODELS FOR PEMFC

A. Model I

First, Model I is based on [6] and provides the equations to model a single cell of the PEMFC. The cell voltage is given by subtracting the losses from a theoretical maximum Nernst voltage as shown:

$$v_{fc} = E - (v_{act} + v_{ohm} + v_{conc}) \qquad (1)$$

The single cell output voltage has a theoretical Nernst voltage, which takes into account the temperature, and partial pressures of the gases. The chemical and thermodynamic losses of potential through activation voltage, ohmic voltage and concentration voltage are then subtracted from the Nernst voltage. The Nernst voltage, which gives the thermodynamic potential of the reactions [7], is:

$$E = 1.229 - 0.85 \times 10^{-3}(T - 298.15) + 4.3085 \times 10^{-5} T \left[\log(p_{H_2}) + \tfrac{1}{2}\log(p_{O_2})\right] \qquad (2)$$

The fuel cell temperature is T while p_{H_2} is the partial pressure of hydrogen and p_{O_2} is the partial pressure of oxygen. The Gibbs standard free energy for the reaction $-\Delta G$ at temperature of 258C is around -237.3 kJ/mol and leads to a standard reversible potential of 1.229V (E_o). This is the highest value that can be obtained at this standard temperature. As the temperatures rises, this potential drops by $0.85 \times 10^{-3} V$ for each degree and is captured in the Nernst voltage equation. The value of E_o is found using the following formula:

$$E_o = \frac{-\Delta G}{nF} \qquad (3)$$

Where F is Faradays constant and n is the number of electrons per molecule of hydrogen. The activation losses arise due to the energy required to break the bonds and the sluggish rate of reactions at the electrode surface [22]. Model I describes the activation losses as: =

$$v_{act} = v_0 + v_a(1 - e^{-c_1 i}) \qquad (4)$$

In this case v_a, v_o and c_1 are constants that need to be estimated. The next set of parameters are found in the ohmic potential losses. The resistance of electron flow through conducting electrodes and the resistance of ion flow through the membrane gives rise to ohmic losses defined as:

$$v_{ohm} = i \, R_{ohm} \qquad (5)$$

The ohmic loss is proportional to the ohmic resistance density in the membrane and electrode. This resistance is calculated through the use of membrane thickness t_m and the membrane conductivity σ_m.

$$R_{ohm} = \frac{t_m}{\sigma_m} \qquad (6)$$

The membrane conductivity is derived using the membrane water content, temperature and several constants:

$$\sigma_m = (b_1 \lambda_m - b_2) exp\left(b_3 \left(\frac{1}{303} - \frac{1}{T}\right)\right) \qquad (7)$$

In the above equation, and overall in the ohmic losses equation there are three parameters b_1, b_2 and b_3 which need to be estimated. The parameter λ_m denotes the membrane water content with a lower limit of 0 and upper limit of 14 [6]. The concentration losses arise due to concentration gradients as the reactants get used up at the cell surfaces and the slow transport of reactants to and from the reaction sites [23].

$$v_{conc} = i \left(c_2 \frac{i}{i_{max}}\right)^{c_3} \qquad (8)$$

In total, there are eight parameters that need to be estimated. There are three constants, c_2, c_3 and i_{max} which need to be retrieved in order to retrieve the concentration losses. The parameter i_{max} is the maximum current density, which causes an abrupt loss in voltage. There are three parameters arising from activation losses (v_a, v_o and c_1). For ohmic losses, a single R_{ohm} parameter is retrieved.

B. Model II

Model II is based on [21] for the single cell voltage model of the PEMFC, which considers the same overall equation of losses as is in equation (1). The Nernst voltage equation is the same as in equation (2) for Model I while the activation voltage loss is given by:

$$V_{act} = a_1 + b_4 \log(i) \qquad (9)$$

Where the parameter a_1 and Tafel slope b_4 are defined using the transfer coefficient α.

$$a_1 = -\frac{RT}{\alpha F} \log(i_o) \quad and \quad b_4 = \frac{RT}{\alpha F} = 0.025 \qquad (10)$$

Where R is the universal gas constant and is i_o the exchange current density which depends on catalyst area and reactant partial pressures [24]. Under ohmic losses, all forms of resistance such as ionic, electronic and contact resistance may be added up to form a single parameter called the total cell internal resistance R_i. This is the same as ohmic resistance stated in [6].

The concentration losses are defined using the limiting current density i_L and can be stated as:

$$V_{conc} = \frac{RT}{nF} \log \left(\frac{i_L}{i_L - i}\right) \qquad (11)$$

In Model II five parameters need to be estimated. These are E, a_1, α, R_{ohm} and i_L.

C. Model III

Model III is based on the work of [21]. The Nernst voltage is expressed in many forms however; the equations can be derived to be the same as in equation (2) of Model I.

The activation losses are defined using the Tafel equation in terms of the exchange current density i_o and a parameter b_5.

$$V_{act} = b_5 \log\left(\frac{i}{i_o}\right) \tag{12}$$

Model II states that the ohmic losses arise mainly from the ionic resistances in the electrolyte and the ohmic loss equation is the same as in Model I. The concentration losses can be stated as:

$$V_{conc} = \alpha_1 i^k \log\left(1 - \frac{i}{i_L}\right) \tag{13}$$

Where the α_1 is the amplification constant, k is the mass transport constant and i_L is the limiting current of the cell. For Model III the cell voltage requires the estimation of these three parameters in the concentration losses. The other parameters are b_5 and i_o. In this was the cell voltage (equation 1) in Model III can be expanded as:

$$v_{fc} = E - b_5 \log\left(\frac{i}{i_o}\right) - i\, R_{ohm} - \alpha_1 i^k \log\left(1 - \frac{i}{i_L}\right) \tag{14}$$

The Nernst voltage E is found through equation (2). The Nernst equation in all three models requires the calculation of gas species partial pressures. Model III calculates the partial pressures empirically using the saturation pressure of water P_{H2O}, which is stated as (P_{sat}) in other models

$$p_{H2} = 0.5\left(\frac{P_{H2}}{exp\left(1.653 * \frac{i}{T_k^{1.334}}\right)} - P_{H_2}o\right) \tag{15}$$

$$p_{O2} = \left(\frac{P_{air}}{exp\left(4.192 \frac{i}{T_k^{1.334}}\right)} - P_{H_2}o\right) \tag{16}$$

Model III uses the following coefficients for saturation pressure calculation.

$$\log_{10} P_{sat} = -2.1794 + 0.02953\, T_c - 9.1837 \times 10^{-5} T_c^{2} + 1.4454 \times 10^{-7} T_c^{3} \tag{17}$$

Here T_c is the temperature in degrees Celsius. Model I [6] derives the partial pressure based on the mass of the gases m, the volumes (V_{ca} for cathode and V_{an} for anode), temperature and the respective gas constants R.

$$p_{O_2} = \frac{m_{O_2} R_{O_2} T}{V_{ca}} \tag{18}$$

$$p_{H_2} = \frac{m_{H_2} R_{H_2} T}{V_{an}} \tag{19}$$

Model I utilizes the saturation pressure equation proposed in [25] to determine the mass values for partial pressure calculations.

$$\log_{10} P_{sat} = -1.69 \times 10^{-10} T^4 + 3.85 \times 10^{-7} T^3 - 3.39 \times 10^{-4} T^2 + 0.143 T - 20.92 \tag{20}$$

III. Methodology

The three models were created in MATLAB® software and a nonlinear least squares method using the Levenberg-Marquardt algorithm was used to estimate the parameters. These were retrieved and compared to those available in literature for the above 3 models. Afterwards experimental data was used to estimate the parameters by using the polarization curve of the experimental data points. For the non-linear least squares application, a set of current-voltage data points $\{(i_1, v_1) \ldots, (i_n, v_n)\}$ have been extracted from the available polarization curves obtained by the parameters. The inputs i_l $(l=1, \ldots n)$, the current density, would form the input vector $\boldsymbol{i}$, the outputs v_l $(l=1, \ldots n)$, the cell voltage, would form the output vector $\boldsymbol{v}$. The polarization curve based on the estimated parameters has been then compared with the previous one and the error computed both for each parameters and for all polarization curve. The estimated polarization curve was denoted another set of voltage values, given by the column vector $\boldsymbol{v_{est}}$ as follows:

$$\boldsymbol{v_{est}} = f(\boldsymbol{i}, \boldsymbol{\theta}) \tag{21}$$

Where the vector $\boldsymbol{\theta}$ is made up of the parameters to be identified, whose number changes according to the model. For example in the case of Model III:

$$\boldsymbol{\theta} = (E, b_5, i_o, R_{ohm}, k, i_L)^T \tag{22}$$

The error in estimation can be given by the scalar E given by the Euclidean norm of the error between the estimated and the actual output voltages:

$$E = \|\boldsymbol{v} - \boldsymbol{v_{est}}\|_2 \tag{23}$$

The error is then minimized by using a second order method, like the Levenberg Marquardt one.

IV. Simulation Results

Keep In the following, the three models have been compared with same input conditions. The input pressure for both hydrogen and oxygen (air) was modelled at 1 bar and the temperature of 25°C. The parameters in each model were estimated using the Levenberg Marquardt Non Linear Least Squares Method and compared to the original parameter supplied by the authors of the models in literature. The percentage error for each parameter is provided together with the error E given by (23).

A. Model I

Fig. 1 shows the activation, ohmic and concentration loss zones in Model I with increasing current density.

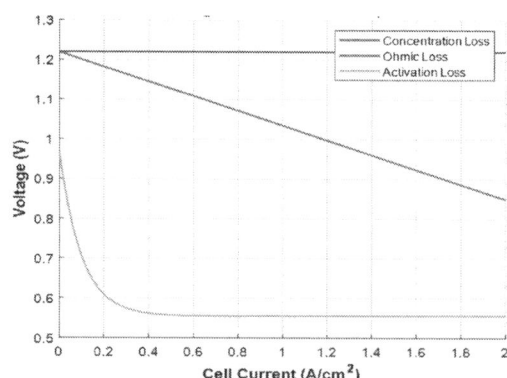

Fig. 1 Voltage losses in Model 1

978-1-5386-7688-2/19 $31.00 © 2019 IEEE

The parameters used in Fig. 1 are shown in table 1, together with the error on the parameter estimation and the norm of the error. Note the Nernst voltage is constant as expected. Activation losses appear more pronounced at lower current densities. Following the parameter estimation with least squares, Table 1 provides a summary of the estimated parameters in Model I. All parameters have been estimated on the basis of the norm of cost function less than 10^{-6}.

TABLE 1. ESTIMATED AND ACTUAL PARAMETERS FOR MODEL 1 (NORM OF ERROR= 10.2)

Parameter	C_2/i_{max}	C_3	R_{ohm}	v_o	v_a	c_1
Original	0.09	2	0.15	0.13	0.3	10
Estimated	0.089	1.93	0.15	0.131	0.3	9.97
Error %	1.1	3.5	0	0	0	1

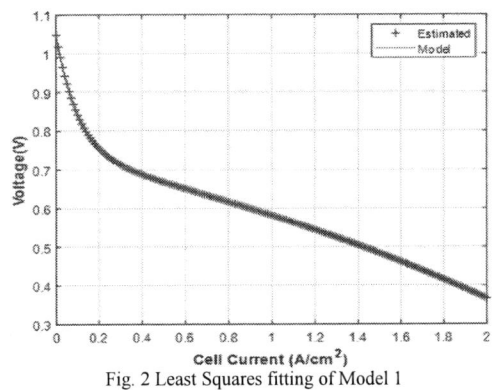

Fig. 2 Least Squares fitting of Model 1

The estimated parameters were used to simulate the polarization curve within the same operating conditions and the comparison is presented in Fig. 2. Despite the under – estimation of the two parameters in Table 1, the estimated polarization curve data fits onto the model curve.

B. Model II

Model II displays similar trends in voltage losses however; the activation losses are more gradual with a sharp loss at lower current densities as seen in Fig. 3. The parameters used to get the results in Fig. 3 are shown in Table 2. In all three models, the ohmic losses behave linearly with current density owing to the linear nature of its equation in all models.

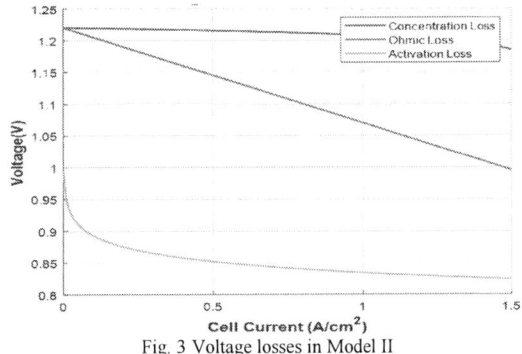

Fig. 3 Voltage losses in Model II

The estimated polarization data points are shown on the model curve in Fig. 4 following least squares parameter estimation. While the estimated and model points are in agreement, it is useful to note that activation losses at lower current are not well defined. The initial sudden loss on the polarization curve indicates activation losses.

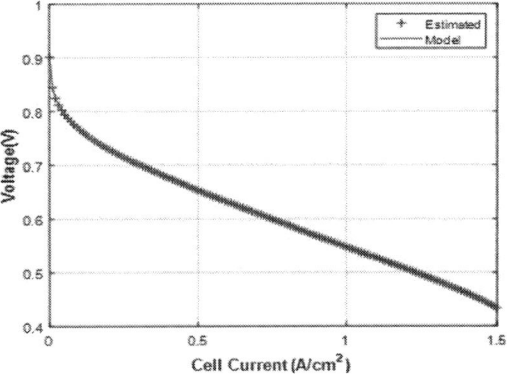

Fig. 4 Least Squares fitting of Model II

TABLE 2. ESTIMATED AND ACTUAL PARAMETERS FOR MODEL II (NORM OF ERROR =1.52)

Parameter	R_{ohm}	i_L	i_o
Original	0.15	1.6	$3e^{-7}$
Estimated	0.153	1.53	$3.3e^{-7}$
Error %	0	4.3	0

C. Model III

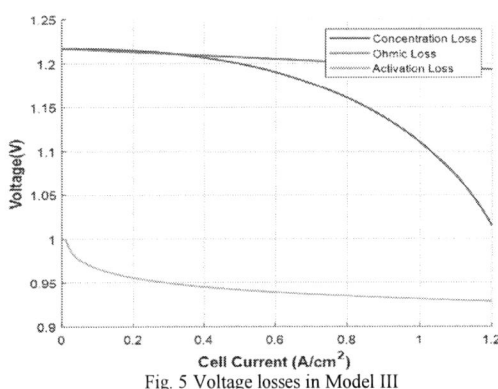

Fig. 5 Voltage losses in Model III

In Model III, the activation losses behave very similar to Model II given the Tafel equations were identical. The parameters used to get the results in Fig. 5 are shown in Table 3. The presence of the mass transport coefficient k as a power of the current density ensures a more pronounced concentration loss as the current density increases.

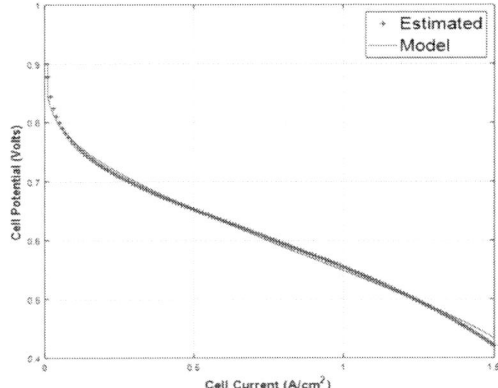

Fig. 6 Least Squares fitting for Model III

978-1-5386-7688-2/19 $31.00 © 2019 IEEE

Fig. 6 shows the data points plotted using the estimated parameters. Ohmic losses appear linearly as in all three models and the Nernst voltage is constant across the current densities.

TABLE 3 ESTIMATED AND ACTUAL PARAMETERS FOR MODEL III (NORM OF ERROR =1.57)

Parameter	R_{ohm}	b_5	i_o	α	K	i_L
Original	0.19	0.03	1.2e-7	0.85	0.87	1.4
Estimated	0.19	0.01	1e-7	0.7	1.1	1.29
Error %	0	66.6	16	17	26	7.8

V. EXPERIMENTAL RESULTS

The PEMFC which has been used for the experimental validation has a membrane with an active area of 62 cm²; a gas diffusion layer with a thickness is equal to 0.42 mm and a graphic block in the anode and the cathode side. This PEMFC is shown in Fig. 7. More details about the component characteristics are mentioned in Table 4.

Fig. 4 The stack PEM fuel cell

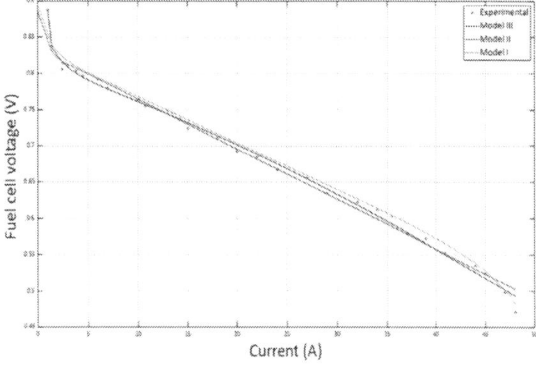

Fig. 5 Experimental and model polarization curves

TABLE 4. SPECIFICATION OF FUEL CELL

Description	Values
Stoichiometry air	1.5
Stoichiometry hydrogen	3
Number of the cell	42
Surface area	62 Cm²
T° max	65 °C

The results of all three models were also compared to experimental polarization curves of the PEMFC presented in [26] as shown in Fig. 8. While the overlapping polzarization curves of the three different models are mostly in agreement, the experimental results are better approximated in the linear ohmic losses region. Fig. 9 shows the absolute error of each model with parameter estimation compared to the experimental results as shown in Fig. 8. Despite the reasonable estimation of the transfer coefficients α_a, α_c and α_l for the activation voltage, the estimated data is not in agreement for the activation losses of the model. This could be due to the number of other variables like saturation and electrode area which need to be estimated separately for more accurate modelling of activation losses on the polarization curves.

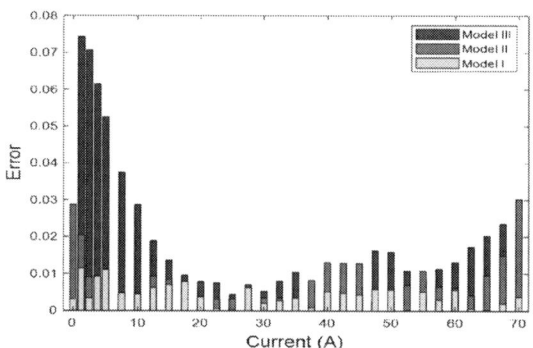

Fig. 6 Error in each model with experimental results

At lower current densities, Model III has higher absolute error. This could be the result of poor parameter estimation for activation losses, which normally occur at lower current densities. For higher current densities, the losses in the polarization curves are mostly linear due to ohmic losses and errors in this region of current density are almost randomly distributed and very small across the three models.

TABLE 5. NOMENCLATURE

Parameter	Definition	Value
T	Fuel cell temperature	°C
i	cell current density	A/cm²
P_{sat}	Water saturation pressure	bar
P_{H_2}	Hydrogen partial pressure	bar
P_{O_2}	Oxygen partial pressure	bar
p_{H_2}	Hydrogen pressure	bar
E	Nernst input voltage	V
P_{ca}	Cathode pressure	bar
t_m	Membrane thickness	μm
λ_m	Membrane water content	-
i_{max}	Maximum Current Density	A/cm²
A_{fc}	Active area of the fuel cell	cm²
F	96487	C
R	8.31	J/molK
α	Transfer Coefficient	1
n	Number of moles of electrons	2

VI. CONCLUSION

Three well-known PEMFC single cell models have been compared and modelled by using the nonlinear least squares method. These three models were discussed the approximation presented both in simulation and on experimental data. The three models provided very similar outputs with slight variation only. The major difference was in the modelling of activation losses at low current densities. The error is modelling is linked to other hidden variables that cannot be identified by external measurement of voltages and currents. Future work in this study will involve the use of other methods such as Total least Squares (TLS) and Genetic Algorithm (GA) optimization to estimate these parameters. Further work also needs to be done on the accurate modeling of activation losses and accurate estimation of parameters related to activation losses. A promising technique which is under investigation by the authors is the Allan Variance analysis, which does not need the construction of an iterative regressor [27] [28].

REFERENCES

[1] A. Accetta, M. Cirrincione, A. Mohammadi, M. Pucci, and K. Ram, "Model Predictive Control Based Air Management of a PEM Fuel Cell System," presented at the International Conference on Emerging and Renewable Energy: Generation and Automation, Tunisia, 2018.

[2] G. Hoogers, *Fuel cell technology handbook*. CRC press, 2002.

[3] J. Balakrishnan, "Fuel cell technology," in *2007 Third International Conference on Information and Automation for Sustainability*, 2007, pp. 159–164.

[4] E. Breaz, F. Gao, B. Blunier, and R. Tirnovan, "Mathematical modeling of proton exchange membrane fuel cell with integrated humidifier for mobile applications," in *2012 IEEE Transportation Electrification Conference and Expo (ITEC)*, 2012, pp. 1–6.

[5] J. Samir, C. Adnen, B. S. Sami, and V. E. Balas, "An efficient design of Fuel Cell Electric Vehicle with Ultra-Battery separated by an energy management system," in *2016 7th International Conference on Sciences of Electronics, Technologies of Information and Telecommunications (SETIT)*, 2016, pp. 29–33.

[6] "Control of Fuel Cell Power Systems - Principles, Modeling, Analysis and Feedback Design | Jay T. Pukrushpan | Springer." [Online]. Available: https://www.springer.com/gp/book/9781852338169. [Accessed: 02-May-2019].

[7] J. C. Amphlett, R. M. Baumert, R. F. Mann, B. A. Peppley, P. R. Roberge, and T. J. Harris, "Performance Modeling of the Ballard Mark IV Solid Polymer Electrolyte Fuel Cell II . Empirical Model Development," *J. Electrochem. Soc.*, vol. 142, no. 1, pp. 9–15, Jan. 1995.

[8] A. Askarzadeh and A. Rezazadeh, "An Innovative Global Harmony Search Algorithm for Parameter Identification of a PEM Fuel Cell Model," *IEEE Trans. Ind. Electron.*, vol. 59, no. 9, pp. 3473–3480, Sep. 2012.

[9] A. Forrai, H. Funato, Y. Yanagita, and Y. Kato, "Fuel-cell parameter estimation and diagnostics," *IEEE Trans. Energy Convers.*, vol. 20, no. 3, pp. 668–675, Sep. 2005.

[10] K. Priya, T. Sudhakar Babu, K. Balasubramanian, K. Sathish Kumar, and N. Rajasekar, "A novel approach for fuel cell parameter estimation using simple Genetic Algorithm," *Sustain. Energy Technol. Assess.*, vol. 12, pp. 46–52, Dec. 2015.

[11] A. Sari, C. Espanet, and D. Hissel, "Particle swarm optimization applied to the co-design of a fuel cell air circuit," *J. Power Sources*, vol. 179, no. 1, pp. 121–131, Apr. 2008.

[12] Fagiolini, A., Babboni, F. and Bicchi, A., "Dynamic distributed intrusion detection for secure multi-robot systems"", IEEE International Conference on Robotics and Automation (pp. 2723-2728). IEEE, 2009.

[13] Q. Li, W. Chen, Y. Wang, S. Liu, and J. Jia, "Parameter Identification for PEM Fuel-Cell Mechanism Model Based on Effective Informed Adaptive Particle Swarm Optimization," *IEEE Trans. Ind. Electron.*, vol. 58, no. 6, pp. 2410–2419, Jun. 2011.

[14] M. Ye, X. Wang, and Y. Xu, "Parameter identification for proton exchange membrane fuel cell model using particle swarm

optimization," *Int. J. Hydrog. Energy*, vol. 34, no. 2, pp. 981–989, Jan. 2009.

[15] K. Priya, K. Sathishkumar, and N. Rajasekar, "A comprehensive review on parameter estimation techniques for Proton Exchange Membrane fuel cell modelling," *Renew. Sustain. Energy Rev.*, vol. 93, pp. 121–144, Oct. 2018.

[16] W. Wu and H.-T. Chen, "Identification and Control of a Fuel Cell System in the Presence of Time-Varying Disturbances," *Ind. Eng. Chem. Res.*, vol. 54, no. 28, pp. 7141–7147, Jul. 2015.

[17] M. V. Moreira and G. E. da Silva, "A practical model for evaluating the performance of proton exchange membrane fuel cells," *Renew. Energy*, vol. 34, no. 7, pp. 1734–1741, Jul. 2009.

[18] N. I. Jannif, G. Cirrincione, M. Cirrincione, A. Mohammadi, and G. Vitale, "Experimental application of least-squares technique for estimation of double layer super capacitor parameters," in *2017 20th International Conference on Electrical Machines and Systems (ICEMS)*, 2017, pp. 1–5.

[19] H.-W. Wu, "A review of recent development: Transport and performance modeling of PEM fuel cells," *Appl. Energy*, vol. 165, pp. 81–106, Mar. 2016.

[20] F. Barbir, *PEM fuel cells: theory and practice*. Academic Press, 2012.

[21] C. Spiegel, *PEM fuel cell modeling and simulation using MATLAB*. Elsevier, 2011.

[22] K. Sedghisigarchi and A. Feliachi, "Dynamic and transient analysis of power distribution systems with fuel Cells-part I: fuel-cell dynamic model," *IEEE Trans. Energy Convers.*, vol. 19, no. 2, pp. 423–428, Jun. 2004.

[23] J. Hua, L. Xu, X. Lin, and M. Ouyang, "Modeling and experimental study of PEM fuel cell transient response for automotive applications," *Tsinghua Sci. Technol.*, vol. 14, no. 5, pp. 639–645, Oct. 2009.

[24] Y. Zhan, J. Zhu, Y. Guo, and Hua Wang, "Performance analysis and improvement of a proton exchange membrane fuel cell using comprehensive intelligent control," in *2008 International Conference on Electrical Machines and Systems*, 2008, pp. 2378–2383.

[25] T. V. Nguyen and R. E. White, "A Water and Heat Management Model for Proton☐Exchange☐Membrane Fuel Cells," *J. Electrochem. Soc.*, vol. 140, no. 8, pp. 2178–2186, Aug. 1993.

[26] A. Mohammadi, G. Cirrincione, A. Djerdir, and D. Khaburi, "A novel approach for modeling the internal behavior of a PEMFC by using electrical circuits," *Int. J. Hydrog. Energy*, vol. 43, no. 25, pp. 11539–11549, Jun. 2018.

[27] N. El-Sheimy, H. Hou and X Niu, "Analysis and Modelling of Inertial Sensors Using Allan Variance", IEEE Transactions on Instrumentation and Measurement, vol. 57, n. 1, pp. 140-149, 2008.

[28] D'Alessandro, A., Vitale, G., Scudero, S., D'Anna, R., Costanza, A., Fagiolini, A. and Greco, L., "Characterization of MEMS accelerometer self-noise by means of PSD and Allan Variance analysis", IEEE International workshop on advances in sensors and interfaces (IWASI) (pp. 159-164). IEEE, 2017.

978-1-5386-7688-2/19 $31.00 © 2019 IEEE

Modeling and Analysis of a Renewable-energy-powered Greenhouse

Yerbol Akhmetov
Department of Electrical and Computer Engineering
Nazarbayev University
Nur-Sultan, Kazakhstan
yerbol.akhmetov@nu.edu.kz

Mehdi Bagheri
Department of Electrical and Computer Engineering
Nazarbayev University
Nur-Sultan, Kazakhstan
mehdi.bagheri@nu.edu.kz

G. B. Gharehpetian
Department of Electricla Engineering
Amirkabir University of Technology
Tehran, Iran
grptian@aut.ac.ir

Abstract—The role of renewable energy powered greenhouses in the development of sustainable agriculture is crucial. Such self-powered standalone greenhouses promise improvement in agriculture in outland regions with incongruous farming conditions and unavailable distribution system, and propose healthy and available food for customers without seasonal interruptions. This study presents the modeling and design of a renewable-energy-powered greenhouse that will be situated in Almaty, Kazakhstan. The geographical location of the greenhouse poses challenges related to the availability of sunlight and substantial annual thermal variation. The greenhouse model presented in this study includes parabolic trough collector (PTC), an earth-to-air heat exchanger (EAHE), and micro combined heat and power (mCHP) as input heating sources and photovoltaic cells (PV) as an electricity generation unit for greenhouse operation. The paper proposes thermal RC modeling of the greenhouse and its components. The results of the simulation demonstrate that the PV cells can generate up to 4.5 kW peak power in the daytime, while the greenhouse air temperature can be sustained between 19 °C and 21 °C.

Keywords—*Micro combined heat and power; Earth-to-air heat exchanger; Greenhouse; Parabolic trough collector; Photovoltaic panels; Renewable energy*

I. INTRODUCTION

The greenhouse is a prominent feature of modern agriculture. It has a significant role in sustaining farming in outland areas with harsh environments, such as arid regions and territories with cold winters. With the rapid development in renewable energy applications, designing and operating greenhouses based on renewable energy becomes attractive [1]. Such a trend towards the development of renewable-energy-based greenhouses is appealing not only because of the sustainability they will enable but also because of autonomy, which is another preferred attribute of outland greenhouses.

There are various techniques for greenhouse energy storage, namely, high-power density batteries, underground heat storage, phase change materials, and hydrogen electrolyze-based energy storage [2]. Furthermore, the sources of renewable energy can be wind, hydropower, geothermal, and solar. However, last two resources are more feasible for this application, because solar energy is more accessible and viable for small-scale electricity generation, and geothermal sources can provide continuous heating to greenhouses. For instance, a design of geothermally heated greenhouse which employs a ground-source heat pump and earth-to-air heat exchanger is proposed in [2] and [3], respectively. Typically, the greenhouse with such heating system has a 6-7 °C higher temperature than a greenhouse without it during the night in winter [3].

On the other side, greenhouses with solar energy-based heating systems are presented in [4]. In [5], solar air collectors located in the roof of the structure were employed for heating and thermal storage. Additionally, solar energy can be utilized for heating liquids which can be stored and circulated through a greenhouse for heat exchange [6]. The study in [6] presented a parabolic trough collector connected to a pump and an oil tank for greenhouse heating. The results obtained from a simulation demonstrated that such greenhouse heating has desirable potential in agriculture.

Generally, thermal modeling is a primary tool for evaluating the energy efficiency of a greenhouse. It includes formulating and solving energy balance equations using the heat transfer between the greenhouse and its surrounding environment [7]. The magnitude of the energy transfer is dependent on the geometrical shape, dimensions, and materials of the greenhouse room, and the climatic conditions of the environment [1]. Additionally, it is essential to become familiar with the geographical location of the greenhouse and the seasonal thermal variation and solar radiation of the region. This assists in identifying the approximate amount of energy required for the greenhouse and power output and efficiency of the solar panels.

Apart from the thermal assessment and modeling of heat sources, photovoltaic (PV) electricity generation has huge relevance to the development of sustainable greenhouses. One such design of PV cells was presented in [8], where an integrated photovoltaic-thermoelectric generator (PV/TEG) was designed for an automated greenhouse. Moreover, the thermal analysis of a PV cell is important since the efficiency of solar panels is dependent on their temperature [8].

This research proposes a renewable-energy-powered greenhouse model suitable for operating in outland regions, where the possibility of supplying the greenhouse with distribution system is minimal. The model includes heating elements, such as a parabolic trough collector (PTC), an earth-to-air heat exchanger (EAHE), and micro combined heat and power (mCHP), and photovoltaic cells (PV). In contrast to other research studies in this field where only one or two of the renewable energy sources are considered simultaneously, this work is focused on integrating several of the most efficient and feasible energy supply elements into single design. Such approach can improve the energy efficiency of the greenhouses and increase their autonomy. The location of the modeled greenhouse is the Almaty region, where the annual temperature varies between -10 °C and 32 °C [9]. However, this study should be considered as a design guide for greenhouse construction in outland areas with harsh conditions.

II. METHODOLOGY

A. Modeling of Greenhouse

Prior to discussing the thermal model of the greenhouse, it is important to consider the overall system operation and

978-1-5386-7688-2/19 $31.00 © 2019 IEEE

its physical dimensions. A CAD model of the proposed greenhouse is presented in Fig. 1. It is a medium-sized greenhouse with floor dimensions of 6 m × 4 m. The total height of the greenhouse is 3 m, while the heights of the north and south walls are 2 m. The north wall is designed to be a brick wall as suggested in [3] since the location of the greenhouse is in the Northern Hemisphere. In contrast to [3], the north roof is made of thick insulation material similar to the north wall. The inclination angle of the north roof is 45°, while the length of the roof is 1 m. It is designed in such a way that it does not block sunlight from falling onto plants at noon-time in the Almaty region. For different locations, the north roof can be redesigned using the local latitude and the solar inclination angle. Other walls, such as the south, east, west, and south roofs are made of a glass cover.

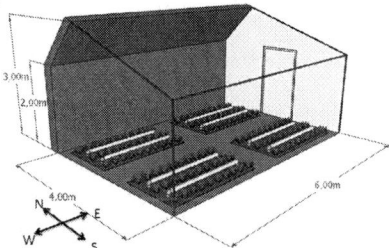

Fig. 1. 3D CAD model of the greenhouse

Apart from the main structure, the proposed greenhouse has several other key components, such as EAHE, PTC, mCHP, and PV panels. A block diagram of the complete system is provided in Fig. 2. This figure presents the heating, electricity generation, energy storage and control units. The greenhouse heating is accomplished by the EAHE, PTC, and mCHP. The first two methods are the main heating elements, while mCHP is a supplementary unit for keeping the temperature at a desired level if the EAHE and PTC are unable to provide sufficient heat. In addition, electric heaters can be employed as backup heating elements. Electricity generation is performed by PV cells located nearby the greenhouse. The generated electricity is accumulated in a battery bank, whereas the excess heat can be stored in a water tank. Finally, a controller is used for maintaining the temperature of the greenhouse in an acceptable range.

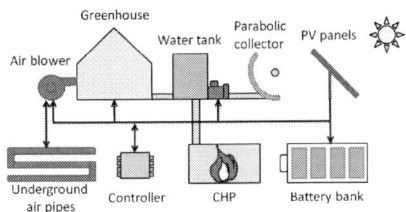

Fig. 2. Block diagram of the greenhouse

B. Thermal RC Modeling of the Greenhouse

The most essential part of the greenhouse design is thermal modeling of the structure. This assists to identify the energy requirements of the greenhouse and heat losses. Fig. 3 demonstrates an RC equivalent thermal model of the proposed greenhouse. In the RC thermal model, temperature, heat flow, thermal resistance and capacitance are equivalent to the voltage, current, electrical resistance, and capacitance, respectively [10].

The RC equivalent thermal model of the greenhouse is purposefully divided into five zones. Each zone represents heat flow in different part of the greenhouse. For instance, the red zone describes the thermal resistance of the glass cover due to convection in the south, east, and west walls and the south roof. In addition, the model includes infiltration loss due to the heat leakage through the cover and ventilation. The purple zone represents the heat flow in the north wall and north roof. It is different from the heat transfer through the glass cover since the north wall and roof are made of a different material, which has a heat storage capability. The green zone reports the heat transfer in the greenhouse floor interfacing with air. The blue region characterizes the thermal capacitance and solar heat gain of the air in the greenhouse. Lastly, the yellow part models the heat gain from the EAHE, electric heater, and water pipes connected to PTC and mCHP.

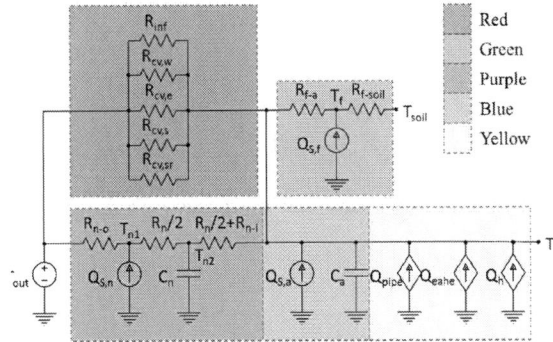

Fig. 3. Thermal RC modeling of the greenhouse

In Fig. 3, C_n and C_a are the specific heats of north wall and greenhouse air, respectively; $R_{cv,i}$ is the convective thermal resistance of cover i, R_{inf} is the infiltration resistance, R_{n-o}, R_{n-l}, R_{f-a}, and R_{f-soil} are the convective resistance between north wall and outside, north wall and inside, greenhouse floor and air, and floor and deeper soil, respectively; R_n is the conductive thermal resistance of north wall, $Q_{S,l}$ is the solar heat falling to surface i, Q_{pipe}, Q_{eahe}, and Q_h are the heat gain from PTC pipes, EAHE, and mCHP, T_{in} is the temperature of greenhouse air.

To observe the greenhouse internal temperature change due to external factors such as solar radiation and outside temperature, a set of differential equations needs to be solved. These differential equations are derived from the RC thermal model. By applying Kirchhoff's current law to the RC thermal model presented in Fig. 3, a set of equations can be written that characterize the relationship between the temperatures of different parts of the greenhouse. The equation for the north wall and roof presented by the purple zone in Fig. 3 are obtained using node voltage analysis:

$$C_n \frac{dT_{n2}}{dt} + \frac{T_{n2} - T_{in}}{R_n/2 + R_{n-i}} - Q_{S,n} + \frac{T_{n1} - T_{out}}{R_{n-o}} = 0 \quad (1)$$

These equations describe the heat transfer between the outer and inner sides of the north wall and roof. The heat transfer through the floor of the greenhouse is expressed as:

$$-Q_{S,f} + \frac{T_f - T_{in}}{R_{f-a}} + \frac{T_f - T_{soil}}{R_{f-soil}} = 0 \quad (2)$$

The greenhouse air, which interfaces all other parts, is obtained as:

$$C_a \frac{dT_{in}}{dt} - Q_{S,a} + \frac{T_{in} - T_{out}}{R_{cover}} + \frac{T_{in} - T_f}{R_{f-a}} + \frac{T_{in} - T_{n2}}{R_n / 2 + R_{n-i}}$$
$$- Q_{pipe} - Q_{eahc} - Q_h = 0 \qquad (3)$$

Taking into account these equations and the differential equation characterizing the temperature of the water inside the PTC pipe, in total three differential equations need to be solved simultaneously.

The thermal resistances of the greenhouse parts are calculated using flowing equations:

$$R_{cv,i} = \frac{1}{A_i h_i} \qquad (4)$$

$$R_{inf} = \frac{1}{0.33NV} \qquad (5)$$

$$R_n = \frac{d_n}{A_n k_n} \qquad (6)$$

where A is the surface area of the surface, h is the convective heat transfer coefficient, d is the thickness of the material, k is the material thermal conductivity, V and N are the volume of the greenhouse and number of air changes per hour. The passive heat gain from the sun can be estimated similar to [3]. For instance, the solar heat gain of the north wall and roof is calculated as:

$$Q_{S,n} = \alpha_n (1 - r_n)(A_n I_n + A_{nr} I_{nr}) \qquad (7)$$

where α_n and r_n are the absorptivity and reflectivity of north wall, I_n and I_{nr} are the solar radiation density incident on north wall and roof, respectively.

The computation of the solar radiation incident on the surface is an important step in both estimating solar heart gain and evaluating the PV efficiency. The magnitude of the solar radiation falling on the earth is dependent on several factors, such as the day of the year, time, geographical location and tilt angle of the object. The presented diurnal solar insolation evaluation is based on the method proposed in [2].

$$I_{tot} = R_b I_b + R_d I_d + R_r I_g \qquad (8)$$

where R_b, R_d, and R_r are the tilt factors for direct, diffuse and reflected solar radiation, respectively; while I_b, I_d, and I_g are the corresponding solar radiation densities. The tilt factors and radiation densities are computed using tilt, hour, and declination angles, day of the Julian calendar and local latitude as shown in details in [2].

1) Earth-to-air Heat Exchanger

One of the heating elements of the greenhouse is the earth-to-air heat exchanger. It is a system that is used for heat transfer from the ground to the air of the greenhouse by employing underground pipes. In the winter season, the temperature of the ground is usually higher than the ambient temperature. According to [11], the ground temperature in winter at 3 m depth can be 19 °C. Thus, installing air pipes at depth can be employed to extract ground heat. The equation characterizing the heat flow is given by [3]:

$$Q_{EAHE} = \dot{m}_a c_a (T_{soil} - T_{in})\left(1 - e^{-2\pi r_u h_u L_u / (\dot{m}_a c_a)}\right) \qquad (9)$$

where c_a is the specific heat of greenhouse air, $\dot{m}_a$ is the air mass flow rate of EAHE, r_u, L_u, and h_u are the radius and length of EAHE pipes and convective heat transfer coefficient between soil and air inside the pipe.

2) Parabolic Trough Collector

Another heat source of the proposed greenhouse is the parabolic trough collector. The PTC is designed to collect sunlight into a small beam, which heats fluid inside of the pipe. The heated fluid is stored in the tank and some portion is pumped into pipes in the greenhouse to heat the air whenever required. In this study, a water-based PTC is used for greenhouse heating. An RC thermal model of the PTC is demonstrated in Fig. 4. It is based on the differential equations developed by [6] and it represents a detailed model of the PTC previously denoted as Q_{pipe} in Fig. 3.

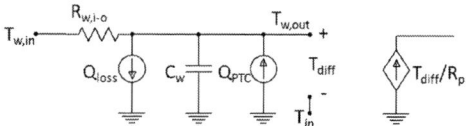

Fig. 4. RC thermal model of the PTC

In Fig. 4, $R_{w,i-o}$ and R_p are the thermal resistance between inlet and outlet of PTC and conductive resistance of pipe; c_w is the specific heat of water. The heat loss Q_{loss} through the pipes and heat gain of the PTC Q_{PTC} are obtained as [12]:

$$Q_{loss} = \frac{0.26(T_{w,out} - T_{out})}{\rho_w A_w} \qquad (10)$$

$$Q_{PTC} = \frac{nGI}{\rho_w A_w} \qquad (11)$$

Using Fig. 4, (10) and (11), the differential equation for the PTC can be obtained as:

$$c_w \frac{dT_{w,out}}{dt} = \frac{nGI}{\rho_w A_w} - \frac{qc_w(T_{w,out} - T_{w,in})}{A_w n_o D} - \frac{0.26(T_{w,out} - T_{out})}{\rho_w A_w} \qquad (12)$$

3) Micro Combined Heat and Power

Combined heat and power is a joint system designed for both heat and electricity generation [13]. Natural gas burned in the gas turbine produces electricity through the generator, and the exhaust heat and CO_2 can be injected into the greenhouse.

$$Q_{exh} = p(T_{out}) P_{fuel} + q(T_{out})\delta \qquad (13)$$

where T_{out} is the ambient temperature, P_{fuel} is the amount of required fuel found from look-up tables, and δ is the binary operation state of the gas turbine [13]. Calculation of $p(T_{out})$ and $q(T_{out})$ can be found in [13] in details.

Interfacing the exhaust unit with a boiler can heat the water inside the tank. In this way, the PTC and mCHP can be combined into a reliable and efficient hybrid system. However, the mCHP in this study is considered as a backup unit, which supplies the remaining required heat whenever the EAHE and PTC do not provide sufficient energy to keep the temperature in livable region for each plant.

4) Photovoltaic Electricity Generation

Electricity generation using photovoltaic cells is one of the features of the proposed greenhouse. The output power of PV cells is dependent not only on the solar radiation intensity, but also on the temperature of the cells. This temperature can be determined by knowing the ambient temperature T_{out}, nominal operating cell temperature T_{NOCT}, nominal power density of the sunlight $I_{T,NOCT}$ and other parameters, such as the current solar radiation intensity I_T as presented in [2]. Using the computed cell temperature T_c, temperature coefficient of PV cell B, reference efficiency η_r and reference temperature T_r, the efficiency of the PV cells can be estimated for an outside temperature variation.

$$\eta_{PV} = \eta_r \left[1 - B \left(T_c - T_r \right) \right] \qquad (14)$$

The total power output of the PV is then estimated using the area of the PV cells, solar radiation, and the efficiency.

$$P_{PV} = A_{PV} I_T \eta_{PV} \qquad (15)$$

III. SIMULATION STUDY AND RESULTS

The temperature dynamics of the greenhouse air are one of the primary factors for effective plant growth. By solving the obtained differential equations, the time variation in the internal temperature of the greenhouse, water temperature inside the PTC pipe, and temperature of the north wall can be identified. The input parameters to these equations are computed using physical dimensions and material constants. The thermal conductivity and specific heat constants of the greenhouse building materials are provided in [14], other constants are provided in [2]. After applying these parameters, a nonlinear system of equations is obtained:

$$\frac{dT_{n2}}{dt} = 10^{-4} \left(9.41 Q_{S,n} + 491 T_{in} - 857 T_{n2} + 366 T_{out} \right)$$

$$\frac{dT_{in}}{dt} = 10^{-4} \left(9.88 Q_{S,a} + 8.33 Q_{S,f} + 407 T_{n2} \right) - 20.9 T_{in} \qquad (16)$$
$$+ 0.987 T_{out} + 19.9 T_{w,out} + 0.274$$

$$\frac{dT_{w,out}}{dt} = 10^{-5} \left(39.9 I_{tot} + 4.95 T_{out} \right) - 265 T_{w,out} + 5305$$

A simulation is performed by taking the hourly variation of the outside temperature T_{out} of Almaty city on April 3, 2018 from a weather forecasting program [16]. An hourly temperature data are interpolated using a cubic spline interpolator for higher resolution (Fig. 5). The output power of the PV cells is shown in Fig. 6.

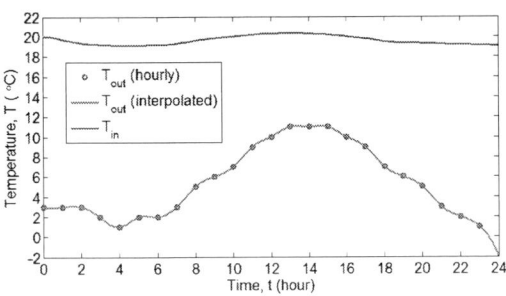

Fig. 5 . External and internal temperatures

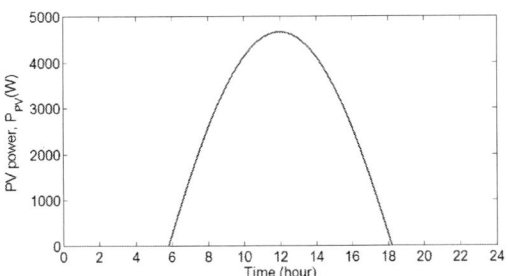

Fig. 6. Output power of the PV cells

The initial conditions for the temperature of the greenhouse air, north wall, and water in the outlet of the PTC pipes are 20 °C each. To solve the system of differential equations, MATLAB with symbolic toolbox was utilized. The simultaneous solution of these equations produced the diurnal temperature change inside of the greenhouse (see Fig. 5). From Fig. 5, it can be noticed that the temperature starts to drop from the initial 20 °C and remains approximately constant at 19 °C from 6 am to 8 am. After the sunrise, the greenhouse air temperature increases and repeats the trend of the solar radiation intensity.

In addition to passive heat obtained from the sun, the greenhouse receives approximately 4.5 kW peak power from the PV cells as shown in Fig. 6. Fig.7 and 8 show the total amount of heat obtained from the PTC and EAHE on an hourly basis, respectively. The transfer from the EAHE is negative since the temperature of the ground is 19 °C, which is less than the greenhouse air temperature. However, it is negligible compared to the heat gain from the PTC. These results show that the implementation of renewable-energy-powered greenhouses is feasible.

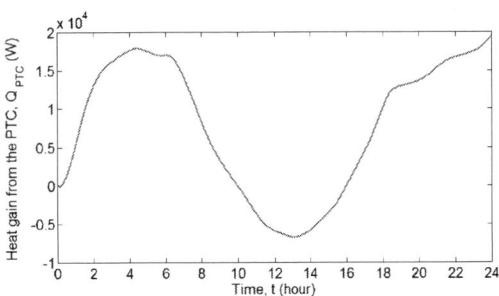

Fig. 7. Heat gain from the PTC

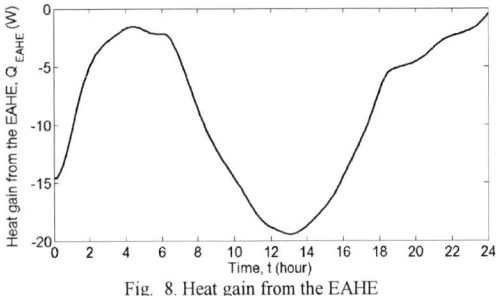

Fig. 8. Heat gain from the EAHE

IV. CONCLUSIONS

The design of a renewable-energy-powered greenhouse was presented in this study. An RC thermal model of the greenhouse and its components were discussed. A simulation study over 24 hours using a real ambient temperature profile from Almaty (in Eurasia) was conducted. The simulation results based on the derived equations demonstrated that the EAHE and PTC can be reasonable sources of heat for the greenhouse. Nevertheless, it is recommended to improve the PTC model in the future, since it introduces limitations into the design. One such limitation is that the model assumes that the inlet temperature of the water to the PTC is constant, while in practice it depends on not only the ambient temperature but also the greenhouse air temperature. For future steps, the integration of the mCHP model and heat storage systems into the design needs to be considered. Finally, these models and simulation results need to be verified by a physical prototype.

ACKNOWLEDGMENT

The authors acknowledge financial support of this study by the Faculty Development Competitive Research Grant (Project No. SOE2018018) and also Social Policy Grant of Nazarbayev University.

REFERENCES

[1] R. H. E. Hassanien, M. Li, and W. D. Lin, "Advanced applications of solar energy in agricultural greenhouses," *Renewable and Sustainable Energy Reviews*, vol. 54, pp. 989-1001, Feb. 2016.

[2] A. S. Anifantis, A. Colantoni, and S. Pascuzzi, "Thermal energy assessment of a small scale photovoltaic, hydrogen and geothermal stand-alone system for greenhouse heating," *Renewable energy*, vol. 103, pp. 115-127, Apr. 2017.

[3] M. K. Ghosal, G. N. Tiwari, and N. S. L. Srivastava, "Thermal modeling of a greenhouse with an integrated earth to air heat exchanger: an experimental validation," *Energy and Buildings*, vol. 36, no. 3, pp. 219-227, Mar. 2004.

[4] S. Tiwari, and G. N. Tiwari, "Thermal analysis of photovoltaic-thermal (PVT) single slope roof integrated greenhouse solar dryer," *Solar Energy*, vol. 138, pp. 128-136, Nov. 2016.

[5] K. A. Joudi, and A. A. Farhan, "Greenhouse heating by solar air heaters on the roof," *Renewable Energy*, vol. 72, pp. 406-414, Dec. 2014.

[6] R. O. Grigoriu, A. Voda, N. Arghira, V. Calofir and S. S. Iliescu, "Temperature control of a greenhouse heated by renewable energy sources," *2015 Intl Aegean Conference on Electrical Machines & Power Electronics (ACEMP), 2015 Intl Conference on Optimization of Electrical & Electronic Equipment (OPTIM) and 2015 Intl Symposium on Advanced Electromechanical Motion Systems (ELECTROMOTION)*, pp. 494-499, 2015.

[7] I. M. Al-Helal, S. A. Waheeb, A. A. Ibrahim, M. R. Shady, and A. M. Abdel-Ghany, "Modified thermal model to predict the natural ventilation of greenhouses," *Energy and Buildings*, vol. 99, pp. 1-8, Jul. 2015.

[8] M. R. Ariffin, S. Shafie, W. Z. W. Hassan, N. Azis and M. E. Ya'acob, "Conceptual design of hybrid photovoltaic-thermoelectric generator (PV/TEG) for Automated Greenhouse system," *2017 IEEE 15th Student Conference on Research and Development (SCOReD)*, Putrajaya, 2017, pp. 309-314.

[9] WorldWeatherOnline.com. 2018. "Almaty Monthly Climate Averages" [Online]. Available: https://www.worldweatheronline.com/almaty-weather-averages/almaty-city/kz.aspx [Accessed: 04- Jan- 2019].

[10] W. Mai and C. Y. Chung, "Model predictive control based on thermal dynamic building model in the demand-side management," *2016 IEEE Power and Energy Society General Meeting (PESGM)*, Boston, MA, pp. 1-5, 2016.

[11] G. Florides, S. Kalogirou, "Measurement of ground temperature at various depths," Apr 2016, [online] Available: https://www.researchgate.net/publication/30500372_Measurements_of_Ground_Temperature_at_Various_Depths

[12] M. Gunther, M. Joemann, S. Csambor, "Parabolic Trough Technology," *Advanced CSP Teaching Materials*, pp. 1-106, 2010.

[13] T. Sun, J. Lu, Z. Li, D. Lubkeman and N. Lu, "Modeling Combined Heat and Power Systems for Microgrid Applications," in *IEEE Transactions on Smart Grid*.

[14] Engineering Toolbox. 2018. "Specific Heat of common Substances" [Online]. Available: https://www.engineeringtoolbox.com/specific-heat-capacity-d_391.html [Accessed: 07- Jan- 2019].

[15] Engineering Toolbox. 2018. "Thermal Conductivity of common Materials and Gases" [Online]. Available: https://www.engineeringtoolbox.com/thermal-conductivity-d_429.html [Accessed: 07- Jan- 2019].

[16] BBC Weather. 2018. "Almaty" [Online]. Available: https://www.bbc.com/weather/1526384 [Accessed: 05- Jan- 2019].

Modeling and Siting of wind farms using Support Vector Regression (SVR)

Meysam Asadi
Department of Electrical Engineering
Tabriz Branch
Islamic Azad University
Tabriz, Iran
Email: stu.Meysam.asadi@iaut.ac.ir

Kazem Pourhossein
Department of Electrical Engineering
Tabriz Branch
Islamic Azad University
Tabriz, Iran
Email: K.pourhossein@iaut.ac.ir

Abstract — Nowadays, the advancement of technology is directly related to energy demand. The largest share of energy production comes from fossil fuels but they will be replaced by renewable energy due to their environmental damage and their limited availability. Wind energy is one of the most important sources of renewable energy. Despite merits of wind energy, dependence between production efficiency and location of the wind power plant is its main drawback. Therefore, proper sitting of wind power plants is very important in its design stage. In this paper, a global and robust model has been developed to locate wind sites using support vector regression.

Keywords — wind power plant, Site selection, Geographical information system (GIS), Support Vector Regression (SVR), Analytic hierarchy process (AHP)

I. INTRODUCTION

Energy is considered as an important factor in achieving countries' economic and social development. As a result, global growth and energy demand are dependent on each other [1]. Renewable energies play an important role in our future energy system. When considering the greenhouse gas emissions and the limitation of fossil resources, as well as the political instability of countries exporting fossil fuels, we will further appreciate the importance of renewable energies. Renewable energies eventually replace fossil fuels, because they are less harmful to the environment and economical [2,3]. Wind energy is now one of the most important renewable energy resources; it is a commonly used and cost-effective energy generated [4].

The locations with the highest wind power density are not always suitable locations to construct wind power plants, because there are various criteria for choosing a site to construct a wind power plant. They can be categorized into Technical, economic, environmental and social factors [5].

Wind turbines also have environmental and social hazards alongside their benefits. Environmental and social hazards can include: bird collisions, noise generation, visual impact, safety issues and electromagnetic interference [6]. Therefore, the location of wind power plants in addition to having high wind power density should also have the appropriate economic criteria. appropriate economic criteria can include the slope with the small angle, proximity to power lines and roads [7]. The sitting of the wind power plant is the most important decision in the development of a wind power plant [4].

II. METHODOLOGY

A. Support Vector Regression (SVR)

Support Vector Machine (SVM) is one of the machine learning algorithms presented by Cortes and Vapnik in 1995. SVM based on statistical learning theory and the principle of structural risk minimization. The popularity of this algorithm is due to its high speed in solving to non-linear data and avoiding overfitting during the training process [8]. SVM can be used for various purposes, such as images retrieval, fault diagnosis, and regression problems [8].

This method is transforming the non-linear input space to a high-dimensional feature space in order to find the hyperplane to distinguish between data. The support vector regression (SVR) is the same support vector machine (SVM) used for approximation of regression functions [8]. The structure of the SVR network is shown in Fig. 1 [9].

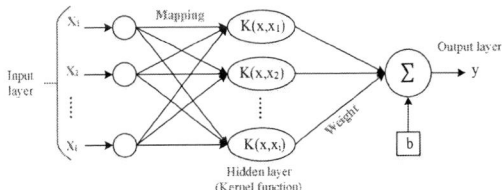

Fig. 1. Structure of the support vector regression (SVR)

The general equation for a linear regression model is as follows [10,11]:

$$f(x) = \langle w, b \rangle + b \tag{1}$$

Where x is the set of input patterns, and w is the the unknown weight vector, and also b is the bias term. In previous models, such as multiple linear regression, we used the reduction of quadratic errors to calculate weight vector while in the ε-SVR model used the optimizing the absolute error. The primary purpose of ε-SVR is to develop a function in where all errors are below a predefined value (ε), but with the best generalization capacity possible (flatness model). With minimizing the norm of the weight vector ‖w‖ can be obtained the flatness model. Moreover, the constraints guarantee that each error is less than (ε), otherwise, if the error value is greater than (ε), used the cost function. These conditions are imposed as follows:

978-1-5386-7688-2/19 $31.00 © 2019 IEEE

$$minimize \quad \frac{1}{2}\|w\|^2 + C\sum_{i=1}^{N}(\xi_i + \xi_i^*)$$

(2)

$$subject \quad \begin{cases} y_i - (\langle w, x_i \rangle + b) \le \varepsilon + \xi_i \\ (\langle w, x_i \rangle + b) - y_i \le \varepsilon + \xi_i^* \\ \xi_i, \xi_i^* \ge 0 \end{cases}$$

Where ξ_i^* and ξ_i are the slack variables. The next change is contained to measure the value of loss through the slack variables. Here we use the loss function for errors that are greater than (ε). The loss function $\xi_{|\varepsilon|}$ lies:

$$\xi_{|\varepsilon|} = \begin{cases} 0 & if \ |y_i - \hat{y}_i| < \varepsilon \\ |y_i - \hat{y}_i| - \varepsilon & otherwise \end{cases}$$

(3)

Where y_i and $\hat{y}_i$ are the measured and the predicted value, while ε is a user-defined parameter. And C is the penalization parameter. Eq.(2) will be solved using standard dual optimization through lagrange multipliers. After we solve it:

$$f(x) = \sum_{i=1}^{n}(\alpha_i - \alpha_i^*)\langle x_i, x \rangle + b$$

(4)

where α_i^* and α_i are Lagrange multipliers. To solve this equation, you can used quadratic programming (QP) techniques.

The SVR method is used to solve non-linear regressions problems using a kernel function. In which the input space is mapped into a higher-dimensional feature space, to be fitted by a linear model. This transformation is mapping $x \rightarrow \phi(x)$, using a non-linear function ϕ. The kernel functions are as follows:

$$K(x_i, x) = \langle \phi(x_i), \phi(x) \rangle$$

(5)

Finally, the generalization of Eq. (4) for nonlinear problems remains:

$$f(x) = \sum_{i=1}^{n}(\alpha_i - \alpha_i^*) k\langle x_i, x \rangle + b$$

(6)

B. Geographical Information System (GIS)

GIS generally is defined as a tool for analyzing, consulting, editing, developing, manipulation and storing geographical information including maps and special information. The spatial data can be represented in the GIS in two types, including raster data and vector data. Raster models are shown with a mesh or grid of rectangles that called pixels or cell; all have the same size. Each cell gives information about the geographic location and cell size indicates the accuracy or resolution of the map. In a vector model, the geographic features in GIS are represented by a point, line, or polygon [12,13]. Comparison of Raster and Vector maps is shown in Fig. 2

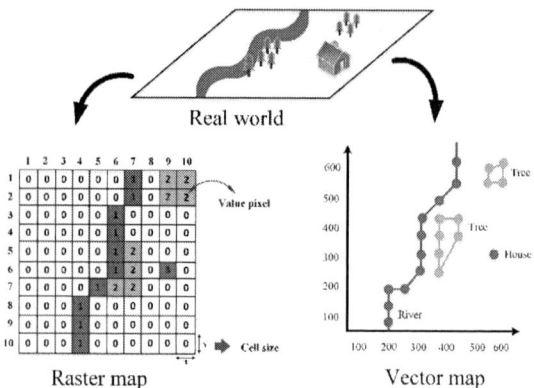

Fig. 2. Comparison of raster and vector map

III. RESULTS

A. Case Study (IRAN)

Islamic Republic of Iran is located in West Asia. The total area of the country is 1,648,000 km2 and has more than 80 million inhabitants. The climate of Iran is hot and dry also has long summers and short, cold winters. This country surrounded by Azerbaijan, Turkmenistan, Armenia, Pakistan, Afghanistan, Iraq and Turkey [14].

B. Wind power in IRAN

The total potential of the country for producing wind energy is estimated at 100,000 GW. In some provinces of Iran, wind power plants have been constructed, the most famous of which is the Manjil Power Plant. Currently, there are 111 wind turbines in Manjil and Rudbar with a capacity of 61 MW. The total wind power energy capacity of Iran's sites is estimated to be around 6500 MW, while the nominal capacity of installed sites is not more than 3,400 megawatts per year. According to reports, wind-assisted power generation will reach 100,000 MW by 2020 [14]. In Fig. 3 and 4, the wind speed map and wind speed distribution diagram are shown respectively.

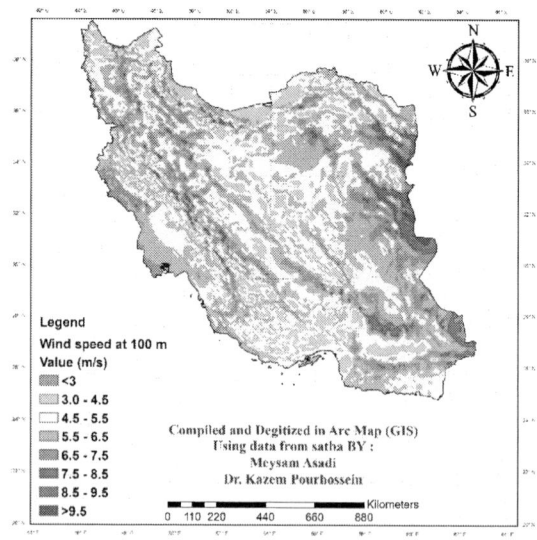

Fig. 3. Wind speed map in 100m [15]

TABLE II. CLASSIFICATION OF MAIN CRITERIA

Class	Wind power density (W/m²) (A)	Distance from Power line (m) (B)	Slope (%) (C)	Distance from road (m) (D)	Distance from cities (m) (E)	Distance from protected area (m) (F)
1	0-300	10500-9000	35-30	3500-3000	0-2000	0-1000
2	300-400	9000-7500	30-25	3000-2500	2000-3000	1000-2000
3	400-550	7500-6000	25-20	2500-2000	3000-4000	2000-3000
4	550-650	6000-4500	20-15	2000-1500	4000-5000	3000-4000
5	650-800	4500-3000	15-10	1500-1000	5000-6000	4000-5000
6	800-1100	3000-1500	10-5	1500-500	6000-7000	5000-6000
7	1100-2500	1500-0	5-0	500-0	7000-8000	6000-7000

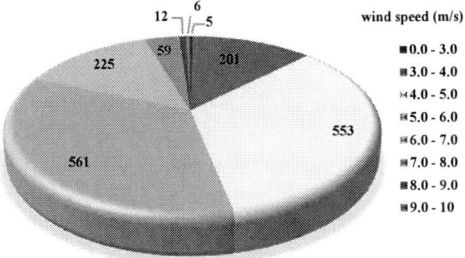

Fig. 4. Distribution of annual average wind speeds in IRAN at 100m. (Thousand-km2).

C. Sampling data of case study

After the overview of previous studies, six criteria were considered as important factors in sitting the suitable location for the construction of the wind power plant. These criteria included wind power density as technical criteria, slope and distance from power line and roads as economic criterias, distance from the city and protected areas as social-environmental criterias. The sampling of the criteria for case study takes place in three steps. In the first step, all criteria are converted to raster layers in GIS so that the database can be created. In the second step, the raster layers are networked and finally, of the networked raster layers are sampled.

The purpose of networking raster layer is that the sampled data is an average of the total space that defined the sampling resolution.

The sampling resolution in this paper is (1 km in 1km). In table I, the sampled data from the case study is shown.

TABLE I. SAMPLES OF 1KM×1KM RESOLUTION FOR IRAN

No	WPD (W/m²)	D.F power line (m)	Slope (%)	D.F road (m)	D.F cities (m)	D.F protected area(m)
1	24.25	640.60	0	0	43008.80	277.41
2	598.04	1390.58	0	0	43615.19	395.14
3	598.19	2347.94	0	0	44347.89	765.41
4	598.33	3310.56	0	0	45101.60	1396.35
5	24.25	3912.44	0	0	45595.89	1923.30
6	24.25	351.66	0	1651.85	38660.10	373.94
7	23.26	244.10	0	1436.18	39141.30	381.25
.	.	.	.	.	.	.
.	.	.	.	.	.	.
.	.	.	.	.	.	.
1629222	543.85	43687.10	7.17	908.55	45323	94828.89
1629223	554.29	44101.10	0	227.09	45889.80	94398.10
1629224	554.55	44289.80	0	666.55	46231.30	93896

D. Evaluation the case study with SVR

In this step, a global and robust regression is trained to evaluate the case study. that includes the following steps:

1) Generate the train data (input and target) to traind the SVR

2) Designing the SVR

3) The evaluate the case study using the trained SVR

1) Generate the train data to traind the SVR

The purpose of this paper is not to modeling a particular method, but modeling a regression that has a global and robust performance. The global train data is needed to modeling this regression. The train data included input and target data. Global train data is obtained as follows: Firstly, each of the criteria is classified into seven classes in the global range. Then, the permutation of all classes is considered as input data. Finally, the scores of each permutation are defined as the target data. The classification of criteria in terms of global range is shown in Table II.

After classified the criteriai in the global range, the permutation of all criteria in each class is considered. The number of permutations (input data) is $8^6 = 262144$. Where in "6" is number of criteria and "8" is number of bounds for classes. The first and last permutation indicates the unsuitable (Eq.7) and most suitable (Eq.8) location mode for the construction of a wind power plant.

Unsuitable location mode:
$$A_1 B_1 C_1 D_1 E_1 F_1 \rightarrow 0 - 10500 - 35 - 3500 - 0 - 0 \quad (7)$$

Most Suitable location mode:
$$A_8 B_8 C_8 D_8 E_8 F_8 \rightarrow 2500 - 0 - 0 - 0 - 8000 - 7000 \quad (8)$$

After obtaining all permutations, the score of each permutation is calculated, which is defined as Target data. The score of each permutation (Target data) is equal to the sum of the weighted normalized bounds of criteria. Target data falls between 0 and 1. Score 0 and 1 represent the unsuitable and most suitable location for the construction of a wind power plant, respectively.

All criteria are not equally important.for this reason, the weight of each criteria is calculated using the AHP method. The AHP method is one of the multi criteria decision making (MCDM) methods. In Fig.5, weights of all criteria are shown.

A little part of prepared input and target data (global train data), based on permutation, is presented in Table III.

978-1-5386-7688-2/19 $31.00 © 2019 IEEE

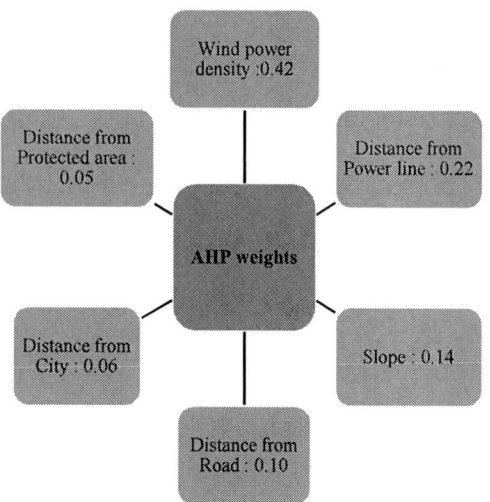

Fig. 5.   Extracted weights for all criteria used in this paper

TABLE III.   A LITTLE PART OF PREPARED GLOBAL TRAIN DATA BASED ON PERMUTATION

Perm number	Input Data						Target Data
	WPD	D.F Powerline	Slope	D.F roads	D.F cities	D.F.PA	
1	0	10500	35	3500	0	0	0
2	0	10500	35	3500	0	1000	0.006
3	0	10500	35	3500	0	2000	0.013
4	0	10500	35	3500	0	3000	0.020
5	0	10500	35	3500	0	4000	0.027
6	0	10500	35	3500	0	5000	0.034
7	0	10500	35	3500	0	6000	0.041
8	0	10500	35	3500	0	70000	0.048
.	.	.	.	.	.	.	.
.	.	.	.	.	.	.	.
.	.	.	.	.	.	.	.
262137	2500	0	0	0	8000	0	0.951
262138	2500	0	0	0	8000	1000	0.958
262139	2500	0	0	0	8000	2000	0.965
262140	2500	0	0	0	8000	3000	0.972
262141	2500	0	0	0	8000	4000	0.979
262142	2500	0	0	0	8000	5000	0.986
262143	2500	0	0	0	8000	6000	0.993
262144	2500	0	0	0	8000	70000	1

2) Design of SVR

Now based on generated data (Table III), SVR can be trained. The trained SVR has the ability to evaluate the extracted data from the case study as global and robust. In training process, the train data has been divided into two parts; training and test data. In training process just used the training data. Test data is used to evaluate the performance of the SVR. The result of training process for test data shown in Fig.6.

Fig.6-a depict error histogram and Fig.6-b presents excellent correlation between target data and output of SVR. μ and σ are mean and standard deviation of error, respectively. μ and σ calculated using Eq.(9,10)

$$\mu = \frac{1}{N} \sum e_i \qquad (9)$$

$$\sigma = \sqrt{\frac{1}{N} \sum_{i=1}^{N} (e_i - \bar{e})} \qquad (10)$$

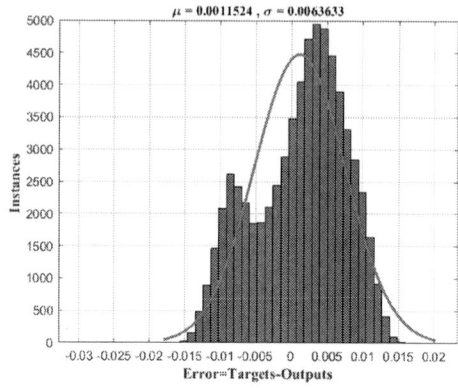

(a) Error histogram plot

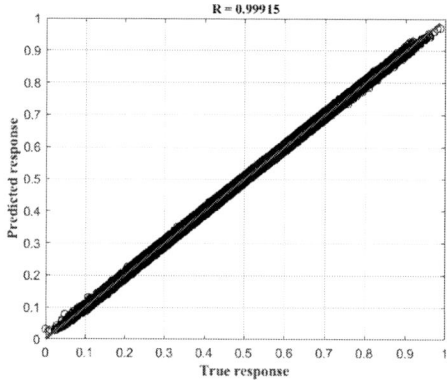

(b) Plot Predicted vs. Actual Response

Fig. 6.   Result of training proceses for test data

3) The evaluate the case study using the trained SVR

Now the trained SVR can be used to assign score to any candidate site location to assess selected area to locate wind power plants.

The sampled data of case study (Table I) are entered into the trained SVR for prioritizing. The sampled data after preprocessing in three different phases are entered into the trained SVR.

Phase 1: Prioritization by considering technical criteria

In this phase, the case study is only prioritized by considering the technical criteria. Wind power density is the only technical criteria that is used to sitting the wind power plant in this paper. For prioritizing the case study only on the basis of technical criteria, it is sufficient that the other criteria in the sampled data are equal to the constant value. The results of the prioritization of the case study on the basis of the technical criteria as the raster layer are shown in Fig. 7.

Phase 2: Prioritization by considering technical and economic criteria

In this phase, in addition to the technical criteria, economic criteria are considered. Economic criteria include the slope and distance from the power line and roads. The results of the prioritization of the case study on the basis of the technical and economic criteria shown in Fig. 8.

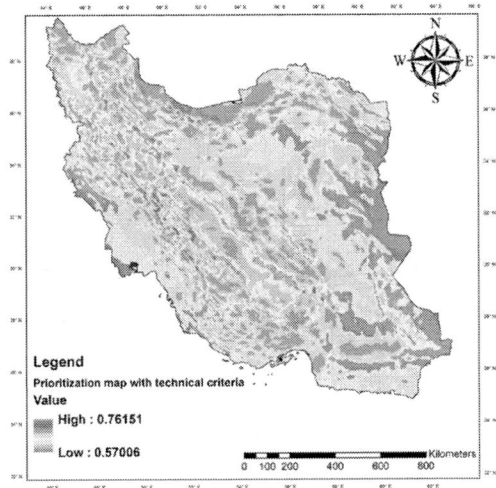

Fig. 7. Prioritization map only Consider Technical criteria

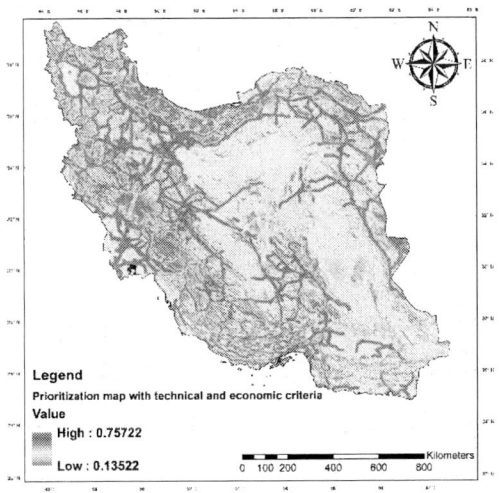

Fig. 8. Prioritization map only Consider Technical and economic criteria

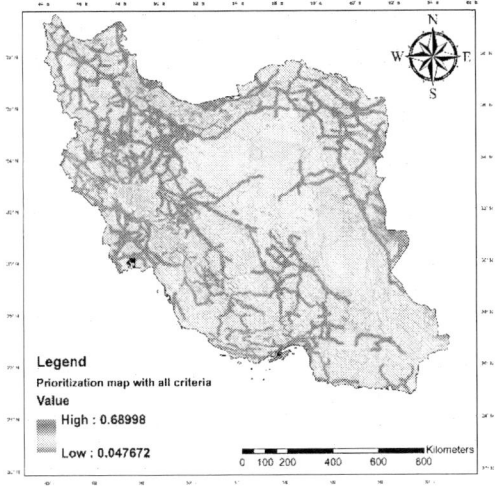

Fig. 9. Prioritization map only Consider all criteria

Phase 3: Prioritization by considering all criteria

Finally, the case study is prioritized with all the criteria. The final phase provided the most complete and precise prioritization map for the construction of a wind power plant (Fig.9).

The high-score areas in the prioritization map with all criteria are shown in Fig.10.

In Fig. 10, several sites have been proposed as the suitable location for the construction of a wind power plant. The score of criteria for proposed sites (table IV) shows that the trained SVR has a high ability to locate wind power.

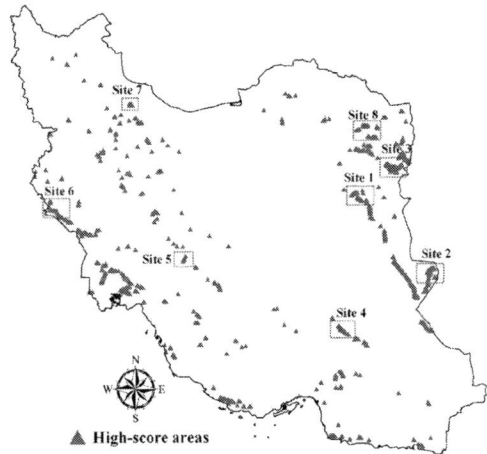

Fig. 10. Suitable (High - score) areas for the construction of wind farm

TABLE IV.   Specifications of Selected High Score Area

Site	WPD (W/m²)	D.F power line (metr)	Slope (%)	D.F. roads (metr)	D.F. city (metr)	D.F. protected area (metr)	Average score in total area
Site 1	483	701	5.24	447	21634	162574	0.5836
Site 2	755	987	2.26	4091	7212	11562	0.5837
Site 3	985	1657	5.47	2722	11595	128886	0.6061
Site 4	306	384	3.29	307	36168	116802	0.5769
Site 5	397	369	6.16	359	24440	56782	0.5773
Site 6	385	460	3.88	358	20088	44228	0.5825
Site 7	922	854	8.13	2133	3703	218223	0.5893
Site 8	434	773	4.10	404	19845	79094	0.5811

IV. Conclusion

One of the most important challenges in developing wind energy is Suitable wind site selection for construction of wind power plants. This paper trained a model for siting of wind power plants using support vector regression. The trained model is global and robust. Application of the proposed method on the selected area shows very good site selection performance of SVR. Values of decision criteria for the determined locations of selected area indicate that they have suitable condition for construction of wind power plants.

978-1-5386-7688-2/19 $31.00 © 2019 IEEE

REFERENCE

[1] Şener, Ş. E. C., Sharp, J. L., & Anctil, A. (2018). Factors impacting diverging paths of renewable energy: A review. Renewable and Sustainable Energy Reviews, 81, 2335-2342.

[2] Rostami, R., Khoshnava, S. M., Lamit, H., Streimikiene, D., & Mardani, A. (2017). An overview of Afghanistan's trends toward renewable and sustainable energies. *Renewable and Sustainable Energy Reviews, 76*, 1440-1464.

[3] Elum, Z. A., & Momodu, A. S. (2017). Climate change mitigation and renewable energy for sustainable development in Nigeria: A discourse approach. *Renewable and Sustainable Energy Reviews, 76*, 72-80.

[4] Noorollahi, Y., Yousefi, H., & Mohammadi, M. (2016). Multi-criteria decision support system for wind farm site selection using GIS. *Sustainable Energy Technologies and Assessments, 13*, 38-50.

[5] Van Haaren, R., & Fthenakis, V. (2011). GIS-based wind farm site selection using spatial multi-criteria analysis (SMCA): Evaluating the case for New York State. *Renewable and sustainable energy reviews, 15*(7), 3332-3340.

[6] Aydin, N. Y., Kentel, E., & Duzgun, S. (2010). GIS-based environmental assessment of wind energy systems for spatial planning: A case study from Western Turkey. *Renewable and Sustainable Energy Reviews, 14*(1), 364-373.

[7] Mellino, S., Ripa, M., Zucaro, A., & Ulgiati, S. (2014). An emergy–GIS approach to the evaluation of renewable resource flows: a case study of Campania Region, Italy. *Ecological Modelling, 271*, 103-112.

[8] Zendehboudi, A., Baseer, M. A., & Saidur, R. (2018). Application of support vector machine models for forecasting solar and wind energy resources: A review. Journal of cleaner production, 199, 272-285.

[9] Chen, Y., & Tan, H. (2017). Short-term prediction of electric demand in building sector via hybrid support vector regression. *Applied energy, 204*, 1363-1374.

[10] Antonanzas-Torres, F., Urraca, R., Antonanzas, J., Fernandez-Ceniceros, J., & Martinez-de-Pison, F. J. (2015). Generation of daily global solar irradiation with support vector machines for regression. Energy conversion and management, 96, 277-286.

[11] Haykin, S. S., Haykin, S. S., Haykin, S. S., Elektroingenieur, K., & Haykin, S. S. (2009). Neural networks and learning machines (Vol. 3). Upper Saddle River: Pearson education.

[12] Tahri, M., Hakdaoui, M., & Maanan, M. (2015). The evaluation of solar farm locations applying Geographic Information System and Multi-Criteria Decision-Making methods: Case study in southern Morocco. *Renewable and Sustainable Energy Reviews, 51*, 1354-1362.

[13] Sánchez-Lozano, J. M., Teruel-Solano, J., Soto-Elvira, P. L., & Garcia-Cascales, M. S. (2013). Geographical Information Systems (GIS) and Multi-Criteria Decision Making (MCDM) methods for the evaluation of solar farms locations: Case study in south-eastern Spain. *Renewable and Sustainable Energy Reviews, 24*, 544-556.

[14] Taghavifar, H., & Mardani, A. (2014). A comparative trend in forecasting ability of artificial neural networks and regressive support vector machine methodologies for energy dissipation modeling of off-road vehicles. Energy, 66, 569-576.

[15] Renewable Energy and Energy Efficiency Organization Country of Iran (SATBA). Wind speed map of iran, https://www.satba.gov.ir/en/album/237/wind/; 2016.

MPPT Based Adaptive Control Algorithm for Small Scale Wind Energy Conversion Systems with PMSG

M. C. AKKAYA
Department of Electrical Engineering
Istanbul Technical University
34469, Istanbul, Turkey
akkayamu@itu.edu.tr

A. POLAT
Department of Electrical Engineering
Istanbul Technical University
34469, Istanbul, Turkey
polata@itu.edu.tr

L. T. ERGENE
Department of Electrical Engineering
Istanbul Technical University
34469, Istanbul, Turkey
ergenel@itu.edu.tr

Abstract— **In this paper, fixed step and adaptive control algorithms by using Hill Climb Search (HCS) method one of the Maximum Power Point Tracking (MPPT) methods are proposed respectively for small scale permanent magnet synchronous generator (PMSG) wind energy conversion systems (WECS). Basically, the controller tracks the maximum power point (MPP) by measurements of current or the ratio of change in power and voltage depending on different control algorithms. To succeed that, controller is dynamically adjusting the duty ratio of boost converter connected to output of WECS by considering needed measurements. The improvements or change observed in different control algorithms are demonstrated with the comparisons.**

Keywords—PMSG, WECS, Adaptive, Perturb & Observe (P&O), MPPT

I. INTRODUCTION

The process of converting the wind energy into mechanical power and converting the mechanical power into electrical power is called as wind energy conversion. Wind energy conversion systems (WECS) basically consist of two fundamental components: wind turbine which converts wind energy into mechanical energy and generator which converts mechanical energy into electrical energy. WECS are divided into four subgroups depending on whether speed characteristic is fixed or variable and gearbox exists or not. In this paper, variable speed (VS) WECS without gearbox is preferred. The mechanical inertia of WECS should be taken into consideration for varying wind speeds. Because of all the rotating mechanical parts of WECS having masses and distances to the rotating point, high moment of inertia is produced and it requires more force to catch the speed of a lighter system in the same environmental conditions. The starting the rotation of turbine is an issue to be overcome in the huge WECS, however this is not a big problem in small scale wind turbines because of having lighter mechanical rotating parts.

One of the brightest application areas of WECS is small scale ones because small scale WECS have become very popular with the considerably reduction in the investment costs. Small scale WECS are widely used in various applications such as supplying electricity to households, farms, rural areas and so on.

Electrical machines used in variable-speed wind power generation systems can be divided into three main types: DC generators, induction generators and synchronous generators. The synchronous generators have many advantages over induction machines and DC generators such as having higher efficiency, working at constant speed, independent of load characteristics, chance of having longer air gaps and so on. Besides, PMSGs having self-excitation bring many benefits such as elimination of the rotor copper losses and external power supply dependency. PMSGs also do not have brushes which require regular maintenance [1]. As a result, PMSGs are highly preferred because of having certain specifications such as not requiring an external excitation circuit and being more efficient and reliable [2].

Wind turbines must operate at optimum rotational speed to reach the MPPs to improve the captured wind power in WECS, which are considerably based on the wind speed, that is why MPPT control is highly needed in VS WECS [3]. The DC-DC converter is controlled by adjusting the duty ratio to obtain the optimal generator speed. Three phase diode rectifier and a boost converter are generally used for small scale application WECS due to its simplicity, low cost, high reliability, and easy control [4]. MPPT control techniques are classified basically into three categories: Tip speed ratio (TSR), power signal feedback (PSF) and hill climb search (HCS).

TSR control which depends on a wind speed estimation is suggested to follow MPPs [5]. The optimal rotor speed is estimated with the knowledge of optimal TSR and estimated wind speed because the main goal is controlling the output power transferred to the load in inverter in TSR.

In PSF control the reference power command, which is symbolised as P_{ref}, is produced then it is applied to the grid side converter control system to reach MPP extraction. Wind speed is used as an input in this method to produce reference power signal [6]. To obtain reference power for PSF based on MPPT control of PMSG WECS, the turbine power equation can be used [7].

The control algorithm of HCS proposes directly adjusting the DC-DC converter duty ratio (d) depending on the results of the comparison of sequential wind generator output power measurements are suggested in [8] as MPPT control [6]. This can be done either as fixed speed and fixed step or variable speed and variable step. The change in d is dependent on step size and a constant value to specify the speed and accuracy of convergence to MPP but at instantly fluctuating wind speeds the P&O method whose performance is based mainly on step size has a slow response and a low efficiency [3]. Therefore, adjustment of the step size, tracking speed, perturbation direction and tracking ability are important parameters should be taken into consideration to improve the P&O method for

978-1-5386-7688-2/19 $31.00 © 2019 IEEE

each wind change [9]. If the step size is adapted by changing step size according to the change in power in each cycle this problems of conventional P&O method can be solved. The MPPT method proposed in [10] uses a variable tracking step [6]. There are many suggestions of adaptive control algorithm for MPPT control in literature [11, 12]. A DC-DC boost converter topology is used for MPPT control of PMSG WECS in [13]. The power as input and the torque as controller output are used for the HCS MPPT control in [14]. The advantages of both the tracking method based on the optimum power versus speed characteristic and the HCS are combined and presented as MPPT control method [15]. In this paper, developed variable tracking step adaptive HCS MPPT control algorithm is proposed.

II. IMPLEMENTATION OF THE WECS

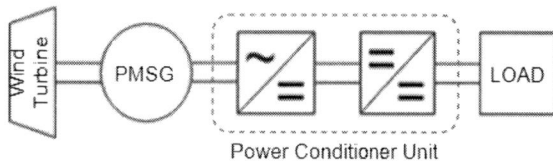

Fig. 1. General model of WECS

PMSG WECS consists of three main components: wind turbine, PMSG and power conditioner unit (PCU) as indicated in Fig. 1.

A. Wind Turbine Model

The power P_m generated by a wind turbine is given by

$$P_m = 0.5 c_p \rho A V^3 \qquad (1)$$

where c_p is the power coefficient depends on λ tip speed ratio, ρ is the air density, A is the cross sectional area (swept area by turbine) and V is the wind speed. c_p is nonlinear function of tip speed ratio and following parameters and given by

$$c_p = c_1 \left(\frac{c_2}{\lambda_i} - c_3\beta - c_4 \right) e^{\frac{-c_5}{\lambda_i}} + c_6 \lambda \qquad (2)$$

$$\frac{1}{\lambda_i} = \frac{1}{\lambda + 0.08\beta} - \frac{0.035}{\beta^3 + 1} \qquad (3)$$

where R is the radius of the turbine, ω is the generator angular speed and β is the blade pitch angle. The coefficients are given in Table 1.

TABLE I. C_P COEFFICIENTS

Coefficients	Value
C_1	0.5176
C_2	116
C_3	0.4
C_4	5
C_5	21
C_6	0.0068

The tip speed ratio λ is defined as

$$\lambda = \frac{R\omega}{V} \qquad (4)$$

The measured relationship between c_p and λ is shown in Fig. 2. It can be deduced from Fig. 2 that there is an optimum λ value to obtain maximum c_p value, so maximum power value.

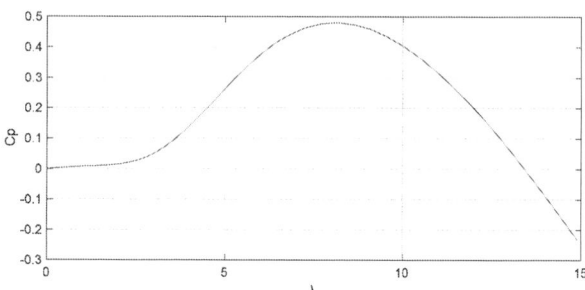

Fig. 2. Power coefficient function based on tip speed ratio

For every wind speed there is an optimum λ value which indicates the MPP, so it can be seen in Fig. 3. that where the MPPs are in different wind speeds equal to where the maximum c_p values are for each wind speed.

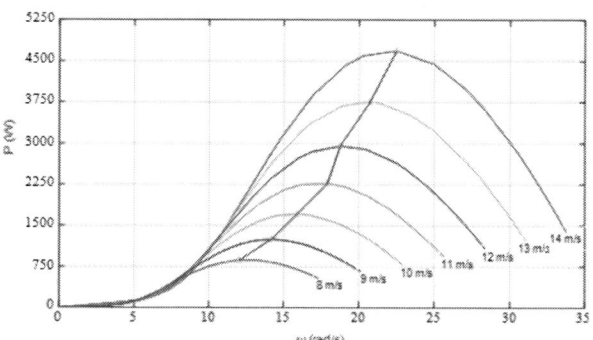

Fig. 3. Output power and generators

As it is shown in the Fig. 3. that every wind speeds have different power curves and MPPs can be obtained in optimal generator speeds which correspond to peak points.

The simulation of turbine model on MATLAB/Simulink is shown in Fig. 4.

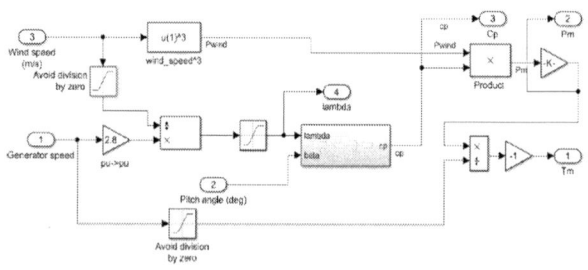

Fig. 4. Simulink model of the wind turbine

978-1-5386-7688-2/19 $31.00 © 2019 IEEE   518

B. PMSG

In this study 3 phase PMSG is used as generator, so its mechanical and electrical torque equations are expressed as

$$T_m = \frac{P_m}{\omega_m} \qquad (4)$$

$$T_e = \frac{P_e}{\omega_e} = \frac{2}{p} x \frac{P_e}{\omega_m} \qquad (5)$$

where T_m is mechanical torque, P_m mechanical power, ω_m mechanical angular speed, T_e electrical torque, P_e electrical power, ω_e electrical angular speed, p number of poles.

C. Power Conditioner Unit

In variable wind speeds PMSG coupled with wind turbine generates 3-phase AC voltage in variable magnitude and frequency, and it is needed to regulate this voltage signal. This three phase electricity is rectified by three phase bridge diode rectifier. The controlled bridge rectifier could be used instead of diodes but bridge diode rectifier has lower cost and higher reliability, that is why it is preferred. The rectified DC voltage is boosted by boost converter topology connected the output of bridge rectifier. It is followed by boost converter and afterwards in some applications by voltage source inverter (VSI). Inverters stabilize the DC voltage and convert it to AC form and enable to feed AC microgrids. VSI is directly connected to output terminals of bridge rectifier without using boost converter in some studies [16]. MPPT is realized by changing d value of inverter in that case.

Boost Converter

In this study converter connects bridge rectifier, which has variable DC voltage, to DC microgrid which has fixed DC voltage of 400 V. The output power is controlled via output current of the boost topology, because of the output voltage is fixed. Boost converter which is shown in Fig. 5 limits the output voltage and obtains MPP by setting d, thereby rotational speed of PMSG driven by wind turbine.

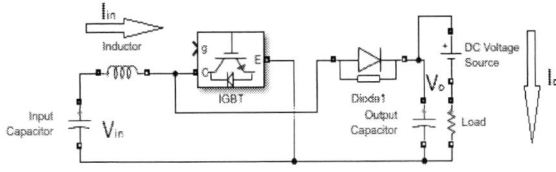

Fig. 5. Simulink model of boost converter

The main goal of controlling the boost converter current is controlling the rotational speed of PMSG to keep the tip speed ratio at the optimal operating point for all wind speeds [17]. The relationship between input current (I_{in}) and output current (I_o) of converter is given as

$$I_o = (1-d)I_{in} \qquad (6)$$

As it is shown in the formula, in order to control output power, so output current, controlling input current is rather enough in a simple way. The parameters of boost converter connected to output of rectifier is presented in Table 2.

TABLE II.   BOOST CONVERTER DESIGN PARAMETERS

Input Capacitor		5 mF
Inductor		700 uH
IGBT	Resistance (R_{on})	1mΩ
	Snubber Resistance (R_s)	100kΩ
	Switching Frequency (f_{sw})	1 kHz
Diode	Resistance (R_{on})	1mΩ
	Voltage Drop (V_{fd})	0.8 V
	Snubber Resistance (R_s)	500 Ω
	Snubber Cap. (C_s)	250 nF
Output Capacitor		10 mF
Load		10 Ω
DC Voltage Source		347 V

III. MPPT CONTROLLER DESIGN

To track MPPs two different control types are simulated: fixed step tracking in fixed and variable wind speed, adaptive (variable step) tracking in fixed and variable wind speed.

A. Fixed Step

In fixed step control, the algorithm tracks the MPP with a constant step size. The output current value of boost converter is compared to previous one for each cycle and if the change in power is positive, d is increased with the constant step, by the time the change in power is negative, whenever the sign changes controller starts to give the previous d without adjusting.

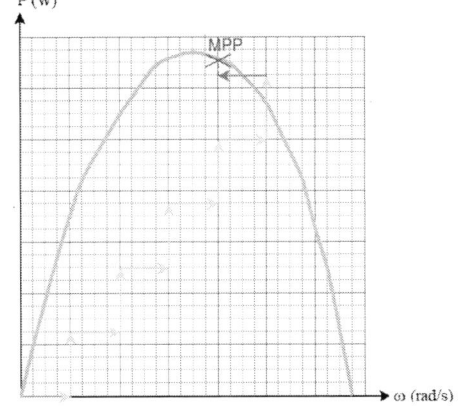

Fig. 6. Characteristic graph of power and angular speed

B. Adaptive

In developed adaptive control algorithm, the step size is dynamically adjusted according to ratio of changes in power and voltage. For variable wind speed, due to using anemometer being a low cost and practical solution, in order to increase the precision and performance of algorithm the wind speed information is used. The captured maximum power by wind turbine is expressed as

$$Pm_{opt} = \frac{0.5\pi\rho c_{p\max}R^5}{\lambda^3_{opt}}\omega^3_{opt} \qquad (7)$$

where Pm_{opt} is maximum power, $c_{p\max}$ max. value of c_p, λ_{opt} point which corresponds the $c_{p\max}$ and ω_{opt} rotational speed of PMSG. The steady state phase voltage of PMSG is stated as

$$V = E - I\sqrt{R_s^2 + (p\omega L_s)^2} \qquad (8)$$

where V is phase RMS voltage, E electromotive force(EMF), I phase RMS current, p pole pair, R_s and L_s winding resistance and inductance respectively. In PMSG, E is a function of ω [17] and the output voltage of bridge rectifier can be expressed as

$$V_{dc} = \frac{3\sqrt{6}}{\pi}V = \frac{3\sqrt{6}}{\pi}\left(E - I\sqrt{R_s^2 + (p\omega L_s)^2}\right) \qquad (9)$$

As seen in (9) there is a relationship between V_{dc} and ω, though it is not a straight line due to resistance and inductance. Depending on this relationship the power curve for one wind speed as shown in Fig. 3. can be examined in three categories: At left side of MPP, at MPP and at right side of MPP. In the left side of MPP the slope will be positive, so

$$\frac{dP_{dc}}{d\omega} > 0 \qquad (10)$$

In the right side of MPP the slope will be negative, then

$$\frac{dP_{dc}}{d\omega} < 0 \qquad (11)$$

At MPP

$$\frac{dP_{dc}}{d\omega} = 0 \qquad (12)$$

Based on the relationship between the variables the chain rule can be specified:

$$\frac{dP_{dc}}{d\omega} = \frac{dP_{dc}}{dD}\frac{dD}{dV}\frac{dV}{d\omega} \qquad (13)$$

Because of $\dfrac{dV}{d\omega}$ being not zero we can arrange and simplify the formula (15) as

$$\frac{dP_{dc}}{d\omega} \cong \frac{dP_{dc}}{dV} \qquad (14)$$

In this study an adaptive control algorithm based on (14) is developed. The flowchart of adaptive algorithm is shown in Fig. 7.

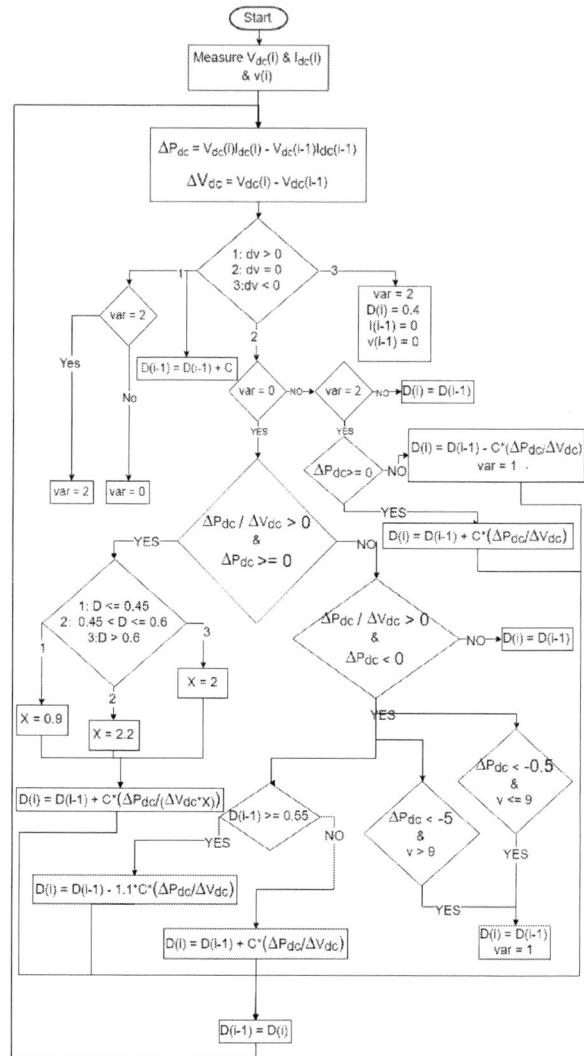

Fig. 7. Flowchart of adaptive control algorithm

In this algorithm, MPP is dynamically tracked, but when the change in power drops in an amount that can be ignored, the controller starts to give same d, so it achieves MPP and does not change d anymore. Additionally, depending on current value of d, the step size is adjusted. There is a trade-off the bigger step the faster response and the smaller step the higher precision, so it should be optimized. In this paper some constants are defined to adjust the change of speed in some d intervals for example if d is smaller than 0.45 algorithm tracks MPP faster and if d is bigger than 0.6, algorithm tracks MPP slower because of generally MPPs being concentrated in the range of 0.45 and 0.6.

IV. RESULT - DISCUSSION

The simulation check is implemented in order to compare and assess the performances of fixed step and adaptive control. The simulation study is performed on MATLAB/Simulink.

Simulation Results

Wind turbine and PMSG parameters are presented in Table 3.

TABLE III.   WIND TURBINE AND PSMG PARAMETERS

Simulation Parameters		Parameters			
		Wind Turbine		**PMSG (3φ-Salient)**	
Simulation time (sec)	10	Rated Pmech (kW)	6	Stator phase resistance (R_s) (Ω)	0.425
Simulation type	Con	Base wind speed (m/s)	12	Inductances (L_d, L_q) (H)	8.4e-3
Sample time (sec)	2e-3	Pitch angle(deg)	0	Flux linkage	0.433
SIMULINK R2019A		Base rotational speed	0.9	Pole pairs	10

General simulation model is stated in Fig. 8.

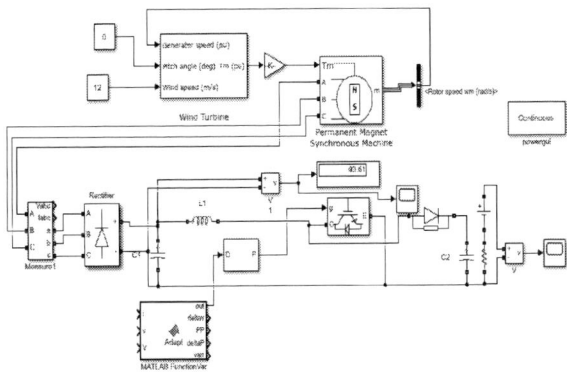

Fig. 8.   General Model of PMSG WECS and MPPT control

The variable wind speed characteristic, adjustment of duty ratio according to variable wind speeds and the performance of adaptive control algorithm in terms of output power are shown in Fig. 9, 10 and 11, respectively.

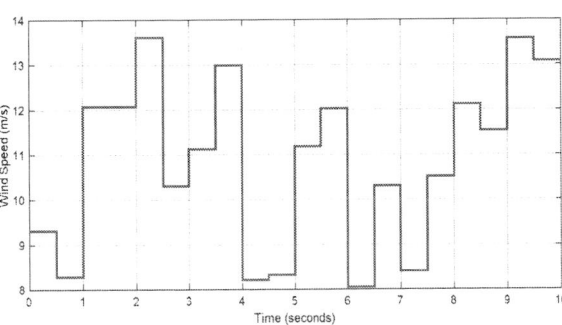

Fig. 9.   Wind speed characteristic

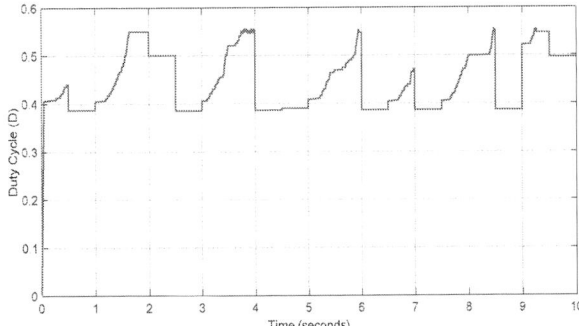

Fig. 10. Adjustment of duty ratio

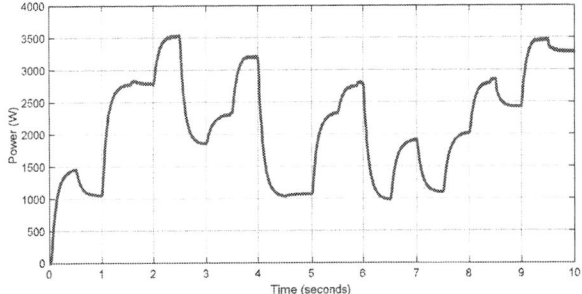

Fig. 11. Output power curves of ADP control

The comparison of MPPs obtained by fixed or conventional (CON) and adaptive (ADP) P&O methods in different wind speeds is presented in Fig. 12 and Fig. 13, respectively.

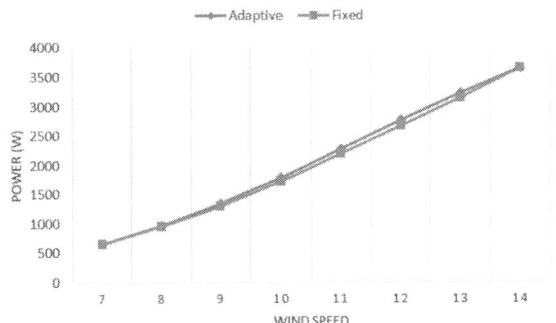

Fig. 12. Comparison of ADP and CON P&O control

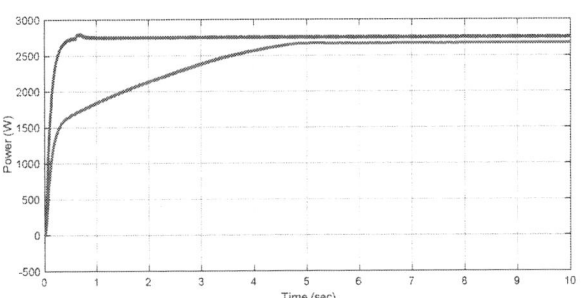

Fig. 13. Comparison of ADP and CON P&O control

The difference in tracking MPPs of CON and ADP is shown in Fig 10 for base wind speed 12 m/s. As seen in Fig. 10 ADP achieves the MPP nearly at 0.5 sec, while CON obtains MPP approximately 4.5 sec later than ADP. It should be emphasized that adaptive control tracks MPP faster than conventional as expected.

V.   CONCLUSION

In small scale WECS, a low cost and high efficiency are the two most important parameters. So in this paper an adaptive MPPT control algorithm which has low cost and higher efficiency than conventional P&O method is proposed. In real life wind energy is intermittent and wind speed can change unpredictably, that is why this control algorithm is checked via simulation and achieved MPPs for variable speeds that may change up to 0.5 sec. Based on the simulation results, the proposed method is determined to be better than the P&O method and has a higher efficiency.

REFERENCES

[1] Lee, J., & Kim, Y.,"Sensorless fuzzy-logic-based maximum power point tracking control for a small-scale wind power generation systems with a switched-mode rectifier." IET Renewable Power Generation, 10(2), 194-202, 2016. doi:10.1049/iet-rpg.2015.0250

[2] Sefidgar, H., & Gholamian, S. A.,"Fuzzy Logic Control of Wind Turbine System Connection to PM Synchronous Generator for Maximum Power Point Tracking",International Journal of Intelligent Systems and Applications, 6(7), 29-35. doi:10.5815/ijisa.2014.07.04

[3] Putri, R., Pujiantara, M., Priyadi, A., Ise, T. and Purnomo, M., "Maximum power extraction improvement using sensorless controller based on adaptive perturb and observe algorithm for PMSG wind turbine application",IET Electric Power Applications, 12(4), pp.455-462, 2018.

[4] Zou, Y., Elbuluk, M.E., Sozer, Y.: "Stability analysis of maximum power point tracking (MPPT) method in wind power systems", IEEE Trans. Ind. Appl., 2013, 49, (3), pp. 1129–1136

[5] H. Li, K. L. Shi and P. G. McLaren, "Neural-network-based sensorless maximum wind energy capture with compensated power coefficient," IEEE Trans. Ind. Appl., vol. 41, no. 6, pp. 1548-1556, Nov./Dec. 2005.

[6] Fundamental and Advanced Topics in Wind Power. (2011). InTech.

[7] A. B. Raju, B. G. Fernandes, and K. Chatterjee, "A UPF power conditioner with maximum power point tracker for grid connected variable speed wind energy conversion system," proc. of 1st International Conf. on Power Electronics Systems and Applications (PESA 2004), Bombay, India, 9-11 Nov., 2004, pp. 107-112.

[8] E. Koutroulis and K. Kalaitzakis, "Design of a maximum power tracking system for wind-energy-conversion applications," IEEE Transactions on Industrial Electronics, vol. 53, no. 2, April 2006, pp. 486-494.

[9] Kumar, D., Chatterjee, K.: "A review of conventional and advanced MPPT algorithms for wind energy systems", Renew. Sustain. Energy Rev., 2016, 55, pp. 957–970

[10] J. Yaoqin, Y. Zhongqing, and C. Binggang, "A new maximum power point tracking control scheme for wind generation," in Proc. International Conference on Power System Technology 2002 (PowerCon 2002). 13-17 Oct., 2002. pp.144-148.

[11] J. Hui and A. Bakhshai, "A new adaptive control algorithm for maximum power point tracking for wind energy conversion systems," in Proc. IEEE PESC 2008, Rhodes, 15-19 June, 2008. pp. 4003-4007.

[12] J. Hui and A. Bakhshai, "Adaptive algorithm for fast maximum power point tracking in wind energy systems," in Proc. IEEE IECON 2008, Orlando, USA, 10-13 No. 2008, pp. 2119-2124.

[13] M. G. Molina and P. E. Mercado, "A new control strategy of variable speed wind turbine generator for three-phase grid-connected applications," in Proc. IEEE/PES Transmission and Distribution Conference and Exposition: Latin America, 2008, Bogota, 13-15 Aug., 2008, pp. 1-8.

[14] J. M. Kwon, J. H. Kim, S. H. Kwak, and H. H. Lee, "Optimal power extraction algorithm for DTC in wind power generation systems," in Proc. IEEE International Conf. On Sustainable Energy Technology, (ICEST 2008), Singapore, 24-27 Nov., 2008 pp. 639 – 643.

[15] C. Patsios, A. Chaniotis, and A. Kladas, "A Hybrid Maximum Power Point Tracking System for Grid-Connected Variable Speed Wind-Generators," in Proc. IEEE PESC 2008, Rhodes, 15-19 June, 2008, pp.1749-1754.

[16] Xia, Y., Ahmed, K. and Williams, B., "Wind Turbine Power Coefficient Analysis of a New Maximum Power Point Tracking Technique.", 2019.

[17] Chen, Y., Xiong, G., Li, W., Qian, J. and Hu, B., "The design of boost circuit in small wind generation system" - IEEE Conference Publication, 2019. [online] Ieeexplore.ieee.org. Available at: https://ieeexplore.ieee.org/document/6611440/ [Accessed 12 May 2019].

[18] Chen, J., Lin, T., Wen, C., et al.: "Design of a unified power controller for variable-speed fixed-pitch wind energy conversion system", IEEE Trans. Ind. Electron., 2016, 63, (8), pp. 4899–4908

Technical track on Mechatronics, Industrial Automation and Control

978-1-5386-7688-2/19 $31.00 © 2019 IEEE

A Novel Development of Acoustic SLAM

Joseph O'Reilly
Science & Engineering. Anglia Ruskin University
Cambridge, Cambridgeshire
joseph.oreilly@pgr.anglia.ac.uk

Silvia Cirstea
Science & Engineering. Anglia Ruskin University
Member, IEEE
Cambridge, Cambridgeshire
silvia.cirstea@anglia.ac.uk

Marcian Cirstea
Science & Engineering. Anglia Ruskin University
Senior Member, IEEE
Cambridge, Cambridgeshire
marcian.cirstea@anglia.ac.uk

Jin Zhang
Science & Engineering. Anglia Ruskin University
Member, IET
Cambridge, Cambridgeshire
jin.zhang@anglia.ac.uk

Abstract— This paper will explore and develop on the novel idea of using acoustics to map and navigate indoor environments. The system requirements, modelling and evaluation are addressed, alongside the design and development process, testing methods, desired outcomes and practical applications. Previous work carried out in this field demonstrates that it is possible to use first order echoes to map a room. The current paper is reporting on initial research to further develop such algorithms into a simultaneous localization and mapping algorithm, having the capability to not only map rooms with sound but to also navigate rooms as well. Such novel system is intended to help visually impaired people to navigate rooms by making use of sounds and their echoes, thus 'listening' their way into navigating through a room. The paper overviews the approach taken towards developing a navigation algorithm using sound, as well as the associated modelling, simulation and testing strategies enabling the desired outcomes of this type of system.

Index Terms— Kalman filters, Autonomous systems, Acoustical engineering, Navigation, Simultaneous localization and mapping, Acoustic signal processing.

I. Introduction

THIS paper covers aspects related to the development process, testing methods, desired outcomes and final uses for the novel idea of using acoustics to map and navigate indoor environments. Some interesting questions to be addressed relate to the ability of autonomous navigation and mapping to be achieved using only audible sound. If so, another question to be answered is if this can be done by only using environmental noise.

It is known that some visually impaired people are capable of navigating rooms by listening to the echoes created by clicking their fingers, so, naturally one wonders if an intelligent system dedicated for such purpose (a robot) could do the same, thus helping the visually impaired to navigate.

A survey of previous work done shows that an interesting question was posed: "Can one hear the shape of a drum?" [1]; the research determined that drums of different shape don't necessarily resonate at different frequencies. Other works carried out [2] developed this theory and applied it to an indoor room, demonstrating that it is possible to use first order echoes to map a room. The present work follows up in further developing this algorithm into a simultaneous localization and mapping algorithms enabling the capability to not only map rooms with sound but also to navigate rooms as well.

The interaction between ambient sound and objects in the environment, manifested through reverberation and echoes, is an important source of information about the composition of the scene, insufficiently exploited in autonomous navigation systems, which can solve the problem of navigation in low light conditions. Show the application of acoustic imaging for mapping a stationary room using a distributed array of five microphones which capture the room impulse response (RIR). This contains the times of arrival of the reverberations from walls and other features using an artificial intelligence (AI) techniques called multidimensional scaling (MDS). This demonstrates that it has been possible to estimate the most likely association between a given reverberation and a particular wall and thus reconstruct the map of the room.

Research in the area of autonomous navigation has developed complex AI systems that combine sensors like cameras, Light Imaging Detection And Ranging (LIDAR) and Inertial Measuring Units (IMUs) with Simultaneous Localization and Mapping (SLAM) algorithms [3] SLAM was developed to guide robots [4] by estimating the map of an unknown environment and, at the same time, updating the position of the robot in relation to this environment, based on probabilistic estimation of feature points (e.g. walls, obstacles) from time-varying inputs of exteroceptive sensors attached to the robot [5]. Sensors like LIDAR and cameras have the advantage of being accurate and of high resolution, however, they come with their downsides: LIDAR is a very expensive sensor and poses health and safety problems in operation; cameras, while becoming less costly, require high processing power as well as having low signal to noise ratios in low light environments.

By using acoustic imaging techniques and a SLAM algorithm adapted for feature detection from sound for navigation in a low light environment, a novel navigation system can be developed. This would have the potential to be used as an assisted living aid for the visually impaired or in robotic navigation for fire-fighting and exploration of dark spaces.

In this paper Section II and III focus on the two major research areas of this novel development. Section II covers the methods of gathering information from the environment with sound and how this information is used to map the environment. Section III overviews the different methodologies and algorithms used in SLAM systems with a focus on indoor environments using microphones as the primary sensor. The development process, testing breakdown and more complex developments is shown in section IV, with the preliminary testing results of RIR input methods is shown in section V.

978-1-5386-7688-2/19 $31.00 © 2019 IEEE

II. ACOUSTIC MAPPING

Sound holds a lot of information of its environment, enough information to accurately estimate the shape of a room. Dokmanic et all.[2] has proved this by developing an algorithm that uses the information from a RIR recorded by an array of microphones to locate the walls in which the echoes from the RIR reflected off of, thus mapping the room. This section will highlight the key elements of RIR and Dokmanic's algorithm.

A. Room impulse response

An impulse response is one of the most important tests one could use to learn the characteristics of a linear system. For this purpose, the room is a linear system. This is where the output of a system is recorded when a delta function is applied at the input. The impulse response of a room is commonly used in acoustic analysis to find acoustic "dead spots", early to late ratios and reverberation time. The reverberation characteristic of a room is commonly used to computationally recreate reverberation for musical application but for acoustic mapping this is the main characteristic which enables the location of the walls of the environment.

There are multiple methods for gathering a RIR: Impulse, Chirp, Linear/Logarithmic sweep, Maximum-Length Signal (MLS) and Inverse Repeated Signal (IRS). These have been compared and tested in many papers [6], [7] and [8]. However, the testing parameters from these works have never explored the situation where the receiver, source or environment is moving. This is due to it not being necessary for a typical RIR use case but for acoustic SLAM, knowing how these methods are affected by a dynamic situation is of high importance.

Impulse is the standard linear system input for an impulse response. The problem with using this method for RIR is that it generates very little energy, which is necessary to acquire clear reverberation in the output. To combat this, it is common to use an impulse train, with multiple impulses, and then calculate the average. Due to these issues, impulse input is not generally used for RIR.

Chirp generates a liner frequency sweep over a short period of time (10-100ms) which is then repeated every 2-3secconds. By using a range of frequencies it reduces the possibility of results being effected by the frequency responses of different rooms. By fast Fourier transform (FFT) deconvolution of the input from the output the RIR can be calculated. This is done by taking the FFT of the input divide it by the FFT of the output and then taking the inverse fast Fourier transform (IFFT) of the result.

Linear/Logarithmic sweep is like a chirp but instead of an impulse train, a longer continuous signal is used. The main advantage of this method over a chirp is that more power is being produced, which in turn results in a better signal to noise ratio.

MLS is close to a white noise signal that has a constant magnitude and a uniform random phase spectra. By using a MLS, the RIR can be found with circular cross-correlation between the output and MLS input, as the generated signal will be unique in its environment. MLS has a long history as a primary input method for RIR as it has many advantages, which are described in [7]. One issue of MLS is that it can produce phantom echoes when faced with non-linear distortion. This issue could be catastrophic, as the primary purpose of the RIR in acoustic mapping is to find the echoes. IRS uses two MLS sequences, one positive and one negative, and alternates between the two. In doing so it doubles the measurement time but increases the immunity to distortion.

B. Mapping a room with sound

Dokmanic et all's.[2] method for mapping the walls of a room is to use an array of 4 or more microphones to record the RIR from a sweeping sine wave generated by a source with an unknown location. The initial audio impulse and all echoes from the RIR are labeled with their time of arrival and magnitude for each individual microphone in the array. The echo information is compared between each microphone to find the direction each echo came from. The labelling process is carried out by taking advantage of Euclidean Distance Matrices (EDM) arithmetic's: each microphone's distance of placement apart from each other is placed in an EDM. The pairing impulses can then be found by entering the distance between the microphone and each recorded impulse virtual source into a new row of the EDM and then by using EDM arithmetic's, false combinations can be found.

Once the directions of all the echoes are known, the room's shape can be constructed. When noise is factored in, there can be a situation where no combinations of the echoes source creates an EDM. In this situation, the probability of how close each arrangement is to a true EDM value is calculated. For a single room measurement, the one which is closest to a true value is used, but in regards to using this method with SLAM it will be beneficial to use a statistical filter for the probability of each EDM arrangement which updates for each sample. Statistical filters are commonly used in SLAM for each sample of sensor data but in this unique situation the sensor information could have multiple outcomes per sample; this will be discussed further in Section III and IV.

It has been shown in Dokmanic's work that this is a reliable method for mapping the walls in a room due to its high precision without even needing to know the location of the source sound or the receiver locations. This means that in a practical environment where the system is constantly moving, recalibration of the receivers can be done purely in the software. This also leads onto another possible improvement where multiple sources from different locations in the environment could be layered on top of each other, allowing a system to not only need to generate a sound to produce a map but also compare multiple generated maps to find the highest probability of creating an accurate map.

III. SIMULTANEOUS LOCALIZATION AND MAPPING

For truly autonomous robotic navigation to be achieved, when an unknown environment is presented, with an unknown position in this environment, a robot must be capable of mapping its soundings with its onboard sensors and accurately estimating its location on this map. This is known as the SLAM problem statement. This allows a robot to be deployed in situations where the area is unknown and the user cannot see the robot, such as fire rescue in high smoke density buildings, small tunneled cave systems and assistance for the visually impaired.

978-1-5386-7688-2/19 $31.00 © 2019 IEEE

Due to the nature of robotics, different situations require different solutions for navigation: the environment, (indoors, outdoors, underwater, airborne) the sensors used (laser, camera, IMU, audio) and the moving system (wheeled, flying, walking, etc.) all affect the SLAM algorithm.

The scope of this paper is focused on the mapping of the environment before it is used with SLAM. So, this paper gives an overview of SLAM methods which are relevant to an indoor environment using microphones as the primary input sensor for gathering information of the environment.

A. General Methodology

The main three elements of SLAM are: the map representation, the data processing and the sensors used. Each of these building blocks are essential for the development of a SLAM algorithm. For this system, the primary sensor will be an array of microphones that gathers the location of a room's walls through sound. The main development process for acoustic SLAM will be finding the best fit of map representation and data processing.

B. Mapping Methods

The mapping method is how the SLAM algorithm information is visually processed. There are many methods to do this: Topological, Semantic, Appearance and Hybrid which have been compared and expanded upon in various works [9], [10]. Due to the vast information of these methods this paper covers the main two methods: 1) feature mapping and 2) grid mapping.

1) Feature mapping

The standard model observation method used in SLAM is landmark/feature detection. This involves labeling landmarks in the environment and using them as reference points for calculating the position of the system. Normally, this method is used with camera based systems, where landmarks in the environment are easily determined with shape, color and depth.

The more unique features that can be recorded, the higher the accuracy of the map. For range-based sensors (laser, ultrasonic, echo), only edge detection and free standing objects can be used as landmarks, which results in clear landmarks but usually less of them compared to cameras.

2) Grid mapping

Rather than mapping features and landmark locations grid based SLAM uses grid maps. This approach splits the environment into ridged independent cells which are recorded as either occupied space or free space. This generates a floor plan like map, where occupied cells are black pixels and free space is white. The main disadvantage of grid based SLAM is that computationally it takes up more memory as the entire map needs to be recorded, compared to feature based SLAM where only the feature locations needs to be recorded. The main advantage of this method though is that feature detection is not required, meaning the system can work directly with the raw sensor data. This is preferable for niche sensors methods where the amount of features are limited.

C. Data Processing methods

The data processing methods refer to how the system calculates its most probable location in the map. The two most common methods are Bayesian based filters and partial based filters, both methods being quite old. There are also some interesting new AI based method which aim to overcome some of the major issues the older methods have.

Bayesian filter framework

For feature mapping techniques, it is common to use a Bayesian filter framework for state estimation. Bayes' theorem describes the probability of an event, based on prior knowledge of conditions; this creates the framework for state estimation.

This breaks down the SLAM problem in to its probabilistic form, by calculating the probability of the system's current location (defined as 'x_k' where x is the locations vector and k is the time period) when all observed landmarks (defined as '$z_{1:k}$', where 'z' is an array of landmark location vectors) and movements of the system (defined as '$u_{1:k}$', where 'u_k' is the system's current movement vector) are known.

$$bel(x_k) = P(x_k | z_{1:k}, u_{1:k}) \qquad (1)$$

This can be defined by a simple probability equation, where the current location is dependent on all past and current observed landmarks and all past and current vector movements of the system. Using statistical methods, shown in [11], the generic SLAM algorithm can be represented through two equations: the prediction step and the correction step.

The prediction step (2) calculates the probability of the systems next possession based on the previous position of the system. This requires the motion model, (3) which models the uncertainty in the optometry data by finding the probability of the current position of the system in relation to the recorded distance traveled and the previous location of the system.

$$\overline{bel}(x_k) = \int P(x_k | x_{k-1}, u_k) bel(x_{k-1}) \, dx_{k-1} \qquad (2)$$
$$P(x_k | x_{k-1}, u_k) \qquad (3)$$

The correction step (4) calculates the probability of the systems current possession based on the sensor data of the environment. This requires the observation model (5), which calculates the probability of a landmark being observed in relation to where the system currently believes it is based on the motion model.

$$bel(x_k) = \eta P(z_k | x_k) \overline{bel}(x_k) \qquad (4)$$
$$P(z_k | x_k) \qquad (5)$$

The most common filters this method is based on are the Kalman filter (KF) [11] and the less computationally intensive Extended Kalman filter (EKF) [12]. KF-SLAM is the optimal estimator when two assumptions are made: all models are linear and all distributions are Gaussian. EKF-SLAM is highly influenced by incorrect observations of landmarks. This is often a problem when the vehicle returns to a previously visited location after traveling a large distance, this is known as the "loop-closure" problem.

2) Particle Filter Framework

Particle filters represent a probability distribution as a line of particles (samples) where the density of the particles represents the probability. When there is a new observation, the points are weighted by the observation probability, then the distribution is resampled and a new distribution is made. Fast-SLAM [13] is currently one of the most used SLAM algorithms, which is favored over the EKF-SLAM one due to the fact that the filter allows non-linear distributions, making it less restrictive when used in a realistic environment.

3) Other Filter Frameworks

The main issue with KF-SLAM and Fast-SLAM is that they assume the environment is static, which is not realistic. Novel approaches to overcome this issue involve using neural-networks and AI, one of them being generalized motion (GEM)-SLAM [14]. GEM-SLAM shows a large improvement in accuracy when there are false readings and uncertainty caused by dynamic objects in the environment.

IV. DEVELOPMENT OF A NOVEL ACOUSTIC SLAM METHOD

This paper presents aspects of the development for an autonomous echo-location system based on acoustic imaging and acoustic SLAM. Such systems can be used for guided navigation in dynamic scenes affected by environmental noise.

A. Stage 1: Initial developments

There are two interconnecting challenges in the initial development of this novel mapping system: I) the acoustic mapping technique has to produce the highest quality environment information for the SLAM algorithm, and II) the SLAM algorithm being used must take as much advantage as possible of the provided information set. Even though these aspects are dependent on each other, they can be addressed independently due to how they are interconnected. As the initial development will be using Dokmanic's algorithm [2], the environment data being handed over to the SLAM algorithm will be in a constant format even if the acoustic recording methods change.

As shown in section II, there are many different input methods for a RIR and for a robust system the one which gives the clearest reverberation whilst having an acceptable signal to noise ratio (SNR) in a realistic environment needs to be found. Using an array of microphones, a range of RIR input methods will be tested. Room shape, room size, input volume, environmental noise and level of clutter are the main elements which need to be tested and how well each method dose will be based on the SNR of the output and the accuracy of mapping. When the optimal RIR method is found/developed, there are multiple situations that need to be tested which go beyond the scope of Dokmanic's work. Useful information like environment material (wood, glass, carpet), open and closed doors or specific object detection could be found inside the RIR which would be useful to take advantage of if the computational cost is not too high. Dokmanic's algorithm creates a 3D graph with the walls, source and receivers plotted as vectors. It would then seam

almost natural to use a graph or grid based method but then, as a 3D map will be developed, the computational costs of these methods greatly increase. Feature based SLAM is less costly but the lack of information may result in unreliable mapping. The SLAM algorithm for the initial testing will need to accurately map with the information from the acoustic mapping algorithm and also be low powered enough to be implemented onto a portable microcontroller. The filter and mapping technique both need to be explored and tested to see which combination produces the most accurate map whilst using the least power.

The acoustic method and SLAM algorithm from these tests will develop the initial system where they will be implemented into a standalone hardware board which will control a wheeled robot. At this point the system will be capable enough to develop a map of its environment with echolocation and when the system is moved it will be able to locate itself. At this point, the system was only been tested by gathering information stationary but in a realistic setting it is assumed that a navigation robot is capable of SLAM whilst moving. This is the next step of development.

B. Stage 2: System developed for realistic environments

The previously mentioned acoustic mapping methods have only been tested using a stationary environment. Sound acts differently when its source is moving and when it interacts with moving objects, this is known as the Doppler Effect. This means that the previously tested RIR input methods, which gave an acceptable stationary result, need to be tested whilst the source and receivers are moving at different speeds, directions and rotations. In turn, this means that an IMU needs to be implemented onto the standalone hardware board to accurately measure the movement of the system. The change in accuracy for these movement elements will be used to develop a margin of error that the SLAM algorithm can use for the probability filter.

Once the system is at a point where it can accurately map whilst it is moving, it will then need to be able to navigate a realistic environment where there are moving objects within it. Firstly, this stage is dependent on how the current system is affected by moving objects, as unlike other sensors the echo location algorithm only maps walls. The next investigation stage would be to test how the system is affected by different size and speeds of moving objects whilst the system is navigating.

If the system is noticeably affected, then there are two development approaches that can be made to solve this problem: develop the acoustic mapping to detect and filter moving objects, or use a more complex SLAM algorithm that can support a dynamic environment. As stated in section III, the majority of SLAM algorithms assume the environment is stationary; this is not realistic. There are SLAM algorithms that tackle this issue [16], though more complex algorithms require higher computational power. The testing process will be repeated from stage 1 for these more complex SLAM algorithms compared with the already functioning algorithm and the further developed algorithm which filters moving objects. This test compares the accuracy and power consumption in a dynamic environment.

978-1-5386-7688-2/19 $31.00 © 2019 IEEE

C. Stage 3: Environmental sound mapping

An interesting factor of previous works [2] is that the sound source location is not need to locate the walls. A sound created in the environment naturally could be used to map the room if it can be successfully isolated. This would mean that the system could navigate purely based on environmental noises being used as inputs for RIR, applying no additional impact into its environment. If a single environmental sound can successfully map a room, it could also be possible to use multiple RIR each using different sounds being computed in parallel, which in theory would increase the accuracy of mapping.

The major problem with this theory is that all methods of RIR use deconvolution, which requires the input signal to be known. This means a new RIR method will need to be developed, which does not have any prior knowledge of the impulse or can estimate what the input signal is before deconvolution.

V. INITIAL EVALUATION AND TESTING OF THE METHOD

A. Evaluation principles

The first stage of testing compares commonly used RIR input methods (impulse, chirp, logarithmic sweep, MLS and popping a balloon) in regards to their SNR in different indoor room environments (Fig 2.) and at different input volumes. Each of the RIR, except popping of a balloon, was generated in Matlab and transmitted by using a Digital Audio Workstation (DAW). The DAW also recorded the four microphones in sync and exported the signal information to be manipulated in Matlab.

For clear results, all initial testing was done with high quality hardware, before the system is developed into a lower power system. An array of 4 microphones was used as audio sensor to record the RIR. It is important that the microphones used are capable of omnidirectional recording to allow no bias to any direction of sound. High-quality microphones require a high amount of power, referred to as phantom power, which is usually provided by the audio interface. The microphones used in these tests were four AKG C414-XLS condenser microphones, used in omnidirectional mode. When using high quality microphones, they often require in line power, amplification and to be converted from analogue to digital to be processed by a computer, which the standard input/output (I/O) ports of a computer cannot provide. Audio interfaces power the microphones and convert their analogue signal into digital for a computer to take advantage of it. The individual important factors of the audio interface are the amount of microphone inputs and the comparability and quality of the audio drivers it can use. The Audio interface used was the ALESIS MULTIMIX 8 USB FX, which has four microphone inputs and uses a common high quality audio driver audio stream input/output (ASIO). A directional Yamaha loud speaker produces the RIR input signal. Both directional speakers and omnidirectional speakers can be used but omnidirectional speakers produce a more realistic spreading sound due to sending sound in all directions. Although these types of speakers are expensive and not necessary for the preliminary testing, both the microphones, input of the audio interface and output of the audio interface all have amplifier settings which are all kept at 0dB.

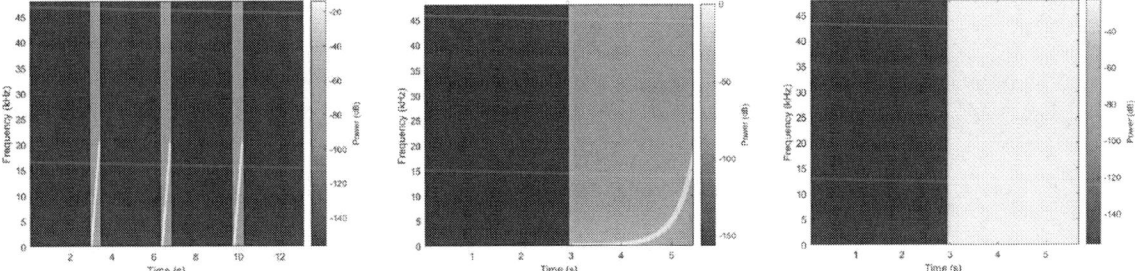

Fig 1. 20Hz to 20kHz linear chirp train, 20Hz to 20kHz logarithmic sweep and MLS RIR input signals used for testing.

Fig 2. Three of the rooms used for testing, each with different shapes and volumes of clutter.

978-1-5386-7688-2/19 $31.00 © 2019 IEEE

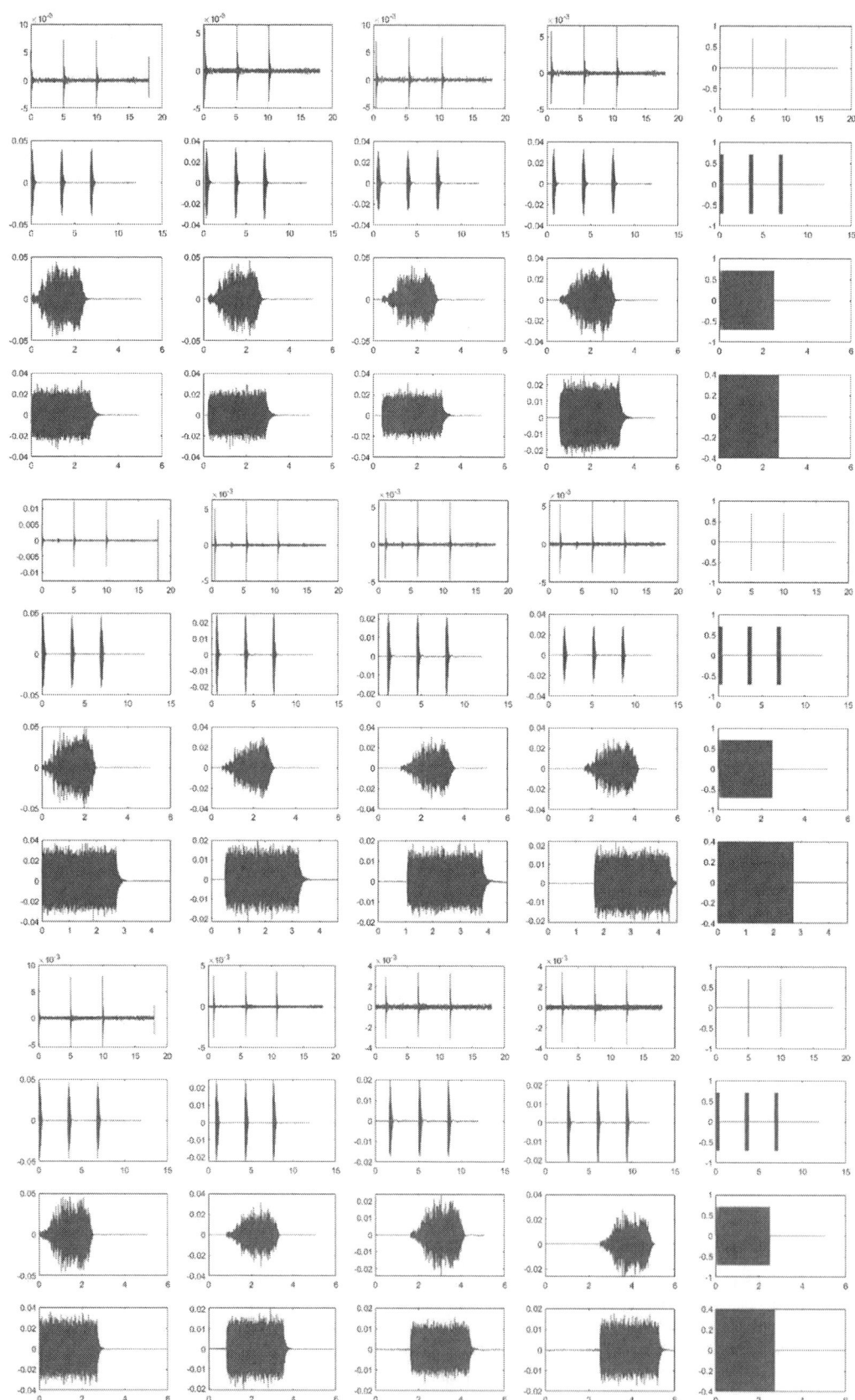

Fig 3. Each rooms microphone (1-4) and input RIR signal for each input method in the time domain.

B. Test results

The RIR signals generated in Matlab (Fig.1) where tested and recorded in three separate rooms (Fig.2). The RIR for each room is shown in Fig.3. From these results there are some clear observations that can be made. The low input power limitation of the impulse signal is clearly visible. The signal output is lower by a magnitude of 10 meaning the ambient noise of the system is noticeable compared to the other methods.

Fig 3 recordings is when there is no reduction in the power of the input signal. These methods were also tested at -12dB and -30dB , which was applied at the DAW output. As the signal power is reduced the SNR of each RIR decreases as expected, at -30dB the impulse RIR error rate is too high to obtain any useful information.

In respect for the system to function without knowledge of the input signal the MLS method seems unlikely to be useful to develop upon. This is due to the deconvolution being impossible without previous knowledge of the unique signal.

VI. Conclusion

Autonomous robotic navigation is an ever developing field, constantly evolving and improving on itself. There is a vast difference in environments, where robots are required to navigate, certain environments requiring specific solutions. Acoustic SLAM takes advantage of the amount of information the sound holds about its environment which is an important tool when designing navigation systems.

The approach taken towards developing a novel navigation algorithm using sound has been covered in this paper, with an overview of the two main elements required to develop it, acoustic mapping and SLAM, and the testing methodology that is being undergone to develop it. With the inclusion of the preliminary testing results of the RIR input in different environments.

When finalized, the completed system would be capable of navigating without needing to generate a sound, creating no impact in the environment. This means that this system will be able to function alongside people without any risk. Not only would this system be safe but it would also be capable to map a 360° three-dimensional room with a single microphone array, which no other sensor that is currently being used with SLAM can do.

In the longer term, and beyond the scope of this project, the concept should lead to a practical solution that would be easy to use and require little or no external infrastructure, to make it ubiquitous available. In addition, such a solution should have minimal impact on the surroundings, be environmentally friendly, robust and cost-effective (both to acquire and use). Other further developments would include integrating acoustic SLAM with other commonly used sensors to increase accuracy for generalized robotic navigation. Sound holds a lot of information about the environment which can be used to support LIDAR and camera based systems, providing more information at a low system cost.

References

[1] K.Mac, "Can't one really hear the shape of a drum?," vol. 57, no. 4, p. 465, Jul. 1966.

[2] I. Dokmanić, R. Parhizkar, A. Walther, Y. M. Lu, and M. Vetterli, "Acoustic echoes reveal room shape," vol. 110, no. 30, pp. 12186–12191, Jul. 2013.

[3] Evers C, Naylor PA. "Acoustic SLAM." IEEE/ACM Transactions on Audio, Speech, and Language Processing. 26(9):1484-98, Sep. 2018;

[4] Dissanayake MG, Newman P, Clark S, Durrant-Whyte HF, Csorba M. "A solution to the simultaneous localization and map building (SLAM) problem." IEEE Transactions on robotics and automation. 17(3):229-41 Jun. 2001.

[5] A. J. Davison, "Mobile Robot Navigation Using Active Vision," Ph.D. dissertation, Dept. Science Eng., Oxford Univ., Oxford, UK, 1998.

[6] A. Carini, S. Cecchi, L. Romoli, A. Carini, S. Cecchi, and L. Romoli, "Robust Room Impulse Response Measurement Using Perfect Sequences for Legendre Nonlinear Filters," vol. 24, no. 11, pp. 1969–1982, Nov. 2016.

[7] Stan GB, Embrechts JJ, Archambeau D. "Comparison of different impulse response measurement techniques." Journal of the Audio Engineering Society. 50(4):249-62. 2002;

[8] Farina A. "Simultaneous measurement of impulse response and distortion with a swept-sine technique." InAudio Engineering Society Convention, 108, Feb. 2000.

[9] Canclini A, Antonacci F, Sarti A, Tubaro S. "Acoustic source localization with distributed asynchronous microphone networks." IEEE Transactions on Audio, Speech, and Language Processing. 21(2):439-43. Feb 2013.

[10] Norcross S, Bradley JS. "Comparison of room impulse response measurement methods." Canadian Acoustics. Sep. 1994.

[11] Welch G, Bishop G. "An introduction to the Kalman filter." 1995.

[12] Ljung L. "Asymptotic behavior of the extended Kalman filter as a parameter estimator for linear systems." IEEE Transactions on Automatic Control. Feb.1979.

[13] Montemerlo M, Thrun S, Koller D, Wegbreit B. "FastSLAM: A factored solution to the simultaneous localization and mapping problem." Jul. 2002.

[14] C. Evers and P. Naylor, "Optimized Self-Localization for SLAMin DynamicScenes Using Probability Hypothesis Density Filters," vol. 26, no. 9, pp. 1484–1498, Sep. 2018.

Alternative Approximation Method for Time Delays in an IMC Scheme

Cristina I. Muresan, Isabela R. Birs, Cosmin Darab, Ovidiu Prodan, Robin De Keyser

Abstract—Internal Model Control (IMC) algorithms have proven to be quite efficient in designing PID type controllers for time delay systems. However, to compute the equivalent PID controller in a feedback loop, using the IMC tuning approach, implies the approximation of the process time delay. Two of the most widely used approximation methods (first order or series and Padé) are used in this paper and compared to a new approximation method, the Non Rational Transfer Function (NRTF) approximation. In the current paper, the NRTF approximation is presented as an alternative method to the series/ Padé approximation techniques for time delay systems in an IMC closed loop design. It is shown that for first order and second order processes, given a certain ratio of the IMC filter time constant and process time delay, the NRTF method could produce better closed loop dynamics, compared to the series/ Padé approximations. Several numerical examples are presented, as well as a case study.

I. INTRODUCTION

Internal Model Control (IMC) is one of the most versatile control strategies [1,2]. The tuning of the IMC controller is among the simplest approaches, with only one tuning parameter, the IMC filter time constant [2, 3]. The IMC controller was first introduced in 1982 [4] and has since been developed to enhance its closed loop performance, both in terms of reference tracking and, especially, in terms of disturbance rejection [3]. Model predictive control and Diophantine equations have been used in the design of improved disturbance rejection IMC control [1]. Other examples include the design of the IMC filter such that it cancels the slow (dominant) process pole [5], the use of the Ziegler-Nichols methodology in the tuning [6] or constructing one or more asymptotic canceling constraints for disturbance rejection [7]. In [8], it is assumed that disturbances have a stochastic nature and are centered around some frequency. The IMC filter is then designed to compensate for the dead time effects, by using this information. Diophantine equations are used to determine the IMC filter. An optimal IMC filter is proposed for the special case of lag dominant processes in [9]. A novel control scheme is presented in [10], designed to improve the anti-interference ability and robustness for the dead time process. The design is based on analyzing the relationship between IMC and disturbance observer control and a disturbance filter is included to realize the active anti-interference ability the presented control scheme.

Successful implementations of the IMC control strategy on real applications are presented in [1,11,12]. An additional advantage of IMC tuning consists in its easy implementation as an equivalent PID controller [13-15]. Issues with the implementation of the IMC controller arise when dealing with time delay systems. The equivalent form of a classical PID controller can be obtained, provided that the time delay is approximated [16]. Quite frequently, this approximation is achieved using a series or Padé expansion of the delay term. In this paper, these approximation methods are evaluated for various types of systems: first order to higher order, lag or delay dominant processes. A new approximation method is introduced, called the Non Rational Transfer Function (NRTF) approximation [17] approach and compared to the previous ones. So far, the NRTF method has been used successfully in implementing fractional order systems [17]. As it has been recently shown for fractional order IMC controllers, the NRTF approach is suitable for a large class of processes and offers better results, in general, compared to the series and Padé approximation methods [18]. In this paper, it will be shown that the NRTF approximation can be considered as an alternative to the series and Padé methods. Several case studies are considered, as well as an experimental validation.

The paper is structured as follows. Section II briefly presents the IMC paradigm, considering the three time delay approximation methods for computing the equivalent PID controllers. Section III presents the numerical results regarding various systems, whereas the validation on a Quanser six tanks process of the proposed NRTF approximation is given in Section IV. Comparisons and advantages of using the proposed approximation method are highlighted. Section V presents the concluding remarks.

II. EQUIVALENT PID CONTROLLERS FOR TIME DELAY PROCESSES BASED ON INTERNAL MODEL CONTROL

The IMC closed loop scheme is given in Fig. 1, where $H_{IMC}(s)$ stands for the IMC controller, $H_p(s)$ is the process, $H_m(s)$ is the process model, $H_c(s)$ is the equivalent controller.

Robin De Keyser is with the DySC research group on Dynamical Systems and Control, Ghent University, Technologiepark 125, B9052, Ghent, Belgium (E-mail: Robain.DeKeyser@UGent.be)

Cristina I. Muresan and Isabela R. Birs are with the Department of Automation, Technical University of Cluj-Napoca, Memorandumului Street, no. 28, 400114 Cluj-Napoca, Romania, (E-mail: Cristina.Muresan@aut.utcluj.ro, Isabela.Birs@aut.utcluj.ro)

Cosmin Darab is with Department of Electroenergetics and Management, Technical University of Cluj-Napoca, Memorandumului Street, no. 28, 400114 Cluj-Napoca, Romania (Cosmin.Darab@enm.utcluj.ro)

Ovidiu Prodan is with Department of Civil Engineering, Technical University of Cluj-Napoca, Memorandumului Street, no. 28, 400114 Cluj-Napoca, Romania (Ovidiu.Prodan@mecon.utcluj.ro)

978-1-5386-7688-2/19 $31.00 © 2019 IEEE

The following transfer function is further used to approximate the dynamics of a stable dead-time process:

$$H_P(s) = \frac{k}{(Ts+1)^n} e^{-s\tau} \qquad (1)$$

where s is the Laplace variable, τ is the dead-time, n is the process order and k is the process gain.

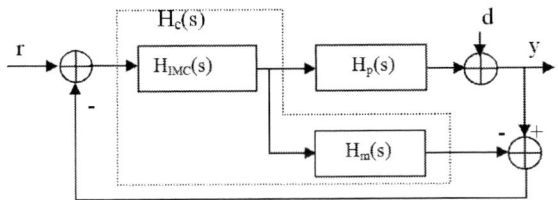

Fig. 1. IMC closed loop control scheme

In a *series (first order) approximation* of the time delay, with $e^{-\tau s} \cong 1 - \tau s$, the IMC controller is given by:

$$H_{IMC}(s) = \frac{(Ts+1)^n}{k} \frac{1}{(\lambda s+1)^n} \qquad (2)$$

where λ is the IMC filter time constant. The equivalent controller is computed as:

$$H_c(s) = \frac{H_{IMC}(s)}{1 - H_{IMC}(s)H_m(s)} = \frac{(Ts+1)^n}{k\left[(\lambda s+1)^n - 1 + \tau s\right]} \qquad (3)$$

For first order processes, $n=1$ and the equivalent controller is a simple PI:

$$H_c(s) = \frac{Ts+1}{k(\lambda+\tau)s} \qquad (4)$$

In a *Padé approximation* of the time delay, with

$e^{-\tau s} \cong \dfrac{1 - \dfrac{\tau}{2}s}{1 + \dfrac{\tau}{2}s}$, the IMC is determined as:

$$H_{IMC}(s) = \frac{(Ts+1)^n}{k} \frac{1+\dfrac{\tau}{2}s}{(\lambda s+1)^{n+1}} \qquad (5)$$

for n≥2. The equivalent controller is computed as:

$$H_c(s) = \frac{(Ts+1)^n\left(1+\dfrac{\tau}{2}s\right)}{k\left[(\lambda s+1)^{n+1} - 1 + \dfrac{\tau}{2}s\right]} \qquad (6)$$

For first order systems, using the Padé approximation, the IMC controller in (5) can be considered as:

$$H_{IMC}(s) = \frac{Ts+1}{k} \frac{1+\dfrac{\tau}{2}s}{\lambda s+1} \qquad (7)$$

leading to an equivalent controller in a PID form:

$$H_c(s) = \frac{(Ts+1)\left(1+\dfrac{\tau}{2}s\right)}{k\left(\lambda+\dfrac{\tau}{2}\right)s} \qquad (8)$$

For the novel *NRTF approximation* method in [17], the IMC controller is determined as:

$$H_{IMC}(s) = \frac{(Ts+1)^n}{k} \frac{1}{(\lambda s+1)^n} \qquad (9)$$

and the equivalent controller is:

$$H_c(s) = \frac{(Ts+1)^n}{k(\lambda s+1)^n - e^{-\tau s}} \qquad (10)$$

Notice that in this last case, the time delay has not been approximated and is used as such in the expression of the equivalent controller in (10).

In what follows the discrete-time approximation of the equivalent controllers in (3), (6) and (10) is used to produce the numerical, as well as the experimental results. The equivalent controllers in (3) and (6) are discretized using classical discretization methods, while for (10), the novel discretization method described in [17] is used. In all cases, the same sampling time is used.

III. THE NRTF APPROXIMATION METHOD

The proposed discrete-time approximation method used in this paper has been developed and presented by the authors in [17]. This method can be applied to any non-rational transfer functions.

The method consists in four steps.

<u>Step 1</u>: Discretize the Laplace operator using a suitable generating function, such as:

$$w(z^{-1}) = \frac{1+\delta}{T_s} \frac{1-z^{-1}}{1+\delta z^{-1}} \qquad (11)$$

with $\delta \in [0 \div 1]$ and T_s-the sampling period. The parameter δ is a shaping knob [17]: a larger value of δ decreases the phase error, while a lower value ensures a lower magnitude error. For $\delta=0$, the Euler discretization rule is obtained, while $\delta=1$ leads to the Tustin discretization rule. This first step produces a discrete time system, $G(z^{-1})$, by replacing the Laplace operator s with the generating function in (11). As the approximation of the equivalent controller in (10) will be performed in a certain bounded frequency interval (0, ω_h), the maximum frequency ω_h needs to be defined at this stage, according to the Nyquist sampling theorem, as $T_s = \dfrac{\pi}{\omega_h}$.

<u>Step 2</u>: Calculate the frequency response of the discrete time equivalent controller in (10). To compute the frequency

response, the Laplace operator s has to be replaced with $j\omega$, where $\omega = \dfrac{2\pi}{T_s N_s}\begin{bmatrix} 0 & 1 & 2 & ... & \dfrac{N_s}{2} \end{bmatrix}$. Then, the frequency response of the discrete-time system is computed according to $z^{-1} = e^{-T_s s}$. The parameter N_s is also a tuning knob. The larger its value, the better the approximation in the low frequency range. This second step produces a vector of frequency response values of the discrete time transfer function.

Step 3: Calculate the impulse response of the discrete time system, based on the inverse Fast Fourier Transform (FFT) algorithm. This step results in a vector of N_s impulse response values:

$$g[n] = \frac{1}{N_s}\sum_{k=0}^{N_s-1} G[k]e^{+j\frac{2\pi}{N_s}nk}, \quad n = 0,1,2,...,N_s-1 \qquad (12)$$

Step 4: Determine a rational discrete time transfer function having a similar impulse response as obtained from the inverse FFT. The Steiglitz-McBride approach is used to achieve this. The order N of the approximation has to be specified. This step results in a rational discrete time transfer function:

$$H(z^{-1}) = \frac{c_0 + c_1 z^{-1} + ... + c_N z^{-N}}{1 + d_1 z^{-1} + ... + d_N z^{-N}} \qquad (13)$$

IV. NUMERICAL RESULTS

The three approximation methods for the equivalent controller in an IMC methodology are evaluated numerically considering lag/delay dominant first order and second order systems. Although not specifically mentioned, each of the considered cases can represent various industrial systems. A simple unit step reference signal is considered in all cases, as a basis for closed loop comparisons.

Consider the first order plus dead time system (FOPDT):

$$H_p(s) = \frac{1}{3s+1}e^{-3s} \qquad (14)$$

for which an IMC controller is designed with $\lambda = 1.5$. Notice that $\tau/T = 1$ and $\lambda/\tau = 0.5$. The closed loop simulation results are indicated in Fig. 2 a), for the outputs and Fig. 2b) for the inputs.

Notice the increased overshoot, settling time and control effort in the Padé approximation. This is due to the lack of robustness of this method, in cases where $\lambda < 0.8\tau$ [2]. Overall, the NRTF method ensures the best closed loop performance, with the lowest overshoot and fastest settling time, combined with an acceptable control effort. The disturbance rejection simulation results are indicated in Fig.

3a) and b). In this cases also, the NRTF ensures better closed loop results.

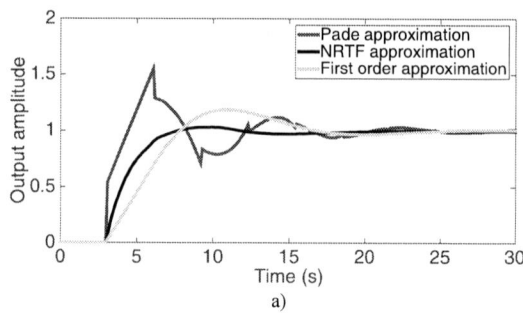

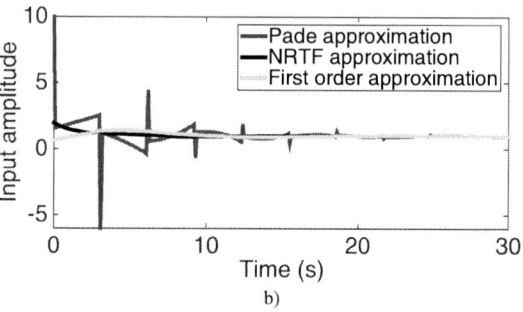

Fig. 2 Reference tracking results for process (14): a) output and b) input signals

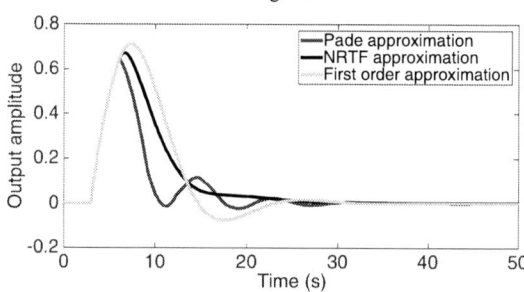

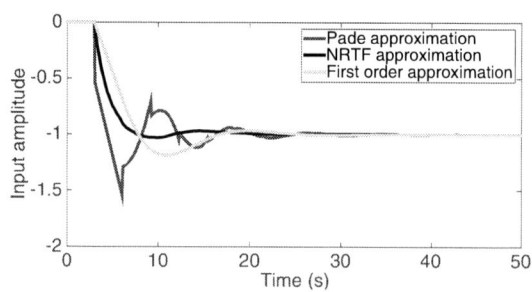

Fig. 3. Disturbance rejection results for process (14): a) output and b) input signals

Consider the first order lag dominant system:

$$H_p(s) = \frac{2}{5s+1}e^{-0.5s} \qquad (15)$$

for which an IMC controller is designed with $\lambda = 2$. Notice that $\tau/T = 0.1$ and $\lambda/\tau = 4$. The closed loop simulation results are indicated in Fig. 4a), for the outputs and Fig. 4b) for the inputs. Fig. 5a) and b) shows the disturbance rejection results.

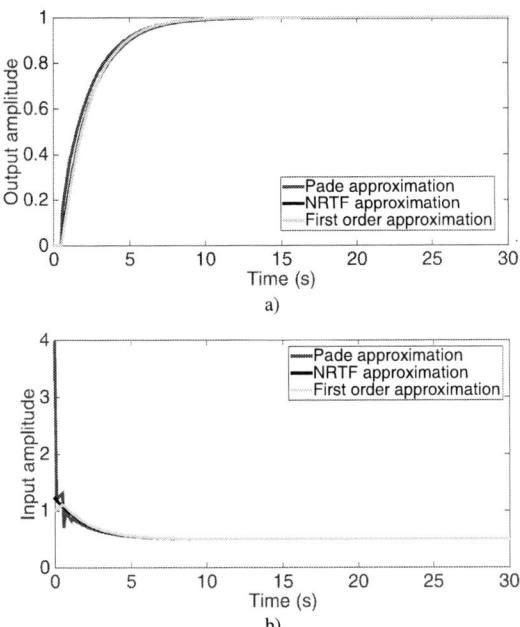

a)

b)

Fig. 4. Reference tracking results for process (15): a) output and b) input signals

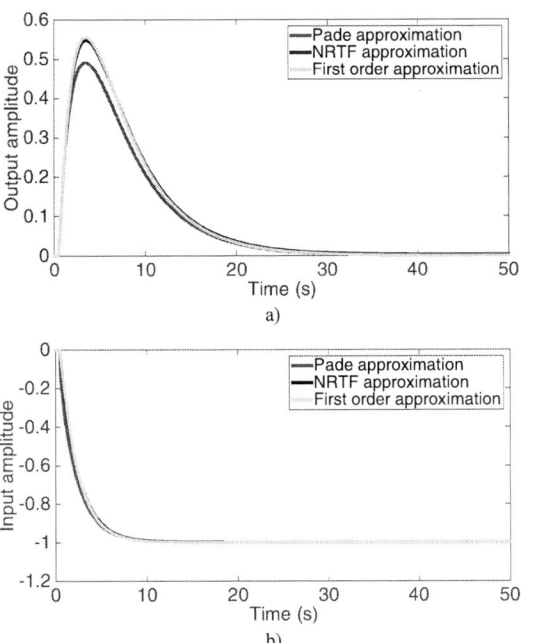

a)

b)

Fig. 5. Disturbance rejection results for process (15): a) output and b) input signals

Notice that for strong lag dominant systems, all methods produce similar results, provided λ/τ large. However, for the Padé approximation there is a significantly larger control effort required. Compare these results with those in Fig. 6, where the IMC controller was designed for the same process in (15), but with $\lambda=0.3$; hence, $\tau/T=0.1$ and $\lambda/\tau=0.6$. For ratios $\lambda/\tau<1$, the NRTF approximation method ensures better

closed loop results compared to the series or Padé approximations.

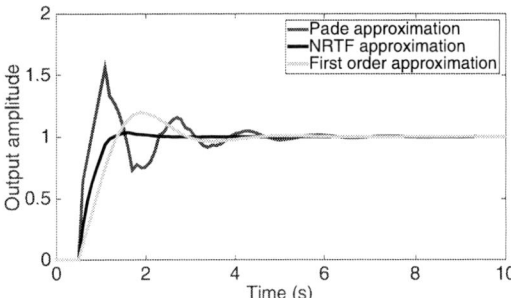

Fig. 6. Reference tracking results for process (15) with $\lambda=0.3$

Finally, consider a delay dominant FOPDT system:

$$H_p(s) = \frac{1}{4s+1}e^{-8s} \qquad (16)$$

for which an IMC controller is designed with $\lambda=1$. Notice that $\tau/T=2$ and $\lambda/\tau=0.125$. The closed loop simulation results are indicated in Fig. 7. Notice that the Padé approximation produces unstable closed loop dynamics because of the low ratio λ/τ [2]. Fig. 8a) and b) show the simulation results regarding reference tracking, while Fig. 9a) and b) present the disturbance rejection results for the NRTF and series approximation approaches. For large ratios λ/τ, the series and Padé approximations approach the NRTF, in terms of closed loop performance. Notice that for strong delay dominant systems, the NRTF method provides for a lower overshoot and faster settling time. Also, it ensures a better disturbance rejection.

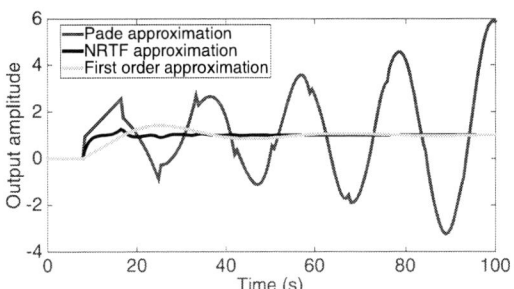

Fig. 7. Unstable closed loop dynamics in Padé approximation

Consider the poorly damped second order system, with damping ratio 0.1:

$$H_p(s) = \frac{1}{s^2+0.8s+16}e^{-4s} \qquad (17)$$

for which an IMC controller is designed with $\lambda=1$. Notice that $\tau/t_s=0.294$, where t_s is the process settling time, and $\lambda/\tau=0.25$. The closed loop simulation results are indicated in Fig. 10a), for the outputs and Fig. 10b) for the inputs. Fig. 11a) and b) shows the disturbance rejection results. Notice that for poorly damped delay dominant systems, the NRTF produces the best results. Decreasing the value for the IMC filter time constant λ and observing the results leads to the conclusion that the NRTF method is most suitable for approximating the time delay in the case of poorly damped systems.

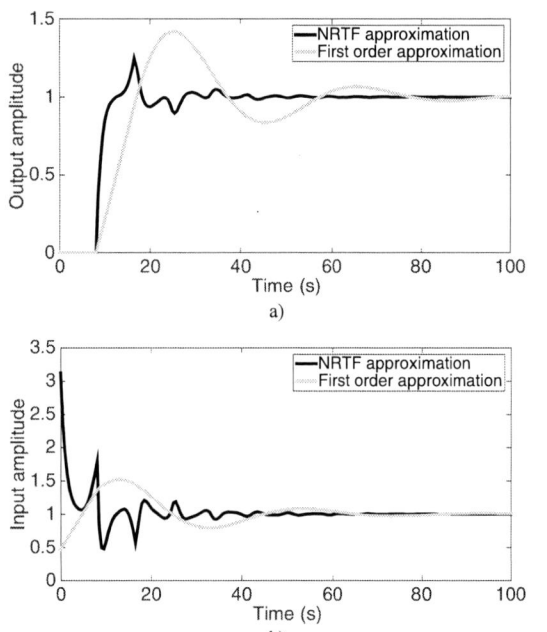

Fig. 8. Reference tracking results for process (16): a) output and b) input signals

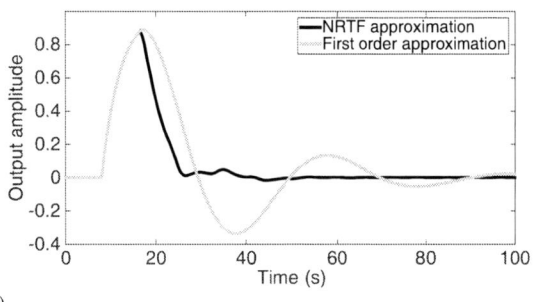

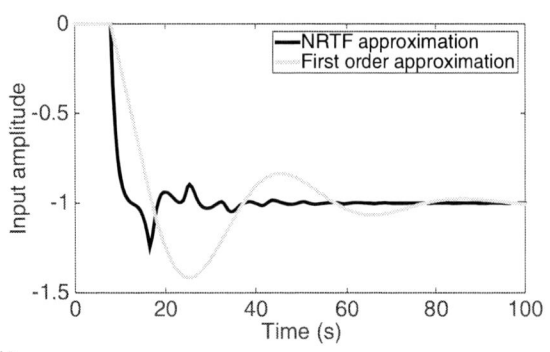

Fig. 9. Disturbance rejection results for process (16): a) output and b) input signals

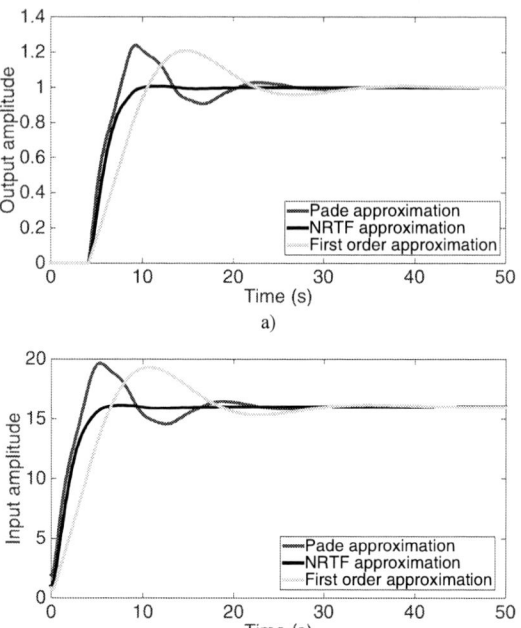

Fig. 10. Reference tracking results for process (17): a) output and b) input signals

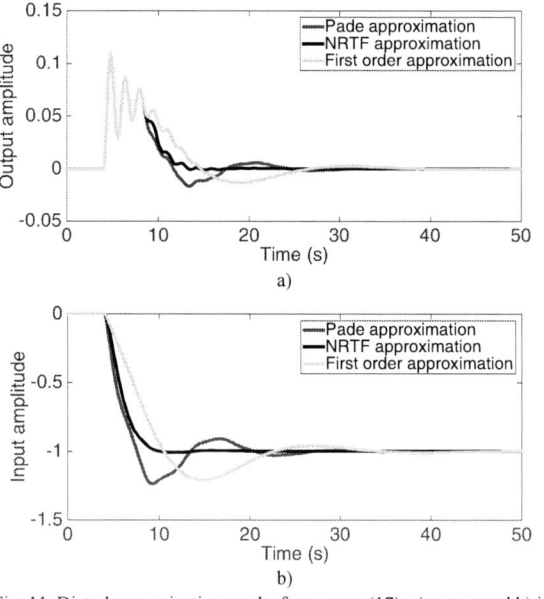

Fig. 11. Disturbance rejection results for process (17): a) output and b) input signals

Consider the delay dominant underdamped system (damping ratio=0.8), such as:

$$H_p(s) = \frac{1}{s^2 + 6.4s + 16} e^{-4s} \qquad (18)$$

with $\lambda=1$, $\tau/t_s=0.8$ and $\lambda/\tau=0.25$. In this case also, the NRTF method provides for the best closed loop results, both in terms of reference tracking and disturbance rejection. See Fig. 12a) and b), for the disturbance rejection case. In conclusions, for underdamped/poorly damped systems, with $\lambda/\tau<0.5$, the NRTF is the most suitable time delay approximation method.

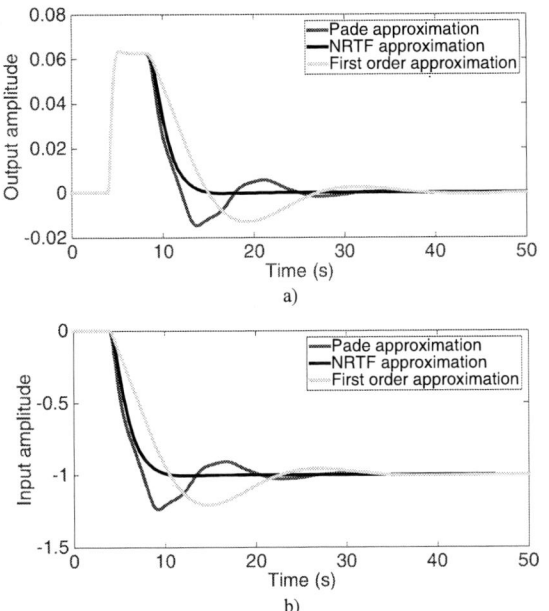

Fig. 12. Disturbance rejection results for process (18): a) output and b) input signals

Consider the underdamped second order system:

$$H_p(s) = \frac{1}{s^2 + 0.8s + 16} e^{-0.5s} \qquad (19)$$

similar with (17), but with a significantly smaller delay. For (19) an IMC controller is designed with $\lambda=1$. Notice that $\tau/t_s=0.35$ and $\lambda/\tau=2$. The closed loop simulation results are indicated in Fig. 13a), for the outputs and Fig. 13b) for the inputs. Fig. 14a) and b) shows the disturbance rejection results. Reconsider now (17), where $\lambda=4$, thus $\tau/t_s=0.294$ but $\lambda/\tau=1$. The closed loop simulation result for reference tracking is given in Fig. 15. Notice that for second order systems with large λ/τ ratios ($\lambda/\tau>0.5$), the three approximations produce similar results. A similar conclusion was also valid for first order lag dominant systems, with $\lambda/\tau>1$.

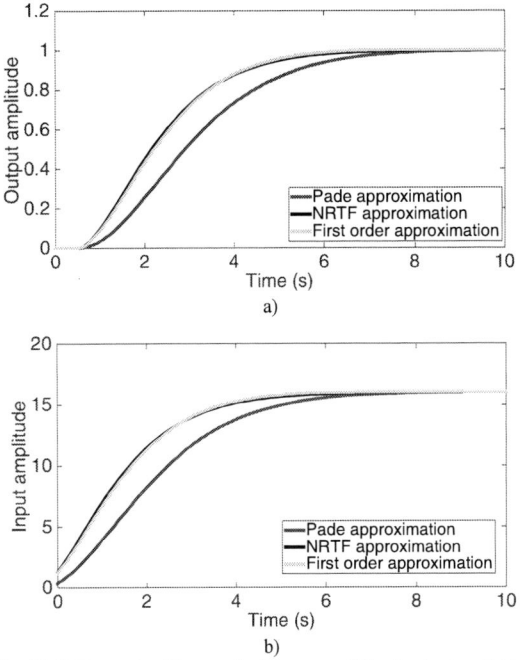

Fig. 13. Reference tracking results for process (19): a) output and b) input signals

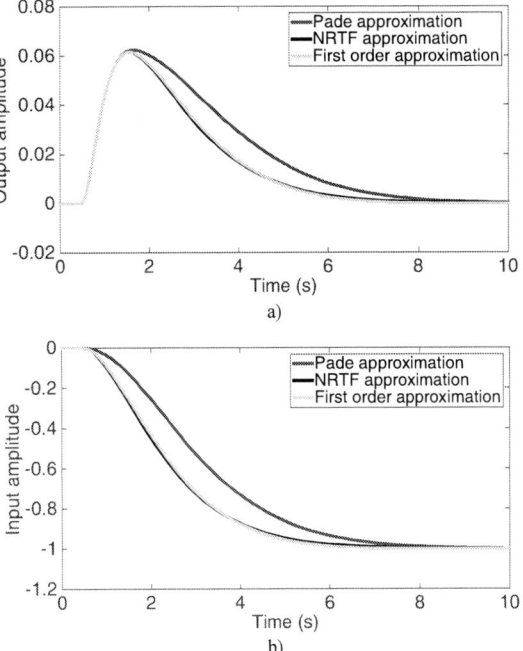

Fig. 14. Disturbance rejection results for process (19): a) output and b) input signals

978-1-5386-7688-2/19 $31.00 © 2019 IEEE

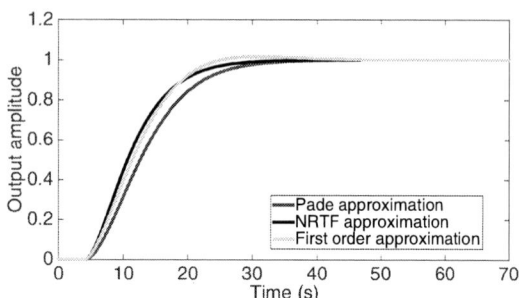

Fig. 15. Reference tracking results for process (17) with $\lambda=4$

V. VALIDATION ON A QUANSER 6 TANKS PROCESS

The NRTF approximation method is validated next on an IMC controller design for the Quanser 6 tanks process, presented in Fig. 16. The description of the unit has been previously presented in [8] and [19]. A simulator for this process has been also developed [19].

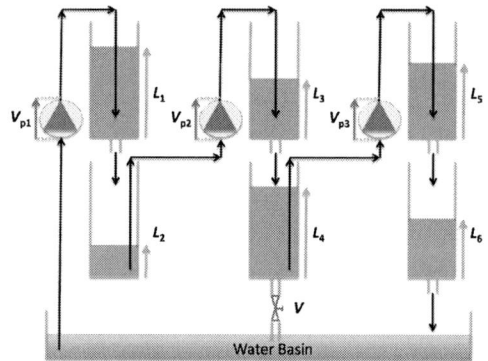

Figure 16. Configuration of the 6 tanks process

The following SOPDT model for the is obtained on the simulator in [19] using a new identification method [20]:

$$\hat{P}(s) = \frac{0.75e^{-22s}}{390s^2+39.5s+1} \tag{20}$$

which will be used to compute the IMC controller. To tune the IMC controller, a filter time constant $\lambda=25$ is selected [20].

The IMC controller using a series approximation is computed as follows:

$$H_{IMC\_series}(s) = \frac{390s^2+39.5s+1}{0.75(1+25s)^2} \tag{21}$$

with the equivalent controller obtained as:

$$H_{c\_series}(s) = \frac{390s^2+39.5s+1}{0.75s(625s+72)} \tag{22}$$

The IMC controller using a Padé approximation is computed as follows:

$$H_{IMC\_Pade}(s) = \frac{(390s^2+39.5s+1)(1+11s)}{0.75(1+25s)^3} \tag{23}$$

with the equivalent controller obtained as:

$$H_{c\_Pade}(s) = \frac{(390s^2+39.5s+1)(1+11s)}{0.75s(15625s^2+1875s+86)} \tag{24}$$

The IMC controller using a NRTF approximation is computed as follows:

$$H_{IMC\_NRTF}(s) = \frac{390s^2+39.5s+1}{0.75(1+25s)^2} \tag{25}$$

with the equivalent controller obtained as:

$$H_{c\_NRTF}(s) = \frac{390s^2+39.5s+1}{0.75(625s^2+50s+1-e^{-27s})} \tag{26}$$

To implement the controllers a sampling period of 1 second has been used:

$$H_{c\_series}(z) = \frac{0.827\,z^2 - 1.572\,z + 0.7474}{z^2 - 1.891\,z + 0.8911} \tag{27}$$

$$H_{c\_Pade}(z) = \frac{0.3791\,z^3 - 1.067\,z^2 + 1.001\,z - 0.3128}{z^3 - 2.882\,z^2 + 2.769\,z - 0.8869} \tag{28}$$

$$H_{c\_NRTF}(z) =$$
$$\frac{0.8412\,z^5 - 3.893\,z^4 + 7.242\,z^3 - 6.772\,z^2 + 3.182\,z - 0.6014}{z^5 - 4.649\,z^4 + 8.687\,z^3 - 8.159\,z^2 + 3.852\,z - 0.7317} \tag{29}$$

For the NRTF approximation, the parameters are: N=5, $\delta=0.9$ [17].

The simulation results are presented in Fig. 17a) and b), including reference tracking and disturbance rejection (a disturbance is applied at 400s). Notice that the results are similar, however there is a slight improvement with respect to the NRTF approximation, which ensures a faster settling time and no overshoot. This is consistent with the previous conclusions: for second order systems with large λ/τ ratios ($\lambda/\tau > 0.5$), the three approximations produce similar results. In the case of the Quanser 6 tanks process, $\lambda/\tau=1.13$.

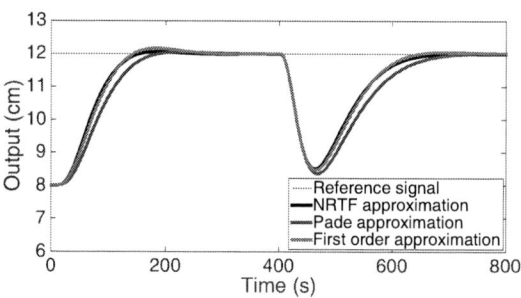

a)

b)

Fig. 17. Closed loop results on the Quanser 6 tanks processes simulator

VI. Conclusion

Internal Model Control (IMC) algorithms have proven to

be quite efficient in designing PID type controllers for time delay systems. Several methods to approximate the time delay have been developed, in order to compute the equivalent controller of a feedback loop. Two of the most widely used approximation methods (first order and Padé) are used in this paper and compared to a new approximation method, the NRTF approach, which has been successfully used so far in approximating fractional order systems.

Based on a series of numerical validations, as well as a case study, the following conclusions were made.

In the case of first order strong lag dominant systems, all methods produce similar results, provided λ/τ large ($\lambda/\tau>1$). However, for the Padé approximation there is a significantly larger control effort required. For ratios $\lambda/\tau<1$, the NRTF approximation method ensures better closed loop results compared to the series or Padé approximations.

For first order delay dominant systems, the series and Padé approximations approach the NRTF, in terms of closed loop performance, provided that the IMC filter time constant is selected such as λ/τ is significantly large. Overall, for strong delay dominant systems, the NRTF method provides for a lower overshoot and faster settling time. Also, it ensures a better disturbance rejection, especially for $\lambda/\tau<1$.

For second order systems, with $\lambda/\tau<0.5$, the NRTF produces the best results. For second order systems with large λ/τ ratios ($\lambda/\tau>0.5$), the three approximations produce similar results.

Further research includes using the proposed alternative approximation method for implementing, testing and experimental validation of the IMC control scheme on various processes.

Acknowledgment

This work was financed by a grant of the Romanian National Authority for Scientific Research and Innovation, CNCS/CCCDI-UEFISCDI, project number PN-III-P1-1.1-TE-2016-1396, TE 65/2018.

References

[1] R. De Keyser, C. Copot, A. Hernandez, C. Ionescu, "Discrete-time internal model control with disturbance and vibration rejection", *J. Vib. Control*, Vol. 23, no. 1, pp. 3–15, 2017.

[2] D.E. Rivera, M. Morari, S. Skogestad, "Internal model control. 4. PID controller design," *Ind. Eng. Chem. Process Des. Dev.*, vol. 25, pp. 252-265, 1986.

[3] S. Saxena, Y.V. Hote, "Advances in Internal Model Control Technique: A Review and Future Prospects," *IETE Tech. Rev.*, vol. 29, no. 6, pp. 461-472, 2012.

[4] C.E. Garcia, M. Morari, "Internal model controls 1. A unifying review and some new results", *Ind. Eng. Chem. Process Des. Dev.*, vol. 21, pp. 308–323, 1982, DOI: 10.1021/i200017a016.

[5] P.V. Gopi Krishna Rao, M.V. Subramanyam, K. Satyaprasad, "Design of Cascaded IMC-PID Controller with Improved Filter for Disturbance Rejection", *International Journal of Applied Science and Engineering*, vol. 12, no. 2, pp. 127-141, 2014.

[6] M. Shamsuzzoha, "IMC based robust PID controller tuning for disturbance rejection," *J. Cent. South Univ.*, vol. 23, no.3, pp. 581–597, 2016.

[7] T. Liu, F. Gao, "New insight into internal model control filter design for load disturbance rejection," *IET Control Theory A.*, vol. 4, no. 3, pp. 448 – 460, 2010.

[8] R. De Keyser, C.I. Muresan, C.M. Ionescu, "Efficient Disturbance Rejection for Dead-Time Processes using Internal Model Control," *The 2018 SICE International Symposium on Control Systems (SICE ISCS)*, 2018, DOI:10.23919/SICEISCS.2018.8330178.

[9] M. Shamsuzzoha, M. Lee, "IMC−PID Controller Design for Improved Disturbance Rejection of Time-Delayed Processes," *Ind. Eng. Chem. Res.*, vol. 46, no. 7, pp. 2077–2091, 2007.

[10] Q. Jin, L. Liu, "Design of active disturbance rejection internal model control strategy for SISO system with time delay process," *J. Cent. South Univ.*, vol. 22, no. 5, pp. 1725–1736, 2015.

[11] B.W. Bequette, "Process Control: Modelling, Design and Simulation", NJ, USA: Prentice-Hall, 2003.

[12] M. Morari, E. Zafiriou, "Robust Process Control", Eaglewood Cliffs, NJ, USA: Prentice-Hall, 1989.

[13] O. Eris, S. Kurtulan, "A new PI tuning rule for first order plus dead-time systems", *The IEEE Afircon*, vol. 11, pp. 1–4, 2011, doi: 10.1109/AFRCON.2011.6072078.

[14] T. Liu, F. Gao, F, "New insight into internal model control filter design for load disturbance rejection", *IET Control Theory A.*, vol. 4, no. 3, pp. 448–460. doi: 10.1049/iet-cta.2008.0472, 2010.

[15] S. Skogestad, "Simple analytic rules for model reduction and PID controller tuning", *J. Process Contr.*, vol. 13, pp. 291–309, doi: 10.1016/S0959-1524(02)00062-8, 2003.

[16] C.I. Muresan, A. Dutta, E.H. Dulf, Z. Pinar, A. Maxim, C.M. Ionescu, "Tuning algorithms for fractional order internal model controllers for time delay processes", *Int. J. Control*, vol. 89, no. 3, pp. 579-593, DOI: 10.1080/00207179.2015.1086027, 2016.

[17] R. De Keyser, C.I. Muresan, C.M. Ionescu, "An efficient algorithm for low-order discrete-time implementation of fractional order transfer functions", *ISA Trans.*, vol. 74, pp. 229-238, DOI: 10.1016/j.isatra.2018.01.026, 2018.

[18] C.I. Muresan, I.R. Birs, O. Prodan, R. De Keyser, "Approximation Methods for FO-IMC Controllers for Time Delay Systems", presented at The 2nd International Conference on Electrical Engineering and Green Energy, Roma, Italy, June 28-30, 2019, Paper G1011.

[19] R. De Keyser, C.I. Muresan, C.I., "Experimental Validation of an Efficient Disturbance Rejection Method for Dead-Time Processes using Internal Model Control", The 24th IEEE Conference on Emerging Technologies and Factory Automation, Zaragoza, Spain, September 10-13, 2019, accepted

[20] R. De Keyser, C.I. Muresan, "Robust Estimation of a SOPDT Model from Highly Corrupted Step Response Data," The European Control Conference, Naples, Italy, June 25-28, 2019, accepted

Design of Programmable, High-Fidelity Haptic Paddle

1st Seyit Yiğit Sızlayan
Electrical and Electronics Engineering
Middle East Technical University
Ankara, Turkey
sizlayan.yigit@metu.edu.tr

2nd Mustafa Mert Ankaralı
Electrical and Electronics Engineering
Middle East Technical University
Ankara, Turkey
mertan@metu.edu.tr

Abstract—In this study, we propose the design of a 1-DOF haptic interface for kinesthetic learning applications and educational platforms in some engineering and science courses. The proposed system is a directly-driven, high fidelity, easy-to-build, easy-to-program haptic device and yet still affordable for use in education. The major advantage of the system is that there is no transmission between the motor and human (i.e. direct-drive) motion which greatly reduces the mechanical complexity, improves robustness, reduces some calibration stages, and reduces some of the important costs. The main drawback associated with a direct-drive actuator interface is that effective motion measurement resolution is very low compared to other systems that utilize capstan drive like mechanisms. To overcome this problem, we used an affordable electronic gyroscope to measure angular velocities directly and integrate gyro and encoder measurements using a statistical fusion filter. The core component of the electronic hardware is a 13$ embedded microcontroller board which assured us of 1KHz hard-real-time measurements and programmable interfaces. The human sense of touch has dynamic characteristic whose bandwidth changes according to action and highest bandwidth with the highest stimulus intensity in an action like slippage has a bandwidth of 1KHz. Because of that, a tactile device being 1KHz hard-real-rime is important for many haptic applications. All these features can be used on various virtual tasks as a kinesthetic learning medium. Such a high fidelity and yet affordable haptic interface is suitable for in-class activities and improves the ability of perception of the students especially on physical systems.

Index Terms—Kinesthetic learning, Haptics, Kalman Filter, Sensor fusion, Gyroscope, Encoder, Virtual environment

I. INTRODUCTION

A. Motivation

Humans receive the information from the outer world through mainly three modalities: visual, auditory, touch (kinesthetic and tactile) according to Felder and Silverman [1]. Learning processes in the human brain heavily uses the information coming from these sources. Kinesthetic feedback provides information and a level of perception that cannot be provided by visual or auditory sensing. Humans interact with physical concepts such as weight, acceleration, toughness/softness, etc. Though kinesthetic feedback, in short, kinesthetic feedback is very relevant to engineering education. Visuals and auditory information can be supplied in our modern (and even nonmodern) classrooms well. However, the higher education system used to ignore possible educational

activities the information that comes with kinesthetic cues. Laboratory experiments used to provide such cues in a limited sense. Fortunately, with the rise of robotics and haptic technologies are penetrating into the education system.

According to the paper of Han and Black [2], to understand and gain intuition regarding the physical systems, people tend to use their imagination. If they have never seen those physical systems before, it is harder to mentally grasp the mechanism of the system. It is hard to verbally explain how the force transmitted by a gear can be increased while transferring the rotary motion from one place to another. It is also not easy to show that force on a screen by a visual aid. Visual and auditory cues remain very abstract about underlying principles in such examples. The most effective way to explain such physical phenomena is by touching. For these reasons, we developed an affordable, direct-drive, haptic device to be used in scientific research studies and as a modular educational platform.

Currently, the most expensive element is the DC motor used in the design. Even though such a high-quality motor is very useful for performing scientific experiments, it can be easily replaced with a low-cost version to further reduce the costs. Such a device is a great tool for teaching concepts like such as spring-damper-mass systems (Both linear and angular), air friction, gear trains.

There are similar haptic paddles or single-axis kinesthetic devices like knobs in the literature. The most comparable kinesthetic devices are the force-feedback haptic paddle developed by Martinez et.al [3], 2-DOF paddle developed by Wong et.al [4] and kinesthetic knob developed by Gillespie et. al [5], haptic paddle developed for human-robot interaction experiments by Gassert et. al [6], haptic paddle developed to understand the rhythmic human behaviors by Ankarali et. al [7] and lastly the paddle developed for educational purposes by Rose et. al [8].

Their common point is the transmission mechanism between the motor and the handgrip, specifically the capstan-drive. Even though it has its own advantages, it causes many deficiencies when it comes to kinesthetic application, namely the need for calibration, mechanical complexity, torque non-linearity and reflected moment of inertia.

Additionally, except [3] which aims the affordability, all the other examples use expensive data acquisition boards, expensive servo driver boards, fast computers and expensive sensors to both measure the position and velocity of the handgrip, to be able to apply a torque to generate the haptic cues in real-time. The main drawback of such systems is that they cannot ensure 1KHz hard-real-time. This property is an important requirement for kinesthetic devices as stated in [9].

In the case of [3], the motor is driven by a generic PWM-controlled motor driver IC which cannot control the torque.

B. Our Contribution

In this study, our contribution is to develop a robust, high fidelity and yet relatively affordable kinesthetic device. We believe that all the devices summarized in I-A have some sacrifices in respect. In the context of robustness, we mean the operational robustness which is mainly related to the mechanical infrastructure. On the other hand, from the context of high fidelity, we mean low measurement and actuation noise and uncertainty. The engineering community is leaning towards direct-drive solutions [10] in many different application domains despite having some obvious deficiencies. Direct-drive coupling reduces some problems associated with mechanical concepts such as hysteresis, friction, backlash, reflected inertia, non-linearities, etc. which are unavoidable in different coupling mechanisms. direct-drive motor control allows us to apply high bandwidth, low-noise, and linear torque control.

The major deficiency of direct-drive (for haptic systems mainly) is its low positional resolution when an encoder (magnetic or optical) is used for position measurement. The reason is that the angular displacement of the motor shaft is significantly smaller than other alternatives. In haptic and kinesthetic applications, it is very critical to measure and track the human motion correctly. For that reason, non-direct drive mechanisms are dominant in haptic systems and technologies.

To compensate for the low encoder resolution associated with a drive solution, we integrated an electronic gyroscope to measure the angular velocity transparently and accurately. We later applied a fusion filter, based on Kalman filter, to integrate the encoder and gyroscope measurements in a statistical fashion.

The motor we have used (Maxon Motors, RE40-150W) has a 500CPR optical encoder (Avago Systems, HEDS-5540). This is a typical resolution for haptic systems that have capstan-drive-like transmission mechanisms. For this reason, our encoder resolution can be categorized as moderate for haptic applications. It is, of course, possible to use higher resolution encoders (up to 3000 CPR) however this would result in a highly expensive solution to be used in in-class applications. Note that the permanent magnet DC motor used in this system is also not an affordable solution, but it can be replaced with cheaper motors easily with the proposed solution.

As electronic parts, many high-cost haptic systems include single or multiple data acquisition card/cards (DAQs) for reading measurements from the sensors (encoders, tachometers, etc.) and then controlling the motor accordingly. However, we choose to handle the data acquisition and control using a microcontroller, which reduces the cost of the whole system substantially. We have chosen Texas Instrument's Tiva-C 123GXL as our controller which has I2C and quadrature encoder interfaces. Those interfaces are finite state machines that can handle the communication (I2C interface) and measurement (Quadrature Encoder Interface) without sacrificing any CPU time and helps us to meet the strict real-time requirements. The microcontroller board we have chosen is 13$ on the manufacturer's website. Yet, it offers 80MHz processing power with floating point acceleration, two quadrature encoder, and four I2C interfaces. All the measurements, calculations and communication and motor control processes are done on this board.

In our system, the microcontroller measures the angular velocity from the gyroscope and angular position from the encoder with 1KHz sampling rate. We fuse this information using a fusion filter base on the Kalman filter to obtain a full state estimation. This filter both filters quantization noises in the encoder as well as averts the gyroscope drift. These smoothed position and velocity measurements are used in the calculation of the haptic forces. Our software infrastructure allows the user to reach the measurements and do the hardware-in-the-loop simulations in the C++ language. The computed force feedback command is used to drive the analog current source circuit which allows us to control the torque generated by the motor. The signal diagram of the system is given in Figure 1.

The paper starts with the comparison between direct-drive and capstan-drive in terms of haptics. Secondly, we explained why using a gyroscope is advantageous measuring the angular velocity and how we fused this angular velocity information with the quantized position information coming from the encoder. Thirdly, we clarified the motor-driving circuit which takes the advantage of current source topology using an Opamp. Then, we gave the details of the software running on the microcontroller and concluded the paper.

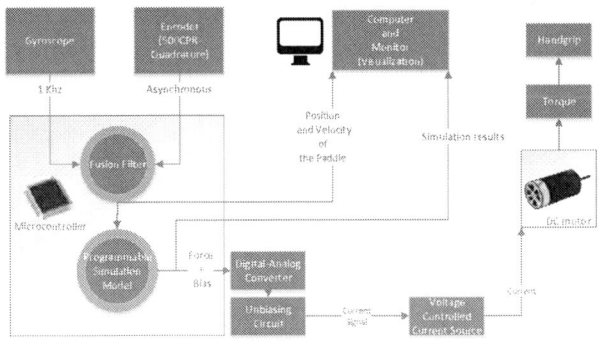

Fig. 1: Signal Diagram of the Haptic Paddle

II. COMPARISON OF DIRECT DRIVE AND CAPSTAN DRIVE MECHANISM FOR HAPTIC APPLICATIONS

In our design, we preferred direct-drive mostly to ensure torque linearity, transparency, and minimal mechanical

978-1-5386-7688-2/19 $31.00 © 2019 IEEE   541

complexity. There are several reasons why direct-drive is not very common in similar applications. We will compare the advantages and disadvantages of the direct-drive against capstan-drive. Then, we are going to explain how we have resolved the disadvantages.

Assembly: In capstan drives, the torque generated by a motor is transferred to the hand-grip using a belt. Since the tension of a belt affects the performance of the transmission significantly and the initial point can be affected by the slips, capstan drive mechanism is disadvantageous. The directly driven handgrip can be assembled just by tightening two screws.

Calibration: Since capstan drive system is affected by slips on the belt, to ensure accuracy they require calibration at the beginning of usage. Directly driven systems are not affected by slips. Another calibration need is due to reflected rotor inertia.

Torque magnitude and motor size: Capstan drive mechanism can increase the torque significantly while the direct drive mechanism cannot. This results in the need for more current and larger motors to generate the same amount of torque in direct drive systems. We had to choose a larger motor to for this reason.

Torque resolution and accuracy: Every transmission mechanism has nonlinearities like dead bands or saturation regions. Capstan drive also has these deficiencies. This situation diminishes the accuracy and linearity of the torque transmitted to the handgrip. However, since the handgrip is assembled on the shaft of the motor, the torque resolution of the system is the same as what motor has. In addition to that, the torque can easily be controlled linearly by controlling the current supplied to the motor.

Moment of inertia: Since our main purpose is to stimulate the sense of touch, to ensure not to irritate the user or make them believe they are just feeling the force which is just generated by the simulation, the handgrip should be as free as possible, namely they should be close to "neutrally buoyant" [11]. Since the moment of inertia of the motor is transferred to the handgrip with a multiple square of the transmission constant, when a person begins to move the handgrip, he/she will feel this multiplied moment of the inertia. However, directly driven hand-grips reflects only the inertia of the motor. To make it clearer comparison, the kinesthetic knob built by Gillespie et.al [5] uses 7.1:1 transmission ratio and the motor they are using has 60-80 gcm^2 rotor inertia. The effective inertia felt by hand just by due to the motor is 3024.6 – 4032.8 gcm^2. On the other hand, the motor we are using has just 137 gcm^2 of rotor inertia. To cancel such inertia, in the kinesthetic device built in [11], Okamura et.al needed to determine a bias current. However, since the movement of a human hand is unpredictable, that bias current needs to be changed online.

Positional resolution: The positional resolution of the capstan drive is superior to the positional resolution of the direct drive. As stated in the [11], the moving part of the kinesthetic devices mostly covers an arc of +/-35°. This range is encoded by a quadrature encoder mounted on the shaft of the motor and position is represented by a number. By measuring the

time difference between two count events of the encoder, the velocity of the handgrip is also found.

In the capstan drive case, the shaft of the motor moves on a larger arc due to transmission. For example, in the kinesthetic knob application given in [5], due to 7.1:1 transmission rate, 70° change in the position of the handgrip results in 497° positional change in the motor shaft. The knob has 7270 counts/revolution(CPR) effective encoder resolution. 70° of range corresponds to observable 1412 different positions on that device. In our direct drive case using 2000CPR encoder (500CPR encoder giving quadrature output), 70° of range corresponds only to 388 positions. Such positional resolution loss has significant impacts on the educational dynamical system simulation applications because such applications require accurate position and velocity measurements. Consequently, to overcome this deficiency, we have used an electrical gyroscope (Invensense Inc., MPU6050) to measure the angular velocity accurately. Then, by using a fusion filter, we have integrated the low-resolution angular position measurements coming from the encoder and the high-resolution angular velocity measurements. This method has both decreased the effect of the positional quantization error and allowed us to measure the velocity accurately without latency. Otherwise, we might need to estimate the velocity from low-resolution position measurement.

III. Advantages of Using Gyroscope for Angular Velocity Measurements

To be able to design a proper kinesthetic device that can be used for teaching physical systems; in addition to position, one needs to use the velocity, too. Kinesthetic devices aiming in-class activities [3], [5], [11] use the low-pass filtered backward difference of the position measurements to estimate the velocity of the handgrip. Due to quantization errors, even they are using a capstan drive, the backward difference comes with unwanted high-frequency components. To simplest way to get rid of them is using an IIR or FIR filters. However, the resulting group delay makes this method inappropriate for kinesthetic applications. When one filters noisy data with an Nth order filter, this causes N/2 samples of delay on the signal. Such delay would cause irritation on the user in kinesthetic applications. The noisiness of the backward-difference of the position measurements and group delay of a filtered signal can be seen from Figure 2. One can easily say that using gyroscope is highly advantageous against the velocity estimation from the encoder.

Another application [6] which aims to be used by undergraduate and graduate students, uses a tachometer to measure the velocity. But the tachometers also give noisy outputs and more expensive than the electronic gyroscopes. Another disadvantage of the tachometer is that it needs to be attached to the shaft or the handgrip. Both ways result in mechanical complexity. In comparison, we attached the gyroscope at the tip of the shaft of the motor to prevent the need for a translation matrix. To ensure alignment, we have used a 3D printed handgrip. Therefore, it requires two screws to attach

978-1-5386-7688-2/19 $31.00 © 2019 IEEE

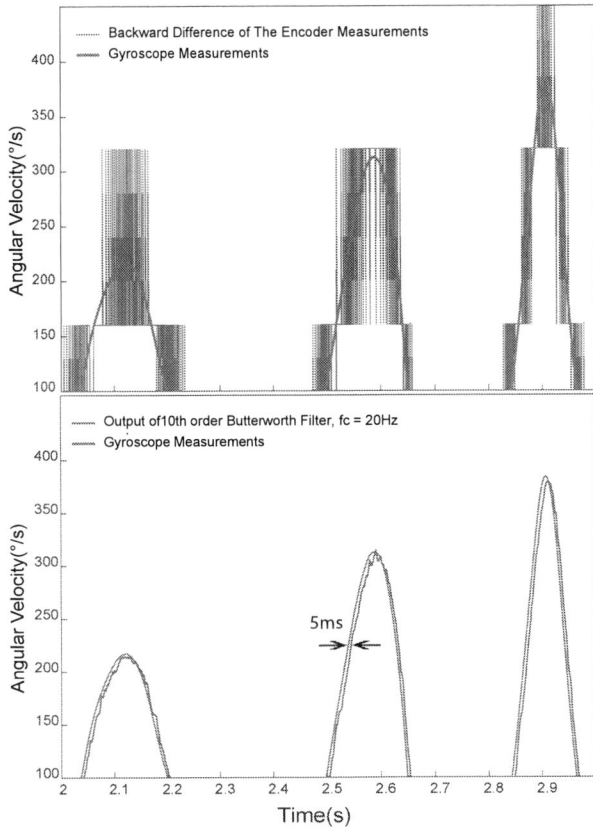

Fig. 2: Comparison of backward-differentiated position measurement, its filtered version with Butterworth filter, f_c and gyroscope measurement

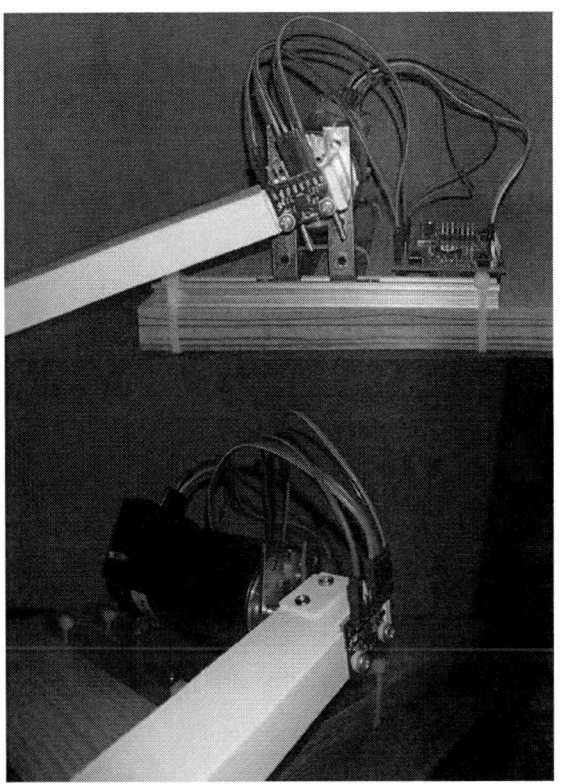

Fig. 3: Coaxial placement of the gyroscope with shaft of the motor, on the handgrip

the gyroscope to the device as can be seen from Figure 3. The 3D CAD models are available on our Github web page [12].

IV. Fusion of Electronic Gyroscope and Optical Encoder using Kalman Filter

In the system, 500CPR rotational incremental encoder is used whose positional resolution is $0.18°$ per pulse. Since our system is direct drive and working range of the paddle is around $70°$, we can measure 388 different positions while a device using capstan drive can measure 1412 different positions on that arc. The effect of the quantization can be seen from the Figure 4.

To compensate for this disadvantage and to be able to measure the angular velocity accurately, we have used a gyroscope that can measure the angular velocities up to $2000°/s$. To increase the resolution and decrease the errors of quantization, we have filtered the angular position using these gyroscope readings using 1-State Kalman estimator. The estimator equations are as follows.

Prediction Equations:

$$\theta_{k|k-1} = \theta_{k-1|k-1} + \Delta t \dot{\theta}_k + \omega_k \tag{1}$$

$$P_{k|k-1} = P_{k-1|k-1} + Q_m \tag{2}$$

Correction Equations:

$\rightarrow$ If the encoder measurement is the same as previous measurement (new encoder measurement has not came):

$$\theta_{k|k} = \theta_{k|k-1} \tag{3}$$

$$P_{k|k-1} = P_{k|k-1} \tag{4}$$

$\rightarrow$ If the encoder measurement has changed (new encoder measurement has come):

$$e_k = \theta_e - \theta_{k|k-1}$$

$$K = \frac{P_{k|k-1}}{P_{k|k-1} + R_e} \tag{5}$$

$$\theta_{k|k} = \theta_{k|k-1} + K_e \tag{6}$$

$$P_{k|k} = (1 - K)P_{k|k-1}k \tag{7}$$

where

θ_k: Angular position of the handgrip, and eventually the shaft of the motor

$\dot{\theta}_k$: Gyroscope measurement and angular velocity of the paddle
θ_e: Optical encoder measurement
Q_m: Model variance
R_e: Noise variance of the encoder measurement

As you can see the gyroscope measurements are used as an input to the system. Since all of the gyroscope measurements have 16-bit resolution, its quantization and measurement noises are negligibly smaller than the one of the optical encoder. Because of that, to make the filter more suitable for the microcontroller, we have used the $1st$ order Kalman filter. This resulted in all of the matrix inversions to simple divisions. For the encoder, the instances where the new measurement has come is unpredictable. Because of that, we have read the samples with 1KHz sampling-rate. This is another factor simplifying the equations because otherwise, we have to measure the time between ticks of the encoder. It is a hard task because the velocity is changing all the time. Another advantage of reading encoder periodically is that this makes the errors in the readings of the measurement closer to the normal distribution. In theory, the quantization of the encoder is modeled as a uniform distribution [13]. But this reading method makes it closer to the normal distribution which is more suitable to the Kalman filter than uniform distribution. Another effect making the error distribution closer to the normal is the manufacturing errors on the encoder wheel. Hence, we have assumed that the error of the encoder is normal with R_e variance. To estimate both R_e and Q_m, we have used the formula given in [14]. We have taken 5 different measurements from the paddle while a user is moving the handgrip randomly. We have preferred this way of movement because the input of the system will be human.

$$(\hat{Q}_m, \hat{R}_e) =$$
$$argmin_{Q_m, R_e}(\sum_{k=1}^{N} log(det(\Sigma_i)) + e_i^T \Sigma_i^{-1} e_i) \quad (8)$$

This likelihood formula is optimized, and the error stabilized when $\hat{Q}_m = 0.00056$ and $\hat{R}_e = 0.0027$. The theoretical value of the uniformly distributed measurement noise variance is also 0.0027. In this case, theoretical and practical values are matched, therefore we have decided to stop numerical optimization.

By using these values, we have implemented the filter on the microcontroller then took some measurement data. One measurement example is given in Figure 4.

One can easily see the effect of quantization from Figure 4. To be able to understand the advantages of using gyroscope for angular velocity measurement, we have also compared the power spectra of the gyroscope and the backward-differentiated encoder measurements.

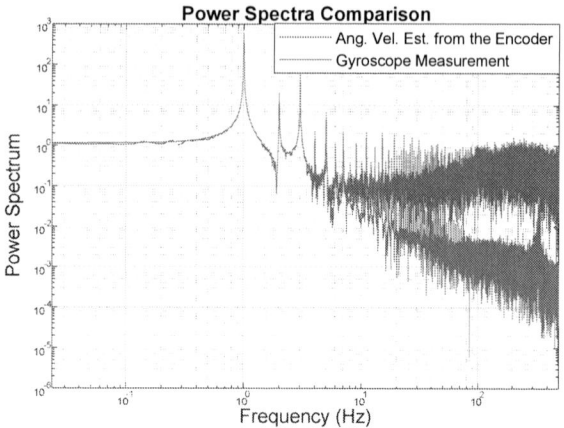

Fig. 5: Comparison of the power spectral densities of the encoder measurements and output of the fusion filter

Figure 5 shows us that the high frequency components coming from quantization errors are reduced more than 100-fold when the gyroscope is used.

V. DC Motor Torque Controller

The kinesthetic devices in [3] and [8] are using PWM-driven motor driver circuits which rapidly opens/closes the motor to drive. By such a method, the only way to control the torque is measuring the current and regulating the duty-cycle of the PWM signal and the direction accordingly using a controller like PI controllers. But as in [6], driving kinesthetic devices by the Opamp circuit is preferred. Opamp circuits have a very important advantage against the PWM-driven circuits: their bandwidths. Because of the very high bandwidth advantage of the Opamps, we have also preferred the Opamp circuit to drive our motor. The main difference from [6] is we do not have a DAQ board which can supply negative voltages to drive the motor. To get very-high-bandwidth torque and direction control, we have designed a circuit using single supply Opamp circuit design techniques. To be able to feed the Opamp, we have used "virtual grounding". In this circuit topology, the main supply voltage is halved by a voltage divider and then buffered by another Opamp. The resulting halved voltage becomes the "virtual" ground of the circuit, making the true ground the negative voltage rail. Since we are supplying a DC motor from this Opamp, we have used the same Opamp used in the current source circuit.(OPA549). The diagram of virtual grounding circuit is given in Figure 6.a.

Another issue designing the circuit was the need for DAC which can supply a negative voltage to drive the motor in the opposite direction. Again, we did not use such an expensive board for that. To get negative voltage from conventional I^2C driven DAC IC (MCP4725, costs 1$), we have added a 1.55V (Half of the supply voltage of the microcontroller) of DC bias value from the software after calculating the force on the microcontroller. Resulting output from the DAC is a DC-biased analog voltage. The DC bias is subtracted using a low-

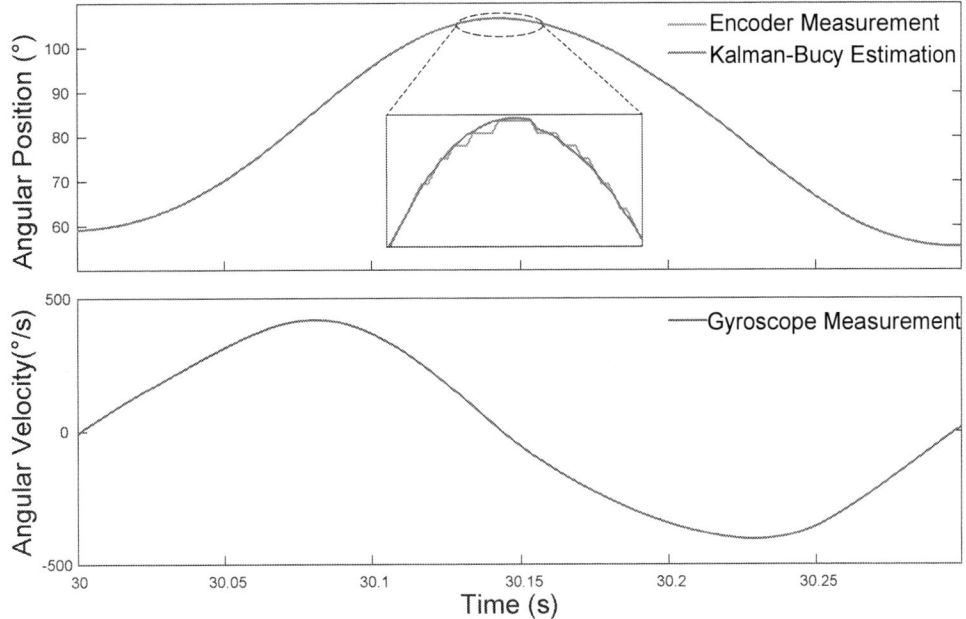

Fig. 4: Combination of electronic gyroscope and optical encoder to reduce the effect of quantization

power difference amplifier (LM358D) circuit and the unbiased signal is given to the main current source Opamp (OPA549). The schematic of the design is given in Figures 6.b and 6.c. All the schematics and the oscilloscope analyses can be found our Github page [12].

VI. PROGRAMMABILITY OF THE KINESTHETIC DEVICE

For the community to be able to use the design and software easily, we have designed the software re-programmable. We have used free-to-use IDE of the Texas Instruments, Code Composer Studio 7 which really helped us a lot during debugging.

The software bases on 1Khz interrupt signal driven by the gyroscope. Since we are using the quadrature encoder interface of the microcontroller, it consumes no CPU time to measure the position. During interrupt subroutine, we simply read the angular velocity from the gyroscope and angular position from the peripheral register. Then we run the fusion filter on them to get the position estimates. This process takes 220us.

After calculating the estimate, we call the "calculateTheForce" function which returns the calculated force value using these measurements. The position estimate, the velocity measurement, and force variable are available globally, so one can just need to implement that function to regulate the force applied by the DC motor. The value of the force just needs to below 2000, numerically. This is because of the resolution limitation of the DAC, it has 12-bit resolution. When the force is calculated, DC bias is added for the single-supply motor driver design. To simulate the spring-mass-damper-like dynamical system, we have also added a software arrays which take mass position or simulation matrices. The software starts waiting for these arrays. The values calculated on the microcontroller can also be transferred to the computer using a RS232-USB converter. We are currently using 2Mbits/s of speed to transfer 8x4-byte variables.

All these "read measurements", "filter measurements", "do the calculations and simulations" then "transfer the values to the computer" cycle takes around 750us when we are running spring-mass-damper simulation on the microcontroller board. The source codes of the software are available on our Github web page [12].

VII. CONCLUSION

In this work, we have developed a force-feedback, single-axis kinesthetic, direct-drive device for educational purposes. The device can provide kinesthetic cues for students to understand the basic concepts of dynamical systems. The device also has low mechanical complexity, does not require calibration for the usage and requires a simple 12V AC to DC adapter. These aspects make the device highly useful for in-class activities.

The developed paddle uses an electronic gyroscope to measure the angular velocity with high-fidelity and diminishes the effect of quantization errors caused by the low positional resolution of quadrature encoder using a fusion filter based on Kalman filter.

The main actuator of the device is a permanent magnet DC motor and it is driven by high-bandwidth, single-supply Opamp circuit. The circuit allows the software to control the torque applied to the handgrip with latency less than 1ms.

Lastly, the software is running on the affordable microcontroller board with 1KHz hard-real-time. This paddle is open-source, easy-to-use and easy-to-program. We believe that this device will be useful for the teachers to develop their own experimental dynamics setups and it will help the students to mentally grasp the concepts of dynamics with less effort.

VIII. Grants

This article is based on work supported by a Scientific Research Projects Coordination Unit of Middle East Technical University grant (GAP-301-2018-2702) to M. M. Ankaralı.

References

[1] R. M. Felder, "Learning and Teaching Styles In Engineering Education (updated 2002)," *Engineering Education*, vol. 78, no. June, pp. 674–681, 1988.

[2] I. Han and J. B. Black, "Incorporating haptic feedback in simulation for learning physics," *Computers and Education*, vol. 57, no. 4, pp. 2281–2290, 2011. [Online]. Available: http://dx.doi.org/10.1016/j.compedu.2011.06.012

[3] M. O. Martinez, T. K. Morimoto, A. T. Taylor, A. C. Barron, J. D. Pultorak, J. Wang, A. Calasanz-Kaiser, R. L. Davis, P. Blikstein, and A. M. Okamura, "3-D printed haptic devices for educational applications," *IEEE Haptics Symposium, HAPTICS*, vol. 2016-April, no. Figure 1, pp. 126–133, 2016.

[4] C. E. Wong and A. M. Okamura, "The snaptic paddle: A modular haptic device," *Proceedings - 1st Joint Eurohaptics Conference and Symposium on Haptic Interfaces for Virtual Environment and Teleoperator Systems; World Haptics Conference, WHC 2005*, pp. 537–538, 2005.

[5] R. B. Gillespie, M. B. Hoffinan, and J. Freudenberg, "Haptic interface for hands-on instruction in system dynamics and embedded control," *Proceedings - 11th Symposium on Haptic Interfaces for Virtual Environment and Teleoperator Systems, HAPTICS 2003*, no. June 2014, pp. 410–415, 2003.

[6] R. Gassert, J. C. Metzger, K. Leuenberger, W. L. Popp, M. R. Tucker, B. Vigaru, R. Zimmermann, and O. Lambercy, "Physical student-robot interaction with the ETHZ haptic paddle," *IEEE Transactions on Education*, vol. 56, no. 1, pp. 9–17, 2013.

[7] M. M. Ankaralı, H. Tutkun Sen, A. De, A. M. Okamura, and N. J. Cowan, "Haptic feedback enhances rhythmic motor control by reducing variability, not improving convergence rate," *Journal of Neurophysiology*, vol. 111, no. 6, pp. 1286–1299, 2014.

[8] C. G. Rose, J. A. French, and M. K. O'Malley, "Design and characterization of a haptic paddle for dynamics education," *IEEE Haptics Symposium, HAPTICS*, pp. 265–270, 2014.

[9] C. Hatzfeld and A. T. Kern, *Engineering Haptic Devices A Beginner's Guide*, 2009.

[10] S. Seok, A. Wang, M. Y. Chuah, D. Otten, J. Lang, and S. Kim, "Design principles for highly efficient quadrupeds and implementation on the MIT Cheetah robot," *Proceedings - IEEE International Conference on Robotics and Automation*, pp. 3307–3312, 2013.

[11] A. M. Okamura, C. Richard, and M. R. Cutkosky, "Feeling is Believing : Using a Force-Feedback," *Journal of Engineering Education*, no. July, pp. 345–349, 2002.

[12] S. Y. Sızlayan and M. M. Ankaralı, "Atlas Haptic," 2019. [Online]. Available: https://www.github.com/sysizlayan/Atlas-Haptic

[13] N. Colonnese and A. M. Okamura, "Propagation of joint space quantization error to operational space coordinates and their derivatives," *IEEE International Conference on Intelligent Robots and Systems*, vol. 2017-Septe, pp. 2054–2061, 2017.

[14] V. A. Bavdekar, A. P. Deshpande, and S. C. Patwardhan, "Identification of process and measurement noise covariance for state and parameter estimation using extended Kalman filter," *Journal of Process Control*, vol. 21, no. 4, pp. 585–601, 2011. [Online]. Available: http://dx.doi.org/10.1016/j.jprocont.2011.01.001

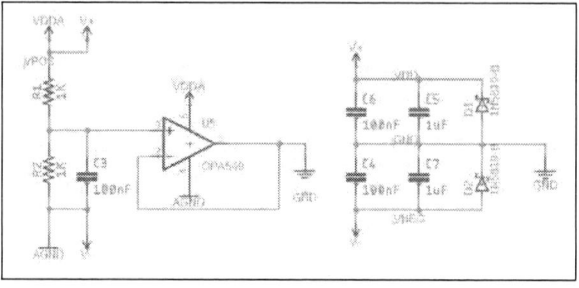

(a) Virtual grounding Circuit

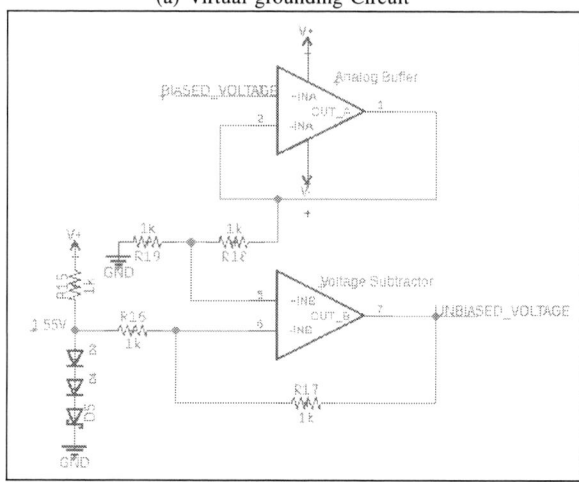

(b) Unbiasing circuit

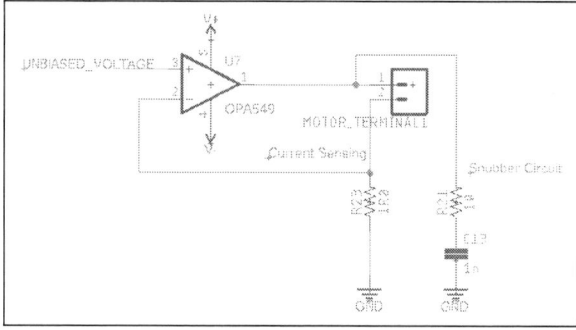

(c) Current source circuit

Fig. 6: Diagram of the DC motor driver circuit

Microcontroller-Based Motion Control for DC Motor Driven Robot Link

1st Mustafa M. Mustafa
Department of Electrical Engineering
Salahaddin University-Erbil
Erbil, Kurdistan Region of Iraq
mustafa.atrushi@su.edu.krd

2nd Ibrahim Hamarash
Department of Computer Science and Engineering
University of Kurdistan, Hawler
Erbil, Kurdistan Region of Iraq
ibrahim.hamad@ukh.edu.krd

Abstract—**This paper presents the design and implementation of a controller for a robot link which is driven by a brushed DC motor and a quadrature encoder. DC motor driven a robot link with an encoder is a challenge for small and micro robotic systems, experimentally accurate measures of encoder resolution, position and speed are necessary parameters while they are not easily achieved. The proposed work is an empirical study of two control technique: i) open-loop control and ii) feedback/closed-loop control. Both control techniques have been implemented using programmable system on chip microcontroller. The algorithms have been compared in term of speed, stability and accuracy. Results show that these algorithms can be implemented in small-scale robots effectively.**

Index Terms—**DC motor, microcontroller, quadrature encoder, robot link**

I. INTRODUCTION

A need common to many robot applications is the accurate following of a specific path by the end-effector of the robot, which translates to accurate positioning of the individual links that constitute the robot. The individual links are driven by joint actuators which are are rotary electric motors. In other words, the motion of the joint produced by the actuators determines the position and orientation of the end-effector at any time [1, 2]. Therefore, most robots are commonly controlled in joint space to perform motion control of the end-effector. Thus, motion control of motors is crucial. Large industrial robots typically use brushless servo motors while small ones use brushed DC motors. The brushed DC motors are widely used because of the high torque, speed controllability over a wide range, well-behaved torque-speed characteristics, and adaptability to various types of control methods [3].

By studding the literature pertaining to motion control of electric motors, it is observed that many control techniques and approaches have been proposed. In [4], the control strategy is developed on the basis of the instantaneous voltage equation of a dc motor and some conditions have been introduced to reduce the position error to be zero without position and speed sensors. The authors, in [5], design and implement integrator backstepping controllers for a brush dc motor driving a one-link robot manipulator. They show that the controllers ensure "good" load position tracking despite parametric uncertainty throughout the DC motor. Artificial intelligence (AI) tools, such as expert system, fuzzy logic, and neural network,

which are presented in [6], could be used for controlling DC motors, like the presented work in [7]. Furthermore, the most common approaches for controlling DC motors are proportional (P), proportional-integral (PI), proportional-derivative (PD), and proportional-integral-derivative (PID) control. These approaches are widely used in industrial control systems and a variety of other applications. All details regarding analysis, design and tuning of these approaches could be found in [8, 9]. In [10], the performance of a PID controller has been presented and investigated to control a DC motor system based on safe experimentation dynamics. The controller, that is obtained in [11], minimizes the error of the angular velocity for a DC motor through data-driven sigmoid-based PI control.

In this study we show a laboratory modeling and motion control for a robot link that is driven by a brushed DC motor. In the implementation, we use a brushed DC motor attached to a quadrature encoder, a driver (IC chip), and a programmable system on chip microcontroller with its open source software supported by C Program. At the beginning we characterize the DC motor in term of position and speed. Also, the encoder is analyzed practically to know how it could cooperate with the motor. After that, we apply two forms of motion control technique to the DC motor which drives a link. The first is an open-loop control technique and the second is a closed-loop. The results of both techniques' algorithms are compared in term of simplicity, speed, stability, and accuracy. Finally, we implemented a proportional control algorithm to show the effect of the proportional gain (k_p) on the speed, stability, and accuracy of the link motion. The contribution of this work is to use the available motor and microcontroller to build a successful controller for a robot motion. This implementation supports robotics teaching and research at colleges where a robot laboratory is not available, since it is simple in building, practical, and it is very efficient. Thus, students and researchers can build their own robot controller and gain insights without need to get access to advanced laboratories.

The rest of this paper is organized as: Section II characterizes the proposed DC motor system in term of encoder resolution and motor speed. The motion control of the DC motor is presented in Section III by applying two control techniques: an open-loop control and a feedback/closed-loop control. For the feedback/closed-loop control technique, three approaches

978-1-5386-7688-2/19 $31.00 © 2019 IEEE

are tested: On-off, proportional, and PID approaches. The experimental results, for both control techniques, are investigated and discussed in the same section. The Conclusion is given in Section IV.

II. CHARACTERIZING THE DC MOTOR

In the proposed design we use a micro metal gearmotor, which is a 6 V brushed DC motor with a 28:1 metal gearbox. A magnetic encoder is attached to the extended motor shaft that uses dual-channel hall effect sensor board and a magnetic disc that can be used to add quadrature encoding to the gearmotor with the extended back shaft. The encoder board senses the rotation of the magnetic disc and provides a resolution of counts per revolution (CPR) of the motor shaft when counting both edges for both channels. The sensors provide digital outputs that can be connected directly to a microcontroller or other digital circuit. Manufacturers summarize products properties in data sheets. We consider this motor is a surplus and somehow the characteristics of the motor have been changed, therefore we need to measure the properties of the motor and the attached encoder. For this purpose, we use the microcontroller and its integrated design environment.

A. Encoder Resolution

Resolution is the smallest change in the physical property that the sensor can detect. Since the encoder counter is typically sampled by the microcontroller, and the microcontroller has other responsibilities besides sampling encoder counts, the resolution of the encoder has to be considered. The encoder has a defined number of cycles that are generated for each 360-degree revolution of the shaft. These cycles/counts are monitored, by a counter or motion controller, and converted to counts for position or velocity control. The microcontroller will count whenever the state of the switches changes and record it, so it is possible to know the number of counts per one revolution of the disc and by this way the angle of the motor shaft, that is attached to the encoder, is measurable. The cycle per revolution of the back and front shafts, gear ratio and motor resolution are presented in Table I. We get the gearbox ratio experimentally through dividing the count per revolution of the back shaft (CPR_1) by the the count per revolution of the front shaft (CPR_2). Depending on the CPR_2, the actual resolution is calculated simply by dividing the angle of one revolution by CPR_2 as presented in (1):

$$Resolution = \frac{360°}{CPR_2} \quad (1)$$

This value tells us the sensor that is used by the encoder is digital. This means that rotation angle of the motor shaft that is less than the resolution, could not be considered.

B. Speed

After that we calculated the CPR_2, now we can measure the angular velocity (ω) of the motor in revolution per minute (rpm) as presented in (2).

$$\omega = \frac{count}{CPR_2} 60 \quad (2)$$

TABLE I
DATA OF THE MOTOR WITH ATTACHED ENCODER

CPR_1	CPR_2	Gear Ratio	Motor Resolution
28	784	28	˜0.46°/count

TABLE II
RECORDED DATA OF THREE SPEED TESTS

D	V	ω_1	ω_2	ω_3	mean	St. Dev
100	4.5	274	271	270	271.67	2.081666
80	3.6	199	198	199	198.67	0.57735
60	2.7	131	130	131	130.67	0.57735
40	1.8	62	63	63	62.67	0.57735

We make the microcontroller to record the count of the encoder for one second when the motor is running. Then, we divide this count by the CPR_2 and multiply the result by 60 to get the speed in rpm. ω is measured when the motor is free of load, so this is considered as no-load speed. The speed is calculated when the supply voltage of the motor is 4.5 volts that is taken from the microcontroller. From the mathematical model of the motor system, which is derived in detail in [12], the angular velocity is represented as

$$\omega = K_v V \quad (3)$$

where K_v and V denote the motor constant and voltage supply of the motor respectively. The value of K_v could be used to measure the angular velocity of the motor at any supplied voltage, but this is inaccurate because the internal friction of the motor is not considered. Therefore, we use pulse width modulation (PWM) technique to control the speed of the motor and to measure K_v at different levels of voltage. We configure the microcontroller to do a speed test for multiple times, by changing the duty cycle (D) of the PWM under certain frequency. The recorded data by the microcontroller is shown in Table II, which are D, voltage V and three set of speed/angular velocity ω_1, ω_2 and ω_3.

We utilize the *mean* value and the standard deviation *St. Dev* of the three data set of the speed to derive the relationship between the voltage and the speed. The data is plotted in Fig. 1 and the equation from this data is represented as

$$\omega = 77.22V - 77.33 \quad (4)$$

The value 77.22 is the slope of the line that is change in ω divided by change in V, so according to the data of our test it is K_v. Since the internal friction of the motor is not considered, we notice a difference between the value of K_v, which is determined straightly from (3), and the value which is calculated in our test. The second value is more accurate because it is calculated based on the data that is collected through three speed tests respectively. By doing multiple tests, the motor will get warmer and it will affect the static friction of the motor, thus the effect of the static friction will appear in the data. The motor stops when the D goes below 40%, this means that the amount of power provided at this ratio is not enough to overcome the static friction of the motor.

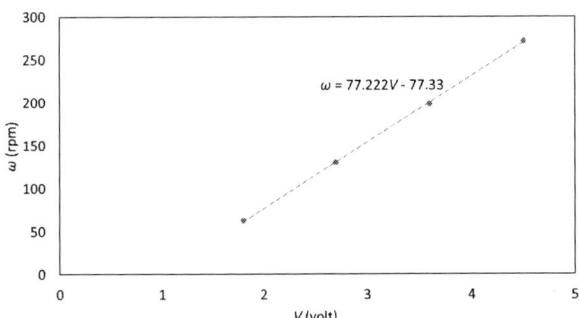

Fig. 1. Voltage vs speed.

III. Motion Control

Since we aim to use the DC motor to actuate an individual link of a robot, we need to investigate the motion control of the motor. There are many control techniques and methodologies that can be applied to control the motion of the motor. In this work we follow two common methods of position control which are open-loop and feedback/closed-loop. For the the feedback/closed-loop control we apply three control approaches. The experimental results, for each test, show how the system behaves in term of speed, stability, and accuracy.

A. Open-loop Control

A motor controller that does not receive any feedback information from the motor is an open-loop control. The block diagram for this control, in Fig. 2, presents that the controller does not use sensors. The provided voltage makes the motor to run at a certain velocity and by integrating the velocity we can get the direction. Therefore, this controller is simple to implement with low cost and it is stable.

The disadvantage of open-loop control is the inaccuracy. In order to implement this controller, we need to program the microcontroller to follow the procedure in Fig. 3.

From characterizing the motor, we knew the speed of the motor at each D. We applied 4.5 volts to the motor that gave us 271 rpm as an angular velocity. We will suppose that we want the motor to turn 90° which is a quarter of one revolution. The wait time is calculated as

$$\frac{1}{271}\frac{min}{rev}\,0.25\,rev\,\frac{60s}{min}\,1000\,ms = 55.35\,ms \quad (5)$$

where rev, min, s, ms denote the revolution, minute, second, and millisecond. Since the maximum operation voltage is 6 volts, D will be calculated as

$$D = \frac{4.5}{6}100\% = 75\% \quad (6)$$

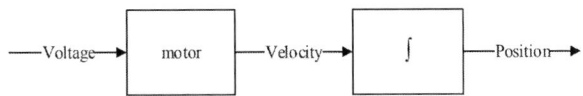

Fig. 2. Open-loop control block diagram.

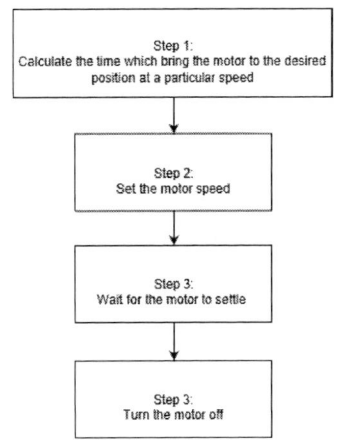

Fig. 3. Open-loop control procedure.

TABLE III
DATA OF EMPIRICAL ACCURACY TEST OF THE OPEN-LOOP CONTROL

Cycle	Target position	Current position	Error
1	90°	97°	7°
2	180°	195°	15°
3	270°	294°	24°
4	0°	32°	32°

From these calculated values, we programmed the micro-controller to provide the motor with 4.5 volts for 55.35 millisecond and to make it stop for a while to settle the motor. During the test, we notice that the motor runs fast, smooth and stable. Although, we applied an external force to the motor, but the control algorithm is not affected. This shows inability of the open-loop control algorithm to reject the disturbance. Therefore, this algorithm is considered as an inaccurate control algorithm. We run the motor for 4 cycles and we aim to get a 90° turn for each cycle. As more turns are made by the motor, it starts to drift which indicates that the error accumulates. The data of this experiments are recorded in Table III.

B. Feedback/Closed-loop Control

Fig. 4 shows the block diagram of the proposed feedback/closed-loop control system for the DC motor. The system consists of a driver for driving the motor in both forward and reverse direction. The control algorithm, which is implemented by the microcontroller, affects the driver directly. The motor position is sensed by the quadrature encoder and fed back to the microcontroller unit. The microcontroller compares the sensed position signal with the desired position. The error signal, that the microcontroller produces as a result of comparison, is given as an input to the driver to drive the motor. We test this error signal in three approaches as a feedback/closed-loop control. The first approach is On-off control, the second is proportional control, and the third is PID control. On-Off control is a simple type of feedback/closed-loop control technique. To test this method, we should follow the flow chart shown in Fig. 5. The target/set-position in the

978-1-5386-7688-2/19 $31.00 © 2019 IEEE

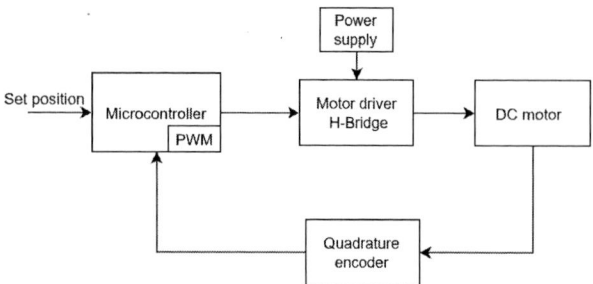

Fig. 4. Feedback/Closed-loop control block diagram.

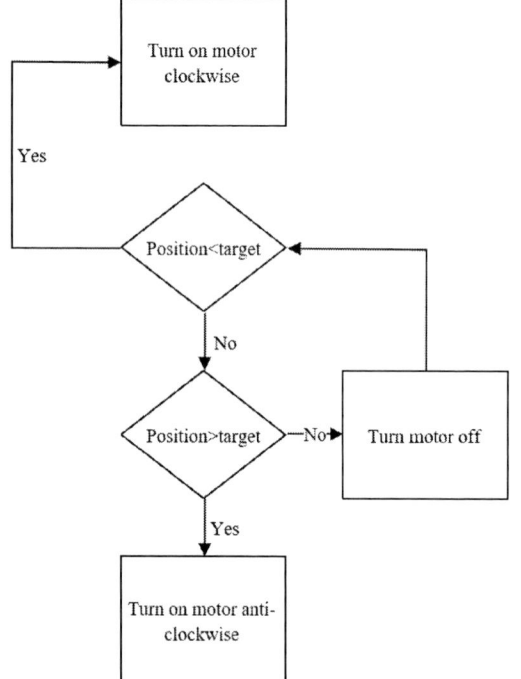

Fig. 5. On-off control algorithm.

flow chart is the desired position that we want the motor to be. We set the desired position depending on the count value range of the encoder that is calculated previously. 200 as a random count value has been chosen for the test. In this algorithm test, we see that the motor runs more accurate than the open-loop control, as well as it rejects disturbance because there are sensor and loops in the algorithm. The problem of this controller is the instability because it does not consider the value of the error. In addition to that, the sensor that is used by the encoder, in our system, is a digital sensor therefore, the motor will never get stable.

Proportional control as the second approach of the feedback/closed-loop control technique is presented. In order to implement this controller, we define a position error between the actual and desired position. The microcontroller is programmed to multiply this error by a given proportional gain

(k_p) to make the motor decreases its speed as it gets closer to the target position. In other words, the proportional control will reduces the error to zero by changing D of the PWM signal which feeds the motor through the motor driver. Thus, the value of k_p makes significant effect on the speed, stability, and accuracy of the motor. We present this effect by the response of the motor system which has been drawn in Fig. 6 based on the collected data from a test that we made. In this test we set the desired position of the motor on 1000 count and then we manually try four values for k_p. We adjust k_p values by setting it where in the output decreases to zero. When the measured value is stabilized, we set the specified value and gradually reduce k_p. We avoid the periodic oscillations in the system by increasing k_p so that error is minimal periodic oscillations decrease to the limit. For each value, we observe its effect on the speed, stability and accuracy. Unlike the On-off approach, this approach drives the motor to the desired position in a stable manner.

Since the microcontroller can handle complicated algorithms, and based on the desired task that the motor will do, it is possible to add the integral and derivative gains to implement a PID control. Moreover, it is possible to utilize the microcontroller in order to tune the gain values of the PID control according to the available tunning techniques. Therefore, we develop the proportional control approach by adding the derivative and integral gains. We achieve this development by replacing the control input algorithm, in the microcontroller, with the general mathematical form of the PID control [9]:

$$u(t) = k_p e(t) + k_i \int_0^t e(t)dt + k_d \frac{de(t)}{dt} \tag{7}$$

where $u(t)$ denote the control input, $e(t)$ is the error between the current and desired position, k_p, k_i, and k_d denote the proportional, integral, and derivative gains respectively. The mathematical equation of PID in (7) is a controller input expressed in continuous time or in the analog domain. In order to make this control input/signal generated by the microcontroller, we need to implement it digitally in software. The discrete time of the PID will be expressed as:

$$u[n] = k_p e[n] + k_i \sum_{k=0}^n e[K]T + k_d \frac{e[n] - e[n-1]}{T} \tag{8}$$

In (8), T is the sample time or the time that PID function gets called by the microcontroller, K is the order of the sample, and n refers to the number of samples. By applying this algorithm to the last test and by using Ziegler–Nichols to tune the gains, the response of the system gets improved as shown in Fig. 7.

The objective of this test is to know the trade-off between the speed, stability, and accuracy as well as the pros and cos of each approach. From the specifications of the response, the rise time t_r and peak time t_p variables are about speed. While the settling time t_s and overshoot measure the stability and the steady state error determine the accuracy. These values, which are calculated from the proportional approach test, are presented in Table IV. The results show that as k_p becomes

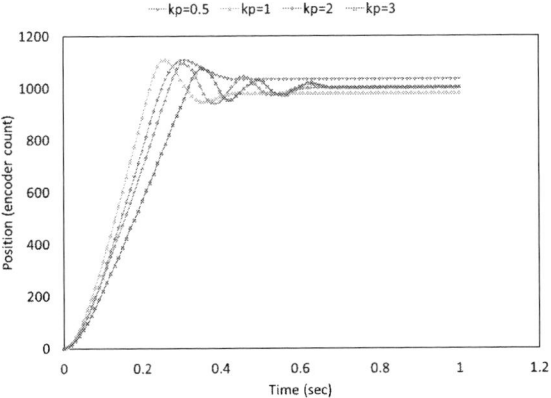

Fig. 6. Motor position response for proportional control approach.

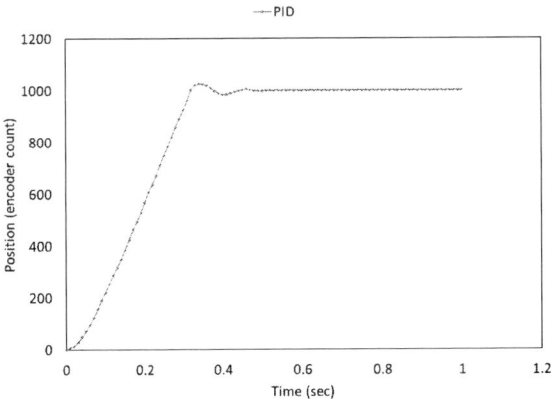

Fig. 7. Motor position response for PID approach.

larger the steady state become smaller which tells us that the accuracy is inversely proportional to k_p. Table V shows the same data for the PID approach. We have noticed that the speed and stability are improved when we apply PID control comparing to the best case in the proportional control. However, for our system the proportional control: i) keeps the overshoot under the certain percentage, ii) it guarantees that the rising time of the output is below the certain limit, and iii) achieve zero static error. Therefore, it satisfies the objective of the DC motor, which is targeting the desired position.

From the implementation and the experimental results, we should point out the pros and cos of the open-loop and the feedback/closed-loop control techniques. We notice that the open-loop control is fast, stable, and it does not use feedback signal so there is no need for sensors. Therefore, it reduces the cost of the control system. In the other hand, the open-loop control does not reject any disturbance which affects the reliability of the technique. Unlike the open-loop control, the feedback/closed-loop control technique needs sensors to provide the feedback signal. The speed, stability, and accuracy

TABLE IV
COLLECTED PARAMETERS OF PROPORTIONAL CONTROL APPROACH

	Speed		Stability		Accuracy
k_p	t_r	t_p	t_s	*Overshoot*	*Steady State Error*
0.5	0.25	0.31	0.43	10.9%	35
1	0.22	0.26	0.44	10.9%	19
2	0.27	0.31	0.63	9.3%	2
3	0.32	0.35	0.68	7.5%	0

TABLE V
COLLECTED PARAMETERS OF PID CONTROL APPROACH

Speed		Stability		Accuracy
t_r	t_p	t_s	*Overshoot*	*Steady State Error*
0.32	0.33	0.51	2.5%	0

depends on the approach that is used. As we showed, the proportional and PID approaches are much better than on-off approach in term of speed, stability, and accuracy. Moreover, regardless the approach, feedback/closed-loop technique rejects the disturbance.

IV. CONCLUSION

In this study, we examined two different motion control techniques for a dc motor with a quadrature encoder driven a robot link using a programmable system on chip microcontroller. The system has been implemented and the controllers have been tested. Experimental results show that the open-loop control technique is easy to implement, make the motor turns smooth, fast with lack of accuracy plus inability to reject external disturbance. In the other hand, the feedback/closed-loop control technique is more accurate, and it easily rejects the external disturbance. The performance of the system has been significantly improved when we apply Proportional then PID control on the cost of complexity. The experimental results from the position control response present the effect of the proportional gain on the speed, stability and accuracy. Moreover, tuning k_p, k_i, and k_d for PID control using Ziegler–Nichols proves that the functionality of the controller and its impact on motion control are satisfactory. The future development of this work could be done by applying an intelligent technique to determine which one of the mentioned techniques should be implemented. The decision should be made by considering the characteristics and the tasks of the DC motor.

REFERENCES

[1] M. Gopal, *Control systems: principles and design*. Tata McGraw-Hill Education, 2002.

[2] C. D. Crane III and J. Duffy, *Kinematic analysis of robot manipulators*. Cambridge University Press, 2008.

[3] B. Siciliano and O. Khatib, *Springer handbook of robotics*. Springer, 2016.

[4] N. Matsui and M. Shigyo, "Brushless dc motor control without position and speed sensors," *IEEE Transactions on Industry Applications*, vol. 28, no. 1, pp. 120–127, 1992.

978-1-5386-7688-2/19 $31.00 © 2019 IEEE

[5] D. M. Dawson, J. J. Carroll, and M. Schneider, "Integrator backstepping control of a brush dc motor turning a robotic load," *IEEE Transactions on Control Systems Technology*, vol. 2, no. 3, pp. 233–244, 1994.

[6] B. K. Bose, "Expert system, fuzzy logic, and neural network applications in power electronics and motion control," *Proceedings of the IEEE*, vol. 82, no. 8, pp. 1303–1323, 1994.

[7] K. Premkumar and B. Manikandan, "Stability and performance analysis of anfis tuned pid based speed controller for brushless dc motor," *Current Signal Transduction Therapy*, vol. 13, no. 1, pp. 19–30, 2018.

[8] A. O'Dwyer, *Handbook of PI and PID controller tuning rules*. Imperial College Press, 2009.

[9] K. H. Ang, G. Chong, and Y. Li, "Pid control system analysis, design, and technology," *IEEE transactions on control systems technology*, vol. 13, no. 4, pp. 559–576, 2005.

[10] M. R. Ghazali, M. A. Ahmad, and R. M. T. R. Ismail, "Data-driven pid control for dc/dc buck-boost converter-inverter-dc motor based on safe experimentation dynamics," in *2018 IEEE Conference on Systems, Process and Control (ICSPC)*, Dec 2018, pp. 89–93.

[11] M. A. Ahmad and R. M. T. R. Ismail, "A data-driven sigmoid-based pi controller for buck-converter powered dc motor," in *2017 IEEE Symposium on Computer Applications Industrial Electronics (ISCAIE)*, April 2017, pp. 81–86.

[12] V. H. García-Rodríguez, R. Silva-Ortigoza, E. Hernández-Márquez, J. R. García-Sánchez, and H. Taud, "Dc/dc boost converter–inverter as driver for a dc motor: Modeling and experimental verification," *Energies*, vol. 11, no. 8, 2018. [Online]. Available: https://www.mdpi.com/1996-1073/11/8/2044

Patient-Specific Imaginary Motor Movement Classification of EEG Signals and Control of Robotic Arm

Özer Can DEVECİOĞLU
Faculty of Engineering and
Computer Science
Izmir University of Economics
Izmir, Turkey
ozercandevecioglu@hotmail.com

Burak YAMAN
Faculty of Engineering and
Computer Science
Izmir University of Economics
Izmir, Turkey
yamanburakk96@gmail.com

Özle MEŞEKOPARAN
Faculty of Engineering and
Computer Science
Izmir University of Economics
Izmir, Turkey
ozlemesekoparan@hotmail.com

Can ÇAKIR
Faculty of Engineering and
Computer Science
Izmir University of Economics
Izmir, Turkey
can_cakir66@windowslive.com

Türker İNCE
Faculty of Engineering and
Computer Science
Izmir University of Economics
Izmir, Turkey
turker.ince@ieu.edu.tr

Abstract— **This paper presents development of a robotic arm that can mimic the mobility of human arm with the recorded EEG (electroencephalogram) signals from the brain. This is a project designed for people who have lost the ability to control their arms. The robot arm will work with the signals that will be produced by the EEG instrument simultaneously when the user thinks that he or she will perform certain movements. These movements can be classified using Deep Multilayer Perceptron (Deep MLP) classifier designed in Python with Keras library. Deep MLP network is trained using Backpropagation algorithm for each patient by using own data. Experimental results demonstrate the proposed method can classify different imaginary movements with high accuracy.**

Keywords- Deep Multilayer Perceptron, Biomedical Signal Processing, EEG Signals, Imaginary Motor Movement

I. INTRODUCTION

As a result of various accidents some people have to continue their lives disabled. As we can see in Figure 1, these disabilities often affect physically instead of vision or hearing disabilities.

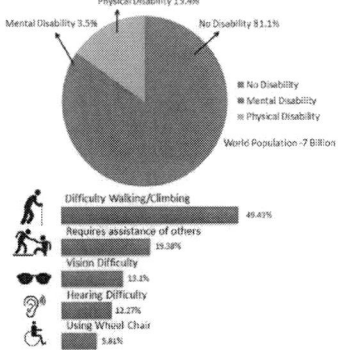

Figure 1. Common Types of Disability [1]

These disabilities may be caused by accidents or illnesses (such as ALS and paralysis). Unfortunately, these events may limit the mobility of these people and affect their freedom. As a result of an unfortunate accident, the individuals can isolate themselves from the society. With this paper they can regain the mobility of their arms and they return partially to their old lives.

This paper can also be useful for industrial applications. In factories or laboratories there are a lot of people who doing dangerous work. Cutting off metal parts, exposure to harmful rays or working with hazardous chemical components can be examples for these dangerous tasks. Instead of a person, this paper has the potential to carry out the same work with a distance. While the robot arm mimics movements in thoughts, the worker can observe their dangerous work from the safest distance.

When we realize any movement, emotion or a thought, even when dream to realization them occurs, some electrical activities are formed in our brains. Monitoring these activities is called as electroencephalography (EEG).

EEG signals can be classified in six types: alpha, beta, theta, delta, gamma and mu waves. Among these waves mu waves are corresponds to imaginary motor reflexes. In other words, the wave occurs when we plan to put a thought into action. We want to our robotic arm work like an actual arm. As a consequence, we used imaginary motor reflexes so that when we want to do something our system understand and control the robotic arm.

In the recent times, the brain-computer interface (BCI) is an advanced technique for connecting a human brain and an external device. With a BCI system, the desired commands can be determined by classifying the task-related neural activities as measured by electroencephalogram (EEG). As a result, people with disabilities will be able to expand their capabilities using BCI systems. In 2016, a method for accurate classification of motor images based on a Bayesian extreme learning machine (BELM) was used by Yu Wang [2]. By combining the ELM and Bayesian extraction, BELM obtained the smallest output weights with automatically predicted flattening to mitigate the excess equipment during the calibration procedure. In conclusion, in ELM 76.9% and in BELM 77.8% classification accuracy were obtained this study. Another study by Hyeong-Jun Park [3], eigenvector centralization feature selection method, wavelet pack separation joint spatial order and kernel extreme learning machine were used to increase the performance of BCIs of motor images and to prevent problems of attachment. In addition, the calculation speed was developed using the kernel extreme learning machine. He used WPD-CSP to extract properties from the data. To avoid over-adaptation, he used the eigenvector centrality feature selection (ECFS) method to select more important features for classification. The classifier used the Extreme learning machine (ELM) and the support vector machine (SVM) with the linear core and the core with radial basis function (RBF). Also, they combined ECFS with WPD-CSP to improve the performance of MI BCIs. In a result, WPD-CSP (%79.28) with ECFS shows much better performance compared with conventional CSP (%76.28) and WPD-CSP (%76.92) without feature selection. In 2017, M. Dharani [4] made studies on the classification of EEG signals (motor imagery) measured under planning and relaxed condition by using advanced learning limiters. Semi-supervised ELM (SS-ELM) and unsupervised ELM (US-ELM) were used for the EEG signal classification task. Both of these algorithms can fit into a unified framework and address multi-class classification or multi-clustering. The overall test efficiency of SSELM is 73% and the USELM is 71.978% respectively. So SSELM is a better classification technique than USELM. Therefore, these two networks show a good classification efficiency and can be used in real-world applications. In 2017, Schubert R. Carvalho [5] presented a new approach to the classification of achieving goals before the movement began. In this approach, he combined the distinguishing power of EEG images with the ability of a deep learning technique called the Convolutional Neural Network (CNN), which was suitable for learning patterns from direct image data and flexible enough to overcome local distortions. In this study, he wanted to use Virtual Reality (VR) to create an experimental setup that allowed the subject to reach their own self in the 3D space to allow for more natural reach movements. The results showed a significant increase in both classification performance and early detection in most experiments. The classification performance in the general training set was 44.16%. Finally, he thinks that the results provide an opportunity for a new field in the use of decoded EEG images for VR applications.

The goal of several researches was on EEG signal classification with different techniques. Their work become inspiration for us. Our target was improving the accuracy of classification with using Deep MLP classification. We tried to improve the evaluation of the classifier performance accuracy with using two different datasets (see section 2.1)

II. METHODOLOGY

A. Dataset

In this paper, two different Datasets were used;
1) Project BCI Dataset [6]

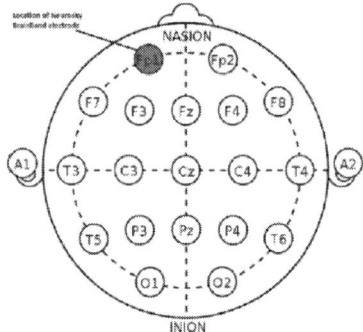

Figure 2. Locations of Electrodes

This dataset includes recordings of 21-year-old, right handed male subject with no known medical conditions. The EEG signals consists of actual random movements of left and right hand acquired with eyes closed. Each row represents one electrode. The order of the electrodes is FP1 FP2 F3 F4 C3 C4 P3 P4 O1 O2 F7 F8 T3 T4 T5 T6 FZ CZ PZ (Figure 2). The recording was done at 500Hz using Neurofax EEG System which uses a daisy chain montage. The data was exported with a common reference using Eemagine EEG. AC Lines in this country work at 50 Hz. This dataset contains lots of imaginary and real movements, for research area of paper imaginary movements are used;

- 1 trial imagined left hand backward movement
- 1 trial imagined right hand backward movement
- 1 trial imagined left hand forward movement
  1 trial imagined right hand forward movement

2) BCI Competition III Dataset V [7][8]

This dataset contains data from 3 normal subjects during 4 non-feedback sessions. The subjects sat in a normal chair, relaxed arms resting on their legs. There are 3 tasks:

- Imagination of repetitive self-paced left-hand movements
- Imagination of repetitive self-paced right-hand movements
- Generation of words beginning with the same random letter

All 4 sessions of a given subject were acquired on the same day, each lasting 4 minutes with 5-10 minutes breaks in between them. The subject performed a given task for about 15 seconds and then switched randomly to another task at the operator's request. EEG data is not splitted into trials since the subjects are continuously performing any of the mental tasks. EEG signals were recorded with a Biosemi system using a cap with 32 integrated electrodes located at standard positions of the International 10-20 system. The sampling rate was 512 Hz. Signals were acquired at full DC. No artifact rejection or correction was employed.

B. Deep MLP classification

To develop a neural network for classifying EEG Signals, we used a deep MLP network. The Deep MLP network is similar to MLP. However, they consist more hidden layers and neurons. Multi-Layer Perceptron is a feed-forward artificial neural network which consists of a number of neurons connected by linking weights. The architecture of MLP is illustrated in Figure-3

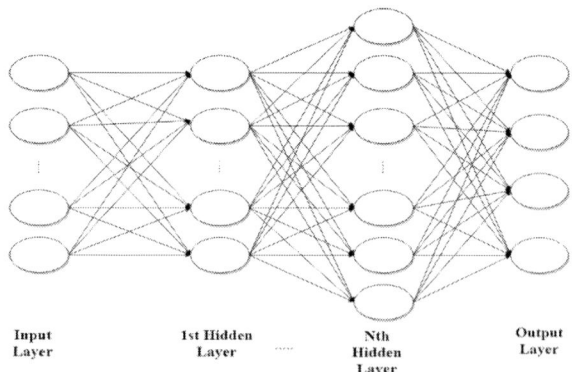

Figure 3. General architecture of Multilayer Perceptron

In MLP, the neurons use non-linear activation functions that is designed to model the behavior of the neurons in the human brain. Multi-layer perceptron has a linear activation function in all its neuron and uses backpropagation for its training. The activation function maps the weighted inputs to the output of the neuron. With activation function, we can calculate the output of any neuron in the MLP. Main operation of Multilayer Perceptron can be expressed as;

$$y = \left(\sum_{i=1}^{n} w_i x_i + b\right) = \emptyset(w^T + b) \quad (1)$$

where **w** denotes the vector of weights, **x** is the vector of inputs, b is the bias and $\emptyset$ is the activation function, the for the ith, neuron and y is the output.

An MLP is made up of a set of nodes which forms the input layer, one or more hidden layers, and an output layer. In contrast, non-linear function is used for activating hidden layers, linear functions can be used between the hidden layers.

The training process is used Backpropagation Algorithm. Backpropagation can conveniently be used to compute the remaining network parameters, weights (w) and bias (b) of each output layer neuron [9]. This algorithm occurs by continuous adjustment of the weights of the connections after each processing. This adjustment is based on the error in output. In the training process of MLP, backpropagation of output error is computed using the derivative of the mean-squared error (MSE) with respect to parameters of each neuron, so that we can perform gradient descent method to minimize their contributions and hence the overall error in an iterative manner. For an input vector i, and its corresponding target and output vectors, d_{zi} and y_{zi}, respectively, *MSE* in the output layer for the input i, Ei, can be expressed as follows:

$$E_i(\text{U}) = \frac{1}{n_i} \sum_{z=1}^{n_i} (d_{zi} - y_{zi})^2 \quad (2)$$

We applied Deep MLP on Python with Keras which is a python library that developed to make implementing deep learning models as fast and easy as possible for research and development [10].

C. Approach

The proposed EEE classification framework is shown in Figure 4. When imaginary movement happens, electrodes which located different positions on head, reads various electrical changes. Instead of dividing what electrode readings to samples and labeling, we classify the movements with different approach. We train the system with unit time values of all electrode readings when action happens. This is a spatial approach. We determine the movement with the spatial changes in electrodes.

Before data classification, we need to standardize data as a preprocessing. For standardizing data, standard score is used. Standard score of x can be defined as;

$$z = \frac{x - \mu}{\sigma} \quad (3)$$

where μ is the mean of dataset and δ is the standard deviation of dataset.

We constructed deep MLP with 2 hidden layers. First layer contains 40 neuron and the other one consists of 60 neurons. System runs with 100 epochs and we get output with sigmoid. We run the system with 10-fold cross validation.

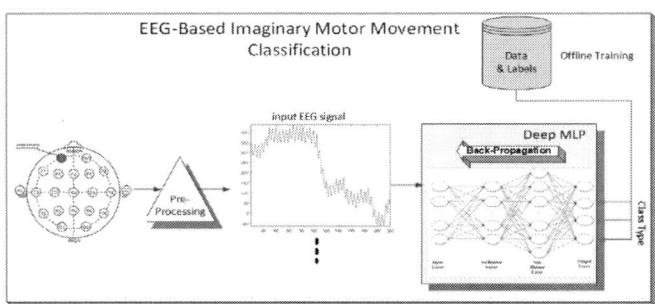

Figure 4. Proposed Classification Framework

The details of drawings of a robotic arm are taken from a robot design website [11]. Parts of robotic arm were produced with 3-D Printer. We produced the parts from PLA material with 30% density. The other parts of the robotic arm are servo motors and strains. Servo motors are the main source of movement in the system. All joints are connected to servo motors with strings. Visualization of robotic arm on computer can be seen on Figure 5. Hardware Implementation of robotic arm can be seen on Figure 6.

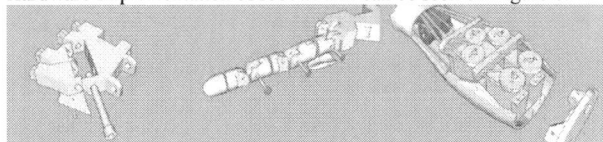

Figure 5. From right to left drawings of hand, finger and forearm can be seen.

Figure 6. Final Images of Robotic Arm with Different Angles

For controlling robotic arm, we used Raspberry Pi 3 as a master device and Arduino UNO as slave device. Raspberry Pi 3 is used for classifying the EEG signals with Deep MLP. Arduino UNO adjusts movements of servo motors with coming signals from Raspberry pi 3.

III. EXPERIMENTAL RESULTS

All the experiments reported in this paper were run on a 2.2GHz Intel Core i7-8750H with 8 GB of RAM and NVIDIA GeForce GTX 1050Ti graphic card. Both train and test sets were processed by CUDA kernels. On average, the training time of one epoch in MLP model is about 1 ms. In figure 6. we can see the process of learning.

From Table 1 and Table 2 deep MLP classifier was tested on two different datasets. Project BCI Dataset contains 1 subject and

978-1-5386-7688-2/19 $31.00 © 2019 IEEE

BCI Competition III Dataset V includes 3 different subjects. Our system is a patient-specific system. Therefore, each patient's system is trained with each recording.

We have achieved some promising results in our experiments on Python. We have trained the system with 90% of the data. We calculated error as a binary cross entropy. Binary cross entropy can be defined as;

$$H_p(q) = -\frac{1}{(N_{pos}+N_{neg})}\left[\sum_{i=1}^{N_{pos}}\log(p(y_i)) + \sum_{i=1}^{N_{neg}}(1-p(y_i))\right] \quad (4)$$

We calculated the test results by comparing the index of the predicted output's maximum value in row with known test output's index of maximum value in row. We run system 10 times with randomly distributed to test and train. Because of the with 10-fold cross validation we have tested all the data. Outputs can be seen in Table 1 and Table 2.

Trials	Accuracy
1	0.9909
2	0.9986
3	0.9951
4	0.9958
5	0.9989
6	0.9961
7	0.9989
8	0.9961
9	0.9989
10	0.9989
Average	0.9968

Table 1. Train and test accuracy rates of the Deep MLP with using Keras for Project BCI Dataset with 10-fold cross validation.

Trials	Accuracy		
	Subject-1	Subject-2	Subject-3
1	0.9811	0.9853	0.9781
2	0.9816	0.9759	0.9765
3	0.9866	0.9824	0.9665
4	0.9850	0.9839	0.9807
5	0.9789	0.9835	0.9765
6	0.9595	0.9895	0.9767
7	0.9804	0.9622	0.9686
8	0.9764	0.9705	0.9732
9	0.9839	0.9840	0.9825
10	0.9779	0.9648	0.9803
Average	0.9792	0.9782	0.9760

Table 2. Train and test accuracy rates of the Deep MLP of 3 Subjects with using Keras for BCI Competition III Dataset V with 10-fold cross validation.

IV. CONCLUSION

In this work, we implemented and tested a new EEG signals-based framework for imaginary motor movement of a robotic arm using Deep MLP classifier.

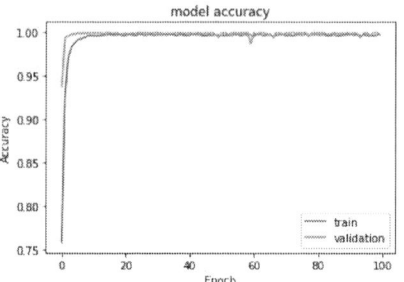

Figure 6. Train and validation accuracy of our model in the 1st trial with 100 epoch.

In the past, there have been numerous approaches to classify the EEG signals. There were other works who used the same dataset with us such as [12], [18], [19] and [7]. Comparison of their classification accuracy rates can be seen in Table 3.

Classification Methods	Average Classification Accuracy
Deep MLP (Our Method)	0.9760
S-dFasArt+RP+vs [17]	0.7607
ANN [13]	0.7996
SVM [14]	0.7095
BCI Competition III Winner [7]	0.6865

Table 3. Comparison of accuracy rates of different classification techniques using the same dataset.

Our project may be a light of hope for the reintegration of people who have not been able to continue their lives as before. Since our project will be controlled by imaginary motor reflex, it has the potential to replace the human arm directly. In near future, more movement classes will be added to our work and Convolutional Neural Networks (CNNs) will be applied and their performance will be compared.

REFERENCES

[1] http://www.disabled-world.com

[2] Yu Zhang, Motor imagery EEG classification via Bayesian extreme learning machine

[3] Hyeung-jun Park, Motor imagery EEG classification with optimal subset of wavelet based common spatial pattern and kernel extreme learning machine

[4] M.Dharani, Motor imagery signal classification using Semi-Supervised and Unsupervised Extreme Learning Machines

[5] Schubert R. Carvalho, A Deep Learning Approach for Classification of Reaching Targets from EEG Images

[6] https://sites.google.com/site/projectbci/

[7] http://bbci.de/competition/iii/

[8] Millán, J. del R.. On the need for on-line learning in brain-computer interfaces *Proc. Int. Joint Conf. on Neural Networks.*, 2004. https://skymind.ai/wiki/multilayer-perceptron

[9] Y. Chauvin and D. E. Rumelhart, "Back Propagation: Theory, Architectures, and Applications", Lawrence Erlbaum Associates Publishers, UK, 1995. https://keras.io/

[10] https://keras.io/

[11] http://inmoov.fr/

[12] J.-M. Cano-Izquierdo, J. Ibarrola, and M. Almonacid Improving motor imagery classification with a new BCI design using neuro-fuzzy S-dFasArt

[13]] Sylvia Bhattacharya, Rami J. Haddad, Mohammad Ahad, A Multiuser EEG Based Imaginary Motion Classification Using Neural Networks

[14] R. Aler, I. Galván, and J. Valls, Transition detection for brain computer interface classification

South African Power Distribution Network Load Forecasting Using Hybrid AI Techniques: ANFIS and OP-ELM

Sibonelo Motepe
Faculty of Engineering and the Built Environment
University of Johannesburg
Johannesburg, South Africa
djscvii@gmail.com

Ali N. Hasan
Faculty of Engineering and the Built Environment
University of Johannesburg
Johannesburg, South Africa
alin@uj.ac.za

Bhekisipho Twala
Faculty of Engineering
University of South Africa
Johannesburg, South Africa
twalab@unisa.ac.za

Riaan Stopforth
Stopforth Mechatronics, Robotics and Research Lab, School of Engineering
University of KwaZulu Natal
Durban, South Africa
stopforth.research@gmail.com

Nancy Alajarmeh
Department of Computer and Information Technology
Tafila Technical University
Tafila, Jordan
najarmeh@ttu.edu.jo

Abstract— South Africa has been a late participant in the previous industrial revolutions. With the fourth industrial revolution upon us, South Africa cannot afford to be left behind. Artificial Intelligence (AI) and big data, which are at the center of this revolution, are improving humans lives. Load forecasting has been shown to have benefits in power systems maintenance and operation. The study of load foresting in South African (SA) power distribution networks using AI is limited. This paper presents a comparative study of hybrid AI techniques, adaptive neuro-fuzzy inference systems (ANFIS) and OP-ELM, in South African distribution load forecasting. This is achieved with a case study on a real South African large power consuming substation using three performance measures, root mean square error (RMSE), mean absolute error (MAE) and symmetric mean absolute percentage error (sMAPE). The paper also investigates the impact of cleaning up loading data and the inclusion of temperature on the two techniques' models' performance. OP-ELM achieved the lowest error in comparison to ANFIS, achieving an sMAPE of 3.83%, MAE of 5.32% and RMSE of 6.52%. Hybrid AI techniques can thus be used to forecast load in South African distribution networks. This application of AI can lead to costs savings for S.A. power utilities.

Keywords—Adaptive Neuro-Fuzzy Inference System, Optimally Prunned Extreme Learning Machines, Load Forecasting, Power System Distribution,

I. INTRODUCTION

The fourth industrial revolution is upon the world and no nation can afford to be left behind. South Africa has been a late participant in the three key industrial revolutions [1] [2]. At the heart of the first industrial revolution was water and steam for mechanical production. The second industrial revolution was mass production stemming from the introduction of the electricity powered assembly line. The transistor was at the heart of the third industrial revolution. At the heart of the fourth industrial revolution is artificial intelligence (AI) and big data [1] [3]. Since AI's inception in the mid-1900s, different AI techniques have been developed and applied in various fields to simplify humans' life [4] [5] [6] [7]. In [5] deep learning was applied to detect domestic

violence victims from Facebook posts . The application of AI in Africa and South Africa is also on the rise [8] [9] [10] [11] [12]. AI application in South African load forecasting is still in its infancy [13]. The application of AI in forecasting distribution networks' loads in South Africa is almost non-existent [13]. South Africa is facing many challenges, from social-economic challenges to energy challenges [1] [8] [14] [15]. Advanced tools such as AI can help decimate big data to help develop solutions and recommendations that will assist South Africa. Load forecasting has been shown to be useful in operations, planning and maintenance [16] [17].

Adaptive neuro-fuzzy inference systems (ANFIS) have been used and shown to be more superior than conventional AI techniques in load forecasting [13] [18] [19] [20]. ANFIS is a hybrid of artificial neural networks (ANN) and fuzzy logic (FL). The combination of two or more techniques is usually to leverage the strengths of the techniques and overcome their shortcomings. ANFIS combines ANN and FL to overcome fuzzy logic's inability to learn and neural networks' lack of knowledge representablity and explainability. Optimally pruned extreme learning machines (OP-ELM) were developed by adding optimal pruning to the standard extreme learning machines [21]. This hybrid form overcomes the challenge of approximation of underlying dynamics when the data set has correlated or irrelevant variables. ELM and OP-ELM were found to outperform the multi-layer perceptron (MLP), ANFIS and support vector regression (SVR) in forecasting South Africa' the total energy demand [22]. In this study, the authors only used the mean square error as a performance measure.

The contributions of this paper are: (1) Presenting a comparative study of hybrid AI techniques, ANFIS and OP-ELM, in South African distribution load forecasting, (2) Presenting a case study on a real South African large power consuming substation using three performance measures, Root Mean Square Error (RMSE), Mean Absolute Error (MAE) and Symmetric Mean Absolute Percentage Error (sMAPE), and (3) Studying the impact of temperature and cleaning loading data on the load forecast accuracy. The paper is structured as follows: Section II gives an overview on load

978-1-5386-7688-2/19 $31.00 © 2019 IEEE

forecasting, Section III presents the two hybrid techniques used in this study, Section IV presents the experiment setup and data, Section V gives the results, and Section VI concludes the paper.

II. LOAD FORECASTING

Electricity is important in different aspects of life, such as modern communication, development of public services such as education and health care, driving industrialization and many others [23]. However, electricity demand in most African countries exceeds supply [23]. South Africa experienced rolling load-shedding in 2007, 2013 and 2018. This illustrated the importance of capacity planning. Access to electricity in developing, sub-Saharan African countries is still a major challenge [24]. Access to electricity is crucial in combating poverty [24]. South Africa plans to achieve universal access by 2025/2016 [25]. Load forecasting can assist with operational and maintenance planning at the power system distribution level. Load forecasting can lead to a reduced number of customer being interrupted when conducting maintenance and/or connecting new customers. The forecast can also help with ensuring that contractors and internal maintenance personnel are dispatched optimally. The effect of load forecast uncertainty on bulk electrical systems has been studied [26]. The study showed that load forecast uncertainty impacts systems with deficient generation (Gx) and/or transmission (Tx) systems than systems with strong Gx and Tx systems. In [27] fuzzy inference systems (FIS) were applied to estimate load in a low voltage European substation. The FIS models were compared to artificial neural networks (ANN) and, FIS models were found to outperform ANN models. Deep learning techniques have also been applied in load forecasting [28] [29].

III. ARTIFICIAL INTELLIGENCE TECHNIQUES USED

A. ANFIS

ANFIS is a hybrid technique that combines ANN and FIS. This technique has been widely used in power systems [13] [30] [31]. In [30] ANFIS was used for photovoltaic (PV) maximum power point tracking. ANFIS is based on the Takagi-Sugeno (TS) method, which is more popular in data-driven modeling [13]. Equation (1) gives the relationship between output and input in the TS technique.

$$y = \sum_{i=1}^{N} \gamma_i(z) \left(a_i^T z + c_i \right) \tag{1}$$

With

$$\gamma_i(z) = \frac{\prod_{j=1}^{p} \exp(-(z_i - g_j^i)^2(z)/2\delta_{ij}^2)}{\sum_{i=1}^{N} \prod_{j=1}^{p} \exp(-(z_i - g_j^i)^2(z)/2\delta_{ij}^2)} \tag{2}$$

where z is the input, y is the output, g is the center of the Gaussian function, δ is the variance. The learning process combines the gradient descent method and the least square estimator. The process can be summarised as follows:

Step 1: Find the optimal number of rules.

Step 2: Partition the input space equally, with functions width and slopes allowing sufficient overlap.

Step 3: Conduct forward pass to determine rule consequent parameters from neuron outputs determined using input data.

Step 4: Conduct the backward pass to update the antecedent parameters using error signals.

B. OP-ELM

Huang et al. proposed extreme learning machines, a single layer neural network, in the mid-2000s. ELMs have been applied in a number of areas [32] [33] [34] [35]. OP-ELM was proposed to overcome underlying approximation estimation challenges when the data set contains irrelevant or correlated variables [36]. The OP-ELM prunes these variables by marginalising the ELM built network's irrelevant neurons. ELM has been shown to outperform ANN, Radial Basis Function Neural Networks (RBFNN) and Back-Propagation Neural Networks (BPNN) in market clearance price forecasting [37]. OP-ELM was illustrated to outperform ANN, ANFIS, Support Vector Machine (SVM), Autoregressive Moving Average (ARMA) and basic ELM. If you have a training set z_i, where $i=1,.....n$ and t_i is the target, the ELMs objective is to minimize the error function. The ELM can thus be given by (3):

$$\sum_{j=1}^{k} f(w_j, b_j, x_i)\beta_j = t_i \tag{3}$$

where w_j is the input weight that connects the input and the jth neuron in the hidden layer, and β_j is the output weight connecting j^{th} hidden layer's neuron and the output, and b_j is the bias. A w_j, β_j and b_j exists such that $\sum_{j=1}^{k} f(w_j, b_j, x_i)\beta_j = y_i$, if the ELM model can estimate the data with an error of zero. We can then write (3) as:

$$H\beta = T \tag{4}$$

with

$$H = \begin{bmatrix} f(w_1, b_1, x_1) & \cdots & f(w_k, b_k, x_1) \\ \vdots & \cdots & \vdots \\ f(w_1, b_1, x_n) & \cdots & f(w_k, b_k, x_n) \end{bmatrix}_{n \times k} \tag{5}$$

where H is the output matrix of the hidden layer. If H is a square matrix, it can randomly be assigned values and β can be calculated by matrix inversion using (4). If H is invertible, the training is equivalent to solving a least square problem. Here β is determined using the matrix generalized inversion, known as the Moore-Penrose. The weight can be given by (6).

$$\beta = H^*T = (HH^T)^{-1}HT^T \tag{6}$$

where H^* is the Moore-Penrose generalized inverse of the matrix H. The OP-ELM algorithm can be summarized in the following 5 steps:

Step 1: Randomly assign input weights and bias.

Step 2: Determine H.

Step 3: Determine β.

Step 4: Rank the neurons using Multiresponse sparse regression (MRSR).

Step 5: Use the leave-one-out (LOO) validation method to choose an optimal neurons number.

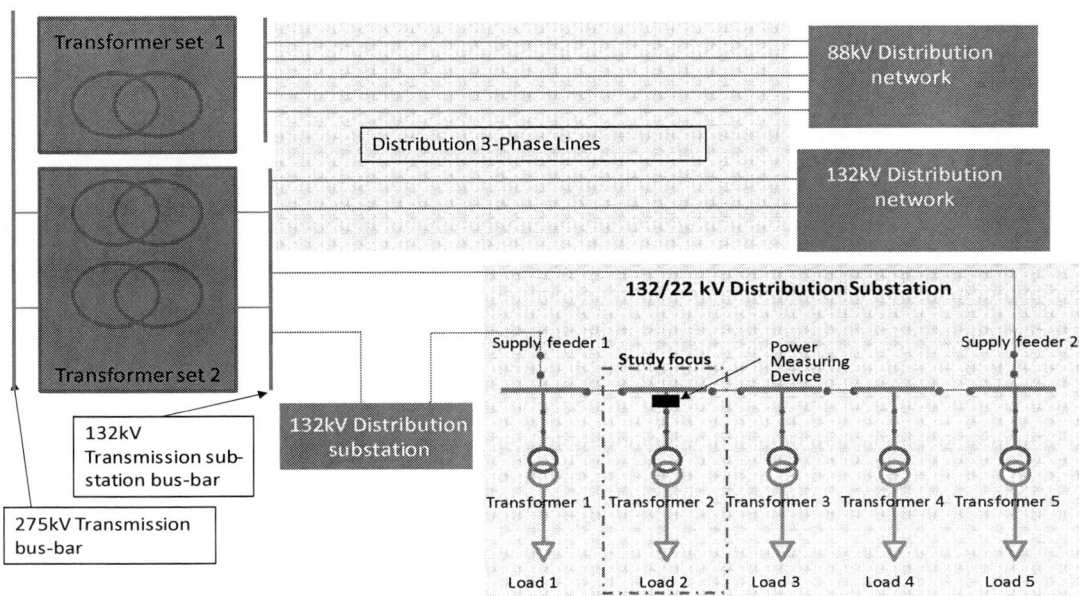

Fig. 1. Network setup where the distribution network under study is located

IV. EXPERIMENT SETUP

A. System setup

The substation under study is part of a 132 kV ring-network. The distribution network is shown in Fig. 1. The substation is supplied by a 275 kV transmission substation, which also supplies two other networks at 132 kV and an 88 kV, respectively. The substation has one supply feeder directly from the transmission substation and another feeder from the 132 kV busbars of the other substation in the ring network. There are 5 × 40 MVA, 132/22 kV transformers in the substation under study. The transformers provide bulk power supply points to five different loads belonging to one customer. The customer has their own internal distribution network with a configuration which combines some loads and separates some of the loads. The second supply point's load is forecasted in this study. The loading profiles for the first and second transformer were similar, and the third and fourth transformer's load profiles were also similar to each other. The load profiles of these two sets of transformers were different from each other. This difference can be a result of each set of transformers supplying the same load based on the internal customer configuration. The choice for the substation portion to study in this paper was arbitrary. The fifth transformer was seen to support either one of the two transformer sets. The power was measured on the transformer high voltage (HV) side.

B. Distribution network loading data

The power measurements are logged to a central database. The data is stored as 30 minutes average loading. This data can be queried from the database and downloaded for local storage. The supplied loading data were from January 2012 until September 2015. The apparent power was used for this study. For model training, the data was normalised to be between 0 and 1 by using (7).

$$x_{normalised} = \frac{x - x_{minimum}}{x_{maximum} - x_{minimum}} \qquad (7)$$

where x is the input variable, $x_{normamlised}$ is the normalised input variable, $x_{maximium}$ and $x_{minimum}$ are the dataset's maximum and minimum values, respectively. The loading data plot is given in Fig. 2. The data was cleaned up to remove dips caused by zero readings. Since the data clean up can take time and may be different for different systems, the study separated the experiments into those with cleaned data and those with the non-cleaned data. This was to investigate the impact of not cleaning up the loading data on the load forecasting accuracy. Weather temperature data were obtained from the South African weather services. These were also normalised using (7). The experiments were further divided into those with temperature as an input variable and those without temperature as a part of the input variables. The experiments were conducted for the winter period as the utility's technical staff stated the winter period as favourable for maintenance due to reduced rainfall and thunderstorms. Table I gives the two groups of input variables, with t as the point of time for the forecast.

TABLE I. EXPERIMENT INPUT VARIABLES

Input variables group	Input variables
Group 1	Loading data at *t-2* years Temperature at *t-2* years Load corresponding time of day Peak or non-peak period indicator Loading at *t-1* year Temperature at *t-1* year Loading at *t-2* weeks Temperature *t-2* weeks
Group 2	Loading *t-2* years Corresponding time of day Peak or non-peak period indicator Loading *t-1* year Loading *t-2* weeks

V. EXPERIMENT RESULTS

A. Performance Measures Utilised

Three error measurements were used to measure the models' performance. These error measurements were the symmetric mean absolute percentage error (sMAPE), mean absolute error (MAE) and root mean square error (RMSE). The error measurements used are given by (8) to (10).

$$sMAPE = \frac{2}{n} \sum_{k=1}^{N} \frac{|F_k - T_k|}{|F_k| + |T_k|} \qquad (8)$$

$$MAE = \frac{\sum_{k=1}^{N} |F_k - T_k|}{n} \qquad (9)$$

$$RMSE = \sqrt{\frac{\sum_{k=1}^{N} (F_k - T_k)^2}{n}} \qquad (10)$$

where F_k is the kth load forecast, T_k is the kth target load and n is the number of forecasts. The results are summarized in the Table II and Table III. The sMAPE equation in (8) gives an error between 0% and 200%, after multiplication by the 100% factor. To get a value between 0% and 100% the "2" in (8) is replaced by the 100% factor.

B. Results Discussion

Experiments were conducted for both techniques. ANFIS models were trained using different training parameters on a trial and error basis. Four key parameters were selected, based on the number of membership function and model behavior observed, to conduct all the ANFIS experiments. The OP-ELM models were trained with the number of hidden nodes being varied. The exact number of hidden units is determined by the LOO algorithm and sometimes an exact specific number of hidden units cannot be obtained. Hence an exact number of hidden units could not be obtained for all the equivalent experiments in in the four setups. An example was not being able to obtain a number of hidden nodes greater than 80 with cleaned loading data and without temperature as an input variable. After training, the models were tested with input and target data that were different from the training data. The results showed that the lowest errors were obtained when using the non-cleaned data to develop both techniques' models. ANFIS obtained the lowest error when the

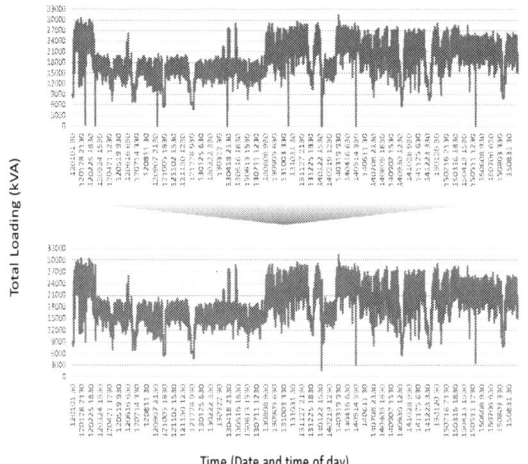

Fig. 2. Plots of substation's non-cleaned and cleaned loading data

temperature was not used in the experiments. The OP-ELM

model that obtained the lowest errors was the one trained and tested with temperature as an input parameter. OP-ELM models were seen to have an increase in error when their number of hidden nodes increased. This observation was however not the case when experimenting with temperature as an input variable and cleaned loading data. In this case, the error increased with an increase in the number of hidden nodes and at 158 hidden nodes the error was seen to be lower than with 108 hidden nodes. The OP-ELM model with 8 hidden achieved the lowest errors in all experiments, with an sMAPE of 3.83%, MAE of 5.32% and RMSE of 6.52%. These errors

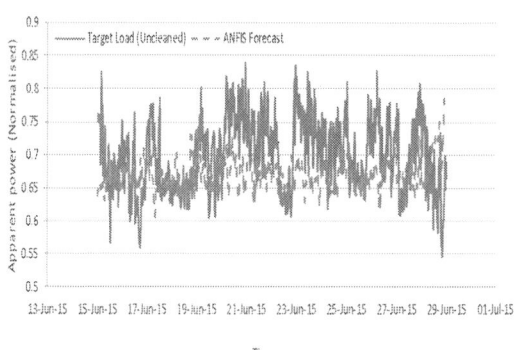

Fig. 3. Normalized ANFIS 2 week-ahead load forecast versus test target load for the lowest attained test error results

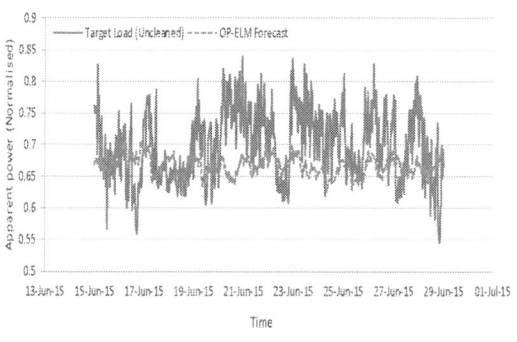

Fig. 4. Normalized OP-ELM 2 week-ahead load forecast versus test target load for the lowest attained test error results

were lower than the lowest errors achieved by the ANFIS models. ANFIS achieved the lowest errors of 3.94%, 5.48% and 6.73% for the sMAPE, MAE, and RMSE, respectively. These results were obtained for a two-week ahead load forecast. The plots of the load forecasting results that achieved the lowest the two-week ahead load forecast (15th to 29th June 2015) versus the target load are presented in Fig. 3 and Fig. 4, respectively, for the ANFIS model and OP-ELM model.

VI. CONCLUSIONS

The objective of this paper was achieved as follows: (1) A comparative study of hybrid AI techniques, ANFIS and OP-ELM, in South African distribution load forecasting was presented, (2) A case was presented using a real South African large power consuming substation and three performance measures, root mean square error (RMSE), mean absolute error (MAE) and symmetric mean absolute percentage error (sMAPE), and (3) The impact of temperature and cleaning loading data, on the load forecasting accuracy, was studied.

The OP-ELM model with 8 hidden nodes achieved the lowest errors in all experiments with an sMAPE of 3.83%, MAE of 5.32% and RMSE of 6.52% in comparison to the lowest errors achieved by the ANFIS models. ANFIS achieved the lowest errors of 3.94%, 5.48% and 6.73% for the sMAPE, MAE, and RMSE, respectively. These results were obtained for a two-week ahead load forecast. This work can be extended to investigate the impact of demand side management, load shedding and electricity price uncertainty on the load forecasting performance of hybrid AI techniques for SA's large power consumers.

TABLE II.   ANFIS LOAD FORECASTING TEST ERRORS

Input variables	Model Tuning parameters	Performance With Non-Cleaned Data			Performance With Cleaned Up Data		
		sMAPE	MAE	RMSE	sMAPE	MAE	RMSE
Group 1	1	0.121348	0.081983	0.104547	0.12329	0.083074	0.105467
	2	0.09584	0.065327	0.080201	0.105026	0.070778	0.087629
	3	0.08605	0.059333	0.073461	0.089798	0.061756	0.076607
	4	0.126163	0.086611	0.114288	0.128889	0.086846	0.108592
Group 2	1	0.086554	0.05962	0.075169	0.09473	0.064364	0.080462
	2	0.233798	0.156265	0.221828	0.236399	0.157662	0.211377
	3	**0.078875**	**0.0548**	**0.067278**	0.088012	0.060583	0.073518
	4	0.088155	0.06063	0.07585	0.0921	0.062804	0.078134

TABLE III.   OP-ELM LOAD FORECASTING TEST ERRORS

Input variables	Number of Hidden Nodes	Performance With Non-Cleaned Data			Number of Hidden Nodes	Performance With Cleaned Up Data		
		sMAPE	MAE	RMSE		sMAPE	MAE	RMSE
Group 1	**8**	**0.076639**	**0.053231**	**0.065293**	8	0.090344	0.057076	0.077942
	58	0.088769	0.061137	0.076092	58	0.112712	0.076688	0.148648
	103	0.100233	0.06852	0.089016	108	0.123649	0.078600	0.115015
	158	0.107616	0.072069	0.093743	158	0.111106	0.073698	0.097400
Group 2	10	0.077614	0.053866	0.065600	10	0.0827701	0.056119	0.068664
	55	0.091838	0.061893	0.097532	50	0.092224	0.061758	0.075695
	80	0.100575	0.067402	0.096671	70	0.101389	0.067256	0.089437
	105	0.114489	0.073351	0.116556	80	0.120470	0.077187	0.138783

ACKNOWLEDGEMENT

The authors would like to thank the SA Weather Services for providing them with weather data. The authors acknowledge the National Research Foundation, Eskom TESP and The DST ROSSA program, who partially funded this research.

REFERENCES

[1] T. Marwala and E. Hurwitz, Artificial Intelligence and Economic Theory: Skynet in the Market, Springer International Publishing AG, 2017.

[3] J. Bloem, M. Van Doorn, S. Duivestein, D. Excoffier, R. Maas and E. Van Ommeren, "The Fourth Industrial Revolution: Things Tighten," Sogeti VINT, 2014.

[4] S. Albarqouni, C. Baur, F. Achilles, V. Belagiannis, S. Demirci and N. Navab, "AggNet: Deep Learning From Crowds for Mitosis Detection in Breast Cancer Histology Images," IEEE Transactions on medical imaging, vol. 35, no. 5, pp. 1313-1321, 2016.

[5] S. Subramani, H. Wang, H. Q. Vu and G. Li, "Domestic Violence Crisis Identification From Facebook Posts Based on Deep Learning," IEEE Access, vol. 6, pp. 54075-54085, 2018.

[6] K. Choi, G. Fazekas, K. Cho and M. Sandler, "The Effects of Noisy Labels on Deep Convolutional Neural Networks for Music Tagging," IEEE Transactions on Emerging Topics in Computational Intelligence, vol. 2, no. 2, pp. 139-149, 2018.

[7] S. Motepe, B. Twala and R. Stopforth, "Determining South African Distribution Power System Big Data Integrity Using Fuzzy Logic," in Pattern Recognition Association of South Africa and Robotics and Mechatronics International Conference (PRASA-RobMech), 2017.

[8] A. N. Hasan, B. Twala, K. Ouahada and T. Marwala, "Energy usage optimisation in South African mines,"," Archives of Mining Sciences, vol. 59, no. 1, pp. 53-69, 2014.

[9] A. N. Hasan, "Improving single classifiers prediction accuracy for underground water pump station in a gold mine using ensemble techniques," in IEEE International Conference on Computer as a Tool (EUROCON), 2015.

[10] A.N. Hasan, P.S.P. Eboule and B. Twala, "The Use of Machine Learning Techniques to Classify Power Transmission Line Fault Types and Locations," in International Conference on Optimization of Electrical and Electronic Equipment (OPTIM) & Intl Aegean Conference on Electrical Machines and Power Electronics (ACEMP, 2017.

[11] E. M. Malatji, J. Zhang and X. Xia, "A multiple objective optimisation model for building energy efficiency investment decision," Energy and Buildings, vol. 61, pp. 81-87, 2013.

[12] E.M. Malatji, B. Twala and N. Mbuli, "Optimal Placement Model of TCSC in Power System Network Considering the Budget Available," in 5th International Electrical Engineering Congress, 2017.

[13] S. Motepe, A. N. Hasan and R. Stopforth, "South African Distribution Networks Load Forecasting Using ANFIS," in IEEE Power Electronics Drivers and Energy Systems (PEDES), 2018.

[14] S. Friedman, "Enrich the rich; Damn the Poor," The Thinker: A Pan-African Quarterly for Thought Leaders, pp. 8-11, 2016.

[15] B. Khanyile, "Labour broking in Southern Africa and other legislative interventions in labour relations," The Thinker: A Pan-African Quarterly for Thought Leaders, pp. 58-59, 2016.

[16] L. Marwala and B. Twala, "Forecasting electricity consumption in South Africa:ARMA, Neural networks and Neurro-fuzzy," in International joint conference on Neural Networks, 2014.

978-1-5386-7688-2/19 $31.00 © 2019 IEEE

[17] E. Ceperic, V. Ceperic and A. Baric, "A strategy for short-term load forecasting by support vector regression machines," IEEE Transaction on Power Systems, vol. 28, no. 4, pp. 4356 - 4364, 2003.

[18] W. Yuill, R. Kgokong, S. Chowdhury and S.P. Chowdhury, "Application of adaptive neuro-fuzzy inference system (ANFIS) based short term load forecasting in South African power networks," in 45th International Universities Power Engineering Conference UPEC2010, 2010.

[19] W. Yuill, R. Kgorong, S. Chowdhury and S.P. Chowdhury, "Management of short term load forecasting in South African power networks," in International conference on power system technology, 2010.

[20] L. Marwala and B. Twala, "Univariate modeling of electricity consumption in South Africa: neural networks and neuro-fuzzy systems," in IEEE International conference on systems, man and cybernetics, 2013.

[21] Y. Miche, A. Sorjamaa and A. Lendasse, "OP-ELM: Theory, experiments and a toolbox," in 18th International Conference on Artificial Neural Networks (ICANN), 2018.

[22] L. Marwala and B. Twala, "Electricity Load Forecasting Using an Ensemble of Optimally-Pruned and Basic Extreme Learning Machines," in 9th IEEE-GCC Conference and Exhibition (GCCCE), 2017.

[23] H. Ahlborg, F. Boräng, S. C. Jagers and P. Söderholm, "Provision of electricity to African households: The importance of democracy and institutional quality," Energy Policy, vol. 87, pp. 125-135, 2015.

[24] P. A. Trotter, M. C. McManus and R. Maconachie, "Electricity planning and implementation in sub-Saharan Africa: A systematic review," Renewable and Sustainable Energy Reviews, vol. 74, pp. 1189-1209, 2017.

[25] South African Government, "Integrated national electrification programme," [Online]. Available: http://www.gov.za/aboutgovernment/government-programmes/inep. [Accessed June 2017].

[26] R. Billinton and D. Huang, "Effects of Load Forecast Uncertainty on Bulk Electric System Reliability Evaluation," IEEE Transactions on power systems, vol. 23, no. 2, pp. 418-425, 2008.

[27] T. Konjic, V. Miranda and I. Kapetanovic, "Fuzzy Inference Systems Applied to LV Substation Load Estimation," IEEE Transactions on power systems, vol. 20, no. 2, pp. 742-749, 2005.

[28] R.F. Berriel, A.T. Lopes, A. Rodrigues, F.M. Varejão and T. Oliveira, "Monthly energy consumption forecast a deep learning approach," in International Joint Conference on Neural Networks (IJCNN), 2017.

[29] G. M. U. Din and A.K. Marnerides, "Short term power load forecasting using deep neural networks," in International Conference on Computing, using deep neural networks, 2017.

[30] A. Ali and A. N. Hasan, "Optimization of PV model using fuzzy-neural network for DC-DC converter systems," in 9th International Renewable Energy Congress (IREC), 2018.

[31] A. M. Farayola, A. N. Hasan and A. Ali, "Curve fitting polynomial technique compared to ANFIS technique for maximum power point tracking," in 8th International Renewable Energy Congress (IREC), 2017.

[32] F. Wang, Z. Zhao, X. Li, F. Yu and H. Zhang, "Stock volatility prediction using multi-kernel learning based extreme learning machine," in International Joint Conference on Neural Networks (IJCNN), 2014.

[33] A.H. Nizar, Z.Y. Dong and Y. Wang, "Power utility nontechnical loss analysis with extreme learning machine method," IEEE Transactions on Power Systems, vol. 23, no. 3, pp. 946-955, 2008.

[34] C. Wan, Z. Xu, P. Pinson, Z.Y Dong and K.P. Wong, "Probabilistic Forecasting of Wind Power Generation Using Extreme Learning Machine," IEEE Transactions on Power Systems, vol. 29, no. 3, pp. 1033-1-44, 2014.

[35] X. Chen, Z. Y. Dong, K. Meng, Y. Xu, K.P. Wong and H.W. Ngan, "Electricity Price Forecasting With Extreme Learning Machine and Bootstrapping," IEEE Transactions on Power Systems, vol. 27, no. 4, pp. 2055-2062, 2012.

[36] Y. Miche, A. Sorjamaa and A. Lendasse, "OP-ELM: Theory, experiments," in 18th International Conference on Artificial Neural, 2008.

[37] X. Chen, Z. Y. Dong, K. Meng, Y. Xu, K.P. Wong and H.W. Ngan, "Electricity Price Forecasting With Extreme Learning Machine and Bootstrapping," IEEE Transactions on Power Systems, vol. 27, no. 4, pp. 2055-1061, 2012.

978-1-5386-7688-2/19 $31.00 © 2019 IEEE

Supercapacitor Parameter Identification Using Grey Wolf Optimization and Its Comparison to Conventional Trust Region Reflection Optimization

Ravneel Prasad, Utkal Mehta, Kajal Kothari, Maurizio Cirrincione and Ali Mohammadi

School of Engineering and Physics,
The University of the South Pacific,
Laucala campus, Fiji.
e-mail: s11108164@student.usp.ac.fj; utkal.mehta@usp.ac.fj; s11151029@student.usp.ac.fj;
maurizio.cirrincione@usp.ac.fj; ali.mohammadi@usp.ac.fj

Abstract—This work focuses on two parameter identification techniques using the traditional Trust Region Reflective (TRR) and the bio inspired Grey Wolf algorithm (GWO). The meta-heuristic parameter identification technique is not widely used in the field of supercapacitor (SC) parameter identificaton since many researchers prefer to make use of classical search method. This work investigates the use of meta-heuristic based parameter identification for SC. Both the algorithms are applied on real data of SCs of different values and brands.This work also extends study on SC's impedance modeling from step input voltages and can estimate fractional impedance model parameters from time response data directly. Comparative results from classical and fractional impedance models are also shown which can extract the actual behavior of supercapacitor. The proposed identified fractional impedance parameters show less error in time domain. The SC model used in the article considers initial conditions (non-zero condition) on which identification is validated experimentally for 0.47F, 1F and 1.5F SCs and result is discussed to form a conclusion.

Index Terms—Supercapacitors; Fractional-order model; Impedance, Trust Region Reflective; Least Square; Grey Wolf Algorithm.

I. INTRODUCTION

Supercapacitors (SCs), also known as Ultracapacitors and Electric Double Layer Capacitors (ELDCs), are type of energy storage component that have large capacitance with quick charge and discharge times. Lately, it has been well accepted that SC can play an important role in renewable energy resources. Application of supercapcitors have gained immense interest in numerous fields due to its superior performance over conventional electrical storage devices. SCs utilize a higher surface area, which is greatly attributed to the porous like structure of electrodes and thinner dielectrics to achieve greater capacitance [1]. Due to this they have greater power densities, typically 10 times than a battery and 100 times greater energy density then most conventional capacitors [2]. Longevity in terms of life cycle is another key feature of the SCs over the batteries but less than the conventional capacitors [3].

Recent decade has shown a significant growth on SC usage in terms of the energy storage and power delivery in numerous applications. It fulfils requirement of not only to store energy but also supply short energy pulses for a short duration [4]. This includes in, electric hybrid cars [5] (e.g. regenerative breaking), renewable energy systems [6] and biomedical sensors [7] to other electronic products. Large scale SCs are also been used as power quality regulators in electrical grids [8].

In order to improve energy storage performances where SCs are present, modeling and identification of impedance parameters of these non-traditional capacitors are key issues to study. Recently, it has been noted in [9], [10], [11] that fractional impedances can describe the behavior better for these capacitors than traditional integer models. Moreover fractional models are able to capture the variability of its parameters achieving the transient charging and discharging region (CDR) [12]. All these studies still reveal a discrepancy between simulated and experimental results in CDR with further error propagation in subsequent CDR. This is the reason why the parameter estimation is a key issue: an improvement in the estimation of the SC parameters would lead up to significant reduction of the error. This paper, in this context, investigates the use of meta-heuristic approach of GWO and compares it to the traditional least squares technique, the so called TRR (Trust Region Reflective) least square. Both the methods are used on fractional SC model with non null initial conditions. Moreover it also shows a comparison with integer model.

II. SUPERCAPACITOR MODEL

An RC (Resistor-Capacitor) model based on the behavior of porous electrode model has been examined in this paper which has been earlier explored in [11]. The structure of the model is shown in Fig.1. This has been used more widely due to it being simple and fitting well with the experimental data at high and low frequencies. This model employs parameterized RC network to imitate the electrical properties of SCs. It comprises of a series resistor, R_s, and a constant phase element (CPE) which is mostly a fractional capacitor, C_α with an order α ranging from 0 to 1. The impedance of CPE can be written as $Z_{CPE} = 1/(C_\alpha s^\alpha)$. The total impedance, $Z(s)$, expression

978-1-5386-7688-2/19 $31.00 © 2019 IEEE

for such porous model from Fig.1 can be written simply by:

$$Z(s) = R_s + \frac{1}{C_\alpha s^\alpha} \tag{1}$$

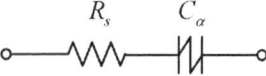

Fig. 1. Simple porous model of a supercapacitor

In order to derive the charging and discharging fractional order equations for SC, Caputo's derivative of an arbitrary function $f(t)$ and its Laplace transform in (2) and (3) respectively is used [13].

$$_c\mathcal{D}_t^\alpha f(t) = \frac{1}{\Gamma(n-\alpha)} \int_0^t \frac{f^{(n)}(\tau)}{(t-\tau)^{\alpha-n+1}} d\tau \tag{2}$$

where $n - 1 < \alpha < n$ and $n \in N$

$$L\left[_c\mathcal{D}_t^\alpha f(t)\right] = s^\alpha F(s) - \sum_{k=0}^{n-1} s^{\alpha-k-1} f^k(0) \tag{3}$$

Assuming charging and discharging circuits for SC as shown in Fig.2 with input voltage, V_{in}, and output transient voltage, V_o. The time domain output response from step input change was derived using Mittag-Leffler function in [11] and yield following expression in time domain for the charging circuit.

$$
\begin{aligned}
V_o(t) &= \frac{V_{in}}{C_\alpha(R+R_s)} t^\alpha E_{\alpha,\alpha+1}\left(\frac{-t^\alpha}{C_\alpha(R+R_s)}\right) \\
&+ \frac{V_{in}R_s + V(0)R}{R+R_s} E_{\alpha,1}\left(\frac{-t^\alpha}{C_\alpha(R+R_s)}\right)
\end{aligned} \tag{4}
$$

$V(0)$ in this particular equation indicates the initial voltage for the SC.

Similarly, the time-domain expression for discharging circuit has been derived with initial peak voltage V_p. The electrical circuit shown in Fig.2 (b) behaves as follow.

$$V_o(t) = V_p \frac{R}{R+R_s} E_{\alpha,1}\left(\frac{-t^\alpha}{C_\alpha(R+R_s)}\right) \tag{5}$$

Using the expressions (4) and (5) for charging and discharging of a SC, one can study the transient behavior and can estimate fractional impedance parameters (C_α, R_s, α). How-

ever, it is worth considering that the charging and discharging parameters may vary. This variation is likely due to the internal reactions that take place because of thermal expansion [14], dielectric relaxation theory [15], aging process [4] and other phenomena associated to internal structure of SC. Impact of internal characteristics has been critical in non-linear behavior of SC.

III. PARAMETER ESTIMATION

Identification of fractional impedance parameters in (1) described by parameter vector $x = (R_s, C_\alpha, \alpha)$ was computed using MATLAB. The constrained problem for solving these parameters was described by Sum of Square Error (SSE) of the experimental data, y_{actual}, and fit or estimated data, $\hat{y}(x)$, at point x as given in (6).

$$
\begin{aligned}
\min_x \|F(x)\|_2^2 &= \min_x \sum_i \left(f(x)_i\right)^2 \\
where: & \\
f(x) &= y_{actual} - \hat{y}(x)
\end{aligned} \tag{6}
$$

In order to obtain the optimized minimum for (6), two different optimization routines have been applied to estimate fractional information from time-domain responses. First technique uses the numerically solved *Trust Region Reflective Least Square Algorithm*, TRR [16] which is based on the concept of reduction in the reflected trust region search space. Second is a swarm meta-heuristic namely *Grey Wolf Optimiser*, GWO [17] which is based on the hunting style of grey wolves. These two methods were selected for identification of the parameters mainly due to the wide spread usage of the conventional technique of TRR and its ability to present better accuracy compared to Gaussian-Newton and Levenberg-Marquardt conventional methods [18]. On the other hand GWO is one of the emerging meta-heuristic technique and hasn't been used in optimization for SC parameters before. Infact, there is very little usage of biological inspired algorithms for SC parameter identification.

The parameter vector was bounded in both the cases to avoid the negative resistance which is not possible to realize in real world. The α value was also bounded to ensure that negative value is not achieved which would indicate an inductor rather than a capacitor.

A. Trust Region Reflective (TRR) Least Square Algorithm

Coleman and Li et. al in [16], [19], [20], presented non-linear minimization algorithm with bounded constrains which was based on the trust region optimization method. It incorporated the interior reflective Newton algorithm. In this study the technique is applied to determine the parameters through the objective function $f(x)$ which is approximated by a quadratic function $\psi(x)$. This is the representation of the function $f(x)$ reflected around the current point x in the neighborhood N. The neighborhood N is known as the trust region and variable x is a vector that is bounded by a lower bound vector lb and an upper bound vector ub. The initial idea for the solution of the trust region is to solve for a trial step p which is by minimization of the region N as depicted by (7), a trust

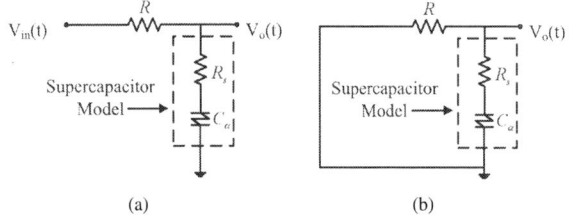

Fig. 2. The supercapacitor circuit during (a) charging, (b) discharging mode.

region sub-problem. In order for this step to be successful and the trust region to remain for the next step, the condition of $f(x) > f(x + p)$ has to be satisfied. If the condition is not met, the step is unsuccessful and the trust region N has to be reduced for the next step.

$$\min_p \psi(\text{p}) = \min_p(\tfrac{1}{2}p^T M p + p^T g) \\ \text{such that} ||Dp|| \le \Delta_t \tag{7}$$

where M is the approximated Hessian Matrix, H, of $f(x)$, g is the gradient of $f(x)$, t is the iteration number, Δ_t is the trust region radius greater than zero and $||.||$ is the second norm.

$$M = H + B \tag{8}$$

$$B = D diag(g(x_t)) J_v(x_t) D \tag{9}$$

$$D = diag(|v(x)|^{-\frac{1}{2}}) \tag{10}$$

where the vector $v(x)$ is defined by:

- If $g_i < 0$ and $ub_i < \infty$ then $v_i = x_i - ub_i$
- If $g_i \ge 0$ and $lb_i < -\infty$ then $v_i = x_i - lb_i$
- If $g_i < 0$ and $ub_i = \infty$ then $v_i = -1$
- If $g_i \ge 0$ and $lb_i = -\infty$ then $v_i = 1$

J_v is a Jacobian matrix of vector $|v|$ and has a diagonal values of $-1, 0$ and 1. If $g_i = 0$ or either of vectors ub or lb is infinite, then the value of J_v is equal to zero. Moreover, if all the values of the bounds lb and ub are finite than the Jacobian matrix is defined by (11).

$$J_v(x_t) = diag(\text{sgn}(g(x_t))) \tag{11}$$

The trust region radius is updated based on (12)

$$r_t = \frac{f(x_t + p_t) - f(x_t) + \tfrac{1}{2}p_t^T B_t p_t}{\psi(p_l)} \tag{12}$$

- if $r_t \ge \sigma_1$, $\Delta_{t+1} \in (0, \gamma_1 \Delta_t]$

- if $\sigma_1 < r_t < \sigma_2$, $\Delta_{t+1} \in [\gamma_1 \Delta_t, \Delta_t]$

- if $r_t \ge \sigma_2$ then,
  if $\Delta_t > \mu_1$, $\Delta_{t+1} \in$ either $[\gamma_1 \Delta_t, \Delta_t]$ or $[\Delta_t, \gamma_2 \Delta_t]$ else, $\Delta_{t+1} \in$ either $[\Delta_t, \min(\gamma_2 \Delta_t, \mu_2)]$

where variables σ_1, σ_1, γ_1, γ_2, μ_1 and μ_2 are constants satisfying $0 < \sigma_1 < \sigma_2 < 1$, $0 < \mu_1 < \mu_2$ and $\gamma_1 < 1 < \gamma_2$.

For a much quicker solution to (7), vector p can be restricted in a two dimensional subspace P. Two direction vectors, p_1 (the direction of the gradient g) and p_2 (the negative direction of curvature or the approximate Newton direction , $p_2^T H p_2 < 0$) span the subspace P. The vector p_2 can be found by applying the preconditioned conjugate gradient method as defined by:

$$H \cdot p_2 = -g \tag{13}$$

$$p_2^T \cdot H \cdot p_2 < 0 \tag{14}$$

After determining the subspace P, the minimization process can be completed, obtaining the value for x which gives the minimum value of $f(x)$. TRR algorithm was implemented in MATLAB using the *lsqcurvefit*.

B. Grey Wolf Optimization (GWO)

This particular method was proposed by Mirjalili et.al in 2014 [17] which was inspired by the grey wolves (Canis Lupus). The algorithm tries to mimic the hunting style and the leadership hierarchy of the grey wolves. The hunting behavior includes the following:

- Tracking, chasing and approaching the prey.
- Pursuing, encircling and harassing the prey until it is motionless.
- Attack the prey

1) The social hierarchy: The hierarchy consists of four levels which are the alpha (ω_3), beta (ω_2), delta (ω_1) and omega (ω_0) ranking from highest to lowest respectively. This hierarchy structure is translated in to the mathematical model where by the fittest solution is considered as ω_3. While the second and the third best solutions are considered as ω_2 and ω_1 respectively. Apart from this, all the other candidate solutions are assumed to be ω_0. These phenomena is seen in the GWO algorithm where by the hunting (optimization) of the solution is guided by these four ranks.

2) Encircling Prey: The encircling behavior of the wolves can be mathematically modeled through (15) and (16).

$$\vec{D} = |\vec{C}.\overrightarrow{Q_p(t)} - \overrightarrow{Q(t)}| \tag{15}$$

$$\overrightarrow{Q(t+1)} = \overrightarrow{Q_p(t)} - \vec{A}.\vec{D} \tag{16}$$

where $\overrightarrow{Q_p}$ is prey's position vector, $\vec{A}$ and $\vec{C}$ are the coefficient vectors and $\vec{Q}$ is a grey wolf's position vector.

The coefficient vectors are given by the following:

$$\vec{A} = 2\vec{a}.\vec{r} - \vec{a} \tag{17}$$

$$\vec{C} = 2.\vec{r_2} \tag{18}$$

where the components of $\vec{a}$ is linearly decreased over the course of iteration from 2 to 0 and vectors r_1 and r_2 are random numbers between the interval [0,1].

3) Exploration(Search for Prey): The wolves diverge from the pack to search for the prey while they converge to attack the prey. This divergence is mathematically modeled by utilizing $\vec{A}$ with random values which is either less than -1 or greater than 1 in order for the search agent to oblige to diverge from the prey. This highlights the exploration phase and allows global search by the GWO algorithm. That is, $|A| > 1$ forces the divergence of the grey wolves from the prey hoping for discovery of a fitter prey.

$\vec{C}$ is another component that favors the exploration in GWO. This is due to the vector C containing random values in [0,2]. In order to stochastically emphasize ($C \ge 1$) or to de-emphasize ($C < 1$) the prey's effect in defining the distance in (15), this component provides with random weights for prey. This adds a more random behavior to the GWO algorithm throughout the optimization, giving preference to exploration and avoidance of local optima. It also worth taking note that unlike $\vec{A}$, $\vec{C}$ does not decrease linearly. $\vec{C}$ is deliberately required to provide random values so that emphasis is given to exploration during initial and final iterations. This component is very useful when it comes to local optima stagnation, especially during the final iterations.

4) Hunting: The hunt is usually commanded by the alpha wolves, and the beta and delta wolves participating occasionally. In order to simulate the hunting behavior of the grey wolves mathematically, it is assumed the best candidate solution is the alpha, while beta and delta being the next best solution respectively. Therefore, the best three solutions so far are saved and the other search agents oblige to update their positions based on the best search agent. The following formulae are utilized with regards to this:

$$\overrightarrow{D_{\omega_3}} = |\overrightarrow{C_1}.\overrightarrow{Q_{\omega_3}} - \vec{Q}|,$$
$$\overrightarrow{D_{\omega_2}} = |\overrightarrow{C_2}.Q_{\omega_2} - \vec{Q}|,$$
$$\overrightarrow{D_{\omega_1}} = |\overrightarrow{C_3}.\overrightarrow{Q_{\omega_1}} - \vec{Q}| \qquad (19)$$

$$\overrightarrow{Q_1} = \overrightarrow{Q_{\omega_3}} - \overrightarrow{A_1}.(\overrightarrow{D_{\omega_3}}),$$
$$\overrightarrow{Q_2} = \overrightarrow{Q_{\omega_2}} - \overrightarrow{A_2}.(\overrightarrow{D_{\omega_2}}),$$
$$\overrightarrow{Q_3} = \overrightarrow{Q_{\omega_1}} - \overrightarrow{A_3}.(\overrightarrow{D_{\omega_1}}) \qquad (20)$$

$$\overrightarrow{Q(t+1)} = \frac{\overrightarrow{Q_1} + \overrightarrow{Q_2} + \overrightarrow{Q_3}}{3} \qquad (21)$$

5) Exploitation (Attacking Prey): The mathematical representation of approaching the prey is done by decreasing the value of $\vec{a}$. Hence, $\vec{A}$ is a random value within the interval $[-a, a]$. When $|A| < 1$ or within the range [-1,1], the wolves are forced to attack the prey.

In summary, the GWO algorithm starts by initializing a random population grey wolves (candidate solutions). During the iterations, the best probable position of the prey is estimated by the alpha, beta and delta wolves and based on each iteration the candidate solutions are updated. The emphasis on exploration or exploitation is denoted by decreasing the parameter a from 2 to 0 respectively. Candidate solutions tend to converge towards the prey if $|A| < 1$, otherwise ($|A| \geq 1$) the solutions diverge from the prey. Finally, the algorithm terminates upon meeting the criterion.

IV. EXPERIMENTAL RESULTS AND DISCUSSIONS

The optimization routines for TRR and GWO are carried out until maximum of 1000 iterations are completed or stopping criteria. The experiments was conducted on three different brands of SCs, namely: Eaton, AVX and Kemet. Three different values of capacitors, 0.47F, 1F and 1.5F were tested from each of the brands. The total number of data points used for each of the SC were 251. A test circuit based on Fig.2 was constructed. The datasets were collected for 30 seconds each to replicate the abrupt change in the voltage where by the voltage to be stored to or discharged from the SC is for short span of time.

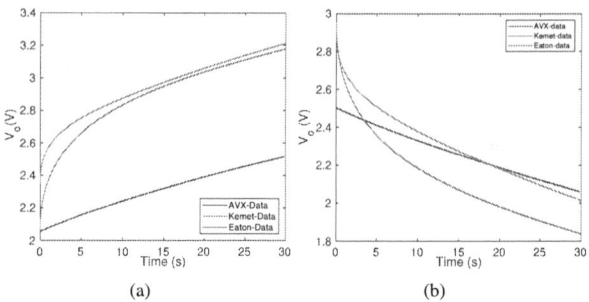

Fig. 3. (a) Charging curve (b) Discharging curve from 0.47F supercapacitor

Initial voltage has been considered not to be zero. This is due to the more realistic result and approach when considering a non zero initial voltage. AVX 1.0F SC was considered for test of initial voltage as non-zero value of 2.23V. The parameters obtained for a zero initial condition was: $C_\alpha = 0.4F$ $R_s = 218.1\Omega$ and $\alpha = 0.9$ where as for non-zero initial condition was: $C_\alpha = 0.8F$ $R_s = 10$ mΩ and $\alpha = 0.9$. This indicates that a high resistance value and a lower capacitance value is obtained when zero initial condition is considered. In contrast to the manufacturers datasheet [21], the series resistance should be 0.7Ω at max when DC voltage is applied which coincides with the SC low resistance property. Thus, it is more appropriate to use a non-zero initial state. However, it is also worth considering that there can not be a direct comparison with the values from manufacturer's datasheet as these values are nominal. All results in this article is with the case of a non-zero initial condition.

From the experimental figures as seen in Fig.3(a), there are different behavior for the same SC value when subjected to same step input voltage of 5V. The circuit for the SC charging is referred to Fig.2(a). The shape of the charging curve differs from brand to brand meaning different manufacturing standards also play a part in the behavior of the SCs. One significant observation was that various charging curve shapes might need different order to accurately fit the fractional model. Same could be depicted from the Fig.3(b), which is the discharge graph of the SC based on the circuit given in Fig.2(b).

Aiming to estimate the fractional impedance parameters using optimization algorithms and experimental data, both the integer order ($\alpha = 1$) and the fractional order models' ($0 < \alpha < 1$) parameters were identified and depicted in Tables I,II,III and IV. All the values provided in the Tables for C is in Farads (F) and R is in Ohms (Ω). The error in identification is calculated using the SSE given by (6) but due to it being relatively small, Sum of Absolute Errors (SAE) as indicated by (22) is also used for a clearer comparison and acting as another benchmark for comparison of the two algorithms. It is clearly noted that the data fitting error generated through the integer model was significantly high compared to that of a fractional order model. This can be graphically seen in Fig.4. It is also clear that the fractional-order model is able to represent

TABLE I
IDENTIFIED INTEGER MODEL PARAMETERS ($\alpha = 1$) FROM CHARGING DATA

Supercapacitors		GWO				TRR			
Brand	Capacitance	C_α (F)	R_s (Ω)	SAE	SSE ($\times 10^{-3}$)	C_α (F)	R_s (Ω)	SAE	SSE ($\times 10^{-3}$)
AVX	0.47F	0.65	1.73	1.42	11.17	0.65	1.73	1.42	11.17
	1F	1.06	1.09	0.66	2.53	1.06	1.09	0.66	2.53
	1.5F	1.54	0.68	0.41	1.00	1.54	0.68	0.41	1.00
Eaton	0.47F	0.26	43.56	12.20	1096.15	0.26	43.55	12.21	1096.15
	1F	0.70	19.47	4.63	162.84	0.70	19.47	4.63	162.84
	1.5F	0.81	15.68	5.06	183.92	0.81	15.67	5.06	183.92
Kemet	0.47F	0.33	30.05	5.66	288.89	0.33	30.04	5.66	288.89
	1F	0.95	4.39	0.81	7.37	0.95	4.39	0.81	7.37
	1.5F	1.18	0.84	0.46	1.22	1.18	0.83	0.46	1.22

TABLE II
IDENTIFIED INTEGER MODEL PARAMETERS ($\alpha = 1$) FROM DISCHARGING DATA

Supercapacitors		GWO				TRR			
Brand	Capacitance	C_α (F)	R_s (Ω)	SAE	SSE ($\times 10^{-3}$)	C_α (F)	R_s (Ω)	SAE	SSE ($\times 10^{-3}$)
AVX	0.47F	0.57	1.41	0.89	4.52	0.57	1.41	0.89	4.52
	1F	0.92	0.77	0.48	1.33	0.92	0.77	0.48	1.33
	1.5F	1.30	0.58	0.31	0.54	1.30	0.58	0.31	0.54
Eaton	0.47F	0.26	43.15	12.39	1124.41	0.26	43.14	12.40	1124.41
	1F	0.73	17.05	4.63	161.66	0.73	17.05	4.63	161.66
	1.5F	0.74	17.30	5.06	183.53	0.74	17.30	5.06	183.53
Kemet	0.47F	0.35	25.91	5.63	284.47	0.35	25.91	5.63	284.47
	1F	0.86	5.84	0.81	6.91	0.86	5.84	0.81	6.91
	1.5F	1.02	0.98	0.38	0.85	1.02	0.98	0.38	0.85

TABLE III
IDENTIFIED FRACTIONAL-ORDER MODEL PARAMETERS FROM CHARGING DATA

Supercapacitors		GWO					TRR				
Brand	Capacitance	C_α (F)	R_s (Ω)	α	SAE	SSE ($\times 10^{-3}$)	C_α (F)	R_s (Ω)	α	SAE	SSE ($\times 10^{-3}$)
AVX	0.47F	0.46	0.00	0.89	0.36	0.74	0.46	0.00	0.89	0.36	0.74
	1F	0.76	0.00	0.90	0.20	0.29	0.76	0.01	0.90	0.20	0.29
	1.5F	1.14	0.01	0.91	0.09	0.05	1.15	0.03	0.91	0.09	0.05
Eaton	0.47F	0.04	0.01	0.46	0.43	1.07	0.04	0.00	0.46	0.43	1.07
	1F	0.08	0.23	0.43	0.54	1.63	0.08	0.00	0.42	0.54	1.62
	1.5F	0.10	0.00	0.43	0.26	0.52	0.10	0.00	0.43	0.26	0.52
Kemet	0.47F	0.08	10.00	0.60	1.95	27.89	0.08	9.98	0.60	1.95	27.89
	1F	0.64	2.93	0.88	0.45	2.39	0.63	2.90	0.87	0.46	2.39
	1.5F	0.92	0.14	0.92	0.09	0.06	0.92	0.13	0.92	0.09	0.06

a dynamic system more accurately compared to the integer counterpart. For comparison, SAEs generated by the integer and fractional order models using the TRR algorithm for each SC are shown in Fig. 5 (a) and (b) for charging and discharging respectively. This further highlighted the fact that fractional order model generated lesser error. The use of fractional based model is more beneficial in reducing the propagating CDR error for a system with subsequent charging and discharging cycles.

$$\text{SAE} = \sum_i |(f(x)_i)| \tag{22}$$

Moreover, results from tables also depict that the parameters during the charging of the SC are different from the discharging ones. This is somewhat attributed to the SCs complex internal structure, electrochemical processes and different phenomena related to it. The transient behavior also describes the fractional impedance parameters which needs to be investigate with two sets of parameters, charging and discharging. Fig.6 shows the Eaton 1.5F SC charging and discharging behavior

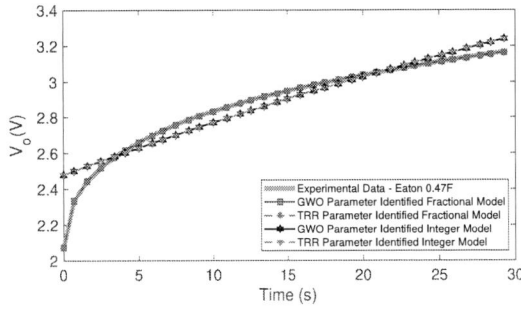

Fig. 4. The charge curve of a 0.47F Eaton Super-Capacitor when the input of 5V is supplied

models' responses obtained from charging and discharging data. The parameters obtained through charging data only replicates the behavior of the charging cycle when the input voltage is 5V step and deviates during the discharging cycle when the input voltage is 0V. Same happens for the discharging which only matches the discharging cycle data. So, there

TABLE IV
IDENTIFIED FRACTIONAL-ORDER MODEL PARAMETERS FROM DISCHARGING DATA

Supercapacitors		GWO					TRR				
Brand	Capacitance	C_α (F)	R_s (Ω)	α	SAE	SSE ($\times 10^{-3}$)	C_α (F)	R_s (Ω)	α	SAE	SSE ($\times 10^{-3}$)
AVX	0.47F	0.44	0.00	0.92	0.16	0.18	0.44	0.00	0.92	0.16	0.18
	1F	0.74	0.00	0.93	0.18	0.26	0.74	0.00	0.93	0.18	0.26
	1.5F	1.04	0.00	0.93	0.08	0.04	1.04	0.00	0.93	0.08	0.04
Eaton	0.47F	0.04	0.00	0.46	0.50	1.54	0.04	0.00	0.46	0.50	1.54
	1F	0.10	0.34	0.45	0.64	2.34	0.10	0.45	0.45	0.64	2.34
	1.5F	0.09	0.00	0.43	0.38	1.12	0.09	0.00	0.43	0.38	1.12
Kemet	0.47F	0.10	8.88	0.63	1.98	28.49	0.09	8.80	0.62	1.99	28.48
	1F	0.51	3.68	0.84	0.46	2.20	0.51	3.68	0.84	0.46	2.20
	1.5F	0.81	0.25	0.93	0.07	0.04	0.81	0.23	0.93	0.07	0.04

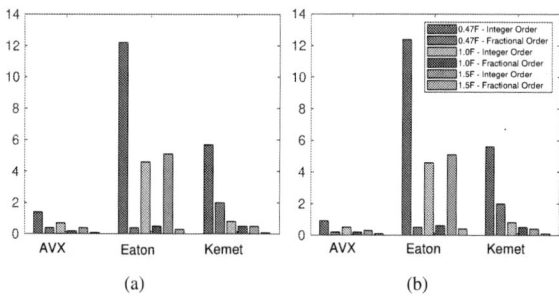

Fig. 5. Error comparison for (a) charging model, (b) discharging model.

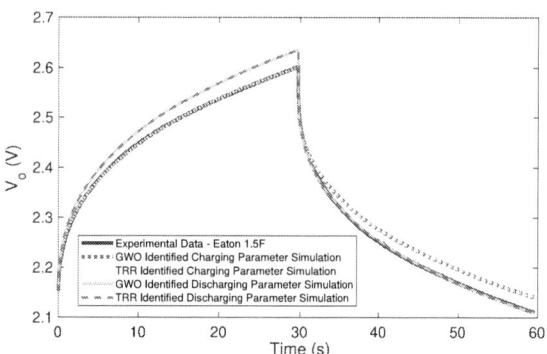

Fig. 6. The charge and discharge curve of a 1.5F Eaton Super-Capacitor

$$\log\ MSE = \log \left(\frac{1}{n} \sum_{i=1}^{n} (f(x)_i)^2 \right) \qquad (23)$$

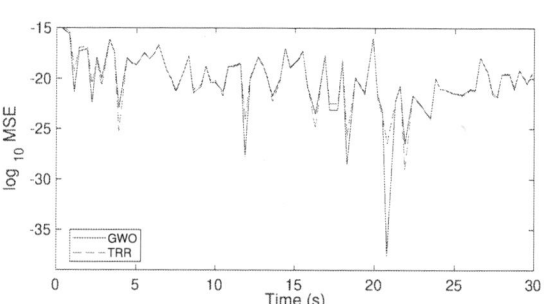

Fig. 7. Logarithmic MSE plot for 1.5F Kemet-Charging time

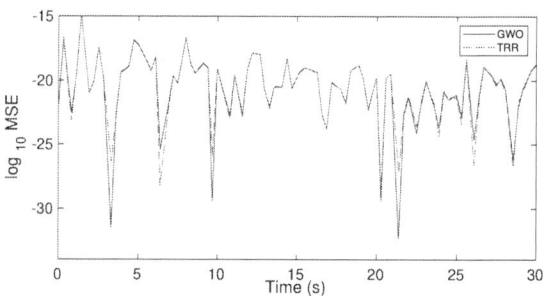

Fig. 8. Logarithmic MSE plot for 1.5F AVX-Discharging time

is a need of expanded fractional-order model for accurately depicting the transient characteristics in SC. The charge-discharge situation and dependence of fractional impedance parameters give an indication for future investigation, that best fit the experimental behavior with model response.

Based on the results generated, the two parameter identification techniques proposed in this paper offer a similar result. Due to minimal difference between identified models, the logarithmic Mean Square of Error (MSE) based on equation (23) is graphed to depict the difference between the two identified models and are clearly differentiated from Figs. 7 and 8. This indicates that better approximation could be made using the evolving meta-heuristic techniques compared to that of conventional technique.

Further evaluation based on convergence, highlights the fact that TRR has a much faster convergence compared to GWO. Faster convergence is likely indication of a faster simulation time. As seen from the Fig. 9, TRR takes less number of iterations to reach a minimum SSE error compared to GWO.

V. CONCLUSIONS

Fractional-order impedance parameters of SC have been approximated using Trust Region Reflective Optimizer and Grey Wolf Optimization Algorithm and then these methods have been compared to each other. The results show that GWO yields similar performance to conventional TRR. The estimation has also been validated experimentally with different SC

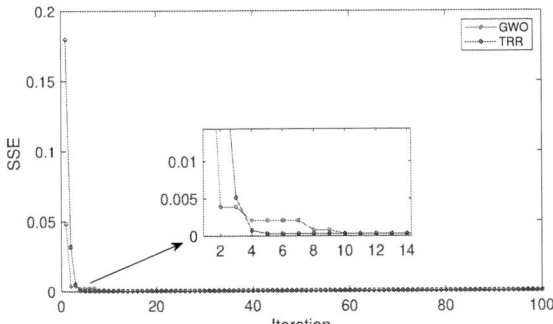

Fig. 9. Convergence Curve for Charging data of 1F AVX

brands which have values 0.47F, 1F and 1.5F measured when a voltage step input was applied to the system. It was also identified that charging and discharging parameters differed from each other for the same SC due to system properties such as relaxation phenomenon attached to it. Furthermore, it was discovered that consideration of initial voltage can better reflect the low resistance property of SC. Considering initial conditions, a simple R-C electrical impedance model was compared for its integer and fractional order behavior. Based on the investigation, it is established that fractional order of a model gives extra flexibility in modeling and meta-heuristic optimization technique is likely to perform equal or better compared to traditional methods for SC. However, due to randomized behavior of meta-heuristic optimization, convergence is slow and may take a longer simulation time. The variation of fractional impedance parameters with different inputs and initial voltages for a same valued SC could be the scope for future study.

REFERENCES

[1] L. Zhang, X. Hu, Z. Wang, F. Sun, and D. G. Dorrell, "A review of super-capacitor modeling, estimation, and applications: A control/management perspective," *Renewable and Sustainable Energy Reviews*, vol. 81, pp. 1868–1878, Jan. 2018.

[2] A. Javaid, "Activated carbon fiber for energy storage," in *Activated Carbon Fiber and Textiles*. Elsevier, 2017, pp. 281–303.

[3] B. E. Conway, *Electrochemical Supercapacitors: Scientific Fundamentals and Technological Applications*. Springer US, 1999.

[4] A. Oukaour, N. Omar, H. Gualous, A. Rachid, P. Van Den Bossche, and J. Van Mierlo, "Electrical double-layer capacitors diagnosis using least square estimation method," *Electric Power Systems Research*, vol. 117, pp. 69–75, 2014.

[5] J. Cao and A. Emadi, "A new battery/UltraCapacitor hybrid energy storage system for electric, hybrid, and plug-in hybrid electric vehicles," *IEEE Transactions on Power Electronics*, vol. 27, no. 1, pp. 122–132, Jan. 2012.

[6] M. Pucci, G. Vitale, G. Cirrincione, and M. Cirrincione, "Parameter identification of a double-layer-capacitor 2-branch model by a least-squares method," in *IECON 2013-39th Annual Conference of the IEEE Industrial Electronics Society*. IEEE, 2013, pp. 6770–6776.

[7] A. M. van Voorden, L. M. R. Elizondo, G. C. Paap, J. Verboomen, and L. van der Sluis, "The application of super capacitors to relieve battery-storage systems in autonomous renewable energy systems," in *2007 IEEE Lausanne Power Tech*. IEEE, Jul. 2007.

[8] J. B. Goodenough, H. D. Abruna, and M. V. Buchanan, "Basic research needs for electrical energy storage. report of the basic energy sciences workshop on electrical energy storage, april 2-4, 2007," 4 2007.

[9] R. Martin, J. J. Quintana, A. Ramos, and I. de la Nuez, "Modeling electrochemical double layer capacitor, from classical to fractional impedance," in *MELECON 2008 - The 14th IEEE Mediterranean Electrotechnical Conference*. IEEE, May 2008.

[10] N. Bertrand, J. Sabatier, O. Briat, and J.-M. Vinassa, "Fractional non-linear modelling of ultracapacitors," *Communications in Nonlinear Science and Numerical Simulation*, vol. 15, no. 5, pp. 1327–1337, May 2010.

[11] T. J. Freeborn, B. Maundy, and A. S. Elwakil, "Measurement of supercapacitor fractional-order model parameters from voltage-excited step response," *IEEE Journal on Emerging and Selected Topics in Circuits and Systems*, vol. 3, no. 3, pp. 367–376, Sep. 2013.

[12] T. J. Freeborn and A. Elwakil, "Variability of supercapacitor fractional-order parameters extracted from discharging behavior using least squares optimization," pp. 1–4, 2017.

[13] I. Petras, *Fractional-Order Nonlinear Systems: Modeling, Analysis and Simulation*. Springer, 2011.

[14] R. Negroiu, P. Svasta, C. Ionescu, and A. Vasile, "Investigation of Supercapacitor's Impedance Based on Spectroscopic Measurements."

[15] N. Ber, J. Sabatier, O. Briat, and J.-M. Vinassa, "Embedded fractional nonlinear supercapacitor model and its parametric estimation method," *IEEE Transactions on Industrial Electronics*, vol. 57, no. 12, pp. 3991–4000, Dec. 2010.

[16] T. F. Coleman and Y. Li, "An interior trust region approach for nonlinear minimization subject to bounds," *SIAM Journal on optimization*, vol. 6, no. 2, pp. 418–445, 1996.

[17] S. Mirjalili, S. M. Mirjalili, and A. Lewis, "Grey wolf optimizer," *Advances in engineering software*, vol. 69, pp. 46–61, 2014.

[18] M. Ahsan and M. A. Choudhry, "System identification of an airship using trust region reflective least squares algorithm," *International Journal of Control, Automation and Systems*, vol. 15, no. 3, pp. 1384–1393, 2017.

[19] T. F. Coleman and Y. Li, "A reflective newton method for minimizing a quadratic function subject to bounds on some of the variables," *SIAM Journal on Optimization*, vol. 6, no. 4, pp. 1040–1058, 1996.

[20] M. A. Branch, T. F. Coleman, and Y. Li, "A subspace, interior, and conjugate gradient method for large-scale bound-constrained minimization problems," *SIAM Journal on Scientific Computing*, vol. 21, no. 1, pp. 1–23, 1999.

[21] *SCM Series - Series Connected SuperCapacitor Modules*, AVX Corporation. [Online]. Available: http://datasheets.avx.com/AVX-SCM.pdf

Tuning PID Controller Using Hybrid Genetic Algorithm Particle Swarm Optimization Method for AVR System

Faouzi Aboura

Laboratoire des Systèmes Electriques
&Industriels (LSEI)
Université des Sciences et Technologie
Houari Boumediéne
Bab Ezzouar 16123 , Algeria
faboura@usthb.dz

Abstract—The proportional-integral-derivative (PID) controller is widely used in industrial applications, one of these important application is the Automatic Voltage Regulator (AVR), due to the necessity of using controller to avoid instability of the system. In our paper a comparison between algorithms Particle Swarm Optimization (PSO) and Genetic Algorithm (GA) and Hybrid Genetic Algorithm Particle Swarm Optimization (HGAPSO) is proposed with their characteristics and performance analysis to find an optimum parameters of the PID controller, a new objective function is also proposed to take into account the relation between the performance criteria's .

Keywords—PID controller, AVR system, objective function, optimization, GA, PSO, HGAPSO.

I. INTRODUCTION

The Proportional-Integral-Derivative (PID) controller has been widely used in industrial process, the evidence of their popularity lies in the fact that even today, 90% of the industry employs PID controller .This kind of controller is popular due to simple structure, reliable in operation and robust in performances. One key factor for their success is that they act in the processes under control in a manner closely similar to human's natural responses to outside stimuli, that is the combined effects of spontaneity (proportional action), post training (integral action) and projection into future (derivative action),[1]. This controller is used in our case in the automatic voltage regulator (AVR) that is used with the synchronous generator in order to maintain constant terminal voltage of the generator under normal operating conditions at various load levels [2]. The AVR controls the terminal voltage by adjusting the exciter voltage of the generator. The tuning parameters of PID controller is not easy especially with a commonly Ziegler-Nichols method that is the most common practical control method,[3].It is difficult to achieve the best performance of the system by using Ziegler-Nichols method, and the designer has to depend on his experience for obtaining the best performance. For the past three decades lots of research has been reported on in the intelligent controllers, evolutionary based approaches have received the increasing attention of engineers dealing with problems. GA (Genetic Algorithm) and PSO (Particle Swarm Optimization) have been used in [4]-[8], to enhance optimal tuning of PID controller for AVR (Automatic Voltage Regulator), other methods have been applied recently [9-10]. GA mainly depends on the concept of survival of the fittest. The PSO is motivated by the social behavior of bird flocking and fish schooling. In our paper we used the two previous methods and a new method called based Hybrid Genetic Algorithm Particle Swarm Optimization (HGAPSO) to integrate the advantages of GA and PSO algorithms to find optimal parameters for controller in AVR system, we also analyze performances for each of the three optimization methods.

II. LINEARIZED MODEL OF AN AVR SYSTEM

The AVR system is composed by four main components, namely amplifier, exciter, generator, and sensor as shown in Fig.1, each of these have its own transfer function given in TABLE I.

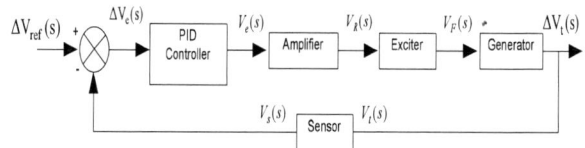

Fig. 1. Block diagram of AVR system with PID controller.

TABLE I.   TRANSFER FUNCTION OF AVR COMPONENTS

Component	Transfer Function
Amplifier	$\dfrac{10}{1+0.1s}$
Exciter	$\dfrac{10}{1+0.5s}$
Generator	$\dfrac{0.7}{1+s}$
Sensor	$\dfrac{1}{1+0.001s}$

The PID controller is in parallel type and his transfer function is:

$$C(s) = K_p + \frac{K_i}{s} + K_d s \qquad (1)$$

-The transfer function of the amplifier model is:

978-1-5386-7688-2/19 $31.00 © 2019 IEEE   570

$$\frac{V_R(s)}{V_e(s)} = \frac{K_A}{1 + \tau_A s} = \frac{10}{1 + 0.1s} \qquad (2)$$

Where K_A are in the range of 10 to 400, and in our work is taken equal to 10, the amplifier time constant τ_A is very small ranging from 0.02 to 0.1 s, and in our work is taken equal to 0.1.

-The transfer function of the exciter model is:

$$\frac{V_F(s)}{V_R(s)} = \frac{K_E}{1 + \tau_E s} = \frac{10}{1 + 0.5s} \qquad (3)$$

Where the gain K_E are in the range of 10 to 400, and in our work is taken equal to 10, the single time constant τ_E is from 0.5 to 1.0 s, and in our work is taken equal to 0.5.

-The transfer function of the generator model is:

$$\frac{V_t(s)}{V_F(s)} = \frac{K_G}{1 + \tau_G s} = \frac{0.7}{1 + s} \qquad (4)$$

Where K_G are in the range of 0.7 to 1.0, and in our work is taken equal to 0.7 and constant τ_G vary in function of load from 1.0 to 2.0 s, and in our work is taken equal to 1.0.

-The transfer function of the sensor model is:

$$\frac{V_s(s)}{V_t(s)} = \frac{K_R}{1 + \tau_R s} = \frac{1}{1 + 0.001s} \qquad (5)$$

The sensor is modeled by a simple first-order transfer function, where $K_R = 1$ and $\tau_R = 0.001$.

III. PERFORMANCE ESTIMATION OF PID CONTROLLER

In our paper we propose a new objective function to integrate the integrated of time-weighted-squared-error (ITSE) in the objective function given in [8] ,the new objective function integrate performance criteria in the time domain and include, the overshoot M_p ,rise time t_r ,settling time t_s and steady-state error E_{ss}.

This new form a objective or fitness function is based on the fact to combine to similar parameter in term of magnitude and nature of parameter.

The new objective function used is:

$$f = \sqrt{((M_p * 100)^2 + (t_s + t_r)^2 + (ITSE + E_{ss})^2} \qquad (6)$$

IV. DESCRIPTION AND IMPLEMENTATION OF PARTICLE SWARM OPTIMIZATION

Particle swarm optimization (PSO) is firstly introduced by Kennedy and Eberhart [11] in 1995. This algorithm searches the space of an objective function by adjusting the trajectories of individual agents, called particles, as these trajectories form piecewise paths in a quasi-stochastic manner, [12]. PSO is initialized with a group of random particles (solutions), k_p, k_i, k_d and then searches for optima by updating generations.

The basic flowchart of particle swarm optimization is presented in Fig. 2.

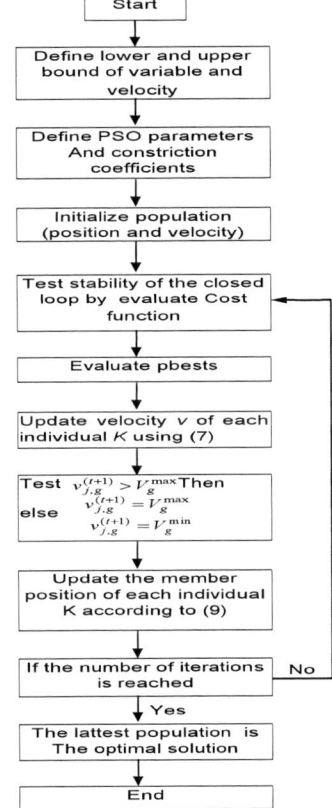

Fig. 2. Algorithm of particle swarm optimization.

The simulation step of the particle swarm optimization (PSO) is shown as follows.

Step 1) Specify the lower and upper bounds of the three controller parameters and initialize randomly the individuals of the population including searching points, velocities, *pbests*, and *gbest* .Acceleration constants c_1 and c_2 were set to be 2.05. These constants represent the weighting of the stochastic acceleration terms that pull each particle toward and positions.

Step 2) For each initial individual of the population, test the closed-loop system stability and calculate the values of the four performance criteria in the time domain, namely M_p, E_{ss}, t_r, and t_s.

Step 3) Calculate the evaluation value of each individual in the population using the evaluation function given by (6).

Step 4) Compare each individual's evaluation value with its pbests . The best evaluation value among the *pbests* is denoted as *gbest* .

Step 5) Modify the member velocity v of each individual K according to (7)

$$v_{j,g}^{(t+1)} = w.v_{j,g}^{(t)} + c_1^* rand()^* (pbest_{j,g} - x_{j,g}^{(t)}) +$$
$$c_2^* rand()^* (gbest_{j,g} - x_{j,g}^{(t)}) \qquad (7)$$
$$j = 1,2,..........,n$$
$$g = 1,2,3$$

Where the value of w that represent inertia weight is set by (8).

$$w = w_{max} - \frac{w_{max} - w_{min}}{iter_{max}} \times iter \qquad (8)$$

Step 6) If $v_{j,g}^{(t+1)} > V_g^{max}$,then $v_{j,g}^{(t+1)} = V_g^{max}$ and if $v_{j,g}^{(t+1)} < V_g^{min}$,then $v_{j,g}^{(t+1)} = V_g^{min}$

Step 7) Modify the member position of each individual K according to (9)

$$k_{j,g}^{(t+1)} = k_{j,g}^{(t)} + v_{j,g}^{(t+1)}$$
$$k_g^{min} \leq k_{j,g}^{(t+1)} \leq k_g^{max} \qquad (9)$$

Where k_g^{min} and k_g^{max} represent the lower and upper bounds, respectively, of member, g of the individual K.

Step 8) If the number of iterations reaches the maximum, then go to Step 9. Otherwise, go to Step 2.

Step 9) The individual that generates the latest $gbest$ is the optimal controller parameter.

V. DESCRIPTION AND IMPLEMENTATION OF GENETIC ALGORITHM

The Genetic Algorithm (GA) is firstly introduced by John Holland in 1975 [13]. GA is a class of stochastic algorithm based on the biological evolution in the natural world. The technological process employs three operators: 1) selection and reproduction, 2) crossover and 3) mutation.
The basic flowchart of genetic algorithm is presented in Fig. 3.

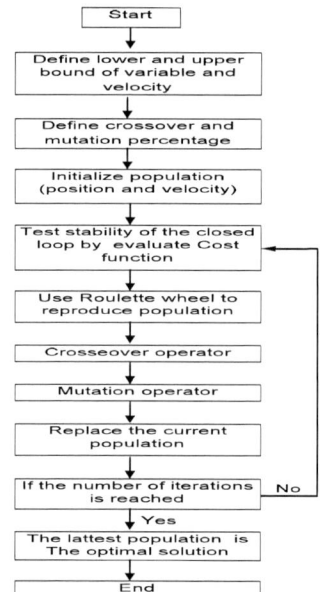

Fig. 3. Algorithm of Genetic Algorithm optimization.

The simulation step of the genetic algorithm optimization (GA) is shown as follows.

Step 1) Specify the lower and upper bounds of the three controller parameters and define crossover and mutation percentage and initialize randomly the individuals of the population. In our case crossover percentage $P_c = 0.70$, mutation percentage $P_m = 0.10$.

Step 2) For each initial individual of the population, test the closed-loop system stability and calculate the values of the four performance criteria in the time domain, namely M_p, E_{ss}, t_r, and t_s.

Step 3) Calculate the evaluation value of each individual in the population using the evaluation function given by (6).

Step 4) Reproduce population using a probabilistic roulette wheel method.

Step 5) Implement crossover operation on the reproduced chromosomes and execute mutation operation.

Step 6) If the number of iterations reaches the maximum, then go to Step 7. Otherwise, go to Step 2.

Step 7) The individual that generates the latest population is the optimal controller parameter solution.

VI. DESCRIPTION AND IMPLEMENTATION OF HYBRID GENETIC ALGORITHM PARTICLE SWARM OPTIMIZATION

The HGAPSO algorithm combines the advantages of swarm intelligence of the PSO algorithm and the natural selection mechanism of the genetic algorithm in order to increase the number of highly evaluated agents at each iteration step.
The basic flowchart of HGAPSO used in this study is given in Fig. 4.

First, multiple solutions are generated randomly as initial population and objective function values are evaluated for each solution. After the evaluation is done, the population is divided into two subpopulations. One of these subpopulations is updated by the GA operation, while the other is updated by the PSO operation.
New solutions created by each operation are combined in the next generation, and non-dominated solutions in the combined population are archived. The archive data is shared between the GA and PSO, i.e., non-dominated solutions created by the PSO can be used as parents in GA, while non-dominated solutions created by GA can be used as global guides in PSO.

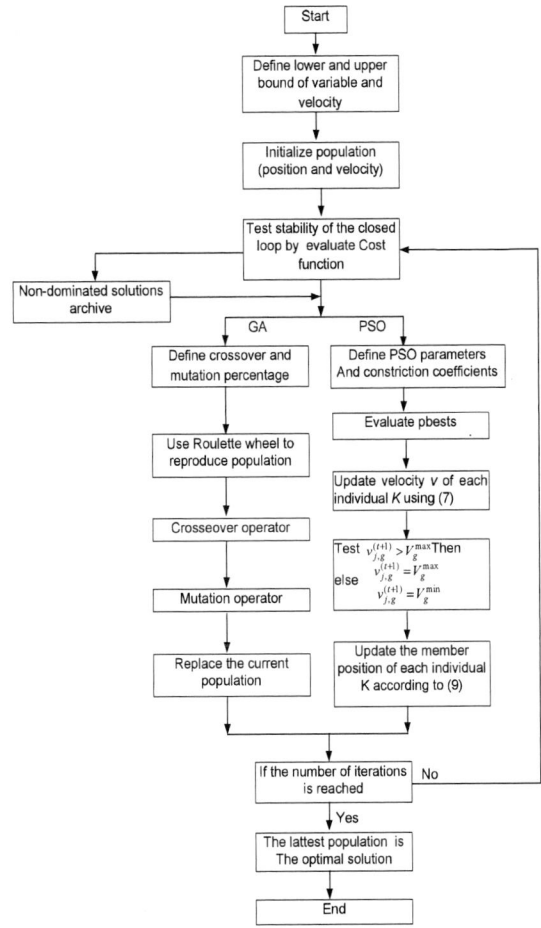

Fig. 4. Algorithm of Hybrid Genetic Algorithm Particle Swarm Optimization .

VII. NUMERICAL RESULTS AND DISCUSSION

In our paper the maximum number of iterations for the three methods is 100, and the number of population is 100.
The curves corresponding to the fitness cost as a function of iterations number obtained for the three optimization techniques are presented in Fig. 5(a)–(c).

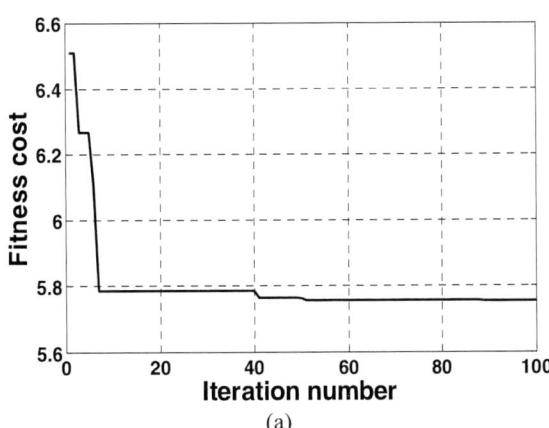

(a)

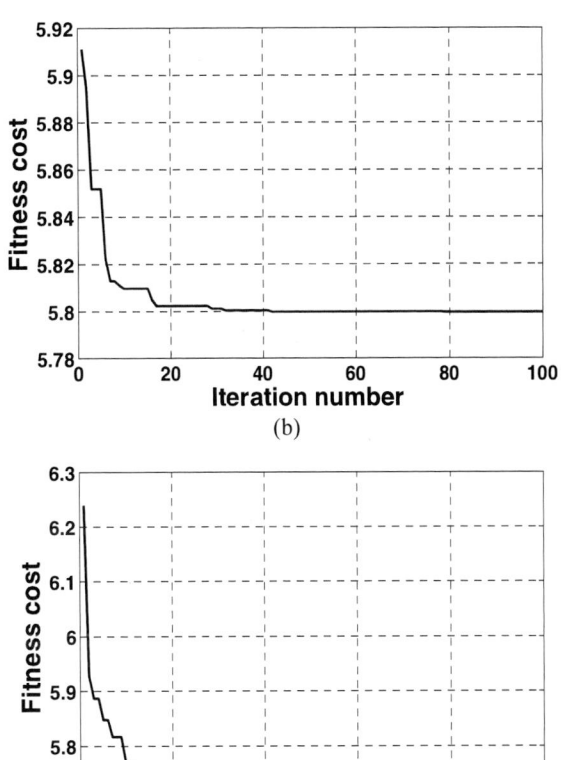

(b)

(c)

Fig. 5. Fitness cost as a function of iterations number. (a) Genetic Algorithm (GA). (b) Particle Swarm Optimization (PSO). (c) Hybrid Genetic Algorithm Particle Swarm Optimization (HGAPSO).

From Fig. 5(a)–(c) ,first we compare GA and PSO ,we can deduce that the fitness cost of GA is lower than PSO ,but the PSO is faster than the GA and reach the minimum cost at

the 40 iteration number, when the GA reach its minimum cost at the 50 iteration number.
The HGAPSO have the advantages described above of the time to reaching the minimum cost value and also the value of this cost value. This algorithm is fastest method and has minimum cost; all the values of the controller parameters and performances are summarized to the Table.2.

TABLE II. PARAMETERS OF PID CONTROLLER AND PERFORMANCES FOR EACH OPTIMIZATION METHODS.

Methods	GA	PSO	HGAPSO
Fitness cost	5.7553	5.7994	5.7507
Kp	3.1613	4.5920	3.3047
Ki	0.0189	0.2020	0.1053
Kd	0.6246	0.9507	0.6548
Ess	0.0170	0.0154	0.0108
Mp	0.0563	0.0018	0.0062
ts	2.9315	0.9238	0.7385
tr	4.9984	4.9984	4.9983

978-1-5386-7688-2/19 $31.00 © 2019 IEEE

The figure.6 represents the terminal voltage step response of the AVR system with the three controllers parameters obtained by the GA, PSO and HGAPSO.

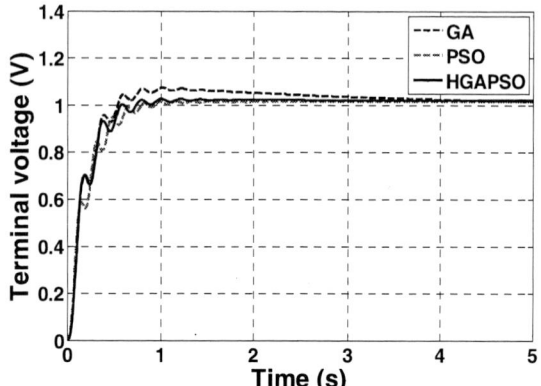

Fig. 6. Terminal voltage step response of the AVR system with the three optimization methods.

VIII. CONCLUSION

This paper study tuning the proportional-integral-derivative (PID) controller with three optimization methods GA, PSO, and a Hybrid version of these two algorithms called HGAPSO. We developed also a new objective function based in the combination of the similar parameter in term of magnitude and nature.

We obtained in our work for AVR system one of the most benefit of the GA method that consist have fitness cost lower than PSO and also that PSO method is faster than the GA, these two characteristics are included in hybrid HGAPSO. Although HGAPSO algorithm is more stable and effective, we prepare further analysis in its internal parameters to study their effect on efficiency.

REFERENCES

[1] K. M. Elbayomy, J. Zongxia and Z. Huaqing, "PID controller optimization by GA and its performances on the electro-hydraulic servo control system," Chinese Journal of Aeronautics, 21(4), 378–384, 2008.

[2] A. H. M. S. Ula and A. R. Hasan, "Design and implementation of a personal computer based automatic voltage regulator for a synchronous generator," IEEE Trans. Energy Convers., vol. 7, no. 1, pp. 125–131, Mar. 1992.

[3] J. G. Ziegler and N. B. Nichols, "Optimum Settings for Automatic Controllers," Transactions of the ASME, Vol. 64, 1942, pp. 759-768.

[4] R. A. Krohling and J. P. Rey, "Design of optimal disturbance rejection PID controllers using genetic algorithms," IEEE Trans. Evol. Comput., vol. 5, no. 1, pp. 78–82, Feb. 2001.

[5] J. Zhang, J. Zhuang, H. Du, and S. Wang, "Self-organizing genetic algorithm based tuning of PID controllers," Inform. Sci., vol. 179, no.7, pp. 1007–1018, Mar. 2009.

[6] D. Devaraj and B. Selvabala, "Real-coded genetic algorithm and fuzzy logic approach for real-time tuning of proportional-integral-derivative controller in automatic voltage regulator system," IET Generation Transmission Distribution, vol. 3, no. 7, pp. 641–649, 2009.

[7] V. Mukherjee and S. P. Ghoshal, "Intelligent particle swarm optimized fuzzy PID controller for AVR system," Electr. Power Syst. Res., vol. 77, no. 12, pp. 1689–1698, 2007.

[8] Z. Gaing , "A particle swarm optimization approach for optimum design of PID controller in AVR system," IEEE Trans. Energy Convers., 2004, 19, (2), pp. 384–391.

[9] S. Ekinci and B. Baran Hekimoğlu, "Improved Kidney-Inspired Algorithm Approach for Tuning of PID Controller in AVR System," IEEE Access., 2019, 7, pp. 39935-39947.

[10] N. Nahas, M. Abouheaf, A.M. Sharaf and W. Gueaieb, "A Self-Adjusting Adaptive AVR-LFC Scheme for Synchronous Generators," IEEE Trans.on Power Systems (Early Access)., 2019, pp. 1-1.

[11] J. Kennedy and R. Eberhart, "Particle swarm optimization," in Proc. IEEE Int. Conf. Neural Networks, 1995, pp.1942-1948.

[12] X-She .Yang, "Engineering optimization, an introduction with metaheuristic applications", Published by John Wiley & Sons, Inc., Hoboken, New Jersey, (2010).

[13] J.H.Holland, "Adaptation in Natural Artificial System", Published by Ann Arbor, the University of Michigan Press, 1975.

978-1-5386-7688-2/19 $31.00 © 2019 IEEE

Using Deep Learning Techniques for South African Power Distribution Networks Load Forecasting

Sibonelo Motepe
Faculty of Engineering and the Built Environment
University of Johannesburg
Johannesburg, South Africa
djscvii@gmail.com

Ali N. Hasan
Faculty of Engineering and the Built Environment
University of Johannesburg
Johannesburg, South Africa
alin@uj.ac.za

Bhekisipho Twala
Faculty of Engineering
University of South Africa
Johannesburg, South Africa
twalab@unisa.ac.za

Riaan Stopforth
Stopforth Mechatronics, Robotics and Research Lab, School of Engineering
University of South Africa
Durban, South Africa
stopforth.research@gmail.com

Abstract— Load forecasting has many benefits for utilities. Artificial Intelligence (AI) has been seen to be effective in load forecasting. Deep Learning AI techniques have been found to perform better than traditional AI techniques. The study of deep learning techniques application in South African load forecasting is in its infancy. This paper presents a study of a South African distribution substation using two deep learning techniques, deep belief networks (DBN) and long short-term memory (LSTM), to address this. The impact of temperature and cleaning up the loading data is also studied. It was found that an LSTM model achieved the lowest errors with a symmetric mean absolute percentage error (sMAPE) of 3.3%, an), mean absolute error (MAE) of 4.6% and root mean square error (RMSE) of 5.5%. These errors were achieved with non-cleaned data with temperature not used as a variable for the training the model. Deep learning techniques can thus be used without weather parameters to forecast distribution substation data with low errors.

Keywords—Deep Belief Networks, Long Short-Term Memory Load Forecasting, Power System Distribution, Deep Learning

I. INTRODUCTION

Load forecasting has many benefits for utilities. Very short-term load forecasting was studied, using recurrent NARX-neural network (Nonlinear Autoregressive Model with Exogenous Input), for Brazillian distribution substations load forecasting. This study was to help achieve reliable and economic power systems operations [1]. Long short-term memory (LSTM) recurrent neural networks (RNN) have been applied in long-term load forecasting to assist with utility capital budgeting [2]. With the rise of computational power and the availability of big data, the use of deep learning techniques has increased [3]. The study of AI application in South African load forecasting, even though in its infancy, is of interest. This interest is due to the multiple challenges that South Africa is faced with in the electricity supply space, which require advanced techniques to help resolve. In 2007, 2013 and recently in 2018, South Africa (SA) experienced load shedding. In [4] the study of SA power consumption was studied in an attempt to address future demand planning. In [5] adaptive neuro-fuzzy inference system (ANFIS) was used to study SA distribution load forecasting. These type of studies are important in distribution maintenance planning and can help reduce customer power down time while driving the

South African electrification programme. Deep learning techniques have been shown to be superior to standard artificial intelligence techniques in various applications, including load forecasting [6]. The study of deep learning techniques in South African load forecasting is a new concept and needs to be explored. The application of deep learning techniques in South African distribution network is also in its infancy.

The most common deep learning techniques are the RNN and convolutional neural networks (CNN), with deep belief network (DBN) also gaining popularity [7] [8]. In natural language processing and load forecasting, RNN-LSTM has been shown to outperform CNN [7] [9]. LSTM's performance is also on par with, and in other cases better than, state of the art AI techniques [6]. LSTM has been applied in domestic violence detection, load forecasting, spot electricity prices forecasting, and many other applications [7] [2] [10]. LSTM has been applied in long-term load forecasting using hourly data as opposed to monthly or annual data to improve accuracy [11]. Belgium's short-term load forecasting was conducted using LSTM. Here LSTM was compared to and found to outperform an Elman neural network [12]. was forecasted in the short-term using LSTM has been used to forecast residential loads of intelligent, flexible and interactive power systems with high renewable generation sources penetration [13]. In [14] LSTM was also applied for load forecasting in smart grids. Researchers have also started looking at other ensemble techniques of RNN in load forecasting [15] [16]. Ensemble/hybrid techniques are usually deployed to achieve improved performance from a combination of techniques as opposed to individual techniques.

DBNs deviate mostly from RNN and CNN by utilising unsupervised learning in their learning process. The utilisation of partially-unsupervised learning is a new concept in SA load forecasting. *Motepe et al.* used DBN to forecast load in a South African distribution substation that redistributes power to different loads [17]. DBN has been utilised in many application such as thermal face recognition and load forecasting [18] [1]. In [19] DBN was combined with K-shape clustering to improve load forecasting. Models with a combination of K-shape clustering and DBN were found to outperform models with neural networks combined with K-shape clustering. DBN has also been applied to predict wind

978-1-5386-7688-2/19 $31.00 © 2019 IEEE

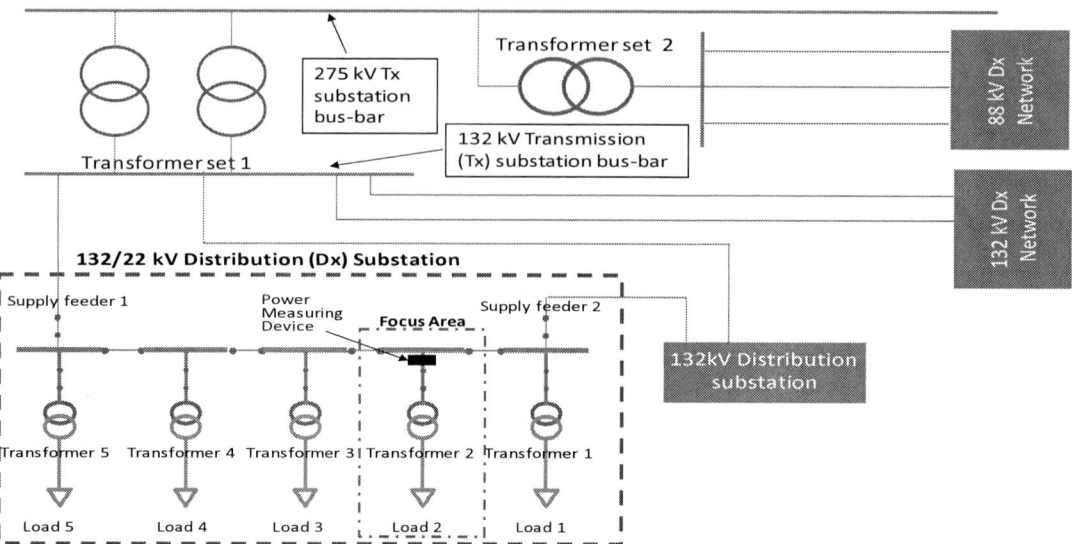

Fig. 1. Network setup where the distribution network under study is located

power from wind farms [20]. Researchers have used different training algorithms to improve DBN for load forecasting [21]. Levenberg-Marquardt (LM) algorithm was found to give the best DBN performance in comparison to gradient descent and back-propagation on medium scale data. The one-step secant algorithm was recommended for large-scale data due to LM algorithm's unacceptable computation time. Researchers have also used DBN combined with LSTM for solar power forecasting [22]. Weather parameters are usually used as training variables in load forecasting studies [23]. These parameters can sometimes not be available for use in training AI model for load forecasting.

This paper addresses the mentioned challenges by: (1) Studying the application of deep learning techniques in South African distribution load forecasting, (2) Investigating the impact of temperature on load forecasting accuracy, for cases when weather data is not available. The authors' hypothesis is that deep learning techniques can be deployed in South African distribution load forecasting with low load forecasting errors and can, therefore, help utilities plan maintenance better. The paper is structured as follows: Section II presents a background on load forecasting, Section III gives an overview of the two deep learning techniques utilized in this study, i.e. deep belief networks and long short-term memory, Section IV presents the experimental setup and results, and Section V concludes the paper.

II. DEEP BELIEF NETWORKS AND LONG SHORT-TERM MEMORY RECURRENT NEURAL NETWORKS

A. Restricted Boltzmann Machines – Deep Belief Network

Deep belief networks (DBN) were proposed by Geoffrey Hinton, with Restricted Boltzmann Machine (RBM) as a building block [24]. An RBM is defined as a generative stochastic neural network model that learns the probability distribution over the inputs [19]. An RBM's layer of neurons is binary-valued and the hidden neurons layer is Boolean [25]. A DBN is trained in two steps:

Step 1: Unsupervised learning using contrastive divergence to reduce the set of features.

Step 2: Supervised training to train an appended layer to pre-trained network in step 1.

The DBN has symmetrical and bidirectional connections between the layers, but no connection of neurons in the same layer [25]. The joint distribution over the hidden and visible units is defined as:

$$P(n,h) = \frac{e^{-E(n,h)}}{\sum_n \sum_h e^{-E(n,h)}} \quad (1)$$

with $E(n,h)$ representing the energy function, expressed as:

$$E(n,h) = -\sum_{j=1}^{k_n} \alpha_j n_j - \sum_{i=1}^{k_h} \beta_i h_i - \sum_{j=1}^{k_n} \sum_{i=1}^{k_h} h_i W_{ij} n_j \quad (2)$$

Here W_{ij} is the weight matrix of the links between the visible and the hidden unit, n_j and h_i, respectively. The visible and hidden layer's respective biases are given by α_j and β_i. The hidden and visible unit conditional probabilities are given by (3) to (6). Equation (3) and (4) are for cases when the hidden and visible units are conditionally independent. Equation (5) and (6) are for when the hidden and visible unit's values are limited between 0 and 1. *Sigmoid()* presents the logistic sigmoid function.

$$p(n|h) = \prod_j p(n_j|h) \quad (3)$$

$$p(h|n) = \prod_i p(h_i|n) \quad (4)$$

$$p(n_j = 1|h) = sigmoid(\alpha_j + \sum_{i=1}^{k_h} W_{ij} h_i) \quad (5)$$

$$p(h_j = 1|n) = sigmoid(\beta_i + \sum_{j=1}^{k_n} W_{ij} n_j) \quad (6)$$

B. Long Short-Term Memory Recurrent Neural Networks

Long Short-Term Memory (LSTM) Recurrent Neural Network (RNN) is a trusted forecasting and sequential data modelling technique [26]. LSTM was introduced to overcome the simple RNN's optimization challenge in capturing long term temporal dependencies. The LSTM has a gating mechanism to overcome the vanishing gradient challenge that RNN faces in its normal state. The LSTM memory cell has an intermediate state, h_{t-1}, and cell state, s_{t-1}, which interacts with the input, z_t, and the output from the

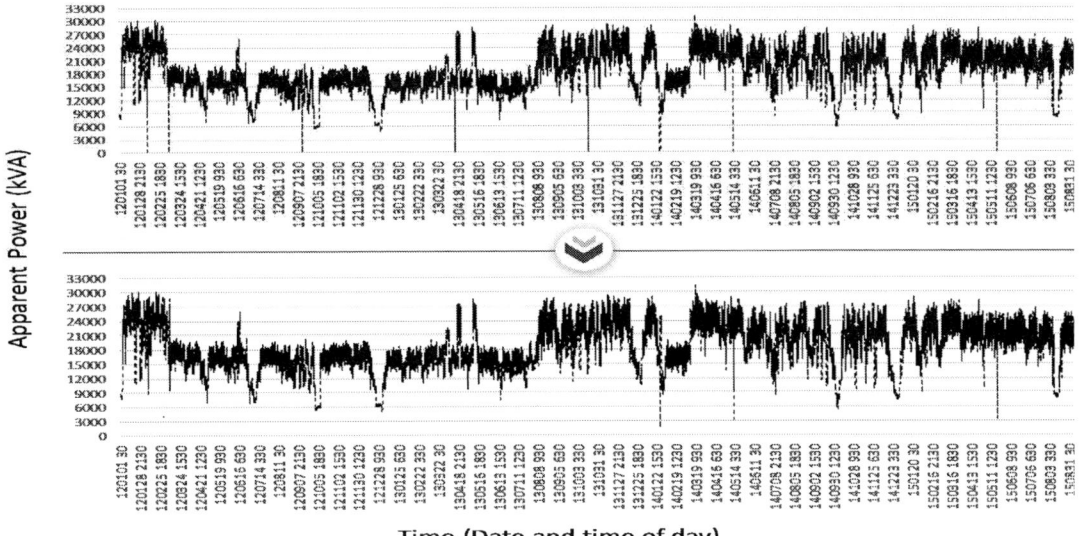

Fig. 2. A plot of uncleaned and cleaned loading data, respectively, for the period between January 2012 and September 2015

previous step to maintain, update and erase internal state vectors. Here t is the time step. The input (i_t), forget (f_t) and output (o_t) gates are respectively given by (7) to (9).

$$i_t = \sigma(W_{iz}z_t + W_{ih}h_{t-1} + b_i) \qquad (7)$$

$$f_t = \sigma(W_{fz}z_t + W_{fh}h_{t-1} + b_f) \qquad (8)$$

$$o_t = \sigma(W_{oz}z_t + W_{oh}h_{t-1} + b_o) \qquad (9)$$

The input node (g_t) is given by (10). The memory cell state (s_t) and the state (h_t) at t are given by (11) and (12) respectively.

$$g_t = \emptyset(W_{gz}z_t + W_{gh}h_{t-1} + b_g) \qquad (10)$$

$$s_t = g_t \odot i_t + s_{t-1} \odot f_t \qquad (11)$$

$$h_t = \emptyset(s_t) \odot o_t \qquad (12)$$

With W_{fz}, W_{fh}, W_{iz}, W_{ih}, W_{gz}, W_{gh}, W_{oz} and W_{oh} being the network's activation functions' corresponding inputs weight matrices, σ the sigmoid function, $\emptyset$ the tanh function and $\odot$ the element-wise multiplication.

III. EXPERIMENTS

A. Experiment setup

The study was conducted using a South African utility 132/88 kV, 5×40 MVA substation's loading data. The transmission (Tx) and distribution (Dx) setup is shown in Fig. 1. The substation under study is a part of a ring network that has two distribution substations. The stations are supplied by a transmission substation, with an interconnecting feeder between the substations. In the station under study, each of the transformers supplies power to a separate customer's point of supply. The customer has their own internal distribution configuration. Transformers 1 and 2 have the same load profile. Transformers 3 and 4 also have the same load profile. Transformer 2's loading data were selected arbitrarily for use

in the experiments. The apparent power was used for training and testing. The data were cleaned up to remove the dips in the load. The cleaned and uncleaned data plots are presented in Fig. 2. The experiments were then separated into those with cleaned up loading and non-cleaned up data. The data was logged from the transformer's 132 kV side. Temperature data was obtained from the South African weather services. Both the loading data and the temperature data were normalized to be between 0 and 1 using (13):

$$x_{norm} = \frac{x - x_{min}}{x_{max} - x_{min}} \qquad (13)$$

where x is the input variable, x_{norm} is the normalized input variable, x_{max} and x_{min} are the maximum and minimum input variable values, respectively. The data were further divided into two more groups for the experiments. These were experiments where temperature is not used as an input variable in the model training and testing, and the other experiments where it is used. Table I gives the input variables used in the experiments.

TABLE I.   EXPERIMENT INPUT VARIABLES

Input variables group	Input variables
Group A	Loading t-2 years Corresponding time of day Peak or non-peak period indicator Loading t-1 year Loading t-2 weeks
Group B	Loading data t-2 years Temperature t-2 years Load corresponding time of day Peak or non-peak period indicator Loading t-1 year Temperature t-1 year Loading t-2 weeks Temperature t-2 weeks

B. Models' Performance Results Discussion

The experiments were conducted for the winter period. This is due to South Africa having less rainfall and thunderstorms in this period, which make it more ideal to execute maintenance. The performance of each model was measured using three error measurements the symmetric mean absolute percentage error (sMAPE), mean absolute error (MAE) and root mean square error (RMSE) given respectively by (14) to (16).

$$sMAPE = \frac{2}{N}\sum_{k=1}^{N}\frac{|Y_k - T_k|}{|Y_k| + |T_k|} \tag{14}$$

$$MAE = \frac{\sum_{k=1}^{N}|Y_k - T_k|}{N} \tag{15}$$

$$RMSE = \sqrt{\frac{\sum_{k=1}^{N}(Y_k - T_k)^2}{N}} \tag{16}$$

In (14) to (16) Y_k is the forecasted variable's k^{th} value, T_k is the target variable k^{th} value and N is the total number of forecasts. The results for a 2-week ahead DBN and LSTM load forecasts are summarized in Table II and Table III, respectively. The number of the hidden units and hidden layers were adjusted in the four different experimental setups for the DBN and LSTM, respectively. It was observed that the lowest errors were attained by an LSTM model trained and tested with uncleaned data, without temperature as an input variable. The LSTM models achieved lower errors at a higher number of hidden layers. The DBN models were seen to have a reduction in errors with an increase in the number of hidden units. This reduction in errors was observed to stop after a certain number of hidden units is reached. The errors then increased with the number of hidden units until they stagnated. The lowest achieved errors for the DBN were an sMAPE of 0.074553, MAE of 0.051886 and an RMSE of 0.06325. This performance was also achieved with uncleaned loading data

and without temperature as an input variable, i.e. with input variable group A. The lowest achieved errors were an sMAPE of 0.065859, MAE of 0.04598 and an RMSE of 0.055058 with an LSTM model.

IV. CONCLUSIONS

The paper presented the study of deep learning techniques application in South African distribution network's load forecasting. A real South African utility's distribution substation apparent power was used for a case study. Two sophisticated deep learning techniques, deep belief networks (DBN) and long short-term memory, were used in this study. The impact of temperature and loading data clean-up was also studied. The models were tested for a 2-week ahead load forecast. An LSTM model achieved the lowest errors achieved. This model's errors were an sMAPE of 3.3%, an MAE of 4.6% and an RMSE of 5.5 %. DBN achieved the lowest errors of sMAPE of 3.7 %, an MAE of 5.2 % and RMSE of 6.3%. These errors were achieved with non-cleaned loading data and without temperature as an input variable in the model training. Deep learning techniques can thus be used without weather parameters to forecast distribution substations' loads with low errors. Accurate load forecasts can help utilities optimally utilise their resources and save costs through improved maintenance planning. This work can be extended to transmission networks and to other large power users. The latter part is to also to be able to generalise the findings in this research.

TABLE II. DBN LOAD FORECASTING TEST ERRORS

Input variables	Number Hidden Units	Performance With Raw Data			Performance With Cleaned Up Data		
		sMAPE	MAE	RMSE	sMAPE	MAE	RMSE
Group A	4	0.084358	0.058398	0.071928	0.096254	0.064928	0.079892
	8	**0.074553**	**0.051886**	**0.06325**	0.081501	0.055427	0.067506
	9	0.355518	0.300051	0.305125	0.376716	0.314994	0.320321
	10	0.355518	0.300051	0.305125	0.376716	0.314994	0.320321
	11	0.355518	0.300051	0.305125	0.376715	0.314993	0.320319
	12	0.355518	0.300051	0.305125	0.376716	0.314994	0.320321
	13	0.355518	0.300051	0.305125	0.376716	0.314994	0.320321
	14	0.355518	0.300051	0.305125	0.376716	0.314994	0.320321
	15	0.355518	0.300051	0.305125	0.376716	0.314994	0.320321
	16	0.355518	0.300051	0.305125	0.376716	0.314994	0.320321
	32	0.355518	0.300051	0.305125	0.376716	0.314994	0.320321
Group B	4	0.098332	0.067465	0.082154	0.082468	0.05608	0.068781
	8	0.076923	0.053481	0.065309	0.083146	0.056521	0.069206
	9	0.077413	0.053808	0.065873	0.081904	0.055714	0.068124
	10	0.076501	0.053202	0.065006	0.080669	0.054858	0.067393
	11	0.07527	0.052338	0.064215	0.376715	0.314993	0.320319
	12	0.077017	0.053483	0.06562	0.376716	0.314994	0.320321
	13	0.355518	0.300051	0.305125	0.083201	0.056469	0.069171
	14	0.355518	0.300051	0.305125	0.082861	0.05626	0.068955
	15	0.355518	0.300051	0.305125	0.376716	0.314994	0.32032
	16	0.355393	0.299922	0.305005	0.37661	0.314884	0.320219
	32	0.355518	0.300051	0.305125	0.376716	0.314994	0.320321

TABLE III. LSTM Load Forecasting Test Errors

Input variables	Hidden layers	Performance With Raw Data			Performance With Cleaned Up Data		
		sMAPE	MAE	RMSE	sMAPE	MAE	RMSE
Group A	60	0.077626	0.053888	0.06557	0.076185	0.051969	0.063520
	67	0.075899	0.052738	0.064054	0.077460	0.052804	0.064516
	336	0.067377	0.047037	0.057051	0.072048	0.049197	0.059428
	470	0.067405	0.047036	0.0567	0.073190	0.049974	0.060671
	538	0.072826	0.050744	0.062387	0.076781	0.052359	0.064038
	672	**0.065859**	**0.04598**	**0.055058**	0.070594	0.048218	0.057900
Group B	60	0.068902	0.04807	0.059121	0.074273	0.050665	0.062063
	67	0.069627	0.048554	0.059077	0.074493	0.050818	0.062484
	336	0.083265	0.057628	0.070237	0.071403	0.048768	0.059263
	470	0.067387	0.047039	0.057631	0.084471	0.057327	0.070240
	538	0.071501	0.049845	0.061517	0.076295	0.052030	0.064442
	672	0.066452	0.046385	0.056338	0.071717	0.048970	0.060011

ACKNOWLEDGEMENT

The authors would like to thank the South African Weather Services for providing them with weather data.

REFERENCES

[1] L. C. M. de Andrade, M. Oleskovicz, A. Q. Santos, D. V. Coury and R. A. S. Fernandes, "Very short-term load forecasting based on NARX recurrent neural networks," in 2014 IEEE PES General Meeting | Conference & Exposition, 2014.

[2] R. K. Agrawal, F. Muchahary and M. M. Tripathi, "Long term load forecasting with hourly predictions based on long-short-term-memory networks," in IEEE Texas Power and Energy Conference (TPEC), 2018.

[3] X.W Chen and X Lin, "Big data deep learning: Challenges and perspectives," IEEE Access, vol. 2, pp. 514-525, 2014.

[5] S. Motepe, Ali N. Hasan and R. Stopforth, "South African Distribution Networks Load Forecasting Using ANFIS," in IEEE Power Electronics Drivers and Energy Systems (PEDES) Conference, 2018.

[6] A. Narayan and K.W. Hipel, "Long short term memory networks for short-term electric load forecasting," in IEEE International conference on systems, man and cybernetics (SMC), 2017.

[7] S. Subramani, H. Wang, H. Q. Vu and G. Li, "Domestic Violence Crisis Identification From Facebook Posts Based on Deep Learning," IEEE Access, vol. 6, pp. 54075-54085, 2018.

[8] E. Mocanu, P. H. Nguyen, M. Gibescu, E. M. Larsen and P. Pinson, "Demand forecasting at low aggregation levels using Factored Conditional Restricted Boltzmann Machine," in Power Systems Computation Conference (PSCC), 2016.

[9] R.F. Berriel, A.T. Lopes, A. Rodrigues, F.M. Varejão and T. Oliveira-Santos, "Monthly energy consumption forecast a deep learning approach," in International Joint Conference on Neural Networks (IJCNN), 2017.

[10] J. Lago, F.D. Ridder and B. Schutter, "Forecasting spot electricity prices: Deep learning approaches and empirical comparison of traditional algorithms," Applied Energy, vol. 221, p. 386–405, 2018.

[11] R. K. Agrawal, F. Muchahary and M. M. Tripathi, "Long term load forecasting with hourly predictions based on long-short-term-memory networks," in 2018 IEEE Texas Power and Energy Conference (TPEC), College Station, TX, USA, 2018.

[12] C. Liu, Z. Jin, J. Gu and C. Qiu, "Short-term load forecasting using a long short-term memory network," in 2017 IEEE PES Innovative Smart Grid Technologies Conference Europe (ISGT-Europe), Torino, Italy, 2017.

[13] W. Kong, Z. Y. Dong, Y. Jia, D. J. Hill, Y. Xu and Y. Zhang, "Short-Term Residential Load Forecasting Based on LSTM Recurrent Neural Network," IEEE Transactions on Smart Grid, vol. 10, no. 1, pp. 841-851, 2019.

[14] J. Zheng, C. Xu, Z. Zhang and X. Li, "Electric load forecasting in smart grids using Long-Short-Term-Memory based Recurrent Neural Network," in 2017 51st Annual Conference on Information Sciences and Systems (CISS), Baltimore, MD, USA, 2017.

[15] L. C. M. de Andrade, M. Oleskovicz and R. A. S. Fernandes, "Very Short-Term Load Forecasting Based on NARX Recurrent Neural Networks Recurrent Neural Networks," in 2014 IEEE PES General Meeting | Conference & Exposition, National Harbor, MD, USA, 2014.

[16] G. M. Khan, F. Zafari and S. A. Mahmud, "Very Short Term Load Forecasting Using Cartesian Genetic Programming Evolved Recurrent Neural Networks (CGPRNN)," in 2013 12th International Conference on Machine Learning and Applications, Miami, FL, USA, 2013.

[17] S. Motepe, A.N. Hasan, B. Twala and R. Stopforth, "Power Distribution Networks Load Forecasting Using Deep Belief Networks: The South African Case," in 2019 IEEE Jordan International Joint Conference on Electrical Engineering and Information Technology (JEEIT), Amman, Jordan.

[18] E. Hurwitz, A. N. Hasan and C. Orji, "Soft biometric thermal face recognition using FWT and LDA feature extraction method with RBM DBN and FFNN classifier algorithms," in Fourth International Conference on Image Information Processing (ICIIP), 2017.

[19] F. Fahiman, S.M. Erfani, S. Rajasegarar, M. Palaniswami and C. Leckie, "Improving load forecasting based on deep and k-shape clustering," in International Joint Conference on Neural Networks (IJCNN), 2017.

[20] Y. Tao, H. Chen and C. Qiu, "Wind power prediction and pattern feature based on deep learning method," in 2014 IEEE PES Asia-Pacific Power and Energy Engineering Conference (APPEEC), Hong Kong, China, 2014.

[21] X. Zhang, R. Wang, T. Zhang and Y. Zha, "Short-term load forecasting based on a improved deep belief network," in 2016 International Conference on Smart Grid and Clean Energy Technologies (ICSGCE), Chengdu, China, 2016.

[22] A. Gensler, J. Henze, B. Sick and N. Raabe, "Deep Learning for Solar Power Forecasting – An Approach Using Autoencoder and LSTM Neural Networks," in IEEE International Conference on Systems, Man, and Cybernetics (SMC), 2016.

[23] K. Grolinger, M. A. M. Capretz and L. Seewald, "Energy Consumption Prediction with Big Data: Balancing Prediction Accuracy and Computational Resources," in IEEE International Congress on Big Data, 2016.

[24] G.E. Hinton, S. Osindero and Y.W. Teh, "A fast learning algorithm for deep belief nets," in Neural Computation, 2006.

[25] Y. He, J. Deng and H. Li, "Short-term power load forecasting with deep belief network and copula models," in 9th International

Conference on Intelligent Human-Machine Systems and Cybernetics, 2017.

[26] S. Hochreiter and J. Schmidhuber, "Long short-term memory," Neural computation, vol. 9, no. 8, p. 1735–1780, 1997.

Automotive Power Conversion - Motors, Power Electronics, Batteries, and Chargers

978-1-5386-7688-2/19 $31.00 © 2019 IEEE

Design of a Controller for Torsional Vibrations of an Electric Vehicle Powertrain

Mustafa Karamuk and Salih Baris Ozturk, *Senior Member, IEEE*
Department of Electrical and Electronics Engineering
Istanbul Okan University
34959 Istanbul, Turkey
E-mails: mkaramuk@stu.okan.edu.tr, baris.ozturk@okan.edu.tr

Abstract—Electric powertrains are subject to vibrations due to weak damping and fast torque response of electric motor. In this study, a PD controller is developed to damp the vibrations. By using PD controller, two solution methods are developed. First solution is to use the wheel speed sensors available in all four wheels. Second solution is to use a reduced order observer to estimate the wheel speed. Differentiation of motor speed and wheel speed is input to the PD controller. Output of the PD controller is the torque damping feedback to the torque reference. Simulation model is developed including a flux vector controlled permanent magnet synchronous motor, powertrain model and vibration controller. The results demonstrate the validity and effectiveness of the proposed scheme.

Index Terms—Electric powertrain vibrations, powertrain model, reduced order observer, PD controller.

I. INTRODUCTION

A. Torsional Vibration Problem in Electric Vehicles

The torsional vibration problem is already known in vehicles propelled by the combustion engines. Overview of the problem and solution methods are given in [1]–[3]. The severity of the problem is worse in battery electric and also hybrid electric vehicles. The first root cause is the lack of components like clutch and torque converter which provide damping in the powertrain system. The second root cause is the fast torque rise time of the electric motors, which is in milliseconds range, excite the torsional resonances. On the other hand, the torque rise time of the combustion engines is in seconds range [4]. The torsional vibration problem in electric powertrains already exist in the industrial applications such as rolling mills, servo motor control or similar multi mass systems. In such systems, depending on the motor and load inertia, and shaft stiffness, torsional resonance frequency exists [5], [6]. The modeling methodology and solutions applied in industrial applications can be adapted into electric vehicle (EV) and also hybrid electric vehicle (HEV) applications. When choosing the right solutions in vehicle applications, the system parameters such as powertrain topology, powertrain parameters (inertia and stiffness of the driveline components, gear ratio), load disturbance severity, vibration damping capability of the method and vehicle performance metrics should be analyzed, simultaneously. Because, the applied method may be too complex or some system parameters may represent variation among the vehicles.

B. Review of State of the Art Methods

The methods known in the state of the art can be summarized as follows

1) Passive damping methods:

a) Applying a filter to the motor torque reference:

2) Active damping methods:

a) Oscillations controller including a Kalman filter for torque estimation: [2]

b) State feedback control including a reduced order observer: [7]

c) LQR (linear quadratic regulator) control including a reduced order observer: [8]

d) Modal state observer including the oscillations characteristics of the drivetrain: [9], [10]

e) Feedforward and feedback compensation: [11]

Passive damping methods reduce the resonance amplitude and shift the resonance frequency. On the other hand, active methods are more common in electric vehicles. Most of the traction inverters are equipped with active damping functionality.

C. Electric Powertrain Topologies

Torsional vibration problem should be analyzed with respect to the electric powertrain topology. There are mainly following topologies, each of which has its own torsional characteristics:

1) Low speed and high torque motors with single speed transmission: These type of motors have relatively high rotor inertia and require low gear ratio.

2) High speed and low torque motors with two speed transmission: High speed (motor speed > 10.000 rpm range) motors have relatively low motor inertia than motors in group (1). Two gear ratios create two different torsional resonance frequencies [9].

3) Direct drive motors: These type of motors have higher inertia and require no gearbox and eliminate the gear backlash. Major concern is the geometrical integration of the motor due to the higher diameter.

4) Electric axle topologies:

a) Single electric motor integrated on the center of the axle: [12]

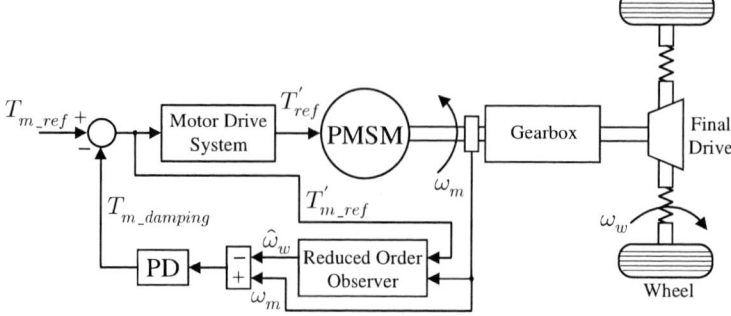

Fig. 1. Block diagram of the PD controller with the proposed scheme with reduced order observer.

b) Near-wheel, in-wheel topologies with two motors in the same axle: [12], [13].

Considering the variation of motor inertia, gear ratios and axle stiffness for each of the topologies given above, different torsional resonance frequency is obtained which can be calculated by [8]

$$f_r = \frac{1}{2\pi} \sqrt{\frac{k_s(J_w + J_m i_g^2)}{J_m J_w i_g^2}} \qquad (1)$$

where k_s is the shaft stiffness coefficient; i_g is given as the product of gear ratio and final drive ratio; J_ω is the equivalent inertia of the rotating parts of the vehicle, which is mainly the wheel inertia. J_m is the equivalent inertia of the motor and transmission.

In this study, central drive topology with single speed transmission is selected for the analysis.

D. Sources of Vibrations

The focus of the paper is the torsional vibration. However, there are several sources of vibration which have influence on damping performance of torsional vibration controller. The following vibrations are additional load disturbances to the damping controller:

1) Internal motor failures: Vibration of electric motor in case of internal failure such as failure of bearing or stator winding insulation. Such a failure is an unexpected and unmodelled disturbance to the damping controller.

2) Motor design: Motor torque ripples due to the motor design. Synchronous reluctance motors, interior permanent magnet motors, permanent magnet assisted synchronous reluctance motors inherently produce higher torque ripples than induction motor.

3) Controller design: Motor control issues such as over modulation at field weakening is also a source of torque ripple and vibration. The amplitude of the torque ripple is dependent on the implementation of the motor controller and motor design as well.

4) Gear backlash: Backlash is required to prevent gears from jamming. However, backlash causes coupling and de-coupling between the motor and load at gearbox deadband region. This problem is known as gear clunk noise. Describing function is typically used as the analytical approach to model the backlash effect [14].

5) Torque transients during gear shifting of two speed transmissions: Minimization of torque transients should be performed by gear shifting controller [15].

6) Road imperfections: Irregular road patterns create load vibrations.

II. DESIGN OF THE VIBRATION DAMPING CONTROLLER

The system configuration is given in Fig. 1. The damping controller is implemented in motor drive unit. The torque reference is sent from electric vehicle control unit to the motor drive unit. In order to damp the vibration effectively, it is important to send the torque reference with high sampling rate such as 1 ms via CAN bus.

A. Mathematical Modeling of the Electric Powertrain

The powertrain system in the analysis has central drive topology in which the electric motor is directly coupled to the axle by a single speed transmission. This system can be represented as a two-mass system, as shown in Fig. 2. Backlash is not included to the transmission model for simplicity. This system is expressed mathematically by three set of state equations, which are the equations of motor, transmission and vehicle driveline, respectively [2], [7], [8]. The state equations of the two-mass system can be described as follows [6], [7]

$$
\underbrace{\begin{bmatrix} \dot{\omega}_m \\ \dot{\theta} \\ \dot{\omega}_w \end{bmatrix}}_{\dot{x}}
=
\underbrace{\begin{bmatrix} -\frac{c_s}{i_g^2 J_m} & \frac{k_s}{i_g J_m} & \frac{c_s}{i_g J_m} \\ -\frac{1}{i_g} & 0 & -1 \\ \frac{c_s}{i_g J_w} & \frac{k_s}{J_w} & -\frac{c_s}{i_g J_w} \end{bmatrix}}_{A}
\underbrace{\begin{bmatrix} \omega_m \\ \theta \\ \omega_w \end{bmatrix}}_{x}
+
$$

$$
\underbrace{\begin{bmatrix} \frac{1}{J_m} \\ 0 \\ 0 \end{bmatrix}}_{B} T_m
+
\underbrace{\begin{bmatrix} 0 \\ 0 \\ -\frac{1}{J_m} \end{bmatrix}}_{D} T_L
\qquad (2)
$$

where c_s is damping coefficient of the shaft; θ is the torsional angle of the shaft; and T_L is the load torque.

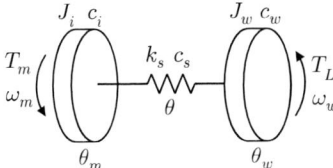

Fig. 2. Electric powertrain model as two mass system [8].

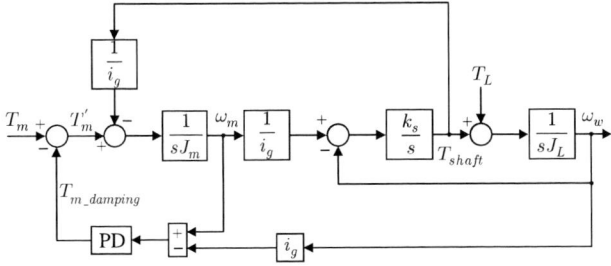

Fig. 3. Simplified model of electric powertrain as two-mass system.

1) Modeling of the PD Controller as Vibration Damper:
The motor speed oscillations occur at higher amplitude and frequency compared to the wheel speed. The oscillations are weakened on the wheel side due to the rotating inertial components of the vehicle. In real life, there are measurement noise and road noise effect on the wheel speed measurement signal. However, in the analysis, it is assumed that the wheel speed signal is filtered. In [9] and [10], the difference between wheel speed and motor shaft speed is used as the input for damping controller. This information has been implemented by using various type of controllers in previous studies. In this study, the difference between wheel speed and motor shaft speed is used as feedback input to a PD (proportional and derivative) controller. The reason of choosing the PD controller and comparison of the step response output with a PID controller are shown in Section II.B. The output of the PD controller is the damping torque applied to the torque reference of the motor. For simplicity, since damping factor c_s described in (2) is small, it is ignored in the model given in Fig. 3. The proposed solution methods can be implemented in two ways as described below:

2) The method including a wheel speed measurement:
Wheel speed sensor is used assuming that the sensor provides high accuracy and measurement noise is filtered. Observer is not needed in this case.

3) The method including a reduced order observer for wheel speed estimation: A reduced order observer is designed to estimate the wheel speed information.

B. Design of the PD Controller as Vibration Damper

The model of a two-mass system is shown in Fig. 3. Transfer functions from ω_m/T_m and ω_w/ω_m are derived by using this model as

$$G_m(s) = \frac{\omega_m}{T_m} = \frac{J_w s^2 + k_s}{J_m J_w s^3 + k_s(J_m + \frac{J_w}{i_g^2})s} \quad (3)$$

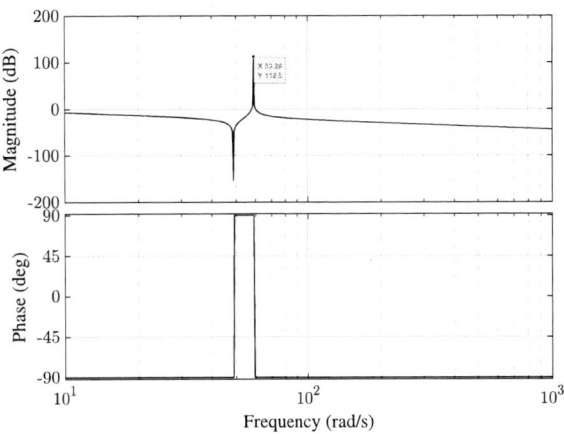

Fig. 4. Bode diagram of the open loop transfer function $G_m(s)$.

$$G_\omega(s) = \frac{\omega_w}{\omega_m} = \frac{k_s}{i_g(J_w s^2 + k_s)} \quad (4)$$

where $G_m(s)$ is the transfer function from motor speed to motor torque and $G_w(s)$ is the transfer function from wheel speed to motor speed. Closed loop transfer function including the PD controller at feedback loop is obtained by using the control diagram given in Fig. 3. It should be noted that the derivative term in transfer function which is given in (5) is an ideal derivative without a filter. State space simulations including the PD control in MATLAB/Simulink® model is implemented with a filtered derivative component.

$$G_{sys}(s) = \frac{G_m(s)}{(1 + (K_p + sK_d))G_m(s)(1 - G_w(s))}. \quad (5)$$

Open loop transfer function $G_m(s)$ has one pole at zero and complex roots on $j\omega$ axis located at as $(0 + j59.25)$ and $(0 - j59.254)$. This is marginally stable system [6], [16].

Open loop bode diagram indicates a resonance frequency at 9.43 Hz and an anti-resonance frequency at 7.79 Hz. Transfer function $G_m(s)$ has a pole at zero. Therefore, a damping control by a PID does not improve the system response because of an additional integrator. As will be seen in the step response analysis, PID increases the oscillation and reduces the system gain gradually. This has been proved both by step response and state space simulations. Step response of transfer function $G_m(s)$ without a damping control generates oscillations due to the presence of undamped complex roots at $j\omega$ axis.

As seen in closed loop Bode diagram in Fig. 5, magnitude of resonance is damped. The closed loop Bode diagram is shown to illustrate the magnitude change of the closed loop system [17]. The anti-resonance peak indicates an increase of the system's stiffness at this frequency. It does not increase the system amplitude as the resonance does. Moreover, system response is reduced at this point [18]. The purpose of the design is more focused on decreasing the resonance amplitude rather than improving the system response at anti-resonance point.

978-1-5386-7688-2/19 $31.00 © 2019 IEEE

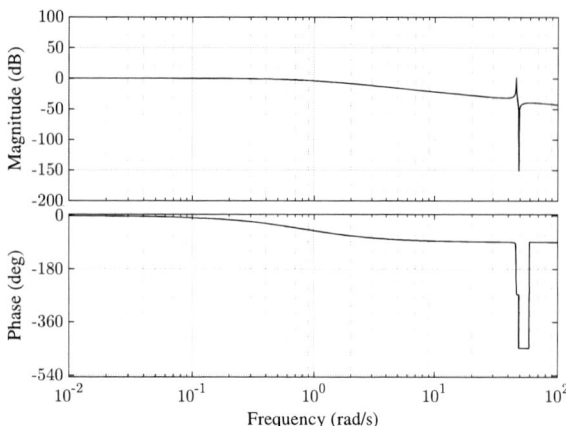

Fig. 5. Bode diagram of the closed loop transfer function $G_{sys}(s)$.

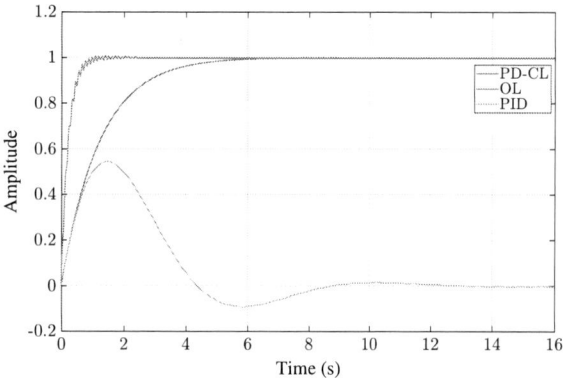

Fig. 6. Step response of the open loop, closed loop with PD and PID controller.

The open loop step response without any controller is given in Fig. 6. The transfer function $G_m(s)$ is actually not an open loop. The vehicle driver controls the vehicle speed by changing the torque reference via acceleration pedal. Therefore, there is a feedback loop acting on the motor speed and wheel speed. Due to the open loop poles at $j\omega$ axis, speed oscillations occur.

Step response of the closed loop transfer function $G_{sys}(s)$ including the PD controller in the loop is shown in Fig. 6 where $K_d = 1.2$ and $K_p = 1.2$. The open loop system has already complex poles at $j\omega$ axis and exhibits slow response with marginal stability. Therefore, a slow response with PD controller is acceptable considering the compensation performance of the vibrations. The damping controller can be tuned faster however, in this case, the requested torque by the vehicle driver will be modified by the damping controller. Therefore, slower response is preferred to prevent any modification of the vehicle driver's torque request.

The system with PD control has almost an exact pole zero cancellation and therefore yields a first order step response. On the other hand, the system with PID control has oscillatory response because of the uncancelled poles near zero. Moreover,

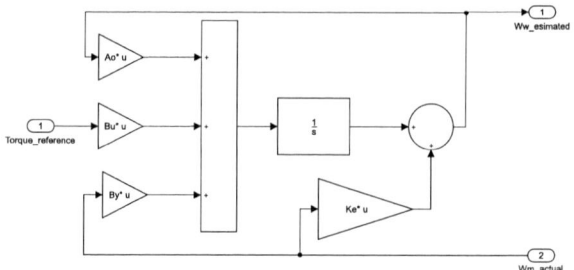

Fig. 7. Simulink model of the reduced order observer for wheel speed estimation.

it sets the system output to zero, as seen in Fig. 6.

C. Design of the Reduced Order Observer

A reduced order Luenberger observer is designed to estimate the wheel speed. The methods given in [5]–[8], [16] are applied as the design methodology. As the first step, observability is checked in MATLAB®. Rank of the observability matrix given in (6) should be equal to the number of states which is three. MATLAB® command $obsv(A, C)$ calculates the observability matrix. It is equal to the number of states.

$$\left[C^* \quad A^* C^* \quad ... \quad (A^*)^{n-1} C^* \right]. \tag{6}$$

The reduced order observer equations are given as

$$[x_a \; ; \; x_b] = [A_{aa} \; A_{ab} \; ; \; A_{ba} \; A_{bb}]. \tag{7}$$

The submatrices of the reduced order observer as a function of system parameters are given as

$$A_{aa} = \left[-\frac{c_s}{i_g{}^2 J_m} \right], A_{ab} = \left[-\frac{k_s}{J_m} - \frac{c_s}{(i_g J_m)} \right] \tag{8}$$

$$A_{ba} = \left[-\frac{1}{i_g} ; \frac{c_s}{(i_g J_w)} \right], A_{bb} = \left[0 \quad -1; \frac{k_s}{J_w} - \frac{c_s}{J_w} \right]. \tag{9}$$

The inputs to the reduced order observer are motor torque reference and motor actual shaft speed. It estimates the wheel speed of the vehicle. The location of the observer poles are set as

$$L = [-100 + 550i \quad -100 - 550i]. \tag{10}$$

Gain of the observer is calculated by the Ackermann formula. MATLAB® command below calculates the Ackermann gain as $K_e = acker(Abb', Aab', L)'$. MATLAB/Simulink® model of the reduced order observer is given in Fig. 7.

III. SIMULATION RESULTS WITH STATE SPACE MODEL

Simulation results are given for the following cases:

1) The electric powertrain is modeled by state equations and PD controller is activated or de-activated in the control loop: The torque reference of the motor is applied as a step function. The motor drive unit is not included in this case. The results are given for both cases where the wheel speed is measured or estimated by the reduced order observer.

978-1-5386-7688-2/19 $31.00 © 2019 IEEE

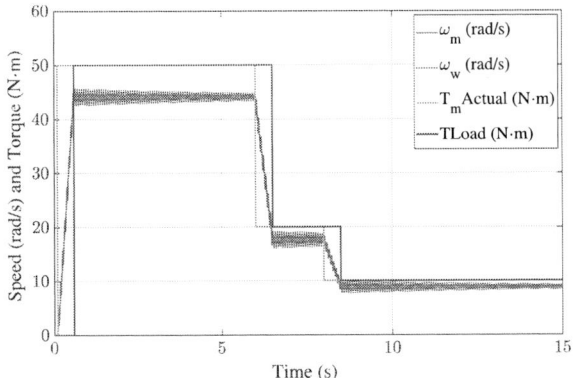

Fig. 8. Motor speed, estimated wheel speed, motor torque and load torque when PD controller is not activated.

TABLE I
PARAMETERS OF THE ELECTRIC POWERTRAIN

Parameters	Value
Motor and gearbox inertia (kg.m^2)	0.15
Wheel inertia (kg.m^2)	2.5
Vehicle mass (kg)	1800
Driveline shaft stiffness coefficient (N·m.rad^{-1})	6000
Damping coefficient of the driveline system	0.5
Gear ratio	3:1
Final drive ratio	2:1

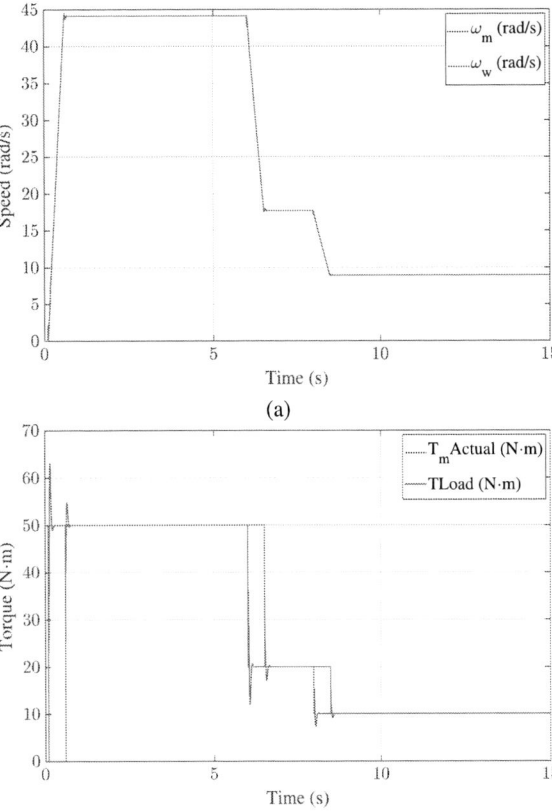

Fig. 9. a) Motor speed and measured wheel speed when PD controller is activated, b) Motor torque and load torque.

2) The vibration damping controller is applied including a vector controlled interior permanent magnet (IPM) synchronous motor drive system: The torque reference of the motor is sent from the vector controlled IPM motor drive model. Torque damping feedback signal is applied to the torque reference of the vector control system. $q-$axis current controller of the vector control system calculates the current reference including the torque damping feedback. The simulation parameters of the electric powertrain system are shown in Table I.

A. Simulation Results Including the State Space Model of Electric Powertrain and Motor Torque Reference as Input

In this case, the results are given with and without PD controller. The system damping is set to a low value to illustrate the effectiveness of the controller. Except the simulation case with observer including IPM motor, PD controller gains are set as $K_d = 1.2$ and $K_p = 1.2$.

1) Simulation results with the wheel speed sensor measurement: In all simulation cases given in Figs. 8 and 9, the load torque is applied with 0.5 s of delay after the activation of the motor torque.

2) Simulation results with the wheel speed estimated by the reduced order observer: They are given in Figs. 10(a) through 10(c).

B. Simulation Results Including a Vector Controlled Interior Permanent Magnet (IPM) Motor and State Space Model of Powertrain Model

In this section, the PD controller is simulated including a vector controlled IPM motor using the MATLAB/Simulink® model given in Fig. 11. The torque

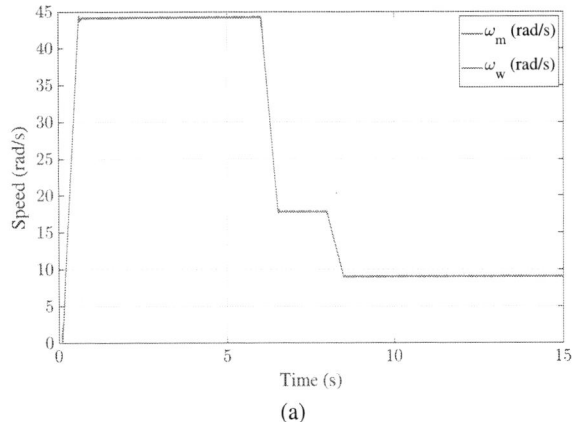

(a)

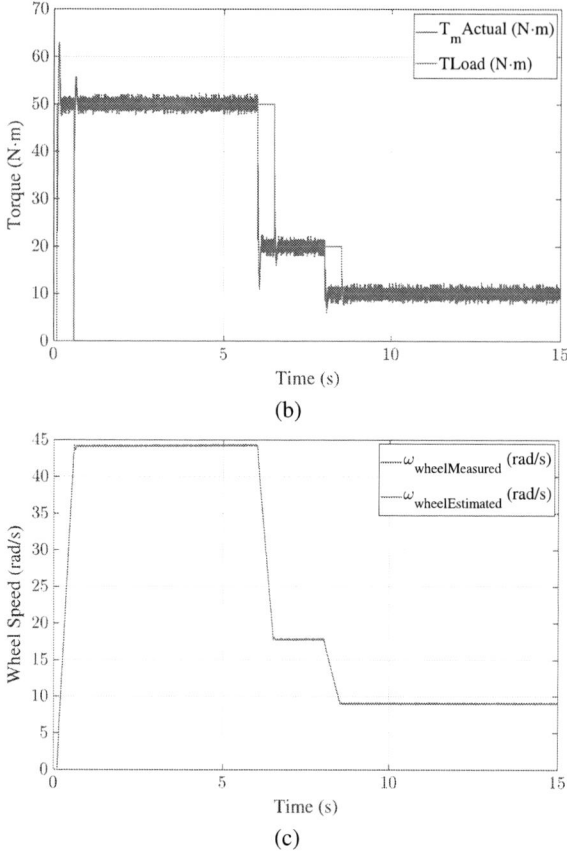

(b)

(c)

Fig. 10. a) Motor speed and estimated wheel speed when PD controller is activated, b) Motor torque and load torque, c) Comparison of measured and estimated wheel speeds.

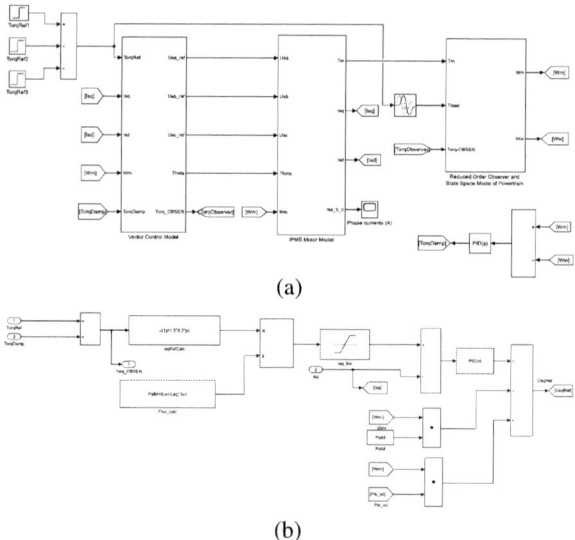

(a)

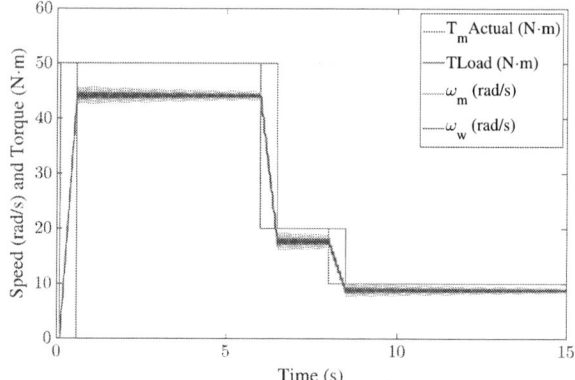

(b)

Fig. 11. a) Vector control model of IPM motor including the vibration controller, b) Calculation of motor torque reference including the vibration damping feedback.

Fig. 12. Motor speed and wheel speed when PD controller is not activated.

damping feedback is applied to the torque reference of the vector control model. IPM motor has 40 kW continuous and 80 kW peak power. Input DC voltage of the inverter is 400 V_{dc}. As seen in simulations, PD controller damps the speed oscillations, effectively. Simulations are performed for both cases including the measured and estimated wheel speeds.

1) Simulation results with the wheel speed sensor measurement and the vector controlled IPM Motor: In this case, the simulations when PD controller is not activated is presented in Fig. 12, whereas the simulation results when PD controller is activated are presented in Figs. 13 and 14.

2) Simulation results with the estimated wheel speed by reduced order observer and the vector controlled IPM Motor: When simulating the model with the reduced order observer for wheel speed estimation and PD controller as the torque damping, the realizable gains of PD controller is reduced. The gains are set as $K_d = 0.6$ and $K_p = 0.6$. The reason is that the complete system has reduced gain margin due to the added complex poles of the reduced order observer. The simulation results when PD controller is activated with the estimated wheel speed by reduced order observer for the vector controlled IPM motor are given in Figs. 15 and 16. Observer

poles are set at the same location which are given as

$$L = [-50 + 700i \quad -50 - 700i]. \quad (11)$$

IV. CONCLUSION

In this study, the proposed PD controller as the active vibration controller has been modeled, tuned and simulated in MATLAB/Simulink®. The simulation results verify that the method works principally. The simulation results including a vector controlled IPM motor are consistent with state space model simulations. Test cases including the wheel speed measurement and estimation by a Luenberger reduced order observer are also consistent between state space and vector-controlled motor model. These results verify the principle of the proposed method.

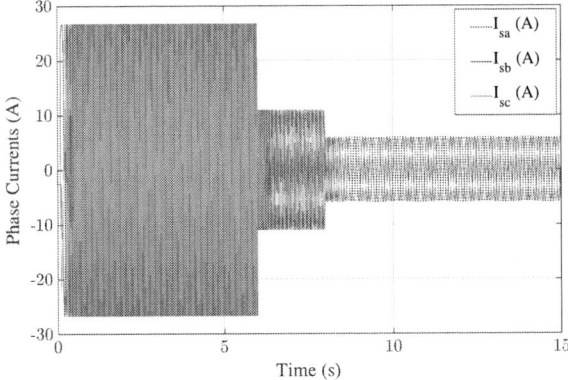

Fig. 13. Three phase motor currents when PD controller is activated.

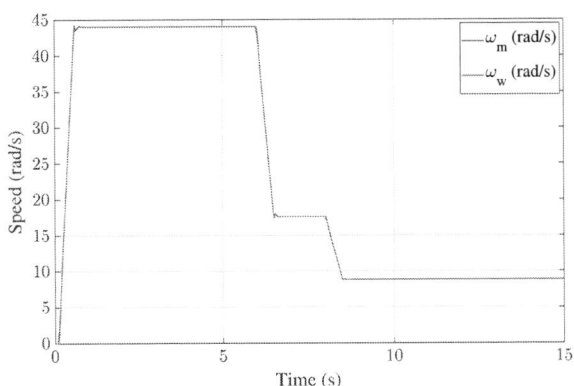

Fig. 14. Motor speed and wheel speed when PD controller is activated.

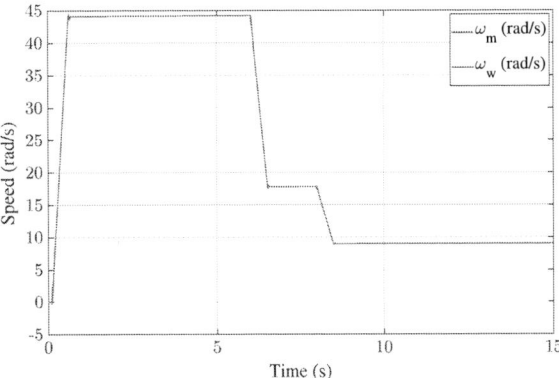

Fig. 15. Motor speed and wheel speed when PD controller is activated (reduced order observer is used for wheel speed estimation).

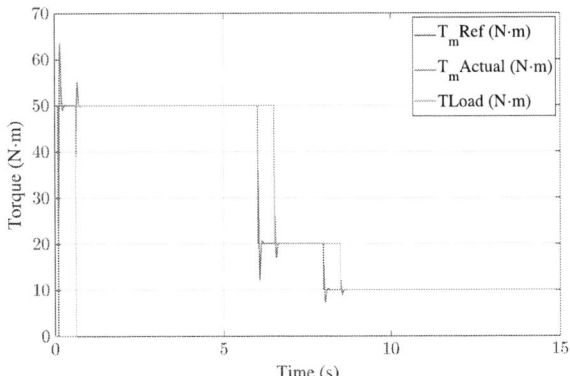

Fig. 16. Reference and actual motor torque and load torque when PD controller is activated (reduced order observer is used for wheel speed estimation).

REFERENCES

[1] Y. Hori, "A review of torsional vibration control methods and a proposal of disturbance observer-based new techniques," *IFAC Proceedings*, vol. 129, no. 1, pp. 990–995, Jun-Jul. 1996.

[2] N. Amann, Y. Bocker, and F. Brenner, "Active damping of drive train oscillations for an electrically driven vehicle," *IEEE/ASME Trans. Mechatron.*, vol. 9, no. 4, pp. 697–700, Dec. 2004.

[3] S. N. Vukosavic, M. R. Stojic, "Supression of torsional oscillations in a high-performance speed servo drive," *IEEE Trans. Ind. Electron.*, vol. 45, no. 1, pp. 108–117, Feb. 1998.

[4] M. Karamuk, "A survey on electric vehicle powertrain systems," in Proc. *IEEE-ACEMP*, Istanbul, 2011, pp. 315–324.

[5] S. Song, J. Ji, S. Sul, and M. Park, "Torsional vibration suppression control in 2-mass system by state feedback speed controller," in Proc. *Int. Conf. on Control and Applications*, Vancouver, BC, Canada, 13–16 Sept. 1993, vol. 1.

[6] R. Zhang, Y. Yang, Z. Chen, and T. Tong, "Torsional vibration suppression control in the main drive system of rolling mill state feedback speed controller based on extended state observer," in Proc. *IEEE Int. Conf. on Control and Autom.*, Guangzhou, May/June 2007, pp. 2172–2177.

[7] Z. M. Zhong and Q. Wei, "Modeling and torsional vibration control based on state feedback for electric vehicle powertrain," *Applied Mechanics and Materials*, vol. 341–342, pp. 411–417, 2013.

[8] H. Fu, C. Feng, Z. Xue, "Vibration suppression control of an integrated powertrain of electric and hybrid vehicles using LQR controller and reduced-order observer," in Proc. *IEEE Conf. and Expo Transp. Electrif. Asia-Pac.*, Beijing, 2014. pp. 1–6.

[9] G. Gotting, R. W.De Doncker, "Active drive control of electric vehicles using a modal state observer," in Proc. *IEEE 35th Annu. Power Electron. Specialists Conf. (PESC)*, Aachen, Germany, 2004, pp. 4585–4590 Vol.6.

[10] G. Gotting, "Dynamische Antriebsregelung von Elektrostrassenfahzugen unter Berücksichtigung eines schwingungsfaehigen Antriebsstrangs," Ph.D. Dissertation, RWTH Aachen University, 2004.

[11] T. Karikomi, K. Itou, T. Okubo, and S. Fujimoto, "Development of the shaking vibration control for electric vehicles," in Proc. *SICE-ICASE Int. Joint Conf.*, Busan, Oct. 2006, pp. 2434–2439.

[12] M. Felden, P. Butterling, P. Jeck, L. Eckstein, and K. Haymer, "Electric vehicle drive trains: From the specification sheet to the drive-train concept," in Proc. *14th Int. Power Electron. and Motion Control Conf. (EPE-PEMC)*, Ohrid, 2010, pp. S11-9-S11-16.

[13] T. Beauduin, S. Yamada, H. Fujimoto, T. Kanou, and E. Katsuyama, "Control-oriented modelling and experimental modal analysis of electric vehicles with geared in-wheel motors," in Proc. *IEEE Int. Conf. on Advanced Intelligent Mechatronics (AIM)*, Munich, 2017, pp. 541–546.

[14] M. Yang, W. Zheng, K. Yang, and D. Xu, "Suppression of Mechanical Resonance Using Torque Disturbance Observer for Two-Inertia System with Backlash," in Proc. *IEEE 9th Int. Conf. on Power Electronics and ECCE Asia (ICPE-ECCE Asia)*, Seoul, 2015, pp. 1860–1866.

[15] C. Lin, S. Sun, and W. Jiang,"Active anti-jerking control of shifting for electric vehicle driveline," *Energy Procedia*, vol. 104, pp.348–353, Dec. 2016.

[16] Ogata, *Modern Control Engineering*. 5th ed. Upper Saddle River, N.J: Prentice Hall, 2010.

[17] G. Franklin, J. D. Powell, and A. Emami-Naeini, *Feedback Control of Dynamic Systems*. 8th ed. Upper Saddle River, N.J: Prentice Hall, 2018.

[18] G.Ellis, R.D.Lorenz, "Resonant load control methods for industrial servo drives," in Proc. *IEEE 35th IAS Annu. Meeting and World Conf. on Ind. Appl. of Elect. Energy*, Rome, Italy, 2000, pp. 1438–1445, vol. 3.

Development of Fuzzy Logic Based Energy Management Control Algorithm for a Plug-In HEV with Fixed Routed

1st Hazal Sölek
Anadolu ISUZU R&D Dept.
Istanbul Technical University
Istanbul, Turkey
hazal.solek@isuzu.com.tr

2nd Kenan Müderrisoğlu
Anadolu ISUZU R&D Dept.
Istanbul Technical University
Istanbul, Turkey
kenan.muderrisoglu@isuzu.com.tr

3rd Cem Armutlu
Anadolu ISUZU R&D Dept.
Istanbul, Turkey
cem.armutlu@isuzu.com.tr

4th Murat Yılmaz
Istanbul Technical University
Istanbul, Turkey
myilmaz@itu.edu.tr

Abstract— **Environmental pollution is increasing due to the increasing number of internal combustion engines. Depending on this situation, there has been a trend towards alternative energy vehicle technologies. Serial hybrid electric vehicles are also one of these alternative energy vehicles. The decision of the control method to increase efficiency in hybrid electric vehicles have great importance. Algorithm of the energy management system has been developed for a series hybrid electric garbage truck. The developed algorithm consists of two parts working as online and offline. The offline algorithm uses the vehicle's previous average consumption data to operate the vehicle in electric mode within residential areas. The online algorithm provides instant control of all subsystems of the vehicle, such as the traction system and generator set system. In the online control algorithm, control mode selection and generator set speed selection functions are made by fuzzy logic controller, considering the non-linear characteristic of the system. The developed controller was tested in the MATLAB/Simulink environment and was able to complete the garbage collection areas in electric mode. By using developed EMS algorithm 27.3% fuel saving is achieved compared to the conventional vehicle.**

Keywords— battery, state of charge, energy management system, fuzzy logic control, TruckMaker, series plug in hybrid electric truck

I. INTRODUCTION

Nowadays, importance is given to alternative energy-based vehicles and alternatives to fossil fuels are increased in the light of environmental awareness and technological developments. Also the increasing environmental awareness along with global warming has accelerated efforts to increase efficiency in conventional vehicles and reduce fossil fuel dependence. Most known types of vehicles developed for this purpose are electric vehicles (EV), hybrid vehicles and fuel cell vehicles. Interest in alternative energy vehicles has been increasing in recent years. Between 2014 and 2015, the sales of electric and hybrid vehicles increased by 70%. Lin et al. [1] conducted a study on people's willingness to pay for the use of alternative energy buses in 4 major cities of China. In this study, it has been seen that 80% of people agree to pay extra to support the use of renewable energy buses.

Alternative energy vehicles are still under development. Therefore, as with any technology in development, alternative energy vehicle technologies also have weaknesses. For example, battery technology is still under development. The energy density of the batteries is much lower than fossil fuels. This results in lower range and higher battery weight for the same range. Since fuel cell vehicles do not have the necessary filling station infrastructure, the initial investment cost of the vehicles is very high [2].

The use of hybrid electric vehicles (HEV), which have been in a transition position between conventional vehicles and alternative energy vehicles, has been increasing in recent years. There are many types of hybrid vehicles that are designed to achieve the highest efficiency and the lowest possible emission level from energy source to wheel in a given driving profile [2]. It can be seen a fuel save ratio about 15% in micro hybrid systems [3-5], 30% in light hybrid vehicles, maximum 25% due to energy conversion in serial hybrid systems and more than 40% in parallel hybrid systems [3].

Especially, hybrid systems which can be used in self-charging and long range are noteworthy in the commercial vehicle industry. These systems can be recharged externally as plug-in or they can provide energy conversion within their own. Internal combustion engines (ICE) are widely preferred in generator systems to generate power for charging the battery. The contribution of these low efficient engines with minimum working time and maximum possible efficiency to the hybrid system is very important. For this reason, the development of energy management algorithms is very important for hybrid systems where several sub-systems are working together.

Energy management system (EMS) is an area where the use of automatic control science is important. Nowadays, many methods are tried and each method is a reference to another. Control structures applied in HEVs are basically divided into two parts. These are online and offline methods. Online methods are divided into two as rule-based and optimization-based methods [6].

Offline hybrid control algorithms include global optimization methods. These are algorithms based on optimization theory that adjust the optimized mathematical models with emission constraints, fuel economy or system state variables. Global optimization methods have the highest yield rates compared to other control methods, but for the correct and efficient use of this method, all driving cycle information is needed. Real-time implementation of this situation is very difficult. In addition, due to the nature of these algorithms, the processing loads are high. Considering the processing load and real-time operation, these methods are mainly suited for analyzing and evaluating the impact of other energy management strategies in offline applications [7]. Linear programming, dynamic programming, particle swarm optimization and genetic algorithms are among the global optimization methods frequently used in hybrid energy management.

The basic idea based on rule-based algorithms, which are subgroups of online controllers, is to limit the operating range of the parameters by setting threshold parameters and

978-1-5386-7688-2/19 $31.00 © 2019 IEEE

to adjust the operating status of the components according to the real-time parameters of the vehicle and the predetermined rules. The simple structure has many advantages, such as a small amount of calculation and high application efficiency, and is very suitable for real-time energy management. However, the settings of threshold parameters usually require experience, so optimization of vehicle fuel economy cannot be guaranteed without a fine calibration [8]. Fuzzy logic based controllers (FLC) have been developed to compensate for this. FLCs, a subdivision of rule-based controllers, do not specify absolute limits, unlike conventional deterministic controllers. The main advantages of the fuzzy logic approach are component variability and compliance with measurement sounds. In this way, fuzzy logic algorithms have higher efficiency values than classical rule based algorithms [9,10].

Online optimization-based control methods reduce the burden of computing by reducing global optimization problems as a succession of local optimization problems. This method eliminates the need for future time driving information so that controllers can be used in real time. Equivalent fuel consumption reduction method (ECMS, Equivalent Consumption Minimization Strategy), model predictive control, artificial neural networks are evaluated within this group.

In this study, an EMS is designed for a hybrid garbage truck. FLC has been found to be suitable because of its ease of application and real time operation advantages. The developed control algorithm is divided into two parts as online and offline controllers. Using the previous consumption data of the vehicle, the offline controller generates the reference SOC signal used to determine the operating mode required to complete the garbage collection areas in electric mode. The online controller generates instant signals such as control mode and generator set power required to control the vehicle.

II. ENERGY MANAGEMENT SYSTEM

The designed serial hybrid plug-in truck is a fixed-route vehicle for its intended use. Structure of designed plug in serial HEV is show in Fig. 1. The vehicle used as garbage truck. So it has certain fixed routes and it is used by the stated drivers at the predetermined hours. In case of the route, driver and operating hours variable are defined before the vehicle moving, thereby the total energy consumed by the vehicle can be predicted with great accuracy. The route is the first parameter that directly affects consumption. The time parameter contains traffic density information. It gives information about the speed profile of the vehicle and affects the consumption. The drive parameter is important to determine the driving tendency of the person whom driving the vehicle. For example, a driver using a vehicle in a high acceleration profile has higher fuel consumption, whereas a driver with a low acceleration profile completes the same driving cycle with less energy consume.

The developed algorithm stores the consumption information of the vehicle depending on the route, driver and time variables. Instead of being stored separately for each drive, this data is stored the data as the average of all drives for the same conditions. In other words, at the end of each ride, the consumption data stored in the vehicle's memory are added to the consumption data of the last drive and averaged. This reduces storage space.

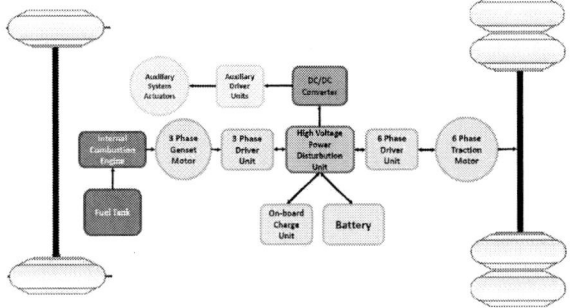

Figure 1 Structure of plug in HEV

When the vehicle is started by the driver, the identity of the driver, route and time parameters are requested by the controller which is developed. Then, according to the information entered, EMS reads the previous average consumption information from the memory. This consumption data is stored separately for every 4 km of the road. In this way, a separate operating mode for each part of the path to be determined by increasing the accuracy of the algorithm. After reading the consumption data, the vehicle's fuel tank and battery charge status are checked and the driver is informed about whether the energy in the vehicle is enough for completing the route. In the next step of the algorithm, the road segments are selected where the vehicle must operate in electrical mode with zero emission values. As the vehicle has fixed route, these parts are predetermined. One of the main objectives of the algorithm is to reach the battery charge level to 90% and above before the vehicle is in the electric mode. For this purpose, it is calculated by the algorithm which value should be at the end of each path part until the battery charge status at the time the vehicle is operated, until it reaches the area where it will operate with the electric mode. The steps mentioned up to this step are calculated offline. Reference battery state of charge information, which is the output of offline steps, is the one of the input of the online control algorithm of the vehicle.

In the online part of the developed controller; the operating mode of the vehicle, the driver power request, the power to be generated by the generator set, and the momentum of the generator set are calculated as instant parameters. In this section, the power demanded by the driver is calculated by usage of the driver's accelerator pedal and the rotational speed of the traction motor. The driver's operating mode is then determined using the driver's instantaneous power request, battery state of charge, battery discharge power, and reference battery charge status. If the specified mode is one of the hybrid modes, the speed and torque value of the generator set are determined. Despite the specified operating mode, if the reference SOC value can not be reached, it is ensured that the side system loads are interfered and their performance is restricted to achieve the required charging power. Fig. 2 shows the flow chart diagram of developed control algorithm.

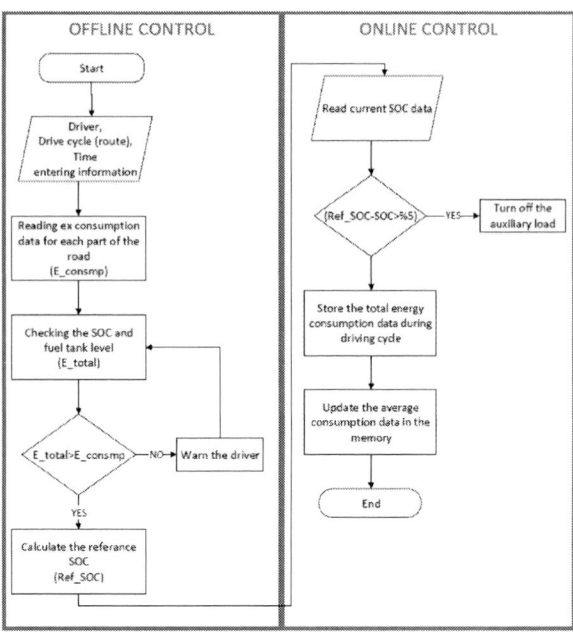

Figure 2 The basic flow diagram of the energy control algorithm

Vehicle operation mode is determined by the FLC. The inputs of this controller shown in Fig. 3 are respectively; the power demand of the driver, the discharge power of the battery, State-of-charge (SOC) and the SOC difference between reference SOC and the actual SOC.

The driver power request input has six membership functions, the battery discharge power and SOC inputs have five each, and the SOC difference input has 3 membership functions.

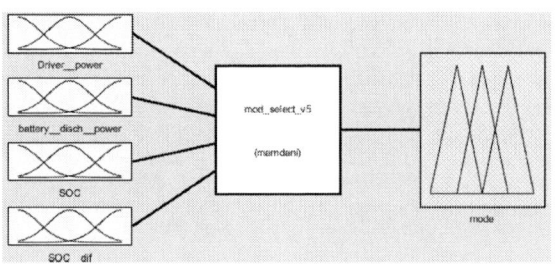

Figure 3 Fuzzy logic mode selection controller

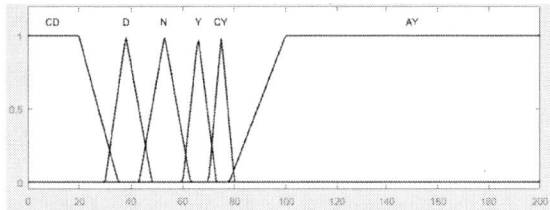

Figure 4 Driver power request membership functions

The membership functions of the driver power request shown in Fig. 4. If the power request is between 0-35 kW, it belongs to the CD membership function. If the power request is between 30-48 kW, it belongs to the D membership function. If the power request is between 43-63 kW, it belongs to the N membership function. If the power request is between 60-73 kW, it belongs to the Y membership function. If the power request is between 70-80

kW, it belongs to the CY membership function. If the power request is between 78-200 kW, it belongs to the AY membership function.

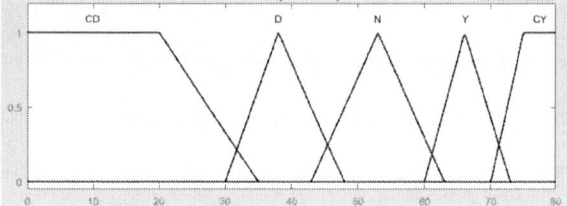

Figure 5 Battery discharge power membership functions

The membership functions of the power request of the battery discharge shown in Fig. 5. If the discharge power input is between 0-35 kW, it belongs to the CD membership function. If the discharge power input is between 30-48 kW, it belongs to the D membership function. If the discharge power input is between 43-63 kW, it belongs to the N membership function. If the discharge input is between 60-73 kW, it belongs to the Y membership function. If the discharge input is between 70-80 kW, it belongs to the CY membership function. If the discharge input is between 78-200 kW, it belongs to the AY membership function.

SOC membership functions are shown in Fig. 6. If SOC is between 0-20%, it belongs to the CD membership function. If SOC is between 15-35%, it belongs to the D membership function. If SOC is between 30-80%, it belongs to the N membership function. If SOC is between 70-90%, it belongs to the Y membership function. If SOC is between 85-100%, it belongs to the CY membership function.

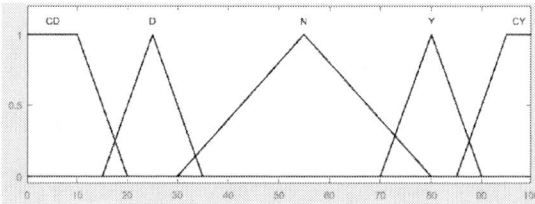

Figure 6 SOC membership functions

The driver power request, battery discharge power and SOC inputs can determine whether the vehicle will operate in a mode regardless of the difference of the SOC input. The reference SOC value is able to direct the system to two different modes according to the sign of the input SOC value. If the value is between -100-0, the decision is electric mode, but if it is between 0-100, the result indicates the hybrid mode. In case the input is close to 0, this input loses its effect, the vehicle is run in the mode decided by the other 3 inputs. This situation is also used for safety purposes when the vehicle goes off route. The membership functions of SOC difference input are shown in Fig. 7.

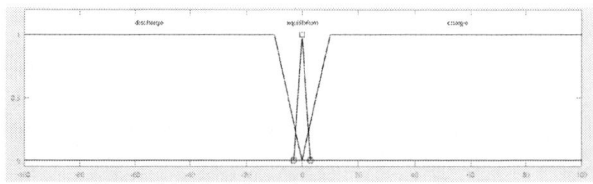

Figure 7 SOC differece's memebership fuctions

The operating mode, which is the output value of the controller shown in Fig. 8, consists of 3 different membership functions. Mode 2 is the electric operation mode of the vehicle. In this mode the generator set is turned off. Mode 1 is the mode for driving support. When the driver power demands above the battery discharge power, the generator set generates extra power to support driving. Mode 0 is the charging mode of the vehicle. The generator set here generates power to charge the battery and, if necessary, to support driving.

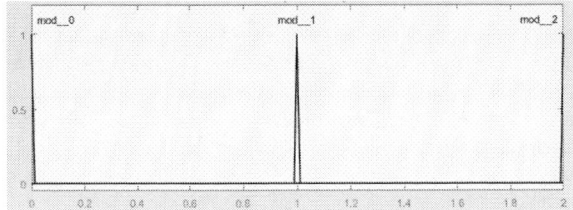

Figure 8 Control mode membership functions

After determining the mode in which the vehicle will run, if the vehicle will run in one of the hybrid modes, the speed at which the generator set will operate is determined. This operation is done by the fuzzy logic controller shown in Fig. 9. The controller has generator set power input and speed output. The membership functions in the controller are based on the efficiency map of the ICE. After determining the speed at which the generator set will operate, the torque value that will be generated for calculating the required output power is calculated. In this way, generator set control has been performed.

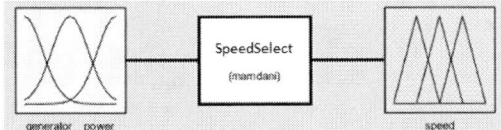

Figure 9 Speed detection controller of the set of the fuzzy logic generator

In the developed hybrid control algorithm, many functions such as driver power request calculation, generator set power requirement calculation and battery charge voltage determination are based on mathematical operations. The outputs of these functions form the inputs of fuzzy logic mode selection and generator speed detection controllers. The relationship between fuzzy logic controllers is shown in Fig. 10. As seen in the figure, FLCs do not directly affect each other's work. The algorithm firstly decides the mode of operation of the vehicle with the fuzzy logic mode selection controller. Then, if the operating mode is one of the hybrid modes, the fuzzy logic speed controller determines the generator speed signal according to generator set power request. However, if this controller selects the electrical operation mode, the generator speed variable is set to 0 rpm.

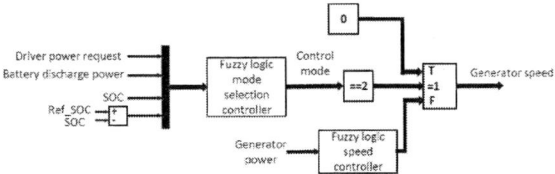

Figure 10 Fuzzy logic controllers

When the reference SOC calculated by the offline controller cannot be reached at the time of travel, the auxiliary system loads are controlled to provide additional power to reach the targeted SOC ratio of the vehicle. This power reduces the performance of the auxiliary system loads, allowing the battery to be charged with the remaining energy. The most important auxiliary system loads here are from the HVAC system of the vehicle. It is the group heater-air conditioner with the most energy consumption from basic auxiliary system loads consisting of heater-air conditioner, battery cooling unit and cooling unit of the power electronic components. If necessary, the units in this group are closed firstly. If sufficient SOC level cannot be reached by switching off the heater-air conditioning group, the performance of the battery cooling and power electronics cooling units are controlled. In these units, performance limitation shall be made to the extent that will not compromise system security.

III. SIMULATION RESULTS

MATLAB/Simulink and TruckMaker programs were used to test the developed controller. The TruckMaker program was used to generate the average consumption data required for the offline controller. The Isuzu NPR10 model vehicle was modeled and the average energy consumption data was obtained by running in the specified routes in different speed profiles. Consumption data were then used to test the controller developed in MATLAB/Simulink environment.

The developed EMS is tested in Simulink environment. For a realistic test simulation, a daily route that has the 56 km total length of a garbage truck is modelled. Modelled route is shown in Fig. 11. In the first shift, the vehicle starts moving from the garage and goes to Acıbadem and Göztepe respectively. After collecting garbage around here, it goes to the garbage dump center. Then he returns to the garage and waits until the second shift. In the second shift, the car goes to the Kadıköy from the garage and collects garbage in Kadıköy. Then the cycle is continued as the garbage dump center and the garage respectively.

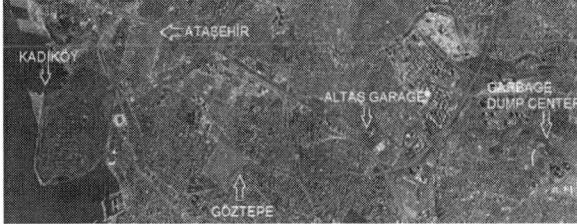

Figure 11 Daily route of garbage truck

The main aim of the controller is to complete two shifts of the garbage truck with a single external night charge. For this purpose, where the vehicle is outside garbage collection areas, the battery is charged internally according to the reference SOC value calculated by the offline controller.

The control algorithm is designed to control the vehicle with the highest possible efficiency while at the same time allowing the vehicle to complete it's daily course. This process is intended to guarantee the movement of the vehicle in electric mode at the garbage collection areas. For this reason, the system, which with the offline controller and the online controller work together, is simulated first. In this case, the mode determination of the vehicle and the control of the generator set are performed by FLCs. Then, the situation in which the offline controller is not used has been tested. In this case, the vehicle determined the optimum mode of operation and completed the route accordingly. The developed two different controllers have been discussed from the angle of fuel consumption, the completion of the specified garbage collection areas in electric mode, the battery operating range.

A. Results with offline controller

The offline controller determines the appropriate operating mode for a each part of route by using the previous driving information of the vehicle. Criteria in the mode determination process are that the vehicle can complete the day with a single charge and is in electric mode in the garbage collection areas.

In order to produce the previous consumption data required for the operation of the offline controller, simulations are run in the same route with different speed profiles have been performed in the TruckMaker program. For these simulations, hybrid NPR 10 type truck, which will be used with the algorithm, is modeled. Simulation consumption data is averaged and this value is entered as data to the offline controller.

The speed-time graph of the truck is shown in Fig. 12. The truck has a low average speed in the garbage collection area. In addition, it is seen that the vehicle have stopped frequently in this area for garbage collection. Outside the garbage collection area, the vehicle has a higher speed profile and less stops.

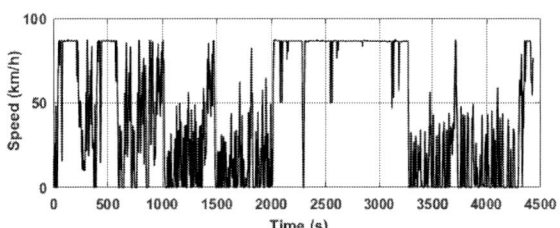

Figure 12 Vehicle speed-time graph

The control mode-time graph is shown in Fig. 13. Vehicle starts to run in mode 0 and charge the battery until it arrives to garbage collection area. It operates in electric mode when it is in first garbage collection area. After that, it run in mode 0 before arrives to second garbage collection area. Second garbage collection area is also completed in electric mode. End of the this area battery SOC level is reach to low limit as shown in Fig. 15. For this reason, control mode switch to mode 0. After SOC level reach to acceptable level, control mode switch to mode 1 to support to drive.

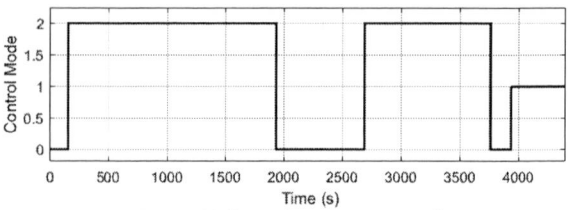

Figure 13 Control mode-time graph

The battery power-time graph is shown in Fig. 14. The batteries are discharged in areas where battery power is positive. In cases where battery power is negative, the battery is charged by the generator set. Due to the structure of the battery, the maximum discharge power is 80 kW and the maximum charging power is limited to 50 kW. According to the control mode-time graph shown in Fig. 13, the vehicle is in electric mode when the battery power is positive.

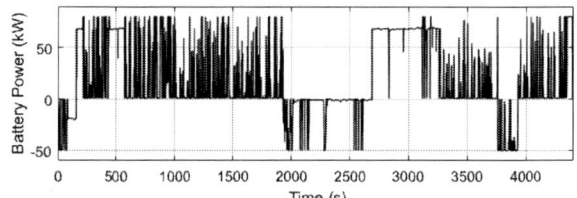

Figure 14 Battery power-time graph

The SOC-time graph is shown in Fig. 15. Initially starting with 95% SOC, the vehicle was charged to 98% SOC until it reached the first garbage collection area. The vehicle then completed the first garbage collection zone in electrical mode. When this area is completed, the battery charge level has decreased to 59%. After leaving the first garbage collection area, the garbage dump center and the car to the garage then increased the battery charge status to 65% until the garbage collection area again in the second shift. At the end of the second garbage collection area, the battery state of charge was reduced to 23%. Since the battery SOC is very close to the lower limit of 20%, firstly the control algorithm has started to charge the battery until it reaches the garbage dump point. After the SOC reached 30% of the level considered adequate, the vehicle entered the support mode. When the vehicle went to the garage and completed the course, the battery occupancy level decreased to 20% again. During the driving cycle, the battery charge status remains within the safe working zone. The vehicle has also completed all garbage collection points in electric mode as targeted.

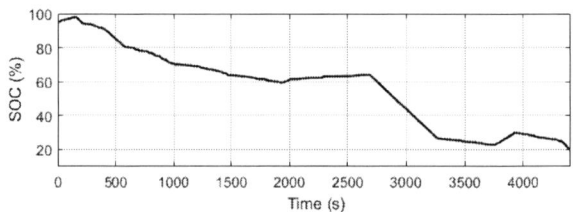

Figure 15 SOC-time graph

978-1-5386-7688-2/19 $31.00 © 2019 IEEE

B. Results without offline controller

In the absence of the offline controller, the online controller determines the mode of operation of the vehicle in accordance with the instantaneous power request, battery discharge power and SOC data. It provides the ideal solution according to instant information. But lack of knowledge about the entire route, it can not guarantee run the vehicle in the electric mode of operation in garbage collection area.

The second simulation was initiated under the same conditions as the first simulation. The route and speed profile during the simulation are the same. On the other hand, depending on the change of vehicle control mode, vehicle performance, distance taken in electric mode and fuel consumption vary.

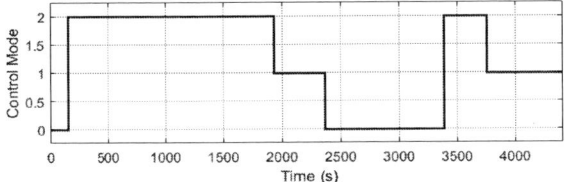

Figure 16 Control mode-time graph without offline controller

Fig. 16 shows the control mode-time graph. When the graph is examined, the first garbage collection area is completed between 211-1928 seconds in electric mode. However, the second garbage collection area cannot be completed in electric mode between 2690-3755 seconds. At the end of the first garbage collection area, the vehicle was put into support mode when the SOC level reached 60% sufficient to exit the electric mode. Switching from support mode to charging mode is achieved when the SOC level reaches the 30% limit, defined as the start of charging in the online controller. The changeover of the vehicle back to electrical mode is achieved when the SOC reaches 40% of the acceptable level. The vehicle has been driven out of electric mode again due to increased drive power demand. The generator set has been activated to support driving, with the drive power request exceeding the highest battery discharge power 80 kW at 3770th second.

Fig. 17 shows the SOC-time graph without offline controller. As shown in figure, vehicle did not exceed the safety operation area limits.

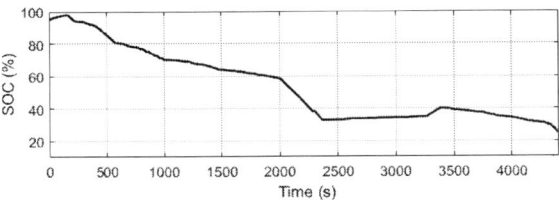

Figure 17 SOC-time graph without offline controller

IV. CONCLUSION

The performance analysis of the controller developed for a hybrid garbage truck was performed. It has been seen that the function of working in electric mode which is one of the main targets of the developed control algorithm is realized in order to reach zero emission and minimum noise level in the garbage collection areas. A conventional NPR10 model truck

was simulated in the TruckMaker program, where it consumed 12.3 liters of fuel to complete the same route. Also, this result was confirmed by road tests.

Fig. 18 illustrates the fuel consumption results of the developed series plug-in HEV with and without offline controller. The vehicle consumes an average of 45.2 kWh of energy for the traction system. As shown in the figure, when the offline controller is in operation, the vehicle completes 32 km of the road in electric mode and consumes 8.9 liters of fuel in total. Compared to the conventional vehicle, a 27.3% fuel saving is achieved. This result is at expected levels compared to the literature. In the absence of the offline controller, the vehicle completed a range of 20 km in electric mode and consumed 10.5 liters of fuel. The vehicle consumed 14.6% less fuel than the conventional vehicle. Reduced electrical range and fuel economy in the absence of an offline controller is due to the fact that the online controller determines the control mode based on instant requests. The offline controller forced the vehicle to run in electric mode in garbage collection areas, increasing the range received in this mode and saving fuel.

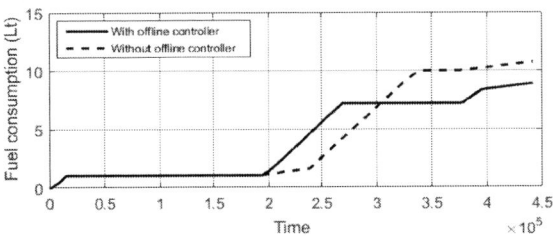

Figure 18 Fuel consumption graph

REFERENCES

[1] Lin, B., Tan , R., "Are people willing to pay more for new energy bus fares?", Energy 130 (2017) 365–372. doi:10.1016/j.energy. 2017.04.153.

[2] Saju, C., Lydia, M., "A comprehensive overview of hybrid electric vehicle: Powertrain configurations, powertrain control techniques and electronic control units", Proceedings of the 2nd International Conference on Inventive Communication and Computational Technologies (ICICCT), 2018.

[3] Das, S., Tan, C., Yatim, A., "Fuel cell hybrid electric vehicles: A review on power conditioning units and topologies", Renewable and Sustainable Energy Reviews 76, 2017.

[4] Tie, SF., Tan, CW., "A review of energy sources and energy management system in electric vehicles". Renew Sustain Energy Rev 20:82–102, 2013.

[5] Micro-Hybrids To Grow Fast: More Than Start-Stop, Less Than Mild Hybrid. Green Car Reports; 2014

[6] Enang, W., Bannister, C., "Modelling and control of hybrid electric vehicles (A comprehensive review)", Renewable and Sustainable Energy Reviews 74 2017.

[7] Ming, L., Ying, Y., Liang, L., "Energy management strategy of a plug-in parallel hybrid electric vehicle using fuzzy control", The 8th International Conference on Applied Energy – ICAE2016.

[8] Peng, J., He, H., Xiong, R., "Rule based energy management strategy for a series–parallel plug-in hybrid electric bus optimized by dynamic programming" Applied Energy 185 1633–1643, 2017

[9] Huang, Y., Wang, H., Khajepour, A., Li, B., "A review of power management strategies and component sizing methods for hybrid vehicles", Renewable and Sustainable Energy Reviews 96 132–144, 2018

[10] Rizzoni, G., Pisu, P., Calo, E., "Control Strategies For Parallel Hybrid Electric Vehicles", IFAC Advances in Automotive Control, Salerno, Italy, 2004

Electric Multipurpose Vehicle Power Take-Off: Overview, Load Cycles and Actuation via Synchronous Reluctance Machine

Branko Ban
Dept. of Electric Machines and Automation
FER; University of Zagreb;
Unska 3, 10000 Zagreb, Croatia
branko.ban@outlook.com

Stjepan Stipetić
Dept. of Electric Machines and Automation
FER; University of Zagreb;
Unska 3, 10000 Zagreb, Croatia
stjepan.stipetic@fer.hr

Abstract—While electric vehicle (EV) technology has been established in passenger vehicle sector, commercial multipurpose vehicle (MPV) penetration is strictly related to the niche end-markets, like medium-duty, short-haul and last mile applications. Some of the examples of short-haul applications are electric multipurpose vehicles (eMPV) like refuse, hook-loader or vacuum trucks. This paper will concentrate on eMPVs, which apart from electric traction, have to actuate additional body systems by the means of electric power take-off (ePTO).

This paper provides load cycle approximations of ePTO load cycles (refuse, concrete mixer, hook-loader and vacuum trucks). These can be used for continuous and peak requirement calculation and optimal ePTO machine sizing. Taking into consideration that the machine needs to be affordable and reliable, synchronous reluctance machine (SyRM) ePTO is proposed for future evaluation.

Keywords—electric machine; traction; synchronous reluctance; optimization; comparison; refuse truck; concrete mixer; vacuum truck; hook-loader truck; design; power take-off; commercial; multipurpose; electric; vehicle; rare-earth free; neodymium.

TABLE I. Abbreviation list

Abbreviation:	Description:
ePTO	Electric power take-off
EV	Electric Vehicle
ICE	Internal combustion engine
IPM	Interior permanent magnet
OEM	Original equipment manufacturer
MPV	Multipurpose vehicle
eMPV	Electric multipurpose vehicle
PTO	Power take off
SyRM	Synchronous reluctance machine

I. INTRODUCTION

In order to reduce emissions, air pollution, resource waste, and traffic noise, government legislation is pushing towards the increase of electric vehicle (EV) production [1]. In addition to legal requirements, shippers and consumers demand cleaner and safer vehicles. This has caused a tectonic change in the automotive industry, both knowledge, and production wise, forcing the industry into a quick development curve. Additionally, the increasing number of start-ups is likely to upset the economics for the traditional original equipment manufacturers (OEM) and suppliers.

Passenger vehicles like Tesla and BMW are well ahead in technology development, which can be used in commercial vehicle industry as leverage (particularly in battery segment) to penetrate niche end-markets like medium-duty, short-haul and last mile applications. Long-haul applications are likely to take more time and be more sensitive to economics.

Short-haul and last mile applications seem to be best suited for immediate EV adoption, with large fleets such as Amazon, UPS and FedEx already making sizable orders for local EV delivery vehicles. Furthermore, the purchasing decisions on these smaller multipurpose vehicles (MPV) face fewer uncertainties than in the case of larger heavy-duty vehicles.

Some of the examples of short-haul applications are electric multipurpose vehicles (eMPV) like refuse, hook-loader or vacuum trucks. This paper concentrates on eMPVs, which apart from electric traction, have to actuate additional body systems (usually powered by some sort of hydraulic pump). Traditionally, this actuation is done via diesel engine or gearbox mounted output shaft, referred as power take-off (PTO). In case of electric trucks, to reduce space claim, weight, and price, the additional electric machine (ePTO) can be the interface to the hydraulic pump shaft. Due to the hydraulic pump and diesel engine operational characteristics, PTO operation is limited to narrow speed range with high torque.

When switching to eMPV, diesel engine does not exists, which creates a big challenge for single size ePTO design. Additionally, hydraulic system load characteristics vary between the applications. This paper will concentrate on providing comprehensive background in eMPV applications including PTO performance requirements and load cycles.

978-1-5386-7688-2/19 $31.00 © 2019 IEEE

II. PTO PERFORMANCE AND REQUIREMENTS

A. General PTO facts

To power the special equipment (cranes, refuse systems, refrigerators, etc.), the MPV must be fitted with an extra means of a power supply, a power take-off (PTO). One or more PTOs transfer power from the engine to drive attachments or load handling equipment. The PTO provides a mechanical link (output shaft) towards load (usually some sort of hydraulic system) with most of the systems having power demand < 90 kW [2]. Historically, the PTO output shaft has been a part of the combustion engine or transmission (Fig. 1a-d). With recent MPV electrification trends [3], [4], PTO will most likely be an extra electric machine mounted on the vehicle chassis (Fig. 1e-h). The ePTO machine will be powered via inverter attached to the traction battery.

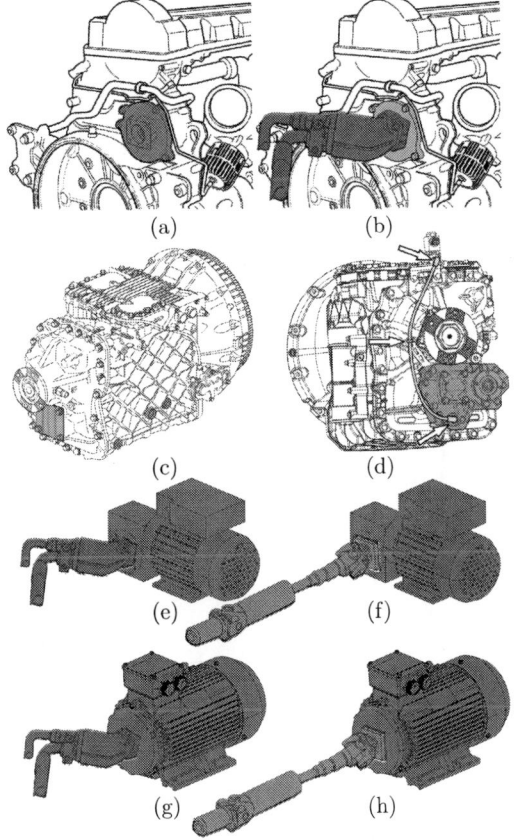

Fig. 1: a) Diesel engine rear PTO mount; b) diesel engine with mounted hydraulic pump; c) Transmission with direct PTO mount; d) transmission with geared PTO [5]; geared ePTO propelling hydraulic pump e) and universal joint shaft f); direct ePTO propelling hydraulic pump g) and universal joint shaft h).

B. Hydraulic pumps

In the applications covered by this paper, the PTO actuates a hydraulic pump (the most common type of PTO attachment). This enables the transmission of mechanical force through the hydraulic system, to any location around the vehicle.

Depending on the application, the hydraulic flow can be finally transformed to linear motion via the piston (e.g. refuse compression), or to torque via a hydraulic motor (e.g. concrete mixer drum rotation) [6].

The main parameters for the pump selection are the required system flow Q_{pump} in l/min and pressure p_{pump} in bars. Pump shaft speed n_{pump} depends on the PTO output speed, which varies depending on the ICE or gearbox gear ratio. For the applications listed in figure 3, PTO output speed can be defined by the referent speed range described by (1) [7].

$$n_{PTO} \in [1500, 3000]\,\text{rpm} \tag{1}$$

Furthermore, the pump type is system related. The most common types are bent axis piston pump 2a), and variable displacement pump 2b). Bent axis piston pump has fixed displacement, meaning that output flow is proportional to the shaft revolution (more flow requires higher shaft speed). These pumps are considered simple and robust, but the disadvantage is a higher level of vibrations in comparison to other alternatives [8]. Variable displacement pumps, as their name indicates, have variable displacement, regulated via special pressure valve, which changes the angle of the piston swash plate. This feature enables more freedom in output flow regulation at different pump shaft speeds. The disadvantage is increased pump weight and system complexity (due to external pressure sensor) [8].

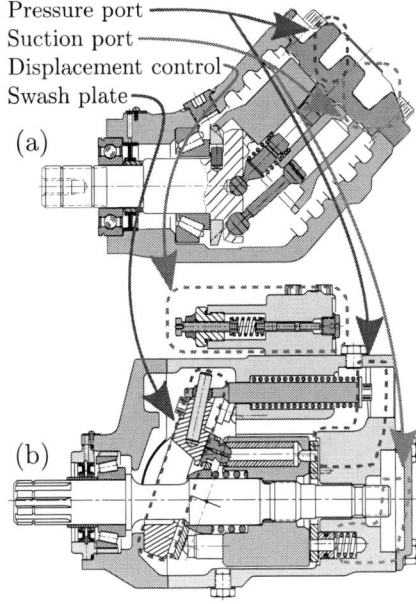

Fig. 2: Example cross-section of truck hydraulic pumps: a) fixed displacement bent axis pump and b) variable displacement pump [8].

C. ICE powered MPV and eMPV PTO requirements

The commercial success of eMPV will depend on the operational range and adaptation to different special body systems. In theory, these systems should have the same functionality as an internal combustion engine (ICE) powered PTO. This assumes the same hydraulic pump size and operational speed range as in (1). Reference [2] provides PTO power range tables for different MPVs. With the assumption of pump speed range (1), and data from [2], peak torque versus power operational areas are calculated in Fig. 3.

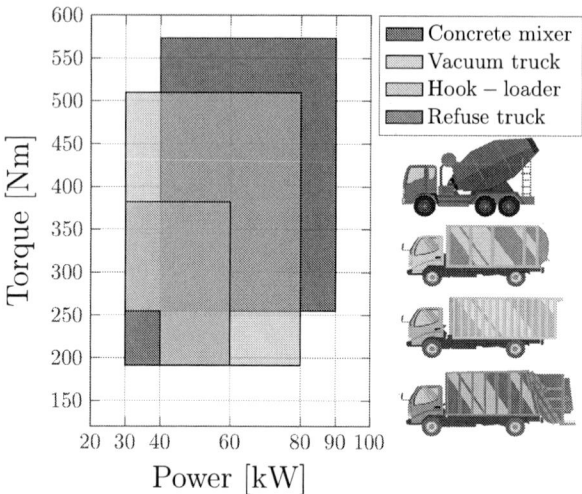

Fig. 3: Torque and power operational areas of some eMPVs.

The most critical MPV types for PTO sizing are: concrete mixer, vacuum truck, hook-loader and refuse truck. Fig. 3 shows that PTO performance requirements are approx. 30-90 kW and 190-570 Nm.

III. PTO LOAD CYCLE FOR DIFFERENT EMPV

A. Refuse truck

Fig. 4: Refuse truck example [9]

Refuse collection applications have a high degree of utilization and are equipped with complex hydraulic circuits. This makes big demands on the reliability of the PTO and requires quiet operation in urban areas [5].

The load cycle has been approximated based on the data from [7] and table II [10]. The cycle consists of three subcycles. Subcycle 1: the opening of the refuse doors and lowering of the bin mounting system, bin loading, and closing of the refuse doors. The vehicle then drives to the next bin and the subcycle 1 is repeated. As the bins are being loaded, the refuse volume (V_{loaded}) is increasing until the truck container is full (V_{truck}). The next step is subcycle 2, in which refuse volume is compacted proportionally to the compaction factor (k_c). The compaction reduces the volume of the loaded refuse, and subcycle 1 repetition starts again. These two subcycles are repeated until no more compactions are possible. The vehicle then begins subcycle 3: driving to the landfill, dumping of the refuse and driving back for another bin collection. Refuse cycle calculation diagram is described in figure 5, subcycle illustrations and real-time cycle are shown in figure 10a.

TABLE II. Refuse truck parameters

Description	Variable	Value	Unit
Max. truck refuse volume	V_{truck}	40	m^3
Max. refuse bin volume	V_{bin}	2	m^3
Compaction factor	k_c	0,5	-
Number of compactions	N_c	> 0	-

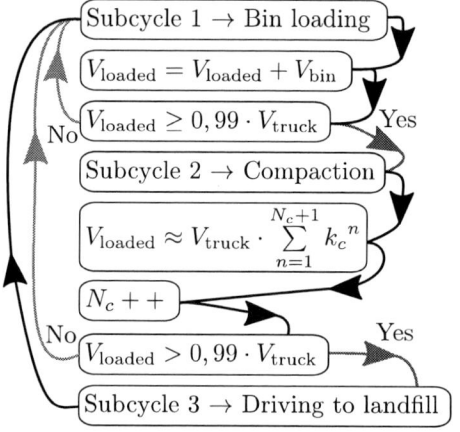

Fig. 5: Refuse cycle calculation diagram.

B. Concrete mixer

Fig. 6: Example concrete mixer truck [11]

Concrete mixers (Fig. 6) come in different sizes differentiated by the payload volume. The calculated load cycle is based on 8 m³ concrete drum volume. This is the most demanding eMPV ePTO application because concrete must be constantly mixed (even while driving), with relatively high torque demand across the PTO load cycle.

The load cycle consists of three subcycles approximated from the data available at [6], [12], [13]. Subcycle 1 describes initial concrete mixing, subcycle 2 is driving to the construction site, while subcycle 3 represents concrete unloading and driving back for another fill. Subcycle illustrations and real-time cycle are shown in figure 10b.

C. Hook-loader

Fig. 7: Hook-loader truck example [14].

Hook-loader (Fig. 7) is a type of MPV with a mounted hydraulic hook-lift hoist. This attachment enables quick swapping of flatbeds, dumpster bodies, and similar containers. Hook-loaders are mostly used for the transportation of materials in the logistic, waste, scrap and demolition industries.

The load cycle consists of two subcycles approximated from [14]. Subcycle 1 describes pushing the container off the the vehicle body, unloading via hook and driving for another container loading. Subcycle 2 represents hook positioning, loading the container and driving to the destination. Subcycle illustrations and real-time cycle are shown in figure 10c.

D. Vacuum truck

Fig. 8: Vacuum truck example [15].

Vacuum truck (Fig. 8) or tanker is a tank hauling MPV with a mounted vacuum pump. The pump is designed for pneumatic suction/unloading of liquids, sludges, slurries, or similar. The payload is transported to the treatment or disposal site (e.g. sewage treatment plant). Data used for load cycle calculation is listed in table III [16].

TABLE III. Vacuum pump parameters

Description	Variable	Value	Unit
Vacuum pressure	V_{vac}	96	kPa
Suction rate	Q_{vac}	40	m³/min
Tank volume	V_{tank}	25	m³
Suction power	P_{vac}	64	kW

The load cycle consists of two subcycles. Subcycle 1 describes the initial tank filling and driving to the disposal site. Subcycle 2 represents tank unloading and driving to another fill site. Subcycle illustrations and real-time cycle is shown in figure 10d.

E. MPV PTO load comparison

Considering the load duration (energy usage), the concrete mixer is the most demanding application and will be setting the continuous (crucial variable for thermal design) and peak ePTO performance requirements. Refuse truck load duration comes second with a distinctive dynamic profile caused by frequent refuse bin loading. This will cause mechanical stress to ePTO bearings, leading to the conclusion that this cycle should be used for rotor cross-section fatigue analysis. Hook-loader and vacuum truck dynamics and load duration are much shorter and do not affect the setting of peak and continuous performance requirements.

IV. ePTO MACHINE SIZING CONSIDERATIONS

A. Peak torque envelope

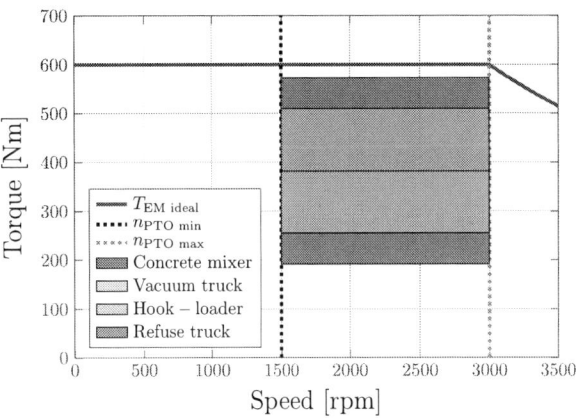

Fig. 9: Torque-speed diagram of PTO operation area for different eMPV and ideal ePTO peak torque envelope (blue line).

Fig. 9 illustrates MPV PTO operational areas in a form of torque-speed graph. The blue line represents the ideal peak torque envelope of ePTO machine.

978-1-5386-7688-2/19 $31.00 © 2019 IEEE

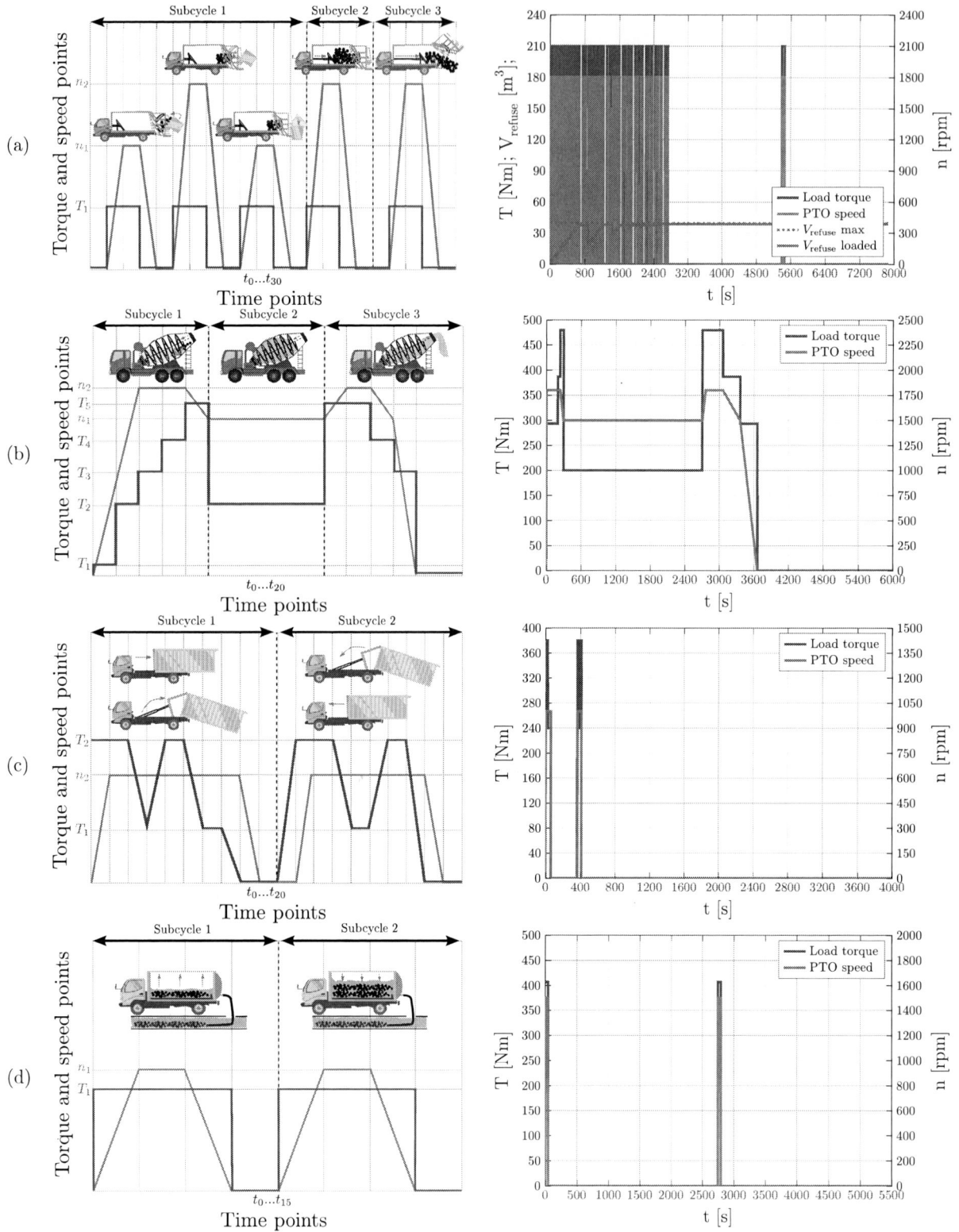

Fig. 10: Illustration of load subcycles (left), and corresponding real-time plots (right) for; a) Refuse truck; b) Concrete mixer; c) Hook-loader truck; d) Vacuum truck.

978-1-5386-7688-2/19 $31.00 © 2019 IEEE

TABLE IV. eMPV ePTO load cycles

	Refuse truck*			Concrete mixer			Vacuum truck			Hook-loader		
	t [s]	T [Nm]	n [rpm]	t [s]	T [Nm]	n [rpm]	t [s]	T [Nm]	n [rpm]	t [s]	T [Nm]	n [rpm]
t_0	0,0	210,0	0,0	0,0	25,0	0,0	0,0	407,0	0,0	0,0	380,0	0,0
t_1	1,0	210,0	1500,0	1,0	25,0	450,0	5,0	407,0	1500,0	0,5	380,0	1000,0
t_2	3,0	210,0	1500,0	1,5	200,0	900,0	5,5	407,0	1500,0	2,0	380,0	1000,0
t_3	3,5	0,0	1500,0	2,0	200,0	1800,0	42,5	407,0	1500,0	20,0	240,0	1000,0
t_4	4,5	0,0	0,0	2,5	293,3	1800,0	43,0	407,0	1500,0	21,0	380,0	1000,0
t_5	6,5	0,0	0,0	200,0	293,3	1800,0	47,5	407,0	0,0	23,0	380,0	1000,0
t_6	7,0	210,0	0,0	200,5	386,7	1800,0	48,0	0,0	0,0	24,0	240,0	1000,0
t_7	8,0	210,0	1800,0	240,0	386,7	1800,0	2747,5	0,0	0,0	60,0	240,0	1000,0
t_8	12,0	210,0	1800,0	240,5	480,0	1800,0	2748,0	407,0	0,0	61,0	0,0	1000,0
t_9	12,5	0,0	1800,0	300,0	480,0	1500,0	2752,5	407,0	1500,0	61,5	0,0	0,0
t_{10}	13,5	0,0	0,0	300,5	200,0	1500,0	2753,0	407,0	1500,0	361,5	0,0	0,0
t_{11}	15,5	0,0	0,0	2700,0	200,0	1500,0	2790,0	407,0	1500,0	365,5	380,0	0,0
t_{12}	16,0	210,0	0,0	2700,5	480,0	1500,0	2790,5	407,0	1500,0	366,0	380,0	1000,0
t_{13}	17,0	210,0	1500,0	2760,0	480,0	1800,0	2795,0	407,0	0,0	371,0	380,0	1000,0
t_{14}	19,0	210,0	1500,0	3060,0	480,0	1800,0	2795,5	0,0	0,0	375,0	240,0	1000,0
t_{15}	19,5	0,0	1500,0	3060,5	386,7	1800,0	5495,0	0,0	0,0	400,0	240,0	1000,0
t_{16}	20,5	0,0	0,0	3360,0	386,7	1500,0	-	-	-	401,0	380,0	1000,0
t_{17}	80,5	0,0	0,0	3360,5	293,3	1500,0	-	-	-	410,0	380,0	1000,0
t_{18}	81,0	210,0	0,0	3660,0	293,3	0,0	-	-	-	410,5	0,0	1000,0
t_{19}	83,0	210,0	1800,0	3660,5	0,0	0,0	-	-	-	411,0	0,0	0,0
t_{20}	143,0	210,0	1800,0	6060,0	0,0	0,0	-	-	-	4011,0	0,0	0,0
t_{21}	143,5	0,0	1800,0	-	-	-	-	-	-	-	-	-
t_{22}	145,5	0,0	0,0	-	-	-	-	-	-	-	-	-
t_{23}	205,5	0,0	0,0	-	-	-	-	-	-	-	-	-
t_{24}	2605,5	0,0	0,0	-	-	-	-	-	-	-	-	-
t_{25}	2606,0	210,0	0,0	-	-	-	-	-	-	-	-	-
t_{26}	2608,0	210,0	1800,0	-	-	-	-	-	-	-	-	-
t_{27}	2668,0	210,0	1800,0	-	-	-	-	-	-	-	-	-
t_{28}	2668,5	0,0	1800,0	-	-	-	-	-	-	-	-	-
t_{29}	2670,5	0,0	0,0	-	-	-	-	-	-	-	-	-
t_{30}	5070,5	0,0	0,0	-	-	-	-	-	-	-	-	-

* Refuse truck load points are valid only for illustrated subcycle in Fig. 10a) (left), real time load can be calculated following the Fig. 5 diagram.

B. SyRM as a potential ePTO

Currently, due to the inherently high torque and power density, interior rare earth permanent magnet synchronous machines (IPM) are predominantly used for automotive traction. Although the performance benefits are undisputed, the use of rare earth permanent magnet (PM) materials, such as neodymium or dysprosium, has raised concerns in a number of areas. In 2011 and 2012, China reportedly threatened to cut off international supplies of these materials [17], leading to dramatic, though short-term, increase in the material price, increasing as much as 3000% in case of dysprosium [18]. These volatilities have forced the automotive industry to search for the alternative electric machine design, which will either use none or minimal amount of rare earth material.

The alternatives which have not been in recent industry focus are synchronous reluctance machines (SyRM), and its derivatives [19]. These solutions rely on high reluctance torque, thus theoretically needing no PM material in the rotor structure. They have relatively low material costs, low rotor losses and are considered as robust [20], [21]. On the other hand, lack of the permanent magnetic field in the rotor is penalized with lower torque density, lower power factor, and higher torque ripple [22], [23].

SyRM is considered to be a direct competitor of induction machines which are often used in railway and ship propulsion. This has been confirmed by [24], concluding that both IM

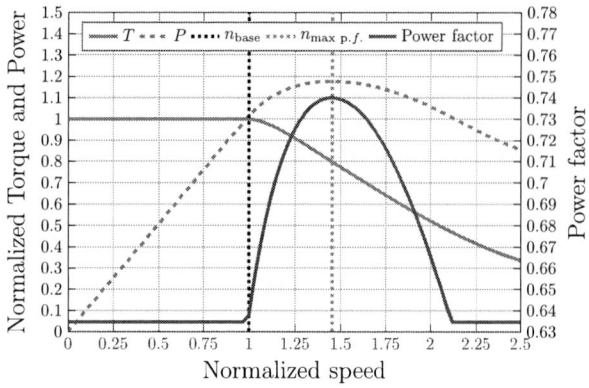

Fig. 11: Example normalized SyRM performance and power factor.

and SyRM have similar electric, magnetic, and thermal performance, with SyRM having lower rotor losses. Germishuizen et al. emphasize that SyRM compared to IM have lower stator winding temperature rise, which allows an increase of the power rating by 5 - 10% [25].

Considering price, overload capability and production simplicity, SyRM can be a viable ePTO solution. Fig. 11 illustrates the normalized performance and power factor achievable by 4 pole, inverter regulated SyRM. It is important to note that the power factor is quite low below the base speed. To increase inverter utilization and increase peak power, ePTO machine should be designed for operation in the area between base speed n_{base} and maximum power factor speed $n_{\text{max p.f.}}$. The following chapter will assume that SyRM ePTO $n_{\text{base}} = n_{\text{PTO min}} = 1500$ rpm and $n_{\text{max p.f.}} = n_{\text{PTO max}} = 3000$ rpm.

C. Direct or geared ePTO machine considerations and future work

Figure (Fig. 12, 13, blue) indicates the ideal ePTO output torque should be 600 Nm. This requirement can be achieved in two ways.

The first approach is high torque (and low speed) machine which does not require a gearbox and can be more easily packaged (Fig. 1g-h). The disadvantage is increased machine volume and weight.

The second approach is a high speed (and low torque) machine with integrated gearbox (Fig. 1e-f). The benefits of this approach are minimized machine volume and weight and increased rotational inertia (traditionally, ICE PTO operates in speed control mode with one or two fixed reference speeds, high rotational inertia comes from ICE crankshaft). The disadvantages are increased system complexity and more complicated packaging. To cover both approaches, two SyRM variants will be investigated in future work.

High torque (low speed) variant performance is illustrated in Fig. 12. It is important to note that output base speed is lower than ideal requirement (Fig. 12, blue). This compromise will maximize SyRM power factor and output power in n_{PTO} (1) speed range. The disadvantage is limitation in pump selection

options, mainly in concrete mixer application. It must be noted that even with this performance adjustment, ePTO can successfully complete concrete mixer cycle.

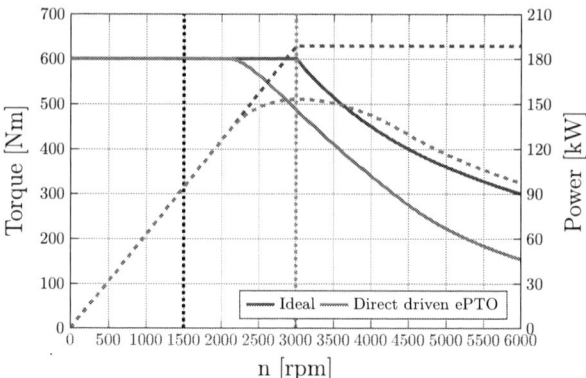

Fig. 12: Ideal (blue) and directly driven ePTO performance.

High speed (low torque) variant (Fig. 13, green), will be coupled to reduction gearbox with transfer ratio $i = 2$, leading to output performance profile (Fig. 13, red). Gear ratio has been selected so that the geared ePTO (Fig. 13, red) has the same output performance as the direct ePTO (Fig. 12, red).

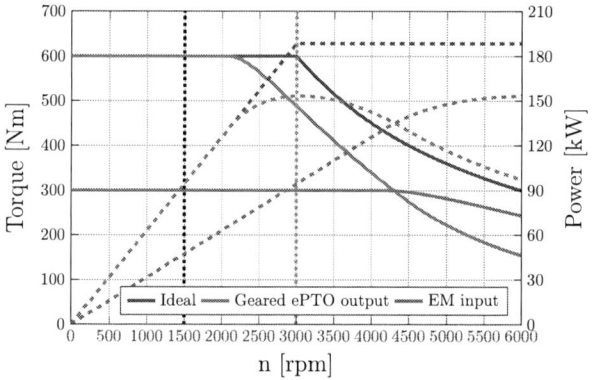

Fig. 13: Ideal performance (blue), input performance before gearbox (green), and ePTO output performance including gearbox (red).

V. CONCLUSION

This paper has covered PTO theory, applications, and recent MPV industry status with emphasis on eMPV e-PTO.

Special attention has been addressed to PTO load cycle definition and calculation. As a next step, this data will be used for the calculation of continuous and peak performance requirements as inputs for optimal ePTO machine design.

Considering robustness, price and reliability, a synchronous reluctance machine (SyRM) has been selected as target ePTO technology. Future work will concentrate on the validation of two concepts, directly driven, or gearbox integrated SyRM design.

REFERENCES

[1] European Environment Agency, "Electric Vehicles in Europe," p. 60, 2016. [Online]. Available: https://www.eea.europa.eu/publications/electric-vehicles-in-europe

[2] VOLVO, "Power take-offs: Fields of application and Calculation guide." 2018. [Online]. Available: http://productinfo.vtc.volvo.se/files/pdf/lo/PowerTake-off(PTO)_Eng_08_580114.pdf

[3] ——, "Volvo Trucks - Premiere for our first all-electric truck," 2018. [Online]. Available: https://www.youtube.com/watch?v=zAbbulKiX-o

[4] ——, "Premiere for Volvo Trucks first all-electric truck," 2018. [Online]. Available: https://www.volvogroup.com/content/dam/volvo/volvo-group/markets/global/en-en/news/2018/apr/180412-volvo-all-electric-truck--en-2018-04-12-08-30-54.pdf

[5] ——, "Body builder instructions: PTO and pumps," 2018. [Online]. Available: https://www.volvotrucks.ca/en-ca/parts-and-services/service/body-builder/manuals/

[6] M. H. Bae, T. Y. Bae, and D. J. Kim, "The Strength Analysis of Differential Planetary Gears of Gearbox for Concrete Mixer Truck," *IOP Conference Series: Materials Science and Engineering*, vol. 317, no. 1, pp. 0–8, 2018. [Online]. Available: https://iopscience.iop.org/article/10.1088/1757-899X/317/1/012010/pdf

[7] VOLVO Trucks Canada, "mDrive PTO Pump Speed Calculator - Volvo Trucks Canada," 2019. [Online]. Available: https://bit.ly/2IhGBZ8

[8] Parker Hannifin, "Series GPA, GP1, F1, T1, F2, F3, VP1, Fixed and Variable Displacement Pumps, Motors and Accessories," p. 80, 2019. [Online]. Available: https://www.parker.com/Literature/PMDE/Catalogs/Truck_Hydraulics/MSG30-8200_UK.pdf

[9] ShinMaywa Industries, "Refuse Compactor (Garbage Compactor Truck)." [Online]. Available: https://www.shinmaywa.co.jp/truck/english/products/refuse_compactor.html

[10] Veolia, "Veolia commertial waste containers," 2019. [Online]. Available: http://veolia.co.uk/birmingham/sites/g/files/dvc501/f/assets/documents/2014/10/Commercial_Waste_Containers.pdf

[11] Construction Equipment, "Putzmeister Redesigns Pro Series, Bridge Maxx Mixers," 2019. [Online]. Available: https://www.constructionequipment.com/putzmeister-redesigns-pro-series-bridge-maxx-mixers

[12] C. Ferraris, "Concrete mixing methods and concrete mixers: State of the art," *Journal of Research of the National Institute of Standards and Technology*, vol. 106, no. 2, p. 391, 2012.

[13] J. Yang, H. Zeng, T. Zhu, and Q. An, "Study on the dynamic performance of concrete mixer's mixing drum," *Mechanical Sciences*, vol. 8, no. 1, pp. 165–178, 2017.

[14] HIAB, "HIAB multilift systems," 2019. [Online]. Available: https://www.hiab.com/

[15] PUMPER, "8 Tips for Vacuum Truck Shoppers," 2019. [Online]. Available: https://www.pumper.com/online_exclusives/2014/09/8_tips_for_vacuum_truck_shoppers

[16] K. vaccum trucks, "Mobile Vac," 2019. [Online]. Available: http://www.kanematsu-eng.jp/english/products/mobile-vac.html

[17] K. Bourzac, "The Rare-Earth Crisis," 2011. [Online]. Available: https://www.technologyreview.com/s/423730/the-rare-earth-crisis/

[18] J. Rowlatt, "Rare earths: Neither rare, nor earths," 2014. [Online]. Available: http://www.bbc.com/news/magazine-26687605

[19] U. Departent of Energy, "Annual Progress Report Electric Drive Technologies Program," Tech. Rep. July, 2011.

[20] J. R. Riba, C. López-Torres, L. Romeral, and A. Garcia, "Rare-earth-free propulsion motors for electric vehicles: A technology review," *Renewable and Sustainable Energy Reviews*, vol. 57, pp. 367–379, 2016.

[21] S. Estenlund, M. Alaküla, and A. Reinap, "PM-less machine topologies for EV traction: A literature review," *2016 International Conference on Electrical Systems for Aircraft, Railway, Ship Propulsion and Road Vehicles and International Transportation Electrification Conference, ESARS-ITEC 2016*, 2016.

[22] G. Pellegrino, T. M. Jahns, N. Bianchi, W. L. Soong, and F. Cupertino, *The Rediscovery of Synchronous Reluctance and Ferrite Permanent Magnet Motors Tutorial Course Notes.* Springer, 2016.

[23] P. Duck, J. Jurgens, and B. Ponick, "Calculation of Synchronous Reluctance Machines Used as Traction Drives," *2015 IEEE Vehicle Power and Propulsion Conference, VPPC 2015 - Proceedings*, pp. 0–4, 2015.

[24] S. M. De Pancorbo, G. Ugalde, J. Poza, and A. Egea, "Comparative study between induction motor and Synchronous Reluctance Motor for electrical railway traction applications," *2015 5th International Conference on Electric Drives Production, EDPC 2015 - Proceedings*, pp. 2–6, 2015.

[25] J. J. Germishuizen, F. S. Van Der Merwe, K. Van Der Westhuizen, and M. J. Kamper, "Performance comparison of reluctance synchronous and induction traction drives for electrical multiple units," *IEEE Transactions on Industry Applications*, pp. 316–323, 2000.

Branko Ban was born in Šibenik (Croatia) in 1991. He completed Bachelor (2012.) and Master (2015.) studies in Electrical Engineering at the University of Zagreb (Faculty of electrical engineering and computing) and Chalmers University of Technology. Since 2015. he is a part of the ALTEN consultancy team working in the automotive sector as electric machine specialist in areas related to design, advanced engineering, quality assurance, and root cause analysis. He is currently enrolled in the University of Zagreb PhD program covering next-generation electric machines.

Stjepan Stipetić was born in Ogulin (Croatia) in 1985. He received Dipl.Eng. and PhD degrees in electrical engineering from the University of Zagreb, Croatia, in 2008 and 2014, respectively. Currently, he is an Assistant Professor at the University of Zagreb Faculty of Electrical Engineering and Computing, Department of Electrical Machines Drives and Automation, Croatia where his research activities are related to design, modelling, analysis and optimization of electrical machines.

978-1-5386-7688-2/19 $31.00 © 2019 IEEE

State-of-Charge Estimation of Li-ion Battery Cell using Support Vector Regression and Gradient Boosting Techniques

Eymen Ipek
Istanbul Technical University
Department of Electrical Engineering
Istanbul, Turkey
ipekey@itu.edu.tr

M. Kerem Eren
Istanbul Technical University
Department of Electrical Engineering
Istanbul, Turkey
erenma@itu.edu.tr

Murat Yilmaz
Istanbul Technical University
Department of Electrical Engineering
Istanbul, Turkey
myilmaz@itu.edu.tr

Abstract— Increasing demand of li-ion batteries brings the need of high accuracy estimation and control of SOC. Different conventional approaches exist to estimate SOC such as open circuit voltage measurement, coulomb counting, electrical model or electrochemical model. Development of data science brings machine learning techniques into SOC estimation of Li-ion batteries. There are number of works which are presented to prove application of machine learning techniques in li-ion battery state estimation. In this paper, two different machine learning algorithms are implemented to estimate SOC of Li-Iron-Phosphate battery cell experimental test data. Support Vector Regression (SVR) and XGBoost are used to estimate SOC. SVR is Support Vector Machine (SVM) based regression method which is used frequently in data science applications. Also, XGBoost is a novel approach for gradient boosting technique which has parallel computation and decreased training time. Radial Basis Function (RBF) kernel of SVR is used to estimate SOC and evaluated to improve results in this study. SVR and XGBoost are compared in terms of ease of implementation, performance, accuracy and duration. Between 97%-99% coefficient of determination is achieved during the estimations by adapting different parameters.

Keywords—electric vehicles, li-ion batteries, state of charge estimation, machine learning, support vector regression, gradient boosting

I. Introduction

The world industry for the vehicles is now changing with the support from the governments that are describing the transition to electric vehicles from the fossil fuel powered vehicles. The main target here are to increase the efficiency, the desire of having healthy and quality city life with emission-free and less noise environment. In that point, electrification is an inevitable result for transportation to decrease CO_2 emissions in the cities and to slow down global warming.

Not only in the automotive industry but also in storage and grid applications, the Li-Ion battery chemistries are the dominant type of chemistry which has superiorities to the other chemistries. Since the first launched Li-Ion battery in 1991 by Sony [1], there have been many technological developments to make Li-Ion batteries the most preferable and energy dense battery. Although the specific energy and power values for EV applications are quite prominent, these values are relatively higher for Li-ion batteries compared to other types of batteries. One of the most important reason is that lithium element has the lowest reduction potential which makes the lithium element with the highest electroactivity. In addition, the lithium element is among the lightest elements in the 3rd place which also contributes to the specific energy and

power values [2]. As stated in [3], the state-of-the-art specific gravimetric energy ratios are in the range of 250-300 Wh/kg and this value is expected to be 400-500 Wh / kg in 2030. Although the volume density of the battery pack is also quite prominent for the design phase, 400-500 Wh/L is considered as a state-of-the-art value for a battery cell [3].

Considering the Li-ion batteries, many different types of chemicals have been considered a key solution for the applications such as lithium cobalt oxide (LCO), lithium iron phosphate (LFP), lithium manganese oxide (LMO), lithium nickel manganese cobalt oxide (NMC), lithium nickel cobalt aluminum oxide (NCA), and lithium titanate oxide (LTO). All of the elements mentioned above serve different properties and the ratios within the chemical affect these properties. For instance; nickel-based batteries have much more energy density [2]. However, the nickel ions have an approach to substitute Li ions which may cause a degradation of the battery capacity because it results to block the lithium diffusion pathways [4]. Apart from the aforementioned commercial types of batteries, there are also many different chemistry types which are promising candidates to be selected in the future as a dominant chemistry such as solid state batteries without liquid electrolyte which can be stable with extreme operations [5], Li-Air battery which oxygen from air is used as an electrode which means very high specific gravimetric density [6], Li sulfur battery which has carbon based sulfur cathode with higher energy density [7].

Since Li-ion battery systems have different security problems such as fire or explosion due to reactive chemicals, they must be equipped with advanced battery management systems. The battery management system (BMS) should be capable of measure the safety relevant input from the battery such as temperature, current which is drawn from the battery and battery states such as state-of-charge (SOC) and state-of-health (SOH).

The measurement of SOC is one of the most important parameters of the BMS and serves as a key to ensure that all cells have a stable voltage. The balancing which can be done with the help of accurate SOC measurement may be the biggest factor in prolonging the life of the battery pack and preventing the damage which can be faced during lifetime such as burning, explosion and swelling.

There are some basic methods for SOC estimation for Li-ion batteries which can be categorized into different chapters. Voltage, current and temperature parameters are generally used as an input to the SOC estimation.

Direct methods [8] include ampere-hour counting, open-circuit-voltage (OCV) method, impedance and internal

978-1-5386-7688-2/19 $31.00 © 2019 IEEE

resistance measured method. While the ampere hour method is used as the SOC measurement method in the industry, the initial value cannot be known by this method and the errors that may occur during the accumulation of the instant calculated values can be listed as the negativity of this method [8]. At the same time, the fact that this method only relies on current measurement which can be held by the sensors, so the measurement errors can also affect the SOC value. As the OCV mapping method is dependent on the voltage measurement, the measurement must be with high precision. Due to the charge-discharge characteristic of the batteries, a certain amount of time must be passed after the charge-discharge cycle of the cells to reach its real voltage value. The other backward of this approach is that, as the different C-rates are applied for charge-discharge cycle, the OCV-SOC curve shifts upward or downward depending on the desired current value and those important points make the OCV mapping method unfit to be used as a base SOC method only in the battery system [8].

Model-based estimation methods include electrochemical model and electrical circuit model. In the electrical circuit model, it is aimed to model the cell characteristics using the basic electrical elements such as the resistances used to represent inner resistance and diffusion resistance and the capacitances represents the capacitive diffusion effects in a cell [8]. Higher number of parallel RC pairs which represent diffusion effects can be beneficial to improve the efficiency [9]. In electrochemical model, the chemical equations are used to create a model regarding the diffusion, intercalation and the kinetics in a cell [8].

In addition to the direct and model-based estimation methods, adaptive filter-based methods using Kalman filters and adaptive artificial intelligence-based methods using genetic algorithms and artificial neural networks can also be used as different SOC estimation methods. With the development of artificial intelligence methods especially increasing machine learning methods can offer solutions that can be used for SOC estimation.

The above-mentioned measurement methods of SOC can also be used together to increase the accuracy. For instance, in [10], a hybrid method which combines adaptive and Kalman observers is proposed to estimate the SOC. Thus, deficiencies of each model can be developed with another model.

Recently, machine learning applications have been used for SOC estimation applications. Using the Extreme Learning Machine, the SOC estimation of Li-Iron-Phosphate was studied and found to be faster than SVM [11]. In another study, SVM was compared with Artificial Neural Network (ANN), and train and test data were separated by 90% and 10%, and the predictive performance of the model was demonstrated. SOC estimation of 18650 cylindrical Li-ion cells with a maximum error of 4.17% [12]. Similar study was performed using SVM and high accuracy was obtained with low calculation times [13]. Also, Michel and Heiries used a hybrid model in cooperate with Kalman filter to improve the electrical model [14].

Machine learning algorithms (MLA) are the methods based on artificial neural networks, which are used to extract a model using previous data. With this approach, they reduce the effort during the creation of the model. Traditional methods require the establishment of a model of electrical or electrochemical and validation of this model. Therefore, it

may require long efforts during the establishment of the model. MLA, on the other hand, are the applications that have little effort to model with the proper test data. However, problems such as overfitting or irrelevant data collection should be avoided. The use of machine learning applications for SOC estimation is becoming popular nowadays and with the development of algorithms it becomes possible to use it in practical applications.

Since the application is easy and it gives high accuracy results with accurate data, the use of machine learning applications for SOC estimation is seen in the literature recently. The Support Vector Machine (SVM) algorithm used in the past also in hybrid form with various conventional methods. Besides, it can be used alone when well trained.

Although it provides relatively accurate results for linear and non-linear systems, incorrect results may be obtained if the kernel functions of the SVM algorithm are not correctly defined. For this reason, SVM must be set up correctly when using. With the development of data science, new methods have been evolved and their applications have been used for SOC estimation. XGBoost is a recent gradient boost algorithm. XGBoost is a decision tree algorithm that works faster than SVM with parallel calculation.

In the scope of this work, SVM and Gradient Boosting algorithms were studied in a comparative way. Usage of these algorithms for SOC estimation has investigated and it has been made possible to apply these algorithms using test data. The SVM algorithm for SOC estimation was used for regression. Results were obtained according to different gamma and C parameters. In addition, the same test data were estimated with XGBoost, a Gradient Boosting algorithm. The results were compared with the test data.

II. SUPPORT VECTOR MACHINE FOR SOC ESTIMATION

SVM has started to be considered as a decision platform based on linear approach before 1980s. It started to be developed by increasing tendency of Neural Network algorithms and it became a non-linear featured decision tree for several applications until 1990s [15].

Modern algorithms have improved also SVM and it has been developed as an optimization algorithm which prevents local minimum points via neural networks [15]. SVM algorithm is used to solve classification problems frequently in the machine learning area. It defines boundary vectors which are imaginary between data groups. There is an infinite possibility of these vectors' locations; therefore, SVM optimizes these vector positions based on the optimum place to data points which should be equally close to data boundaries.

Since SVM is an optimization algorithm for these support vectors, it can be also used for regression problems. Thus, the same concept is valid for regression: it tries to find a minimum error by maximizing the margin and separating the boundaries.

Fig. 1 and Fig. 2 describes how linear and non-linear regression is done by SVM algorithm. By using available data, SVM tries to optimize norm vectors to find a suitable equation. This is the relation between input vectors, weight of the feature and kernel functions for non-linear SVM algorithm. As shown in Fig. 2, SVM tries to linearize the function by converting inputs to functions.

978-1-5386-7688-2/19 $31.00 © 2019 IEEE

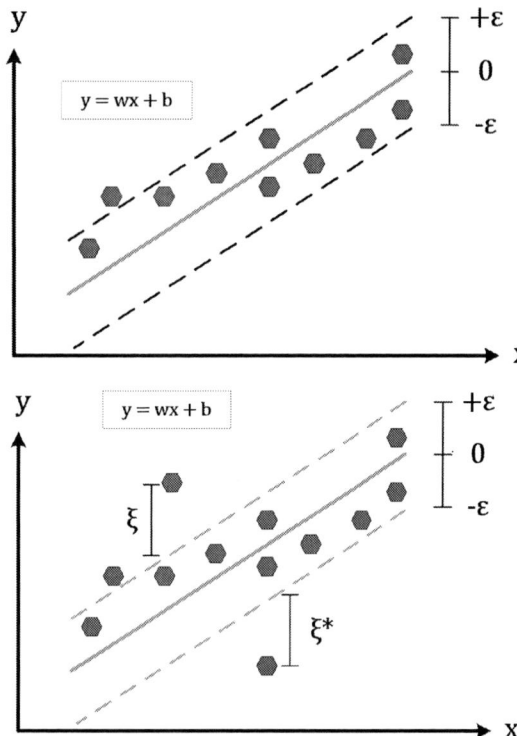

Fig. 1. Linear regression with SVM

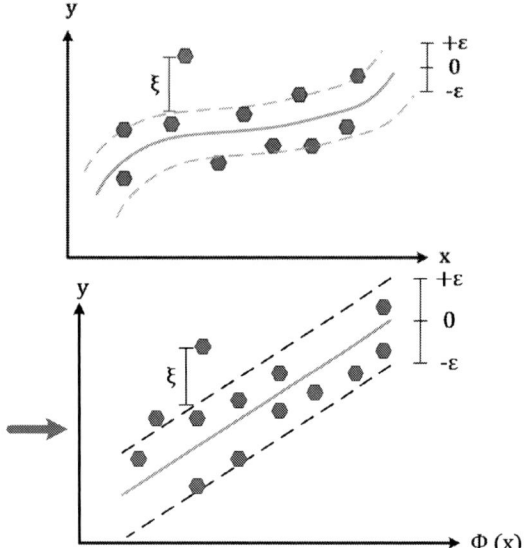

Fig. 2. Non-linear regression with SVM

Based on error equations, SVM tries to optimize norm vector [16]:

$$y_i - wx_i - b \leq \varepsilon + \xi_i \qquad (1)$$

$$wx_i + b - y_i \leq \varepsilon + \xi_i^* \qquad (2)$$

$$min \frac{1}{2}\|w\|^2 + C \sum_{i=0}^{N}(\xi_i + \xi_i^*) \qquad (3)$$

Where;
x: input vector
b: bias
y: predicted output value
$C > 0$: error and flatness trade-off constant
w, σ = weight the feature (or w_i for each)
ξ = optimal hyperplane or maximal margin

$$\xi_i, \xi_i^* \geq 0 \qquad (4)$$

Linear SVM:

$$y = \sum_{i=1}^{N}(a_i + a_i^*) \cdot \langle x_i, x \rangle + b \qquad (5)$$

Non-linear SVM:

$$y = \sum_{i=1}^{N}(a_i + a_i^*) \cdot \langle \Phi(x_i), \Phi(x) \rangle + b \qquad (6)$$

$$y = \sum_{i=1}^{N}(a_i + a_i^*) \cdot K\langle x_i, x \rangle + b \qquad (7)$$

In the modern applications, based on linear and non-linear equations of SVM are evolved. Kernel functions are created by using polar coordinates to improve classification and regression performance of SVM algorithm:

Polynomial:

$$k(x_i, x_j) = (x_i \cdot x_j)^d \qquad (8)$$

Gaussian Radial Basis Function (RBF) [16]:

$$k(x_i, x_j) = exp\left(-\frac{\|x_i - x_j\|^2}{2\sigma}\right) \qquad (9)$$

In order to apply these functions for SOC estimation, A123 Nanophosphate M1 26650 cylindrical li-ion cell test data is used based on University of Maryland CALCE Battery Research Group experimental test data [17,18]. The experimental data is investigated and re-organized to perform proper training of the machine. Besides, irrelevant data is excluded from dataset and required features are specified to perform SOC estimation.

Fig. 3 exhibits the experimental test data from University of Maryland CALCE Battery Research Group which is conducted under low current. This means, internal resistance effect on the OCV can be assumed as low. Fig. 4 presents OCV-SOC relation based on manufacturer datasheet. These available data are used as a reference to train and test the machine learning models after some adjustment on the data structure. Since, Python is powerful language with open source development environment, out of these datasets, required training and test data are created in Python language and its development environment. Different functions of the development environment are used to get results with the errors and computational durations. The improvements on the kernel function parameters are made based on the estimation results and the errors. Linear, polynomial and RBF functions are tested to see results.

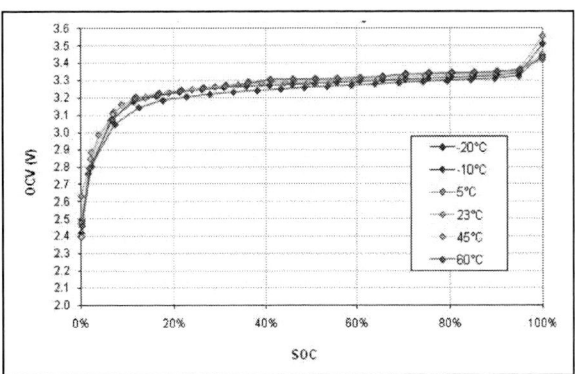

Fig. 4. A123 M1 26650 Cell SOC-OCV Curve [19]

SVM algorithm is implemented by using Python libraries. First, all data is read and then train and test data are separated. Next, the model is trained by train data. RBF kernel function of the SVR algorithm is used for estimation. Approximately, 97% coefficient of determination which is called R2 score is achieved by test results and predicted SOC values after some trials on the parameter tuning.

Gamma and C parameters are tuned during getting results. Since gamma describes how is the effect of a single training data on the estimation performance, if it has a lower value, the boundaries will be smoother, and the results will be slightly off-line. Otherwise, high gamma value will affect sharpness of boundaries. If the data has fault values, the result can be affected by these fault values easily in case of using high gamma value. Besides, C is the trade-off between flatness of the model and the training error which tries to fit data perfectly.

Fig. 5 presents the first feasible tuning results. To improve prediction results SVR, parameters are adjusted, and the outputs are recorded. Increasing gamma values is resulted better predictions. Also, increasing C values is resulted better end-data predictions. However, it is observed that increasing complexity of the equation is resulted high computational effort and; therefore, longer time to train data and get results.

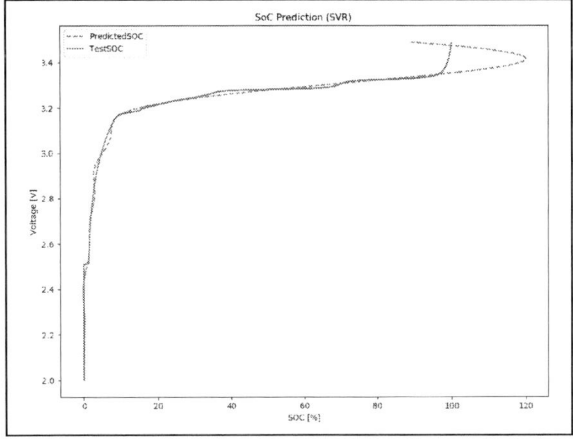

Fig. 5. SVR RBF with C=10, Gamma=15

Fig. 6 and Fig. 7 exhibits the results of the tuning on the C and gamma parameters of RBF kernel function. Out of the results, the determination coefficient is increased by increasing gamma values. Besides, increasing gamma parameter causes computational duration increment.

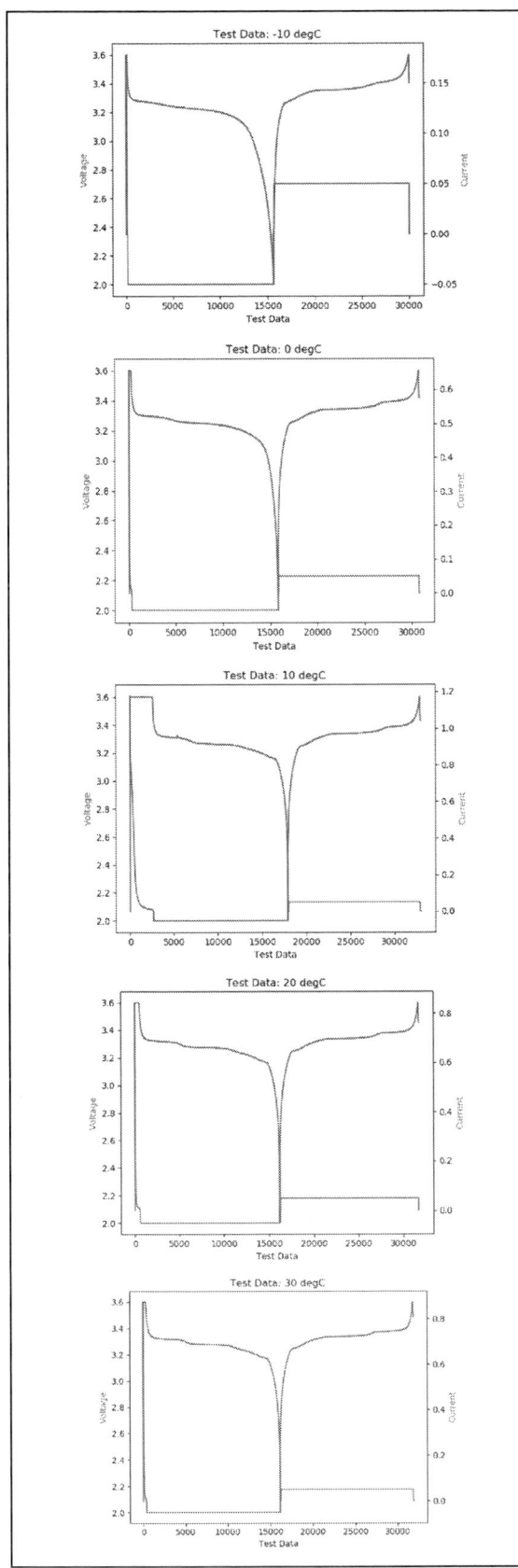

Fig. 3. Experimental test data of A123 M1 26650 [17,18]

978-1-5386-7688-2/19 $31.00 © 2019 IEEE   607

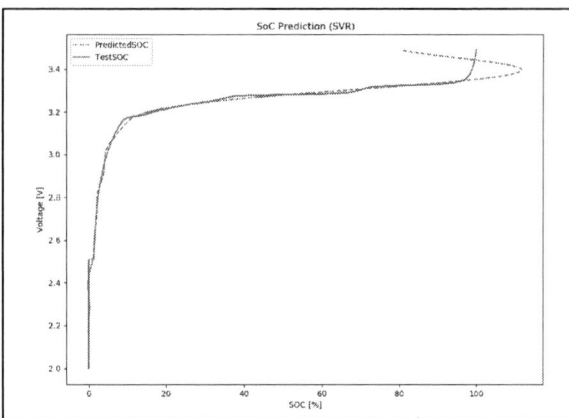

Fig. 6. SVR RBF with C=1200, Gamma=15

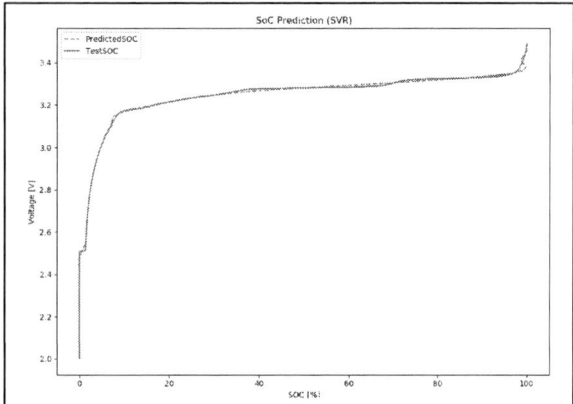

Fig. 7. SVR RBF with C=1200, Gamma=85

Results exhibits that training data also has a big impact on predicted data as well as kernel parameters. Therefore, training data must be used carefully to create proper model for SOC estimation with the proper kernel parameters.

Since SOC estimation accuracy is crucial for li-ion battery applications in EVs, a huge amount of data must be used with proper model training to achieve acceptable results. However, for the applications that SOC estimation accuracy is not that important, less data can be used for training the model by describing a certain SOC window where the accuracy of the model is acceptable.

III. XGBOOST FOR SOC ESTIMATION

Gradient boosting is an also frequently used machine learning algorithm based on decision trees. It uses weak learners to transform them strong learners. Therefore, the ultimate algorithm is an orchestra which includes several different instruments. Basically, gradient boosting algorithms try to improve predictions by optimizing of several arbitrary loss functions which are differentiable.

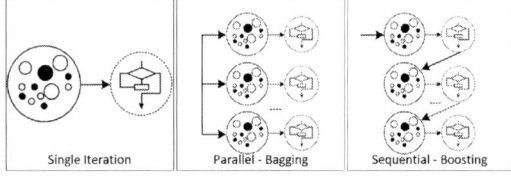

Fig. 8. Explanation of iteration, bagging and boosting

One of the most recent gradient boosting algorithm is XGBoost. Since, evaluation of XGBoost is based on decision trees, it creates possible solutions which relies on certain conditions. Similar to bagging, combination of different models are created and random decision trees are collected. To achieve error minimization, the dominancy of models which have high performing behavior is boosted (increased). Also, to prevent overfitting missing regularization is done and parallel processing is made.

XGBoost has several advantages. Due to it is a scalable and it scales to distributed examples, it does not rely on just one machine learning model. Besides, XGBoost applies quick model exploration by parallel and distributed computing. It has continuous predictions by fast learner iterations. Thus, error minimization performance of XGBoost is high enough and even with small amount of data, the algorithm can predict with very high accuracy.

SOC estimation is also performed by using the same training data. XGBoost is used to train machine and to get results. Fig. 9 represents the results of SOC estimation by using XGBoost. Due to its parallel computation feature, the duration to do calculations is 0.37 seconds by using XGBoost algorithm on the same data.

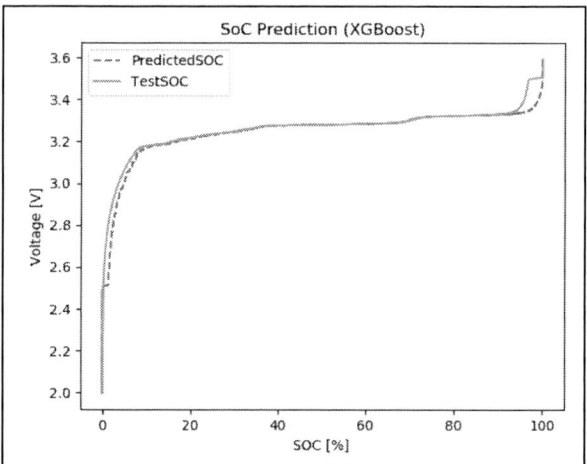

Fig. 9. XGBoost for SOC estimation

IV. COMPARISON OF SVR AND XGBOOST

Since SVR algorithm is a kind of linear separator, it needs a kernel function to separate non-linear data correctly. This kernel function performance affects the accuracy of the results. RBF kernel function allows that the regression can be made with less error. However, decreasing error rate may result long duration of training times especially considering large amount of data. Despite of ease of implementation, SVR has data separating nature. Thus, it can be hard to implement into online systems. On the other hand, XGBoost is a decision tree-based algorithm with parallel computation which has lower computation times and higher accuracies. Increased number of different variables makes XGBoost more effective and applicable to SOC estimation systems than SVR considering number of datasets and parameters.

Above it is tried to explain that SVR is needed to configure the correct kernel function with the feasible parameter sets to achieve acceptable machine learning based SOC estimation model. On the other side, XGBoost is easy to set and it is applicable to SOC estimation applications with several input parameters such as voltage, current, temperature even aging.

Table I shows the performance results of the different evaluations on SOC estimation by using SVR and XGBoost respectively. The results are presented considering duration of computation, root-mean-square-error (RMSE) and R2 score.

TABLE I: PERFORMANCE COMPARISON OF SVR AND XGBOOST

Algorithm	C	Gamma	Duration	RMSE	R2 Score
SVR	10	15	11.36 sec	23.659	0.97123
SVR	1200	15	30.08 sec	19.296	0.97653
SVR	1200	85	1 min 28 sec	16.057	0.98047
XGBoost			0.37 sec	1.322	0.99833

It is observed that XGBoost is more feasible to use for SOC estimation applications due to its quick computation and higher error rate based on this study. To increase SVR performance, two different parameters are tuned which is resulted from increasing computational duration. It can be clearly seen that increasing C parameter with the same gamma values does not affect R2 score substantially. However, it decreased the root mean square error (RMSE) as approximately 18%. The other improvement is done by increasing gamma with the same C value which caused a decrease in RMSE and R2 score with an increase in the computation time. On the other hand, it is noticed that XGBoost estimates with very low RMSE and high R2 score comparing SVR even below one second.

V. CONCLUSION

In this study, the experimental test data of a cylindrical li-ion cell is examined and arranged according to the application of machine learning. Then, SVR is applied to the test data and the results are tried to be improved by changing algorithm parameters C and gamma. Similarly, a different machine learning algorithm is applied to the same test data and the results are compared in terms of error rates and computational duration. As a result of the comparison, it is observed that the SVM algorithm has to do a lot of calculation to minimize the error especially when it is applied for the whole data range. Therefore, it consumes long time to do calculations. Because the multiplicity of test parameters can be evaluated according to different data and parallel decision trees accelerate the operations, it can be interpreted that XGBoost algorithm can generate faster and high accuracy models for SOC estimation applications. Comparing conventional SOC estimation techniques, machine learning algorithms offer easy to implement way with the drawbacks which can be overcome such as need of test data, overfitting and parameter tuning.

As future work, it is intended to extend proposed estimation techniques to include other parameters such as current, temperature and aging. Another improvement would be increasing estimation accuracy and decreasing errors by also avoiding overfitting.

ACKNOWLEDGMENT

We thank University of Maryland CALCE Battery Research Group and contributors for sharing test results of several different li-ion cells.

REFERENCES

[1] Lithium ion rechargeable batteries technical handbook. Sony Corporation. <https://www.4project.co.il/documents/doc_286_2661.pdf).

[2] N. Nitta, F. Wu, J. T. Lee and G. Yushin, "Li-ion battery materials: present and future," Elsevier Materials Today vol. 18, pp. 252-264, 2015.

[3] C. Cluzel, C. Douglas, "Cost and performance of EV batteries," The Committee on Climate Change, Element Energy Limited, 2012.

[4] A. Rougier, P. Gravereau and C. Delmas, "Optimization of the Composition of the Li1 − z Ni1 + z O 2 Electrode Materials: Structural, Magnetic, and Electrochemical Studies," J. Electrochem. Soc. 1996, 143 (4), p. 1168.

[5] C.R. Stoldt and S.-H. Lee, "All-Solid-State Lithium Metal Batteries for Next Generation Energy Storage," Transducers & Eurosensors XXVII: The 17th Int. Conf. on Solid-State Sensors, Actuators and Microsystems, 2013, pp. 2819-2822.

[6] J. Lee, S. Cao, N. Choi, M. Liu, K. Lee and J. Cho, "Metal–Air Batteries with High Energy Density: Li–Air versus Zn–Air," Advanced Energy Materials, 2011, pp. 34-50.

[7] A. Manthiram, S. Chung, and C. Zu, "Lithium–Sulfur Batteries: Progress and Prospects," Advanced Energy Materials, Feb. 2015, 27 (12), pp. 1980-2006.

[8] J. Rivera-Barrera, N. Munoz-Galeano, H. O. Sarmiento-Maldonado, "SOC Estimation for Lithium-Ion Batteries: Review and Future Challenges," Electronics, vol. 6, 102, 2017.

[9] S. Li, C. Liao, and L. Wang, "Research Progress of Equivalent Circuit Models for SOC Estimation of Batteries in Electric Vehicles," 2014 IEEE Conf. and Expo Transportation Electrification Asia-Pacific (ITEC Asia-Pacific), 2014, pp. 1-6.

[10] A. Mejdoubi, A. Oukaour, H. Chaoui, H. Gualous, J. Sabor, Y. Slamani, "State-of-Charge and State-of-Health Lithium-Ion Batteries' Diagnosis According to Surface Temperature Variation," IEEE Trans. Ind. Electron., vol. 63 (4) , April 2016, pp. 2391-2402.

[11] Z. Wang, D. Yang, "State-of-charge estimation of lithium iron phosphate battery using extreme learning machine," 6th International Conf. on Power Electronics Systems and Applications (PESA), 2015.

[12] F. Liu, T. Liu, Y. Fu, "An Improved SoC Estimation Algorithm Based on Artificial Neural Network," 8th International Symposium on Computational Intelligence and Design, 2015.

[13] J. C. A. Anton, P. J. G. Nieto, C. B. Viejo, and J. A. V. Vilan, "Support Vector Machines Used to Estimate the Battery State of Charge," IEEE Trans. On Power Electron., vol. 28, no. 12, 2013, pp. 5919 - 5926.

[14] P.-H. Michel, V. Heiries, "An Adaptive Sigma Point Kalman Filter Hybridized by Support Vector Machine Algorithm for Battery SoC and SoH Estimation," IEEE 81st Vehicular Technology Conference (VTC Spring), 2015.

[15] R. Berwick, "An Idiot's Guide to Support Vector Machines (SVMs)," http://web.mit.edu/6.034/wwwbob/svm-notes-long-08.pdf , Access date: 22.04.2019.

[16] A. J. Smola, B. Scholkopf, "A tutorial on support vector regression", Kluwer Academic Publishers, 2003.

[17] Y. Xing, W. He, M. Pecht and K. L. Tsui, "State of Charge Estimation of Lithium-Ion Batteries Using the Open-Circuit Voltage at Various Ambient Temperatures," Applied Energy, 113, pp.106-115, 2014.

[18] W. He, N. Williard, C. Chen, M. Pecht, "State of Charge Estimation for Li-Ion Batteries Using Neural Network Modeling and Unscented Kalman Filter-based Error Cancellation," Int. Journal of Electrical Power & Energy Systems, vol. 62, pp.783-791, 2014.

[19] A123, "Cylindrical Battery Pack Design, Validation, and Assembly Guide," User Documentation, 2013.

978-1-5386-7688-2/19 $31.00 © 2019 IEEE

Tracking controller design of a RF matching box with plasma load varying

Li, Yen-Fang
Dept. of Elec. Eng.,
Ming-Hsin University of Science and Technology,
Hsin-Chu 30401, Taiwan
Email:yfli@must.edu.tw

Hsieh, Ming-Heng
Dept. of Elec. Eng.,
Ming-Hsin University of Science and Technology,
Hsin-Chu 30401, Taiwan
Email: f0928887921@gmail.com

Liou, Ren-Sian
Dept. of Elec. Eng.,
Ming-Hsin University of Science and Technology,
Hsin-Chu 30401, Taiwan
Email: joe43113@gmail.com

Abstract—**Based on a good RF power supply, the plasma density control is the most important work for getting a good product success rate. The RF power is transmitted from RF generator to the plasma chamber through a RF matching box. The matching box is used to match the impedance between the transmitter and receiver by tuning a series adjustable capacitor and a parallel one. Unfortunately, only an unsatisfied result is given since the matching is not an easy work for the traditional matching box with an analog controller. In this paper, an auto-tracking and rapidly matching controller will be developed by analyzing the behaviors of reflectance while the capacitances of the matching components vary. Since the reflectance is a complex nonlinear function in terms of the matching capacitances. These is not a linear controller can be design to tune the matching capacitances. Some simulation examples are presented to analyze the variation of magnitude and phase of the reflectance while the matching capacitances are changing. A control strategy is proposed to build the tracking controller to tune two matching capacitances in the matching box while the impedance of plasma load is varying. Also, some simulation examples are presented to illustrate the proposed control method and verify the performance of the tracking controller. A good agreement is gotten to examine the tuning performance for the proposed tracking controller.**

Keywords- RF matching box, capacitance tuning, reflectance detection, tracking controller, load varying

I. INTRODUCTION

Over the past decade, the plasma processing technology has been widely applied in many advanced process for the industry of DVD, CD coatings, glass coating processes, and semiconductor manufacturing. Especially in the semiconductor manufacturing, RF power is used to generate plasma in a vacuum chamber to etch and deposit doped layers for building up the structures of a wafer. By the high-speed ion impacting, the process controllability of the nowadays semiconductor manufacturing has been greatly enhanced.

An AC power source is used to supply the RF power, namely RF generator, to drive the plasma load and generate the required plasma. In the plasma process, the essential problem is to control the power consuming of the plasma chamber. This power is delivered from a RF power generator to chamber to generate constantly plasma through a matching box. A high quality manufactured process requires the stable plasma supply. It leads the plasma power control to be an important work for producing good quality products.

The work was supported by the Ministry of Science and Technology of the Republic of China under Grant MOST 105-2622-E-159-002 -CC3 and MOST 107-2221-E-159-005.

Practically, the aforementioned industrial applications often use inductively-coupled plasma. The impedance of plasma load is an inductive load. At standpoint of the RF source, the plasma load is an inductive load. Hence, the matching circuit must to match the inductive load to a resistance load for the RF generator. Not only an inductive load matching problem but also the load will vary due to the produced process. This is the requirement for the matching box controlling.

For a RF system working, the maximum real power can be delivered to the plasma load based on the output impedance of power source is equal to complex conjugate of the load impedance. Hence, usually a matching network, such as L-matching network or LC-matching network [14], is established to match the impedance between the power source and the plasma load for delivering the maximum AC power from RF generator to plasma chamber. Unfortunately, in the case of plasma system, the equivalent impedance of the load may vary greatly during the plasma producing and also depending upon the used gas and the gas pressure in the plasma chamber. Any change in the plasma impedance is reflected back to the input of the matching circuit, so the RF system must be re-matched by tuning the matching network such as series capacitor or shunt capacitor in the L-matching network to keep in an ideal load for the RF generator.

In general, the matching circuit is a two-port network composed with adjustable capacitors or adjustable inductors. There are many literatures on the analysis of this matching network. [1]-[5] analyzed the circuit impedance characteristics by series and parallel method. [6][7] analyzed the circuit characteristics by the two-port network theorem. They focused on how to accurately estimate the impedance of the load to perform matching component design. Through the S-parameter or ABCD transmission parameters measuring, the impedance of the plasma chamber has been calculated [6]. But the measurement must do in the off-line state, and the on-line close-loop matching control cannot be achieved. Some literatures showed the method by installing voltage and current probes (VI probe) in the chamber to actually measure the load impedance condition [8][9]. Further, the matching network control problem was studied for the RF heating system [10]. However, while the circuit is matched, the RF power circuit is almost resonant, and the voltage at the load terminal is extremely high. Taking the operating condition of 13.56 MHz 1 kW as an example, the voltage at the load terminal is up to AC 4kV, which the probes are installed to detect the feedback signal for the control circuit. The feedback components are not only expensive but also extremely dangerous. In contrast, the voltage at the input of the matching network is only about

AC 200V. It is more practical to use this point as the detected point or sampled point for sensing the matching condition.

Typically, the industrial RF generator is designed with 50Ω output impedance hence it can deliver the maximum RF power into a 50Ω load and decrease the output power accordingly while the reflection coefficient Γ of the load increases from zero to one. In other words, the matching network will transfer the plasma impedance for any case to 50Ω in the standpoint of the input side. Generally, the most frequently used matching networks are the L network, Π network, and T network. The advantage of the L network is that when given an impedance matching problem, provided that this lies within the capabilities of the network a unique solution may be found. This is due to the L network having only two variables, i.e. a series capacitor and a shunt capacitor, allowing the network to be tuned easily for specified load impedances [11].

In general, we can distinguish two categories of adaptive control algorithms. The first category includes deterministic algorithms [12] that make use of searching and optimization and requiring an elaborate baseband processing. Usually, these computation-based algorithms need a set of the factory calibration data to store in lookup tables. The second category includes functional algorithms that commonly use the detected information for direct control of variable network elements [13] [14] while making use of the fundamental impedance transformation properties of LC networks.

In this paper, with the L type matching network, a tracking control method is proposed by controlling the series and parallel capacitance for the inductively-coupled plasma. The signals of magnitude and phase of the power reflectance are detected to be the feedback messages for the tracking controller. Since the proposed method sampled the feedback signals at input side of the matching box, lower price will be paid contrast to the method that probes the voltage and current in the plasma chamber. Moreover, a control strategy is proposed to tune the two adjustable capacitances in the matching box while the impedance of plasma load is varying. Also, simulation examples are presented to illustrate the proposed control method and verify the performance of the tracking controller. Good agreements from the results of simulation give verification for the analysis and design of the tracking controller.

II. PRELIMINARY AND PROBLEM FORMULATION

A basic RF plasma system includes with three subsystems: a RF generator for the RF power source, a plasma chamber to achieve the plasma reaction, and a matcher to balance the impedance between the RF generator and chamber load for maximizing the power delivery.

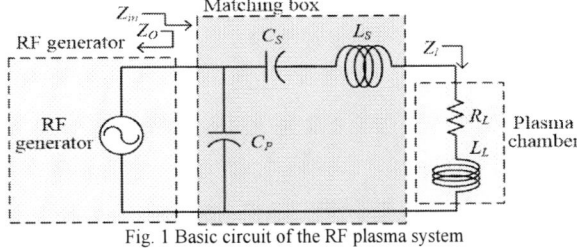

Fig. 1 Basic circuit of the RF plasma system

As Fig. 1, a basic RF plasma system is shown in Fig. 1, where Z_o denotes the output impedance of RF generator, R_L

and L_L denote the equivalent resistance and inductance of the chamber load. The components C_P, C_S, and L_S make up the matching circuit namely matching box which is in charge of to balance the impedance of two sides. Also, Z_{in} shows the input impedance of the matching box.

The matching box is built to maximize the power delivery from RF generator to chamber load. Usually, it is formed only by L-C circuit for lossless transmission. The basic concept is to tune the L-C circuit, i.e. adjusts the inductor or capacitors, to make the impedance Z_{in} conjugate to the impedance of Z_O, that is $Z_{in}=Z_O*$, even when the load varies. Our goal is to design the controlling approach and controller to tune the L-C circuit of the matching box automatically to execute the impedance matching process.

In generally, the parallel capacitor C_P and series capacitor C_S are replaced by adjustable capacitors. Normally, the output impedance of the RF generator is 50Ω. In other words, the controller should be in charge of to tune the two adjustable capacitors C_P and C_S to make the input impedance of the matcher with $|Z_{in}| = 50Ω$ and phase angle being 0°. Fig. 2 shows the redrawn RF circuit.

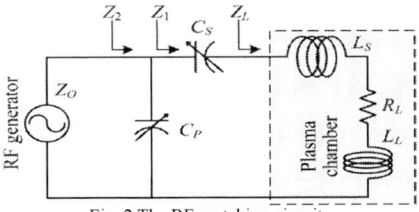

Fig. 2 The RF matching circuit

III. THE MATCHING PROCESS - SMITH'S CHART ANALYSIS

***Method* 1**: Smith's chart analyzed method

In Fig. 1 and Fig. 2, by the transmission principle, the voltage on each point of the transmission line can be expressed as $V(x,t)=V^+(x,t)+V^-(x,t)$, where $V^+(x,t)$ denotes the transmission wave and $V^-(x,t)$ denotes the reflection wave. Before the matching circuit is added, the reflectance Γ_z is defined as follows

$$\Gamma_Z = \Gamma_r + i\Gamma_i \equiv \frac{V^-(x,t)}{V^+(x,t)} = \cdots = \frac{Z_L - Z_O}{Z_L + Z_O}, \quad (1)$$

where, Γ_r and Γ_i denote the real part and imaginary part of the reflectance respectively. Also, it can be rewritten as

$$\Gamma_Z = \Gamma_r + i\Gamma_i = \frac{Z_O(z-1)}{Z_O(z+1)} = \frac{r+ix-1}{r+ix+1} = \frac{(r-1)+ix}{(r+1)+ix}, \quad (2)$$

where, z is the normalized impedance in Z_O base i.e. $z \equiv Z_L/Z_O = r+ix$. Further, we obtain

$$(r-1)+ix = (r+1)\Gamma_r - x\Gamma_i + i((r+1)\Gamma_i + x\Gamma_r). \quad (3)$$

By separating the real and imaginary part in eq.(3), it can be gotten as follows

$$\begin{cases} (r-1) = (r+1)\Gamma_r - x\Gamma_i \\ x = (r+1)\Gamma_i + x\Gamma_r \end{cases}, \quad (4)$$

$$x = (r+1)\Gamma_i + x\Gamma_r = \frac{(r+1)\Gamma_i}{1-\Gamma_r}, \quad$$

$$(r-1) = (r+1)\Gamma_r - x\Gamma_i = (r+1)\Gamma_r - \frac{(r+1)\Gamma_i^2}{1-\Gamma_r}, \quad (5)$$

By a routine manipulation, eq.(5) can be rewritten as follows

$$(\Gamma_r - \frac{r}{r+1})^2 + \Gamma_i^2 = (\frac{1}{r+1})^2, \tag{6}$$

Also, the imaginary part of eq.(4) can be manipulated in the following

$$(r+1) = \frac{2-x\Gamma_i}{(1-\Gamma_r)}, \quad x = \Gamma_i \frac{2-x\Gamma_i}{(1-\Gamma_r)} + x\Gamma_r ,$$

$$(1-\Gamma_r)^2 = \frac{1}{x^2} - \frac{1}{x^2} + 2\frac{1}{x}\Gamma_i \cdot \Gamma_i^2 ,$$

$$(\Gamma_r - 1)^2 + (\Gamma_r - \frac{1}{x})^2 = (\frac{1}{x})^2 , \tag{7}$$

Eq.(6) and (7) imply two circles in Γ_r and Γ_i axes coordinate system, $i.e.$ Smith's chart. Hence, the moving locus of the reflectance Γ_z versus to the load impedance variation can be expressed as the equations of two circles which are shown as eq.(8) and plotted as Fig. 3(a).

$$\begin{cases} (\Gamma_r - \frac{r}{r+1})^2 + \Gamma_i^2 = (\frac{1}{r+1})^2, & \text{center}=(\frac{r}{r+1},0), \text{ radius}=\frac{1}{r+1}, \\ (\Gamma_r - 1)^2 + (\Gamma_i - \frac{1}{x})^2 = (\frac{1}{x})^2, & \text{center}=(1,\frac{1}{x}), \text{ radius}=\frac{1}{|x|}. \end{cases} \tag{8}$$

The origin of Smith's chart in Fig. 3 (a) implies the normalized load impedance is equal to 1, i.e. $Z_L = Z_O$ the matching condition. Briefly, the matching process is to push the load impedance from point z to the origin of Smith's chart in Fig. 3(a).

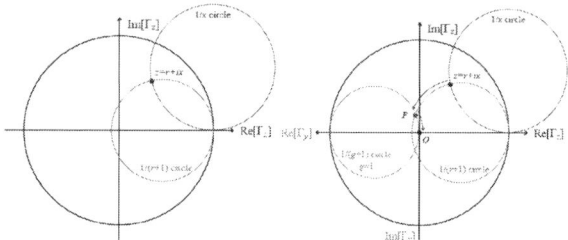

(a) Smith's chart of the load impedance   (b) basic matching process
Fig. 3 The reflectance Γ_z in Smith's chart

Now we need to get matching by tuning the two adjustable capacitors C_P and C_S. While the series capacitor C_S is added, the impedance in load side can express as eq.(9).

$$z_1 = \frac{Z_1}{Z_O} = (Z_L - j\frac{1}{\omega C_S})/Z_O = r + jx_1, \tag{9}$$

Here, the real part is unchanging and the imaginary part will vary according to the added capacitance. Also, the trajectory while the series capacitor C_S is tuning can be plotted in Fig. 3(b). There the locus will move from point z to point P along with the real part circle.

Referring to Fig. 1, Y_{in} and Y_o denote the admittance on each viewpoint of a transmission line and $y=g+ib=1/z$ denotes the normalized admittance. Similarly, the reflectance of admittance on each point of a transmission line, namely Γ_y, can be shown in the following:

$$\Gamma_y = \Gamma_r^y + i\Gamma_i^y = \frac{Y_{in} - Y_0}{Y_{in} + Y_0} = \frac{y-1}{y+1} = \frac{1-z}{1+z} = -\Gamma_z = -(\Gamma_r + i\Gamma_i). \tag{10}$$

Similarly, the real part and imaginary part of the reflectance of admittance, i.e. the counterpart of the reflectance of impedance, will be drawn in Smith's chart as follows.

$$\begin{cases} (\Gamma_r^y - \frac{g}{g+1})^2 + \Gamma_i^{y2} = (\frac{1}{g+1})^2, & \text{center}=(\frac{g}{g+1},0), \text{ radius}=\frac{1}{g+1}, \\ (\Gamma_r^y - 1)^2 + (\Gamma_i^y - \frac{1}{b})^2 = (\frac{1}{b})^2, & \text{center}=(1,\frac{1}{b}), \text{ radius}=\frac{1}{|b|}. \end{cases} \tag{11}$$

As Fig. 3(b), the matching condition is tuning the series capacitor C_S to reach point P. In other words, it will be the solution of the two following equations while the locus reaches point P. There are:

$$\Gamma_y = \Gamma_r^y + i\Gamma_i^y = \frac{Y_{in} - Y_0}{Y_{in} + Y_0} = \frac{y-1}{y+1} = \frac{1-z}{1+z} = -\Gamma_z = -(\Gamma_r + i\Gamma_i).$$

With eq.(10) in hand, we get the solution in the following.

$$\begin{cases} \Gamma_r = \frac{r-1}{3r+1}, \\ \Gamma_i = \pm\frac{2\sqrt{r-r^2}}{3r+1}. \end{cases} \tag{12}$$

Also another condition about the imaginary part of eq.(8) should be satisfied, for this reason the following result can be gotten by a routine manipulation.

$$x' = x - (\omega C_S Z_O)^{-1} = \sqrt{r-r^2} ,$$

$$C_S^{-1} = (x - \sqrt{r-r^2})\omega Z_O . \tag{13}$$

Hence, the locus about a reflectance of admittance can show in Smith's chart similar to the reflectance of impedance. Now, the parallel capacitor C_P will be added. Referring to Fig. 2, Y_2 shows the admittance that corresponds to Z_2. The normalized admittance will be

$$y_2 = \frac{Y_2}{Y_O} = (Y_1 + j\omega C_P)/Y_O = z_1^{-1} + jy_{C_P}, . \tag{14}$$

where, y_{CP} shows the normalized admittance of parallel capacitor C_P. By tuning the parallel capacitors C_P, the trajectory can move from point P to point O in Fig. 3(b) along with the real part admittance circle since the real part or the conductance is fixed. Eventually, the matching process theoretically by tuning the two adjustable capacitors C_P and C_S is shown as Fig. 3 (b) from point z to point P then to point O.

Before the parallel capacitor C_P is added, the admittance at point P, i.e. y_P, is

$$y_P = \frac{1}{z_P} = \frac{1}{r+i\sqrt{r-r^2}} = \frac{r-i\sqrt{r-r^2}}{r} = 1-i\sqrt{r^{-1}-1}. \tag{15}$$

Here, the parallel capacitors C_P is in charge of to eliminate the imaginary part of y_P in eq.(15). That leads to get the result for solving the capacitors C_P that shows in eq.(16).

$$C_P = (\omega Z_O)^{-1}\sqrt{r^{-1}-1}. \tag{16}$$

***Method* 2**: The impedance analyzed method

Referring to Fig. 2, let $L'=(L_L+L_S)$ and assume that the system reaches match, we get

$$Z_1(s) = R_L + sL' + \frac{1}{sC_S} , \quad Y_2(s)=Z_2^{-1}(s)=sC_P + (R_L + sL' + \frac{1}{sC_S})^{-1},$$

$$Y_2(j\omega) = j\omega C_P + (\frac{-L'C_S\omega^2 + j\omega R_L C_S + 1}{j\omega C_S})^{-1} = \frac{1}{Z_O},$$

$$(\frac{-L'C_S\omega^2 + j\omega R_L C_S + 1}{j\omega C_S})^{-1} = \frac{1 - j\omega C_P Z_O}{Z_O}, \tag{17}$$

Hence, by a routine manipulation, it can be obtained.

$$\begin{cases} R_L = \dfrac{Z_O}{1+(\omega C_P Z_O)^2} \\ (\omega L' - \dfrac{1}{\omega C_S}) = \dfrac{\omega C_P Z_O^2}{1+(\omega C_P Z_O)^2} \end{cases}, \quad (18)$$

In order to simplify the calculation, a normalized expression is derived. Let $r_L = R_L/Z_O$, $x_L = \omega L'/Z_O$, $x_{Cp} = (\omega C_P Z_O)^{-1}$, and $x_{Cs} = (\omega C_S Z_O)^{-1}$ and if $r_L < 1$, eq.(18) can be rewritten as follows.

$$x_{Cp} = \sqrt{\frac{r_L}{1-r_L}}, \quad x_{Cs} = x_{L'} - \frac{x_{Cp}}{x_{Cp}^2+1} \quad (19)$$

Hence, by a routine manipulation, it can be obtained.

$$\begin{cases} C_P = \dfrac{1}{\omega Z_O} \times \sqrt{(\dfrac{1-r_L}{r_L})} \\ C_S = \dfrac{1}{\omega Z_O} \times \left(x_{L'} - \dfrac{x_{Cp}}{x_{Cp}^2+1} \right)^{-1} = \dfrac{1}{\omega Z_O} \times \left(x_{L'} - \sqrt{r_L - r_L^2} \right)^{-1} \end{cases}. \quad (20)$$

To illustrate the methods, the following examples are used to check the methods mutually.

Example 1 :

A RF plasma system with a RF power generator that the output frequency $f = 13.56$ MHz and output impedance $Z_O = 50$ Ω, and the impedance of the plasma load $Z_L = 38.46 + i57.69$ Ω.

Solution by Method 1:

By normalizing the load impedance, get the parameters $r = r_L = 0.7692$ and $x = x_L = 1.1538$. By eq.(13) and eq.(15), it can be calculated that the system matching conditions are:

$$C_S = \frac{1}{(1.1538 - 0.4213)\omega Z_O} = 320.486 \ pF,$$

$$C_P = \frac{\sqrt{r^{-1}-1}}{\omega Z_O} = \frac{0.54777}{\omega Z_O} = 128.585 \ pF.$$

Solution by Method 2 :

The normalized load impedance is $r_L = 0.7692$ and $x_L = 1.1538$. By eq.(20), the system matching conditions also can be gotten which are $C_P = 128.585$ pF and $C_S = 320.486$ pF same as the above solutions.

IV. THE MATCHING SYSTEM MODELING AND CONTROL

Eq.(20) shows the solution for the matching conditions if the load impedance in plasma chamber has been measured and fixed. In fact, the load impedance isn't a constant value and it will change according to the process situation. For this reason, the two capacitances in the matching box should be tuned to adapt the matched requirement while the load impedance changes. A tracking controller is necessary for tuning the two capacitances.

The control target is the delivered power occurring in plasma chamber from RF generator. The load impedance conditions in plasma chamber can be actually measured by installing the VI probe in the chamber. Since the voltage at the load terminal is up to AC 4kV typically, it implies that the signal feedback by measuring the impedance is not only expensive but also extremely dangerous.

Whereas the detection of the impedance about the plasma load is difficult and high payment, the feedback sensing point should be designed at lower voltage side such as the input side of the matching box for practically applications. By probing the magnitude and phase of the power reflectance, the situations about the power delivery will be inspected. At this viewpoint, the considering reflectance will be the input point of the matching box, and the controlled targets are the two adjustable capacitance as the aforementioned discussion. Referring to Fig. 2, the impedance can be rewritten in the following.

$$Z_2(j\omega) = \frac{(1-L'C_S\omega^2)+jR_LC_S\omega}{j\omega(-L'C_SC_p\omega^2+jR_LC_SC_p\omega+C_S+C_P)}$$
$$= \frac{\alpha(\omega)+j\beta(\omega)}{A(\omega)}, \quad (21)$$

where,

$A(\omega) = ((R_LC_SC_p\omega)^2 + (L'C_SC_p\omega^2-(C_S+C_P))^2)\omega,$

$\alpha(\omega) = ((L'C_S\omega^2-1)R_LC_SC_p - (L'C_SC_p\omega^2-(C_S+C_P))R_LC_S)\omega,$

$\beta(\omega) = (1-L'C_S\omega^2)(L'C_SC_p\omega^2-(C_S+C_P)) - C_p(R_LC_S\omega)^2.$

Let $\alpha'(\omega) = \alpha(\omega) \times (Z_O A(\omega))^{-1}$ and $\beta'(\omega) = \beta(\omega) \times (Z_O A(\omega))^{-1}$. The magnitude M and phase θ of the reflectance at input side of the matching box can be gotten as follows.

$$\Gamma_Z = \frac{z_2-1}{z_2+1} = \frac{\alpha'-1+j\beta'}{\alpha'+1+j\beta'} = M\angle\theta,$$

$$M = \left| \frac{(\alpha'^2+\beta'^2-1)+j(2\beta')}{(\alpha'+1)^2+\beta'^2} \right|, \quad \theta = \tan^{-1}\left(\frac{2\beta'}{\alpha'^2+\beta'^2-1} \right). \quad (22)$$

Eq.(22) show the reflectance at input side of the matching box. Obviously, they are complex nonlinear functions in terms of the two matching components C_S and C_P. Hence, there is not a linear controller for tuning the reflectance by the two matching components. Also, the two functions are mutual coupling with C_S and C_P, which implies that the two capacitances can't be controlled independently.

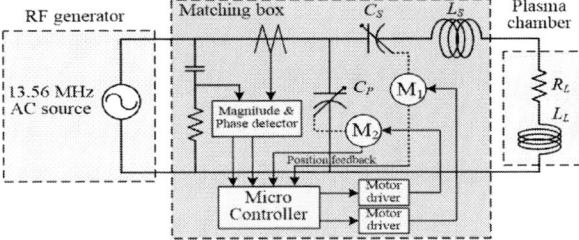

Fig. 4 The block diagram of control system for the RF matching box

In the paper, the matching system control structure is recommended as the block diagram in Fig. 4. The adjustable capacitor C_S and C_P are driven by two motors individually. The detector is designed to measure the magnitude and phase of the reflectance at point of the input side of matching box. It should be noted that the detected point is the low voltage side of the matching circuit. That implies the detecting circuit can be established with lower cost contrast to the on-line load impedance measurement in the chamber. For the sake of focusing on matching box control, here the detecting circuit about the reflectance measurement is omitted.

V. THE REFLECTANCE VARIATION SIMULATION

Obviously, eq.(22) is complex nonlinear functions in terms of the two matching capacitances C_S and C_P. There isn't unique solution to solve the capacitances C_S and C_P from eq. (22). Moreover, the two functions are mutual coupling with C_S and C_P. Hence, the tuning control of the two matching components is difficult to achieve only by sensing the magnitude and phase messages directly. To observe the magnitude and phase variation of the reflectance while the plasma impedance is changing or drifting, a simulation example is given as follows.

Example 3: (Theoretical simulation by Matlab)

A given RF system with the RF power source frequency f =13.56 MHz, the output impedance Z_O=50 Ω, and the load impedance of the plasma chamber Z_L=2+i46 Ω.

Solving and simulation :

Normalize the load impedance, get r_L= 0.04, x_L=0.92. By eq.(20), it can be gotten that the matching conditions theoretically are C_P=1150 pF and C_S=324.21 pF. For observing the behaviors while a matching component is drifting, each matching component has been set by changing from 70% to 120% individually with respect to the theoretically matching condition, *i.e.* C_{S0}, C_{P0}, and L'_0. Using eq.(22), the variation of the magnitude and phase of the reflectance can be simulated according to the matching components drift. Also, the magnitude of reflectance is scaled without normalizing and shifting to 50Ω for easy examining the behaviors.

(1) C_S and C_P changing 70%~120% (normalized by C_{S0} and C_{P0})

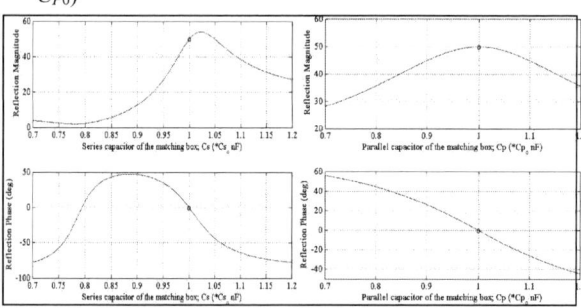

Fig. 5 M and θ of the reflectance varying versus C_S and C_P changing

(2) C_S & C_P changing 50%~200%

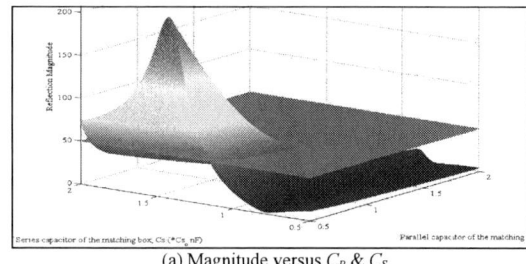

(a) Magnitude versus C_P & C_S

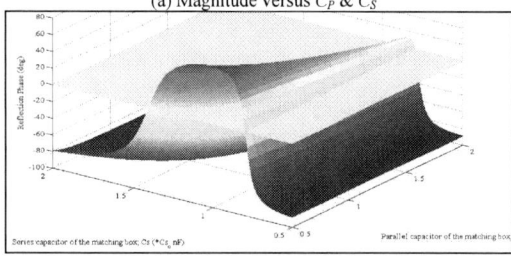

(b) Phase versus C_P & C_S

Fig. 6 3D meshed figure for M & θ varying versus C_S & C_P changing

VI. SIMULATION OF THE TRACKING CONTROLLER

Comparing the simulation results in Fig. 5 ~ Fig. 6 summaries can be concluded in the following:

1. By checking Fig. 5, the reflectance is more sensitive with the series capacitance C_S than the parallel capacitance C_P.

2. By way of examining the 3D mesh plots in Fig. 6, the peak values of the magnitude and phase are influenced by the parallel capacitance C_P obviously.

3. The reflectance variation is similar to the influence of the series capacitance C_S while the load inductance changes.

4. In the phase plot, the optimal matching point occurs at zero degree and falling slope region with respect to the series capacitance C_S increase.

5. In the phase plot, the optimal matching point occurs at zero degree and falling slope region with respect to the parallel capacitance C_P increase.

6. In the magnitude plot, the optimal matching point occurs at 50Ω and rising slope region with respect to the series capacitance C_S increase and also with respect to the series capacitance C_P increase.

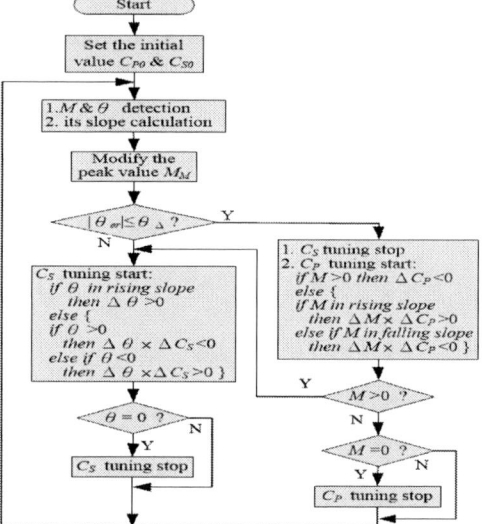

Fig. 7 The flow chart of the capacitances tuning process

By observing the behaviors in above figures and the summaries, the controlled strategy for the tracking controller is proposed as the follow chart in Fig. 7. According to the proposed controlled strategy in Fig. 7, the tracking controller can be built with a digital controller such as a SOC-based micro-processor. Here, a simulation example for this controller is shown with the considering system in *Example* 3. By setting the initial capacitances of the two matching components with 2*C_{P0} and 0.2*C_{S0}, the tuning processes of the two capacitances is checked in Fig. 8.

The simulation results verify the two capacitances are tuning according to the feedback, *i.e.* magnitude and phase of reflectance from the detector. Also, the reflectance is shown in Fig. 8 to examine the matching performance.

Certainly, the simulation results can be extended to the case for the impedance changing of plasma load. Only a routine work, the simulation can be done as the same procedure and it is omitted here. It will be similar to the results in *Example* 3.

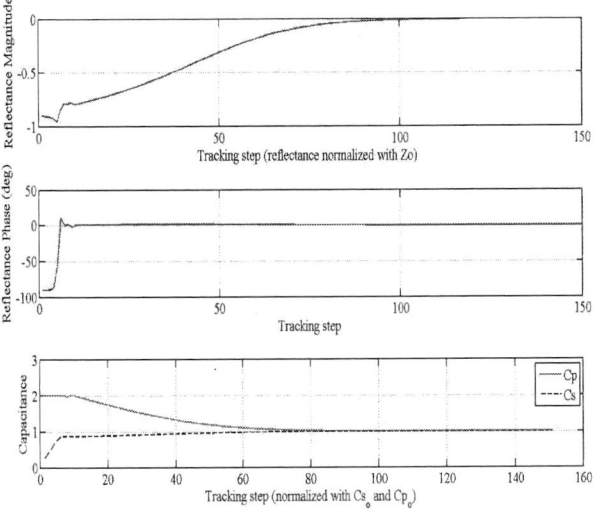

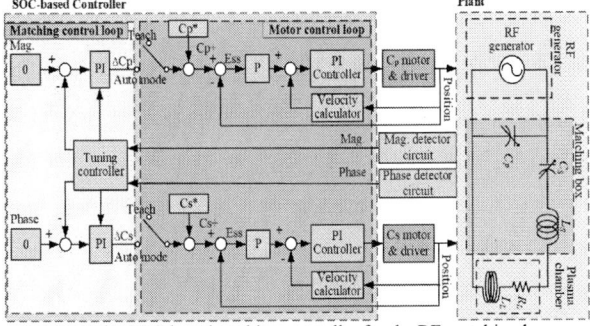

Fig. 8 The tuning process of the tracking controller

The practical control system is proposed as the block diagram in Fig. 9. A SOC-based controller is used to implement the proposed tracking controller. The tracking strategies or the control rules in Fig. 7 are built in the matching loop in Fig. 9, namely tuning controller.

Fig. 9 The SOC-based tracking controller for the RF matching box

VII. CONCLUSION

The matching box is used to match the impedance between the transmitter and receiver by tuning a series adjustable capacitor and a parallel one. Since the reflectance is a complex nonlinear function in terms of the matching capacitances. These isn't a linear controller can be design to tune the matching capacitances. By analyzing the varying behaviors of the magnitude and phase of the reflectance versus the two matching capacitances vary, the controlled strategy has been summarized to build the tuning controller. For the simulation results, a good agreement has gotten to examine the tuning processes and matching performance for the proposed tracking controller. According to the proposed controlled strategy in Fig. 7, the tracking controller can be built with a digital controller such as a SOC-based microprocessor. Also, based on a SOC-based controller, the auto-tracking controller has implemented to verify the tune rules.

REFERENCES

[1] Logan, J. S., Mazza, N. M., and Davidse, P. D., "Electrical characterization of radio-frequency sputtering gas discharge", *J. Vac. Sci. & Tech.*, vol. 6, 1968.

[2] Dragan B.llic, "Impedance measurement as a diagnostic for plasma reactor", *Rev. Sci. Instrum.*, vol. 52, no. 10, Oct 1981.

[3] Roosmalen, A. J. van, "Plasma parameter estimation from RF impedance measurements in a dry etching system", *Appl. Ohys. Lett.*, vol. 42, no. 5, March 1983.

[4] Bakker, L. P., Kroesen, G. M. W., and Hoog, F. J. de, "RF discharge impedance measurements using a new method to determine the sray impedance", *IEEE Transactions on Plasma science*, vol. 27, , no. 3, pp. 759-765, June 1999.

[5] Payling, R., Bonnot, O., Fretel, E., Rogerieux, O., Aeberhard, M., Michler, J., Nelis, T., Hansen, U., Hartmann, A., Belenguer, P., and Guillot, P., "Modelling the RF source in GDOES", *J. Anal. At. Spectrom.*, vol. 18, pp. 656-664, 2003.

[6] Andries, B., Ravel, G., and Peccoud, L., "Electrical characterization of radio-frequency parallel plate capacitively coupled discharge", *J. Vac. Sci. & Tech., A*, vol. 7, no. 4, July/Aug 1989.

[7] Wüst, K., Groh, K. H., and Löb, H. W., "Impedance measurements at rf plasmas in noble gases by use of the four-terminal network formalism", *Rev. Sci. Instrum.*, vol. 63, no. 4, April 1992.

[8] Butterbaugh, J. W., Baston, L. D., and Sawin, H. H., "Measurement and analysis of radio frequency glow discharge electrical Impedance and network power loss", *J. Vac. Sci. & Tech., A*, vol. 8, no. 2, Mar/Apr. 1990.

[9] Sobolewski, M. A., "Electrical characterization of radio-frequency discharges in the Gaseous Electronics Conference Reference cell", *J. Vac. Sci. & Tech., A*, vol. 10, no. 6), Nov/Dec 1992.

[10] Cottee, C. J. and Duncan, S. R., "Design of Matching Circuit Controllers for Radio-Frequency Heating", *IEEE Transactions on Control Systems Technology*, vol. 11, Issue 1, pp. 91-100, 2003.

[11] Thompson, M., and Fidler, J. K., "Determination of the impedance matching domain of impedance matching networks," *IEEE Trans. Circuits Syst. I, Reg. Papers*, vol. 51, no. 10, pp. 2098-2106, Oct. 2004.

[12] Thompson, M. and Fidler, J. F., "Fast antenna tuning using transputer based simulated annealing," *Electron. Lett.*, vol. 36, no. 7, pp. 603–604, Mar. 2000.

[13] Mileusnic, M., Petrociv, P., and Todorovic, J., "Design and implementation of fast antenna tuners for HF radio systems," in *Proc. Int. Inf., Commun. Signal Process. Conf.*, Sep. 1997, vol. 3, pp. 1722–1726.

[14] Bezooijen, A., Jongh, M. A., Straten, F., Mahmoudi, R., and Roermund, H. M., "Adaptive impedance-matching techniques for controlling L networks," *IEEE Trans. Circuits Syst. I, Reg. Papers*, vol. 57, no. 2, pp. 495-505, Feb. 2010.

978-1-5386-7688-2/19 $31.00 © 2019 IEEE

AUTHOR INDEX

Aarniovuori, L. 114
Abdelaziz, Ahmed Sayed Abdelaal 366
Aboura, Faouzi 570
Agroui, Kamel 402
Aguilar, Fabio Gómez-Estern 359
Aimer, Ameur Fethi 216, 254
Akhmetov, Yerbol 506
Akkaya, M. C. 517
Alajarmeh, Nancy 557
Allaoui, Tayeb 486
Andresen, Jan 191
Ankarali, Mustafa Mert 540
Armutlu, Cem 590
Asadi, Meysam 481, 511
Askar, Murat 143
Avci, Emre 383
Aydemir, M. Timur 389
Babau, Radu 77
Bagheri, Mehdi 506
Bakbak, Ali 347
Ban, Branko 596
Bastovansky, R. 240
Bauer, Pavol 8
Bayram, Duygu 155
Bazaz, Mohammad Abid 310
Bendiabdellah, Azeddine 205, 254
Benhala, Bachir 417
Benouzza, Noureddine 205, 227
Berkani, Abderrahmane 486
Bidgoli, Farzaneh Sabbaghian 85
Bigdeli, Mehdi 341
Bilir, Bulent 143
Birs, Isabela R. 532
Blaabjerg, Frede 402
Boldea, I. 65
Boldea, Ion 29, 77, 260
Bostanci, Emine 147
Boudinar, Ahmed Hamida 216, 254
Bouguerra, Sara 402, 431
Boukhelifa, Akkila 473
Boulaam, Karima 473
Boumediene, Bachir 459, 486
Bourahla, Mohamed 211
Çakir, Can 553
Calin, Marius Daniel 166
Cardona, Juan Diego Nieto 359
Cetin, Sevilay 315
Ceylan, Doga 297
Chibani, Youcef 274
Choi, Jaeho 409, 466
Choi, Sungmin 280
Chun, Tae-Won 439
Cirrincione, M. 372
Cirrincione, Maurizio 221, 500, 563

Cirstea, Marcian 525
Cirstea, Silvia 525
Collins, Michael 71
Cornea, Octavian 289
Darab, Cosmin 532
De Keyser, Robin 532
Devecioglu, Özer Can 553
Di Benedetto, M. 372
Dida, Abdelkader Hadj 211
Ding, Hao 172
Dityagraha, R Dimas 466
Enokizono, Masato 41, 178
Ercan, Tuncay 493
Eren, Levent 143
Eren, M. Kerem 604
Ergene, L. T. 186, 517
Ertan, H. Bülent 5, 211, 486
Ertan, Hulusi Bülent 395
Eshwiage, Ahmed 431
Fricke, Torben 93
Gassab, Oussama 402
Gharehpetian, G. B. 506
Gherabi, Zakaria 205, 227
Gherman, Tudor 322, 329
Gholaminejad, Azadeh 85
Gireada, Mihaita 289
Gordillo, Francisco 359
Haddad, Abdel Gafoor 139
Hamarash, Ibrahim 547
Hamdani, Samir 274
Hanigovszki, Norbert 51
Hariyanto, Nanang 409, 466
Hasan, Ali N. 557, 575
Helerea, Elena 166
Henini, Noureddine 227
Henke, Markus 58
Himmelstoss, Felix A. 304
Ho, Anh-Vu 439
Hofmann, Lutz 132
Hollmann, Jan 246
Hsieh, Ming-Heng 610
Hulea, Dan 289
Hur, Kyeon 453
Huynh, Anh-Tuan 439
Imecs, Maria 351
Ince, Türker 143, 553
Ipek, Eymen 604
Jabbari, Mahdi 341
Jahns, Thomas M. 16
Jahns, Tom 10
Jun, Hee-Deuk 233
Jung, Sang-Young 233
Kalla, Matthias 132
Karamuk, Mustafa 18, 583

AUTHOR INDEX

Keysan, Ozan 297
Khan, Hadhiq 310
Khelfi, Hamid 274
Khodja, Mohammed-El-Amine 254
Kim, Chan-Ki 453
Kim, Rae-Eun 233
Kim, Sang-Min 453
Kim, Youn-Hwan 233
Koçak, Ibrahim 395
Kocan, S. 240
Korolova, Olga 132
Kothari, Kajal 563
Koulali, Mostefa 459
Koura, Mohamed Boudiaf 216
Kumar, D. M. 372
Kumar, Ravinesh 500
Kutay, Mahir 493
Laghrouche, Salah 221
Lange, Sebastian 108
Lee, Jae Suk 280
Lee, Seung-Hwan 280
Li, Yen-Fang 610
Lidozzi, A. 372
Lindh, P. 114
Liou, Ren-Sian 610
Liu, Mingda 172
Liu, Yong-Chao 221
Locatt, Howard 71
Loraz, Ismail Onur 389
Madan, Alican 147
Makys, P. 240
Mankour, Mohamed 459
Marignetti, F. 65
Martin, Adrian Daniel 77
Matehkolaei, Mohammad Jafari 120, 447
Mehta, Utkal 563
Mertens, Axel 191
Mese, Erkan 347
Mesekoparan, Özle 553
Mohammadi, Ali 500, 563
Mokhtari, Hossein 120, 447
Moon, Jae-Won 233
Moradlou, Majid 341
Moradzadeh, Arash 199, 268
Motepe, Sibonelo 557, 575
Mousavi, Sadra 155
Mudaliar, H. K. 372
Müderrisoglu, Kenan 590
Mukhopadhyay, Shayok 139, 366
Muntean, Nicolae 260, 289
Muresan, Cristina I. 532
Mustafa, Mustafa M. 547
Nahvi, Shahkar Ahmad 310
Naidu, Karteek 500

N'Diaye, Abdoul 221
Neacsu, Dorin O. 423
Negadi, Karim 459, 486
Nemoianu, Iosif-Vasile 166
Neufeld, Alexander 132
O'Reilly, Joseph 525
Ozcan, I. Halil 143
Ozturk, Salih Baris 583
Pahalvandust, Mohammad 336
Paltanea, Gheorghe 166
Paltanea, Veronica Manescu 166
Pellegrino, Gianmario 57
Petreus, Dorin 322, 329
Pfost, Martin 108
Polat, A. 186, 517
Ponick, Bernd 93, 132, 191, 246
Popa, Ana 29
Poshtan, Javad 85
Poshtan, Majid 85, 120
Pourhossein, Kazem 199, 268, 481, 511
Pourkeivannour, Siamak 297, 459, 486
Prasad, Ravneel 563
Prodan, Ovidiu 532
Pyrhönen, J. 114
Rae, Michael 71
Rafajdus, P. 240
Rafajdus, Pavol 100
Ram, Krishnil R 500
Rehman, Habibur 139
Rehman, Habib-Ur 366
Rhyu, Se-Hyun 233
Rizqiawan, Arwindra 409
Sangwongwanich, Ariya 402
Sarikaya, T. A. 186
Sarlioglu, Bülent 60, 172
Schwarz, Babette 93
Seker, Serhat 155
Sharkh, Suleiman M. 431
Sim, Soo-Yeon 453
Sizlayan, Seyit Yigit 540
Smaili, Attalah 459
Soda, Naoya 178
Sölek, Hazal 590
Stipetic, Stjepan 596
Stopforth, Riaan 557, 575
Stulrajter, Marek 100
Sufianto, Abdul Muiz 409
Sumega, Martin 100
Szabo, Csaba 351
Szekely, Norbert 351
Szoke, Eniko 351
Teodorescu, Remus 322, 329
Torac, I. 65
Toumi, Djilali 205, 227

AUTHOR INDEX

Trinh, Ngoc-Tu...161
Tsuchida, Yuji..178
Tutelea, L. N...65
Tutelea, Lucian.........................29, 77, 260
Twala, Bhekisipho.............................557, 575
Ucar, Mehmet...383
Ueno, Shohei..178
Ugur, Abdulkerim..377
Usman, Ahmad Mustapha..............................493
Usman, Hafiz M. 139
Varecha, Patrik..100
Vidal-Naquet, Fabien.....................................161
Vip, Stephan-Akash..246
Vitan, Liviu-Danut...260
Votzi, Helmut L..304
Wakabayashi, Daisuke 178
Yaman, Burak...553
Yazdani, Mohammad Rouhollah....................336
Yenil, Veli..315
Yilmaz, Murat...........................377, 590, 604
Zacharias, Vlad..351
Zaheer, M. 114
Zaskalicky, Pavel...126
Zhang, Jin..525
Zhang, Ning..71
Zossak, Simon 100

IEEE
445 Hoes Lane
Piscataway, NJ 08854-4141

ISBN 978-1-5386-7688-2